# IMPORTANT:

## HERE IS YOUR REGISTRATION CODE TO ACCESS
## YOUR PREMIUM McGRAW-HILL ONLINE RESOURCES.

For key premium online resources you need THIS CODE to gain access. Once the code is entered, you will be able to use the Web resources for the length of your course.

If your course is using **WebCT** or **Blackboard**, you'll be able to use this code to access the McGraw-Hill content within your instructor's online course.

Access is provided if you have purchased a new book. If the registration code is missing from this book, the registration screen on our Website, and within your WebCT or Blackboard course, will tell you how to obtain your new code.

## Registering for McGraw-Hill Online Resources

**TO gain access to your MCGraw-HiLL web resources simply follow the steps below:**

1. USE YOUR WEB BROWSER TO GO TO: **www.mhhe.com/roberts**

2. CLICK ON **FIRST TIME USER**.

3. ENTER THE REGISTRATION CODE* PRINTED ON THE TEAR-OFF BOOKMARK ON THE RIGHT.

4. AFTER YOU HAVE ENTERED YOUR REGISTRATION CODE, CLICK **REGISTER**.

5. FOLLOW THE INSTRUCTIONS TO SET-UP YOUR PERSONAL UserID AND PASSWORD.

6. WRITE YOUR UserID AND PASSWORD DOWN FOR FUTURE REFERENCE. KEEP IT IN A SAFE PLACE.

**TO GAIN ACCESS** to the McGraw-Hill content in your instructor's **WebCT** or **Blackboard** course simply log in to the course with the UserID and Password provided by your instructor. Enter the registration code exactly as it appears in the box to the right when prompted by the system. You will only need to use the code the first time you click on McGraw-Hill content.

**Thank you, and welcome to your MCGraw-HiLL online Resources!**

**REGISTRATION CODE**

**G089-P6WH-K3BU-06MN-OAUF**

0-07-291976-0   T/A   ROBERTS: SIGNALS AND SYSTEMS

# Signals and Systems
*Analysis Using Transform Methods and MATLAB*®

**Michael J. Roberts**

*Professor, Department of Electrical and Computer Engineering*
*University of Tennessee*

Boston   Burr Ridge, IL   Dubuque, IA   Madison, WI   New York   San Francisco   St. Louis
Bangkok   Bogotá   Caracas   Kuala Lumpur   Lisbon   London   Madrid   Mexico City
Milan   Montreal   New Delhi   Santiago   Seoul   Singapore   Sydney   Taipei   Toronto

# Higher Education

SIGNALS AND SYSTEMS: ANALYSIS USING TRANSFORM METHODS AND MATLAB®

Publisher: *Elizabeth A. Jones*
Senior sponsoring editor: *Carlise Paulson*
Developmental editor: *Melinda Dougharty*
Marketing manager: *Dawn R. Bercier*
Project manager: *Sheila M. Frank*
Lead production supervisor: *Sandy Ludovissy*
Media project manager: *Sandra M. Schnee*
Senior media technology producer: *Phillip Meek*
Senior coordinator of freelance design: *Michelle D. Whitaker*
Cover designer: *Christopher Reese*
Cover image: *© Getty Images/Scott Montgomery*
Lead photo research coordinator: *Carrie K. Burger*
Compositor: *Interactive Composition Corporation*
Typeface: *10.5/12 Times Roman*
Printer: *Courier Westford*

**Library of Congress Cataloging-in-Publication Data**

Roberts, M. J. (Michael J.)
  Signals and systems : analysis using transform methods and MATLAB® / M. J. Roberts. — 1st ed.
     p.   cm.
  Includes index.
  ISBN 0–07–249942–7
  1. Signal processing.   2. System analysis.   3. MATLAB®.   I. Title.

TK5102.9.R63   2004
621.382'2—dc21

2003046424
CIP

www.mhhe.com

To my wife Barbara for giving me the time and space to complete this effort
and to the memory of my parents, Bertie Ellen Pinkerton and Jesse Watts Roberts,
for their early emphasis on the importance of education.

# CONTENTS

iv

## MOTIVATION

I wrote this book because I love the mathematical beauty of signals and systems analysis. As in most areas of science and engineering the most important and useful theories are the ones, like Newton's laws, Maxwell's equations, and Einstein's theory of relativity, that capture the essence, and therefore the beauty, of physical phenomena. I don't know how many hours I have spent writing this material, although it must be at least several thousand, but I think it would be difficult, if not impossible, for anyone to do this much work without a real passionate commitment to it.

## AUDIENCE

I have written class notes for my junior-level classes in this area for many years and in 2000 decided that these notes had reached a level of maturity such that I could contemplate publishing them more widely. This book, which grew out of those class notes, is intended to cover a two-semester course sequence in the basics of signal and system analysis during the junior year. It could also be used in a senior-level course in these topics, although in most engineering curricula this material is covered in the junior year. It can also be used (as I have used it) as a book for a quick one-semester master's-level review of transform methods as applied to linear systems.

## OVERVIEW

The book begins with mathematical methods for describing signals and systems, in both continuous and discrete time. I introduce the idea of a transform with the Fourier series, and from that base move to the Fourier transform as an extension of the Fourier series to aperiodic signals. There is a chapter on applications of Fourier analysis, including filters and communication systems. After covering Fourier methods, I use them in explaining the implications of sampling and in the analysis of the correlation between two signals and the energy and power spectral density of signals. I introduce the Laplace transform both as a generalization of the continuous-time Fourier transform for unbounded signals and unstable systems and as a powerful tool in system analysis because of its very close association with the eigenvalues and eigenfunctions of continuous-time linear systems. Then I present applications of the Laplace transform in circuit analysis, feedback systems, and multiple-input, multiple-output systems. I take a similar path for discrete-time systems using the $z$ transform. In the last chapter I spend significant time on approximating continuous-time systems with discrete-time systems with extensive coverage of digital filter design methods. Throughout the book I present examples and introduce MATLAB functions and operations to implement the methods presented. A chapter-by-chapter summary follows.

## CHAPTER SUMMARIES

Chapter 1 is an introduction to the general concepts involved in signal and system analysis without any mathematical rigor. It is intended to motivate the student by demonstrating the ubiquity of signals and systems in everyday life and the importance of understanding them.

Chapter 2 is an exploration of methods of mathematically describing signals of various kinds. It begins with familiar functions, continuous-time (CT) sinusoids and exponentials, and then extends the range of signal-describing functions to include CT singularity functions (switching functions) and other functions that are related to them through convolution and/or Fourier transformation. Like most, if not all, signals and systems textbooks, I define the unit step, the signum, the unit impulse, the unit ramp, and the unit sinc function. In addition to these I define the unit rectangle, the unit triangle, and the unit comb function (a periodic sequence of unit impulses). I find them very convenient and useful because of the compact notation that results. The unit comb function, along with convolution, provides an especially compact way of mathematically describing arbitrary periodic signals.

After introducing the new CT signal functions, I cover the common types of signal transformations (amplitude scaling, time shifting, time scaling, differentiation, and integration) and apply them to the signal functions. Then I cover some characteristics of signals that make them invariant to certain transformations (evenness, oddness, and periodicity) and some of the implications of these signal characteristics in signal analysis.

The next major section of Chapter 2 is coverage of discrete-time (DT) signals following a path analogous to that followed in CT signals. I introduce the DT sinusoid and exponential and comment on the problems of determining the period of a DT sinusoid. This is the student's first exposure to some of the implications of sampling. I define some DT signal functions analogous to CT singularity functions. Then I explore amplitude scaling, time shifting, time scaling, differencing, and accumulation for DT signal functions pointing out the unique implications and problems that occur, especially when time-scaling DT functions.

The last section of Chapter 2 is on signal energy and power. I define both for CT and DT signals and comment on the need for both by defining and discussing energy signals and power signals.

Chapter 3 is an introduction to the mathematical description of systems. First I cover the most common forms of classification of systems (homogeneity, additivity, linearity, time invariance, causality, memory, static nonlinearity, and invertibility). By example I present various types of systems which have, or do not have, these properties and how to prove various properties from the mathematical description of the system.

The next major section of Chapter 3 is the introduction of impulse response and convolution as components in the systematic analysis of the response of linear, time-invariant DT systems. I present the mathematical properties of convolution and a graphical method of understanding what the convolution-sum formula says. I also show how the properties of convolution can be used to combine subsystems which are connected in cascade or parallel into one system and what the impulse response of the overall system must be. This section is followed by an analogous coverage of CT convolution. This order of coverage seems best because the students are better able to understand how to find the impulse response of a DT system than the impulse response of a CT system. Also DT convolution is easier to conceive because there are no limit concepts involved. The last section of Chapter 3 is on the relations between block diagrams of systems and the system equations.

Chapter 4, on the Fourier series, is the beginning of the student's exposure to transform methods. I begin by graphically introducing the concept that any CT signal with engineering usefulness can be expressed over a finite time by a linear combination of CT sinusoids, real or complex. Then I show that periodic signals can be expressed for all time as a linear combination of sinusoids. Then I formally derive the Fourier series using the concept of orthogonality (without the name at this point) to show where the signal description as a function of discrete harmonic number (the harmonic function) comes from. I mention the Dirichlet conditions to let the student know that the CT Fourier series applies to all practical CT signals, but not to all imaginable CT signals.

There is a major section on simply following the mathematical process of finding the harmonic function for a time function, with many graphical illustrations, beginning with a single sinusoid and progressing to multiple sinusoids and nonsinusoidal functions. Along the way the concepts of orthogonality and correlation arise naturally and are briefly discussed.

The next few sections in Chapter 4 are an exploration of the properties of the Fourier series. I have tried to make the Fourier series notation and properties as similar as possible and analogous to the Fourier transform which comes later. That is, the harmonic function forms a Fourier series pair with the time function. As is conventional in most signals and systems textbooks, I have used a notation for all the transform methods in which lowercase letters are used for time-domain quantities and uppercase letters are used for their transforms (in this case their harmonic functions). This supports the understanding of the interrelationship among the Fourier methods. I have taken an ecumenical approach to two different notational conventions that are commonly seen in books on signals and systems, control systems, digital signal processing, communication systems, and other applications of Fourier methods such as image processing and Fourier optics; the use of either cyclic frequency $f$ or radian frequency $\omega$. I use both and emphasize that the two are simply related through a change of variable. I think this better prepares students for seeing both forms in other books in their college and professional careers. Also I emphasize some aspects of the Fourier series, especially with regard to using different representation periods, because this is an important idea which later appears in Chapter 7 on sampling and the discrete Fourier transform (DFT). I encourage students to use tables and properties to find harmonic functions, and this practice prepares them for a similar process in finding Fourier transforms and later Laplace and $z$ transforms. I also have a section on the convergence of the Fourier series illustrating the Gibbs phenomenon at function discontinuities.

The next major section of Chapter 4 covers the same basic concepts as the first section, but as applied to DT signals. I emphasize the important differences caused by the differences between continuous- and discrete-time signals, especially the finite summation range of the DT Fourier series as opposed to the (generally) infinite summation range in the CT Fourier series. I also point out the importance of the fact that the DT Fourier series relates a finite set of numbers to another finite set of numbers, making it amenable to direct numerical machine computation. Later, in Chapter 7, I show the strong similarity between the DT Fourier series and the discrete Fourier transform (DFT).

Chapter 5 extends the concepts of the Fourier series to aperiodic signals and introduces the Fourier transform. I introduce the concept by examining what happens to a CT Fourier series as the period of the signal approaches infinity and then define and derive the CT Fourier transform as a generalization of the CT Fourier series. Following that I derive all the important properties of the CT Fourier transform. The next major section covers the DT Fourier transform, introducing and deriving it in an analogous way. There are numerous examples of the properties of both the CT and DT Fourier transform.

The last major section is a comparison of the four Fourier methods. This section is important because it reemphasizes many of the concepts of (1) continuous time and discrete time, and (2) sampling in time and sampling in frequency (which will be important in Chapter 7 which covers sampling and the discrete Fourier transform). I emphasize particularly the duality between sampling in one domain and periodic repetition in the other domain and the information equivalence of a sampled and an impulse-sampled signal.

Chapter 6 is devoted to the application of Fourier methods to two kinds of system analysis for which it is particularly well suited, filters and communication systems. I define the ideal filter and return to the concept of causality to show that the ideal filter cannot be realized as a physical system. This is an example of a design in the frequency domain which cannot be achieved, but can be approached, in the time domain. Then I discuss and analyze some simple practical passive and active filters and demonstrate that they are causal systems. Bode diagrams are introduced as a method of quick analysis of cascaded systems. Then I introduce the simplest forms of modulation and show how Fourier analysis greatly simplifies understanding them. I also explore the concepts of phase and group delay and demonstrate them with a modulated signal. In the next major section I apply the same modulation principles to DT signals and systems in an analogous way. In the last sections I briefly discuss the use of filters to reduce noise in communication systems and the operation of a spectrum analyzer.

Chapter 7 is the first exploration of the correspondence between a CT signal and a DT signal formed by sampling it. The first section covers how sampling is usually done in real systems using a sample-and-hold and an analog-to-digital converter. The second section starts by asking the question of how many samples are enough to describe a CT signal. Then the question is answered by deriving Shannon's sampling theorem, first using the DT Fourier transform to describe a DT signal formed by sampling a CT signal. Then I impulse-sample a signal to show the correspondence between a sampled signal and an impulse-sampled signal and use the CT Fourier transform to show the same result. Then I discuss interpolation methods, theoretical and practical; the special properties of band-limited periodic signals; and finally the discrete Fourier transform, relating it to the DT Fourier series. I do a complete development of the DFT starting with a CT signal and then time-sampling, windowing, and frequency-sampling it to form two signals, each completely described by a finite set of numbers and exactly related by the DFT. Then I show how the DFT can be used to approximate the CT Fourier transform of an energy signal or a periodic signal. The next major section is a sequence of examples of the use of and properties of the DFT, and the last major section discusses the fast Fourier transform and shows why it is a very efficient algorithm for computing the DFT.

Chapter 8 is on correlation, energy spectral density, and power spectral density. These topics are not often covered in a signals and systems textbook. They are traditionally introduced in (or after) a course on random processes. I introduce the ideas here from the point of view of analyzing the similarity of two signals. Correlation concepts are important in system identification and matched filtering. I demonstrate correlation, and then autocorrelation, using both random and deterministic signals, but in the exercises I only ask the students to analyze the correlation or autocorrelation of deterministic signals. Since energy spectral density and power spectral density are the Fourier transforms of the autocorrelation of energy and power signals, I include those topics also.

Chapter 9 introduces the Laplace transform. I approach the Laplace transform from two points of view, as a generalization of the Fourier transform to a larger class of signals and as a result which naturally follows from the excitation of a linear, time-invariant system by a complex exponential signal. I begin by defining the bilateral Laplace transform and discussing the significance of the region of convergence. Then I define the unilateral Laplace transform and use it for most of the rest of the chapter before returning to the bilateral form at the end. I derive all the important properties of the Laplace transform and fully explore the method of partial-fraction expansion for finding inverse transforms. Then examples of solving differential equations with initial conditions are presented. Lastly I return to the bilateral form and show that bilateral transforms can be found by using unilateral transform tables.

Chapter 10 covers various applications of the Laplace transform, including block-diagram representation of systems in the complex-frequency domain, system stability, system interconnections, feedback systems including root-locus and gain and phase margin, block-diagram reduction, system responses to standard signals, frequency response, Butterworth filters, and lastly standard realizations and state-space methods for CT systems.

Chapter 11 introduces the $z$ transform. The development parallels the development of the Laplace transform except it is applied to DT signals and systems. I initially define a bilateral transform and discuss the region of convergence and then define a unilateral transform. I derive all the important properties and demonstrate the inverse transform using partial-fraction expansion and the solution of difference equations with initial conditions. Then I return to the bilateral transform showing that they can be found using unilateral tables. I also show the relationship between the Laplace and $z$ transforms, an important idea in the approximation of CT systems by DT systems in Chapter 12.

Chapter 12 is the last chapter and deals with applications of the $z$ transform. The main topics are approximating CT systems with DT systems, especially digital filter design as an approximation to optimal analog filters, responses to standard signals, system interconnections, standard system realizations, and state-space methods.

There are several appendices ranging from one page on how to find least common multiples to many pages describing the major commands and operations available in MATLAB. Appendices E, F, and G are tables of the Fourier methods, the Laplace transform, and the $z$ transform, respectively. Appendices H, I, and J are on complex numbers and variables, differential and difference equations, and vectors and matrices, respectively, topics which are generally considered as background for a signals and systems course. These appendices are written like book chapters with exercises at the end and can be used for purposes of review if the students in a particular class need it.

## CONTINUITY

The book is fairly integrated in its approach, and each chapter builds on earlier chapters. However, in a two-semester sequence spanning the entire book, the following topics could be omitted without loss of continuity.

System characteristics (except for linearity and time invariance).

The response of linear, time-invariant systems to periodic excitation using the Fourier series.

Some applications of Fourier methods including discrete-time filters, some of the modulation techniques, phase and group delay, and spectral analysis.

Sampling methods (as opposed to sampling theory).

Sampling discrete-time signals.

The fast Fourier transform algorithm.

All the discussions of correlation, energy spectral density, and power spectral density in Chapter 8.

The return to the bilateral Laplace transform at the end of Chapter 9.

Certain topics on Laplace transform applications like specific stability analysis methods, block-diagram reduction, and Butterworth filters.

Standard realizations of systems and state-space analysis, continuous-time and/or discrete-time.

The return to the bilateral $z$ transform at the end of Chapter 11.

Digital filter design with MATLAB.

## REVIEWS AND EDITING

I often tell my students that if they really want to learn a subject well, they should agree to teach a course in that subject. The process of standing up in front of a group of very intelligent people and presenting material is a strong discipline for learning the material (if the presenter is at all disturbed by public humiliation). After writing this book, I can amend that statement to say that if one wants to learn a subject *very* well, he or she should agree to write a textbook on it. The process of review is a somewhat similar discipline although not quite as public. The public part comes after the book is published. This book owes a lot to the reviewers, especially those who really took time and criticized and suggested improvements. I am indebted to them.

I am also indebted to the many students who have endured my classes over the years. I believe that our relationship is more symbiotic than they realize. That is, they learn signals and systems analysis from me and I learn how to teach signals and systems analysis from them. I cannot count the number of times I have been asked a very perceptive question by a student that revealed not only that the students were not understanding a concept but that I did not understand it as well as I had previously thought.

## WRITING STYLE

Every author thinks he or she has found a better way to present material so that students can grasp it, and I am no different. I have taught this material for many years and through the experience of grading tests have found what students generally do and do not grasp. I have spent countless hours in my office one on one with students explaining these concepts to them, and, through that experience, I have found out what needs to be said. In my writing I have tried to simply speak directly to the reader in a straightforward conversational way, trying to avoid off-putting formality and, to the extent possible, anticipating the usual misconceptions and revealing the fallacies in them. Transform methods are not an obvious idea, and, at first exposure, students can easily get bogged down in a bewildering morass of abstractions and lose sight of the goal which is to analyze a system's response to signals. I have tried (as every author does) to find the magic combination of accessibility and mathematical rigor because both are important. I think my writing is clear and direct, but you, the reader, will be the final judge of whether or not that is true.

## EXERCISES

The book contains over 500 exercises. Each chapter has a group of exercises with answers provided and a second group of exercises without answers. The first group is intended more or less as a set of drill exercises, and the second group as a set of more challenging exercises.

## CONCLUDING REMARKS

Although I have tried hard to make this book a good one, no book in its first edition is perfect, and mine will not disprove that assertion. I have discovered what all authors must discover, that even though they feel they really understand the concepts and how to do all the exercises, with this much text and this many exercises mistakes are inevitable. Therefore, I welcome any and all criticism, corrections, and suggestions. All comments, including ones I disagree with and ones which disagree with others, will have a constructive impact on the next edition because they will point out a problem. If something does not seem right to you, it probably will bother others also, and it is my task, as an author, to find a way to solve that problem. So I encourage you to be direct and clear in any remarks about what you believe should be changed and not to hesitate to mention any errors you may find, from the most trivial to the most significant.

I wish to thank the following reviewers for their invaluable help in making the book better.

Ali Amini, *California State University, Northridge*
Vijayakumar Bhagavatula, *Carnegie Mellon University*
Jose B. Cruz, Jr., *Ohio State University*
Thomas S. Denney, Jr., *Auburn University*
Frank Gross, *Florida A&M University*
John Y. Hung, *Auburn University*
Aziz Inan, *University of Portland*
James S. Kang, *Cal Poly Pomona*
Thomas Kincaid, *Boston University*
Wojtek J. Kolodziej, *Oregon State University*
Darryl Morrell, *Arizona State University*
Farzad Pourboghrat, *Southern Illinois University*
Lee Swindlehurst, *Brigham Young University*

**Michael J. Roberts, Professor**
**Electrical and Computer Engineering**
**University of Tennessee at Knoxville**

### CHAPTER 1

# Introduction

## 1.1 SIGNALS AND SYSTEMS DEFINED

It is always best to begin at the beginning. Since this text is about signals and systems, the first question to answer is, What are they? Any time-varying physical phenomenon which is intended to convey information is a *signal*. Examples of signals are the human voice, a dog's bark, a lion's roar, bird songs, smoke signals, drums, sign language, Morse code, and traffic signals. Examples of modern high-speed signals are the voltages on telephone wires, the electric fields emanating from radio or television transmitters, and variations of light intensity in an optical fiber on a telephone or computer network. *Noise,* which is sometimes called a random signal, is like a signal in that it is a time-varying physical phenomenon, but unlike a signal it usually does not carry useful information and is almost always considered undesirable.

Signals are processed or operated on by systems. When one or more excitation signals are applied at one or more system inputs, the system produces one or more response signals at its outputs. Figure 1.1 shows a diagram of a single-input, single-output system.

In a communication system, a *transmitter* is a device that produces a signal and a *receiver* is a device which acquires the signal. A *channel* is the path a signal and/or noise take from a transmitter and/or noise source to a receiver (Figure 1.2). The transmitter, channel, and receiver are all systems, which are components or subsystems of the overall system. Other types of systems also process signals which are analyzed using signal analysis. Scientific instruments which measure a physical phenomenon (temperature, pressure, speed, etc.) convert that phenomenon to a voltage or current, a signal. Commercial building control systems, industrial plant process control systems, avionics in airplanes, ignition and fuel pumping control in automobiles, etc., are all systems which process signals.

The definition of the term *system* even encompasses things one might not think of as systems, for example, the stock market, government, weather, and the human body.

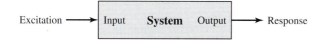

**Figure 1.1**
Block diagram of a simple system.

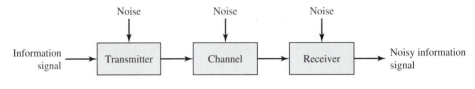

**Figure 1.2**
A communication system.

They all respond to excitations. Some systems are readily analyzed in detail, some can be analyzed approximately, but some are so complicated or difficult to measure that we hardly know enough to understand them or control them.

## 1.2  TYPES OF SIGNALS

There are several broad classifications of signals: continuous-time, discrete-time, continuous-value, discrete-value, random, and nonrandom. A *continuous-time* signal is one which is defined at every instant of time over some time interval. Another common name for a continuous-time signal is an *analog* signal. The name analog comes from the fact that in many systems the variation of the analog signal with time is *analog*ous to some physical phenomenon which is being measured or monitored.

The process of *sampling* a signal is to take values from it at discrete points in time and then to use only the samples to represent the original continuous-time signal. The set of samples taken from a continuous-time signal is one example of a *discrete-time* signal. A discrete-time signal can also be created by an inherently discrete-time *system* which produces signal values only at discrete times. A discrete-time signal has defined values only at discrete points in time and not between them. In Chapter 7 we will investigate under what conditions a discrete-time signal produced by sampling a continuous-time signal can be considered an adequate representation of the continuous-time signal from which it came.

A continuous-value signal is one which may have a value anywhere within a continuum of allowed values. The continuum may have a finite or infinite extent. A *continuum* is a set of values with no "space" between allowed values; two allowed values can be arbitrarily close together. The set of real numbers is a continuum with infinite extent. The set of real numbers between zero and one is a continuum with finite extent. Each of these examples is a set with infinitely many members.

A discrete-value signal can only have values taken from a discrete set of values. A *discrete* set of values is a set for which there is a finite space between allowed values. Another way of saying this is that the magnitude of the difference between any two values in the set is greater than some positive number. The set of integers is an example of a discrete set of values. Discrete-time signals are usually transmitted as digital signals. The term *digital signal* applies to the transmission of a sequence of values of a discrete-time signal in the form of *digits* in some encoded form (usually binary). The term digital is also sometimes used loosely to refer to a discrete-value signal which only has two possible values.

A *random* signal is one whose values cannot be predicted exactly and cannot be described by any mathematical function. A nonrandom signal, which is also called a *deterministic* signal, is one which can be mathematically described, at least approximately. As previously stated, a common name for a random signal is *noise*. Figures 1.3 through 1.5 are examples of different kinds of signals.

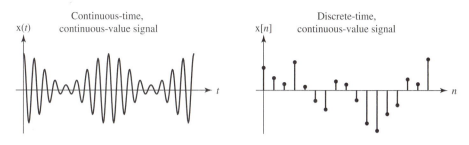

**Figure 1.3**
Examples of continuous-time and discrete-time signals.

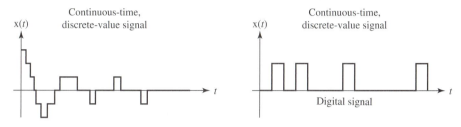

**Figure 1.4**
Examples of continuous-time and digital signals.

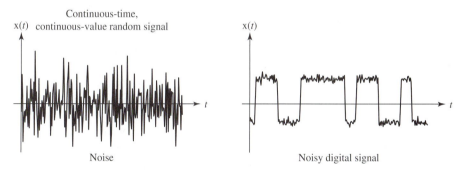

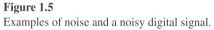

**Figure 1.5**
Examples of noise and a noisy digital signal.

In practical signal processing it is very common to acquire a signal for processing using a computer and following these steps: sampling, quantizing, and encoding (Figure 1.6). The original signal is typically a continuous-value, continuous-time signal. The process of sampling acquires its values at discrete times, and that sequence of values constitutes a continuous-value, discrete-time signal. The process of quantization approximates each sample as the nearest member of a finite set of discrete values, producing a discrete-value, discrete-time signal. Then each signal value in the set of discrete values at discrete times is converted to a sequence of rectangular pulses which encode that member of the set of discrete values into a binary number, creating a discrete-value, continuous-time signal, commonly called a digital signal. [It should be noted here that the steps illustrated in Figure 1.6 are usually carried out by a single device, an *analog-to-digital converter (ADC),* and that the signals at the intermediate

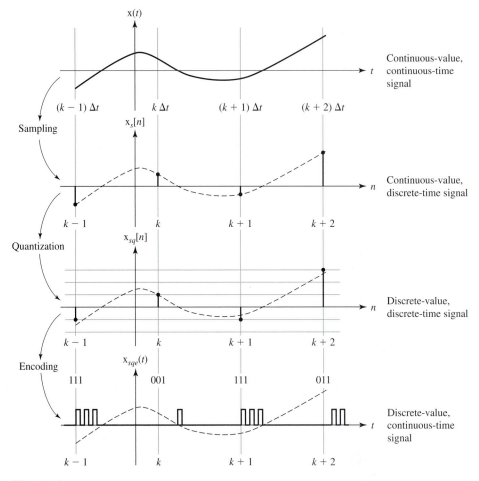

**Figure 1.6**
Sampling, quantization, and encoding of a signal to illustrate various signal types.

steps are not available external to the ADC. In fact, those signals may not even *exist* inside the ADC because of the way it accomplishes the conversion. Nevertheless, the operation of an ADC is often usefully analyzed as though the individual steps of sampling, quantizing, and encoding, were taken in sequence.]

One very common use of binary digital signals is for sending text messages using the American Standard Code for Information Interchange (ASCII). The letters of the alphabet, the digits 0 to 9, some punctuation characters, and several nonprinting control characters, for a total of 128 characters, are all encoded into a sequence of 7 binary bits. In asynchronous serial transmission of ASCII messages, the 7 bits are sent sequentially, preceded by a *start* bit and followed by one or two *stop* bits which are used for synchronization purposes. In some cases an extra bit called a *parity bit* is also transmitted. Parity bits are used to detect transmission errors. Typically in direct-wired connections between digital equipment the bits are represented by voltage levels, a *high* voltage [typically around +5 volts (V)] for a 1 and a *low* voltage level (typically around 0 V) for a 0. In an asynchronous transmission using one start and one stop bit and no parity, sending the message "SIGNAL," the voltage versus time would look as illustrated in Figure 1.7.

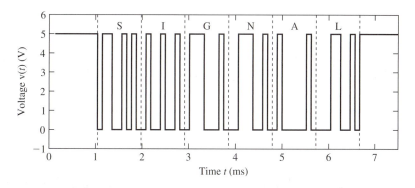

**Figure 1.7**
Asynchronous serial binary ASCII-encoded voltage signal for the message
"SIGNAL."

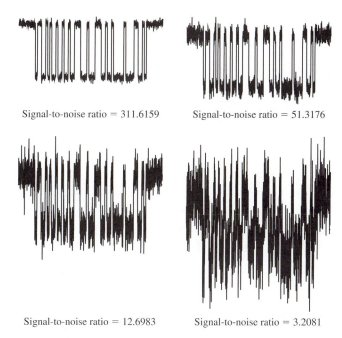

Signal-to-noise ratio = 311.6159    Signal-to-noise ratio = 51.3176

Signal-to-noise ratio = 12.6983    Signal-to-noise ratio = 3.2081

**Figure 1.8**
Noisy digital ASCII signal.

Digital signals are increasingly important in modern signal analysis because of the spread of digital systems. Digital signals have the advantage of better immunity to noise. That can be demonstrated by illustrating the digital signal in Figure 1.7 with different levels of noise added (Figure 1.8). Even though in the worst case in Figure 1.8 the noise has made the 1s and 0s in the binary signal difficult to see with the naked eye, practically all the 1s and 0s from the binary bit stream can be detected correctly, and, therefore, the received binary signal after detection is still practically perfect. In binary signal communication the bits can be detected very cleanly until the noise gets very large, as illustrated in Figure 1.9. The detection of bit values in a stream of bits is usually done by comparing the signal value at a predetermined bit time with a threshold. If it is above the threshold, it is declared a 1, and if it is below the threshold, it is declared a 0. In Figure 1.9, the $x$'s mark the signal value at the detection time, and when

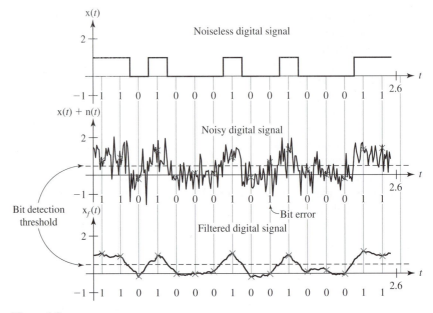

**Figure 1.9**
Use of a filter to reduce bit error rate in a digital signal.

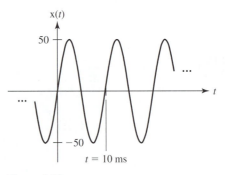

**Figure 1.10**
A continuous-time signal described by a
mathematical function.

this technique is applied to the noisy digital signal, one of the bits is incorrectly detected. But when the signal is filtered, all the bits are correctly detected. Even though the filtered digital signal does not look very clean in comparison with the noiseless digital signal, the bits can still be detected with a very low probability of error. This is the basic reason that digital signals have better noise immunity than analog signals.

The first problem in analysis of signals is finding a way to describe them mathematically. The first signals we will study will be deterministic, continuous-time signals. Some continuous-time signals can be described mathematically by simple continuous functions of time. For example, a signal x might be described by a function of continuous time $t$,

$$x(t) = 50 \sin(200\pi t).$$     **(1.1)**

This is an exact description of the signal at every instant of time. This signal can also be described graphically by plotting a graph of the continuous-time function describing the signal (Figure 1.10).

Many continuous-time signals that are important in signal and system analysis are not as easy to describe mathematically. Consider the signal graphed in Figure 1.11. Waveforms similar to this actually occur in various types of instrumentation and communication systems. With the definition of some signal functions and an operation called *convolution,* this signal can be compactly described, analyzed, and manipulated mathematically. Continuous-time signals that can be described by mathematical functions can be transformed into another domain called the *frequency domain* through the *continuous-time Fourier transform (CTFT).* Although it may not be clear to the reader at this time what transformation means, transformation of a signal to the frequency domain is a very important tool in signal analysis which allows certain characteristics of the signal to be more clearly observed and more easily manipulated than in the time domain. Without frequency-domain analysis, design and analysis of many systems would be considerably more difficult.

Discrete-time signals are only defined at discrete points in time and not between. Figure 1.12 illustrates some discrete-time signals. All these signals appear to be deterministic except the one in the lower right-hand corner which appears random.

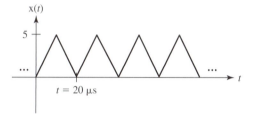

**Figure 1.11**
Another continuous-time signal.

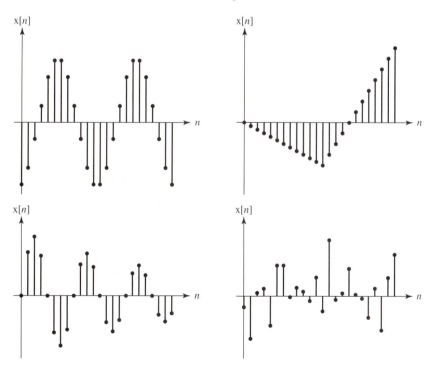

**Figure 1.12**
Some discrete-time signals.

(The word *appear* is used because one can never say for sure whether a signal is or is not random by observing a finite-time record of it.) In Chapter 7 we will explore the relationship between continuous-time signals and discrete-time signals which have been formed by sampling continuous-time signals.

Random signals cannot be described exactly by a mathematical function. In fact, there is no general way to completely describe random signals except to actually enumerate every value of the random signal. For continuous-time random signals this is impossible, even for a finite time, because there are infinitely many values of a continuous-time signal in a finite time interval. It is possible to describe a discrete-time random signal exactly over a finite time interval. The description would be a finite-length sequence of numbers. But, even though the discrete-time random signal is described over a finite time, that does not imply that the signal is known or described over any *other* finite time.

Even though random signals cannot be exactly described by mathematical functions they can be approximately described. Figure 1.13 illustrates four different random continuous-time signals. Even though these signals are all random, they obviously have different characteristics. They vary over different ranges, some change value rapidly while others change value more slowly, their values are distributed differently, they have different average values, etc.

The analysis of random signals and their interaction with systems must be done by using what are called *descriptors* of the random signals. Descriptors approximately describe certain important aspects of random signals but can never give an exact description. The descriptors of the four random signals in Figure 1.13 would express some of their important differences in general behavior. Using descriptors is the best way of describing and analyzing random signals. Even though descriptors cannot describe a signal exactly, they can in most practical cases be very effective in accomplishing goals in system design.

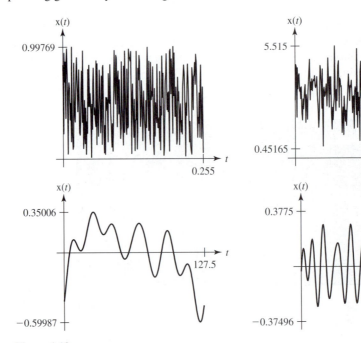

**Figure 1.13**
Four random continuous-time signals.

In this text we will see many examples of the operation of systems on exact deterministic signals because that is an important part of system design which lends insight into system operation. But ultimately most systems are designed to handle certain *types* of signals, rather than any exact signal or signals.

So far all the signals we have considered have been mathematically described by functions of time. There is another important class of signals which are functions of *space* instead of time, images. Most of the theories of signals, the information they convey, and how they are processed by systems in this text will be based on the signals which are a variation of a physical phenomenon with time. But the theories and methods so developed also apply, with only minor modifications, to the processing of images. Time signals are described by the variation of a physical phenomenon as a function of a single independent variable, time. Spatial signals, or images, are described by the variation of a physical phenomenon as a function of two orthogonal, independent, spatial variables, conventionally referred to as *x* and *y*. The physical phenomenon described is most commonly light or something that affects the transmission or reflection of light, but the techniques of image processing are also applicable to anything that can be mathematically described by a function of two independent variables.

Historically, image-processing techniques have lagged behind signal-processing techniques because in practice the amount of information that has to be processed to gather the information from a typical image is much larger than the amount of information required to get the information from a typical time signal. But now, with the great increases in computer power and the huge reductions in computer costs, image processing is increasingly a practical technique in many situations. Most image processing is digital and performed by computers. Some simple image-processing operations can be done directly with optics, and those can, of course be done at very high speeds (at the speed of light!). But direct optical image processing is very limited in its flexibility compared with digital image processing performed by computers.

Figure 1.14 shows two images. The one on the left is an unprocessed X-ray image of a carry-on bag at an airport checkpoint. The one on the right is the same image except processed by some image filtering operations to reveal the presence of a weapon. This text will not go into image processing in any depth but will use occasional examples of image processing to illustrate concepts in signal processing.

**Figure 1.14**
An example of image processing to reveal information.
[*Original X-ray image and processed version provided by M. A. Abidi, Imaging, Robotics and Intelligent Systems (IRIS) Laboratory of the Department of Electrical and Computer Engineering at the University of Tennessee, Knoxville.*]

An understanding of how signals carry information, how systems process signals, how multiple signals may be carried on one channel simultaneously, and how noise interferes with the transmission of information by signals is fundamental to multiple areas of engineering. Techniques for the analysis of signals and noise as they are processed by systems are the subject of this text. This material can be considered almost as an applied mathematics text more than a text covering the actual building of useful devices, but an understanding of this material is very important for the successful design of useful devices. The material which follows builds from some fundamental definitions and concepts to a full range of analysis techniques for continuous-time and discrete-time signals in systems.

## 1.3  A SIGNAL AND SYSTEM EXAMPLE

As an example of signal and system analysis, let's look at a signal and system that everyone is familiar with, sound, and a system which produces and/or measures sound. Sound is simply what the ear senses, and the human ear is sensitive to acoustic pressure waves in a limited frequency range, typically between about 15 hertz (Hz) and 20 kilohertz (kHz) with some sensitivity variation in that range. In the material which follows are some graphs of air-pressure variations which produce some common sounds. These sounds were recorded by a system consisting of a microphone which converts air-pressure variation into a continuous-time voltage signal, electronic circuitry which processes the continuous-time voltage signal, and an analog-to-digital converter which changes the continuous-time voltage signal to a digital signal in the form of a sequence of binary numbers which are then stored in computer memory (Figure 1.15).

Consider the pressure variation graphed in Figure 1.16. It is the continuous-time pressure signal that produces the sound of the word *signal* spoken by an adult male (the author). Analysis of sounds is a large subject in itself, but some things about the relationship between this graph of air-pressure variation and what a human hears as the word *signal* can be seen by looking at the graph. There are three identifiable signal bursts: burst 1 from 0 to about 0.12 seconds(s), burst 2 from about 0.12 to about 0.19 s, and burst 3 from about 0.22 to about 0.4 s. Burst 1 is the *s* in the word *signal*. It is a hissing sound and has a different character than the other bursts which are voiced sounds. Burst 2 is the *i* sound. The region between bursts 2 and 3 is the double consonant *gn*. Consonants are simply the things in speech that divide vowel sounds. The *g* stops the *i* sound, and the *n* begins the *a* sound. Burst 3 is the *a* sound terminated by the *l* consonant stop. An *l* is not quite as abrupt a stop as some other consonants, like *b* or *p*, for example, so the sound tends to trail off rather than stopping quickly. Notice that

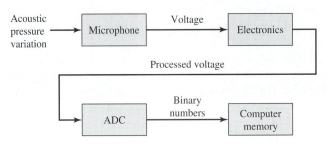

**Figure 1.15**
A sound recording system.

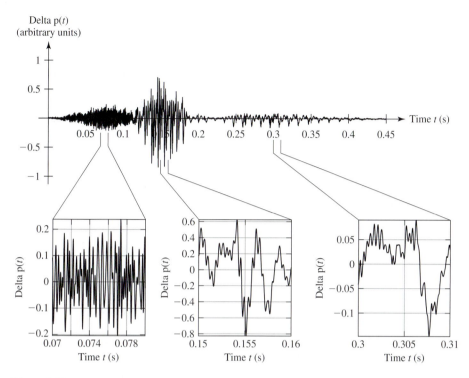

**Figure 1.16**
The word *signal* spoken by an adult male voice.

the variation of air pressure is generally faster for the *s* than for the *i* or the *a*. In signal analysis we would say that it has more high-frequency content. In the blowup diagram of the *s* sound one can see that the air-pressure variation looks almost random. The *i* and *a* sounds are different in that they vary more slowly and seem to be more regular or predictable (although not exactly predictable). The *i* and *a* are formed by vibrations of the vocal cords and therefore exhibit an approximately oscillatory behavior. This is described by saying that the *i* and *a* are tonal, or voiced, and the *s* is not. The word *tonal* simply means having the basic quality of a single tone or pitch or frequency. This description is not mathematically precise but is useful qualitatively.

Another way of looking at a signal is in what is known as the *frequency domain,* by examining the frequencies, or pitches, that are present in the signal. A common way of illustrating the variation of signal power with frequency is the *power spectral density* which will be introduced in Chapter 8. Figure 1.17 shows the three bursts (*s*, *i*, and *a*) from the word *signal* and their associated power spectral densities [the $G_s(f)$ functions].

Power spectral density is just another mathematical tool for analyzing a signal. It does not contain any new information, but sometimes it can reveal things that are difficult to see otherwise. (In fact, power spectral density contains *less* information than the original signal, but, because of the way the information is displayed, power spectral density can reveal information that is hard to recognize otherwise.) In this case, the power spectral density of the *s* sound is fairly widely distributed in frequency, whereas the power spectral densities of the *i* and *a* sounds are fairly narrowly distributed in the lowest frequencies (those nearest zero in the power spectral density plots). There is

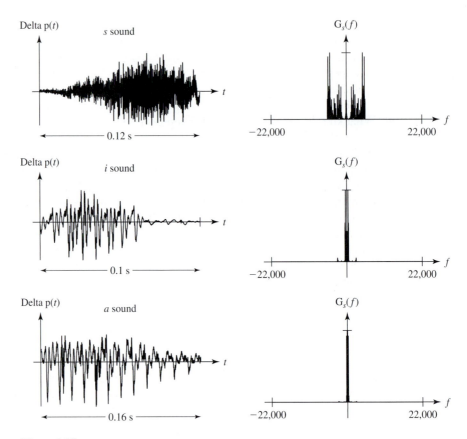

**Figure 1.17**
Three sounds in the word *signal* and their associated power spectral densities.

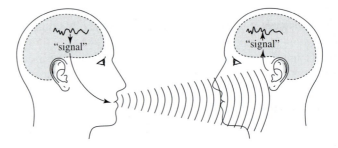

**Figure 1.18**
Communication between two people involving signals and signal processing by systems.

more power in the *s* sound at higher frequencies than in the *i* and *a* sounds. The *s* sound has an *edge,* or hissing quality, caused by the high frequencies in the *s* sound.

The signal in Figure 1.16 carries information. Consider what happens in conversation when one person says the word *signal* and another hears it (Figure 1.18). The speaker thinks first of the concept of a signal. His brain quickly converts the concept to the word *signal*. Then his brain sends nerve impulses to his vocal cords and diaphragm to create the air movement and vibration and tongue and lip movements to produce the sound. This sound then propagates through the air between the speaker and the listener. The sound strikes the listener's eardrum, and the vibrations are converted to nerve impulses which the listener's brain converts to first the sound, then the word, and then the concept. Therefore, ordinary conversation is accomplished through a system of some considerable sophistication.

How does the listener's brain know that the complicated pattern in Figure 1.16 is the word *signal?* What kind of signal processing is happening? The listener is certainly not aware of all the detailed air-pressure variations but instead hears sounds which are caused by the air-pressure variation. Somehow the eardrum and brain convert the complicated air-pressure pattern into a few simple features. That conversion is similar to what we will do when we convert signals into the frequency domain. The individual features of a word are called phonemes. *Phonemes* are a set of combinations of time and pitches or frequencies which generally characterize the sounds people are capable of making in conversation in all languages. The eardrum and brain are trained to recognize phonemes and use them to determine the word being spoken. A listener can do this for many different speakers, male and female, young and old, soft and loud, even with strong accents (up to a point). The process of recognizing a sound by reducing it to a small set of features reduces the amount of information the brain has to process down to a manageable level. Signal processing and analysis in the technical sense do the same thing but in a more mathematically precise and defined way.

Two very common problems in all signal and system analysis are noise and interference. *Noise* is an undesirable *random* signal. *Interference* is an undesirable *nonrandom* signal. Noise and interference both tend to obscure the information in a signal. In Figure 1.19 are examples of the signal from Figure 1.16 with different levels of noise added. As the noise power increases, there is a gradual degradation in the intelligibility

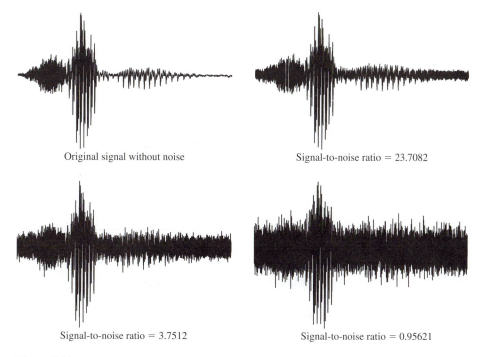

Original signal without noise     Signal-to-noise ratio = 23.7082

Signal-to-noise ratio = 3.7512     Signal-to-noise ratio = 0.95621

**Figure 1.19**
Sound of the word *signal* with different levels of noise added.

of the signal, and at some level of noise the signal becomes unintelligible. A measure of the quality of a received signal corrupted by noise is the ratio of the signal power to the noise power, commonly called the signal-to-noise ratio (SNR).

Figures 1.20 and 1.21 show two more examples of sounds. Figure 1.20 shows that the sound of whistling a constant tone produces an almost perfect sinusoidal variation of air pressure with time.

Figure 1.21 shows that the sound of a paper clip dropped onto a desktop is characterized by certain discrete events, bursts of air-pressure variation. Each burst

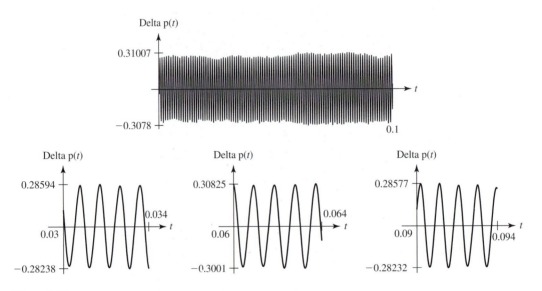

**Figure 1.20**
The sound of a whistled single tone.

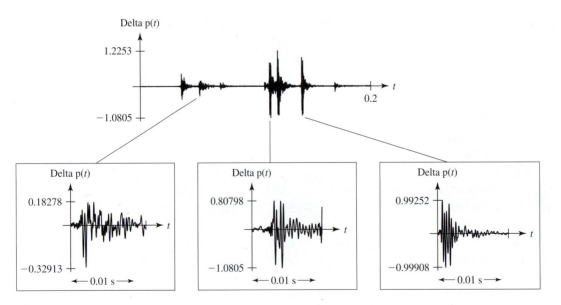

**Figure 1.21**
The sound of a paper clip dropped onto the top surface of a wooden desk.

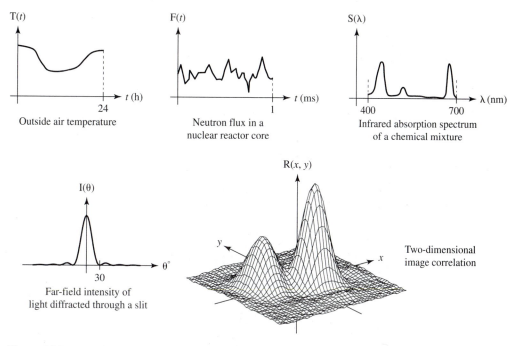

**Figure 1.22**

Examples of signals which are functions of one or more continuous independent variables.

corresponds to a strike of the paper clip on the desk as it bounces multiple times very quickly. From the graph it is apparent that the paper clip is at rest within about one-eighth of a second, but in that time it strikes the surface about 7 times. Each time it strikes it creates a pressure variation pattern that looks about the same except for its amplitude. There is a ringing after each strike that is caused by the mechanical vibrations of the paper clip.

Sounds are not the only signals, of course. Any physical phenomenon that is measured or observed is a signal. Also, although the majority of signals we will consider in this text will be functions of time, a signal can be a function of some other independent variable, like frequency, wavelength, and distance. Figures 1.22 and 1.23 illustrate some other kinds of signals.

Just as sounds are not the only signals, conversation between two people is not the only system. Examples of other systems are

■ An automobile suspension for which the road surface is the excitation and the position of the chassis relative to the road is the response

■ A chemical mixing vat for which streams of chemicals are the excitation and the mixture of chemicals is the response

■ A building environmental control system for which the exterior temperature is the excitation and the interior temperature is the response

■ A cup anemometer for which the wind is the excitation and the rotation of the cups is the response

■ A chemical spectroscopy system for which white light is the excitation and the spectrum of transmitted light is the response

■ A telephone network for which voices and data are the excitations and reproductions of those voices and data at a distant location are the responses

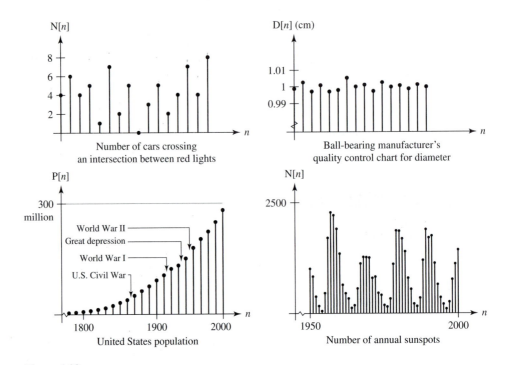

**Figure 1.23**
Examples of signals which are functions of a discrete independent variable.

- Earth's atmosphere for which energy from the sun is the excitation and the weather is the response (ocean temperature, wind, clouds, humidity, etc.)
- A thermocouple for which the temperature gradient along its length is the excitation and the voltage developed at one end is the response
- A computer mouse for which its movement is the excitation and the cursor position on the screen is the response
- A trumpet for which the vibration of the player's lips and the positions of the valves are the excitations and the tone emanating from the bell is the response

The list is endless. Any physical entity can be thought of as a system, because if we excite it with physical energy, it has a physical response.

## 1.4  USE OF MATLAB

Throughout the text, examples will be presented showing how many signal-analysis calculations can be done using MATLAB. MATLAB is a high-level mathematical tool available on many types of computers. It has been designed with signal processing and system analysis in mind. There is an introduction to MATLAB in Appendix B.

# Mathematical Description of Signals

## 2.1 INTRODUCTION AND GOALS

Over the years, signal and system analysts have observed many signals and have observed that they can be classified into groups with certain types of similar behavior. Figure 2.1 shows some examples of the types of signals that occur in real systems.

In signal and system analysis, signals are described (to the extent possible) by mathematical functions. The *signal* is the actual physical phenomenon which carries information, and the *function* is a mathematical description of the signal. Although, strictly speaking, the two concepts are distinct, the relation between a signal and the mathematical function that describes it is so intimate that the two terms, signal and function, are used almost interchangeably in signal and system analysis.

Some of the functions that describe real signals should already be familiar: exponentials and sinusoids. These occur frequently in signal and system analysis and are often used to describe signals. One set of functions has been defined to describe the effects on signals of switching operations that often occur in systems. Some other functions arise in the development of certain system-analysis techniques, which will be introduced in later chapters. They will all be defined here and used as they are needed in later chapters. These functions are all carefully chosen to be simply related to each other and to be easily transformed by a well-chosen set of transformation operations. They are prototype functions, which have simple definitions and are easily remembered.

There are two distinct types of signals and systems: continuous-time and discrete-time. Both terms and applicable functions and transformation operations will be defined. Also the types of symmetries and patterns that most frequently occur in real signals will be defined and their effects on signal analysis explored.

### CHAPTER GOALS

1. To define some mathematical functions that can be used to describe various types of signals

2. To develop methods of transforming and combining those functions in useful ways to represent real signals

3. To recognize certain symmetries and patterns and use them to simplify signal and system analysis

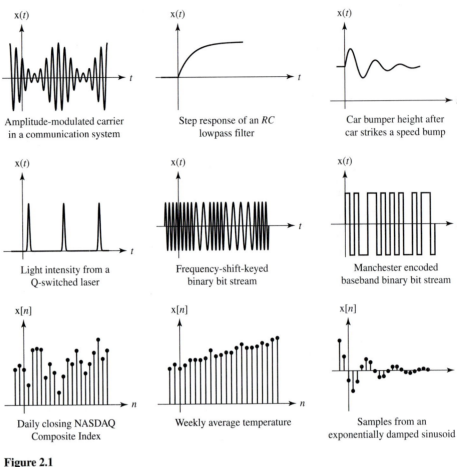

**Figure 2.1**
Examples of signals in real systems.

## 2.2 CONTINUOUS-TIME VERSUS DISCRETE-TIME FUNCTIONS

### CONTINUOUS-TIME FUNCTIONS

Most of the reader's experience with mathematical functions has probably been with functions of the form $g(x)$ where the independent variable $x$ can, in general, have any real value in a continuum of real values. If the independent variable is time $t$ and can have any real value, the function $g(t)$ is called a *continuous-time (CT)* function because it is defined on a continuum of points in time. Figure 2.2 illustrates some CT functions. Observe that Figure 2.2(b) illustrates a function with a discontinuous first derivative and Figure 2.2(d) illustrates a discontinuous function. At a discontinuity, the limit of the function value as we approach the discontinuity from above is not the same as the limit as we approach the same point from below. Stated mathematically, if the time $t = t_0$ is a point of discontinuity of a function $g(t)$, then

$$\lim_{\varepsilon \to 0} g(t + \varepsilon) \neq \lim_{\varepsilon \to 0} g(t - \varepsilon). \tag{2.1}$$

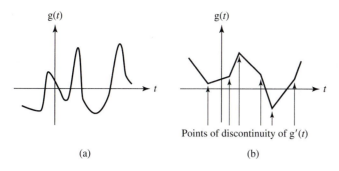

Points of discontinuity of g′(t)

(a)                              (b)

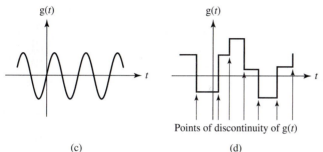

Points of discontinuity of g(t)

(c)                              (d)

**Figure 2.2**
Examples of CT functions.

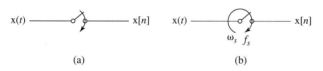

(a)                              (b)

**Figure 2.3**
(a) An ideal sampler, and (b) an ideal sampler sampling
uniformly.

But all four functions, Figure 2.2(a) to (d), are continuous-time functions because their values are defined on a continuum of times $t$. Therefore, the terms *continuous* and *continuous time* mean different things. At any time $t$ on a continuous function, the function value is the same as the limit of the function approaching that same time $t$ from above or below. A CT function is defined on a continuum of times, but it is not necessarily continuous at every point in time.

## SAMPLING AND DISCRETE TIME

Of great importance in signal and system analysis are functions defined only at discrete points in time and not between them. These are *discrete-time (DT)* functions which describe discrete-time signals. One very common example of DT signals are those obtained by sampling CT signals. *Sampling* means acquiring the values of a signal at discrete points in time. One way to visualize sampling is through the example of a voltage signal and a switch used as an ideal sampler [Figure 2.3(a)]. The switch closes for an infinitesimal time at defined discrete points in time. Only the values of the CT signal $x(t)$ at those discrete times are assigned to the DT signal $x[n]$. If there is a fixed time $T_s$

between samples (which is by far the most common situation in practice), the sampling is called *uniform* sampling in which the sampling times are integer multiples of the sampling interval $T_s$. The specification of the time of the sample $nT_s$ can be replaced by simply specifying the number $n$ which indexes the sample. This type of operation can be envisioned by imagining that the switch simply rotates at a constant cyclic velocity $f_s$, in cycles per second, or at a constant angular velocity $\omega_s$, in radians per second, as in Figure 2.3(b). Then the time between samples is

$$T_s = \frac{1}{f_s} = \frac{2\pi}{\omega_s}. \tag{2.2}$$

We will use a commonly accepted simplified notation for DT functions, g[$n$], which, at every point of continuity of g($t$), is the same as g($nT_s$), and in which $n$ can only have integer values. The square brackets [ ] enclosing the argument indicate a DT function, as contrasted with the parentheses ( ), which indicate a CT function. The independent variable $n$ is commonly called a discrete-time variable because it indexes discrete points in time, even though it is dimensionless and does not have units of seconds like $t$ and $T_s$ do. Since DT functions are only defined for integer values of $n$, the value of an expression like g[2.7] is simply undefined.

The values of g($t$) that are acquired at the sampling instants are g($nT_s$). This formulation of the relation between a CT function and its sample values works well except for the special case in which the sampling time $nT_s$ falls on a discontinuity of g($t$). In that case we will adopt the convention that, at a discontinuity, the sample value will be defined by

$$g[n] = \lim_{\varepsilon \to 0} g(nT_s + \varepsilon), \qquad \varepsilon > 0. \tag{2.3}$$

In words, at a discontinuity, the proper sample value is the limit as $t$ approaches $nT_s$ *from above.*

Functions which are inherently discrete time are indicated by notation of the form g[$n$], where the square brackets indicate that the function has a defined value only if $n$ is an integer. Functions which are defined for continuous arguments can also be given discrete time as an argument, for example, $\sin(2\pi f_0 nT_s)$. We can form a DT function from a CT function by sampling, for example, g[$n$] = $\sin(2\pi f_0 nT_s)$. Then, although the sine is defined for any argument value in the complex plane, the function g[$n$] is only defined for real integer values of $n$. That is, g[7.8] is undefined even though $\sin(2\pi f_0(7.8)T_s)$ is defined.

> If we were to define a function as g($n$) = $\sin(2\pi f_0 nT_s)$, the parentheses in g($n$) would indicate that any value of $n$ would be acceptable, integer or otherwise. Although this is legal, it is not a good idea because, at least in this text, we are using the symbol $t$ for continuous time and the symbol $n$ for discrete time and the notation g($n$), although mathematically defined, would be confusing.

DT functions do not necessarily come from samples of CT functions. There are many signals and systems which are inherently discrete time. The classic example is a financial system in which interest on a savings account is credited at discrete times (at the end of every day, week, month, or year). The value of the account is fixed during the time between discrete points and only changes at discrete points in time. In all inherently DT systems nothing happens *between* discrete points in time. Events occur only *at* discrete points in time.

In engineering practice the most important examples of DT systems are those which involve the use of *sequential-state machines,* the most common example being a computer. Computers are driven by a clock. The clock generates pulses at regular intervals in time, and at the end of each clock cycle the computer has executed an instruction and changed from one logical state to the next. Of course, at the integrated microcircuit level, physical events occur between clock pulses. But that is of concern only to the designers of the integrated circuit. Only the sequential states of the computer are actually significant to computer users. It must now be obvious that the computer has become a very important tool in engineering and business (and many other endeavors), so understanding how DT signals are processed by sequential-state machines is very important, especially to engineers. Figure 2.4 illustrates some DT functions which could describe DT signals. The type of plot used in Figure 2.4 is called a *stem* plot in which a dot indicates the functional value and the stems always connect the dot to the discrete-time *n* axis. This is a widely used method of plotting DT functions. MATLAB has a command, `stem`, which can be used to generate stem plots.

In signal and system analysis, by far the most important signals are time-varying phenomena and are directly described by functions of time, either continuous or discrete. That is why the shorthand terminology, CT or DT, has been introduced. Functions of time are just special cases of functions of a continuous or discrete independent variable which could represent something other than time. In Chapters 4 and 5 we will be using transforms to describe signals, and, after transformation, the independent variable will no longer be time. For example, we will be dealing with functions of harmonic number, a discrete independent variable, and functions of frequency, a continuous independent variable. All the mathematical characteristics which apply to functions of continuous or discrete time also apply to functions of other continuous or discrete independent variables.

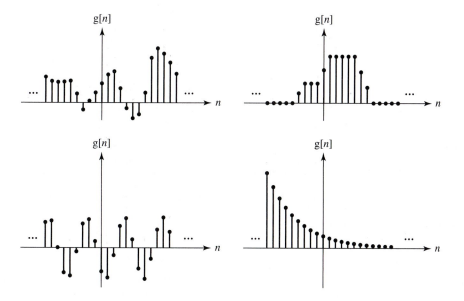

**Figure 2.4**
Examples of DT functions.

## 2.3 CONTINUOUS-TIME SIGNAL FUNCTIONS

### COMPLEX EXPONENTIALS AND SINUSOIDS

Some of the most commonly used mathematical functions for describing signals (see Figure 2.5) should already be familiar: CT sinusoids

$$g(t) = A \cos\left(\frac{2\pi t}{T_0} + \theta\right) = A \cos(2\pi f_0 t + \theta) = A \cos(\omega_0 t + \theta) \qquad (2.4)$$

and exponential functions

$$g(t) = A e^{(\sigma_0 + j\omega_0)t} = A e^{\sigma_0 t}[\cos(\omega_0 t) + jB \sin(\omega_0 t)] \qquad (2.5)$$

where  $A$ = amplitude of sinusoid or complex exponential

$T_0$ = real fundamental period of sinusoid

$f_0$ = real fundamental frequency of sinusoid, Hz

$\omega_0$ = real fundamental frequency of a sinusoid, radians per second (rad/s)

$t$ = continuous time

$\sigma_0$ = real damping rate

In Figure 2.5 the units indicate what kind of physical signal is being described. Very often in system analysis, when only one kind of signal is being followed through a system, the units are omitted for the sake of brevity.

In signal and system analysis, sinusoids are expressed in either of two ways: the cyclic frequency $f$ form, $A \cos(2\pi f_0 t + \theta)$, and the radian frequency $\omega$ form, $A \cos(\omega_0 t + \theta)$. There are advantages and disadvantages to either form. The advantages of the $f$ form are

1.  The fundamental period $T_0$ and the fundamental cyclic frequency $f_0$ are simple reciprocals of each other.

2.  In communication system analysis, a spectrum analyzer is often used, and its display scale is usually calibrated in hertz not radians per second. Therefore, $f$ is the directly observed variable.

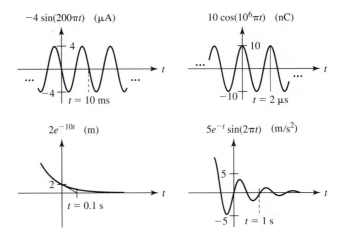

**Figure 2.5**
Examples of signals described by sines, cosines, and exponentials.

**3.** The definition of the Fourier transform and certain transforms and transform relationships will be simpler in the $f$ form than in the $\omega$ form.

The advantages of the $\omega$ form are

**1.** Resonant frequencies of real systems, expressed directly in terms of physical parameters are more simply expressed in the $\omega$ form than in the $f$ form. For example, in an *LC* oscillator, the resonant frequency is related to the inductance and capacitance by

$$\omega_0^2 = \frac{1}{LC} = (2\pi f_0)^2 . \qquad (2.6)$$

An *LC* oscillator is a circuit in which an inductance and a capacitance resonate and that resonance controls its frequency of oscillation. Also, the relation between time constants and corresponding critical frequencies in a system are more simply related through the $\omega$ form than through the $f$ form. For example, the half-power corner frequency of an *RC* lowpass filter is related to *R* and *C* by

$$\omega_c = \frac{1}{RC} = 2\pi f_c . \qquad (2.7)$$

**2.** The Laplace transform (Chapter 9) is defined in a form that is more simply related to the $\omega$ form than to the $f$ form.

**3.** Some Fourier transforms are simpler in the $\omega$ form.

CT sinusoids and exponentials are important in signal and systems analysis because they arise naturally in the solutions of the differential equations which describe CT system dynamics. As we will see in Chapters 4 and 5 in the study of the Fourier series and Fourier transform, even if signals are not sinusoids or exponentials they can be expressed in terms of sinusoids or exponentials.

## FUNCTIONS WITH DISCONTINUITIES

CT sines, cosines, and exponentials are all continuous and differentiable at every point in time. But there are many other types of important CT signals that occur in practical systems that are not continuous or differentiable at every point in time. One very common operation in systems is to switch on or switch off a signal at some specified time. Some examples of signals that are switched on or off are shown in Figure 2.6. Each of these signals has a point at which it is either discontinuous or its first derivative is discontinuous. The functional descriptions of the signals in Figure 2.6 are complete and accurate but are in an awkward form. Signals of this type can be better described mathematically by multiplying a function which is continuous and differentiable for all time by another function which is zero before some time and one after that time, or one before and zero after.

## SINGULARITY FUNCTIONS AND RELATED FUNCTIONS

In signal and system analysis there is a set of functions that are related to each other through integrals and derivatives which can be used to mathematically describe signals which have discontinuities or discontinuous derivatives. They are called *singularity functions*. These functions, and functions which are closely related to them through some common system operations, are the subject of this section.

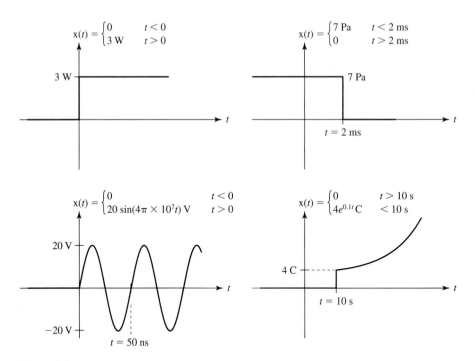

**Figure 2.6**
Examples of signals that are switched on or off at some specified time.

**The Unit Step Function**   Before defining the unit step function it is important to establish a principle of signal and system analysis. Consider the function,

$$g(t) = \begin{cases} A & t < t_0 \\ B & t > t_0 \end{cases}, \qquad A \neq B \tag{2.8}$$

(Figure 2.7). This is a CT function that has a defined value at every point in time *except* $t = t_0$. We can approach $t = t_0$ from below as closely as we want, and the function value is $A$ until we get to $t = t_0$. We can approach $t = t_0$ from above, and the function value is $B$ until we get to $t = t_0$. But *at* $t = t_0$ the value is undefined. We could, of course, assign $g(t)$ a value at $t = t_0$, but that would not change the fact that $g(t)$ is discontinuous there.

Now suppose we redefine $g(t)$ as

$$g(t) = \begin{cases} A & t < t_0 \\ \dfrac{A + B}{2} & t = t_0, \\ B & t > t_0 \end{cases} \qquad A \neq B \tag{2.9}$$

and define another function $h(t)$ as

$$h(t) = \begin{cases} A & t \leq t_0 \\ B & t > t_0 \end{cases}, \qquad A \neq B. \tag{2.10}$$

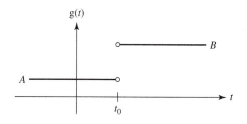

**Figure 2.7**
A discontinuous function.

It is obvious that $g(t)$ and $h(t)$ are unequal because their values are different at the point $t = t_0$. But the definite integrals of these two functions over any interval are the same; not approximately the same, *exactly the same*. That is,

$$\int_{\alpha}^{\beta} g(t)\, dt = \int_{\alpha}^{\beta} h(t)\, dt \qquad (2.11)$$

for *any* $\alpha$ and $\beta$, including $\alpha < t_0 < \beta$. This can be shown by writing the integral of $g(t)$ as

$$\int_{\alpha}^{\beta} g(t)\, dt = \int_{\alpha}^{t_0 - \varepsilon} g(t)\, dt + \int_{t_0 - \varepsilon}^{t_0 + \varepsilon} g(t)\, dt + \int_{t_0 + \varepsilon}^{\beta} g(t)\, dt. \qquad (2.12)$$

In the limit as $\varepsilon$ approaches zero, the integral $\int_{t_0 - \varepsilon}^{t_0 + \varepsilon} g(t)\, dt$ approaches zero because the function value is finite and the area under it in the interval $t_0 - \varepsilon < t_0 < t_0 + \varepsilon$ approaches zero in that limit. Similarly the integral $\int_{t_0 - \varepsilon}^{t_0 + \varepsilon} h(t)\, dt$ approaches zero, even though the function values of $g(t)$ and $h(t)$ are *different* at $t = t_0$. We will show later that one implication of this result is that if $g(t)$ and $h(t)$ describe signals applied as an excitation to any real physical system, the response of the system to $g(t)$ and $h(t)$ is *exactly* the same. Also, when we come to transforms later, it will turn out that a finite difference between two signals at one point (or any finite number of points) has no real consequence; the transforms are the same.

This discussion was for two specific functions. We can now generalize and say that any two functions which have finite values everywhere and differ in value only at a finite number of isolated points are equivalent in their effect on any real physical system. The responses of any real physical system to excitation by the two signals are identical.

Now we define the CT *unit step* function as

$$u(t) = \begin{cases} 1 & t > 0 \\ \frac{1}{2} & t = 0 \\ 0 & t < 0 \end{cases} \qquad (2.13)$$

(Figure 2.8). The left-hand graph in Figure 2.8 is drawn according to the strict mathematical definition. On the right is the more common way of drawing the function. The graph on the right is more common in engineering practice because no real physical phenomenon can change a finite amount in zero time. A graph versus time of any actual signal which is approximated by a unit step would look like the graph on the

**Figure 2.8**
The CT unit step function.

right. This function is called the unit step because the height of the step change in function value is one unit in the system of units used to describe the signal.

Some authors define the unit step by

$$u(t) = \begin{cases} 1 & t \geq 0 \\ 0 & t < 0 \end{cases} \quad \text{or} \quad u(t) = \begin{cases} 1 & t > 0 \\ 0 & t < 0 \end{cases} \quad \text{or}$$

$$u(t) = \begin{cases} 1 & t > 0 \\ 0 & t \leq 0 \end{cases}$$

For most analysis purposes these definitions are all equivalent. The unit steps defined by these definitions have an identical effect on any real physical system. The definition used in this text is, in a few special situations, more convenient than the other definitions and also corresponds more precisely to the signum function to be presented next.

The unit step is defined and used in signal and system analysis because it can mathematically represent a very common action in real physical systems, fast switching from one state to another. For example, in the circuit of Figure 2.9 the switch moves from one position to the other at time $t = 0$. The voltage applied to the $RC$ network can be described mathematically by

$$v_{RC}(t) = V_b u(t). \tag{2.14}$$

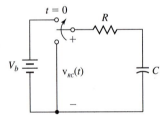

**Figure 2.9**
Circuit with a switch whose action can be represented mathematically by a unit step.

**The Signum Function**   The signum function (Figure 2.10) is closely related to the unit step function. For nonzero arguments, the value of the signum function has a magnitude of one and a sign which is the same as the sign of its argument. For this reason this function is sometimes called the *sign* function. But the name *signum* is more common because of the confusion which can result between the homonyms *sign* and *sine*.

$$\text{sgn}(t) = \left\{ \begin{array}{ll} 1 & t > 0 \\ 0 & t = 0 \\ -1 & t < 0 \end{array} \right\} = 2u(t) - 1 \tag{2.15}$$

**The Unit Ramp Function**   Another type of signal that occurs in systems is one which is switched on at some time and changes linearly after that time or one which changes linearly before some time and is switched off at that time. Figure 2.11 illustrates some examples. Signals of this kind can be described with the use of the *ramp* function. The CT unit ramp function (Figure 2.12) is the integral of the

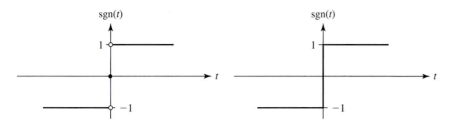

**Figure 2.10**
The CT signum function.

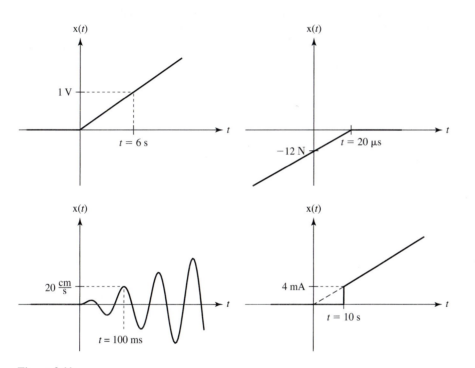

**Figure 2.11**
Functions that change linearly before or after some time or that are multiplied by functions
that change linearly before or after some time.

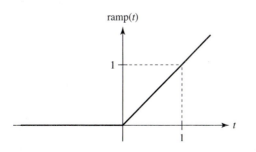

**Figure 2.12**
The CT unit ramp function.

unit step function. It is called the unit ramp function because, for positive $t$, its slope is one.

$$\text{ramp}(t) = \begin{cases} t & t > 0 \\ 0 & t \le 0 \end{cases} = \int_{-\infty}^{t} u(\lambda) \, d\lambda = t \, u(t) \qquad \textbf{(2.16)}$$

The ramp is defined by

$$\text{ramp}(t) = \int_{-\infty}^{t} u(\lambda) \, d\lambda. \qquad \textbf{(2.17)}$$

The symbol $\lambda$ is used in (2.17) as the independent variable of the unit step function and as the variable of integration. But $t$ is used as the independent variable of the ramp function. In words, (2.17) says, to find the value of the ramp function at any arbitrary value of $t$, start with negative infinity as the argument $\lambda$ of the unit step function and move, in $\lambda$, up to where $\lambda = t$, all the while accumulating the area under the unit step function. The total area accumulated from $\lambda = -\infty$ to $\lambda = t$ is the function value of the ramp function with an argument of $t$ (Figure 2.13). For values of $t$ less than zero, no area at all is accumulated. For values of $t$ greater than zero, the area accumulated equals $t$ because it is simply the area of a rectangle with width $t$ and height one.

**The Unit Impulse**    Before we define the unit impulse we should explore an important idea. Consider a unit-area rectangular pulse defined by the function

$$\delta_a(t) = \begin{cases} \dfrac{1}{a} & |t| < \dfrac{a}{2} \\[2mm] 0 & |t| > \dfrac{a}{2} \end{cases} \qquad \textbf{(2.18)}$$

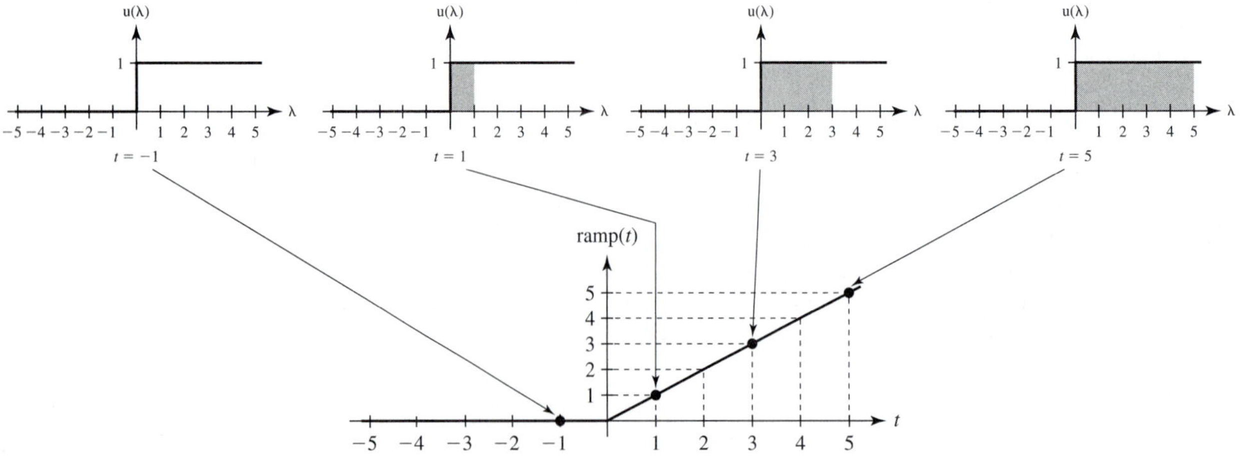

**Figure 2.13**
Illustration of the integral relationship between the CT unit step and the CT unit ramp.

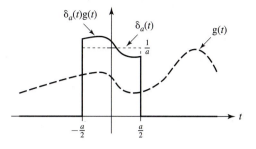

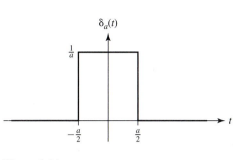

**Figure 2.14**
A unit-area rectangular pulse of width $a$.

**Figure 2.15**
Product of a unit-area rectangular pulse centered
at $t = 0$ and a function g($t$), which is continuous
and finite at $t = 0$.

(Figure 2.14). Let this function multiply another function g($t$), which is finite and continuous at $t = 0$, and find the area $A$ under the product of the two functions,

$$A = \int_{-\infty}^{\infty} \delta_a(t)g(t)\,dt \qquad (2.19)$$

(Figure 2.15). Using the definition of $\delta_a(t)$ we can rewrite the integral as

$$A = \frac{1}{a} \int_{-(a/2)}^{a/2} g(t)\,dt. \qquad (2.20)$$

Now imagine taking the limit of this integral as $a$ approaches zero. In that limit, the two limits of integration approach the same value, zero, from above and below. Since we are evaluating the value of a function over the range covered by the integration, as $a$ approaches zero the value of g($t$) approaches the same value at both limits and everywhere between them because it is continuous and finite at $t = 0$. So, in that limit, the value of g($t$) becomes g(0), a constant, and can be taken out of the integration process. Then

$$\lim_{a \to 0} A = g(0) \lim_{a \to 0} \frac{1}{a} \int_{-(a/2)}^{a/2} dt = g(0) \lim_{a \to 0} \frac{1}{a}(a) = g(0). \qquad (2.21)$$

So, in the limit as $a$ approaches zero, the function $\delta_a(t)$ has the interesting property of extracting the value of any continuous finite function g($t$) at time $t = 0$ when the product of $\delta_a(t)$ and g($t$) is integrated between any two limits which include time $t = 0$.

Now, to make the point that impulses are somewhat different from the functions we are familiar with, let's use a different definition of the function $\delta_a(t)$ and see what happens. Let it now be defined as

$$\delta_a(t) = \begin{cases} \dfrac{1}{a}\left(1 - \dfrac{|t|}{a}\right) & |t| < a \\ 0 & |t| > a \end{cases} \qquad (2.22)$$

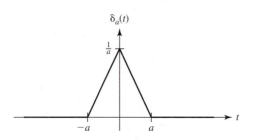

**Figure 2.16**
A unit area triangular pulse with a half width of $a$.

(Figure 2.16). If we make the same argument as before, we get the area

$$A = \int_{-\infty}^{\infty} \delta_a(t)g(t)\,dt = \frac{1}{a} \int_{-a}^{a} \left(1 - \frac{|t|}{a}\right) g(t)\,dt. \qquad (2.23)$$

Taking the limit as $a$ approaches zero,

$$\lim_{a \to 0} A = \lim_{a \to 0} \frac{1}{a} \int_{-a}^{a} \left(1 - \frac{|t|}{a}\right) g(t)\,dt = g(0) \lim_{a \to 0} \frac{2}{a} \int_{0}^{a} \left(1 - \frac{t}{a}\right) dt. \qquad (2.24)$$

Doing the integral and taking the limit we get

$$\lim_{a \to 0} A = g(0) \lim_{a \to 0} \frac{2}{a} \left[t - \frac{t^2}{2a}\right]_0^a = g(0) \lim_{a \to 0} \frac{2}{a} \frac{a}{2} = g(0). \qquad (2.25)$$

This is exactly the same result we got with the previous definition of $\delta_a(t)$. Therefore, the two different definitions of $\delta_a(t)$ have exactly the same effect in the limit as $a$ approaches zero. The important point here is that it is not the *shape* of the function that is important in the limit, but its *area*. In each case, $\delta_a(t)$ is a function with an area of one, independent of the value of $a$. (In the limit as $a$ approaches zero these functions do not have a shape because there is no time in which to develop one.) There are many other definitions of $\delta_a(t)$ that could be used with exactly the same effect in the limit.

The unit impulse $\delta(t)$ can now be defined by the property that when it is multiplied by any function $g(t)$, which is finite and continuous at $t = 0$, and the product is integrated between limits which include $t = 0$, the result is

$$\boxed{g(0) = \int_{-\infty}^{\infty} \delta(t)g(t)\,dt}. \qquad (2.26)$$

In other words,

$$\int_{-\infty}^{\infty} \delta(t)g(t)\,dt = \lim_{a \to 0} \int_{-\infty}^{\infty} \delta_a(t)g(t)\,dt \qquad (2.27)$$

where $\delta_a(t)$ is any of many functions which have the characteristics just described. The notation $\delta(t)$ is simply a convenient shorthand notation which allows us to avoid having to constantly take a limit when using impulses.

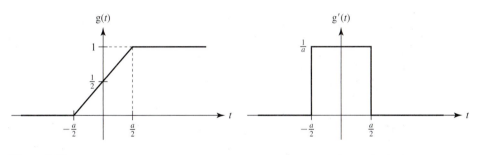

**Figure 2.17**
Functions which approach the unit step and unit impulse.

The unit step function is the derivative of the unit ramp function. One way of introducing the unit impulse is to define it as the derivative of the unit step function. Strictly speaking, the derivative of the unit step $u(t)$ is undefined at $t = 0$. But consider a function $g(t)$ of time and its time derivative $g'(t)$ in Figure 2.17. The derivative of $g(t)$ exists for all $t$ except at $t = -(a/2)$ and at $t = +a/2$. In the limit as $a$ approaches zero, the function $g(t)$ approaches the unit step function. In that same limit, the width of the function $g'(t)$ approaches zero while its area remains the same, one. So $g'(t)$ is a short-duration pulse whose area is always one, the same as the initial definition of $\delta_a(t)$ previously given with the same implications. The limit as $a$ approaches zero of $g'(t)$ is called the *generalized* derivative of $u(t)$. Therefore, the unit impulse is the generalized derivative of the unit step.

The generalized derivative of any function $g(t)$, with a discontinuity at time $t = t_0$, is defined as

$$\frac{d}{dt}(g(t)) = \frac{d}{dt}(g(t))_{t \neq t_0} + \lim_{\varepsilon \to 0}[g(t + \varepsilon) - g(t - \varepsilon)]\delta(t - t_0), \qquad \varepsilon > 0. \qquad \textbf{(2.28)}$$

Since the unit impulse is the generalized derivative of the unit step, it must follow that the unit step is the integral of the unit impulse,

$$u(t) = \int_{-\infty}^{t} \delta(\lambda) \, d\lambda, \qquad \textbf{(2.29)}$$

which is the same as the relationship between the unit ramp and the unit step. Since the derivative of the unit step $u(t)$ is zero everywhere except at $t = 0$, the unit impulse is zero everywhere except at $t = 0$. Since the unit step is the integral of the unit impulse, a definite integral of the unit impulse whose integration range includes $t = 0$ must have the value of one. These two facts are often used to define the unit impulse.

$$\delta(t) = 0 \qquad t \neq 0 \qquad \text{and} \qquad \int_{t_1}^{t_2} \delta(t) \, dt = \begin{cases} 1 & t_1 < 0 < t_2 \\ 0 & \text{otherwise} \end{cases} \qquad \textbf{(2.30)}$$

The area under an impulse is called its *strength,* or sometimes its *weight.* An impulse with a strength of one is called a unit impulse. The exact definition and characteristics of the impulse require a plunge into generalized function or distribution theory. For our purposes it will suffice to consider a unit impulse simply to be a pulse of unit area whose width is so small that making it any smaller would not significantly change any signals in the system to which it is applied.

The impulse cannot be graphed in the same way as other functions because its amplitude is undefined when its argument is zero. The usual convention for graphing an impulse is to use a vertical arrow. Sometimes the strength of the impulse is written beside it in parentheses, and sometimes the height of the arrow indicates the strength of the impulse. In Figure 2.18 are illustrated some ways of representing impulses graphically.

A common mathematical operation that occurs in signal and system analysis is the product of an impulse with another function, of the form

$$h(t) = g(t)A\delta(t - t_0) \tag{2.31}$$

where the impulse $A\delta(t - t_0)$ has a strength of $A$ and occurs at time $t = t_0$. Using the same argument we used in the introduction to the impulse, consider that the impulse $A\delta(t - t_0)$ is the limit of a pulse with area $A$, centered at time $t = t_0$, with width $a$, as $a$ approaches zero (Figure 2.19). The product is then a pulse whose height at the midpoint is $Ag(t_0)/a$ and whose width is $a$. In the limit, as $a$ approaches zero the pulse becomes an impulse and the strength of that impulse is $Ag(t_0)$. Therefore,

$$h(t) = g(t)A\delta(t - t_0) = Ag(t_0)\delta(t - t_0). \tag{2.32}$$

Equation (2.32) is sometimes called the *equivalence* property of the impulse.

Another important property of the unit impulse that follows naturally from the equivalence property is its so-called sampling property,

$$\int_{-\infty}^{\infty} g(t)\delta(t - t_0)\, dt = g(t_0). \tag{2.33}$$

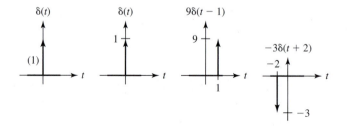

**Figure 2.18**
Graphical representations of impulses.

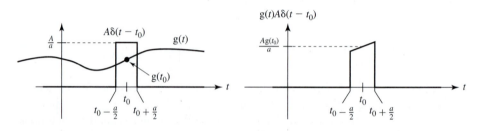

**Figure 2.19**
Product of a function $g(t)$ and a rectangle function which becomes an impulse as its width approaches zero.

This is easily seen by observing that, according to the equivalence property, the product $g(t)\delta(t - t_0)$ is equal to $g(t_0)\delta(t - t_0)$. Since $t_0$ is one particular value of $t$, it is a constant and $g(t_0)$ is also a constant and

$$\int_{-\infty}^{\infty} g(t)\delta(t - t_0)\, dt = g(t_0) \underbrace{\int_{-\infty}^{\infty} \delta(t - t_0)\, dt}_{=1} = g(t_0). \qquad (2.34)$$

Equation (2.34) is called the *sampling* property of the impulse because it samples the value of the function $g(t)$, at time $t = t_0$. [It is also sometimes called the *sifting* property because it sifts out the value of $g(t)$, at time, $t = t_0$.]

Another important property of the impulse function is its scaling property

$$\delta(a(t - t_0)) = \frac{1}{|a|}\delta(t - t_0). \qquad (2.35)$$

This can be proven through a change of variable in the integral definition and separate consideration of positive and negative values for $a$ (see Exercise 39).

**The Unit Comb**   Another useful function is the unit comb function (Figure 2.21). The *unit comb* function is a uniformly spaced sequence of unit impulses.

$$\boxed{\text{comb}(t) = \sum_{n=-\infty}^{\infty} \delta(t - n), \qquad n \text{ is an integer}} \qquad (2.36)$$

This is a unit comb function because the strength of each impulse, the spacing between impulses, and the average value of the function are all one.

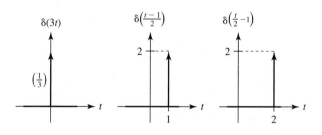

**Figure 2.20**
Examples of the effect of the scaling property of impulses.

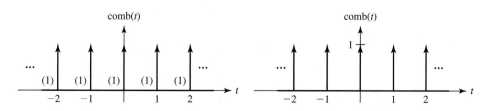

**Figure 2.21**
The unit comb function.

The impulse and comb functions may seem very abstract and unrealistic. The impulse will appear later as a result of a fundamental operation of linear system analysis, the convolution integral. Although, as a practical matter, a true impulse is impossible to generate, the mathematical impulse is very useful in signal and system analysis, as is the comb, being a periodic repetition of impulses. Using the comb and the convolution operation we can mathematically represent, in a succinct notation, many kinds of useful signals which would be more awkward to represent in another way.

**Singularity Functions**   We have seen in the previous few sections that the unit step, unit impulse, and unit ramp are related through integral and generalized derivative relationships. These functions are the most important members of the family of functions called *singularity functions.* In some system literature these functions are indicated by the coordinated notation $u_k(t)$, where the value of $k$ determines the function. For example,

$$u_0(t) = \delta(t), \qquad u_{-1}(t) = u(t), \qquad \text{and} \qquad u_{-2}(t) = \text{ramp}(t). \qquad \textbf{(2.37)}$$

In this coordinated notation, the subscript $k$ indicates how many times an impulse is differentiated to obtain the function in question and a negative value of $k$ indicates that integration is done instead of differentiation. The unit doublet $u_1(t)$ is defined as the generalized derivative of the unit impulse, the unit triplet $u_2(t)$ is defined as the generalized derivative of the unit doublet, etc. Even though the unit doublet and triplet and higher generalized derivatives are even less practical than the unit impulse, they are sometimes useful in signal and system theory.

Just as the impulse has a sampling property, the doublet does also. Consider the integral of the product of a doublet with a function $g(t)$ that is continuous and finite at the location of the doublet $t = t_0$,

$$I = \int_{-\infty}^{\infty} g(t)u_1(t - t_0)\, dt. \qquad \textbf{(2.38)}$$

We can use integration by parts to evaluate this integral. In

$$\int u\, dv = uv - \int v\, du \qquad \textbf{(2.39)}$$

let $u = g(t)$ and $dv = u_1(t - t_0)\, dt$. Then

$$I = g(t)u_0(t - t_0)\Big|_{-\infty}^{\infty} - \int_{-\infty}^{\infty} u_0(t - t_0)g'(t)\, dt. \qquad \textbf{(2.40)}$$

Since $u_0(t) = \delta(t)$, $g(t)u_0(t)|_{-\infty}^{\infty} = 0$ and, using the sampling property of the impulse,

$$I = -g'(t_0). \qquad \textbf{(2.41)}$$

Analogous sampling properties can be derived for higher-order singularities.

**The Unit Rectangle Function**   A very common type of signal occurring in systems is one in which a signal $x(t)$ is switched on at some time and then back off at a later time. It is very convenient to define another function especially for describing this type of signal. Use of this function shortens the notation when describing some complicated signals. The unit rectangle function (Figure 2.22) is defined to serve this purpose. It is a *unit* rectangle function because its width, height, and area are all one.

$$\operatorname{rect}(t) = \begin{cases} 1 & |t| < \frac{1}{2} \\ \frac{1}{2} & |t| = \frac{1}{2} \\ 0 & |t| > \frac{1}{2} \end{cases} \tag{2.42}$$

The unit rectangle function can be thought of as a *gate* function. When the unit rectangle function multiplies another function, the result is zero outside the nonzero range of the rectangle function and is equal to the other function inside the nonzero range of the rectangle function. The rectangle "opens a gate," allowing the other function through and then closes the gate again.

**The Unit Triangle Function**   The unit triangle function is defined in Figure 2.23. In Chapter 3 we will see that it has a close relationship with the unit rectangle function through the convolution operation to be introduced later. It is called a *unit* triangle because its height and area are both one (but its base width is not).

$$\operatorname{tri}(t) = \begin{cases} 1 - |t| & |t| < 1 \\ 0 & |t| \geq 1 \end{cases} \tag{2.43}$$

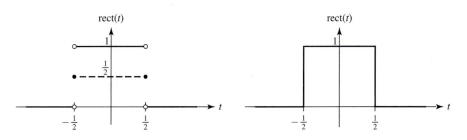

**Figure 2.22**
The CT unit rectangle function.

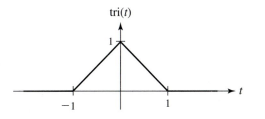

**Figure 2.23**
The CT unit triangle function.

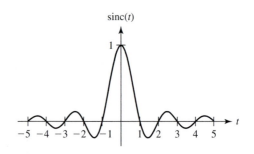

**Figure 2.24**
The CT unit sinc function.

**The Unit Sinc Function**   The unit sinc function (Figure 2.24) is also related to the unit rectangle function. It is the Fourier transform of the CT unit rectangle function. The Fourier transform will be introduced in Chapter 5. The unit sinc function is called a *unit* function because its height and area are both one. (In Chapter 5 we will see a way to find the area of this function.)

$$\text{sinc}(t) = \frac{\sin(\pi t)}{\pi t} \qquad\qquad (2.44)$$

Be aware that the definition of the sinc function is generally, but not universally, accepted as

$$\text{sinc}(t) = \frac{\sin(\pi t)}{\pi t}.$$

In some textbooks and reference books the sinc function is defined as

$$\text{sinc}(t) = \frac{\sin(t)}{t}.$$

In other references this second form is called the Sa function,

$$\text{Sa}(t) = \frac{\sin(t)}{t}.$$

How the sinc function is defined is not really critical. As long as one definition is accepted and used consistently, the results of the signal and system analysis will be useful.

One common question that arises when one first encounters the sinc function is how to determine the value of $\text{sinc}(0)$. When the independent variable $t$ in $\sin(\pi t)/\pi t$ has a value of zero, both the numerator $\sin(\pi t)$ and the denominator $\pi t$ evaluate to zero, leaving us with an indeterminate form. The solution to this problem is, of course, to use L'Hôpital's rule. Then

$$\lim_{t \to 0} \text{sinc}(t) = \lim_{t \to 0} \frac{\sin(\pi t)}{\pi t} = \lim_{t \to 0} \frac{\pi \cos(\pi t)}{\pi} = 1. \qquad\qquad (2.45)$$

So $\text{sinc}(t)$ is continuous at $t = 0$.

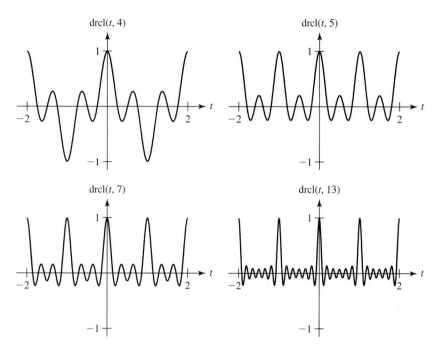

**Figure 2.25**
The Dirichlet function for $N = 4, 5, 7$, and 13.

**The Dirichlet Function**    A function that is related to the sinc function is the *Dirichlet* function (Figure 2.25) defined by

$$\boxed{\operatorname{drcl}(t, N) = \frac{\sin(\pi N t)}{N \sin(\pi t)}}. \qquad (2.46)$$

For $N$ odd, the similarity to a sinc function is obvious; the Dirichlet function is a sum of uniformly spaced sinc functions. The numerator $\sin(N\pi t)$ is zero when $t$ is any integer multiple of $1/N$. Therefore, the Dirichlet function is zero at those points, *unless* the denominator is also zero. The denominator $N \sin(\pi t)$ is zero for every integer value of $t$. Therefore, we must again use L'Hôpital's rule to evaluate the Dirichlet function at integer values of $t$.

$$\lim_{t \to m} \operatorname{drcl}(t, N) = \lim_{t \to m} \frac{\sin(N\pi t)}{N \sin(\pi t)} = \lim_{t \to m} \frac{N\pi \cos(N\pi t)}{N\pi \cos(\pi t)} = \pm 1 \qquad (2.47)$$

where $m$ is an integer. If $N$ is even, the extrema of the Dirichlet function alternate between $+1$ and $-1$. If $N$ is odd, the extrema are all $+1$. A version of the Dirichlet function is a part of the MATLAB signal toolbox with the function name `diric`. It is defined as

$$\operatorname{diric}(x, N) = \frac{\sin(Nx/2)}{N \sin(x/2)}. \qquad (2.48)$$

Therefore,

$$\text{drcl}(t, N) = \text{diric}(2\pi t, N). \tag{2.49}$$

The Dirichlet function appears in Chapters 4 and 5 in the material on the discrete-time Fourier series and discrete-time Fourier transform.

## MATLAB .m FILES FOR SOME SINGULARITY FUNCTIONS AND RELATED FUNCTIONS

Some of the functions introduced in this chapter exist as callable intrinsic functions in some computer languages and mathematical tools. For example, in MATLAB the sgn function is given the name `sign` and has exactly the same definition. In MATLAB the sinc function is defined exactly as done in this text.

We can create our own functions in MATLAB which, after definition, become functions we can call upon just like the built-in functions `cos`, `sin`, `exp`, etc. MATLAB functions are defined by creating a `.m` file, which is a file whose name has the extension `.m`. For example, we could create a file which finds the length of the hypotenuse of a right triangle given the lengths of the other two sides.

```
%       Function to compute the length of the hypotenuse of a
%       right triangle given the lengths of the other two sides
%
%       a - The length of one side
%       b - The length of the other side
%       c - The length of the hypotenuse

function c = hyp(a,b)
        c = sqrt(a^2 + b^2) ;
```

The first seven lines are comment lines which are not executed but serve to document how the function is used. The first executable line must begin with the keyword `function`. The rest of the first line is in the form,

$$\text{result} = \text{name}(\text{arg1}, \text{arg2}, \ldots),$$

where *result* is the name of the variable which contains the returned value which can be a scalar, a vector, or a matrix (or even a cell array or a structure array, the discussion of which are beyond the scope of this text); *name* is the name of the function; and *arg1, arg2,* ..., are the parameters or arguments passed to the function. The arguments can also be scalars, vectors, or matrices (or cell arrays or structure arrays). The name of the file containing the function definition must be *name*`.m`.

Here is a listing of some MATLAB functions to implement the functions just discussed.

```
%       Unit step function defined as 0 for input argument values
%       less than zero, ½ for input argument values equal to zero,
%       and one for input argument values greater than zero. Works
%       for vectors and scalars equally well.

function y = u(t)
        zero = (t == 0) ; pos = (t>0) ; y = zero/2 + pos ;
```

```
%       Function to compute the ramp function defined as zero for
%       values of the argument less than zero and the value of
%       the argument for arguments greater than or equal to zero.
%       Works for vectors and scalars equally well.

function y = ramp(t)
       y = t.*(t >= 0) ;


%       Rectangle function. Uses the definition of the rectangle
%       function in terms of the unit step function. Works on
%       vectors or scalars equally well.

function y = rect(t)
       y = u(t+0.5) - u(t-0.5) ;


%       Function to compute the triangle function. Uses the definition
%       of the triangle function in terms of the ramp function. Works
%       for vectors and scalars equally well.

function y = tri(t)
       y = ramp(t+1) - 2*ramp(t) + ramp(t-1) ;


%       Function to compute sinc(t) defined as sin(pi*t)/(pi*t).
%       Works for vectors or scalars equally well. This function
%       may be intrinsic in some versions of MATLAB.

function y = sinc(t)
       zero = (t==0) ;     % Indicate the locations of zeros in t.
       num = (~zero).*sin(pi*t) + zero ; den = (~zero).*(pi*t) + zero ;
       y = num./den ;


%       Function to compute values of the Dirichlet function.
%       Works for vectors or scalars equally well.
%
%       x = sin(N*pi*t)/(N*sin(pi*t))

function x = drcl(t,N)
       x = diric(2*pi*t,N) ;
```

We did not include the CT unit impulse function in our listing of function .m files. There is a good reason for this. It is not a function in the ordinary sense. A function accepts an argument and returns a related value. The impulse does that at all points *except* where its argument is zero. At that point its value is not defined. Therefore, it is not possible for MATLAB to return a value at that point like it does for ordinary functions. In Section 2.9 we will define a DT impulse which does have defined values at all allowable values of its argument and it *will* have a MATLAB function .m file description.

## 2.4 FUNCTIONS AND COMBINATIONS OF FUNCTIONS

Recall from basic mathematics that a function accepts a number from its domain and creates and returns another number from its range which is mathematically related. Standard functional notation for CT functions is in the form g($t$) in which g is the function name and everything inside the parentheses is called the *argument* of the function. The argument is an expression written in terms of the *independent variable*. In the case of g($t$), $t$ is the independent variable and the expression is the simplest possible expression in terms of $t$, itself. A function in the form g($t$) creates and returns a value for g for every value of $t$ it accepts. For example, in the function

$$g(t) = 2 + 4t^2 \tag{2.50}$$

for any particular value of $t$, there is a corresponding value of g. If $t$ is 1, then g is 6. That is indicated by the notation, g(1) = 6.

The argument of a function need not be simply the independent variable. It can be any mathematical expression written in terms of the independent variable, including another function of the independent variable. For example, if g($t$) = $5e^{-2t}$, what is g($t$ + 3)? We simply replace $t$ by $t$ + 3 everywhere on both sides of g($t$) = $5e^{-2t}$ to get g($t$ + 3) = $5e^{-2(t+3)}$. Observe carefully that we do not get $5e^{-2t+3}$. The reasoning is that since $t$ was multiplied by −2 in the exponent of $e$, then the entire expression $t$ + 3 must also be multiplied by −2 in the new exponent of $e$. In other words, whatever was done with $t$ in the function g($t$) must be done with the entire expression involving $t$ in any other function, g(*expression involving t*). The following equations are some examples of functions defined with $t$ as the argument and then written with expressions involving $t$.

If

$$g(t) = 3 + t^2 - 2t^3 \tag{2.51}$$

then

$$g(2t) = 3 + (2t)^2 - 2(2t)^3 = 3 + 4t^2 - 16t^3 \tag{2.52}$$

and

$$g(1 - t) = 3 + (1 - t)^2 - 2(1 - t)^3 = 2 + 4t - 5t^2 + 2t^3 \tag{2.53}$$

If

$$g(t) = 10 \cos(20\pi t) \tag{2.54}$$

then

$$g\left(\frac{t}{4}\right) = 10 \cos\left(20\pi\frac{t}{4}\right) = 10 \cos(5\pi t) \tag{2.55}$$

and

$$g(e^t) = 10 \cos(20\pi e^t) \tag{2.56}$$

Even though g may be defined with g($t$) on the left side of the equation and expressions involving $t$ on the right side, the function g does not always have to have its argument written in terms of an independent variable named $t$. For example, if g($t$) = $5e^{-10t}$, then g($2x$) = $5e^{-20x}$ and g($z$ − 1) = $5e^{10}e^{-10z}$. When g($t$) is defined by indicating what should be done with $t$ to create the value of g($t$), all that really means is any time you see g(*expression*), do to *expression* exactly what was done to $t$ in the definition.

When a function is invoked by passing an argument to it, the first thing MATLAB does is numerically evaluate the argument and then it computes the function value. For example, the MATLAB instruction,

$$g = \cos(2*pi*f0*t) \ ; ,$$

first finds numerical values for $t$ and $f0$, forms their product, and then multiplies the result by 2 and by $pi$ ($\pi$). Then the number that results is sent to the cosine function to return the cosine value. The cosine returns a value by interpreting the number sent to it as an angle in radians. If $t$ is a vector or matrix of times, instead of a scalar, each element of the vector or matrix $t$ is multiplied by $f0$, 2, and $pi$ ($\pi$) and then the resulting vector or matrix of numbers is sent to the cosine function for evaluation. The cosine returns a vector or matrix of cosines of the numbers sent to it in the vector or matrix, again interpreting the numbers as angles in radians. Therefore, MATLAB functions do exactly what is described here for arguments which are functions of the independent variable; they accept numbers and return other numbers.

## COMBINATIONS OF FUNCTIONS

In some cases a single mathematical function may completely describe a signal, a sinusoid, for example. But often one function is not enough for an accurate description. One operation which allows versatility in the mathematical representation of arbitrary signals is that of combining two or more functions. The combinations can be sums, differences, products, and/or quotients of functions. Figure 2.26 shows some examples of sums, products, and quotients of functions. (We do not include differences because they are so similar to sums.)

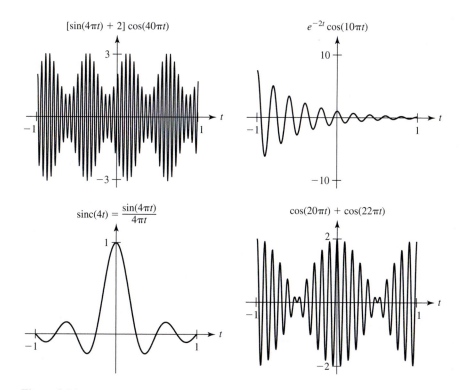

**Figure 2.26**
Examples of sums, products, and quotients of functions.

**EXAMPLE 2.1**

Using MATLAB, graph these function combinations,

$$x_1(t) = e^{-t}\sin(20\pi t) + e^{-(t/2)}\sin(19\pi t) \tag{2.57}$$

$$x_2(t) = \text{sinc}(t)\cos(20\pi t). \tag{2.58}$$

■ **Solution**

```
%  Program to plot some demonstrations of CT function combinations.

t = 0:1/120:6 ; x1 = exp(-t).*sin(20*pi*t) + exp(-t/2).*sin(19*pi*t) ;
subplot(2,1,1) ; p = plot(t,x1,'k') ; set(p,'LineWidth',2) ;
xlabel('\itt') ; ylabel('x_1({\itt})') ;
t = -4:1/60:4 ; x2 = sinc(t).*cos(20*pi*t) ;
subplot(2,1,2) ; p = plot(t,x2,'k') ; set(p,'LineWidth',2) ;
xlabel('\itt') ; ylabel('x_2({\itt})') ;
```

The plots which result are shown in Figure 2.27.

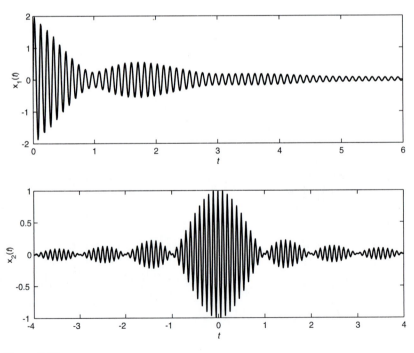

**Figure 2.27**
MATLAB graphical result.

## 2.5 CONTINUOUS-TIME SCALING AND SHIFTING TRANSFORMATIONS

It is important in signal and system analysis to be able to describe signals both analytically and graphically and to be able to relate the two different kinds of descriptions to each other. So let's now look at some graphical descriptions of functions and deduce how they will look when the function is transformed. Let g($t$) be defined by the graph in Figure 2.28. Since the graph only extends over the range $-5 < t < 5$, we don't know what the function does outside that range without auxiliary information. To keep it simple let's say that g($t$) = 0, $|t| > 5$.

### AMPLITUDE SCALING

Consider first the simplest functional transformation, multiplying the function by a constant. This transformation can be indicated by the notation

$$g(t) \rightarrow Ag(t). \tag{2.59}$$

For any arbitrary $t$, this transformation multiplies the returned value g($t$) by $A$. Thus the transformation g($t$) $\rightarrow$ $A$g($t$) multiplies g($t$) at every value of $t$ by $A$. This type of functional transformation is called *amplitude scaling*. Figure 2.29 shows two examples of amplitude-scaling of the function g($t$) that was defined in Figure 2.28.

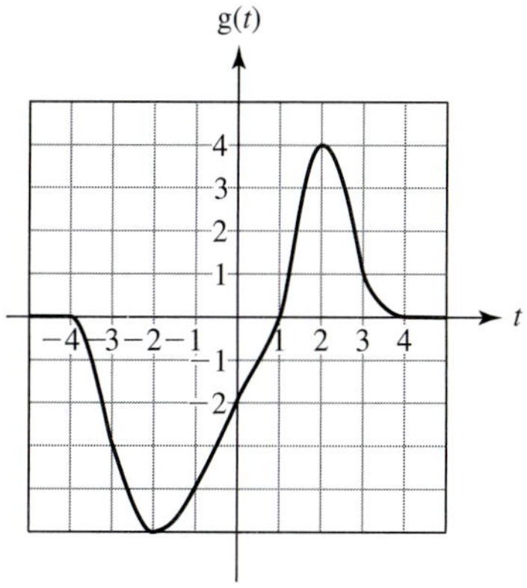

**Figure 2.28**
Graphical definition of a CT function g($t$).

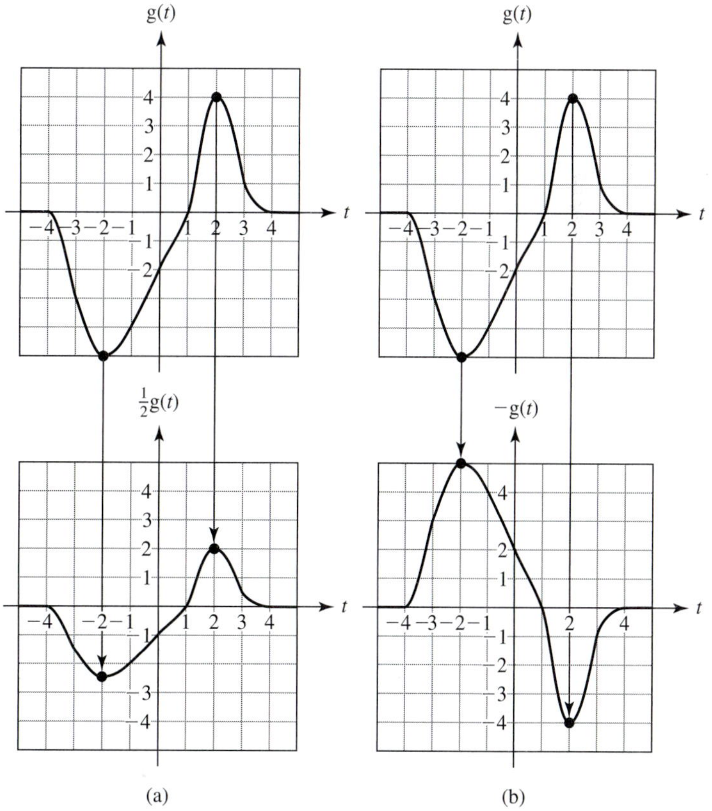

**Figure 2.29**
Two examples of amplitude scaling.

(a)

(b)

**Table 2.1** Selected values of $g(t-1)$

| $t$ | $t-1$ | $g(t-1)$ |
|---|---|---|
| $-5$ | $-6$ | $0$ |
| $-4$ | $-5$ | $0$ |
| $-3$ | $-4$ | $0$ |
| $-2$ | $-3$ | $-3$ |
| $-1$ | $-2$ | $-5$ |
| $0$ | $-1$ | $-4$ |
| $1$ | $0$ | $-2$ |
| $2$ | $1$ | $0$ |
| $3$ | $2$ | $4$ |
| $4$ | $3$ | $1$ |
| $5$ | $4$ | $0$ |

It is apparent from Figure 2.29(b) that a negative amplitude-scaling factor flips the function, with the $t$ axis as the rotation axis of the flip. If the scaling factor is $-1$ as in this example, flipping is the only action. If the scaling factor is some other factor $A$, and $A$ is negative, the amplitude-scaling transformation can be thought of as two successive transformations,

$$g(t) \rightarrow -g(t) \rightarrow |A|(-g(t)), \tag{2.60}$$

a flip followed by a positive amplitude scaling. Amplitude scaling is a transformation of the *dependent* variable g. In the next two sections we introduce transformations of the *independent* variable.

## TIME SHIFTING

If the graph in Figure 2.28 defines $g(t)$, what does $g(t-1)$ look like? We can begin to understand how to make this transformation by computing the values of $g(t-1)$ at a few selected points (Table 2.1).

It should now be apparent that replacing $t$ by $t-1$ has the effect of shifting the function one unit to the right (Figure 2.30). The transformation

$$t \rightarrow t-1 \tag{2.61}$$

can be described by saying, for every value of $t$, look back one unit in time, get the value of g at that time and use it as the value to plot for $g(t-1)$ at time $t$. This type of functional transformation is called *time shifting,* or *time translation.*

We can summarize time shifting by saying that the transformation $t \rightarrow t - t_0$, where $t_0$ is any arbitrary constant, has the effect of shifting $g(t)$ to the right by $t_0$ units. (Consistent with the accepted interpretation of negative numbers, if $t_0$ is negative, the shift is to the *left* by $|t_0|$ units.) Figure 2.31 shows some time-shifted step functions.

The rectangle function can be defined as the difference between two unit step functions time-shifted in opposite directions,

$$\text{rect}(t) = \left[ \text{u}\left(t + \frac{1}{2}\right) - \text{u}\left(t - \frac{1}{2}\right) \right], \tag{2.62}$$

and the triangle function can be defined as the sum of three ramp functions, two of which are shifted in time,

$$\text{tri}(t) = \text{ramp}(t + 1) - 2\,\text{ramp}(t) + \text{ramp}(t - 1). \tag{2.63}$$

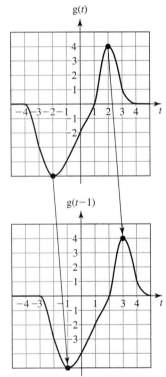

**Figure 2.30**
Graph of $g(t-1)$ in relation to $g(t)$ illustrating the time-shifting functional transformation.

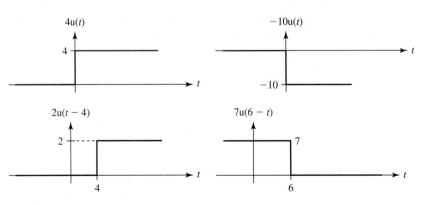

**Figure 2.31**
Transformed step functions.

Time shifting is a transformation of the independent variable. This type of transformation can be done on any independent variable; it need not be time. Our examples here use terminology involving time, but the independent variable could be a spatial dimension. In that case we could call this type of transformation *space shifting*. In Chapter 5 in the sections on transforms, we will have functions of an independent variable (frequency), and this transformation will be called *frequency shifting*. The mathematical significance is the same regardless of the name used to describe the process.

Amplitude scaling and time shifting occur in many real physical systems. For example, in ordinary conversation there is a propagation delay, the time required for a sound wave to propagate from one person's mouth to another person's ear. If that distance is 2 meters (m) and sound travels at about 330 m/s, the propagation delay is about 6 milliseconds (ms), a delay that is not noticeable. But consider an observer watching a pile driver, located 100 m away, drive a pile. The first thing the observer senses is the image of the driver striking the pile. There is a slight delay due to the speed of light from the pile driver to the eye, but it is less than a microsecond. The sound of the driver striking the pile arrives about 0.3 s later, a noticeable delay. This is an example of time shift, in this case, delay. Also the sound of the driver striking the pile is much louder near the driver than at a distance of 100 m, an example of amplitude scaling. Another familiar example is the delay between observing a lightning strike and hearing the thunder it produces.

As a more technological example, consider a satellite communication system. A ground station sends a strong electromagnetic signal to a satellite. When the signal reaches the satellite, the electromagnetic field is much weaker than when it left the ground station and it arrives later because of the propagation delay. If the satellite is geosynchronous, it is about 36,000 kilometers (km) above the surface of the earth, so if the ground station were directly below the satellite, the propagation delay on the uplink would be about 120 ms. For ground stations not directly below the satellite the delay is a little more. If the transmitted signal is $Ax(t)$, the received signal is $Bx(t - t_p)$ where $B$ is typically much smaller than $A$, and $t_p$ is the propagation time. In communications links between locations on earth that are very far apart, more than one uplink and downlink may be required to communicate. If that communication is voice communication between a television anchorperson in New York and a reporter in Calcutta, the delay can easily be 1s, which is noticeable enough to cause significant awkwardness in conversation. Imagine the problem of communicating with the first astronauts on Mars. The minimum one-way delay when Earth and Mars are in their closest proximity is more than 4 minutes (min).

In the case of long-range, two-way communication, time delay is a problem. In other cases it can be quite useful, as in radar and sonar. In this case the time delay between when a pulse is sent out and when a reflection returns indicates the distance to the object from which the pulse reflected, for example, an airplane or a submarine.

## TIME SCALING

Consider next the functional transformation, indicated by

$$t \rightarrow \frac{t}{a}. \tag{2.64}$$

As an example, let's compute selected values of $g(t/2)$ (Table 2.2). This transformation expands the function $g(t)$ horizontally (in $t$) by the factor $a$ in $g(t/a)$ (Figure 2.32). This functional transformation is called *time scaling*.

**Table 2.2** Selected values of $g(t/2)$

| $t$ | $\dfrac{t}{2}$ | $g\left(\dfrac{t}{2}\right)$ |
| --- | --- | --- |
| $-4$ | $-2$ | $-5$ |
| $-2$ | $-1$ | $-4$ |
| $0$ | $0$ | $-2$ |
| $2$ | $1$ | $0$ |
| $4$ | $2$ | $4$ |

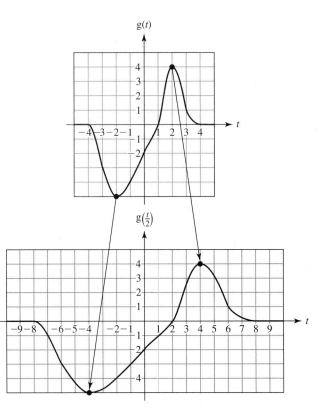

**Figure 2.32**
Graph of g($t/2$) in relation to g($t$) illustrating the time-scaling functional transformation.

**Table 2.3** Selected values of g($-t/2$)

| $t$ | $\dfrac{t}{2}$ | $g\left(\dfrac{t}{2}\right)$ |
|---|---|---|
| −4 | 2 | 4 |
| −2 | 1 | 0 |
| 0 | 0 | −2 |
| 2 | −1 | −4 |
| 4 | −2 | −5 |

Now consider the transformation

$$t \rightarrow -\frac{t}{2}. \tag{2.65}$$

This is identical to the last example except for the sign of the scaling factor which is now −2 instead of 2. The new relationship is shown in Table 2.3 and graphed in Figure 2.33.

We can summarize by saying that the time-scaling functional transformation $t \rightarrow t/a$ expands the function horizontally by a factor of $|a|$ and, if $a < 0$, the function is also time-inverted. *Time inversion* means flipping the curve with the g($t$) axis as the rotation axis of the flip. The case of a negative $a$ can be conceived as two successive transformations, $t \rightarrow -t$ followed by $t \rightarrow t/|a|$. The first step $t \rightarrow -t$ simply time-inverts the function without changing its horizontal scale. The second step $t \rightarrow t/|a|$ then time-scales the function that is already time-inverted by the positive scaling factor $|a|$.

Time scaling can also be indicated by the transformation $t \rightarrow bt$. This is not really new because it is the same as $t \rightarrow t/a$ with $b \rightarrow 1/a$. So all the rules for time scaling still apply with that relation between the two scaling constants $a$ and $b$.

We can imagine an experiment which would demonstrate the phenomenon of time scaling. Suppose we have an analog tape recording of some music. When we play the tape in the usual way, we hear the music as performed. But if we increase the speed of the tape movement across the sensing head, we hear a speeded-up version of the music.

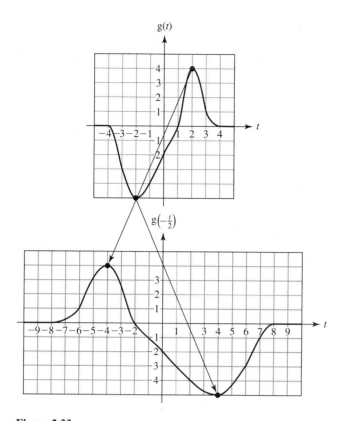

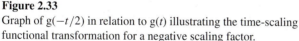

**Figure 2.33**
Graph of g(−t/2) in relation to g(t) illustrating the time-scaling
functional transformation for a negative scaling factor.

All the frequencies in the original recording are now higher and the time of the performance is reduced. If we slow down the tape, the opposite effect occurs. If we reverse the direction of tape travel, we hear the time inverse of the music, a very strange sound. If a human voice is recorded on tape in the usual way and then played back speeded up, it is often described as sounding like a chipmunk, high-pitched and very fast.

A common experience which illustrates the effect of time scaling is the *Doppler* effect. If we stand by the side of a road and a fire truck approaches while sounding its horn, we will notice that both the volume and pitch of the horn seem to change as the fire truck passes. The volume changes because of the proximity of the horn; the closer it is to us the louder it is. But why does the pitch change? The horn is doing exactly the same thing all the time, so it is not the pitch of the sound produced by the horn that changes but rather the pitch of the sound that arrives at our ears. As the fire truck approaches, each successive compression of air caused by the horn occurs a little closer to us than the last one, so it arrives at our ears in a shorter time than the previous compression and that makes the frequency of the sound wave at our ear higher than the frequency emitted by the horn. As the fire truck passes, the opposite effect occurs and the sound of the horn arriving at our ears shifts to a lower frequency. While we are hearing a pitch change, the firefighters on the truck hear a constant horn pitch.

Let the sound heard by the firefighters on the truck be described by g(t). As the fire truck approaches, the sound we hear is A(t)g(at) where A(t) is an increasing function of time which accounts for the volume change and a is a number slightly greater than one. The change in amplitude as a function of time is an effect called *amplitude modulation*

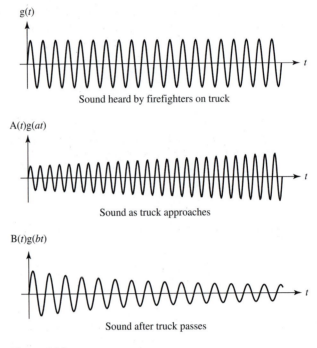

$g(t)$

Sound heard by firefighters on truck

$A(t)g(at)$

Sound as truck approaches

$B(t)g(bt)$

Sound after truck passes

**Figure 2.34**
Illustration of the Doppler effect.

in communication systems. After the fire truck passes, the sound we hear shifts to $B(t)g(bt)$ where $B(t)$ is a decreasing function of time and $b$ is slightly less than one (Figure 2.34). (In Figure 2.34 sinusoids are used to represent the horn sound. This is not precise, but it still serves to illustrate the important points.)

The same phenomenon occurs with light waves. The *red shift* of spectra from distant stars is what first indicated that the universe is expanding. It is called a red shift because when a star is receding from the earth the light we receive on earth experiences a Doppler shift which reduces the frequency of all the light waves emitted by the star. The light from a star has many characteristic variations with frequency because of the composition of the star and the path from the star to the observer. The amount of shift can be determined by comparing the spectral patterns of the light from the star with known spectral patterns from various elements measured on earth in a laboratory.

Time scaling is a transformation of the independent variable. As was true of time shifting, this type of transformation can be done on any independent variable; it need not be time. In Chapter 5 we will do some frequency scaling.

## MULTIPLE TRANSFORMATIONS

All three transformations, amplitude scaling, time scaling, and time shifting, can be applied simultaneously, for example,

$$g(t) \rightarrow Ag\left(\frac{t - t_0}{a}\right). \tag{2.66}$$

To understand the overall effect, it is usually best to break down a transformation like (2.66) into successive simple transformations,

$$g(t) \xrightarrow[\text{scaling } A]{\text{Amplitude}} Ag(t) \xrightarrow{t \rightarrow t/a} Ag\left(\frac{t}{a}\right) \xrightarrow{t \rightarrow t - t_0} Ag\left(\frac{t - t_0}{a}\right). \tag{2.67}$$

Observe here that the order of the transformations is important. For example, if we exchange the order of the time-scaling and time-shifting operations in (2.67), we get

$$g(t) \xrightarrow[\text{scaling } A]{\text{Amplitude}} Ag(t) \xrightarrow{t \to t - t_0} Ag(t - t_0) \xrightarrow{t \to t/a} Ag\left(\frac{t}{a} - t_0\right) \neq Ag\left(\frac{t - t_0}{a}\right). \qquad \textbf{(2.68)}$$

The result of this sequence of transformations is different from the preceding result (unless $a = 1$ or $t_0 = 0$). We could follow this sequence and get the preceding result in (2.67) by using a different time shift if we first observe that

$$Ag\left(\frac{t - t_0}{a}\right) = Ag\left(\frac{t}{a} - \frac{t_0}{a}\right). \qquad \textbf{(2.69)}$$

Then we could time-shift first and time-scale second, yielding

$$g(t) \xrightarrow[\text{scaling } A]{\text{Amplitude}} Ag(t) \xrightarrow{t \to t - t_0/a} Ag\left(t - \frac{t_0}{a}\right) \xrightarrow{t \to t/a} Ag\left(\frac{t}{a} - \frac{t_0}{a}\right) = Ag\left(\frac{t - t_0}{a}\right).$$
$$\textbf{(2.70)}$$

But, even though this works, it is simpler and more logical to use the first sequence, time scaling before time shifting. For a different transformation, a different sequence may be better, for example,

$$Ag(bt - t_0). \qquad \textbf{(2.71)}$$

In this case the sequence of amplitude scaling, time shifting, and then time scaling is the simplest path to a correct transformation.

$$g(t) \xrightarrow[\text{scaling } A]{\text{Amplitude}} Ag(t) \xrightarrow{t \to t - t_0} Ag(t - t_0) \xrightarrow{t \to bt} Ag(bt - t_0). \qquad \textbf{(2.72)}$$

Figures 2.35 and 2.36 illustrate some transformation steps for two functions. In these figures certain points are labeled with letters, beginning with $a$ and proceeding alphabetically. As each functional transformation is performed, corresponding points have the same letter designation.

The functions previously introduced, along with function transformations, allow us to describe a wide variety of signals. For example, a signal that has a decaying exponential shape after some time $t = t_0$ and is zero before that can be represented in the succinct mathematical form,

$$x(t) = Ae^{-\frac{t}{\tau}}u(t - t_0) \qquad \textbf{(2.73)}$$

(Figure 2.37). A signal which has the shape of a negative sine function before time $t = 0$ and a positive sine function after time $t = 0$ can be represented by $x(t) = A \sin(2\pi f_0 t) \operatorname{sgn}(t)$ (Figure 2.38). A signal which is a burst of a sinusoid between times $t = 1$ and $t = 5$ and is zero elsewhere can be represented by

$$x(t) = A \cos(2\pi f_0 t + \theta)\operatorname{rect}\left(\frac{t - 3}{4}\right) \qquad \textbf{(2.74)}$$

(Figure 2.39). Figure 2.40 illustrates time scaling and shifting a comb function.

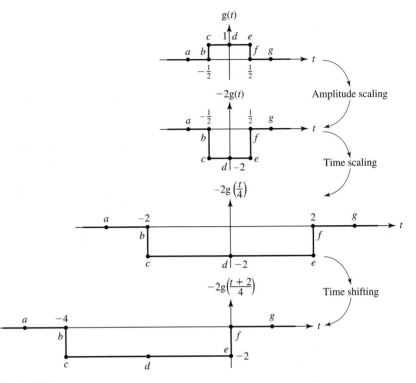

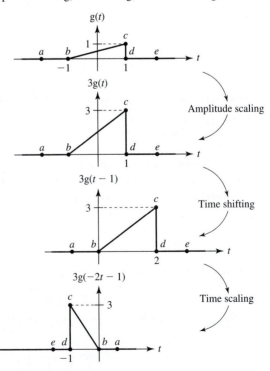

**Figure 2.35**
A sequence of amplitude scaling, time scaling, and time shifting a function.

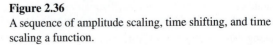

**Figure 2.36**
A sequence of amplitude scaling, time shifting, and time
scaling a function.

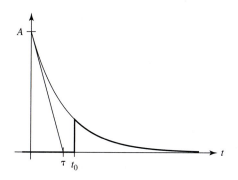

**Figure 2.37**
A decaying exponential "switched" on at time $t = t_0$.

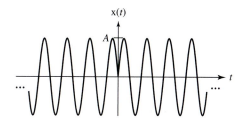

**Figure 2.38**
Product of a sine and a signum function.

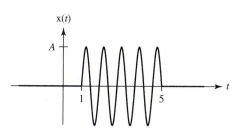

**Figure 2.39**
A sinusoidal "burst."

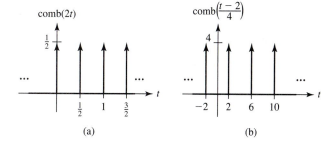

**Figure 2.40**
(a) Time-scaled and (b) time-shifted comb functions.

**EXAMPLE 2.2**

Using MATLAB, plot the function defined by

$$g(t) = \begin{cases} 0 & t < -2 \\ -4 - 2t & -2 < t < 0 \\ -4 + 3t & 0 < t < 4 \\ 16 - 2t & 4 < t < 8 \\ 0 & t > 8 \end{cases}. \tag{2.75}$$

Then plot the transformed functions

$$3g(t + 1), \qquad \frac{1}{2}g(3t), \qquad -2g\left(\frac{t-1}{2}\right).$$

■ **Solution**

We must first choose a range of $t$ over which to plot the function, and a spacing between points in $t$ to yield a curve that closely approximates the actual function. Let's choose a range of $-5 < t < 20$ and a spacing between points of 0.1. Also let's use the function feature of

MATLAB which allows us to define the function $g(t)$ as a separate MATLAB program, a .m file. Then we can simply refer to it when plotting the transformed functions and not have to retype the function description every time. The `g.m` file contains the code,

```
function y = g(t)

    % Calculate the functional variation for each range of t.
    y1 = -4 - 2*t; y2 = -4 + 3*t; y3 = 16 - 2*t;

    % Splice together the different functional variations in
    % their respective ranges of validity.
    y = y1.*(-2<t & t<=0) + y2.*(0<t & t<=4) + y3.*(4<t & t<=8);
```

The MATLAB program contains the following code,

```
%  Program to plot the function, g(t) and then to
%  plot 3*g(t+1), g(3*t)/2, and -2*g((t-1)/2).

tmin = -4 ; tmax = 20;      %  Set the time range for the plot.
dt = 0.1;                   %  Set the time between points.
t = tmin:dt:tmax;           %  Set the vector of times for the plot.
g0 = g(t);                  %  Compute the original g(t).
g1 = 3*g(t+1);             %  Compute the first transformation.
g2 = g(3*t)/2;             %  Compute the second transformation.
g3 = -2*g((t-1)/2);        %  Compute the third transformation.
```

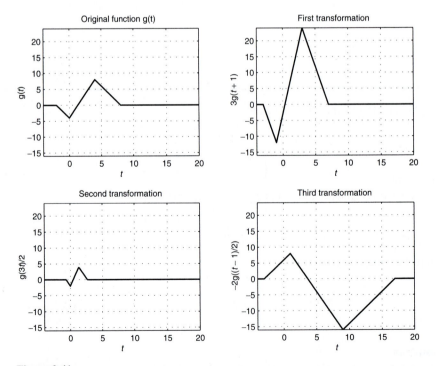

**Figure 2.41**
MATLAB graphs of transformed functions.

```
%   Find the maximum and minimum g values in all the transformed
%   functions and use them to scale all plots the same.

gmax = max([max(g0), max(g1), max(g2), max(g3)]);
gmin = min([min(g0), min(g1), min(g2), min(g3)]);

%   Plot all four functions in a 2 by 2 arrangement.
%   Plot them all on the same scale using the axis command.
%   Plot grid lines, using the grid command, to aid in reading values.

subplot(2,2,1); p = plot(t,g0,'k'); set(p,'LineWidth',2);
xlabel('t'); ylabel('g(t)'); title('Original Function, g(t)');
axis([tmin,tmax,gmin,gmax]); grid;
subplot(2,2,2); p = plot(t,g1,'k'); set(p,'LineWidth',2);
xlabel('t'); ylabel('3g(t+1)'); title ('First Transformation');
axis([tmin,tmax,gmin,gmax]); grid;
subplot(2,2,3); p = plot(t,g2,'k'); set(p,'LineWidth',2);
xlabel('t'); ylabel('g(3t)/2'); title ('Second Transformation');
axis([tmin,tmax,gmin,gmax]); grid;
subplot(2,2,4); p = plot(t,g3,'k'); set(p,'LineWidth',2);
xlabel('t'); ylabel('-2g((t-1)/2)');title('Third Transformation');
axis([tmin,tmax,gmin,gmax]); grid;
```

The graphical results are displayed in Figure 2.41. ∎

In Figures 2.42 and 2.43 are more examples of amplitude-scaled, time-shifted, and time-scaled versions of the functions just introduced.

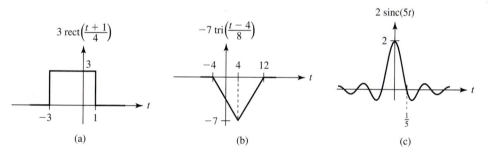

**Figure 2.42**
Examples of (a) amplitude-scaled, (b) time-shifted, and (c) time-scaled functions.

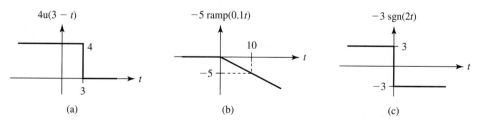

**Figure 2.43**
More examples of (a) amplitude-scaled, (b) time-shifted, and (c) time-scaled functions introduced in this chapter.

## 2.6 DIFFERENTIATION AND INTEGRATION

Another very common type of transformation of a function used to represent a signal is to differentiate or integrate it because integration and differentiation are common signal-processing operations in real systems. The *derivative* of a function at any time $t$ is its slope *at* that time, and the integral of a function at any time $t$ is the accumulated area under the function *up to* that time. Figure 2.44 illustrates some CT functions and their derivatives. Note that the zero crossings of all the derivatives have been indicated by light vertical lines which lead exactly to the maxima and minima of the corresponding function, points at which the slope of the function is zero.

Integration is a little more problematic than differentiation. Any function's derivative is unambiguously determinable (if it exists). However, its integral is not unambiguously determinable without some more information. This is inherent in one of the first principles learned in integral calculus. If a function $g(x)$ has a derivative $g'(x)$, then the function, $g(x) + K$ ($K$ is a constant) has exactly the same derivative $g'(x)$, regardless of the value of the constant $K$. Turning the logic around, since integration is the opposite of differentiation, what shall we say is the integral of $g'(x)$? It could be $g(x)$, but it could also be $g(x) + K$, with $K$ having any arbitrary value.

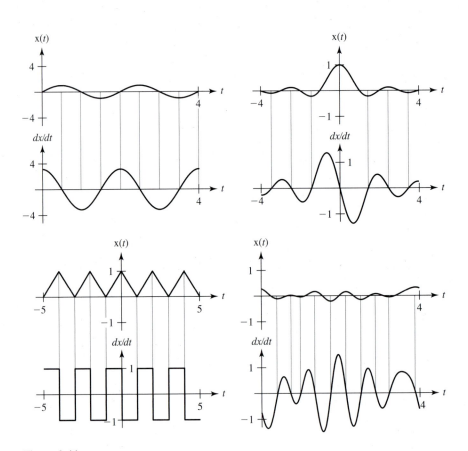

**Figure 2.44**
Some CT functions and their derivatives.

Often, in practice, we know that a function of time is zero, before some initial time $t = t_0$. Then we know that the integral from negative infinity to time $t = t_0$ is zero,

$$\int_{-\infty}^{t_0} g(t) \, dt = 0.$$  **(2.76)**

Then the integral of that function from any time $t_1 < t_0$ to any time $t > t_0$ is unambiguous. It can only be the area under the function from time $t = t_0$ to time $t$,

$$\int_{t_1}^{t} g(\lambda) \, d\lambda = \underbrace{\int_{t_1}^{t_0} g(\lambda) \, d\lambda}_{=0} + \int_{t_0}^{t} g(\lambda) \, d\lambda = \int_{t_0}^{t} g(\lambda) \, d\lambda.$$  **(2.77)**

Figure 2.45 illustrates some functions and their integrals. Two of the functions are zero before time $t = 0$, and the integrals illustrated assume a lower limit on the integral less than zero, thereby producing a single, unambiguous result. The other two are illustrated with multiple possible integrals, differing from each other only by constants.

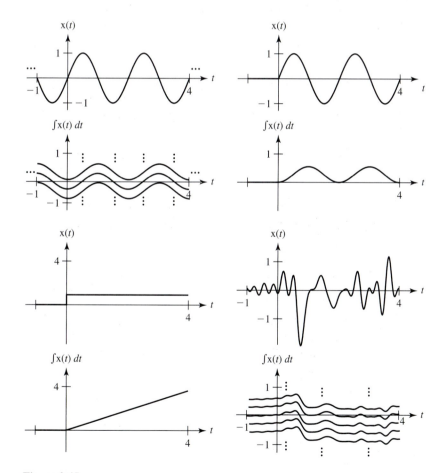

**Figure 2.45**
Some functions and their integrals.

They all have the same derivative and are all equally valid candidates for the integral in the absence of auxiliary information.

MATLAB can do some differentiation and integration with the `diff` and `int` commands. For example,

```
»diff('sin(2*pi*t)')
ans =
2*cos(2*pi*t)*pi
»int('sin(2*pi*t)')
ans =
-1/2*cos(2*pi*t)/pi
```

The `int` command does what is called an *antiderivative*. That simply means the integral with an assumed constant of integration of zero. See the MATLAB help files for more details.

## 2.7  CONTINUOUS-TIME EVEN AND ODD FUNCTIONS

Some functions have the property that when they undergo certain types of transformations they do not actually change. They are said to be *invariant under* that transformation. An *even* function is one which is invariant under the transformation $t \to -t$, and an *odd* function is one which is invariant under the transformation $g(t) \to -g(-t)$. That is, an even function $g(t)$ is one for which $g(t) = g(-t)$, and an odd function is one for which $g(t) = -g(-t)$.

A simple way of visualizing even and odd functions is to imagine that the ordinate axis [the $g(t)$ axis] is a mirror. For even functions, the part of $g(t)$ for $t > 0$ and the part of $g(t)$ for $t < 0$ are mirror images of each other. For an odd function, the same two parts of the function are *negative* mirror images of each other. Figures 2.46 and 2.47 show some examples of even and odd CT functions.

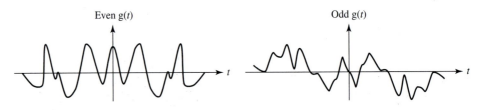

**Figure 2.46**
Examples of even and odd CT functions.

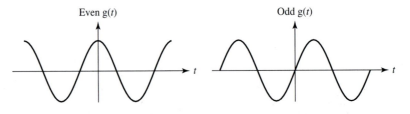

**Figure 2.47**
Very common and useful even and odd CT functions.

Some functions are even, some are odd, and some are neither even nor odd. But any function g($t$), even if it is neither even nor odd, can be expressed as a sum of its even and odd parts as $g(t) = g_e(t) + g_o(t)$. In other words, every function is composed of an even part plus an odd part. The even and odd parts of a function g($t$) are

$$g_e(t) = \frac{g(t) + g(-t)}{2} \qquad g_o(t) = \frac{g(t) - g(-t)}{2}. \qquad \textbf{(2.78)}$$

Suppose, for example, that g($t$) is an even function. Then

$$g_e(t) = \frac{g(t) + g(-t)}{2} = g(t) \qquad \text{and} \qquad g_o(t) = \frac{g(t) - g(-t)}{2} = 0, \qquad \textbf{(2.79)}$$

which says that an even function has an odd part of zero. If the function g($t$) is odd,

$$g_e(t) = \frac{g(t) + g(-t)}{2} = 0 \qquad \text{and} \qquad g_o(t) = \frac{g(t) - g(-t)}{2} = g(t). \qquad \textbf{(2.80)}$$

If the odd part of a function is zero, the function is even, and if the even part of a function is zero, the function is odd.

**EXAMPLE 2.3**

What are the even and odd parts of the function $g(t) = 4\cos(3\pi t)$?

■ **Solution**
They are

$$g_e(t) = \frac{g(t) + g(-t)}{2} = \frac{4\cos(3\pi t) + 4\cos(-3\pi t)}{2} = \frac{8\cos(3\pi t)}{2} = 4\cos(3\pi t)$$

$$g_o(t) = \frac{4\cos(3\pi t) - 4\cos(-3\pi t)}{2} = 0$$

because the cosine is an even function.                                     ■

## SUMS, PRODUCTS, DIFFERENCES, AND QUOTIENTS

Consider two functions $g_1(t)$ and $g_2(t)$. Let both be even functions. Then

$$g_1(t) = g_1(-t) \qquad \text{and} \qquad g_2(t) = g_2(-t). \qquad \textbf{(2.81)}$$

Now let

$$g(t) = g_1(t) + g_2(t). \qquad \textbf{(2.82)}$$

Then

$$g(-t) = g_1(-t) + g_2(-t) \qquad \textbf{(2.83)}$$

and, using the evenness of $g_1(t)$ and $g_2(t)$,

$$g(-t) = g_1(t) + g_2(t) = g(t), \qquad \textbf{(2.84)}$$

proving that the sum of two even functions is also even. Now let

$$g(t) = g_1(t)g_2(t). \qquad \textbf{(2.85)}$$

Then

$$g(-t) = g_1(-t)g_2(-t) = g_1(t)g_2(t) = g(t), \qquad (2.86)$$

proving that the product of two even functions is also even.

Now let $g_1(t)$ and $g_2(t)$ both be odd. Then

$$g(-t) = g_1(-t) + g_2(-t) = -g_1(t) - g_2(t) = -g(t), \qquad (2.87)$$

proving that the sum of two odd functions is odd. Then let

$$g(-t) = g_1(-t)g_2(-t) = [-g_1(t)][-g_2(t)] = g_1(t)g_2(t) = g(t), \qquad (2.88)$$

proving that the product of two odd functions is *even*.

By similar reasoning we can show that if two functions are even, their difference and quotient are also even. If two functions are odd, their difference is odd but their quotient is even. If one function is even and the other is odd, their product and quotient are odd.

The most important even and odd functions in signal analysis are cosines and sines. Cosines are even, and sines are odd. Figures 2.48 through 2.51 show some examples of products of even and odd CT functions.

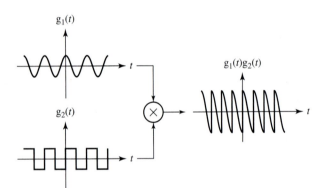

**Figure 2.48**
Product of even and odd CT functions.

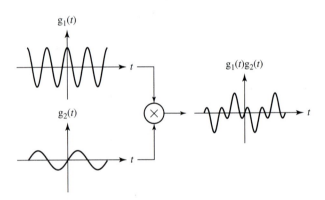

**Figure 2.49**
Product of even and odd CT functions.

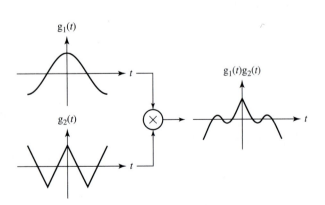

**Figure 2.50**
Product of two even CT functions.

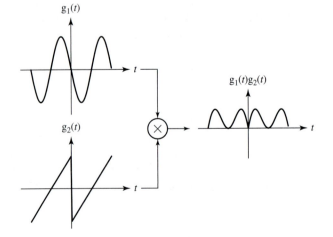

**Figure 2.51**
Product of two odd CT functions.

Let g($t$) be an even function. Then

$$g(t) = g(-t). \tag{2.89}$$

Using the chain rule of differentiation, the derivative of g($t$) is

$$g'(t) = -g'(-t), \tag{2.90}$$

an odd function. So the derivative of any even function is an odd function. Similarly, the derivative of any odd function is an even function. We can turn the arguments around to say that the integral of any even function is an odd function *plus* a constant of integration, and the integral of any odd function is an even function, plus a constant of integration. That is, except for a possible additive constant, the integrals of even and odd functions are, respectively, odd and even.

The integrals of even and odd CT functions can be simplified in certain common cases. If g($t$) is an even function and $a$ is a real constant,

$$\int_{-a}^{a} g(t)\, dt = \int_{-a}^{0} g(t)\, dt + \int_{0}^{a} g(t)\, dt = -\int_{0}^{-a} g(t)\, dt + \int_{0}^{a} g(t)\, dt. \tag{2.91}$$

Making the change of variable $\lambda = -t$ in the first integral term on the right side of (2.91), and then using the fact that g($\lambda$) = g($-\lambda$) for an even function, it is easy to show that

$$\int_{-a}^{a} g(t)\, dt = 2\int_{0}^{a} g(t)\, dt, \tag{2.92}$$

which should be geometrically obvious by looking at the graph of the function [Figure 2.52(a)]. By similar reasoning, if g($t$) is an odd function,

$$\int_{-a}^{a} g(t)\, dt = 0, \tag{2.93}$$

which should also be geometrically obvious [Figure 2.52 (b)].

MATLAB has several built-in functions, in addition to trigonometric and exponential functions, which can be used to generate waveforms of various types (Figure 2.53). The examples in Figure 2.53 were generated using the following MATLAB script file.

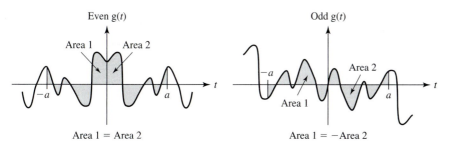

**Figure 2.52**
Integrals of an even function and an odd function.

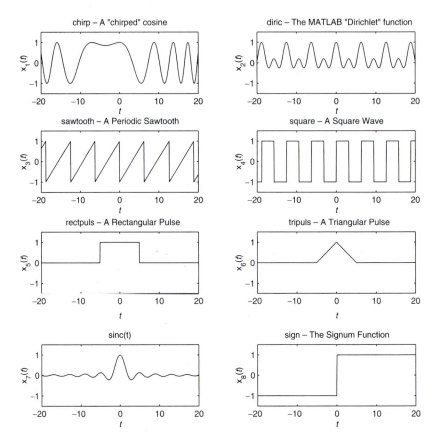

**Figure 2.53**
Examples of waveforms that can be generated using built-in MATLAB functions.

```
%       Program to illustrate some of MATLAB's built-in functions
close all ;
t = -20:1/20:20 ;
x1 = chirp(t,1/20,20,1/3) ; subplot(4,2,1) ; p = plot(t,x1,'k') ;
axis([-20,20,-1.5,1.5]) ; title('chirp - A "chirped" cosine') ;
xlabel ('\itt') ; ylabel('x_1({\itt})') ;
x2 = diric(t,5) ; subplot(4,2,2) ; p = plot(t,x2,'k') ;
axis([-20,20,-1.5,1.5]) ;
title('diric - The MATLAB "Dirichlet" function') ;
xlabel('\itt') ; ylabel('x_2({\itt})') ;
x3 = sawtooth(t) ; subplot(4,2,3) ; p = plot(t,x3,'k') ;
axis([-20,20,-1.5,1.5]) ; title('sawtooth - A Periodic Sawtooth') ;
xlabel('\itt') ; ylabel('x_3({\itt})') ;
x4 = square(t) ; subplot(4,2,4) ; p = plot(t,x4,'k') ;
axis([-20,20,-1.5,1.5]) ; title('square - A Square Wave') ;
xlabel('\itt') ; ylabel('x_4({\itt})') ;
```

```
x5 = rectpuls(t/10) ; subplot(4,2,5) ; p = plot(t,x5,'k') ;
axis([-20,20,-1.5,1.5]) ; title('rectpuls - A Rectangular Pulse Wave') ;
xlabel('\itt') ; ylabel('x_5({\itt})') ;

x6 = tripuls(t/10) ; subplot(4,2,6) ; p = plot(t,x6,'k') ;
axis([-20,20,-1.5,1.5]) ; title('tripuls - A Triangular Pulse Wave') ;
xlabel('\itt') ; ylabel('x_6({\itt})') ;

x7 = sinc(t/2) ; subplot(4,2,7) ; p = plot(t,x7,'k') ;
axis([-20,20,-1.5,1.5]) ; title('sinc(t)') ;
xlabel('\itt') ; ylabel('x_7({\itt})') ;

x8 = sign(t/2) ; subplot(4,2,8) ; p = plot(t,x8,'k') ;
axis([-20,20,-1.5,1.5]) ; title('sign - The Signum Function') ;
xlabel('\itt') ; ylabel('x_8({\itt})') ;
```

We can form products of these functions showing that products of even functions are even, products of odd functions are even, and mixed products of even and odd functions are odd (Figure 2.54).

```
x24 = x2.*x4 ;
subplot(2,2,1) ; plot(t,x24,'k') ;
axis([-20,20,-1.5,1.5]) ; title('x_2*x_4 - Even*Odd') ;
xlabel('\itt') ; ylabel('x_2_4({\itt})') ;

x34 = x3.*x4 ;
subplot(2,2,2) ; plot(t,x34,'k') ;
axis([-20,20,-1.5,1.5]) ; title('x_3*x_4 - Odd*Odd') ;
xlabel('\itt') ; ylabel('x_3_4({\itt})') ;

x26 = x2.*x6 ;
subplot(2,2,3) ; plot(t,x26,'k') ;
axis([-20,20,-1.5,1.5]) ; title('x_2*x_6 - Even*Even') ;
xlabel('\itt') ; ylabel('x_2_6({\itt})') ;

x37 = x3.*x7 ;
subplot(2,2,4) ; plot(t,x37,'k') ;
axis([-20,20,-1.5,1.5]) ; title('x_3*x_7 - Odd*Even') ;
xlabel('\itt') ; ylabel('x3_7({\itt})') ;
```

Signal $x_1(t)$ in Figure 2.53 is neither even nor odd. But we find its even and odd parts using MATLAB (Figure 2.55).

```
x1e = (x1 + x1(end:-1:1))/2 ; x1o = (x1 - x1(end:-1:1))/2 ;
subplot(2,1,1) ; plot(t,x1e,'k') ;
axis([-20,20,-1.5,1.5]) ; title('Even Part of x_1') ;
xlabel('\itt') ; ylabel('x_1_e({\itt})') ;
subplot(2,1,2) ; plot(t,x1o,'k') ;
axis([-20,20,-1.5,1.5]) ; title('Odd Part of x_1') ;
xlabel('\itt') ; ylabel('x_1_o({\itt})') ;
```

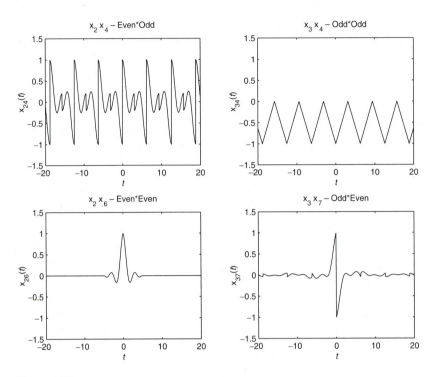

**Figure 2.54**
Products of even and odd functions.

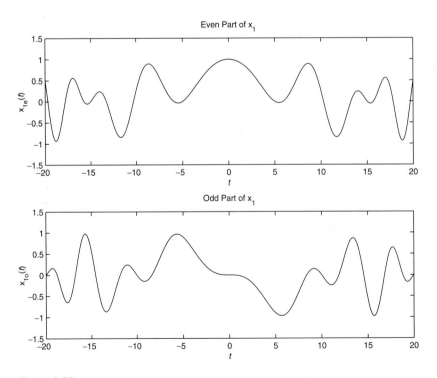

**Figure 2.55**
The even and odd parts of $x_1(t)$ of Figure 2.50.

MATLAB can also find the numerical value of an integral of an arbitrary function over an arbitrary time using the numerical integration function `quad`. For example, the MATLAB code,

```
ns = (1:20)' ; areas = [ ] ;
for n = 1:20 ;
      area = quad('sinc',-n,n) ; areas = [areas ; area] ;
end
disp([ns,areas]) ;
```

calculates the area under the `sinc` function between two symmetrical limits, $-n$ and $n$, for $n$ going from 1 to 20, and produces the result,

```
 1.0000    1.1790
 2.0000    0.9028
 3.0000    1.0662
 4.0000    0.9499
 5.0000    1.0402
 6.0000    0.9664
 7.0000    1.0288
 8.0000    0.9499
 9.0000    1.0225
10.0000    0.9798
11.0000    1.0184
12.0000    0.9831
13.0000    1.0156
14.0000    0.9855
15.0000    1.0135
16.0000    0.9499
17.0000    1.0119
18.0000    0.9887
19.0000    1.0107
20.0000    0.9899
```

where the left column is the value of $n$ and the right column is the corresponding area.

## 2.8 CONTINUOUS-TIME PERIODIC FUNCTIONS

A periodic function is one which has been repeating an exact pattern for an infinite time and will continue to repeat that exact pattern for an infinite time. That is, a periodic function $g(t)$ is one for which

$$g(t) = g(t + nT), \tag{2.94}$$

for any integer value of $n$, where $T$ is a *period* of the function. Another way of saying that a function is periodic is to say that it is invariant under the transformation

$$t \to t + nT. \tag{2.95}$$

The function repeats every $T$ s. Of course, it also repeats every $2T$, $3T$, and $nT$ s ($n$ an integer). Therefore, $2T$, $3T$, and $nT$ are all periods of the function because the function repeats over any of those intervals. The *minimum positive* interval over which a function repeats is called the *fundamental period* $T_0$. The *fundamental frequency* $f_0$ of

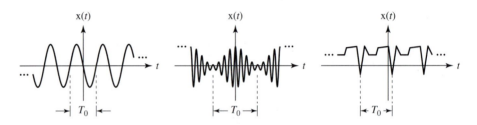

**Figure 2.56**
Examples of periodic CT functions with fundamental period $T_0$.

a periodic function is the reciprocal of the fundamental period $f_0 = 1/T_0$. This is the fundamental *cyclic* frequency, which is the number of cycles (periods) per second. The fundamental *radian* frequency is $\omega_0 = 2\pi f_0 = 2\pi/T_0$, which is the number of radians per second. Both frequency types are used in signal and system analysis.

Some common examples of CT periodic functions are real or complex sinusoids. We will see in Chapter 3 that other, more complicated, types of periodic functions with different periodically repeating shapes can be generated and mathematically described. Figure 2.56 shows some examples of periodic CT functions.

A function which is not periodic is called an *aperiodic* function. In real systems, a signal is never actually periodic because, presumably, it was turned on at some finite time in the past and will be turned off at some finite time in the future. However, it is often the case that a signal has been repeating for a very long time before the time we want to analyze the signal and will repeat for a very long time after that time. In many such cases approximating the signal by a periodic function introduces negligible error. Examples of signals that would be properly approximated by periodic functions would be rectified sinusoids in an AC to DC converter, horizontal sync signals in a television or a computer monitor, the angular shaft position of a generator in a power plant, a carrier in a radio transmitter before modulation, the firing pattern of spark plugs in an automobile traveling at constant speed, the vibration of a quartz crystal in a wristwatch, and the angular position of a pendulum on a grandfather clock. Many natural phenomena are, for all practical purposes, periodic: most planet, satellite, and comet orbital positions; the phases of the moon; the electric field emitted by a cesium atom at resonance; the migration patterns of birds; sunspot activity; and the caribou mating season. Therefore, periodic phenomena play a large part both in the natural world and in the realm of artificial systems.

## EXAMPLE 2.4

Determine which of the following functions or signals are periodic and, for those that are, determine the fundamental period.

a.  $g(t) = 7\sin(400\pi t)$                                                             **(2.96)**

b.  $g(t) = 3 + t^2$     **(2.97)**

c.  $g(t) = 3\tan(4t)$     **(2.98)**

d.  $g(t) = 10\sin(12\pi t) + 4\cos(18\pi t)$     **(2.99)**

e.  $g(t) = 10\sin(12\pi t) + 4\cos(18t)$     **(2.100)**

f.  $g(t) = 10\cos(\pi t)\sin(4\pi t)$     **(2.101)**

■ **Solution**

a. The sine function repeats when its total argument is increased or decreased by any integer multiple of $2\pi$ rad. Therefore,

$$\sin(400\pi t \pm 2n\pi) = \sin[400\pi(t \pm nT_0)] \qquad (2.102)$$

Setting the arguments equal in (2.102),

$$400\pi t \pm 2n\pi = 400\pi(t \pm nT_0) \qquad (2.103)$$

$$\pm 2n\pi = \pm 400\pi n T_0 \qquad (2.104)$$

$$T_0 = \frac{1}{200}. \qquad (2.105)$$

An alternate way of finding the fundamental period is to realize that $7\sin(400\pi t)$ is in the form $A\sin(2\pi f_0 t)$ or $A\sin(\omega_0 t)$, where $f_0$ is the cyclic fundamental frequency of the sinusoid and $\omega_0$ is the fundamental radian frequency. In this case, $f_0 = 200$ and $\omega_0 = 400\pi$. Since period is the reciprocal of cyclic frequency, $T_0 = 1/200$.

b. This function is a parabola. As its argument $t$ increases or decreases from zero, the function value increases monotonically (always in the same direction). No function which increases monotonically can be periodic because if a fixed amount is added to the argument $t$, the function must be larger or smaller than for the current $t$. This function is not periodic.

c. The tangent function repeats every $\pi$ rad of its total argument. Therefore,

$$3\tan[4(t \pm nT_0)] = 3\tan(4t \pm n\pi) \qquad (2.106)$$

$$T_0 = \frac{\pi}{4} \qquad (2.107)$$

d. This function is a little harder to deal with. It is the sum of two functions which are both periodic. But is the sum also periodic? That depends. If a time can be found inside which both functions have an integer number of periods, then the sum will repeat with that period. The fundamental period of the first one is $\frac{1}{6}$ s. The fundamental period of the second one is $\frac{1}{9}$ s. What is the shortest time in which both these signals have an integer number of periods? If both functions repeat exactly an integer number of times in some minimum time interval, then they will repeat exactly an integer number of times again in the next time interval of the same length. That time is then the fundamental period of the overall function, the minimum positive time in which it repeats. This is a common problem in mathematics. We are looking for the *least common multiple* of the two fundamental periods; that is, the smallest number into which both numbers divide an integer number of times. In this case the least common multiple is $\frac{1}{3}$ s. (See Appendix C • for a systematic method for finding least common multiples.) There are two fundamental periods of the first function and three fundamental periods of the second function in that time. Therefore, the fundamental period of the overall function is $\frac{1}{3}$ s (Figure 2.57). Similarly, the fundamental frequency of the sum of the two signals is the *greatest common divisor* of the frequencies of the two signals.

　　Figure 2.58 is another example of the fundamental period of the sum of two periodic functions with different fundamental periods.

e. This function is exactly like that of part (d) (2.99) except that a $\pi$ is missing in the second argument. The two fundamental periods are now $\frac{1}{6}$ and $\pi/9$ s. What is the least common multiple of these two times? It is infinite because $\pi$ is irrational. That is, this function,

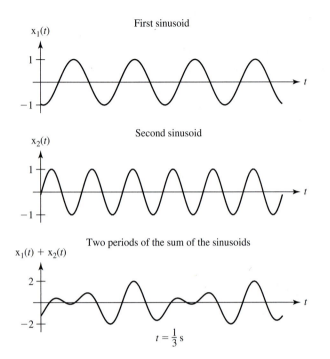

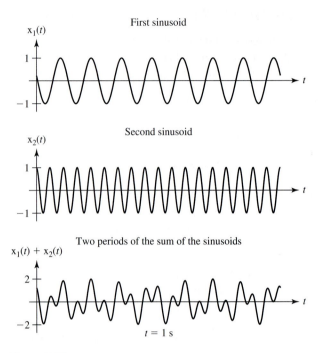

**Figure 2.57**
Signals with frequencies of 6 and 9 Hz and their sum.

**Figure 2.58**
Signals with frequencies of 4 and 9 Hz and their sum.

although made up of the sum of two periodic functions, is not itself periodic because it does not repeat exactly in a finite time. (It is sometimes referred to as *almost periodic* because it almost repeats in a finite time. But, strictly speaking, it is aperiodic.)

f.    This function is the product of two functions instead of the sum, but the arguments for finding the fundamental period are the same as for a sum of two functions. That is, this product repeats in a time which is the least common multiple of the fundamental periods of the two functions, in this case 2 s. Although the least common multiple of the fundamental periods in the product of two functions is a period, it may not be the *fundamental* period of the product. For example, let the function be

$$g(t) = A \cos(2\pi f_0 t) \sin(2\pi f_0 t). \qquad (2.108)$$

Since the two functions in the product on the right side of (2.108) have the same fundamental period and frequency, the least common multiple of the two fundamental periods is $T_0 = 1/f_0$. But the fundamental period of the product is actually $T_0 = 1/2 f_0$. This can be seen by applying the trigonometric identity

$$\sin(x)\cos(y) = \frac{1}{2}[\sin(x - y) + \sin(x + y)] \qquad (2.109)$$

to (2.108) to yield

$$g(t) = \frac{A}{2}\sin(4\pi f_0 t) \qquad (2.110)$$

whose period is $1/2 f_0$. So the least common multiple of the periods of two multiplied functions is a period, but it is not necessarily the fundamental period.  ∎

## 2.9 DISCRETE-TIME SIGNAL FUNCTIONS

Exponentials and sinusoids are as important in DT signal and system analysis as in CT signal and system analysis. DT exponentials and sinusoids can be defined in a manner analogous to their CT counterparts as

$$g[n] = Ae^{\beta n} \quad \text{or} \quad g[n] = A\alpha^n, \ \alpha = e^{\beta} \tag{2.111}$$

and

$$g[n] = A\cos\left(\frac{2\pi n}{N_0} + \theta\right) \quad \text{or} \quad A\cos(2\pi F_0 n + \theta) \quad \text{or}$$

$$g[n] = A\cos(\Omega_0 n + \theta) \tag{2.112}$$

where $\alpha, \beta$ = complex constants

$\quad A$ = real constant

$\quad \theta$ = real phase shift, rad

$\quad N_0$ = real number

and $F_0$ and $\Omega_0$ are related to $N_0$ through $1/N_0 = F_0 = \Omega_0/2\pi$, where $n$ is the previously defined discrete time.

There are some important differences between CT and DT sinusoids. The first is the fact that if we create a DT sinusoid by sampling a CT sinusoid, their periods may not be the same and, in fact, the DT sinusoid may not even be periodic. Let a DT sinusoid

$$g[n] = A\cos(2\pi K n + \theta) \tag{2.113}$$

be related to a CT sinusoid

$$g(t) = A\cos(2\pi f_0 t + \theta) \tag{2.114}$$

through

$$g[n] = g(nT_s). \tag{2.115}$$

Then, for (2.115) to be correct,

$$K = f_0 T_s = \frac{f_0}{f_s}. \tag{2.116}$$

The requirement on a DT sinusoid that it be periodic is that, for some discrete time $n$ and some integer $m$,

$$2\pi K n = 2\pi m. \tag{2.117}$$

Solving,

$$K = \frac{m}{n}. \tag{2.118}$$

In words, (2.118) says that $K$ must be a rational number (a ratio of integers). Since sampling forces the relationship $K = f_0/f_s$, this requirement also means that for a DT sinusoid to be periodic, the ratio of the fundamental frequency of the CT sinusoid to the sampling rate must be rational for a DT sinusoid formed by sampling a CT sinusoid. For example, what is the fundamental period of the following DT sinusoid?

$$g[n] = 4\cos\left(\frac{72\pi}{19}n + \theta\right) = 4\cos\left(2\pi\underbrace{\frac{36}{19}}_{K}n + \theta\right) \tag{2.119}$$

The smallest positive discrete time $n$ which solves $Kn = m$, where $m$ is an integer, is $n = 19$. So the fundamental period is 19. If $K$ is a rational number and is expressed as a ratio of integers in the form

$$K = \frac{p}{q},$$    (2.120)

and if the fraction has been reduced to its simplest form by canceling common factors in $p$ and $q$, then the fundamental period of the DT sinusoid is $q$, *not*

$$\frac{1}{K} = \frac{q}{p}$$    (2.121)

*unless* $p = 1$. Compare this result with the fundamental period of the CT sinusoid,

$$g(t) = 4\cos\left(\frac{72\pi}{19}t + \theta\right),$$    (2.122)

whose period $T_0$ is $\frac{19}{36}$, not 19. Figure 2.59 is a plot of some DT sinusoids with their fundamental periods indicated.

In the analysis of DT signals a handy relation to remember is that in the form $g[n] = A\cos(2\pi n/N_0 + \theta)$, if $N_0$ is a positive integer, it is also the fundamental period of $g[n]$.

One other aspect of DT sinusoids which will be very important in Chapter 7 in the consideration of sampling is that two DT sinusoids

$$g_1[n] = A\cos(2\pi K_1 n + \theta) \qquad \text{and} \qquad g_2[n] = A\cos(2\pi K_2 n + \theta)$$    (2.123)

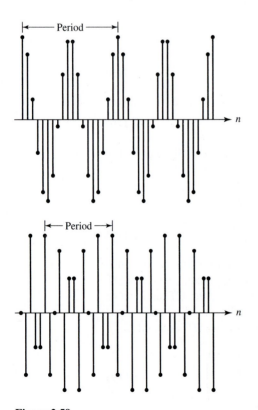
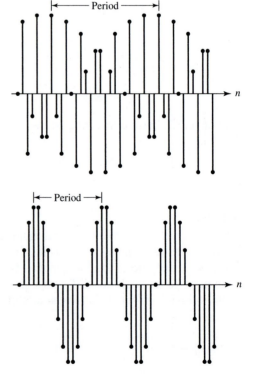

**Figure 2.59**
Four DT sinusoids.

can be identical, even if $K_1$ and $K_2$ are different. For example, the two DT sinusoids

$$g_1[n] = \cos\left(\frac{2\pi}{5}n\right) \qquad \text{and} \qquad g_2[n] = \cos\left(\frac{12\pi}{5}n\right) \qquad \textbf{(2.124)}$$

are described by different-looking analytical expressions, but when we plot them versus discrete time $n$, they look identical (Figure 2.60). The dashed lines in Figure 2.60 are the CT functions

$$g_1(t) = \cos\left(\frac{2\pi}{5}t\right) \qquad \text{and} \qquad g_2(t) = \cos\left(\frac{12\pi}{5}t\right). \qquad \textbf{(2.125)}$$

The CT functions are obviously different but the DT functions are not. The reason the two DT functions are identical can be seen by rewriting $g_2[n]$ in the form

$$g_2[n] = \cos\left(\frac{2\pi}{5}n + \frac{10\pi}{5}n\right) = \cos\left(\frac{2\pi}{5}n + 2\pi n\right). \qquad \textbf{(2.126)}$$

Then, using the principle that if any integer multiple of $2\pi$ is added to the angle of a sinusoid the value is not changed,

$$g_2[n] = \cos\left(\frac{2\pi}{5}n + 2\pi n\right) = \cos\left(\frac{2\pi}{5}n\right) = g_1[n], \qquad \textbf{(2.127)}$$

because discrete time $n$ is always an integer.

DT exponentials can have a variety of functional behaviors depending on the value of $\alpha$ in $A\alpha^n$. Figures 2.61 and 2.62 summarize several cases of the functional form of an exponential when $\alpha$ has different values.

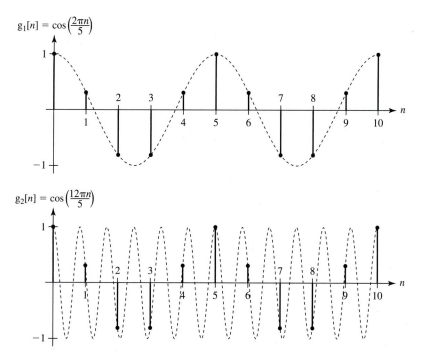

**Figure 2.60**
Two DT cosines with different $K$'s but the same functional behavior.

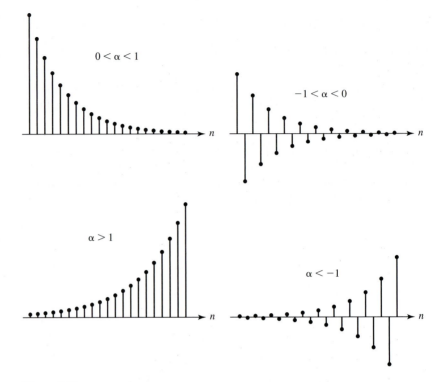

**Figure 2.61**
Behavior of $A\alpha^n$ for different real $\alpha$'s.

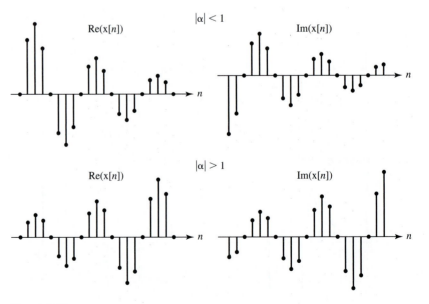

**Figure 2.62**
Behavior of $A\alpha^n$ for different complex $\alpha$'s.

## DISCRETE-TIME SINGULARITY FUNCTIONS

**The Unit Impulse**   The DT unit impulse (Figure 2.63) is defined by

$$\delta[n] = \begin{cases} 1 & n = 0 \\ 0 & n \neq 0 \end{cases}. \qquad (2.128)$$

The DT unit impulse suffers from none of the mathematical peculiarities that the CT unit impulse has. (This function is sometimes referred to as the *Kronecker delta* function.) The DT unit impulse does not have a property corresponding to the scaling property of the CT unit impulse. Therefore,

$$\delta[n] = \delta[an] \qquad (2.129)$$

for any nonzero, finite integer value of $a$. But the DT impulse does have a sampling property. It is

$$\sum_{n=-\infty}^{\infty} A\delta[n - n_0]x[n] = Ax[n_0]. \qquad (2.130)$$

This is easily seen by realizing that, since the impulse is only nonzero where its argument is zero, the summation over all $n$ is a summation of terms which are all zero except the one for which $n = n_0$. When $n = n_0$, $x[n] = x[n_0]$ and that result is simply multiplied by the scale factor $A$.

**The Unit Sequence**   The DT function which corresponds to the unit step is the *unit sequence* function (Figure 2.64).

$$u[n] = \begin{cases} 1 & n \geq 0 \\ 0 & n < 0 \end{cases} \qquad (2.131)$$

For this function there is no disagreement or ambiguity about its value at $n = 0$, it is one, and every author agrees. The unit sequence can be generated by sampling the unit step function in the manner described in Section 2.2,

$$u[n] = \lim_{\varepsilon \to 0} u(t + nT_s + \varepsilon), \qquad \varepsilon > 0. \qquad (2.132)$$

So, at the discontinuity in the CT step function the sample value is one and the DT unit sequence has the value $u[0] = 1$.

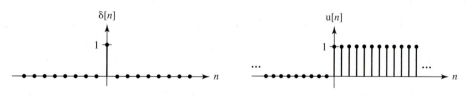

**Figure 2.63**
The DT unit impulse function.

**Figure 2.64**
The unit sequence function.

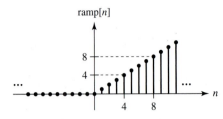

**Figure 2.65**
The DT unit ramp function.

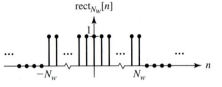

**Figure 2.66**
The DT rectangle function.

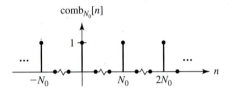

**Figure 2.67**
The DT comb function.

**The Unit Ramp**   The DT function corresponding to the CT unit ramp is defined in Figure 2.65.

$$\text{ramp}[n] = \begin{cases} n & n \geq 0 \\ 0 & n < 0 \end{cases} \tag{2.133}$$

**The Rectangle Function**   A DT rectangle is shown in Figure 2.66 and defined as

$$\text{rect}_{N_w}[n] = \begin{cases} 1 & , & |n| \leq N_w \\ 0 & , & |n| > N_w \end{cases} = u[n + N_w] - u[n - N_w - 1]. \tag{2.134}$$

where $N_w$ is an integer $\geq 0$. Because of the difference in the effects of time scaling between CT and DT functions, it is more convenient to define a general rectangle whose width is characterized by a parameter $N_w$ than to make a direct analog to the CT unit rectangle.

**The Comb Function**   The DT comb function (Figure 2.67) is defined by

$$\text{comb}_{N_0}[n] = \sum_{m=-\infty}^{\infty} \delta[n - m N_0]. \tag{2.135}$$

These DT functions can be implemented in MATLAB by the following .m files.

```
%    Function to generate the discrete-time impulse function defined as one
%    for input integer arguments equal to zero and zero otherwise.
%    Returns "NaN" for noninteger arguments. Works for vectors and
%    scalars equally well.
%
%    function y = impDT(n)
```

```
function y = impDT(n)
      y = double(n == 0) ;            %     Impulse is one where argument
                                      %     is zero and zero otherwise.
      ss = find(round(n)~=n) ;        %     Find noninteger values of n.
      y(ss) = NaN ;                   %     Set corresponding outputs to
                                      %     NaN.


%     Unit sequence function defined as zero for input integer argument
%     values less than zero, and one for input integer argument values
%     equal to or greater than zero. Returns NaN for noninteger
%     arguments. Works for vectors and scalars equally well.
%
%     function y = uDT(n)

function y = uDT(n)
      y = double(n>=0) ;              %     Set output to one for
                                      %     nonnegative arguments
      ss = find(round(n)~=n) ;        %     Find all noninteger n's.
      y(ss) = NaN ;                   %     Set the corresponding outputs
                                      %     all to NaN.


%     Unit discrete-time ramp function defined as zero for input integer
%     argument values equal to or less than zero, and n for input
%     integer argument values greater than zero. Returns NaN for
%     noninteger arguments. Works for vectors and scalars equally well.
%
%     function y = rampDT(n)

function y = rampDT(n)
      pos = double(n>0) ; y = n.*pos ;  %   Set output to n for
                                        %   positive n.
      ss = find(round(n)~=n) ;          %   Find all noninteger n's.
      y(ss) = NaN ;                     %   Set the corresponding
                                        %   outputs all to NaN.


%     Discrete-time rectangle function defined as one for input integer
%     argument values equal to or less than Nw in magnitude and zero
%     for other integer argument values. Nw must be an integer.
%     Returns NaN for noninteger input values.
%
%     y = rectDT(Nw,n)

function y = rectDT(Nw,n)
      if Nw == round(Nw),
            y = double(abs(n)<=abs(Nw)) ;   %     Set output to one if
                                            %     |n|<=|Nw|and zero
                                            %     otherwise.
            ss = find(round(n)~=n) ;        %     Find all noninteger
                                            %     n's.
            y(ss) = NaN ;                   %     Set the corresponding
                                            %     outputs all to NaN.
```

```
          else
              disp('In rectDT, width parameter, Nw, is not an integer') ;
          end

%     Discrete-time comb function defined as 1 for input integer
%     argument values equal to integer multiples of N0, and zero
%     otherwise. N0 must be an integer. Returns NaN for noninteger
%     input values. Works for vectors and scalars equally well.
%
%     function y = combDT(N0,n)

function y = combDT(N0,n)
          if N0 == round(N0),
              y = double(n/N0 == round(n/N0)) ;   %  Set output to one for
                                                  %  for all n's which are
                                                  %  integer multiples of
                                                  %  N0 and zero otherwise.
              ss = find(round(n)~=n) ;            %  Find all noninteger
                                                  %  n's.
              y(ss) = NaN ;                       %  Set the corresponding
                                                  %  outputs all to NaN.
          else
              disp('In combDT, period parameter, N0, is not an integer') ;
          end
```

The following are examples of the use of these DT MATLAB functions.

```
»impDT(3)
ans =
      0
»impDT(0)
ans =
      1
»impDT(1.5)
ans =
    NaN
»impDT(-4)
ans =
      0
»uDT(0)
ans =
      1
»uDT(-6)
ans =
      0
»uDT(pi)
ans =
    NaN
»rampDT(0)
ans =
      0
```

```
»rampDT(10)
ans =
     10
»rampDT(-10)
ans =
     0
»rampDT(11.5)
ans =
     NaN
»rectDT(3,0)
ans =
     1
»rectDT(3,5)
ans =
     0
»rectDT(3.3,1)
In rectDT, width parameter, W, is not an integer
»rectDT(5,-7)
ans =
     0
»combDT(5,0)
ans =
     1
»combDT(5,2)
ans =
     0
»combDT(1.2,0)
In combDT, period parameter, N0, is not an integer
»combDT(8,-8)
ans =
     1
```

## 2.10 DISCRETE-TIME SCALING AND SHIFTING TRANSFORMATIONS

The general principles that govern functional transformations of CT functions also apply to DT functions, but with some interesting differences caused by the fundamental differences between DT and CT functions. Just as a CT function does, a DT function accepts a number and returns another number. The general principle that the *expression* in g[*expression*] is treated in exactly the same way that discrete time $n$ is treated in the definition g[$n$] still holds. Amplitude scaling for DT functions is exactly the same as it is for CT functions.

### TIME SHIFTING

Let a DT function g[$n$] be defined by the graph in Figure 2.68. Now let $n \rightarrow n + 3$. Time shifting is essentially the same for DT and CT functions, except that for DT functions the shift must be an integer; otherwise the shifted function would have undefined values (Figure 2.69).

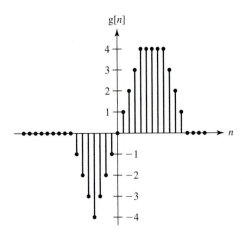

**Figure 2.68**
Graphical definition of a DT function g[n],
where g[n] = 0 and |n| > 15.

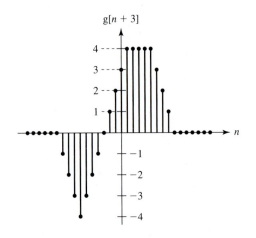

**Figure 2.69**
Graph of g[n + 3] illustrating the time-shifting
functional transformation.

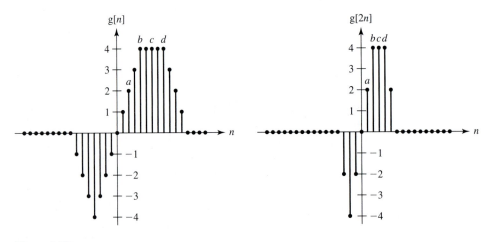

**Figure 2.70**
Time compression for a DT function.

## TIME SCALING

Amplitude scaling and time shifting for DT and CT functions are very similar, but time scaling for DT functions is different than for CT functions. There are two cases to examine, time compression and time expansion. Time compression is accomplished by a transformation of the form $n \rightarrow Kn$, where $K$ is an integer whose magnitude is greater than one. Time compression for DT functions is similar to time compression for CT functions in that the function occurs faster in time. But in the case of DT functions there is another effect called *decimation* that has meaning only for DT functions.

Consider the time scaling $n \rightarrow 2n$, illustrated in Figure 2.70. As is obvious from this figure, for each integer $n$ in g[2n], the functional argument value $2n$ must be an even integer. Therefore, for this scaling by a factor of two, the odd integer values of the originally defined g[n] are never needed to find values for g[2n]. The function has been decimated by a factor of two because the plot of g[2n] only uses every other value of the defined function g[n]. For larger scaling constants, the decimation factor is obviously higher. Decimation does not happen in scaling CT functions because a continuum of *at* values maps into a corresponding continuum of *t* values without any missing values. The fundamental difference between CT and DT functions can be expressed by observing that the domain of a CT function is all real numbers, an *uncountable* infinity of times, but the domain of DT functions is all integers, a *countable* infinity of discrete times.

The other time-scaling case, time expansion, is even stranger. For example, if we want to graph g[$n/2$] for each integer value of *n*, we must assign a value to g[$n/2$] by finding the corresponding value in the original function definition. But when $n = 1, n/2 = \frac{1}{2}$, and g[$\frac{1}{2}$] is not defined. The value of the transformed function g[$n/K$] is undefined unless $n/K$ is an integer. We can simply leave those values undefined, or we can interpolate between them using the values of g[$n/K$] at the next higher and next lower values of *n* at which $n/K$ is an integer. (*Interpolation* is a process of computing functional values between two known values according to some formula.) Since interpolation begs the question of what interpolation formula to use, we will simply leave g[$n/K$] undefined if $n/K$ is not an integer.

DT functions can be combined in the same way as CT functions through addition, subtraction, multiplication, and division. But there are two issues that arise which do not occur when combining CT functions.

As we have just seen, it is possible, through functional transformation, to create a DT function which is undefined at some particular discrete times. If a DT function is undefined at any particular discrete time, then any combination of that function and any other DT function will also be undefined at that discrete time.

Suppose two DT functions are created by sampling two CT functions, and the sampling rates are different for the two functions. We can combine the two DT functions by combining their values at corresponding discrete times. But in doing so we are combining samples taken at different actual times from the original CT functions. Even though there are no mathematical rules preventing this kind of combination of DT functions, it is not clear what useful meaning the result would have. Usually in discrete-time signal analysis, all signals have the same discrete time. That is, at any particular value of *discrete* time *n*, all the signals have values acquired at the same *actual* time *t*. Figures 2.71 and 2.72 illustrate some DT function combinations.

When writing MATLAB .m files to implement a DT function, a predefined constant NaN comes in very handy. The name NaN is an acronym for "not a number" and simply indicates an undefined value. For example, we can define a CT polynomial function poly.

```
function       x = poly(t)
      x = 3*t.^2 - t + 8 ;
```

(MATLAB uses parentheses exclusively for function arguments, even when we define the function as one with discrete-time behavior. Square brackets are used to enclose vectors or matrices. So in MATLAB, even DT functions are written with parentheses. The distinction comes in the body of the function's .m file and is not immediately clear when the function is invoked in a script file.) As written, this MATLAB function computes a defined numerical value of x for every t sent to it. We can now modify this function to make it into a DT function.

```
function       x = polyDT(n)
      x = 3*n.^2 - n + 8 ;
      nonInt = find(round(n)~=n) ;    % Find all noninteger n's.
      x(nonInt) = NaN ;               % Set the corresponding x's to NaN.
```

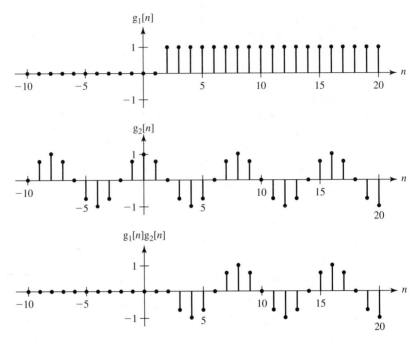

**Figure 2.71**
Product of two DT functions.

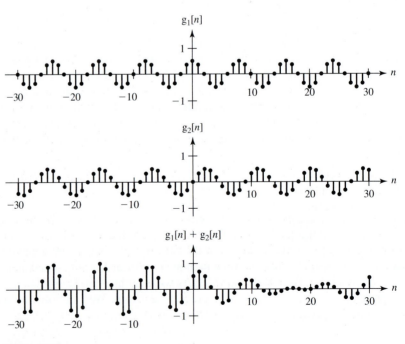

**Figure 2.72**
Sum of two DT functions.

When the `stem` command is used in MATLAB to plot a DT function with some undefined values, the defined values are plotted and the undefined values are simply omitted, just as they should be.

**EXAMPLE 2.5**

Using MATLAB, graph the DT function

$$g[n] = 10e^{-n/4} \sin\left(\frac{3\pi n}{16}\right) u[n]. \qquad (2.136)$$

Then graph the functions g[2n] and g[n/3].

### ■ Solution

In some respects, DT functions are actually easier to program in MATLAB than CT functions because MATLAB is inherently oriented toward calculation of functional values at discrete values of the independent variable. For DT functions there is no need to decide how close together to make the time points to make the plot look continuous because the function is *not* continuous. A good way to handle graphing the function and the time-scaled functions is to define the original function as a `.m` file. But we need to ensure that the function definition includes its discrete-time behavior, namely, that for noninteger values of discrete time the function is undefined. MATLAB handles undefined results by assigning to them the special value NaN. The three-character sequence NaN is predefined by MATLAB to represent results that are undefined mathematically. The user of MATLAB can redefine the value of the variable NaN, but it is strongly recommended that it not be redefined because that could cause significant confusion. The only other programming problem is how to handle the two different functional descriptions in the two different ranges of *n*. We can do that nicely with logical and relational operators as demonstrated below in `g.m`.

```
function y = g(n),

ss = find(round(n) ~= n) ;        % Find all noninteger n's.
n(ss) = NaN ;                     % Set them all to NaN.
y = 10*exp(-n/4).*sin(3*pi*n/16) ; % Compute the function without the
                                  % specification of zero value for
                                  % negative discrete times.
y = y.*uDT(n) ;                   % Set the negative-time part of
                                  % the function to zero.
```

We still must decide over what range of discrete times to plot the function. Since it is zero for negative times, we should represent that time range with at least a few points to show that it suddenly turns on at time zero. Then, for positive times it has the shape of an exponentially decaying sinusoid. Therefore, if we graph a few time constants of the exponential decay, the function will be practically zero after that time. So the time range should be something like $-5 < n < 16$ to graph a reasonable representation of the original function. But the time-expanded function g[n/3] will be wider in discrete time and require more discrete time to see the functional behavior. Therefore, to really see all the functions on the same DT scale for comparison, let's make the range of discrete times $-5 < n < 48$.

The only other programming problem is how to handle the two different functional descriptions in the two different ranges of *n*. We can do that with logical and relational operators.

```
%    Graphing a discrete-time function and compressed and expanded
%    transformations of it.
```

```
%-------------------------------------------------------------------
%    Compute values of the original function and the transformed
%    versions in this section
%-------------------------------------------------------------------

n = -5:48 ;                              % Set the discrete times for
                                         % function computation.

g0 = g(n) ;                              % Compute the original function
                                         % values.

g1 = g(2*n) ;                            % Compute the compressed function
                                         % values.

g2 = g(n/3) ;                            % Compute the expanded function.
                                         % values.


%-------------------------------------------------------------------
% Display the original and transformed functions graphically
% in this section.
%-------------------------------------------------------------------

%
%    Plot the original function.
%
subplot(3,1,1) ;                         % Plot first of three plots stacked
                                         % vertically.

p = stem(n,g0,'k','filled') ;            % "Stem" plot the original function.
set(p,'LineWidth',2,'MarkerSize',4) ;    % Set the line weight and dot
                                         % size.

ylabel('g[n]') ;                          % Label the original function axis.
title('Example 2.5') ;                   % Title the plots.
%
%    Plot the time-compressed function.
%
subplot(3,1,2) ;                         % Plot second of three plots
                                         % stacked vertically.

p = stem(n,g1,'k','filled') ;            % "Stem" plot the compressed
% function.
set(p,'LineWidth',2,'MarkerSize',4) ;    % Set the line weight and dot
                                         % size.

ylabel('g[2n]') ;                        % Label the compressed function
                                         % axis.

%
%    Plot the time-expanded function.
%
subplot (3,1,3) ;                        % Plot third of three plots
                                         % stacked vertically.

p = stem(n,g2,'k','filled') ;            % "Stem" plot the expanded
                                         % function.

set(p,'LineWidth',2,'MarkerSize',4) ;    % Set the line weight and dot
                                         % size.

xlabel('Discrete time, n') ;             % Label the expanded function axis.
ylabel('g[n/3]') ;                       % Label the discrete-time axis.
```

The MATLAB plots are shown in Figure 2.73.

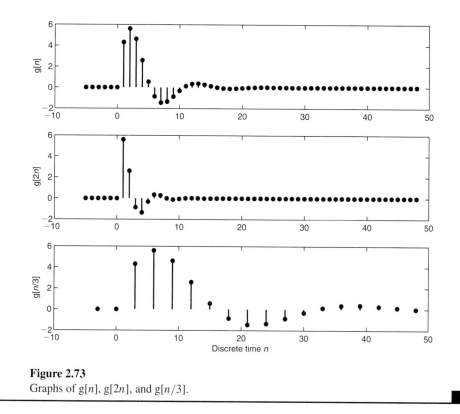

**Figure 2.73**
Graphs of g[n], g[2n], and g[n/3].

## 2.11 DIFFERENCING AND ACCUMULATION

Just as differentiation and integration are important transformations for CT functions, differencing and accumulation are important transformations for DT functions. The first forward difference of a DT function g[n] is defined by

$$\Delta g[n] = g[n+1] - g[n]. \qquad (2.137)$$

(See Appendix I for more on differencing and difference equations.) The first backward difference of a DT function is $g[n] - g[n-1]$, which is the first forward difference of $g[n-1]$,

$$\Delta g[n-1] = g[n] - g[n-1]. \qquad (2.138)$$

Figure 2.74 illustrates some DT functions and their first forward or backward differences. If you imagine a DT function as being created by sampling a CT function, you can see that the differencing operation yields a result which looks like samples of the derivative of that CT function (to within a scale factor).

The discrete-time counterpart of integration is accumulation (or summation), and the same ambiguity problem which occurs in the integration of a CT function exists for DT functions also. That is, even though the first forward or backward difference of a DT function is unambiguous, the accumulation of a DT function is not unique or unambiguous. Multiple DT functions can have the same forward or backward difference, but, just as in integration, these functions which all have the same difference can only differ from each other by an additive constant. Let h[n] be the first forward difference

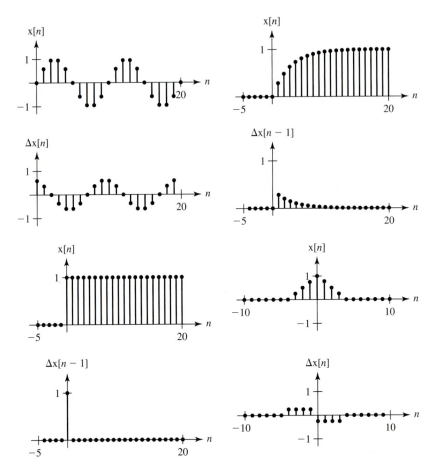

**Figure 2.74**
Some DT functions and their forward or backward differences.

of $g[n]$,

$$h[n] = \Delta(g[n-1]).$$ **(2.139)**

Then we can find $g[n]$ from $h[n]$ by accumulating both sides,

$$g[n] = \sum_{m=-\infty}^{n} h[m] = g[n_0] + \sum_{m=n_0+1}^{n} h[m].$$ **(2.140)**

This can be proven to be correct by substituting $\Delta(g[m-1])$ for $h[m]$,

$$g[n] = g[n_0] + \sum_{m=n_0+1}^{n} \Delta(g[m-1]) = g[n_0] + \sum_{m=n_0+1}^{n} (g[m] - g[m-1])$$

$$g[n] = g[n_0] + g[n_0+1] - g[n_0] + g[n_0+2]$$ **(2.141)**
$$\quad - g[n_0+1] + \cdots + g[n] - g[n-1].$$

$$g[n] = g[n]$$

Figure 2.75 illustrates some DT functions $h[n]$ and their accumulations $g[n]$. In each of the plots in the figure the accumulation was done based on the assumption that all function values of $h[n]$ before the time range plotted are zero.

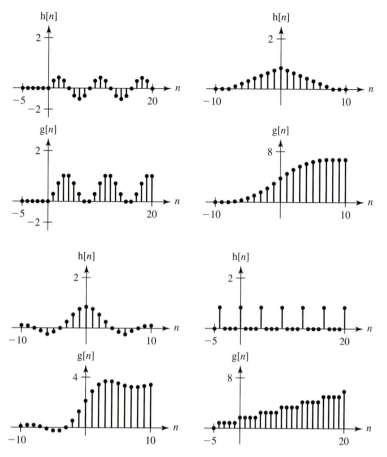

**Figure 2.75**
Some DT functions h[n] and their accumulations g[n].

In a manner analogous to the integral-derivative relationship between the CT unit step and the CT unit impulse, the unit sequence is the accumulation of the unit impulse,

$$u[n] = \sum_{m=-\infty}^{n} \delta[m], \qquad (2.142)$$

and the unit impulse is the first backward difference of the unit sequence,

$$\delta[n] = u[n] - u[n - 1] = \Delta(u[n - 1]). \qquad (2.143)$$

Also in a manner similar to the integral definition of the CT unit ramp function, the DT unit ramp is defined as the accumulation of a unit sequence function delayed by one in discrete time,

$$\text{ramp}[n] = \sum_{m=-\infty}^{n} u[m - 1] = \sum_{m=-\infty}^{n-1} u[m], \qquad (2.144)$$

and the unit sequence is the first *forward* difference of the unit ramp,

$$u[n] = \text{ramp}[n + 1] - \text{ramp}[n] = \Delta(\text{ramp}[n]). \qquad (2.145)$$

We can define a *family* of DT singularity functions with analogous characteristics to the CT doublet, triplet, etc. For example, we can define the first backward difference of the DT unit impulse to be a DT unit doublet,

$$u_1[n] = \delta[n] - \delta[n-1]. \tag{2.146}$$

The DT unit doublet samples the first backward difference of a function.

$$\sum_{n=-\infty}^{\infty} g[n]u_1[n-n_0] = \sum_{n=-\infty}^{\infty} g[n](\delta[n-n_0] - \delta[n-n_0-1])$$

$$= g[n_0] - g[n_0+1] = -(g[n_0+1] - g[n_0]) \tag{2.147}$$

This is the *negative* of the first backward difference of $g[n]$ at $n = n_0 + 1$, which is the same as the negative of the first *forward* difference of $g[n]$ at $n = n_0$. This is analogous to the sampling property of the CT unit doublet. It samples the negative of the first derivative of a CT function. Other characteristics of the DT singularity functions are analogous to those of the corresponding CT singularity functions.

MATLAB can compute differences of DT functions using the `diff` function. The `diff` function accepts a vector as its argument and returns a vector of forward differences whose length is one less than the length of the vector it accepted. MATLAB can also compute the accumulation of a DT function using the `cumsum` function. The `cumsum` function accepts a vector as its argument and returns a vector of equal length which is the accumulation of the elements in the argument vector. For example,

```
»a=1:10
a =
     1     2     3     4     5     6     7     8     9    10
»diff(a)
ans =
     1     1     1     1     1     1     1     1     1
»cumsum(a)
ans =
     1     3     6    10    15    21    28    36    45    55
»b = randn(1,5)
b =
     1.1909    1.1892   -0.0376    0.3273    0.1746
»diff(b)
ans =
    -0.0018   -1.2268    0.3649   -0.1527
»cumsum(b)
ans =
     1.1909    2.3801    2.3424    2.6697    2.8444
```

Of course, `cumsum` implicitly assumes that the value of the accumulation is zero before the first element in the vector.

## EXAMPLE 2.6

Using MATLAB, find the accumulation of the DT function

$$x[n] = \cos\left(\frac{2\pi n}{18}\right) \tag{2.148}$$

from $n = 0$ to 36 under the assumption that the accumulation before time $n = 0$ is zero.

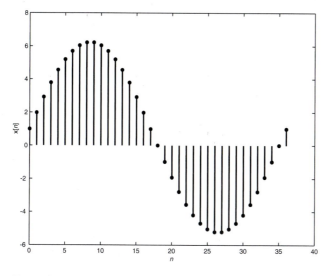

**Figure 2.76**
Accumulation of a DT cosine.

```
%     Program to demonstrate accumulation of a DT function over a finite
%     time using the cumsum function.

n = 0:36 ; x = cos(2*pi*n/36) ;
p = stem(n,cumsum(x),'k','filled') ;
set(p,'LineWidth',2,'MarkerSize',4) ;
```

Notice that this cosine accumulation (see Figure 2.76) looks a lot like (but not exactly like) a DT sine function. That occurs because the accumulation process is analogous to the integration process for CT functions and the integral of a cosine is a sine. ∎

## 2.12 DISCRETE-TIME EVEN AND ODD FUNCTIONS

Like CT functions, DT functions can also be classified by the properties of evenness and oddness. The defining relationships are completely analogous to those for CT functions. If $g[n] = g[-n]$, then $g[n]$ is even and if $g[n] = -g[-n]$, $g[n]$ is odd. Figure 2.77 shows some examples of DT even and odd functions.

The even and odd parts of a DT function $g[n]$ are found exactly the same way as for CT functions.

$$g_e[n] = \frac{g[n] + g[-n]}{2} \quad \text{and} \quad g_o[n] = \frac{g[n] - g[-n]}{2} \quad \textbf{(2.149)}$$

An even function has an odd part which is zero, and an odd function has an even part which is zero.

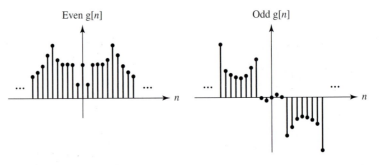

**Figure 2.77**
Examples of even and odd DT functions.

---

EXAMPLE 2.7

What are the even and odd parts of the function $g[n] = \sin\left(\dfrac{2\pi n}{7}\right)(1 + n^2)$?

■ **Solution**

$$
g_e[n] = \frac{\sin\left(\dfrac{2\pi n}{7}\right)(1 + n^2) + \sin\left(-\dfrac{2\pi n}{7}\right)(1 + (-n)^2)}{2}
$$

$$
= \frac{\sin\left(\dfrac{2\pi n}{7}\right)(1 + n^2) - \sin\left(\dfrac{2\pi n}{7}\right)(1 + n^2)}{2} = 0 \qquad \textbf{(2.150)}
$$

$$
g_o[n] = \frac{\sin\left(\dfrac{2\pi n}{7}\right)(1 + n^2) - \sin\left(-\dfrac{2\pi n}{7}\right)(1 + (-n)^2)}{2}
$$

$$
= \sin\left(\dfrac{2\pi n}{7}\right)(1 + n^2) \qquad \textbf{(2.151)}
$$

The function $g[n]$ is odd. ■

---

## SUMS, PRODUCTS, DIFFERENCES, AND QUOTIENTS

All the properties of combinations of functions that apply to CT functions also apply to DT functions. If two functions are even, their sum, difference, product, and quotient are even also. If two functions are odd, their sum and difference are odd but their product and quotient are even. If one function is even and the other is odd, their product and quotient are odd. Figures 2.78 through 2.80 show some examples of products of DT even and odd functions.

## ACCUMULATION

Integration of CT functions is analogous to accumulation of DT functions. Properties hold for accumulations of DT functions that are similar to (but not identical to) those for integrals of CT functions. If $g[n]$ is an even function and $N$ is a positive

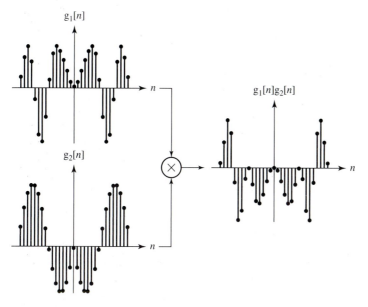

**Figure 2.78**
Product of two even DT functions.

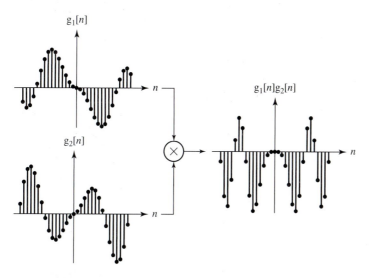

**Figure 2.79**
Product of two odd DT functions.

integer,

$$\sum_{n=-N}^{N} g[n] = g[0] + 2\sum_{n=1}^{N} g[n] \qquad \textbf{(2.152)}$$

and, if g[n] is an odd function,

$$\sum_{n=-N}^{N} g[n] = 0 \qquad \textbf{(2.153)}$$

(Figure 2.81).

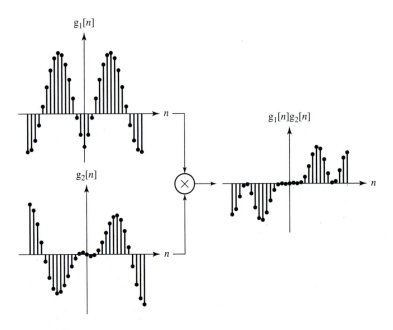

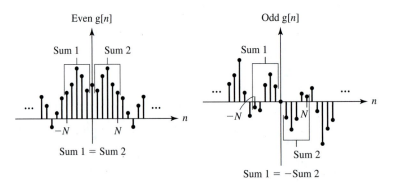

**Figure 2.80**
Product of an even and an odd DT function.

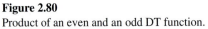

**Figure 2.81**
Accumulations of even and odd DT functions.

# 2.13 DISCRETE-TIME PERIODIC FUNCTIONS

A DT periodic function is one which is invariant under the transformation

$$n \rightarrow n + mN \tag{2.154}$$

where $N$ = any period of the function
$\quad m$ = any integer
$\quad N_0$ = fundamental period

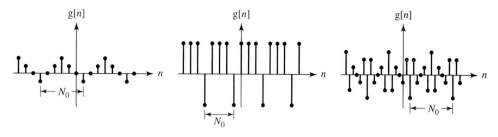

**Figure 2.82**
Examples of periodic functions with fundamental period $N_0$.

The fundamental period is the minimum positive discrete time in which the function repeats. Figure 2.82 shows some examples of DT periodic functions.

The fundamental discrete-time frequency is $F_0 = 1/N_0$ in cycles or $\Omega_0 = 2\pi/N_0$ in radians. Observe that the units of DT frequency are not hertz or radians per second because the units of discrete time are not seconds. Just as discrete time is not really time but rather an integer number of sampling times, DT frequency is not really frequency but rather the number of cycles or radians occurring between two consecutive discrete times.

**EXAMPLE 2.8**

Graph the DT function

$$g[n] = 2\cos\left(\frac{9\pi n}{4}\right) - 3\sin\left(\frac{6\pi n}{5}\right) \tag{2.155}$$

over the range $-50 \le n \le 50$. From the graph determine the period.

■ **Solution**
See Figure 2.83.

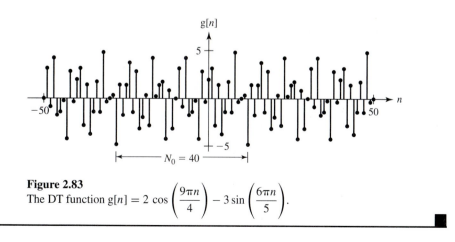

**Figure 2.83**
The DT function $g[n] = 2\cos\left(\frac{9\pi n}{4}\right) - 3\sin\left(\frac{6\pi n}{5}\right)$.

## 2.14 SIGNAL ENERGY AND POWER

It is important at this point to establish some terminology describing the energy and power of signals. In the study of signals in systems, the signals are often treated as mathematical abstractions. Often the physical significance of the signal is either unknown or ignored for the sake of simplicity of analysis. Typical signals in electrical systems would be voltages or currents but could be charge or electric field or some other physical quantity. In other types of systems a signal could be a force, a temperature, a chemical concentration, a neutron flux, etc. Because of the many different kinds of physical signals that can be operated on by systems, sometimes many different kinds of signals in a *single* system, the term *signal energy* has been defined. Signal energy (as opposed to just energy) of a CT signal is defined as the area under the square of the magnitude of the signal. If x(t) is a CT signal, its signal energy is

$$E_x = \int_{-\infty}^{\infty} |x(t)|^2 \, dt. \tag{2.156}$$

Therefore, the units of signal energy depend on the units of the signal. If the signal unit is the volt, (V), the signal energy of that signal is expressed in $V^2 \cdot s$. Signal energy is defined this way to be *proportional to* the actual physical energy delivered by a signal but not necessarily *equal to* that physical energy. In the case of a voltage signal v(t) across a resistor R, the actual energy delivered to the resistor by the voltage would be

$$\text{Energy} = \int_{-\infty}^{\infty} \frac{|v(t)|^2}{R} \, dt = \frac{1}{R} \int_{-\infty}^{\infty} |v(t)|^2 \, dt = \frac{E_v}{R}. \tag{2.157}$$

By this definition, signal energy is proportional to actual energy and the proportionality constant, in this case, is R. For a different kind of signal, the proportionality constant would be different. In many kinds of system analysis the use of signal energy is more convenient than the use of actual physical energy.

Signal energy for a DT signal is defined in an analogous way as

$$E_x = \sum_{n=-\infty}^{\infty} |x[n]|^2, \tag{2.158}$$

and its units are simply the square of the units of the signal itself.

**EXAMPLE 2.9**

Find the signal energy of

$$x(t) = 3 \, \text{tri}\left(\frac{t}{4}\right). \tag{2.159}$$

■ **Solution**

From the definition,

$$E_x = \int_{-\infty}^{\infty} |x(t)|^2 \, dt = \int_{-\infty}^{\infty} \left| 3 \, \text{tri}\left(\frac{t}{4}\right) \right|^2 \, dt = 9 \int_{-\infty}^{\infty} \text{tri}^2\left(\frac{t}{4}\right) \, dt. \tag{2.160}$$

Using the definition of the triangle function

$$\text{tri}(t) = \begin{cases} 1 - |t| & |t| < 1 \\ 0 & |t| \geq 1 \end{cases}, \tag{2.161}$$

tri$(t/4)$ is defined by

$$\text{tri}\left(\frac{t}{4}\right) = \begin{cases} 1 - \left|\dfrac{t}{4}\right| & \left|\dfrac{t}{4}\right| < 1 \text{ or } |t| < 4 \\ 0 & \left|\dfrac{t}{4}\right| \geq 1 \text{ or } |t| \geq 4 \end{cases} \tag{2.162}$$

and

$$E_x = 9 \int_{-4}^{4} \left(1 - \left|\frac{t}{4}\right|\right)^2 dt. \tag{2.163}$$

Since the integrand of (2.163) is an even function,

$$E_x = 18 \int_{0}^{4} \left(1 - \left|\frac{t}{4}\right|\right)^2 dt = 18 \int_{0}^{4} \left(1 - \frac{t}{4}\right)^2 dt = 18 \int_{0}^{4} \left(1 - \frac{t}{2} + \frac{t^2}{16}\right) dt$$

$$= 18 \left[t - \frac{t^2}{4} + \frac{t^3}{48}\right]_0^4 = 24 \tag{2.164}$$

■

**EXAMPLE 2.10**

Find the signal energy of

$$x[n] = \left(\frac{1}{2}\right)^n u[n]. \tag{2.165}$$

**■ Solution**

From the definition of DT-signal energy,

$$E_x = \sum_{n=-\infty}^{\infty} |x[n]|^2 = \sum_{n=-\infty}^{\infty} \left|\left(\frac{1}{2}\right)^n u[n]\right|^2 = \sum_{n=0}^{\infty} \left|\left(\frac{1}{2}\right)^n\right|^2 = \sum_{n=0}^{\infty} \left(\frac{1}{2}\right)^{2n}$$

$$= 1 + \frac{1}{2^2} + \frac{1}{2^4} + \cdots. \tag{2.166}$$

This infinite series in (2.166) can be rewritten as

$$E_x = 1 + \frac{1}{4} + \frac{1}{4^2} + \cdots. \tag{2.167}$$

In (2.167) we can use the formula for the summation of a power series

$$\frac{1}{1-x} = 1 + x + x^2 + \cdots, \quad |x| < 1 \tag{2.168}$$

to get

$$E_x = \frac{1}{1 - \frac{1}{4}} = \frac{4}{3}. \tag{2.169}$$

For many signals encountered in signal and system analysis, neither the integral

$$E_x = \int_{-\infty}^{\infty} |\mathrm{x}(t)|^2 \, dt \tag{2.170}$$

nor the summation

$$E_x = \sum_{n=-\infty}^{\infty} |\mathrm{x}[n]|^2 \tag{2.171}$$

converge because the signal energy is infinite. This usually occurs because the signal is not time-limited. (*Time limited* means that the signal is nonzero over only a finite time.) An example of a CT signal with infinite energy would be the sinusoidal signal

$$\mathrm{x}(t) = A \cos(2\pi f_0 t). \tag{2.172}$$

The signal energy is infinite because, over an infinite time interval, the area under the square of this signal is infinite. The unit sequence is an example of a DT signal with infinite energy. For signals of this type, it is usually more convenient to deal with the average signal power of the signal instead of the signal energy. The average signal power of a CT signal is defined by

$$\boxed{P_x = \lim_{T \to \infty} \frac{1}{T} \int_{-(T/2)}^{T/2} |\mathrm{x}(t)|^2 \, dt}. \tag{2.173}$$

In this definition of average signal power, the integral is the signal energy of the signal over a time $T$, and that is then divided by $T$ yielding the average signal power over time $T$. Then, as $T$ approaches infinity, this average signal power becomes the average signal power over all time. For DT signals the definition of signal power is

$$\boxed{P_x = \lim_{N \to \infty} \frac{1}{2N} \sum_{n=-N}^{N-1} |\mathrm{x}[n]|^2}, \tag{2.174}$$

which is the average signal power over all discrete time.

For periodic signals, the average signal power calculation may be simpler. The average value of any periodic function is the average over any period. Therefore, since the square of a periodic function is also periodic, for periodic CT signals,

$$P_x = \frac{1}{T} \int_{t_0}^{t_0+T} |\mathrm{x}(t)|^2 \, dt = \frac{1}{T} \int_T |\mathrm{x}(t)|^2 \, dt \tag{2.175}$$

where the notation $\int_T$ means the same thing as $\int_{t_0}^{t_0+T}$ for any arbitrary choice of $t_0$, where $T$ can be any period (usually the fundamental period). For DT signals

$$P_x = \frac{1}{N} \sum_{n=k}^{k+N-1} |\mathrm{x}[n]|^2 = \frac{1}{N} \sum_{n=\langle N \rangle} |\mathrm{x}[n]|^2, \qquad \textbf{(2.176)}$$

where $k$ is any integer and the notation $\sum_{n=\langle N \rangle}$ means the summation over any range $N$ in length where $N$ can be any period (usually the fundamental period).

---

**EXAMPLE 2.11**

Find the signal power of

$$\mathrm{x}(t) = A \cos(2\pi f_0 t + \theta). \qquad \textbf{(2.177)}$$

■ **Solution**

From the definition of signal power for a periodic signal,

$$P_x = \frac{1}{T} \int_T |A \cos(2\pi f_0 t + \theta)|^2 \, dt = \frac{A^2}{T_0} \int_{-(T_0/2)}^{T_0/2} \cos^2\left(\frac{2\pi}{T_0}t + \theta\right) dt. \qquad \textbf{(2.178)}$$

Using the trigonometric identity

$$\cos(x)\cos(y) = \frac{1}{2}[\cos(x-y) + \cos(x+y)] \qquad \textbf{(2.179)}$$

in (2.178) we get

$$P_x = \frac{A^2}{2T_0} \int_{-(T_0/2)}^{T_0/2} \left[1 + \cos\left(\frac{4\pi}{T_0}t + 2\theta\right)\right] dt = \frac{A^2}{2T_0} \int_{-(T_0/2)}^{T_0/2} dt$$

$$+ \frac{A^2}{2T_0} \underbrace{\int_{-(T_0/2)}^{T_0/2} \cos\left(\frac{4\pi}{T_0}t + 2\theta\right) dt}_{=0}. \qquad \textbf{(2.180)}$$

The second integral on the right side of (2.180) is zero because it is the integral of a sinusoid over exactly two fundamental periods. Therefore, the signal power is

$$P_x = \frac{A^2}{2} \qquad \textbf{(2.181)}$$

Notice that this result is independent of the phase $\theta$ and the frequency $f_0$. It depends only on the amplitude $A$.

---

Signals which have finite signal energy are referred to as *energy signals* and signals which have infinite signal energy but finite average signal power are referred to as *power signals*. No real physical signal can actually have infinite energy or infinite average power because there is not enough energy or power in the universe available. But we often analyze signals that, according to their strict mathematical definition, have infinite energy, a sinusoid for example. How relevant can an analysis be if it is done with signals that cannot physically exist? Very relevant! The reason mathematical sinusoids have infinite signal energy is that they have always existed and will always exist. Of course, real signals that we use in systems and call sinusoids never have that exact quality. They all had to begin at some finite time and, presumably, they will all end at some later finite time and, therefore, are actually time limited and have finite signal energy. But in much system analysis we perform a steady-state analysis of a system in which all signals are treated as though they are periodic. Therefore, the analysis is still very relevant and useful because it is a good approximation to reality. There are other signals with infinite signal energy or infinite average signal power. All periodic signals are power signals [except for the trivial signal, $x(t) = 0$] because they all endure for an infinite time. Figure 2.84 shows examples of CT and DT energy and power signals.

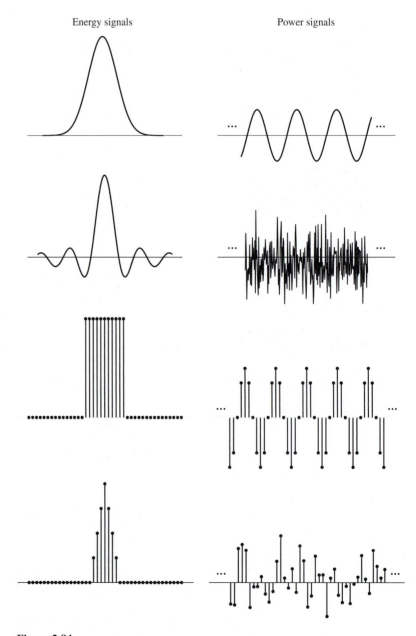

**Figure 2.84**
Examples of CT and DT energy and power signals.

**EXAMPLE 2.12**

Using MATLAB find the signal energy or power of the following signals:

a. $x(t) = \mathrm{tri}\left(\dfrac{t-3}{10}\right)$

b. $x[n] = e^{-|n/10|}\sin\left(\dfrac{2\pi n}{4}\right)$

c. A periodic signal of fundamental period 10, described over one period by $x(t) = -3t$, $-5 < t < 5$.

d. $x[n] = 4\,\mathrm{comb}_5[n] - 7\,\mathrm{comb}_7[n]$

Then compare the results with analytical calculations.

### ■ Solution

```
%       Program to compute the signal energy or power of some example signals.
%       (a)
dt = 0.1 ; t = -7:dt:13 ;           %   Set up a vector of times at which to
                                    %   compute the function. Time interval
                                    %   is 0.1.
x = tri((t-3)/10) ;                 %   Compute the function values and
                                    %   their squares.
xsq = x.^2 ;
Ex = trapz(t,xsq) ;                 %   Use trapezoidal-rule numerical
                                    %   integration to find the area under
                                    %   the function squared and display the
                                    %   result.
disp(['(a) Ex = ',num2str(Ex)]) ;
%       (b)
n = -100:100 ;                      %   Set up a vector of discrete times at
                                    %   which to compute the value of the
                                    %   function.
%   Compute the value of the function and its square.
x = exp(-abs(n/10)).*sin(2*pi*n/4) ; xsq = x.^2 ;
Ex = sum(xsq) ;                     %   Use the sum function in MATLAB to
                                    %   find the total energy and display
                                    %   the result.
disp(['(b) Ex = ',num2str(Ex)]) ;
%       (c)
T0 = 10 ;                           %   The fundamental period is 10.
dt = 0.1 ; t = -5:dt:5 ;            %   Set up a vector of times at which to
                                    %   compute the function. Time interval
                                    %   is 0.1.
x = -3*t ; xsq = x.^2 ;             %   Compute the function values and
                                    %   their squares over one fundamental
                                    %   period.
```

```
Px = trapz(t,xsq)/T0 ;              %      Use trapezoidal rule numerical
                                    %      integration to find the area under
                                    %      the function squared, divide the
                                    %      period, and display the result.
disp(['(c) Px = ',num2str(Px)]) ;
%          (d)
N0 = 35 ;                           %      The fundamental period is 35.
n = 0:N0-1 ;                        %      Set up a vector of discrete times
                                    %      over one period at which to compute
                                    %      the value of the function.
%       Compute the value of the function and its square.
x = 4*combN0n(5,n) - 7*combN0n(7,n) ; xsq = x.^2 ;
Px = sum(xsq)/N0 ;                  %      Use the sum function in MATLAB to
                                    %      find the average power and display
                                    %      the result.
disp(['(d) Px = ',num2str(Px)]) ;
```

The output of this program is

(a)  Ex = 6.667
(b)  Ex = 4.9668
(c)  Px = 75.015
(d)  Px = 8.6

The analytical computations are as follows:

a.  $$E_x = \int_{-\infty}^{\infty} |x(t)|^2\, dt = \int_{-\infty}^{\infty} \left|\text{tri}\left(\frac{t-3}{10}\right)\right|^2 dt = \int_{-\infty}^{\infty} \left|\text{tri}\left(\frac{\lambda}{10}\right)\right|^2 d\lambda = 2\int_{0}^{10}\left(1 - \frac{\lambda}{10}\right)^2 d\lambda$$

$$= 2\int_{0}^{10}\left(1 - \frac{\lambda}{5} + \frac{\lambda^2}{100}\right) d\lambda = 2\left[\lambda - \frac{\lambda^2}{10} + \frac{\lambda^3}{300}\right]_{0}^{10} = \frac{20}{3} \cong 6.667 \qquad \text{Check.}$$

b.  $$E_r = \sum_{n=-\infty}^{\infty} |x[n]|^2 = \sum_{n=-\infty}^{\infty} \left|e^{-|n/10|} \sin\left(\frac{2\pi n}{4}\right)\right|^2$$

$$= \sum_{n=0}^{\infty} \left|e^{-(n/10)} \sin\left(\frac{2\pi n}{4}\right)\right|^2 + \sum_{n=-\infty}^{0} \left|e^{n/10} \sin\left(\frac{2\pi n}{4}\right)\right|^2 - \underbrace{|x[0]|^2}_{=0}$$

$$= \sum_{n=0}^{\infty} e^{-(n/5)} \sin^2\left(\frac{2\pi n}{4}\right) + \sum_{n=-\infty}^{0} e^{n/5} \sin^2\left(\frac{2\pi n}{4}\right)$$

$$= \frac{1}{2}\sum_{n=0}^{\infty} e^{-(n/5)}(1 - \cos(\pi n)) + \frac{1}{2}\sum_{n=-\infty}^{0} e^{n/5}(1 - \cos(\pi n))$$

$$= \frac{1}{2}\sum_{n=0}^{\infty} e^{-(n/5)}(1 - \cos(\pi n)) + \frac{1}{2}\sum_{n=0}^{\infty} e^{-(n/5)}(1 - \cos(\pi n))$$

$$= \sum_{n=0}^{\infty} e^{-(n/5)}(1 - \cos(\pi n))$$

$$= \sum_{n=0}^{\infty} \left( e^{-(n/5)} - e^{-(n/5)} \frac{e^{j\pi n} + e^{-j\pi n}}{2} \right)$$

$$= \sum_{n=0}^{\infty} e^{-(n/5)} - \frac{1}{2} \left[ \sum_{n=0}^{\infty} e^{[j\pi - (1/5)]n} + \sum_{n=0}^{\infty} e^{[-j\pi - (1/5)]n} \right]$$

Using the formula for the sum of a geometric series,

$$\sum_{n=0}^{\infty} r^n = \frac{1}{1-r}, \qquad |r| < 1$$

$$E_x = \frac{1}{1 - e^{-(1/5)}} - \frac{1}{2} \left[ \frac{1}{1 - e^{[j\pi - (1/5)]}} + \frac{1}{1 - e^{[-j\pi - (1/5)]}} \right]$$

$$= \frac{1}{1 - e^{-(1/5)}} - \frac{1}{2} \left[ \frac{2 - e^{[-j\pi - (1/5)]} - e^{[j\pi - (1/5)]}}{1 - e^{[j\pi - (1/5)]} - e^{[-j\pi - (1/5)]} + e^{-(2/5)}} \right]$$

$$= \frac{1}{1 - e^{-(1/5)}} - \frac{1}{2} \left[ \frac{2 - e^{-(1/5)}(e^{-j\pi} + e^{j\pi})}{1 - 2e^{-(1/5)} \cos(\pi) + e^{-(2/5)}} \right]$$

$$= \frac{1}{1 - e^{-(1/5)}} - \frac{1 + e^{-(1/5)}}{1 + 2e^{-(1/5)} + e^{-(2/5)}}$$

$$= \frac{1}{0.1813} - \frac{1.8187}{1 + 1.637 + 0.67} = 4.966 \qquad \text{Check.}$$

c. $$P_x = \frac{1}{10} \int_{-5}^{5} (-3t)^2 \, dt = \frac{1}{5} \int_{0}^{5} 9t^2 \, dt = \frac{1}{5}(3t^3)_0^5 = \frac{375}{5} = 75 \qquad \text{Check.}$$

d. $$P_x = \frac{1}{N_0} \sum_{n=(N_0)} |x[n]|^2 = \frac{1}{N_0} \sum_{n=0}^{N_0-1} |x[n]|^2 = \frac{1}{35} \sum_{n=0}^{34} |4 \, \text{comb}_5[n] - 7 \, \text{comb}_7[n]|^2$$

The impulses in the two comb functions only coincide at integer multiples of 35. Therefore, in this summation range they coincide only at $n = 0$. The net impulse strength at $n = 0$ is therefore $-3$. All the other impulses occur alone, and the sum of the squares is the same as the square of the sum. Therefore,

$$P_x = \frac{1}{35} \left( \underbrace{(-3)^2}_{n=0} + \underbrace{4^2}_{n=5} + \underbrace{(-7)^2}_{n=7} + \underbrace{4^2}_{n=10} + \underbrace{(-7)^2}_{n=14} + \underbrace{4^2}_{n=15} + \underbrace{4^2}_{n=20} + \underbrace{(-7)^2}_{n=21} + \underbrace{4^2}_{n=25} + \underbrace{(-7)^2}_{n=28} + \underbrace{4^2}_{n=30} \right)$$

$$= \frac{9 + 6 \times 4^2 + 4 \times (-7)^2}{35} = \frac{9 + 96 + 196}{35} = 8.6 \qquad \text{Check.}$$

## 2.15  SUMMARY OF IMPORTANT POINTS

1. The terms *continuous* and *continuous-time* mean different things.
2. A DT signal can be formed from a CT signal by sampling.
3. A DT function is not defined at noninteger values of discrete time.
4. Two CT signals which differ only at a finite number of points have exactly the same effect on any real physical CT system.
5. A CT impulse, although very useful in signal and system analysis, is not a function in the ordinary sense.
6. The order in which multiple functional transformations are done is significant.
7. DT signals formed by sampling periodic CT signals may have a different period or may even be aperiodic.
8. Two different-looking analytical descriptions of DT functions may, in fact, be identical.
9. A time-shifted version of a DT function is only defined for integer shifts in discrete time.
10. Time-scaling a DT function can produce decimation or undefined values, phenomena which do not occur when time-scaling CT functions.
11. *Signal energy* is, in general, not the same thing as the actual energy delivered by a signal.
12. A signal with finite signal energy is called an *energy* signal, and a signal with finite average power is called a *power* signal.

## EXERCISES WITH ANSWERS

1.  If $g(t) = 7e^{-2t-3}$, write out and simplify each function.

    *a.*  $g(3)$  $\qquad\qquad$ *b.*  $g(2 - t)$

    *c.*  $g\left(\dfrac{t}{10} + 4\right)$  $\qquad$ *d.*  $g(jt)$

    *e.*  $\dfrac{g(jt) + g(-jt)}{2}$

    *f.*  $\dfrac{g((jt-3)/2) + g((-jt-3)/2)}{2}$

**Answers:**
$7\cos(t),\quad 7e^{-7+2t},\quad 7e^{-j2t-3},\quad 7e^{-(t/5)-11},\quad 7e^{-3}\cos(2t),\quad 7e^{-9}$

2.  If $g(x) = x^2 - 4x + 4$, write out and simplify each function.

    *a.*  $g(z)$  $\qquad\qquad$ *b.*  $g(u + v)$

    *c.*  $g(e^{jt})$  $\qquad\qquad$ *d.*  $g(g(t))$

    *e.*  $g(2)$

**Answers:**
$(e^{jt} - 2)^2,\quad z^2 - 4z + 4,\quad 0,\quad u^2 + v^2 + 2uv - 4u - 4v + 4,$
$t^4 - 8t^3 + 20t^2 - 16t + 4$

3. What would be the numerical value of g in each of the following MATLAB instructions?

   *a.*   `t = 3 ; g = sin(t) ;`

   *b.*   `x = 1:5 ; g = cos(pi*x) ;`

   *c.*   `f = -1:0.5:1 ; w = 2*pi*f ; g = 1./(1+j*w') ;`

**Answers:**

$$0.1411, \quad [-1, 1, -1, 1, -1], \quad \begin{bmatrix} 0.0247 + j0.155 \\ 0.0920 + j0.289 \\ 1 \\ 0.0920 - j0.289 \\ 0.0247 - j0.155 \end{bmatrix}$$

4. Let two functions be defined by

$$x_1(t) = \begin{cases} 1 & \sin(20\pi t) \geq 0 \\ -1 & \sin(20\pi t) < 0 \end{cases} \quad \text{and} \quad x_2(t) = \begin{cases} t & \sin(2\pi t) \geq 0 \\ -t & \sin(2\pi t) < 0 \end{cases}.$$

Graph the product of these two functions versus time over the time range $-2 < t < 2$.

**Answer:**

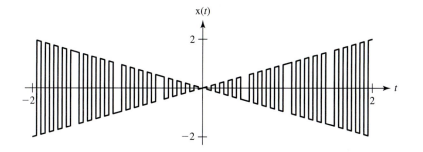

5. For each function g(t), sketch $g(-t)$, $-g(t)$, $g(t-1)$, and $g(2t)$.

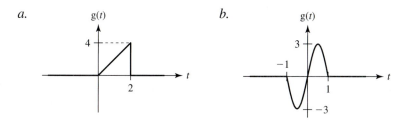

**Answers:**

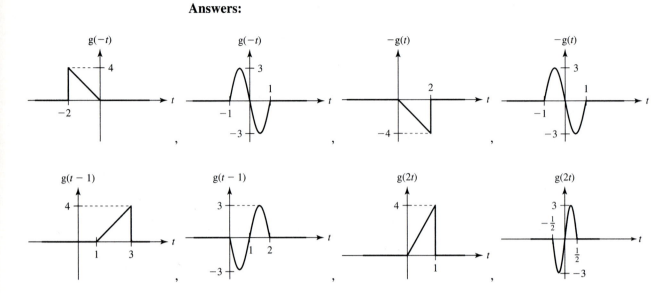

, , , ,

, , ,

6.   A function G($f$) is defined by

$$G(f) = e^{-j2\pi f} \, \text{rect}\left(\frac{f}{2}\right).$$

Graph the magnitude and phase of G($f - 10$) + G($f + 10$) over the range $-20 < f < 20$.

**Answer:**

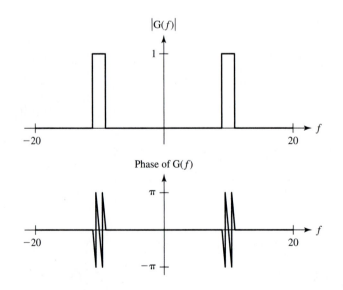

7.   Sketch the derivatives of these signals.
  *a.*   x($t$) = sinc($t$)
  *b.*   x($t$) = $(1 - e^{-t})$u($t$)

**Answers:**

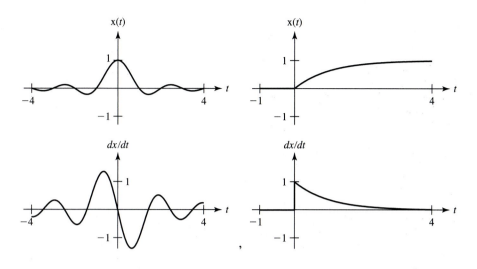

,

8.  Sketch the integral from negative infinity to time $t$ of these functions which are zero for all time before time $t = 0$.

    *a.*                                    *b.*

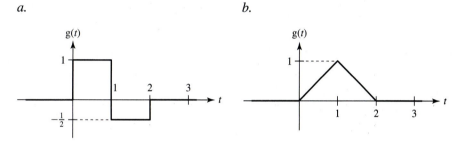

**Answers:**

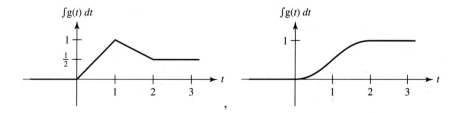

,

9.  Find the even and odd parts of these functions.

    *a.*  $g(t) = 2t^2 - 3t + 6$          *b.*  $g(t) = 20\cos\left(40\pi t - \dfrac{\pi}{4}\right)$

    *c.*  $g(t) = \dfrac{2t^2 - 3t + 6}{1 + t}$          *d.*  $g(t) = \text{sinc}(t)$

    *e.*  $g(t) = t(2 - t^2)(1 + 4t^2)$     *f.*  $g(t) = t(2 - t)(1 + 4t)$

**Answers:**

$$t(2 - 4t^2), \quad \frac{20}{\sqrt{2}}\cos(40\pi t), \quad 0, \quad -t\frac{2t^2 + 9}{1 - t^2}, \quad 7t^2, \quad 0,$$

$$\frac{20}{\sqrt{2}}\sin(40\pi t), \quad 2t^2 + 6, \quad t(2 - t^2)(1 + 4t^2), \quad \frac{6 + 5t^2}{1 - t^2}, \quad \text{sinc}(t), \quad -3t$$

**10.**   Sketch the even and odd parts of these functions.

*a.*                                                     *b.*

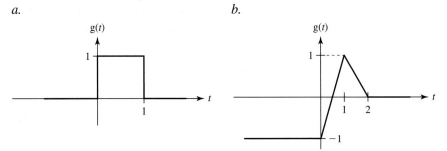

**Answers:**

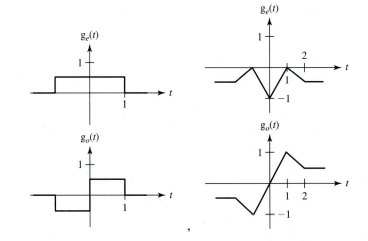

**11.**   Sketch the indicated product or quotient $g(t)$ of these functions.

*a.*                                                     *b.*

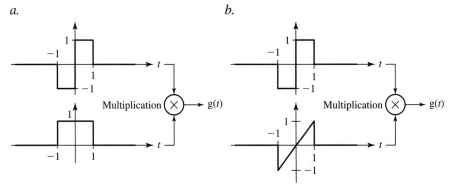

c.
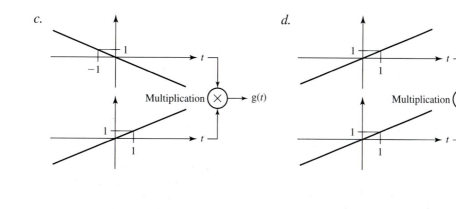

d.

e.
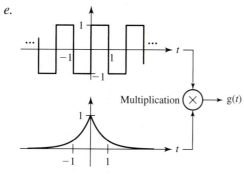

f.
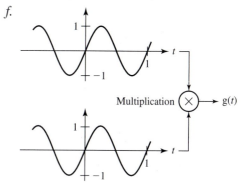

g.
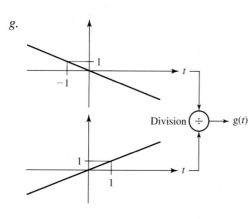

h.
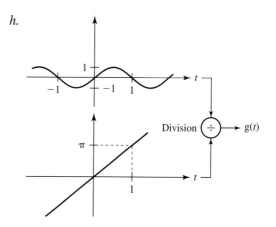

**Answers:**

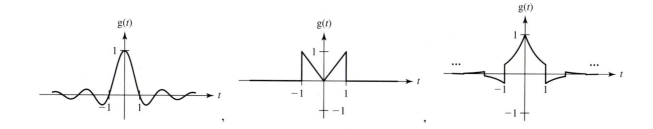

,                    ,

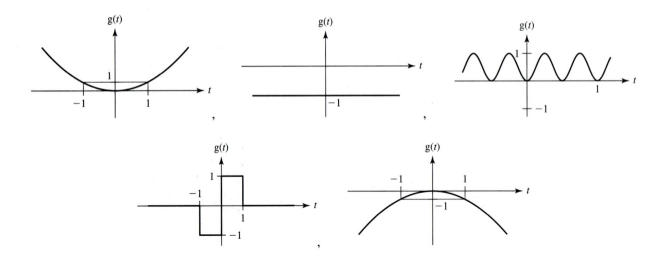

,                                    ,

,

**12.** Use the properties of integrals of even and odd functions to evaluate these integrals in the quickest way.

a. $\displaystyle\int_{-1}^{1} (2 + t)\, dt$

b. $\displaystyle\int_{-(1/20)}^{1/20} [4\,\cos(10\pi t) + 8\,\sin(5\pi t)]\, dt$

c. $\displaystyle\int_{-(1/20)}^{1/20} 4t\,\cos(10\pi t)\, dt$

d. $\displaystyle\int_{-(1/10)}^{1/10} t\,\sin(10\pi t)\, dt$

e. $\displaystyle\int_{-1}^{1} e^{-|t|}\, dt$

f. $\displaystyle\int_{-1}^{1} t e^{-|t|}\, dt$

**Answers:**

$0,\quad \dfrac{8}{10\pi},\quad \dfrac{1}{50\pi},\quad 0,\quad 1.264,\quad 4$

**13.** Find the fundamental period and fundamental frequency of each of these functions.

a. $g(t) = 10\,\cos(50\pi t)$

b. $g(t) = 10\,\cos\left(50\pi t + \dfrac{\pi}{4}\right)$

c. $g(t) = \cos(50\pi t) + \sin(15\pi t)$

d. $g(t) = \cos(2\pi t) + \sin(3\pi t) + \cos\left(5\pi t - \dfrac{3\pi}{4}\right)$

**Answers:**

$2\ \text{s},\quad \dfrac{1}{25}\ \text{s},\quad 2.5\ \text{Hz},\quad \dfrac{1}{25}\ \text{s},\quad \dfrac{1}{2}\ \text{Hz},\quad 0.4\ \text{s},\quad 25\ \text{Hz},\quad 25\ \text{Hz}$

**14.** Find the fundamental period and fundamental frequency of g(t).

*a.*    *b.*

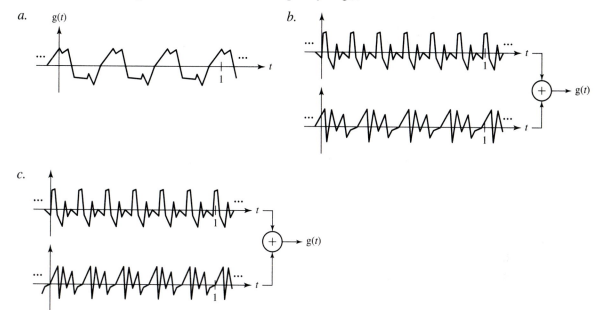

*c.*

**Answers:**

$$1 \text{ Hz}, \quad 2 \text{ Hz}, \quad \frac{1}{2} \text{ s}, \quad 1 \text{ s}, \quad \frac{1}{3} \text{ s}, \quad 3 \text{ Hz}$$

**15.** Plot these DT functions.

*a.* $\quad x[n] = 4 \cos\left(\dfrac{2\pi n}{12}\right) - 3 \sin\left(\dfrac{2\pi(n-2)}{8}\right) \qquad -24 \le n < 24$

*b.* $\quad x[n] = 3ne^{-|n/5|} \qquad -20 \le n < 20$

*c.* $\quad x[n] = 21\left(\dfrac{n}{2}\right)^2 + 14n^3 \qquad -5 \le n < 5$

**Answers:**

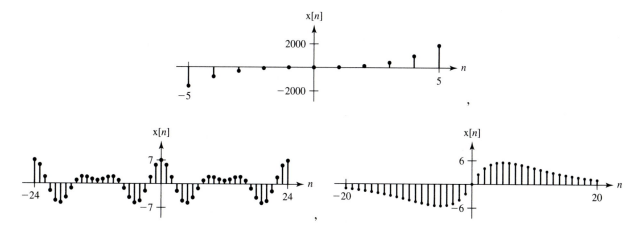

16. Let $x_1[n] = 5\cos(2\pi n/8)$ and $x_2[n] = -8e^{-(n/6)^2}$. Plot the following combinations of those two signals over the DT range, $-20 \le n < 20$. If a signal has some defined and some undefined values, just plot the defined values.

a. $x[n] = x_1[n]x_2[n]$

b. $x[n] = 4x_1[n] + 2x_2[n]$

c. $x[n] = x_1[2n]x_2[3n]$

d. $x[n] = \dfrac{x_1[2n]}{x_2[-n]}$

e. $x[n] = 2x_1\left[\dfrac{n}{2}\right] + 4x_2\left[\dfrac{n}{3}\right]$

**Answers:**

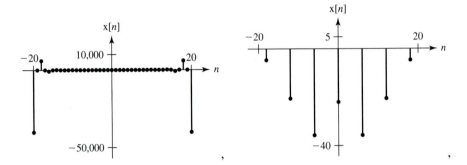

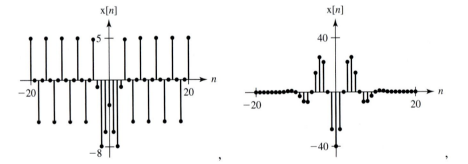

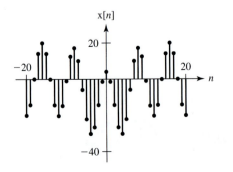

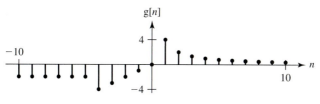

**Figure E17**

**17.** A function g[n] (see Figure E17) is defined by

$$g[n] = \begin{cases} -2, & n < -4 \\ n, & -4 \le n < 1. \\ \dfrac{4}{n}, & 1 \le n \end{cases}$$

Sketch

a.  g[−n],     b.  g[2 − n],     c.  g[2n],     d.  g[n/2]

**Answers:**

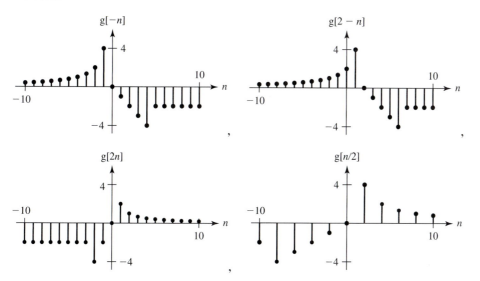

**18.** Sketch the backward differences of these DT functions.

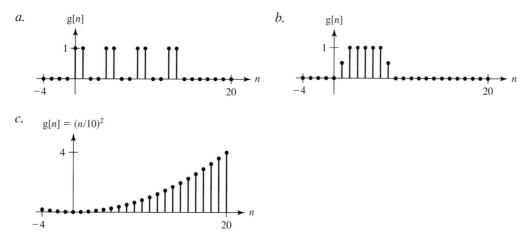

**Answers:**

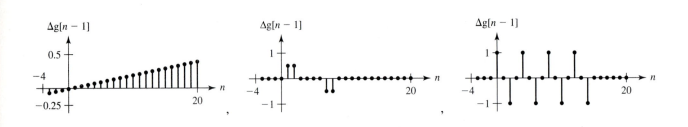

19.   Sketch the accumulation g[n] from negative infinity to n of each of these DT functions h[n].

   *a.*   $h[n] = \delta[n]$       *b.*   $h[n] = u[n]$

   *c.*   $h[n] = \cos\left(\dfrac{2\pi n}{16}\right)u[n]$       *d.*   $h[n] = \cos\left(\dfrac{2\pi n}{8}\right)u[n]$

   *e.*   $h[n] = \cos\left(\dfrac{2\pi n}{16}\right)u[n + 8]$

**Answers:**

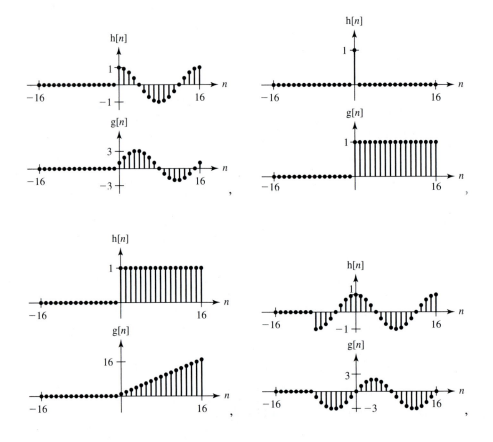

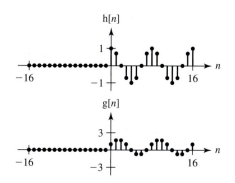

20.  Find and sketch the even and odd parts of these functions.

    *a.*  $g[n] = u[n] - u[n-4]$        *b.*  $g[n] = e^{-(n/4)}u[n]$

    *c.*  $g[n] = \cos\left(\dfrac{2\pi n}{4}\right)$        *d.*  $g[n] = \sin\left(\dfrac{2\pi n}{4}\right)u[n]$

**Answers:**

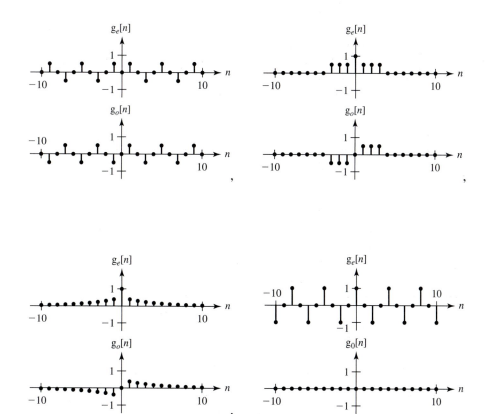

**21.** Sketch g[n].

*a.*

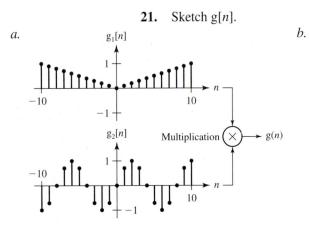

*b.*

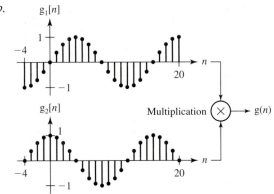

*c.*

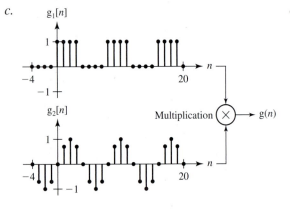

*d.*

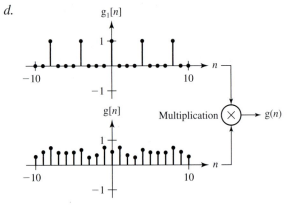

**Answers:**

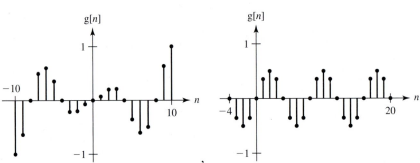

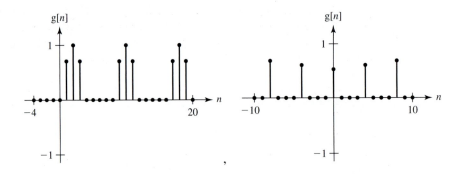

**22.** Find the fundamental DT period $N_0$ and frequency $F_0$ of these functions.

*a.* $g[n] = \cos\left(\dfrac{2\pi n}{10}\right)$ 

*b.* $g[n] = \cos\left(\dfrac{\pi n}{10}\right)$

*c.* $g[n] = \cos\left(\dfrac{2\pi n}{5}\right) + \cos\left(\dfrac{2\pi n}{7}\right)$ 

*d.* $g[n] = e^{j(2\pi n/20)} + e^{-j(2\pi n/20)}$

*e.* $g[n] = e^{-j(2\pi n/3)} + e^{-j(2\pi n/4)}$

**Answers:**

$N_0 = 10, \quad F_0 = \dfrac{1}{10}, \quad N_0 = 35, \quad F_0 = \dfrac{1}{35}, \quad N_0 = 20, \quad F_0 = \dfrac{1}{20},$

$N_0 = 12, \quad F_0 = \dfrac{1}{12}, \quad N_0 = 20, \quad F_0 = \dfrac{1}{20}$

**23.** Graph the following functions and determine from the graphs the fundamental period of each one (if it is periodic).

*a.* $g[n] = 5\,\sin\left(\dfrac{2\pi n}{4}\right) + 8\,\cos\left(\dfrac{2\pi n}{6}\right)$

*b.* $g[n] = 5\,\sin\left(\dfrac{7\pi n}{12}\right) + 8\,\cos\left(\dfrac{14\pi n}{8}\right)$

*c.* $g[n] = \mathrm{Re}\left(e^{j\pi n} + e^{-j(\pi n/3)}\right)$ 

*d.* $g[n] = \mathrm{Re}\left(e^{jn} + e^{-j(n/3)}\right)$

**Answers:**

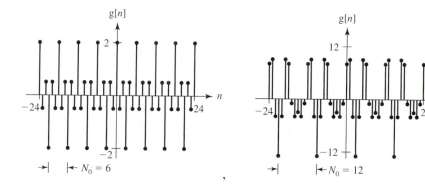

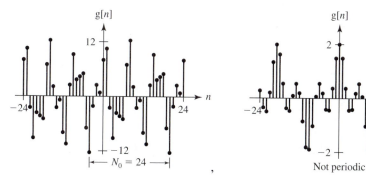

**24.** Find the signal energy of these signals.

a. $x(t) = 2 \operatorname{rect}(t)$              b.   $x(t) = A(u(t) - u(t - 10))$

c. $x(t) = u(t) - u(10 - t)$         d.   $x(t) = \operatorname{rect}(t) \cos(2\pi t)$

e. $x(t) = \operatorname{rect}(t) \cos(4\pi t)$      f.   $x(t) = \operatorname{rect}(t) \sin(2\pi t)$

g. $x[n] = A \operatorname{rect}_{N_0}[n]$           h.   $x[n] = A\delta[n]$

i. $x[n] = \operatorname{comb}_{N_0}[n]$           j.   $x[n] = \operatorname{ramp}[n]$

k. $x[n] = \operatorname{ramp}[n] - 2 \operatorname{ramp}[n - 4] + \operatorname{ramp}[n - 8]$

**Answers:**

$\infty, \quad 44, \quad \infty, \quad (2N_0 + 1)A^2, \quad \dfrac{1}{2}, \quad \infty, \quad 10A^2, \quad \dfrac{1}{2}, \quad 4, \quad A^2, \quad \dfrac{1}{2}$

**25.** Find the signal power of these signals.

a. $x(t) = A$                                    b.   $x(t) = u(t)$

c. $x(t) = A \cos(2\pi f_0 t + \theta)$         d.   $x(t) = A \displaystyle\sum_{n=-\infty}^{\infty} \operatorname{rect}(t - 2n)$

e. $x(t) = 2A\left[-\dfrac{1}{2} + \displaystyle\sum_{n=-\infty}^{\infty} \operatorname{rect}(t - 2n)\right]$     f.   $x[n] = A$

g. $x[n] = u[n]$                                 h.   $x[n] = A \displaystyle\sum_{m=-\infty}^{\infty} \operatorname{rect}_2[n - 8m]$

i. $x[n] = \operatorname{comb}_{N_0}[n]$          j.   $x[n] = \operatorname{ramp}[n]$

**Answers:**

$A^2, \quad \dfrac{5A^2}{8}, \quad \dfrac{A^2}{2}, \quad A^2, \quad \infty, \quad \dfrac{1}{2}, \quad A^2, \quad \dfrac{1}{N_0}, \quad \dfrac{A^2}{2}, \quad \dfrac{1}{2}$

# EXERCISES WITHOUT ANSWERS

**26.** Using MATLAB, plot the CT signal $x(t) = \sin(2\pi t)$ over the time range $0 < t < 10$, with the following choices of the time resolution $\Delta t$ of the plot. Explain why the plots look the way they do.

a. $\dfrac{1}{24}$          b. $\dfrac{1}{12}$          c. $\dfrac{1}{4}$          d. $\dfrac{1}{2}$

e. $\dfrac{2}{3}$          f. $\dfrac{5}{6}$          g.  1

**27.** Given the function definitions, find the function values.

a. $g(t) = 100 \sin\left(200\pi t + \dfrac{\pi}{4}\right)$      $g(0.001)$

b. $g(t) = 13 - 4t + 6t^2$      $g(2)$

c. $g(t) = -5e^{-2t}e^{-j2\pi t}$      $g\left(\dfrac{1}{4}\right)$

**28.** Sketch these CT exponential and trigonometric functions.

a. $g(t) = 10 \cos(100\pi t)$          b.   $g(t) = 40 \cos(60\pi t) + 20 \sin(60\pi t)$

c. $g(t) = 5e^{-(t/10)}$                    d.   $g(t) = 5e^{-(t/2)} \cos(2\pi t)$

**29.** Sketch these CT singularity and related functions.

a. $g(t) = 2u(4 - t)$

b. $g(t) = u(2t)$

c. $g(t) = 5 \, \text{sgn}(t - 4)$

d. $g(t) = 1 + \text{sgn}(4 - t)$

e. $g(t) = 5 \, \text{ramp}(t + 1)$

f. $g(t) = -3 \, \text{ramp}(2t)$

g. $g(t) = 2\delta(t + 3)$

h. $g(t) = 6\delta(3t + 9)$

i. $g(t) = -4\delta(2(t - 1))$

j. $g(t) = 2 \, \text{comb}\left(t - \dfrac{1}{2}\right)$

k. $g(t) = 8 \, \text{comb}(4t)$

l. $g(t) = -3 \, \text{comb}\left(\dfrac{t + 1}{2}\right)$

m. $g(t) = 2 \, \text{rect}\left(\dfrac{t}{3}\right)$

n. $g(t) = 4 \, \text{rect}\left(\dfrac{t + 1}{2}\right)$

o. $g(t) = \text{tri}(4t)$

p. $g(t) = -6 \, \text{tri}\left(\dfrac{t - 1}{2}\right)$

q. $g(t) = 5 \, \text{sinc}\left(\dfrac{t}{2}\right)$

r. $g(t) = -\text{sinc}(2(t + 1))$

s. $g(t) = -10 \, \text{drcl}(t, \, 4)$

t. $g(t) = 5 \, \text{drcl}\left(\dfrac{t}{4}, \, 7\right)$

u. $g(t) = -3 \, \text{rect}(t - 2)$

v. $g(t) = 0.1 \, \text{rect}\left(\dfrac{t - 3}{4}\right)$

w. $g(t) = -4 \, \text{tri}\left(\dfrac{3 + t}{2}\right)$

x. $g(t) = 4 \, \text{sinc}(5(t - 3))$

y. $g(t) = 4 \, \text{sinc}(5t - 3)$

**30.** Sketch these CT functions.

a. $g(t) = u(t) - u(t - 1)$

b. $g(t) = \text{rect}\left(t - \dfrac{1}{2}\right)$

c. $g(t) = -4 \, \text{ramp}(t)u(t - 2)$

d. $g(t) = \text{sgn}(t) \, \sin(2\pi t)$

e. $g(t) = 5e^{-(t/4)}u(t)$

f. $g(t) = \text{rect}(t) \, \cos(2\pi t)$

g. $g(t) = -6 \, \text{rect}(t) \, \cos(3\pi t)$

h. $g(t) = \text{rect}(t) \, \text{tri}(t)$

i. $g(t) = \text{rect}(t) \, \text{tri}\left(t + \dfrac{1}{2}\right)$

j. $g(t) = u\left(t + \dfrac{1}{2}\right) \text{ramp}\left(\dfrac{1}{2} - t\right)$

k. $g(t) = \text{tri}^2(t)$

l. $g(t) = \text{sinc}^2(t)$

m. $g(t) = |\text{sinc}(t)|$

n. $g(t) = \dfrac{d}{dt}(\text{tri}(t))$

o. $g(t) = \text{rect}\left(t + \dfrac{1}{2}\right) - \text{rect}\left(t - \dfrac{1}{2}\right)$

p. $g(t) = \displaystyle\int_{-\infty}^{t} [\delta(\lambda + 1) - 2\delta(\lambda) + \delta(\lambda - 1)] \, d\lambda$

q. $g(t) = 3 \, \text{tri}\left(\dfrac{2t}{3}\right) + 3 \, \text{rect}\left(\dfrac{t}{3}\right)$

$r.$  $g(t) = 6\, \text{tri}\left(\dfrac{t}{3}\right) \text{rect}\left(\dfrac{t}{3}\right)$        $s.$  $g(t) = 4\, \text{sinc}(2t)\, \text{sgn}(-t)$

$t.$  $g(t) = 2\, \text{ramp}(t)\, \text{rect}\left(\dfrac{t-1}{2}\right)$        $u.$  $g(t) = 4\, \text{tri}\left(\dfrac{t-2}{2}\right) u(2-t)$

$v.$  $g(t) = 3\, \text{rect}\left(\dfrac{t}{4}\right) - 6\, \text{rect}\left(\dfrac{t}{2}\right)$        $w.$  $g(t) = 10\, \text{drcl}\left(\dfrac{t}{4}, 5\right) \text{rect}\left(\dfrac{t}{8}\right)$

**31.**  Using MATLAB, for each given function plot the original function and the transformed function.

   $a.$  $g(t) = 10\, \cos(20\pi t)\, \text{tri}(t)$    $5g(2t)$    vs. $t$

   $b.$  $g(t) = \begin{cases} -2 & t < -1 \\ 2t & -1 < t < 1 \\ 3 - t^2 & 1 < t < 3 \\ -6 & t > 3 \end{cases}$    $-3g(4 - t)$    vs. $t$

   $c.$  $g(t) = \text{Re}\left(e^{j\pi t} + e^{j1.1\pi t}\right)$    $g\left(\dfrac{t}{4}\right)$    vs. $t$

   $d.$  $G(f) = \left| \dfrac{5}{f^2 - j2 + 3} \right|$    $|G(10(f - 10)) + G(10(f + 10))|$    vs. $f$

**32.**  Let two signals be defined by

$$x_1(t) = \begin{cases} 1 & \cos(2\pi t) \geq 1 \\ 0 & \cos(2\pi t) < 1 \end{cases} \quad \text{and} \quad x_2(t) = \sin\left(\dfrac{2\pi t}{10}\right).$$

Plot these products over the time range $-5 < t < 5$.

   $a.$  $x_1(2t)x_2(-t)$        $b.$  $x_1\left(\dfrac{t}{5}\right) x_2(20t)$

   $c.$  $x_1\left(\dfrac{t}{5}\right) x_2(20(t + 1))$        $d.$  $x_1\left(\dfrac{t-2}{5}\right) x_2(20t)$

**33.**  Given the graphical definition of a function, graph the indicated transformation(s).

   $a.$  $g(t) = 0, t < -2$ or $t > 6$. Graph $t \to 2t$ and $g(t) \to -3g(-t)$.

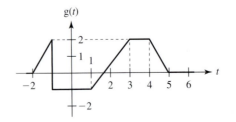

b. g(t) is periodic with fundamental period 4. Graph $t \rightarrow t + 4$ and
$$g(t) \rightarrow -2g\left(\frac{t-1}{2}\right).$$

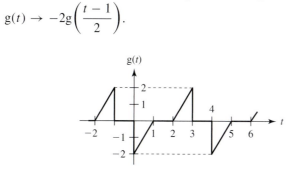

**34.** For each graphed pair of functions determine what transformation has been performed and write a correct analytical expression for the transformation. For part (b), assuming g(t) is periodic with fundamental period 2, find two different transformations which yield the same result.

a.

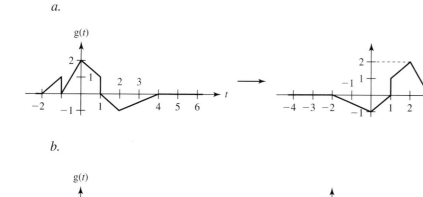

b.

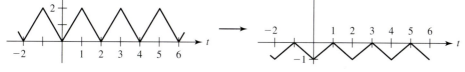

**35.** Sketch the magnitude and phase of each function versus $f$.

a. $G(f) = \text{sinc}(f)\, e^{-j(\pi f/8)}$     b. $G(f) = \dfrac{jf}{1 + j(f/10)}$

c. $G(f) = \left[\text{rect}\left(\dfrac{f - 1000}{100}\right) + \text{rect}\left(\dfrac{f + 1000}{100}\right)\right] e^{-j(\pi f/500)}$

d. $G(f) = \dfrac{1}{250 - f^2 + j3f}$

e. $G(f) = \text{comb}(100\, f)\, \text{sinc}(25\, f)\, e^{j(\pi f/50)}$

**36.** Graph versus $f$, in the range $-4 < f < 4$, the magnitude and phase of each function.

a. $X(f) = \text{sinc}(f)$     b. $X(f) = 2\,\text{sinc}(f)\, e^{-j4\pi f}$

   c.   $X(f) = 5 \text{ rect}(2f) e^{+j2\pi f}$        d.   $X(f) = 10 \text{ sinc}^2\left(\dfrac{f}{4}\right)$

   e.   $X(f) = j5\delta(f+2) - j5\delta(f-2)$     f.   $X(f) = 2 \text{ comb}(4f) e^{-j\pi f}$

**37.** Sketch the even and odd parts of these CT signals.

   a.   $x(t) = \text{rect}(t-1)$             b.   $x(t) = \text{tri}\left(t - \dfrac{3}{4}\right) + \text{tri}\left(t + \dfrac{3}{4}\right)$

   c.   $x(t) = 4 \text{ sinc}\left(\dfrac{t-1}{2}\right)$        d.   $x(t) = 2 \sin\left(4\pi t - \dfrac{\pi}{4}\right) \text{rect}(t)$

**38.** Let the CT unit impulse function be represented by the limit

$$\delta(x) = \lim_{a\to 0} \frac{1}{a} \text{ tri}\left(\frac{x}{a}\right), \qquad a > 0.$$

The function, $(1/a) \text{ tri}(x/a)$ has an area of one regardless of the value of $a$.

   a.   What is the area of the function $\delta(4x) = \lim_{a\to 0} \dfrac{1}{a} \text{ tri}\left(\dfrac{4x}{a}\right)$?

   b.   What is the area of the function $\delta(-6x) = \lim_{a\to 0} \dfrac{1}{a} \text{ tri}\left(\dfrac{-6x}{a}\right)$?

   c.   What is the area of the function $\delta(bx) = \lim_{a\to 0} \dfrac{1}{a} \text{ tri}\left(\dfrac{bx}{a}\right)$ for $b$ positive and for $b$ negative?

**39.** Using a change of variable and the definition of the unit impulse, prove that

$$\delta(a(t - t_0)) = \frac{1}{|a|}\delta(t - t_0).$$

**40.** Using the results of Exercise 39, show that

   a.   $\text{comb}(ax) = \dfrac{1}{|a|} \displaystyle\sum_{n=-\infty}^{\infty} \delta\left(x - \dfrac{n}{a}\right).$

   b.   The average value of $\text{comb}(ax)$ is one, independent of the value of $a$.

   c.   A comb function of the form, $(1/a) \text{ comb}(t/a)$ is a sequence of *unit* impulses spaced $a$ units apart.

   d.   Even though $\delta(at) = (1/|a|)\delta(t)$, $\text{comb}(ax) \neq (1/|a|) \text{ comb}(x)$.

**41.** Sketch the generalized derivative of $g(t) = 3 \sin(\pi t/2) \text{ rect}(t)$.

**42.** Sketch the following CT functions.

   a.   $g(t) = 3\delta(3t) + 6\delta(4(t-2))$        b.   $g(t) = 2 \text{ comb}\left(-\dfrac{t}{5}\right)$

   c.   $g(t) = \text{comb}(t) \text{ rect}\left(\dfrac{t}{11}\right)$       d.   $g(t) = 5 \text{ sinc}\left(\dfrac{t}{4}\right)\left[\dfrac{1}{2} \text{ comb}\left(\dfrac{t}{2}\right)\right]$

   e.   $g(t) = \dfrac{1}{2} \displaystyle\int_{-\infty}^{t} \left[\text{comb}\left(\dfrac{\lambda}{2}\right) - \text{comb}\left(\dfrac{\lambda-1}{2}\right)\right] d\lambda$

**43.** What is the numerical value of each of the following integrals?

a. $\displaystyle\int_{-\infty}^{\infty} \delta(t) \cos(48\pi t)\, dt$

b. $\displaystyle\int_{-\infty}^{\infty} \delta(t-5) \cos(\pi t)\, dt$

c. $\displaystyle\int_{0}^{20} \delta(t-8)\, \mathrm{tri}\left(\frac{t}{32}\right) dt$

d. $\displaystyle\int_{0}^{20} \delta(t-8)\, \mathrm{rect}\left(\frac{t}{16}\right) dt$

e. $\displaystyle\int_{-2}^{2} \delta(t-1.5)\, \mathrm{sinc}(t)\, dt$

f. $\displaystyle\int_{-2}^{2} \delta(t-1.5)\, \mathrm{sinc}(4t)\, dt$

**44.** What is the numerical value of each of the following integrals?

a. $\displaystyle\int_{-\infty}^{\infty} \mathrm{comb}(t) \cos(48\pi t)\, dt$

b. $\displaystyle\int_{-\infty}^{\infty} \mathrm{comb}(t) \sin(2\pi t)\, dt$

c. $\displaystyle\int_{0}^{20} \mathrm{comb}\left(\frac{t-2}{4}\right) \mathrm{rect}(t)\, dt$

d. $\displaystyle\int_{-2}^{2} \mathrm{comb}(t)\, \mathrm{sinc}(t)\, dt$

**45.** Sketch the derivatives of these CT functions.

a. $g(t) = \sin(2\pi t)\, \mathrm{sgn}(t)$

b. $g(t) = 2\, \mathrm{tri}\left(\dfrac{t}{2}\right) - 1$

c. $g(t) = |\cos(2\pi t)|$

**46.** Sketch the derivatives of these CT functions. Compare the average values of the magnitudes of the derivatives.

a.

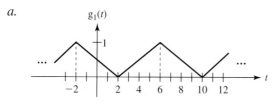

b.

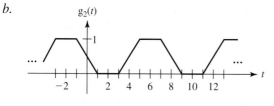

c.

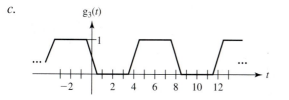

**47.** A function g(t) has this description: It is zero for $t < -5$. It has a slope of $-2$ in the range $-5 < t < -2$. It has the shape of a sine wave of unit amplitude and with a frequency of $\frac{1}{4}$ Hz plus a constant in the range $-2 < t < 2$. For $t > 2$ it decays exponentially toward zero with a time constant of 2 s. It is continuous everywhere. Write an exact mathematical description of this function.

   *a.* Graph g(t) in the range $-10 < t < 10$.

   *b.* Graph g(2t) in the range $-10 < t < 10$.

   *c.* Graph 2g(3 − t) in the range $-10 < t < 10$.

   *d.* Graph $-2g\left(\dfrac{t+1}{2}\right)$ in the range $-10 < t < 10$.

**48.** Find the even and odd parts of each of these CT functions.

   *a.* $g(t) = 10\ \sin(20\pi t)$          *b.* $g(t) = 20t^3$

   *c.* $g(t) = 8 + 7t^2$              *d.* $g(t) = 1 + t$

   *e.* $g(t) = 6t$                   *f.* $g(t) = 4t\ \cos(10\pi t)$

   *g.* $g(t) = \dfrac{\cos(\pi t)}{\pi t}$          *h.* $g(t) = 12 + \dfrac{\sin(4\pi t)}{4\pi t}$

   *i.* $g(t) = (8 + 7t)\ \cos(32\pi t)$    *j.* $g(t) = (8 + 7t^2)\ \sin(32\pi t)$

**49.** Is there a function that is both even and odd simultaneously? Discuss.

**50.** Find and sketch the even and odd parts of the CT signal x(t), shown in Figure E50.

**51.** For each of the following CT signals decide whether it is periodic and, if it is, find the fundamental period.

   *a.* $g(t) = 28\ \sin(400\pi t)$      *b.* $g(t) = 14 + 40\ \cos(60\pi t)$

   *c.* $g(t) = 5t - 2\ \cos(5000\pi t)$   *d.* $g(t) = 28\ \sin(400\pi t) + 12\ \cos(500\pi t)$

   *e.* $g(t) = 10\ \sin(5t) - 4\ \cos(7t)$  *f.* $g(t) = 4\ \sin(3t) + 3\ \sin(\sqrt{3}t)$

**52.** The voltage illustrated in Figure E52 occurs in an analog-to-digital converter. Write a mathematical description of it.

**53.** A signal occurring in a television set is illustrated in Figure E53. Write a mathematical description of it.

**54.** The signal illustrated in Figure E54 is part of a binary-phase-shift-keyed (BPSK) binary data transmission. Write a mathematical description of it.

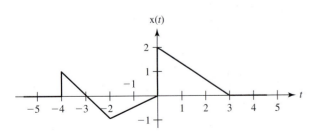

**Figure E50**

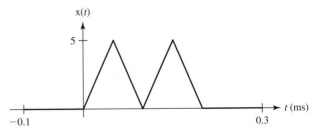

**Figure E52**
Signal occurring in an analog-to-digital converter.

**Figure E53**
Signal occurring in a television set.

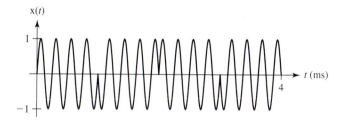

**Figure E54**
BPSK signal.

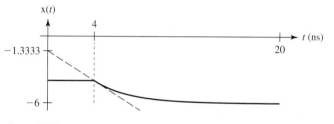

**Figure E55**
Transient response of an *RC* filter.

**55.** The signal illustrated in Figure E55 is the response of an *RC* lowpass filter to a sudden change in excitation. Write a mathematical description of it.

**56.** Find the signal energy of each of these signals:

a. $x(t) = 2 \, \text{rect}(-t)$  

b. $x(t) = \text{rect}(8t)$

c. $x(t) = 3 \, \text{rect}\left(\dfrac{t}{4}\right)$  

d. $x(t) = \text{tri}(2t)$

e. $x(t) = 3 \, \text{tri}\left(\dfrac{t}{4}\right)$  

f. $x(t) = 2 \sin(200\pi t)$

g. $x(t) = \delta(t)$    (*Hint:* First find the signal energy of a signal which approaches an impulse in some limit, and then take the limit.)

h. $x(t) = \dfrac{d}{dt}(\text{rect}(t))$  

i. $x(t) = \displaystyle\int_{-\infty}^{t} \text{rect}(\lambda) \, d\lambda$

j. $x(t) = e^{(-1-j8\pi)t} \, u(t)$

**57.** Find the average signal power of each of these signals:

  a.  $x(t) = 2 \sin(200\pi t)$          b.  $x(t) = \text{comb}(t)$

  c.  $x(t) = e^{j100\pi t}$

**58.** Sketch these DT exponential and trigonometric functions.

  a.  $g[n] = -4 \cos\left(\dfrac{2\pi n}{10}\right)$          b.  $g[n] = -4 \cos(2.2\pi n)$

  c.  $g[n] = -4 \cos(1.8\pi n)$          d.  $g[n] = 2 \cos\left(\dfrac{2\pi n}{6}\right) - 3 \sin\left(\dfrac{2\pi n}{6}\right)$

  e.  $g[n] = \left(\dfrac{3}{4}\right)^n$          f.  $g[n] = 2\,(0.9)^n \sin\left(\dfrac{2\pi n}{4}\right)$

**59.** Sketch these DT singularity functions.

  a.  $g[n] = 2u[n+2]$          b.  $g[n] = u[5n]$

  c.  $g[n] = -2 \text{ ramp}[-n]$          d.  $g[n] = 10 \text{ ramp}\left[\dfrac{n}{2}\right]$

  e.  $g[n] = 7\delta[n-1]$          f.  $g[n] = 7\delta[2(n-1)]$

  g.  $g[n] = -4\delta\left[\dfrac{2}{3}n\right]$          h.  $g[n] = -4\delta\left[\dfrac{2}{3}n - 1\right]$

  i.  $g[n] = 8 \text{ comb}_4[n]$          j.  $g[n] = 8 \text{ comb}_4[2n]$

  k.  $g[n] = \text{rect}_4[n]$          l.  $g[n] = 2 \text{ rect}_5\left[\dfrac{n}{3}\right]$

  m.  $g[n] = \text{tri}\left(\dfrac{n}{5}\right)$          n.  $g[n] = -\text{sinc}\left(\dfrac{n}{4}\right)$

  o.  $g[n] = \text{sinc}\left(\dfrac{n+1}{4}\right)$          p.  $g[n] = \text{drcl}\left(\dfrac{n}{10}, 9\right)$

**60.** Sketch these combinations of DT functions.

  a.  $g[n] = u[n] + u[-n]$          b.  $g[n] = u[n] - u[-n]$

  c.  $g[n] = \cos\left(\dfrac{2\pi n}{12}\right) \text{comb}_3[n]$

  d.  $g[n] = \cos\left(\dfrac{2\pi n}{12}\right) \text{comb}_3\left[\dfrac{n}{2}\right]$

  e.  $g[n] = \cos\left(\dfrac{2\pi(n+1)}{12}\right)u[n+1] - \cos\left(\dfrac{2\pi n}{12}\right)u[n]$

  f.  $g[n] = \displaystyle\sum_{m=0}^{n} \cos\left(\dfrac{2\pi m}{12}\right)u[m]$

  g.  $g[n] = \displaystyle\sum_{m=0}^{n} (\text{comb}_4[m] - \text{comb}_4[m-2])$

  h.  $g[n] = \displaystyle\sum_{m=0}^{n} (\text{comb}_4[m] + \text{comb}_3[m])\text{rect}_4[m]$

    *i.*   $g[n] = \text{comb}_2[n + 1] - \text{comb}_2[n]$

    *j.*   $g[n] = \displaystyle\sum_{m=-\infty}^{n+1} \delta[m] - \sum_{m=-\infty}^{n} \delta[m]$

**61.**  Sketch the magnitude and phase of each function versus $k$.

    *a.*  $G[k] = 20 \sin\left(\dfrac{2\pi k}{8}\right) e^{-j(\pi k/4)}$    *b.*  $G[k] = 20 \cos\left(\dfrac{2\pi k}{8}\right) \text{sinc}\left(\dfrac{k}{40}\right)$

    *c.*  $G[k] = (\delta[k + 8] - 2\delta[k + 4] + \delta[k] - 2\delta[k - 4] + \delta[k - 8])\, e^{j(\pi k/8)}$

**62.**  Given the function definitions, find the function values.

    *a.*  $g[n] = \dfrac{3n + 6}{10} e^{-2n}$    $g[3]$

    *b.*  $g[n] = \text{Re}\left(\left(\dfrac{1 + j}{\sqrt{2}}\right)^n\right)$    $g[5]$

    *c.*  $g[n] = (j2\pi n)^2 + j10\pi n - 4$    $g[4]$

**63.**  Using MATLAB, for each function, plot the original function and the transformed function.

    *a.*  $g[n] = \begin{cases} 5 & n \le 0 \\ 5 - 3n & 0 < n \le 4 \\ -23 + n^2 & 4 < n \le 8 \\ 41 & n > 8 \end{cases}$    $g[3n]$    vs. $n$

    *b.*  $g[n] = 10 \cos\left(\dfrac{2\pi n}{20}\right) \cos\left(\dfrac{2\pi n}{4}\right)$    $4g[2(n + 1)]$    vs. $n$

    *c.*  $g[n] = \left|8e^{j(2\pi n/16)} u[n]\right|$    $g\left[\dfrac{n}{2}\right]$    vs. $n$

**64.**  Given the graphical definition of a function $g[n]$, graph the indicated function(s) $h[n]$.

    *a.*  $g[n] = 0$, $|n| > 8$. Graph $h[n] = g[2n - 4]$.

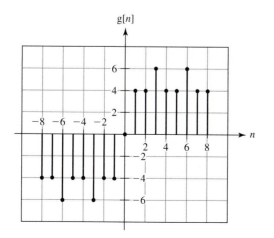

b.   $g[n] = 0,\ |n| > 8$. Graph $h[n] = g[n/2]$.

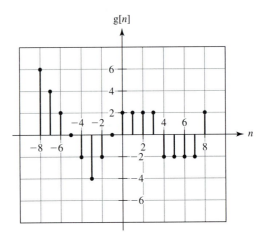

c.   $g[n]$ is periodic with fundamental period 8. Graph $h[n] = g[n/2]$.

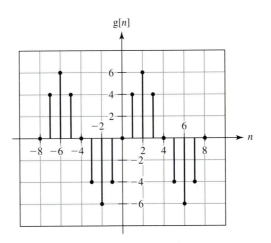

**65.**   Sketch the accumulation from negative infinity to $n$ of each of these DT functions.

a.   $g[n] = \cos(2\pi n)\, u[n]$

b.   $g[n] = \cos(4\pi n)\, u[n]$

**66.**   Find and sketch the magnitude and phase of the even and odd parts of this discrete-$k$ function.

$$G[k] = \frac{10}{1 - j4k}$$

**67.**   Find and sketch the even and odd parts of the DT function shown in Figure E67.

**68.**   Using MATLAB, plot each of these DT functions. If a function is periodic, find the period analytically and verify the period from the plot.

a.   $g[n] = \sin\left(\frac{3\pi n}{2}\right)$

b.   $g[n] = \sin\left(\frac{2\pi n}{3}\right) + \cos\left(\frac{10\pi n}{3}\right)$

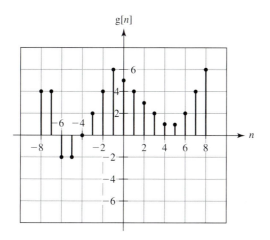

**Figure E67**

c.  $g[n] = 5\cos\left(\dfrac{2\pi n}{8}\right) + 3\sin\left(\dfrac{2\pi n}{5}\right)$

d.  $g[n] = 10\cos\left(\dfrac{n}{4}\right)$

e.  $g[n] = -3\cos\left(\dfrac{2\pi n}{7}\right)\sin\left(\dfrac{2\pi n}{6}\right)$    (*Hint:* A trigonometric identity will be useful here.)

**69.**  Sketch the following DT functions.

a.  $g[n] = 5\delta[n-2] + 3\delta[n+1]$

b.  $g[n] = 5\delta[2n] + 3\delta[4(n-2)]$

c.  $g[n] = 5(u[n-1] - u[4-n])$

d.  $g[n] = 8\,\text{rect}_4[n+1]$

e.  $g[n] = 8\cos\left(\dfrac{2\pi n}{7}\right)$

f.  $g[n] = -10e^{n/4}u[n]$

g.  $g[n] = -10(1.284)^n u[n]$

h.  $g[n] = \left|\left(\dfrac{j}{4}\right)^n u[n]\right|$

i.  $g[n] = \text{ramp}[n+2] - 2\,\text{ramp}[n] + \text{ramp}[n-2]$

j.  $g[n] = \text{rect}_2[n]\,\text{comb}_2[n]$

k.  $g[n] = \text{rect}_2[n]\,\text{comb}_2[n+1]$

l.  $g[n] = 3\sin\left(\dfrac{2\pi n}{3}\right)\text{rect}_4[n]$

m.  $g[n] = 5\cos\left(\dfrac{2\pi n}{8}\right)u\left[\dfrac{n}{2}\right]$

**70.**  Graph versus $k$, in the range, $-10 < k < 10$, the magnitude and phase of each function.

a.  $X[k] = \text{sinc}\left(\dfrac{k}{2}\right)$

b.  $X[k] = \text{sinc}\left(\dfrac{k}{2}\right)e^{-j(2\pi k/4)}$

c.  $X[k] = \text{rect}_3[k]\,e^{-j(2\pi k/3)}$

d.  $X[k] = \dfrac{1}{1 + jk/2}$

e.  $X[k] = \dfrac{jk}{1 + jk/2}$

f.  $X[k] = \text{comb}_2[k]e^{-j(2\pi k/4)}$

**71.** Sketch the even and odd parts of these signals.

    *a.*  $x[n] = \text{rect}_5[n + 2]$                         *b.*  $x[n] = \text{comb}_3[n - 1]$

    *c.*  $x[n] = 15 \cos\left(\dfrac{2\pi n}{9} + \dfrac{\pi}{4}\right)$         *d.*  $x[n] = \sin\left(\dfrac{2\pi n}{4}\right) \text{rect}_5[n - 1]$

**72.** What is the numerical value of each of the following accumulations?

    *a.*  $\displaystyle\sum_{n=0}^{10} \text{ramp}[n]$                           *b.*  $\displaystyle\sum_{n=0}^{6} \dfrac{1}{2^n}$

    *c.*  $\displaystyle\sum_{n=-\infty}^{\infty} \dfrac{u[n]}{2^n}$                      *d.*  $\displaystyle\sum_{n=-10}^{10} \text{comb}_3[n]$

    *e.*  $\displaystyle\sum_{n=-10}^{10} \text{comb}_3[2n]$                    *f.*  $\displaystyle\sum_{n=-\infty}^{\infty} \text{sinc}(n)$

**73.** Find the signal energy of each of these signals:

    *a.*  $x[n] = 5\,\text{rect}_4[n]$                       *b.*  $x[n] = 2\delta[n] + 5\delta[n - 3]$

    *c.*  $x[n] = \dfrac{u[n]}{n}$                           *d.*  $x[n] = \left(-\dfrac{1}{3}\right)^n u[n]$

    *e.*  $x[n] = \cos\left(\dfrac{\pi n}{3}\right)(u[n] - u[n - 6])$

**74.** Find the average signal power of each of these signals:

    *a.*  $x[n] = u[n]$      *b.*  $x[n] = (-1)^n$      *c.*  $x[n] = A\cos(2\pi F_0 n + \theta)$

    *d.*  $x[n] = \begin{cases} A & n = \cdots, 0, 1, 2, 3, 8, 9, 10, 11, 16, 17, 18, 19, \cdots \\ 0 & n = \cdots, 4, 5, 6, 7, 12, 13, 14, 15, 20, 21, 22, 23, \cdots \end{cases}$

    *e.*  $x[n] = e^{-j(\pi n/2)}$

### CHAPTER 3

# Description and Analysis of Systems

## 3.1 INTRODUCTION AND GOALS

The words *signal* and *system* were defined very generally in Chapter 1. Analysis of systems is a discipline that has been developed by engineers. Engineers are educated by learning mathematics (differential calculus, complex variables, vectors, differential equations, etc.) and science (physics, chemistry, biology, etc.). This education is important because an *engineer* uses mathematical theories and tools developed by *mathematicians* and applies them to knowledge of the physical world that has been discovered by *scientists* to design things that do something useful for society. The things an engineer designs are systems, but, as indicated in Chapter 1, the definition of a system is broader than that. The term *system* is so broad and abstract it is difficult to define. A system can be almost anything.

One way to define a system is as anything that performs a function. That is, it operates on something and produces something else. Another way to define a system is as anything that responds when stimulated or excited. A system can be an electrical system, a mechanical system, a biological system, a computer system, an economic system, a political system, etc. Systems designed by engineers are *artificial* systems, and systems which have developed organically over a period of time through evolution and the rise of civilization are *natural* systems. Some systems can be analyzed very thoroughly and completely through mathematics. Others may be so complicated that mathematical analysis is extremely difficult. And still others are just not well understood because of the difficulty in measuring their characteristics. Although the definition of the term system is very broad, in engineering the term usually refers to an artificial system which is excited by certain signals and responds with other signals.

Many systems were developed in earlier times by craftspeople who designed and improved their systems by experience and observation, apparently with the use of only the simplest mathematics. One of the most important distinctions between engineers and craftspeople is in the engineer's use of higher mathematics, especially calculus, to describe and analyze systems.

## CHAPTER GOALS

1. To introduce nomenclature which describes important system characteristics

2. To develop techniques for classifying systems according to their characteristics

3. To develop systematic methods of finding the responses to arbitrary excitations of a very important type of system

## BLOCK DIAGRAMS AND SYSTEM TERMINOLOGY

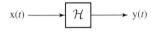

**Figure 3.1**
A single-input, single-output system.

Although systems can be of many different kinds, they have some features in common. A system operates on signals at one or more inputs to produce signals at one or more outputs. In system analysis it is very useful to represent systems by block diagrams. A very simple system with one input and one output would be represented as in Figure 3.1. In this case the signal at the input $x(t)$ is operated on by the operator $\mathcal{H}$ to produce the signal at the output $y(t)$. The operator $\mathcal{H}$ could perform just about any general operation imaginable. Common terminology in system analysis is that if one or more *excitation* signals are applied at one or more inputs, *response* signals appear at one or more outputs. That is, a signal applied at an input is an excitation signal (or just an excitation) and a signal appearing at an output is a response signal (or just a response). Other equivalent names are *input* signal for excitation and *output* signal for response.

> In this text we will consistently refer to a signal at an input as an excitation or input signal and a signal at an output as a response or output signal. Some other authors simply use the term input for both the location where the excitation is applied and for the excitation itself, and they use the term output for both the location where the response appears and the response itself. That is, they leave off the *signal* from both input signal and output signal. Although that leaves some possibility of ambiguity, usually the meaning is clear in context.

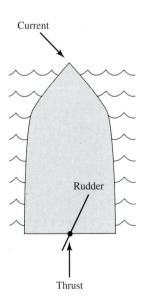

**Figure 3.2**
A simplified diagram of a boat.

One example of a system would be a boat steered by a rudder. The thrust developed by the propeller, the rudder position, and the current of the water are excitations of this system, and the heading and speed of the boat are responses (Figure 3.2). Notice that this statement says that the heading and speed of the boat are responses. It does not say that the heading and speed are *the* responses (which might imply that there are not any others). Practically every system has multiple responses, some significant and some insignificant. In the case of the boat, the heading and speed of the boat are significant but the vibration of the boat structure, the sounds created by the water splashing on the sides, the wake created behind the boat (both in the water and in the air), the rocking and/or tipping of the boat, and a myriad of other physical phenomena are probably not significant (unless they are very large) and would probably be ignored in a simplified analysis of this system.

An automobile suspension is excited by the surface topology of the road as the car travels over it, and the position of the chassis relative to the road is a significant response (Figure 3.3). When we set a thermostat in a room, the setting is an input signal to the heating and cooling system and a response of the system is the temperature inside the room.

A whole class of systems, measurement instruments, are single-input, single-output systems. The excitation is the physical phenomenon being measured, and the response

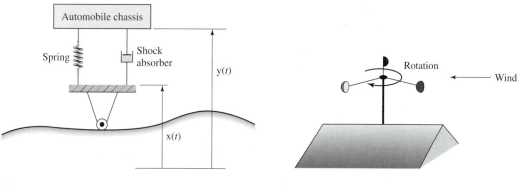

**Figure 3.3**
Simplified model of an automobile suspension system.

**Figure 3.4**
Cup anemometer.

is the instrument's indication of the value of that physical phenomenon. A good example is a cup anemometer. The wind excites the anemometer, and the angular velocity of the anemometer is the significant response (Figure 3.4).

An example of something that is not ordinarily thought of as a system is a suspension bridge. It is not usually thought of as a system because there is no obvious or deliberate excitation that produces a desired response. The ideal bridge would be one that doesn't respond at all because we want it to just stay exactly where it is and not move. But a suspension bridge *is* excited by the traffic that rolls across it, the wind that blows onto it, and the water currents that push on its support structure, and *it does move*. A very dramatic example that suspension bridges respond when excited was the failure of the Tacoma Narrows bridge in the State of Washington. On one very windy day the bridge responded to the wind by oscillating wildly and eventually actually tore itself apart. This is a very dramatic example of why good analysis is important. The conditions under which the bridge would respond so strongly should have been discovered in the design and modeling process so that the design could have been changed to avoid this disaster.

A single biological cell in a plant or animal is a system of astonishing complexity, especially considering the cell's size. The human body is a system comprising a huge number of cells and is, therefore, a very complicated system. But it can be modeled as a much simpler system in some cases to calculate an isolated effect. In pharmacokinetics the human body is often modeled as a single compartment, a volume containing liquid. The taking of a drug is an excitation, and the concentration of drug in the body is the significant response. Rates of infusion and excretion of the drug determine the drug concentration.

A manufacturing plant is a system. The input signal is an order from a customer. The response is to fill the order from inventory and to replenish inventory when it gets too low.

A distillation column is a system. It is fed by one or more streams of chemicals and maintained at an optimal temperature profile. The distillation product or products are removed from the column at various levels. Most industrial distillation columns are operated continuously and, therefore, are continuous-time systems.

Musical instruments are systems. A clarinet is excited by the air from the player's mouth causing the reed to vibrate. The body of the clarinet serves as a resonant cavity to filter the noise made by the vibrating reed, thereby producing a more pleasant sound. The effective length of the cavity, and, therefore, its resonant frequency, is determined by which holes are closed or opened as the player moves the keys. A pipe organ is excited by airflow into its pipes, and the responses are the resonant tones determined by the lengths of the pipes. Stringed instruments are excited by the movement of the bow on the strings, and the resultant vibration creates the sounds we hear. The string resonances are determined by the positions of the player's fingers on the strings. Even the human voice is a system excited by the vibrations of the vocal folds in the throat which are filtered by the resonances in the oral cavity.

A system is often described and analyzed as an assembly of components. A *component* is a smaller, simpler system, usually one which is standard in some sense and whose characteristics are already known. Just what is considered a component as opposed to a system is a matter of opinion and depends on the situation being described. To a circuit designer, components are resistors, capacitors, inductors, operational amplifiers, etc., and systems are power amplifiers, ADCs, modulators, filters, etc. To a communication system designer, components are amplifiers, modulators, filters, antennas, etc., and systems are microwave links, fiber-optic trunk lines, telephone central offices, etc. To an automobile designer, components are wheels, engines, bumpers, lights, seats, etc., and the system is the automobile. In large, complicated systems like commercial airliners, telephone networks, supertankers, power plants, etc., there are many levels of hierarchy of components and systems.

By knowing how to mathematically describe and characterize all the components in a system and how the components interact with each other, an engineer can predict, using mathematics, how a system will work, without actually building it and testing it. A system made up of components is diagrammed in Figure 3.5. The process of describing a system and analyzing it without building it is often called *modeling*. An engineer works with a mathematical model of a system. This capability is especially important in designing large, expensive systems like commercial aircraft, suspension bridges, supertankers, and communication networks. So the study of systems is the study of how interconnected components function as a coordinated whole.

In signals and systems there are common references to two general types of systems, open-loop and closed-loop. An *open-loop* system is one which simply responds to an input signal. A *closed-loop* system is one which responds to an input signal but also senses the output signal and alters the input signal to modify the output signal to satisfy some system requirement. Any measuring instrument is an open-loop system. The response simply indicates what the excitation is without altering it. A human driving a car is a good example of a closed-loop system. The driver signals the car to move at a certain speed and in a certain direction by pressing the accelerator or brake and by turning the steering wheel. As the car moves down a road the driver is constantly sensing with her eyes the speed and position of the car relative to the road and

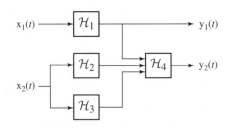

**Figure 3.5**
A system composed of four interconnected components, with two inputs and two outputs.

the other cars. Based on what the driver senses she modifies the input signals to maintain the desired direction of the car and keep it at a safe speed and position on the road.

## DISCRETE-TIME VERSUS CONTINUOUS-TIME SYSTEMS

Just as signals can either be continuous-time or discrete-time, systems can also. A CT system operates on a CT excitation to produce a CT response. A DT system operates on a DT excitation to produce a DT response (Figure 3.6). We will analyze these two types of systems alternately in Chapters 4, 5, 6, 9, 10, 11, and 12, indicating the similarities and differences between them. In Chapter 12, we will explore systems in which there are both CT and DT signals and in Chapter 7 we will develop ways of understanding the relation between DT signals which are formed by sampling CT signals and the CT signals from which they came.

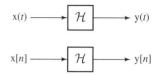

**Figure 3.6**
CT and DT system block diagrams.

## 3.2 SYSTEM CHARACTERISTICS

To build an understanding of large, generalized systems, let us begin with examples of some very simple systems which will illustrate some important properties of more general systems. Circuits are familiar to electrical engineers. Circuits are electrical systems. A very common circuit is the *RC* lowpass filter, a single-input, single-output system, illustrated in Figure 3.7. The voltage at the input $v_{in}(t)$ is the excitation of the system, and the voltage at the output $v_{out}(t)$ is the response of the system. The input voltage signal is applied to the left-hand pair of terminals, which is sometimes called a *port* in circuit theory, and the output voltage signal appears at the right-hand port. This system consists of two components familiar to electrical engineers, a resistor and a capacitor. The mathematical voltage-current relations for resistors and capacitors are well known and are illustrated in Figure 3.8.

Assume that the circuit of Figure 3.7 is quiescent before time $t = 0$, and the input voltage signal $v_{in}(t)$ changes suddenly from 0 to $A$ volts at time $t = 0$. *Quiescent* means the circuit has no stored energy. In this case that means the capacitor is initially uncharged. In system terminology, this system is initially *at rest* and the response of such a system is known as the *zero-state* response because the initial state of the system is that there is zero energy stored in it. In this case the zero-state condition is 0 V across the capacitor because that is the only energy-storage element in the circuit.

Using the unit step function, the input voltage signal $v_{in}(t)$ in Figure 3.7 can be written as $v_{in}(t) = Au(t)$. Then we can write a differential equation describing the

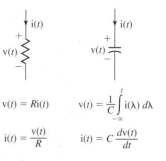

**Figure 3.8**
Mathematical voltage-current relationships for a resistor and a capacitor.

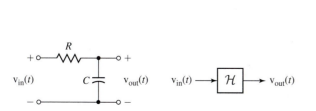

**Figure 3.7**
An *RC* lowpass filter: a single-input, single-output system.

circuit,

$$RCv'_{out}(t) + v_{out}(t) = v_{in}(t). \tag{3.1}$$

The solution of this differential equation is the sum of the transient (homogeneous) and steady-state (particular) solutions. The transient solution $v_{out,tr}(t)$ for times $t > 0$ is

$$v_{out,tr}(t) = Ke^{-(t/RC)} \qquad t > 0, \tag{3.2}$$

where, so far, $K$ is unknown. The steady-state solution depends on the functional form of $v_{in}(t)$. In this case, since the input voltage signal is constant for time $t > 0$ and no constant current can flow through a capacitor, the steady-state solution is simply $v_{out,ss}(t) = A$ and the total solution is

$$v_{out}(t) = v_{out,tr}(t) + v_{out,ss}(t) = Ke^{-(t/RC)} + A \qquad t > 0 \tag{3.3}$$

or

$$v_{out}(t) = \left(Ke^{-(t/RC)} + A\right)u(t). \tag{3.4}$$

The constant $K$ can be found by observing that the initial output voltage signal $v_{out}(0^+)$ is zero because $v_{out}(0^-)$ is zero and it cannot change instantaneously in response to a finite excitation. Therefore,

$$v_{out}(0^+) = K + A = 0 \Rightarrow K = -A. \tag{3.5}$$

Now the output voltage signal can be written as

$$v_{out}(t) = A\left(1 - e^{-(t/RC)}\right)u(t) \tag{3.6}$$

and is illustrated in Figure 3.9.

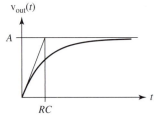

**Figure 3.9**
*RC* lowpass filter response to a unit step excitation.

If the capacitor had had an initial nonzero charge, the solution would have been different. The other standard type of system response commonly referred to in system theory is the so-called zero-input (signal) response of a system. This is the response of a system whose initial state is not zero but whose excitation is zero. The zero-input response (the capacitor voltage) after time $t = 0$ in this circuit would be

$$v_{out}(t) = v_C(0^+)e^{-(t/RC)} \qquad t \geq 0. \tag{3.7}$$

If we had both a nonzero initial state and a nonzero excitation, we could also find the solution by solving the differential equation with a different initial condition and the solution after time $t = 0$, would be

$$v_{out}(t) = A\left(1 - e^{-(t/RC)}\right) + v_C(0^+)e^{-(t/RC)} \qquad t \geq 0 \tag{3.8}$$

or

$$v_{out}(t) = A + [v_C(0^+) - A]e^{-(t/RC)} \qquad t \geq 0. \tag{3.9}$$

Notice that this solution for time $t \geq 0$ is the sum of the zero-input and zero-state responses.

It should be noted here that if the excitation were truly a voltage step of height $A$, that would imply that the excitation was defined for all negative time to be 0 V. If we assume that the circuit has been connected with this excitation between the input terminals for an infinite time (since $t = -\infty$), the initial capacitor voltage at time $t = 0$ would have to be zero unless some external energy source injected the charge onto the capacitor at time $t = 0$. If the circuit is connected as shown for all time, the solution for a nonzero initial state must be assumed to be under the condition that there is an initial capacitor charge injected by some process occurring at time $t = 0$. The solution could be for either of two situations. One possibility is that the excitation is not connected for all time before time $t = 0$ and is suddenly connected at that time to a constant voltage $A$. The other situation is that the circuit is connected for all time but that some external energy source deposits a charge on the capacitor at time $t = 0$. In either case we must change the system either by injecting a charge or by closing a switch as illustrated in Figure 3.10. If the initial capacitor voltage is zero in both circuits of Figure 3.10, the response is the same.

As an example of a DT system, consider the system in Figure 3.11. The $D$ block in the system block diagram is a delay component whose output signal is its input signal, delayed by one, in discrete time. So the system is characterized by the difference equation

$$y[n] = x[n] + \frac{4}{5}y[n - 1]. \tag{3.10}$$

The homogeneous solution is $y_h[n] = K\left(\frac{4}{5}\right)^n$. Let the excitation be the unit sequence. Then the particular solution is $y_p[n] = 5$, and the total solution is $y[n] = K\left(\frac{4}{5}\right)^n + 5$. If the system is initially at rest, this solution is the zero-state solution and the total solution is

$$y[n] = \begin{cases} 5 - 4\left(\dfrac{4}{5}\right)^n & n \geq 0 \\ 0 & n < 0 \end{cases} \tag{3.11}$$

or

$$y[n] = \left[5 - 4\left(\frac{4}{5}\right)^n\right] u[n] \tag{3.12}$$

(Figure 3.12).

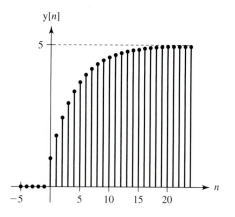

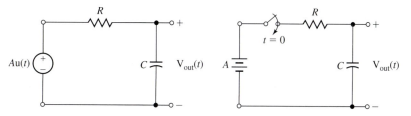

**Figure 3.10**
Two ways of suddenly applying $A$ V to the $RC$ lowpass filter.

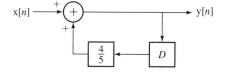

**Figure 3.11**
A DT system.

**Figure 3.12**
DT system response to a unit sequence excitation.

The similarity between the shape of the CT $RC$ lowpass filter's response to a unit step excitation and the envelope of the DT system response to a unit sequence excitation is not an accident. The DT system is a very simple digital lowpass filter (digital filters will be discussed further in Chapters 6 and 12).

## HOMOGENEITY

If we were to double the input voltage signal of the $RC$ lowpass filter to $v_{in}(t) = 2Au(t)$, the factor $2A$ would carry through and the output voltage signal would double to

$$v_{out}(t) = 2A\left(1 - e^{-(t/RC)}\right)u(t). \tag{3.13}$$

The quality of this system that makes that true is called *homogeneity*.

> In a *homogeneous* system, multiplying the excitation by any constant (including complex constants) multiplies the response by the same constant.

Figure 3.13 illustrates, in a block diagram sense, what homogeneity means. If we were to double the excitation of the DT system in the figure through the transformation $x[n] \to 2x[n]$, its response would also double, $y[n] \to 2y[n]$. Therefore, it is also a homogeneous system. The property of homogeneity can also be indicated by the shorthand notation

$$x[n] \xrightarrow{\mathcal{H}} y[n] \Rightarrow Kx[n] \xrightarrow{\mathcal{H}} Ky[n], \tag{3.14}$$

where $x[n] \xrightarrow{\mathcal{H}} y[n]$ means "the excitation x of system $\mathcal{H}$ produces the response y" and $K$ can be any complex constant.

A very simple example of a system which is not homogeneous is a system characterized by the relationship

$$y(t) = x(t) + 1. \tag{3.15}$$

For an excitation x of 1, the response y is 2, and for an excitation x of 2, the response y is 3. The excitation was doubled, but the response was not. The thing that makes this system inhomogeneous is the presence of the 1 on the right side of the equation which was not considered to be part of the excitation. This system has a nonzero zero-input response. Notice that, if we were to redefine the excitation to be $x_{new}(t) = x(t) + 1$ instead of just $x(t)$, we would have $y(t) = x_{new}(t)$, and doubling $x_{new}(t)$ would double the response and the system would then be homogeneous under this new definition of the excitation.

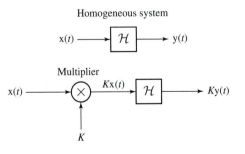

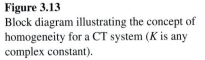

**Figure 3.13**
Block diagram illustrating the concept of homogeneity for a CT system ($K$ is any complex constant).

## TIME INVARIANCE

Suppose the excitation of the DT system in Figure 3.11 were delayed by some time $n_0$. That is, let the input signal be changed to $x[n] = u[n - n_0]$. What happens to the response? Going through the solution process again we would find that the response is $[5 - 4 \left(\frac{4}{5}\right)^{(n-n_0)}]u[n - n_0]$, which is exactly the original response except with $n$ replaced by $n - n_0$. Saying it in another way, the excitation $x_1[n] = u[n]$ produced the response $y_1[n]$, and the excitation $x_2[n] = u[n - n_0]$ produced the response $y_2[n] = y_1[n - n_0]$. Delaying the excitation delayed the response by the same amount without changing the functional form of the response. This quality is called *time invariance*.

> If an arbitrary excitation $x[n]$ of a system causes a response $y[n]$, and an excitation of the system $x[n - n_0]$ causes a response $y[n - n_0]$, for any arbitrary $n_0$, the system is said to be *time invariant*.

Figure 3.14 illustrates the concept of time invariance. The time-invariance property for CT systems is analogous,

$$x(t) \xrightarrow{\mathcal{H}} y(t) \Rightarrow x(t - t_0) \xrightarrow{\mathcal{H}} y(t - t_0). \tag{3.16}$$

A very simple example of a system which is not time-invariant would be one described by

$$y[n] = x[2n]. \tag{3.17}$$

Let $x_1[n] = g[n]$ and let $x_2[n] = g[n - 1]$, where $g[n]$ is the signal illustrated in Figure 3.15, and let the response to $x_1[n]$ be $y_1[n]$ and the response to $x_2[n]$ be $y_2[n]$. These excitations and responses are illustrated in Figure 3.16. Since the excitation $x_2[n]$ is the same as the excitation $x_1[n]$, except delayed by one discrete time unit, for the system to be time-invariant the response $y_2[n]$ must be the same as the response $y_1[n]$, except delayed by one unit, but it is not. Therefore, this system is not time-invariant, it is *time-variant*.

A simple example of a time-variant CT system is a thermistor driven by a current. A thermistor is a device whose resistance is a function of its temperature. The resistance of a thermistor falls as its temperature rises. Let the current through the thermistor be the excitation, and let the voltage across the thermistor be the response. If a constant current is suddenly applied, there is a voltage response determined by Ohm's law and power is dissipated in the thermistor. This power dissipation causes the temperature of the thermistor to rise which, in turn, causes the resistance of the thermistor to fall. As the resistance falls, the power dissipation also falls, causing the temperature to rise

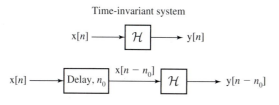

**Figure 3.14**
Block diagram illustrating the concept of time invariance for a DT system.

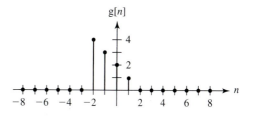

**Figure 3.15**
A DT excitation.

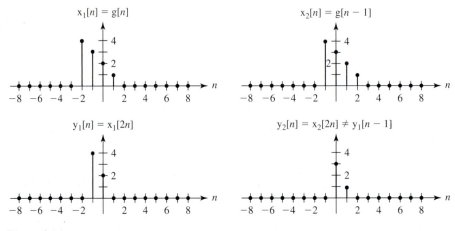

**Figure 3.16**
Responses of the system described by $y[n] = x[2n]$ to two different excitations.

more slowly and the resistance of the thermistor to fall more slowly. At some resistance level the heating effect of the power dissipated in the thermistor and the cooling effects of heat loss from the thermistor will be equal and the voltage will stabilize. Since the relationship between excitation and response (the resistance) is a function of time, this is a time-variant system.

## ADDITIVITY

Let the input voltage signal to the $RC$ lowpass filter be the sum of two voltages, $v_{in}(t) = v_{in1}(t) + v_{in2}(t)$. For a moment let $v_{in2}(t) = 0$ and let the solution for $v_{in1}(t)$ acting alone be $v_{out1}(t)$. The differential equation describing that situation is

$$RCv'_{out1}(t) + v_{out1}(t) = v_{in1}(t). \tag{3.18}$$

Similarly, if $v_{in2}(t)$ acts alone,

$$RCv'_{out2}(t) + v_{out2}(t) = v_{in2}(t). \tag{3.19}$$

Adding (3.18) and (3.19),

$$RC[v'_{out1}(t) + v'_{out2}(t)] + v_{out1}(t) + v_{out2}(t) = v_{in1}(t) + v_{in2}(t) = v_{in}(t) \tag{3.20}$$

The equation describing the application of both input voltage signals simultaneously is

$$RCv'_{out}(t) + v_{out}(t) = v_{in}(t). \tag{3.21}$$

Combining (3.20) and (3.21),

$$RC[v'_{out1}(t) + v'_{out2}(t)] + v_{out1}(t) + v_{out2}(t) = RCv'_{out}(t) + v_{out}(t), \tag{3.22}$$

which implies that

$$v_{out1}(t) + v_{out2}(t) = v_{out}(t). \tag{3.23}$$

This result depends on the fact that the derivative of a sum of two functions equals the sum of the derivatives of those two functions. If the excitation is the sum of two excitations, the solution of *this* differential equation is the sum of the responses to those

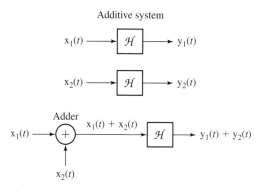

Additive system

**Figure 3.17**
Block diagram illustrating the concept of additivity
in a CT system.

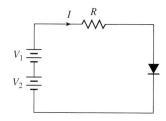
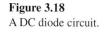

**Figure 3.18**
A DC diode circuit.

excitations acting alone. That is, if $v_{in}(t) = v_{in1}(t) + v_{in2}(t)$, then $v_{out}(t) = v_{out1}(t) + v_{out2}(t)$. A system in which added excitations yield added responses is called *additive* (Figure 3.17).

> If an arbitrary excitation $x_1$ produces a response $y_1$, and an arbitrary excitation $x_2$ produces a response $y_2$, and the excitation $x_1 + x_2$ always produces the response $y_1 + y_2$, the system is additive.

The additivity property for DT systems is analogous,

$$x_1[n] \xrightarrow{\mathcal{H}} y_1[n] \quad \text{and} \quad x_2[n] \xrightarrow{\mathcal{H}} y_2[n] \Rightarrow (x_1[n] + x_2[n]) \xrightarrow{\mathcal{H}} (y_1[n] + y_2[n]).$$
$$(3.24)$$

A very common example of a nonadditive system is a simple DC diode circuit (Figure 3.18). Let the input voltage signal of the circuit $V$ be the series connection of two DC voltage sources $V_1$ and $V_2$, making the overall input voltage signal the sum of the two individual input voltage signals. Let the overall response be the current $I$, and let the individual current responses to the individual voltage sources acting alone be $I_1$ and $I_2$. To make the result obvious let $V_1 > 0$ and $V_1 = -V_2$. The response to $V_1$ acting alone is a positive current $I_1$. The response to $V_2$ acting alone is an extremely small negative (ideally zero) current $I_2$. The sum of the two excitations is zero, but the sum of the two responses is not. So this is not an additive system.

## LINEARITY AND SUPERPOSITION

Any system which is both homogeneous and additive is called a *linear* system.

> In any linear system, if an excitation $x_1[n]$ causes a response $y_1[n]$ and an excitation $x_2[n]$ causes a response $y_2[n]$, then an excitation
>
> $$x[n] = \alpha x_1[n] + \beta x_2[n] \qquad (3.25)$$
>
> will cause the response
>
> $$y[n] = \alpha y_1[n] + \beta y_2[n]. \qquad (3.26)$$

This characteristic of linear systems is called *superposition*. This term comes from the verb *superpose*. The "pose" part of superpose means to put something into a certain position and the "super" part means "on top of." Together then superpose means to place something on top of something else. That is what is done when we add one excitation to another, and, in a linear system, the overall response is one of the responses "on top of" (added to) the other.

By far the most common type of system analyzed in practical system design and analysis is the linear, time-invariant system. If a system is both linear and time-invariant, it is called an *LTI* system. Analysis of LTI systems forms the majority of the material in this text.

Superposition is the basis of a powerful technique for finding the response of a linear system with an arbitrary excitation. The salient characteristic of equations which describe linear systems is that the dependent variable and its integrals and derivatives, or summations and differences, appear only to the first power. To illustrate this rule, consider a system described by the differential equation

$$a y''(t) + b y^2(t) = x(t) \tag{3.27}$$

where $x(t)$ is the excitation and $y(t)$ is the response. If the excitation were changed to $x_{\text{new}}(t) = x_1(t) + x_2(t)$, the system would then be described by

$$a y''_{\text{new}}(t) + b y^2_{\text{new}}(t) = x_{\text{new}}(t). \tag{3.28}$$

The equations describing the system for the two individual excitations $x_1(t)$ and $x_2(t)$ acting alone would be

$$a y''_1(t) + b y^2_1(t) = x_1(t) \qquad \text{and} \qquad a y''_2(t) + b y^2_2(t) = x_2(t). \tag{3.29}$$

The sum of the two equations in (3.29) is

$$a\big[y''_1(t) + y''_2(t)\big] + b\big[y^2_1(t) + y^2_2(t)\big] = x_1(t) + x_2(t) = x_{\text{new}}(t) \tag{3.30}$$

which is (in general) *not* equal to

$$a[y_1(t) + y_2(t)]'' + b[y_1(t) + y_2(t)]^2 = x_1(t) + x_2(t) = x_{\text{new}}(t). \tag{3.31}$$

The difference is caused by the $y^2(t)$ term, which is not consistent with a differential equation that describes a linear system. Therefore, in this system, superposition does *not* apply.

Another very simple example of a differential equation that does not describe a linear system is

$$a y'(t) + b y(t) + c = x(t) \qquad c \neq 0. \tag{3.32}$$

The presence of the constant $c$ causes this system to be inhomogeneous because, if the forcing function $x(t)$ is zero, the response is not zero. That is, the system's zero-input response is not zero. We could rewrite this equation as

$$a y'(t) + b y(t) = x(t) - c. \tag{3.33}$$

Written this way the equation looks like a typical differential equation describing a system whose response is $y(t)$ and which is excited either by one excitation $x(t) - c$ or by two excitations $x(t)$ and $-c$. This change of point of view does not change the system; it only changes the way we define the excitation of the system. Now if the excitation $x(t) - c$ is multiplied by a constant, the response is multiplied by the same constant. Or if we apply the two excitations $x(t)$ and $-c$ individually and multiply them by constants, the individual responses are multiplied by the same constants. Now homogeneity is satisfied, and the system can be considered linear. Thus the way we identify a system's excitation or excitations affects how we classify it.

Instead of redefining the excitation, we could redefine the response as

$$y_{new}(t) = y(t) + \frac{c}{b}. \tag{3.34}$$

Then the differential equation, written in terms of the new response, would be

$$ay'_{new}(t) + b\left[y_{new}(t) - \frac{c}{b}\right] + c = x(t) \tag{3.35}$$

or, simplifying,

$$ay'_{new}(t) + by_{new}(t) = x(t). \tag{3.36}$$

This equation describes a linear system. Again we did not really change the system itself, just the way we described it mathematically.

A time-invariant system is one which is described by differential or difference equations in which the coefficients of the dependent variable and all its integrals and derivatives or summations and differences are constants. The coefficients are not functions of time.

A very common analysis technique in signal and system analysis is to use the methods of linear systems to analyze nonlinear systems. This process is called *linearizing* the system. Of course, the analysis is not exact because the system is not actually linear. But many nonlinear systems can be usefully analyzed by linear-system methods if the excitations and responses are small enough. As an example consider a pendulum (Figure 3.19). Assume that the mass is supported by a massless rigid rod of length $L$. If a force is applied to the mass $m$, it responds by moving. At any position in its motion the vector sum of the forces acting on the mass tangential to the direction of motion is equal to the product of the mass and the acceleration in that same direction. That is,

$$x(t) - mg\sin(\theta(t)) = mL\theta''(t) \tag{3.37}$$

or

$$mL\theta''(t) + mg\sin(\theta(t)) = x(t) \tag{3.38}$$

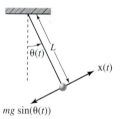

**Figure 3.19**
A pendulum.

where $m$ = mass at end of pendulum
$\quad x(t)$ = force applied to mass tangential to direction of motion
$\quad L$ = length of pendulum
$\quad g$ = gravitational constant
$\quad \theta(t)$ = angular position of pendulum

In this system, $x(t)$ is the excitation and $\theta(t)$ is the response. Equation (3.38) is nonlinear. But if $\theta(t)$ is small enough, $\sin(\theta(t))$ can be closely approximated by $\theta(t)$. In that approximation,

$$mL\theta''(t) + mg\theta(t) \cong x(t) \tag{3.39}$$

and this is a linear equation. So, for small perturbations from the rest position, this system can be usefully analyzed by (3.39).

## STABILITY

In the *RC* lowpass filter example, the excitation, a step of voltage, was bounded. If a signal is *bounded,* this means its absolute value is less than some finite value for all time,

$$|x(t)| < \infty \qquad -\infty < t < \infty. \tag{3.40}$$

The response of the *RC* lowpass filter to this bounded signal was also bounded.

> Any system for which the response is bounded when the excitation is bounded is called a *bounded-input–bounded-output* (BIBO) stable system.

A good example of a system which is not stable is the DT financial system of accruing compound interest. If a principle amount *P* of money is deposited in a fixed-income investment at an interest rate *r* per annum compounded annually, the amount A[*n*], which is the value of the investment *n* years later, is

$$A[n] = P(1 + r)^n. \tag{3.41}$$

The amount A[*n*] grows without bound as discrete time *n* passes. Does that mean our banking system is unstable?

## INCREMENTAL LINEARITY

As an example of another type of CT system, consider a mechanical system consisting of a linear spring suspending a mass *m* (Figure 3.20) which is being acted upon by an external force (the excitation) x(*t*) applied at time *t* = 0. Let the position of the end of the unstretched spring in Figure 3.20(a) be the reference for vertical position. When the system is in equilibrium (before the external force is applied), the top of the mass *m* is at the equilibrium position $y_e$. If the mass is acted upon by the excitation x(*t*), it will respond by moving. The equation of motion is based on the fundamental mechanical principle that the vector sum of the forces on a body equals the product of the mass of the body and the vector acceleration of the body,

$$\sum_{i=1}^{N} \vec{F}_i = m\vec{a}. \tag{3.42}$$

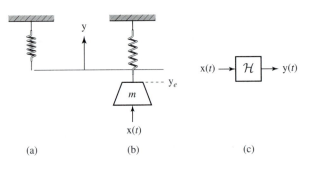

(a)                      (b)                      (c)

**Figure 3.20**
A mechanical system: (a) an unstretched spring, (b) the spring supporting a mass, and (c) a system block diagram of this single-input, single-output system.

In this case all the vectors act in the same direction and the vector equation (3.42) reduces to the scalar equation

$$-K_s y(t) - mg + x(t) = my''(t) \qquad \textbf{(3.43)}$$

or

$$my''(t) + K_s y(t) + mg = x(t) \qquad \textbf{(3.44)}$$

(if the spring is lossless), where $K_s$ is the spring constant and $g$ is the gravitational constant. If the spring has loss, there will also be a force directly proportional to velocity which opposes motion and the equation of motion will be

$$my''(t) + K_v y'(t) + K_s y(t) + mg = x(t) \qquad \textbf{(3.45)}$$

where $K_v$ is the proportionality constant between velocity and force. Equation (3.45) is a *second-order* equation because the highest derivative is a second derivative. Although it is a linear, constant-coefficient, differential equation, (3.45) *does not* describe an LTI system. That fact is easily verified by solving (3.45) for $x(t) = 0$,

$$my''(t) + K_v y'(t) + K_s y(t) + mg = 0. \qquad \textbf{(3.46)}$$

Since the excitation is zero, the system does not respond at all to the excitation, all the derivatives of the response are zero, and (3.46) reduces to

$$K_s y(t) + mg = 0 \Rightarrow y(t) = -\frac{mg}{K_s}. \qquad \textbf{(3.47)}$$

We have a nonzero response for a zero excitation and that violates the principle of homogeneity. Therefore, the system, as described by this equation, is not linear, even though the equation is commonly classified as a *linear* differential equation. This points out one annoying little difference between linear differential equations and equations which describe linear systems. This is another example of a system whose zero-input response is not zero.

It is natural to think that, even though the system is not linear, it has many characteristics similar to linear systems. In fact it is what is termed *incrementally linear*. This system can be modeled as an LTI system with an extra signal $y_0(t)$, the zero-input response, added to its response (Figure 3.21). In this case the zero-input response is

$$y_0(t) = -\frac{mg}{K_s}. \qquad \textbf{(3.48)}$$

In the upper half of Figure 3.21 the overall system is characterized by the operator $\mathcal{H}$. In the lower half the system is broken into two parts, an LTI system characterized by the operator $\mathcal{H}_{\text{LTI}}$, and the addition of the zero-input response $y_0(t)$. That is, an incrementally linear system is a system whose response is the sum of a zero-input response and the response of an LTI system to the excitation. If it were not for the addition of the zero-input response, the system would be LTI. The designation *incrementally linear* comes from the fact that changes in the excitation cause proportional changes in the response. That is, the increment in the response is proportional to the increment in the excitation. It is important here to comment on terminology. All LTI systems are also incrementally linear because incremental changes in their excitations cause proportional incremental changes in their responses. Thus, LTI systems form a *subset* of incrementally linear systems.

Incrementally linear system

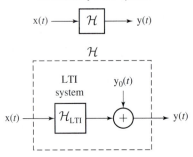

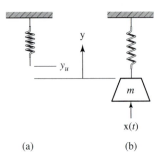

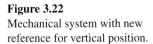

(a)                    (b)

**Figure 3.21**
The relation between the incrementally
linear system and an LTI system.

**Figure 3.22**
Mechanical system with new
reference for vertical position.

This system has a nonzero zero-input response. That is what makes it incrementally linear instead of just linear. Systems are often characterized by their zero-input response and their zero-state response. As mentioned earlier, the zero-input response is the response of a system when the excitation is zero. The zero-state response is the response of a system when the initial state is zero; that is, the system is initially at rest. If a system can be described by a linear, constant-coefficient differential equation, its response is the sum of its zero-input response and its zero-state response.

In a manner similar to the previous discussion on linearity, we can convert the mechanical system description to an LTI system description. If we were to redefine the zero reference for vertical position to $y_e$, the equilibrium position of the top of the mass (Figure 3.22), then the equation of motion would become

$$m[y(t) - y_u]'' + K_v y'(t) + K_s y(t) + mg = x(t), \qquad (3.49)$$

where $y_u$ is the position of the end of the unstretched spring with respect to the new reference position. Now the value of y just before time $t = 0$ *is* zero which implies that

$$K_s y_u - mg = 0, \qquad (3.50)$$

and the equation of motion becomes

$$my''(t) + K_v y'(t) + K_s y(t) = x(t). \qquad (3.51)$$

This system description is linear because the zero-input response is zero. The system itself did not change, but the way we described it did and changed it from an incrementally linear system description to a linear one.

We could also return to the original differential equation with the original reference,

$$my''(t) + K_v y'(t) + K_s y(t) + mg = x(t), \qquad (3.52)$$

and rewrite it as

$$my''(t) + K_v y'(t) + K_s y(t) = x(t) - mg. \qquad (3.53)$$

Now, written this way, we could interpret the equation to mean that there are either two excitations, the force $x(t)$ and the force due to gravity $-mg$, or one excitation $x(t) - mg$, the *net* force on the mass. This is another way of describing the system as linear rather than as incrementally linear.

# CAUSALITY

In the analysis of the three systems we have considered so far, we observe that each system responds only during or after the time the excitation is applied. This should seem obvious and natural. How could a system respond to an excitation that has not yet been applied? It seems obvious because we live in a physical world in which responses of real physical systems always occur while excitations are applied or after. But, as we shall later discover in considering ideal filters, some system design approaches may lead to a system design for which the response begins *before* the excitation is applied. Such a system cannot actually be built.

The fact that a real system response only occurs while or after the excitation is applied is a result of the commonsense idea of cause and effect. An effect has a cause, and the effect occurs during or after the application of the cause.

> Any system for which the response occurs only during or after the time in which the excitation is applied is called a *causal* system.

All physical systems are causal because they are unable to look into the future and anticipate an excitation that will be applied later.

Even though all real physical systems must be causal in the strict sense that a response must occur only during or after the excitation which caused it, there are real signal-processing systems which are sometimes described, in a superficial sense, as noncausal. These are data-processing systems in which signals are recorded and then processed off-line at a later time to produce a computed response. Since the whole history of the excitation has been recorded, the computed response at some designated time in the data stream can be based on future values of the already-recorded excitation (Figure 3.23). But, since the whole data-processing operation occurs after the excitations have been recorded, this kind of system is still causal in the strict sense.

The term *causal* is also commonly (albeit somewhat imprecisely) applied to signals. A causal signal is one which is zero before time $t = 0$ or $n = 0$. The use of this

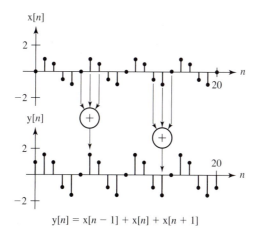

$$y[n] = x[n-1] + x[n] + x[n+1]$$

**Figure 3.23**
A so-called noncausal filter calculating responses from a record of excitations.

terminology for signals comes from the fact that if an excitation which is zero before time $t = 0$ or $n = 0$ is applied to a causal system, the response is also zero before time $t = 0$ or $n = 0$. By this definition, the response would be a causal signal because it is the response of a causal system to a causal excitation. The term *anticausal* is sometimes used to describe signals which are zero *after* time $t = 0$ or $n = 0$.

## MEMORY

The responses of the three systems we have considered so far depend on the present excitation and the past excitation. In the *RC* lowpass filter, the charge on the capacitor is determined by the current that has flowed through it in the past. By this mechanism it remembers something about its past. The DT system has a delay element in it which remembers the last value of the response. The dynamic behavior of the mechanical system at any time depends on stored energy in the spring which is determined by the past history of the forces applied to it. These systems remember their past excitations and use that memory, along with their present excitations, to determine their present responses.

There are systems for which the present value of the response depends only on the present value of the excitation. A resistive voltage divider is a good example (Figure 3.24). This type of system is said to have no memory or is called a *memoryless* or static system.

> If any system's response at an arbitrary time $t = t_0$, $y(t_0)$, depends only on the excitation at time $t = t_0$, $x(t_0)$, and not on the value of the excitation or response at any other time, the system has no memory and is called a *static* system.

The term *dynamic* is commonly used for a system with memory. Figure 3.25 is an example of a DT system without memory. The response at any discrete time $n$ depends only on the excitations at discrete time $n$.

## STATIC NONLINEARITY

We have already seen one example of a nonlinear system, the incrementally linear system. The incrementally linear system is nonlinear because it violates the principle of homogeneity. The nonlinearity is not an intrinsic result of nonlinearity of the components themselves, but rather a result of the fact that the zero-input response of the system is not zero.

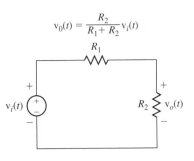

$$v_0(t) = \frac{R_2}{R_1 + R_2} v_i(t)$$

**Figure 3.24**
A resistive voltage divider.

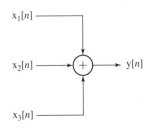

**Figure 3.25**
Memoryless DT system.

The more common meaning of nonlinear system in practice is a system in which, even with a zero-input response of zero, the response is still a nonlinear function of the excitation. This is often the result of components in the system which have static nonlinearities. A static nonlinearity is one in which the nonlinearity is not a result of memory but rather of a component which has a static or memoryless relation between the excitation and response which is a nonlinear function. Examples of statically nonlinear components include diodes, transistors, and multipliers. These components are nonlinear because if the excitation is changed by some factor, the response may change by a different factor.

The difference between linear and nonlinear components of this type can be seen by plotting the relationship between the excitation and response. For a linear resistor, the relation is determined by Ohm's law

$$v(t) = Ri(t).$$

A graph of voltage versus current is linear (Figure 3.26).

A diode is a good example of a statically nonlinear component. Its voltage-current relationship is

$$i(t) = I_s\left(e^{qv(t)/kT} - 1\right), \tag{3.54}$$

where $I_s$ = reverse saturation current

$q$ = charge on an electron

$k$ = Boltzmann's constant

$T$ = absolute temperature

as illustrated in Figure 3.27.

Another example of a statically nonlinear component is an analog multiplier used as a squarer. An analog multiplier has two inputs and one output, and the output signal is the product of the signals applied at the two inputs. It is memoryless, or static, because the present output signal depends only on the present input signals and not on any past output signal or input signal (Figure 3.28). The output signal $y(t)$ is the product of the input signals $x_1(t)$ and $x_2(t)$. If $x_1(t)$ and $x_2(t)$ are the same signal $x(t)$, then

$$y(t) = x^2(t). \tag{3.55}$$

This is a statically nonlinear relationship because if the excitation is multiplied by some factor $A$, the response is multiplied by the factor $A^2$; and that is a violation of the principle of homogeneity.

A very common example of a static nonlinearity is the phenomenon of saturation in real (as opposed to *ideal*) operational amplifiers. An operational amplifier has two

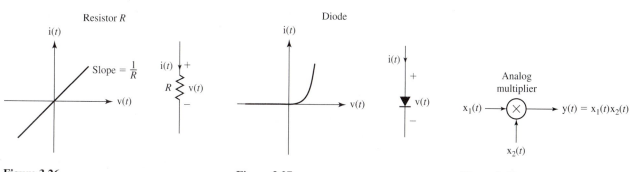

**Figure 3.26**
Voltage-current relationship for a resistor.

**Figure 3.27**
Voltage-current relationship for a diode.

**Figure 3.28**
An analog multiplier.

inputs, the inverting input and the noninverting input, and one output. When input voltage signals are applied to the inputs, the output voltage signal of the operational amplifier is a fixed multiple of the difference between the two input voltage signals, up to a point. For small signals, the relationship is

$$v_{out}(t) = A[v_{in+}(t) - v_{in-}(t)]. \tag{3.56}$$

But the output voltage signal is constrained by the power supply voltages and can only approach those voltages, not exceed them. Therefore, if the difference between the input voltage signals is large enough that the output voltage signal calculated from

$$v_{out}(t) = A[v_{in+}(t) - v_{in-}(t)] \tag{3.57}$$

would cause it to be outside the range $-V_{ps}$ to $+V_{ps}$ (ps = power supply), the operational amplifier will saturate. The output voltage signal will go that far and no farther. When the operational amplifier is saturated, the relationship between the excitations and the response becomes statically nonlinear. That is illustrated in Figure 3.29.

Even if a system is statically nonlinear, linear system analysis techniques may still be useful in analyzing it. As a simple example of using linear system analysis on nonlinear systems consider a circuit containing a voltage source, a resistor, and a diode in series (Figure 3.30). By Kirchhoff's voltage law the voltage across the diode is equal to the voltage across the series combination of the voltage source and the resistor,

$$v_D(t) = v_s(t) - Ri(t). \tag{3.58}$$

Equation (3.58) can be solved for the current in the form

$$i(t) = \frac{v_s(t) - v_D(t)}{R}. \tag{3.59}$$

The current is also described by the diode equation

$$i(t) = I_s\left(e^{qv_D(t)/kT} - 1\right). \tag{3.60}$$

This is a system of two equations, one of which is nonlinear. The equations can be solved graphically by drawing the two V-I diagrams and finding the intersection (Figure 3.31).

The solution in Figure 3.31 is drawn as though the voltage and current $v_s(t)$ and $i(t)$ are constants. Suppose $v_s(t)$ consists of a constant $v_{s,DC}$, plus a small time-varying part $v_{s,AC}(t)$,

$$v_s(t) = v_{s,DC} + v_{s,AC}(t). \tag{3.61}$$

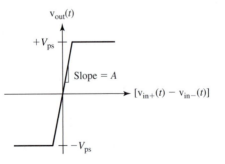

**Figure 3.29**
Input-output signal relationship for a saturating operational amplifier.

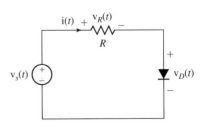

**Figure 3.30**
Diode circuit to demonstrate linear analysis of a nonlinear system.

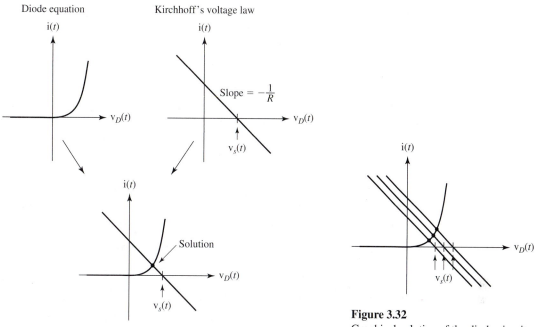

**Figure 3.31**
Graphical solution of the two diode circuit equations.

**Figure 3.32**
Graphical solution of the diode circuit equations as the source voltage changes with time.

Then the graphical solution for the voltage and current would track the point of intersection of the two curves as $v_s(t)$ changes with time (Figure 3.32). If the AC part of the source voltage is small enough, the part of the diode curve traversed by the intersection point is very short and almost linear. Equation (3.58) can be rewritten in a form that illustrates how to find an approximate solution analytically.

$$v_D(t) = v_{s,\mathrm{DC}} + v_{s,\mathrm{AC}}(t) - Ri(t) \tag{3.62}$$

From the diode equation,

$$i(t) = I_s\left(e^{qv_D(t)/kT} - 1\right). \tag{3.63}$$

Combining (3.62) and (3.63),

$$v_D(t) = v_{s,\mathrm{DC}} + v_{s,\mathrm{AC}}(t) - RI_s\left(e^{qv_D(t)/kT} - 1\right). \tag{3.64}$$

From the graphical solution we see that the diode voltage also consists of a constant plus a small variation,

$$v_D(t) = v_{D,\mathrm{DC}} + v_{D,\mathrm{AC}}(t). \tag{3.65}$$

Therefore,

$$v_D(t) = v_{s,\mathrm{DC}} + v_{s,\mathrm{AC}}(t) - RI_s\left(e^{q(v_{D,\mathrm{DC}}+v_{D,\mathrm{AC}}(t))/kT} - 1\right) \tag{3.66}$$

or

$$v_D(t) = v_{s,\mathrm{DC}} + v_{s,\mathrm{AC}}(t) - RI_s\left(e^{qv_{D,\mathrm{DC}}/kT}\,e^{qv_{D,\mathrm{AC}}(t)/kT} - 1\right). \tag{3.67}$$

We can now linearize (3.67) by first expressing the exponential function in its series form,

$$e^{qv_{D,\mathrm{AC}}(t)/kT} = 1 + \frac{qv_{D,\mathrm{AC}}(t)}{kT} + \frac{1}{2}\left(\frac{qv_{D,\mathrm{AC}}(t)}{kT}\right)^2 + \frac{1}{6}\left(\frac{qv_{D,\mathrm{AC}}(t)}{kT}\right)^3 + \cdots \tag{3.68}$$

and then, assuming that the variation of the diode voltage is small enough, approximating the exponential by the first two terms of the series,

$$e^{qv_{D,\text{AC}}(t)/kT} \cong 1 + \frac{qv_{D,\text{AC}}(t)}{kT}. \tag{3.69}$$

Then (3.67) becomes

$$v_D(t) \cong v_{s,\text{DC}} + v_{s,\text{AC}}(t) - RI_s \left( e^{qv_{D,\text{DC}}/kT} \left[ 1 + \frac{qv_{D,\text{AC}}(t)}{kT} \right] - 1 \right) \tag{3.70}$$

or

$$v_{D,\text{DC}} + v_{D,\text{AC}}(t) \cong v_{s,\text{DC}} - RI_s e^{qv_{D,\text{DC}}/kT} + RI_s + v_{s,\text{AC}}(t) - e^{qv_{D,\text{DC}}/kT} \frac{qRI_s}{KT} v_{D,\text{AC}}(t). \tag{3.71}$$

We can equate the constant parts and time-varying parts of (3.71) separately,

$$v_{D,\text{DC}} \cong v_{s,\text{DC}} - RI_s e^{qv_{D,\text{DC}}/kT} + RI_s \tag{3.72}$$

and

$$v_{D,\text{AC}}(t) \cong v_{s,\text{AC}}(t) - e^{qv_{D,\text{DC}}/kT} \frac{qRI_s}{kT} v_{D,\text{AC}}(t). \tag{3.73}$$

The equation for the constant part of the diode voltage (called the *bias* voltage) is nonlinear and must still be solved by a graphical or numerical technique. Once the diode's bias voltage is found, the second equation becomes a linear equation. That is, the equation for the AC or time-varying part of the voltages and currents is approximately linear and can be solved by linear-system analysis techniques. So one nonlinear dynamic system analysis has been converted into a static nonlinear analysis plus an approximate linear dynamic analysis. This kind of technique is the basis for what is called *small signal analysis* in electronics.

## INVERTIBILITY

In the analysis of the systems we usually find the response of the system, given an excitation. But we can often find the excitation, given the response, if the system is invertible.

> A system is said to be *invertible* if unique excitations produce unique responses.

If unique excitations produce unique responses, then it is possible, in principle at least, given the response, to associate it with the excitation that produced it. Most practical systems are invertible, at least in principle.

An example of a DT system that is invertible is one described by the accumulation operation

$$y[n] = \sum_{m=-\infty}^{n} x[m]. \tag{3.74}$$

The inverse of this relationship is the *first backward difference*, defined by

$$y[n] - y[n-1]. \tag{3.75}$$

Applying the first backward difference to the system equation (3.74),

$$y[n] - y[n-1] = \sum_{m=-\infty}^{n} x[m] - \sum_{m=-\infty}^{n-1} x[m] = x[n]. \tag{3.76}$$

Therefore, the excitation $x[n]$ is simply the first backward difference of the response $y[n] - y[n-1]$.

An example of a system that is not invertible is a static system whose excitation-response functional relationship is

$$y(t) = \sin(x(t)). \tag{3.77}$$

For any given value of excitation $x(t)$, it is possible to uniquely determine the value of the response $y(t)$. Knowledge of the excitation uniquely determines the response. However, if we attempt to find the excitation, given the response, by rearranging the functional relationship (3.77) into

$$x(t) = \sin^{-1}(y(t)), \tag{3.78}$$

we encounter a problem. The inverse sine function is multiple-valued. Therefore, knowledge of the response does not uniquely determine the excitation. This system violates the principle of invertibility because different excitations can produce the same response. For example, if, at some time $t = t_0$, $x(t_0) = \pi/4$, then $y(t_0) = \sqrt{2}/2$. But if, at time $t = t_0$, $x(t_0)$ had a different value, $x(t_0) = 3\pi/4$, then $y(t_0)$ would have the same value, $y(t_0) = \sqrt{2}/2$. Therefore, by observing only the response we would have no idea which excitation value had caused it.

Another example of a system that is not invertible is one that is very familiar to electronic circuit designers, the full-wave rectifier (Figure 3.33). Assume that the transformer is an ideal, 1:2-turns-ratio transformer and that the diodes are ideal so that there is no voltage drop across them in forward bias and no current through them in reverse bias. Then the output voltage signal $v_o(t)$ and input signal $v_i(t)$ are related by

$$v_o(t) = |v_i(t)|. \tag{3.79}$$

Suppose that at some particular time the output voltage signal is $+1$ V. The input voltage signal at that time could be $+1$ or $-1$ V. There is no way of knowing which of these two input voltage signals is the excitation just by observing the output voltage signal. Therefore, we could not be assured of correctly reconstructing the excitation from the response. The response is uniquely determined by the excitation, but the excitation is not uniquely determined by the response. Therefore, this system is not invertible.

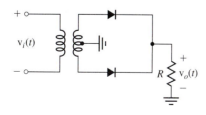

**Figure 3.33**
A full-wave rectifier.

## 3.3  EIGENFUNCTIONS OF LTI SYSTEMS

### CONTINUOUS-TIME SYSTEMS

As an example of a second-order system consider the *RLC* circuit in Figure 3.34. Assume the circuit is initially in its zero state (no stored energy in the inductor or capacitor) and that the input voltage signal is $v_{in}(t) = Au(t)$. Then the sum of voltages around the closed loop yields

$$LCv''_{out}(t) + RCv'_{out}(t) + v_{out}(t) = Au(t) \tag{3.80}$$

and the solution for the output voltage signal is

$$v_{out}(t) = K_1 e^{\left[-(R/2L)+\sqrt{(R/2L)^2-(1/LC)}\right]t} + K_2 e^{\left[-(R/2L)-\sqrt{(R/2L)^2-(1/LC)}\right]t} + A \tag{3.81}$$

and $K_1$ are $K_2$ arbitrary constants.

This solution is considerably more complicated than the solution for the *RC* lowpass filter was. Now there are two exponential terms, and each of them has a much more complicated exponent. Notice also that the exponent involves a square root of a quantity which could be negative. Therefore, the exponent could be complex. For this reason, the eigenfunction $e^{\lambda t}$ is called a *complex exponential*. The solutions to ordinary linear differential equations with constant coefficients are always linear combinations of complex exponentials.

> A *linear combination* of numbers, variables, or functions is simply a sum of products of the numbers, variables, or functions and a set of constant coefficients. For example, a linear combination of $N$ complex exponentials would be $K_1 e^{\lambda_1 t} + K_2 e^{\lambda_2 t} + \cdots + K_N e^{\lambda_N t}$, where the $K$'s are constants.

Complex exponentials are very important in signal and system analysis and will be a recurring theme in this text. In the *RLC* circuit, if the exponents are real, the response is simply the sum of two real exponentials. The more interesting case is complex exponents. The exponents are complex if

$$\left(\frac{R}{2L}\right)^2 - \frac{1}{LC} < 0. \tag{3.82}$$

In this case the solution can be written in terms of two standard parameters of second-order systems, the undamped resonant radian frequency $\omega_0$ and the damping rate $\alpha$, as

$$v_{out}(t) = K_1 e^{\left(-\alpha+\sqrt{\alpha^2-\omega_0^2}\right)t} + K_2 e^{\left(-\alpha-\sqrt{\alpha^2-\omega_0^2}\right)t} + A \tag{3.83}$$

where

$$\omega_0^2 = \frac{1}{LC} \quad \text{and} \quad \alpha = \frac{R}{2L}. \tag{3.84}$$

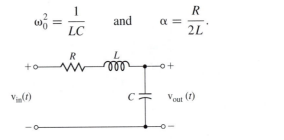

**Figure 3.34**
An *RLC* circuit.

When condition (3.82) is satisfied, the system is said to be *underdamped* and the response can be written as

$$v_{out}(t) = K_1 e^{\left(-\alpha + j\sqrt{\omega_0^2 - \alpha^2}\right)t} + K_2 e^{\left(-\alpha - j\sqrt{\omega_0^2 - \alpha^2}\right)t} + A. \qquad (3.85)$$

The exponents are complex conjugates of each other. [They must be for $v_{out}(t)$ to be a real-valued function.]

Applying initial conditions, the output voltage signal is

$$v_{out}(t) = A \begin{bmatrix} \dfrac{1}{2}\left(-1 + j\dfrac{\alpha}{\sqrt{\omega_0^2 - \alpha^2}}\right) e^{\left(-\alpha + j\sqrt{\omega_0^2 - \alpha^2}\right)t} \\[2ex] + \dfrac{1}{2}\left(-1 - j\dfrac{\alpha}{\sqrt{\omega_0^2 - \alpha^2}}\right) e^{\left(-\alpha - j\sqrt{\omega_0^2 - \alpha^2}\right)t} + 1 \end{bmatrix} \qquad (3.86)$$

This response, (3.86), appears to be a complex response to a real system with real excitation. But, even though the coefficients and exponents are complex, the overall solution is real because the output voltage signal can be reduced to

$$v_{out}(t) = A \left\{ 1 - e^{-\alpha t}\left[ \frac{\alpha}{\sqrt{\omega_0^2 - \alpha^2}}\sin\left(\sqrt{\omega_0^2 - \alpha^2}\,t\right) + \cos\left(\sqrt{\omega_0^2 - \alpha^2}\,t\right) \right] \right\}. \qquad (3.87)$$

This solution is in the form of a *damped* sinusoid, a sinusoid multiplied by a decaying exponential. The undamped resonant frequency $f_0 = \omega_0/2\pi$ is the frequency at which the response voltage would oscillate if the damping factor were zero. The rate at which the sinusoid is damped is determined by the damping factor $\alpha$. Any system described by a second-order linear differential equation could be analyzed by an analogous procedure.

An important special case of linear system analysis is an LTI system excited by a complex sinusoid. Let the input voltage signal of the *RLC* circuit now be

$$v_{in}(t) = Ae^{j2\pi f_0 t}. \qquad (3.88)$$

It is important to realize that this excitation is described exactly for all time. Not only is the excitation going to be a complex sinusoid from now on, it *has always been* a complex sinusoid. Since the excitation began at an infinite time in the past, any transients that may have occurred have long since died away (if the system is stable, as this *RLC* circuit is). Thus, the only solution that is left at this time is the steady-state solution. The steady-state response is the particular solution of the describing differential equation. Since all the derivatives of the complex sinusoid are also complex sinusoids, the particular solution of (3.88) is simply

$$v_{out,p}(t) = Be^{j2\pi f_0 t} \qquad (3.89)$$

where $B$ is yet to be determined. That is, if this LTI system is excited by a complex sinusoid, the response is also a complex sinusoid, at the same frequency, but with a different multiplying constant (in general). In general, for *any* LTI system, if its excitation is a complex exponential, its response is that same complex exponential multiplied by a complex constant.

The steady-state solution can be found by the method of undetermined coefficients. Substituting the form of the solution into the differential equation (3.80),

$$(j2\pi f_0)^2 LCBe^{j2\pi f_0 t} + j2\pi f_0 RCBe^{j2\pi f_0 t} + Be^{j2\pi f_0 t} = Ae^{j2\pi f_0 t} \qquad (3.90)$$

and solving,

$$B = \frac{A}{(j2\pi f_0)^2 LC + j2\pi f_0 RC + 1}. \tag{3.91}$$

Using the principle of superposition for LTI systems, if the excitation is an arbitrary function, which is a linear combination of complex sinusoids of various frequencies, then the response is also a linear combination of complex sinusoids at those same frequencies. This idea is the basis for the methods of Fourier series and Fourier transform analysis, to be introduced in Chapters 4 and 5, which express arbitrary system excitations and responses as linear combinations of complex sinusoids to solve for system responses.

### DISCRETE-TIME SYSTEMS

Discrete-time LTI systems are described by linear, constant-coefficient difference equations. The eigenfunctions of these equations are functions of the form $\alpha^n$, where $\alpha$ is a complex constant (Appendix I). Suppose a discrete-time LTI system is described by the difference equation

$$2y[n] + 2y[n-1] + y[n-2] = x[n]. \tag{3.92}$$

If $\alpha^n$ is an eigenfunction, then the solution of (3.92) must be in the form

$$y[n] = A\alpha^n \tag{3.93}$$

and the homogeneous equation becomes

$$2A\alpha^n + 2A\alpha^{n-1} + A\alpha^{n-2} = 0. \tag{3.94}$$

We can divide (3.94) through by $A\alpha^{n-2}$ leaving

$$2\alpha^2 + 2\alpha + 1 = 0, \tag{3.95}$$

and the solution for $\alpha$ is

$$\alpha = -\frac{1}{2} \pm \frac{j}{2}. \tag{3.96}$$

The solution of the homogeneous equation (3.94) is then of the form,

$$y_h[n] = A_1 \left(\frac{-1+j}{2}\right)^n + A_2 \left(\frac{-1-j}{2}\right)^n, \tag{3.97}$$

which can also be written as

$$y_h[n] = A_1 \left(\frac{e^{j(3\pi/4)}}{2}\right)^n + A_2 \left(\frac{e^{-j(3\pi/4)}}{2}\right)^n = 2^{-n}\left(A_1 e^{j(3\pi/4)n} + A_2 e^{-j(3\pi/4)n}\right). \tag{3.98}$$

Just as we found for CT systems, the eigenfunctions of DT systems are DT complex exponentials, and if the system is excited by a DT complex exponential, its response is also a DT complex exponential.

## 3.4 ANALOGIES

Compare the equations describing the mechanical system in Figure 3.22 and the *RLC* circuit in Figure 3.34,

$$my''(t) + K_v y'(t) + K_s y(t) = x(t) \tag{3.99}$$

and

$$LCv''_{\text{out}}(t) + RCv'_{\text{out}}(t) + v_{\text{out}}(t) = v_{\text{in}}(t). \tag{3.100}$$

Rewriting (3.100) as

$$Lv''_{\text{out}}(t) + Rv'_{\text{out}}(t) + \frac{1}{C}v_{\text{out}}(t) = \frac{1}{C}v_{\text{in}}(t) \tag{3.101}$$

we see that the two equations have the same form. We can now make some *analogies* between them.

$$m \leftrightarrow L \qquad y \leftrightarrow v_{\text{out}} \qquad K_v \leftrightarrow R \qquad K_s \leftrightarrow \frac{1}{C} \qquad x(t) \leftrightarrow \frac{1}{C}v_{\text{in}}(t) \tag{3.102}$$

The mechanical and the electrical system are *analogous*. The describing equations are of the same form, and if we can solve one, we can solve the other. System analysis encompasses both because they are both LTI systems. A problem-solving technique that was once very popular is the *analog computer*. An analog computer solves system problems by analogy by simulating system properties with voltage, current, capacitance, inductance, resistance, etc. The advantage of this technique is that the dynamics of a large, expensive system can be modeled in electronic hardware for a small fraction of the cost of actually building the system. Analog computing has faded away as digital computing has become more powerful and economical. Now most system modeling is done with digital computation instead of analog simulation. But that does not mean that analogies are no longer important. In the generalized study of systems, observing and understanding analogies between systems of widely varying types enriches and deepens our understanding of all systems.

## 3.5 THE CONVOLUTION SUM

We have seen techniques for finding the solutions to differential or difference equations which describe systems. The total solution is the sum of the homogeneous and particular solutions. The homogeneous solution is a linear combination of eigenfunctions. The particular solution depends on the form of the forcing function. Although these methods work, there is a more systematic way of finding how systems respond to excitations and it lends insight into important system properties. It is called *convolution*. Convolution will be introduced in this section for DT systems. It will then be extended in Section 3.6 to CT systems.

### UNIT IMPULSE RESPONSE

The convolution technique for finding the response of a discrete-time LTI system is based on a simple idea. No matter how complicated an excitation signal is, it is simply a sequence of DT impulses. And, for LTI systems, we can find the response of the system to one impulse at a time and then add all those responses to form the actual overall response. The responses to those impulses all have the same functional form (because the system is time-invariant) except shifted in time. The responses are of different sizes because the impulses are of different sizes and the sizes of the responses are proportional to the sizes of the excitation impulses (because the system is homogeneous). Therefore, if we can find the response of an LTI system to a unit impulse excitation occurring at time $n = 0$, we can easily find the response to any other excitation. Therefore, use of the convolution technique begins with the assumption that the response to a unit impulse excitation occurring at time $n = 0$ has already been found, and we will call that response, h[n], the *impulse response*.

Finding the impulse response of the most common types of discrete-time LTI systems is relatively simple (at least in principle). Consider first a system described by a difference equation of the form

$$a_n y[n] + a_{n-1} y[n-1] + \cdots + a_{n-D} y[n-D] = x[n]. \tag{3.103}$$

This is not the most general form of difference equation describing a discrete-time LTI system, but it is a good place to start because, from the analysis of this system, we can easily extend to finding the impulse responses of more general systems. This system is causal and LTI and, to find the impulse response, we let the excitation $x[n]$ be a unit impulse at time $n = 0$, and that is the *only* excitation of the system. Therefore, the system has never been excited by anything before that time and the response $y[n]$ has been zero for all negative time

$$y[n] = 0 \qquad n < 0. \tag{3.104}$$

For all times *after* time $n = 0$, the system excitation is also zero. The solution of the difference equation after time $n = 0$ is the homogeneous solution because the excitation is zero and there is no forced response after time $n = 0$. All we need, to find the homogeneous solution after time $n = 0$, are $D$ initial conditions we can use to evaluate the $D$ arbitrary constants in the homogeneous solution. We need an initial condition for each order of the difference equation. We can always find these initial conditions by recursion. This difference equation can always be put into a recursion form in which the present response is a linear combination of the present excitation and previous responses,

$$y[n] = \frac{x[n] - a_{n-1} y[n-1] - \cdots - a_{n-D} y[n-D]}{a_n}. \tag{3.105}$$

Then we can find an exact homogeneous solution $y_h[n]$, which is valid for all times $n \geq 0$. That solution, together with the fact that $y[n] = 0$, $n < 0$, forms the total solution, which we will call the *impulse response* $h[n]$. In a very real sense, the application of an impulse to a system simply establishes some initial conditions and the system relaxes back to its former equilibrium after that (if it is stable).

Now consider a more general system described by a difference equation of the form

$$\begin{aligned} a_n y[n] + a_{n-1} y[n-1] + \cdots + a_{n-D} y[n-D] = b_n x[n] + b_{n-1} x[n-1] \\ + \cdots + b_{n-N} x[n-N]. \end{aligned} \tag{3.106}$$

Since the system is LTI, we can find the impulse response by first finding the impulse responses to systems described by the difference equations

$$\begin{aligned} a_n y[n] + a_{n-1} y[n-1] + \cdots + a_{n-D} y[n-D] = b_n x[n] \\ a_n y[n] + a_{n-1} y[n-1] + \cdots + a_{n-D} y[n-D] = b_{n-1} x[n-1] \\ \vdots \qquad\qquad\qquad\qquad \vdots \\ a_n y[n] + a_{n-1} y[n-1] + \cdots + a_{n-D} y[n-D] = b_{n-N} x[n-N] \end{aligned} \tag{3.107}$$

and then adding all those responses. Since all the equations are the same except for the strength and time of occurrence of the impulse, the overall impulse response is simply the sum of a set of impulse responses weighted and delayed appropriately. The impulse response of the general system must be

$$h[n] = b_n h_1[n] + b_{n-1} h_1[n-1] + \cdots + b_{n-N} h_1[n-N] \tag{3.108}$$

where $h_1[n]$ is the impulse response found earlier.

**EXAMPLE 3.1**

Find the impulse response h[$n$] of the system described by the difference equation

$$8y[n] + 6y[n-1] = x[n]. \tag{3.109}$$

■ **Solution**

This equation describes a causal system, so

$$h[n] = 0 \qquad n < 0. \tag{3.110}$$

We can find the first response to a unit impulse at time $n = 0$ by recursion,

| $n$ | x[$n$] | y[$n$] |
|-----|--------|--------|
| 0   | 1      | $\frac{1}{8}$ |

For all times $n \geq 0$, the solution is the homogeneous solution of the form

$$y_h[n] = K_h \left( -\frac{3}{4} \right)^n \tag{3.111}$$

Applying initial conditions,

$$y_h[0] = \frac{1}{8} = K_h. \tag{3.112}$$

Then, the impulse response of the system is

$$h[n] = \frac{1}{8} \left( -\frac{3}{4} \right)^n u[n]. \tag{3.113}$$

## CONVOLUTION

To demonstrate the convolution idea, suppose that a discrete-time LTI system is excited by a signal x[$n$] = δ[$n$] + δ[$n - 1$], and that its impulse response is

$$h[n] = e^{-(n/4)}u[n] = (0.7788)^n u[n] \tag{3.114}$$

(Figure 3.35). The excitation for any DT system is made up of a sequence of impulses with different strengths, occurring at different times. Therefore, invoking linearity and time invariance, the response of a discrete-time LTI system will be the sum of all the individual responses to the individual excitation impulses. Since we know the response of the system to a single unit impulse occurring at discrete time $n = 0$, we can find the responses to the individual impulses in the excitation by shifting and scaling the unit impulse response appropriately.

In the example excitation, the first nonzero impulse in the excitation occurs at time $n = 0$ and its strength is one. Therefore, the system will respond to this with exactly its impulse response. The second nonzero impulse in the excitation occurs at time $n = 1$, and its strength is also one. The response of the system to this single impulse is the impulse response, except delayed by one in discrete time. So, by the additivity property of LTI systems, the overall system response to the excitation x[$n$] = δ[$n$] + δ[$n - 1$] is

$$y[n] = e^{-(n/4)}u[n] + e^{-((n-1)/4)}u[n-1] \tag{3.115}$$

(Figure 3.35).

Suppose the excitation is now changed to x[$n$] = 2δ[$n$]. Then, since the system is LTI and the excitation is an impulse of strength two, occurring at time $n = 0$, by the

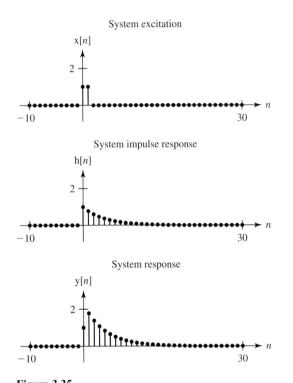

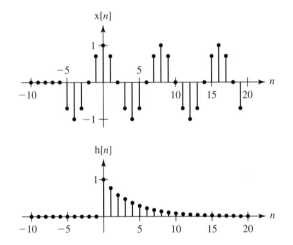

**Figure 3.35**
System excitation x[n], system impulse response h[n], and system response y[n].

**Figure 3.36**
A suddenly applied sinusoidal excitation and the system impulse response.

homogeneity property of LTI systems, the system response is twice the impulse response or

$$y[n] = 2e^{-(n/4)}u[n]. \tag{3.116}$$

Now let the excitation be the one illustrated in Figure 3.36, while the impulse response remains the same. The responses to the first four nonzero DT impulses in the excitation are plotted in Figure 3.37. Figure 3.38 shows the next four impulse responses. When we add all the responses to all the impulses in the excitation, we get the total system response to the total system excitation (Figure 3.39).

We have seen graphically what happens; now it is time to see analytically what happens. The total system response can be written as

$$y[n] = \cdots x[-5]h[n + 5] + \cdots + x[0]h[n] + \cdots + x[2]h[n - 2] + \cdots \tag{3.117}$$

or

$$y[n] = \sum_{m=-\infty}^{\infty} x[m]h[n - m]. \tag{3.118}$$

The result (3.118) is called the *convolution sum* expression for the system response. In words, it says that the value of the response y at any discrete time n can be found by summing all the products of the excitation x at discrete times m with the impulse response h at discrete times n − m, for m ranging from negative to positive infinity. Now, in order to find a system response we only need to know the system's impulse

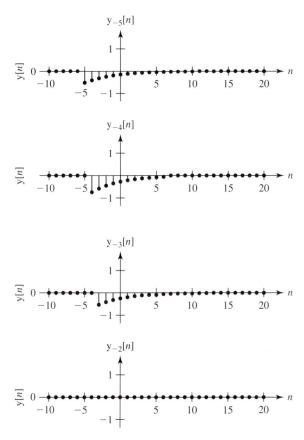

**Figure 3.37**
System responses to the impulses x[−5], x[−4], x[−3], and x[−2] in the excitation.

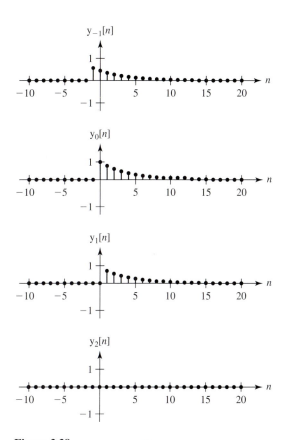

**Figure 3.38**
System responses to the impulses x[−1], x[0], x[1], and x[2] in the excitation.

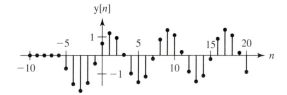

**Figure 3.39**
The total system response.

response h, and we can find its response to any arbitrary excitation x. Another way of saying this is that, for an LTI system, the impulse response of the system is a *complete description* of how it responds to excitations. So we can imagine first testing a system by applying an impulse to it and recording the response. Once we have it, we can compute the response to any desired excitation. This is a very powerful technique. In DT system analysis we only have to solve the difference equation for the system once for the simplest possible nonzero excitation, a unit impulse, and then, for any forcing function, we can find the response by convolution.

It is not quite precise to say the impulse response h[n] is the system response to a unit impulse. Notice that we have derived an expression for the response of the system to an excitation x[n] in the form,

$$y[n] = \sum_{m=-\infty}^{\infty} x[m]h[n-m]. \tag{3.119}$$

Suppose the system is a voltage amplifier. That is, it is excited by a voltage, and its response is also a voltage. Then the response h[n] to a unit impulse voltage would be a voltage, and the impulse response would, therefore, have units of volts. The excitation also has units of volts. Therefore, according to (3.119), y[n], being a sum of products of x[n] and h[n], would have units of volts squared. But we know that is incorrect. Therefore, to be precise, h[n] is the response to a unit impulse at time $n = 0$ *divided by* the units of the excitation. This does not change the numerical value of the impulse response, only its units.

## EXAMPLE 3.2

Show that the convolution sum of the impulse response with the excitation yields a response which solves the original difference equation for the equation in Example 3.1,

$$8y[n] + 6y[n-1] = x[n], \tag{3.120}$$

whose impulse response was found to be

$$h[n] = \frac{1}{8}\left(-\frac{3}{4}\right)^n u[n]. \tag{3.121}$$

### ■ Solution

Expressing the response as a convolution sum,

$$y[n] = \sum_{m=-\infty}^{\infty} x[m]h[n-m] = \sum_{m=-\infty}^{\infty} x[m]\frac{1}{8}\left(-\frac{3}{4}\right)^{n-m} u[n-m]. \tag{3.122}$$

Substituting the response into the difference equation,

$$8\sum_{m=-\infty}^{\infty} x[m]\frac{1}{8}\left(-\frac{3}{4}\right)^{n-m} u[n-m] + 6\sum_{m=-\infty}^{\infty} x[m]\frac{1}{8}\left(-\frac{3}{4}\right)^{n-1-m} u[n-1-m] = x[n]. \tag{3.123}$$

Combining summations,

$$\frac{1}{8}\left(-\frac{3}{4}\right)^n \sum_{m=-\infty}^{\infty} \left(-\frac{3}{4}\right)^{-m} \left(8u[n-m] + 6\left(-\frac{3}{4}\right)^{-1} u[n-1-m]\right) x[m] = x[n]. \tag{3.124}$$

Factoring and simplifying, we recognize a difference between two unit sequences to be a unit impulse,

$$\frac{1}{8}\left(-\frac{3}{4}\right)^n \sum_{m=-\infty}^{\infty} \left(-\frac{3}{4}\right)^{-m} 8\left(\underbrace{u[n-m] - u[n-1-m]}_{\delta[n-m]}\right) x[m] = x[n]. \tag{3.125}$$

Then, using the sampling property of the unit impulse, the two sides of the equation are both equal to x[n] and the difference equation is satisfied.

$$\left(-\frac{3}{4}\right)^n \sum_{m=-\infty}^{\infty} \left(-\frac{3}{4}\right)^{-m} \delta[n-m]x[m] = \left(-\frac{3}{4}\right)^n \left(-\frac{3}{4}\right)^{-n} x[n] = x[n] \qquad \textbf{(3.126)}$$

■

Although the convolution operation is completely defined by (3.118), it is helpful to explore some graphical concepts that aid in actually performing convolution. The two functions which are multiplied and then summed over $-\infty < m < \infty$ are x[m] and h[n − m]. To illustrate the idea of graphical DT convolution let the two functions x[n] and h[n] be the simple functions illustrated in Figure 3.40. Since the summation index in (3.118) is m, the function h[n − m] should be considered a function of m for purposes of performing the summation in (3.118). With that point of view, we can imagine that h[n − m] is created by two transformations, $m \to -m$ which changes h[m] to h[−m] and then $m \to m − n$ which changes h[−m] to h[−(m − n)] = h[n − m]. The first transformation forms the discrete-time inverse of h[m], and the second transformation shifts the already-time-inverted function n units to the right (Figure 3.41).

Now, realizing that the convolution result is $y[n] = \sum_{m=-\infty}^{\infty} x[m]h[n-m]$, the process of graphing the convolution result y[n] versus n is to pick a value of n and do the operation $\sum_{m=-\infty}^{\infty} x[m]h[n-m]$ for that n, graph that single numerical result for y[n] at that n, and then repeat the whole process for each n. Every time a new n is chosen, the function h[n − m] shifts to a new position, x[m] stays right where it is because there is no n in x[m], and the summation $\sum_{m=-\infty}^{\infty} x[m]h[n-m]$ is simply the sum of the products of the values of x[m] and h[n − m] for that choice of n. Figure 3.42 is an illustration of this process. For all values of n not represented in Figure 3.42, y[n] = 0, so we can now graph y[n] as illustrated in Figure 3.43.

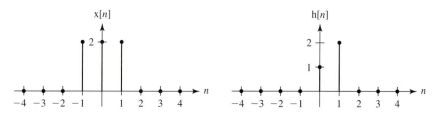

**Figure 3.40**
Two DT functions.

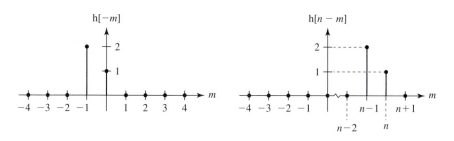

**Figure 3.41**
h[−m] and h[n − m] versus m.

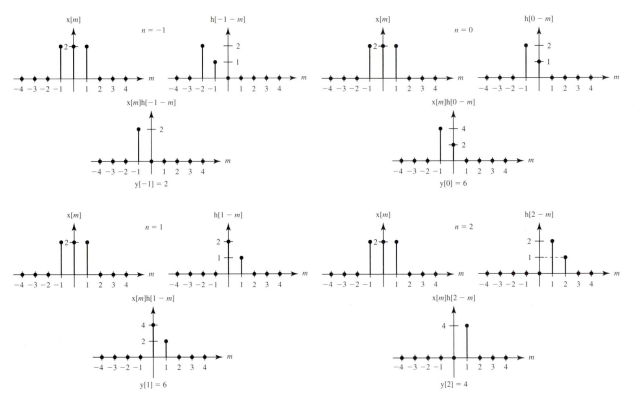

**Figure 3.42**
y[$n$] for $n = -1, 0, 1,$ and 2.

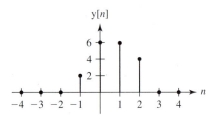

**Figure 3.43**
Graph of y[$n$].

## CONVOLUTION PROPERTIES

Convolution, as a mathematical operation, is indicated by the operator "$*$". For example,

$$y[n] = x[n] * h[n] = \sum_{m=-\infty}^{\infty} x[m]h[n-m]. \qquad (3.127)$$

> Try not to confuse the convolution operator "$*$" with the indicator for the complex conjugate of a complex number or function, "$*$". For example, x[$n$] $*$ h[$n$] is x[$n$] convolved with h[$n$], but x[$n$]$*$h[$n$] is the product of the complex conjugate of x[$n$] and h[$n$]. Usually the difference is clear in context.

Consider first the special case

$$x[n] = \delta[n]. \tag{3.128}$$

Since the excitation is now the unit impulse occurring at discrete time $n = 0$, we know that the response should be the impulse response. Using the definition of convolution (3.127), we get

$$y[n] = \delta[n] * h[n] = \sum_{m=-\infty}^{\infty} \delta[m]h[n - m]. \tag{3.129}$$

Look at the summand $\delta[m]h[n - m]$ in (3.129). This product has exactly one nonzero value, and it occurs where $m = 0$ because the impulse is nonzero only at that discrete time. During the summation process, all the $m$'s except $m = 0$ contribute nothing to the summation. Therefore, we can write the convolution (3.129) as

$$\delta[n] * h[n] = \underbrace{\delta[0]}_{=1} h[n - 0] = h[n] \tag{3.130}$$

confirming mathematically that the response is indeed the impulse response. In a more general mathematical sense, this result says that any function convolved with an unshifted unit impulse is unchanged,

$$g[n] * \delta[n] = g[n] \tag{3.131}$$

for any $g[n]$.

If the excitation, $x[n]$, of a discrete-time LTI system is multiplied by a constant $A$, the response is multiplied by the same constant.

$$(Ax[n]) * h[n] = \sum_{m=-\infty}^{\infty} Ax[m]h[n - m] = A \sum_{m=-\infty}^{\infty} x[m]h[n - m] = A(x[n] * h[n])$$
$$\tag{3.132}$$

This is just a restatement of the fact that the system is LTI and, therefore, homogeneous.

If the excitation is shifted in time by some amount $n_0$, we get the response

$$x[n - n_0] * h[n] = \sum_{m=-\infty}^{\infty} x[m - n_0]h[n - m]. \tag{3.133}$$

Making the change of variable $q = m - n_0$, we can rewrite the summation in (3.133) as

$$\sum_{q=-\infty}^{\infty} x[q]h[n - q - n_0] = \sum_{q=-\infty}^{\infty} x[q]h[(n - n_0) - q]. \tag{3.134}$$

Now we can see that if $y[n]$ is given by

$$y[n] = \sum_{m=-\infty}^{\infty} x[m]h[n - m] \tag{3.135}$$

and we replace $n$ by $n - n_0$ everywhere $n$ occurs, then we get the result

$$y[n - n_0] = \sum_{m=-\infty}^{\infty} x[m]h[(n - n_0) - m], \tag{3.136}$$

which is identical to (3.135) (except for the symbol, $m$ or $q$, used for the summation index which doesn't change the summation result). Therefore, we can say that

$$x[n - n_0] * h[n] = y[n - n_0], \tag{3.137}$$

which in words says that when we shift the excitation in time by some amount $n_0$, we shift the response by the same amount. The linearity and time-shifting properties of convolution with an impulse can be stated in the succinct form

$$\mathrm{x}[n] * A\delta[n - n_0] = A\mathrm{x}[n - n_0] \qquad (3.138)$$

(Figure 3.44).

Notice that if we define a function

$$\mathrm{g}[n] = \mathrm{g}_0[n] * \delta[n], \qquad (3.139)$$

that a time-shifted version of $\mathrm{g}[n]$, $\mathrm{g}[n - n_0]$, can be expressed in either of two alternate forms,

$$\mathrm{g}[n - n_0] = \mathrm{g}_0[n - n_0] * \delta[n] \quad \text{or} \quad \mathrm{g}[n - n_0] = \mathrm{g}_0[n] * \delta[n - n_0], \qquad (3.140)$$

but *not* in the form $\mathrm{g}_0[n - n_0] * \delta[n - n_0]$. Instead

$$\mathrm{g}[n - 2n_0] = \mathrm{g}_0[n - n_0] * \delta[n - n_0]. \qquad (3.141)$$

This property is true not only when convolving with impulses, but with *any* function. Using the definition of convolution

$$\mathrm{y}[n] = \mathrm{x}[n] * \mathrm{h}[n] = \sum_{m=-\infty}^{\infty} \mathrm{x}[m]\mathrm{h}[n - m] = \sum_{m=-\infty}^{\infty} \mathrm{x}[n - m]\mathrm{h}[m] \qquad (3.142)$$

we can form

$$\mathrm{y}[n - n_0] = \sum_{m=-\infty}^{\infty} \mathrm{x}[m]\mathrm{h}[n - n_0 - m] = \sum_{m=-\infty}^{\infty} \mathrm{x}[n - n_0 - m]\mathrm{h}[m] \qquad (3.143)$$

or

$$\mathrm{y}[n - n_0] = \mathrm{x}[n] * \mathrm{h}[n - n_0] = \mathrm{x}[n - n_0] * \mathrm{h}[n]. \qquad (3.144)$$

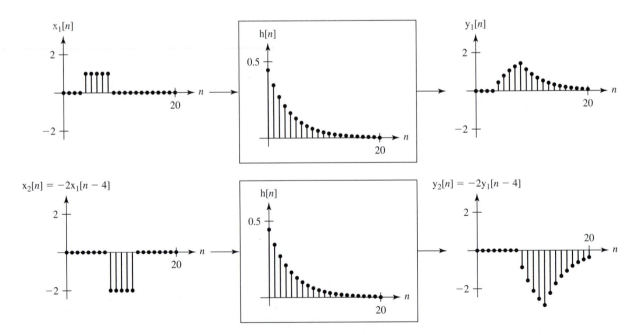

**Figure 3.44**
Scaling and shifting properties of convolution.

In words, if an overall function is the convolution of two component functions, then shifting either, but not both, of the two component functions shifts the overall function by the same amount.

Convolution is commutative, associative, and distributive. These properties are proven in Appendix D. The properties, stated mathematically, are

Commutativity: $\quad\quad x[n] * y[n] = y[n] * x[n]$ $\quad\quad\quad\quad\quad\quad\quad\quad$ **(3.145)**

Associativity: $\quad\quad (x[n] * y[n]) * z[n] = x[n] * (y[n] * z[n])$ $\quad\quad$ **(3.146)**

Distributivity: $\quad\quad (x[n] + y[n]) * z[n] = x[n] * y[n] + x[n] * z[n]$ $\quad$ **(3.147)**

We can prove two more interesting properties of the convolution sum. Let two functions be denoted $x[n]$ and $h[n]$. Then their convolution sum is

$$y[n] = x[n] * h[n]. \tag{3.148}$$

Then, using (3.140), the first backward difference of their convolution sum is

$$y[n] - y[n-1] = x[n] * h[n] - x[n] * h[n-1] \tag{3.149}$$

or

$$y[n] - y[n-1] = \sum_{m=-\infty}^{\infty} x[m]h[n-m] - \sum_{m=-\infty}^{\infty} x[m]h[n-m-1]. \tag{3.150}$$

Combining summations,

$$y[n] - y[n-1] = \sum_{m=-\infty}^{\infty} x[m](h[n-m] - h[n-m-1]) \tag{3.151}$$

or

$$y[n] - y[n-1] = x[n] * (h[n] - h[n-1]). \tag{3.152}$$

In words, the first difference of a convolution of two functions is the convolution of either one of the functions with the first difference of the other function. This is the differencing property of the convolution sum.

Let the sum of all the impulses in the functions y, x, and h be $S_y$, $S_x$, and $S_h$, respectively. Then

$$y[n] = \sum_{m=-\infty}^{\infty} x[m]h[n-m] \tag{3.153}$$

and

$$S_y = \sum_{n=-\infty}^{\infty} y[n] = \sum_{n=-\infty}^{\infty} \sum_{m=-\infty}^{\infty} x[m]h[n-m]. \tag{3.154}$$

Interchanging the order of summation,

$$S_y = \underbrace{\sum_{m=-\infty}^{\infty} x[m]}_{=S_x} \underbrace{\sum_{n=-\infty}^{\infty} h[n-m]}_{=S_h} = S_x S_h. \tag{3.155}$$

In words, the sum of the impulses in a convolution sum of two DT functions is the product of the sums of the impulses in the two individual functions. This is the sum property of the convolution sum.

We have seen that any DT function convolved with a DT unit impulse at time $n = 0$ is unchanged. If we convolve a function with the first backward difference of that impulse, a DT unit doublet, we get

$$g[n] * u_1[n] = \sum_{m=-\infty}^{\infty} g[m]u_1[n - m] = \sum_{m=-\infty}^{\infty} g[m](\delta[n - m] - \delta[n - m - 1])$$

**(3.156)**

or

$$g[n] * u_1[n] = g[n] - g[n - 1].$$

**(3.157)**

So convolution with an unshifted unit doublet yields the first backward difference of a function. Similarly, higher-order singularity functions produce higher-order differences when convolved with a function.

MATLAB has a command `conv` which does convolution. The syntax is

```
y = conv(x,h)
```

where $x$ and $h$ are vectors of values of discrete-time signals and $y$ is the vector containing the values of the convolution of $x$ with $h$. Of course, MATLAB cannot actually compute an infinite sum as indicated by (3.153). Therefore, MATLAB can only convolve time-limited signals, and the vectors, $x$ and $h$ should contain all the nonzero values of the signals they represent. (They can also contain extra zero values, if desired.) The length of $y$ is one less than the sum of the lengths of $x$ and $h$. If the time of the first element in $x$ is $n_{x0}$ and the time of the first element in $h$ is $n_{h0}$, the time of the first element in $y$ is $n_{x0} + n_{h0}$. As an example, suppose

$$x[n] = \text{rect}_2[n - 3] \qquad \text{and} \qquad h[n] = \text{tri}\left(\frac{n - 6}{4}\right).$$

**(3.158)**

Then $x[n]$ is time-limited to the range $1 \le n \le 5$ and $h[n]$ is time-limited to the range $3 \le n \le 9$. Therefore, any vector describing $x[n]$ should be at least five elements long, and any vector describing $h[n]$ should be at least seven elements long. Let's put in some extra zeros, compute the convolution, and graph the two signals and their convolutions using the following MATLAB code.

```
nx = -2:8 ; nh = 0:12 ;        % Set time vectors for x and h.
x = rectDT(2,nx-3) ;           % Compute values of x.
h = tri((nh-6)/4) ;            % Compute values of h.
y = conv(x,h) ;                % Compute the convolution of x with h.
%
%       Generate a discrete-time vector for y.
%
ny = (nx(1) + nh(1)) + (0:(length(nx) + length(nh) - 2)) ;
%
%       Plot the results.
%
subplot(3,1,1) ; stem(nx,x,'k','filled') ;
xlabel('n') ; ylabel('x') ; axis([-2,20,0,4]) ;
subplot(3,1,2) ; stem(nh,h,'k','filled') ;
xlabel('n') ; ylabel('h') ; axis([-2,20,0,4]) ;
subplot(3,1,3) ; stem(ny,y,'k','filled') ;
xlabel('n') ; ylabel('y') ; axis([-2,20,0,4]) ;
```

The three signals are illustrated in Figure 3.45.

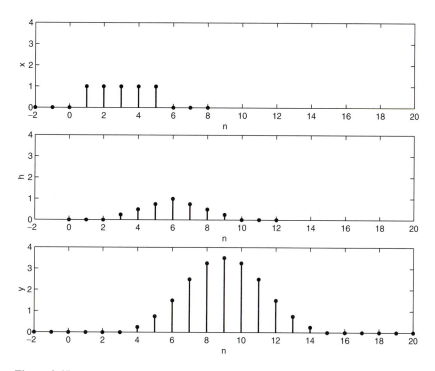

**Figure 3.45**
Excitation, impulse response, and response of a DT system found using the MATLAB `conv` command.

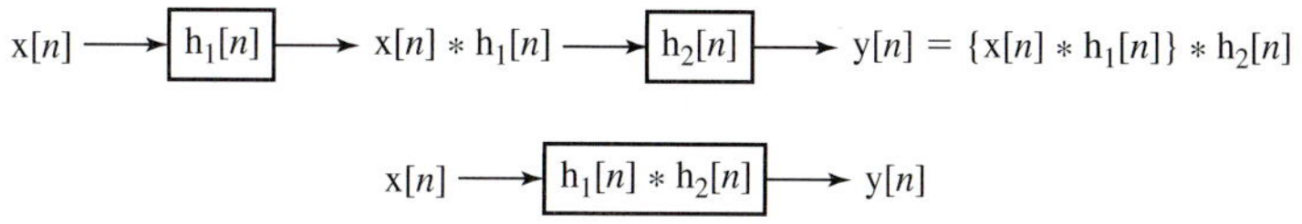

$$x[n] \longrightarrow \boxed{h_1[n]} \longrightarrow x[n] * h_1[n] \longrightarrow \boxed{h_2[n]} \longrightarrow y[n] = \{x[n] * h_1[n]\} * h_2[n]$$

$$x[n] \longrightarrow \boxed{h_1[n] * h_2[n]} \longrightarrow y[n]$$

**Figure 3.46**
Cascade connection of systems.

## SYSTEM INTERCONNECTIONS

There are some common ways that systems are connected to form larger systems. Two of the most common are the cascade connection and the parallel connection. The cascade connection (Figure 3.46) is the connection of the output of one system to the input of another. If the excitation of the first system is $x[n]$, then its response is $x[n] * h_1[n]$, which is also the excitation of the second system. Therefore, the response of the second system is $\{x[n] * h_1[n]\} * h_2[n]$. Using the associativity of convolution we then say that the impulse response of the overall system is

$$h[n] = h_1[n] * h_2[n]. \tag{3.159}$$

This can be extended to a cascade connection of any number of systems. In words, the impulse response of a cascade connection of LTI systems is the convolution of all the individual system impulse responses.

The parallel connection of systems is illustrated in Figure 3.47. The excitation of both systems is $x[n]$. The response of the top system is $x[n] * h_1[n]$, and the response of the bottom system is $x[n] * h_2[n]$. Therefore, the overall system response (using the

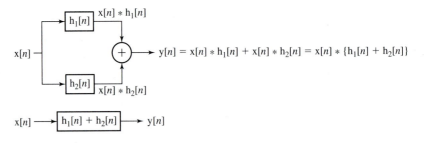

**Figure 3.47**
Parallel connection of systems.

distributivity of convolution) is

$$x[n] * h_1[n] + x[n] * h_2[n] = x[n] * (h_1[n] + h_2[n]). \qquad \textbf{(3.160)}$$

The impulse response of the overall system is $h_1[n] + h_2[n]$. This can be extended to a parallel connection of any number of systems. In words, the impulse response of a parallel connection of LTI systems is the sum of all the individual system impulse responses.

## STABILITY AND IMPULSE RESPONSE

Stability was generally defined in Section 3.2 by saying that a stable system has a bounded response to a bounded excitation. We can now find a way to determine whether a system is stable by examining its impulse response. The response $y[n]$ of a DT system to an excitation $x[n]$ is

$$y[n] = \sum_{m=-\infty}^{\infty} x[m]h[n-m]. \qquad \textbf{(3.161)}$$

If the excitation $x[n]$ is bounded, we can say that $|x[m]| < B < \infty$, for all $m$. The response magnitude is

$$|y[n]| = \left| \sum_{m=-\infty}^{\infty} x[m]h[n-m] \right|. \qquad \textbf{(3.162)}$$

Using the principle that the magnitude of any sum of terms is less than or equal to the sum of the magnitudes of the terms,

$$|y[n]| \leq \sum_{m=-\infty}^{\infty} |x[m]||h[n-m]|. \qquad \textbf{(3.163)}$$

Since $x[m]$ is less than $B$ in magnitude for any $m$,

$$|y[n]| \leq \sum_{m=-\infty}^{\infty} |x[m]||h[n-m]| \leq \sum_{m=-\infty}^{\infty} B|h[n-m]| \qquad \textbf{(3.164)}$$

or

$$|y[n]| \leq B \sum_{m=-\infty}^{\infty} |h[n-m]|. \qquad \textbf{(3.165)}$$

Therefore, the response $y[n]$ is bounded if $\sum_{m=-\infty}^{\infty} |h[n-m]|$ is bounded.

A DT system is stable if its impulse response is absolutely summable.

## RESPONSES OF SYSTEMS TO STANDARD SIGNALS

It is useful, in preparation for later work, to examine the form of DT system responses to some standard signals, the unit sequence, a complex exponential, and a sinusoid.

The response of any system is the convolution of the excitation with the impulse response,

$$y[n] = x[n] * h[n] = \sum_{m=-\infty}^{\infty} x[m]h[n-m]. \tag{3.166}$$

Let the excitation be a unit sequence, and let the response to a unit sequence be designated $h_{-1}[n]$. Then

$$h_{-1}[n] = u[n] * h[n] = \sum_{m=-\infty}^{\infty} u[m]h[n-m]. \tag{3.167}$$

The unit sequence is defined as the accumulation of the unit impulse

$$u[n] = \sum_{m=-\infty}^{n} \delta[m]. \tag{3.168}$$

Combining (3.167) and (3.168),

$$h_{-1}[n] = u[n] * h[n] = \sum_{m=-\infty}^{\infty} \sum_{q=-\infty}^{m} \delta[q]h[n-m]. \tag{3.169}$$

The summand $\delta[q]h[n-m]$ in (3.169) has a nonzero value only where $q = 0$. The inner summation with respect to $q$ goes from negative infinity to $m$, where $m$ is the index of summation of the outer summation. Therefore, for all values of $m$ less than zero, the response is zero and the summation can be simplified to

$$h_{-1}[n] = u[n] * h[n] = \sum_{m=0}^{\infty} h[n-m]. \tag{3.170}$$

Now, making the change of variable $q = n - m$,

$$h_{-1}[n] = \sum_{q=n}^{-\infty} h[q] = \sum_{q=-\infty}^{n} h[q]. \tag{3.171}$$

In words, the response at any discrete time $n$ of an LTI system excited by a unit sequence is the accumulation of the impulse response. Therefore, we can say that just as the unit sequence is the accumulation of the impulse response, the unit sequence response is the accumulation of the unit impulse response. That is the reason for the notation $h_{-1}[n]$. It follows the same logic as the notation for singularity functions. The subscript indicates the number of differentiations, or in this case, differences. In this case there is a $-1$ difference, or one accumulation, in going from the impulse response to the unit sequence response. In fact, this relationship holds not just for impulse and unit sequence excitations, but for *any* excitation. If any excitation is changed to its accumulation, the response also changes to its accumulation. We can also turn these relationships around and say that since the first backward difference is the inverse of accumulation, if the excitation is changed to its first backward difference, the response is also changed to its first backward difference.

The complex exponential excitation is the second important standard signal. It is important because it will be the basis of the transform methods in all the chapters to follow. It has the form

$$x[n] = z^n \tag{3.172}$$

where $z$ can be any complex constant. The response of the system can be written in either of the two convolution forms

$$y[n] = \sum_{m=-\infty}^{\infty} z^m h[n-m] = \sum_{m=-\infty}^{\infty} z^{n-m} h[m]. \tag{3.173}$$

Rearranging the second form in (3.173),

$$y[n] = z^n \underbrace{\sum_{m=-\infty}^{\infty} h[m] z^{-m}}_{\text{complex constant}}. \tag{3.174}$$

This result shows that the functional form of the response is the same as the functional form of the excitation, a constant, which can be complex, multiplied by $z^n$. In Chapter 11, the summation

$$\sum_{n=-\infty}^{\infty} h[n] z^{-n} \tag{3.175}$$

will be called the *z transform* of $h[n]$ and will be one of the transform methods for LTI system analysis, in this case, for DT systems. The commonly used transform methods will be introduced in Chapters 4, 5, 9, and 11.

We can specialize the complex exponential excitation to a complex sinusoidal excitation by letting $z = e^{j\Omega} = e^{j2\pi F}$. If the excitation is $e^{j2\pi Fn}$, then the response is

$$y[n] = e^{j2\pi Fn} \sum_{m=-\infty}^{\infty} h[m] e^{-j2\pi Fm}. \tag{3.176}$$

If the excitation is $e^{-j2\pi Fn}$, the response is

$$y[n] = e^{-j2\pi Fn} \sum_{m=-\infty}^{\infty} h[m] e^{j2\pi Fm}. \tag{3.177}$$

Invoking linearity, if the excitation is $(e^{j2\pi Fn} + e^{-j2\pi Fn})/2$ (a cosine) the response is

$$y[n] = \frac{1}{2}\left(e^{j2\pi Fn} \sum_{m=-\infty}^{\infty} h[m] e^{-j2\pi Fm} + e^{-j2\pi Fn} \sum_{m=-\infty}^{\infty} h[m] e^{j2\pi Fm}\right) \tag{3.178}$$

or

$$y[n] = \frac{1}{2}\left(e^{j2\pi Fn} \sum_{m=-\infty}^{\infty} h[m](\cos(2\pi Fm) - j\sin(2\pi Fm))\right.$$
$$\left. + e^{-j2\pi Fn} \sum_{m=-\infty}^{\infty} h[m](\cos(2\pi Fm) + j\sin(2\pi Fm))\right) \tag{3.179}$$

or

$$y[n] = \frac{1}{2}\left((e^{j2\pi Fn} + e^{-j2\pi Fn}) \sum_{m=-\infty}^{\infty} h[m]\cos(2\pi Fm)\right.$$
$$\left. + (je^{-j2\pi Fn} - je^{j2\pi Fn}) \sum_{m=-\infty}^{\infty} h[m]\sin(2\pi Fm)\right) \tag{3.180}$$

$$y[n] = \cos(2\pi Fn) \underbrace{\sum_{m=-\infty}^{\infty} h[m]\cos(2\pi Fm)}_{\text{real constant}} + \sin(2\pi Fn) \underbrace{\sum_{m=-\infty}^{\infty} h[m]\sin(2\pi Fm)}_{\text{real constant}} \tag{3.181}$$

So we conclude that a real-valued sinusoidal excitation produces a real-valued sinusoidal response at the same frequency as the excitation, just as a complex exponential excitation produces a complex exponential response (since a real sinusoid is just a linear combination of complex sinusoids and a complex sinusoid is just a special case of a complex exponential).

**EXAMPLE 3.3**

The DT system of Figure 3.11 is replicated in Figure 3.48. We could find the impulse directly using the methods presented earlier, but, in this case, since we have already found its response to a unit sequence excitation

$$y[n] = \left[ 5 - 4 \left( \frac{4}{5} \right)^n \right] u[n], \qquad (3.182)$$

we can find the impulse response as the first backward difference of the unit sequence response,

$$h[n] = y[n] - y[n - 1]$$

$$= \left[ 5 - 4 \left( \frac{4}{5} \right)^n \right] u[n] - \left[ 5 - 4 \left( \frac{4}{5} \right)^{n-1} \right] u[n - 1]$$

$$= 5 \underbrace{(u[n] - u[n - 1])}_{= \delta[n]} - 4 \left( \frac{4}{5} \right)^{n-1} \left( \frac{4}{5} u[n] - u[n - 1] \right)$$

$$= \underbrace{5\delta[n] - 4 \left( \frac{4}{5} \right)^n \delta[n]}_{= \delta[n]} + \left( \frac{4}{5} \right)^n u[n - 1]$$

$$= \left( \frac{4}{5} \right)^n u[n]. \qquad (3.183)$$

This impulse response can also be found from the recursion

$$y[n] = x[n] + \frac{4}{5} y[n - 1]. \qquad (3.184)$$

Table 3.1 compares the two results.

**Table 3.1**

| $n$ | $\delta[n]$ | $y[n] = \delta[n] + \frac{4}{5} y[n-1]$ | $h[n] = \left( \frac{4}{5} \right)^n u[n]$ |
|---|---|---|---|
| 0 | 1 | 1 | 1 |
| 1 | 0 | $\frac{4}{5}$ | $\frac{4}{5}$ |
| 2 | 0 | $\frac{16}{25}$ | $\frac{16}{25}$ |
| 3 | 0 | $\frac{64}{125}$ | $\frac{64}{125}$ |
| ⋮ | ⋮ | ⋮ | ⋮ |

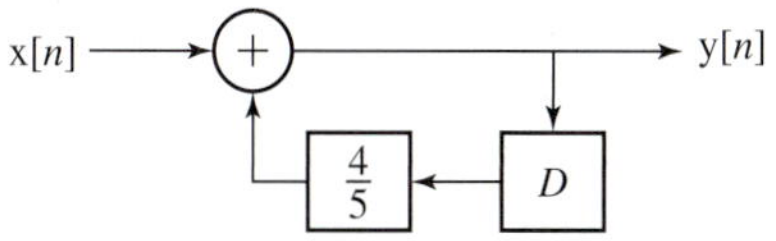

**Figure 3.48**
A DT system.

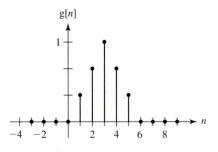

**Figure 3.49**
Excitation of the DT system.

Now that we have the impulse response, we can find the response to any excitation by convolution. Let the excitation be that illustrated in Figure 3.49. All that remains is to perform the convolution. We can do that using this MATLAB program:

```
%       Program to demonstrate discrete-time convolution
nx = -5:15 ;          %       Set a discrete-time vector for the excitation.

x = tri((n-3)/3) ;   %       Generate the excitation vector.

nh = 0:20 ;           %       Set a discrete-time vector for the impulse
                      %       response.

%       Generate the impulse response vector.
h = ((4/5).^nh).*uDT(nh) ;

%       Compute the beginning and ending discrete times for the system
%       response vector from the discrete-time vectors for the excitation
%       and the impulse response.
nymin = nx(1) + nh(1) ; nymax = nx(length(nx)) + length(nh) ;
ny = nymin:nymax-1 ;

%       Generate the system response vector by convolving the excitation
%       with the impulse response.
y = conv(x,h) ;

%       Plot the excitation, impulse response, and system response, all
%       on the same time scale for comparison.

%       Plot the excitation.
subplot(3,1,1) ; p = stem(nx,x,'k','filled') ;
set(p,'LineWidth',2,'MarkerSize',4) ;
axis([nymin,nymax,0,3]) ;
xlabel('n') ; ylabel('x[n]') ;

%       Plot the impulse response.
subplot(3,1,2) ; p = stem(nh,h,'k','filled') ;
set(p,'LineWidth',2,'MarkerSize',4) ;
axis([nymin,nymax,0,3]) ;
xlabel('n') ; ylabel('h[n]') ;
```

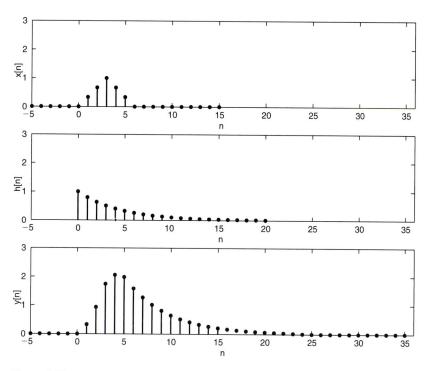

**Figure 3.50**
Excitation, impulse response, and system response.

```
%       Plot the system response.
subplot(3,1,3) ; p = stem(ny,y,'k','filled') ;
set(p,'LineWidth',2,'MarkerSize',4) ;
axis([nymin,nymax,0,3]) ;
xlabel('n') ; ylabel('y[n]') ;
```

The three signals as plotted by MATLAB are illustrated in Figure 3.50.

**EXAMPLE 3.4**

For a system described by

$$8y[n] + 4y[n-1] + y[n-2] = x[n] \tag{3.185}$$

find the response to a unit amplitude complex sinusoidal excitation at a DT cyclic frequency $F$, and then plot the amplitude of the steady-state complex sinusoidal response versus DT cyclic frequency $F$ and versus DT radian frequency $\Omega$.

■ **Solution**

The difference equation describing the system with a unit amplitude complex sinusoidal excitation at DT frequency $F$ is

$$8y[n] + 4y[n-1] + y[n-2] = e^{j2\pi Fn} \tag{3.186}$$

Since we are only interested in the steady-state solution, we can just find the particular solution to (3.186) of the form

$$y_p[n] = Ke^{j2\pi Fn} \tag{3.187}$$

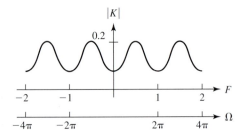

**Figure 3.51**
Amplitude of response complex sinusoid
versus DT frequency.

where $K$ is a complex constant which is yet to be determined. Substituting (3.187) into (3.186),

$$8Ke^{j2\pi Fn} + 4Ke^{j2\pi F(n-1)} + Ke^{j2\pi F(n-2)} = e^{j2\pi Fn}. \tag{3.188}$$

Dividing through by $e^{j2\pi F(n-2)}$,

$$8Ke^{j4\pi F} + 4Ke^{j2\pi F} + K = e^{j4\pi F}. \tag{3.189}$$

Solving for $K$,

$$K = \frac{e^{j4\pi F}}{8e^{j4\pi F} + 4e^{j2\pi F} + 1}. \tag{3.190}$$

Since $\Omega = 2\pi F$, $K$ as a function of $\Omega$ is

$$K = \frac{e^{j2\Omega}}{8e^{j2\Omega} + 4e^{j2\Omega} + 1}. \tag{3.191}$$

Figure 3.51 is a plot of $|K|$ versus $F$ and $\Omega$. ∎

## 3.6  THE CONVOLUTION INTEGRAL

### IMPULSE RESPONSE

In a manner analogous to the development of the convolution sum for DT signals we will now develop the convolution integral for CT signals. The basic conceptual process is the same, but it is a little more involved. We express the excitation as the sum of simple functions, all of the same type; find the response to a standard function; and then find the responses to all the simple functions and add them. The simple functions will turn out to be CT impulses. Therefore, as was true for DT systems, in order to use convolution to find a system response we first need the system's impulse response. Example 3.5 illustrates some methods for finding the impulse response from the system's differential equation.

**EXAMPLE 3.5**

Find the impulse response of a system characterized by the differential equation

$$y'(t) + ay(t) = x(t), \tag{3.192}$$

where $x(t)$ is the excitation and $y(t)$ is the response.

■ **Solution**

**Method 1**

Since the excitation is the unit impulse at time $t = 0$, we know that the impulse response before time $t = 0$ is zero,

$$h(t) = 0 \qquad t < 0. \tag{3.193}$$

The homogeneous solution for times $t > 0$ is

$$y_h(t) = K e^{-at}, \tag{3.194}$$

and this is the form of the impulse response for times $t > 0$, because in that time range the excitation is zero. We now know the form of the impulse response before and after time $t = 0$. All that is left is to find out what happens *at* time $t = 0$. The differential equation (3.192) must be satisfied at all times. We can determine what happens at time $t = 0$ by integrating both sides of (3.192) from $t = 0^-$ to $t = 0^+$,

$$y(0^+) - y(0^-) + a \int_{0^-}^{0^+} y(t)\, dt = \int_{0^-}^{0^+} \delta(t)\, dt = 1. \tag{3.195}$$

If $y(t)$ does not have an impulse or higher-order singularity at time $t = 0$, then

$$\int_{0^-}^{0^+} y(t)\, dt = 0. \tag{3.196}$$

If $y(t)$ does have an impulse or higher-order singularity at time $t = 0$, then

$$\int_{0^-}^{0^+} y(t)\, dt \neq 0. \tag{3.197}$$

If $y(t)$ has an impulse or higher-order singularity at time $t = 0$, then $y'(t)$ must be a doublet or higher-order singularity. Since there is no doublet or higher-order singularity on the right side of (3.192), the equation cannot be satisfied. Therefore, we know that there is no impulse or higher-order singularity in $y(t)$ at time $t = 0$ and, thus, $\int_{0^-}^{0^+} y(t)\, dt = 0$. Then

$$y(0^+) - y(0^-) = 1. \tag{3.198}$$

Since the system is initially at rest before the impulse is applied, we know that $y(0^-) = 0$ and, therefore, that $y(0^+) = 1$. This is the needed initial condition to find an exact numerical form of the homogeneous solution which applies after time $t = 0$. Applying that initial condition,

$$h(0^+) = 1 = K \tag{3.199}$$

and the total solution is

$$h(t) = e^{-at} u(t). \tag{3.200}$$

Let's verify this solution by substituting it into the differential equation

$$y'(t) + a y(t) = e^{-at} \delta(t) - a e^{-at} u(t) + a e^{-at} u(t) = \delta(t) \tag{3.201}$$

or

$$e^{-at} \delta(t) = e^{0} \delta(t) = \delta(t). \tag{3.202}$$

In words, the function $h(t) = e^{-at}u(t)$ has the property that its derivative plus $a$ times itself, $h'(t) + ah(t)$, is zero before time $t = 0$, is also zero after time $t = 0$, and has exactly the right-size step discontinuity at time $t = 0$ to equate $h'(t) + ah(t)$ to a unit impulse at that time. Therefore, for any time $t$ the differential equation $h'(t) + ah(t) = \delta(t)$ is satisfied by $h(t) = e^{-at}u(t)$, and it must be the impulse response.

## Method 2

Another way to find the impulse response is to take the approach of finding the response of the system to a rectangular pulse of width $w$ and height $1/w$, beginning at time $t = 0$, and, after finding the solution, letting $w$ approach zero. As $w$ approaches zero, the rectangular pulse approaches an impulse at time $t = 0$, and, therefore, the response approaches the impulse response.

Using the principle of linearity, the response to the pulse is the sum of the response to a step of height $1/w$ at time $t = 0$, and the response to a step of height $t - (1/w)$ at time $t = w$. The total solution for time $t > 0$ to a step excitation is

$$y(t) = Ke^{-at} + \frac{1}{aw}. \tag{3.203}$$

Referring to (3.192),

$$y'(t) + ay(t) = x(t). \tag{3.204}$$

If $y(t)$ had a discontinuity at $t = 0$, then $y'(t)$ would have an impulse at $t = 0$. Therefore, since $x(t)$ does not contain an impulse, $y(t)$ must be continuous at $t = 0$, otherwise (3.204) could not be correct. Since $y(t)$ has been zero for all negative time, it must also be zero at $t = 0^+$. Then

$$y(0^+) = 0 = Ke^0 + \frac{1}{aw} \Rightarrow K = -\frac{1}{aw} \tag{3.205}$$

and

$$y(t) = \frac{1 - e^{-at}}{aw} \qquad t > 0. \tag{3.206}$$

Combining this with the fact that $y(t) = 0$ for $t < 0$, we get the solution for all time,

$$y_1(t) = \frac{1 - e^{-at}}{aw}u(t). \tag{3.207}$$

Using linearity and time invariance, the response to the second step is

$$y_2(t) = -\frac{1 - e^{-a(t-w)}}{aw}u(t - w). \tag{3.208}$$

Therefore, the response to the pulse is

$$y(t) = \frac{(1 - e^{-at})u(t) - (1 - e^{-a(t-w)})u(t - w)}{aw}. \tag{3.209}$$

Then, letting $w$ approach zero,

$$h(t) = \lim_{w \to 0} y(t) = \lim_{w \to 0} \frac{(1 - e^{-at})u(t) - (1 - e^{-a(t-w)})u(t - w)}{aw}. \tag{3.210}$$

This is an indeterminate form, so we must use L'Hôpital's rule to evaluate it.

$$\lim_{w \to 0} y(t) = \lim_{w \to 0} \frac{\frac{d}{dw}\left((1 - e^{-at})u(t) - \left(1 - e^{-a(t-w)}\right)u(t - w)\right)}{(d/dw)(aw)}$$

$$= \lim_{w \to 0} \frac{-(d/dw)\left(\left(1 - e^{-a(t-w)}\right)u(t - w)\right)}{a}$$

$$= -\lim_{w \to 0} \frac{\left(1 - e^{-a(t-w)}\right)(-\delta(t - w)) - ae^{-a(t-w)}u(t - w)}{a}$$

$$= -\frac{(1 - e^{-at})(-\delta(t)) - ae^{-at}u(t)}{a} = -\frac{-ae^{-at}u(t)}{a} = e^{-at}u(t) \quad \textbf{(3.211)}$$

Therefore, the impulse response is

$$h(t) = e^{-at}u(t) \quad \textbf{(3.212)}$$

as before.

### Method 3

The third (and last) method is to find the unit step response instead of the unit impulse response and then differentiate it to find the unit impulse response. From method 2, we know that the unit step response is

$$y(t) = \frac{1 - e^{-at}}{a}u(t). \quad \textbf{(3.213)}$$

Therefore,

$$h(t) = \frac{d}{dt}(y(t)) = \frac{d}{dt}\left(\frac{1 - e^{-at}}{a}u(t)\right) = \frac{1}{a}\left[\underbrace{(1 - e^{-at})\delta(t)}_{= 0} + ae^{-at}u(t)\right] \quad \textbf{(3.214)}$$

$$h(t) = e^{-at}u(t) \quad \textbf{(3.215)}$$

as before.

◼

The principles in Example 3.5 can be generalized to apply to finding the impulse response of a system described by a differential equation of the general form

$$a_n y^{(n)}(t) + a_{n-1} y^{(n-1)}(t) + \cdots + a_1 y'(t) + a_0 y(t)$$
$$= b_m x^{(m)}(t) + b_{m-1} x^{(m-1)}(t) + \cdots + b_1 x'(t) + b_0 x(t). \quad \textbf{(3.216)}$$

The response $y(t)$ to an impulse excitation must have a functional form such that (1) when it is differentiated multiple times, up to the $n$th derivative, all those derivatives must match a corresponding derivative of the impulse up to the $m$th derivative at time $t = 0$, and (2) the linear combination of all the derivatives of $y(t)$ on the left-hand side of (3.216) must add to zero for any time $t \neq 0$. Requirement 2 is met by a solution of the form $y_h(t)u(t)$, where $y_h(t)$ is the homogeneous solution of (3.216). To meet requirement 1 we may need to add another function or functions to $y_h(t)u(t)$. Consider three cases.

*Case 1: $m < n$.* In this case the derivatives of $y_h(t)u(t)$ provide all the singularity functions necessary to match the impulse and derivatives of the impulse on the right side and no other terms need to be added.

*Case 2: m = n*. Here we only need to add an impulse term $K_0\delta(t)$ and solve for $K_0$ by matching coefficients of impulses on both sides.

*Case 3: m > n*. In this case, the $n$th derivative of the function we add to $y_h(t)u(t)$ must have a term that matches the $m$th derivative of the unit impulse. So the function we add must be of the form $K_{m-n}u_{m-n}(t) + K_{m-n-1}u_{m-n-1}(t) + \cdots + \underbrace{K_0 u_0(t)}_{=\delta(t)}$, and the $K$'s will be determined by matching coefficients of corresponding terms on the two sides. All the other derivatives of the impulse will be accounted for by differentiating the solution form $y_h(t)u(t)$, multiple times. (It should be mentioned that this case is rare in practice.)

---

### EXAMPLE 3.6

Find the impulse response of a system described by

$$y'(t) + ay(t) = x'(t). \tag{3.217}$$

#### ■ Solution

In this case, the highest derivative is the same for the excitation and response, and the general form of the impulse response must be

$$y(t) = Ke^{-at}u(t) + K_0\delta(t). \tag{3.218}$$

Substituting (3.218) into (3.217) we get

$$Ke^{-at}\delta(t) - aKe^{-at}u(t) + K_0 u_1(t) + a[Ke^{-at}u(t) + K_0\delta(t)] = u_1(t) \tag{3.219}$$

or

$$Ke^{-at}\delta(t) + K_0 u_1(t) + aK_0\delta(t) = u_1(t). \tag{3.220}$$

The coefficients of the impulse and doublet must match independently on both sides. This requirement leads to the two equations

$$K_0 u_1(t) = u_1(t) \Rightarrow K_0 = 1 \tag{3.221}$$

$$Ke^{-at}\delta(t) + aK_0\delta(t) = 0 \Rightarrow K = -aK_0 = -a \tag{3.222}$$

Therefore,

$$y(t) = \delta(t) - ae^{-at}u(t). \tag{3.223}$$

Checking the solution, by substituting it into (3.217),

$$u_1(t) - ae^{-at}\delta(t) + a^2 e^{-at}u(t) + a[\delta(t) - ae^{-at}u(t)] = u_1(t) \tag{3.224}$$

or

$$u_1(t) = u_1(t). \text{ Check.} \tag{3.225}$$

---

## CONVOLUTION

Now, assuming that the impulse response is known, we can proceed to a development of a method for finding the response to a general excitation using convolution. Suppose the CT excitation $x(t)$ to some CT system is an arbitrary waveform as illustrated in Figure 3.52. How would we find the response? We could find an approximate response

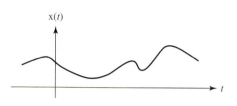

**Figure 3.52**
An arbitrary excitation.

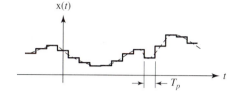

**Figure 3.53**
Contiguous-pulse approximation to an
arbitrary excitation.

by approximating this excitation as a sequence of contiguous rectangular pulses, all of the same width $T_p$ (Figure 3.53).

Now we can (approximately) find the response to the original excitation as the sum of the responses to all those pulses, acting individually. Since all the pulses are rectangular and of the same width, the only differences between pulses are when they occur and how tall they arc. So the pulse responses all have the same form except delayed by some amount, to account for time of occurrence, and multiplied by a weighting constant, to account for the height. We can make the approximation as good as necessary just by using more pulses of smaller width. In summary, just as for DT systems, the problem of finding the response of an LTI system to an arbitrary excitation becomes the problem of adding responses of a known functional form, but weighted and delayed appropriately.

Using the CT rectangle function, the description of the approximation to the arbitrary excitation can now be written analytically. The height of a pulse is the value of the excitation at the time the center of the pulse occurs. Then the approximation to the excitation can be written as

$$x(t) \cong \cdots + x(-T_p) \operatorname{rect}\left(\frac{t + T_p}{T_p}\right) + x(0) \operatorname{rect}\left(\frac{t}{T_p}\right) + x(T_p) \operatorname{rect}\left(\frac{t - T_p}{T_p}\right) + \cdots$$

$$\textbf{(3.226)}$$

or

$$x(t) \cong \sum_{n=-\infty}^{\infty} x(nT_p) \operatorname{rect}\left(\frac{t - nT_p}{T_p}\right). \qquad \textbf{(3.227)}$$

Let the response to a single pulse excitation of width $T_p$, centered at time $t = 0$, with unit area (an unshifted unit pulse) be a function $h_p(t)$ called the *unit pulse response*. The mathematical form of the unit pulse is

$$\frac{1}{T_p} \operatorname{rect}\left(\frac{t}{T_p}\right). \qquad \textbf{(3.228)}$$

The actual pulses in (3.226) have the form

$$x(nT_p) \operatorname{rect}\left(\frac{t - nT_p}{T_p}\right). \qquad \textbf{(3.229)}$$

Therefore, (3.227) could be written in terms of shifted unit pulses as

$$x(t) \cong \sum_{n=-\infty}^{\infty} T_p x(nT_p) \underbrace{\frac{1}{T_p} \operatorname{rect}\left(\frac{t - nT_p}{T_p}\right)}_{\text{shifted unit pulse}}. \qquad \textbf{(3.230)}$$

Invoking the linearity and time invariance of LTI systems, the response to each of these actual excitation pulses must be the unit pulse response $h_p(t)$, amplitude-scaled by the factor $T_p x(nT_p)$ and time-shifted from the time origin the same amount as the excitation pulse. Then the approximation to the response can be written as

$$y(t) \cong \sum_{n=-\infty}^{\infty} T_p x(nT_p) h_p(t - nT_p). \tag{3.231}$$

As was noted in the development of DT convolution, notice that if the unit pulse response is simply taken as the response when a unit pulse, with the same units as the excitation, is applied as the excitation, the units don't work out right because

Response units = excitation units × unit pulse response units × time.

Therefore, to make the units work out right the unit pulse response has to be defined as the response when an excitation unit pulse is applied, divided by the integral of the excitation unit pulse over all time. Since the integral of the unit pulse is defined as one, that simply means changing units. For example, let the excitation and response of the system both be volts. When a unit pulse of voltage is applied, the response is a voltage, but the unit pulse response must be that response voltage divided by the integral of the excitation unit voltage pulse which is 1 volt-second (V-s). Therefore, the units of the unit pulse response are inverse seconds, $s^{-1}$. In most system analysis this consideration of the units of the impulse response is not a critical point, but in some cases it can cause confusion.

As an illustration, let the unit pulse response be that of the $RC$ lowpass filter introduced above (Figure 3.54),

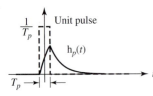

**Figure 3.54**
Unit pulse response of an $RC$ lowpass filter.

$$h_p(t) = \left(\frac{1 - e^{-([t+(T_p/2)]/RC)}}{T_p}\right) u\left(t + \frac{T_p}{2}\right) - \left(\frac{1 - e^{-([t+(T_p/2)]/RC)}}{T_p}\right) u\left(t - \frac{T_p}{2}\right). \tag{3.232}$$

Then let the excitation be a triangular pulse. The system exact and approximate excitations and responses are illustrated in Figure 3.55 (with $T_p = \frac{1}{8}$).

Recall from basic calculus that a real integral of a real variable is defined as the limit of a summation,

$$\int_a^b g(x)\, dx = \lim_{\Delta x \to 0} \sum_{n=a/\Delta x}^{b/\Delta x} g(n\,\Delta x)\,\Delta x. \tag{3.233}$$

We will apply (3.233) to the summations of pulse and pulse responses, (3.230) and (3.231), respectively, in the limit as the pulse width approaches zero. As the pulse width $T_p$ becomes smaller, the excitation and response approximations become better. In the limit as $T_p$ approaches zero, the summation becomes an integral and the approximations become exact. In that same limit, the unit *pulse*

$$\frac{1}{T_p} \operatorname{rect}\left(\frac{t}{T_p}\right) \tag{3.234}$$

approaches a unit *impulse*. Now a limit can be taken as the pulse width $T_p$ approaches zero. As $T_p$ approaches zero, the points in time $nT_p$ become closer and closer together.

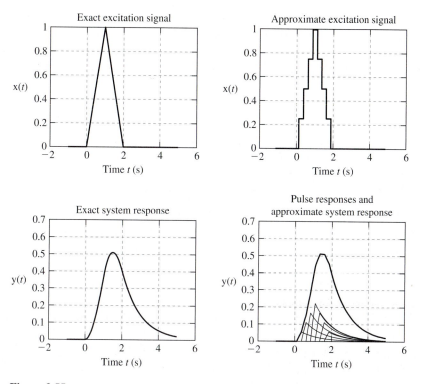

**Figure 3.55**
Exact and approximate system responses to an arbitrary excitation, illustrating superposition.

In the limit, the discrete time shifts $nT_p$ merge into a continuum of time shifts. It is convenient (and conventional) to call that new continuous time shift $\tau$. Changing the name of the time shift amount $nT_p$ to $\tau$ and taking the limit as $T_p$ approaches zero, the width of the pulse $T_p$ approaches a differential $d\tau$ and

$$\mathrm{x}(t) \cong \sum_{n=-\infty}^{\infty} \underbrace{T_p}_{\substack{\downarrow \\ d\tau}} \ \underbrace{\mathrm{x}(nT_p)}_{\substack{\downarrow \\ (\tau)}} \ \underbrace{\frac{1}{T_p} \operatorname{rect}\!\left(\frac{t-nT_p}{T_p}\right)}_{\substack{\downarrow \\ \delta(t-\tau)}} \tag{3.235}$$

$$\underbrace{\phantom{\sum}}_{\int}$$

and

$$\mathrm{y}(t) \cong \sum_{n=-\infty}^{\infty} \underbrace{T_p}_{\substack{\downarrow \\ d\tau}} \ \underbrace{\mathrm{x}(nT_p)}_{\substack{\downarrow \\ (\tau)}} \ \underbrace{\mathrm{h}(t-nT_p)}_{\substack{\downarrow \\ \mathrm{h}(t-\tau)}} \tag{3.236}$$

$$\underbrace{\phantom{\sum}}_{\int}$$

and, in the limit, these summations become integrals of the forms

$$\boxed{\ \mathrm{x}(t) = \int_{-\infty}^{\infty} \mathrm{x}(\tau)\delta(t-\tau)\, d\tau \ } \tag{3.237}$$

and

$$y(t) = \int_{-\infty}^{\infty} x(\tau)h(t - \tau)\, d\tau \qquad \textbf{(3.238)}$$

where the unit pulse response $h_p(t)$ approaches the unit impulse response $h(t)$ (more commonly called just the *impulse response*) of the system. The integral in (3.237) is easily verified by application of the sampling property of the impulse. The integral in (3.238) is called the *convolution* integral.

Compare the convolution integral for CT signals with the convolution sum for DT signals,

$$y(t) = \int_{-\infty}^{\infty} x(\tau)h(t - \tau)\, d\tau \qquad \text{and} \qquad y[n] = \sum_{m=-\infty}^{\infty} x[m]h[n - m]. \qquad \textbf{(3.239)}$$

In each case, one of the two signals is time-inverted and shifted and then multiplied by the other. Then, for CT signals, the product is integrated to find the total area under the product. For DT signals the product is summed to find the total value of the product. Since an integral can be conceived as a limit of a summation, the analogy between the two processes is complete.

The impulse response of an LTI system is a very important descriptor of the way it responds to excitations because once it is determined, the response to any arbitrary excitation can be found. The effect of convolution can be illustrated in a block diagram (Figure 3.56) just the same as for DT signals and systems.

## CONVOLUTION PROPERTIES

It is important at this point to develop a mathematical understanding of the convolution integral. The general mathematical form of the convolution integral is

$$x(t) * h(t) = \int_{-\infty}^{\infty} x(\tau)h(t - \tau)\, d\tau. \qquad \textbf{(3.240)}$$

A graphical example of the steps involved is very helpful in a conceptual understanding of CT convolution. Suppose that $h(t)$ and $x(t)$ are the functions in Figure 3.57. These functions are not typical of linear system excitations and responses but will serve to demonstrate the process of convolution. The integrand in the first form of the convolution integral is $x(\tau)h(t - \tau)$. What is $h(t - \tau)$? It is a function of two variables $t$ and $\tau$. Since the variable of integration in the convolution integral is $\tau$, we should consider $h(t - \tau)$ to be a function of $\tau$ in order to see how to do the integral. We can

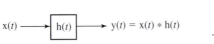

**Figure 3.56**
Block diagram illustration of
convolution.

**Figure 3.57**
Two functions to be convolved.

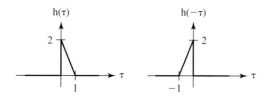

**Figure 3.58**
$h(\tau)$ and $h(-\tau)$ graphed versus $\tau$.

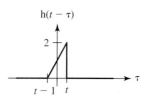

**Figure 3.59**
$h(t - \tau)$ graphed versus $\tau$.

start by graphing $h(\tau)$ and then $h(-\tau)$ (Figure 3.58). The addition of the $t$ in $h(t - \tau)$ just shifts the function $t$ units to the right (Figure 3.59).

The transformation from $h(\tau)$ to $h(t - \tau)$ can be described as two successive transformations,

$$h(\tau) \xrightarrow{\tau \to -\tau} h(-\tau) \xrightarrow{\tau \to \tau - t} h(-(\tau - t)) = h(t - \tau). \qquad (3.241)$$

To verify that the function illustrated in Figure 3.59 is correct, if we substitute $t$ for $\tau$ in $h(t - \tau)$ we have $h(0)$. From the first definition of the function $h(t)$ we see that $t = 0$ is the point of discontinuity where $h(t)$ goes from 0 to 1 (or 1 to 0 depending on which side we approach from). That is the same point on $h(t - \tau)$. Do the same for $\tau = t - 1$ and see if it works.

One confusion that is common is to look at the integral and not understand what the process of integrating from $\tau = -\infty$ to $\tau = +\infty$ means. Since $t$ is not the variable of integration, it is a constant during the integration process. But it is the variable in the final function which results from the convolution!

Think of the process as two general procedures. First pick a value for $t$, and do the integration and get a result. Then pick another value of $t$ and repeat the process. Each integration yields one point on the curve describing the final function. Stated in other words, each point on the $y(t)$ curve corresponding to some particular value of $t$ will be found by finding the total area under the product $x(\tau)h(t - \tau)$.

Now try to visualize the product $x(\tau)h(t - \tau)$. The product depends on what $t$ is. For most values of $t$, the nonzero portions of the two functions do not overlap and the product is zero. (This is not typical of real impulse responses because they usually are not time-limited. Real impulse responses of stable systems usually begin at some time and approach zero as $t$ approaches infinity.) But for some times $t$, their nonzero portions do overlap and there is nonzero area under their product curve. To illustrate these two cases, consider $t = 5$ and $t = 0$. When $t = 5$, the nonzero portions of $x(\tau)$ and $h(5 - \tau)$ do not overlap and the product is zero everywhere (Figure 3.60). When $t = 0$, the nonzero portions of $x(\tau)$ and $h(5 - \tau)$ *do* overlap and the product is not zero everywhere (Figure 3.61).

For $-1 < t < 0$ the convolution of the two functions is twice the area of the h function (which is 1) minus the area of a triangle of width $-t$ and height $-4t$ (Figure 3.62). Therefore, the convolution function value over this range of $t$ is

$$y(t) = 2 - \frac{1}{2}(-t)(-4t) = 2(1 - t^2) \qquad -1 < t < 0. \qquad (3.242)$$

For $0 < t < 1$ the convolution of the two functions is the constant 2.

$$y(t) = 2 \qquad 0 < t < 1. \qquad (3.243)$$

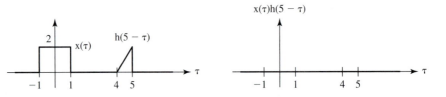

**Figure 3.60**
Impulse response, excitation signal, and their product when $t = 5$.

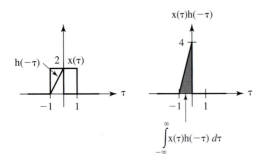

**Figure 3.61**
Impulse response, input signal, and their product
when $t = 0$.

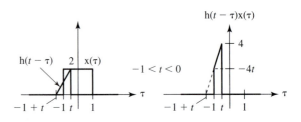

**Figure 3.62**
Product of $h(t - \tau)$ and $x(\tau)$ for $-1 < t < 0$.

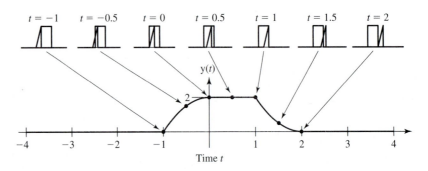

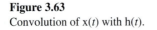

**Figure 3.63**
Convolution of $x(t)$ with $h(t)$.

For $1 < t < 2$, the convolution of the two functions is the area of a triangle whose base
width is $(2 - t)$ and whose height is $(8 - 4t)$ or

$$y(t) = \frac{1}{2}(2 - t)(8 - 4t) = 2(2 - t)^2 \qquad 1 < t < 2. \qquad \textbf{(3.244)}$$

The final function $y(t)$ is illustrated in Figure 3.63.

As a more practical exercise let us now find the unit step response of an $RC$
lowpass filter using convolution. We already know the answer from prior analysis,

$$v_{\text{out}}(t) = \left(1 - e^{-(t/RC)}\right)u(t). \qquad \textbf{(3.245)}$$

First we need to find the impulse response. The response to a unit pulse was given in an example in the derivation of convolution,

$$h(t) = \left( \frac{1 - e^{(-[t + (T_p/2)]/RC)}}{T_p} \right) u\left( t + \frac{T_p}{2} \right) - \left( \frac{1 - e^{(-[t - (T_p/2)]/RC)}}{T_p} \right) u\left( t - \frac{T_p}{2} \right).$$

$$(3.246)$$

The impulse response is the limit of this function as the pulse width $T_p$ goes to zero. That limit is

$$h(t) = \frac{e^{-(t/RC)}}{RC} u(t) \tag{3.247}$$

as illustrated in Figure 3.64. Then the response $v_{out}(t)$ to a unit step excitation $v_{in}(t)$ is $v_{out}(t) = v_{in}(t) * h(t)$ or

$$v_{out}(t) = \int_{-\infty}^{\infty} v_{in}(\tau) h(t - \tau) \, d\tau = \int_{-\infty}^{\infty} u(\tau) \frac{e^{-[(t-\tau)/RC]}}{RC} u(t - \tau) \, d\tau. \tag{3.248}$$

We can immediately simplify the integral some by observing that the first unit step function $u(\tau)$ makes the integrand zero for negative $\tau$. Therefore,

$$v_{out}(t) = \int_{0}^{\infty} \frac{e^{-[(t-\tau)/RC]}}{RC} u(t - \tau) \, d\tau. \tag{3.249}$$

Now we must consider the effect of the other unit step function, $u(t - \tau)$. Since we are integrating over a range of $\tau$ from zero to infinity, if $t$ is negative, for any $\tau$ in that range this unit step has a value of zero (Figure 3.65). Therefore, for negative $t$,

$$v_{out}(t) = 0 \qquad t < 0. \tag{3.250}$$

For positive $t$, the unit step $u(t - \tau)$ will be one for $\tau < t$ and zero for $\tau > t$. Therefore, for positive $t$,

$$v_{out}(t) = \int_{0}^{t} \frac{e^{-[(t-\tau)/RC]}}{RC} \, d\tau = \left[ e^{-[(t-\tau)/RC]} \right]_{0}^{t} = 1 - e^{-(t/RC)} \qquad t > 0. \tag{3.251}$$

Combining (3.250) and (3.251),

$$v_{out}(t) = \left( 1 - e^{-(t/RC)} \right) u(t). \tag{3.252}$$

**Figure 3.64**
The impulse response and excitation of the $RC$ lowpass filter.

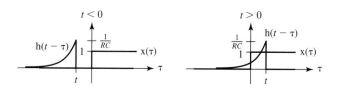

**Figure 3.65**
The relation between the two functions which form the product in the convolution integrand for $t$ negative and $t$ positive.

An operation that appears frequently in CT system analysis is the convolution of a CT signal with an impulse,

$$x(t) * A\delta(t - t_0) = \int_{-\infty}^{\infty} x(\tau)A\delta(t - \tau - t_0) \, d\tau. \tag{3.253}$$

We can use the sampling property of the impulse to evaluate the integral. The variable of integration is $\tau$. The impulse occurs in $\tau$ where

$$t - \tau - t_0 = 0 \quad \text{or} \quad \tau = t - t_0. \tag{3.254}$$

Therefore,

$$x(t) * A\delta(t - t_0) = Ax(t - t_0). \tag{3.255}$$

This is a very important result and will show up many times in the exercises and later material. Just as we found with DT convolution, if we define a function

$$g(t) = g_0(t) * \delta(t), \tag{3.256}$$

then a time-shifted version of $g(t)$, $g(t - t_0)$, can be expressed in either of the two alternate forms,

$$g(t - t_0) = g_0(t - t_0) * \delta(t) \quad \text{or} \quad g(t - t_0) = g_0(t) * \delta(t - t_0), \tag{3.257}$$

but *not* in the form,

$$g_0(t - t_0) * \delta(t - t_0). \tag{3.258}$$

Instead,

$$g(t - 2t_0) = g_0(t - t_0) * \delta(t - t_0). \tag{3.259}$$

This property is true not only when convolving with impulses, but with any function. The same properties, commutativity, associativity, and distributivity can be shown for CT convolution just as for DT convolution (Appendix D). Figure 3.66 illustrates convolution with impulses.

If $y(t) = x(t) * h(t)$, then $y'(t) = x'(t) * h(t) = x(t) * h'(t)$. This is the differentiation property of the convolution integral, and the proof is similar to the corresponding

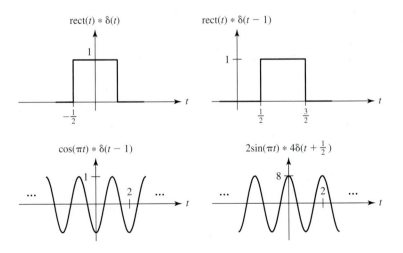

**Figure 3.66**
Examples of convolution with impulses.

proof for the first backward difference of a convolution sum. If $y(t) = x(t) * h(t)$, then the total area under y is the product of the total areas under x and h. This is the area property of the convolution integral. It is analogous to the sum property of the convolution sum and the proof is similar.

There is one more property of the convolution integral which has no counterpart for the convolution sum, the scaling property. Let $z(t) = x(at) * h(at)$, $a > 0$. Then

$$z(t) = \int_{-\infty}^{\infty} x(a\tau)h(a(t - \tau)) \, d\tau. \tag{3.260}$$

Then, making the change of variable, $\lambda = a\tau \Rightarrow d\tau = d\lambda/a$, for $a > 0$, we get

$$z(t) = \frac{1}{a} \int_{-\infty}^{\infty} x(\lambda)h(at - \lambda) \, d\lambda. \tag{3.261}$$

Since

$$y(t) = x(t) * h(t) = \int_{-\infty}^{\infty} x(\tau)h(t - \tau) \, d\tau, \tag{3.262}$$

comparing (3.261) and (3.262) it follows that

$$z(t) = \frac{1}{a}y(at) \qquad \text{and} \qquad \frac{1}{a}y(at) = x(at) * h(at). \tag{3.263}$$

If we do a similar proof for $a < 0$, we get $-(1/a)y(at) = x(at) * h(at)$. Therefore, in general, if

$$y(t) = x(t) * h(t) \tag{3.264}$$

then

$$y(at) = |a|x(at) * h(at). \tag{3.265}$$

This is the scaling property of the convolution integral.

## AN EXPLORATION OF IMPULSE PROPERTIES USING CONVOLUTION

An important property of convolution is that when we convolve any function $g(t)$ with a unit impulse $\delta(t)$, the result is just $g(t)$. Convolution of $\delta(t)$ with any CT function leaves it unchanged. This holds true for the impulse itself,

$$\delta(t) * \delta(t) = \delta(t). \tag{3.266}$$

We can see that by the following argument. We earlier described the unit impulse as the limit of a unit-area rectangular pulse as the width approaches zero,

$$\delta(t) = \lim_{a \to 0} \frac{1}{a} \text{rect}\left(\frac{t}{a}\right). \tag{3.267}$$

If we convolve $(1/a) \text{rect}(t/a)$ with itself, we get

$$\frac{1}{a} \text{rect}\left(\frac{t}{a}\right) * \frac{1}{a} \text{rect}\left(\frac{t}{a}\right) = \frac{1}{a} \text{tri}\left(\frac{t}{a}\right) \tag{3.268}$$

(Figure 3.67).

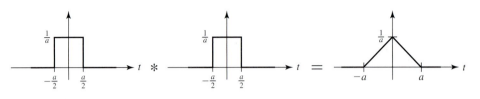

**Figure 3.67**
Convolution of a rectangle with itself.

If we now take the limit of both sides of (3.268) as $a$ approaches zero,

$$\lim_{a \to 0} \left[ \frac{1}{a} \operatorname{rect}\left(\frac{t}{a}\right) * \frac{1}{a} \operatorname{rect}\left(\frac{t}{a}\right) \right] = \lim_{a \to 0} \frac{1}{a} \operatorname{tri}\left(\frac{t}{a}\right). \tag{3.269}$$

From (3.266) and (3.267) the left side of (3.269) must be $\delta(t) * \delta(t)$. In Chapter 2 we showed that

$$\delta(t) = \lim_{a \to 0} \frac{1}{a} \operatorname{tri}\left(\frac{t}{a}\right). \tag{3.270}$$

Therefore, $\delta(t) * \delta(t) = \delta(t)$.

As demonstrated in Chapter 2 when the unit impulse was defined, this shows that a unit impulse can be conceived as the limit of a unit-area rectangle as the width approaches zero *or* as the limit of a unit-area triangle as the width approaches zero. In the limit the two functions are equivalent. This is one of the qualities that makes the impulse different from normal functions. We could carry the argument further to show that a unit triangle convolved with itself is also an impulse in that same limit. This result reemphasizes an important concept; the shape of an impulse is not important, only its strength. In fact, as these functions approach zero width, the whole notion of shape becomes undefined. There is no time in which to develop a shape. Other functions also approach a unit impulse as their widths approach zero. For example,

$$\delta(t) = \lim_{a \to 0} \frac{1}{a} \operatorname{sinc}\left(\frac{t}{a}\right) = \lim_{a \to 0} \frac{1}{a\sqrt{2\pi}} e^{-(t^2/2a^2)}. \tag{3.271}$$

Later, in the study of the Fourier transform we will show that an impulse can even be defined by

$$\delta(t) = \int_{-\infty}^{\infty} e^{-j2\pi ft} \, df. \tag{3.272}$$

The fact that all these different-looking descriptions of a unit impulse are all correct emphasizes the point that an impulse is different from an ordinary function. An ordinary function is defined by its value at every value of its independent variable. An impulse can be defined many ways. One way to define an impulse is by the effect it has when it is convolved with other functions. This leads back to a property of the unit impulse proven earlier in this section, Eq. 3.255.

$$g(t) * \delta(t) = g(t). \tag{3.273}$$

This is yet another way of defining an impulse. The other definitions follow from this one. For example, let $g(t) = 1$. Then

$$\delta(t) * g(t) = 1 = \int_{-\infty}^{\infty} \delta(\tau) \underbrace{g(t-\tau)}_{=1} \, d\tau = \int_{-\infty}^{\infty} \delta(\tau) \, d\tau, \qquad \textbf{(3.274)}$$

proving that the strength of a unit impulse is one.

Consider a system whose response is the derivative of its excitation,

$$y(t) = x'(t). \qquad \textbf{(3.275)}$$

Then its impulse response is the derivative of the unit impulse which was earlier identified as the unit doublet,

$$h(t) = u_1(t). \qquad \textbf{(3.276)}$$

For any general excitation, the response is the impulse response convolved with the excitation,

$$y(t) = x'(t) = x(t) * u_1(t). \qquad \textbf{(3.277)}$$

So just as $g(t) * \delta(t) = g(t)$ can be used to define a unit impulse, $g(t) * u_1(t) = g'(t)$ can be used to define a unit doublet. This concept can be extended to all the singularity functions

$$g(t) * u_2(t) = g''(t), \qquad g(t) * u_3(t) = g'''(t), \qquad \text{etc.} \qquad \textbf{(3.278)}$$

Using the principle that the second derivative is the first derivative of the first derivative we can write

$$g''(t) = (g'(t))' = (g(t) * u_1(t))' = g(t) * u_1(t) * u_1(t) \qquad \textbf{(3.279)}$$

and, using (3.278),

$$g(t) * u_2(t) = g(t) * u_1(t) * u_1(t). \qquad \textbf{(3.280)}$$

So it follows that

$$u_2(t) = u_1(t) * u_1(t). \qquad \textbf{(3.281)}$$

This can be extended to a general principle for the convolution of any of the singularity functions,

$$u_{m+n}(t) = u_m(t) * u_n(t). \qquad \textbf{(3.282)}$$

Applying this to the unit step we would get

$$u_{-1}(t) * u_{-1}(t) = u_{-2}(t) = \text{ramp}(t). \qquad \textbf{(3.283)}$$

From these properties we can form a general principle. If a system has a known impulse response $h(t)$, then the response to the singularity function $u_n(t)$ is the $n$th derivative of $h(t)$ and, if $n$ is negative, that simply indicates that integration is to be done instead of differentiation.

## SYSTEM INTERCONNECTIONS

The results for cascade and parallel system interconnections from Section 3.5 on DT signal and system analysis apply also to CT signal and system analysis (Figures 3.68 and 3.69).

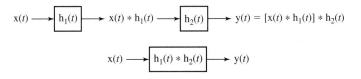

**Figure 3.68**
Cascade connection of two systems.

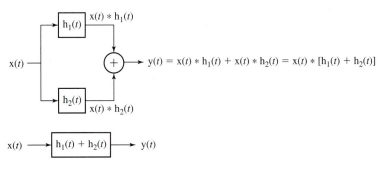

**Figure 3.69**
Parallel connection of two systems.

## STABILITY AND IMPULSE RESPONSE

In a manner analogous to that used to find the stability criterion for DT systems, we can show the relationship between impulse response and stability for a CT system. The response $y(t)$ of a CT system to an excitation $x(t)$ is

$$y(t) = \int_{-\infty}^{\infty} x(\tau)h(t - \tau)\, d\tau. \tag{3.284}$$

If the excitation $x(t)$ is bounded, we can say that $|x(\tau)| < B < \infty$ for all $\tau$. The response magnitude is

$$|y(t)| = \left| \int_{-\infty}^{\infty} x(\tau)h(t - \tau)\, d\tau \right|. \tag{3.285}$$

Using the principle that the magnitude of an integral of a product is less than or equal to the integral of the magnitude of a product,

$$|y(t)| \leq \int_{-\infty}^{\infty} |x(\tau)||h(t - \tau)|\, d\tau. \tag{3.286}$$

Since $x(\tau)$ is less than $B$ in magnitude for any $\tau$,

$$|y(t)| \leq B \int_{-\infty}^{\infty} |h(t - \tau)|\, d\tau. \tag{3.287}$$

Therefore, the response $y(t)$ is bounded if $\int_{-\infty}^{\infty} |h(t - \tau)|\, d\tau$ is bounded, and, therefore, a CT system is stable if its impulse response is absolutely integrable.

## RESPONSES OF SYSTEMS TO STANDARD SIGNALS

Just as we did for DT systems we need to find the responses of CT systems to the unit step, the complex exponential, and the real sinusoid. The response of a CT system to a unit step is (using the same style of notation introduced for the unit sequence response),

$$h_{-1}(t) = h(t) * u(t) = \int_{-\infty}^{\infty} h(\tau)u(t - \tau)\, d\tau = \int_{-\infty}^{t} h(\tau)\, d\tau, \qquad \textbf{(3.288)}$$

which proves that the response of a CT system to a unit step is the integral of its response to a unit impulse. More generally, the response of a CT system to the integral of *any* excitation is the integral of the response to that excitation. The converse holds true also. If we differentiate the excitation, we also differentiate the response (Figure 3.70).

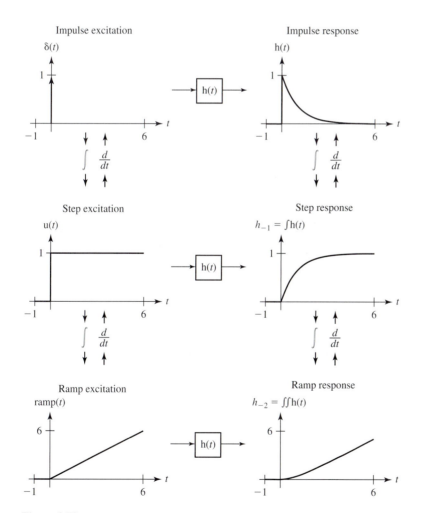

**Figure 3.70**
Relations between integrals and derivatives of excitations and responses for an LTI system.

The response of a system to a complex exponential $e^{st}$, where $s$ is any complex constant is

$$y(t) = h(t) * e^{st} = \int_{-\infty}^{\infty} h(\tau)e^{s(t-\tau)} \, d\tau = e^{st} \underbrace{\int_{-\infty}^{\infty} h(\tau)e^{-s\tau} d\tau}_{\text{complex constant}}, \qquad \textbf{(3.289)}$$

proving that the response of an LTI system to a complex exponential is a complex exponential with the same functional form except multiplied by a complex constant which depends on the impulse response and the constant $s$. In Chapter 9 the integral

$$\int_{-\infty}^{\infty} h(t)e^{-st} \, dt \qquad \textbf{(3.290)}$$

will be called the *Laplace* transform of $h(t)$. This result is another basis for transform analysis methods in LTI systems.

We can specialize the result for complex exponentials to complex sinusoids by letting

$$s = j\omega = j2\pi f, \qquad \textbf{(3.291)}$$

where $\omega$ and $f$ are real. Then, if the excitation is $e^{j\omega t}$, the response is

$$y(t) = e^{j\omega t} \int_{-\infty}^{\infty} h(\tau)e^{-j\omega\tau} \, d\tau, \qquad \textbf{(3.292)}$$

and if the excitation is $e^{-j\omega t}$, it follows that

$$y(t) = e^{-j\omega t} \int_{-\infty}^{\infty} h(\tau)e^{j\omega\tau} \, d\tau. \qquad \textbf{(3.293)}$$

Invoking linearity, if the excitation is $(e^{j\omega t} + e^{-j\omega t})/2$, the response will be

$$y(t) = \frac{1}{2}\left[ e^{j\omega t} \int_{-\infty}^{\infty} h(\tau)e^{-j\omega\tau} \, d\tau + e^{-j\omega t} \int_{-\infty}^{\infty} h(\tau)e^{j\omega\tau} \, d\tau \right] \qquad \textbf{(3.294)}$$

or

$$y(t) = \frac{1}{2}\left\{ e^{j\omega t} \int_{-\infty}^{\infty} h(\tau)[\cos(\omega\tau) - j\sin(\omega\tau)] \, d\tau \right.$$
$$\left. + e^{-j\omega t} \int_{-\infty}^{\infty} h(\tau)[\cos(\omega\tau) + j\sin(\omega\tau)] \, d\tau \right\} \qquad \textbf{(3.295)}$$

or

$$y(t) = \frac{1}{2}\left[ (e^{j\omega t} + e^{-j\omega t}) \int_{-\infty}^{\infty} h(\tau)\cos(\omega\tau) \, d\tau + j(e^{-j\omega t} - e^{j\omega t}) \int_{-\infty}^{\infty} h(\tau)\sin(\omega\tau) \, d\tau \right]$$
$$\textbf{(3.296)}$$

or

$$y(t) = \cos(\omega t) \underbrace{\int_{-\infty}^{\infty} h(\tau)\cos(\omega\tau) \, d\tau}_{\text{real constant}} - \sin(\omega t) \underbrace{\int_{-\infty}^{\infty} h(\tau)\sin(\omega\tau) \, d\tau}_{\text{real constant}}. \qquad \textbf{(3.297)}$$

This shows that a continuous-time LTI system excited by a real sinusoid has a response which is also a real sinusoid but with a different amplitude and phase, in general.

## 3.7  BLOCK DIAGRAM SIMULATION OF DIFFERENTIAL OR DIFFERENCE EQUATIONS

Differential and difference equations describe the dynamics of most CT and DT systems. An important aspect of signal and system analysis is to develop a view of how signals in a system are related to each other and how the system form is related to the describing equation. A good way of doing that is by drawing a block diagram or system simulation diagram from the equation.

Consider a DT system described by the equation

$$y[n] + 3y[n-1] - 2y[n-2] = x[n].\qquad\textbf{(3.298)}$$

It can be rearranged into the recursion form

$$y[n] = x[n] - 3y[n-1] + 2y[n-2].\qquad\textbf{(3.299)}$$

In words, this equation says that the present value of the response y can be computed by adding the present value of the excitation x, the immediate past value of y multiplied by $-3$, and the value of y immediately before that multiplied by two. This relationship can be diagrammed in the form illustrated in Figure 3.71. In this block diagram the $D$ represents a delay of one, in discrete time (for a DT signal created by sampling one sample time $T_s$). This type of representation of a system aids understanding of the system dynamics and can also be used to actually build the system, or a simulation of the system, using delays, gains, and adders. Notice that, since delays are involved, the system must remember some past values of the signals. Therefore, this is a system with memory, a dynamic system.

Differential equations can be converted into system block diagrams also. Let a system be characterized by the differential equation

$$2y''(t) + 5y'(t) + 4y(t) = x(t).\qquad\textbf{(3.300)}$$

Following an analogy with the previous discrete-time example, (3.300) can be rewritten as

$$y(t) = \frac{x(t)}{4} - \frac{5}{4}y'(t) - \frac{1}{2}y''(t).\qquad\textbf{(3.301)}$$

The block diagram for the system is illustrated in Figure 3.72. Although this diagram is correct, it is not the preferred way of simulating the system. It uses two differentiators

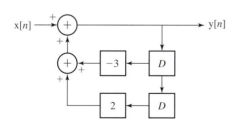

**Figure 3.71**

Block diagram representation of the recursion relation $y[n] = x[n] - 3y[n-1] + 2y[n-2]$.

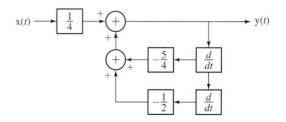

**Figure 3.72**

Block diagram representation of the differential equation

$$y(t) = \frac{x(t)}{4} - \frac{5}{4}y'(t) - \frac{1}{2}y''(t).$$

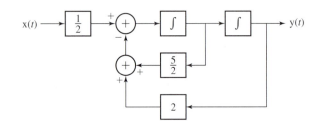

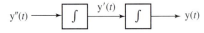

**Figure 3.73**
A system with a signal and its first two derivatives.

**Figure 3.74**
Block diagram simulation of the system.

and, as a practical matter, differentiators are more troublesome in real systems because of their emphasis of high-frequency noise. As a practical matter, the integrator is usually a much better component to use in system simulations. We can form the system block diagram by realizing that if the response of an integrator is $g(t)$, then its excitation must be $g'(t)$. We can rearrange (3.301) into

$$y''(t) = \frac{x(t)}{2} - 2y(t) - \frac{5}{2}y'(t). \qquad (3.302)$$

In this equation we have $y(t)$, $y'(t)$, and $y''(t)$. We can start the block diagram simulation with two integrators (Figure 3.73). Then, from (3.302), we see that to form $y''(t)$ we need $x(t)/2 - 2y(t) - \frac{5}{2}y'(t)$. We can form that as illustrated in Figure 3.74. Since each integral accumulates information from the past behavior of the signal it is integrating, this CT system has memory.

Block diagrams are especially useful in the analysis of more complicated systems with multiple inputs and outputs. But it is best to wait to explore more complicated systems until we have developed the transform analysis methods in Chapters 4, 5, 9, and 11.

## 3.8  SUMMARY OF IMPORTANT POINTS

1. A system which is both homogeneous and additive is linear.
2. A system which is both linear and time-invariant is called an LTI system.
3. The total response of any LTI system is the sum of its zero-input and zero-state responses.
4. Often, nonlinear systems can be analyzed with linear system techniques through an approximation called *linearization.*
5. A system is said to be BIBO-stable if bounded excitations always produce bounded responses.
6. All real physical systems are causal, although some may be conveniently and superficially described as noncausal.
7. The response of a system excited by its eigenfunction is of the same functional form as the excitation. This fact forms the basis for transform methods to be introduced in Chapters 4, 5, 9, and 11.
8. Systems of quite different physical structure or nature may be described by differential or difference equations of the same mathematical form.
9. Every LTI system is completely characterized by its impulse response.
10. The response of an LTI system to an arbitrary excitation can be found by convolving the excitation with its impulse response.

11. The impulse response of a cascade connection of LTI systems is the convolution of the individual impulse responses.
12. The impulse response of a parallel connection of LTI systems is the sum of the individual impulse responses.
13. A system is BIBO stable if its impulse response is absolutely summable or absolutely integrable.
14. LTI systems can be represented by block diagrams, and this type of representation is useful both in synthesizing systems and in understanding their dynamic behavior.

## EXERCISES WITH ANSWERS

1. Show that a system with excitation $x(t)$ and response $y(t)$, described by

$$y(t) = u(x(t)),$$

   is nonlinear, time-invariant, stable, and noninvertible.

2. Show that a system with excitation $x(t)$ and response $y(t)$, described by

$$y(t) = x(t - 5) - x(3 - t),$$

   is linear, noncausal, and noninvertible.

3. Show that a system with excitation $x(t)$ and response $y(t)$, described by

$$y(t) = x\left(\frac{t}{2}\right),$$

   is linear, time-variant, and noncausal.

4. Show that a system with excitation $x(t)$ and response $y(t)$, described by

$$y(t) = \cos(2\pi t)x(t),$$

   is time-variant, BIBO stable, static, and noninvertible.

5. Show that a system whose response is the magnitude of its excitation is nonlinear, BIBO stable, causal, and noninvertible.

6. Show that the system in Figure E6 is linear, time-invariant, BIBO-unstable, and dynamic.

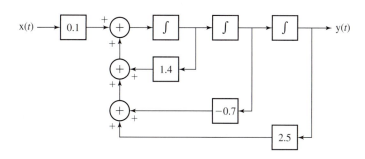

**Figure E6**
A CT system.

7. Show that the system of Figure E7 is nonlinear, BIBO-stable, static, and noninvertible. (The output signal of an analog multiplier is the product of its two input signals.)

8. Show that a system with excitation $x[n]$ and response $y[n]$, described by

$$y[n] = nx[n],$$

is linear, time-variant, and static.

9. Show that the system of Figure E9 is linear, time-invariant, BIBO-unstable, and dynamic.

10. Show that a system with excitation $x[n]$ and response $y[n]$, described by

$$y[n] = \text{rect}(x[n]),$$

is nonlinear, time-invariant, and noninvertible.

11. Show that the system of Figure E11 is nonlinear, time-invariant, static, and invertible.

12. Show that the system of Figure E12 is time-invariant, BIBO-stable, and causal.

13. Find the impulse responses of these systems.

   a.  $y[n] = x[n] - x[n-1]$
   b.  $25y[n] + 6y[n-1] + y[n-2] = x[n]$
   c.  $4y[n] - 5y[n-1] + y[n-2] = x[n]$
   d.  $2y[n] + 6y[n-2] = x[n] - x[n-2]$

**Answers:**

$$\left[\frac{1}{3} - \frac{1}{12}\left(\frac{1}{4}\right)^n\right]u[n], \quad \delta[n] - \delta[n-1], \quad \frac{(\sqrt{3})^n}{2}\cos\left(\frac{\pi n}{2}\right)\left(u[n] + \frac{1}{3}u[n-2]\right),$$

$$h[n] = \frac{\cos(2.214n + 0.644)}{20(5)^n}$$

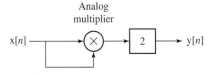

**Figure E7**
A CT system.

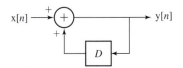

**Figure E9**
A DT system.

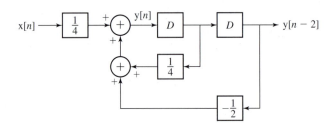

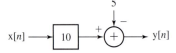

**Figure E11**
A DT system.

**Figure E12**
A DT system.

**14.** Sketch g[n]. To the extent possible find analytical solutions. Where possible, compare analytical solutions with the results of using the MATLAB command `conv` to do the convolution.

a.  $g[n] = u[n] * u[n]$

b.  $g[n] = u[n + 2] * \text{rect}_3[n]$

c.  $g[n] = \text{rect}_2[n] * \text{rect}_2[n]$

d.  $g[n] = \text{rect}_2[n] * \text{rect}_4[n]$

e.  $g[n] = 3\delta[n - 4] * \left(\dfrac{3}{4}\right)^n u[n]$

f.  $g[n] = 2\,\text{rect}_4[n] * \left(\dfrac{7}{8}\right)^n u[n]$

g.  $g[n] = \text{rect}_3[n] * \text{comb}_{14}[n]$

**Answers:**

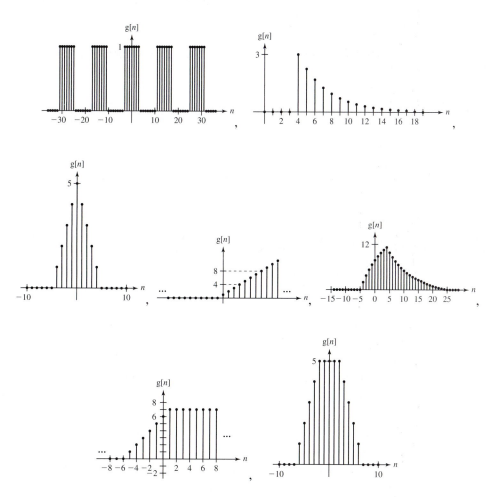

**15.** Given the excitations x[n] and the impulse responses h[n], find closed-form expressions for and plot the system responses y[n].

a.  $x[n] = e^{j(2\pi n/32)}$    $h[n] = (0.95)^n u[n]$
    (Plot the real and imaginary parts of the response.)

b.  $x[n] = \sin\left(\dfrac{2\pi n}{32}\right)$    $h[n] = (0.95)^n u[n]$

**Answers:**

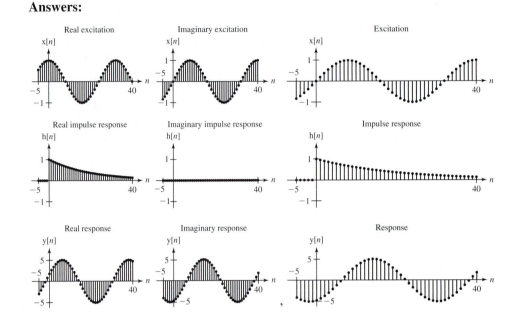

16. Given the excitations x[n] and the impulse responses h[n], use MATLAB to plot the system responses y[n].

   *a.*   $x[n] = u[n] - u[n - 8]$

   $$h[n] = \sin\left(\frac{2\pi n}{8}\right)(u[n] - u[n - 8])$$

   *b.*   $x[n] = \sin\left(\frac{2\pi n}{8}\right)(u[n] - u[n - 8])$

   $$h[n] = -\sin\left(\frac{2\pi n}{8}\right)(u[n] - u[n - 8])$$

**Answers:**

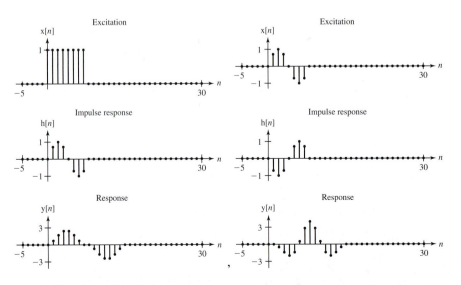

**17.** Which of these systems are BIBO stable?

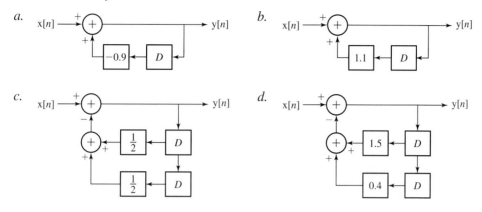

a.

b.

c.

d.

**Answers:**

Two stable and two unstable.

**18.** Find and plot the unit sequence responses of these systems.

a.

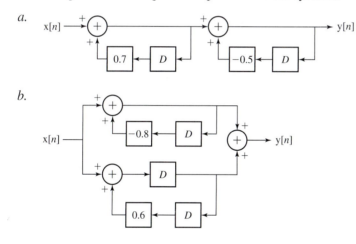

b.

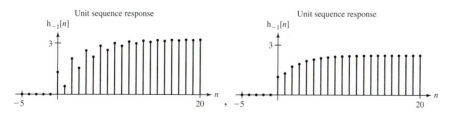

**Answers:**

Unit sequence response

Unit sequence response

**19.** Find the impulse responses of these systems.

    a.  $y'(t) + 5y(t) = x(t)$        b.  $y''(t) + 6y'(t) + 4y(t) = x(t)$

    c.  $2y'(t) + 3y(t) = x'(t)$       d.  $4y'(t) + 9y(t) = 2x(t) + x'(t)$

**Answers:**

$$h(t) = -\frac{1}{16}e^{-(9/4)t}u(t) + \frac{1}{4}\delta(t), \quad e^{-5t}u(t), \quad -\frac{3}{4}e^{-(3/2)t}u(t) + \frac{1}{2}\delta(t),$$

$$0.2237(e^{-0.76t} - e^{-5.23t})u(t)$$

**20.** Sketch g(t).

 a. $g(t) = \text{rect}(t) * \text{rect}(t)$

 b. $g(t) = \text{rect}(t) * \text{rect}\left(\dfrac{t}{2}\right)$

 c. $g(t) = \text{rect}(t - 1) * \text{rect}\left(\dfrac{t}{2}\right)$

 d. $g(t) = [\text{rect}(t - 5) + \text{rect}(t + 5)] * [\text{rect}(t - 4) + \text{rect}(t + 4)]$

**Answers:**

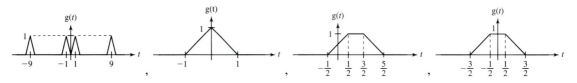

,

**21.** Sketch these functions.

 a. $g(t) = \text{rect}(4t)$      b. $g(t) = \text{rect}(4t) * 4\delta(t)$

 c. $g(t) = \text{rect}(4t) * 4\delta(t - 2)$      d. $g(t) = \text{rect}(4t) * 4\delta(2t)$

 e. $g(t) = \text{rect}(4t) * \text{comb}(t)$      f. $g(t) = \text{rect}(4t) * \text{comb}(t - 1)$

 g. $g(t) = \text{rect}(4t) * \text{comb}(2t)$      h. $g(t) = \text{rect}(t) * \text{comb}(2t)$

**Answers:**

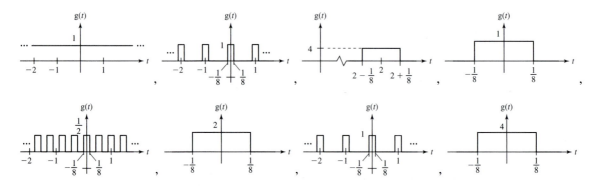

,

**22.** Plot these convolutions.

 a. $g(t) = \text{rect}\left(\dfrac{t}{2}\right) * [\delta(t + 2) - \delta(t + 1)]$

 b. $g(t) = \text{rect}(t) * \text{tri}(t)$

 c. $g(t) = e^{-t}\text{u}(t) * e^{-t}\text{u}(t)$

 d. $g(t) = \left[\text{tri}\left(2\left(t + \dfrac{1}{2}\right)\right) - \text{tri}\left(2\left(t - \dfrac{1}{2}\right)\right)\right] * \dfrac{1}{2}\,\text{comb}\left(\dfrac{t}{2}\right)$

 e. $g(t) = \left[\text{tri}\left(2\left(t + \dfrac{1}{2}\right)\right) - \text{tri}\left(2\left(t - \dfrac{1}{2}\right)\right)\right] * \text{comb}(t)$

**Answers:**

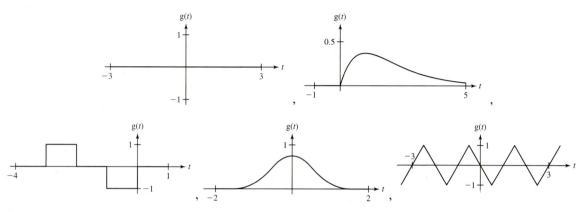

**23.** A system has an impulse response, $h(t) = 4e^{-4t}u(t)$. Find and plot the response of the system to the excitation $x(t) = \text{rect}\left(2\left(t - \frac{1}{4}\right)\right)$.

**Answer:**

**24.** Change the system impulse response in Exercise 23 to $h(t) = \delta(t) - 4e^{-4t}u(t)$ and find and plot the response to the same excitation $x(t) = \text{rect}\left(2\left(t - \frac{1}{4}\right)\right)$.

**Answer:**

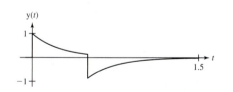

**25.** Find the impulse responses of the two systems in Figure E25. Are these systems BIBO stable?

**Answers:**
One BIBO stable, one BIBO unstable.

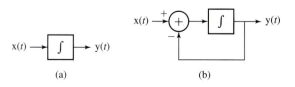

**Figure E25**
Two single-integrator systems.

**26.** Find the impulse response of the system in Figure E26. Is this system BIBO stable?

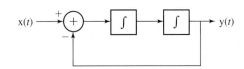

**Figure E26**
A double-integrator system.

**Answer:**

BIBO unstable.

**27.** In the circuit of Figure E27 the input signal voltage is $v_i(t)$ and the output signal voltage is $v_o(t)$.

     *a.*   Find the impulse response in terms of $R$ and $L$.

     *b.*   If $R = 10 \text{ k}\Omega$ and $L = 100 \text{ }\mu\text{H}$, graph the unit step response.

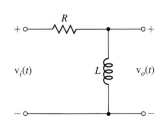

**Figure E27**
An *RL* circuit.

**Answers:**

$$\delta(t) - \frac{R}{L}e^{-(R/L)t}u(t),$$

**28.** Find the impulse response of the system in Figure E28 and evaluate its BIBO stability.

**Answers:**

$4.589e^{0.05t}\sin(0.2179t)u(t)$, not BIBO stable

**29.** Find the impulse response of the system in Figure E29 and evaluate its BIBO stability.

**Answers:**

$8.482e^{-(t/3)}\sin(0.1179t)u(t)$, BIBO stable

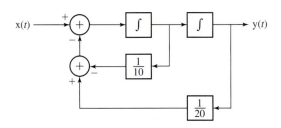

**Figure E28**
A two-integrator system.

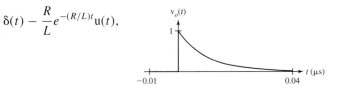

**Figure E29**
A two-integrator system.

**30.** Plot the amplitudes of the responses of the systems of Exercise 19 to the excitation $e^{j\omega t}$ as a function of radian frequency $\omega$.

**Answers:**

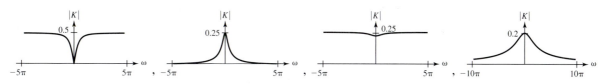

**31.** Plot the responses of the systems of Exercise 19 to a unit step excitation.

**Answers:**

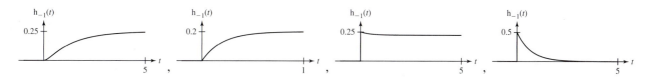

# EXERCISES WITHOUT ANSWERS

**32.** A CT system is described by the block diagram in Figure E32. Classify the system as to homogeneity, additivity, linearity, time invariance, stability, causality, memory, and invertibility.

**33.** A CT system has a response that is the cube of its excitation. Classify the system as to linearity, time invariance, stability, causality, memory, and invertibility.

**34.** A CT system is described by the differential equation

$$ty'(t) - 8y(t) = x(t).$$

Classify the system as to linearity, time invariance, and stability.

**35.** A CT system is described by the equation

$$y(t) = \int_{-\infty}^{t/3} x(\lambda)\, d\lambda.$$

Classify the system as to time invariance, stability, and invertibility.

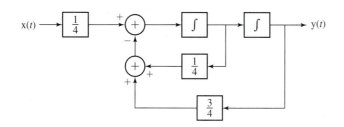

**Figure E32**
A CT system.

**36.** A CT system is described by the equation

$$y(t) = \int_{-\infty}^{t+3} x(\lambda) \, d\lambda.$$

Classify the system as to linearity, causality, and invertibility.

**37.** Show that the system described by $y(t) = \text{Re}(x(t))$ is additive but not homogeneous. (Remember, if the excitation is multiplied by any complex constant and the system is homogeneous, the response must be multiplied by that same complex constant.)

**38.** Graph the magnitude and phase of the complex-sinusoidal response of the system described by

$$y'(t) + 2y(t) = e^{-j2\pi f t}$$

as a function of cyclic frequency $f$.

**39.** A DT system is described by

$$y[n] = \sum_{m=-\infty}^{n+1} x[m].$$

Classify this system as to time invariance, BIBO stability, and invertibility.

**40.** A DT system is described by

$$ny[n] - 8y[n-1] = x[n].$$

Classify this system as to time invariance, BIBO stability, and invertibility.

**41.** A DT system is described by

$$y[n] = \sqrt{x[n]}.$$

Classify this system as to linearity, BIBO stability, memory, and invertibility.

**42.** Graph the magnitude and phase of the complex-sinusoidal response of the system described by

$$y[n] + \frac{1}{2}y[n-1] = e^{-j\Omega n}$$

as a function of $\Omega$.

**43.** Find the impulse response $h[n]$ of the system in Figure E43.

**44.** Find the impulse responses of these systems.

a. $3y[n] + 4y[n-1] + y[n-2] = x[n] + x[n-1]$

b. $\frac{5}{2}y[n] + 6y[n-1] + 10y[n-2] = x[n]$

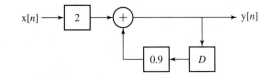

**Figure E43**

DT system block diagram.

**45.** Plot g[n]. Use the MATLAB `conv` function if needed.

   *a.* $g[n] = \text{rect}_1[n] * \sin\left(\dfrac{2\pi n}{9}\right)$

   *b.* $g[n] = \text{rect}_2[n] * \sin\left(\dfrac{2\pi n}{9}\right)$

   *c.* $g[n] = \text{rect}_4[n] * \sin\left(\dfrac{2\pi n}{9}\right)$

   *d.* $g[n] = \text{rect}_3[n] * \text{rect}_3[n] * \text{comb}_{14}[n]$

   *e.* $g[n] = \text{rect}_3[n] * \text{rect}_3[n] * \text{comb}_7[n]$

   *f.* $g[n] = 2\cos\left(\dfrac{2\pi}{7}n\right) * \left(\dfrac{7}{8}\right)^n u[n]$

   *g.* $g[n] = \dfrac{\text{sinc}(n/4)}{2\sqrt{2}} * \dfrac{\text{sinc}(n/4)}{2\sqrt{2}}$

**46.** Find the impulse responses of the subsystems in Figure E46 and then convolve them to find the impulse response of the cascade connection of the two subsystems. You may find this formula for the summation of a geometric series useful,

$$\sum_{n=0}^{N-1} \alpha^n = \begin{cases} N & \alpha = 1 \\ \dfrac{1 - \alpha^N}{1 - \alpha} & \alpha \neq 1 \end{cases}.$$

**47.** For the system of Exercise 43, let the excitation x[n] be a unit amplitude complex sinusoid of DT cyclic frequency $F$. Plot the amplitude of the response complex sinusoid versus $F$ over the range $-1 < F < 1$.

**48.** In the second-order DT system given in Figure E48, what is the relationship between $a$, $b$, and $c$ that ensures that the system is stable?

**49.** Given the excitations x[n] and the impulse responses h[n], find closed-form expressions for and plot the system responses y[n].

   *a.* $x[n] = u[n]$     $h[n] = n\left(\dfrac{7}{8}\right)^n u[n]$

   $$\left( \text{Hint: Differentiate } \sum_{n=0}^{N-1} r^n = \begin{cases} \dfrac{1 - r^N}{1 - r} & r \neq 1 \\ N & r = 1 \end{cases} \text{ with respect to } r. \right)$$

   *b.* $x[n] = u[n]$     $h[n] = \dfrac{4}{7}\delta[n] - \left(-\dfrac{3}{4}\right)^n u[n]$

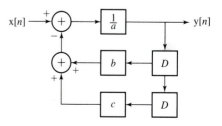

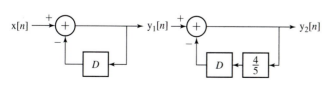

**Figure E46**
Two cascaded subsystems.

**Figure E48**

**50.** A CT function is nonzero over a range of its argument from 0 to 4. It is convolved with a function which is nonzero over a range of its argument from $-3$ to $-1$. What is the nonzero range of the convolution of the two?

**51.** What function convolved with $-2\cos(t)$ would produce $6\sin(t)$?

**52.** Sketch these functions.

a. $g(t) = 3\cos(10\pi t) * 4\delta\left(t + \dfrac{1}{10}\right)$

b. $g(t) = \text{tri}(2t) * \text{comb}(t)$

c. $g(t) = [\text{tri}(2t) - \text{rect}(t-1)] * \text{comb}\left(\dfrac{t}{2}\right)$

d. $g(t) = \left[\text{tri}\left(\dfrac{t}{4}\right)\text{comb}(t)\right] * \text{comb}\left(\dfrac{t}{8}\right)$

e. $g(t) = \text{sinc}(4t) * \dfrac{1}{2}\text{comb}\left(\dfrac{t}{2}\right)$

f. $g(t) = e^{-2t}u(t) * \dfrac{1}{4}\left[\text{comb}\left(\dfrac{t}{4}\right) - \text{comb}\left(\dfrac{t-2}{4}\right)\right]$

g. $g(t) = \left[\text{sinc}(t)\,\text{rect}\left(\dfrac{t}{2}\right)\right] * \dfrac{1}{2}\text{comb}\left(\dfrac{t}{2}\right)$

h. $g(t) = \left[\text{sinc}(2t) * \dfrac{1}{2}\text{comb}\left(\dfrac{t}{2}\right)\right]\text{rect}\left(\dfrac{t}{4}\right)$

**53.** Find the signal power of these signals.

a. $x(t) = \text{rect}(t) * \text{comb}\left(\dfrac{t}{4}\right)$    b. $x(t) = \text{tri}(t) * \text{comb}\left(\dfrac{t}{4}\right)$

**54.** A rectangular voltage pulse which begins at $t = 0$, is 2 s wide, and has a height of 0.5 V drives an $RC$ lowpass filter in which $R = 10$ kilo-ohms (k$\Omega$) and $C = 100$ microfarads ($\mu$F).

a. Sketch the voltage across the capacitor versus time.

b. Change the pulse duration to 0.2 s and the pulse height to 5 V and repeat.

c. Change the pulse duration to 2 milliseconds (ms) and the pulse height to 500 V and repeat.

d. Change the pulse duration to 2 $\mu$s and the pulse height to 500 kV and repeat. Based on these results what do you think would happen if you let the input voltage be a unit impulse?

**55.** Write the differential equation for the voltage $v_C(t)$ in the circuit in Figure E55 for time $t > 0$, and then find an expression for the current $i(t)$ for time $t > 0$.

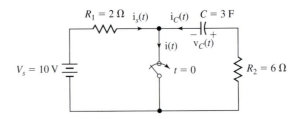

**Figure E55**

**56.** The water tank in Figure E56 is filled by an inflow $x(t)$ and is emptied by an outflow $y(t)$. The outflow is controlled by a valve which offers resistance $R$ to the flow of water out of the tank. The water depth in the tank is $d(t)$ and the surface area of the water is $A$, independent of depth (cylindrical tank). The outflow is related to the water depth (head) by

$$y(t) = \frac{d(t)}{R}.$$

The tank is 1.5 meters (m) high with a 1-m diameter and the valve resistance is 10 seconds per square meter ($s/m^2$).

   *a.*  Write the differential equation for the water depth in terms of the tank dimensions and valve resistance.

   *b.*  If the inflow is 0.05 $m^3/s$, at what water depth will the inflow and outflow rates be equal, making the water depth constant?

   *c.*  Find an expression for the depth of water versus time after 1 $m^3$ of water is dumped into an empty tank.

   *d.*  If the tank is initially empty at time $t = 0$, and the inflow is a constant 0.2 $m^3/s$ after time $t = 0$, at what time will the tank start to overflow?

**57.** The suspension of a car can be modeled by the mass-spring-dashpot system of Figure E57. Let the mass $m$ of the car be 1500 kilograms (kg), let the spring constant $K_s$ be 75,000 newtons per meter (N/m), and let the shock absorber (dashpot) viscosity coefficient $K_d$ be 20,000 N·s/m. At a certain spring length $d_0$, the spring is unstretched and uncompressed and exerts no force. Let that length be 0.6 m.

   *a.*  What is the distance $y(t) - x(t)$ when the car is at rest?

   *b.*  Define a new variable $z(t) = y(t) - x(t) -$ constant such that, when the system is at rest, $z(t) = 0$, and write an equation in z and x which describes an LTI system. Then find the impulse response.

   *c.*  The effect of the car striking a curb can be modeled by letting the road surface height change discontinuously by the height of the curb $h_c$. Let $h_c = 0.15$ m. Graph $z(t)$ versus time after the car strikes a curb.

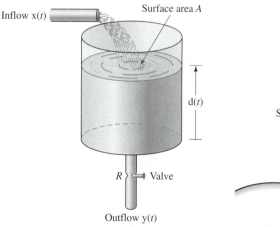

**Figure E56**
Water tank with inflow and outflow.

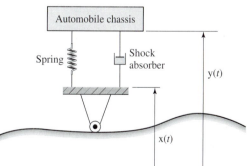

**Figure E57**
Car suspension model.

58. As derived in the text, a simple pendulum is approximately described for small angles θ by the differential equation

$$mL\theta''(t) + mg\theta(t) \cong \mathrm{x}(t)$$

where $m$ = mass of pendulum

$L$ = length of massless rigid rod supporting mass

$\theta$ = angular deviation of pendulum from vertical

    *a.* Find the general form of the impulse response of this system.

    *b.* If the mass is 2 kg and the rod length is 0.5 m, at what cyclic frequency will the pendulum oscillate?

59. Pharmacokinetics is the study of how drugs are absorbed into, distributed through, metabolized by, and excreted from the human body. Some drug processes can be approximately modeled by a one-compartment model of the body in which $V$ is the volume of the compartment, $C(t)$ is the drug concentration in that compartment, $k_e$ is a rate constant for excretion of the drug from the compartment, and $k_0$ is the infusion rate at which the drug enters the compartment.

    *a.* Write a differential equation in which the infusion rate is the excitation and the drug concentration is the response.

    *b.* Let the parameter values be $k_e = 0.4\,\mathrm{h}^{-1}$, $V = 20$ liters (L), and $k_0 = 200$ milligrams per hour (mg/h). If the initial drug concentration is $C(0) = 10$ mg/L, plot the drug concentration as a function of time (in hours) for the first 10 h of infusion. Find the solution as the sum of the zero-input response and zero-state response.

60. At the beginning of the year 2000, the country Freedonia had a population $p$ of 100 million people. The birth rate is 4 percent per annum and the death rate is 2 percent per annum, compounded daily. That is, the births and deaths occur every day at a uniform fraction of the current population and the next day the number of births and deaths changes because the population changed the previous day. For example, every day the number of people who die is the fraction 0.02/365 of the total population at the end of the previous day (neglect leap-year effects). Every day 275 immigrants enter Freedonia.

    *a.* Write a difference equation for the population at the beginning of the $n$th day after January 1, 2000, with the immigration rate as the excitation of the system.

    *b.* By finding the zero-input and zero-state responses of the system determine what the population of Freedonia will be at the beginning of the year 2050.

61. A car rolling on a hill can be modeled as shown in Figure E61. The excitation is the force f($t$) for which a positive value represents accelerating the car forward with the motor and a negative value represents slowing the car by braking action. As it rolls, the car experiences drag due to various frictional phenomena which can be approximately modeled by a coefficient $k_f$ which multiplies the car's velocity to produce a force which tends to slow the car when it moves in either direction. The mass of the car is $m$, and gravity acts on it at all times tending to make it roll down the hill in the absence of other forces. Let the mass $m$ of the car be 1000 kg, let the friction coefficient $k_f$ be 5 N-s/m, and let the angle θ be $\pi/12$.

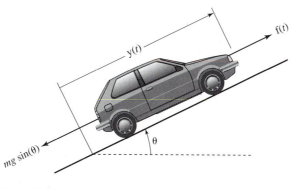

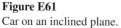

**Figure E61**
Car on an inclined plane.

a. Write a differential equation for this system with the force $f(t)$ as the excitation and the position of the car $y(t)$ as the response.

b. If the nose of the car is initially at position $y(0) = 0$, with an initial velocity $[y'(t)]_{t=0} = 10$ m/s, and no applied acceleration or braking force, graph the velocity of the car $y'(t)$ for positive time.

c. If a constant force $f(t)$ of 200 N is applied to the car, what is its terminal velocity?

62. A block of aluminum is heated to a temperature of 100 degrees Celsius (°C). It is then dropped into a flowing stream of water which is held at a constant temperature of 10°C. After 10 s, the temperature of the ball is 60°C. (Aluminum is such a good heat conductor that its temperature is essentially uniform throughout its volume during the cooling process.) The rate of cooling is proportional to the temperature difference between the ball and the water.

a. Write a differential equation for this system with the temperature of the water as the excitation and the temperature of the block as the response.

b. Compute the time constant of the system.

c. Find the impulse response of the system and, from it, the step response.

d. If the same block is cooled to 0°C and dropped into a flowing stream of water at 80°C at time $t = 0$, at what time will the temperature of the block reach 75°C?

63. A well-stirred vat has been fed for a long time by two streams of liquid, fresh water at 0.2 m³/s and concentrated blue dye at 0.1 m³/s. The vat contains 10 m³ of this mixture, and the mixture is being drawn from the vat at a rate of 0.3 m³/s to maintain a constant volume. The blue dye is suddenly changed to red dye at the same flow rate. At what time after the switch does the mixture drawn from the vat contain a ratio of red to blue dye of 99:1?

64. Some large auditoriums have a noticeable echo or reverberation. While a little reverberation is desirable, too much is undesirable. Let the response of an auditorium to an acoustic impulse of sound be

$$\text{h}(t) = \sum_{n=0}^{\infty} e^{-n} \delta \left( t - \frac{n}{5} \right).$$

We would like to design a signal-processing system that will remove the effects of reverberation. In Chapters 5 and 9 on transform theory we will be able to show that the compensating system that can remove the reverberations has an impulse response of the form

$$h_c(t) = \sum_{n=0}^{\infty} g[n]\delta\left(t - \frac{n}{5}\right).$$

Find the function $g[n]$.

**65.**  Show that the area property and the scaling property of the convolution integral are in agreement by finding the area of $x(at) * h(at)$ and comparing it with the area of $x(t) * h(t)$.

**66.**  The convolution of a function $g(t)$ with a doublet can be written as

$$g(t) * u_1(t) = \int_{-\infty}^{\infty} g(\tau)u_1(t - \tau)\, d\tau.$$

Integrate by parts to show that $g(t) * u_1(t) = g'(t)$.

**67.**  Derive the sampling property for a unit triplet. That is, find an expression for the integral

$$\int_{-\infty}^{\infty} g(t)u_2(t)\, dt$$

which is analogous to the sampling property of the unit doublet $-g'(t) = \int_{-\infty}^{\infty} g(t)u_1(t)\, dt$.

**68.**  Sketch block diagrams of the systems described by these equations. For the differential equation use only integrators in the block diagrams.

*a.*   $y''(t) + 3y'(t) + 2y(t) = x(t)$

*b.*   $6y[n] + 4y[n - 1] - 2y[n - 2] + y[n - 3] = x[n]$

### CHAPTER 4

# The Fourier Series

## 4.1 INTRODUCTION AND GOALS

In Chapter 3 we developed a technique, convolution, for finding the response of an LTI system to an arbitrary excitation. The basic idea of convolution is to break up or decompose a signal into a sum of elementary functions. Then we find the response of the system to each of those elementary functions individually and add the responses to get the overall response. In the case of convolution, the elementary functions are impulses, and convolution is a process of combining shifted and weighted impulse responses to form the overall response. This method works in LTI systems because of the properties of linearity and time invariance.

In this chapter we will decompose a signal in a different way. We will express it as a sum of real or complex sinusoids instead of as a sum of impulses. Real and complex sinusoids are linear combinations of special cases of the eigenfunctions of LTI systems, complex exponentials. The responses of LTI systems to sinusoids are also sinusoids of the same frequency but with, in general, different amplitude and phase. Expressing signals in this way leads to the frequency-domain concept in which differential or difference equations are converted to algebraic equations and systems can be analyzed by methods involving systems of algebraic equations with complex coefficients instead of systems of differential or difference equations. Looking at signals this way also lends new insight into the nature of systems and, for certain types of systems, greatly simplifies designing and analyzing them.

Analyzing signals as linear combinations of sinusoids is not as strange as it may sound. The human ear does something very similar. When we hear a sound, what is the actual response of the brain? As presented in Chapter 1, the ear senses a time variation of air pressure. This variation might be a single tone like the sound of a person whistling. When we hear a whistled tone, we are not aware of the (very fast) oscillation of air pressure with time. Rather we are aware of three important characteristics of the sound, its *pitch,* which is a synonym for frequency; its *intensity* or amplitude; and its *duration*. The ear–brain system effectively parameterizes the signal into three simple descriptive parameters, pitch, intensity and duration, and does not attempt to follow the rapidly changing (and very repetitive) air pressure in detail. In doing so the ear–brain system has distilled the information in the signal down to its essence. The mathematical analysis of signals as linear combinations of sinusoids does something similar but in a more mathematically precise way.

1. To develop methods of expressing CT and DT signals as linear combinations of sinusoids, real or complex

2. To explore the general properties of these ways of expressing signals

3. To apply these methods to finding the responses of CT and DT systems to arbitrary excitations

## 4.2 THE CONTINUOUS-TIME FOURIER SERIES (CTFS)

### LINEARITY AND COMPLEX EXPONENTIAL EXCITATION

A very important result from Chapter 3 is that if an LTI system is excited by a complex sinusoid, the response is also a complex sinusoid, with the same frequency but generally a different multiplying constant. This occurs because the complex exponential is the eigenfunction of the difference or differential equations describing LTI systems and a complex sinusoid is just a special case of a complex exponential.

Another important result from Chapter 3 is that if an LTI system is excited by a sum of signals, the overall response is the sum of the responses to each of the signals individually. If we could find a way to express arbitrary excitation signals as linear combinations of complex sinusoids, we could use superposition to find the response of any LTI system to any arbitrary excitation simply by summing the responses to the individual complex sinusoids (Figure 4.1). It may seem strange to try to express a *real* function as a linear combination of *complex* sinusoids, but we already know that real cosines and real sines can be expressed as

$$\cos(x) = \frac{e^{jx} + e^{-jx}}{2} \qquad \text{and} \qquad \sin(x) = \frac{e^{jx} - e^{-jx}}{j2} \qquad \textbf{(4.1)}$$

(Figure 4.2).

We will soon see that any periodic function with engineering usefulness can be expressed as a combination of complex sinusoids through the Fourier series, which is

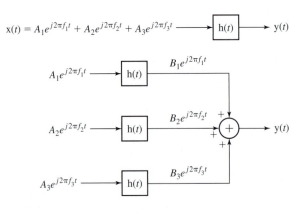

**Figure 4.1**
Equivalence of the responses of an LTI system to an excitation and to a linear combination of complex sinusoids which is equivalent to the excitation.

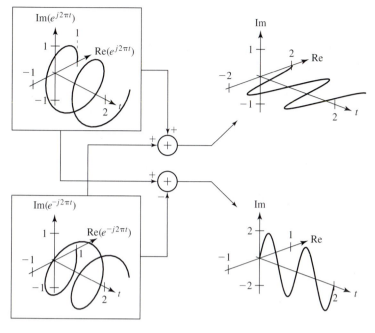

**Figure 4.2**
Addition and subtraction of $e^{j2\pi t}$ and $e^{-j2\pi t}$ to form $2\cos(2\pi t)$ and $j2\sin(2\pi t)$.

the subject of this chapter. (The phrase "periodic function with engineering usefulness" means a function describing a signal that can actually occur in a real, physical system. It is possible to imagine periodic mathematical functions that cannot be expressed as a Fourier series, but such functions have no known engineering use.) In Chapter 5 we will extend the Fourier series to the Fourier transform to represent aperiodic functions.

Consider an arbitrary original signal that we would like to represent as a linear combination of sinusoids over a finite range of time from an initial time $t_0$ to a final time $t_0 + T_F$ as illustrated at the top of Figure 4.3. Let $f_F = 1/T_F$ be called the *fundamental frequency* of this kind of representation of the signal. Then it is possible, as illustrated in Figure 4.3, to add a constant and sines and cosines at integer multiples of the fundamental frequency with the right amplitudes to represent the original signal in that finite time interval. (How we choose those amplitudes is the subject of Section 4.3.) The last representation in the lower right-hand corner of Figure 4.3 is not exact because we have only presented the results for sines and cosines at up to four times the fundamental frequency. We will soon show that if we kept adding correctly chosen sines and cosines at higher integer multiples of the fundamental frequency indefinitely, the representation would approach the original signal in the time interval $t_0 < t < t_0 + T_F$.

Notice that, in the example in Figure 4.3, the representation approaches the original signal only in the time interval $t_0 < t < t_0 + T_F$, not outside it. Notice also that the representation is periodic with fundamental period $T_F$, because all the sines and cosines used to form it have an integer number of periods in that time. Therefore, if the original signal happened to be periodic with fundamental period $T_0$, and if we were to choose $T_F = T_0$, the representation would be correct for *all* time (Figure 4.4).

The representation of a signal in the form of a linear combination of complex sinusoids is called the *Fourier series* in honor of Jean Baptiste Joseph Fourier, a French mathematician of the late eighteenth and early nineteenth centuries. (The name Fourier is commonly pronounced "fore-yay" because of the similarity to the English word *four*, but the proper French pronunciation is "foor-yay" where "foor" rhymes with *tour*.)

Fourier lived in a time of great turmoil in France, the French Revolution and the reign of Napolean Bonaparte. Fourier served as secretary of the Paris Academy of Science. In studying the propagation of heat in solids, Fourier developed the Fourier series and the Fourier integral. When he first presented his work to the great French mathematicians of the time, Laplace, LaGrange, and LaCroix, they were intrigued by his theories, but they (especially LaGrange) thought they lacked mathematical rigor. The publication of his paper at that time was denied. Some years later Dirichlet put the theories on a firmer foundation explaining exactly what functions could, and could not, be expressed by a Fourier series. Then Fourier published his theories in what is now a classic text, *Theorie analytique de la chaleur*.

Jean Baptiste Joseph Fourier, 3/21/1768–5/16/1830.

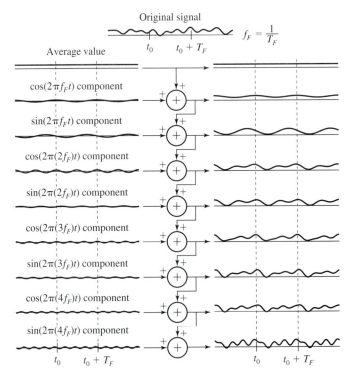

**Figure 4.3**
Illustration of the concept of representing an arbitrary signal as a linear combination of sinusoids.

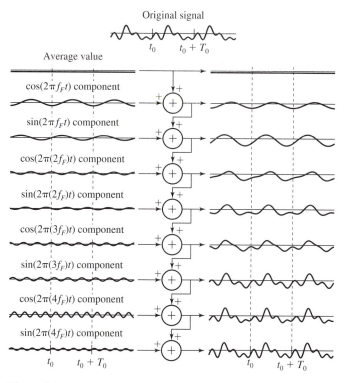

**Figure 4.4**
Illustration of the concept of representing a periodic signal, for all
time, as a linear combination of sinusoids by choosing $T_F = T_0$.

(It would also be correct for all time if we chose $T_F$ to be *any* period, not just the fundamental period.)

In a Fourier series representation of a signal, the higher-frequency sines and cosines have frequencies that are integer multiples of the fundamental frequency. The multiple is called the *harmonic number* and will be designated by $k$. So, for example, the function $\cos(2\pi(kf_F)t)$ is the $k$th-harmonic cosine, and its frequency is $kf_F$. If the signal to be represented is $x(t)$, then the amplitude of the $k$th-harmonic sine will be designated $X_s[k]$ and the amplitude of the $k$th-harmonic cosine will be designated $X_c[k]$. So the amplitudes of the sines and cosines are both functions of a discrete independent variable, not discrete time $n$, but rather discrete harmonic number $k$. The functions $X_s[k]$ and $X_c[k]$, together with the constant term, will be called the *harmonic functions* of the Fourier series of the original signal, in this case the trigonometric Fourier series.

The demonstration of the Fourier series idea in Figure 4.3 uses a constant, sines, and cosines to represent the original function, the trigonometric Fourier series. For our purposes, it is important as a prelude to later work to see the equivalence of another (and more important) form of Fourier series, the *complex* form of the Fourier series. As indicated earlier, every sine and cosine in the trigonometric form of the Fourier series could be replaced by its complex-sinusoid equivalent

$$\cos(2\pi(kf_F)t) = \frac{e^{j2\pi(kf_F)t} + e^{-j2\pi(kf_F)t}}{2} \qquad \text{and}$$

$$\sin(2\pi(kf_F)t) = \frac{e^{j2\pi(kf_F)t} - e^{-j2\pi(kf_F)t}}{j2}. \tag{4.2}$$

So for every sine and cosine harmonic component of the Fourier series, there is a pair of complex sinusoids that can replace it. If we add the sine and cosine with amplitudes $X_s[k]$ and $X_c[k]$, at any particular harmonic $k$, we get

$$X_c[k] \cos(2\pi(k f_F)t) + X_s[k] \sin(2\pi(k f_F)t)$$

$$= X_c[k]\frac{e^{j2\pi(k f_F)t} + e^{-j2\pi(k f_F)t}}{2} + X_s[k]\frac{e^{j2\pi(k f_F)t} - e^{-j2\pi(k f_F)t}}{j2}. \qquad \textbf{(4.3)}$$

We can combine like complex-sinusoid terms on the right-hand side to form

$$X_c[k] \cos(2\pi(k f_F)t) + X_s[k] \sin(2\pi(k f_F)t)$$

$$= \frac{1}{2}\left\{(X_c[k] - jX_s[k])e^{j2\pi(k f_F)t} + (X_c[k] + jX_s[k])e^{j2\pi((-k) f_F)t}\right\}. \qquad \textbf{(4.4)}$$

Now if we define

$$X[k] = \frac{X_c[k] - jX_s[k]}{2} \qquad \text{and} \qquad X[-k] = \frac{X_c[k] + jX_s[k]}{2}, \qquad \textbf{(4.5)}$$

we can write

$$X_c[k] \cos(2\pi(k f_F)t) + X_s[k] \sin(2\pi(k f_F)t) = X[k]e^{j2\pi(k f_F)t} + X[-k]e^{j2\pi((-k) f_F)t} \qquad \textbf{(4.6)}$$

and we have the amplitudes $X[k]$ of the complex sinusoids $e^{j2\pi(k f_F)t}$, at all positive and negative integer multiples of the fundamental frequency, and the sum of all these complex sinusoids adds to the original function just like the sines and cosines did.

We still have the constant term to consider. To include the constant term in the general formulation of complex sinusoids, we can let it be the zeroth harmonic of the fundamental. Letting $k$ be zero, the complex sinusoid $e^{j2\pi(k f_F)t}$ is just the number one, and if we multiply it by a correctly chosen weighting factor $X[0]$, we can complete the complex-Fourier-series representation. It will turn out in the material to follow, that the same general formula for finding $X[k]$ for any nonzero $k$ can also be used, without modification, to find $X[0]$, and that $X[0]$ is simply the average value in the time interval $t_0 < t < t_0 + T_F$ of the function to be represented. That fact makes the complex Fourier series more efficient and compact than the trigonometric Fourier series. Figure 4.5 illustrates how the Fourier series converges to the original signal as $k$ is increased for two other original signals.

## DEFINITION OF THE CONTINUOUS-TIME FOURIER SERIES

**Derivation of the Continuous-Time Fourier Series**   Let's assume, provisionally, that a signal $x(t)$ can be expressed, over a time interval $t_0 < t < t_0 + T_F$, as a linear combination of complex sinusoids in the form

$$x_F(t) = \sum_{k=-\infty}^{\infty} X[k]e^{j2\pi(k f_F)t}, \qquad \textbf{(4.7)}$$

where $f_F = 1/T_F$. Then

$$x(t) = x_F(t) \qquad t_0 < t < t_0 + T_F. \qquad \textbf{(4.8)}$$

Later we will see under which conditions this assumption is a good one. Notice that we are representing the original signal $x(t)$ as a linear combination of complex sinusoids

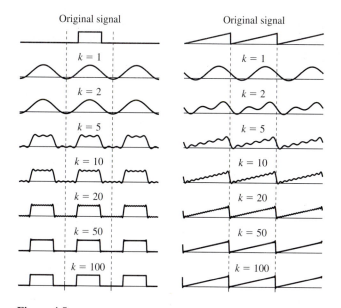

**Figure 4.5**
Examples of Fourier series representations.

over a finite time range $t_0 < t < t_0 + T_F$, *not* over all time. This is the most general application of the continuous-time Fourier series (CTFS). Later we will restrict its application to periodic signals and show that a CTFS can represent a periodic signal for all time. The representation of a periodic signal for all time is, by far, the most common application of the CTFS.

The problem of finding the CTFS representation becomes one of finding the proper function $X[k]$ to make valid the equality

$$x(t) = x_F(t) \qquad t_0 < t < t_0 + T_F. \tag{4.9}$$

If $x_F(t)$ is to be the same as $x(t)$ in the interval $t_0 < t < t_0 + T_F$, then

$$x(t) = \sum_{k=-\infty}^{\infty} X[k] e^{j2\pi(kf_F)t} \qquad t_0 < t < t_0 + T_F. \tag{4.10}$$

We can multiply (4.10) through by $e^{-j2\pi(qf_F)t}$ ($q$ is an integer) yielding

$$x(t)e^{-j2\pi(qf_F)t} = \sum_{k=-\infty}^{\infty} X[k] e^{j2\pi(kf_F)t} e^{-j2\pi(qf_F)t}$$

$$= \sum_{k=-\infty}^{\infty} X[k] e^{j2\pi((k-q)f_F)t} \qquad t_0 < t < t_0 + T_F. \tag{4.11}$$

If we now integrate both sides of (4.11) over the time interval $t_0 < t < t_0 + T_F$, we get

$$\int_{t_0}^{t_0+T_F} x(t)e^{-j2\pi(qf_F)t}\, dt = \int_{t_0}^{t_0+T_F} \left[ \sum_{k=-\infty}^{\infty} X[k] e^{j2\pi((k-q)f_F)t} \right] dt. \tag{4.12}$$

Since $k$ and $t$ are independent variables, the integral of the sum on the right side of (4.12) is equivalent to a sum of integrals. Therefore, (4.12) can be written as

$$\int_{t_0}^{t_0+T_F} x(t)e^{-j2\pi(qf_F)t}\,dt = \sum_{k=-\infty}^{\infty} X[k] \int_{t_0}^{t_0+T_F} e^{j2\pi((k-q)f_F)t}\,dt. \qquad (4.13)$$

The complex-sinusoid integrand on the right side of (4.13) can be expressed, through Euler's identity, as the sum of a cosine and a sine,

$$\int_{t_0}^{t_0+T_F} x(t)e^{-j2\pi(qf_F)t}\,dt$$

$$= \sum_{k=-\infty}^{\infty} X[k] \int_{t_0}^{t_0+T_F} [\cos(2\pi((k-q)f_F)t) - j\sin(2\pi((k-q)f_F)t)]\,dt. \qquad (4.14)$$

Since $k$ and $q$ are both integers, so is $k-q$. Then, for the case $k \neq q$, we are integrating sinusoidal functions over exactly $k-q \neq 0$ fundamental periods (because $f_F = 1/T_F$). The integral of any sinusoid over any period (any integer multiple of the fundamental period) is zero (Figure 4.6). Therefore, for $k \neq q$,

$$\int_{t_0}^{t_0+T_F} [\cos(2\pi((k-q)f_F)t) - j\sin(2\pi((k-q)f_F)t)]\,dt = 0. \qquad (4.15)$$

The case $k = q$ is unique because the arguments of the sine and cosine are identically zero for any $t$. If $k = q$, the integral reduces to

$$\int_{t_0}^{t_0+T_F} [\cos(0) - j\sin(0)]\,dt = \int_{t_0}^{t_0+T_F} dt = T_F. \qquad (4.16)$$

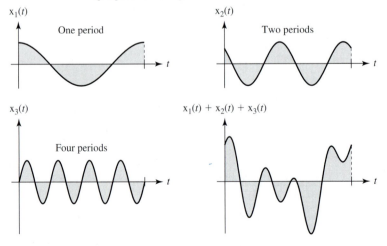

**Figure 4.6**
Graphical illustration of the fact that the integral of any sinusoid (or sum of sinusoids) over any period is zero.

Therefore, the summation

$$\sum_{k=-\infty}^{\infty} X[k] \int_{t_0}^{t_0+T_F} e^{j2\pi((k-q)f_F)t} \, dt \tag{4.17}$$

reduces to $X[q]T_F$ and

$$\int_{t_0}^{t_0+T_F} x(t)e^{-j2\pi(qf_F)t} \, dt = X[q]T_F. \tag{4.18}$$

Solving for $X[q]$,

$$X[q] = \frac{1}{T_F} \int_{t_0}^{t_0+T_F} x(t)e^{-j2\pi(qf_F)t} \, dt. \tag{4.19}$$

If this is a correct expression for $X[q]$, then $X[k]$ in the original Fourier series expression (4.7) must be

$$X[k] = \frac{1}{T_F} \int_{t_0}^{t_0+T_F} x(t)e^{-j2\pi(kf_F)t} \, dt. \tag{4.20}$$

From this derivation we conclude that if the integral in (4.20) converges, a signal $x(t)$ can be represented exactly, in the time interval $t_0 < t < t_0 + T_F$, by

$$\boxed{x_F(t) = \sum_{k=-\infty}^{\infty} X[k]e^{j2\pi(kf_F)t}}, \tag{4.21}$$

where

$$\boxed{X[k] = \frac{1}{T_F} \int_{t_0}^{t_0+T_F} x(t)e^{-j2\pi(kf_F)t} \, dt}. \tag{4.22}$$

If the integral does not converge, a CTFS of the signal cannot be found.

In the CTFS representation of a function,

$$x_F(t) = \sum_{k=-\infty}^{\infty} X[k]e^{j2\pi(kf_F)t}, \tag{4.23}$$

where $k$ is the harmonic number. The CTFS is written with $kf_F$ enclosed in parentheses to emphasize the fact that the frequency of each complex sinusoid is $k$ times the fundamental frequency of the Fourier series representation. For example, the frequency $2f_F$ is the second harmonic of the fundamental frequency $f_F$. In CTFS expressions the signals are always represented by linear combinations of complex sinusoids at the fundamental frequency of the Fourier series representation and its harmonics.

Representing the CT function $x(t)$ by a CTFS is a transformation of the way we mathematically represent the function. The function is the same because if we were to plot points of $x(t)$ and $\sum_{k=-\infty}^{\infty} X[k]e^{j2\pi(kf_F)t}$ on a graph in the time range

$t_0 < t < t_0 + T_F$, the two functional representations would yield exactly the same values at points of continuity of x($t$). A common way of expressing the relation between a CT function and its CTFS harmonic function is to say that they form a *transform pair*. That is often indicated in the shorthand notation

$$x(t) \xleftrightarrow{\;\mathcal{FS}\;} X[k] \tag{4.24}$$

where the function on the left represents the signal in the time domain because its independent variable is time $t$ and the function on the right represents the transformation of the signal to a harmonic-number domain because the independent variable is harmonic number $k$.

The process of forming a signal $x_F(t)$ as the sum of a series of complex sinusoids is sometimes called *synthesis*. We can synthesize a signal from its component parts, the individual complex sinusoids. The process of finding the harmonic function of the Fourier series $X[k]$ is sometimes called *analysis*. We analyze the signal x($t$) by breaking it up into its component parts.

**Limitations on Functions Representable by a Continuous-Time Fourier Series**
As indicated in the previous section, if the integral of a signal x($t$) over the time interval $t_0 < t < t_0 + T_F$ diverges, a CTFS cannot be found for the signal. There are two other conditions on the applicability of the CTFS, which, together with the condition on the convergence of the integral, are called the *Dirichlet* conditions. The Dirichlet conditions are

1.  The signal must be absolutely integrable over the time $t_0 < t < t_0 + T_F$, that is,

$$\int_{t_0}^{t_0+T_F} |x(t)|\, dt < \infty. \tag{4.25}$$

2.  The signal must have a finite number of maxima and minima in the time $t_0 < t < t_0 + T_F$.

3.  The signal must have a finite number of discontinuities, all of finite size, in the time $t_0 < t < t_0 + T_F$.

There are hypothetical signals for which the Dirichlet conditions are not met, but they have no known engineering use.

**The Trigonometric Continuous-Time Fourier Series**   Assuming a CTFS can be found, we can say that

$$x(t) = x_F(t) \qquad t_0 < t < t_0 + T_F, \tag{4.26}$$

where

$$x_F(t) = \sum_{k=-\infty}^{\infty} X[k] e^{j2\pi(kf_F)t}. \tag{4.27}$$

This holds for any signal, real or complex, which satisfies the Dirichlet conditions. It is useful to explore the characteristics of the complex conjugate of $x_F(t)$. If we conjugate both sides of (4.27), we get

$$x_F^*(t) = \sum_{k=-\infty}^{\infty} X^*[k] e^{-j2\pi(kf_F)t} = \sum_{k=\infty}^{-\infty} X^*[-k] e^{j2\pi(kf_F)t}$$

$$= \sum_{k=-\infty}^{\infty} X^*[-k] e^{-j2\pi(-kf_F)t}. \tag{4.28}$$

In words, this says that to find the CTFS harmonic function $X[k]$ for the complex conjugate of a signal, conjugate it and change the sign of $k$. The transformation is

$$X[k] \rightarrow X^*[-k], \tag{4.29}$$

and then for any $x(t)$, $x^*(t) \xleftrightarrow{\ \mathcal{FS}\ } X^*[-k]$. In the very important special case in which $x(t)$ is a real-valued function, $x(t) = x^*(t)$ and, therefore, $x_F(t) = x_F^*(t)$. That means that the two representations

$$x_F(t) = \sum_{k=-\infty}^{\infty} X[k]e^{j2\pi(kf_F)t} \quad \text{and} \quad x_F^*(t) = \sum_{k=-\infty}^{\infty} X^*[-k]e^{j2\pi(kf_F)t} \tag{4.30}$$

must be equal and, therefore, that $X[k] = X^*[-k]$, implying that, for real-valued signals and for any $k$, $X[k]$ and $X[-k]$ are complex conjugates. That being the case, for real signals we can express the CTFS in the form

$$x_F(t) = X[0] + \sum_{k=1}^{\infty} \left[ X[k]e^{j2\pi(kf_F)t} + X^*[k]e^{-j2\pi(kf_F)t} \right] \tag{4.31}$$

or

$$x_F(t) = X[0] + \sum_{k=1}^{\infty} \left[ \mathrm{Re}(X[k])e^{j2\pi(kf_F)t} + \mathrm{Re}(X[k])e^{-j2\pi(kf_F)t} \right.$$
$$\left. + j\,\mathrm{Im}(X[k])e^{j2\pi(kf_F)t} - j\,\mathrm{Im}(X[k])e^{-j2\pi(kf_F)t} \right]. \tag{4.32}$$

where $X[0]$ is

$$X[0] = \frac{1}{T_F} \int_{t_0}^{t_0+T_F} x(t)\,dt, \tag{4.33}$$

which is simply the average value of the signal $x(t)$ over the time interval $t_0 < t < t_0 + T_F$. Therefore, if $x(t)$ is real-valued, $X[0]$ is a real number. Using the trigonometric relations

$$\cos(x) = \frac{e^{jx} + e^{-jx}}{2} \quad \text{and} \quad \sin(x) = \frac{e^{jx} - e^{-jx}}{j2} \tag{4.34}$$

in (4.32), we get

$$x_F(t) = X[0] + \sum_{k=1}^{\infty} [2\,\mathrm{Re}(X[k])\cos(2\pi(kf_F)t) - 2\,\mathrm{Im}(X[k])\sin(2\pi(kf_F)t)]. \tag{4.35}$$

This is now a representation of the real-valued signal $x(t)$ in terms of a linear combination of a real constant and real-valued cosines and sines. It is relatively easy to show that

$$x_F(t) = X_c[0] + \sum_{k=1}^{\infty} [X_c[k]\cos(2\pi(kf_F)t) + X_s[k]\sin(2\pi(kf_F)t)] \tag{4.36}$$

where $X_c[0] = X[0]$,

$$X_c[k] = 2\,\mathrm{Re}(X[k]) = \frac{2}{T_F} \int_{t_0}^{t_0+T_F} x(t)\cos(2\pi(kf_F)t)\,dt \tag{4.37}$$

and

$$X_s[k] = -2\,\mathrm{Im}(X[k]) = \frac{2}{T_F} \int_{t_0}^{t_0+T_F} x(t)\sin(2\pi(kf_F)t)\,dt. \tag{4.38}$$

This is known as the *trigonometric* form of the CTFS harmonic function for real-valued signals. Therefore, the relationships between the complex and trigonometric forms are

$$X_c[0] = X[0]$$
$$X_s[0] = 0$$
$$X_c[k] = X[k] + X^*[k]$$
$$X_s[k] = j(X[k] - X^*[k]) \qquad k = 1, 2, 3, \ldots \tag{4.39}$$

and

$$X[0] = X_c[0]$$
$$X[k] = \frac{X_c[k] - jX_s[k]}{2} \qquad k = 1, 2, 3, \ldots \tag{4.40}$$
$$X[-k] = X^*[k] = \frac{X_c[k] + jX_s[k]}{2}$$

The trigonometric form is the one actually used by Fourier, but the complex sinusoidal form is more important in the modern study of signals and systems. The relationships between the complex and trigonometric forms of the CTFS are closely related because of Euler's identity

$$e^{jx} = \cos(x) + j\sin(x), \tag{4.41}$$

which indicates that when we find a complex sinusoid in a CTFS representation of a signal, we are, by implication, simultaneously finding a cosine and a sine.

**Periodicity of Continuous-Time Fourier Series Representations**   The representation $x_F(t)$ of a function $x(t)$ as a complex CTFS is of the form

$$x_F(t) = \sum_{k=-\infty}^{\infty} X[k]e^{j2\pi(kf_F)t}. \tag{4.42}$$

If we increment the time $t$ by an integer multiple $q$, of the time $T_F$, in (4.42), we get

$$x_F(t + qT_F) = \sum_{k=-\infty}^{\infty} X[k]e^{j2\pi(kf_F)(t+qT_F)} = \sum_{k=-\infty}^{\infty} X[k]e^{j2\pi(kf_F)t}e^{j2\pi(kqf_F)T_F}. \tag{4.43}$$

But $f_F T_F = 1$ and, therefore,

$$x_F(t + qT_F) = \sum_{k=-\infty}^{\infty} X[k]e^{j2\pi(kf_F)t}\underbrace{e^{j2kq\pi}}_{=1} = \sum_{k=-\infty}^{\infty} X[k]e^{j2\pi(kf_F)t}$$

$$= x_F(t) \tag{4.44}$$

proving that $x_F(t)$ is periodic with fundamental period $T_F$. The time range $t_0 < t < t_0 + T_F$ of the CTFS representation of $x(t)$ is one fundamental period of $x_F(t)$ and is equal to $x(t)$ at points of continuity in that time interval.

The preceding proofs were of the equality of the two signals $x(t)$ and $x_F(t)$ in the time interval $t_0 < t < t_0 + T_F$ and the periodicity of $x_F(t)$. A natural question is whether $x(t)$ and $x_F(t)$ are also equal *outside* the time interval $t_0 < t < t_0 + T_F$. That depends on the nature of $x(t)$ and the choice of the time interval length $T_F$. Since $x_F(t)$ is periodic with fundamental period $T_F$ seconds, if $x(t)$ is periodic with fundamental period $T_0$ and if $T_F$ is an integer multiple of $T_0$, then $x(t)$ and $x_F(t)$ are equal for

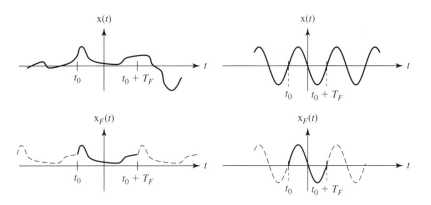

**Figure 4.7**
Signals represented over a finite interval by a CTFS.

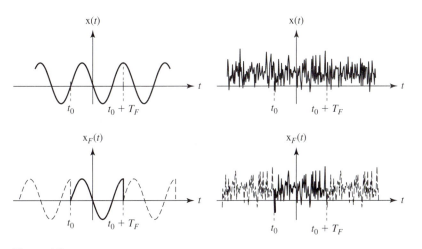

**Figure 4.8**
More signals represented over a finite interval by a CTFS.

all time at points of continuity of $x(t)$. Stated mathematically, if $x(t)$ is periodic with fundamental period $T_0$ and

$$T_F = mT_0 \qquad m \text{ is an integer,} \qquad \textbf{(4.45)}$$

then $x(t) = x_F(t)$, at points of continuity of $x(t)$, for all $t$.

Figures 4.7 and 4.8 show how various kinds of signals are represented by a CTFS over a finite time. (The dashed lines are periodic continuations of the CTFS representation.)

## 4.3 CALCULATION OF THE CONTINUOUS-TIME FOURIER SERIES

### SINUSOIDAL SIGNALS

To begin understanding the process of finding the CTFS and to develop a deeper understanding of the relationship between a function and its CTFS, start with a very simple example signal, a cosine with an amplitude of 2 at a frequency of 200 Hz,

$$x(t) = 2\cos(400\pi t). \qquad \textbf{(4.46)}$$

As a practical matter, it would be a waste of time to go through the steps of finding a CTFS expression for this signal since it is already in the form of a trigonometric CTFS consisting of exactly one cosine. But it is useful as a pedagogic device to develop an understanding of how the CTFS is found and what it really indicates about a signal. It also introduces by example the concepts of *orthogonality* and *correlation,* which will be important in Chapters 4 and 8, respectively.

We will express this signal as a CTFS over an interval $0 < t < 5$ ms, which is exactly one fundamental period of this signal ($T_F = T_0$). We can find either the trigonometric or complex form of the harmonic functions. To find the trigonometric form we evaluate integrals of the forms

$$X_c[k] = \frac{2}{T_F} \int_{t_0}^{t_0+T_F} x(t) \cos(2\pi(kf_F)t) \, dt \qquad (4.47)$$

and

$$X_s[k] = \frac{2}{T_F} \int_{t_0}^{t_0+T_F} x(t) \sin(2\pi(kf_F)t) \, dt, \qquad (4.48)$$

and to find the complex form we evaluate an integral of the form

$$X[k] = \frac{1}{T_F} \int_{t_0}^{t_0+T_F} x(t) e^{-j2\pi(kf_F)t} \, dt. \qquad (4.49)$$

The integral (4.49) can be written as

$$X[k] = \underbrace{\frac{1}{T_F} \int_{t_0}^{t_0+T_F} x(t) \cos(2\pi(kf_F)t) \, dt}_{\frac{X_c[k]}{2}} - j \underbrace{\frac{1}{T_F} \int_{t_0}^{t_0+T_F} x(t) \sin(2\pi(kf_F)t) \, dt}_{\frac{X_s[k]}{2}} \qquad (4.50)$$

again illustrating the close relationship between the trigonometric and complex forms. The trigonometric harmonic functions are found by evaluating integrals of real functions. Because it is easier to graphically illustrate real functions than complex functions, we will find the trigonometric harmonic functions and then relate them to the corresponding complex harmonic functions. First let's find $X_c[1]$ and $X_s[1]$ (Figure 4.9).

The top two graphs in Figure 4.9 show the signal (solid curves) and the cosine and sine (dashed curves) by which it is multiplied to form the integrands in the formulas for $X_c[1]$ and $X_s[1]$,

$$X_c[1] = \frac{2}{T_F} \int_{t_0}^{t_0+T_F} \underbrace{x(t) \cos(2\pi f_F t)}_{\uparrow} \, dt \qquad \text{and}$$

$$X_s[1] = \frac{2}{T_F} \int_{t_0}^{t_0+T_F} \underbrace{x(t) \sin(2\pi f_F t)}_{\uparrow} \, dt. \qquad (4.51)$$

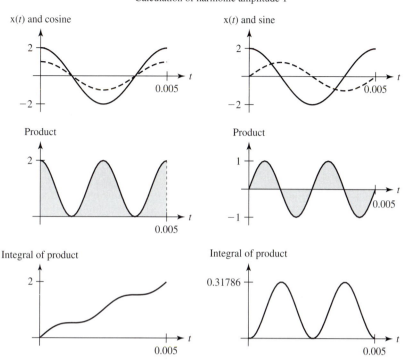

**Figure 4.9**
Graphical illustration of the steps involved in calculating $X_c[1]$ and $X_s[1]$ for a cosine.

The middle graphs show the product of the signal and the cosine and sine, respectively, with the area under the curve shaded to emphasize the integration process of finding the area under the product. The bottom graphs show the accumulated area under the product as the integration proceeds from $t_0$ to $t_0 + T_F$. They are graphs of

$$\frac{2}{T_F} \int_{t_0}^{t_0+\Delta t} x(t) \cos(2\pi f_F t)\, dt \qquad \text{and} \qquad \frac{2}{T_F} \int_{t_0}^{t_0+\Delta t} x(t) \sin(2\pi f_F t)\, dt. \qquad \textbf{(4.52)}$$

In these integrals, when $\Delta t$ becomes the same as $T_F$, we have arrived at the values of the $X_c[1]$ and $X_s[1]$. The cosine integral arrives at a value of two indicating the amplitude of the fundamental cosine component in the CTFS is two. Since the signal is a cosine of amplitude, two, at the fundamental frequency, this is obviously correct. The sine integral arrives at a value of zero even though the sine is also at the fundamental frequency. This is also correct because the signal has no sine components at all.

Now we will find the second-harmonic cosine and sine amplitudes by the same technique (Figure 4.10). Now both harmonic amplitudes are zero as they should be since there is neither a second-harmonic cosine nor sine in the signal.

Analytically we can find the CTFS using either of the two forms, complex or trigonometric. Let's use the complex form to analytically evaluate the CTFS.

$$X[k] = \frac{1}{T_F} \int_{t_0}^{t_0+T_F} x(t)e^{-j2\pi(kf_F)t}\, dt = 200 \int_{0}^{1/200} 2\cos(400\pi t)e^{-j400k\pi t}\, dt \qquad \textbf{(4.53)}$$

Calculation of harmonic amplitude 2

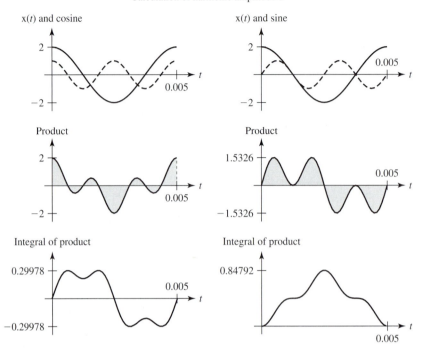

**Figure 4.10**
Graphical illustration of the steps involved in calculating $X_c[2]$ and $X_s[2]$ for a cosine.

Using

$$\cos(x) = \frac{e^{jx} + e^{-jx}}{2} \tag{4.54}$$

we can rewrite (4.53) as

$$X[k] = 200 \int_0^{1/200} (e^{j400\pi t} + e^{-j400\pi t})e^{-j400k\pi t}\, dt$$

$$= 200 \int_0^{1/200} e^{-j400\pi(k-1)t} + e^{-j400\pi(k+1)t}\, dt \tag{4.55}$$

or

$$X[k] = \frac{j}{2\pi}\left[\frac{e^{-j2\pi(k-1)} - 1}{k - 1} + \frac{e^{-j2\pi(k+1)} - 1}{k + 1}\right]. \tag{4.56}$$

Unless $k = 1$ or $k = -1$, this expression evaluates to zero. For $k = 1$ the expression evaluates (using L'Hôpital's rule) to the real number, one, and for $k = -1$ it also evaluates to the real number one. We can then express the complex CTFS harmonic function as

$$X[k] = \delta[k - 1] + \delta[k + 1]. \tag{4.57}$$

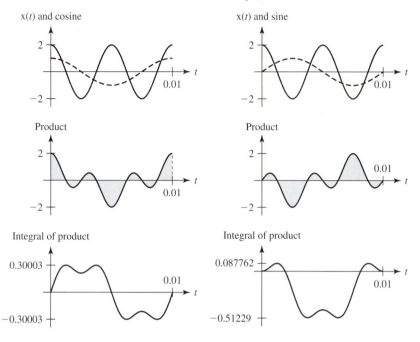

Calculation of harmonic amplitude 1

**Figure 4.11**
Graphical illustration of the steps involved in calculating $X_c[1]$ and $X_s[1]$ for a cosine at the second harmonic of the fundamental CTFS frequency.

For $k > 0$, the corresponding trigonometric harmonic functions are

$$X_c[k] = X[k] + X^*[k] = 2\delta[k - 1] \qquad \text{and}$$
$$X_s[k] = j(X[k] - X^*[k]) = 0 \qquad k > 0 \qquad \qquad \textbf{(4.58)}$$

indicating that there is a cosine of amplitude, two, at the first harmonic (the fundamental) and no other harmonic components. If we were to find the amplitudes of other harmonics by the same graphical steps as just illustrated, we would find that they are all zero, which is in agreement with the analytical result.

Now let us find the CTFS of this same signal over a different interval, $0 < t < 10$ ms. This choice of interval makes the fundamental frequency of the signal $x(t)$ and the fundamental frequency of the CTFS representation of the signal $x_F(t)$, different. The fundamental frequency of $x(t)$, $f_0$, is 200 Hz; the fundamental frequency of $x_F(t)$, $f_F = 1/T_F$, is 100 Hz. First let's find $X_c[1]$ and $X_s[1]$ (Figure 4.11).

In this case both integrals arrive at a value of zero, indicating that $X_c[1] = 0$ and $X_s[1] = 0$ (and, therefore, that $X[1] = X[-1] = 0$). This should be expected because $X_c[1]$ and $X_s[1]$ tell us how much of a sinusoid at the fundamental frequency $f_F$ is in the signal. Since the signal consists only of a cosine at the second harmonic of $f_F$, $f_0$, the signal does not contain *any* sinusoid at the fundamental frequency $f_F$.

Now repeat the process for $X_c[2]$ and $X_s[2]$ (Figure 4.12). From a graphical point of view, what makes $X_c[2]$ nonzero? It is nonzero because the signal and the second harmonic cosine have the same frequency and "rise and fall" together and, since they

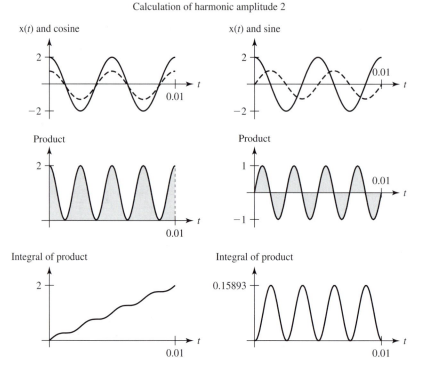

Calculation of harmonic amplitude 2

**Figure 4.12**
Graphical illustration of the steps involved in calculating $X_c[2]$ and $X_s[2]$ for a cosine at the second harmonic of the fundamental CTFS frequency.

always have the same sign, when their product is taken, the product is always positive and the integral accumulates positive area. $X_s[2]$ is zero because the product is sometimes positive and sometimes negative, with equal areas under the product which cancel each other. This occurs because of the 90° phase shift between the cosine and sine functions.

When a signal has a certain frequency component in it which is a harmonic of the fundamental frequency $f_F$ and it is multiplied by a sinusoid of the same harmonic of $f_F$, which is in phase with it, and integrated over any period, the integral is nonzero. When a signal has a certain frequency component in it and it is multiplied by a sinusoid of a different harmonic of $f_F$, or the same harmonic but 90° out of phase with it, and integrated over any period, the integral is zero. This is the way the formulas determine the amplitudes of the components.

The process of finding the CTFS can be conceived graphically as looking for a certain sinusoidal shape which is in the signal. In this simple example the shape is obvious because the signal *is* a sinusoid. (In Chapter 8, this process of finding how much of one signal is contained within another signal will be generalized to a process called *correlation*.) When the integral of the product of two functions over some interval is zero (as occurred in the second example for $|k| \neq 2$), the two functions are said to be *orthogonal* on that interval. In this case, the two functions are $2\cos(400\pi t)$ and $e^{-j200k\pi t}$, the interval is $0 < t < 10$ ms, and these two functions are orthogonal if $|k| \neq 2$. This is just one example of orthogonality of functions which happens to occur in Fourier analysis. The general study of orthogonal functions encompasses many other kinds of functions.

Analytically, finding the complex CTFS of $x(t) = 2\cos(400\pi t)$ with $T_F = \frac{1}{10}$,

$$X[k] = \frac{j}{2\pi} \left[ \frac{e^{-j2\pi(k-2)} - 1}{k-2} + \frac{e^{-j2\pi(k+2)} - 1}{k+2} \right]. \tag{4.59}$$

Unless $k = 2$ or $-2$, this expression evaluates to zero. That indicates that the only harmonic component present is the cosine at the second harmonic of the CTFS fundamental frequency $f_F$. This agrees with the graphical analysis.

We have found two different harmonic functions for the same signal. This was done to illustrate that the harmonic function alone is not sufficient to determine a signal; we must also know the fundamental period or frequency. One signal may have any number of harmonic functions by choosing different representation times $T_F$. By the same token, two different signals can have the same harmonic function. For example, the signal

$$x(t) = A\cos(2\pi f_1 t) \tag{4.60}$$

represented by a CTFS over the interval $0 < t < 1/f_1$ and the signal

$$x(t) = A\cos(2\pi f_2 t) \tag{4.61}$$

represented by a CTFS over the interval $0 < t < 1/f_2$, $f_1 \neq f_2$, both have the same complex CTFS harmonic function

$$X[k] = \frac{A}{2} \left( \delta[k-1] + \delta[k+1] \right). \tag{4.62}$$

To illustrate another point, let's use these same analysis methods on a slightly more complicated signal $x(t)$, which is the sum of a constant and two sinusoids,

$$x(t) = \frac{1}{2} - \frac{3}{4}\cos(20\pi t) + \frac{1}{2}\sin(30\pi t) \tag{4.63}$$

on the time interval $-100\text{ ms} < t < 100\text{ ms}$, which is exactly one fundamental period of this signal. This time interval is exactly one fundamental period of the signal $x(t)$, but it is exactly two fundamental periods of $\frac{3}{4}\cos(20\pi t)$ and exactly three fundamental periods of $\frac{1}{2}\sin(30\pi t)$. The graphical calculations of some of the harmonic amplitudes are illustrated in Figures 4.13 and 4.14.

Even though the signal is more complicated than that in the previous example, the process of finding each sinusoidal component in the signal by evaluating the area under the product over one fundamental period still yields exactly the right harmonic amplitudes without any interference from the constant and the other sinusoid present in the signal. This occurs again because of orthogonality. The other sinusoids are orthogonal to the sinusoidal harmonic being sought, and their contribution to the integral is therefore zero. (Also notice that, in this case, the fundamental frequency of the signal is 5 Hz but the CTFS harmonic function is zero at the fundamental. That is, $X[1] = 0$.)

Analytically we can show why a constant or a sinusoid at another harmonic does not interfere with the calculation of the amplitude of any harmonic. Let's find the

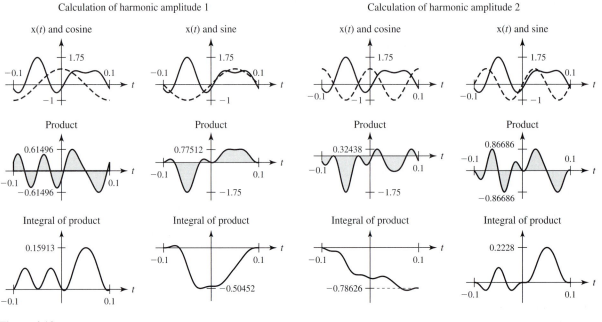

**Figure 4.13**
Graphical illustration of the steps involved in calculating the trigonometric harmonic amplitudes for the fundamental and second harmonic.

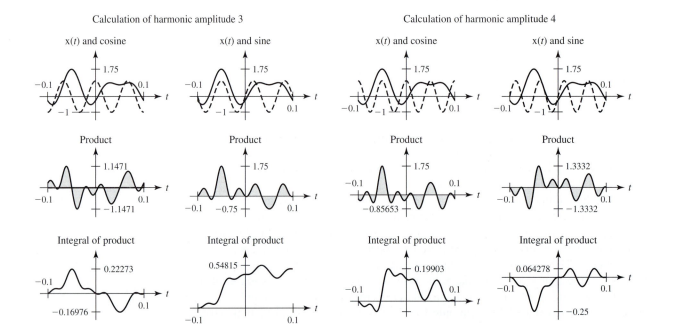

**Figure 4.14**
Graphical illustration of the steps involved in calculating the trigonometric harmonic amplitudes for the third and fourth harmonics.

trigonometric CTFS harmonic function directly this time using

$$X_c[0] = \frac{1}{T_F} \int_{t_0}^{t_0+T_F} x(t)\, dt$$

$$X_c[k] = \frac{2}{T_F} \int_{t_0}^{t_0+T_F} x(t) \cos(2\pi(kf_F)t)\, dt \qquad \textbf{(4.64)}$$

$$X_s[k] = \frac{2}{T_F} \int_{t_0}^{t_0+T_F} x(t) \sin(2\pi(kf_F)t)\, dt.$$

First find $X_c[0]$.

$$X_c[0] = 5 \int_{-(1/10)}^{1/10} \left[ \frac{1}{2} - \frac{3}{4}\cos(20\pi t) + \frac{1}{2}\sin(30\pi t) \right] dt = \frac{1}{2} \qquad \textbf{(4.65)}$$

Therefore, $X_c[0]$ is exactly equal to the constant term $\frac{1}{2}$. This should not be surprising since we are integrating three functions, two of which are sinusoids, over a time interval which contains exactly two fundamental periods of one of the sinusoids and exactly three fundamental periods of the other. The integral of any sinusoid over any period is always zero, so the only term left to contribute to the average value is the constant term $\frac{1}{2}$ whose average value is, of course, $\frac{1}{2}$.

Next let's find an expression for $X_c[k]$.

$$X_c[k] = \frac{2}{T_F} \int_{t_0}^{t_0+T_F} x(t) \cos(2\pi(kf_F)t)\, dt$$

$$= 10 \int_{-(1/10)}^{1/10} \left[ \frac{1}{2} - \frac{3}{4}\cos(20\pi t) + \frac{1}{2}\sin(30\pi t) \right] \cos(10k\pi t)\, dt$$

$$= \underbrace{5 \int_{-(1/10)}^{1/10} \cos(10k\pi t)\, dt}_{I_1} - \underbrace{\frac{15}{2} \int_{-(1/10)}^{1/10} \cos(20\pi t)\cos(10k\pi t)\, dt}_{I_2}$$

$$+ \underbrace{5 \int_{-(1/10)}^{1/10} \sin(30\pi t)\cos(10k\pi t)\, dt}_{I_3}$$

The first integral $I_1$ is zero, because it is the integral of a cosine over exactly two fundamental periods. Another way of saying this is that the constant $\frac{1}{2}$ and the function $\cos(10k\pi t)$ are orthogonal on the interval $-\frac{1}{10} < t < \frac{1}{10}$. We can evaluate integral $I_2$ by using the trigonometric identity

$$\cos(x)\cos(y) = \frac{1}{2}[\cos(x - y) + \cos(x + y)], \qquad \textbf{(4.66)}$$

yielding

$$I_2 = \frac{1}{2} \int\limits_{-(1/10)}^{1/10} [\cos(10\pi(2-k)t) + \cos(10\pi(2+k)t)] \, dt. \qquad (4.67)$$

As we have seen before, except in the two cases $k = \pm 2$, this is an integral over a period of two sinusoids and is therefore zero. Since we are finding the trigonometric CTFS, only the $k = +2$ case is of interest. In that case,

$$I_2 = \frac{1}{2} \int\limits_{-(1/10)}^{1/10} [1 + \cos(40\pi t)] \, dt = \frac{1}{10}. \qquad (4.68)$$

Again, another way of saying why the integral is zero when $|k| \neq 2$ is that the two functions $\cos(20\pi t)$ and $\cos(10k\pi t)$ are orthogonal on the interval $-\frac{1}{10} < t < \frac{1}{10}$ unless $k = \pm 2$. Integral $I_3$ is zero for all $k$ because its integrand is odd and it is integrated over limits which are symmetrical about $t = 0$. It is also zero because $\sin(30\pi t)$ and $\cos(10k\pi t)$ are orthogonal on the interval $-\frac{1}{10} < t < \frac{1}{10}$ for any integer value of $k$. Therefore,

$$X_c[2] = -\frac{15}{2} \times \frac{1}{10} = -\frac{3}{4} \qquad \text{and} \qquad X_c[k] = 0 \qquad k \neq 2 \qquad (4.69)$$

or

$$X_c[k] = -\frac{3}{4}\delta[k-2]. \qquad (4.70)$$

$X_c[2]$ is exactly the amplitude of the cosine term in the original expression of x $(t)$, with no interference from the sine term or the constant term. In a similar way the harmonic amplitudes of the sine terms in the CTFS are found to be

$$X_s[3] = \frac{1}{2} \qquad \text{and} \qquad X_s[k] = 0 \qquad k \neq 3 \qquad (4.71)$$

or

$$X_s[k] = \frac{1}{2}\delta[k-3]. \qquad (4.72)$$

As has been illustrated by this example, an important feature of the CTFS functions, complex or real sinusoids, is that they are all mutually orthogonal on the interval $t_0 < t < t_0 + T_F$. That is why no interference between them occurs when representing any period of a periodic signal with the CTFS.

This lack of interference from other sinusoidal components leads to an important property of the CTFS. When we have a signal which is composed of a sum of component signals, we can find the CTFS of each component signal separately and then add the CTFSs to form the CTFS of the overall signal. That is, the harmonic function of a sum of signals is the sum of the harmonic functions of the signals (Figure 4.15).

It is important to emphasize here that when finding the harmonic functions of the component signals, the same representation time $T_F$ and, therefore, the same fundamental frequency $f_F$ must be used for all of them. Otherwise the relation between harmonic number and sinusoid frequency is different for each component signal and the sum of the CTFS representations adds signals at different frequencies for the same harmonic number.

$$x(t) = x_1(t) + x_2(t) + \cdots$$

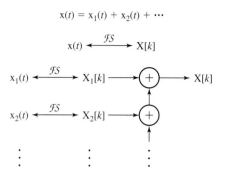

**Figure 4.15**
The harmonic function of a sum is the sum
of the harmonic functions.

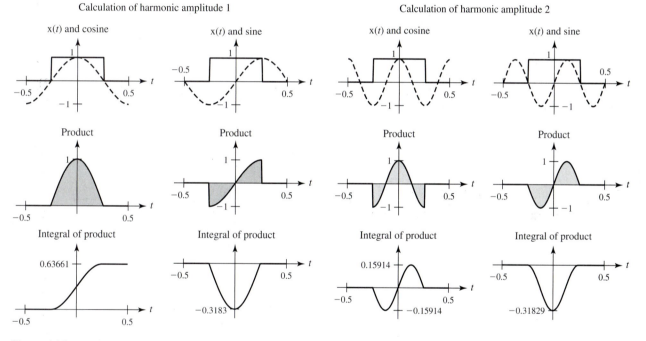

**Figure 4.16**
Graphical illustration of the steps involved in calculating the trigonometric harmonic amplitudes of the fundamental and
second harmonic for a square wave at a frequency of 1 Hz with a 50 percent duty cycle.

## NONSINUSOIDAL SIGNALS

Now, to expand our horizons, let's find the CTFS description of a periodic signal $x(t)$
that is not initially described in terms of sinusoids, a 50%-duty-cycle square wave with
an amplitude of one, and a fundamental period $T_0 = 1$ (Figures 4.16 and 4.17),

$$x(t) = \text{rect}(2t) * \text{comb}(t). \qquad (4.73)$$

We can find $X[k]$ analytically also.

$$X[k] = \frac{1}{T_F} \int_{t_0}^{t_0+T_F} x(t)e^{-j2\pi(kf_F)t}\, dt = \int_{-(1/4)}^{1/4} e^{-j2k\pi t}\, dt = 2\int_{0}^{1/4} \cos(2k\pi t)\, dt \qquad (4.74)$$

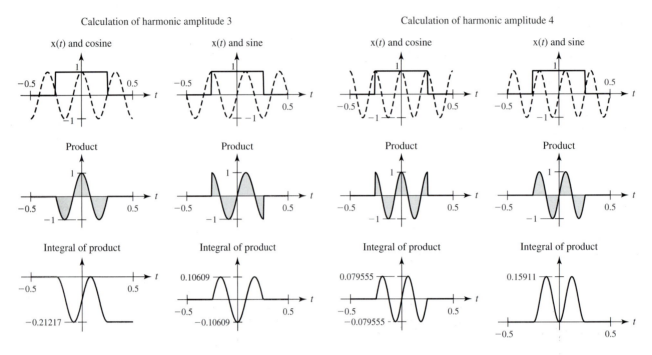

**Figure 4.17**
Graphical illustration of the steps involved in calculating the trigonometric harmonic amplitudes of the third and fourth harmonics for a square wave at a frequency of 1 Hz with a 50 percent duty cycle.

or

$$X[k] = 2 \left[ \frac{\sin(2k\pi t)}{2k\pi} \right]_0^{1/4} = \frac{1}{2} \frac{\sin(k\pi/2)}{k\pi/2} = \frac{1}{2} \operatorname{sinc}(k/2) \qquad \textbf{(4.75)}$$

It follows that

$$X_c[0] = \frac{1}{2} \quad \text{and} \quad X_c[k] = \operatorname{sinc}\left(\frac{k}{2}\right) \qquad X_s[k] = 0 \qquad k > 0. \qquad \textbf{(4.76)}$$

In this case, in contrast with previous ones, we have infinitely many nonzero CTFS harmonic function values. A very useful way of presenting the CTFS harmonic function is through a graph of its magnitude and phase versus harmonic number, or versus frequency (which is harmonic number times fundamental frequency) (Figure 4.18).

## THE CONTINUOUS-TIME FOURIER SERIES OF PERIODIC SIGNALS OVER A NONINTEGER NUMBER OF FUNDAMENTAL PERIODS

Now let us consider a different case. Suppose now that we find the CTFS for a sinusoid but over an interval $T_F$, which is not an integer multiple of its fundamental period $T_0$. Let the signal be the same as in the first example,

$$x(t) = 2\cos(400\pi t), \qquad \textbf{(4.77)}$$

but now let the representation interval be 7.5 ms long instead of 10 ms long. The graphical calculations of the first three cosine and sine harmonic amplitudes are illustrated in Figures 4.19 and 4.20.

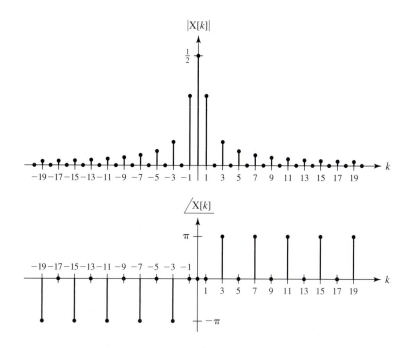

**Figure 4.18**
Magnitude and phase of the complex CTFS harmonic function for a
unit-amplitude, 50 percent duty-cycle, square wave versus harmonic
number.

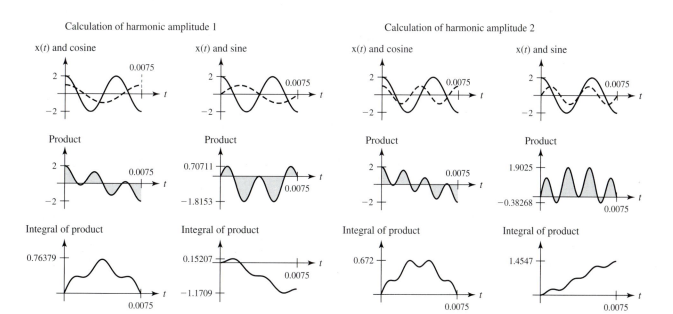

**Figure 4.19**
Graphical illustration of the steps involved in calculating $X_c[1]$ and $X_s[1]$ and $X_c[2]$ and $X_s[2]$ for a cosine at a frequency
of 1.5 times the fundamental frequency.

Calculation of harmonic amplitude 3

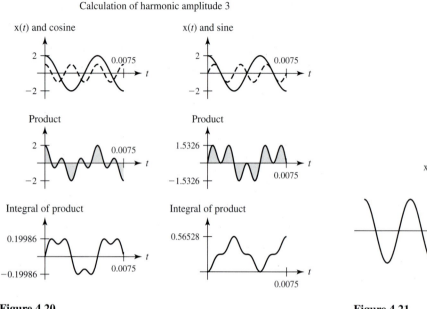

Figure 4.20
Illustration of the steps involved in calculating $X_c[3]$ and $X_s[3]$
for a cosine at a frequency of 1.5 times the fundamental frequency.

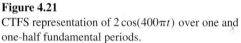

Figure 4.21
CTFS representation of $2\cos(400\pi t)$ over one and
one-half fundamental periods.

What's going on here? We know the signal $x(t)$ is a cosine, yet all the cosine harmonic amplitudes of $x_F(t)$ are zero and all the sine harmonic amplitudes are nonzero. The reconstruction of the signal from sinusoidal components is done with sinusoids at harmonics of $f_F$. If the original signal is periodic and its fundamental frequency $f_0$ is not a harmonic of $f_F$, none of the available harmonic frequencies $kf_F$ is the same as that fundamental frequency $f_0$. The CTFS gives a result which equals the original signal in the interval $t_0 < t < t_0 + T_F$ but not necessarily anywhere else. The problem is the difference between the fundamental periods of $x(t)$ and $x_F(t)$. As shown above, $x_F(t)$ repeats periodically with fundamental period $T_F$. If we repeat the part of this $x(t)$ lying in the interval $0 < t < 7.5$ ms, periodically we get the signal in Figure 4.21 which is quite different from the original signal, although it matches perfectly in the interval $0 < t < 7.5$ ms. This illustrates the reason the harmonic amplitudes come out in such an apparently strange way. We are reconstructing the cosine $x(t)$ over a finite interval using cosines and sines at harmonics of $f_F$ not $f_0$. In this case, $x_F(t)$ is an odd function even though $x(t)$ is an even function.

It is natural to wonder at this point why anyone would want to do a CTFS in this form since $X[k]$ is so much more complicated and the CTFS expression only equals the original signal in a finite time interval. This is certainly an awkward and inelegant analysis of the signal. But it is included here to illustrate a similar phenomenon called *leakage* which will arise in the application of the discrete Fourier transform to the analysis of sampled CT signals which will be introduced in Chapter 7.

Even though, in general, the CTFS can represent any signal satisfying the Dirichlet conditions over a finite time interval, the most common use of the CTFS in signal and system analysis is to exactly represent, over all time, a periodic signal which satisfies the Dirichlet conditions. As previously indicated, the representation is exact for all time if $T_F = mT_0$, where $m$ is an integer, and the most common choice for $m$ is $m = 1$.

# THE CONTINUOUS-TIME FOURIER SERIES OF PERIODIC SIGNALS OVER AN INTEGER NUMBER OF FUNDAMENTAL PERIODS

Consider the complex CTFS representation of any periodic signal $x(t)$ over exactly one fundamental period $T_0$.

$$X[k] = \frac{1}{T_0} \int_{t_0}^{t_0+T_0} x(t)e^{-j2\pi(kf_0)t} \, dt$$

$$= \frac{1}{T_0} \left[ \int_{t_0}^{t_0+T_0} x(t) \cos(2\pi(kf_0)t) \, dt - j \int_{t_0}^{t_0+T_0} x(t) \sin(2\pi(kf_0)t) \, dt \right] \qquad (4.78)$$

Since, by assumption, $x(t)$ is periodic with fundamental period $T_0$ and, by definition, $e^{-j2\pi(kf_0)t}$ is periodic with fundamental period $T_0$, the products $x(t)e^{-j2\pi(kf_0)t}$, $x(t)\cos(2\pi(kf_0)t)$, and $x(t)\sin(2\pi(kf_0)t)$ all repeat exactly in time $T_0$. That being the case, the value of the integral is independent of the choice of the lower limit $t_0$ because it is always the area under a period of the product of the signal and the sinusoid (complex or real) (Figure 4.22).

When the CTFS is used to represent a signal over exactly one fundamental period (and, by implication, over all time), we can simplify the notation to

$$X[k] = \frac{1}{T_0} \int_{T_0} x(t)e^{-j2\pi(kf_0)t} \, dt, \qquad (4.79)$$

where $\int_{T_0} = \int_{t_0}^{t_0+T_0}$ for any arbitrary $t_0$ means an integral over any time interval of length, $T_0$. Summarizing for the case of CTFS representations over exactly one fundamental period $T_0 = 1/f_0$,

$$x(t) = \sum_{k=-\infty}^{\infty} X[k]e^{j2\pi(kf_0)t} \xleftarrow{\ \mathcal{FS}\ } X[k] = \frac{1}{T_0} \int_{T_0} x(t)e^{-j2\pi(kf_0)t} \, dt. \qquad (4.80)$$

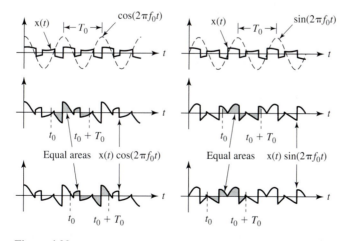

**Figure 4.22**
Graphical demonstration that for periodic signals exactly described by a CTFS, the CTFS is independent of the lower limit $t_0$ in the integral formula for the coefficients.

All these arguments apply equally well to finding the CTFS over any period, that is, over any integer multiple of the fundamental period. If the number of fundamental periods is $m$, then

$$x(t) = \sum_{k=-\infty}^{\infty} X[k]e^{j2\pi((k/m)f_0)t} \xrightarrow{\;\mathcal{FS}\;} X[k] = \frac{1}{mT_0}\int_{mT_0} x(t)e^{-j2\pi((k/m)f_0)t}\,dt. \quad \textbf{(4.81)}$$

The CTFS is an infinite summation of sinusoids. In general, for exact equality between an arbitrary original signal and its CTFS representation, infinitely many terms must be used. (There are signals for which the equality is achieved with a finite number of terms and they are called *band-limited* signals, but, in general, infinitely many terms are required.) If a partial-sum approximation

$$x_N(t) = \sum_{k=-N}^{N} X[k]e^{j2\pi(kf_F)n} \quad \textbf{(4.82)}$$

is made to a signal $x(t)$ by using only the first $N$ harmonics of the CTFS, the difference between $x(t)$ and $x_N(t)$ is the approximation error

$$e_N(t) = x_N(t) - x(t). \quad \textbf{(4.83)}$$

It can be shown that, for any value of $N$, the mean-squared approximation error

$$\overline{e_N^2(t)} = \overline{[x_N(t) - x(t)]^2} = \frac{1}{T_0}\int_{t_0}^{t_0+T_0}[x_N(t) - x(t)]^2\,dt \quad \textbf{(4.84)}$$

could not be any smaller if we chose any other $X[k]$ as the CTFS harmonic function of $x(t)$. The discrete-harmonic-number function $X[k]$ determined by the Fourier series formulas is optimal in that sense.

## THE CTFS OF EVEN AND ODD PERIODIC SIGNALS

Consider the case of representing a periodic even signal $x(t)$ with fundamental period $T_0$ for all time with a complex CTFS whose fundamental period is $T_F = T_0$. The CTFS harmonic function is

$$X[k] = \frac{1}{T_0}\int_{T_0} x(t)e^{-j2\pi(kf_0)t}\,dt. \quad \textbf{(4.85)}$$

For periodic signals this integral over exactly one fundamental period is independent of the starting point. Therefore, we can rewrite the integral as

$$X[k] = \frac{1}{T_0}\int_{-(T_0/2)}^{T_0/2} x(t)e^{-j2\pi(kf_0)t}\,dt$$

$$= \frac{1}{T_0}\left[\int_{-(T_0/2)}^{T_0/2} \underbrace{\underbrace{x(t)}_{\text{even}}\underbrace{\cos(2\pi(kf_0)t)}_{\text{even}}}_{\text{even}}\,dt - j\int_{-(T_0/2)}^{T_0/2} \underbrace{\underbrace{x(t)}_{\text{even}}\underbrace{\sin(2\pi(kf_0)t)}_{\text{odd}}}_{\text{odd}}\,dt\right]. \quad \textbf{(4.86)}$$

Using the fact that an odd function integrated over symmetrical limits about zero is zero, X[k] must be real and, since

$$X_c[0] = X[0]$$
$$\left. \begin{array}{l} X_c[k] = X[k] + X^*[k] \\[6pt] X_s[k] = j(X[k] - X^*[k]) \end{array} \right\} \quad k = 1, 2, 3, \ldots, \tag{4.87}$$

$X_s[k]$ must be zero for all $k$. By a similar argument, for a periodic odd function $X[k]$ must be imaginary and $X_c[k]$ must be zero for all $k$ (including $k = 0$).

## CYCLIC FREQUENCY AND RADIAN FREQUENCY FORMS

The CTFS is often expressed as a function of *radian* frequency $\omega$, instead of *cyclic* frequency $f$. Since $\omega = 2\pi f$, the CTFS representations of a signal $x_F(t)$ in terms of radian frequency would be

$$x_F(t) = \sum_{k=-\infty}^{\infty} X[k] e^{j(k\omega_F)t} \tag{4.88}$$

or

$$x_F(t) = X_c[0] + \sum_{k=1}^{\infty} [X_c[k] \cos((k\omega_F)t) + X_s[k] \sin((k\omega_F)t)] \tag{4.89}$$

where

$$X_c[0] = \frac{1}{T_F} \int_{t_0}^{t_0+T_F} x(t)\, dt$$

$$X_c[k] = \frac{2}{T_F} \int_{t_0}^{t_0+T_F} x(t) \cos((k\omega_F)t)\, dt, \qquad k = 1, 2, 3, \ldots$$

$$X_s[k] = \frac{2}{T_F} \int_{t_0}^{t_0+T_F} x(t) \sin((k\omega_F)t)\, dt, \qquad k = 1, 2, 3, \ldots$$

$$X[k] = \frac{1}{T_F} \int_{t_0}^{t_0+T_F} x(t) e^{-j(k\omega_F)t}\, dt, \qquad k = \ldots, = 3, -2, -1, 0, 1, 2, 3, \ldots \tag{4.90}$$

and $\omega_F = 2\pi f_F = 2\pi/T_F$. Note that, regardless of whether cyclic or radian frequency is used, $X[k]$ is the same.

In signal and system analysis, both the $f$ form and the $\omega$ form of the CTFS are used. In communication system analysis, Fourier optics, and image processing the $f$ form is usually used. In control system analysis the $\omega$ form is usually used. There are good reasons for either choice, depending on how the CTFS is used. As we progress through the transform methods in succeeding chapters, we will use whichever variable, $f$ or $\omega$, is more convenient and natural in any particular application always realizing that conversion from one form to the other can be done using the relationship $\omega = 2\pi f$.

## THE CONTINUOUS-TIME FOURIER SERIES OF A RANDOM SIGNAL

Let's consider one last example of a signal for which we might want to find the CTFS (Figure 4.23). This signal presents some problems. It is not at all obvious how to describe it, other than graphically. It is not sinusoidal, or any other obvious mathematical functional form. Up to this time in our study of the CTFS, in order to find a CTFS of a signal, we needed a mathematical description of it. But just because we cannot describe a signal mathematically does not mean it does not have a CTFS description. Most real signals which we might want to analyze in practice do not have a known exact mathematical description. We could, in principle, find it graphically as illustrated in Figure 4.24 for the fundamental.

If we examine the graph of the signal in Figure 4.24 closely, it may be apparent that there seems to be a tendency toward being periodic with 10 cycles in the time

**Figure 4.23**
A random signal.

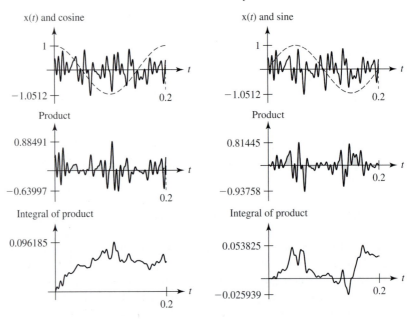

**Figure 4.24**
Graphical illustration of the steps involved in calculating $X_c[1]$ and $X_s[1]$ for a random signal.

shown. Calculating the 10th harmonic we get the results illustrated in Figure 4.25. The amplitude of the 10th sine harmonic component $X_s[10]$ is larger than the others, confirming our observation. A graph of the magnitude of the first 30 harmonics points this out in a clear way (Figure 4.26).

So there is a way to find the Fourier series coefficients of a signal for which we have no mathematical description. But there must be a better way! There is, but a full exploration of the better way will have to wait until we have considered the implications of sampling a signal in Chapter 7.

In everyday, practical Fourier analysis of signals, most CTFSs (and other Fourier transforms we will soon consider) are actually found, at least approximately, by using the discrete Fourier transform (Chapter 7) which depends on having a description of the signal in the form of a set of samples from it instead of a mathematical description of

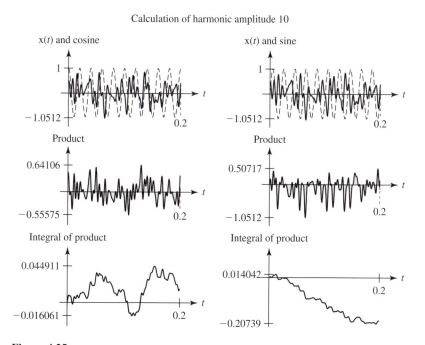

**Figure 4.25**
Graphical illustration of the steps involved in calculating $X_c[10]$ and $X_s[10]$ for a random signal.

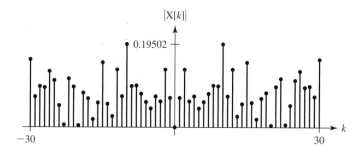

**Figure 4.26**
Magnitude of the harmonic amplitudes of the random signal.

it. The analytical form of the CTFS presented here has great value in that it can be used to help us develop an understanding of the frequency content of signals of certain shapes and the effects of LTI systems on those signals. The discrete Fourier transform has great value in that it can be used to help us determine the frequency content of real signals acquired experimentally which usually do not have a known mathematical description.

## 4.4 PROPERTIES OF THE CONTINUOUS-TIME FOURIER SERIES

Let the CTFS harmonic function of a periodic signal $x(t)$ with fundamental period $T_{0x}$ be $X[k]$ and let the CTFS harmonic function of a periodic signal $y(t)$ with fundamental period $T_{0y}$ be $Y[k]$. We can find the CTFS harmonic function for each of these two signals represented over exactly its fundamental period. The CTFS harmonic function is found from the general formula

$$X[k] = \frac{1}{T_F} \int_{t_0}^{t_0+T_F} x(t)e^{-j2\pi(kf_F)t} \, dt. \tag{4.91}$$

By assumption, for $x(t)$, $T_F = T_{0x}$, $f_F = f_{0x}$, and

$$X[k] = \frac{1}{T_{0x}} \int_{T_{0x}} x(t)e^{-j2\pi(kf_{0x})t} \, dt \tag{4.92}$$

and

$$x(t) = \sum_{k=-\infty}^{\infty} X[k]e^{j2\pi(kf_{0x})t}. \tag{4.93}$$

Similarly, for $y(t)$, $T_F = T_{0y}$, $f_F = f_{0y}$, and

$$Y[k] = \frac{1}{T_{0y}} \int_{T_{0y}} y(t)e^{-j2\pi(kf_{0y})t} \, dt \tag{4.94}$$

and

$$y(t) = \sum_{k=-\infty}^{\infty} Y[k]e^{j2\pi(kf_{0y})t}. \tag{4.95}$$

Then, using those signals as examples, in the following sections various properties of the CTFS of related signals will be introduced.

### LINEARITY

Let $z(t) = \alpha x(t) + \beta y(t)$. If $T_{0x} = T_{0y} = T_{0z} = T_0$, then the CTFS harmonic function of $z(t)$ is

$$Z[k] = \frac{1}{T_0} \int_{T_0} z(t)e^{-j2\pi(kf_0)t} \, dt = \frac{1}{T_0} \int_{T_0} [\alpha x(t) + \beta y(t)]e^{-j2\pi(kf_0)t} \, dt \tag{4.96}$$

and

$$z(t) = \sum_{n=-\infty}^{\infty} Z[k]e^{j2\pi(kf_0)t} = \sum_{n=-\infty}^{\infty} (\alpha X[k] + \beta Y[k])e^{j2\pi(kf_0)t} \tag{4.97}$$

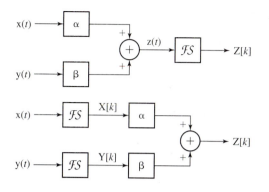

**Figure 4.27**
Linearity property of the Fourier series.

and

$$Z[k] = \alpha X[k] + \beta Y[k] \qquad (4.98)$$

or

$$\boxed{\alpha x(t) + \beta y(t) \xleftrightarrow{\;\mathcal{FS}\;} \alpha X[k] + \beta Y[k].} \qquad (4.99)$$

(If the fundamental periods of x and y are not the same, we must use the change-of-period property (4.139) to find the harmonic function for z.)

From this property we can say that if we think of the process of finding the CTFS harmonic function of a signal as a *system* whose excitation is $x(t)$ and whose response is $X[k]$, the system is linear (Figure 4.27). This is a very important property of the Fourier series because it is used so much in analysis. It is such a natural property that we usually don't even think about it when we use it (if the fundamental periods of x and y are the same).

## TIME SHIFTING

Let $z(t) = x(t - t_0)$ and $T_{0x} = T_{0z} = T_0$. Then

$$Z[k] = \frac{1}{T_0} \int_{T_0} z(t) e^{-j2\pi(kf_0)t} \, dt = \frac{1}{T_0} \int_{T_0} x(t - t_0) e^{-j2\pi(kf_0)t} \, dt. \qquad (4.100)$$

Making the change of variable $\lambda = t - t_0 \Rightarrow d\lambda = dt$,

$$Z[k] = \frac{1}{T_0} \int_{T_0} x(\lambda) e^{-j2\pi(kf_0)(\lambda + t_0)} \, d\lambda = e^{-j2\pi(kf_0)t_0} \frac{1}{T_0} \int_{T_0} x(\lambda) e^{-j2\pi(kf_0)\lambda} \, d\lambda \qquad (4.101)$$

$$Z[k] = e^{-j2\pi(kf_0)t_0} X[k] \qquad (4.102)$$

or

$$\boxed{x(t - t_0) \xleftrightarrow{\;\mathcal{FS}\;} e^{-j2\pi(kf_0)t_0} X[k]} \qquad (4.103)$$

$$\boxed{x(t - t_0) \xleftrightarrow{\;\mathcal{FS}\;} e^{-j(k\omega_0)t_0} X[k]}. \qquad (4.104)$$

The time-shifting property indicates that a shift of a function in time corresponds to multiplication of the CTFS harmonic function by a complex function of harmonic number $k$. To illustrate why that makes physical sense, consider a very simple time function

$$x(t) = \cos(2\pi t). \tag{4.105}$$

Its CTFS harmonic function (over exactly one fundamental period) is

$$X[k] = \frac{1}{2}(\delta[k+1] + \delta[k-1]). \tag{4.106}$$

Now let's shift the function $x(t)$ by a time delay $t_0$ to form a new function $z(t)$. The time-shifting property indicates that the CTFS harmonic function of $z(t)$ should be

$$Z[k] = \frac{e^{-j2\pi k t_0}}{2}(\delta[k+1] + \delta[k-1]) = \frac{1}{2}(e^{+j2\pi t_0}\delta[k+1] + e^{-j2\pi t_0}\delta[k-1]) \tag{4.107}$$

or

$$Z[k] = \frac{1}{2}\{[\cos(2\pi t_0) + j\sin(2\pi t_0)]\delta[k+1] + [\cos(2\pi t_0) - j\sin(2\pi t_0)]\delta[k-1]\}. \tag{4.108}$$

To verify that (4.108) makes sense, first let $t_0 = 0$. Then

$$z(t) = x(t) = \cos(2\pi t) \quad \text{and} \quad Z[k] = \frac{1}{2}(\delta[k+1] + \delta[k-1]), \tag{4.109}$$

which is as it should be. Then let $t_0 = \frac{1}{4}$.

$$z(t) = x\left(t - \frac{1}{4}\right) = \cos\left(2\pi\left(t - \frac{1}{4}\right)\right) \tag{4.110}$$

and

$$Z[k] = \frac{1}{2}\left(e^{+j(\pi/2)}\delta[k+1] + e^{-j(\pi/2)}\delta[k-1]\right) \tag{4.111}$$

or

$$Z[k] = \frac{j}{2}(\delta[k+1] - \delta[k-1]). \tag{4.112}$$

Then the CTFS representation of the time-shifted signal would be

$$z_F(t) = \sum_{k=-\infty}^{\infty} \frac{j}{2}(\delta[k+1] - \delta[k-1])e^{j2\pi(kf_0)t} = \frac{j}{2}(e^{-j2\pi f_0 t} - e^{j2\pi f_0 t}) \tag{4.113}$$

or

$$z_F(t) = \sin(2\pi t). \tag{4.114}$$

This is correct because a delay in a 1-Hz cosine of $\frac{1}{4}$ s produces a 1-Hz sine. So the multiplication of the CTFS harmonic function by the complex constant $e^{-j2\pi(kf_0)t_0}$ adjusts the phase of every sinusoidal component of the CTFS representation so that the time delay of each sinusoidal component is exactly $t_0$. At higher harmonics of the fundamental, the same time delay corresponds to a greater phase shift. A time delay of $\frac{1}{4}$ second is equivalent to a phase shift of $\pi/2$ radians at the fundamental frequency of 1 Hz, but it is equivalent to a phase shift of $\pi$ radians at the second harmonic frequency of 2 Hz and $2\pi$ radians at the fourth harmonic frequency of 4 Hz, etc. This describes a linear dependence of phase shift on harmonic number, which is exactly what the factor $e^{-j2\pi(kf_0)t_0}$ does to the CTFS harmonic function.

## FREQUENCY SHIFTING

Let $z(t) = e^{j2\pi(k_0 f_0)t} x(t)$, where $k_0$ is an integer, and let $T_{0x} = T_{0z} = T_0$. Then

$$Z[k] = \frac{1}{T_0} \int_{T_0} z(t) e^{-j2\pi(kf_0)t} \, dt = \frac{1}{T_0} \int_{T_0} e^{j2\pi(k_0 f_0)t} x(t) e^{-j2\pi(kf_0)t} \, dt \qquad \textbf{(4.115)}$$

$$Z[k] = \frac{1}{T_0} \int_{T_0} x(t) e^{-j2\pi((k-k_0)f_0)t} \, dt \qquad \textbf{(4.116)}$$

$$Z[k] = X[k - k_0] \qquad \textbf{(4.117)}$$

and

$$\boxed{e^{j2\pi(k_0 f_0)t} x(t) \overset{\mathcal{FS}}{\longleftrightarrow} X[k - k_0]} \qquad \textbf{(4.118)}$$

$$\boxed{e^{j(k_0 \omega_0)t} x(t) \overset{\mathcal{FS}}{\longleftrightarrow} X[k - k_0]}. \qquad \textbf{(4.119)}$$

(This property is commonly called the *frequency-shifting* property even though it should more precisely be called the *harmonic-number–shifting* property since $k$ is harmonic number. But, since the product of harmonic number and fundamental frequency is frequency, the name still has meaning and it is easier to say.) Notice the duality between the time-shifting and frequency-shifting properties. Shifting in one domain corresponds to multiplication by a complex exponential in the other domain.

A particularly important application of the frequency-shifting property occurs when a CTFS harmonic function is frequency-shifted both up and down by the same amount and added together. Let

$$Z[k] = X[k - k_0] + X[k + k_0]. \qquad \textbf{(4.120)}$$

Then, using the linearity and frequency-shifting properties,

$$z(t) = x(t) e^{j2\pi(k_0 f_0)t} + x(t) e^{-j2\pi(k_0 f_0)t} = 2x(t) \cos(2\pi(k_0 f_0)t). \qquad \textbf{(4.121)}$$

So shifting up and down by equal amounts in frequency (harmonic number) is equivalent to multiplication by a cosine of that frequency (harmonic number) in the time domain. This operation is called *modulation* and is very important in communication systems (Figure 4.28).

## TIME REVERSAL

Let $z(t) = x(-t)$ and let $T_{0x} = T_{0z} = T_0$. If

$$x(t) = \sum_{k=-\infty}^{\infty} X[k] e^{j2\pi(kf_0)t}, \qquad \textbf{(4.122)}$$

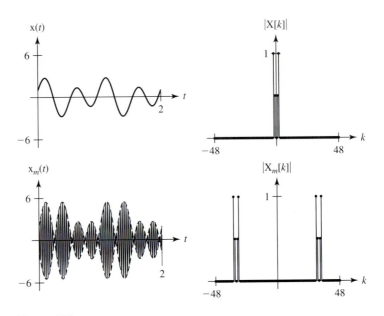

**Figure 4.28**
A periodic signal and that signal multiplied by a cosine of amplitude 2, at 32 times its fundamental frequency and the CTFS harmonic functions of both.

then

$$x(-t) = \sum_{k=-\infty}^{\infty} X[k]e^{j2\pi(kf_0)(-t)} = \sum_{k=-\infty}^{\infty} X[k]e^{j2\pi(-kf_0)t}. \tag{4.123}$$

Let $q = -k$; then

$$x(-t) = \sum_{q=\infty}^{-\infty} X[-q]e^{j2\pi(qf_0)t} \tag{4.124}$$

and, since the order of summation does not matter,

$$x(-t) = \sum_{q=-\infty}^{\infty} X[-q]e^{j2\pi(qf_0)t}. \tag{4.125}$$

Therefore, since

$$z(t) = \sum_{k=-\infty}^{\infty} Z[k]e^{j2\pi(kf_0)t}, \tag{4.126}$$

we can say that

$$\sum_{k=-\infty}^{\infty} Z[k]e^{j2\pi(kf_0)t} = \sum_{k=-\infty}^{\infty} X[-k]e^{j2\pi(kf_0)t} \tag{4.127}$$

and

$$Z[k] = X[-k] \tag{4.128}$$

or

$$\boxed{x(-t) \xleftarrow{\;\mathcal{FS}\;} X[-k]}. \tag{4.129}$$

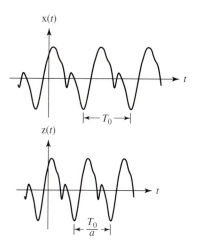

**Figure 4.29**
A signal $x(t)$ and a time-scaled
version $z(t)$ of that signal.

So time reversal of a time function corresponds to harmonic-number reversal in its CTFS harmonic function.

## TIME SCALING

Let $z(t) = x(at)$, with $a > 0$, and let $T_{0x} = T_0$ (Figure 4.29). The first thing to realize is that if $x(t)$ is periodic with fundamental period $T_{0x} = T_0$, then $z(t)$ is periodic with fundamental period $T_{0z} = T_0/a$ and fundamental frequency $af_0$.

***Case I.*** $z(t)$ represented by a CTFS over the fundamental period of $z(t)$, $T_0/a = T_F$. The CTFS harmonic function will be

$$Z[k] = \frac{a}{T_0} \int_{t_0}^{t_0+(T_0/a)} z(t)e^{-j2\pi(kaf_0)t}\,dt = \frac{a}{T_0} \int_{t_0}^{t_0+(T_0/a)} x(at)e^{-j2\pi(kaf_0)t}\,dt. \qquad \textbf{(4.130)}$$

We can make the change of variable $\lambda = at \Rightarrow d\lambda = a\,dt$ in (4.130).

$$Z[k] = \frac{a}{T_0}\frac{1}{a} \int_{at_0}^{at_0+T_0} x(\lambda)e^{-j2\pi(kaf_0)(\lambda/a)}\,d\lambda = \frac{1}{T_0} \int_{at_0}^{at_0+T_0} x(\lambda)e^{-j2\pi(kf_0)\lambda}\,d\lambda \qquad \textbf{(4.131)}$$

Since the starting point $t_0$ is arbitrary,

$$Z[k] = \frac{1}{T_0}\int_{T_0} x(\lambda)e^{-j2\pi(kf_0)\lambda}\,d\lambda = X[k] \qquad \textbf{(4.132)}$$

and the CTFS harmonic function describing $z(t)$ over its fundamental period $T_0/a$ is the same as the CTFS harmonic function describing $x(t)$ over its fundamental period $T_0$.

Even though the CTFS harmonic functions of $x(t)$ and $z(t)$ are the same, the CTFS representations themselves are not because the fundamental frequencies are different. The representations are

$$x(t) = \sum_{k=-\infty}^{\infty} X[k]e^{j2\pi(kf_0)t} \qquad \text{and} \qquad z(t) = x(at) = \sum_{k=-\infty}^{\infty} Z[k]e^{j2\pi(akf_0)t}. \qquad \textbf{(4.133)}$$

*Case II.*   $z(t)$ represented by a CTFS over the fundamental period of $x(t)$, $T_0 = T_F$. The CTFS harmonic function will be

$$Z[k] = \frac{1}{T_0} \int_{t_0}^{t_0+T_0} z(t)e^{-j2\pi(kf_0)t}\, dt = \frac{1}{T_0} \int_{t_0}^{t_0+T_0} x(at)e^{-j2\pi(kf_0)t}\, dt. \tag{4.134}$$

Let $\lambda = at \Rightarrow d\lambda = a\,dt$. Then

$$Z[k] = \frac{1}{aT_0} \int_{at_0}^{at_0+aT_0} x(\lambda)e^{-j2\pi[k(f_0/a)]\lambda}\, d\lambda. \tag{4.135}$$

If $a$ is not an integer, the relationship between the two harmonic functions $Z[k]$ and $X[k]$ cannot be simplified further.

Let $a$ be an integer. The signal $x(\lambda)$ is made up of frequency components at integer multiples of its fundamental frequency $f_0$. Therefore, for ratios $k/a$ that are not integers $x(\lambda)$ and $e^{-j2\pi[k(f_0/a)]\lambda}$ are orthogonal on the interval $at_0 < \lambda < at_0 + aT_0$ and $Z[k] = 0$. For ratios $k/a$ that are integers the integral over $a$ periods is $a$ times the integral over one period,

$$Z[k] = a\left(\frac{1}{aT_0} \int_{at_0}^{at_0+T_0} x(\lambda)e^{-j2\pi[k(f_0/a)]\lambda}\, d\lambda\right) = X\left[\frac{k}{a}\right] \tag{4.136}$$

where $k/a$ is an integer.

Summarizing, where $a$ is an integer,

$$Z[k] = \begin{cases} X\left[\dfrac{k}{a}\right] & \dfrac{k}{a} \text{ is an integer} \\[2mm] 0 & \text{otherwise} \end{cases}. \tag{4.137}$$

Typical harmonic functions for $a = 2$ are shown in Figure 4.30.

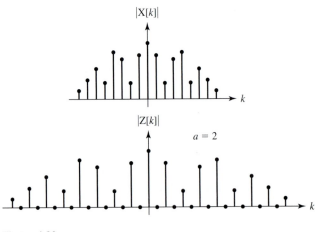

**Figure 4.30**
Comparison of $|X[k]|$ and $|Z[k]|$ for $a = 2$.

## CHANGE OF REPRESENTATION PERIOD

If the CTFS harmonic function of $x(t)$ over its fundamental period $T_{x0} = T_0$ is $X[k]$, we can find the CTFS harmonic function $X_m[k]$ of $x(t)$ over a time $T_F = mT_0$, which is a positive integer multiple $m$ of that period. The new fundamental CTFS frequency is then $f_F = f_0/m$ and

$$X_m[k] = \frac{1}{mT_0} \int_{mT_0} x(t) e^{-j2\pi[(k/m)f_0]t} \, dt. \tag{4.138}$$

This is exactly the same as the result for time scaling by a positive integer in the previous section, and the result is that

$$X_m[k] = \begin{cases} X\left[\dfrac{k}{m}\right] & \dfrac{k}{m} \text{ is an integer} \\ 0 & \text{otherwise} \end{cases}. \tag{4.139}$$

**EXAMPLE 4.1**

For a signal

$$x(t) = \frac{1}{2} \text{rect}\left(\frac{t}{2}\right) * \frac{1}{4} \text{comb}\left(\frac{t}{4}\right) \tag{4.140}$$

find the harmonic function and CTFS representation over one fundamental period. Then find the harmonic function and CTFS representation over three fundamental periods.

■ **Solution**

In the table of Fourier pairs, Appendix E, we find the harmonic function for a rectangular wave

$$\text{rect}\left(\frac{t}{w}\right) * \frac{1}{T_0} \text{comb}\left(\frac{t}{T_0}\right) \xleftrightarrow{\mathcal{FS}} \frac{w}{T_0} \text{sinc}\left(\frac{w}{T_0}k\right), \tag{4.141}$$

where $T_0$ is the fundamental period of $x(t)$. Therefore, the harmonic function corresponding to the fundamental period is

$$X[k] = \frac{1}{4} \text{sinc}\left(\frac{k}{2}\right) \tag{4.142}$$

and the CTFS representation of the signal is

$$x(t) = \sum_{k=-\infty}^{\infty} X[k] e^{j2\pi(kf_0)t} = \frac{1}{4} \sum_{k=-\infty}^{\infty} \text{sinc}\left(\frac{k}{2}\right) e^{j(\pi k/2)t}. \tag{4.143}$$

If we now find the harmonic function corresponding to the period $3T_0$, using the change-of-period property we get

$$X_3[k] = \begin{cases} \dfrac{1}{4} \text{sinc}\left(\dfrac{k}{6}\right) & \dfrac{k}{3} \text{ is an integer} \\ 0 & \text{otherwise} \end{cases}. \tag{4.144}$$

This can be written more compactly as

$$X_3[k] = \frac{1}{4} \text{sinc}\left(\frac{k}{6}\right) \text{comb}_3[k]. \tag{4.145}$$

The CTFS representation is

$$x(t) = \sum_{k=-\infty}^{\infty} X_m[k]e^{j2\pi((kf_0)/m)t} = \frac{1}{4} \sum_{k=-\infty}^{\infty} \text{sinc}\left(\frac{k}{6}\right) \text{comb}_3[k]e^{j(\pi k/6)t}. \qquad \textbf{(4.146)}$$

These CTFS representations are equal because for the complex exponential function for every $k$ in the summation $\sum_{k=-\infty}^{\infty} \text{sinc}(k/2)e^{j(\pi k/2)t}$, there is a corresponding complex exponential function, for $3k$, in $\sum_{k=-\infty}^{\infty} \text{sinc}(k/6)\text{scomb}_3[k]e^{j(\pi k/6)t}$ which is exactly the same function, and all the in-between values of $k$ in $\sum_{k=-\infty}^{\infty} \text{sinc}(k/6) \text{comb}_3[k]e^{j(\pi k/6)t}$ have no effect because of the comb function. That is,

$$\frac{1}{4} \sum_{k=-\infty}^{\infty} \text{sinc}(k/2)e^{j(\pi k/2)t} = \frac{1}{4} \sum_{k=-\infty}^{\infty} \text{sinc}(k/6)\text{comb}_3[k]e^{j(\pi k/6)t} \qquad \textbf{(4.147)}$$

which must be true if both CTFS representations are to be correct. ■

## TIME DIFFERENTIATION

Let $z(t) = (d/dt)(x(t))$ and let $T_{0x} = T_{0z} = T_0$. Then we can represent $z(t)$ by

$$z(t) = \frac{d}{dt}(x(t)) = \frac{d}{dt}\left(\sum_{n=-\infty}^{\infty} X[k]e^{j2\pi(kf_0)t}\right) = \sum_{n=-\infty}^{\infty} j2\pi(kf_0)X[k]e^{j2\pi(kf_0)t}. \qquad \textbf{(4.148)}$$

Then, if

$$z(t) = \sum_{k=-\infty}^{\infty} Z[k]e^{j2\pi(kf_0)t}, \qquad \textbf{(4.149)}$$

it follows that

$$\sum_{k=-\infty}^{\infty} Z[k]e^{j2\pi(kf_0)t} = j2\pi(kf_0) \sum_{k=-\infty}^{\infty} X[k]e^{j2\pi(kf_0)t} \qquad \textbf{(4.150)}$$

and

$$Z[k] = j2\pi(kf_0)X[k] \qquad \textbf{(4.151)}$$

and

$$\boxed{\frac{d}{dt}(x(t)) \xleftarrow{\;\;\mathcal{FS}\;\;} j2\pi(kf_0)X[k]} \qquad \textbf{(4.152)}$$

$$\boxed{\frac{d}{dt}(x(t)) \xleftarrow{\;\;\mathcal{FS}\;\;} j(k\omega_0)X[k]}. \qquad \textbf{(4.153)}$$

So differentiation of a time function corresponds to a multiplication of its CTFS harmonic function by an imaginary number whose value is a linear function of the harmonic number. This is a handy feature because it converts differentiation in the time domain into multiplication by a complex number in the frequency (harmonic-number) domain. This feature in Fourier and Laplace transform methods (presented in Chapters 5 and 9) is one reason they are so powerful in the solution of systems described by differential equations. Differential equations are converted to algebraic equations by the transformation process.

## TIME INTEGRATION

Let $z(t) = \int_{-\infty}^{t} x(\lambda)\, d\lambda$. We must consider two cases separately, $X[0] = 0$ and $X[0] \neq 0$. If $X[0] \neq 0$, then, even though $x(t)$ is periodic, $z(t)$ is not and we cannot represent it exactly for all time with a CTFS (Figure 4.31). If $X[0] = 0$, then we can represent $z(t)$ by

$$z(t) = \int_{-\infty}^{t} x(\lambda)\, d\lambda = \int_{-\infty}^{t} \sum_{k=-\infty}^{\infty} X[k] e^{j2\pi(kf_0)\lambda}\, d\lambda = \sum_{k=-\infty}^{\infty} X[k] \int_{-\infty}^{t} e^{j2\pi(kf_0)\lambda}\, d\lambda$$

$$= \sum_{k=-\infty}^{\infty} X[k] \frac{e^{j2\pi(kf_0)t}}{j2\pi(kf_0)} \qquad\qquad (4.154)$$

Then, if

$$z(t) = \sum_{k=-\infty}^{\infty} Z[k] e^{j2\pi(kf_0)t}, \qquad\qquad (4.155)$$

it follows that

$$Z[k] = \frac{X[k]}{j2\pi(kf_0)} \qquad\qquad (4.156)$$

and

$$\boxed{\int_{-\infty}^{t} x(\lambda)\, d\lambda \overset{\mathcal{FS}}{\longleftrightarrow} \frac{X[k]}{j2\pi(kf_0)} \qquad \text{if } X[0] = 0} \qquad\qquad (4.157)$$

$$\boxed{\int_{-\infty}^{t} x(\lambda)\, d\lambda \overset{\mathcal{FS}}{\longleftrightarrow} \frac{X[k]}{j(k\omega_0)} \qquad \text{if } X[0] = 0} . \qquad\qquad (4.158)$$

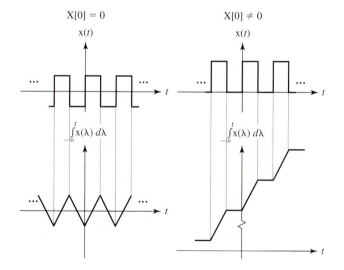

**Figure 4.31**
Effect of a nonzero average value on the integral of a periodic function.

Just as differentiation corresponds to multiplication by an imaginary number proportional to harmonic number, integration (of a function whose average value is zero) corresponds to division by an imaginary number proportional to harmonic number.

## MULTIPLICATION–CONVOLUTION DUALITY

Let $z(t) = x(t)y(t)$ and let $T_{0x} = T_{0y} = T_0$. Then

$$Z[k] = \frac{1}{T_0} \int_{T_0} z(t) e^{-j2\pi(kf_0)t} \, dt = \frac{1}{T_0} \int_{T_0} x(t)y(t) e^{-j2\pi(kf_0)t} \, dt. \tag{4.159}$$

Then, using

$$y(t) = \sum_{k=-\infty}^{\infty} Y[k] e^{j2\pi(kf_0)t} = \sum_{q=-\infty}^{\infty} Y[q] e^{j2\pi(qf_0)t} \tag{4.160}$$

in (4.159),

$$Z[k] = \frac{1}{T_0} \int_{T_0} x(t) \left( \sum_{q=-\infty}^{\infty} Y[q] e^{j2\pi(qf_0)t} \right) e^{-j2\pi(kf_0)t} \, dt. \tag{4.161}$$

Reversing the order of integration and summation,

$$Z[k] = \frac{1}{T_0} \sum_{q=-\infty}^{\infty} Y[q] \int_{T_0} x(t) e^{j2\pi(qf_0)t} e^{-j2\pi(kf_0)t} \, dt \tag{4.162}$$

or

$$Z[k] = \sum_{q=-\infty}^{\infty} Y[q] \underbrace{\frac{1}{T_0} \int_{T_0} x(t) e^{-j2\pi((k-q)f_0)t} \, dt}_{=X[k-q]}. \tag{4.163}$$

Then

$$Z[k] = \sum_{q=-\infty}^{\infty} Y[q]X[k-q] \tag{4.164}$$

or

$$\boxed{x(t)y(t) \xleftrightarrow{\;\mathcal{FS}\;} \sum_{q=-\infty}^{\infty} Y[q]X[k-q] = X[k] * Y[k]}. \tag{4.165}$$

This result, $\sum_{q=-\infty}^{\infty} Y[q]X[k-q]$, is a convolution sum. So multiplying CT signals corresponds to convolving their CTFS harmonic functions.

If $T_{0x} \neq T_{0y}$, then we must first find a period $T_0$ which is common to both x and y (the smallest value being the least common multiple of the two fundamental periods). Then, if we find the harmonic functions of x and y using this common period and the change-of-period property (4.139), (4.165) applies to these harmonic functions with that common period.

Now let $Z[k] = X[k]Y[k]$ and $T_{0x} = T_{0y} = T_{0z} = T_0$. Then

$$z(t) = \sum_{k=-\infty}^{\infty} X[k]Y[k]e^{j2\pi(kf_0)t}$$

$$= \sum_{k=-\infty}^{\infty} \frac{1}{T_0} \int_{T_0} x(\tau)e^{-j2\pi(kf_0)\tau} \, d\tau \, Y[k]e^{j2\pi(kf_0)t}$$

$$= \frac{1}{T_0} \int_{T_0} x(\tau) \, d\tau \underbrace{\sum_{k=-\infty}^{\infty} Y[k]e^{j2\pi(kf_0)(t-\tau)}}_{y(t-\tau)}$$

$$= \frac{1}{T_0} \int_{T_0} x(\tau)y(t-\tau) \, d\tau. \tag{4.166}$$

This integral looks just like a convolution integral except that it covers the range $t_0 < \tau < t_0 + T_0$ instead of $-\infty < \tau < \infty$ and is divided by $T_0$. This integral operation is called *periodic convolution* and is symbolized by

$$x(t) \circledast y(t) = \int_{T_0} x(\tau)y(t-\tau) \, d\tau. \tag{4.167}$$

Therefore,

$$z(t) = \frac{1}{T_0} x(t) \circledast y(t). \tag{4.168}$$

Since $x(t)$ is periodic, it can be expressed as the periodic extension of an aperiodic function $x_{ap}(t)$,

$$x(t) = \sum_{m=-\infty}^{\infty} x_{ap}(t - mT_0) = x_{ap}(t) * \frac{1}{T_0} \text{comb}\left(\frac{t}{T_0}\right). \tag{4.169}$$

[The function $x_{ap}(t)$ is not unique. It can be any function which satisfies (4.169).] Then

$$x(t) \circledast y(t) = \int_{T_0} \left[ \sum_{m=-\infty}^{\infty} x_{ap}(\tau - mT_0) \right] y(t - \tau) \, d\tau \tag{4.170}$$

$$x(t) \circledast y(t) = \sum_{m=-\infty}^{\infty} \int_{t_0}^{t_0+T_0} x_{ap}(\tau - mT_0)y(t - \tau) \, d\tau \tag{4.171}$$

Let $\lambda = \tau - mT_0$. Then $d\lambda = d\tau$ and

$$x(t) \circledast y(t) = \sum_{m=-\infty}^{\infty} \int_{t_0+mT_0}^{t_0+(m+1)T_0} x_{ap}(\lambda)y(t - (\lambda + mT_0)) \, d\lambda. \tag{4.172}$$

Since $y(t)$ is periodic, with fundamental period $T_0$,

$$y(t - (\lambda + mT_0)) = y(t - mT_0 - \lambda) = y(t - \lambda) \qquad \textbf{(4.173)}$$

and the summation of integrals $\sum_{m=-\infty}^{\infty} \int_{t_0+mT_0}^{t_0+mT_0+T_0}$ is equivalent to the single integral over infinite limits $\int_{-\infty}^{\infty}$, we conclude that

$$x(t) \circledast y(t) = \int_{-\infty}^{\infty} x_{ap}(\lambda) y(t - \lambda)\, d\lambda = x_{ap}(t) * y(t). \qquad \textbf{(4.174)}$$

So the periodic convolution of two functions $x(t)$ and $y(t)$ each with fundamental period $T_0$ can be expressed as an aperiodic convolution of $y(t)$ with a function $x_{ap}(t)$, which, when periodically repeated with the same fundamental period $T_0$, equals $x(t)$. The periodic convolution of two periodic functions corresponds to the product of their CTFS harmonic function representations and the fundamental period $T_0$,

$$\boxed{x(t) \circledast y(t) \xleftrightarrow{\mathcal{FS}} T_0 X[k] Y[k]} \qquad \textbf{(4.175)}$$

If $T_{0x} \neq T_{0y}$, then we must find a common period $T_0$ for the two signals. If a common period can be found, then (4.175) applies to the harmonic functions found using that common period.

## CONJUGATION

Let $z(t) = x^*(t)$ and $T_{0x} = T_{0z} = T_0$. Then

$$\sum_{k=-\infty}^{\infty} Z[k] e^{j2\pi(kf_0)t} = \left( \sum_{k=-\infty}^{\infty} X[k] e^{j2\pi(kf_0)t} \right)^*$$

$$= \sum_{k=-\infty}^{\infty} X^*[k] e^{-j2\pi(kf_0)t} = \sum_{k=\infty}^{-\infty} X^*[-k] e^{j2\pi(kf_0)t} \qquad \textbf{(4.176)}$$

and, since changing the order of summation does not change the sum,

$$\sum_{k=-\infty}^{\infty} Z[k] e^{j2\pi(kf_0)t} = \sum_{k=-\infty}^{\infty} X^*[-k] e^{j2\pi(kf_0)t} \qquad \textbf{(4.177)}$$

and

$$Z[k] = X^*[-k] \qquad \textbf{(4.178)}$$

and

$$\boxed{x^*(t) \xleftrightarrow{\mathcal{FS}} X^*[-k]} \qquad \textbf{(4.179)}$$

as was shown earlier.

## PARSEVAL'S THEOREM

The signal energy in any single fundamental period $T_{0x} = T_0$ of any periodic signal $x(t)$ is

$$E_{x,T_0} = \int_{T_0} |x(t)|^2 \, dt = \int_{T_0} \left| \sum_{k=-\infty}^{\infty} X[k] e^{j2\pi(kf_0)t} \right|^2 dt$$

$$= \int_{T_0} \left( \sum_{k=-\infty}^{\infty} X[k] e^{j2\pi(kf_0)t} \right) \left( \sum_{q=-\infty}^{\infty} X[q] e^{j2\pi(qf_0)t} \right)^* dt$$

$$= \int_{T_0} \left( \sum_{k=-\infty}^{\infty} \sum_{q=-\infty}^{\infty} X[k] e^{j2\pi(kf_0)t} X^*[q] e^{-j2\pi(qf_0)t} \right) dt$$

$$= \int_{T_0} \left( \sum_{k=-\infty}^{\infty} \sum_{q=-\infty}^{\infty} X[k] X^*[q] e^{j2\pi((k-q)f_0)t} \right) dt$$

$$= \int_{T_0} \left( \sum_{k=-\infty}^{\infty} X[k] X^*[k] + \underset{k\neq q}{\sum_{k=-\infty}^{\infty} \sum_{q=-\infty}^{\infty}} X[k] X^*[q] e^{j2\pi((k-q)f_0)t} \right) dt$$

$$= \int_{T_0} \sum_{k=-\infty}^{\infty} |X[k]|^2 \, dt + \underbrace{\int_{T_0} \underset{k\neq q}{\sum_{k=-\infty}^{\infty} \sum_{q=-\infty}^{\infty}} X[k] X^*[q] e^{j2\pi((k-q)f_0)t} \, dt}_{=0, \, k\neq q}$$

$$= T_0 \sum_{k=-\infty}^{\infty} |X[k]|^2. \qquad (4.180)$$

Therefore, for any periodic signal $x(t)$,

$$\boxed{\frac{1}{T_0} \int_{T_0} |x(t)|^2 \, dt = \sum_{k=-\infty}^{\infty} |X[k]|^2} \cdot \qquad (4.181)$$

The quantity on the left side of (4.181) is the average power of the signal $x(t)$; therefore, the quantity on the right must be also. Therefore, (4.181) just says that the average power of a periodic signal is the sum of the average powers in its harmonic components.

## SUMMARY OF CTFS PROPERTIES

Linearity $\qquad \alpha x(t) + \beta y(t) \xleftrightarrow{\mathcal{FS}} \alpha X[k] + \beta Y[k]$

Time shifting $\qquad x(t - t_0) \xleftrightarrow{\mathcal{FS}} e^{-j2\pi(kf_0)t_0} X[k]$

Time reversal $\qquad x(-t) \xleftrightarrow{\mathcal{FS}} X[-k]$

Time scaling $\qquad$ If $z(t) = x(at)$, $a > 0$, then

$\qquad\qquad\qquad$ *a.* On the period $T_F = T_0/a$:

$$Z[k] = X[k]$$

$$\text{and} \quad z(t) = x(at) = \sum_{k=-\infty}^{\infty} Z[k] e^{j2\pi(akf_0)t}.$$

b. On the period $T_F = T_0$:

$$Z[k] = \begin{cases} X\left[\dfrac{k}{a}\right] & \dfrac{k}{a} \text{ is an integer} \\ 0 & \text{otherwise} \end{cases}$$

and   $z(t) = x(at) = \displaystyle\sum_{k=-\infty}^{\infty} Z[k]e^{j2\pi(kf_0)t}$ .

**Change of representation period**   On the period $T_F = mT_0$, where $m$ is a positive integer,

$$X_m[k] = \begin{cases} X\left[\dfrac{k}{m}\right] & \dfrac{k}{m} \text{ is an integer} \\ 0 & \text{otherwise} \end{cases}$$

**Time differentiation**   $\dfrac{d}{dt}(x(t)) \overset{\mathcal{FS}}{\longleftrightarrow} j2\pi(kf_0)X[k]$

**Time integration**   $\displaystyle\int_{-\infty}^{t} x(\lambda)\,d\lambda \overset{\mathcal{FS}}{\longleftrightarrow} \dfrac{X[k]}{j2\pi(kf_0)}$   if $X[0] = 0$

**Multiplication–convolution duality**   $x(t)y(t) \overset{\mathcal{FS}}{\longleftrightarrow} \displaystyle\sum_{q=-\infty}^{\infty} Y[q]X[k-q] = X[k] * Y[k]$

$x(t) \circledast y(t) = \displaystyle\int_{T_0} x(\tau)y(t-\tau)\,d\tau \overset{\mathcal{FS}}{\longleftrightarrow} T_0 X[k]Y[k]$

**Conjugation**   $x^*(t) \overset{\mathcal{FS}}{\longleftrightarrow} X^*[-k]$

**Parseval's theorem**   $\dfrac{1}{T_0}\displaystyle\int_{T_0} |x(t)|^2\,dt = \displaystyle\sum_{k=-\infty}^{\infty} |X[k]|^2$

Appendix E is a table which includes CTFS pairs for a few basic signals.

## 4.5  USE OF TABLES AND PROPERTIES

By using a table of CTFS pairs, one can, in many cases, avoid the use of the integral definition to find a CTFS harmonic function.

**EXAMPLE 4.2**

Find the CTFS harmonic function of $x(t) = \cos(50\pi t - (\pi/4))$ with $T_F = T_0$ .

■ **Solution**

From the CTFS pairs in Appendix E we find

$$\cos(2\pi(mf_0)t) \overset{\mathcal{FS}}{\longleftrightarrow} \frac{1}{2}(\delta[k-m] + \delta[k+m]). \tag{4.182}$$

We can use (4.182) along with the time-shifting property

$$x(t - t_0) \xleftrightarrow{\;\mathcal{FS}\;} e^{-j2\pi(kf_0)t_0}X[k] \qquad \textbf{(4.183)}$$

to find this CTFS harmonic function. First recognize that $x(t)$ can be written as

$$x(t) = \cos\left(50\pi\left(t - \frac{1}{200}\right)\right). \qquad \textbf{(4.184)}$$

Then, since $f_0 = 25$ [expanding the CTFS over one fundamental period of $x(t)$],

$$\cos(50\pi t) \xleftrightarrow{\;\mathcal{FS}\;} \frac{1}{2}(\delta[k - 1] + \delta[k + 1]). \qquad \textbf{(4.185)}$$

Applying the time-shifting property,

$$\cos\left(50\pi\left(t - \frac{1}{200}\right)\right) \xleftrightarrow{\;\mathcal{FS}\;} e^{-j(\pi k/4)}\frac{1}{2}(\delta[k - 1] + \delta[k + 1]) \qquad \textbf{(4.186)}$$

or

$$\cos\left(50\pi\left(t - \frac{1}{200}\right)\right) \xleftrightarrow{\;\mathcal{FS}\;} \frac{1}{2}\left(\delta[k - 1]e^{-j(\pi/4)} + \delta[k + 1]e^{+j(\pi/4)}\right) \qquad \textbf{(4.187)}$$

or

$$\cos\left(50\pi\left(t - \frac{1}{200}\right)\right) \xleftrightarrow{\;\mathcal{FS}\;} \frac{1}{2\sqrt{2}}\{(1 - j)\delta[k - 1] + (1 + j)\delta[k + 1]\} \qquad \textbf{(4.188)}$$

(Figure 4.32). ∎

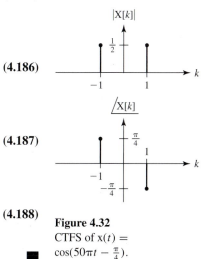

**Figure 4.32**
CTFS of $x(t) = \cos(50\pi t - \frac{\pi}{4})$.

**EXAMPLE 4.3**

Find the CTFS harmonic function of $x(t) = 5\cos(10\pi t)\cos(10{,}000\pi t)$ with $T_F = T_0$.

■ **Solution**
We can use

$$\cos(2\pi(mf_0)t) \xleftrightarrow{\;\mathcal{FS}\;} \frac{1}{2}(\delta[k - m] + \delta[k + m]) \qquad \textbf{(4.189)}$$

again along with the multiplication–convolution duality property

$$x(t)y(t) \xleftrightarrow{\;\mathcal{FS}\;} \sum_{q=-\infty}^{\infty} Y[q]X[k - q] = X[k] * Y[k] \qquad \textbf{(4.190)}$$

to find this CTFS harmonic function. The fundamental frequency of $x(t)$ is $f_0 = 5$.

$$5\cos(10\pi t) \xleftrightarrow{\;\mathcal{FS}\;} \frac{5}{2}(\delta[k - 1] + \delta[k + 1]) \qquad \textbf{(4.191)}$$

$$\cos(10{,}000\pi t) \xleftrightarrow{\;\mathcal{FS}\;} \frac{1}{2}(\delta[k - 1000] + \delta[k + 1000]). \qquad \textbf{(4.192)}$$

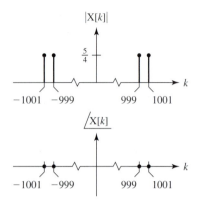

**Figure 4.33**
CTFS harmonic function of
$x(t) = 5\cos(10\pi t)\cos(10{,}000\pi t)$.

So, using the multiplication–convolution duality property,

$5\cos(10\pi t)\sin(10{,}000\pi t)$

$$\xleftarrow{\mathcal{FS}} \frac{5}{4}(\delta[k - 1000] + \delta[k + 1000]) * (\delta[k - 1] + \delta[k + 1])$$

$$\xleftarrow{\mathcal{FS}} \frac{5}{4}(\delta[k - 999] + \delta[k - 1001] + \delta[k + 999] + \delta[k + 1001]) \qquad (4.193)$$

(Figure 4.33). ∎

## EXAMPLE 4.4

Find the CTFS harmonic function of an even CT rectangular wave with a peak-to-peak ampli-tude of 15, a fundamental frequency of 100 Hz, a duty cycle of 10 percent, and an average value of zero with $T_F = T_0$.

■ **Solution**

First we need to describe this signal mathematically. It can be described by a rectangle convolved with a comb minus a constant to make the average value zero. The rectangle sets the width of each pulse, and the comb sets the fundamental period. The fundamental period is 10 ms, and 10 percent duty cycle means that the pulse width is 10 percent of the fundamental period, or 1 ms. Since the function is even, the rectangle is unshifted. So the mathematical description is

$$x(t) = [15\,\text{rect}(1000t) * 100\,\text{comb}(100t)] - \frac{3}{2}. \qquad (4.194)$$

From the CTFS pairs in Appendix E we can use

$$\text{rect}\left(\frac{t}{w}\right) * \frac{1}{T_0}\,\text{comb}\left(\frac{t}{T_0}\right) \xleftarrow{\mathcal{FS}} \frac{w}{T_0}\,\text{sinc}((kf_0)w) \qquad (4.195)$$

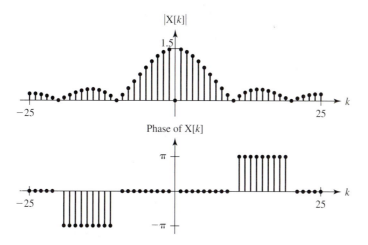

**Figure 4.34**
CTFS harmonic function of $x(t) = [15\,\text{rect}(1000t) * 100\,\text{comb}(100t)] - \frac{3}{2}$.

and

$$1 \overset{\mathcal{FS}}{\longleftrightarrow} \delta[k] \tag{4.196}$$

$$\text{rect}(1000t) * 100\,\text{comb}(100t) \overset{\mathcal{FS}}{\longleftrightarrow} \frac{1}{10}\text{sinc}(0.1k) \tag{4.197}$$

$$15\,\text{rect}(1000t) * 100\,\text{comb}(100t) \overset{\mathcal{FS}}{\longleftrightarrow} \frac{3}{2}\text{sinc}(0.1k) \tag{4.198}$$

$$15\,\text{rect}(1000t) * 100\,\text{comb}(100t) - \frac{3}{2} \overset{\mathcal{FS}}{\longleftrightarrow} \frac{3}{2}\text{sinc}(0.1k) - \frac{3}{2}\delta[k] \tag{4.199}$$

(Figure 4.34). [Notice that there is no impulse at $k = 0$ because the average value of $x(t)$ is zero.]

■

**EXAMPLE 4.5**

Find a function whose CTFS harmonic function representation is

$$X[k] = 2\,\text{sinc}^2\left(\frac{k}{5}\right) \tag{4.200}$$

assuming that the CTFS represents the signal for all time and that $T_F = T_0$.

**■ Solution**
Looking at the CTFS pairs in Appendix E, we find the pair

$$\text{tri}\left(\frac{t}{w}\right) * \frac{1}{T_0}\text{comb}\left(\frac{t}{T_0}\right) \overset{\mathcal{FS}}{\longleftrightarrow} \frac{w}{T_0}\text{sinc}^2\left(\frac{w}{T_0}k\right). \tag{4.201}$$

There is a problem here. We can identify the value of the ratio $w/T_0$ to be $\frac{1}{5}$, but we cannot identify the values of $w$ or $T_0$ individually. If we multiply both sides of (4.201) by 10 we get the correct expression for $X[k]$ implying that $x[n]$ is $10\,\text{tri}\,(t/w) * (1/T_0)\,\text{comb}\,(t/T_0)$ but still not determining what $w$ or $T_0$ are individually. So $T_0$ could still have any arbitrary value. This illustrates an important aspect of the CTFS previously mentioned. The CTFS harmonic function is a function of harmonic number $k$, and the harmonic number is a multiple of the fundamental frequency. So knowledge of the fundamental frequency is necessary to represent any signal by the CTFS representation

$$x(t) = \sum_{k=-\infty}^{\infty} X[k]e^{+j2\pi(kf_0)t}.$$     **(4.202)**

We cannot find an exact representation of $x(t)$ until we know the fundamental frequency. So let's do the next best thing and find a representation in terms of $f_0$ or $T_0$ without knowing what it is. Since $w/T_0 = \frac{1}{5}$,

$$x(t) = \frac{5}{T_0} \text{tri}\left(\frac{5t}{T_0}\right) * \frac{1}{T_0} \text{comb}\left(\frac{t}{T_0}\right) = 5 f_0 \text{tri}(5 f_0 t) * f_0 \text{comb}(f_0 t).$$     **(4.203)**

## 4.6  BAND-LIMITED SIGNALS

In general, infinitely many terms are required in a CTFS representation of a signal for exact equality according to

$$x(t) = \sum_{k=-\infty}^{\infty} X[k]e^{j2\pi(kf_0)t}.$$     **(4.204)**

But there are signals for which a finite number of terms yields an exact equality. For such signals, for $k > k_{\max} < \infty$, $X[k]$ is zero. As mentioned previously, such signals are said to be band-limited. The term *band-limited* comes from the concept of a band of frequencies (meaning a range of frequencies) in a signal. If the band of frequencies is limited (finite), the signal is band-limited. The characteristics of band-limited signals will become important later when we study sampling and the discrete Fourier transform.

## 4.7  CONVERGENCE OF THE CONTINUOUS-TIME FOURIER SERIES

### CONTINUOUS SIGNALS

In this section we will examine how the CTFS summation approaches the signal it represents as the number of terms used in the sum approaches infinity. We do this by examining the partial sum

$$x_N(t) = \sum_{k=-N}^{N} X[k]e^{j2\pi(kf_0)t}$$     **(4.205)**

or

$$x_N(t) = X_c[0] + \sum_{k=1}^{N}[X_c[k]\cos(2\pi(kf_0)t) + X_s[k]\sin(2\pi(kf_0)t)]$$     **(4.206)**

for successively higher values of $N$. As a first example consider the CTFS representation of the continuous periodic signal in Figure 4.35.

$$x(t) = A \text{tri}\left(\frac{2t}{T_0}\right) * \frac{1}{T_0} \text{comb}\left(\frac{t}{T_0}\right)$$     **(4.207)**

Letting $T_F = T_0$ to make the CTFS representation be valid for all time, we find the complex CTFS harmonic function

$$X[k] = \frac{A}{2} \text{sinc}^2\left(\frac{k}{2}\right),$$     **(4.208)**

and the approximations to $x(t)$ for $N = 1, 3, 5,$ and $59$ are illustrated in Figure 4.36. At $N = 59$ (and probably at lower values of $N$), it is impossible to distinguish the CTFS partial sum representation from the original signal by observing a plot on this scale.

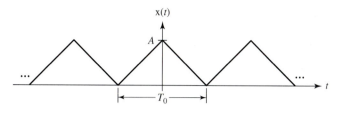

**Figure 4.35**
A continuous signal to be represented by a CTFS.

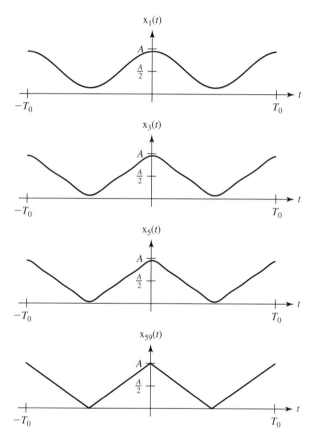

**Figure 4.36**
Successively closer approximations to a triangle wave.

## SIGNALS WITH DISCONTINUITIES AND THE GIBBS PHENOMENON

Now consider a periodic CT signal with discontinuities

$$x(t) = A \operatorname{rect}\left(\frac{2(t - (T_0/4))}{T_0}\right) * \frac{1}{T_0} \operatorname{comb}\left(\frac{t}{T_0}\right) \tag{4.209}$$

(Figure 4.37). Letting $T_F = T_0$ to make the CTFS representation be valid for all time, we get the complex CTFS harmonic function

$$X[k] = \frac{A}{2} \operatorname{sinc}\left(\frac{k}{2}\right) e^{-j(\pi/2)(kf_0)}, \tag{4.210}$$

and the approximations to $x(t)$ for $N = 1, 3, 5$, and 59 are illustrated in Figure 4.38.

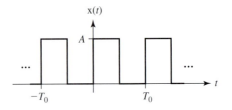

**Figure 4.37**
A discontinuous CT signal to be represented
by a CTFS.

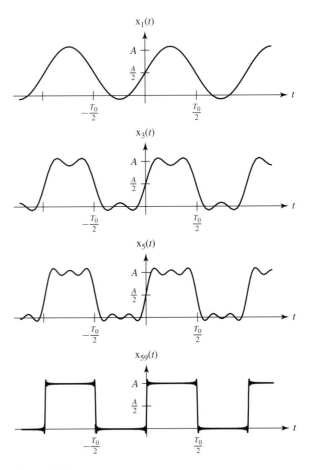

**Figure 4.38**
Successively closer approximations to a square wave.

Although the mathematical derivation indicates that the original signal and its CTFS representation are equal everywhere, it is natural to wonder whether that is true after looking at Figure 4.38. There is an obvious "overshoot" or "ripple" near the discontinuities which does not appear to become smaller as $N$ increases. It is in fact true that the maximum vertical overshoot near a discontinuity does not decrease with $N$, even as $N$ approaches infinity. This overshoot is called the *Gibbs phenomenon* in honor of Josiah Gibbs who first mathematically described it. But notice that the ripple is also confined ever more closely in the vicinity of the discontinuity as $N$ increases. In the limit as $N$ ap-

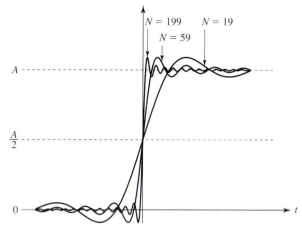

**Figure 4.39**
Illustration of the Gibbs phenomenon for increasing values of *N*.

proaches infinity the *height* of the overshoot is constant, but its *width* approaches zero. So in the limit as *N* approaches infinity the signal power of the CTFS representation converges to the same value as the signal power of the original signal because the zero-width overshoot contains no signal energy. Also, at any particular value of *t* (except exactly at a discontinuity) the value of the CTFS representation approaches the value of the original signal as *N* approaches infinity. At a discontinuity the functional value of the CTFS representation is always the average of the two limits of the original function approached from above and from below, for any *N*. Figure 4.39 is a magnified view of the CTFS representation at a discontinuity for three different values of *N*. Since the two signals have exactly the same signal energy in any finite time interval, their effect on any real physical system is the same and they can be considered equal without error.

The CTFS just found and graphed in Figure 4.39 goes through exactly the midpoint of each discontinuity of x($t$), independent of the choice of the fundamental period. If the fundamental period $T_0$ is allowed to approach infinity, the signal

$$x(t) = A \operatorname{rect}\left(\frac{2[t - (T_0/4)]}{T_0}\right) * \frac{1}{T_0} \operatorname{comb}\left(\frac{t}{T_0}\right) \qquad \textbf{(4.211)}$$

approaches the unit step and the CTFS still goes through the midpoint of the only discontinuity left, the one at $t = 0$. This is why the unit step is defined in this text with the value at zero to be one-half, $u(0) = \frac{1}{2}$. Defined this way, the unit step is a simple transformation of the signum function at *all* points,

$$\operatorname{sgn}(t) = 2u(t) - 1 \qquad \textbf{(4.212)}$$

In Section 6.3, in the material on ideal filters, we will see another reason why it is convenient to define the unit step this way.

## 4.8 THE DISCRETE-TIME FOURIER SERIES (DTFS)

### MATHEMATICAL DEVELOPMENT

We next explore the applicability of the Fourier series in DT signal and system analysis. The same advantages of expressing a signal as a linear combination of complex sinusoids applies to DT signal analysis. The development will be parallel to the development of the CTFS for CT signals.

Let's begin by assuming provisionally, as we did in the development of the CTFS, that a DT signal x[n] can be represented over a finite discrete-time interval $n_0 \le n < n_0 + N_F$ by a summation of complex sinusoids $x_F[n]$ of the form

$$x_F[n] = \sum_{k=-\infty}^{\infty} X[k]e^{j2\pi(kF_F)n} = \sum_{k=-\infty}^{\infty} X[k]e^{j2\pi(nk/N_F)} \qquad \textbf{(4.213)}$$

or

$$x_F[n] = \sum_{k=-\infty}^{\infty} X[k]e^{j(k\Omega_F)n} = \sum_{k=-\infty}^{\infty} X[k]e^{j2\pi(nk/N_F)}, \qquad \textbf{(4.214)}$$

where $N_F =$ integer

$F_F = 1/N_F =$ discrete-time cyclic frequency

$\Omega_F = \frac{2\pi}{N_F} =$ discrete-time radian frequency

As indicated in Chapter 2, neither $F_F$ nor $\Omega_F$ is actually frequency in the usual sense because $n$ is not continuous time in seconds, but discrete time, which is dimensionless. Therefore, $F_F$ is not frequency in hertz and $\Omega_F$ is not frequency in radians per second, but they are both angular change per unit discrete time measured in either cycles or radians. Therefore, the units of $F_F$ are cycles, and the units of $\Omega_F$ are radians. (The radian is defined as a ratio of lengths and is therefore dimensionless. A cycle is just $2\pi$ radians, so it is also dimensionless.) Also note that, when using the DTFS to represent a signal, since $N_F$ is an integer, $F_F$ is restricted to values that are reciprocals of integers and $\Omega_F$ is restricted to values that are $2\pi$ divided by an integer.

The first difference to notice between the DT and CT cases is the fact that the summation in (4.213) or (4.214) need not go to infinity because, after some value of $k$, the exponentials $e^{j2\pi(kF_F)n}$ start to repeat. This is a consequence of the fact that $n$ and $k$ are integers and

$$e^{j2\pi((k+N_F)F_F)n} = e^{j2\pi(kF_F)n}e^{j2\pi(N_F F_F)n} \qquad \textbf{(4.215)}$$

and, since $F_F = 1/N_F$,

$$e^{j2\pi((k+N_F)F_F)n} = e^{j2\pi(kF_F)n} \underbrace{e^{j2\pi n}}_{=1} = e^{j2\pi(kF_F)n}. \qquad \textbf{(4.216)}$$

So if we add $N_F$ to $k$, we don't change the exponential function. In fact, if we add any integer multiple of $N_F$ to $k$, we don't change the exponential function. Any set of $k$'s which cover the range $k_0 \le k < k_0 + N_F$, where $k_0$ is arbitrary, generates a complete set of exponential functions. Adding any more would be redundant. Therefore, the DTFS can be written without redundancy as

$$x_F[n] = \sum_{k=k_0}^{k_0+N_F-1} X[k]e^{j2\pi(kF_F)n}, \qquad \textbf{(4.217)}$$

where $k_0$ is arbitrary. Since $k_0$ is arbitrary, we can use the notation

$$x_F[n] = \sum_{k=\langle N_F \rangle} X[k]e^{j2\pi(kF_F)n} \qquad \textbf{(4.218)}$$

where $\sum_{k=\langle N_F \rangle}$ means summation over any range of consecutive $k$'s exactly $N_F$ in length.

If the equality between x[n] and $x_F[n]$ is to hold, we must be able to say that

$$x[n] = \sum_{k=\langle N_F \rangle} X[k]e^{j2\pi(kF_F)n} \qquad n_0 \le n < n_0 + N_F. \qquad \textbf{(4.219)}$$

We can use a method similar to the one we used in the derivation of the CTFS by multiplying both sides of (4.219) by $e^{-j2\pi(qF_F)n}$, where $q$ is an integer. Then we get

$$x[n]e^{-j2\pi(qF_F)n} = \sum_{k=\langle N_F \rangle} X[k]e^{-j2\pi(qF_F)n}e^{j2\pi(kF_F)n} \qquad n_0 \le n < n_0 + N_F \qquad \textbf{(4.220)}$$

or

$$x[n]e^{-j2\pi(qF_F)n} = \sum_{k=\langle N_F \rangle} X[k]e^{j2\pi((k-q)F_F)n} \qquad n_0 \le n < n_0 + N_F. \qquad \textbf{(4.221)}$$

Following the analogy with the development of the CTFS, sum both sides of (4.221) over the range $n_0 \le n < n_0 + N_F$,

$$\sum_{n=n_0}^{n_0+N_F-1} x[n]e^{-j2\pi(qF_F)n} = \sum_{n=n_0}^{n_0+N_F-1} \sum_{k=\langle N_F \rangle} X[k]e^{j2\pi((k-q)F_F)n} \qquad n_0 \le n < n_0 + N_F.$$

$$\textbf{(4.222)}$$

Reversing the order of summation on the right side of (4.222),

$$\sum_{n=n_0}^{n_0+N_F-1} x[n]e^{-j2\pi(qF_F)n} = \sum_{k=\langle N_F \rangle} X[k] \sum_{n=n_0}^{n_0+N_F-1} e^{j2\pi((k-q)F_F)n} \qquad n_0 \le n < n_0 + N_F$$

$$\textbf{(4.223)}$$

or

$$\sum_{n=n_0}^{n_0+N_F-1} x[n]e^{-j2\pi(qF_F)n} = \sum_{k=\langle N_F \rangle} X[k] \sum_{n=n_0}^{n_0+N_F-1} e^{j2\pi((k-q)/N_F)n} \qquad n_0 \le n < n_0 + N_F.$$

$$\textbf{(4.224)}$$

Now, examine the summation $\sum_{n=n_0}^{n_0+N_F-1} e^{j2\pi((k-q)/N_F)n}$. If $k \ne q$, the complex sinusoids $e^{j2\pi((k-q)/N_F)n}$ all have a magnitude of one and are arrayed at equal angular intervals on the unit circle in the complex plane (Figure 4.40). The angular interval between complex sinusoids is the fraction $(k-q)/N_F$ of a complete $2\pi$-radian cycle.

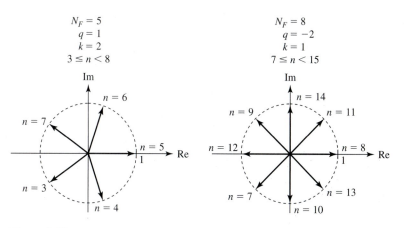

**Figure 4.40**
Two examples of complex sinusoids spaced at equal angular intervals in the complex plane.

Since the summation over $n$ covers the range $n_0 \leq n < n_0 + N_F$, the total angular range covered by the summation is an integer multiple of $2\pi$ radians. When plotted as in Figure 4.40, we see that the sum of all those complex sinusoids is zero.

There is also an analytical approach to this result using a very handy formula for the summation of a finite geometric series,

$$\sum_{n=0}^{N-1} r^n = \begin{cases} N & r = 1 \\ \dfrac{1 - r^N}{1 - r} & r \neq 1 \end{cases}, \tag{4.225}$$

where $r$ can be complex. Applying this formula to $\sum_{n=n_0}^{n_0+N_F-1} e^{j2\pi[(k-q)/N_F]n}$, first make the change of variable $m = n - n_0$. Then

$$\sum_{m=0}^{N_F-1} e^{j2\pi((k-q)/N_F)(m+n_0)} = e^{j2\pi((k-q)/N_F)n_0} \sum_{m=0}^{N_F-1} \left(e^{j2\pi((k-q)/N_F)}\right)^m \tag{4.226}$$

$$e^{j2\pi((k-q)/N_F)n_0} \sum_{m=0}^{N_F-1} \left(e^{j2\pi((k-q)/N_F)}\right)^m = e^{j2\pi((k-q)/N_F)n_0} \frac{1 - \left(e^{j2\pi((k-q)/N_F)}\right)^{N_F}}{1 - e^{j2\pi(k-q)/N_F}}$$

$$= e^{j2\pi((k-q)/N_F)n_0} \underbrace{\frac{\overbrace{1 - e^{j2\pi(k-q)}}^{=0}}{1 - e^{j2\pi(k-q)/N_F}}}_{\substack{=0,\, k=q \\ \neq 0,\, k \neq q}} \tag{4.227}$$

and, since $k - q$ is an integer,

$$e^{j2\pi((k-q)/N_F)n_0} \sum_{m=0}^{N_F-1} \left(e^{j2\pi((k-q)/N_F)}\right)^m = 0 \qquad k \neq q. \tag{4.228}$$

Therefore, only the case $k = q$ yields a nonzero result. When $k = q$,

$$e^{j2\pi((k-q)/N_F)n} = e^0 = 1. \tag{4.229}$$

Then we can simplify (4.224) to

$$\sum_{n=n_0}^{n_0+N_F-1} x[n]e^{-j2\pi(qF_F)n} = \sum_{k=\langle N_F \rangle} X[q] = N_F X[q] \qquad n_0 \leq n < n_0 + N_F, \tag{4.230}$$

and, solving for $X[q]$, we get

$$X[q] = \frac{1}{N_F} \sum_{n=n_0}^{n_0+N_F-1} x[n]e^{-j2\pi(qF_F)n}. \tag{4.231}$$

Therefore, in the original statement of the DTFS representation of a DT signal,

$$\boxed{x_F[n] = \sum_{k=\langle N_F \rangle} X[k]e^{j2\pi(kF_F)n}}, \tag{4.232}$$

$X[k]$ is given by

$$\boxed{X[k] = \frac{1}{N_F} \sum_{n=n_0}^{n_0+N_F-1} x[n]e^{-j2\pi(kF_F)n}}. \tag{4.233}$$

Using the relationship $\Omega_F = 2\pi F_F$, the DTFS can also be written in terms of DT radian frequency as

$$x_F[n] = \sum_{k=\langle N_F \rangle} X[k]e^{j(k\Omega_F)n} \quad \text{and} \quad X[k] = \frac{1}{N_F} \sum_{n=n_0}^{n_0+N_F-1} x[n]e^{-j(k\Omega_F)n}. \quad \textbf{(4.234)}$$

The harmonic function $X[k]$ is the same in either form.

Just as the CTFS representation of a signal is periodic with fundamental period $T_F$, the DTFS representation $x_F[n]$ of a signal $x[n]$ is periodic with fundamental period $N_F$. This is easily shown by replacing $n$ by $n + N_F$.

$$x[n + N_F] = \sum_{k=\langle N_0 \rangle} X[k]e^{j2\pi(kF_F)(n+N_F)} = \sum_{k=\langle N_0 \rangle} X[k]e^{j2\pi(kF_F)n} \underbrace{e^{j2\pi(kF_F)N_F}}_{=1}$$

$$= \sum_{k=\langle N_0 \rangle} X[k]e^{j2\pi(kF_F)n} = x[n] \quad \textbf{(4.235)}$$

If, as with the CTFS, we consider the very common case in which we represent the signal $x[n]$ over exactly one fundamental period $N_0$, then $N_F = N_0$ and $F_F = F_0 = 1/N_0$, $\Omega_F = \Omega_0 = 2\pi/N_0$, $x_F[n] = x[n]$ for *all* $n$, and the relationships reduce to the forms

$$x[n] = \sum_{k=\langle N_0 \rangle} X[k]e^{j2\pi(kF_0)n} \xleftrightarrow{\mathcal{FS}} X[k] = \frac{1}{N_0} \sum_{n=\langle N_0 \rangle} x[n]e^{-j2\pi(kF_0)n} \quad \textbf{(4.236)}$$

or

$$x[n] = \sum_{k=\langle N_0 \rangle} X[k]e^{j(k\Omega_0)n} \xleftrightarrow{\mathcal{FS}} X[k] = \frac{1}{N_0} \sum_{n=\langle N_0 \rangle} x[n]e^{-j(k\Omega_0)n}. \quad \textbf{(4.237)}$$

The CTFS representation of a signal $x_F(t)$ is periodic in time with fundamental period $T_F$. The CTFS harmonic function $X[k]$ is not periodic in $k$ [unless $x_F(t)$ consists only of impulses at uniform time intervals]. The DTFS representation of a signal $x_F[n]$ is periodic in discrete time with fundamental period $N_F$, and the DTFS harmonic function $X[k]$ is periodic in harmonic number $k$ with fundamental period $N_F$. Therefore, in the most common case in which we represent a periodic signal $x[n]$ over exactly one fundamental period $N_0$, $x[n]$ and $X[k]$ are both periodic with fundamental period $N_0$.

All the information about a periodic signal $x[n]$ with fundamental period $N_0$ is contained in $N_0$ real numbers, the values of the signal over exactly one fundamental period, because, if we know all the $x[n]$ in any period and we know the period, we can reconstruct the whole signal simply by repeating those numbers in all the other periods. Also, all the information about the periodic signal $x[n]$ with fundamental period $N_0$ is contained in a different set of $N_0$ numbers, the values of the $X[k]$ over exactly its fundamental period, which is also $N_0$, because, if we know those numbers, we can reconstruct the whole signal by using the relationship,

$$x[n] = \sum_{k=\langle N_0 \rangle} X[k]e^{j2\pi(kF_0)n} \quad \textbf{(4.238)}$$

(Figure 4.41).

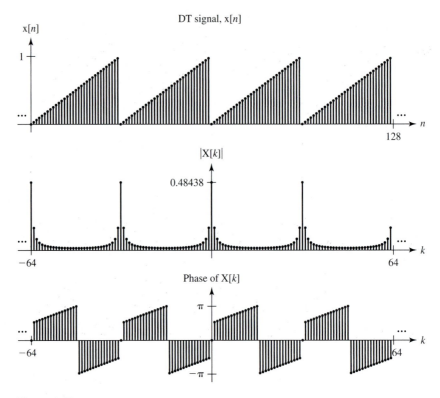

**Figure 4.41**
A periodic DT signal and its periodic DTFS.

This symmetry leads to an important idea. The information about a signal is conserved as we transform it from a function of discrete time $n$ to its equivalent representation as a function of discrete harmonic number $k$. This idea will be significant when we explore sampling and the discrete Fourier transform in Chapter 7.

## EXAMPLE 4.6

Find the DTFS harmonic function of the DT signal $x[n] = \text{rect}_2[n] * \text{comb}_8[n]$ over exactly one fundamental period (Figure 4.42).

**■ Solution**

The fundamental period $N_0$ is 8. The DTFS harmonic function is found from

$$X[k] = \frac{1}{N_0} \sum_{n=\langle N_0 \rangle} x[n] e^{-j2\pi(kF_0)n}. \tag{4.239}$$

The range of summation can be any range $m \le n < m + 8$, where $m$ is any integer. Let the range of the summation be $-4 \le n < 4$. Then

$$X[k] = \frac{1}{N_0} \sum_{n=-4}^{3} x[n] e^{-j2\pi(kF_0)n} = \frac{1}{8} \sum_{n=-2}^{2} e^{-j2\pi(nk/8)}. \tag{4.240}$$

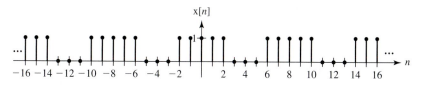

**Figure 4.42**
A DT signal, $x[n] = \text{rect}_2[n] * \text{comb}_8[n]$.

We can find the DTFS harmonic function by simply letting $k$ be each of the integers in a range $q \leq k < q + 8$, where $q$ is any integer, one at a time, and adding the terms in the summation for each $k$. But it would be nice to have a compact closed-form expression for $X[k]$ as a function of $k$. As it turns out that can be done by again using the handy relation

$$\sum_{n=0}^{N-1} r^n = \begin{cases} N & r = 1 \\ \dfrac{1 - r^N}{1 - r} & r \neq 1 \end{cases} \qquad (4.241)$$

for summing a finite geometric series. First make the change of variable $m = n + 2$ in (4.240). Then

$$X[k] = \frac{1}{8} \sum_{m=0}^{4} e^{-j2\pi[(m-2)k/8]} = \frac{1}{8} e^{j(k/2)\pi} \sum_{m=0}^{4} \left(e^{-j2\pi(k/8)}\right)^m \qquad (4.242)$$

and, using (4.241),

$$X[k] = \frac{1}{8} e^{j(k/2)\pi} \frac{1 - e^{-j(5k/4)\pi}}{1 - e^{-j(k/4)\pi}} = \frac{1}{8} e^{j(k/2)\pi} \frac{e^{-j(5k/8)\pi}}{e^{-j(k/8)\pi}} \frac{e^{+j(5k/8)\pi} - e^{-j(5k/8)\pi}}{e^{+j(k/8)\pi} - e^{-j(k/8)\pi}}$$

$$= \left(\frac{1}{8}\right) \frac{\sin((5k/8)\pi)}{\sin((k/8)\pi)}. \qquad (4.243)$$

Recall that the Dirichlet function is defined by

$$\text{drcl}(t, N) = \frac{\sin(N\pi t)}{N \sin(\pi t)}. \qquad (4.244)$$

Using that definition,

$$X[k] = \frac{5}{8} \text{drcl}\left(\frac{k}{8}, 5\right). \qquad (4.245)$$

This is a Dirichlet function with an $N$ of 5, an odd number, so the extrema of $\text{drcl}(k/8, 5)$ are all $+1$ at integer values of $k/8$. Therefore, when $k$ is an integer multiple of 8, $X[k]$ is $\frac{5}{8}$. The top graph in Figure 4.43 is a plot of the magnitude of the DTFS harmonic function versus harmonic number $k$.

Now, to illustrate a concept that will be important in understanding the Fourier transform in Chapter 5, let's redo this example except with a greater discrete time between the rectangular pulses. Let

$$x[n] = \text{rect}_2[n] * \text{comb}_{32}[n]. \qquad (4.246)$$

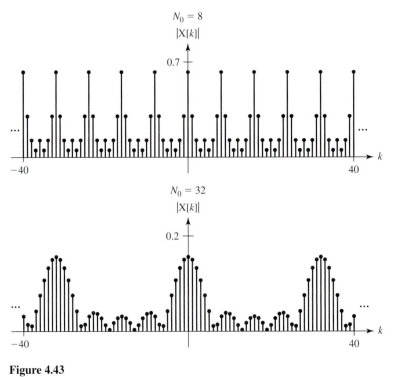

**Figure 4.43**
$X[k]$ for $x[n] = \text{rect}_2[n] * \text{comb}_8[n]$ and $x[n] = \text{rect}_2[n] * \text{comb}_{32}[n]$.

This signal is the same except that the fundamental period is now 32 instead of 8. The calculation of the DTFS harmonic function is basically the same, and the result is

$$X[k] = \frac{1}{32} \frac{\sin((5k/32)\pi)}{\sin((k/32)\pi)} = \frac{5}{32} \text{drcl}\left(\frac{k}{32}, 5\right). \tag{4.247}$$

The DTFS harmonic function is graphed in the bottom plot of Figure 4.43. ∎

---

## EXAMPLE 4.7

Find the DTFS harmonic function of $x[n] = A\cos(2\pi F_0 n) = A\cos(2\pi n/N_0)$.

■ **Solution**
This harmonic function is in the table of Fourier pairs, but let's derive it from the definition.

$$X[k] = \frac{1}{N_0} \sum_{n=\langle N_0 \rangle} A\cos\left(\frac{2\pi n}{N_0}\right) e^{-j2\pi(kn/N_0)} = \frac{A}{N_0} \sum_{n=\langle N_0 \rangle} \frac{e^{j(2\pi n/N_0)} + e^{-j(2\pi n/N_0)}}{2} e^{-j2\pi(kn/N_0)}$$

$$= \frac{A}{2N_0} \sum_{n=\langle N_0 \rangle} \left(e^{j(2\pi n/N_0)(1-k)} + e^{j(2\pi n/N_0)(-1-k)}\right)$$

$$= \frac{A}{2N_0} \left[\sum_{n=\langle N_0 \rangle} \left(e^{j(2\pi/N_0)(1-k)}\right)^n + \sum_{n=\langle N_0 \rangle} \left(e^{j(2\pi/N_0)(-1-k)}\right)^n\right] \tag{4.248}$$

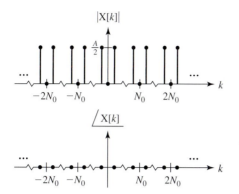

**Figure 4.44**
Magnitude and phase of $X[k]$.

Then, using

$$\sum_{n=0}^{N-1} r^n = \frac{1 - r^N}{1 - r} \qquad r \neq 1 \tag{4.249}$$

$$X[k] = \frac{A}{2N_0} \left[ \frac{1 - \left(e^{j(2\pi/N_0)(1-k)}\right)^{N_0}}{1 - e^{j(2\pi/N_0)(1-k)}} + \frac{1 - \left(e^{j(2\pi/N_0)(-1-k)}\right)^{N_0}}{1 - e^{j(2\pi/N_0)(-1-k)}} \right]$$

$$= \frac{A}{2N_0} \left( \frac{1 - e^{j2\pi(1-k)}}{1 - e^{j(2\pi/N_0)(1-k)}} + \frac{1 - e^{j2\pi(-1-k)}}{1 - e^{j(2\pi/N_0)(-1-k)}} \right)$$

$$= \frac{A}{2N_0} \left[ \frac{e^{j\pi(1-k)}}{e^{j(\pi/N_0)(1-k)}} \frac{\sin[\pi(k-1)]}{\sin[(\pi/N_0)(k-1)]} + \frac{e^{j\pi(-1-k)}}{e^{j(\pi/N_0)(-1-k)}} \frac{\sin[\pi(k+1)]}{\sin[(\pi/N_0)(k+1)]} \right] \tag{4.250}$$

or

$$X[k] = \frac{A}{2} \left[ \frac{e^{j\pi(1-k)}}{e^{j(\pi/N_0)(1-k)}} \operatorname{drcl}\left(\frac{k-1}{N_0}, N_0\right) + \frac{e^{j\pi(-1-k)}}{e^{j(\pi/N_0)(-1-k)}} \operatorname{drcl}\left(\frac{k+1}{N_0}, N_0\right) \right]. \tag{4.251}$$

This rather complicated looking result is actually quite simple. Its value is zero for every $k$ except those $k$'s for which

$$\frac{k-1}{N_0} = q \qquad \text{or} \qquad \frac{k+1}{N_0} = q, \tag{4.252}$$

where $q$ is an integer, and at those values of $k$ the value of $X[k]$ is $A/2$. In summary,

$$X[k] = \begin{cases} \dfrac{A}{2} & k = qN_0 \pm 1 \\ 0 & \text{otherwise} \end{cases} = \frac{A}{2} \left( \operatorname{comb}_{N_0}[k-1] + \operatorname{comb}_{N_0}[k+1] \right) \tag{4.253}$$

(Figure 4.44). ∎

## 4.9 PROPERTIES OF THE DISCRETE-TIME FOURIER SERIES

Let the DTFS harmonic function of a periodic signal $x[n]$ with fundamental period $N_{0x}$ be $X[k]$, and let the DTFS harmonic function of a periodic signal $y[n]$ with fundamental period $N_{0y}$ be $Y[k]$. We can find the DTFS harmonic function for each of these two signals over its fundamental period. The DTFS harmonic function is found from the transformation relationships

$$x[n] = \sum_{k=\langle N_{0x} \rangle} X[k] e^{j2\pi(kF_{0x})n} \xleftrightarrow{\mathcal{FS}} X[k] = \frac{1}{N_{0x}} \sum_{n=\langle N_{0x} \rangle} x[n] e^{-j2\pi(kF_{0x})n} \qquad \textbf{(4.254)}$$

or

$$y[n] = \sum_{k=\langle N_{0y} \rangle} Y[k] e^{j2\pi(kF_{0y})n} \xleftrightarrow{\mathcal{FS}} Y[k] = \frac{1}{N_{0y}} \sum_{n=\langle N_{0y} \rangle} y[n] e^{-j2\pi(kF_{0y})n}. \qquad \textbf{(4.255)}$$

Then, using those signals as examples, in the following sections various properties of the DTFS of related signals will be introduced. Where the proofs are similar to the equivalent proofs for the CTFS only the result will be presented.

## LINEARITY

This property is identical to the linearity property for the CTFS and the proof is similar.

$$\boxed{\alpha x[n] + \beta y[n] \xleftrightarrow{\mathcal{FS}} \alpha X[k] + \beta Y[k]}. \qquad \textbf{(4.256)}$$

## TIME SHIFTING

Let $z[n] = x[n - n_0]$ and $N_{x0} = N_{z0} = N_0$. Then

$$Z[k] = \frac{1}{N_0} \sum_{n=\langle N_0 \rangle} z[n] e^{-j2\pi(kF_0)n} = \frac{1}{N_0} \sum_{n=\langle N_0 \rangle} x[n - n_0] e^{-j2\pi(kF_0)n}. \qquad \textbf{(4.257)}$$

Now let $q = n - n_0$ in the x summation. Then, since $n$ covers a range of $N_0$, $q$ does also and

$$Z[k] = \frac{1}{N_0} \sum_{n=\langle N_0 \rangle} z[n] e^{-j2\pi(kF_0)n} = \frac{1}{N_0} \sum_{q=\langle N_0 \rangle} x[q] e^{-j2\pi(kF_0)(q+n_0)} \qquad \textbf{(4.258)}$$

$$Z[k] = \frac{1}{N_0} \sum_{n=\langle N_0 \rangle} z[n] e^{-j2\pi(kF_0)n} = e^{-j2\pi(kF_0)n_0} \underbrace{\frac{1}{N_0} \sum_{q=\langle N_0 \rangle} x[q] e^{-j2\pi(kF_0)q}}_{=X[k]} \qquad \textbf{(4.259)}$$

$$Z[k] = e^{-j2\pi(kF_0)n_0} X[k] \qquad \textbf{(4.260)}$$

and

$$\boxed{x[n - n_0] \xleftrightarrow{\mathcal{FS}} e^{-j2\pi(kF_0)n_0} X[k]} \qquad \textbf{(4.261)}$$

or

$$\boxed{x[n - n_0] \xleftrightarrow{\mathcal{FS}} e^{-j(k\Omega_0)n_0} X[k]}. \qquad \textbf{(4.262)}$$

This property is very similar to the equivalent property for the CTFS except that "time" is now discrete time.

## FREQUENCY SHIFTING

This property is similar to the frequency-shifting property for the CTFS and the proof is similar.

$$e^{j2\pi(k_0 F_0)n}\mathrm{x}[n] \xleftrightarrow{\ \mathcal{FS}\ } \mathrm{X}[k - k_0] \qquad (4.263)$$

or

$$e^{j(k_0\Omega_0)n}\mathrm{x}[n] \xleftrightarrow{\ \mathcal{FS}\ } \mathrm{X}[k - k_0] \qquad (4.264)$$

## CONJUGATION

This property is similar to the conjugation property for the CTFS and the proof is similar.

$$\mathrm{x}^*[n] \xleftrightarrow{\ \mathcal{FS}\ } \mathrm{X}^*[-k] \qquad (4.265)$$

## TIME REVERSAL

This property is similar to the time-reversal property for the CTFS and the proof is similar.

$$\mathrm{x}[-n] \xleftrightarrow{\ \mathcal{FS}\ } \mathrm{X}[-k] \qquad (4.266)$$

## TIME SCALING

This property is quite different for DT functions than it was for CT functions. Let $z[n] = \mathrm{x}[an]$, $a > 0$. If $a$ is not an integer, then some values of $z[n]$ will be undefined and a DTFS cannot be found for it. If $a$ is an integer, then $z[n]$ is a decimated version of $\mathrm{x}[n]$ and some of the values of $\mathrm{x}[n]$ do not appear in $z[n]$. In that case, there cannot be a unique relationship between the harmonic functions of $\mathrm{x}[n]$ and $z[n]$ through the transformation $n \to an$. However, there is a related operation for which the relationship between $\mathrm{x}[n]$ and $z[n]$ is unique. Let $m$ be a positive integer and let

$$z[n] = \begin{cases} \mathrm{x}\left[\dfrac{n}{m}\right] & \dfrac{n}{m} \text{ is an integer} \\ 0 & \text{otherwise} \end{cases} . \qquad (4.267)$$

That is, $z[n]$ is a time-expanded version of $\mathrm{x}[n]$ formed by placing $m - 1$ zeros between adjacent values of $\mathrm{x}[n]$ (Figure 4.45).

If the fundamental period of $\mathrm{x}[n]$ is $N_{0x} = N_0$, the fundamental period of $z[n]$ is $N_{0z} = mN_0$. Then the DTFS harmonic function for $z[n]$ with a representation period of $mN_0$ is

$$Z[k] = \frac{1}{mN_0} \sum_{n=\langle mN_0 \rangle} z[n] e^{-j2\pi(nk/mN_0)}. \qquad (4.268)$$

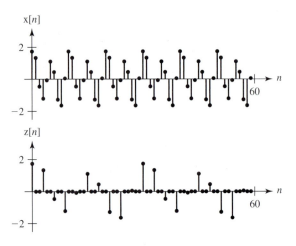

**Figure 4.45**
A DT function and an expanded version formed by
inserting zeros between values.

Since all the values of z are zero when $n/m$ is not an integer,

$$Z[k] = \frac{1}{mN_0} \sum_{\substack{n=\langle mN_0\rangle \\ n/m \text{ is an integer}}} z[n]e^{-j2\pi(nk/mN_0)} \qquad (4.269)$$

Let $p = n/m$, $n/m$ an integer. Then

$$Z[k] = \frac{1}{mN_0} \sum_{p=\langle N_0\rangle} z[mp]e^{-j2\pi(kp/N_0)} \qquad (4.270)$$

and $z[mp] = x[p]$ with all the other values of $z[n]$ equal to zero. Therefore,

$$\boxed{Z[k] = \frac{1}{mN_0} \sum_{p=\langle N_0\rangle} x[p]e^{-j2\pi(kp/N_0)} = \frac{1}{m}X[k]} \qquad (4.271)$$

This result, (4.271), says that the harmonic function of z is the same as the harmonic
function for x except divided by $m$. But that does not mean that the DTFS representation of z is the same as the DTFS representation of x except divided by $m$ because the
periods of the two signals are not the same. The DTFS representation of z is

$$\boxed{z[n] = \sum_{k=\langle mN_0\rangle} Z[k]e^{j2\pi(kn/mN_0)}} \qquad (4.272)$$

## CHANGE OF PERIOD

If we know that the DTFS harmonic function of $x[n]$ over the representation period $N_{0x} = N_0$ is $X[k]$, we can find the harmonic function of $x[n]$, $X_q[k]$, over the

representation period $qN_0$, where $q$ is a positive integer. It is

$$X_q[k] = \frac{1}{qN_0} \sum_{n=\langle qN_0 \rangle} x[n]e^{-j2\pi(nk/qN_0)} . \qquad (4.273)$$

The DT function $x[n]$ has a fundamental period $N_0$ and, therefore, consists of DT sinusoids at integer multiples of its fundamental frequency $1/N_0$. The DT function $e^{-j2\pi(nk/qN_0)}$ has a fundamental period $qN_0$ and fundamental frequency $1/qN_0$. Therefore, on the DT interval, $n_0 \leq n < n_0 + qN_0$, the two DT functions $x[n]$ and $e^{-j2\pi(nk/qN_0)}$ are orthogonal unless $k/q$ is an integer. Therefore, when $k/q$ is not an integer, $X_q[k] = 0$. When $k/q$ is an integer,

$$X_q[k] = q\left(\frac{1}{qN_0}\sum_{n=\langle N_0 \rangle} x[n]e^{-j2\pi(nk/qN_0)}\right) = \frac{1}{N_0}\sum_{n=\langle N_0 \rangle} x[n]e^{-j2\pi(nk/qN_0)} = X\left[\frac{k}{q}\right].$$

$$(4.274)$$

Summarizing,

$$X_q[k] = \begin{cases} X\left[\dfrac{k}{q}\right] & \dfrac{k}{q} \text{ is an integer} \\ 0 & \text{otherwise} \end{cases} \qquad (4.275)$$

(Figure 4.46).

Examining the two DTFS harmonic functions we see that using two periods instead of one does not add any information because the signal is exactly the same in

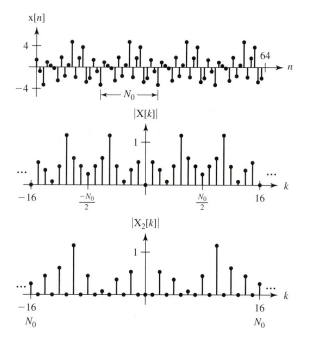

**Figure 4.46**
A DT signal and the magnitudes of its DTFS harmonic function with $N_F = N_0$ and with $N_F = 2N_0$.

each period. By using a representation period that is longer we get information about any lower-frequency harmonics that might be in the signal. In the example in Figure 4.46, the signal is periodic with period $N_0$; therefore, its fundamental frequency is $1/N_0$. When we use a representation period of $2N_0$, we see the harmonic amplitudes at integer multiples of $1/2N_0$. But in this signal the harmonics are all at integer multiples of $1/N_0$ which are the same as the even multiples of $1/2N_0$. Therefore, there are not any harmonics at odd multiples of $1/2N_0$. We get extra harmonic information, but all the extra harmonics have zero amplitude.

---

**EXAMPLE 4.8**

---

Find the DTFS harmonic function for

$$x[n] = \cos\left(\frac{2\pi n}{3}\right) + \text{comb}_5[n].$$   (4.276)

■ **Solution**

This is a signal consisting of the sum of two periodic DT signals whose individual fundamental periods are 3 and 5. The fundamental period of the sum of these two signals is the least common multiple of 3 and 5, 15. Using the linearity principle we can find the DTFS harmonic function of $x[n]$, by adding the DTFS harmonic functions of $\cos(2\pi n/3)$ and $\text{comb}_5[n]$. But in order to be able to add them we must find them over the same representation time which must also be a period of each of them. The fundamental period of $x[n]$ is 15. Therefore, any positive integer multiple of 15 would be appropriate. We will use 15 as the representation time.

From the Fourier pairs in Appendix E,

$$\cos\left(\frac{2\pi n}{N_0}\right) \overset{\mathcal{FS}}{\longleftrightarrow} \frac{1}{2}(\text{comb}_{N_0}[k-1] + \text{comb}_{N_0}[k+1])$$   (4.277)

and

$$\text{comb}_{N_0}[n] \overset{\mathcal{FS}}{\longleftrightarrow} \frac{1}{N_0}.$$   (4.278)

Let the cosine CTFS harmonic function over its fundamental period be

$$X_{\cos}[k] = \frac{1}{2}(\text{comb}_3[k-1] + \text{comb}_3[k+1]).$$   (4.279)

Therefore, using the change-of-period property to find the DTFS harmonic function over five of its fundamental periods we get

$$X_{\cos,5}[k] = \begin{cases} \frac{1}{2}\left(\text{comb}_3\left[\frac{k}{5}-1\right] + \text{comb}_3\left[\frac{k}{5}+1\right]\right) & \frac{k}{5} \text{ is an integer} \\ 0 & \text{otherwise} \end{cases}$$   (4.280)

From the definition of the DT comb function,

$$\text{comb}_{N_0}\left[\frac{k}{5}-1\right] = \text{comb}_{N_0}\left[\frac{k-5}{5}\right] = \text{comb}_{N_0}[k-5].$$   (4.281)

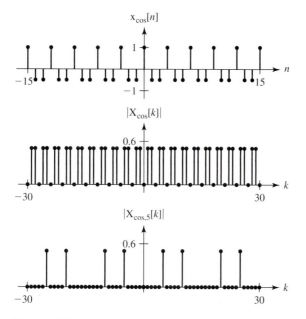

**Figure 4.47**
The cosine signal and its DTFS harmonic functions.

Therefore,

$$
X_{cos,5}[k] = \begin{cases} \dfrac{1}{2}\left(\mathrm{comb}_3[k-5] + \mathrm{comb}_3[k+5]\right) & \dfrac{k}{5} \text{ is an integer} \\ 0 & \text{otherwise} \end{cases} \tag{4.282}
$$

which can also be written as

$$
X_{cos,5}[k] = \frac{1}{2}(\mathrm{comb}_{15}[k-5] + \mathrm{comb}_{15}[k+5]) \tag{4.283}
$$

(Figure 4.47).

The DTFS harmonic function of the DT comb over its fundamental period is

$$
X_{comb}[k] = \frac{1}{5}. \tag{4.284}
$$

Then, using the change-of-period property to find the DTFS harmonic function over three of its fundamental periods, we get

$$
X_{comb,3}[k] = \left\{ \begin{array}{ll} \dfrac{1}{5} & \dfrac{k}{3} \text{ is an integer} \\ 0 & \text{otherwise} \end{array} \right\} = \frac{1}{5}\,\mathrm{comb}_3[k] \tag{4.285}
$$

(Figure 4.48).

Then the DTFS harmonic function for $x[n]$ is

$$
X[k] = X_{cos,5}[k] + X_{comb,3}[k] = \frac{1}{2}(\mathrm{comb}_{15}[k-5] + \mathrm{comb}_{15}[k+5]) + \frac{1}{5}\,\mathrm{comb}_3[k] \tag{4.286}
$$

(Figure 4.49).

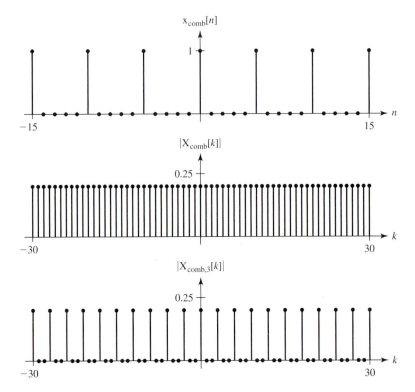

**Figure 4.48**
The comb signal and its DTFS harmonic functions.

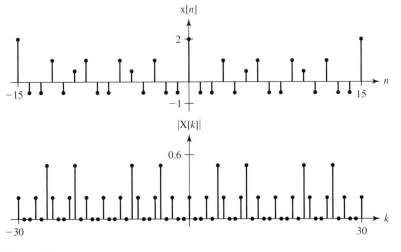

**Figure 4.49**
The overall signal and the magnitude of its DTFS harmonic function.

## MULTIPLICATION–CONVOLUTION DUALITY

Let $z[n] = x[n]y[n]$ and let $N_{x0} = N_{y0} = N_0$. Then

$$Z[k] = \frac{1}{N_0} \sum_{n=\langle N_0 \rangle} z[n]e^{-j2\pi(kF_0)n} = \frac{1}{N_0} \sum_{n=\langle N_0 \rangle} x[n]y[n]e^{-j2\pi(kF_0)n}, \tag{4.287}$$

and, using

$$y[n] = \sum_{q=\langle N_0 \rangle} Y[q] e^{j2\pi(qF_0)n}, \tag{4.288}$$

$$Z[k] = \frac{1}{N_0} \sum_{n=\langle N_0 \rangle} x[n] \sum_{q=\langle N_0 \rangle} Y[q] e^{j2\pi(qF_0)n} e^{-j2\pi(kF_0)n}$$

$$= \frac{1}{N_0} \sum_{n=\langle N_0 \rangle} x[n] \sum_{q=\langle N_0 \rangle} Y[q] e^{-j2\pi((k-q)F_0)n}$$

$$= \sum_{q=\langle N_0 \rangle} Y[q] \underbrace{\frac{1}{N_0} \sum_{n=\langle N_0 \rangle} x[n] e^{-j2\pi((k-q)F_0)n}}_{=X[k-q]}$$

$$= \sum_{q=\langle N_0 \rangle} Y[q] X[k-q] \tag{4.289}$$

This result looks just like a convolution sum except that $q$ extends over a finite range instead of an infinite one. Therefore, by analogy with the CTFS multiplication–convolution duality property, this is a *periodic convolution sum* which can be symbolized by

$$Z[k] = Y[k] \circledast X[k]. \tag{4.290}$$

Therefore,

$$\boxed{x[n]y[n] \xleftrightarrow{\mathcal{FS}} Y[k] \circledast X[k] = \sum_{q=\langle N_0 \rangle} Y[q] X[k-q]}. \tag{4.291}$$

In a manner similar to CT convolution, multiplication of two DT signals corresponds to the convolution sum of their DTFS harmonic functions, but the convolution is now a *periodic* convolution sum. If $N_{x0} \neq N_{y0}$, then, as in the CTFS, we must find a common period for the two signals and use that as the $N_0$ in (4.291).

Now let $Z[k] = Y[k]X[k]$ and $N_{x0} = N_{y0} = N_0$. Then

$$z[n] = \sum_{k=\langle N_0 \rangle} X[k] Y[k] e^{j2\pi(kF_0)n}$$

$$= \sum_{k=\langle N_0 \rangle} \frac{1}{N_0} \sum_{m=\langle N_0 \rangle} x[m] e^{-j2\pi(kF_0)m} Y[k] e^{j2\pi(kF_0)n}$$

$$= \frac{1}{N_0} \sum_{m=\langle N_0 \rangle} x[m] \underbrace{\sum_{k=\langle N_0 \rangle} Y[k] e^{j2\pi(kF_0)(n-m)}}_{=y[n-m]}$$

$$= \frac{1}{N_0} \sum_{m=\langle N_0 \rangle} x[m] y[n-m] \tag{4.292}$$

or

$$\boxed{\text{x}[n] \circledast \text{y}[n] \xleftrightarrow{\;\mathcal{FS}\;} N_0 \text{Y}[k] \text{X}[k]} \;.$$

(4.293)

So for the DTFS there is an elegant symmetry. Multiplication in either domain corresponds to a periodic convolution sum in the other domain (except for a scale factor of $N_0$ in the case of discrete-time periodic convolution). As we have seen before, if $N_{x0} \neq N_{y0}$, then a common period for the two signals must be used both to find the harmonic functions and as $N_0$ in (4.293).

## FIRST BACKWARD DIFFERENCE

This property is analogous to the time-differentiation property of the CTFS. Let $\text{z}[n] = \text{x}[n] - \text{x}[n-1]$ and let $N_{x0} = N_{z0} = N_0$. Then

$$\text{Z}[k] = \frac{1}{N_0} \sum_{n=\langle N_0 \rangle} \text{z}[n] e^{-j2\pi(kF_0)n} = \frac{1}{N_0} \sum_{n=\langle N_0 \rangle} (\text{x}[n] - \text{x}[n-1]) e^{-j2\pi(kF_0)n}$$

$$= \frac{1}{N_0} \left[ \sum_{n=\langle N_0 \rangle} \text{x}[n] e^{-j2\pi(kF_0)n} - \sum_{n=\langle N_0 \rangle} \text{x}[n-1] e^{-j2\pi(kF_0)n} \right].$$

(4.294)

Using the time-shifting property previously derived,

$$\text{Z}[k] = \text{X}[k] - e^{-j2\pi(kF_0)n} \text{X}[k] = \left(1 - e^{-j2\pi(kF_0)}\right) \text{X}[k]$$

(4.295)

$$\boxed{\text{x}[n] - \text{x}[n-1] \xleftrightarrow{\;\mathcal{FS}\;} \left(1 - e^{-j2\pi(kF_0)}\right) \text{X}[k]}$$

(4.296)

or

$$\boxed{\text{x}[n] - \text{x}[n-1] \xleftrightarrow{\;\mathcal{FS}\;} \left(1 - e^{-j(k\Omega_0)}\right) \text{X}[k]} \;.$$

(4.297)

This result is similar to the equivalent result for the CTFS.

## ACCUMULATION

The property is analogous to the time-integration property of the CTFS. Let $\text{z}[n] = \sum_{m=-\infty}^{n} \text{x}[m]$. It is important for this property to consider the effect of the average value of $\text{x}[n]$. We can write the signal $\text{x}[n]$ as

$$\text{x}[n] = \text{x}_0[n] + \text{X}[0],$$

(4.298)

where $\text{x}_0[n]$ is a signal with an average value of zero and $\text{X}[0]$ is the average value of $\text{x}[n]$. Then

$$\text{z}[n] = \sum_{m=-\infty}^{n} \text{x}_0[m] + \sum_{m=-\infty}^{n} \text{X}[0].$$

(4.299)

Since $\text{X}[0]$ is a constant, $\sum_{m=-\infty}^{n} \text{X}[0]$ increases or decreases linearly with $n$, *unless* $\text{X}[0] = 0$. Therefore, if $\text{X}[0] \neq 0$, $\text{z}[n]$ is not periodic and we cannot find its DTFS. If the average value of x[n] is zero, z[n] is periodic and we can find a DTFS for it. Since accumulation is the inverse of the first backward difference,

$$\text{If} \quad z[n] = \sum_{m=-\infty}^{n} x[m], \quad \text{then} \quad x[n] = z[n] - z[n-1]. \qquad \textbf{(4.300)}$$

The first backward difference property proved that $X[k] = (1 - e^{-j2\pi(kF_0)})Z[k]$. Therefore,

$$Z[k] = \frac{X[k]}{1 - e^{-j2\pi(kF_0)}} \qquad k \neq 0 \qquad \textbf{(4.301)}$$

and

$$\boxed{\sum_{m=-\infty}^{n} x[m] \xleftrightarrow{\mathcal{FS}} \frac{X[k]}{1 - e^{-j2\pi(kF_0)}} \qquad k \neq 0} \qquad \textbf{(4.302)}$$

or

$$\boxed{\sum_{m=-\infty}^{n} x[m] \xleftrightarrow{\mathcal{FS}} \frac{X[k]}{1 - e^{-j(k\Omega_0)}} \qquad k \neq 0}. \qquad \textbf{(4.303)}$$

## EVEN AND ODD SIGNALS

These properties are identical to the equivalent properties of the CTFS. If x[n] is an even signal, $x[n] = x[-n]$ and

$$X[k] = X[-k]. \qquad \textbf{(4.304)}$$

If, in addition, x[n] is real-valued, then we already know that $X[k] = X^*[-k]$. Therefore, when x[n] is even and real-valued, $X[k]$ is also even and real-valued.

If $x[n] = -x[-n]$, then the derivation is exactly the same except for a sign and

$$X[k] = -X[-k]. \qquad \textbf{(4.305)}$$

If, in addition, x[n] is real-valued, $X[k] = X^*[-k]$. Therefore, when x[n] is odd and real-valued, if $X[k]$ is to satisfy both

$$X[k] = -X[-k] \qquad \text{and} \qquad X[k] = X^*[-k] \qquad \textbf{(4.306)}$$

simultaneously, it must be odd and *purely imaginary*.

## PARSEVAL'S THEOREM

The total signal energy of x[n] is infinite. The signal energy over one period $N_{x0} = N_0$ is defined as

$$E_{x,N_0} = \sum_{n=\langle N_0 \rangle} |x[n]|^2 = \sum_{n=\langle N_0 \rangle} \left| \sum_{k=\langle N_0 \rangle} X[k] e^{j2\pi(kF_0)n} \right|^2$$

$$= \sum_{n=n_0}^{n_0+N_0-1} \left( \sum_{k=\langle N_0 \rangle} X[k] e^{j2\pi(kF_0)n} \right) \left( \sum_{q=\langle N_0 \rangle} X[q] e^{j2\pi(qF_0)n} \right)^*$$

$$= \sum_{n=\langle N_0 \rangle} \left( \sum_{k=\langle N_0 \rangle} |X[k]|^2 + \sum_{\substack{k=\langle N_0 \rangle \\ k \neq q}} \sum_{q=\langle N_0 \rangle} X[k]e^{j2\pi(kF_0)n} X^*[q]e^{-j2\pi(qF_0)n} \right)$$

$$= \sum_{n=\langle N_0 \rangle} \left( \sum_{k=\langle N_0 \rangle} |X[k]|^2 + \underbrace{\sum_{\substack{k=\langle N_0 \rangle \\ k \neq q}} \sum_{q=\langle N_0 \rangle} X[k]X^*[q]e^{j2\pi((k-q)F_0)n}}_{=0} \right)$$

$$= N_0 \sum_{k=\langle N_0 \rangle} |X[k]|^2. \tag{4.307}$$

Then

$$\boxed{\frac{1}{N_0} \sum_{n=\langle N_0 \rangle} |x[n]|^2 = \sum_{k=\langle N_0 \rangle} |X[k]|^2}, \tag{4.308}$$

which, in words, says that the average signal power of the signal is equal to the sum of the average signal powers in its DTFS harmonics.

## SUMMARY OF DTFS PROPERTIES

Linearity $\qquad\qquad \alpha x[n] + \beta y[n] \overset{\mathcal{FS}}{\longleftrightarrow} \alpha X[k] + \beta Y[k]$

Time shifting $\qquad\quad x[n - n_0] \overset{\mathcal{FS}}{\longleftrightarrow} e^{-j2\pi(kF_0)n_0} X[k]$

$\qquad\qquad\qquad\quad x[n - n_0] \overset{\mathcal{FS}}{\longleftrightarrow} e^{-j(k\Omega_0)n_0} X[k]$

Frequency shifting $\quad e^{j2\pi(k_0 F_0)n} x[n] \overset{\mathcal{FS}}{\longleftrightarrow} X[k - k_0]$

$\qquad\qquad\qquad\quad e^{j(k_0 \Omega_0)n} x[n] \overset{\mathcal{FS}}{\longleftrightarrow} X[k - k_0]$

Conjugation $\qquad\quad x^*[n] \overset{\mathcal{FS}}{\longleftrightarrow} X^*[-k]$

Time reversal $\qquad\; x[-n] \overset{\mathcal{FS}}{\longleftrightarrow} X[-k]$

Time scaling $\qquad$ If $z[n] = \begin{cases} x\left[\dfrac{n}{m}\right] & \dfrac{n}{m} \text{ is an integer} \\ 0 & \text{otherwise} \end{cases}$, then for

$\qquad\qquad\qquad\quad N_F = mN_0,\; Z[k] = \dfrac{1}{m}X[k].$

Change of period $\quad$ For $N_F = qN_0,\; X_q[k] = \begin{cases} X\left[\dfrac{k}{q}\right] & \dfrac{k}{q} \text{ an integer} \\ 0 & \text{otherwise} \end{cases}$

Multiplication–
convolution duality $\quad x[n]y[n] \overset{\mathcal{FS}}{\longleftrightarrow} Y[k] \circledast X[k] = \displaystyle\sum_{q=\langle N_0 \rangle} Y[q]X[k - q]$

$\qquad\qquad\quad x[n] \circledast y[n] = \displaystyle\sum_{m=\langle N_0 \rangle} x[m]y[n - m] \overset{\mathcal{FS}}{\longleftrightarrow} N_0 Y[k]X[k]$

First backward difference

$$x[n] - x[n-1] \xleftrightarrow{\ \mathcal{FS}\ } \left(1 - e^{-j2\pi(kF_0)}\right)X[k]$$

$$x[n] - x[n-1] \xleftrightarrow{\ \mathcal{FS}\ } \left(1 - e^{-j(k\Omega_0)}\right)X[k]$$

Accumulation

$$\sum_{m=-\infty}^{n} x[m] \xleftrightarrow{\ \mathcal{FS}\ } \frac{X[k]}{1 - e^{-j2\pi(kF_0)}} \qquad k \neq 0$$

$$\sum_{m=-\infty}^{n} x[m] \xleftrightarrow{\ \mathcal{FS}\ } \frac{X[k]}{1 - e^{-j(k\Omega_0)}} \qquad k \neq 0$$

Real-valued functions    If $\mathrm{Re}(x[n]) = x[n]$, then $X[k] = X^*[-k]$.

Parseval's theorem    $\dfrac{1}{N_0}\displaystyle\sum_{n=\langle N_0\rangle} |x[n]|^2 = \sum_{k=\langle N_0\rangle} |X[k]|^2$

## 4.10  CONVERGENCE OF THE DISCRETE-TIME FOURIER SERIES

The convergence of the DTFS is simpler than the convergence of the CTFS. Since the summation is finite, exact equality between a function and its DTFS representation is achieved with a finite number of terms $N_0$. A DT signal $x[n]$, similar to the CT square wave used to illustrate CTFS convergence, and its DTFS harmonic function $X[k]$ are illustrated in Figure 4.50 along with the corresponding CT signal and its CTFS harmonic function for comparison. In Figures 4.51 through 4.53 are the partial sums,

$$x_N[n] = \sum_{k=-N}^{N} X[k] e^{j2\pi(nk/N_0)} \qquad N < \frac{N_0}{2} \qquad\qquad \textbf{(4.309)}$$

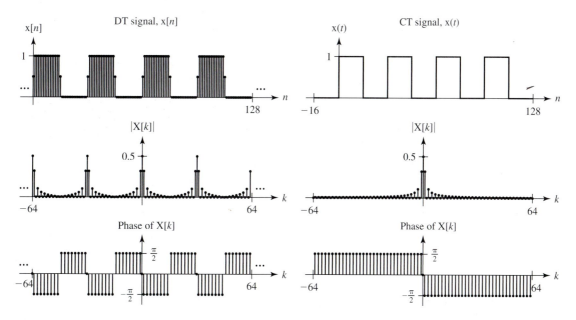

**Figure 4.50**
A DT signal with its DTFS harmonic function and the corresponding CT signal with its CTFS harmonic function.

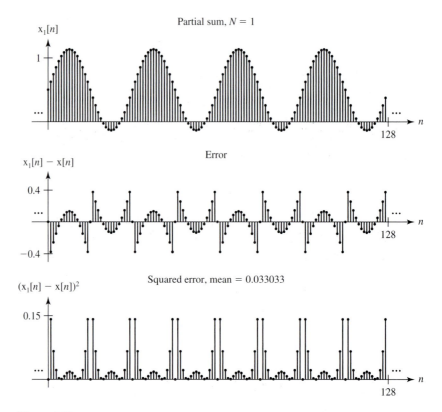

**Figure 4.51**
Partial sum, error, and squared error for $N = 1$.

**Table 4.1** Mean-squared error (MSE) versus $N$

| $N$ | MSE |
|---|---|
| 0 | 0.23438 |
| 1 | 0.03303 |
| 2 | 0.03303 |
| 3 | 0.01181 |
| 4 | 0.01181 |
| 5 | 0.00497 |
| 6 | 0.00497 |
| 7 | 0.00207 |
| 8 | 0.00207 |
| 9 | 0.00076 |
| 10 | 0.00076 |
| 11 | 0.00020 |
| 12 | 0.00020 |
| 13 | 0.00002 |
| 14 | 0.00002 |
| 15 | 0.00000 |

for $N = 1, 3, 5$, the associated errors $x_N[n] - x[n]$, and the squared errors $(x_N[n] - x[n])^2$.

It is obvious from these figures that the partial sums approach the DT function and that the mean-squared error is a monotonically decreasing function of $N$ (Table 4.1). The partial sum for $N = 15$ is exactly the same as for the original DT function. There is no Gibbs phenomenon in the DTFS case. Also the convergence to the mean at a discontinuity, which was shown in the CTFS case, has no meaning in the DT case because *discontinuity* has no meaning. A DT function has no continuity.

The DT signal in Figure 4.49 is *almost* the signal formed by sampling the corresponding CT signal, according to the rule presented in Chapter 3 for sampling in the presence of discontinuities,

$$g[n] = \lim_{\varepsilon \to 0} g(nT_s + \varepsilon) \qquad \varepsilon > 0, \tag{4.310}$$

but not quite. Notice that at discrete times $n = 0 \pm m(N_0/2)$, where $m$ is an integer, the signal has the value $\frac{1}{2}$ instead of zero or one. It is interesting to see what happens if we actually sample the CT signal (Figure 4.54).

Since the two DT functions are not the same, we shouldn't expect the DTFS harmonic functions to be the same. Their magnitudes are almost the same, but the phase is noticeably different. It now no longer simply switches back and forth between $\pi/2$ and $-(\pi/2)$, but instead has a linear dependence on frequency superimposed on it. If

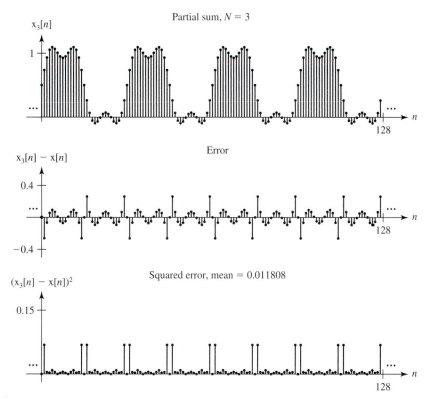

**Figure 4.52**
Partial sum, error, and squared error for $N = 3$.

we examine the first DTFS harmonic, we can see why. For the second DT signal,

$$X[1] = \frac{1}{N_0} \sum_{n=\langle N_0 \rangle} x[n]e^{-j2\pi(nk/N_0)} = \frac{1}{32} \sum_{n=0}^{N_0-1} x[n]e^{-j2\pi(n/32)} = \frac{1}{32} \sum_{n=0}^{15} e^{-j2\pi(n/32)}$$

$$= \frac{1}{32} \frac{1-e^{-j\pi}}{1-e^{-j(\pi/16)}} = \frac{1}{32} \frac{e^{-j(\pi/2)}}{e^{-j(\pi/32)}} \frac{e^{+j(\pi/2)}-e^{-j(\pi/2)}}{e^{+j(\pi/32)}-e^{-j(\pi/32)}}$$

$$= -j0.3188 e^{+j(\pi/32)} = 0.03124 - j0.3173 \tag{4.311}$$

By comparison, this same harmonic for the first DT signal was

$$X[1] = -j0.3173. \tag{4.312}$$

The imaginary parts are the same, but the DTFS harmonic function of the second signal has a small, but nonzero, real part 0.03124 and that causes a phase difference. And remember, a phase difference in the DTFS harmonic function corresponds to a time shift in the DT domain. Figure 4.55 illustrates the relation between the original DT signal and its $N = 1$ partial sum for both DT signals.

For both DT signals, the $N = 1$ partial sum consists of the average value $\frac{1}{2}$, plus a sine whose fundamental period is the same as the square wave fundamental period $N_0 = 32$. For the first DT signal, the fundamental sine wave has its zero crossings at

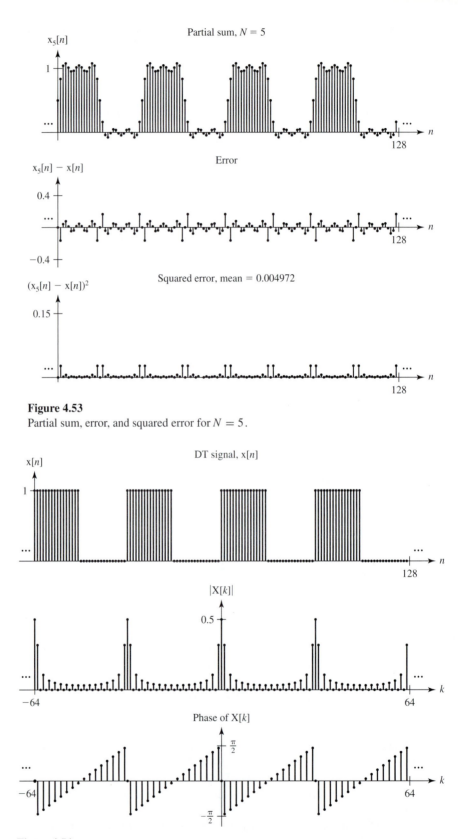

**Figure 4.53**
Partial sum, error, and squared error for $N = 5$.

**Figure 4.54**
An alternate DT signal, formed by sampling the corresponding CT signal, and its DTFS.

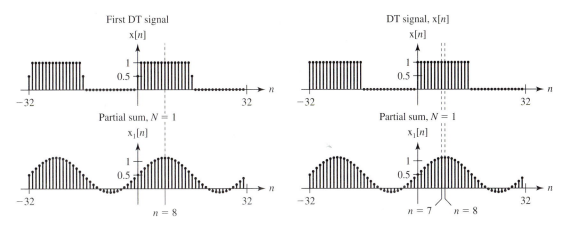

**Figure 4.55**
Comparison of $N = 1$ partial sums for the two DT signals.

exactly the $n = 0$ and $n = 16$ points. It also has its peaks at the $n = 8$ and $n = 24$ points which are exactly the symmetry points of the first and second half-periods of the square wave. It has no phase shift (which corresponds to the phase plots which are the phase of a *cosine*). For the second DT signal, the fundamental sine wave is phase-shifted slightly because the symmetry point of the first half-period of the square wave does not fall exactly on an integer DT value, but rather, halfway between the $n = 7$ and $n = 8$ points, and the sine function has to shift by a corresponding amount. This illustrates one of the many subtleties of sampling. As the sampling rate increases, the difference between these two DT signals becomes progressively less significant.

## 4.11 FREQUENCY RESPONSE OF LTI SYSTEMS WITH PERIODIC EXCITATION

The reason for introducing the Fourier series is that it is a tool for analysis of an LTI system's response to an excitation. Since only periodic signals can be expressed exactly for all time as a Fourier series, the analysis using the Fourier series will be limited to an excitation that is periodic. (That limitation will be removed in Chapter 5 with the introduction of the Fourier transform.) Let us return to the $RC$ lowpass filter as a first example. Recall that the differential equation describing the relationship between the input voltage signal $v_{in}(t)$ and the output voltage signal $v_{out}(t)$ is

$$RC v'_{out}(t) + v_{out}(t) = v_{in}(t). \tag{4.313}$$

Let the input voltage signal $v_{in}(t)$ be a periodic signal expressed as a complex CTFS,

$$v_{in}(t) = \sum_{k=-\infty}^{\infty} V_{in}[k] e^{j2\pi(kf_0)t}, \tag{4.314}$$

where $f_F$ is chosen to be equal to $f_0$, the fundamental frequency of the excitation. Since this is an LTI system, the response can be found by finding the response to each complex sinusoid individually and then summing those responses. The equation for the $k$th input voltage signal complex sinusoid is

$$RCv'_{\text{out},k}(t) + v_{\text{out},k}(t) = v_{\text{in},k}(t) = V_{\text{in}}[k]e^{j2\pi(kf_0)t}. \tag{4.315}$$

The steady-state output voltage signal will be of the same form as the input voltage signal, with the same frequency $kf_0$, but a different complex CTFS (in general). Let the form of the response to the $k$th complex sinusoidal excitation be

$$v_{\text{out},k}(t) = V_{\text{out}}[k]e^{j2\pi(kf_0)t}. \tag{4.316}$$

Then the equation becomes

$$j2k\pi f_0 RC V_{\text{out}}[k]e^{j2\pi(kf_0)t} + V_{\text{out}}[k]e^{j2\pi(kf_0)t} = V_{\text{in}}[k]e^{j2\pi(kf_0)t}. \tag{4.317}$$

It is important, at this point, to observe that by assuming a solution in this form, the differential equation has been changed into an algebraic equation. We have transformed the way we describe the excitation and response and, by doing so, have transformed the differential equation into an algebraic equation. Solving the algebraic equation for $V_{\text{out}}[k]$,

$$V_{\text{out}}[k] = \frac{V_{\text{in}}[k]}{j2k\pi f_0 RC + 1}. \tag{4.318}$$

The steady-state solution for the response is then

$$v_{\text{out}}(t) = \sum_{k=-\infty}^{\infty} V_{\text{out}}[k]e^{j2\pi(kf_0)t} = \sum_{k=-\infty}^{\infty} \frac{V_{\text{in}}[k]}{j2k\pi f_0 RC + 1}e^{j2\pi(kf_0)t}. \tag{4.319}$$

Notice that the relationship between the response CTFS harmonic function $V_{\text{out}}[k]$ and the CTFS harmonic function $V_{\text{in}}[k]$ is a function of the frequency $kf_0$. If the fundamental frequency $f_0$ of the input voltage signal is small compared to $1/2\pi RC$, then for small $k$, $V_{\text{out}}[k]$ and $V_{\text{in}}[k]$ are approximately the same,

$$\lim_{kf_0 \to 0} \frac{V_{\text{out}}[k]}{V_{\text{in}}[k]} = \lim_{kf_0 \to 0} \frac{1}{j2k\pi f_0 RC + 1} = 1. \tag{4.320}$$

The higher $k$ becomes, the smaller $V_{\text{out}}[k]$ becomes in comparison with $V_{\text{in}}[k]$,

$$\lim_{kf_0 \to \infty} \frac{V_{\text{out}}[k]}{V_{\text{in}}[k]} = \lim_{kf_0 \to \infty} \frac{1}{j2k\pi f_0 RC + 1} = 0. \tag{4.321}$$

This circuit tends to deemphasize higher frequencies in the input voltage signal while having very little effect on low frequencies. That is why this circuit is called a *lowpass filter;* it passes low-frequency input voltage signal components. This can be illustrated by graphing the magnitude and phase of the ratio of the two CTFS harmonic functions, $V_{\text{out}}[k]/V_{\text{in}}[k]$ (for a particular choice of $R$, $C$, and $f_0$), as in Figure 4.56. As can be observed in Figure 4.56, the ratio of magnitudes is approximately one for low frequencies and falls toward zero at higher frequencies.

Now compare CTFS analysis with a technique learned in elementary circuit analysis, the *phasor* technique used to solve for steady-state responses in circuits excited by a single sinusoid. The sinusoid is first converted to a complex sinusoid with a magnitude equal to the sinusoid's amplitude and an angle equal to the phase shift of the sinusoid relative to a cosine,

$$A\cos(2\pi f_0 t + \theta) \to Ae^{j2\pi f_0 t}e^{j\theta}. \tag{4.322}$$

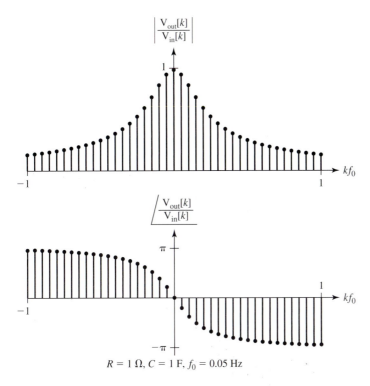

**Figure 4.56**
Magnitude and phase of the ratio $V_{out}[k]/V_{in}[k]$ versus frequency.

Then the variation with $t$ is suppressed because, in steady state, all the terms in the differential equation have the same functional form of time dependence,

$$Ae^{j2\pi f_0 t}e^{j\theta} \to Ae^{j\theta} \qquad \text{or} \qquad A\angle\theta. \tag{4.323}$$

Then the solution is found using the impedances $Z_R$ and $Z_c$ of the resistor and capacitor, respectively, and phasor notation for the excitation and response voltages. The phasor equation for the $RC$ lowpass filter, relating the excitation and response (writing impedance in the conventional way in terms of the radian frequency $\omega$), is

$$V_{out} = \frac{Z_c}{Z_c + Z_R}V_{in} = \frac{1/j\omega C}{(1/j\omega C) + R}V_{in} = \frac{V_{in}}{j\omega RC + 1} = \frac{V_{in}}{j2\pi fRC + 1}. \tag{4.324}$$

The similarity between this result and the CTFS analysis result

$$V_{out}[k] = \frac{V_{in}[k]}{j2\pi(kf_0)RC + 1} = \frac{V_{in}[k]}{j(k\omega_0)RC + 1} \tag{4.325}$$

is obvious. The only real difference is that in phasor analysis the assumption is that the excitation is a single-frequency complex sinusoid. In the CTFS analysis the excitation is a linear combination of sinusoids at integer multiples of the fundamental frequency of the excitation. Therefore, CTFS analysis is equivalent to phasor analysis, done multiple times, once for each component at the harmonic radian frequencies $\omega$ or harmonic cyclic frequencies $f$ present in the excitation.

**EXAMPLE 4.9**

Find the response of the system illustrated in Figure 4.57 to the periodic square-wave DT excitation

$$x[n] = \text{rect}_2[n] * \text{comb}_8[n]. \tag{4.326}$$

■ **Solution**

The difference equation describing this system can be found by combining the two difference equations

$$y[n] = 5x[n] - y_1[n] \tag{4.327}$$

and

$$y_1[n] = x[n] + \frac{4}{5}y_1[n-1] \tag{4.328}$$

into

$$y[n] = 4(x[n] - x[n-1]) + \frac{4}{5}y[n-1]. \tag{4.329}$$

Since the excitation $x[n] = \text{rect}_2[n] * \text{comb}_8[n]$ is periodic, it can be expressed as

$$x[n] = \sum_{k=\langle N_0 \rangle} X[k]e^{j2\pi(kn/N_0)} = \sum_{k=\langle 8 \rangle} X[k]e^{j\pi(kn/4)}, \tag{4.330}$$

where, using

$$\text{rect}_{N_w}[n] * \text{comb}_{N_0}[n] \xleftrightarrow{\mathcal{FS}} \frac{2N_w + 1}{N_0} \text{drcl}\left(\frac{k}{N_0}, 2N_w + 1\right), \tag{4.331}$$

we get

$$X[k] = \frac{5}{8} \text{drcl}\left(\frac{k}{8}, 5\right) = \frac{1}{8} \frac{\sin\left(\frac{5}{8}k\pi\right)}{\sin(k\pi/8)}. \tag{4.332}$$

The response which is also periodic with the same fundamental period can also be expressed as a DTFS in the form

$$y[n] = \sum_{k=\langle 8 \rangle} Y[k]e^{j\pi(kn/4)}. \tag{4.333}$$

Substituting this form for $y[n]$ into (4.329) we get

$$\sum_{k=\langle 8 \rangle} Y[k]e^{j\pi(kn/4)} = 4\left(\sum_{k=\langle 8 \rangle} X[k]e^{j\pi(kn/4)} - \sum_{k=\langle 8 \rangle} X[k]e^{j\pi(k(n-1)/4)}\right) + \frac{4}{5}\sum_{k=\langle 8 \rangle} Y[k]e^{j\pi(k(n-1)/4)}. \tag{4.334}$$

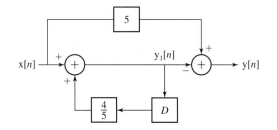

**Figure 4.57**
A DT system.

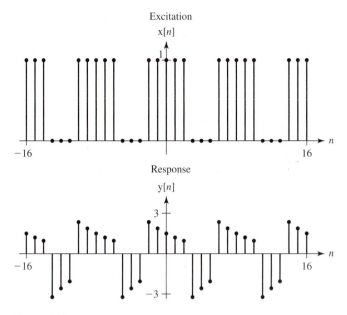

**Figure 4.58**
Excitation and response of the DT system.

Solving for $Y[k]$ we get

$$Y[k] = 4\frac{1 - e^{-j(k\pi/4)}}{1 - \frac{4}{5}e^{-j(k\pi/4)}}X[k] = j8e^{-j(k\pi/8)}\frac{\sin(k\pi/8)}{1 - \frac{4}{5}e^{-j(k\pi/4)}}X[k] \qquad (4.335)$$

or

$$Y[k] = je^{-j(k\pi/8)}\frac{\sin\left(\frac{5}{8}k\pi\right)}{1 - \frac{4}{5}e^{-j(k\pi/4)}}. \qquad (4.336)$$

Therefore, combining (4.333) and (4.336),

$$y[n] = j\sum_{k=(8)} e^{-j(k\pi/8)}\frac{\sin\left(\frac{5}{8}k\pi\right)}{1 - \frac{4}{5}e^{-j(k\pi/4)}}e^{j\pi(kn/4)}. \qquad (4.337)$$

We can choose any interval for $k$ of length eight. For convenience, choose $-4 \le k < 4$. Then

$$y[n] = j\sum_{k=-4}^{3} e^{-j(k\pi/8)}\frac{\sin\left(\frac{5}{8}k\pi\right)}{1 - \frac{4}{5}e^{-j(k\pi/4)}}e^{j\pi(kn/4)}. \qquad (4.338)$$

Using (4.338) we can now plot $x[n]$ and $y[n]$ versus $n$ (Figure 4.58).

## 4.12 SUMMARY OF IMPORTANT POINTS

**1.** The Fourier series expresses a periodic signal as a sum of sinusoids at harmonics of the fundamental frequency of the signal.
**2.** The sinusoids used in the Fourier series to represent a signal are orthogonal to each other.
**3.** A CTFS can be found for any CT signal that satisfies the Dirichlet conditions.
**4.** The complex and trigonometric forms of the Fourier series are related through Euler's identity.

5. For continuous CT signals the Fourier series converges exactly to the signal at every point.
6. For discontinuous CT signals the Fourier series converges exactly to the signal at every point except points of discontinuity. The effects of the actual signal and its Fourier series representation on any real physical system are the same.
7. For DT signals, the convergence of the DTFS is exact at every point.
8. The DTFS of a signal is a finite summation because of the nature of discrete time.
9. If an LTI system is excited by a periodic signal, the response is also a periodic signal with the same fundamental period.
10. The relationship between the periodic excitation and the periodic response of an LTI system characterizes the system.

## EXERCISES WITH ANSWERS

1. Using MATLAB plot each sum of complex sinusoids over the time period indicated.

   a. $x(t) = \dfrac{1}{10} \displaystyle\sum_{k=-30}^{30} \operatorname{sinc}\left(\dfrac{k}{10}\right) e^{j200\pi kt}$      $-15 \text{ ms} < t < 15 \text{ ms}$

   b. $x(t) = \dfrac{j}{4} \displaystyle\sum_{k=-9}^{9} \left[ \operatorname{sinc}\left(\dfrac{k+2}{2}\right) - \operatorname{sinc}\left(\dfrac{k-2}{2}\right) \right] e^{j10\pi kt}$

   $-200 \text{ ms} < t < 200 \text{ ms}$

**Answers:**

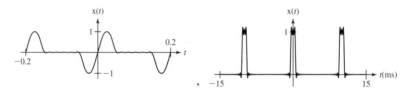

2. Show by direct analytical integration that the integral of the function

   $$g(t) = A\sin(2\pi t)B\sin(4\pi t)$$

   is zero over the interval $-\frac{1}{2} < t < \frac{1}{2}$.

3. Convert the function $g(t) = (1 + j)e^{j4\pi t} + (1 - j)e^{-j4\pi t}$ to an equivalent form in which $j$ does not appear.

**Answer:**
$2\cos(4\pi t) - 2\sin(4\pi t)$

4. Using MATLAB plot these products over the time range indicated and observe in each case that the net area under the product is zero.

   a. $x(t) = -3\sin(16\pi t) \times 2\cos(24\pi t)$      $0 < t < \dfrac{1}{4}$

   b. $x(t) = -3\sin(16\pi t) \times 2\cos(24\pi t)$      $0 < t < 1$

c.  $x(t) = -3 \sin(16\pi t) \times 2 \cos(24\pi t)$      $-\dfrac{1}{16} < t < \dfrac{3}{16}$

d.  $x(t) = x_1(t)x_2(t)$, where $x_1(t)$ is an even, 50%-duty-cycle square wave with a fundamental period of 4 s, an amplitude of 2 and an average value of zero and $x_2(t)$ is an odd, 50%-duty-cycle square wave with a fundamental period of 4 s, an amplitude of 3 and an average value of zero

e.  $x(t) = x_1(t)x_2(t)$, where $x_1(t) = \text{rect}(2t) * \text{comb}(t)$ and $x_2(t) = [\text{rect}(4(t - \frac{1}{8})) * \frac{1}{2} \text{comb}(t/2)] - \frac{1}{8}$

**Answers:**

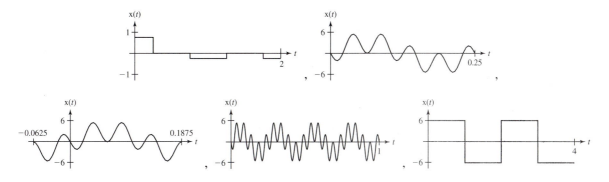

5.  A sine function can be written as

$$\sin(2\pi f_0 t) = \frac{e^{j2\pi f_0 t} - e^{-j2\pi f_0 t}}{j2}.$$

This is a very simple complex CTFS in which the harmonic function is only nonzero at two harmonic numbers, $+1$ and $-1$. Verify that we can write the harmonic function directly as

$$X[k] = \frac{j}{2}(\delta[k + 1] - \delta[k - 1]).$$

Write the equivalent expressions for $\sin(2\pi(-f_0)t)$, and show that the harmonic function is the complex conjugate of the previous one for $\sin(2\pi f_0 t)$.

6.  For each signal, find a complex CTFS which is valid for all time, plot the magnitude and phase of the harmonic function versus harmonic number $k$, and then convert the answers to the trigonometric form of the harmonic function.

a.  $x(t) = 4\,\text{rect}(4t) * \text{comb}(t)$

b.  $x(t) = 4\,\text{rect}(4t) * \dfrac{1}{4}\text{comb}\left(\dfrac{t}{4}\right)$

c.  A periodic signal which is described over one fundamental period by

$$x(t) = \begin{cases} \text{sgn}(t) & |t| < 1 \\ 0 & 1 < |t| < 2 \end{cases}$$

**Answers:**

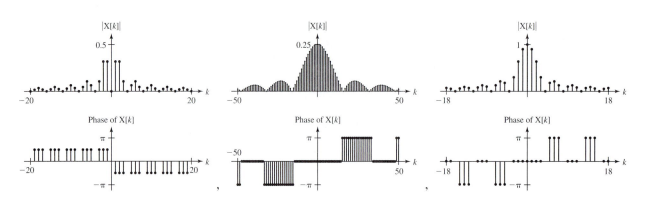

$$X_c[k] = 0, \quad X_s[k] = -2\frac{\cos(\pi k/2) - 1}{\pi k},$$

$$X_c[k] = 2\,\text{sinc}\left(\frac{k}{4}\right), \quad X_s[k] = 0, \quad X_c[k] = \frac{1}{4}\,\text{sinc}\left(\frac{k}{16}\right), \quad X_s[k] = 0$$

7.  Using the CTFS table of transforms in Appendix E and the CTFS properties, find the CTFS harmonic function of each of these periodic signals using the representation period $T_F$ indicated.

   a.  $x(t) = 10\sin(20\pi t) \qquad T_F = \dfrac{1}{10}$

   b.  $x(t) = 2\cos(100\pi(t - 0.005)) \qquad T_F = \dfrac{1}{50}$

   c.  $x(t) = -4\cos(500\pi t) \qquad T_F = \dfrac{1}{50}$

   d.  $x(t) = \dfrac{d}{dt}(e^{-j10\pi t}) \qquad T_F = \dfrac{1}{5}$

   e.  $x(t) = \text{rect}(t) * \text{comb}\left(\dfrac{t}{4}\right) \qquad T_F = 4$

   f.  $x(t) = \text{rect}(t) * \text{comb}(t) \qquad T_F = 1$

   g.  $x(t) = \text{tri}(t) * \text{comb}(t) \qquad T_F = 1$

**Answers:**

$-2(\delta[k - 5] + \delta[k + 5]), \quad \delta[k], \quad \delta[k], \quad j5(\delta[k + 1] - \delta[k - 1]),$

$j(\delta[k + 1] - \delta[k - 1]), \quad -j10\pi\delta[k + 1], \quad \text{sinc}\left(\dfrac{k}{4}\right)$

8.  If a periodic signal $x(t)$ has a fundamental period of 10 s and its harmonic function is

$$X[k] = 4\,\text{sinc}\left(\frac{k}{20}\right)$$

with a representation period of 10 s, what is the harmonic function of $z(t) = x(4t)$ using the same representation period of 10 s?

**Answer:**

$$Z[k] = \begin{cases} 4\,\text{sinc}\left(\dfrac{k}{80}\right) & \dfrac{k}{4} \text{ is an integer} \\ 0 & \text{otherwise} \end{cases}$$

9. A periodic signal $x(t)$ has a fundamental period of 4 ms, and its harmonic function is

$$X[k] = 15(\delta[k-1] + \delta[k+1])$$

with a representation period of 4 ms. Find the integral of $x(t)$.

**Answer:**

$$\frac{3}{50\pi}\sin(500\pi t)$$

10. If $X[k]$ is the harmonic function over one fundamental period of a unit-amplitude, 50%-duty-cycle square wave with an average value of zero and a fundamental period of 1 μs, find an expression consisting of only real-valued functions for the signal whose harmonic function is $X[k-10] + X[k+10]$.

**Answer:**

$2[2\,\text{rect}(2 \times 10^6 t) * 10^6\text{comb}(10^6 t) - 1]\cos(2 \times 10^7 \pi t)$

11. Find the harmonic function for a sine wave of the general form $A\sin(2\pi f_0 t)$. Then, using Parseval's theorem, find its signal power and verify that it is the same as the signal power found directly from the function itself.

**Answer:**

$$\frac{A^2}{2}$$

12. Show for a cosine and a sine that the CTFS harmonic functions have the property

$$X[k] = X^*[-k].$$

13. Find the time functions associated with these harmonic functions assuming $T_F = 1$.

    a. $X[k] = \delta[k-2] + \delta[k] + \delta[k+2]$

    b. $X[k] = 10\,\text{sinc}\left(\dfrac{k}{10}\right)$

**Answers:**

$2\cos(4\pi t) + 1, \quad 100\,\text{rect}(10t) * \text{comb}(t)$

**14.** Find the even and odd parts $x_e(t)$ and $x_o(t)$ of

$$x(t) = 20 \cos\left(40\pi t + \frac{\pi}{6}\right).$$

Then find the harmonic functions $X_e[k]$ and $X_o[k]$ corresponding to them. Using the time-shifting property find the harmonic function $X[k]$ and compare it to the sum of the two harmonic functions $X_e[k]$ and $X_o[k]$.

**Answer:**
$$10(\delta[k-1]e^{j(\pi/6)} + \delta[k+1]e^{-j(\pi/6)})$$

**15.** Using the direct summation formula find and sketch the DTFS harmonic function of $\text{comb}_{N_0}[n]$ with $N_F = N_0$.

**Answer:**

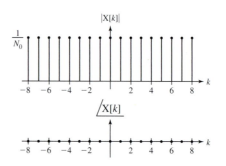

**16.** Using the DTFS table of transforms in Appendix E and the DTFS properties, find the DTFS harmonic function of each of these periodic signals using the representation period $N_F$ indicated.

 *a.*   $x[n] = 6 \cos\left(\dfrac{2\pi n}{32}\right)$       $N_F = 32$

 *b.*   $x[n] = 10 \sin\left(\dfrac{2\pi(n-2)}{12}\right)$       $N_F = 12$

 *c.*   $x[n] = \begin{cases} x_1\left[\dfrac{n}{8}\right] & \dfrac{n}{8} \text{ is an integer} \\ 0 & \text{otherwise} \end{cases}$       $N_F = 48$

  where $x_1[n] = \sin\left(\dfrac{2\pi n}{6}\right)$

 *d.*   $x[n] = e^{j2\pi n}$       $N_F = 6$

 *e.*   $x[n] = \cos\left(\dfrac{2\pi n}{16}\right) - \cos\left(\dfrac{2\pi(n-1)}{16}\right)$       $N_F = 16$

f.  $x[n] = -\sin\left(\dfrac{33\pi n}{32}\right)$    $N_F = 64$

g.  $x[n] = \text{rect}_5[n] * \text{comb}_{11}[n]$    $N_F = 11$

h.  $x[n] = \text{rect}_2[n] * \text{comb}_{21}[n-3]$    $N_F = 21$

**Answers:**

$X[k] = \text{comb}_{11}[k]$,   $X[k] = \dfrac{5}{21}\,\text{drcl}\left(\dfrac{k}{21}, 5\right) e^{-j(2\pi k/7)}$,

$\dfrac{1 - e^{-j(\pi/8)}}{2}\,\text{comb}_{16}[k-1] + \dfrac{1 - e^{j(\pi/8)}}{2}\,\text{comb}_{16}[k+1]$,

$X[k] = 3(\text{comb}_{32}[k-1] + \text{comb}_{32}[k+1])$,

$X[k] = \dfrac{j}{16}(\text{comb}_6[k+1] - \text{comb}_6[k-1])$,

$X[k] = -\dfrac{j}{2}(\text{comb}_{64}[k+33] - \text{comb}_{64}[k-33])$,   $\text{comb}_6[k]$,

$X[k] = j5(\text{comb}_{12}[k+1] - \text{comb}_{12}[k-1])e^{-j(\pi/3)k}$

**17.**  Find the DTFS harmonic function of

$$x[n] = \sum_{m=-\infty}^{n} \text{comb}_3[m] - \text{comb}_3[m-1]$$

with $N_F = N_0 = 3$.

**Answer:**

$X[k] = \dfrac{1}{3}$

**18.**  Find the average signal power of

$$x[n] = \text{rect}_4[n] * \text{comb}_{20}[n]$$

directly in the DT domain and then find its harmonic function $X[k]$ and the signal power in the $k$ domain and show that they are the same.

**Answer:**

$\dfrac{9}{20}$

**19.**  Using the frequency-shifting property of the DTFS find the DT-domain signal $x[n]$ corresponding to the harmonic function

$$X[k] = \dfrac{7}{32}\,\text{drcl}\left(\dfrac{k-16}{32}, 7\right).$$

**Answer:**

$(\text{rect}_3[n] * \text{comb}_{32}[n])(-1)^n$

**20.** Find the DTFS harmonic function for

$$x[n] = \text{rect}_3[n] * \text{comb}_8[n]$$

with the representation period $N_F = 8$. Then, using MATLAB, plot the DTFS representation

$$x_F[n] = \sum_{k=0}^{7} X[k]e^{j2\pi(kn/8)}$$

over the DT range $-8 \le n < 8$. For comparison, plot the function

$$x_{F2}[n] = \sum_{k=13}^{20} X[k]e^{j2\pi(kn/8)}$$

over the same range. The plots should be identical.

$$\text{rect}_3[n] * \text{comb}_8[n] \xleftrightarrow{\mathcal{FS}} \frac{7}{8} \, \text{drcl}\left(\frac{k}{8}, 7\right)$$

**Answer:**

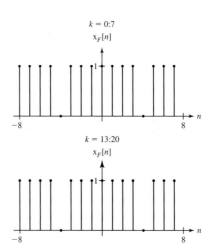

## EXERCISES WITHOUT ANSWERS

**21.** A periodic signal $x(t)$ with a period of 4 s is described over one fundamental period by

$$x(t) = 3 - t \qquad 0 < t < 4.$$

Plot the signal and find its trigonometric CTFS description. Then plot on the same scale approximations to the signal $x_N(t)$ given by

$$x_N(t) = X_c[0] + \sum_{k=1}^{N} X_c[k]\cos(2\pi(kf_F)t) + X_s[k]\sin(2\pi(kf_F)t)$$

for $N = 1$, 2, and 3. (In each case the time scale of the plot should cover at least two fundamental periods of the original signal.)

22. A periodic signal x(t) with a fundamental period of 2 s is described over one fundamental period by

$$
x(t) = \begin{cases} \sin(2\pi t) & |t| < \dfrac{1}{2} \\ 0 & \dfrac{1}{2} < |t| < 1 \end{cases}.
$$

Plot the signal and find its complex CTFS description. Then plot on the same scale approximations to the signal $x_N(t)$ given by

$$
x_N(t) = \sum_{k=-N}^{N} X[k] e^{j2\pi(kf_F)t}
$$

for $N = 1$, 2, and 3. (In each case the time scale of the plot should cover at least two fundamental periods of the original signal.)

23. Find and plot two fundamental periods of the complex CTFS description of $\cos(2\pi t)$

   a. Over the interval $0 < t < 1$

   b. Over the interval $0 < t < 1.5$

24. Using MATLAB, plot the following signals over the time range $-3 < t < 3$.

   a. $x_0(t) = 1$

   b. $x_1(t) = x_0(t) + 2\cos(2\pi t)$

   c. $x_2(t) = x_1(t) + 2\cos(4\pi t)$

   d. $x_{20}(t) = x_{19}(t) + 2\cos(40\pi t)$

   For each part, (a) through (d), numerically evaluate the area of the signal over the time range $-\frac{1}{2} < t < \frac{1}{2}$.

25. Using the CTFS table of transforms in Appendix E and the CTFS properties, find the CTFS harmonic function of each of these periodic signals using the representation period $T_F$ indicated.

   a. $x(t) = 3 \operatorname{rect}\left(2\left(t - \dfrac{1}{4}\right)\right) * \operatorname{comb}(t) \qquad T_F = 1$

   Remember:  If $\qquad g(t) = g_0(t) * \delta(t)$,

   then $\qquad g(t - t_0) = g_0(t - t_0) * \delta(t) = g_0(t) * \delta(t - t_0)$

   and $\qquad g(t - t_0) \neq g_0(t - t_0) * \delta(t - t_0) = g(t - 2t_0)$.

   b. $x(t) = 5[\operatorname{tri}(t - 1) - \operatorname{tri}(t + 1)] * \dfrac{1}{4} \operatorname{comb}\left(\dfrac{t}{4}\right) \qquad T_F = 4$

   c. $x(t) = 3\sin(6\pi t) + 4\cos(8\pi t) \qquad T_F = 1$

   d. $x(t) = 2\cos(24\pi t) - 8\cos(30\pi t) + 6\sin(36\pi t) \qquad T_F = 2$

   e. $x(t) = \displaystyle\int_{-\infty}^{t} \left[\operatorname{comb}(\lambda) - \operatorname{comb}\left(\lambda - \dfrac{1}{2}\right)\right] d\lambda \qquad T_F = 1$

f.   $x(t) = 4\cos(100\pi t)\sin(1000\pi t)$          $T_F = \dfrac{1}{50}$

g.   $x(t) = \left[14\operatorname{rect}\left(\dfrac{t}{8}\right) * \operatorname{comb}\left(\dfrac{t}{12}\right)\right] \circledast \left[7\operatorname{rect}\left(\dfrac{t}{5}\right) * \operatorname{comb}\left(\dfrac{t}{8}\right)\right]$

$T_F = 24$

h.   $x(t) = \left[8\operatorname{rect}\left(\dfrac{t}{2}\right) * \operatorname{comb}\left(\dfrac{t}{5}\right)\right] \circledast \left[-2\operatorname{rect}\left(\dfrac{t}{6}\right) * \operatorname{comb}\left(\dfrac{t}{20}\right)\right]$

$T_F = 20$

26.   A signal $x(t)$ is described over one fundamental period by

$$x(t) = \begin{cases} -A & -\dfrac{T_0}{2} < t < 0 \\ A & 0 < t < \dfrac{T_0}{2} \end{cases}.$$

Find its complex CTFS harmonic function and then, using the integration property, find the CTFS harmonic function of its integral and plot the resulting CTFS representation of the integral.

27.   In some types of communication systems binary data are transmitted using a technique called binary phase-shift keying (BPSK) in which a 1 is represented by a burst of a CT sine wave and a 0 is represented by a burst which is the exact negative of the burst that represents a 1. Let the sine frequency be 1 MHz and let the burst width be 10 periods of the sine wave. Find and plot the CTFS harmonic function for a periodic binary signal consisting of alternating 1s and 0s using its fundamental period as the representation period.

28.   Using the DTFS table of transforms in Appendix E and the DTFS properties, find the harmonic function of each of these periodic signals using the representation period $N_F$ indicated.

a.   $x[n] = e^{-j(\pi n/8)} \circledast \operatorname{comb}_{24}[n]$          $N_F = 48$

b.   $x[n] = (\operatorname{rect}_5[n] * \operatorname{comb}_{24}[n])\sin\left(\dfrac{2\pi n}{6}\right)$          $N_F = 24$

c.   $x[n] = x_1[n] - x_1[n-1]$   where   $x_1[n] = \operatorname{tri}\left(\dfrac{n}{8}\right) * \operatorname{comb}_{20}[n]$

$N_F = 20$

29.   Find the signal power of

$$x[n] = 5\sin\left(\dfrac{14\pi n}{15}\right) - 8\cos\left(\dfrac{26\pi n}{30}\right).$$

30.   Find the DTFS harmonic function $X[k]$ of $x[n] = (\operatorname{rect}_1[n-1] - \operatorname{rect}_1[n-4]) * \operatorname{comb}_6[n]$. Plot the partial sum $x_N[n] = \sum_{k=-N}^{N} X[k]e^{j(\pi nk/3)}$ for $N = 0, 1, 2,$ and then plot the total sum $x[n] = \sum_{k=0}^{5} X[k]e^{j(\pi nk/3)}$.

**31.** Find and sketch the magnitude and phase of the DTFS harmonic function of

$$x[n] = 4\cos\left(\frac{2\pi}{7}n\right) + 3\sin\left(\frac{2\pi}{3}n\right)$$

which is valid for all discrete time.

**32.** The sun shining on the earth is a system in which the radiant power from the sun is the excitation and the atmospheric temperature (among many other things) is a response. A simplified model of the radiant power falling on a typical midlatitude location in North America is that it is periodic with a fundamental period of one year and that every day the radiant power of the sunlight rises linearly from the time the sun rises until the sun is at its zenith and then the radiant power falls linearly until the sun sets. The earth absorbs and stores the radiant energy and reradiates some of the energy into space every night. To keep the model of the excitation as simple as possible assume that the energy loss every night can be modeled as a continuation of the daily linear radiant power pattern except negative at night. There is also a variation with the seasons caused by the tilt of the earth's axis of rotation. This causes the linear rise-and-fall pattern to rise and fall sinusoidally on a much longer time scale as illustrated in Figure E32.

  *a.* Write a mathematical description of the radiant power from the sun.

  *b.* Assume that the earth is a first-order system with a time constant of 0.16 years. What day of the year should be the hottest according to this simplified model?

**33.** The speed and timing of computer computations are controlled by a clock. The clock is a periodic sequence of rectangular pulses, typically with a 50 percent duty cycle. One problem in the design of computer circuit boards is that the clock signal can interfere with other signals on the board by being coupled into adjacent circuits through stray capacitance. Let the computer clock be modeled by a square wave voltage source alternating between 0.4 and 1.6 V at a

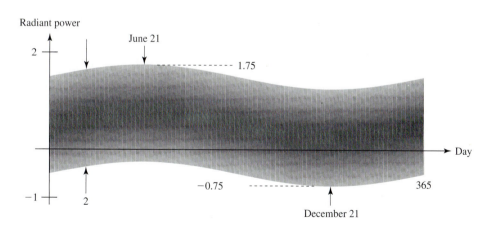

**Figure E32**

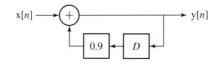

**Figure E34**

frequency of 2 GHz, and let the coupling into an adjacent circuit be modeled by a series combination of a 0.1-picofarad (pF) capacitor and a 50-Ω resistance. Find and plot over two fundamental periods the voltage across the 50-Ω resistance.

**34.** Find and plot versus $F$ the magnitude of the response $y[n]$ to the periodic excitation

$$x[n] = \cos(2\pi F n)$$

in the system shown in Figure E34.

### CHAPTER 5

# The Fourier Transform

## 5.1 INTRODUCTION AND GOALS

The Fourier series is a fine analysis tool, but it is limited. It can describe any signal with engineering usefulness over a finite time and any periodic signal with engineering usefulness over all time as a linear combination of sinusoids. But it cannot describe an aperiodic signal for all time.

In this chapter we will extend the idea of the Fourier series to make it apply to aperiodic signals, developing the Fourier transform. We will see that the Fourier series is just a special case of the Fourier transform. This extension will be done for both continuous-time and discrete-time signals in analogous ways. We will begin to develop techniques of analysis and design of systems in the *frequency domain,* techniques that will be applied more in Chapter 6. Finally we will take an overall view of the four Fourier methods, comparing them, converting between them, and developing various relationships that will be valuable in later chapters, especially Chapter 7 on sampling.

### CHAPTER GOALS

1. To generalize the Fourier series to include aperiodic signals by defining the Fourier transform for both CT and DT signals

2. To establish which types of signals can and cannot be described by a Fourier transform

3. To derive and demonstrate the properties of the Fourier transform for CT and DT signals

4. To demonstrate the interrelationships among all the Fourier methods

## 5.2 THE CONTINUOUS-TIME FOURIER TRANSFORM

### THE TRANSITION FROM THE CONTINUOUS-TIME FOURIER SERIES TO THE CONTINUOUS-TIME FOURIER TRANSFORM

The Fourier series as an analysis tool for LTI systems is very good in many ways, but it has one significant drawback: it can represent, for all time, *only* periodic signals. The

Fourier transform is an extension of the Fourier series to allow the representation of both periodic and aperiodic signals for all time. The salient difference between a periodic signal and an aperiodic signal is that a periodic signal repeats in a finite time $T_0$, called the *fundamental period*. It has been repeating with that fundamental period forever and will continue to do so forever. An aperiodic signal does not have a period. It may repeat a pattern many times within some finite time, but not *over all time*. The transition between the Fourier series and the Fourier transform is accomplished by finding the form of the Fourier series for a periodic signal and then letting the fundamental period approach infinity. If the fundamental period goes to infinity, the signal cannot repeat in a finite time and therefore is no longer periodic. In other words, saying a signal has an infinite fundamental period and saying it is aperiodic are really saying the same thing.

Consider a time-domain signal x($t$) consisting of rectangular pulses of height $A$ and width $w$ with fundamental period $T_0$ (Figure 5.1). This is a specific periodic signal, but it will illustrate the phenomena that occur in letting the fundamental period approach infinity for a general signal. Representing this pulse train with a complex CTFS over exactly one fundamental period ($T_F = T_0$), the CTFS harmonic function X[$k$] of x($t$) is found to be

$$X[k] = \frac{Aw}{T_0} \operatorname{sinc}\left(\frac{kw}{T_0}\right). \tag{5.1}$$

Suppose $w = T_0/2$ (meaning the waveform is at $A$ half the time and at zero the other half). Then

$$X[k] = \frac{A}{2} \operatorname{sinc}\left(\frac{k}{2}\right). \tag{5.2}$$

A graph of X[$k$] versus harmonic number $k$ is shown in Figure 5.2 (with $A = 1$ and $T_0 = 1$).

Now let the fundamental period $T_0$ (and $T_F$) increase from one to five. Then X[0] becomes $\frac{1}{10}$ and the CTFS harmonic function is

$$X[k] = \frac{1}{10} \operatorname{sinc}\left(\frac{k}{10}\right) \tag{5.3}$$

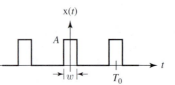

**Figure 5.1**
Rectangular-wave signal.

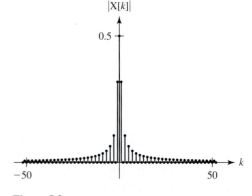

**Figure 5.2**
Magnitude of the CTFS harmonic function of a 50 percent duty-cycle, rectangular-wave signal.

(Figure 5.3). The maximum harmonic amplitude magnitude is 5 times smaller than before because the average value of the function is 5 times smaller than before. As the fundamental period $T_0$ gets larger, the harmonic amplitudes lie on a wider sinc function whose amplitude goes down as $T_0$ increases. In the limit as $T_0$ approaches infinity, the original time-domain waveform $x(t)$ approaches a single rectangular pulse at the origin and the CTFS harmonic function approaches samples from an infinitely wide sinc function with zero amplitude. If we were to multiply $X[k]$ by $T_0$ before plotting it, the amplitude would not go to zero as $T_0$ approached infinity but would stay where it is and simply trace, with higher and higher density, points on a widening sinc function. Also, plotting against $kf_0$ instead of $k$ would make the horizontal scale be frequency instead of harmonic number $k$ and the sinc function would remain the same width on that scale as $T_0$ increases. Making those changes, the last two plots would look like those in Figure 5.4. Call this a *modified* complex CTFS harmonic function for the pulse train. For this modified CTFS harmonic function,

$$T_0 \, X[k] = Aw \, \text{sinc}(w(kf_0)). \tag{5.4}$$

As $T_0$ increases without bound (making the pulse train a single pulse), the discrete variable $kf_0$ approaches a continuous variable which we will call $f$ and the modified CTFS

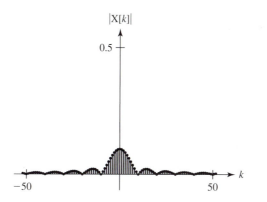

**Figure 5.3**
Magnitude of the CTFS harmonic function for a rectangular-wave signal with reduced duty cycle.

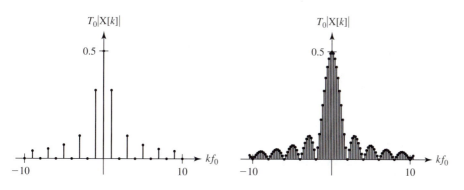

**Figure 5.4**
Magnitudes of the modified CTFS harmonic functions for rectangular-wave signals of 50 and 10 percent duty cycles.

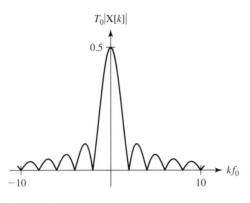

**Figure 5.5**
Limiting form of modified CTFS harmonic
function for a rectangular-wave signal.

harmonic function approaches the function illustrated in Figure 5.5. This modified
CTFS harmonic function will (with some notation changes) be called the *continuous-
time Fourier transform* of that single pulse.

The frequency difference between adjacent CTFS harmonic amplitudes is the
same as the fundamental frequency of the CTFS representation $f_F$, which is related to
the fundamental period of the CTFS representation by

$$f_F = \frac{1}{T_F}. \tag{5.5}$$

To emphasize its relationship to a frequency differential (which it will become in the
limit as the fundamental period goes to infinity), let this spacing be called $\Delta f$. That is,
let $\Delta f = f_F = 1/T_F$. Then the complex CTFS representation of $x(t)$ can be written as

$$x(t) = \sum_{k=-\infty}^{\infty} X[k]e^{j2\pi(k\Delta f)t}. \tag{5.6}$$

Substituting the integral expression for $X[k]$ into (5.6),

$$x(t) = \sum_{k=-\infty}^{\infty} \left[ \frac{1}{T_F} \int_{t_0}^{t_0+T_F} x(\tau)e^{-j2\pi(k\Delta f)\tau}\, d\tau \right] e^{j2\pi(k\Delta f)t}. \tag{5.7}$$

(The variable of integration is $\tau$ to distinguish it from the $t$ in the function $e^{j2\pi(k\Delta f)t}$,
which is outside the integral.) Since the starting point $t_0$ for the integral is arbitrary, let
it be

$$t_0 = -\frac{T_F}{2}. \tag{5.8}$$

Then

$$x(t) = \sum_{k=-\infty}^{\infty} \left[ \int_{-(T_F/2)}^{T_F/2} x(\tau)e^{-j2\pi(k\Delta f)\tau}\, d\tau \right] e^{j2\pi(k\Delta f)t}\, \Delta f \tag{5.9}$$

where $\Delta f$ replaces $1/T_F$. In the limit as $T_F$ approaches infinity, $\Delta f$ approaches the
differential $df$, $k\,\Delta f$ becomes a continuous variable $f$, the integration limits approach

plus and minus infinity, and the summation becomes an integral. That is,

$$x(t) = \lim_{T_F \to \infty} \left\{ \sum_{k=-\infty}^{\infty} \left[ \int_{-(T_F/2)}^{T_F/2} x(\tau)e^{-j2\pi(k\Delta f)\tau} \, d\tau \right] e^{j2\pi(k\Delta f)t} \, \Delta f \right\}$$

$$= \int_{-\infty}^{\infty} \left[ \int_{-\infty}^{\infty} x(\tau)e^{-j2\pi f\tau} \, d\tau \right] e^{j2\pi ft} \, df. \tag{5.10}$$

The bracketed quantity on the right side of (5.10) is called the *continuous-time Fourier transform* (CTFT) of $x(t)$,

$$X(f) = \int_{-\infty}^{\infty} x(t)e^{-j2\pi ft} \, dt. \tag{5.11}$$

That is,

$$x(t) = \int_{-\infty}^{\infty} X(f)e^{j2\pi ft} \, df. \tag{5.12}$$

Don't be bothered by the change of variable name from $\tau$ to $t$ in (5.10) to (5.12), because the name of the variable of integration is not important.

$$X(f) = \int_{-\infty}^{\infty} x(t)e^{-j2\pi ft} \, dt = \int_{-\infty}^{\infty} x(\tau)e^{-j2\pi f\tau} \, d\tau \tag{5.13}$$

## DEFINITION OF THE CONTINUOUS-TIME FOURIER TRANSFORM

The CTFT is defined by

$$\boxed{X(f) = \mathcal{F}(x(t)) = \int_{-\infty}^{\infty} x(t)e^{-j2\pi ft} \, dt}$$

$$\boxed{x(t) = \mathcal{F}^{-1}(X(f)) = \int_{-\infty}^{\infty} X(f)e^{+j2\pi ft} \, df} \tag{5.14}$$

or

$$\boxed{X(j\omega) = \mathcal{F}(x(t)) = \int_{-\infty}^{\infty} x(t)e^{-j\omega t} \, dt}$$

$$\boxed{x(t) = \mathcal{F}^{-1}(X(j\omega)) = \frac{1}{2\pi} \int_{-\infty}^{\infty} X(j\omega)e^{+j\omega t} \, d\omega} \tag{5.15}$$

where the operator $\mathcal{F}$ means "Fourier transform of" and the operator $\mathcal{F}^{-1}$ means "*inverse* Fourier transform of." These are the two most commonly used definitions of the Fourier transform in engineering. The first one (5.14) is written in terms of cyclic frequency $f$ and has the advantage of being very symmetrical. The forward and inverse transforms are almost the same. Only the sign of the exponent and the variable of integration change. This is the form most often used in communication system analysis, Fourier optics, and image processing. The second definition (5.15) is written in terms of the radian frequency variable $\omega$ instead of cyclic frequency $f$. Radian frequency has a somewhat more direct relationship to the time constants and resonant frequencies of real systems, and, as a result, the transforms of some system functions are somewhat simpler using this form. This is the form most often used in control-system analysis. Either definition can be converted to the other by using the relationship $\omega = 2\pi f$.

The signal $x(t)$ is said to be in the *time* domain because its functional argument $t$ represents time, and the transform function $X(f)$ or $X(j\omega)$ is said to be in the *frequency* domain because its functional argument $f$ or $\omega$ represents frequency. Cyclic frequency is the reciprocal of time, and radian frequency is *proportional to* the reciprocal of time. In some other applications of the CTFT in mathematics, physics, and engineering the two independent variables are not time and frequency, but they are always proportional to the reciprocals of each other.

The forward transform

$$X(f) = \mathcal{F}(x(t)) = \int_{-\infty}^{\infty} x(t)e^{-j2\pi ft}\,dt \quad \text{or} \quad X(j\omega) = \mathcal{F}(x(t)) = \int_{-\infty}^{\infty} x(t)e^{-j\omega t}\,dt$$

(5.16)

is sometimes referred to as *analysis* of the signal $x(t)$ because it extracts the component parts of $x(t)$, the complex exponentials $X(f)$ or $X(j\omega)$, at every value of the continuous variables $f$ or $\omega$. The inverse transform

$$x(t) = \mathcal{F}^{-1}(X(f)) = \int_{-\infty}^{\infty} X(f)e^{+j2\pi ft}\,df \qquad \text{or}$$

$$x(t) = \mathcal{F}^{-1}(X(j\omega)) = \frac{1}{2\pi}\int_{-\infty}^{\infty} X(j\omega)e^{+j\omega t}\,d\omega \qquad (5.17)$$

is sometimes called the *synthesis* of the signal $x(t)$ because it recombines the components $X(f)$ or $X(j\omega)$ back into the original signal $x(t)$.

It is natural to wonder at this point what the physical significance of $X(f)$ is. One way to understand it is to find the units of $X(f)$. They depend on what the units of $x(t)$ are. To make the idea concrete suppose for the moment that the units of $x(t)$ are volts (V). The transformation process begins by multiplying $x(t)$ by the complex exponential $e^{-j2\pi ft}$. The exponent of $e$ consists of three dimensionless numbers, $-j$, 2, and $\pi$ along with $f$ and $t$ which are frequency and time, respectively. Frequency has units of hertz or 1/second, and time has units of seconds. Therefore, the exponent of $e$ is dimensionless and so is $e^{-j2\pi ft}$. Then we multiply by $dt$ which has units of seconds. Thus the integration process accumulates the area under the product of $x(t)$ and $e^{-j2\pi ft}$. This area has units of (in this case) volt-seconds. Therefore, $X(f)$ has units of volt-seconds. But it is more physically meaningful to express these units as volts per hertz, since hertz is the same as 1/second and is the unit of the independent variable $f$. Similarly $X(j\omega)$ would have units of volts per radian per second. If the unit of the time-domain signal $x(t)$ is not volts, then the units of the Fourier transform would be $x(t)$ units per hertz or $x(t)$ units per radian per second.

The function $X(f)$ or $X(j\omega)$ is sometimes called the *amplitude spectral density* or just the *spectrum* of $x(t)$. It expresses the variation of the amplitude of complex sinusoids with frequency which, when added, form $x(t)$. The word *spectral* refers, in this case, to the variation with respect to frequency. (Sometimes in other disciplines it may refer to the variation with some other physical entity. For example, in optics, it may refer to a wavelength spectrum. In X-ray spectroscopy it may refer to an energy spectrum.) The word *density* comes from the units, volts per hertz. This is analogous to other densities which are more familiar. For example pressure is the two-dimensional density of force; that is, force per unit area. In the case of $X(f)$ or $X(j\omega)$ it is the one-dimensional density of amplitude; that is, amplitude per unit frequency. In this example the amplitude was voltage, but it could have been current or something else.

There are many reasons to transform signals from the time to the frequency domain, and vice versa. Some very useful and common operations in linear system analysis are more convenient in one domain than the other. The fact that $x(t)$ and $X(f)$ or $X(j\omega)$ are transforms of each other can be indicated by a shorthand notation,

$$x(t) \xleftrightarrow{\mathcal{F}} X(f) \qquad \text{or} \qquad x(t) \xleftrightarrow{\mathcal{F}} X(j\omega), \qquad \textbf{(5.18)}$$

and they are said to form a *Fourier transform pair*.

---

Other authors define other forms of the CTFT. For example,

$$X(f) = \int_{-\infty}^{\infty} x(t)e^{+j2\pi ft}\, dt \qquad x(t) = \int_{-\infty}^{\infty} X(f)e^{-j2\pi ft}\, df \qquad \textbf{(5.19)}$$

or

$$X(j\omega) = \frac{1}{\sqrt{2\pi}} \int_{-\infty}^{\infty} x(t)e^{+j\omega t}\, dt \qquad x(t) = \frac{1}{\sqrt{2\pi}} \int_{-\infty}^{\infty} X(j\omega)e^{-j\omega t}\, d\omega. \qquad \textbf{(5.20)}$$

They are just as mathematically valid as the ones presented here but are not often used in engineering.

---

It is important at this point to comment on conventional notation. In communication system literature, Fourier optics, and image processing, the functional form $X(f)$ is usually used for the transform of $x(t)$. In control-system literature one can find the use of both $X(\omega)$ and $X(j\omega)$. The form $X(\omega)$ has the advantage of being written directly in terms of the independent variable $\omega$. The other form $X(j\omega)$ may seem less efficient or direct, but there is a good reason for it. Later, in the study of the Laplace transform, we will be able to simply replace $j\omega$ by $s$ to form many Laplace transforms without having to change the mathematical meaning of the function X.

Note that, strictly speaking mathematically, the "X" of $X(f)$ and the "X" of $X(j\omega)$ are *not* the same function because one cannot form $X(j\omega)$ by replacing $f$ by $j\omega$ in $X(f)$. Instead $X(j\omega)$ is formed by replacing $f$ by $\omega/2\pi$ in $X(f)$,

$$X(j\omega) = X(f)|_{f \to \omega/2\pi}.$$

In stating this equality the symbol X does not represent the same function on the left side as it does on the right side. This is bad mathematical form because of the confusion that can result. But most of the time this is not an issue because usually either one or the other form of the transform is used exclusively in any particular analysis and the symbol X denotes only that form of the transform and, therefore, a unique function, unambiguously. There is a table of Fourier transform pairs in Appendix E.

## EXAMPLE 5.1

Find the CTFT of $x(t) = \text{rect}(t)$ (see Figure 5.6).

### ■ Solution

From the $f$ form of the CTFT,

$$X(f) = \int_{-\infty}^{\infty} \text{rect}(t)\, e^{-j2\pi ft}\, dt = \int_{-\frac{1}{2}}^{\frac{1}{2}} e^{-j2\pi ft}\, dt = \int_{-\frac{1}{2}}^{\frac{1}{2}} \left[ \underbrace{\cos(2\pi ft)}_{\text{even}} - j\,\underbrace{\sin(2\pi ft)}_{\text{odd}} \right] dt \qquad (5.21)$$

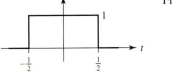
$x(t)$

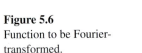

**Figure 5.6**
Function to be Fourier-transformed.

and

$$X(f) = 2 \int_{0}^{\frac{1}{2}} \cos(2\pi ft)\, dt = \left. \frac{1}{\pi f}\sin(2\pi ft) \right|_{0}^{\frac{1}{2}} = \frac{\sin(\pi f)}{\pi f} = \text{sinc}(f). \qquad (5.22)$$

From the $\omega$ form of the CTFT,

$$X(j\omega) = \int_{-\infty}^{\infty} \text{rect}(t)\, e^{-j\omega t}\, dt = \int_{-\frac{1}{2}}^{\frac{1}{2}} e^{-j\omega t}\, dt = \int_{-\frac{1}{2}}^{\frac{1}{2}} \left[ \underbrace{\cos(\omega t)}_{\text{even}} - j\,\underbrace{\sin(\omega t)}_{\text{odd}} \right] dt \qquad (5.23)$$

and

$$X(j\omega) = 2 \int_{0}^{\frac{1}{2}} \cos(\omega t)\, dt = \left. \frac{2}{\omega}\sin(\omega t) \right|_{0}^{\frac{1}{2}} = \frac{2}{\omega}\sin\left(\frac{\omega}{2}\right) = \frac{\sin(\pi(\omega/2\pi))}{\pi(\omega/2\pi)} = \text{sinc}\left(\frac{\omega}{2\pi}\right). \qquad (5.24)$$

and, as stated above for all Fourier transforms, the $f$ and $\omega$ forms of this Fourier transform can each be found from the other by using the relationship $\omega = 2\pi f$. ∎

Fourier transforms are, in general, complex functions of the real variable $f$ or $\omega$. Therefore, they are usually plotted as two graphs, one for the magnitude and another for the phase. The sinc function in Example 5.1 as the CTFT of the rectangle function could, of course, be plotted as a real function (Figure 5.7).

But what about the CTFT of $x(t) = 2\,\text{rect}(t - 2)$? Its CTFT is

$$X(f) = 2 \int_{-\infty}^{\infty} \text{rect}(t - 2)\, e^{-j2\pi ft}\, dt. \qquad (5.25)$$

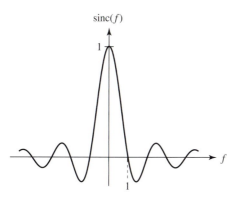

**Figure 5.7**
CTFT of rect($t$) plotted as a real function.

The best way to find this transform (until we get to the time-shifting property of the CTFT) is to make a change of variable in (5.25). Let

$$\lambda = t - 2 \qquad \text{and} \qquad d\lambda = dt. \tag{5.26}$$

Then

$$X(f) = 2 \int_{-\infty}^{\infty} \text{rect}(\lambda)\, e^{-j2\pi f(\lambda+2)}\, d\lambda = 2e^{-j4\pi f} \underbrace{\int_{-\frac{1}{2}}^{\frac{1}{2}} e^{-j2\pi f\lambda}\, d\lambda}_{=\text{sinc}(f)} = 2\,\text{sinc}(f)\, e^{-j4\pi f}. \tag{5.27}$$

Now the CTFT is *definitely not* a real function and must be plotted as magnitude and phase (Figure 5.8).

Notice that amplitude scaling by a factor of 2 changed the magnitude of the CTFT by the same scale factor but not the shape of the magnitude of the CTFT. The time-shift transformation $t \to t - 2$ causes a big change in the phase but does not affect the magnitude. The shape of the CTFT magnitude in each case indicates that the rectangle function is dominated by low frequencies because that is where the magnitude of the CTFT is the largest. At higher frequencies there are still significant components, but their magnitudes generally decrease with frequency. This type of signal might be informally characterized by saying that it has more low-frequency content than high-frequency content.

Consider the remarkable implications of the inverse CTFT integral formula

$$x(t) = \mathcal{F}^{-1}(X(f)) = \int_{-\infty}^{\infty} X(f)e^{+j2\pi ft}\, df. \tag{5.28}$$

In words, this formula says that the summation (integral) of an infinite continuum of weighted infinitesimal-amplitude complex exponentials (which oscillate for all time) is a finite-amplitude, real signal which can be time-limited. *Time limited* means that a signal is nonzero only over a finite time and zero everywhere else. It is amazing to think that by choosing the weighting function $X(f)$ exactly right that all those complex exponentials can exactly cancel each other outside some finite time range and exactly equal the signal inside that same time range. The fact that such seemingly different functions can be made equivalent is the magic of the CTFT. It converts a signal

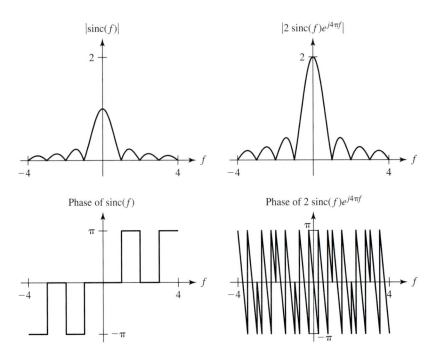

**Figure 5.8**
Magnitudes and phases for the CTFT of rect($t$) and CTFT of 2 rect($t - 2$).

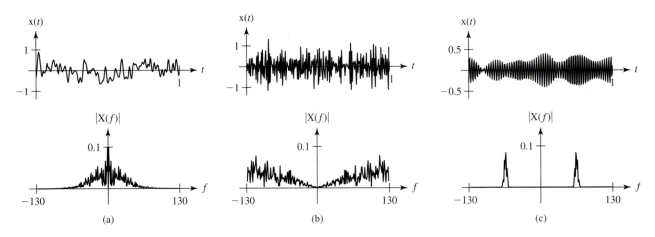

**Figure 5.9**
Examples of signals with different kinds of spectral content.

from one form into a very different form, and sometimes this makes the analysis of the signal much easier because certain features which may be obscure in one form may be obvious in the other form.

It is important to develop an intuitive understanding of the relationship between the shape and time variation of a signal and the shape and frequency variation of its CTFT (Figure 5.9). In Figure 5.9a is a lowpass signal, one with more low-frequency content than anything else. The name *lowpass* comes from the idea that the response of a system, which preferentially allows low frequencies to pass through and attenuates or stops high frequencies, to an excitation with a uniform spectrum (constant over all frequencies) would have such a spectral shape. Figure 5.9b and c show *highpass* and *bandpass* signals, respectively. A highpass system stops or attenuates low frequencies and allows high frequencies through. A bandpass system allows a finite range of frequencies not containing zero to pass through. A *bandstop* system attenuates or stops a finite range of frequencies not containing zero and lets the rest of the frequencies through. (These ideas will be explored in more detail in Chapter 6.)

Lowpass signals are characterized by their smoothness. They are smooth because a low-frequency signal changes value slowly and is therefore smoother than a signal which changes value more rapidly. A highpass signal is characterized by its rapidly changing value. A bandpass signal is characterized by its similarity to a sinusoid. The CTFT of the bandpass signal in Figure 5.8c has two tall, narrow peaks. If the two peaks had been infinitely tall and infinitely narrow, they would have been impulses and the signal would have been a perfect sinusoid. The narrower the peaks become, the closer the signal is to a sinusoid.

## 5.3 CONVERGENCE AND THE GENERALIZED FOURIER TRANSFORM

As an example of a problem using the Fourier transform, let's find the CTFT of a very simple function $x(t) = A$, where $A$ is a constant.

$$X(f) = \int_{-\infty}^{\infty} Ae^{-j2\pi ft}\, dt = A \int_{-\infty}^{\infty} e^{-j2\pi ft}\, dt \qquad \textbf{(5.29)}$$

The integral does not converge. Therefore, strictly speaking, the Fourier transform does not exist. But we can avoid this problem by use of a trick. Let us first find the CTFT of

$$x_\sigma(t) = Ae^{-\sigma|t|} \qquad \sigma > 0. \qquad \textbf{(5.30)}$$

Then we will let $\sigma$ approach zero *after* finding the transform. The factor $e^{-\sigma|t|}$ is a convergence factor which allows us to evaluate the integral (Figure 5.10).

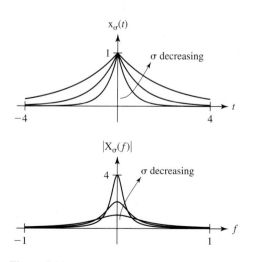

**Figure 5.10**
Effect of the convergence factor $\sigma$.

The transform is

$$X_\sigma(f) = \int_{-\infty}^{\infty} A e^{-\sigma|t|} e^{-j2\pi ft} \, dt = \int_{-\infty}^{0} A e^{\sigma t} e^{-j2\pi ft} \, dt + \int_{0}^{\infty} A e^{-\sigma t} e^{-j2\pi ft} \, dt \quad \textbf{(5.31)}$$

$$X_\sigma(f) = A \left[ \int_{-\infty}^{0} e^{(\sigma - j2\pi f)t} \, dt + \int_{0}^{\infty} e^{(-\sigma - j2\pi f)t} \, dt \right] = A \frac{2\sigma}{\sigma^2 + (2\pi f)^2}. \quad \textbf{(5.32)}$$

Now take the limit as $\sigma$ approaches zero, $\lim_{\sigma \to 0} A(2\sigma/[\sigma^2 + (2\pi f)^2])$. If $f \neq 0$, then

$$\lim_{\sigma \to 0} A \frac{2\sigma}{\sigma^2 + (2\pi f)^2} = 0. \quad \textbf{(5.33)}$$

Therefore, in the limit as $\sigma$ approaches zero, the CTFT of $x_\sigma(t)$, $X_\sigma(f)$, approaches zero for $f \neq 0$. Next let's find the area under the function $A(2\sigma/[\sigma^2 + (2\pi f)^2])$, as $\sigma$ approaches zero.

$$\text{Area} = A \int_{-\infty}^{\infty} \frac{2\sigma}{\sigma^2 + (2\pi f)^2} \, df \quad \textbf{(5.34)}$$

This integral can be evaluated by contour integration in the complex plane (which is beyond the scope of this text) or by looking it up in an integral table. Using

$$\int \frac{dx}{a^2 + (bx)^2} = \frac{1}{ab} \tan^{-1}\left(\frac{bx}{a}\right) \quad \textbf{(5.35)}$$

we get

$$\text{Area} = A \left[ \frac{2\sigma}{2\pi\sigma} \tan^{-1}\left(\frac{2\pi f}{\sigma}\right) \right]_{-\infty}^{\infty} = \frac{A}{\pi}\left(\frac{\pi}{2} + \frac{\pi}{2}\right) = A. \quad \textbf{(5.36)}$$

The area under the function is $A$, independent of $\sigma$. Therefore, in the limit $\sigma \to 0$, the Fourier transform of the constant $A$ is a function which is zero for $f \neq 0$ and has an area of $A$. According to the definition of an impulse, this exactly describes an impulse of strength $A$, occurring at $f = 0$. Therefore, we can form the Fourier transform pair

$$A \xleftrightarrow{\mathcal{F}} A\delta(f). \quad \textbf{(5.37)}$$

Appealing to this limiting process using a convergence factor to find the CTFT of a function yields what is called the *generalized* Fourier transform. It extends the CTFT to a class of very useful functions, constants and periodic functions. By similar reasoning the CTFT transform pairs

$$\cos(2\pi f_0 t) \xleftrightarrow{\mathcal{F}} \frac{1}{2}[\delta(f - f_0) + \delta(f + f_0)] \quad \textbf{(5.38)}$$

and

$$\sin(2\pi f_0 t) \xleftrightarrow{\mathcal{F}} \frac{j}{2}[\delta(f + f_0) - \delta(f - f_0)] \quad \textbf{(5.39)}$$

can be found. By making the substitution, $f = \omega/2\pi$ and using the scaling property of impulses, the equivalent radian-frequency forms of these transforms are found to be

$$A \xleftrightarrow{\;\mathcal{F}\;} 2\pi A \delta(\omega) \tag{5.40}$$

$$\cos(\omega_0 t) \xleftrightarrow{\;\mathcal{F}\;} \pi[\delta(\omega - \omega_0) + \delta(\omega + \omega_0)] \tag{5.41}$$

$$\sin(\omega_0 t) \xleftrightarrow{\;\mathcal{F}\;} j\pi[\delta(\omega + \omega_0) - \delta(\omega - \omega_0)]. \tag{5.42}$$

## 5.4 COMPARISONS BETWEEN THE CONTINUOUS-TIME FOURIER SERIES AND THE CONTINUOUS-TIME FOURIER TRANSFORM

The CTFS is a way of expressing a signal as an infinite summation of sinusoids (real or complex) at discrete frequencies, integer multiples of the fundamental frequency,

$$x(t) = \sum_{k=-\infty}^{\infty} X[k] e^{j2\pi(kf_F)t}. \tag{5.43}$$

The CTFT is a way of expressing a signal as an integral of complex sinusoids, weighted by the Fourier transform of the time-domain signal. This amounts to a summation of complex sinusoids over an infinite continuum of frequencies. The limit of the summation as the frequencies merge into a continuum is an integral,

$$x(t) = \int_{-\infty}^{\infty} X(f) e^{+j2\pi ft} \, df. \tag{5.44}$$

The CTFT can represent aperiodic (and, by extending it to the generalized Fourier transform, also periodic) signals for all time. As illustrated earlier, one way of seeing the relationship between the CTFS and the CTFT is by observing the changes in the CTFS representation of a periodic signal as the fundamental period gets larger. When the period becomes infinite, the signal is no longer periodic because it cannot repeat.

Notice that when a CTFT is done, the result is valid for all $f$ or $\omega$, including negative values. A common source of confusion when first encountering Fourier analysis of signals is the idea of a *negative* frequency. Fourier analysis, at its most basic level, is simply expressing a signal as a summation of sinusoids. Consider a very simple signal in Figure 5.11. It is periodic with fundamental period $T_0$. What is it, or more specifically, what are the sinusoidal components that can be added to produce it? It is obviously sinusoidal. Therefore a single, correctly chosen sinusoid can completely describe it. It is tempting to say that the only mathematical function that exactly describes this signal is

$$x(t) = A \cos\left(\frac{2\pi t}{T_0}\right) = A \cos(2\pi f_0 t), \tag{5.45}$$

where $f_0 = 1/T_0$. But that is not true because the function

$$x(t) = A \cos(2\pi(-f_0)t) \tag{5.46}$$

also exactly describes it. Who is to say what the *frequency* is, $f_0$ or $-f_0$? The functions

$$x(t) = A_1 \cos(2\pi f_0 t) + A_2 \cos(2\pi(-f_0)t) \qquad A_1 + A_2 = A \tag{5.47}$$

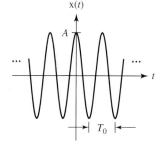

**Figure 5.11**
A sinusoidal signal to be expressed as a CTFS.

and

$$x(t) = A \frac{e^{j2\pi f_0 t} + e^{-j2\pi f_0 t}}{2} \tag{5.48}$$

also exactly describe the signal. If the signal had been a sine instead of a cosine, then either of the two forms $A \sin(2\pi f_0 t)$ and $-A \sin(2\pi(-f_0)t)$ (and others) could have been chosen, with equal mathematical validity.

The conventional trigonometric CTFS is defined by

$$x(t) = X_c[0] + \sum_{k=1}^{\infty} [X_c[k] \cos(2\pi(k f_F)t) + X_s[k] \sin(2\pi(k f_F)t)] \tag{5.49}$$

$$X_c[0] = \frac{1}{T_F} \int_{t_0}^{t_0+T_F} x(t) \, dt \tag{5.50}$$

$$X_c[k] = \frac{2}{T_F} \int_{t_0}^{t_0+T_F} x(t) \cos(2\pi(k f_F)t) \, dt \qquad k = 1, 2, 3, \ldots \tag{5.51}$$

$$X_s[k] = \frac{2}{T_F} \int_{t_0}^{t_0+T_F} x(t) \sin(2\pi(k f_F)t) \, dt \qquad k = 1, 2, 3, \ldots \tag{5.52}$$

The conventional trigonometric CTFS would describe this signal by saying that $X_c[1]$ is $A$ and all other $X_c[k]$'s and $X_s[k]$'s are zero and $X_c[1]$ is the amplitude of $\cos(2\pi f_F t)$, the only nonzero term in the CTFS representation. But this is arbitrary. The CTFS could have been defined, with equal mathematical validity, by

$$x(t) = X_c[k] + \sum_{k=1}^{\infty} [X_c[k] \cos(2\pi(-k f_F)t) - X_s[k] \sin(2\pi(-k f_F)t)] \tag{5.53}$$

$$X_c[0] = \frac{1}{T_F} \int_{t_0}^{t_0+T_F} x(t) \, dt \tag{5.54}$$

$$X_c[k] = \frac{2}{T_F} \int_{t_0}^{t_0+T_F} x(t) \cos(2\pi(-k f_F)t) \, dt \qquad k = 1, 2, 3, \ldots \tag{5.55}$$

$$X_s[k] = -\frac{2}{T_F} \int_{t_0}^{t_0+T_F} x(t) \sin(2\pi(-k f_F)t) \, dt \qquad k = 1, 2, 3, \ldots \tag{5.56}$$

and $X_c[k]$ and $X_s[k]$ would be exactly the same.

One could also set up a system in which the harmonic amplitudes were equally distributed between positive and corresponding negative frequencies, giving half the amplitude to each. That is essentially what the complex CTFS and the CTFT are. Therefore, mathematically, a negative frequency is just as legitimate as a positive one.

The basic problem with negative frequency is, of course, that one cannot imagine something happening "minus 10 times per second." Examine the general sinusoidal signal

$$x(t) = A \cos(2\pi(-f_0)t + \theta). \tag{5.57}$$

It is written to emphasize the idea of a negative frequency $-f_0$. But it can be mathematically rearranged to emphasize another idea,

$$x(t) = A\cos(2\pi f_0(-t) + \theta). \tag{5.58}$$

In this formulation, we can conceive of the signal as being the same as the positive-frequency signal $x(t) = A\cos(2\pi f_0 t - \theta)$, *except* that it is time-inverted. So a negative-frequency sinusoid is equivalent to the corresponding positive-frequency sinusoid but with time reversed. When we proceed along the time-inverted sinusoid, we encounter the same number of cycles per second, so a negative frequency signal has the same oscillatory behavior as a positive frequency signal, but its phase is, in general, different.

Fourier analysis is a mathematical tool for manipulating signals which expresses the signals as summations of sinusoids or complex exponentials. Sinusoids or complex exponentials with negative frequencies are just as mathematically valid as sinusoids or complex exponentials with positive frequencies. It is not obvious at this point, but later the symmetry of using both positive and negative frequencies will be very useful in simplifying the analysis of some complicated systems. Trying to analyze them by attributing the signal amplitude only to positive frequencies, although possible, leads to much undesirable awkwardness in the analysis. (By the way, there are other uses of the CTFT in science and engineering. For example, the diffraction of light is nicely described, in some situations, by a two-dimensional Fourier transform and the inclusion of both positive and negative frequencies is absolutely essential in that type of analysis.)

Now look at the definition of the CTFT,

$$X(f) = \int_{-\infty}^{\infty} x(t)e^{-j2\pi ft}\, dt \qquad x(t) = \int_{-\infty}^{\infty} X(f)e^{j2\pi ft}\, df \tag{5.59}$$

or

$$X(j\omega) = \int_{-\infty}^{\infty} x(t)e^{-j\omega t}\, dt \qquad x(t) = \frac{1}{2\pi}\int_{-\infty}^{\infty} X(j\omega)e^{+j\omega t}\, d\omega \tag{5.60}$$

and compare it to the definition of the complex CTFS,

$$X[k] = \frac{1}{T_F}\int_{t_0}^{t_0+T_F} x(t)e^{-j2\pi(kf_F)t}\, dt \qquad x(t) = \sum_{k=-\infty}^{\infty} X[k]e^{j2\pi(kf_F)t} \tag{5.61}$$

or

$$X[k] = \frac{1}{T_F}\int_{t_0}^{t_0+T_F} x(t)e^{-j(k\omega_F)t}\, dt \qquad x(t) = \sum_{k=-\infty}^{\infty} X[k]e^{j(k\omega_F)t} \tag{5.62}$$

They are analogous in that they both use positive and negative frequencies to synthesize the time signal $x(t)$ from its complex-exponential components. The CTFS analyzes a signal as an infinite summation of complex sinusoids at discrete frequencies, and the CTFT analyzes a signal as an infinite summation of complex sinusoids over a continuum of frequencies (an integral). The CTFS converts a CT signal $x(t)$

into a discrete-harmonic number function $X[k]$. The CTFT converts a CT signal $x(t)$ into a continuous-frequency function $X(f)$. Both the CTFS and the CTFT are transformations of the time-domain signal into a different form which contains the same information.

In the realm of Fourier analysis, signals can be viewed as having only positive frequencies (single-sided spectra) or both positive and negative frequencies (double-sided spectra) with equal validity. (One could even choose to use only negative frequencies, but that view has few, if any, adherents.) There is much inconsistency in the communication system, control system, and signal analysis disciplines. Sometimes single-sided spectra are used; other times double-sided spectra are used. There are some advantages to the mathematical symmetry of double-sided analysis, and there are some advantages to the physical concepts of single-sided analysis. It is important to realize that both are useful and to be able to use both without confusion.

## 5.5  PROPERTIES OF THE CONTINUOUS-TIME FOURIER TRANSFORM

There are several important properties of the CTFT which, along with a table of transform pairs, can, in most practical cases, let us avoid directly applying the integral definition of the CTFT.

If two signals have CTFTs,

$$\mathcal{F}(x(t)) = X(f) \quad \text{or} \quad X(j\omega) \quad \text{and} \quad \mathcal{F}(y(t)) = Y(f) \quad \text{or} \quad Y(j\omega), \quad \textbf{(5.63)}$$

then the following properties apply regardless of the forms of the signals. (Some of the $f$ forms are proven, and all the $\omega$ forms are stated without proof, since the proofs of the $\omega$ forms are similar.)

### LINEARITY

The linearity property is exactly the same as it is for the CTFS and DTFS,

$$\boxed{\alpha x(t) + \beta y(t) \xleftrightarrow{\ \mathcal{F}\ } \alpha X(f) + \beta Y(f)} \qquad \textbf{(5.64)}$$

or

$$\boxed{\alpha x(t) + \beta y(t) \xleftrightarrow{\ \mathcal{F}\ } \alpha X(j\omega) + \beta Y(j\omega)}, \qquad \textbf{(5.65)}$$

and the proofs are similar.

### TIME SHIFTING AND FREQUENCY SHIFTING

Let $t_0$ be any real constant and let

$$z(t) = x(t - t_0). \qquad \textbf{(5.66)}$$

Then the CTFT of $z(t)$ is

$$Z(f) = \int_{-\infty}^{\infty} z(t) e^{-j2\pi ft}\, dt = \int_{-\infty}^{\infty} x(t - t_0) e^{-j2\pi ft}\, dt. \qquad \textbf{(5.67)}$$

Make the change of variable

$$\lambda = t - t_0 \qquad \text{and} \qquad d\lambda = dt. \tag{5.68}$$

Then

$$Z(f) = \int_{-\infty}^{\infty} x(\lambda)e^{-j2\pi f(\lambda+t_0)}\, d\lambda = e^{-j2\pi f t_0}\int_{-\infty}^{\infty} x(\lambda)e^{-j2\pi f\lambda}\, d\lambda = e^{-j2\pi f t_0}X(f) \tag{5.69}$$

and the time-shifting property is

$$\boxed{x(t - t_0) \overset{\mathcal{F}}{\longleftrightarrow} X(f)e^{-j2\pi f t_0}} \tag{5.70}$$

or

$$\boxed{x(t - t_0) \overset{\mathcal{F}}{\longleftrightarrow} X(j\omega)e^{-j\omega t_0}}. \tag{5.71}$$

As an example of why the time-shifting property makes sense, let the time signal be the complex sinusoid

$$x(t) = e^{j2\pi f_0 t}. \tag{5.72}$$

Then

$$x(t - t_0) = e^{j2\pi f_0(t-t_0)} = e^{j2\pi f_0 t}e^{-j2\pi f_0 t_0} \tag{5.73}$$

(Figure 5.12). So shifting this signal in time corresponds to multiplying it by the complex number $e^{-j2\pi f_0 t_0}$. The CTFT expression

$$x(t) = \int_{-\infty}^{\infty} X(f)e^{+j2\pi f t}\, df \tag{5.74}$$

says that any signal that is Fourier-transformable can be expressed as a linear combination of complex sinusoids over a continuum of frequencies $f$, and, if $x(t)$ gets shifted by $t_0$, each of those complex sinusoids gets multiplied by the complex number $e^{-j2\pi f t_0}$.

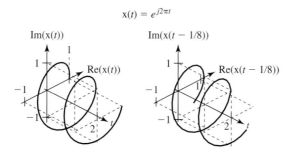

**Figure 5.12**
A complex exponential $x(t) = e^{j2\pi f_0 t}$ and a delayed version $x(t - \frac{1}{8}) = e^{j2\pi f_0(t-(1/8))}$.

The frequency-shifting property can be proven by starting with a frequency-shifted version of $X(f)$, $X(f - f_0)$, and using the inverse CTFT integral. The result is

$$x(t)e^{+j2\pi f_0 t} \overset{\mathcal{F}}{\longleftrightarrow} X(f - f_0) \tag{5.75}$$

or

$$x(t)e^{+j\omega_0 t} \overset{\mathcal{F}}{\longleftrightarrow} X(j(\omega - \omega_0)). \tag{5.76}$$

Notice the similarity between the time-shifting and frequency-shifting properties. They both contain a complex sinusoid multiplier in the other domain. However, the signs of the exponents in the complex sinusoids are different. This occurs because of the signs in the forward and inverse CTFTs,

$$X(f) = \int_{-\infty}^{\infty} x(t)e^{-j2\pi ft}\,dt \qquad x(t) = \int_{-\infty}^{\infty} X(f)e^{+j2\pi ft}\,df. \tag{5.77}$$

The time-shifting property is very handy for finding transforms of signals which are composed of multiple functions which have been time-shifted and added. The frequency-shifting property is fundamental to understanding the effects of modulation in communication systems.

## TIME SCALING AND FREQUENCY SCALING

Let $a$ be any real constant that is not zero and let

$$z(t) = x(at). \tag{5.78}$$

Then the CTFT of $z(t)$ is

$$Z(f) = \int_{-\infty}^{\infty} z(t)e^{-j2\pi ft}\,dt = \int_{-\infty}^{\infty} x(at)e^{-j2\pi ft}\,dt. \tag{5.79}$$

Make the change of variable

$$\lambda = at \qquad \text{and} \qquad d\lambda = a\,dt. \tag{5.80}$$

Then, if $a > 0$,

$$Z(f) = \int_{-\infty}^{\infty} x(\lambda)e^{-j2\pi f(\lambda/a)}\,\frac{d\lambda}{a} = \frac{1}{a}\int_{-\infty}^{\infty} x(\lambda)e^{-j2\pi(f/a)\lambda}\,d\lambda = \frac{1}{a}X\left(\frac{f}{a}\right) \tag{5.81}$$

and if $a < 0$,

$$Z(f) = \int_{\infty}^{-\infty} x(\lambda)e^{-j2\pi f(\lambda/a)}\,\frac{d\lambda}{a} = -\frac{1}{a}\int_{-\infty}^{\infty} x(\lambda)e^{-j2\pi(f/a)\lambda}\,d\lambda = -\frac{1}{a}X\left(\frac{f}{a}\right). \tag{5.82}$$

Therefore, in either case,

$$Z(f) = \frac{1}{|a|}X\left(\frac{f}{a}\right) \tag{5.83}$$

and the time-scaling property is

$$\boxed{x(at) \overset{\mathcal{F}}{\longleftrightarrow} \frac{1}{|a|}X\left(\frac{f}{a}\right)} \quad \text{or} \quad \boxed{x(at) \overset{\mathcal{F}}{\longleftrightarrow} \frac{1}{|a|}X\left(j\frac{\omega}{a}\right)}. \quad \textbf{(5.84)}$$

The frequency-scaling property can be proven in a similar manner and the result is

$$\boxed{\frac{1}{|a|}x\left(\frac{t}{a}\right) \overset{\mathcal{F}}{\longleftrightarrow} X(af)} \quad \text{or} \quad \boxed{\frac{1}{|a|}x\left(\frac{t}{a}\right) \overset{\mathcal{F}}{\longleftrightarrow} X(ja\omega)}. \quad \textbf{(5.85)}$$

The time-scaling property and the frequency-scaling property allow the calculation of the CTFT and inverse CTFT of functions which have been stretched or compressed to fit the time scale or frequency scale of actual signals.

One consequence of the time-scaling and frequency-scaling properties is that a compression in one domain is an expansion in the other domain. One interesting way of illustrating this is through the function

$$x(t) = e^{-\pi t^2}, \quad \textbf{(5.86)}$$

whose CTFT happens to be of the same functional form,

$$e^{-\pi t^2} \overset{\mathcal{F}}{\longleftrightarrow} e^{-\pi f^2}. \quad \textbf{(5.87)}$$

If we now time-scale through the transformation $t \rightarrow t/2$, for example, the transform pair becomes

$$e^{-\pi(t/2)^2} \overset{\mathcal{F}}{\longleftrightarrow} 2e^{-\pi(2f)^2} \quad \textbf{(5.88)}$$

(Figure 5.13). The transformation $t \rightarrow t/2$ is a time expansion, and the corresponding effect in the frequency domain is a frequency compression (accompanied by an

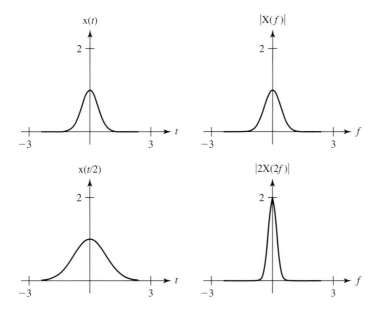

**Figure 5.13**
Time expansion and the corresponding frequency compression.

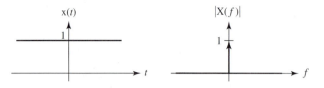

**Figure 5.14**
Constant and impulse as limits of time and frequency scaling
$x(t) = e^{-\pi t^2}$ and its CTFT.

amplitude scale factor). As the time-domain signal is expanded, it falls from one more and more slowly as time departs from zero in either direction and, in the limit as the time expansion factor approaches infinity, it does not fall at all and approaches the constant one. As the time-domain signal is expanded by some factor, its CTFT is frequency-compressed and its height is multiplied by the same factor. In the limit as the time-domain expansion factor approaches infinity, the CTFT becomes an impulse,

$$\lim_{a\to\infty} e^{-\pi(t/a)^2} = 1 \overset{\mathcal{F}}{\longleftrightarrow} \lim_{a\to\infty} \frac{1}{|a|} e^{-\pi(af)^2} = \delta(f) \tag{5.89}$$

(Figure 5.14).

The relation between compression in one domain and expansion in the other is the basis for an idea called the *uncertainty principle* of Fourier analysis. As $a \to \infty$ in (5.89), the time-domain function becomes less localized and the corresponding frequency-domain function becomes more localized. In that limit, the frequency of the signal is infinitely localized to a single frequency $f = 0$, while the time function is uniformly spread over the range $-\infty < t < \infty$ and is, therefore, infinitely unlocalized. If we compress the time function instead, it becomes an impulse at time $t = 0$, and its location is exact while its CTFT becomes spread uniformly over the range $-\infty < f < \infty$ and has no locality at all. We can express the idea in words by saying that as we know the location of one function better and better, we lose knowledge of the location of the other function. The name, uncertainty principle comes, of course, from the principle in quantum mechanics of the same name, and the mathematical description of the uncertainty principle in quantum mechanics can be expressed using the Fourier transform.

## TRANSFORM OF A CONJUGATE

The inverse CTFT of $X(f)$ is

$$\mathcal{F}^{-1}(X(f)) = x(t) = \int_{-\infty}^{\infty} X(f)e^{j2\pi ft}\, df. \tag{5.90}$$

The complex conjugate of $x(t)$ is

$$x^*(t) = \left[\int_{-\infty}^{\infty} X(f)e^{j2\pi ft}\, df\right]^* = \int_{-\infty}^{\infty} X^*(f)e^{-j2\pi ft}\, df. \tag{5.91}$$

Make the change of variable

$$\lambda = -f \qquad \text{and} \qquad d\lambda = -df. \tag{5.92}$$

Then

$$x^*(t) = -\int_{\infty}^{-\infty} X^*(-\lambda)e^{j2\pi\lambda t} \, d\lambda = \underbrace{\int_{-\infty}^{\infty} X^*(-\lambda)e^{j2\pi\lambda t} \, d\lambda}_{\mathcal{F}^{-1}[X^*(-f)]}. \tag{5.93}$$

The conjugation property is

$$\boxed{x^*(t) \overset{\mathcal{F}}{\longleftrightarrow} X^*(-f)} \quad \text{or} \quad \boxed{x^*(t) \overset{\mathcal{F}}{\longleftrightarrow} X^*(-j\omega)}. \tag{5.94}$$

Using this property we can discover another useful characteristic of Fourier transforms of real-valued signals. If $x(t)$ is real-valued, then $x(t) = x^*(t)$. The CTFT of $x(t)$ is $X(f)$, and the CTFT of $x^*(t)$ is $X^*(-f)$. Therefore, if $x(t) = x^*(t)$, $X(f) = X^*(-f)$. In words, if the time-domain signal is real-valued, its CTFT has the property that the behavior for negative frequencies is the complex conjugate of the behavior for positive frequencies. Therefore, if we know the positive-frequency functional form of the CTFT of a real-valued signal, we also know the negative-frequency functional form. This is analogous to the previously observed property that the complex CTFS harmonic amplitudes of a real signal occur in complex-conjugate pairs.

Let $x(t)$ be a real-valued signal. The magnitude of $X(f)$ is

$$|X(f)|^2 = X(f)X^*(f). \tag{5.95}$$

Then using $X(f) = X^*(-f)$ we can show that the magnitude of $X(-f)$ is

$$|X(-f)|^2 = \underbrace{X(-f)}_{X^*(f)}\underbrace{X^*(-f)}_{X(f)} = X(f)X^*(f) = |X(f)|^2, \tag{5.96}$$

proving that the magnitude of the CTFT of a real-valued signal is an even function of frequency. Using $X(f) = X^*(-f)$, we can also show that the phase of the CTFT of a real-valued signal is an odd function of frequency. Often, in practical signal and system analysis the CTFT of a signal is only displayed for positive frequencies because, since $X(f) = X^*(-f)$, if we know the functional behavior for positive frequencies we also know it for negative frequencies.

## MULTIPLICATION–CONVOLUTION DUALITY

Let the convolution of $x(t)$ and $y(t)$ be

$$z(t) = x(t) * y(t) = \int_{-\infty}^{\infty} x(\tau)y(t-\tau) \, d\tau. \tag{5.97}$$

The CTFT of $z(t)$ is

$$Z(f) = \int_{-\infty}^{\infty} z(t)e^{-j2\pi ft} \, dt \tag{5.98}$$

or

$$Z(f) = \int_{-\infty}^{\infty} \left[ \int_{-\infty}^{\infty} x(\tau)y(t-\tau) \, d\tau \right] e^{-j2\pi ft} \, dt. \tag{5.99}$$

Reversing the order of integration in (5.99),

$$Z(f) = \int_{-\infty}^{\infty} x(\tau) \underbrace{\left[ \int_{-\infty}^{\infty} y(t - \tau) e^{-j2\pi ft}\, dt \right]}_{\mathcal{F}[y(t-\tau)]} d\tau. \tag{5.100}$$

Then, using the time-shifting property in (5.100),

$$Z(f) = \int_{-\infty}^{\infty} x(\tau) e^{-j2\pi f\tau} Y(f)\, d\tau \tag{5.101}$$

Since $Y(f)$ is not a function of $\tau$,

$$Z(f) = Y(f) \underbrace{\int_{-\infty}^{\infty} x(\tau) e^{-j2\pi f\tau}\, d\tau}_{\mathcal{F}[x(\tau)]}, \tag{5.102}$$

and finally

$$Z(f) = X(f)Y(f). \tag{5.103}$$

The time-domain convolution property is

$$\boxed{x(t) * y(t) \xleftrightarrow{\mathcal{F}} X(f)Y(f)} \tag{5.104}$$

or

$$\boxed{x(t) * y(t) \xleftrightarrow{\mathcal{F}} X(j\omega)Y(j\omega)}. \tag{5.105}$$

The proof of the frequency-domain convolution property

$$\boxed{x(t)y(t) \xleftrightarrow{\mathcal{F}} X(f) * Y(f)} \tag{5.106}$$

or

$$\boxed{x(t)y(t) \xleftrightarrow{\mathcal{F}} \frac{1}{2\pi} X(j\omega) * Y(j\omega)}. \tag{5.107}$$

is similar.

The convolution property may be the most important property of the Fourier transform (indeed of any transform method) because it converts the basic LTI system time-domain property, *the response is the excitation convolved with the impulse response*, into the simpler frequency-domain property, *the response is the product of the excitation and the transfer function* (Figure 5.15).

*Transfer function* is the name given to the CTFT of the impulse response because it is the function which transfers the excitation to the response. In Chapter 4 we showed that the impulse response of the cascade of two LTI systems is the convolution

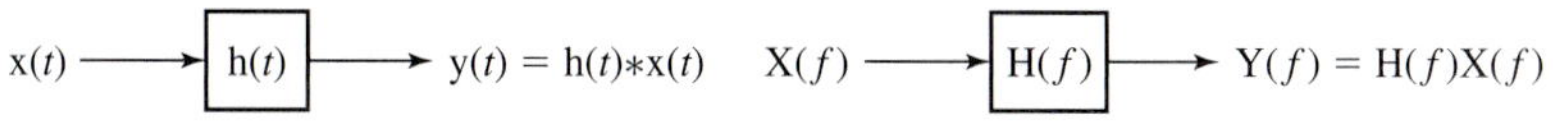

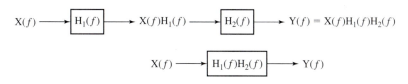

**Figure 5.15**
Equivalence of time-domain convolution and frequency-domain multiplication in LTI system analysis.

**Figure 5.16**
Transfer function of a cascade of two LTI systems.

of their impulse responses. It then follows that the transfer function of the cascade of two systems is the product of their transfer functions (Figure 5.16). [Some authors reserve the name transfer function for the Laplace transform of the impulse response and call H($f$) the *frequency response* of the system. But the mathematical relations are the same no matter what the name is.]

## TIME DIFFERENTIATION

The time-domain function x($t$) can be expressed as

$$x(t) = \int_{-\infty}^{\infty} X(f)e^{j2\pi ft}\, df. \tag{5.108}$$

Differentiating both sides of (5.108), with respect to time,

$$\frac{d}{dt}(x(t)) = \frac{d}{dt}\int_{-\infty}^{\infty} X(f)e^{j2\pi ft}\, df = \int_{-\infty}^{\infty} j2\pi f X(f)e^{j2\pi ft}\, df = \mathcal{F}^{-1}(j2\pi f X(f)). \tag{5.109}$$

Therefore, the time-differentiation property is

$$\boxed{\frac{d}{dt}(x(t)) \xleftrightarrow{\ \mathcal{F}\ } j2\pi f X(f)} \tag{5.110}$$

or

$$\boxed{\frac{d}{dt}(x(t)) \xleftrightarrow{\ \mathcal{F}\ } j\omega X(j\omega)}. \tag{5.111}$$

This property, along with the integration property to be covered later in this section, can be used to convert integrodifferential equations in the time domain into algebraic equations in the frequency domain.

**EXAMPLE 5.2**

Find the CTFT of $x(t) = \text{rect}((t + 1)/2) - \text{rect}((t - 1)/2)$ using the differentiation property of the CTFT and the table entry in Appendix E for the CTFT of the triangle function (Figure 5.17).

■ **Solution**

The function $x(t)$ is the derivative of a triangle function centered at zero with a half-width of two,

$$x(t) = \frac{d}{dt}\left(2 \text{ tri}\left(\frac{t}{2}\right)\right). \tag{5.112}$$

Using the scaling property,

$$2 \text{ tri}\left(\frac{t}{2}\right) \xleftrightarrow{\mathcal{F}} 4 \text{ sinc}^2(2f). \tag{5.113}$$

Then, using the differentiation property,

$$x(t) \xleftrightarrow{\mathcal{F}} j8\pi f \text{ sinc}^2(2f). \tag{5.114}$$

If we find the CTFT of $x(t)$ by using the table entry for the CTFT of a rectangle and the time-scaling and time-shifting properties, we get

$$x(t) \xleftrightarrow{\mathcal{F}} j4 \text{ sinc}(2f) \sin(2\pi f). \tag{5.115}$$

which, using the definition of the sinc function, can be shown to be equivalent to

$$x(t) \xleftrightarrow{\mathcal{F}} j8\pi f \text{ sinc}^2(2f) = j8\pi f \text{ sinc}(2f)\frac{\sin(2\pi f)}{2\pi f} = j4 \text{ sinc}(2f)\sin(2\pi f). \tag{5.116}$$ ■

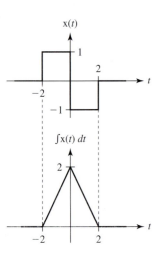

**Figure 5.17**
$x(t)$ and its integral.

## MODULATION

The CTFT of the product of $x(t)$ and a cosine at the frequency $f_0$ is

$$\mathcal{F}(x(t)\cos(2\pi f_0 t)) = X(f) * \left\{\frac{1}{2}[\delta(f - f_0) + \delta(f + f_0)]\right\} \tag{5.117}$$

or

$$\mathcal{F}(x(t)\cos(2\pi f_0 t)) = \frac{1}{2}[X(f - f_0) + X(f + f_0)]. \tag{5.118}$$

The modulation property is

$$\boxed{x(t)\cos(2\pi f_0 t) \xleftrightarrow{\mathcal{F}} \frac{1}{2}[X(f - f_0) + X(f + f_0)]} \tag{5.119}$$

or

$$\boxed{x(t)\cos(\omega_0 t) \xleftrightarrow{\mathcal{F}} \frac{1}{2}[X(j(\omega - \omega_0)) + X(j(\omega + \omega_0))]}. \tag{5.120}$$

This property is really just a specialization of the frequency-shifting property to describe the basic effect in the frequency domain of modulation in the time domain. We will explore this in much more detail in Chapter 6.

## TRANSFORMS OF PERIODIC SIGNALS

If a time signal $x(t)$ is periodic, it can be represented exactly by a complex CTFS (by letting $T_F = T_0$). Therefore, for periodic signals (using the frequency-shifting property),

$$x(t) = \sum_{k=-\infty}^{\infty} X[k] e^{-j2\pi(kf_F)t} \xleftrightarrow{\mathcal{F}} X(f) = \sum_{k=-\infty}^{\infty} X[k]\delta(f - kf_0) \qquad \textbf{(5.121)}$$

or

$$x(t) = \sum_{k=-\infty}^{\infty} X[k] e^{-j(k\omega_F)t} \xleftrightarrow{\mathcal{F}} X(j\omega) = 2\pi \sum_{k=-\infty}^{\infty} X[k]\delta(\omega - k\omega_0) \;. \qquad \textbf{(5.122)}$$

The CTFT of a periodic signal consists only of impulses.

This property of the CTFT illustrates that the CTFS can be considered as just a special case of the CTFT. Once we know how to do the CTFT we can use it to find the CTFS harmonic function of a periodic signal by finding the CTFT in the form $\sum_{k=-\infty}^{\infty} X[k]\delta(f - kf_0)$ and identifying the function $X[k]$ as the CTFS harmonic function.

---

**EXAMPLE 5.3**

Use (5.121) to find the CTFS harmonic function of

$$x(t) = \text{rect}(2t) * \text{comb}(t). \qquad \textbf{(5.123)}$$

■ **Solution**

This is a convolution of two functions. Therefore, the CTFT of $x(t)$ is the product of the CTFTs of the individual functions,

$$X(f) = \frac{1}{2} \text{sinc}\left(\frac{f}{2}\right) \text{comb}(f) = \frac{1}{2} \sum_{k=-\infty}^{\infty} \text{sinc}\left(\frac{k}{2}\right) \delta(f - k) \qquad \textbf{(5.124)}$$

Then, using (5.121), the CTFS harmonic function is

$$X[k] = \frac{1}{2} \text{sinc}\left(\frac{k}{2}\right). \qquad \textbf{(5.125)}$$

■

---

## PARSEVAL'S THEOREM

Even though an energy signal and its CTFT may look quite different, they do have something in common. They have the same total signal energy. In a manner similar to the definition of signal energy for a time-domain signal, the total signal energy of a frequency-domain signal $X(f)$ is defined by

$$E_x = \int_{-\infty}^{\infty} |X(f)|^2 \, df. \qquad \textbf{(5.126)}$$

Parseval's theorem states that the signal energies of an energy signal and its CTFT are equal. (The name of the theorem comes from Marc-Antoine Parseval des Chênes, another French mathematician of the late eighteenth and early nineteenth centuries.) This can be shown by the following logic.

The product of two energy signals in the time domain corresponds to the convolution of their transforms in the frequency domain.

$$\mathcal{F}[x(t)y(t)] = X(f) * Y(f) = \int_{-\infty}^{\infty} X(\phi)Y(f - \phi) \, d\phi. \tag{5.127}$$

From the definition of the CTFT,

$$\mathcal{F}[x(t)y(t)] = \int_{-\infty}^{\infty} x(t)y(t)e^{-j2\pi ft} \, dt. \tag{5.128}$$

Combining (5.127) and (5.128)

$$\int_{-\infty}^{\infty} x(t)y(t)e^{-j2\pi ft} \, dt = \int_{-\infty}^{\infty} X(\phi)Y(f - \phi) \, d\phi. \tag{5.129}$$

This relation holds for any value of $f$. Setting $f = 0$ in (5.129),

$$\int_{-\infty}^{\infty} x(t)y(t) \, dt = \int_{-\infty}^{\infty} X(\phi)Y(-\phi) \, d\phi. \tag{5.130}$$

This is known as the *generalized form* of Parseval's theorem. For the special case in which $y(t) = x^*(t)$, (5.130) becomes

$$\int_{-\infty}^{\infty} x(t)x^*(t) \, dt = \int_{-\infty}^{\infty} |x(t)|^2 \, dt = \int_{-\infty}^{\infty} X(\phi)X(-\phi) \, d\phi$$

$$= \int_{-\infty}^{\infty} X(\phi)X^*(\phi) \, d\phi = \int_{-\infty}^{\infty} |X(\phi)|^2 \, d\phi, \tag{5.131}$$

where we have used

$$yy^* = |y|^2 \quad \text{and} \quad \mathcal{F}[y^*(t)] = Y^*(-f). \tag{5.132}$$

So, ultimately, we have the equality of total signal energy in the time and frequency domains,

$$\boxed{\int_{-\infty}^{\infty} |x(t)|^2 \, dt = \int_{-\infty}^{\infty} |X(f)|^2 \, df}. \tag{5.133}$$

The equivalent result for the $\omega$ form of the CTFT is

$$\boxed{\int_{-\infty}^{\infty} |x(t)|^2 \, dt = \frac{1}{2\pi} \int_{-\infty}^{\infty} |X(j\omega)|^2 \, df}. \tag{5.134}$$

## INTEGRAL DEFINITION OF AN IMPULSE

The definition of the Fourier transform pair,

$$X(f) = \mathcal{F}(x(t)) = \int_{-\infty}^{\infty} x(t)e^{-j2\pi ft} \, dt \qquad x(t) = \mathcal{F}^{-1}(X(f)) = \int_{-\infty}^{\infty} X(f)e^{+j2\pi ft} \, df,$$

$$(5.135)$$

can be used to prove a handy result. Since

$$x(t) = \int_{-\infty}^{\infty} X(f)e^{+j2\pi ft} \, df \qquad (5.136)$$

and (making the change of variable $\tau = t$ in the definition of the forward CTFT)

$$X(f) = \int_{-\infty}^{\infty} x(\tau)e^{-j2\pi f\tau} \, d\tau, \qquad (5.137)$$

we can combine (5.136) and (5.137) to get

$$x(t) = \int_{-\infty}^{\infty} \left[ \int_{-\infty}^{\infty} x(\tau)e^{-j2\pi f\tau} \, d\tau \right] e^{+j2\pi ft} \, df \qquad (5.138)$$

or

$$x(t) = \int_{-\infty}^{\infty} \int_{-\infty}^{\infty} x(\tau)e^{-j2\pi f(t-\tau)} \, d\tau \, df. \qquad (5.139)$$

Rearranging to do the $f$ integration first in (5.139),

$$x(t) = \int_{-\infty}^{\infty} x(\tau) \left[ \int_{-\infty}^{\infty} e^{-j2\pi f(t-\tau)} \, df \right] d\tau. \qquad (5.140)$$

In words, this integral says that if we take any arbitrary signal $x(t)$, replace $t$ by $\tau$, multiply that by a function $\int_{-\infty}^{\infty} e^{-j2\pi f(t-\tau)} \, df$, and then integrate over all $\tau$, we get $x(t)$ back. We could rewrite (5.140) in the form,

$$x(t) = \int_{-\infty}^{\infty} x(\tau)g(t-\tau) \, d\tau, \qquad (5.141)$$

where

$$g(t-\tau) = \int_{-\infty}^{\infty} e^{-j2\pi f(t-\tau)} \, df. \qquad (5.142)$$

Recall the sampling property of the impulse

$$\int_{-\infty}^{\infty} g(t)\delta(t - t_0)\, dt = g(t_0).$$

(5.143)

The only way the equality

$$x(t) = \int_{-\infty}^{\infty} x(\tau)g(t - \tau)\, d\tau$$

(5.144)

can be satisfied is if $g(t - \tau) = \delta(t - \tau)$, that is, if

$$\int_{-\infty}^{\infty} e^{-j2\pi f(t - \tau)}\, df = \delta(t - \tau).$$

(5.145)

This is yet another valid definition of a unit impulse. A more common form of this result is

$$\boxed{\int_{-\infty}^{\infty} e^{-j2\pi xy}\, dy = \delta(x)}.$$

(5.146)

## DUALITY

If the CTFT of $x(t)$ is $X(f)$, what is the CTFT of $X(t)$?

$$X(f) = \int_{-\infty}^{\infty} x(t)e^{-j2\pi ft}\, dt = \int_{-\infty}^{\infty} x(\tau)e^{-j2\pi f\tau}\, d\tau$$

(5.147)

Therefore, replacing $f$ by $t$ in the right side of (5.147),

$$X(t) = \int_{-\infty}^{\infty} x(\tau)e^{-j2\pi t\tau}\, d\tau.$$

(5.148)

The CTFT of (5.148) is

$$\mathcal{F}(X(t)) = \mathcal{F}\left[\int_{-\infty}^{\infty} x(\tau)e^{-j2\pi t\tau}\, d\tau\right] = \int_{-\infty}^{\infty}\left[\int_{-\infty}^{\infty} x(\tau)e^{-j2\pi t\tau}\, d\tau\right]e^{-j2\pi ft}\, dt$$

(5.149)

or

$$\mathcal{F}(X(t)) = \int_{-\infty}^{\infty} x(\tau)\int_{-\infty}^{\infty} e^{-j2\pi(\tau + f)t}\, dt\, d\tau.$$

(5.150)

Using the integral definition of an impulse derived in the previous section,

$$\int_{-\infty}^{\infty} e^{-j2\pi xy}\, dy = \delta(x),$$

(5.151)

we can write

$$
\int_{-\infty}^{\infty} e^{-j2\pi t(\tau + f)} \, dt = \delta(\tau + f). \tag{5.152}
$$

Now, substituting into (5.150),

$$
\mathcal{F}(X(t)) = \int_{-\infty}^{\infty} x(\tau)\delta(\tau + f) \, d\tau = x(-f). \tag{5.153}
$$

The duality property is

$$
\boxed{X(t) \xleftrightarrow{\ \mathcal{F}\ } x(-f) \qquad \text{and} \qquad X(-t) \xleftrightarrow{\ \mathcal{F}\ } x(f)} \tag{5.154}
$$

or

$$
\boxed{X(jt) \xleftrightarrow{\ \mathcal{F}\ } 2\pi x(-\omega) \qquad \text{and} \qquad X(-jt) \xleftrightarrow{\ \mathcal{F}\ } 2\pi x(\omega)}. \tag{5.155}
$$

The proof of

$$
X(-t) \xleftrightarrow{\ \mathcal{F}\ } x(f) \tag{5.156}
$$

is similar.

The basic reason this property exists stems from the definition of the CTFT,

$$
X(f) = \int_{-\infty}^{\infty} x(t)e^{-j2\pi ft} \, dt \qquad \text{and} \qquad x(t) = \int_{-\infty}^{\infty} X(f)e^{+j2\pi ft} \, df. \tag{5.157}
$$

The forward and inverse CTFTs are very similar, differing only in the sign of the exponent of $e$ and the name of the variable of integration. A good example of this is the duality of the rect( ) and sinc( ) functions,

$$
\text{rect}(t) \xleftrightarrow{\ \mathcal{F}\ } \text{sinc}(f) \qquad \text{and} \qquad \text{sinc}(t) \xleftrightarrow{\ \mathcal{F}\ } \text{rect}(f) \tag{5.158}
$$

(Figure 5.18).

## TOTAL-AREA INTEGRAL USING FOURIER TRANSFORMS

Another property of the CTFT which comes directly from the definition is that the total area under a time- or frequency-domain signal can be found by evaluating its CTFT or inverse CTFT with an argument of zero.

$$
\boxed{X(0) = \left[ \int_{-\infty}^{\infty} x(t)e^{-j2\pi ft} \, dt \right]_{f \to 0} = \int_{-\infty}^{\infty} x(t) \, dt} \tag{5.159}
$$

and

$$
\boxed{x(0) = \left[ \int_{-\infty}^{\infty} X(f)e^{+j2\pi ft} \, df \right]_{t \to 0} = \int_{-\infty}^{\infty} X(f) \, df} \tag{5.160}
$$

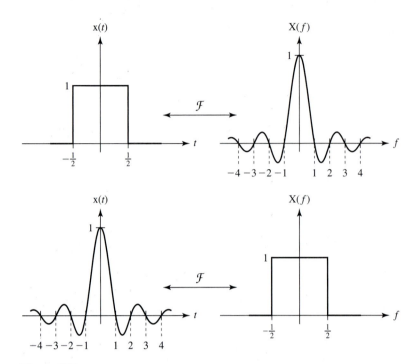

**Figure 5.18**
Duality of rect( ) and sinc( ) functions under Fourier transformation.

or

$$X(0) = \left[ \int_{-\infty}^{\infty} x(t)e^{-j\omega t}\, dt \right]_{\omega \to 0} = \int_{-\infty}^{\infty} x(t)\, dt \qquad (5.161)$$

and

$$x(0) = \left[ \frac{1}{2\pi} \int_{-\infty}^{\infty} X(j\omega)e^{+j\omega t}\, d\omega \right]_{t \to 0} = \frac{1}{2\pi} \int_{-\infty}^{\infty} X(j\omega)\, d\omega. \qquad (5.162)$$

**EXAMPLE 5.4**

Find the total area under the function

$$x(t) = 10 \, \text{sinc} \left( \frac{t+4}{7} \right). \qquad (5.163)$$

■ **Solution**

Ordinarily we would try to directly integrate the function over all $t$,

$$\text{Area} = \int_{-\infty}^{\infty} x(t)\, dt = \int_{-\infty}^{\infty} 10 \, \text{sinc} \left( \frac{t+4}{7} \right)\, dt = \int_{-\infty}^{\infty} 10 \frac{\sin(\pi(t+4)/7)}{\pi(t+4)/7}\, dt. \qquad (5.164)$$

This integral is a variant of a type of mathematical function called a *sine integral* defined by

$$\text{Si}(z) = \int_0^z \frac{\sin(t)}{t} \, dt. \tag{5.165}$$

The sine integral is related to a more general function, the *exponential integral*. The sine integral and exponential integral can be found tabulated in mathematical tables books, and MATLAB has a built-in function to evaluate the exponential integral. However, a plunge into the sine integral is not necessary to solve this problem. We can use (5.159). First we find the CTFT of $x(t)$,

$$X(f) = 70 \, \text{rect}(7f) \, e^{j8\pi f}. \tag{5.166}$$

Then

$$\text{Area} = X(0) = 70. \tag{5.167}$$

■

## INTEGRATION

The proof of this property was left until last because it depends on many of the other properties already proven. It is tempting to try to reverse the differentiation property to prove the integration property. Let

$$x(t) = \frac{d}{dt}(y(t)). \tag{5.168}$$

Then $X(f) = j2\pi f Y(f)$ or $X(j\omega) = j\omega Y(j\omega)$. We can rearrange these results into

$$Y(f) = \frac{X(f)}{j2\pi f} \quad \text{or} \quad Y(j\omega) = \frac{X(j\omega)}{j\omega}. \tag{5.169}$$

Therefore, since $x(t)$ is the derivative of $y(t)$, $y(t)$ must be the integral of $x(t)$, and we could say that the integration property is

$$\int_{-\infty}^t x(\lambda) \, d\lambda \xleftrightarrow{\ \mathcal{FT}\ } \frac{X(f)}{j2\pi f} \quad \text{or} \quad \int_{-\infty}^t x(\lambda) \, d\lambda \xleftrightarrow{\ \mathcal{FT}\ } \frac{X(j\omega)}{j\omega}. \tag{5.170}$$

But what is the value of $(X(f)/j2\pi f)$ at $f = 0$, given that division by zero is undefined? This uncertainty comes about because if $x(t)$ is the derivative of $y(t)$, given any $x(t)$ there are multiple possible answers for what $y(t)$ could be. The addition of any constant to $y(t)$ does not change its derivative $x(t)$. So we need to be more careful in the proof of the integration property.

Before proving the integration property it is useful to prove some other CTFT results. First we will find the CTFT of

$$x(t) = \frac{1}{j\pi t}. \tag{5.171}$$

$$\mathcal{F}(x(t)) = \int_{-\infty}^\infty \frac{e^{-j2\pi ft}}{j\pi t} \, dt = \int_{-\infty}^\infty \underbrace{\frac{\cos(2\pi ft)}{j\pi t}}_{\text{odd}} - \underbrace{\frac{j\sin(2\pi ft)}{j\pi t}}_{\text{even}} \, dt \tag{5.172}$$

$$\mathcal{F}(x(t)) = - \int_{-\infty}^\infty \frac{\sin(2\pi ft)}{\pi t} \, dt \tag{5.173}$$

For $f = 0$,

$$\mathcal{F}(\mathrm{x}(t)) = \lim_{f \to 0} \left( -\int_{-\infty}^{\infty} \frac{\sin(2\pi f t)}{\pi t} \, dt \right) = -\int_{-\infty}^{\infty} \lim_{f \to 0} \frac{\sin(2\pi f t)}{\pi t} \, dt = 0. \qquad (5.174)$$

For $f \neq 0$,

$$\mathcal{F}(\mathrm{x}(t)) = -\int_{-\infty}^{\infty} \frac{\sin(2\pi f t)}{\pi t} \, dt = -2f \int_{-\infty}^{\infty} \frac{\sin(2\pi f t)}{2\pi f t} \, dt = -2f \int_{-\infty}^{\infty} \mathrm{sinc}(2f t) \, dt.$$

$$(5.175)$$

Let $\lambda = 2ft$ and $d\lambda = 2f \, dt$. Then, for $f > 0$,

$$\mathcal{F}[\mathrm{x}(t)] = -2f \int_{-\infty}^{\infty} \mathrm{sinc}(\lambda) \frac{d\lambda}{2f} = -\int_{-\infty}^{\infty} \mathrm{sinc}(\lambda) \, d\lambda. \qquad (5.176)$$

Using the property that the total area under a function is its CTFT evaluated at $f = 0$,

$$\int_{-\infty}^{\infty} \mathrm{sinc}(\lambda) \, d\lambda = \mathcal{F}[\mathrm{sinc}(t)]_{f \to 0}. \qquad (5.177)$$

We already know from Example 5.1 that $\mathcal{F}[\mathrm{rect}(t)] = \mathrm{sinc}(f)$. We can use the duality property to show that $\mathcal{F}[\mathrm{sinc}(t)] = \mathrm{rect}(f)$. Then

$$\int_{-\infty}^{\infty} \mathrm{sinc}(\lambda) \, d\lambda = \mathrm{rect}(0) = 1 \qquad \text{and} \qquad \mathcal{F}(\mathrm{x}(t)) = -1 \qquad f > 0. \qquad (5.178)$$

By a similar argument, for $f < 0$,

$$\mathcal{F}(\mathrm{x}(t)) = -2f \int_{\infty}^{-\infty} \mathrm{sinc}(\lambda) \frac{d\lambda}{2f} = \underbrace{\int_{-\infty}^{\infty} \mathrm{sinc}(\lambda) \, d\lambda}_{=1} = +1 \qquad f < 0. \qquad (5.179)$$

Therefore,

$$\mathcal{F}(\mathrm{x}(t)) = \begin{cases} -1 & f > 0 \\ 0 & f = 0 \\ 1 & f < 0 \end{cases} = -\mathrm{sgn}(f) = \mathrm{sgn}(-f). \qquad (5.180)$$

Using the duality property,

$$\mathcal{F}(\mathrm{sgn}(t)) = \frac{1}{j\pi f}. \qquad (5.181)$$

Using (5.181) it is easy to find the CTFT of a unit step, $\mathrm{u}(t) = \frac{1}{2}[\mathrm{sgn}(t) + 1]$. Using the linearity property,

$$\mathcal{F}(\mathrm{u}(t)) = \frac{1}{2}\{\mathcal{F}[\mathrm{sgn}(t)] + \mathcal{F}(1)\}. \qquad (5.182)$$

Using the integral definition of an impulse previously derived,

$$\mathcal{F}(1) = \int_{-\infty}^{\infty} e^{-j2\pi ft}\, dt = \delta(f). \qquad (5.183)$$

Therefore,

$$\mathcal{F}(u(t)) = \frac{1}{2}\left[\frac{1}{j\pi f} + \delta(f)\right] = \frac{1}{j2\pi f} + \frac{1}{2}\delta(f). \qquad (5.184)$$

If a signal $x(t)$ is convolved with a unit step,

$$x(t) * u(t) = \int_{-\infty}^{\infty} x(\tau)u(t-\tau)\, d\tau = \int_{-\infty}^{t} x(\tau)\, d\tau, \qquad (5.185)$$

proving that the convolution of a signal with a unit step is the same as the cumulative integral of the signal. Now we can prove the integration property of the CTFT. Using the convolution property and the equivalence property of an impulse,

$$\mathcal{F}(x(t) * u(t)) = X(f)\left[\frac{1}{j2\pi f} + \frac{1}{2}\delta(f)\right] = \frac{X(f)}{j2\pi f} + \frac{1}{2}X(0)\,\delta(f). \qquad (5.186)$$

Then, finally, the integration property of the CTFT is

$$\boxed{\int_{-\infty}^{t} x(\lambda)\, d\lambda \xleftrightarrow{\ \mathcal{F}\ } \frac{X(f)}{j2\pi f} + \frac{1}{2}X(0)\,\delta(f)} \qquad (5.187)$$

or

$$\boxed{\int_{-\infty}^{t} x(\lambda)\, d\lambda \xleftrightarrow{\ \mathcal{F}\ } \frac{X(j\omega)}{j\omega} + \pi X(0)\,\delta(\omega)}. \qquad (5.188)$$

The term $\frac{1}{2}X(0)\,\delta(f)$ or $\pi X(0)\,\delta(\omega)$ accounts for the effect of the total area of $x(t)$. This development of the CTFT of the integral of a function is a good example of the use of the properties.

**EXAMPLE 5.5**

Verify the table entry for the CTFT of $x(t) = \mathrm{tri}(t)$ using the integration property of the CTFT.

■ **Solution**

The first derivative of the triangle function is

$$\frac{d}{dt}(x(t)) = \mathrm{rect}\left(t + \frac{1}{2}\right) - \mathrm{rect}\left(t - \frac{1}{2}\right), \qquad (5.189)$$

and the second derivative is

$$\frac{d^2}{dt^2}(x(t)) = \delta(t+1) - 2\delta(t) + \delta(t-1) \qquad (5.190)$$

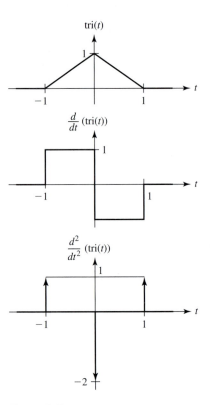

**Figure 5.19**
The unit triangle and its first and second
derivatives.

(Figure 5.19). The Fourier transform of the second derivative is

$$\mathcal{F}\left(\frac{d^2}{dt^2}(\mathrm{x}(t))\right) = e^{j2\pi f} - 2 + e^{-j2\pi f} = 2[\cos(2\pi f) - 1]. \tag{5.191}$$

Using the integration property of the CTFT and the fact that the area under the second derivative is zero, the Fourier transform of the first derivative is

$$\mathcal{F}\left(\frac{d}{dt}(\mathrm{x}(t))\right) = 2\frac{\cos(2\pi f) - 1}{j2\pi f}, \tag{5.192}$$

and, using the integration property again,

$$\mathcal{F}(\mathrm{x}(t)) = \mathcal{F}(\mathrm{tri}(t)) = 2\frac{\cos(2\pi f) - 1}{(j2\pi f)^2}. \tag{5.193}$$

Then, using

$$\sin(x)\sin(y) = \frac{1}{2}[\cos(x - y) - \cos(x + y)], \tag{5.194}$$

it follows that

$$\sin^2(x) = \frac{1}{2}[1 - \cos(2x)] \tag{5.195}$$

and

$$\mathcal{F}(\mathrm{tri}(t)) = 2\frac{\cos(2\pi f) - 1}{(j2\pi f)^2} = 2\frac{-2\,\sin^2(\pi f)}{(j2\pi f)^2} = \frac{\sin^2(\pi f)}{(\pi f)^2} = \mathrm{sinc}^2(f). \tag{5.196}$$

## SUMMARY OF CTFT PROPERTIES

Linearity

$$\alpha x(t) + \beta y(t) \overset{\mathcal{F}}{\longleftrightarrow} \alpha X(f) + \beta Y(f)$$

$$\alpha x(t) + \beta y(t) \overset{\mathcal{F}}{\longleftrightarrow} \alpha X(j\omega) + \beta Y(j\omega)$$

Time shifting

$$x(t - t_0) \overset{\mathcal{F}}{\longleftrightarrow} X(f)e^{-j2\pi f t_0}$$

$$x(t - t_0) \overset{\mathcal{F}}{\longleftrightarrow} X(j\omega)e^{-j\omega t_0}$$

Frequency shifting

$$x(t)e^{+j2\pi f_0 t} \overset{\mathcal{F}}{\longleftrightarrow} X(f - f_0)$$

$$x(t)e^{+j\omega_0 t} \overset{\mathcal{F}}{\longleftrightarrow} X[j(\omega - \omega_0)]$$

Time scaling

$$x(at) \overset{\mathcal{F}}{\longleftrightarrow} \frac{1}{|a|}X\left(\frac{f}{a}\right)$$

$$x(at) \overset{\mathcal{F}}{\longleftrightarrow} \frac{1}{|a|}X\left(j\frac{\omega}{a}\right)$$

Frequency scaling

$$\frac{1}{|a|}x\left(\frac{t}{a}\right) \overset{\mathcal{F}}{\longleftrightarrow} X(af)$$

$$\frac{1}{|a|}x\left(\frac{t}{a}\right) \overset{\mathcal{F}}{\longleftrightarrow} X(ja\omega)$$

Transform of a conjugate

$$x^*(t) \overset{\mathcal{F}}{\longleftrightarrow} X^*(-f)$$

$$x^*(t) \overset{\mathcal{F}}{\longleftrightarrow} X^*(-j\omega)$$

Multiplication–convolution duality

$$x(t) * y(t) \overset{\mathcal{F}}{\longleftrightarrow} X(f)Y(f)$$

$$x(t) * y(t) \overset{\mathcal{F}}{\longleftrightarrow} X(j\omega)Y(j\omega)$$

$$x(t)y(t) \overset{\mathcal{F}}{\longleftrightarrow} X(f) * Y(f)$$

$$x(t)y(t) \overset{\mathcal{F}}{\longleftrightarrow} \frac{1}{2\pi}X(j\omega) * Y(j\omega)$$

Differentiation

$$\frac{d}{dt}(x(t)) \overset{\mathcal{F}}{\longleftrightarrow} j2\pi f X(f)$$

$$\frac{d}{dt}(x(t)) \overset{\mathcal{F}}{\longleftrightarrow} j\omega X(j\omega)$$

Modulation

$$x(t)\cos(2\pi f_0 t) \overset{\mathcal{F}}{\longleftrightarrow} \frac{1}{2}[X(f - f_0) + X(f + f_0)]$$

$$x(t)\cos(\omega_0 t) \overset{\mathcal{F}}{\longleftrightarrow} \frac{1}{2}[X(j(\omega - \omega_0)) + X(j(\omega + \omega_0))]$$

Transforms of periodic signals

$$x(t) = \sum_{k=-\infty}^{\infty} X[k]e^{-j2\pi(kf_F)t} \overset{\mathcal{F}}{\longleftrightarrow} X(f) = \sum_{k=-\infty}^{\infty} X[k]\delta(f - kf_0)$$

$$x(t) = \sum_{k=-\infty}^{\infty} X[k]e^{-j(k\omega_F)t} \overset{\mathcal{F}}{\longleftrightarrow} X(j\omega) = 2\pi\sum_{k=-\infty}^{\infty} X[k]\delta(\omega - k\omega_0)$$

Parseval's theorem
$$\int\limits_{-\infty}^{\infty} |x(t)|^2 \, dt = \int\limits_{-\infty}^{\infty} |X(f)|^2 \, df$$

$$\int\limits_{-\infty}^{\infty} |x(t)|^2 \, dt = \frac{1}{2\pi} \int\limits_{-\infty}^{\infty} |X(j\omega)|^2 \, df$$

Integral definition of an impulse
$$\int\limits_{-\infty}^{\infty} e^{-j2\pi xy} \, dy = \delta(x)$$

Duality
$$X(t) \overset{\mathcal{F}}{\longleftrightarrow} x(-f) \quad \text{and} \quad X(-t) \overset{\mathcal{F}}{\longleftrightarrow} x(f)$$

$$X(jt) \overset{\mathcal{F}}{\longleftrightarrow} 2\pi x(-\omega) \quad \text{and} \quad X(-jt) \overset{\mathcal{F}}{\longleftrightarrow} 2\pi x(\omega)$$

Total-area integral using Fourier transforms

$$X(0) = \left[ \int\limits_{-\infty}^{\infty} x(t) e^{-j2\pi ft} \, dt \right]_{f\to 0} dt$$

$$x(0) = \left[ \int\limits_{-\infty}^{\infty} X(f) e^{+j2\pi ft} \, df \right]_{t\to 0} df$$

$$X(0) = \left[ \int\limits_{-\infty}^{\infty} x(t) e^{-j\omega t} \, dt \right]_{\omega\to 0} dt$$

$$x(0) = \left[ \frac{1}{2\pi} \int\limits_{-\infty}^{\infty} X(j\omega) e^{+j\omega t} \, d\omega \right]_{t\to 0} d\omega$$

Integration
$$\int\limits_{-\infty}^{t} x(\lambda) \, d\lambda \overset{\mathcal{F}}{\longleftrightarrow} \frac{X(f)}{j2\pi f} + \frac{1}{2} X(0) \delta(f)$$

$$\int\limits_{-\infty}^{t} x(\lambda) \, d\lambda \overset{\mathcal{F}}{\longleftrightarrow} \frac{X(j\omega)}{j\omega} + \pi X(0) \delta(\omega)$$

## USE OF TABLES AND PROPERTIES

In this section are some examples which will illustrate the use of the tables in Appendix E and the properties presented at the beginning of this chapter to find the CTFTs of some signals.

**EXAMPLE 5.6**

If $x(t) = 10\sin(t)$, then find the CTFT of

a. $x(t)$    b. $x(t-2)$    c. $x(2(t-1))$    d. $x(2t-1)$

■ **Solution**

a. Using the linearity property and looking up the transform of the general sine form

$$\sin(2\pi f_0 t) \overset{\mathcal{F}}{\longleftrightarrow} \frac{j}{2}[\delta(f+f_0) - \delta(f-f_0)] \tag{5.197}$$

$$\sin(t) \overset{\mathcal{F}}{\longleftrightarrow} \frac{j}{2}\left[\delta\left(f+\frac{1}{2\pi}\right) - \delta\left(f-\frac{1}{2\pi}\right)\right] \tag{5.198}$$

$$10\sin(t) \overset{\mathcal{F}}{\longleftrightarrow} j5\left[\delta\left(f+\frac{1}{2\pi}\right) - \delta\left(f-\frac{1}{2\pi}\right)\right] \tag{5.199}$$

or, in the radian-frequency form,

$$10\sin(t) \overset{\mathcal{F}}{\longleftrightarrow} j10\pi[\delta(\omega+1) - \delta(\omega-1)]. \tag{5.200}$$

This is a case in which the radian-frequency form of the transform is slightly more compact than the cyclic-frequency form.

b. Using the result of part (a)

$$10\sin(t) \overset{\mathcal{F}}{\longleftrightarrow} j5\left[\delta\left(f+\frac{1}{2\pi}\right) - \delta\left(f-\frac{1}{2\pi}\right)\right] \tag{5.201}$$

and the time-shifting property,

$$10\sin(t-2) \overset{\mathcal{F}}{\longleftrightarrow} j5\left[\delta\left(f+\frac{1}{2\pi}\right) - \delta\left(f-\frac{1}{2\pi}\right)\right]e^{-j4\pi f} \tag{5.202}$$

or

$$10\sin(t-2) \overset{\mathcal{F}}{\longleftrightarrow} j10\pi[\delta(\omega+1) - \delta(\omega-1)]e^{-j2\omega}. \tag{5.203}$$

Since the only values of $f$ in $e^{-j4\pi f}$ or $\omega$ in $e^{-j2\omega}$ that really matter are those where the impulses occur, the solution can also be written in the forms

$$10\sin(t-2) \overset{\mathcal{F}}{\longleftrightarrow} j5\left[\delta\left(f+\frac{1}{2\pi}\right)e^{j2} - \delta\left(f-\frac{1}{2\pi}\right)e^{-j2}\right] \tag{5.204}$$

or

$$10\sin(t-2) \overset{\mathcal{F}}{\longleftrightarrow} j10\pi[\delta(\omega+1)e^{j2} - \delta(\omega-1)e^{-j2}]. \tag{5.205}$$

c. From part (a),

$$10\sin(t) \overset{\mathcal{F}}{\longleftrightarrow} j5\left[\delta\left(f+\frac{1}{2\pi}\right) - \delta\left(f-\frac{1}{2\pi}\right)\right]. \tag{5.206}$$

Using the time-scaling property,

$$10\sin(2t) \overset{\mathcal{F}}{\longleftrightarrow} j\frac{5}{2}\left[\delta\left(\frac{f}{2}+\frac{1}{2\pi}\right) - \delta\left(\frac{f}{2}-\frac{1}{2\pi}\right)\right]. \tag{5.207}$$

Then, using the time-shifting property,

$$10\sin(2(t-1)) \overset{\mathcal{F}}{\longleftrightarrow} j\frac{5}{2}\left[\delta\left(\frac{f}{2}+\frac{1}{2\pi}\right) - \delta\left(\frac{f}{2}-\frac{1}{2\pi}\right)\right]e^{-j2\pi f}. \tag{5.208}$$

Finally, using the scaling property of the impulse function,

$$10\sin(2(t-1)) \overset{\mathcal{F}}{\longleftrightarrow} j5\left[\delta\left(f+\frac{1}{\pi}\right) - \delta\left(f-\frac{1}{\pi}\right)\right]e^{-j2\pi f} \tag{5.209}$$

or

$$10\sin(2(t-1)) \overset{\mathcal{F}}{\longleftrightarrow} j5\left[\delta\left(f+\frac{1}{\pi}\right)e^{j2} - \delta\left(f-\frac{1}{\pi}\right)e^{-j2}\right] \tag{5.210}$$

or

$$10\sin(2(t-1)) \overset{\mathcal{F}}{\longleftrightarrow} j10\pi[\delta(\omega+2)e^{j2} - \delta(\omega-2)e^{-j2}]. \tag{5.211}$$

d. From part (a),

$$10\sin(t) \overset{\mathcal{F}}{\longleftrightarrow} j5\left[\delta\left(f+\frac{1}{2\pi}\right) - \delta\left(f-\frac{1}{2\pi}\right)\right]. \tag{5.212}$$

Applying the time-shifting property first,

$$10\sin(t-1) \overset{\mathcal{F}}{\longleftrightarrow} j5\left[\delta\left(f+\frac{1}{2\pi}\right) - \delta\left(f-\frac{1}{2\pi}\right)\right]e^{-j2\pi f}. \tag{5.213}$$

Then, applying the time-scaling property,

$$10\sin(2t-1) \overset{\mathcal{F}}{\longleftrightarrow} j\frac{5}{2}\left[\delta\left(\frac{f}{2}+\frac{1}{2\pi}\right) - \delta\left(\frac{f}{2}-\frac{1}{2\pi}\right)\right]e^{-j\pi f}. \tag{5.214}$$

Finally, using the scaling property of the impulse,

$$10\sin(2t-1) \overset{\mathcal{F}}{\longleftrightarrow} j5\left[\delta\left(f+\frac{1}{\pi}\right) - \delta\left(f-\frac{1}{\pi}\right)\right]e^{-j\pi f} \tag{5.215}$$

or

$$10\sin(2t-1) \overset{\mathcal{F}}{\longleftrightarrow} j5\left[\delta\left(f+\frac{1}{\pi}\right)e^{j} - \delta\left(f-\frac{1}{\pi}\right)e^{-j}\right], \tag{5.216}$$

or, in the radian-frequency form,

$$10\sin(2t-1) \overset{\mathcal{F}}{\longleftrightarrow} j10\pi[\delta(\omega+2)e^{-j} - \delta(\omega-2)e^{j}]. \tag{5.217}$$

■

## EXAMPLE 5.7

If $x(t) = 25\,\text{rect}((t-4)/10)$, find the CTFT of $x(t)$.

**■ Solution**

We can find the CTFT of the unit rectangle function in the table of Fourier transforms.

$$\text{rect}(t) \overset{\mathcal{F}}{\longleftrightarrow} \text{sinc}(f) \qquad \text{or} \qquad \text{rect}(t) \overset{\mathcal{F}}{\longleftrightarrow} \text{sinc}\left(\frac{\omega}{2\pi}\right) \tag{5.218}$$

First apply the linearity property.

$$25\,\text{rect}(t) \overset{\mathcal{F}}{\longleftrightarrow} 25\,\text{sinc}(f) \qquad \text{or} \qquad 25\,\text{rect}(t) \overset{\mathcal{F}}{\longleftrightarrow} 25\,\text{sinc}\left(\frac{\omega}{2\pi}\right) \tag{5.219}$$

Then apply the time-scaling property.

$$25 \operatorname{rect}\left(\frac{t}{10}\right) \overset{\mathcal{F}}{\longleftrightarrow} 10 \times 25 \operatorname{sinc}(10f) \quad \text{or} \quad 25 \operatorname{rect}\left(\frac{t}{10}\right) \overset{\mathcal{F}}{\longleftrightarrow} 10 \times 25 \operatorname{sinc}\left(\frac{5\omega}{\pi}\right)$$

(5.220)

$$25 \operatorname{rect}\left(\frac{t}{10}\right) \overset{\mathcal{F}}{\longleftrightarrow} 250 \operatorname{sinc}(10f) \quad \text{or} \quad 25 \operatorname{rect}\left(\frac{t}{10}\right) \overset{\mathcal{F}}{\longleftrightarrow} 250 \operatorname{sinc}\left(\frac{5\omega}{\pi}\right)$$

(5.221)

Then apply the time-shifting property.

$$25 \operatorname{rect}\left(\frac{t-4}{10}\right) \overset{\mathcal{F}}{\longleftrightarrow} 250 \operatorname{sinc}(10f)e^{-j2\pi f(4)} \quad \text{or}$$

$$25 \operatorname{rect}\left(\frac{t-4}{10}\right) \overset{\mathcal{F}}{\longleftrightarrow} 250 \operatorname{sinc}\left(\frac{5\omega}{\pi}\right)e^{-j\omega(4)}$$

(5.222)

$$25 \operatorname{rect}\left(\frac{t-4}{10}\right) \overset{\mathcal{F}}{\longleftrightarrow} 250 \operatorname{sinc}(10f)e^{-j8\pi f} \quad \text{or}$$

$$25 \operatorname{rect}\left(\frac{t-4}{10}\right) \overset{\mathcal{F}}{\longleftrightarrow} 250 \operatorname{sinc}\left(\frac{5\omega}{\pi}\right)e^{-j4\omega}$$

(5.223)

**EXAMPLE 5.8**

Find the CTFT of the convolution of $10\sin(t)$ with $2\delta(t+4)$.

■ **Solution**

*Method 1.* Do the convolution first and find the CTFT of the result.

$$10\sin(t) * 2\delta(t+4) = 20\sin(t+4)$$

(5.224)

The transform can be done in a manner similar to part (b) of Example 5.6,

$$20\sin(t+4) \overset{\mathcal{F}}{\longleftrightarrow} j10\left[\delta\left(f+\frac{1}{2\pi}\right) - \delta\left(f-\frac{1}{2\pi}\right)\right]e^{j8\pi f}$$

(5.225)

or

$$20\sin(t+4) \overset{\mathcal{F}}{\longleftrightarrow} j20\pi[\delta(\omega+1) - \delta(\omega-1)]e^{j4\omega}.$$

(5.226)

*Method 2.* Do the CTFT first to avoid the convolution.

$$10\sin(t) * 2\delta(t+4) \overset{\mathcal{F}}{\longleftrightarrow} \mathcal{F}(10\sin(t))\,\mathcal{F}(2\delta(t+4)) = 2\,\mathcal{F}(10\sin(t))\,\mathcal{F}(\delta(t))\,e^{j8\pi f}$$

(5.227)

$$10\sin(t) * 2\delta(t+4) \overset{\mathcal{F}}{\longleftrightarrow} j10\left[\delta\left(f+\frac{1}{2\pi}\right) - \delta\left(f-\frac{1}{2\pi}\right)\right]e^{j8\pi f}$$

(5.228)

or

$$10\sin(t) * 2\delta(t+4) \overset{\mathcal{F}}{\longleftrightarrow} \mathcal{F}(10\sin(t))\,\mathcal{F}(2\delta(t+4)) = 2\,\mathcal{F}(10\sin(t))\,\mathcal{F}(\delta(t))\,e^{j4\omega}$$

(5.229)

$$10\sin(t) * 2\delta(t+4) \overset{\mathcal{F}}{\longleftrightarrow} j20\pi[\delta(\omega+1) - \delta(\omega-1)]e^{j4\omega}$$

(5.230)

## 5.6 THE DISCRETE-TIME FOURIER TRANSFORM

The transition from the DTFS to the DTFT is analogous to the transition from the CTFS to the CTFT. We begin with a graphical demonstration of the concepts, and then do a formal analytical derivation.

### GRAPHICAL ILLUSTRATION

Consider first a DT rectangular-wave signal

$$x[n] = \text{rect}_{N_w}[n] * \text{comb}_{N_0}[n] \tag{5.231}$$

(Figure 5.20). The DTFS representing this DT signal $x[n]$ over exactly one fundamental period $N_0$ is given by

$$X[k] = \frac{1}{N_0} \sum_{n=\langle N_0 \rangle} x[n] e^{-j2\pi(kF_0)n} = \frac{1}{N_0} \sum_{n=-N_w}^{N_w} e^{-j2\pi(kn/N_0)}. \tag{5.232}$$

Letting $m = n + N_w$ in (5.232),

$$X[k] = \frac{1}{N_0} \sum_{m=0}^{2N_w} e^{-j2\pi(k(m-N_w)/N_0)} = e^{j2\pi(kN_w/N_0)} \frac{1}{N_0} \sum_{m=0}^{2N_w} e^{-j2\pi(km/N_0)}. \tag{5.233}$$

Then, using

$$\sum_{n=0}^{N-1} r^n = \begin{cases} N & r = 1 \\ \dfrac{1-r^N}{1-r} & r \neq 1 \end{cases} \tag{5.234}$$

in (5.233) we get

$$X[k] = e^{j2\pi(kN_w/N_0)} \frac{1}{N_0} \sum_{m=0}^{2N_w} \left( e^{-j2\pi(k/N_0)} \right)^m = \frac{e^{j2\pi(kN_w/N_0)}}{N_0} \frac{1 - e^{-j2\pi((2N_w+1)k/N_0)}}{1 - e^{-j2\pi(k/N_0)}} \tag{5.235}$$

or

$$X[k] = \frac{e^{+j2\pi(kN_w/N_0)}}{N_0} \frac{e^{-j\pi((2N_w+1)k/N_0)}}{e^{-j\pi(k/N_0)}} \frac{e^{+j\pi((2N_w+1)k/N_0)} - e^{-j\pi((2N_w+1)k/N_0)}}{e^{+j\pi(k/N_0)} - e^{-j\pi(k/N_0)}}$$

$$= \frac{1}{N_0} \frac{\left( e^{+j\pi((2N_w+1)k/N_0)} - e^{-j\pi((2N_w+1)k/N_0)} \right)/j2}{\left( e^{+j\pi(k/N_0)} - e^{-j\pi(k/N_0)} \right)/j2} \tag{5.236}$$

$$X[k] = \frac{1}{N_0} \frac{\sin((2N_w+1)(k\pi/N_0))}{\sin(k\pi/N_0)} = \frac{2N_w+1}{N_0} \text{drcl}\left( \frac{k}{N_0}, 2N_w+1 \right) \tag{5.237}$$

a Dirichlet function with extrema of $(2N_w+1)/N_0$ and a period of $N_0$.

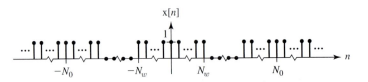

**Figure 5.20**
General DT rectangular-wave signal.

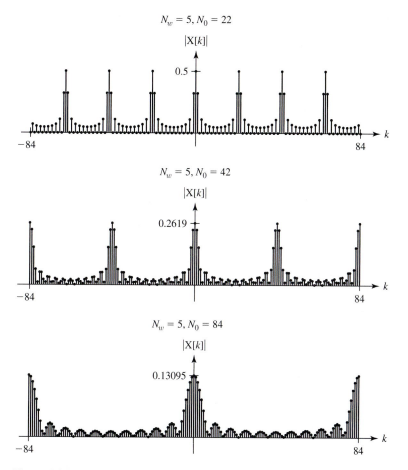

**Figure 5.21**
Effect of the fundamental period $N_0$ on the magnitude of the DTFS harmonic function of a rectangular-wave signal.

To illustrate the effects of different fundamental periods $N_0$, let $W = 5$ and plot the magnitude of $X[k]$ versus $k$ for $N_0 = 22$, $42$, and $84$ (Figure 5.21). The effect on the DTFS harmonic function of making the fundamental period of $x[n]$ longer is similar to the same effect on the CTFS harmonic function of making the fundamental period of $x(t)$ longer. The CTFS harmonic function is not periodic, and the DTFS harmonic function is periodic. But in either case the shape of the *envelope* of the harmonic amplitudes approaches a sinc function which is the CTFT of the rectangle function. As the fundamental period of $x[n]$ is increased, the resolution of the envelope of the DT sinc function shape is increased. In the DTFS case the DT sinc function shape is repeated periodically and that defines a Dirichlet function.

We will next do two normalizations similar to those done for the transition from the CTFS to the CTFT. First, as the fundamental period $N_0$ is increased, the harmonic amplitudes decrease. In the limit as $N_0$ approaches infinity, the harmonic amplitudes all approach zero. This effect can be eliminated by modifying the DTFS harmonic function by multiplying by the fundamental period $N_0$. The modified DTFS harmonic function is

$$N_0 X[k] = \frac{\sin((2N_w + 1)(k\pi/N_0))}{\sin(k\pi/N_0)} = (2N_w + 1) \, \mathrm{drcl}\left(\frac{k}{N_0}, 2N_w + 1\right). \quad \textbf{(5.238)}$$

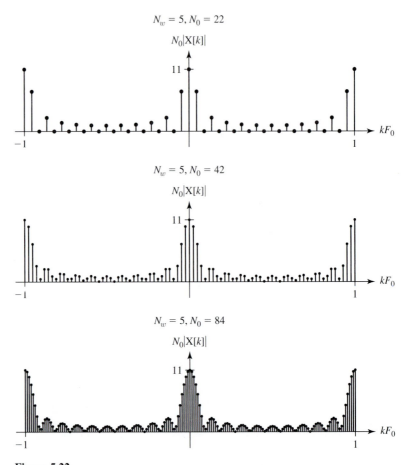

**Figure 5.22**
Magnitude of the modified DTFS harmonic function of a rectangular-wave signal.

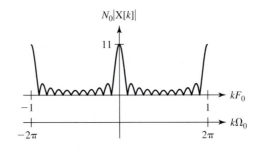

**Figure 5.23**
Limiting modified DTFS harmonic function of a
rectangular-wave signal.

The other effect is that the fundamental period $N_0$ of x[$n$] is also the fundamental period of the DTFS harmonic function, and as it is increased, the width of the plot of one fundamental period of the modified DTFS harmonic function $N_0X[k]$ increases to infinity. We can normalize by plotting the modified DTFS harmonic function versus $k/N_0 = kF_0$ instead of versus $k$. Then the fundamental period of the modified DTFS harmonic function (as plotted) is always one, rather than $N_0$ (Figure 5.22). As $N_0$ approaches infinity, the separation between points of $N_0X[k]$ approaches zero and the discrete DT-frequency plot becomes a continuous DT-frequency plot (Figure 5.23).

## ANALYTICAL DERIVATION

This derivation proceeds along a path analogous to the derivation of the CTFT from the CTFS. To expand the DTFS to aperiodic signals, first let $F_0 = \Delta F = 1/N_0$. Then the DTFS representation of $x[n]$ can be written as

$$x[n] = \sum_{k=\langle N_0 \rangle} X[k] e^{j2\pi(k\Delta F)n}. \tag{5.239}$$

Substituting the summation expression for $X[k]$ in the DTFS definition,

$$x[n] = \sum_{k=\langle N_0 \rangle} \left( \frac{1}{N_0} \sum_{m=\langle N_0 \rangle} x[m] e^{-j2\pi(k\Delta F)m} \right) e^{j2k\pi\Delta Fn} \tag{5.240}$$

or

$$x[n] = \sum_{k=\langle N_0 \rangle} \left( \sum_{m=\langle N_0 \rangle} x[m] e^{-j2\pi(k\Delta F)m} \right) e^{j2\pi(k\Delta F)n} \, \Delta F. \tag{5.241}$$

(The index of summation $n$ in the expression for $X[k]$ has been changed to $m$ to avoid confusion with the $n$ in the expression for $x[n]$). Since the inner summation is over any arbitrary range of $m$ of width $N_0$, let the range be $-(N_0/2) \le m < (N_0/2)$ for $N_0$ even or $-((N_0 - 1)/2) \le m < ((N_0 + 1)/2)$ for $N_0$ odd. The outer summation is over any arbitrary range of $k$ of width $N_0$, so let its range be $k_0 \le k < k_0 + N_0$. Then

$$x[n] = \sum_{k=k_0}^{k_0+N_0-1} \left( \sum_{m=-(N_0/2)}^{(N_0/2)-1} x[m] e^{-j2\pi(k\Delta F)m} \right) e^{j2\pi(k\Delta F)n} \, \Delta F \qquad \text{when } N_0 \text{ is even} \tag{5.242}$$

or

$$x[n] = \sum_{k=k_0}^{k_0+N_0-1} \left( \sum_{m=-(N_0-1)/2}^{(N_0-1)/2} x[m] e^{-j2\pi(k\Delta F)m} \right) e^{j2\pi(k\Delta F)n} \, \Delta F \qquad \text{when } N_0 \text{ is odd.} \tag{5.243}$$

Now let the fundamental period $N_0$ of the DTFS approach infinity. In that limit, $\Delta F$ approaches the differential discrete-time frequency $dF$; $k \, \Delta F$ approaches the continuous, discrete-time frequency $F$; the outer summation approaches an integral in $F = k \, \Delta F$ which covers a range of $k_0 \le k < k_0 + N_0$ which, through $F = k \, \Delta F = k F_0 = k/N_0$, translates to the range $F_0 < F < F_0 + 1$; the inner summation covers an infinite range; and (5.242) and (5.243) both become

$$x[n] = \int_1 \left( \sum_{m=-\infty}^{\infty} x[m] e^{-j2\pi Fm} \right) e^{j2\pi Fn} \, dF. \tag{5.244}$$

The equivalent radian-frequency form is

$$x[n] = \frac{1}{2\pi} \int_{2\pi} \left( \sum_{m=-\infty}^{\infty} x[m] e^{-j\Omega m} \right) e^{j\Omega n} \, d\Omega. \tag{5.245}$$

## DEFINITION OF THE DISCRETE-TIME FOURIER TRANSFORM

The discrete-time Fourier transform is defined by

$$\boxed{x[n] = \int_1 X(F) e^{j2\pi Fn} \, dF \overset{\mathcal{F}}{\longleftrightarrow} X(F) = \sum_{n=-\infty}^{\infty} x[n] e^{-j2\pi Fn}} \tag{5.246}$$

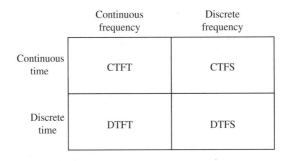

**Figure 5.24**
Fourier methods matrix.

or

$$\boxed{x[n] = \frac{1}{2\pi} \int_{2\pi} X(j\Omega)\, e^{j\Omega n}\, d\Omega \overset{\mathcal{F}}{\longleftrightarrow} X(j\Omega) = \sum_{n=-\infty}^{\infty} x[n]e^{-j\Omega n}}.$$     **(5.247)**

The DTFT completes the four Fourier analysis methods. These four methods form a *matrix* of methods for the four combinations of continuous and discrete time and continuous and discrete frequency (Figure 5.24).

## 5.7  CONVERGENCE OF THE DISCRETE-TIME FOURIER TRANSFORM

The condition for convergence of the DTFT is simply that the summation in

$$X(F) = \sum_{n=-\infty}^{\infty} x[n]e^{-j2\pi Fn} \quad \text{or} \quad X(j\Omega) = \sum_{n=-\infty}^{\infty} x[n]e^{-j\Omega n} \qquad \textbf{(5.248)}$$

actually converges. It will converge if

$$\sum_{n=-\infty}^{\infty} |x[n]| < \infty. \qquad \textbf{(5.249)}$$

If $|X(F)|$ is bounded, the inverse transform

$$x[n] = \int_{1} X(F)e^{j2\pi Fn}\, dF \quad \text{or} \quad x[n] = \frac{1}{2\pi} \int_{2\pi} X(j\Omega)e^{j\Omega n}\, d\Omega \qquad \textbf{(5.250)}$$

will always converge because the integration interval is finite.

## 5.8  PROPERTIES OF THE DISCRETE-TIME FOURIER TRANSFORM

Let $x[n]$ and $y[n]$ be two DT signals whose DTFTs are $X(F)$ and $Y(F)$ or $X(j\Omega)$ and $Y(j\Omega)$. Then the following properties apply.

## LINEARITY

This property is the same as it is for every other Fourier method (also for the Laplace and $z$ transforms to come later).

$$\alpha x[n] + \beta y[n] \overset{\mathcal{F}}{\longleftrightarrow} \alpha X(F) + \beta Y(F) \qquad (5.251)$$

or

$$\alpha x[n] + \beta y[n] \overset{\mathcal{F}}{\longleftrightarrow} \alpha X(j\Omega) + \beta Y(j\Omega) \qquad (5.252)$$

## TIME SHIFTING AND FREQUENCY SHIFTING

These can be proven in a manner similar to the equivalent proofs for the CTFT. The results are

$$x[n - n_0] \overset{\mathcal{F}}{\longleftrightarrow} e^{-j2\pi F n_0} X(F) \qquad (5.253)$$

or

$$x[n - n_0] \overset{\mathcal{F}}{\longleftrightarrow} e^{-j\Omega n_0} X(j\Omega) \qquad (5.254)$$

and

$$e^{j2\pi F_0 n} x[n] \overset{\mathcal{F}}{\longleftrightarrow} X(F - F_0) \qquad (5.255)$$

or

$$e^{j\Omega_0 n} x[n] \overset{\mathcal{F}}{\longleftrightarrow} X(j(\Omega - \Omega_0)). \qquad (5.256)$$

**EXAMPLE 5.9**

Find and sketch the inverse DTFT of

$$X(F) = \left[ \text{rect}\left( 50\left( F - \frac{1}{4}\right)\right) + \text{rect}\left( 50\left( F + \frac{1}{4}\right)\right)\right] * \text{comb}(F) \qquad (5.257)$$

(Figure 5.25).

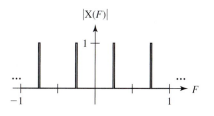

**Figure 5.25**
Magnitude of X($F$).

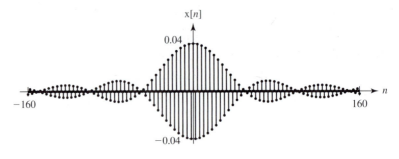

**Figure 5.26**
Inverse DTFT of X($F$).

### ■ Solution

We can start with the table entry,

$$\text{sinc}\left(\frac{n}{w}\right) \xleftrightarrow{\mathcal{F}} w\,\text{rect}(w F) * \text{comb}(F), \tag{5.258}$$

or, in this case,

$$\frac{1}{50}\,\text{sinc}\left(\frac{n}{50}\right) \xleftrightarrow{\mathcal{F}} \text{rect}(50 F) * \text{comb}(F). \tag{5.259}$$

Now apply the frequency-shifting property, $e^{j2\pi F_0 n}\,\text{x}[n] \xleftrightarrow{\mathcal{F}} \text{X}(F - F_0)$,

$$e^{j(\pi n/2)}\frac{1}{50}\,\text{sinc}\left(\frac{n}{50}\right) \xleftrightarrow{\mathcal{F}} \text{rect}\left(50\left(F - \frac{1}{4}\right)\right) * \text{comb}(F) \tag{5.260}$$

and

$$e^{-j(\pi n/2)}\frac{1}{50}\,\text{sinc}\left(\frac{n}{50}\right) \xleftrightarrow{\mathcal{F}} \text{rect}\left(50\left(F + \frac{1}{4}\right)\right) * \text{comb}(F). \tag{5.261}$$

Remember, when two functions are convolved, a shift of either one of them (but not both) shifts the convolution by the same amount. Finally, combining (5.260) and (5.261) and simplifying,

$$\frac{\text{sinc}(n/50)\cos(\pi n/2)}{25} \xleftrightarrow{\mathcal{F}} \left[\text{rect}\left(50\left(F - \frac{1}{4}\right)\right) + \text{rect}\left(50\left(F + \frac{1}{4}\right)\right)\right] * \text{comb}(F) \tag{5.262}$$

(Figure 5.26). ■

## TRANSFORM OF A CONJUGATE

If we conjugate a DT function x[$n$] its DTFT is

$$\mathcal{F}(\text{x}^*[n]) = \sum_{n=-\infty}^{\infty} \text{x}^*[n]e^{-j2\pi F n} = \left(\sum_{n=-\infty}^{\infty} \text{x}[n]e^{+j2\pi F n}\right)^* = \text{X}^*(-F) \tag{5.263}$$

$$\boxed{\text{x}^*[n] \xleftrightarrow{\mathcal{F}} \text{X}^*(-F)} \tag{5.264}$$

or

$$\boxed{\text{x}^*[n] \xleftrightarrow{\mathcal{F}} \text{X}^*(-j\Omega)} \tag{5.265}$$

The implications of this property are the same for DT signals as for CT signals. The magnitude of the DTFT of a real DT function is even and the phase is odd.

## DIFFERENCING AND ACCUMULATION

Differencing and accumulation are analogous to differentiation and integration in the CTFT.

$$\mathcal{F}(x[n] - x[n-1]) = \sum_{n=-\infty}^{\infty} x[n]e^{-j2\pi Fn} - \sum_{n=-\infty}^{\infty} x[n-1]e^{-j2\pi Fn} \qquad (5.266)$$

In the second summation of (5.266) let $m = n - 1$. Then

$$\mathcal{F}(x[n] - x[n-1]) = \sum_{n=-\infty}^{\infty} x[n]e^{-j2\pi Fn} - \sum_{m=-\infty}^{\infty} x[m]e^{-j2\pi F(m+1)} \qquad (5.267)$$

$$\mathcal{F}(x[n] - x[n-1]) = (1 - e^{-j2\pi F}) \sum_{n=-\infty}^{\infty} x[n]e^{-j2\pi Fn} = (1 - e^{-j2\pi F})X(F) \qquad (5.268)$$

$$\boxed{x[n] - x[n-1] \overset{\mathcal{F}}{\longleftrightarrow} (1 - e^{-j2\pi F})X(F)} \qquad (5.269)$$

or

$$\boxed{x[n] - x[n-1] \overset{\mathcal{F}}{\longleftrightarrow} (1 - e^{-j\Omega})X(j\Omega)}. \qquad (5.270)$$

In a manner analogous to the proof of the integration property of the CTFT, the accumulation property of the DTFT can be shown to be

$$\boxed{\sum_{m=-\infty}^{n} x[m] \overset{\mathcal{F}}{\longleftrightarrow} \frac{X(F)}{1 - e^{-j2\pi F}} + \frac{1}{2}X(0)\,\mathrm{comb}(F)} \qquad (5.271)$$

or

$$\boxed{\sum_{m=-\infty}^{n} x[m] \overset{\mathcal{F}}{\longleftrightarrow} \frac{X(j\Omega)}{1 - e^{-j\Omega}} + \frac{1}{2}X(0)\,\mathrm{comb}\left(\frac{\Omega}{2\pi}\right)}. \qquad (5.272)$$

**EXAMPLE 5.10**

Find the DTFT of $x[n] = \mathrm{rect}_{N_w}[n]$ using the integration property and the DTFT of an impulse.

### ■ Solution

The first backward difference of $x[n]$ is

$$x[n] - x[n-1] = \delta[n + N_w] - \delta[n - (N_w + 1)] \qquad (5.273)$$

and

$$\delta[n + N_w] - \delta[n - (N_w + 1)] \overset{\mathcal{F}}{\longleftrightarrow} e^{j2\pi F N_w} - e^{-j2\pi F(N_w+1)} \qquad (5.274)$$

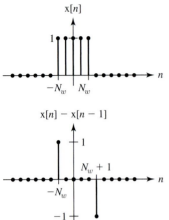

**Figure 5.27**
$\text{rect}_{N_w}[n]$ and its first backward difference.

(Figure 5.27). Then, using the integration property of the DTFT and the fact that the first backward difference of the rectangle has a sum of zero,

$$x[n] = \text{rect}_{N_w}[n] \xleftrightarrow{\mathcal{F}} \frac{e^{j2\pi F N_w} - e^{-j2\pi F(N_w+1)}}{1 - e^{-j2\pi F}} = \frac{e^{-j\pi F}}{e^{-j\pi F}} \frac{e^{j\pi F(2N_w+1)} - e^{-j\pi F(2N_w+1)}}{e^{j\pi F} - e^{-j\pi F}} \quad (5.275)$$

$$x[n] = \text{rect}_{N_w}[n] \xleftrightarrow{\mathcal{F}} \frac{\sin(\pi F(2N_w+1))}{\sin(\pi F)} = (2N_w + 1)\,\text{drcl}(F, 2N_w + 1) \quad (5.276)$$ ■

## TIME REVERSAL

If we time invert a DT function $x[n]$ its DTFT is

$$\mathcal{F}(x[-n]) = \sum_{n=-\infty}^{\infty} x[-n]e^{-j2\pi Fn} \quad (5.277)$$

Let $m = -n$. Then

$$\mathcal{F}(x[-n]) = \sum_{m=\infty}^{-\infty} x[m]e^{+j2\pi Fm} = \sum_{m=-\infty}^{\infty} x[m]e^{-j2\pi(-F)m} = X(-F) \quad (5.278)$$

$$\boxed{x[-n] \xleftrightarrow{\mathcal{F}} X(-F)} \quad (5.279)$$

or

$$\boxed{x[-n] \xleftrightarrow{\mathcal{F}} X(-j\Omega)}. \quad (5.280)$$

## MULTIPLICATION–CONVOLUTION DUALITY

Let

$$z[n] = x[n] * y[n] = \sum_{m=-\infty}^{\infty} x[m]y[n-m]. \quad (5.281)$$

Then

$$Z(F) = \sum_{n=-\infty}^{\infty} z[n]e^{-j2\pi Fn} = \sum_{n=-\infty}^{\infty}\sum_{m=-\infty}^{\infty} x[m]y[n-m]e^{-j2\pi Fn}. \quad (5.282)$$

Reversing the order of summation in (5.282),

$$Z(F) = \sum_{m=-\infty}^{\infty} x[m]\underbrace{\sum_{n=-\infty}^{\infty} y[n-m]e^{-j2\pi Fn}}_{\mathcal{F}(y[n-m])} = \sum_{m=-\infty}^{\infty} x[m]Y(F)e^{-j2\pi Fm} \quad (5.283)$$

$$Z(F) = Y(F)\underbrace{\sum_{m=-\infty}^{\infty} x[m]e^{-j2\pi Fm}}_{\mathcal{F}(x[m])} = Y(F)X(F). \quad (5.284)$$

Therefore,

$$\boxed{x[n] * y[n] \xleftrightarrow{\mathcal{F}} X(F)Y(F)} \quad (5.285)$$

or

$$\boxed{x[n] * y[n] \overset{\mathcal{F}}{\longleftrightarrow} X(j\Omega)Y(j\Omega)}. \qquad \textbf{(5.286)}$$

Let

$$z[n] = x[n]y[n]. \qquad \textbf{(5.287)}$$

Then

$$Z(F) = \sum_{n=-\infty}^{\infty} x[n]y[n]e^{-j2\pi Fn} \qquad \textbf{(5.288)}$$

$$Z(F) = \sum_{n=-\infty}^{\infty} \left( \int_1 X(\lambda)e^{j2\pi\lambda n}\, d\lambda \right) y[n]e^{-j2\pi Fn} = \int_1 X(\lambda) \sum_{n=-\infty}^{\infty} e^{j2\pi\lambda n}\, y[n]e^{-j2\pi Fn}\, d\lambda \qquad \textbf{(5.289)}$$

$$Z(F) = \int_1 X(\lambda) \underbrace{\sum_{n=-\infty}^{\infty} y[n]e^{-j2\pi(F-\lambda)n}}_{Y(F-\lambda)}\, d\lambda = \int_1 X(\lambda)Y(F-\lambda)\, d\lambda. \qquad \textbf{(5.290)}$$

The last integral $\int_1 X(\lambda)Y(F - \lambda)\, d\lambda$ is another instance of periodic convolution. Therefore,

$$\boxed{x[n]y[n] \overset{\mathcal{F}}{\longleftrightarrow} X(F) \circledast Y(F)} \qquad \textbf{(5.291)}$$

or

$$\boxed{x[n]y[n] \overset{\mathcal{F}}{\longleftrightarrow} \frac{1}{2\pi}X(j\Omega) \circledast Y(j\Omega)}. \qquad \textbf{(5.292)}$$

The implications of multiplication–convolution duality for signal and system analysis are the same for DT signals and systems as for CT signals and systems. The response of a system is the convolution of the excitation with the impulse response and the equivalent statement in the DT-frequency domain is that the DTFT of the response of a system is the product of the DTFT of the excitation and the transfer function, which is the DTFT of the impulse response (Figure 5.28). The implications for cascade connections of systems are also the same (Figure 5.29).

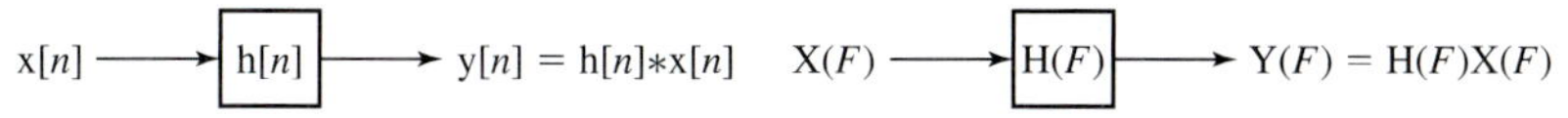

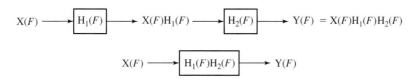

**Figure 5.28**
Equivalence of convolution in the DT domain and multiplication in the DT frequency domain.

$$X(F) \longrightarrow \boxed{H_1(F)} \longrightarrow X(F)H_1(F) \longrightarrow \boxed{H_2(F)} \longrightarrow Y(F) = X(F)H_1(F)H_2(F)$$

$$X(F) \longrightarrow \boxed{H_1(F)H_2(F)} \longrightarrow Y(F)$$

**Figure 5.29**
Cascade connection of DT systems.

## ACCUMULATION DEFINITION OF A COMB FUNCTION

The CTFT leads to an integral definition of an impulse. In a similar manner the DTFT leads to an accumulation definition of a comb. Begin with the definition,

$$X(F) = \sum_{n=-\infty}^{\infty} x[n]e^{-j2\pi Fn} \qquad \text{and} \qquad x[n] = \int_1 X(F)e^{j2\pi Fn} \, dF. \qquad \textbf{(5.293)}$$

Then, in

$$X(F) = \sum_{n=-\infty}^{\infty} x[n]e^{-j2\pi Fn} \qquad \textbf{(5.294)}$$

replace $x[n]$ by its integral equivalent,

$$X(F) = \sum_{n=-\infty}^{\infty} \left[ \int_1 X(\phi)e^{j2\pi\phi n} \, d\phi \right] e^{-j2\pi Fn}$$

$$= \sum_{n=-\infty}^{\infty} \int_{\phi_0}^{\phi_0+1} X(\phi)e^{j2\pi(\phi-F)n} \, d\phi \qquad \textbf{(5.295)}$$

or

$$X(F) = \sum_{n=-\infty}^{\infty} \int_{-\infty}^{\infty} X_p(\phi)e^{j2\pi(\phi-F)n} \, d\phi$$

$$= \sum_{n=-\infty}^{\infty} X_p(F) * e^{-j2\pi Fn} \qquad \textbf{(5.296)}$$

$$X(F) = X_p(F) * \left[ \sum_{n=-\infty}^{\infty} e^{-j2\pi Fn} \right] \qquad \textbf{(5.297)}$$

where

$$X_p(\phi) = \begin{cases} X(\phi) & \phi_0 < \phi < \phi_0 + 1 \\ 0 & \text{otherwise} \end{cases} \qquad \textbf{(5.298)}$$

is any arbitrary single period of $X(F)$. Since $X_p(F)$ is one period of $X(F)$ and the period is one, it follows that

$$X(F) = X_p(F) * \text{comb}(F). \qquad \textbf{(5.299)}$$

Therefore, if (5.297) and (5.299) are both true, that means that

$$\sum_{n=-\infty}^{\infty} e^{-j2\pi Fn} = \text{comb}(F) \qquad \textbf{(5.300)}$$

and, since $\text{comb}(F)$ is an even function,

$$\boxed{\sum_{n=-\infty}^{\infty} e^{j2\pi Fn} = \text{comb}(F)}. \qquad \textbf{(5.301)}$$

EXAMPLE 5.11

Find the DTFT of the DT cosine

$$x[n] = A \, \cos \left( \frac{\pi n}{2} \right). \qquad (5.302)$$

■ **Solution**

Applying the definition,

$$X(F) = \sum_{n=-\infty}^{\infty} x[n] e^{-j2\pi F n} = \sum_{n=-\infty}^{\infty} A \, \cos \left( \frac{\pi n}{2} \right) e^{-j2\pi F n}$$

$$= \frac{A}{2} \sum_{n=-\infty}^{\infty} \left( e^{j(\pi n/2)} + e^{-j(\pi n/2)} \right) e^{-j2\pi F n} \qquad (5.303)$$

$$X(F) = \frac{A}{2} \sum_{n=-\infty}^{\infty} \left[ e^{j2\pi \left( \frac{1}{4} - F \right) n} + e^{j2\pi \left( -\frac{1}{4} - F \right) n} \right] \qquad (5.304)$$

or

$$X(j\Omega) = \frac{A}{2} \sum_{n=-\infty}^{\infty} \left[ e^{j((\pi/2) - \Omega)n} + e^{j(-(\pi/2) - \Omega)n} \right]. \qquad (5.305)$$

Using $\sum_{n=-\infty}^{\infty} e^{j2\pi x n} = \text{comb}(x)$, and the fact that the comb function is even,

$$X(F) = \frac{A}{2} \left[ \text{comb} \left( F - \frac{1}{4} \right) + \text{comb} \left( F + \frac{1}{4} \right) \right] \qquad (5.306)$$

or, using the scaling property of the comb function,

$$X(j\Omega) = A\pi \left[ \text{comb} \left( \Omega - \frac{\pi}{2} \right) + \text{comb} \left( \Omega + \frac{\pi}{2} \right) \right] \qquad (5.307)$$

(Figure 5.30).

Since x[n] is periodic, we can also find its DTFS harmonic function.

$$X[k] = \frac{1}{N_0} \sum_{n=\langle N_0 \rangle} x[n] e^{-j2\pi(k F_0)n} = \frac{A}{4} \sum_{n=\langle N_0 \rangle} \cos \left( \frac{\pi n}{2} \right) e^{-j\pi(kn/2)} \qquad (5.308)$$

$$X[k] = \frac{A}{4} \left( 1 - e^{-jk\pi} \right) = \frac{A}{4} e^{-j(k\pi/2)} \left( e^{+j(k\pi/2)} - e^{-j(k\pi/2)} \right) \qquad (5.309)$$

$$X[k] = j \frac{A}{2} e^{-j(k\pi/2)} \, \sin \left( \frac{k\pi}{2} \right) \qquad (5.310)$$

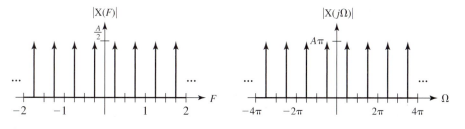

**Figure 5.30**

Magnitude of DTFT of $x[n] = A \cos \left( \frac{\pi n}{2} \right)$.

This expression is zero for even values of $k$ and evaluates to $A/2$ for odd values of $k$,

$$X[k] = \frac{A}{2} \begin{cases} 0 & k \text{ even} \\ 1 & k \text{ odd} \end{cases}. \tag{5.311}$$

These values are exactly the strengths of the impulses in $X(F)$ at $\cdots -\frac{3}{4}, -\frac{1}{4}, \frac{1}{4}, \frac{3}{4}, \cdots$.

This result illustrates that the DTFS is just a special case of the DTFT just as the CTFS was a special case of the CTFT. If a DT signal is periodic, its DTFT consists only of impulses and the strengths of those impulses are the values of the DTFS harmonic function at the harmonics of the fundamental frequency. ∎

## PARSEVAL'S THEOREM

Just as was true for the other Fourier methods there is a Parseval relation for the DTFT. The total signal energy in a signal $x[n]$ is

$$\sum_{n=-\infty}^{\infty} |x[n]|^2 = \sum_{n=-\infty}^{\infty} \left| \int_1 X(F)e^{j2\pi Fn}\, dF \right|^2$$

$$= \sum_{n=-\infty}^{\infty} \left( \int_1 X(F)e^{j2\pi Fn}\, dF \right) \left( \int_1 X(F')e^{j2\pi F'n}\, dF' \right)^* \tag{5.312}$$

$$\sum_{n=-\infty}^{\infty} |x[n]|^2 = \sum_{n=-\infty}^{\infty} \int_1 X(F) \int_1 X^*(F')e^{-j2\pi(F'-F)n}\, dF'\, dF \tag{5.313}$$

We can exchange the order of summation and integration to yield

$$\sum_{n=-\infty}^{\infty} |x[n]|^2 = \int_1 X(F) \int_1 X^*(F') \underbrace{\sum_{n=-\infty}^{\infty} e^{-j2\pi(F'-F)n}}_{=\text{comb}(F'-F)}\, dF'\, dF \tag{5.314}$$

$$\sum_{n=-\infty}^{\infty} |x[n]|^2 = \int_1 X(F) \int_1 X^*(F')\delta(F' - F)\, dF'\, dF \tag{5.315}$$

and

$$\boxed{\sum_{n=-\infty}^{\infty} |x[n]|^2 = \int_1 X(F)X^*(F)\, dF = \int_1 |X(F)|^2\, dF}, \tag{5.316}$$

proving that the total energy over all discrete time $n$ is equal to the total energy in one fundamental period of DT frequency $F$ (that fundamental period being one for any DTFT). The equivalent result for the radian-frequency form of the DTFT is

$$\boxed{\sum_{n=-\infty}^{\infty} |x[n]|^2 = \frac{1}{2\pi} \int_{2\pi} |X(j\Omega)|^2\, d\Omega}. \tag{5.317}$$

**EXAMPLE 5.12**

Find the signal energy of

$$x[n] = \frac{1}{5} \operatorname{sinc}\left(\frac{n}{100}\right). \tag{5.318}$$

■ **Solution**

The signal energy of a DT signal is defined as

$$E_x = \sum_{n=-\infty}^{\infty} |x[n]|^2. \tag{5.319}$$

But we can avoid doing that summation by using Parseval's theorem. The DTFT of $x[n]$ can be found by starting with the table entry

$$\operatorname{sinc}\left(\frac{n}{w}\right) \xleftrightarrow{\mathcal{F}} w \operatorname{rect}(w F) * \operatorname{comb}(F) \tag{5.320}$$

and applying the linearity property to form

$$\frac{1}{5} \operatorname{sinc}\left(\frac{n}{100}\right) \xleftrightarrow{\mathcal{F}} 20 \operatorname{rect}(100 F) * \operatorname{comb}(F). \tag{5.321}$$

Parseval's theorem is

$$\sum_{n=-\infty}^{\infty} |x[n]|^2 = \int_1 |X(F)|^2 \, dF. \tag{5.322}$$

So the signal energy is

$$E_x = \int_1 |20 \operatorname{rect}(100 F) * \operatorname{comb}(F)|^2 \, dF = \int_{-\infty}^{\infty} |20 \operatorname{rect}(100 F)|^2 \, dF \tag{5.323}$$

or

$$E_x = 400 \int_{-\frac{1}{200}}^{\frac{1}{200}} dF = 4. \tag{5.324}$$

## SUMMARY OF DTFT PROPERTIES

| | |
|---|---|
| Linearity | $\alpha x[n] + \beta y[n] \xleftrightarrow{\mathcal{F}} \alpha X(F) + \beta Y(F)$ |
| | $\alpha x[n] + \beta y[n] \xleftrightarrow{\mathcal{F}} \alpha X(j\Omega) + \beta Y(j\Omega)$ |
| Time shifting | $x[n - n_0] \xleftrightarrow{\mathcal{F}} e^{-j2\pi F n_0} X(F)$ |
| | $x[n - n_0] \xleftrightarrow{\mathcal{F}} e^{-j\Omega n_0} X(j\Omega)$ |
| Frequency shifting | $e^{j2\pi F_0 n} x[n] \xleftrightarrow{\mathcal{F}} X(F - F_0)$ |
| | $e^{j\Omega_0 n} x[n] \xleftrightarrow{\mathcal{F}} X(j(\Omega - \Omega_0))$ |
| Transform of a conjugate | $x^*[n] \xleftrightarrow{\mathcal{F}} X^*(-F)$ |
| | $x^*[n] \xleftrightarrow{\mathcal{F}} X^*(-j\Omega)$ |

Differencing
$$x[n] - x[n-1] \overset{\mathcal{F}}{\longleftrightarrow} (1 - e^{-j2\pi F})X(F)$$
$$x[n] - x[n-1] \overset{\mathcal{F}}{\longleftrightarrow} (1 - e^{-j\Omega})X(j\Omega)$$

Accumulation
$$\sum_{m=-\infty}^{n} x[m] \overset{\mathcal{F}}{\longleftrightarrow} \frac{X(F)}{1 - e^{-j2\pi F}} + \frac{1}{2}X(0)\,\text{comb}(F)$$
$$\sum_{m=-\infty}^{n} x[m] \overset{\mathcal{F}}{\longleftrightarrow} \frac{X(j\Omega)}{1 - e^{-j\Omega}} + \frac{1}{2}X(0)\,\text{comb}\left(\frac{\Omega}{2\pi}\right)$$

Time reversal
$$x[-n] \overset{\mathcal{F}}{\longleftrightarrow} X(-F)$$
$$x[-n] \overset{\mathcal{F}}{\longleftrightarrow} X(-j\Omega)$$

Multiplication–convolution duality
$$x[n] * y[n] \overset{\mathcal{F}}{\longleftrightarrow} X(F)Y(F)$$
$$x[n] * y[n] \overset{\mathcal{F}}{\longleftrightarrow} X(j\Omega)Y(j\Omega)$$
$$x[n]y[n] \overset{\mathcal{F}}{\longleftrightarrow} X(F) \circledast Y(F)$$
$$x[n]y[n] \overset{\mathcal{F}}{\longleftrightarrow} \frac{1}{2\pi}X(j\Omega) \circledast Y(j\Omega)$$

Accumulation definition of a comb function
$$\sum_{n=-\infty}^{\infty} e^{j2\pi Fn} = \text{comb}(F)$$

Parseval's theorem
$$\sum_{n=-\infty}^{\infty} |x[n]|^2 = \int_{1} |X(F)|^2 \, dF$$
$$\sum_{n=-\infty}^{\infty} |x[n]|^2 = \frac{1}{2\pi} \int_{2\pi} |X(j\Omega)|^2 \, d\Omega$$

---

## EXAMPLE 5.13

Find the inverse DTFT of $X(F) = \text{rect}(wF) * \text{comb}(F)$, where $w > 1$.

■ **Solution**

$$x[n] = \int_{1} X(F)e^{j2\pi Fn} \, dF \qquad x[n] = \int_{1} \text{rect}(wF) * \text{comb}(F)e^{j2\pi Fn} \, dF \qquad \textbf{(5.325)}$$

Since we can choose to integrate over any interval in $F$ of width one, let's choose the simplest one,

$$x[n] = \int_{-\frac{1}{2}}^{\frac{1}{2}} \text{rect}(wF) * \text{comb}(F)e^{j2\pi Fn} \, dF. \qquad \textbf{(5.326)}$$

In this integration interval, there is exactly one rectangle function of width $1/w$ and

$$x[n] = \int_{-(1/2w)}^{1/2w} e^{j2\pi Fn} \, dF = 2 \int_{0}^{1/2w} \cos(2\pi Fn) \, dF = \frac{\sin(\pi n/w)}{\pi n} = \frac{1}{w}\,\text{sinc}\left(\frac{n}{w}\right). \qquad \textbf{(5.327)}$$

From this result we can also establish the handy DTFT pair (which appears in the table of Fourier pairs in Appendix E),

$$\text{sinc}\left(\frac{n}{w}\right) \xleftrightarrow{\mathcal{F}} w\,\text{rect}(wF) * \text{comb}(F) \qquad w > 1 \tag{5.328}$$

or

$$\text{sinc}\left(\frac{n}{w}\right) \xleftrightarrow{\mathcal{F}} w\sum_{n=-\infty}^{\infty}\text{rect}(w(F-k)) \qquad w > 1 \tag{5.329}$$

or, in radian-frequency form,

$$\text{sinc}\left(\frac{n}{w}\right) \xleftrightarrow{\mathcal{F}} w\,\text{rect}\left(\frac{w\Omega}{2\pi}\right) * \text{comb}\left(\frac{\Omega}{2\pi}\right) \qquad w > 1 \tag{5.330}$$

or

$$\text{sinc}\left(\frac{n}{w}\right) \xleftrightarrow{\mathcal{F}} w\sum_{n=-\infty}^{\infty}\text{rect}\left(\frac{w}{2\pi}(\Omega - 2\pi k)\right) \qquad w > 1. \tag{5.331}$$

## 5.9 RELATIONS AMONG FOURIER METHODS

A careful reader will have noticed that there are many similarities among the Fourier analysis methods, CTFS, DTFS, CTFT, and DTFT. This section explores the relations among them and shows that the information in a CTFS, DTFS, and DTFT exists in an equivalent form using the CTFT. Here is a summary of the defining relations of the four Fourier methods.

*CTFS.* Representation, for all time, of a periodic CT function with fundamental period $T_0$ which satisfies the Dirichlet conditions:

$$X[k] = \frac{1}{T_0}\int_{T_0} x(t)e^{-j2\pi(kf_0)t}\,dt \qquad x(t) = \sum_{k=-\infty}^{\infty} X[k]e^{+j2\pi(kf_0)t} \tag{5.332}$$

or

$$X[k] = \frac{1}{T_0}\int_{T_0} x(t)e^{-j(k\omega_0)t}\,dt \qquad x(t) = \sum_{k=-\infty}^{\infty} X[k]e^{+j(k\omega_0)t} \tag{5.333}$$

*DTFS.* Representation, for all discrete time, of a periodic DT function with fundamental period $N_0$:

$$X[k] = \frac{1}{N_0}\sum_{n=\langle N_0\rangle} x[n]e^{-j2\pi(kF_0)n} \qquad x[n] = \sum_{k=\langle N_0\rangle} X[k]e^{+j2\pi(kF_0)n} \tag{5.334}$$

or

$$X[k] = \frac{1}{N_0}\sum_{n=\langle N_0\rangle} x[n]e^{-j(k\Omega_0)n} \qquad x[n] = \sum_{k=\langle N_0\rangle} X[k]e^{+j(k\Omega_0)n} \tag{5.335}$$

*CTFT.*

$$X(f) = \int_{-\infty}^{\infty} x(t)e^{-j2\pi ft}\, dt \qquad x(t) = \int_{-\infty}^{\infty} X(f)e^{+j2\pi ft}\, df \qquad \textbf{(5.336)}$$

or

$$X(j\omega) = \int_{-\infty}^{\infty} x(t)e^{-j\omega t}\, dt \qquad x(t) = \frac{1}{2\pi}\int_{-\infty}^{\infty} X(j\omega)e^{+j\omega t}\, d\omega \qquad \textbf{(5.337)}$$

*DTFT.*

$$X(F) = \sum_{n=-\infty}^{\infty} x[n]e^{-j2\pi Fn} \qquad x[n] = \int_{1} X(F)e^{+j2\pi Fn}\, dF \qquad \textbf{(5.338)}$$

or

$$X(j\Omega) = \sum_{n=-\infty}^{\infty} x[n]e^{-j\Omega n} \qquad x[n] = \frac{1}{2\pi}\int_{2\pi} X(j\Omega)e^{+j\Omega n}\, d\Omega \qquad \textbf{(5.339)}$$

Table 5.1 presents the definitions and some corresponding properties for the four Fourier methods.

**Table 5.1**   Comparison of Fourier methods

| Discrete frequency | Continuous frequency |
|---|---|
| **Continuous time** | |
| $\displaystyle\sum_{k=-\infty}^{\infty} X[k]e^{j2\pi(kf_0)t} \xleftrightarrow{\mathcal{FS}} \frac{1}{T_0}\int_{T_0} x(t)e^{-j2\pi(kf_0)t}\, dt$ | $\displaystyle\int_{-\infty}^{\infty} X(f)e^{+j2\pi ft}\, df \xleftrightarrow{\mathcal{F}} \int_{-\infty}^{\infty} x(t)e^{-j2\pi ft}\, dt$ |
| $x(t-t_0) \xleftrightarrow{\mathcal{FS}} e^{-j2\pi(kf_0)t_0}X[k]$ | $x(t-t_0) \xleftrightarrow{\mathcal{F}} X(f)e^{-j2\pi ft_0}$ |
| $e^{j2\pi(k_0 f_0)t}x(t) \xleftrightarrow{\mathcal{FS}} X[k-k_0]$ | $x(t)e^{+j2\pi f_0 t} \xleftrightarrow{\mathcal{F}} X(f-f_0)$ |
| $x(t)y(t) \xleftrightarrow{\mathcal{FS}} X[k]*Y[k]$ | $x(t)y(t) \xleftrightarrow{\mathcal{F}} X(f)*Y(f)$ |
| $x(t)\circledast y(t) \xleftrightarrow{\mathcal{FS}} T_0 X[k]Y[k]$ | $x(t)*y(t) \xleftrightarrow{\mathcal{F}} X(f)Y(f)$ |
| $\dfrac{1}{T_0}\displaystyle\int_{T_0} |x(t)|^2\, dt = \sum_{k=-\infty}^{\infty} |X[k]|^2$ | $\displaystyle\int_{-\infty}^{\infty} |x(t)|^2\, dt = \int_{-\infty}^{\infty} |X(f)|^2\, df$ |
| **Discrete time** | |
| $\displaystyle\sum_{k=\langle N_0\rangle} X[k]e^{j2\pi(kn/N_0)} \xleftrightarrow{\mathcal{FS}} \frac{1}{N_0}\sum_{n=\langle N_0\rangle} x[n]e^{-j2\pi(kn/N_0)}$ | $\displaystyle\int_{1} X(F)e^{j2\pi Fn}\, dF \xleftrightarrow{\mathcal{F}} \sum_{n=-\infty}^{\infty} x[n]e^{-j2\pi Fn}$ |
| $x[n-n_0] \xleftrightarrow{\mathcal{FS}} e^{-j2\pi(kn_0/N_0)}X[k]$ | $x[n-n_0] \xleftrightarrow{\mathcal{F}} e^{-j2\pi Fn_0}X(F)$ |
| $e^{j2\pi(k_0 n/N_0)}x[n] \xleftrightarrow{\mathcal{FS}} X[k-k_0]$ | $e^{j2\pi F_0 n}x[n] \xleftrightarrow{\mathcal{F}} X(F-F_0)$ |
| $x[n]y[n] \xleftrightarrow{\mathcal{FS}} Y[k]\circledast X[k]$ | $x[n]y[n] \xleftrightarrow{\mathcal{F}} X(F)\circledast Y(F)$ |
| $x[n]\circledast y[n] \xleftrightarrow{\mathcal{FS}} N_0 Y[k]X[k]$ | $x[n]*y[n] \xleftrightarrow{\mathcal{F}} X(F)Y(F)$ |
| $\dfrac{1}{N_0}\displaystyle\sum_{n=\langle N_0\rangle} |x[n]|^2 = \sum_{k=\langle N_0\rangle} |X[k]|^2$ | $\displaystyle\sum_{n=-\infty}^{\infty} |x[n]|^2 = \int_{1} |X(F)|^2\, dF$ |

Observe that a shift in one domain corresponds to multiplication by a complex function in the other domain. Multiplication of two functions in one domain corresponds to convolution in the other domain. If two functions being convolved are periodic, the convolution is also periodic. If a time signal is aperiodic, Parseval's theorem equates its signal energy to the corresponding signal energy in the frequency domain. If a time signal is periodic, Parseval's theorem equates its average signal power to the average signal power in the frequency domain.

The four Fourier methods are compared graphically in Figure 5.31 for four corresponding signals. We can observe some general characteristics of these transform pairs. If a function is discrete in one domain, it is periodic in the other, and vice versa. In the DTFS corner both functions are discrete and both are periodic. In the CTFT corner both functions are continuous and both are aperiodic.

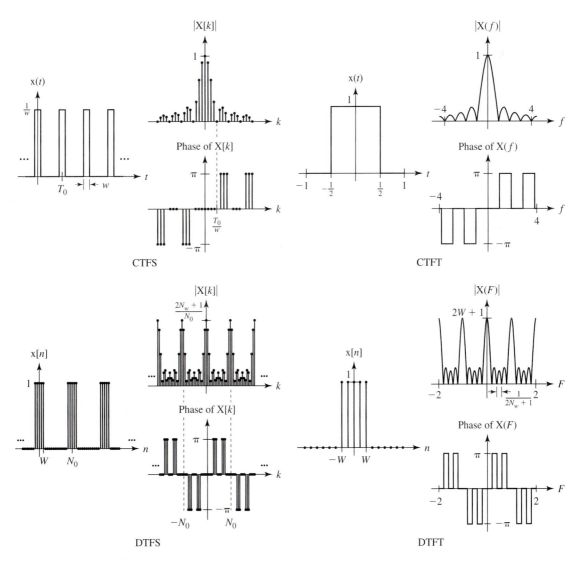

**Figure 5.31**

Graphical comparison of signals and their Fourier transforms.

## CTFT AND CTFS

A periodic CT signal $x(t)$ with fundamental period $T_0 = 1/f_0$ can be represented for all time by a CTFS,

$$x(t) = \sum_{k=-\infty}^{\infty} X[k]e^{j2\pi(kf_0)t} \quad \text{or} \quad x(t) = \sum_{k=-\infty}^{\infty} X[k]e^{j(k\omega_0)t}. \quad \textbf{(5.340)}$$

Using the frequency-shifting property $e^{j2\pi f_0 t}x(t) \overset{\mathcal{F}}{\longleftrightarrow} X(f - f_0)$ and the CTFT transform pair $1 \overset{\mathcal{F}}{\longleftrightarrow} \delta(f)$, we can find the CTFT of $x(t)$, yielding

$$X(f) = \sum_{k=-\infty}^{\infty} X[k]\delta(f - kf_0) \quad \text{or} \quad X(j\omega) = 2\pi \sum_{k=-\infty}^{\infty} X[k]\delta(\omega - k\omega_0). \quad \textbf{(5.341)}$$

Therefore, the CTFT of a periodic CT function is a continuous-frequency function which consists of a sum of impulses, spaced apart by the fundamental frequency of the signal, whose strengths are the same as the CTFS harmonic function at the same harmonic-number multiple of the fundamental frequency. The CTFS is just a special case of the CTFT with some notation changes (Figure 5.32).

This is the first example of the information equivalence of a function $X[k]$ of a discrete independent variable, in this case harmonic number $k$ and a function $X(f)$ or $X(j\omega)$ of a continuous independent variable, in this case frequency. They are equivalent in the sense that $X(f)$ or $X(j\omega)$ is nonzero only at integer multiples $k$ of the fundamental frequency $f_0$ or $\omega_0$ and $X[k]$ is only *defined* at integer values of $k$. Also, the values of $X[k]$ at the integer values of $k$ are the same as the strengths of the impulses in $X(f)$ that occur at $kf_0$. Summarizing, for a periodic CT function $x(t)$,

$$X(f) = \sum_{k=-\infty}^{\infty} X[k]\delta(f - kf_0). \quad \textbf{(5.342)}$$

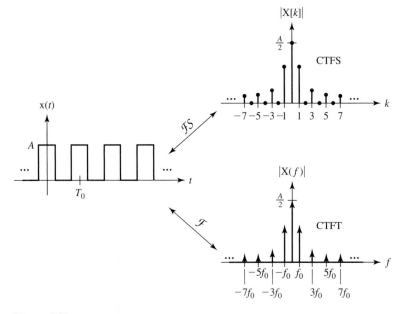

**Figure 5.32**
CTFS harmonic function and CTFT for a square-wave function.

(The functions, X( ) and X[ ] should not be confused here as the same function because, even though they have the same name, one is a function of a continuous independent variable and the other is a function of a discrete independent variable.)

Another important comparison between the CTFS and CTFT is the relation between the CTFT of an aperiodic CT signal and the CTFS harmonic function of a periodic extension of that signal. Let $x(t)$ be an aperiodic function of time. Let $x_p(t)$ be a periodic extension of $x(t)$ with fundamental period $T_p$, defined by

$$x_p(t) = \sum_{n=-\infty}^{\infty} x(t - nT_p) = x(t) * \frac{1}{T_p} \text{comb}\left(\frac{t}{T_p}\right) \tag{5.343}$$

(Figure 5.33). The CTFT of $x(t)$ is $X(f)$. Using the multiplication–convolution duality of the CTFT, the CTFT of $x_p(t)$ is

$$X_p(f) = X(f) \text{comb}(T_p f) = f_p \sum_{k=-\infty}^{\infty} X(kf_p)\delta(f - kf_p) \tag{5.344}$$

where $f_p = 1/T_p$. Now, using (5.342),

$$X_p(f) = \sum_{k=-\infty}^{\infty} X_p[k]\delta(f - kf_p) \tag{5.345}$$

and combining (5.344) and (5.345), we get

$$X_p[k] = f_p X(kf_p). \tag{5.346}$$

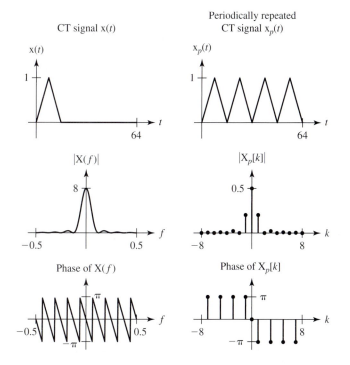

**Figure 5.33**
A CT signal and its CTFT, and the periodic repetition of the CT signal and its CTFS harmonic function.

In words, this says that if an aperiodic CT function is periodically extended to form a periodic function $x_p(t)$ with fundamental period $T_p$, that the values of the CTFS harmonic function $X_p[k]$, of $x_p(t)$ are samples of the CTFT, $X(f)$, of $x(t)$ taken at the CT frequencies $kf_p$ and then multiplied by the fundamental frequency of the CTFS, $f_p$. This forms an equivalence between sampling in the CT-frequency domain and periodic repetition in the CT domain. This idea will be important in the study of sampling in Chapter 7.

**EXAMPLE 5.14**

Using (5.346), find the CTFS harmonic function of

$$x(t) = \text{sinc}\left(\frac{t}{2}\right) * \frac{1}{10} \text{comb}\left(\frac{t}{10}\right). \tag{5.347}$$

■ **Solution**

The fundamental period of this periodically extended aperiodic function is $T_p = 10$. From (5.346), $X_p[k] = f_p X(kf_p)$, where, in this case,

$$X(f) = \mathcal{F}\left(\text{sinc}\left(\frac{t}{2}\right)\right) = 2\,\text{rect}(2f). \tag{5.348}$$

Therefore, for this $x(t)$

$$X[k] = \frac{1}{5}\,\text{rect}\left(\frac{k}{5}\right). \tag{5.349}$$

Then $x(t)$ can be expressed as a CTFS,

$$x(t) = \sum_{k=-\infty}^{\infty} X[k]e^{j2\pi(kf_0)t} = \frac{1}{5} \sum_{k=-\infty}^{\infty} \text{rect}\left(\frac{k}{5}\right) e^{j(\pi k/5)t} \tag{5.350}$$

or

$$x(t) = \frac{1}{5} \sum_{k=-2}^{2} e^{j(\pi k/5)t}. \tag{5.351}$$

This result can be expressed in two interesting alternate forms. First, combining complex sinusoids in conjugate pairs we get

$$x(t) = \frac{1}{5}\left[1 + 2\,\cos\left(\frac{\pi}{5}t\right) + 2\,\cos\left(\frac{2\pi}{5}t\right)\right]. \tag{5.352}$$

Second, we can use the formula for the summation of a finite series

$$\sum_{n=0}^{N-1} r^n = \begin{cases} N & r = 1 \\ \dfrac{1 - r^N}{1 - r} & r \neq 1 \end{cases} \tag{5.353}$$

and the change of variable $q = k + 2$ to get

$$x(t) = \frac{e^{-j(2\pi/5)t}}{5} \sum_{q=0}^{5-1} e^{j(\pi q/5)t} = \frac{e^{-j(2\pi/5)t}}{5} \frac{1 - e^{j\pi t}}{1 - e^{j(\pi/5)t}}$$

$$= \frac{e^{-j(2\pi/5)t}}{5} \frac{e^{j(\pi/2)t}}{e^{j(\pi/10)t}} \frac{e^{-j(\pi/2)t} - e^{j(\pi/2)t}}{e^{-j(\pi/10)t} - e^{j(\pi/10)t}} \tag{5.354}$$

or

$$x(t) = \frac{1}{5} \frac{e^{-j(\pi/2)t} - e^{j(\pi/2)t}}{e^{-j(\pi/10)t} - e^{j(\pi/10)t}} = \frac{1}{5} \frac{\sin((\pi/2)t)}{\sin((\pi/10)t)} = \text{drcl}\left(\frac{t}{10}, 5\right) \qquad (5.355)$$

So the original time-domain function $x(t)$ can be expressed in three very different-looking ways,

$$x(t) = \text{sinc}\left(\frac{t}{2}\right) * \frac{1}{10} \text{comb}\left(\frac{t}{10}\right) = \frac{1}{5}\left[1 + 2\cos\left(\frac{\pi}{5}t\right) + 2\cos\left(\frac{2\pi}{5}t\right)\right]$$

$$= \text{drcl}\left(\frac{t}{10}, 5\right). \qquad (5.356)$$

---

**EXAMPLE 5.15**

Generalize the results of Example 5.14 starting with

$$x(t) = \text{sinc}\left(\frac{t}{w}\right) * \frac{1}{T_0} \text{comb}\left(\frac{t}{T_0}\right) = \text{sinc}\left(\frac{t}{w}\right) * f_0 \text{comb}(f_0 t). \qquad (5.357)$$

### ■ Solution

The fundamental period of this periodically extended aperiodic function is $T_p = T_0$. From (5.346), $X_p[k] = f_p X(kf_p)$, where, in this case,

$$X(f) = \mathcal{F}\left(\text{sinc}\left(\frac{t}{w}\right)\right) = w\,\text{rect}(wf). \qquad (5.358)$$

Therefore,

$$X[k] = wf_0\,\text{rect}(wkf_0). \qquad (5.359)$$

Then $x(t)$ can be expressed as a CTFS,

$$x(t) = \sum_{k=-\infty}^{\infty} X[k]e^{j2\pi(kf_0)t} = wf_0 \sum_{k=-\infty}^{\infty} \text{rect}(wkf_0)\,e^{j2\pi kf_0 t} \qquad (5.360)$$

*Case 1.* $T_0/2w$ is not an integer.

$$x(t) = wf_0 \sum_{k=-M}^{M} e^{j2\pi kf_0 t}, \qquad (5.361)$$

where $M$ is the greatest integer in $T_0/2w$. Combining complex sinusoids in conjugate pairs we get

$$x(t) = wf_0[1 + 2\cos(2\pi f_0 t) + 2\cos(4\pi f_0 t) + \cdots + 2\cos(2M\pi f_0 t)]. \qquad (5.362)$$

We can use the formula for the summation of a finite series,

$$\sum_{n=0}^{N-1} r^n = \begin{cases} N & r = 1 \\ \dfrac{1 - r^N}{1 - r} & r \neq 1 \end{cases} \qquad (5.363)$$

and the change of variable $q = k + M$ to get

$$x(t) = wf_0 \sum_{k=-M}^{M} e^{j2\pi kf_0 t} = wf_0 \sum_{q=0}^{2M} e^{j2\pi(q-M)f_0 t} = wf_0 e^{-j2\pi Mf_0 t} \sum_{q=0}^{2M} e^{j2\pi qf_0 t} \qquad (5.364)$$

or

$$x(t) = wf_0 e^{-j2\pi M f_0 t} \frac{1 - e^{j2\pi(2M+1)f_0 t}}{1 - e^{j2\pi f_0 t}} = wf_0 \frac{e^{-j2\pi M f_0 t} e^{j\pi(2M+1)f_0 t}}{e^{j\pi f_0 t}} \frac{e^{-j\pi(2M+1)f_0 t} - e^{j\pi(2M+1)f_0 t}}{e^{-j\pi f_0 t} - e^{j\pi f_0 t}}$$

(5.365)

or

$$x(t) = wf_0 \frac{\sin(\pi(2M+1)f_0 t)}{\sin(\pi f_0 t)} = wf_0(2M+1)\,\mathrm{drcl}(f_0 t,\ 2M+1)\,.$$ (5.366)

So the original time-domain function $x(t)$ can be expressed in three very different-looking ways,

$$x(t) = \mathrm{sinc}\left(\frac{t}{w}\right) * f_0\,\mathrm{comb}(f_0 t)$$ (5.367)

$$x(t) = wf_0[1 + 2\cos(2\pi f_0 t) + 2\cos(4\pi f_0 t) + \cdots + 2\cos(2M\pi f_0 t)]$$ (5.368)

and

$$x(t) = wf_0(2M+1)\,\mathrm{drcl}(f_0 t, 2M+1).$$ (5.369)

*Case 2.*   $T_0/2w$ is an integer.

$$x(t) = wf_0\left[\sum_{k=-((T_0/2w)-1)}^{(T_0/2w)-1} e^{j2\pi k f_0 t}\right] + wf_0\left(\frac{1}{2}\right)e^{-j2\pi(T_0/2w)f_0 t} + wf_0\left(\frac{1}{2}\right)e^{j2\pi(T_0/2w)f_0 t}$$ (5.370)

$$x(t) = wf_0\left[\sum_{k=-((T_0/2w)-1)}^{(T_0/2w)-1} e^{j2\pi k f_0 t}\right] + \frac{wf_0}{2}\left(e^{-j(\pi t/w)} + e^{j(\pi t/w)}\right)$$

$$= wf_0\left[\cos\left(\frac{\pi t}{w}\right) + \sum_{k=-[(T_0/2w)-1]}^{(T_0/2w)-1} e^{j2\pi k f_0 t}\right]$$ (5.371)

Combining complex sinusoids in conjugate pairs we get

$$x(t) = wf_0\left[1 + 2\cos(2\pi f_0 t) + 2\cos(4\pi f_0 t) + \cdots\right.$$
$$\left. + 2\cos\left(2\left(\frac{T_0}{2w}-1\right)\pi f_0 t\right) + \cos\left(\frac{\pi t}{w}\right)\right].$$ (5.372)

We can use the formula for the summation of a finite series,

$$\sum_{n=0}^{N-1} r^n = \begin{cases} N & r = 1 \\ \dfrac{1-r^N}{1-r} & r \neq 1 \end{cases}$$ (5.373)

and the change of variable $q = k + ((T_0/2w) - 1)$ to get

$$x(t) = wf_0\left[\cos\left(\frac{\pi t}{w}\right) + \sum_{q=0}^{(T_0/w)-2} e^{j2\pi(q-((T_0/2w)-1))f_0 t}\right]$$

$$= wf_0\left[\cos\left(\frac{\pi t}{w}\right) + e^{-j2\pi((T_0/2w)-1)f_0 t}\sum_{q=0}^{(T_0/w)-2} e^{j2\pi q f_0 t}\right]$$ (5.374)

or

$$x(t) = w f_0 \left[ \cos\left(\frac{\pi t}{w}\right) + e^{-j2\pi((T_0/2w)-1)f_0 t} \frac{1 - e^{j2\pi((T_0/w)-1)f_0 t}}{1 - e^{j2\pi f_0 t}} \right] \tag{5.375}$$

$$x(t) = w f_0 \left[ \cos\left(\frac{\pi t}{w}\right) + e^{-j2\pi((T_0/2w)-1)f_0 t} \frac{e^{j\pi((T_0/w)-1)f_0 t}}{e^{j\pi f_0 t}} \frac{e^{-j\pi((T_0/w)-1)f_0 t} - e^{j\pi((T_0/w)-1)f_0 t}}{e^{-j\pi f_0 t} - e^{j\pi f_0 t}} \right] \tag{5.376}$$

$$x(t) = w f_0 \left[ \cos\left(\frac{\pi t}{w}\right) + \frac{\sin(\pi((T_0/w)-1)f_0 t)}{\sin(\pi f_0 t)} \right] \tag{5.377}$$

$$x(t) = w f_0 \left[ \cos\left(\frac{\pi t}{w}\right) + \left(\frac{T_0}{w} - 1\right) \mathrm{drcl}\left(f_0 t, \frac{T_0}{w} - 1\right) \right]. \tag{5.378}$$

So, in this case, the original time-domain function $x(t)$ can be expressed in these three ways,

$$x(t) = \mathrm{sinc}\left(\frac{t}{w}\right) * f_0 \, \mathrm{comb}(f_0 t) \tag{5.379}$$

$$x(t) = w f_0 \left[ 1 + 2 \, \cos(2\pi f_0 t) + 2 \, \cos(4\pi f_0 t) + \cdots \right.$$
$$\left. + 2 \cos\left( 2 \left(\frac{T_0}{2w} - 1\right) \pi f_0 t \right) + \cos\left(\frac{\pi t}{w}\right) \right] \tag{5.380}$$

and

$$x(t) = w f_0 \left[ \cos\left(\frac{\pi t}{w}\right) + \left(\frac{T_0}{w} - 1\right) \mathrm{drcl}\left(f_0 t, \frac{T_0}{w} - 1\right) \right]. \tag{5.381}$$

∎

## CTFT AND DTFT

The CTFT is the Fourier transform of a CT function, and the DTFT is the Fourier transform of a DT function. If we multiply a CT function $x(t)$ by a periodic train of unit impulses spaced $T_s$ seconds apart (a comb function), we create the CT impulse function

$$x_\delta(t) = x(t) \frac{1}{T_s} \mathrm{comb}\left(\frac{t}{T_s}\right) = \sum_{n=-\infty}^{\infty} x(nT_s)\delta(t - nT_s). \tag{5.382}$$

If we now form a DT function $x[n]$ whose values are the values of the original CT function $x(t)$ at integer multiples of $T_s$ and are, therefore, also the strengths of the impulses in the CT impulse function $x_\delta(t)$, we get the relationship,

$$x[n] = x(nT_s). \tag{5.383}$$

Therefore, the two functions, $x[n]$ and $x_\delta(t)$ are completely described by the same set of numbers and contain the same information. If we now find the CTFT of (5.382), we get

$$X_\delta(f) = X_{\mathrm{CTFT}}(f) * \mathrm{comb}(T_s f) = \sum_{n=-\infty}^{\infty} x(nT_s)e^{-j2\pi f n T_s} \tag{5.384}$$

or

$$X_\delta(f) = f_s \sum_{k=-\infty}^{\infty} X_{\mathrm{CTFT}}(f - kf_s) = \sum_{n=-\infty}^{\infty} x[n]e^{-j(2\pi f n/f_s)} \tag{5.385}$$

where $f_s = 1/T_s$. If we make the change of variable $f \to f_s F$ in (5.385), we get

$$X_\delta(f_s F) = f_s \sum_{k=-\infty}^{\infty} X_{\text{CTFT}}(f_s(F - k)) = \underbrace{\sum_{n=-\infty}^{\infty} x[n]e^{-j2\pi n F}}_{=X_{\text{DTFT}}(F)} \tag{5.386}$$

The last expression in (5.386) is exactly the definition of the DTFT of $x[n]$, $X_{\text{DTFT}}(F)$. Summarizing, if $x[n] = x(nT_s)$ and $x_\delta(t) = \sum_{n=-\infty}^{\infty} x[n]\delta(t - nT_s)$, then

$$X_{\text{DTFT}}(F) = X_\delta(f_s F) \tag{5.387}$$

or

$$X_\delta(f) = X_{\text{DTFT}}\left(\frac{f}{f_s}\right). \tag{5.388}$$

Also

$$X_{\text{DTFT}}(F) = f_s \sum_{k=-\infty}^{\infty} X_{\text{CTFT}}(f_s(F - k)). \tag{5.389}$$

The subscripts CTFT and DTFT are needed here because

$$X_{\text{DTFT}}(\ ) \neq X_{\text{CTFT}}(\ ). \tag{5.390}$$

Instead, from (5.389),

$$X_{\text{DTFT}}(F) = f_s \sum_{k=-\infty}^{\infty} X_{\text{CTFT}}(f_s(F - k)). \tag{5.391}$$

The functions $X_{\text{DTFT}}(\ )$ and $X_{\text{CTFT}}(\ )$ are both functions of a continuous independent variable and are mathematically distinct, so they need distinguishable names. Usually the subscripts are not needed in practical analysis because only one of the two functions is used, but here when we are relating Fourier methods it is necessary to make a distinction between the two functions.

Here again we have a correspondence between a function $x[n]$ of a discrete independent variable, in this case discrete time $n$, and an impulse function $x_\delta(t)$ of a continuous independent variable, in this case continuous time $t$. So there is also an information equivalence between the DTFT of the DT function $x[n]$ and the CTFT of the CT function $x_\delta(t)$ (Figure 5.34).

There is also some equivalence between the CTFT of the original CT function $x(t)$ and the DTFT of the DT function $x[n]$ through (5.389). Given $X_{\text{CTFT}}(f)$, we can find $X_{\text{DTFT}}(F)$. However, the reverse of this statement is not always true. Given $X_{\text{DTFT}}(F)$, we cannot always be sure of being able to find $X_{\text{CTFT}}(f)$. The conditions under which $X_{\text{CTFT}}(f)$ can be found from $X_{\text{DTFT}}(F)$ are the subject of Chapter 7.

**DTFT AND DTFS**

The DTFS of a periodic DT function $x[n]$ with fundamental period $N_0 = 1/F_0$ is defined by

$$x[n] = \sum_{k=\langle N_0 \rangle} X[k]e^{j2\pi(kF_0)n} \xleftrightarrow{\ \mathcal{FS}\ } X[k] = \frac{1}{N_0} \sum_{n=\langle N_0 \rangle} x[n]e^{-j2\pi(kF_0)n}. \tag{5.392}$$

Using the frequency-shifting property $e^{j2\pi F_0 n} x[n] \xleftrightarrow{\ \mathcal{F}\ } X(F - F_0)$ and the DTFT transform pair $1 \xleftrightarrow{\ \mathcal{F}\ } \text{comb}(F)$, we can find the DTFT of $x[n]$, yielding

$$X(F) = \sum_{k=\langle N_0 \rangle} X[k] \text{comb}(F - kF_0). \tag{5.393}$$

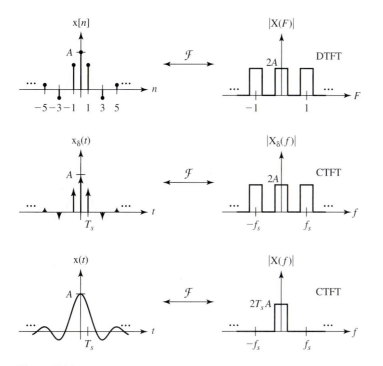

**Figure 5.34**
DTFT of a DT sinc function $x[n]$; CTFT of a CT impulse function $x_\delta(t)$, whose impulse strengths are the values of $x[n]$; and CTFT of the original CT function $x(t)$.

Then

$$X(F) = \sum_{k=\langle N_0 \rangle} X[k] \sum_{q=-\infty}^{\infty} \delta(F - kF_0 - q) = \sum_{k=-\infty}^{\infty} X[k]\delta(F - kF_0). \qquad \textbf{(5.394)}$$

This shows that, for periodic DT functions, the DTFS is simply a special case of the DTFT. If a function $x[n]$ is periodic, its DTFT consists only of impulses occurring at $kF_0$ with strengths $X[k]$ (Figure 5.35).

Summarizing, for a periodic DT function $x[n]$ with fundamental period $N_0 = 1/F_0$,

$$X(F) = \sum_{k=-\infty}^{\infty} X[k]\delta(F - kF_0). \qquad \textbf{(5.395)}$$

Another case that will be important in the exploration of sampling in Chapter 7 is the relationship between the DTFT of an aperiodic DT signal and the DTFS harmonic function of a periodic extension of that signal. Let $x[n]$ be an aperiodic DT function. Its DTFT is $X(F)$. Let $x_p(n)$ be a periodic extension of $x[n]$ with fundamental period $N_p$ such that

$$x_p[n] = \sum_{m=-\infty}^{\infty} x[n - mN_p] = x[n] * \text{comb}_{N_p}[n] \qquad \textbf{(5.396)}$$

(Figure 5.36). Using the multiplication–convolution duality of the DTFT,

$$X_p(F) = X(F) \text{comb}(N_p F) = \frac{1}{N_p} \sum_{k=-\infty}^{\infty} X\left(\frac{k}{N_p}\right) \delta\left(F - \frac{k}{N_p}\right). \qquad \textbf{(5.397)}$$

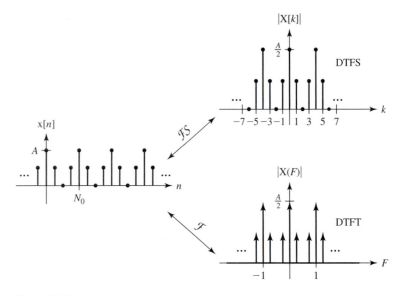

**Figure 5.35**
Harmonic function and DTFT of $x[n] = \frac{A}{2}\left[1 + \cos\left(\frac{2\pi}{4}n\right)\right]$.

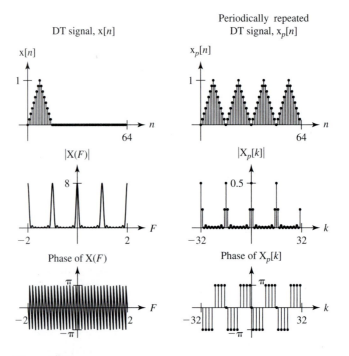

**Figure 5.36**
A DT signal and its DTFT, and the periodic repetition of the
DT signal and its DTFS harmonic function.

Using (5.395),

$$X_p(F) = \sum_{k=-\infty}^{\infty} X_p[k]\delta(F - kF_p) \tag{5.398}$$

and, combining (5.397) and (5.398),

$$X_p[k] = \frac{1}{N_p}X(kF_p). \tag{5.399}$$

In words, this says that if an aperiodic DT signal x[n] is periodically repeated with fundamental period $N_p$ to form a periodic DT signal $x_p[n]$, the values of its DTFS harmonic function $X_p[k]$ can be found from $X(F)$, the DTFT of x[n], evaluated at the discrete DT frequencies $kF_p$, where $F_p = 1/N_p$. This forms an equivalence between sampling in the DT frequency domain and periodic repetition in the DT domain. This will be useful in the study of sampling in Chapter 7.

**EXAMPLE 5.16**

Find the DTFT of the bipolar DT pulse

$$x[n] = \text{rect}_2[n - 2] - \text{rect}_2[n - 7] \tag{5.400}$$

and compare it to the DTFS harmonic function of a periodic extension of this signal with periods $N_p = 10$, 20, and 50, multiplied by the period $N_p$.

■ **Solution**
The DTFT is

$$X(F) = 5\,\text{drcl}(F, 5)(e^{-j4\pi F} - e^{-j14\pi F}) = j10e^{-j9\pi F}\,\text{drcl}(F, 5)\,\sin(5\pi F) \tag{5.401}$$

(Figure 5.37). From (5.399) the DTFS harmonic function of a periodic extension of the signal is

$$X_p[k] = \frac{1}{N_p}X(kF_p). \tag{5.402}$$

The three DTFSs are illustrated in Figure 5.38.

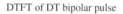
DT bipolar pulse

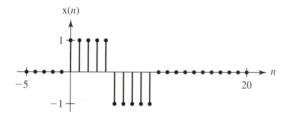

DTFT of DT bipolar pulse

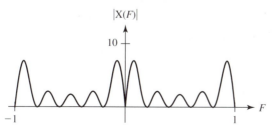

**Figure 5.37**
Bipolar pulse and magnitude of the DT signal's DTFT.

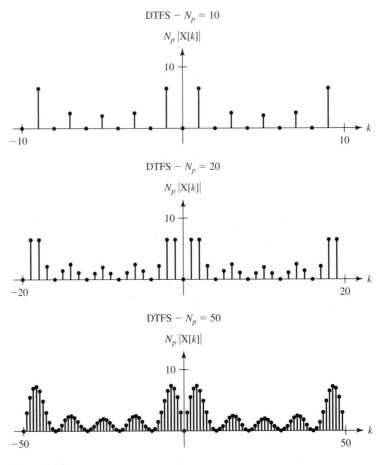

**Figure 5.38**
Magnitude of the DTFS harmonic function of the periodic extension of the
DT bipolar pulse for three different periods.

As the period of the periodic extension of the DT bipolar pulse is increased, the DTFS harmonic function (scaled by the period) approaches the same shape as the DTFT of the original DT bipolar pulse. This is the same kind of relationship illustrated in the development of the DTFT as a generalization of the DTFS.

## METHOD COMPARISON EXAMPLES

Examples 5.17 and 5.18 compare all the Fourier methods.

**EXAMPLE 5.17**

Find the CTFS harmonic function and CTFT of $x(t) = A \cos(2\pi f_0 t)$; the CTFS and CTFT of $x_\delta(t) = A \cos(2\pi f_0 t) f_s \, \text{comb}(f_s t)$, where $f_s = N_0 f_0$ and $N_0$ is an integer; and the DTFS harmonic function and DTFT of $x[n] = A \cos(2\pi n / N_0)$, and observe the relationships among them.

■ **Solution**

The CTFS harmonic function of $x(t)$ (see Figure 5.39) is

$$X[k] = \frac{A}{2}(\delta[k-1] + \delta[k+1]).\tag{5.403}$$

The CTFT of $x(t)$ (see Figure 5.40) is

$$X(f) = \frac{A}{2}[\delta(f - f_0) + \delta(f + f_0)]\tag{5.404}$$

or (see Figure 5.41)

$$X(j\omega) = A\pi[\delta(\omega - \omega_0) + \delta(\omega + \omega_0)].\tag{5.405}$$

The CTFS harmonic function of $x_\delta(t)$ (see Figure 5.42) is

$$X_\delta[k] = \frac{A f_s}{2}(\text{comb}_{N_0}[k-1] + \text{comb}_{N_0}[k+1])\tag{5.406}$$

or

$$X_\delta[k] = \frac{A f_s}{2}\left(\sum_{q=-\infty}^{\infty} \delta[k - qN_0 - 1] + \sum_{q=-\infty}^{\infty} \delta[k - qN_0 + 1]\right).\tag{5.407}$$

The CTFT of $x_\delta(t)$ (see Figure 5.43) is

$$X_\delta(f) = \frac{A}{2}\left[\text{comb}\left(\frac{f}{f_s} - \frac{1}{N_0}\right) + \text{comb}\left(\frac{f}{f_s} + \frac{1}{N_0}\right)\right]\tag{5.408}$$

$$X_\delta(f) = \frac{A f_s}{2}\left[\sum_{q=-\infty}^{\infty} \delta(f - qf_s - f_0) + \sum_{q=-\infty}^{\infty} \delta(f - qf_s + f_0)\right]\tag{5.409}$$

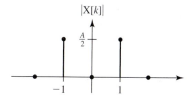

**Figure 5.39**

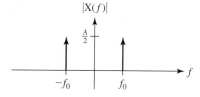

**Figure 5.40**

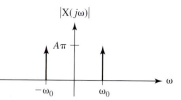

**Figure 5.41**

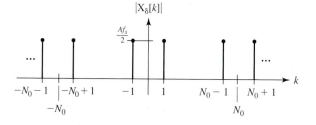

**Figure 5.42**

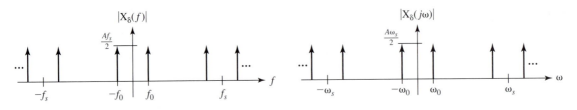

**Figure 5.43**

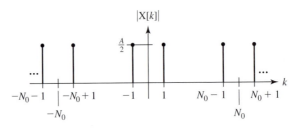

**Figure 5.44**

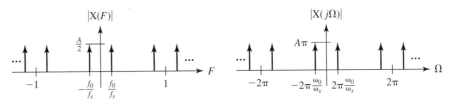

**Figure 5.45**

or

$$X_\delta(j\omega) = \frac{A}{2}\left[ \text{comb}\left(\frac{\omega}{\omega_s} - \frac{1}{N_0}\right) + \text{comb}\left(\frac{\omega}{\omega_s} + \frac{1}{N_0}\right)\right] \qquad (5.410)$$

$$X_\delta(j\omega) = \frac{A\omega_s}{2}\left[ \sum_{q=-\infty}^{\infty} \delta(\omega - q\omega_s - \omega_0) + \sum_{q=-\infty}^{\infty} \delta(\omega - q\omega_s + \omega_0)\right]. \qquad (5.411)$$

The DTFS harmonic function of x[n] (see Figure 5.44) is

$$X[k] = \frac{A}{2}(\text{comb}_{N_0}[k - 1] + \text{comb}_{N_0}[k + 1]) \qquad (5.412)$$

or

$$X[k] = \frac{A}{2}\left( \sum_{q=-\infty}^{\infty} \delta[k - qN_0 - 1] + \sum_{q=-\infty}^{\infty} \delta[k - qN_0 + 1]\right). \qquad (5.413)$$

The DTFT of x[n] (see Figure 5.45) is

$$X(F) = \frac{A}{2}\left[ \text{comb}\left(F - \frac{f_0}{f_s}\right) + \text{comb}\left(F + \frac{f_0}{f_s}\right)\right] \qquad (5.414)$$

$$X(F) = \frac{A}{2}\left[ \sum_{q=-\infty}^{\infty} \delta\left(F - q - \frac{f_0}{f_s}\right) + \sum_{q=-\infty}^{\infty} \delta\left(F - q + \frac{f_0}{f_s}\right)\right] \qquad (5.415)$$

or

$$X(j\Omega) = \frac{A}{2}\left[\text{comb}\left(\frac{\Omega}{2\pi} - \frac{\omega_0}{\omega_s}\right) + \text{comb}\left(\frac{\Omega}{2\pi} + \frac{\omega_0}{\omega_s}\right)\right] \tag{5.416}$$

$$X(j\Omega) = A\pi\left[\sum_{q=-\infty}^{\infty}\delta\left(\Omega - 2\pi\left(\frac{\omega_0}{\omega_s} - q\right)\right) + \sum_{q=-\infty}^{\infty}\delta\left(\Omega + 2\pi\left(\frac{\omega_0}{\omega_s} - q\right)\right)\right]. \tag{5.417}$$

Observations:

1. The values of the CTFS harmonic function of $x(t)$ are the impulse strengths in the CTFT of $x(t)$.
2. The CTFS harmonic function of $x_\delta(t)$ and the DTFS of $x[n]$ are exactly the same except for a factor $f_s$, and they are both periodic repetitions of the CTFS harmonic function of $x(t)$ with fundamental period $N_0$.
3. The CTFT of $x_\delta(t)$ is a periodic repetition of the CTFT of $x(t)$, except for a factor of $f_s$ or $\omega_s$, with fundamental period $f_s$ or $\omega_s$.
4. The CTFT of $x_\delta(t)$ and the DTFT of $x[n]$ are related by $X_\delta(f) = X(f/f_s)$ or $X_\delta(j\omega) = X(j2\pi(\omega/\omega_s))$ as was earlier shown to be true in general.

**EXAMPLE 5.18**

Find the CTFT of $x(t) = A\,\text{sinc}(t/T)$, the CTFT of $x_\delta(t) = A\,\text{sinc}(t/T)\,f_s\,\text{comb}(f_s t)$, and the DTFT of $x[n] = A\,\text{sinc}(n/N_0)$, where $f_s T = N_0$, and $N_0$ is an integer $>1$.

■ **Solution**

The CTFT of $x(t)$ (see Figure 5.46) is

$$X(f) = AT\,\text{rect}(Tf) \qquad \text{or} \qquad X(j\omega) = AT\,\text{rect}\left(\frac{T}{2\pi}\omega\right). \tag{5.418}$$

The CTFT of $x_\delta(t)$ (see Figure 5.47) is

$$X_\delta(f) = ATf_s\sum_{q=-\infty}^{\infty}\text{rect}(T(f - qf_s)) = AN_0\sum_{q=-\infty}^{\infty}\text{rect}(Tf - nN_0) \tag{5.419}$$

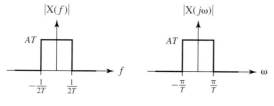

**Figure 5.46**

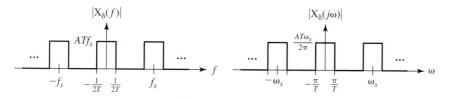

**Figure 5.47**

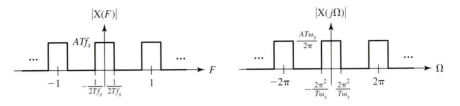

**Figure 5.48**

or

$$X_\delta(j\omega) = \frac{AT\omega_s}{2\pi} \sum_{q=-\infty}^{\infty} \mathrm{rect}\left(\frac{T}{2\pi}(\omega - q\omega_s)\right) = AN_0 \sum_{q=-\infty}^{\infty} \mathrm{rect}\left(\frac{T\omega - 2\pi q N_0}{2\pi}\right).$$
(5.420)

The DTFT of $x[n] = A \, \mathrm{sinc}(n/N_0)$ (see Figure 5.48) is

$$X(F) = AN_0 \sum_{q=-\infty}^{\infty} \mathrm{rect}(N_0(F - q)) = ATf_s \sum_{q=-\infty}^{\infty} \mathrm{rect}(Tf_s(F - q))$$
(5.421)

or

$$X(j\Omega) = AN_0 \sum_{q=-\infty}^{\infty} \mathrm{rect}\left(N_0\left(\frac{\Omega - 2\pi q}{2\pi}\right)\right) = \frac{AT\omega_s}{2\pi} \sum_{q=-\infty}^{\infty} \mathrm{rect}\left(\frac{T\omega_s}{2\pi}\left(\frac{\Omega - 2\pi q}{2\pi}\right)\right).$$
(5.422)

The results of Examples 5.17 and 5.18 show how interconnected all the Fourier methods are and how one can be substituted for another in many analyses. This will be important in the study of sampling.

## 5.10 SUMMARY OF IMPORTANT POINTS

1. The CTFS is a special case of the CTFT, and the DTFS is a special case of the DTFT.
2. A signal with an infinite period is aperiodic.
3. Signals and systems are often more usefully described by their frequency-domain properties than their time-domain properties.
4. The generalized CTFT which allows impulses in the transform includes periodic signals.
5. The more a signal is localized in one domain (time or frequency), the less it is localized in the other domain.
6. Convolution and multiplication of functions are dual operations in the time and frequency domains.
7. The Fourier transform of a periodic signal consists only of impulses.
8. Signal energy is conserved in the Fourier-transformation process.
9. Most Fourier transforms of signals with engineering usefulness can be done most efficiently using tables of transforms and the properties of the transform.
10. The DTFT is always periodic with period one in the $F$ domain or period $2\pi$ in the $\Omega$ domain.
11. For periodic signals there are simple conversions between a Fourier transform and a Fourier series.
12. If a signal is discrete in one domain, it is periodic in the other.
13. A shift in one domain corresponds to a multiplication by a complex exponential in the other domain.

**14.** If a CT signal is sampled to form a DT signal, the DTFT of the DT signal can be found from the CTFT of the CT signal by a change of variable, but the converse is not generally true.

## EXERCISES WITH ANSWERS

**1.** The transition from the CTFS to the CTFT is illustrated by the signal

$$x(t) = \text{rect}\left(\frac{t}{w}\right) * \frac{1}{T_0}\text{comb}\left(\frac{t}{T_0}\right)$$

or

$$x(t) = \sum_{n=-\infty}^{\infty} \text{rect}\left(\frac{t - nT_0}{w}\right).$$

The complex CTFS harmonic function for this signal is given by

$$X[k] = \frac{Aw}{T_0}\text{sinc}\left(\frac{kw}{T_0}\right).$$

Plot the modified CTFS harmonic function

$$T_0\,X[k] = Aw\,\text{sinc}(w(kf_0))$$

for $w = 1$ and $f_0 = 0.5, 0.1,$ and $0.02$ versus $kf_0$ for the range $-8 < kf_0 < 8$.

**Answers:**

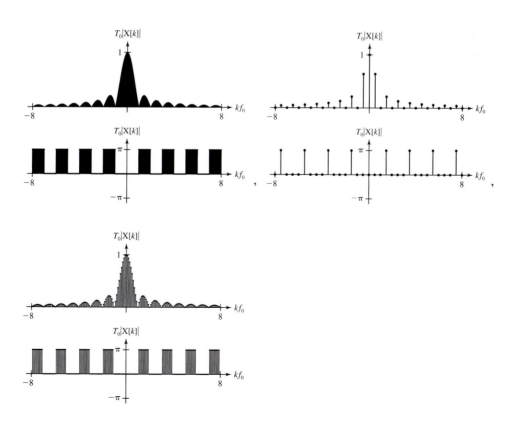

**2.** Suppose a function m($x$) has units of kg/m$^3$ and is a function of spatial position $x$ in meters. Write the mathematical expression for its CTFT, M($y$). What are the units of M and $y$?

**Answers:**

kg/m$^2$,    m$^{-1}$.

**3.** Using the integral definition of the Fourier transform, find the CTFT of these functions.

    *a.*   x($t$) = tri($t$)

    *b.*   $x(t) = \delta\left(t + \dfrac{1}{2}\right) - \delta\left(t - \dfrac{1}{2}\right)$

**Answers:**

$j2\sin(\pi f)$,    sinc$^2(f)$

**4.** In Figure E4 there is one example each of a lowpass, highpass, bandpass, and bandstop signal. Identify them.

**Answers:**

*a.*   bandstop,     *b.*   bandpass,     *c.*   lowpass,     *d.*   highpass

**5.** Starting with the definition of the CTFT find the radian-frequency form of the generalized CTFT of a constant. Then verify that a change of variable $\omega \rightarrow 2\pi f$ yields the correct result in cyclic-frequency form. Check your answer against the Fourier transform table in Appendix E.

**6.** Starting with the definition of the CTFT, find the generalized CTFT of a sine of the form $A\sin(\omega_0 t)$ and check your answer against the Fourier transform table in Appendix E.

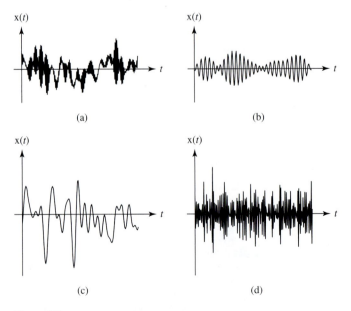

**Figure E4**
Signals with different frequency content.

7. Find the CTFS harmonic function and CTFT of each of these periodic signals and compare the results. After finding the transforms, formulate a general method of converting between the two forms for periodic signals.

    a.  $x(t) = A\cos(2\pi f_0 t)$          b.   $x(t) = \text{comb}(t)$

**Answers:**

1, $\dfrac{A}{2}(\delta(f - f_0) + \delta(f + f_0))$,   $\text{comb}(f)$, $\dfrac{A}{2}(\delta[k - 1] + \delta[k + 1])$,

$X(f) = \displaystyle\sum_{k=-\infty}^{\infty} X[k]\delta(f - kf_0)$

8. Let a signal be defined by

$$x(t) = 2\cos(4\pi t) + 5\cos(15\pi t).$$

Find the CTFTs of $x(t - \frac{1}{40})$ and $x(t + \frac{1}{20})$ and identify the resultant phase shift of each sinusoid in each case. Plot the phase of the CTFT, and draw a straight line through the four phase points which result in each case. What is the general relationship between the slope of that line and the time delay?

**Answers:**

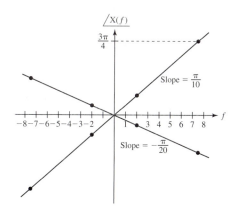

The slope of the line is $-2\pi f$ times the delay.

9. Using the frequency-shifting property, find and plot versus time the inverse CTFT of

$$X(f) = \text{rect}\left(\frac{f - 20}{2}\right) + \text{rect}\left(\frac{f + 20}{2}\right).$$

**Answer:**

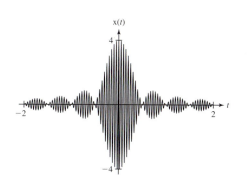

10. Find the CTFT of

$$x(t) = \text{sinc}(t).$$

Then make the transformation $t \to 2t$ in $x(t)$ and find the CTFT of the transformed signal.

**Answers:**

$$\text{rect}(f), \quad \frac{1}{2}\text{rect}\left(\frac{f}{2}\right)$$

11. Using the multiplication–convolution duality of the CTFT, find an expression for $y(t)$ which does not use the convolution operator $*$, and plot $y(t)$.

  a.  $y(t) = \text{rect}(t) * \cos(\pi t)$

  b.  $y(t) = \text{rect}(t) * \cos(2\pi t)$

  c.  $y(t) = \text{sinc}(t) * \text{sinc}\left(\frac{t}{2}\right)$

  d.  $y(t) = \text{sinc}(t) * \text{sinc}^2\left(\frac{t}{2}\right)$

  e.  $y(t) = e^{-t}u(t) * \sin(2\pi t)$

**Answers:**

$$\frac{\cos(2\pi t + 2.984)}{\sqrt{1 + (2\pi)^2}}, \quad \frac{2}{\pi}\cos(\pi t), \quad 0, \quad \text{sinc}\left(\frac{t}{2}\right), \quad \text{sinc}^2\left(\frac{t}{2}\right)$$

12. Using the CTFT of the rectangle function and the differentiation property of the CTFT find the Fourier transform of

$$x(t) = \delta(t - 1) - \delta(t + 1).$$

Check your answer against the CTFT found using the table and the time-shifting property.

**Answer:**

$$-j2\sin(2\pi f)$$

13. Find the CTFS harmonic function and CTFT of these periodic functions and compare answers.

  a.  $x(t) = \text{rect}(t) * \frac{1}{2}\text{comb}\left(\frac{t}{2}\right)$

  b.  $x(t) = \text{tri}(10t) * 4\,\text{comb}(4t)$

**Answers:**

$$\sum_{k=-\infty}^{\infty} \frac{5}{4}\frac{\cos(4\pi k/5) - 1}{(\pi k)^2}\delta(f - 4k), \quad \frac{1}{2}\text{sinc}(f)\sum_{k=-\infty}^{\infty}\delta\left(f - \frac{k}{2}\right), \quad \frac{1}{2}\text{sinc}\left(\frac{k}{2}\right),$$

$$\frac{5}{4}\frac{\cos\left(\frac{4}{5}\pi k\right) - 1}{(\pi k)^2}$$

14. Using Parseval's theorem, find the signal energy of these signals.

  a.  $x(t) = 4\,\text{sinc}\left(\frac{t}{5}\right)$    b.  $x(t) = 2\,\text{sinc}^2(3t)$

**Answers:**

80, $\frac{8}{9}$

**15.** What is the total area under the function, $g(t) = 100 \, \text{sinc}((t-8)/30)$?

**Answer:**

3000

**16.** Using the integration property, find the CTFT of each of these functions and compare with the CTFT found using other properties.

*a.* $g(t) = \begin{cases} 1 & |t| < 1 \\ 2 - |t| & 1 < |t| < 2 \\ 0 & \text{elsewhere} \end{cases}$ 　　*b.* $g(t) = 8 \, \text{rect}\left(\dfrac{t}{3}\right)$

**Answers:**

$24 \, \text{sinc}(3f), \quad 3 \, \text{sinc}(3f) \, \text{sinc}(f)$

**17.** Sketch the magnitudes and phases of the CTFTs of these signals in the $f$ form.

*a.* $x(t) = \delta(t - 2)$

*b.* $x(t) = u(t) - u(t - 1)$

*c.* $x(t) = 5 \, \text{rect}\left(\dfrac{t+2}{4}\right)$

*d.* $x(t) = 25 \, \text{sinc}(10(t-2))$

*e.* $x(t) = 6 \sin(200\pi t)$

*f.* $x(t) = 2e^{-3t}u(3t)$

*g.* $x(t) = 4e^{-3t^2}$

**Answers:**

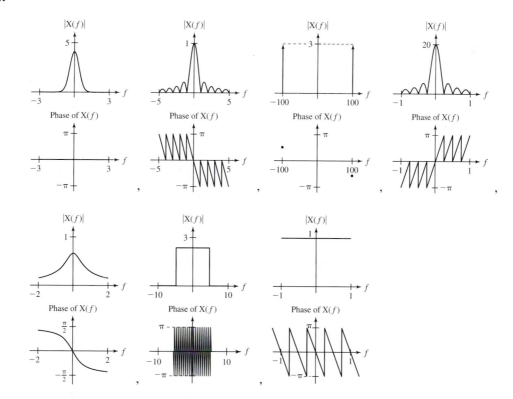

**18.** Sketch the magnitudes and phases of the CTFTs of these signals in the $\omega$ form.

a.   $x(t) = \dfrac{1}{2}\,\mathrm{comb}\left(\dfrac{t}{2}\right)$

b.   $x(t) = \mathrm{sgn}(2t)$

c.   $x(t) = 10\,\mathrm{tri}\left(\dfrac{t-4}{20}\right)$

d.   $x(t) = \dfrac{\mathrm{sinc}^2((t+1)/3)}{10}$

e.   $x(t) = \dfrac{\cos(200\pi t - (\pi/4))}{4}$

f.   $x(t) = 2e^{-3t}u(t)$

g.   $x(t) = 7e^{-5|t|}$

**Answers:**

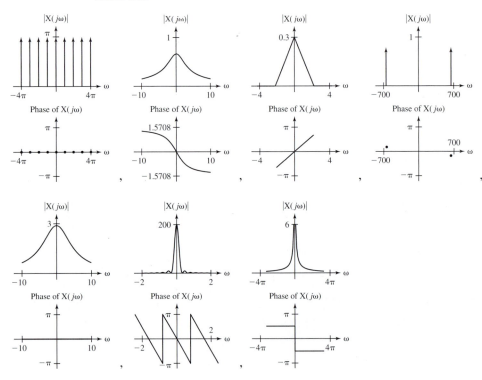

**19.** Sketch the inverse CTFTs of these functions.

a.   $X(f) = -15\,\mathrm{rect}\left(\dfrac{f}{4}\right)$

b.   $X(f) = \dfrac{\mathrm{sinc}(-10f)}{30}$

c.   $X(f) = \dfrac{18}{9+f^2}$

d.   $X(f) = \dfrac{1}{10+jf}$

e.   $X(f) = \dfrac{\delta(f-3)+\delta(f+3)}{6}$

f.   $X(f) = 8\delta(5f)$

g.   $X(f) = -\dfrac{3}{j\pi f}$

**Answers:**

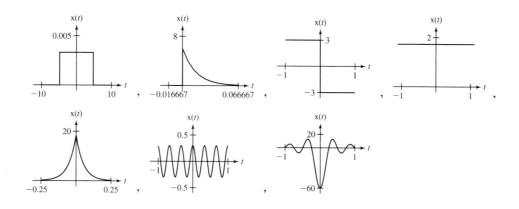

**20.** Sketch the inverse CTFTs of these functions.

a. $X(j\omega) = e^{-4\omega^2}$

b. $X(j\omega) = 7 \operatorname{sinc}^2\left(\dfrac{\omega}{\pi}\right)$

c. $X(j\omega) = j\pi[\delta(\omega + 10\pi) - \delta(\omega - 10\pi)]$

d. $X(j\omega) = \dfrac{\operatorname{comb}(4\omega/\pi)}{5}$

e. $X(j\omega) = \dfrac{5\pi}{j\omega} + 10\pi\delta(\omega)$

f. $X(j\omega) = \dfrac{6}{3 + j\omega}$

g. $X(j\omega) = 20 \operatorname{tri}(8\omega)$

**Answers:**

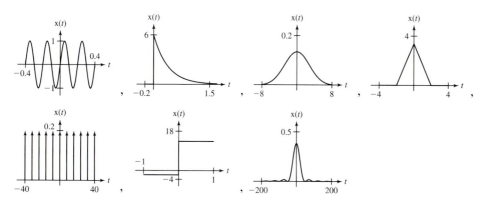

**21.** Find the CTFTs of these signals in either the $f$ or $\omega$ form, whichever is more convenient.

a. $x(t) = 3\cos(10t) + 4\sin(10t)$

b. $x(t) = \operatorname{comb}\left(\dfrac{t}{2}\right) - \operatorname{comb}\left(\dfrac{t-1}{2}\right)$

c. $x(t) = 4\,\mathrm{sinc}(4t) - 2\,\mathrm{sinc}\left(4\left(t - \dfrac{1}{4}\right)\right) - 2\,\mathrm{sinc}\left(4\left(t + \dfrac{1}{4}\right)\right)$

d. $x(t) = \left[2e^{(-1+j2\pi)t} + 2e^{(-1-j2\pi)t}\right]u(t)$

e. $x(t) = 4e^{-(|t|/16)}$

**Answers:**

$(5\pi e^{-j0.927})\delta(\omega - 10) + (5\pi e^{j0.927})\delta(\omega + 10),\quad 4\dfrac{j2\pi f + 1}{(j2\pi f + 1)^2 + (2\pi)^2},$

$\mathrm{rect}\left(\dfrac{\omega}{8\pi}\right) - \mathrm{rect}\left(\dfrac{\omega}{8\pi}\right)\cos\left(\dfrac{\omega}{4}\right),\quad \dfrac{128}{1 + 256\omega^2},\quad j4e^{-j(\omega/2)}\,\mathrm{comb}\left(\dfrac{\omega}{\pi}\right)\sin\left(\dfrac{\omega}{2}\right)$

**22.** Sketch the magnitudes and phases of these functions. Sketch the inverse CTFTs of the functions also.

a. $X(j\omega) = \dfrac{10}{3 + j\omega} - \dfrac{4}{5 + j\omega}$

b. $X(f) = 4\left[\mathrm{sinc}\left(\dfrac{f - 1}{2}\right) + \mathrm{sinc}\left(\dfrac{f + 1}{2}\right)\right]$

c. $X(f) = \dfrac{j}{10}\left[\mathrm{tri}\left(\dfrac{f + 2}{8}\right) - \mathrm{tri}\left(\dfrac{f - 2}{8}\right)\right]$

d. $X(f) = \delta(f + 1050) + \delta(f + 950) + \delta(f - 950) + \delta(f - 1050)$

e. $X(f) = [\delta(f + 1050) + 2\delta(f + 1000) + \delta(f + 950) + \delta(f - 950)$
$\qquad + 2\delta(f - 1000) + \delta(f - 1050)]$

**Answers:**

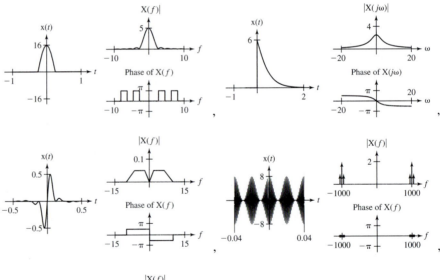

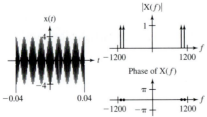

**23.**  Sketch these signals versus time. Sketch the magnitudes and phases of their CTFTs in either the $f$ or $\omega$ form, whichever is more convenient.

a.  $x(t) = \text{rect}(2t) * \text{comb}(t) - \text{rect}(2t) * \text{comb}\left(t - \dfrac{1}{2}\right)$

b.  $x(t) = -1 + 2\,\text{rect}(2t) * \text{comb}(t)$

c.  $x(t) = e^{-(t/4)}u(t) * \sin(2\pi t)$

d.  $x(t) = e^{-\pi t^2} * [\text{rect}(2t) * \text{comb}(t)]$

e.  $x(t) = \text{rect}(t) * [\text{tri}(2t) * \text{comb}(t)]$

f.  $x(t) = \text{sinc}(2.01t) * \text{comb}(t)$

g.  $x(t) = \text{sinc}(1.99t) * \text{comb}(t)$

h.  $x(t) = e^{-t^2} * e^{-t^2}$

**Answers:**

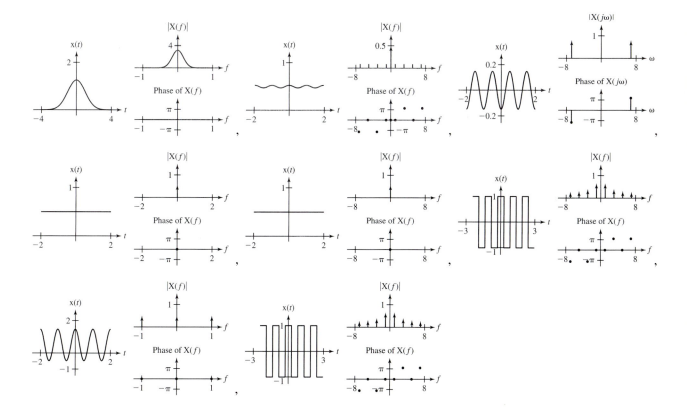

**24.**  Sketch the magnitudes and phases of these functions. Sketch the inverse CTFTs of the functions also.

a.  $X(f) = \text{sinc}\left(\dfrac{f}{100}\right) * [\delta(f - 1000) + \delta(f + 1000)]$

b.  $X(f) = \text{sinc}(10f) * \text{comb}(f)$

**Answers:**

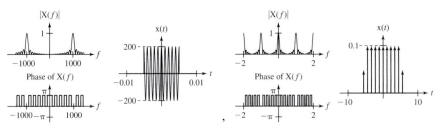

25.    Sketch these signals versus time. Sketch the magnitudes and phases of the CTFTs of these signals in either the $f$ or $\omega$ form, whichever is more convenient. In some cases the time sketch may be conveniently done first. In other cases it may be more convenient to do the time sketch after the CTFT has been found, by finding the inverse CTFT.

a.    $x(t) = e^{-\pi t^2} \sin(20\pi t)$

b.    $x(t) = \cos(400\pi t)\, \text{comb}(100t)$

c.    $x(t) = [1 + \cos(400\pi t)]\cos(4000\pi t)$

d.    $x(t) = [1 + \text{rect}(100t) * 50\, \text{comb}(50t)]\cos(500\pi t)$

e.    $x(t) = \text{rect}\left(\dfrac{t}{7}\right) \text{comb}(t)$

**Answers:**

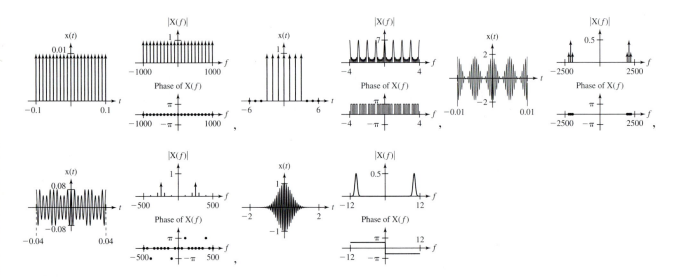

26.    Sketch the magnitudes and phases of these functions. Sketch the inverse CTFTs of the functions also.

a.    $X(f) = \text{sinc}\left(\dfrac{f}{4}\right)\text{comb}(f)$

b.    $X(f) = \left[\text{sinc}\left(\dfrac{f-1}{4}\right) + \text{sinc}\left(\dfrac{f+1}{4}\right)\right]\text{comb}(f)$

c.    $X(f) = \text{sinc}(f)\,\text{sinc}(2f)$

**Answers:**

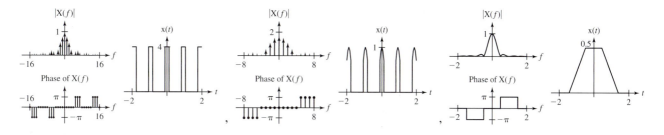

27.  Sketch these signals versus time and the magnitudes and phases of their CTFTs.

a.  $x(t) = \dfrac{d}{dt}[\text{sinc}(t)]$

b.  $x(t) = \dfrac{d}{dt}\left[4\,\text{rect}\left(\dfrac{t}{6}\right)\right]$

c.  $x(t) = \dfrac{d}{dt}[\text{tri}(2t) * \text{comb}(t)]$

**Answers:**

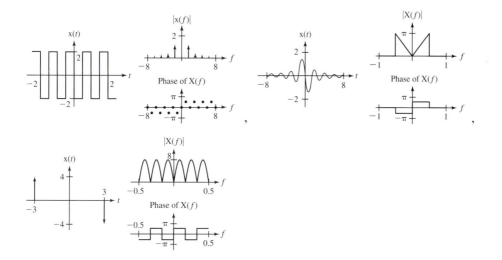

28.  Sketch these signals versus time and the magnitudes and phases of their CTFTs.

a.  $x(t) = \displaystyle\int_{-\infty}^{t} \sin(2\pi\lambda)\, d\lambda$

b.  $x(t) = \displaystyle\int_{-\infty}^{t} \text{rect}(\lambda)\, d\lambda$

c.  $x(t) = \displaystyle\int_{-\infty}^{t} 3\,\text{sinc}(2\lambda)\, d\lambda$

**Answers:**

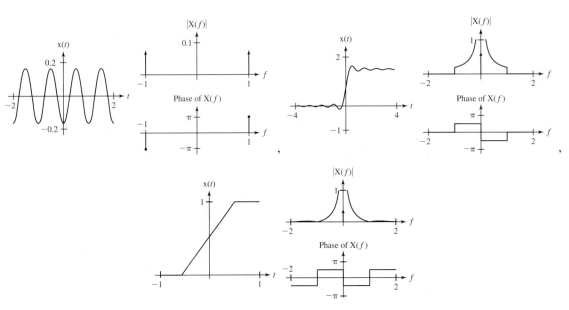

29.  From the summation definition, find the DTFT of

$$x[n] = 10 \ \text{rect}_4[n]$$

and compare with the Fourier transform table in Appendix E.

30.  From the definition, derive a general expression for the $F$ and $\Omega$ forms of the DTFT of functions of the form

$$x[n] = A \sin(2\pi F_0 n) = A \sin(\Omega_0 n).$$

[It should remind you of the CTFT of $x(t) = A \sin(2\pi f_0 t) = A \sin(\omega_0 t)$.] Compare with the Fourier transform table in Appendix E.

31.  A DT signal is defined by

$$x[n] = \text{sinc}\left(\frac{n}{8}\right).$$

Sketch the magnitude and phase of the DTFT of $x[n - 2]$.

**Answer:**

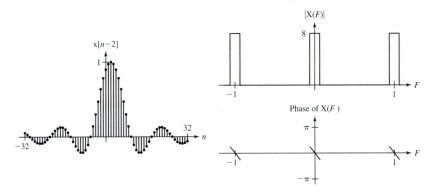

**32.** A DT signal is defined by

$$x[n] = \sin\left(\frac{\pi n}{6}\right).$$

Sketch the magnitude and phase of the DTFT of
*a.* $x[n-3]$        *b.* $x[n+12]$.

**Answers:**

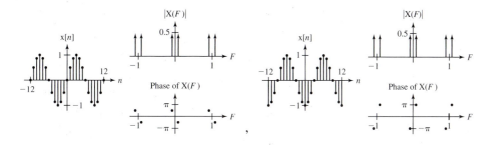

,

**33.** The DTFT of a DT signal is defined by

$$X(j\Omega) = 4\left[\text{rect}\left(\frac{2}{\pi}\left(\Omega - \frac{\pi}{2}\right)\right) + \text{rect}\left(\frac{2}{\pi}\left(\Omega + \frac{\pi}{2}\right)\right)\right] * \text{comb}\left(\frac{\Omega}{2\pi}\right).$$

Sketch $x[n]$.

**Answer:**

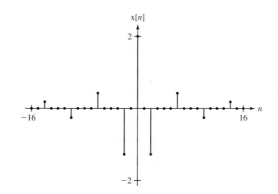

**34.** Sketch the magnitude and phase of the DTFT of

$$x[n] = \text{rect}_4[n] * \cos\left(\frac{2\pi n}{6}\right).$$

Then sketch $x[n]$.

**Answer:**

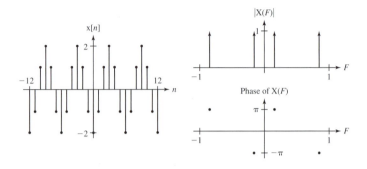

**35.** Sketch the inverse DTFT of

$$X(F) = [\text{rect}(4F) * \text{comb}(F)] \circledast \text{comb}(2F).$$

**Answer:**

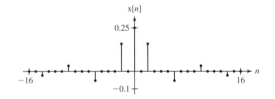

**36.** Using the differencing property of the DTFT and the transform pair

$$\text{tri}\left(\frac{n}{2}\right) \overset{\mathcal{F}}{\longleftrightarrow} 1 + \cos(2\pi F),$$

find the DTFT of $\frac{1}{2}(\delta[n + 1] + \delta[n] - \delta[n - 1] - \delta(n - 2))$. Compare it with the Fourier transform found using the table in Appendix E.

**37.** Using Parseval's theorem, find the signal energy of

$$x[n] = \text{sinc}\left(\frac{n}{10}\right) \sin\left(\frac{2\pi n}{4}\right).$$

**Answer:**

5

**38.** Sketch the magnitude and phase of the CTFT of

$$x_1(t) = \text{rect}(t)$$

and of the CTFS harmonic function of

$$x_2(t) = \text{rect}(t) * \frac{1}{8} \text{comb}\left(\frac{t}{8}\right).$$

For comparison purposes, sketch $X_1(f)$ versus $f$ and $T_0 X_2[k]$ versus $k f_0$ on the same set of axes. [$T_0$ is the fundamental period of $x_2(t)$ and $T_0 = 1/f_0$.]

**Answer:**

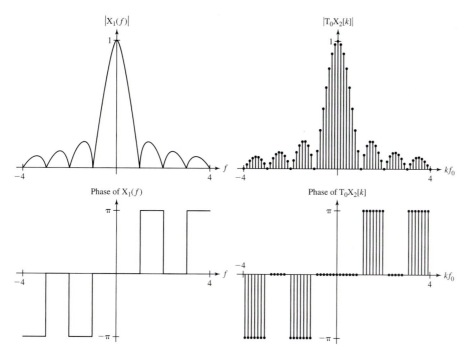

**39.** Sketch the magnitude and phase of the CTFT of

$$x_1(t) = 4\cos(4\pi t)$$

and of the DTFT of

$$x_2[n] = x_1(nT_s)$$

where $T_s = \frac{1}{16}$. For comparison purposes sketch $X_1(f)$ and $T_s X_2(T_s f)$ versus $f$ on the same set of axes.

**Answer:**

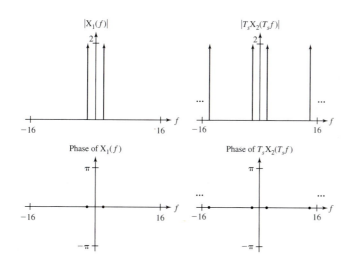

**40.** Sketch the magnitude and phase of the DTFT of

$$x_1[n] = \frac{\text{sinc}(n/16)}{4}$$

and of the DTFS harmonic function of

$$x_2[n] = \frac{\text{sinc}(n/16)}{4} * \text{comb}_{32}[n].$$

For comparison purposes sketch $X_1(F)$ versus $F$ and $N_0 X_2[k]$ versus $kF_0$ on the same set of axes.

**Answers:**

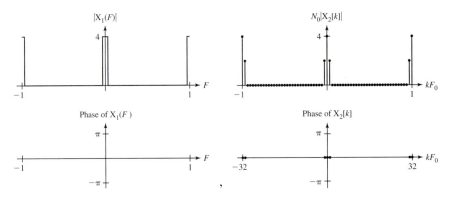

,

# EXERCISES WITHOUT ANSWERS

**41.** A system is excited by a signal,

$$x(t) = 4\,\text{rect}\left(\frac{t}{2}\right)$$

and its response is

$$y(t) = 10\left[\left(1 - e^{-(t+1)}\right)u(t+1) - \left(1 - e^{-(t-1)}\right)u(t-1)\right].$$

What is its impulse response?

**42.** Sketch the magnitudes and phases of the CTFTs of the following functions.

a. $g(t) = 5\delta(4t)$

b. $g(t) = \text{comb}\left(\frac{t+1}{4}\right) - \text{comb}\left(\frac{t-3}{4}\right)$

c. $g(t) = u(2t) + u(t-1)$

d. $g(t) = \text{sgn}(t) - \text{sgn}(-t)$

e. $g(t) = \text{rect}\left(\frac{t+1}{2}\right) + \text{rect}\left(\frac{t-1}{2}\right)$

f. $g(t) = \text{rect}\left(\frac{t}{4}\right)$

g.   $g(t) = 5 \, \mathrm{tri} \left( \dfrac{t}{5} \right) - 2 \, \mathrm{tri} \left( \dfrac{t}{2} \right)$

h.   $g(t) = \dfrac{3}{2} \, \mathrm{rect} \left( \dfrac{t}{8} \right) * \mathrm{rect} \left( \dfrac{t}{2} \right)$

**43.**   Sketch the magnitudes and phases of the CTFTs of the following functions.

   a.   $\mathrm{rect}(4t)$
   b.   $\mathrm{rect}(4t) * 4\delta(t)$
   c.   $\mathrm{rect}(4t) * 4\delta(t-2)$
   d.   $\mathrm{rect}(4t) * 4\delta(2t)$
   e.   $\mathrm{rect}(4t) * \mathrm{comb}(t)$
   f.   $\mathrm{rect}(4t) * \mathrm{comb}(t-1)$
   g.   $\mathrm{rect}(4t) * \mathrm{comb}(2t)$
   h.   $\mathrm{rect}(t) * \mathrm{comb}(2t)$

**44.**   Plot these signals over two fundamental periods centered at $t = 0$.

   a.   $x(t) = 2\cos(20\pi t) + 4\sin(10\pi t) + 3\cos(-20\pi t) - 3\sin(-10\pi t)$
   b.   $x(t) = 5\cos(20\pi t) + 7\sin(10\pi t)$

   Compare the results of parts (a) and (b).

**45.**   A periodic signal has a fundamental period of 4 s.

   a.   What is the lowest positive frequency at which its CTFT could be nonzero?
   b.   What is the next-lowest positive frequency at which its CTFT could be nonzero?

**46.**   Sketch the magnitude and phase of the CTFT of each of the following signals ($\omega$ form):

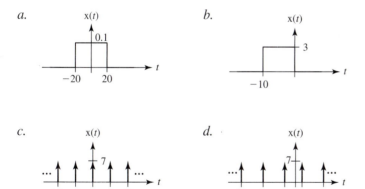

**47.**   Sketch the inverse CTFTs of the following functions:

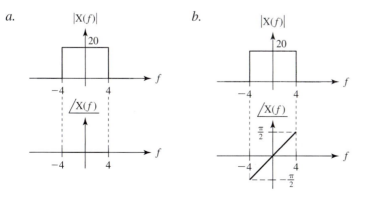

c.    d.

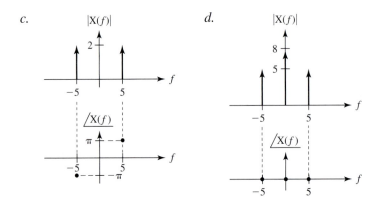

48. Find the inverse CTFT of this real frequency-domain function (Figure E48) and sketch it. (Let $A = 1$, $f_1 = 95$ kHz, and $f_2 = 105$ kHz.)

49. Find the CTFT (either form) of this signal (Figure E49) and sketch its magnitude and phase versus frequency on separate graphs. (Let $A = -B = 1$, $t_1 = 1$, and $t_2 = 2$.) *Hint:* Express this signal as the sum of two functions and use the linearity property.

50. In many communication systems a device called a *mixer* is used. In its simplest form a mixer is simply an analog multiplier. That is, its response signal $y(t)$ is the product of its two excitation signals. If the two excitation signals are

$$x_1(t) = 10 \operatorname{sinc}(20t) \qquad \text{and} \qquad x_2(t) = 5 \cos(2000\pi t),$$

plot the magnitude of the CTFT of $y(t)$, $Y(f)$, and compare it to the magnitude of the CTFT of $x_1(t)$. In simple terms what does a mixer do?

51. Sketch a graph of the convolution of the two functions in each case.

a.    $\operatorname{rect}(t) * \operatorname{rect}(t)$

b.    $\operatorname{rect}\left(t - \dfrac{1}{2}\right) * \operatorname{rect}\left(t + \dfrac{1}{2}\right)$

c.    $\operatorname{tri}(t) * \operatorname{tri}(t - 1)$

d.    $3\delta(t) * 10 \cos(t)$

e.    $10 \operatorname{comb}(t) * \operatorname{rect}(t)$

f.    $5 \operatorname{comb}(t) * \operatorname{tri}(t)$

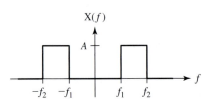

**Figure E48**
A real frequency-domain function.

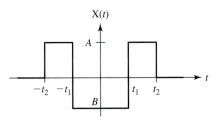

**Figure E49**
A CT function.

**52.** In electronics, one of the first circuits studied is the rectifier. There are two forms, the half-wave rectifier and the full-wave rectifier. The half-wave rectifier cuts off half of an input voltage sinusoidal signal and leaves the other half intact. The full-wave rectifier reverses the polarity of half of the input voltage sinusoidal signal and leaves the other half intact. Let the input voltage sinusoid be a typical household voltage, 120 V rms at 60 Hz, and let both types of rectifiers alter the negative half of the sinusoid while leaving the positive half unchanged. Find and plot the magnitudes of the CTFTs of the output voltage signals of both types of rectifiers (either form).

**53.** Find the DTFT of each of these signals:

*a.* $x[n] = \left(\dfrac{1}{3}\right)^n u[n-1]$

*b.* $x[n] = \sin\left(\dfrac{\pi}{4}n\right)\left(\dfrac{1}{4}\right)^n u[n-2]$

*c.* $x[n] = \operatorname{sinc}\left(\dfrac{2\pi n}{8}\right) * \operatorname{sinc}\left(\dfrac{2\pi(n-4)}{8}\right)$

*d.* $x[n] = \operatorname{sinc}^2\left(\dfrac{2\pi n}{8}\right)$

**54.** Sketch the magnitudes and phases of the DTFTs of the following functions:

*a.* $\operatorname{rect}_2[n]$

*b.* $\operatorname{rect}_2[n] * (-5\delta[n])$

*c.* $\operatorname{rect}_2[n] * 3\delta[n+3]$

*d.* $\operatorname{rect}_2[n] * (-5\delta[4n])$

*e.* $\operatorname{rect}_2[n] * \operatorname{comb}_8[n]$

*f.* $\operatorname{rect}_2[n] * \operatorname{comb}_8[n-3]$

*g.* $\operatorname{rect}_2[n] * \operatorname{comb}_8[2n]$

*h.* $\operatorname{rect}_2[n] * \operatorname{comb}_5[n]$

**55.** Sketch the inverse DTFTs of these functions.

*a.* $X(F) = \operatorname{comb}(F) - \operatorname{comb}\left(F - \dfrac{1}{2}\right)$

*b.* $X(F) = j\operatorname{comb}\left(F + \dfrac{1}{8}\right) - j\operatorname{comb}\left(F - \dfrac{1}{8}\right)$

*c.* $X(F) = \left[\operatorname{sinc}\left(10\left(F - \dfrac{1}{4}\right)\right) + \operatorname{sinc}\left(10\left(F + \dfrac{1}{4}\right)\right)\right] * \operatorname{comb}(F)$

*d.* $X(F) = \left[\delta\left(F - \dfrac{1}{4}\right) + \delta\left(F - \dfrac{3}{16}\right) + \delta\left(F - \dfrac{5}{16}\right)\right] * \operatorname{comb}(2F)$

56.  Using the relationship between the CTFT of a signal and the CTFS of a periodic extension of that signal, find the CTFS of

$$x(t) = \text{rect}\left(\frac{t}{w}\right) * \frac{1}{T_0}\text{comb}\left(\frac{t}{T_0}\right)$$

and compare it with the table entry.

57.  Using the relationship between the DTFT of a signal and the DTFS of a periodic extension of that signal, find the DTFS of

$$\text{rect}_{N_w}[n] * \text{comb}_{N_0}[n]$$

and compare it with the table entry.

# CHAPTER 6

# Fourier Transform Analysis of Signals and Systems

## 6.1 INTRODUCTION AND GOALS

Up to this point in this text the material has been highly mathematical and abstract. We have seen some occasional examples of the use of these signal and system analysis techniques but no really in-depth exploration of their use. We are now at the point at which we have enough analytical tools to attack some important types of signals and systems and demonstrate why Fourier methods are so popular and powerful in the analysis of many systems. Once we have developed a real facility and familiarity with frequency-domain methods, we will understand why many professional engineers spend practically their whole careers "in the frequency domain," creating, designing, and analyzing systems with Fourier methods and other transform methods.

Every LTI system has an impulse response and, through the Fourier transform, also a frequency response. We will analyze systems called *filters* which are designed to have a certain frequency response. We will define the term *ideal filter,* and we will see ways of approximating ideal CT and DT filters. Since frequency response is so important in the analysis of systems, we will develop efficient methods of finding the frequency responses of complicated systems. The last major application examples of Fourier methods are communication systems, which use filters and other frequency-domain techniques.

### CHAPTER GOALS

1. To demonstrate the use of Fourier methods in the analysis of a variety of systems with practical engineering importance such as filters and communication systems

2. To develop an appreciation of the power of signal and system analysis done directly in the frequency domain

## 6.2 FREQUENCY RESPONSE

The real power of the CTFT comes in the generalized analysis of signals and systems in the frequency domain. An LTI system is completely characterized by its impulse response. It is also completely characterized by its transfer function or frequency response, which is the CTFT of its impulse response (Figure 6.1).

As shown in Chapter 5, when two systems are cascaded, the overall system impulse response is the convolution of the two individual impulse responses. Since the frequency-domain counterpart of convolution is multiplication, when two systems are cascaded, the overall transfer function is the product of the two individual transfer functions (Figure 6.2). Since multiplication of complex functions is generally easier than convolution of real functions, signal and system analysis is often more convenient in the frequency domain. The overall system impulse response of parallel-connected systems is the sum of the individual impulse responses. Since the CTFT of a sum of time-domain functions is the sum of the CTFTs of the individual functions, the overall system transfer function of parallel-connected systems is the sum of their transfer functions (Figure 6.3).

So far we have been solving for the response of a known system to a known excitation. It is very common in system analysis *not* to know the exact time-domain behavior of an excitation signal but rather to know its general characteristics in the frequency domain. Figures 6.4 to 6.6 illustrate various kinds of signals and how their signal powers vary with frequency.

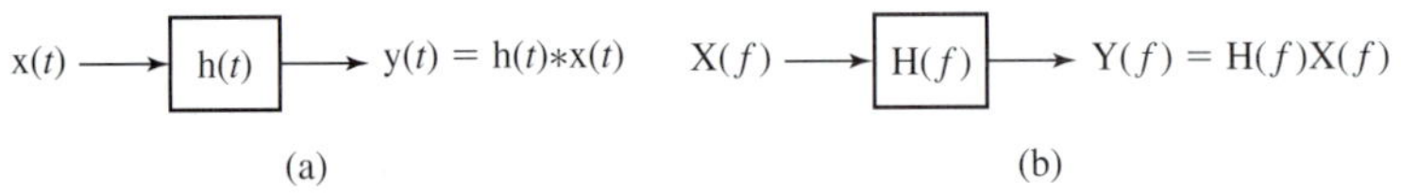

$$x(t) \longrightarrow \boxed{h(t)} \longrightarrow y(t) = h(t)*x(t) \qquad X(f) \longrightarrow \boxed{H(f)} \longrightarrow Y(f) = H(f)X(f)$$

(a)                                         (b)

**Figure 6.1**
(a) Time-domain system block diagram, and (b) frequency-domain system block diagram.

$$X(f) \longrightarrow \boxed{H_1(f)} \longrightarrow X(f)H_1(f) \longrightarrow \boxed{H_2(f)} \longrightarrow Y(f) = X(f)H_1(f)H_2(f)$$

$$X(f) \longrightarrow \boxed{H_1(f)H_2(f)} \longrightarrow Y(f)$$

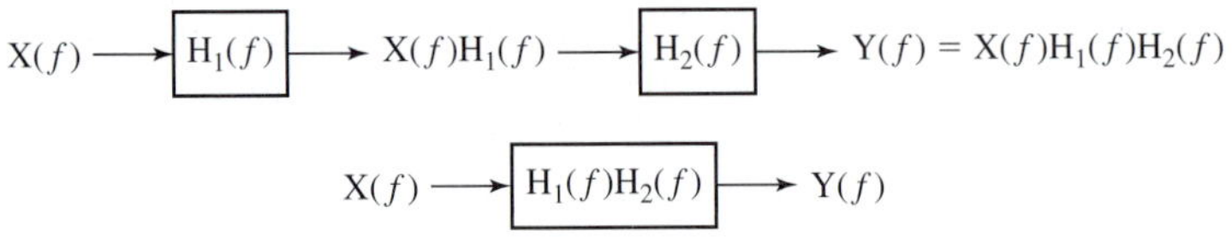

**Figure 6.2**
Cascade connection of systems in the frequency domain.

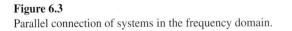

$$Y(f) = X(f)H_1(f) + X(f)H_2(f) = X(f)[H_1(f) + H_2(f)]$$

$$X(f) \longrightarrow \boxed{H_1(f) + H_2(f)} \longrightarrow Y(f)$$

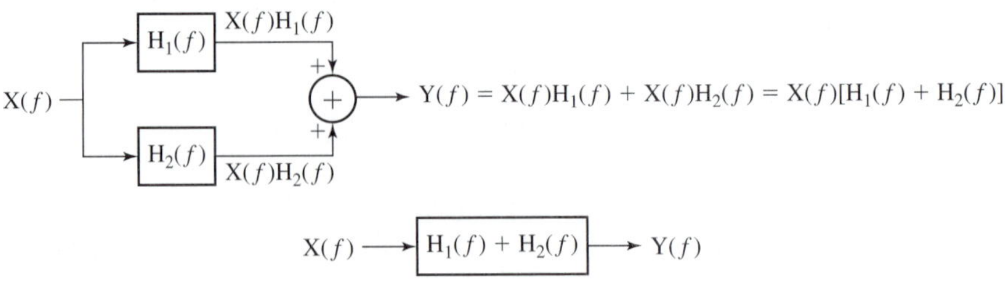

**Figure 6.3**
Parallel connection of systems in the frequency domain.

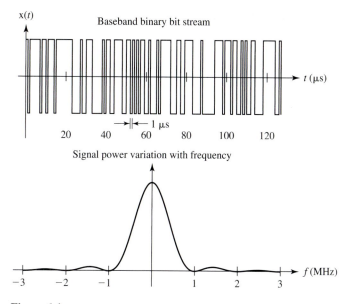

**Figure 6.4**
A baseband binary bit stream and its signal power variation with frequency.

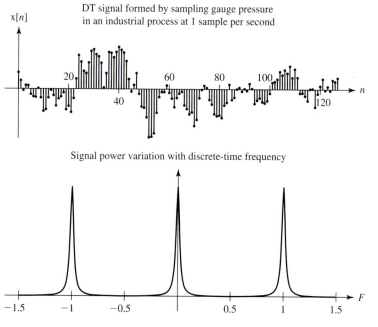

**Figure 6.5**
A DT signal and its signal power variation with DT frequency.

If the signal to be processed is a radio program source like an announcer or music, either the excitation is not known (because it is a live broadcast) or it is known (if the broadcast is a recorded message or music), but its mathematical description would be so complicated that the analysis would be practically impossible. But even though we cannot describe the excitation signal exactly, we do know something about it. We

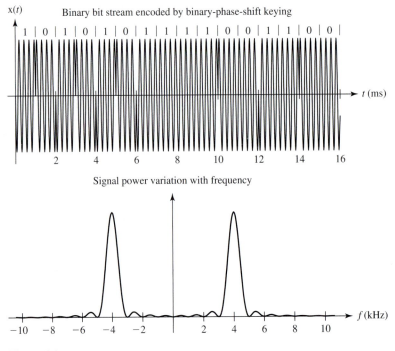

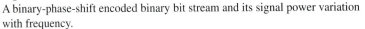

**Figure 6.6**
A binary-phase-shift encoded binary bit stream and its signal power variation
with frequency.

know that people's voices do not create significant signal power outside the range of
30 to 300 Hz and musical instruments do not create significant signal power at frequencies outside the range of about 15 Hz to 20 kHz. If we listened for a while and
measured the signal power, we could probably describe how much signal power we
might expect on average in various frequency ranges.

Another example of an unknown excitation would be a stream of binary data. The
bits come in a sequence that is unknown to the receiver of the data stream and, therefore, might as well be random. An exact description of the signal is unavailable. But
the receiver is usually designed with some knowledge of the signal characteristics, typically the time occupied by one bit and the method used to encode the bits for transmission. With this knowledge it is possible to make very good estimates of how the
power of the signal varies, on average, with frequency. Knowing that, we can design
an appropriate signal processor for that type of signal.

Another example would be an instrumentation system measuring pressure, temperature, flow, etc., in an industrial process. We do not know exactly how these process
parameters vary. But they normally lie within some known range and can vary no
faster than some maximum rate because of the physical limitations of the process.
Again, this knowledge allows us to design a signal-processing system appropriate for
these types of signals.

As an exact analysis problem, analyzing these signals as they are modified by LTI
systems is an impossible task. But these are real engineering problems. We usually design systems to process a certain type of signal, not an exact known signal. We need only
to know enough about the signal to design the system to process it to achieve the desired

goal. Signals like this are usually treated as though they are random. The following questions arise: How do we analyze the response of a system to a signal that is random? How do we design systems to process signals that are random?

Even though a signal may be random, we usually know something about it. We often know its approximate *power spectrum*. We have an approximate description of the signal power of the excitation signal in the frequency domain. It is natural at this point to wonder how we could know the spectrum of a signal since we do not have a mathematical description of it. We could measure it. There are many ways one could, in principle, measure the power spectrum of a signal. One way would be through the use of filters.

## 6.3 IDEAL FILTERS

We have already analyzed a circuit called a lowpass filter and shown why it has that name. In general a filter is a device for separating something desirable from something undesirable. A coffee filter separates the desirable coffee from the undesirable coffee grounds. In signal and system analysis, a filter separates the desirable part of a signal from the undesirable part. Just what is desirable and undesirable depends on what we are trying to accomplish with the signals and systems. The desirable part of a signal could be the part that occurs at a certain time or times, and the parts that occur at other times would then be undesirable. With that interpretation a filter would separate parts of the signal occurring at certain times from parts occurring at other times. A filter could also be defined as a device to separate signal values above and below a certain level or between and outside certain level ranges. But a filter is conventionally defined in signal and system analysis as a device which separates the power of a signal in one frequency range from the power in another frequency range. Devices which do the other functions mentioned are conventionally given other names.

### DISTORTION

The term *lowpass filter* means a device that *passes* signal power at low frequencies and *stops* signal power at high frequencies. An *ideal* lowpass filter would pass *all* signal power at frequencies below some maximum, without distorting the signal at all in that range, and *completely* stop or block all signal power at frequencies above that maximum. It is important here to precisely define what is meant by *distortion*. Distortion is commonly construed in signal and system analysis to mean that the shape of a signal has been changed. This does not mean that if we change the signal we necessarily distort it. Multiplication of the signal by a *gain* constant or a time shift of the signal are changes that are not considered to be distortion.

Suppose a CT signal $x(t)$ and a DT signal $x[n]$ have the shapes illustrated at the top of Figure 6.7. Then the signals at the bottom of Figure 6.7 are undistorted versions of those signals. Figure 6.8 illustrates two types of distortion.

The response of a filter (and of any LTI system) is the convolution of its excitation with its impulse response. Any signal convolved with a unit impulse at the origin is unchanged, $x(t) * \delta(t) = x(t)$. If the impulse has a strength other than one, the signal is multiplied by that strength but the shape is still unchanged, $x(t) * A\delta(t) = Ax(t)$. If the impulse is shifted from the origin, the convolution is shifted also, but without changing the shape, $x(t) * A\delta(t - t_0) = Ax(t - t_0)$. Therefore, the impulse response of a filter which does not distort would be an impulse, possibly with a strength other

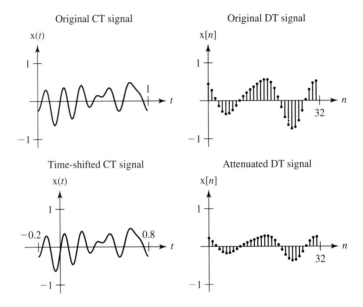

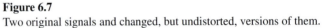

**Figure 6.7**
Two original signals and changed, but undistorted, versions of them.

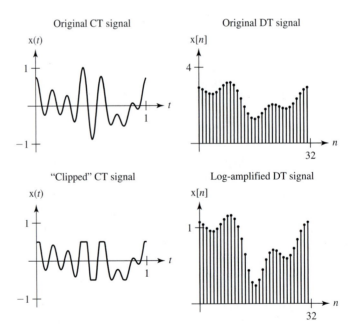

**Figure 6.8**
Two original signals and distorted versions of them.

than one and possibly shifted in time. The most general form of an impulse response of a distortionless system would be

$$h(t) = A\delta(t - t_0) \tag{6.1}$$

for CT systems or

$$h[n] = A\delta[n - n_0] \tag{6.2}$$

for DT systems. The corresponding transfer function would be the Fourier transform of the impulse response,

$$H(f) = Ae^{-j2\pi f t_0} \tag{6.3}$$

or

$$H(F) = Ae^{-j2\pi F n_0}. \tag{6.4}$$

The transfer function can be characterized by its magnitude and phase,

$$|H(f)| = A \qquad \text{or} \qquad |H(F)| = A \tag{6.5}$$

and

$$\angle H(f) = -2\pi f t_0 \qquad \text{or} \qquad \angle H(F) = -2\pi F n_0. \tag{6.6}$$

Therefore, a distortionless system has a transfer function magnitude which is constant with frequency and a transfer function phase which is linear with frequency (Figure 6.9).

The variation of the magnitude and phase of the transfer function of a system plotted versus frequency $f$ or $F$ is called the *frequency response* of the system. The magnitude of the frequency response of a distortionless system is flat (not a function of frequency), and the phase frequency response is linear. In the case of CT systems the phase is linear over the range $-\infty < f < \infty$, and in the case of DT systems, the phase is linear over the range $-\frac{1}{2} < F < \frac{1}{2}$ and repeats periodically outside that range. Since $n_0$ is an integer, the phase of a DT distortionless filter $-2\pi F n_0$ is guaranteed to repeat every time $F$ changes by one.

It should be noted here that a distortionless impulse response or transfer function is a concept that cannot actually be realized in any real physical CT system. No real CT system can possibly have a frequency response that is constant all the way to an infinite frequency. Therefore, the frequency responses of all real physical CT systems must approach zero as frequency approaches infinity.

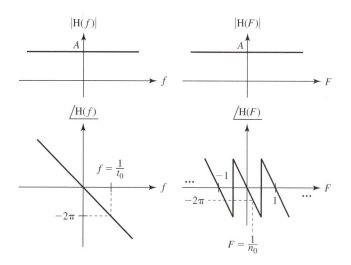

**Figure 6.9**
Magnitude and phase of a distortionless system.

## FILTER CLASSIFICATIONS

For any filter, a range of frequencies over which a filter passes signal power is called a *passband* and a range of frequencies over which a filter blocks signal power is called a *stopband*. Since the purpose of a filter is to remove the undesirable part of a signal and leave the rest, no filter, not even an ideal one, is distortionless because its magnitude is not constant with frequency. But an ideal filter *is* distortionless within its passband. That is, its transfer function magnitude is constant within the passband and its transfer function phase is linear within the passband.

There are four commonly defined types of filters: lowpass, highpass, bandpass, and bandstop. For CT filters,

1. A lowpass filter passes signal power in a range of frequencies $0 < |f| < f_m$ and stops signal power at all other frequencies.
2. A highpass filter stops signal power in a range of frequencies $0 < |f| < f_m$ and passes signal power at all other frequencies.
3. A bandpass filter passes signal power in a range of frequencies $0 < f_1 < |f| < f_2 < \infty$ and stops signal power at all other frequencies.
4. A bandstop filter stops signal power in a range of frequencies $0 < f_1 < |f| < f_2 < \infty$ and passes signal power at all other frequencies.

The descriptions of ideal DT filters are similar in concept but have to be modified slightly because of the fact that all DT systems have periodic transfer functions. For DT filters, in the DT-frequency range $-\frac{1}{2} < F < \frac{1}{2}$,

1. A lowpass filter passes signal power in a range of frequencies $0 < |F| < F_m < \frac{1}{2}$ and stops signal power at all other frequencies.
2. A highpass filter stops signal power in a range of frequencies $0 < |F| < F_m < \frac{1}{2}$ and passes signal power at all other frequencies.
3. A bandpass filter passes signal power in a range of frequencies $0 < F_1 < |F| < F_2 < \frac{1}{2}$ and stops signal power at all other frequencies.
4. A bandstop filter stops signal power in a range of frequencies $0 < F_1 < |F| < F_2 < \frac{1}{2}$ and passes signal power at all other frequencies.

## IDEAL-FILTER FREQUENCY RESPONSES

In Figures 6.10 and 6.11 are the magnitude and phase frequency responses of the four basic types of *ideal* filters. (Notice that the phases of these filters are not indicated in the regions where the magnitudes are zero. The phase is the inverse tangent of the ratio of the imaginary part of the transfer function to the real part. Since both parts are zero, their ratio 0/0 is undefined and so is the phase of the transfer function. It is a common practice in some signal analysis literature to indicate a phase of zero when the magnitude is zero, even though, precisely speaking, it is undefined.)

## BANDWIDTH

It is appropriate here to define a word that is very commonly used in signal analysis, *bandwidth*. The term bandwidth is applied to both signals and filters. It generally means a range of frequencies. This could be the range of frequencies present in a signal or the range of frequencies a filter allows to pass. For historical reasons, it is usually construed to mean a range of frequencies in positive frequency space. For example, an

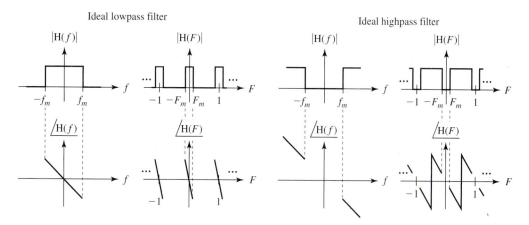

**Figure 6.10**
Magnitude and phase frequency responses of ideal lowpass and highpass filters.

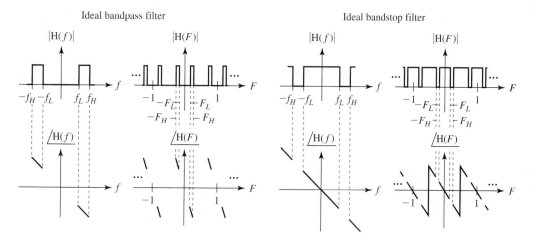

**Figure 6.11**
Magnitude and phase frequency responses of ideal bandpass and bandstop filters.

ideal lowpass filter with corner frequencies of $\pm f_m$, as illustrated in Figure 6.10, is said to have a bandwidth of $f_m$, even though the width of the filter as you look at the plot of the magnitude response is obviously $2 f_m$. The ideal bandpass filter has a bandwidth of $f_H - f_L$ which is the width of the region in positive frequency in which the filter passes a signal.

There are many different kinds of bandwidths, including absolute bandwidth, half-power bandwidth, and null bandwidth (Figure 6.12). Each of them is a range of frequencies but is defined in a different way. For example, if a signal has no signal power at all below some minimum positive frequency and above some maximum positive frequency, its absolute bandwidth is simply the difference between those two frequencies. If a signal has a finite absolute bandwidth, it is said to be strictly band-limited or, more commonly, just band-limited. Most real signals are not band-limited. That is why other definitions of bandwidth are needed.

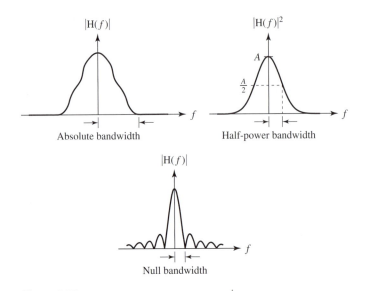

**Figure 6.12**
Examples of bandwidth definitions.

## IMPULSE RESPONSES AND CAUSALITY

Since ideal filters do not pass all frequencies, their impulse responses are not impulses. They are the inverse transforms of the filter transfer functions. The ideal lowpass filter has a transfer function which is mathematically described by a rectangle function for CT systems or a periodically repeated rectangle function for DT systems,

$$\mathrm{H}(f) = A \, \mathrm{rect}\left(\frac{f}{2f_m}\right) e^{-j2\pi f t_0} \tag{6.7}$$

or

$$\mathrm{H}(F) = A \, \mathrm{rect}\left(\frac{F}{2F_m}\right) e^{-j2\pi F n_0} * \mathrm{comb}(F). \tag{6.8}$$

The corresponding impulse responses are the CT and DT sinc functions,

$$h(t) = 2A f_m \, \mathrm{sinc}(2f_m(t - t_0)) \tag{6.9}$$

or

$$h[n] = 2A F_m \, \mathrm{sinc}(2F_m(n - n_0)). \tag{6.10}$$

These descriptions are general in the sense that they involve an arbitrary gain constant $A$ and an arbitrary time delay $t_0$ or $n_0$.

The ideal highpass filter performs an operation which is exactly the opposite of the ideal lowpass filter. Therefore, its transfer function is a constant minus a rectangle or a periodically repeated rectangle,

$$\mathrm{H}(f) = A\left[1 - \mathrm{rect}\left(\frac{f}{2f_m}\right)\right] e^{-j2\pi f t_0} \tag{6.11}$$

or

$$\mathrm{H}(F) = A e^{-j2\pi F n_0}\left[1 - \mathrm{rect}\left(\frac{F}{2F_m}\right) * \mathrm{comb}(F)\right]. \tag{6.12}$$

The corresponding impulse responses are each a CT or DT impulse minus a CT or DT sinc function,

$$h(t) = A\delta(t - t_0) - 2Af_m \operatorname{sinc}(2f_m(t - t_0)) \tag{6.13}$$

or

$$h[n] = A\delta[n - n_0] - 2AF_m \operatorname{sinc}(2F_m(n - n_0)). \tag{6.14}$$

Notice that the ideal CT highpass filter has a frequency response extending all the way to infinity. This is impossible in any real physical system. Therefore, practical approximations to the ideal CT highpass filter block low-frequency signals and allow higher-frequency signals to pass but only up to some very high, not infinite, frequency. *Very high* is a relative term and, as a practical matter, usually means beyond the frequencies of any signals actually expected to occur in the system.

The ideal bandpass filter has a transfer function that can be conveniently described in two equivalent ways. One description is the difference between two unshifted rectangle functions or two periodically repeated unshifted rectangle functions,

$$H(f) = A\left[\operatorname{rect}\left(\frac{f}{2f_H}\right) - \operatorname{rect}\left(\frac{f}{2f_L}\right)\right]e^{-j2\pi f t_0} \tag{6.15}$$

or

$$H(F) = A\left[\operatorname{rect}\left(\frac{F}{2F_H}\right) - \operatorname{rect}\left(\frac{F}{2F_L}\right)\right]e^{-j2\pi F n_0} * \operatorname{comb}(F), \tag{6.16}$$

where $f_L$ or $F_L$ and $f_H$ or $F_H$ are the low- and high-frequency corners, respectively. The other description is the sum of two shifted rectangle functions or the periodic repetition of two shifted rectangle functions,

$$H(f) = A\left[\operatorname{rect}\left(\frac{f - f_0}{\Delta f}\right) + \operatorname{rect}\left(\frac{f + f_0}{\Delta f}\right)\right]e^{-j2\pi f t_0} \tag{6.17}$$

or

$$H(F) = A\left[\operatorname{rect}\left(\frac{F - F_0}{\Delta F}\right) + \operatorname{rect}\left(\frac{F + F_0}{\Delta F}\right)\right]e^{-j2\pi F n_0} * \operatorname{comb}(F), \tag{6.18}$$

where $\Delta f = f_H - f_L$

$$f_0 = \frac{f_H + f_L}{2}$$

$$\Delta F = F_H - F_L$$

$$F_0 = \frac{F_H + F_L}{2}$$

The impulse response of the ideal bandpass filter is the inverse transform of the transfer function and therefore can also be described in two alternate, but equivalent, ways,

$$h(t) = 2Af_H \operatorname{sinc}(2f_H(t - t_0)) - 2Af_L \operatorname{sinc}(2f_L(t - t_0)) \tag{6.19}$$

or

$$h[n] = 2AF_H \operatorname{sinc}(2F_H(n - n_0)) - 2AF_L \operatorname{sinc}(2F_L(n - n_0)) \tag{6.20}$$

or

$$h(t) = 2A \, \Delta f \, \text{sinc}(\Delta f(t - t_0)) \cos(2\pi f_0(t - t_0)) \tag{6.21}$$

or

$$h[n] = 2A \, \Delta F \, \text{sinc}(\Delta F(n - n_0)) \cos(2\pi F_0(n - n_0)). \tag{6.22}$$

The ideal bandstop filter, being the opposite of the ideal bandpass filter, also has a transfer function that can be conveniently described in two equivalent ways. Each of the two alternate description forms is a constant minus the corresponding bandpass form. The first form is

$$H(f) = A \left[ 1 - \text{rect}\left(\frac{f}{2f_H}\right) + \text{rect}\left(\frac{f}{2f_L}\right) \right] e^{-j2\pi f t_0} \tag{6.23}$$

or

$$H(F) = Ae^{-j2\pi F n_0} \left\{ 1 - \left[ \text{rect}\left(\frac{F}{2F_H}\right) - \text{rect}\left(\frac{F}{2F_L}\right) \right] * \text{comb}(F) \right\} \tag{6.24}$$

where $f_L$ or $F_L$ and $f_H$ or $F_H$ are the low- and high-frequency corners, respectively, and the second form is

$$H(f) = A \left[ 1 - \text{rect}\left(\frac{f - f_0}{\Delta f}\right) - \text{rect}\left(\frac{f + f_0}{\Delta f}\right) \right] e^{-j2\pi f t_0} \tag{6.25}$$

or

$$H(F) = Ae^{-j2\pi F n_0} \left\{ 1 - \left[ \text{rect}\left(\frac{F - F_0}{\Delta F}\right) + \text{rect}\left(\frac{F + F_0}{\Delta F}\right) \right] * \text{comb}(F) \right\} \tag{6.26}$$

where $\Delta f = f_H - f_L$

$$f_0 = \frac{f_H + f_L}{2}$$

$$\Delta F = F_H - F_L$$

$$F_0 = \frac{F_H + F_L}{2}$$

and

$$h(t) = A\delta(t - t_0) - 2Af_H \, \text{sinc}(2f_H(t - t_0)) + 2Af_L \, \text{sinc}(2f_L(t - t_0)) \tag{6.27}$$

or

$$h[n] = A\delta[n - n_0] - 2AF_H \, \text{sinc}(2F_H(n - n_0)) + 2AF_L \, \text{sinc}(2F_L(n - n_0)) \tag{6.28}$$

or

$$h(t) = A\delta(t - t_0) - 2A \, \Delta f \, \text{sinc}(\Delta f(t - t_0)) \cos(2\pi f_0(t - t_0)) \tag{6.29}$$

or

$$h[n] = A\delta[n - n_0] - 2A \, \Delta F \, \text{sinc}(\Delta F(n - n_0)) \cos(2\pi F_0(n - n_0)). \tag{6.30}$$

As was true for the CT highpass filter, the ideal CT bandstop filter has a frequency response that extends all the way to infinity. For the same reason, no real physical system can have that frequency response. Its frequency response must approach zero as the frequency approaches infinity.

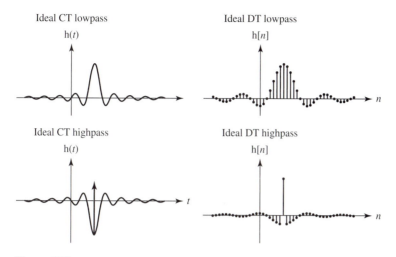

**Figure 6.13**
Typical impulse responses of ideal lowpass and highpass filters.

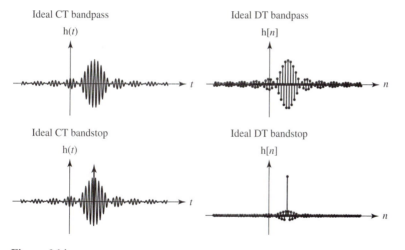

**Figure 6.14**
Typical impulse responses of ideal bandpass and bandstop filters.

In Figures 6.13 and 6.14 are some typical shapes of impulse responses for the four basic types of ideal filters. The ideal lowpass filter makes a transition from allowing signals through at frequencies below its cutoff frequency with no distortion at all and completely blocking frequencies above its cutoff frequency. This transition occurs in an infinitesimal range of frequencies around the cutoff frequency. One might wonder what happens to a signal at *exactly* the cutoff frequency. From a practical point of view this question is not significant because, as we will soon see, the ideal filter cannot actually be built. No real filter can have vertical sides in its frequency response, but it is interesting from a theoretical point of view to see what would happen if we could build it. We can analyze this situation by convolving the impulse response of an ideal lowpass filter with a cosine and see what happens as we change the frequency of the cosine.

The response of a unity-gain, zero-phase-shift, ideal lowpass filter to a unit cosine excitation is the convolution of its impulse response with that cosine,

$$y(t) = 2 f_m \, \text{sinc}(2 f_m t) * \cos(2 \pi f_0 t). \tag{6.31}$$

Using the integral definition of convolution,

$$y(t) = 2 f_m \int_{-\infty}^{\infty} \text{sinc}(2 f_m \tau) \cos(2 \pi f_0 (t - \tau)) \, d\tau \tag{6.32}$$

or, using the definition of the sinc function,

$$y(t) = \int_{-\infty}^{\infty} \frac{\sin(2 \pi f_m \tau)}{\pi \tau} \cos(2 \pi f_0 (t - \tau)) \, d\tau. \tag{6.33}$$

We can use a trigonometric identity for the cosine of a difference angle to write

$$y(t) = \int_{-\infty}^{\infty} \frac{\sin(2 \pi f_m \tau)}{\pi \tau} [\cos(2 \pi f_0 t) \cos(2 \pi f_0 \tau) + \sin(2 \pi f_0 t) \sin(2 \pi f_0 \tau)] \, d\tau. \tag{6.34}$$

The integral of this sum is a sum of integrals, and the second integral is zero because it is the integral of an odd function over symmetrical limits. Therefore,

$$y(t) = \cos(2 \pi f_0 t) \int_{-\infty}^{\infty} \frac{\sin(2 \pi f_m \tau)}{\pi \tau} \cos(2 \pi f_0 \tau) \, d\tau. \tag{6.35}$$

Then, using a trigonometric identity for the product of a sine and a cosine,

$$y(t) = \frac{1}{2} \cos(2 \pi f_0 t) \int_{-\infty}^{\infty} \frac{\sin(2 \pi \tau (f_m - f_0)) + \sin(2 \pi \tau (f_m + f_0))}{\pi \tau} \, d\tau. \tag{6.36}$$

Now, to evaluate the integral, consider three cases.

***Case I***    $f_0 < f_m$.

$$y(t) = \frac{1}{2} \cos(2 \pi f_0 t) \left[ 2(f_m - f_0) \int_{-\infty}^{\infty} \frac{\sin(2 \pi \tau (f_m - f_0))}{2 \pi \tau (f_m - f_0)} \, d\tau \right.$$
$$\left. + 2(f_m + f_0) \int_{-\infty}^{\infty} \frac{\sin(2 \pi \tau (f_m + f_0))}{2 \pi \tau (f_m + f_0)} \, d\tau \right] \tag{6.37}$$

or

$$y(t) = \frac{1}{2} \cos(2 \pi f_0 t) \left[ 2(f_m - f_0) \int_{-\infty}^{\infty} \text{sinc}(2 \tau (f_m - f_0)) \, d\tau \right.$$
$$\left. + 2(f_m + f_0) \int_{-\infty}^{\infty} \text{sinc}(2 \tau (f_m + f_0)) \, d\tau \right]. \tag{6.38}$$

We can use the CTFT to find the area under these sinc functions, and the final result is

$$y(t) = \cos(2\pi f_0 t). \tag{6.39}$$

In this case, the excitation and response are identical.

**Case II**   $f_0 = f_m$.   In this case, $\sin(2\pi\tau(f_m - f_0)) = 0$ and, from (6.37),

$$y(t) = \frac{1}{2}\cos(2\pi f_0 t) \times 2(f_m + f_0) \int\limits_{-\infty}^{\infty} \frac{\sin(2\pi\tau(f_m + f_0))}{2\pi\tau(f_m + f_0)} d\tau = \frac{1}{2}\cos(2\pi f_0 t). \tag{6.40}$$

In this case the response is exactly half the excitation.

**Case III**   $f_0 > f_m$.   In this case,

$$2(f_m - f_0) \int\limits_{-\infty}^{\infty} \frac{\sin(2\pi\tau(f_m - f_0))}{2\pi\tau(f_m - f_0)} d\tau = -2(f_m - f_0) \int\limits_{-\infty}^{\infty} \text{sinc}(2\tau(f_m - f_0)) \, d\tau, \tag{6.41}$$

the two integrals in (6.38) exactly cancel, and

$$y(t) = 0. \tag{6.42}$$

In this case the response is identically zero. We can see that, for the ideal lowpass filter, the definition of the rectangle function as having a value of one-half at its discontinuity fits exactly its frequency response.

As mentioned earlier, one reason ideal filters are called ideal is that they cannot physically exist. The reason is not simply that perfect circuit components with ideal characteristics do not exist (although that would be sufficient). It is more fundamental than that. Consider the impulse responses depicted in Figures 6.13 and 6.14. They are the responses of the filters to a unit impulse applied at time $t = 0$ or $n = 0$. That is what *impulse response* means. Notice that all the impulse responses of these ideal filters have a nonzero response before the impulse is applied at time $t = 0$ or $n = 0$. In fact, all these impulse responses begin at an infinite time before zero. It should be intuitively obvious that a real system cannot look into the future and anticipate the application of the excitation and start responding before it occurs. Ideal filters are all noncausal.

As first discussed in Chapter 3, a system whose response begins before the excitation occurs is said to violate the principle of causality and is termed a *noncausal* system. The words *causality* and *causal* come from the principle of cause and effect in which, for real systems, there can be no effect until after its cause has occurred. All real physical systems are causal. That is, their impulse responses occur only when or after the excitation is applied. In Figures 6.15 and 6.16 are some examples of the impulse responses and magnitude frequency responses of some nonideal, causal filters of the four common filter types.

The term causal is also commonly applied to signals. A causal signal is one which is zero for all time $t < 0$ or $n < 0$. Therefore, a causal signal is one which could be the impulse response of a causal system. An *anticausal* signal is one which is zero for all time $t > 0$ or $n > 0$.

Some effects of a filter can be illustrated by exciting it with a standard signal and observing the response. The standard signal might be a unit step, a square wave, or even a random signal. Another way of seeing the effects of a filter is to excite it with a

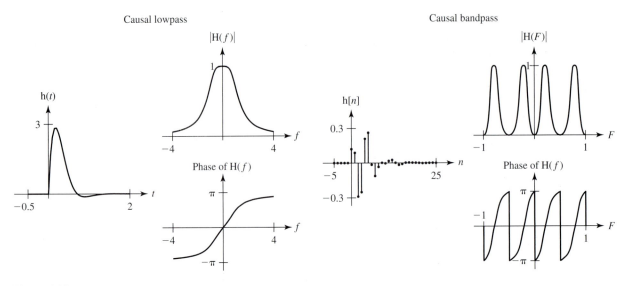

**Figure 6.15**
Impulse responses and frequency responses of causal lowpass and bandpass filters.

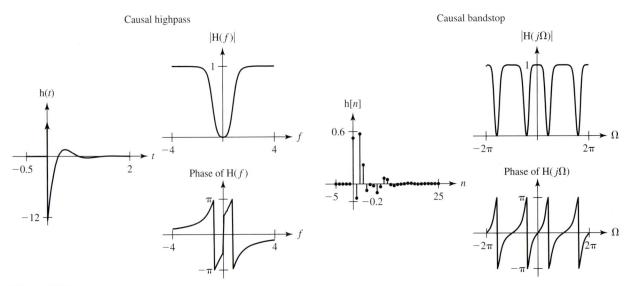

**Figure 6.16**
Impulse responses and frequency responses of causal highpass and bandstop filters.

sequence of sinusoids of different frequencies and observe the amplitudes and phases of the responses. In Figures 6.17 and 6.18 are some examples of the responses of some causal filters to some of these types of excitations.

One interesting way to demonstrate what filters do is to filter an image. An *image* is a two-dimensional signal. Images can be acquired in various ways. A film camera exposes light-sensitive film to a scene through a lens system which puts an optical image of the scene on the film. The photograph could be a color photograph or a black-and-white (monochrome) photograph. This discussion will be confined to monochrome images. A digital camera acquires an image by imaging the scene on a rectangular

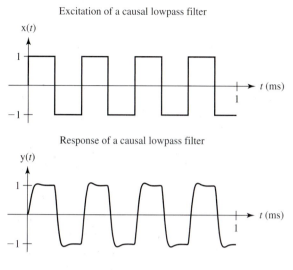

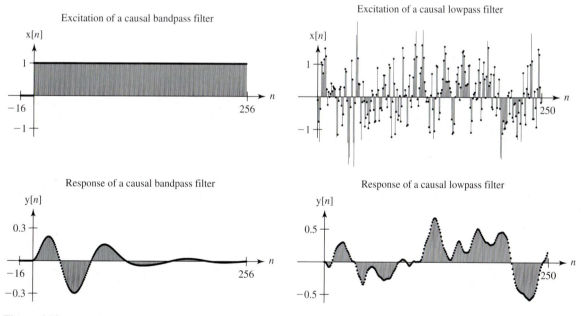

**Figure 6.17**
Excitations and responses of lowpass and highpass CT filters.

**Figure 6.18**
Excitations and responses of bandpass and lowpass DT filters.

array of detectors which convert light energy to electric charge. Each detector sees a very tiny part of the image called a *pixel* (short for *picture element*). The image acquired by the digital camera then consists of an array of numbers, one for each pixel indicating the light intensity at that point (again assuming a monochrome image).

A photograph is a continuous-space function of two spatial coordinates conventionally called *x* and *y*. An acquired digital image is a discrete-space function of two discrete-space coordinates $n_x$ and $n_y$. In principle a photograph could be directly filtered. In fact there are optical techniques which do just that. But by far the most common type of image filtering is done digitally meaning that the image is filtered by a computer using numerical methods.

The techniques used to filter images are very similar to the techniques used to filter time signals, except that they are done in two dimensions. Consider the example image in Figure 6.19. One technique for filtering an image is to take one row of pixels as a one-dimensional signal and filter it just like a discrete-time signal. Figure 6.20 is a plot of the brightness of the pixels in the top row of the image versus horizontal discrete-space $n_x$. If the signal were actually a function of discrete time and we were filtering in real time (meaning we would not have future values available during the filtering process), the lowpass-filtered signal might look like Figure 6.21. After lowpass filtering, all the rows in the image would look smeared in the horizontal direction and unaltered in the vertical direction (Figure 6.22). If we had filtered the columns instead of the rows, the effect would have been as illustrated in Figure 6.23.

One nice thing about image filtering is that causality is not relevant to the filtering process. Usually the entire image is acquired and then processed. Following the analogy between time and space, during horizontal filtering past signal values would lie to the left and future values to the right. In real-time filtering of time signals we cannot use future values because we don't yet know what they are. In image filtering we have the entire image before we begin filtering and, therefore, future values are available. If

**Figure 6.19**
A white-cross image.

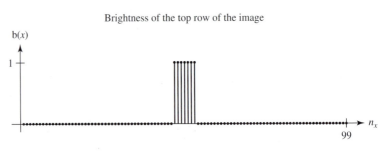

Brightness of the top row of the image

**Figure 6.20**
Brightness of the top row of pixels in a white-cross image.

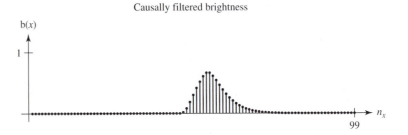

Causally filtered brightness

**Figure 6.21**
Brightness of the top row of pixels after being lowpass-filtered by a causal lowpass filter.

**Figure 6.22**
White-cross image after all rows have
been lowpass-filtered by a causal
lowpass filter.

**Figure 6.23**
White-cross image after all columns
have been lowpass-filtered by a causal
lowpass filter.

Noncausally filtered brightness

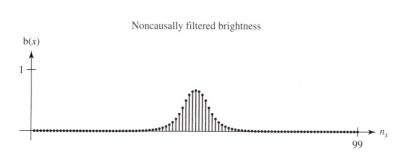

**Figure 6.24**
Brightness of the top row of pixels after being lowpass-filtered by a noncausal
lowpass filter.

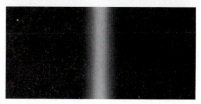

**Figure 6.25**
White-cross image after all rows have
been lowpass-filtered by a noncausal
lowpass filter.

we horizontally filtered the top row of the image with a noncausal lowpass filter, the
effect might look as illustrated in Figure 6.24. If we horizontally lowpass filtered the
entire image with a noncausal lowpass filter, the result would look like Figure 6.25.
The overall effect of this type of filtering can be seen in Figure 6.26 where both the
rows and columns of the image have been filtered by a lowpass filter.

Of course, the filter referred to as noncausal is actually causal because all the
image data are acquired before the filtering process begins. It is only called noncausal
because if a space coordinate were instead time, and we were doing real-time filtering,
the filtering would be noncausal.

## THE POWER SPECTRUM

The whole purpose of launching into the exploration of the idea of a filter was to
explain one way of determining the power spectrum of a signal by measuring it. That

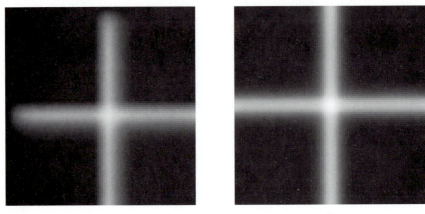

(a)                                            (b)

**Figure 6.26**
White-cross image filtered by a lowpass filter: (a) causal, (b) noncausal.

could be accomplished for CT signals by the system illustrated in Figure 6.27. The excitation signal is routed to multiple bandpass filters, each with the same bandwidth but different center frequencies. Each filter's response is that part of the signal lying in the frequency range of the filter. Then the output signal from each filter is the input signal of a *squarer* and its output signal is the input signal of a *time averager*. A squarer simply takes the square of the signal. This is not a linear operation, so this is not a linear system. The output signal from any squarer is that part of the instantaneous signal power of the original excitation $x(t)$ which lies in the passband of the bandpass filter. Then the time averager simply forms the time-average signal power. Each output response $P_x(f_n)$ is a measure of the signal power of the original excitation $x(t)$ in a narrow band of frequencies centered at $f_n$. Taken together, the P's are an indication of the variation of the signal power with frequency, the power spectrum.

No engineer today would actually build a system like this to measure the power spectrum of a signal. A better way to measure it is to use an instrument called a *spectrum analyzer* which will be introduced in Section 6.10. But this illustration is useful because it reinforces the concept of what a filter does and what the term power spectrum means. In Chapter 8 we will explore a closely related idea, power spectral density.

## NOISE REMOVAL

Every useful signal always has another undesirable signal called *noise* added to it. One very important use of filters is in removing noise from a signal. The sources of noise are many and varied. By careful design, noise can be reduced to a minimum but can never be completely eliminated. As an example of filtering, suppose the signal power is confined to a range of low frequencies and the noise power is spread over a much wider range of frequencies. We can filter the signal plus noise with a lowpass filter and reduce the noise power without having much effect on the signal power (Figure 6.28).

The ratio of the signal power of the desired signal to the signal power of the noise is called the *signal-to-noise ratio* (SNR). Probably the most fundamental consideration in communication system design is to maximize the SNR, and filtering is a very important technique in achieving this.

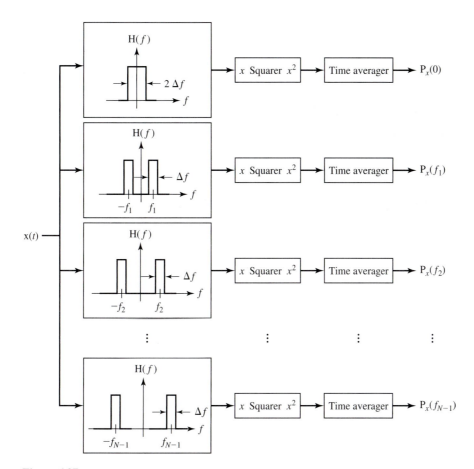

**Figure 6.27**
A system to measure the power spectrum of a signal.

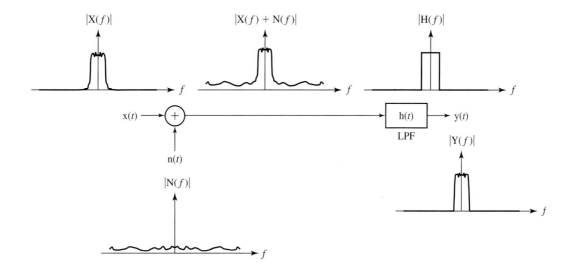

**Figure 6.28**
Partial removal of noise by a lowpass filter.

## 6.4 PRACTICAL PASSIVE FILTERS

### THE *RC* LOWPASS FILTER

Approximations to the ideal lowpass and bandpass filters can be made with certain types of circuits. The simplest approximation to the ideal lowpass filter is the one we have already analyzed more than once, the so-called single-pole *RC* lowpass filter (Figure 6.29). We have found its response to a step and to a sinusoid. Let us now analyze it directly in the frequency domain.

The differential equation describing this circuit is

$$RC v'_{\text{out}}(t) + v_{\text{out}}(t) = v_{\text{in}}(t). \tag{6.43}$$

Fourier-transforming both sides,

$$(j\omega C)R V_{\text{out}}(f) + V_{\text{out}}(f) = V_{\text{in}}(f). \tag{6.44}$$

We can now solve directly for the transfer function,

$$H(j\omega) = \frac{V_{\text{out}}(j\omega)}{V_{\text{in}}(j\omega)} = \frac{1}{(j\omega C)R + 1} \qquad \text{or} \qquad H(f) = \frac{V_{\text{out}}(f)}{V_{\text{in}}(f)} = \frac{1}{(j2\pi f C)R + 1} \tag{6.45}$$

The method commonly used in elementary circuit analysis to solve for the transfer function is based on the phasor and impedance concepts. *Impedance* is a generalization of the idea of resistance to apply to inductors and capacitors. Recall the voltage-current relationships for resistors, capacitors, and inductors (Figure 6.30). If we Fourier-transform these relationships, we get

$$V(j\omega) = R I(j\omega), \qquad V(j\omega) = j\omega L I(j\omega), \qquad \text{and} \qquad I(j\omega) = j\omega C V(j\omega) \tag{6.46}$$

or

$$V(f) = R I(f), \qquad V(f) = j2\pi L I(f), \qquad \text{and} \qquad I(f) = j2\pi f C V(f) \tag{6.47}$$

The impedance concept comes from the similarity of the inductor and capacitor equations to Ohm's law for resistors. If we form the ratios of voltage to current, we get

$$\frac{V(j\omega)}{I(j\omega)} = R, \qquad \frac{V(j\omega)}{I(j\omega)} = j\omega L, \qquad \text{and} \qquad \frac{V(j\omega)}{I(j\omega)} = \frac{1}{j\omega C} \tag{6.48}$$

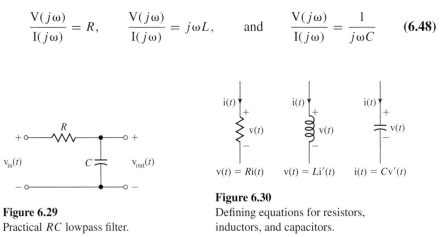

**Figure 6.29**
Practical *RC* lowpass filter.

**Figure 6.30**
Defining equations for resistors, inductors, and capacitors.

or

$$\frac{V(f)}{I(f)} = R, \qquad \frac{V(f)}{I(f)} = j2\pi f L, \qquad \text{and} \qquad \frac{V(f)}{I(f)} = \frac{1}{j2\pi f C}. \qquad \textbf{(6.49)}$$

For resistors this ratio is called *resistance*. In the generalization this ratio is called impedance. Impedance is conventionally symbolized by Z. Using that symbol,

$$Z_R(j\omega) = R, \qquad Z_L(j\omega) = j\omega L, \qquad \text{and} \qquad Z_C(j\omega) = \frac{1}{j\omega C} \qquad \textbf{(6.50)}$$

or

$$Z_R(f) = R, \qquad Z_L(f) = j2\pi f L, \qquad \text{and} \qquad Z_C(f) = \frac{1}{j2\pi f C}. \qquad \textbf{(6.51)}$$

This allows us to apply many of the techniques of resistive circuit analysis to circuits which contain inductors and capacitors and are analyzed in the frequency domain. In the case of the *RC* lowpass filter we can view it as a voltage divider (Figure 6.31). Then we can directly write the transfer function in the frequency domain,

$$H(j\omega) = \frac{V_{\text{out}}(j\omega)}{V_{\text{in}}(j\omega)} = \frac{Z_c(j\omega)}{Z_c(j\omega) + Z_f(j\omega)} = \frac{1/j\omega C}{(1/j\omega C) + R} = \frac{1}{j\omega RC + 1} \qquad \textbf{(6.52)}$$

or

$$H(f) = \frac{1}{j2\pi f RC + 1}, \qquad \textbf{(6.53)}$$

arriving at the same result as before while completely ignoring the time-domain relationships. The magnitude and phase of the *RC* lowpass filter transfer function are illustrated in Figure 6.32.

The impulse response of the *RC* single-pole lowpass filter is the inverse CTFT of its transfer function,

$$h(t) = \frac{e^{-(t/RC)}}{RC} u(t) \qquad \textbf{(6.54)}$$

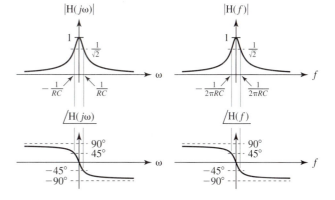

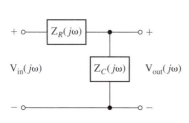

**Figure 6.31**
Impedance voltage divider representation of the *RC* lowpass filter.

**Figure 6.32**
Magnitude and phase frequency responses of an *RC* lowpass filter.

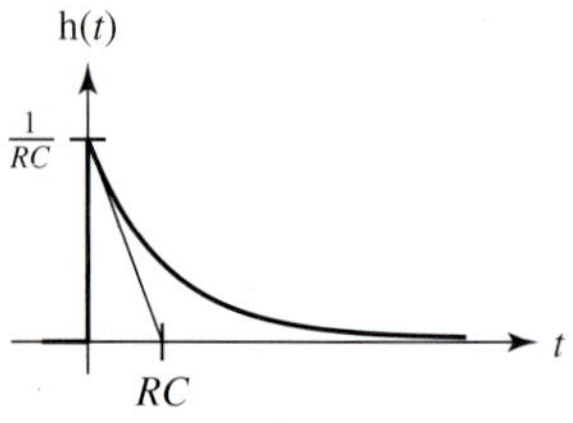

**Figure 6.33**
Impulse response of an $RC$ lowpass filter.

as illustrated in Figure 6.33. For this physically realizable filter the impulse response is zero before time $t = 0$. That is, it is *causal*.

For this circuit the physical operation can be thought of in the frequency domain this way: At very low frequencies (approaching zero) the capacitor's impedance is much greater in magnitude than the resistor's impedance, and, therefore, the voltage division ratio approaches one and the output voltage signal and input voltage signal are about the same. At very high frequencies the capacitor's impedance becomes much smaller in magnitude than the resistor's impedance and the voltage division ratio approaches zero. Thus we can say that low frequencies pass through and high frequencies get stopped. This qualitative analysis of the circuit agrees with the mathematical form of the transfer function,

$$H(j\omega) = \frac{1}{j\omega RC + 1} \quad \text{or} \quad H(f) = \frac{1}{j2\pi f RC + 1}. \tag{6.55}$$

At low frequencies

$$\lim_{\omega \to 0} H(j\omega) = 1 \quad \text{or} \quad \lim_{f \to 0} H(f) = 1, \tag{6.56}$$

and at high frequencies

$$\lim_{\omega \to \infty} H(j\omega) = 0 \quad \text{or} \quad \lim_{f \to \infty} H(f) = 0. \tag{6.57}$$

The $RC$ lowpass filter is lowpass only because the excitation is defined as the voltage at the input and the response is defined as the voltage at the output. If the response had been defined as the current, the nature of the filtering process would change completely. In that case the transfer function would become

$$H(j\omega) = \frac{I(j\omega)}{V_{in}(j\omega)} = \frac{1}{Z_R(j\omega) + Z_c(j\omega)} = \frac{1}{(1/j\omega C) + R} = \frac{j\omega C}{j\omega RC + 1}. \tag{6.58}$$

With this definition of the response, at low frequencies the capacitor impedance is very large, blocking current flow so the response approaches zero. At high frequencies the capacitor impedance approaches zero, so the circuit responds as though it were a short circuit and the current flow is controlled by the resistance $R$. Mathematically the response approaches zero at *low* frequencies and approaches the constant $1/R$ at high frequencies. This defines a *highpass* filter.

$$\lim_{\omega \to 0} H(j\omega) = 0 \quad \text{and} \quad \lim_{\omega \to \infty} H(j\omega) = \frac{1}{R} \tag{6.59}$$

Notice that we are no longer considering any particular response to any particular excitation. The value of the transfer function is that it relates the response to the excitation generally. The transfer function characterizes the system itself not the excitation or response, and most system design is done by knowing the general frequency-domain nature of the expected excitations and desired responses and designing transfer functions to achieve them.

Another (much less common) form of a lowpass filter is illustrated in Figure 6.34.

$$H(j\omega) = \frac{V_{out}(j\omega)}{V_{in}(j\omega)} = \frac{R}{j\omega L + R} \quad \text{or} \quad H(f) = \frac{V_{out}(f)}{V_{in}(f)} = \frac{R}{j2\pi f L + R}. \tag{6.60}$$

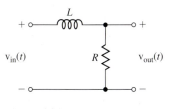

**Figure 6.34**
Alternate form of a practical
lowpass filter.

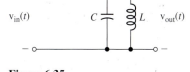

**Figure 6.35**
An $RLC$ practical bandpass filter.

Using the impedance and voltage divider ideas, can you explain in words why this circuit is a lowpass filter?

## THE $RLC$ BANDPASS FILTER

One of the simplest forms of a practical bandpass filter is illustrated in Figure 6.35.

$$H(j\omega) = \frac{V_{out}(j\omega)}{V_{in}(j\omega)} = \frac{j\omega/RC}{(j\omega)^2 + j(\omega/RC) + (1/LC)} \qquad (6.61)$$

or

$$H(f) = \frac{V_{out}(f)}{V_{in}(f)} = \frac{j2\pi f/RC}{(j2\pi f)^2 + j(2\pi f/RC) + (1/LC)}. \qquad (6.62)$$

Even though it may be a little difficult to envision the magnitude of this mathematical expression, consider the following reasoning. At very low frequencies, the capacitor is an open circuit (it might as well not be there) and the inductor is a short circuit (no voltage across it). Therefore, at very low frequencies, the output voltage signal is practically zero. At very high frequencies, the inductor is an open circuit and the capacitor is a short circuit, again making the output voltage signal zero. But at the resonant frequency of the parallel-$LC$ tank circuit, the impedance of that parallel combination of inductor and capacitor goes to infinity and the output voltage signal is the same as the input voltage signal. This frequency is the value of $\omega$ or $f$ at which the real part of the denominator of the transfer function goes to zero,

$$(j\omega)^2 + \frac{1}{LC} = 0 \Rightarrow \omega = \pm\frac{1}{\sqrt{LC}} \quad \text{or} \quad f = \pm\frac{1}{2\pi\sqrt{LC}}. \qquad (6.63)$$

Therefore, the overall behavior of the circuit is to pass frequencies near the resonant frequency and block other frequencies; hence, it is a practical bandpass filter. A plot of the magnitude and phase of the transfer function (Figure 6.36) (for a particular choice of component values) will reveal the bandpass nature of the transfer function.

The impulse response of the $RLC$ bandpass filter is

$$h(t) = \mathcal{F}^{-1}\left(\frac{j\omega/RC}{(j\omega)^2 + j(\omega/RC) + (1/LC)}\right) \qquad (6.64)$$

or

$$h(t) = \frac{1}{RC}\mathcal{F}^{-1}\left(\frac{j\omega}{(j\omega + (1/2RC))^2 + (1/LC) - (1/2RC)^2}\right) \qquad (6.65)$$

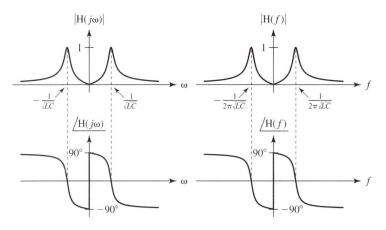

**Figure 6.36**
Magnitude and phase frequency responses of a practical $RLC$ bandpass filter.

or

$$h(t) = \frac{1}{RC} \mathcal{F}^{-1} \left( \frac{j\omega + (1/2RC)}{(j\omega + (1/2RC))^2 + (1/LC) - (1/2RC)^2} \right.$$

$$\left. - \frac{1}{2RC\sqrt{(1/LC) - (1/2RC)^2}} \frac{\sqrt{(1/LC) - (1/2RC)^2}}{(j\omega + (1/2RC))^2 + (1/LC) - (1/2RC)^2} \right).$$

(6.66)

From the Fourier transform tables in Appendix E,

$$e^{-at} \sin(\omega_0 t) \, u(t) \xleftrightarrow{\mathcal{F}} \frac{\omega_0}{(j\omega + a)^2 + \omega_0^2}$$

$$e^{-at} \cos(\omega_0 t) \, u(t) \xleftrightarrow{\mathcal{F}} \frac{j\omega + a}{(j\omega + a)^2 + \omega_0^2}$$

(6.67)

and

$$h(t) = \frac{e^{-(t/2RC)}}{RC} \left( \cos\left( \sqrt{\frac{1}{LC} - \left(\frac{1}{2RC}\right)^2} \, t \right) - \frac{\sin\left(\sqrt{(1/LC) - (1/2RC)^2}\, t\right)}{2RC\sqrt{(1/LC) - (1/2RC)^2}} \right) u(t)$$

(6.68)

or

$$h(t) = 2\zeta\,\omega_0 e^{-\zeta\,\omega_0 t} \left[ \cos\left(\omega_0 \sqrt{1 - \zeta^2}\, t\right) - \frac{\zeta}{\sqrt{1 - \zeta^2}} \sin\left(\omega_0 \sqrt{1 - \zeta^2}\, t\right) \right] u(t) \qquad (6.69)$$

where

$$2\zeta\,\omega_0 = \frac{1}{RC} \qquad \text{and} \qquad \omega_0^2 = \frac{1}{LC} \qquad (6.70)$$

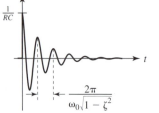

**Figure 6.37**
Impulse response of a practical $RLC$ bandpass filter.

(Figure 6.37). Notice that the impulse response of this physically realizable filter is causal.

All physical systems are filters in the sense that each of them has a response to excitations that has a characteristic variation with frequency. This is what gives a musical instrument and each human voice its characteristic sound. To see how important this is, try playing just the mouthpiece of any wind instrument. The sound is very unpleasant until the instrument is attached, and then it becomes very pleasant (when played by a good musician). The sun periodically heats the earth as it rotates, and the earth acts like a lowpass filter, smoothing out the daily variations and responding with a lagging seasonal variation of temperature. When a diver jumps off the end of a diving board, he excites it, and it responds with a vibration at its characteristic resonant frequency. Industrial foam-rubber ear plugs are designed to allow lower frequencies through, so that people wearing them can converse, but to block intense high-frequency sounds that may damage the ear. In prehistoric times people tended to live in caves because the thermal mass of the rock around them smoothed the seasonal variation of temperature and allowed them to be cooler in the summer and warmer in the winter, another example of lowpass filtering. The list of examples of systems that we are familiar with in daily life that perform filtering operations is endless.

# 6.5 LOG-MAGNITUDE FREQUENCY-RESPONSE PLOTS AND BODE DIAGRAMS

Often linear plots of frequency response like Figures 6.32 and 6.36, although accurate, do not reveal important system behavior. As an example, consider the plots of the frequency responses of two quite different-looking system transfer functions,

$$H_1(f) = \frac{1}{j2\pi f + 1} \quad \text{and} \quad H_2(f) = \frac{30}{30 - 4\pi^2 f^2 + j62\pi f} \quad (6.71)$$

(Figure 6.38). Plotted this way, the two magnitude frequency-response plots look identical, yet we know the transfer functions are different. The phase plots do show

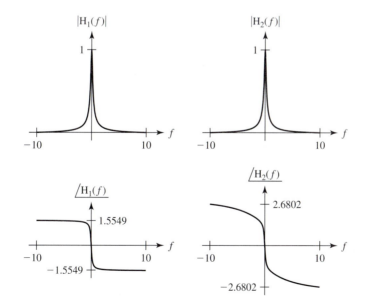

**Figure 6.38**
Comparison of the frequency responses of two apparently different transfer functions.

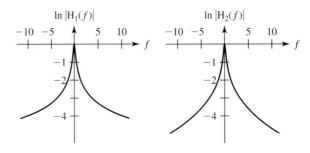

**Figure 6.39**
Log-magnitude plots of the two frequency responses.

some difference, but it is not immediately obvious what aspects of the systems cause the difference. One way of seeing subtle differences between frequency responses is to plot the logarithm of the magnitude of the response instead of the magnitude itself. A logarithm deemphasizes large values and emphasizes small values. Then small differences between frequency responses can be more easily seen (Figure 6.39). In the linear plots, the behavior of the magnitude frequency response looked identical because at very small values, the two plots look the same. In a log-magnitude plot, the difference between the two magnitude frequency responses at very small values is obvious.

Although log-magnitude plots are used some, a more common way of displaying frequency response is the *Bode diagram* or *Bode plot*. Like the log-magnitude plot, the Bode diagram reveals small differences between frequency responses, but it is also a systematic way of quickly sketching or estimating the overall frequency response of a system which may contain multiple cascaded transfer functions. A log-magnitude plot is logarithmic in one dimension; a Bode diagram is logarithmic in both dimensions. A magnitude frequency-response Bode diagram is a plot of the logarithm of the magnitude of a frequency response against a logarithmic frequency scale.

> Since the frequency scale is now logarithmic, only positive frequencies can be plotted. That is not a loss of information since, for transfer functions of real systems, the value of the frequency response at any negative frequency is the complex conjugate of the value at the corresponding positive frequency.

In a Bode diagram, the magnitude of the frequency response is converted to a logarithmic scale using a special unit called the decibel (dB). If the transfer function magnitude is

$$|H(f)| = \left| \frac{Y(f)}{X(f)} \right|, \tag{6.72}$$

then that magnitude, expressed in decibels, is

$$|H_{dB}(f)| = 20 \log_{10} |H(f)| = 20 \log_{10} \left| \frac{Y(f)}{X(f)} \right| = \left| Y_{dB}(f) \right| - \left| X_{dB}(f) \right|. \tag{6.73}$$

The name of the unit decibel comes from the original unit defined by Bell Telephone engineers, the bel (B), named in honor of Alexander Graham Bell, the inventor of the telephone. The bel is defined as the common logarithm (base 10) of a power ratio. For example, if the response signal-power of a system is 100 and the excitation signal-power (expressed in the same units) is 20, the power gain of the system, expressed in bels would be

$$\log_{10}\left(\frac{P_Y}{P_X}\right) = \log_{10}\left(\frac{100}{20}\right) \cong 0.699 \text{ B}.$$

Since *deci* is the standard international prefix for one-tenth, a decibel is one-tenth of a bel, and that same power ratio would be 6.99 dB. So the power gain, expressed in dB would be

$$10\log_{10}\left(\frac{P_Y}{P_X}\right).$$

Since signal power is proportional to the square of the signal itself, the ratio of powers, expressed directly in terms of the signals, would be

$$10\log_{10}\left(\frac{P_Y}{P_X}\right) = 10\log_{10}\left(\frac{Y^2}{X^2}\right) = 10\log_{10}\left[\left(\frac{Y}{X}\right)^2\right] = 20\log_{10}\left(\frac{Y}{X}\right).$$

In a system in which multiple subsystems are cascaded, the overall transfer function is the product of the individual transfer functions, but the overall transfer function expressed in decibels is the sum of the individual transfer functions expressed in decibels because of the logarithmic definition of the decibel.

Returning now to the two different system transfer functions

$$H_1(f) = \frac{1}{j2\pi f + 1} \qquad \text{and} \qquad H_2(f) = \frac{30}{30 - 4\pi^2 f^2 + j62\pi f}, \qquad \textbf{(6.74)}$$

if we make a Bode diagram of each of them, their difference becomes more evident (Figure 6.40). The logarithmic decibel scale makes the behavior of the two magnitude frequency responses at the higher frequencies distinguishable. Plotting them on the same scale emphasizes the difference (Figure 6.41).

Although the fact that differences between low levels of magnitude frequency response can be better seen with a Bode diagram is a good reason to use it, it is by no means the only reason. The fact that system gains in decibels add instead of multiplying when systems are cascaded makes the quick graphical estimation of overall system gain characteristics easier using Bode diagrams than using linear plots.

LTI systems are described by linear differential equations with constant coefficients. The most general form of such an equation is

$$\sum_{k=0}^{D} a_k \frac{d^k}{dt^k} y(t) = \sum_{k=0}^{N} b_k \frac{d^k}{dt^k} x(t), \qquad \textbf{(6.75)}$$

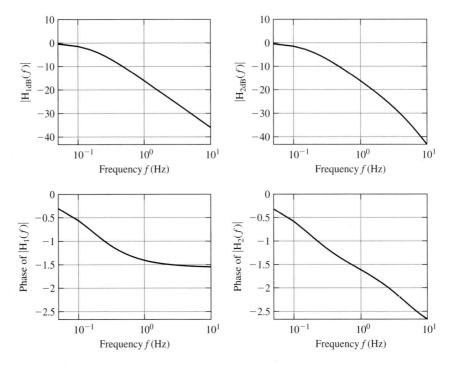

**Figure 6.40**
Bode diagrams of the two example transfer function frequency responses.

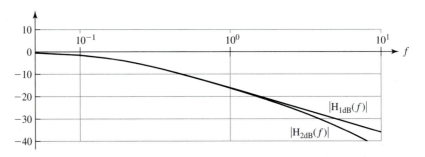

**Figure 6.41**
Bode diagrams of the two example transfer function magnitude frequency responses on the same scale for better comparison.

where $x(t)$ is the excitation and $y(t)$ is the response. Fourier-transforming both sides of the equation we get

$$\sum_{k=0}^{D} a_k (j2\pi f)^k Y(f) = \sum_{k=0}^{N} b_k (j2\pi f)^k X(f) \qquad \text{or}$$

$$\sum_{k=0}^{D} a_k (j\omega)^k Y(j\omega) = \sum_{k=0}^{N} b_k (j\omega)^k X(j\omega). \tag{6.76}$$

This can be rearranged into the transfer function

$$H(f) = \frac{Y(f)}{X(f)} = \frac{\sum_{k=0}^{N} b_k (j2\pi f)^k}{\sum_{k=0}^{D} a_k (j2\pi f)^k} \quad \text{or} \quad H(j\omega) = \frac{Y(j\omega)}{X(j\omega)} = \frac{\sum_{k=0}^{N} b_k (j\omega)^k}{\sum_{k=0}^{D} a_k (j\omega)^k},$$

**(6.77)**

showing that the transfer functions of LTI systems are in the form of a ratio of polynomials in either $f$ or $j\omega$. The transfer function can be expressed in the form

$$H(f) = \frac{b_N (j2\pi f)^N + b_{N-1} (j2\pi f)^{N-1} + \cdots + b_1 (j2\pi f) + b_0}{a_D (j2\pi f)^D + a_{D-1} (j2\pi f)^{D-1} + \cdots + a_1 (j2\pi f) + b_0}$$

**(6.78)**

or

$$H(j\omega) = \frac{b_N (j\omega)^N + b_{N-1} (j\omega)^{N-1} + \cdots + b_1 (j\omega) + b_0}{a_D (j\omega)^D + a_{D-1} (j\omega)^{D-1} + \cdots + a_1 (j\omega) + b_0}.$$

**(6.79)**

The numerator and denominator polynomials can (at least in principle) be factored, putting the transfer function into the form,

$$H(f) = A \frac{(1 - (j2\pi f / z_1))(1 - (j2\pi f / z_2)) \cdots (1 - (j2\pi f / z_N))}{(1 - (j2\pi f / p_1))(1 - (j2\pi f / p_2)) \cdots (1 - (j2\pi f / p_D))}$$

**(6.80)**

or

$$H(j\omega) = A \frac{(1 - (j\omega / z_1))(1 - (j\omega / z_2)) \cdots (1 - (j\omega / z_N))}{(1 - (j\omega / p_1))(1 - (j\omega / p_2)) \cdots (1 - (j\omega / p_D))}.$$

**(6.81)**

(Formulated this way the units of the $p$'s and $z$'s are radians per second rather than hertz. This conforms to accepted conventions for the Laplace transform to be introduced later and is consistent with the notation and conventions in the control system area where Bode plots are used most.) This is a good point at which to define two very common terms in signal and system analysis, pole and zero. A *pole* of a function is a value of its independent variable at which the function value goes to infinity, and a *zero* of a function is a value of its independent variable at which the function goes to zero.

In (6.80) and (6.81) the $k$th pole of H occurs where $j2\pi f = p_k$ or $j\omega = p_k$. So the $p$'s are not the frequencies $f$ or $\omega$, at which the transfer function magnitude goes to infinity but are rather the values of $j2\pi f$ or $j\omega$ at which the transfer function magnitude goes to infinity. So when referring to the $p$'s as poles, we mean the values of $j2\pi f$ or $j\omega$ where the transfer function goes to infinity. (When we come to the Laplace transform later, these $p$'s will be the actual values of the Laplace independent variable $s$ at which a system function goes to infinity.) The same holds true for the zeros of the transfer which occur at $j2\pi f = z_k$ or $j\omega = z_k$.

For real systems the coefficients $a$ and $b$ in the general form of a linear, constant-coefficient differential equation

$$\sum_{k=0}^{D} a_k \frac{d^k}{dt^k} y(t) = \sum_{k=0}^{N} b_k \frac{d^k}{dt^k} x(t)$$

**(6.82)**

are all real. Since these coefficients are also the coefficients in the transfer function in the ratio-of-polynomials form

$$H(f) = \frac{b_N(j2\pi f)^N + b_{N-1}(j2\pi f)^{N-1} + \cdots + b_1(j2\pi f) + b_0}{a_D(j2\pi f)^D + a_{D-1}(j2\pi f)^{D-1} + \cdots + a_1(j2\pi f) + b_0} \tag{6.83}$$

or

$$H(j\omega) = \frac{b_N(j\omega)^N + b_{N-1}(j\omega)^{N-1} + \cdots + b_1(j\omega) + b_0}{a_D(j\omega)^D + a_{D-1}(j\omega)^{D-1} + \cdots + a_1(j\omega) + b_0} \tag{6.84}$$

all the $p$'s and $z$'s in the factored forms

$$H(f) = A\frac{(1 - (j2\pi f/z_1))(1 - (j2\pi f/z_2)) \cdots (1 - (j2\pi f/z_N))}{(1 - (j2\pi f/p_1))(1 - (j2\pi f/p_2)) \cdots (1 - (j2\pi f/p_D))} \tag{6.85}$$

or

$$H(j\omega) = A\frac{(1 - (j\omega/z_1))(1 - (j\omega/z_2)) \cdots (1 - (j\omega/z_N))}{(1 - (j\omega/p_1))(1 - (j\omega/p_2)) \cdots (1 - (j\omega/p_D))} \tag{6.86}$$

must either be real or must occur in complex-conjugate pairs, so that when the factored numerator and denominator are multiplied out to obtain the ratio-of-polynomials form, all the coefficients of the powers of $j2\pi f$ or $j\omega$ are real.

From the factored form, the system transfer function can be considered as the cascade of multiple subsystems, each having a transfer function with one pole or one zero (Figure 6.42). Each simple system will have a Bode diagram and, because the magnitude

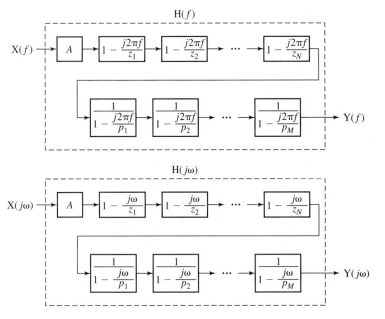

**Figure 6.42**
A system transfer function represented as a cascade of simpler systems.

Bode diagrams are plotted in decibels, which is a logarithmic scale, the overall magnitude Bode diagram is the sum of the individual magnitude Bode diagrams. Phase is plotted linearly as before (against a logarithmic frequency scale), and the overall phase Bode plot is the sum of all the phases contributed by the subsystems.

## COMPONENT DIAGRAMS

**One-Real-Pole System**    Consider the frequency response of a subsystem with a single real pole and no zeros,

$$ \text{H}(f) = \frac{1}{1 - (j2\pi f/p_k)} \qquad \text{or} \qquad \text{H}(j\omega) = \frac{1}{1 - (j\omega/p_k)}. \qquad \textbf{(6.87)} $$

It obviously depends on the value of $p_k$ which must either be real or complex and, if it is complex, it must have a companion $p$ which is its complex conjugate. Consider first the case in which $p_k$ is real and negative. The magnitudes and phases of $\text{H}(f) = 1/(1 - (j2\pi f/p_k))$ and $\text{H}(j\omega) = 1/(1 - (j\omega/p_k))$ versus frequency are plotted in Figure 6.43.

For frequencies $2\pi f = \omega \ll |p_k|$, the magnitude response is approximately 0 dB and the phase response is approximately 0 radians (rad). For frequencies $2\pi f = \omega \gg |p_k|$, the magnitude response approaches a linear slope of $-6$ dB per octave or $-20$ dB per decade and the phase response approaches a constant $-(\pi/2)$ rad. (An octave is a factor of two change in frequency, and a decade is a factor of 10 change in frequency.) These limiting behaviors for extreme frequencies define magnitude and phase asymptotes. The intersection of the two magnitude asymptotes occurs at $2\pi f = |p_k|$ or $\omega = |p_k|$. At that point the actual Bode diagram is 3 dB below the corner formed by the

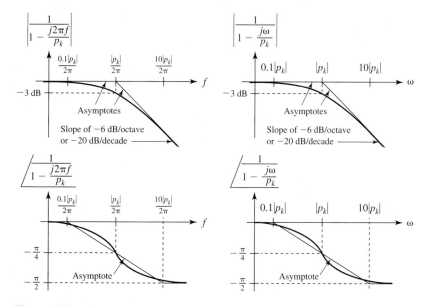

**Figure 6.43**
The magnitude and phase frequency response of a single-real-pole subsystem.

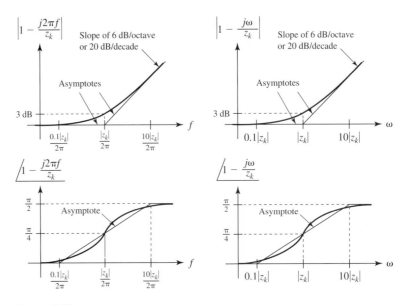

**Figure 6.44**
The magnitude and phase frequency response of a single-real-zero subsystem.

asymptotes. This is the point of largest deviation of the magnitude Bode diagram from its asymptotes. The phase Bode diagram goes through $-(\pi/4)$ rad at the corner frequency and approaches 0 rad below and $-(\pi/2)$ rad above the corner frequency.

**One-Real-Zero System** A similar analysis yields the magnitude and phase Bode plots for a subsystem with a single real negative zero and no poles whose transfer function is of the form

$$\text{H}(f) = 1 - \frac{j2\pi f}{z_k} \qquad \text{or} \qquad \text{H}(j\omega) = 1 - \frac{j\omega}{z_k} \qquad \textbf{(6.88)}$$

(Figure 6.44).

The diagrams are very similar to those for the simple numerator factor except that the magnitude asymptote above the corner frequency has a slope of $+6$ dB per octave or $+20$ dB per decade and the phase approaches $+(\pi/2)$ instead of $-(\pi/2)$ rad. They are basically the single-real-pole Bode diagrams turned upside down.

**Integrators and Differentiators** We must also consider a pole or a zero at zero frequency (Figures 6.45 and 6.46). A system component with a single pole at zero is called an integrator because its transfer function is $\text{H}(f) = 1/j2\pi f$ or $\text{H}(j\omega) = 1/j\omega$. The type of system component in Figure 6.46 is called a differentiator because its transfer function is $j2\pi f$ or $j\omega$.

**Frequency-Independent Gain** The only remaining type of system component is a frequency-independent gain (Figure 6.47). In the figure, the gain constant $A$ is assumed to be positive. That is why the phase is zero. If $A$ is negative, the phase is $\pm\pi$ rad.

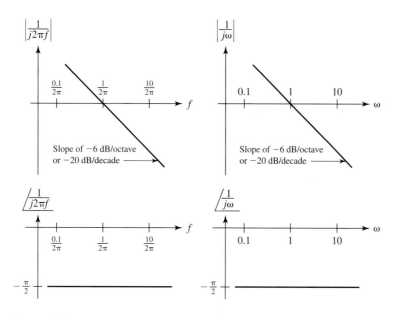

**Figure 6.45**
The magnitude and phase frequency response of a single $p_k$ at zero.

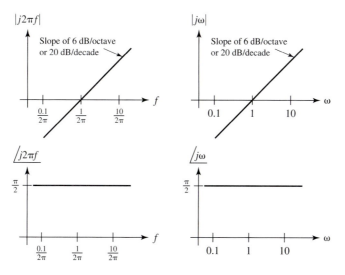

**Figure 6.46**
The magnitude and phase frequency response of a single $z_k$ at zero.

The asymptotes are helpful in drawing the actual Bode diagram, especially in sketching the overall Bode diagram for a more complicated system. The asymptotes can be quickly sketched from a knowledge of a few simple rules and added together. Then the magnitude Bode diagram can be sketched approximately by drawing a smooth curve which approaches the asymptotes and deviates at the corners by $\pm 3$ dB.

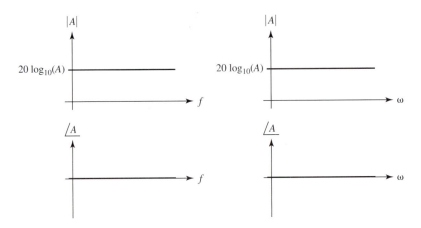

**Figure 6.47**
The magnitude and phase frequency response of a frequency-independent gain $A$.

## EXAMPLE 6.1

Plot the Bode diagram for the voltage transfer function of the circuit in Figure 6.48, where $C_1 = 1$ F, $C_2 = 2$ F, $R_s = 4\ \Omega$, $R_1 = 2\ \Omega$, $R_2 = 3\ \Omega$.

### ■ Solution

Using the impedance concept the transfer function is found to be

$$H(j\omega)$$

$$= R_2 \frac{(j\omega)R_1C_1 + 1}{(j\omega)^2 R_1 R_2 R_s C_1 C_2 + (j\omega)[R_1 R_2(C_1 + C_2) + (R_1C_1 + R_2C_2)R_s] + (R_1 + R_2 + R_s)}.$$
$$(6.89)$$

Substituting numerical values for the components,

$$H(j\omega) = 3\frac{2j\omega + 1}{48(j\omega)^2 + 50(j\omega) + 9}$$

$$= 0.125\frac{j\omega + 0.5}{(j\omega + 0.2316)(j\omega + 0.8104)} \qquad (6.90)$$

$$H(j\omega) = 0.333\frac{1 - (j\omega/(-0.5))}{[1 - (j\omega/(-0.2316))][1 - (j\omega/(-0.8104))]}$$

$$= A\frac{1 - (j\omega/z_1)}{(1 - (j\omega/p_1))(1 - (j\omega/p_2))} \qquad (6.91)$$

where $A = 0.333$, $z_1 = -0.5$, $p_1 = -0.2316$, and $p_2 = -0.8104$.

So this transfer function has two poles, one zero and one frequency-independent gain. We can quickly construct an overall asymptotic Bode diagram by adding the asymptotic Bode diagrams for the four individual components of the overall transfer function (Figure 6.49).

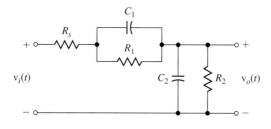

**Figure 6.48**
Circuit.

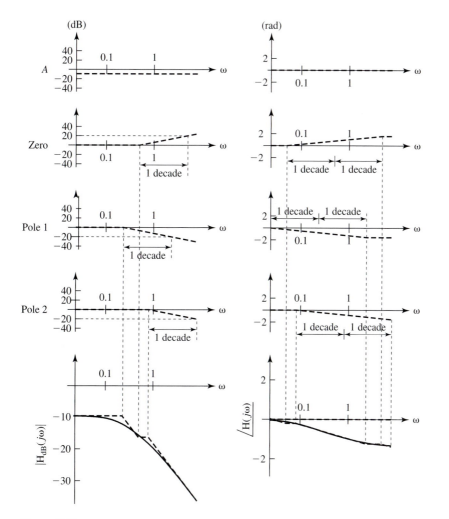

**Figure 6.49**
Individual asymptotic and overall asymptotic and exact Bode magnitude and phase
diagrams for the circuit voltage transfer function.

MATLAB has a function `bode` for plotting Bode diagrams of systems. The syntax is

$$\texttt{bode(sys)} \quad \text{or} \quad \texttt{bode(sys,w)}$$

where `sys` is a MATLAB-system object and `w` is a vector of radian frequencies. (There are also other syntaxes. Type `help bode` for more information.)

## COMPLEX POLE AND ZERO PAIRS

Now consider the case of complex poles and zeros. For real system functions, they always occur in complex-conjugate pairs. So a complex-conjugate pair of poles would form a subsystem transfer function of the form

$$H(j\omega) = \frac{1}{(1 - (j\omega/p_1))(1 - (j\omega/p_2))} = \frac{1}{1 - j\omega((1/p_1) + (1/p_1^*)) + ((j\omega)^2/p_1 p_1^*)}$$

$$= \frac{1}{1 - j\omega(2\text{Re}(p_1)/|p_1|^2) + ((j\omega)^2/|p_1|^2)}. \tag{6.92}$$

From the table of Fourier pairs, we find the pair

$$e^{-\omega_0 \zeta t} \sin\left(\omega_0\sqrt{1 - \zeta^2}\, t\right) u(t) \overset{\mathcal{F}}{\longleftrightarrow} \frac{\omega_0\sqrt{1 - \zeta^2}}{(j\omega)^2 + j\omega(2\zeta\omega_0) + \omega_0^2} \tag{6.93}$$

in the $\omega$ domain, which can be expressed in the form,

$$\omega_0 \frac{e^{-\omega_0 \zeta t} \sin\left(\omega_0\sqrt{1 - \zeta^2}\, t\right)}{\sqrt{1 - \zeta^2}} u(t) \longleftrightarrow \frac{1}{1 + j\omega\left(2\zeta\omega_0/\omega_0^2\right) + \left((j\omega)^2/\omega_0^2\right)} \tag{6.94}$$

whose right side is of the same functional form as

$$H(j\omega) = \frac{1}{1 - j\omega(2\text{Re}(p_1)/|p_1|^2) + ((j\omega)^2/|p_1|^2)}. \tag{6.95}$$

This is a standard form of a second-order underdamped system response where the undamped resonant radian frequency is $\omega_0$ and the damping factor is $\zeta$. Therefore, for this type of subsystem,

$$\omega_0^2 = |p_1|^2 = p_1 p_2 \quad \text{and} \quad \zeta = -\frac{\text{Re}(p_1)}{\omega_0} = -\frac{p_1 + p_2}{2\sqrt{p_1 p_2}}. \tag{6.96}$$

The Bode diagram for this subsystem is illustrated in Figure 6.50.

A complex pair of zeros would form a subsystem transfer function of the form

$$H(j\omega) = \left(1 - \frac{j\omega}{z_1}\right)\left(1 - \frac{j\omega}{z_2}\right) = 1 - j\omega\left(\frac{1}{z_1} + \frac{1}{z_1^*}\right) + \frac{(j\omega)^2}{z_1 z_1^*}$$

$$= 1 - j\omega\frac{2\text{Re}(z_1)}{|z_1|^2} + \frac{(j\omega)^2}{|z_1|^2} \tag{6.97}$$

In this type of subsystem we can identify the undamped resonant radian frequency and the damping factor as

$$\omega_0^2 = |z_1|^2 = z_1 z_2 \quad \text{and} \quad \zeta = -\frac{\text{Re}(z_1)}{\omega_0} = -\frac{z_1 + z_2}{2\sqrt{z_1 z_2}}. \tag{6.98}$$

The Bode diagram for this subsystem is illustrated in Figure 6.51.

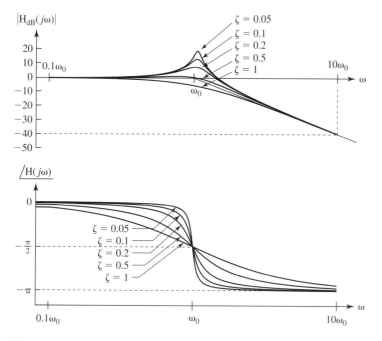

**Figure 6.50**
Magnitude and phase Bode diagram for a second-order complex
pole pair.

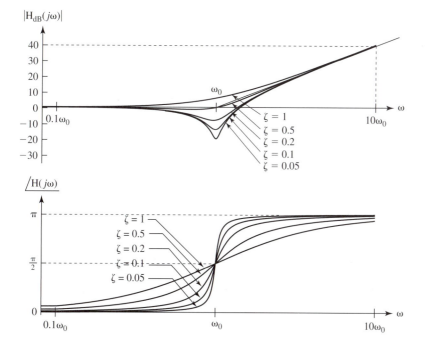

**Figure 6.51**
Magnitude and phase Bode diagram for a second-order complex zero pair.

# 6.6 PRACTICAL ACTIVE FILTERS

All the practical filters we have examined so far have been passive filters. *Passive* means they contained no devices with the capability of having a response with more real power than the excitation. Many modern filters are *active* filters. That is, they contain active devices like transistors and/or operational amplifiers and require an external source of power to operate properly. With the use of active devices the real response power can be greater than the real excitation power. The subject of active filters is a large one, and only the simplest forms of active filters will be introduced here.

> In some passive circuits, there is voltage gain at some frequencies. That is, the output voltage signal can be larger than the input voltage signal. Therefore the signal power of the response, as defined previously, would be greater than the signal power of the excitation. But this is not *real* power gain because that higher output voltage signal is across a higher impedance.

## OPERATIONAL AMPLIFIERS

There are two commonly used forms of operational amplifier circuits, the inverting amplifier form and the noninverting amplifier form (Figure 6.52). The analysis here will use the simplest possible model for the operational amplifier, the *ideal* operational amplifier. An ideal operational amplifier has infinite input impedance, zero output impedance, infinite gain, and infinite bandwidth.

For each type of amplifier there are two impedances $Z_i(f)$ and $Z_f(f)$ which control the gain. The gain of the inverting amplifier can be derived by observing that, since the operational amplifier input impedance is infinite, the current flowing into either input terminal is zero and, therefore,

$$\mathrm{I}_f(f) = \mathrm{I}_i(f). \qquad (6.99)$$

Also, since the output voltage is finite and the operational amplifier gain is infinite, the voltage difference between the two input terminals must be zero. Therefore,

$$\mathrm{I}_i(f) = \frac{\mathrm{V}_i(f)}{\mathrm{Z}_i(f)} \qquad (6.100)$$

and

$$\mathrm{I}_f(f) = -\frac{\mathrm{V}_f(f)}{\mathrm{Z}_f(f)}. \qquad (6.101)$$

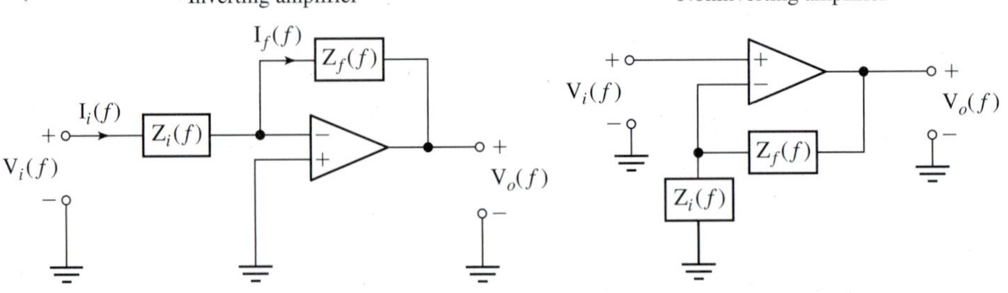

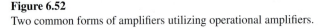

**Figure 6.52**
Two common forms of amplifiers utilizing operational amplifiers.

Equating (6.100) and (6.101) according to (6.99), and solving for the transfer function,

$$\frac{V_o(f)}{V_i(f)} = -\frac{Z_f(f)}{Z_i(f)}.$$  (6.102)

By a similar analysis it can be shown that the noninverting amplifier gain is

$$\frac{V_o(f)}{V_i(f)} = \frac{Z_f(f) + Z_i(f)}{Z_i(f)}.$$  (6.103)

## FILTERS

Probably the most common and simplest form of active filter is the active integrator (Figure 6.53). Using the inverting amplifier gain formula for the transfer function,

$$H(f) = -\frac{Z_f(f)}{Z_i(f)} = -\frac{1/j2\pi f C}{R} = -\frac{1}{j2\pi f RC}.$$  (6.104)

The action of the integrator is easier to see if the transfer function is rearranged to the form

$$V_o(f) = -\frac{1}{RC} \underbrace{\frac{V_i(f)}{j2\pi f}}_{\substack{\text{integral} \\ \text{of } V_i(f)}}.$$  (6.105)

That is, the integrator integrates the signal but, at the same time, multiplies it by $-(1/RC)$. Notice that we did not introduce a practical passive integrator. The passive $RC$ lowpass filter acts much like an integrator for frequencies well above its corner frequency but at a low enough frequency that its response is not like an integrator. So the active device (the operational amplifier in this case) has given the filter designer another degree of freedom in design.

The integrator is easily changed to a lowpass filter by the addition of a single resistor (Figure 6.54). For this circuit,

$$\frac{V_0(f)}{V_i(f)} = -\frac{R_f}{R_s}\frac{1}{j2\pi f C R_f + 1}.$$  (6.106)

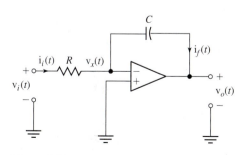

**Figure 6.53**
An active integrator.

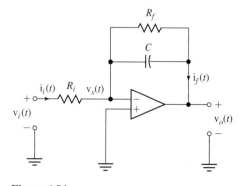

**Figure 6.54**
An active $RC$ lowpass filter.

This transfer function has the same functional form as the passive $RC$ lowpass filter except for the factor $-(R_f/R_s)$. So this is a filter with gain. That is, it filters and amplifies the signal simultaneously. In this case the voltage gain is negative.

**EXAMPLE 6.2**

Plot the Bode magnitude and phase diagrams for the two-stage active filter in Figure 6.55.

■ **Solution**

The transfer function of the first stage is

$$H_1(f) = -\frac{Z_{f1}(f)}{Z_{i1}(f)} = -\frac{R_{f1}}{R_{i1}} \frac{1}{1 + j2\pi f C_{f1} R_{f1}}. \tag{6.107}$$

The transfer function of the second stage is

$$H_2(f) = -\frac{Z_{f1}(f)}{Z_{i1}(f)} = -\frac{j2\pi f R_{f2} C_{i2}}{1 + j2\pi f R_{f2} C_{f2}}. \tag{6.108}$$

Since the output impedance of an ideal operational amplifier is zero, the second stage does not load the first stage, and, therefore, the overall transfer function is simply the product of the two transfer functions,

$$H(f) = \frac{R_{f1}}{R_{i1}} \frac{j2\pi f R_{f2} C_{i2}}{(1 + j2\pi f C_{f1} R_{f1})(1 + j2\pi f R_{f2} C_{f2})}. \tag{6.109}$$

Substituting in parameter values,

$$H(f) = \frac{j1000\,f}{(1000 + j\frac{f}{10})(1000 + jf)} \tag{6.110}$$

(Figure 6.56). This is obviously a practical bandpass filter.

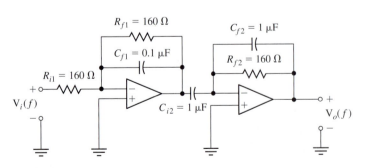

**Figure 6.55**
A two-stage active filter.

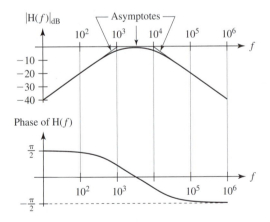

**Figure 6.56**
Bode diagram of the frequency response of the two-stage active filter.

EXAMPLE 6.3

Design an active filter which attenuates signals at 60 Hz and below by more than 40 dB and amplifies signals at 10 kHz and above with a positive gain that deviates from 20 dB by no more than 2 dB.

■ **Solution**

This specifies a highpass filter. The gain must be positive. A positive gain and some highpass filtering can be accomplished by one noninverting amplifier. However, looking at the gain formula for the noninverting amplifier

$$\frac{V_o(f)}{V_i(f)} = \frac{Z_f(f) + Z_i(f)}{Z_i(f)}, \tag{6.111}$$

we see that if the two impedances consist of only resistors and capacitors, the amplifier's gain is never less than one and we need attenuation (a gain less than one) at low frequencies. [If we were to use inductors and capacitors, we could make the magnitude of the sum $Z_f(f) + Z_i(f)$ be less than the magnitude of $Z_i(f)$ at some frequencies and achieve a gain less than one. But we could not make that occur for all frequencies below 60 Hz, and the use of inductors is generally avoided in practical design unless absolutely necessary. There are other practical difficulties with this idea when using real, as opposed to ideal, operational amplifiers.]

If we use one inverting amplifier, we have a negative gain. But we could follow it with another inverting amplifier making the overall gain positive. Gain is the opposite of attenuation. If the attenuation is 60 dB, the gain is −60 dB. If the gain at 60 Hz is −40 dB and the response is that of a single-pole highpass filter, the Bode diagram asymptote on the magnitude frequency response would pass through −20 dB of gain at 600 Hz, 0 dB of gain at 6 kHz, and 20 dB of gain at 60 kHz. But we need 20 dB of gain at 10 kHz, so a single-pole filter is inadequate to meet the specifications. We need a two-pole highpass filter. We can achieve that with a cascade of two single-pole highpass filters, meeting the requirements for attenuation and positive gain simultaneously.

Now we must choose $Z_f(f)$ and $Z_i(f)$ to make the inverting amplifier a highpass filter. Figure 6.54 illustrates an active lowpass filter. That filter is lowpass because the gain is $-(Z_f(f)/Z_i(f))$, $Z_i(f)$ is constant, and $Z_f(f)$ has a larger magnitude at low frequencies than at high frequencies. There is more than one way to make a highpass filter using the same inverting amplifier configuration. We could make the magnitude of $Z_f(f)$ be small at low frequencies and larger at high frequencies. That requires the use of an inductor, but, again for practical reasons, inductors should be avoided unless really needed. We could make $Z_f(f)$ constant and make the magnitude of $Z_i(f)$ large at low frequencies and small at high frequencies. That general goal can be accomplished by either a parallel or series combination of a resistor and a capacitor (Figure 6.57).

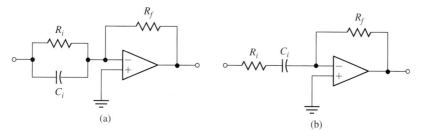

(a)  (b)

**Figure 6.57**
Two ideas for a highpass filter using only capacitors and resistors.

If we just think about the limiting behavior of these two design ideas at very low and very high frequencies, we immediately see that only one of them meets the specifications of this design. The design in Figure 6.57a has a finite gain at very low frequencies and a gain which rises with frequency at higher frequencies, never approaching a constant. The design in Figure 6.57b has a gain that falls with frequency at low frequencies, approaching zero at zero frequency and approaches a constant gain at high frequencies. This latter design can be used to meet our specification.

So now the design is a cascade of two inverting amplifiers (Figure 6.58). At this point we must select the resistor and capacitor values to meet the attenuation and gain requirements. There are many ways of doing that. The design is not unique. We can begin by selecting the resistors to meet the high-frequency gain requirement of 20 dB. That is an overall high-frequency gain of 10 which we can apportion any way we want between the two amplifiers. Let's let the two stage gains be approximately the same. Then the resistor ratios in each stage should be about 3.16. We should choose resistors large enough not to load the outputs of the operational amplifiers but small enough that stray capacitances don't cause problems. Resistors in the range of 500 $\Omega$ to 50 k$\Omega$ are usually good choices. But unless we are willing to pay a lot, we cannot arbitrarily choose a resistor value. Resistors come in standard values, typically in the following sequence:

$$1, 1.2, 1.5, 1.8, 2.2, 2.7, 3.3, 3.9, 4.7, 5.6, 6.8, 8.2 \times 10^n$$

where $n$ sets the *decade* of the resistance value. Some ratios that are very near 3.16 are

$$\frac{3.9}{1.2} = 3.25 \qquad \frac{4.7}{1.5} = 3.13 \qquad \frac{5.6}{1.8} = 3.11 \qquad \frac{6.8}{2.2} = 3.09 \qquad \frac{8.2}{2.7} = 3.03.$$

To set the overall gain very near 10 we can choose the first stage ratio to be $3.9/1.2 = 3.25$ and the second stage ratio to be $6.8/2.2 = 3.09$ and achieve an overall high-frequency gain of 10.043. So we set

$$R_{f1} = 3.9 \text{ k}\Omega \qquad R_{i1} = 1.2 \text{ k}\Omega \qquad R_{f2} = 6.8 \text{ k}\Omega \qquad R_{i2} = 2.2 \text{ k}\Omega. \qquad \textbf{(6.112)}$$

Now we must choose the capacitor values to achieve the attenuation at 60 Hz and below and the gain at 10 kHz and above. To simplify the design let's set the two corner frequencies of the two stages at the same (or very nearly the same) value. With a two-pole low-frequency rolloff of 40 dB per decade and a high-frequency gain of approximately 20 dB, we get a 60-dB difference between the transfer function magnitude at 60 Hz and 10 kHz. If we were to set the gain at 60 Hz to be exactly $-40$ dB, then at 600 Hz we would have approximately 0 dB gain and at 6 kHz we would have a gain of 40 dB and it would be higher at 10 kHz. This does not meet the specification.

We can start at the high-frequency end and set the gain at 10 kHz to be approximately 10, meaning that the corner for the low-frequency rolloff should be well below 10 kHz. If we put it

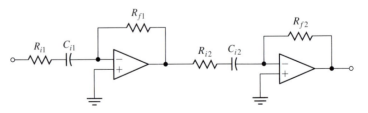

**Figure 6.58**
Cascade of two inverting highpass active filters.

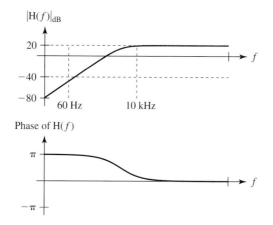

**Figure 6.59**
Bode diagram for two-stage active highpass filter design.

at 1 kHz, the approximate gain at 100 Hz based on asymptotic approximations will be $-20$ dB and at 10 Hz it will be $-60$ dB, and we need $-40$ dB at 60 Hz. But we only get about $-29$ dB at 60 Hz. So we need to put the corner frequency a little higher, say 3 kHz. If we put the corner frequency at 3 kHz, the calculated capacitor values will be $C_{i1} = 46$ nF and $C_{i2} = 24$ nF. Again we cannot arbitrarily choose a capacitor value. Standard capacitor values are typically at the same intervals as standard resistor values.

There is some leeway in the location of the corner frequency, so we probably don't need a really precise value of capacitance. We can choose $C_{i1} = 47$ nF and $C_{i2} = 22$ nF making one a little high and one a little low. This will separate the poles slightly but will still create the desired 40 dB per decade low-frequency rolloff. This looks like a good design, but we need to verify its performance by drawing a Bode diagram (Figure 6.59). It is apparent from the diagram that the attenuation at 60 Hz is adequate. Calculation of the gain at 10 kHz yields about 19.2 dB which also meets specifications.

These results are based on exact values of resistors and capacitors. In reality all resistors and capacitors are typically chosen based on their nominal values, but their actual values may differ from the nominal by a few percent. So any good design should have some tolerance in the specifications to allow for small deviations of component values from the design values. ∎

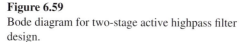

A popular filter design that can be found in many books on electronics or filters is the two-pole, single-stage, constant-$K$ bandpass filter (Figure 6.60). The triangle symbol with the $K$ inside in Figure 6.60 represents an ideal noninverting amplifier with a finite voltage gain $K$, an infinite input impedance, a zero output impedance, and infinite bandwidth (not an operational amplifier). The overall bandpass filter transfer function is

$$H(j\omega) = \frac{V_o(j\omega)}{V_i(j\omega)}$$

$$= \frac{[j\omega(K/(1-K))](1/R_1C_2)}{(j\omega)^2 + j\omega[(1/R_1C_1) + (1/R_2C_2) + (1/R_1C_2(1-K))] + (1/R_1R_2C_1C_2)}$$

$$\tag{6.113}$$

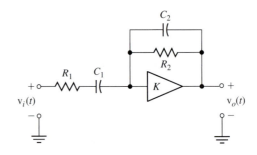

**Figure 6.60**
Constant-$K$ bandpass filter.

which is of the form,

$$H(j\omega) = H_0 \frac{j\alpha\omega_0^2}{(j\omega)^2 + j\omega\alpha\omega_0 + \omega_0^2} = \frac{j\omega A}{(j\omega)^2 + j\omega\alpha\omega_0 + \omega_0^2} \qquad (6.114)$$

where

$$A = \frac{K}{(1 - K)} \frac{1}{R_1 C_2} \qquad (6.115)$$

$$\omega_0^2 = \frac{1}{R_1 R_2 C_1 C_2} \qquad (6.116)$$

$$\alpha = \frac{R_1 C_1 + R_2 C_2 + (R_2 C_1/(1 - K))}{\sqrt{R_1 R_2 C_1 C_2}} \qquad (6.117)$$

and

$$H_0 = \frac{K}{1 + (1 - K)((C_2/C_1) + (R_1/R_2))}. \qquad (6.118)$$

The recommended design procedure is to choose the $Q$ and the resonant frequency $f_0$, to choose $C_1 = C_2 = C$ as some convenient value, and then to calculate

$$R_1 = R_2 = \frac{1}{2\pi f_0 C} \quad \text{and} \quad K = \frac{3Q - 1}{2Q - 1} \quad \text{and} \quad |H_0| = 3Q - 1.$$

Also, it is recommended that $Q$ should be less than 10 for this design. Design a filter of this type with a $Q$ of 5 and a center frequency of 50 kHz.

■ **Solution**

We can pick convenient values of capacitance, so let $C_1 = C_2 = C = 10$ nF. Then $R_1 = R_2 = 318$ Ω, $K = 1.556$, and $|H_0| = 14$. That makes the transfer function

$$H(j\omega) = -\frac{j8.792 \times 10^5 \omega}{9.86 \times 10^{10} - \omega^2 + j6.4 \times 10^4 \omega} \qquad (6.119)$$

or, written as a function of cyclic frequency,

$$H(f) = -\frac{j1.398 \times 10^5 f}{2.5 \times 10^9 - f^2 + j1.02 \times 10^4 f}$$

(Figure 6.61).

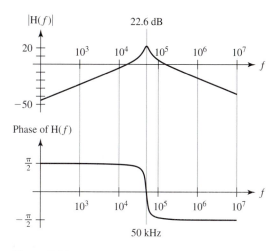

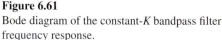

**Figure 6.61**
Bode diagram of the constant-$K$ bandpass filter frequency response.

As in Example 6.3, we cannot choose the component values to be exactly those calculated, but we can come close. We would probably have to use nominal 300-$\Omega$ resistors, and that would alter the frequency response slightly depending on their actual values and the actual values of the capacitors. ∎

Operational amplifiers are often used to form functional blocks which are interconnected to form larger systems with desired frequency-response characteristics. The most commonly used functional block is the integrator. We have already seen in Chapter 3 the integrator used in the configuration of Figure 6.62.

If the response of the integrator is $y(t)$, the excitation of the integrator is $y'(t)$. Then the differential equation describing this system is

$$y'(t) = x(t) - y(t) \tag{6.120}$$

or

$$y'(t) + y(t) = x(t). \tag{6.121}$$

The CTFT of this equation is

$$j\omega Y(j\omega) + Y(j\omega) = X(j\omega). \tag{6.122}$$

Solving (6.122) for the transfer function,

$$H(j\omega) = \frac{Y(j\omega)}{X(j\omega)} = \frac{1}{j\omega + 1}. \tag{6.123}$$

This is the transfer function of a system with a lowpass frequency response.

A system with a highpass frequency response can be formed by a small modification of the system in Figure 6.62 (Figure 6.63). Parallel and cascade connections of lowpass and highpass frequency responses can produce bandpass and bandstop frequency responses.

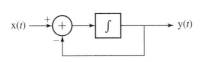

**Figure 6.62**
An integrator used to form a system
with a lowpass frequency response.

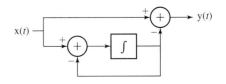

**Figure 6.63**
A system with a highpass frequency
response using an integrator as a
functional block.

## 6.7 DISCRETE-TIME FILTERS

In Chapter 3 there was an example of an LTI discrete-time lowpass filter (Figure 6.64). Its unit sequence response was found to be $[5 - 4(\frac{4}{5})^n]u[n]$ (Figure 6.65). The impulse response of any DT system is the first backward difference of its unit sequence response. In this case that is

$$h[n] = \left[5 - 4\left(\frac{4}{5}\right)^n\right]u[n] - \left[5 - 4\left(\frac{4}{5}\right)^{n-1}\right]u[n-1] \tag{6.124}$$

which reduces to

$$h[n] = \left(\frac{4}{5}\right)^n u[n] \tag{6.125}$$

(Figure 6.66). The transfer function is the DTFT of the impulse response

$$H(F) = \frac{1}{1 - \frac{4}{5}e^{-j2\pi F}} \tag{6.126}$$

(Figure 6.67).

It is instructive to compare the impulse and frequency responses of this DT lowpass filter and the $RC$ continuous-time lowpass filter analyzed in the previous section. The impulse response of the DT lowpass filter looks like a sampled version of the impulse response of the CT lowpass filter (Figure 6.68). Their frequency responses also have some similarities (Figure 6.69).

If we compare the shapes of the magnitudes and phases of these transfer functions over the DT-frequency range $-\frac{1}{2} < F < \frac{1}{2}$, they look very much alike (magnitudes more than phases). But a DT frequency response is always periodic and can never be lowpass in the same sense as the frequency response of the CT lowpass filter. The name lowpass applies accurately to the behavior of the DT frequency response in the range $-\frac{1}{2} < F < \frac{1}{2}$, and that is the only sense in which the designation lowpass is correctly used for DT systems.

Another very common type of DT lowpass filter which will illustrate some principles of DT filter design and analysis is the moving-average filter (Figure 6.70). The difference equation describing this filter is

$$y[n] = \frac{x[n] + x[n-1] + x[n-2] + \cdots + x[n-N]}{N+1}, \tag{6.127}$$

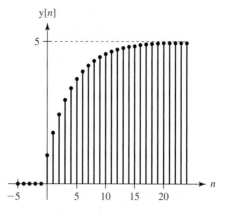

**Figure 6.65**
Unit-sequence response of the DT lowpass filter.

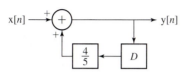

**Figure 6.64**
A DT lowpass filter.

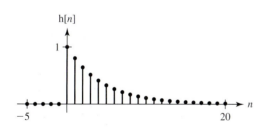

**Figure 6.66**
Impulse response of the DT lowpass filter.

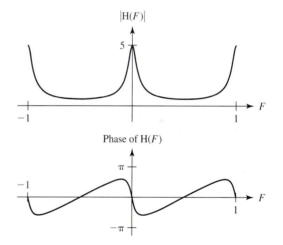

**Figure 6.67**
Frequency response of the DT lowpass filter.

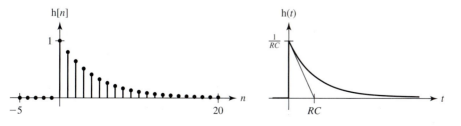

**Figure 6.68**
Comparison of the impulse responses of DT and CT lowpass filters.

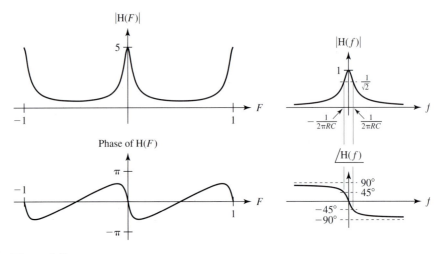

**Figure 6.69**
Frequency responses of DT and CT lowpass filters.

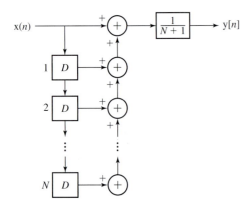

**Figure 6.70**
A DT moving-average filter.

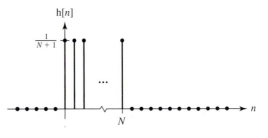

**Figure 6.71**
Impulse response of a moving-average filter.

and its impulse response is

$$h[n] = \frac{\delta[n] + \delta[n-1] + \delta[n-2] + \cdots + \delta[n-N]}{N+1} \qquad (6.128)$$

(Figure 6.71). Its frequency response is

$$H(F) = \frac{1 + e^{-j2\pi F} + e^{-j4\pi F} + \cdots + e^{-j2N\pi F}}{N+1} = \frac{1}{N+1} \sum_{m=0}^{N} e^{-j2\pi mF}, \qquad (6.129)$$

which can be simplified to

$$H(F) = \frac{e^{-j\pi NF}}{N+1} \frac{\sin(\pi(N+1)F)}{\sin(\pi F)} = e^{-j\pi NF} \, \mathrm{drcl}(F, N+1) \qquad (6.130)$$

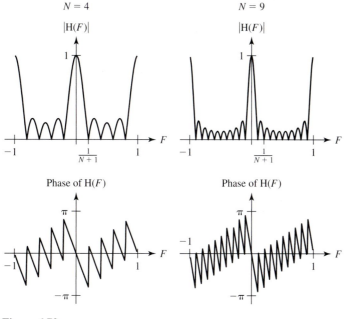

**Figure 6.72**
Frequency response of a moving-average filter for two different
averaging times.

(Figure 6.72). This filter is obviously a lowpass filter in the same general sense as the previous DT lowpass filter, in the DT-frequency range $-\frac{1}{2} < F < \frac{1}{2}$ . And the longer the averaging time is, the more limited is the frequency response of the filter.

The moving-average filter is very easy to implement because it simply adds the current and multiple past values of the excitation to form the response. But it is not the most desirable lowpass filter. Usually we want a filter to approach the ideal lowpass filter in the sense that it passes frequencies in some range with a constant-magnitude and a linear-phase frequency response and that it completely suppresses frequencies outside that range. If we arbitrarily identify the passband of this filter as the range of frequencies between the two first nulls of the magnitude response $-(1/(N + 1)) < F < (1/(N + 1))$, then it does not have a constant magnitude in that range and does not completely suppress frequencies outside that range. It does, however, have a linear phase shift in the passband.

If we want to approach the frequency-domain performance of the ideal lowpass filter, we must design a DT filter with an impulse response that closely approaches the inverse DTFT of the ideal frequency response. We have previously shown that the ideal lowpass filter is noncausal and cannot be physically realized. However, we can closely approach it. The ideal lowpass filter impulse response is illustrated in Figure 6.73.

The problem in realizing this filter physically is the part of the impulse response that occurs before time $n = 0$. If we arrange to delay the impulse response by a large amount, then the part of the impulse response that occurs before time $n = 0$ will become very small and we can truncate it and closely approach the ideal frequency response (Figures 6.74 and 6.75). The magnitude response in the stopband is so small we cannot see its shape when plotted on a linear scale as in Figure 6.75. In cases like

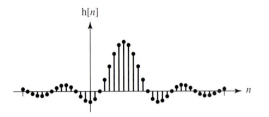

**Figure 6.73**
Ideal DT lowpass filter impulse response.

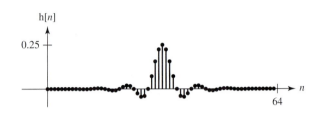

**Figure 6.74**
Almost-ideal DT lowpass filter impulse response.

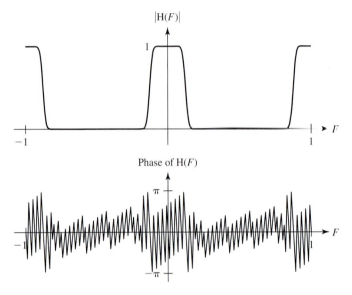

**Figure 6.75**
Almost-ideal DT lowpass filter frequency response.

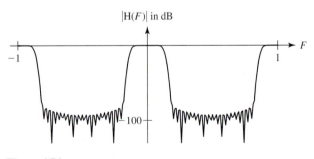

**Figure 6.76**
Almost-ideal DT lowpass filter frequency response plotted on a
decibel scale.

this a log-magnitude plot helps us see what the real attenuation is in the stopband
(Figure 6.76).

This filter has a very nice lowpass filter magnitude response, but it comes at a
price. We must wait for it to respond. The closer a filter approaches the ideal, the
greater time delay there is in the impulse response. This is apparent in the time delay

of the impulse response and the phase shift of the frequency response. The fact that a long delay is required for filters which approach the ideal is true of highpass, bandpass, and bandstop filters and it is true for both CT and DT filters. It is a general principle of filter design that any filter designed to be able to discriminate between two closely spaced frequencies and pass one while stopping the other must observe them for a long time to be able to distinguish one from the other. The closer they are in frequency, the longer the filter must observe them to be able to make the distinction. That is the basic reason for the requirement for a long time delay in the response of a filter which approaches an ideal filter.

One might wonder why we would want to use a DT filter instead of a CT filter. There are several reasons. DT filters are built with three basic elements, a delay device, a multiplier, and an adder. These can be implemented with digital devices. As long as we stay within their intended ranges of operation these devices always do exactly the same thing. That cannot be said of devices such as resistors and capacitors which make up CT filters. A resistor of a certain nominal resistance is never exactly that value, even under ideal conditions. And even if it were at some time, temperature or other environmental effects would change it. The same thing can be said of capacitors, inductors, transistors, etc. So DT filters are more stable and reproducible than CT filters.

It is often difficult to implement a CT filter at very low frequencies because the component sizes become unwieldy; for example, very large capacitor values may be needed. Also, at very low frequencies thermal drift effects on components become a very big problem because they are indistinguishable from signal changes in the same frequency range. DT filters do not have these same problems.

DT filters are often implemented with programmable digital hardware. That means that this type of DT filter can be reprogrammed to perform a different function without changing the hardware. CT filters do not have this flexibility. Also there are some types of DT filters that are so computationally sophisticated that they would be practically impossible to implement as CT filters.

DT signals can be reliably stored for very long times without any degradation on magnetic disk or tape or CD-ROM. CT signals can be stored on analog magnetic tape, but over time the exact values can degrade.

By time-multiplexing DT signals, one DT filter can accommodate multiple signals in a way that seems to be, and effectively is, simultaneous. CT filters cannot do that because to operate correctly they require that the signal always be present.

## 6.8 FILTER SPECIFICATIONS AND FIGURES OF MERIT

Practical filters are often specified or characterized by certain descriptors or figures of merit which quantify how closely they approach the ideal filter behavior. We have seen four types of ideal filters, lowpass, highpass, bandpass, and bandstop. But a filter may be more complicated than one of these prototype forms. A general ideal filter may have multiple passbands and stopbands. It has a constant magnitude and linear phase throughout its passbands, and an absolute zero response throughout its stopbands, and all the transitions between passbands and stopbands typically have zero width (Figure 6.77).

No practical filter can ever achieve the ideal-filter behavior, so it is important to be able to describe in a quantitative manner how close any particular practical filter comes to the ideal in various ways.

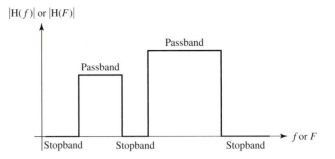

**Figure 6.77**
General ideal-filter magnitude specification.

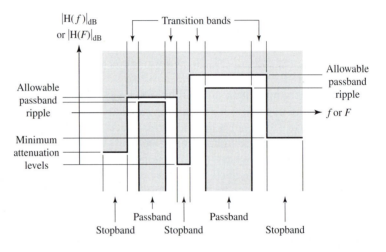

**Figure 6.78**
A typical practical-filter specification.

A typical practical-filter specification usually includes several elements,

1. One or more passbands
2. One or more stopbands
3. Transition bands between passbands and stopbands
4. An allowable ripple in the passbands
5. A minimum required attenuation in the stopbands

Sometimes, in addition, there may be a specification on the phase frequency response in the passbands. A typical practical-filter magnitude specification might look like Figure 6.78. The filter's magnitude frequency response is required to lie entirely between the shaded areas. A filter which meets this specification might look like Figure 6.79.

As stated previously, the ideal filter has a constant magnitude response throughout its passband and an absolute zero response throughout its stopband. One measure of how close a filter comes to the ideal is its passband ripple. *Ripple* is usually defined as the maximum peak-to-peak variation in the passband, usually specified in decibels (Figure 6.80).

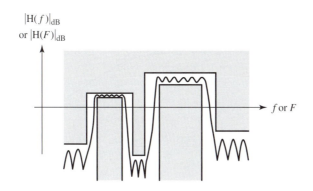

$|H(f)|_{dB}$
or $|H(F)|_{dB}$

$f$ or $F$

**Figure 6.79**
A filter meeting the specification of Figure 6.78.

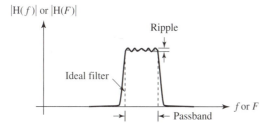

$|H(f)|$ or $|H(F)|$

Ripple

Ideal filter

$f$ or $F$

Passband

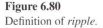

**Figure 6.80**
Definition of *ripple*.

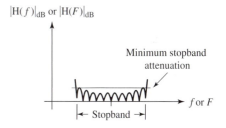

$|H(f)|_{dB}$ or $|H(F)|_{dB}$

Minimum stopband
attenuation

$f$ or $F$

Stopband

**Figure 6.81**
Stopband attenuation.

An ideal filter has infinite attenuation in its stopband. No practical filter can ever achieve that, so an important specification is the minimum attenuation in the stopband (Figure 6.81).

Another commonly used descriptor of filter performance is *rolloff*, the rate at which a filter's magnitude frequency response falls in moving from a passband to a stopband through a transition band. This is, of course, closely associated with the specification of the width of the transition band; the narrower the transition band, the faster the rolloff is required to be. Rolloff is typically specified as some number of decibels per octave or per decade. An octave is a factor of two in frequency, and a decade is a factor of 10 in frequency.

**EXAMPLE 6.5**

Find the ripple, minimum stopband attenuation, and rolloff of a single-pole lowpass $RC$ filter with a $-3$-dB corner frequency of 100 Hz, a passband of $0 < f < 50$, and a stopband of $f > 200$.

**■ Solution**
The transfer function of this filter is

$$H(f) = \frac{1}{1 + j(f/100)} = \frac{100}{100 + jf}. \qquad (6.131)$$

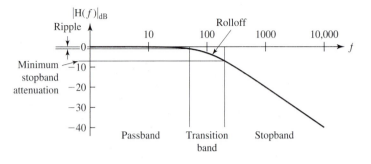

**Figure 6.82**
Bode diagram of the magnitude frequency response of an $RC$ lowpass filter.

The maximum magnitude frequency response occurs at $f = 0$ and $|H(0)| = 1$. The minimum magnitude frequency response occurs at the passband edge, $f = 50$, and $|H(50)| = 0.8944$. Therefore, the ripple in this case is the difference between these two extremes which is 0.97 dB. The minimum signal attenuation in the stopband occurs at the band edge $f = 200$ and is about 7 dB. The rolloff of the filter is the rate at which the filter's magnitude response falls in the transition band. This rolloff is a function of frequency, and the best way to grasp it is to plot the magnitude response as a Bode diagram (Figure 6.82). The rolloff asymptotically approaches 20 dB per decade.

## EXAMPLE 6.6

Design an almost-ideal bandpass DT filter with a passband of $0.2 < F < 0.3$ by truncating the impulse response of an ideal bandpass filter to 64 nonzero DT impulses and then determine its passband ripple and minimum stopband attenuation.

■ **Solution**

An ideal DT bandpass filter with a passband of $0.2 < F < 0.3$ would have a transfer function

$$H(F) = \left\{ \left[ \text{rect}\left( 10\left( F - \frac{1}{4} \right) \right) + \text{rect}\left( 10\left( F + \frac{1}{4} \right) \right) \right] * \text{comb}(F) \right\} e^{-j2\pi F n_0} \qquad \textbf{(6.132)}$$

where $n_0$ is yet to be determined. Using

$$\text{sinc}\left( \frac{n}{w} \right) \overset{\mathcal{F}}{\longleftrightarrow} w\,\text{rect}(wF) * \text{comb}(F) \qquad \textbf{(6.133)}$$

and the time- and frequency-shifting properties of the DTFT, the impulse response of this ideal filter is found to be

$$h[n] = \frac{1}{5}\,\text{sinc}\left( \frac{n - n_0}{10} \right) \cos\left( \frac{\pi(n - n_0)}{2} \right). \qquad \textbf{(6.134)}$$

Since the length of the filter's impulse response is specified to be 64, for symmetry let $n_0 = 32$ (Figure 6.83).

The frequency response of the truncated ideal-impulse-response filter is illustrated in Figure 6.84. Figure 6.85 is a magnified view showing the ripple in the passband of about 7 dB.

The minimum attenuation in the stopbands depends on how each stopband is designated. The ideal filter makes an instantaneous transition from passband to stopband. Any real filter

Ideal-bandpass-filter impulse response

Truncated ideal-bandpass-filter impulse response

**Figure 6.83**
Ideal and truncated impulse responses of an ideal-bandpass DT filter.

Truncated h[*n*] bandpass filter magnitude frequency response

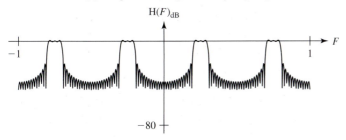

**Figure 6.84**
Magnitude frequency response of the bandpass filter designed by truncating an ideal-filter impulse response.

Truncated h[*n*] BP filter magnitude frequency response

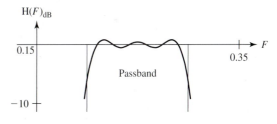

**Figure 6.85**
Passband ripple of the truncated impulse response of the bandpass DT filter.

must have a finite-width transition band. If we take the practical filter passband to be the same as the ideal filter passband, then we must choose some transition bandwidth. A simple choice would be to let the stopband begin at the first local minimum of the frequency-response magnitude outside the passband (Figure 6.86). Using that criterion the minimum stopband attenuation for this design is about 22 dB.

This design can be modified to improve the flatness of the passband and the minimum attenuation in the stopband. If we smooth the truncated impulse response as illustrated in Figure 6.87, we get the magnitude frequency responses of Figures 6.88 and 6.89. The passband response is noticeably smoother and the stopband attenuation is greater than the previous design, but we have lost something; the transition band (as previously defined) is now wider than before.

Truncated h[n] BP filter magnitude frequency response

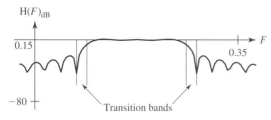

**Figure 6.86**
Stopband attenuation of the filter's truncated impulse response.

Truncated and smoothed h[n] bandpass filter impulse response

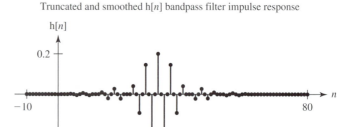

**Figure 6.87**
Truncated and smoothed impulse response.

Truncated and smoothed h[n] BP filter magnitude frequency response

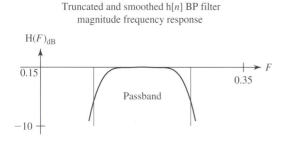

**Figure 6.88**
Passband ripple of the truncated, smoothed impulse response of the filter.

Truncated and smoothed h[n] BP filter magnitude frequency response

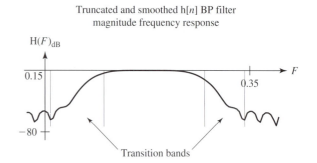

**Figure 6.89**
Stopband attenuation of the truncated, smoothed impulse response of the filter.

The design and analysis of filters like the one in Example 6.6 will be covered in Chapter 12.

## 6.9 COMMUNICATION SYSTEMS

One of the most important applications of the Fourier transform is in the analysis and design of communication systems. Let us approach this concept by analyzing the operation of a radio transmitter and receiver. Why do we have radios? Because they solve the problem of communication between people who are too far apart to communicate directly with sound. There are, of course, many types of communication at a distance. The communication could be one-way as in radio and television or two-way as in telephone, amateur radio, and Internet. The information transferred could be voice, data, images, etc. The communication could be real-time or delayed.

Suppose a person in Miami and a person in Seattle want to converse. The human voice is obviously too weak to be heard at that distance. We could use amplifiers and loudspeakers to increase the acoustic power of the voice, but because the acoustical power dies quite rapidly with distance, we would need an incredibly powerful system to be heard at that distance (Figure 6.90).

If a voice in Miami could be heard in Seattle and vice versa, with acoustic amplification, there might be a few complaints from the people in Orlando and Spokane

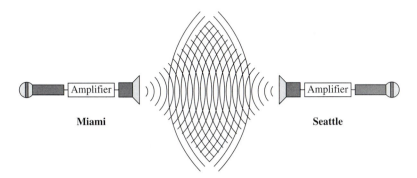

**Figure 6.90**
A crude, naive communication system.

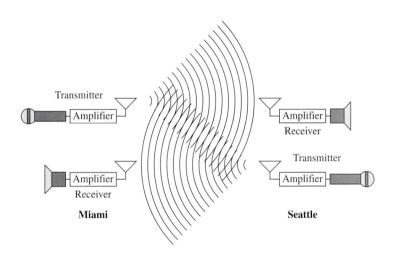

**Figure 6.91**
Communication system using direct acoustic-to-electromagnetic and
electromagnetic-to-acoustic conversion.

about the noise. (There would not be any complaints from people in Miami and Seattle
because they would all have been killed by the acoustic energy.) Also, if the communication is two-way, given the speed of sound in air, the person in Seattle would have
to wait more than 8 h to hear a response to a question she asked of the person in Miami.
If we throw in the problems of millions of people in the United States talking simultaneously and the attendant lack of privacy of their communication, we realize that this
would be an extremely unsatisfactory, and ridiculous, system.

A good solution to many of these problems is to use electromagnetic energy propagation to carry messages between remote locations. Its speed is so much greater than
the speed of sound, that the problem of delay would be solved. But we now have some
other problems to solve. How do we encode an acoustic message in an electromagnetic
signal so the message will propagate at the speed of the electromagnetic wave (the
speed of light)?

The simplest idea is to simply use a microphone to directly convert acoustic energy to electromagnetic energy (Figure 6.91). Then the electromagnetic energy could
drive an amplifier which would drive a transmitter antenna. A receiver antenna at the

remote location could collect some of the transmitted electromagnetic energy, and an amplifier and speaker could reconvert the electromagnetic energy to acoustic energy.

There are two principal problems with this simple approach. First, the frequency spectrum of voice communication is mostly between 30 and 300 Hz and even music program sources do not extend much beyond 10 kHz. An efficient antenna in this frequency range would have to be very long (many miles long). Also the variation of frequency over a range of 10:1 up to maybe 1000:1 in frequency would mean that the signal would be significantly distorted by the variation of antenna efficiency with frequency. Maybe we could build a very long antenna or maybe we could just live with an inefficient one. But the second problem is more significant. On the assumption that many people would want to talk simultaneously (a pretty good assumption), after conversion of the energy back to acoustic form, we still have the problem of hearing everyone talking at once because they are all transmitting at the same time.

Standard telephone systems solve this problem by confining the electromagnetic energy to a cable, either copper or optical fiber. That is, the signals are *spatially* separated by having a dedicated direct connection between the parties. But with modern wireless cellular telephones that solution does not work because the electromagnetic energy is not confined on its path between the handset and the nearest cellular antenna. Another solution would be to assign to each transmitter a unique set of time intervals in which every other transmitter would not transmit. Then, to receive the correct message, the receiver would have to be synchronized to these same times (accounting for propagation delays). This solution is called *time multiplexing*. Time multiplexing is used extensively in telephone systems where the signal is confined to cables or in local cellular areas where the telephone company can control all the timing and the intervals can be made so short that they are not noticed by the people using the system. But time multiplexing has some problems in other communication systems. If the electromagnetic energy propagation is in free space, with multiple independent transmitters and receivers involved in a national communication system, time multiplexing becomes practically impossible. There is a better solution, and understanding it requires the Fourier transform. The solution is called *frequency multiplexing,* and it depends on using a technique called *modulation.*

## MODULATION

**Double-Sideband Suppressed-Carrier Modulation**    Let us represent a signal to be transmitted by x($t$). If we were to multiply this signal by a sinusoid as illustrated in Figure 6.92, we would get a new signal y($t$), which is the product of the original signal and the sinusoid. In the language of communication systems the signal x($t$) *modulates* the carrier $\cos(2\pi f_c t)$. In this case the modulation is called *amplitude modulation* because the amplitude of the carrier is constantly being modified by the signal level of the modulation x($t$) (Figure 6.93).

The modulator response signal is

$$y(t) = x(t)\cos(2\pi f_c t). \tag{6.135}$$

Fourier-transforming both sides,

$$Y(f) = X(f) * \frac{1}{2}[\delta(f - f_c) + \delta(f + f_c)] \tag{6.136}$$

or

$$Y(f) = \frac{1}{2}[X(f - f_c) + X(f + f_c)]. \tag{6.137}$$

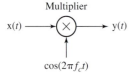

Multiplier

x($t$) —→ ⊗ —→ y($t$)

$\cos(2\pi f_c t)$

**Figure 6.92**
An analog multiplier acting as a modulator.

x(t)

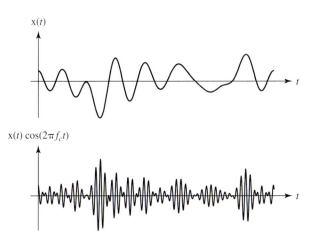

$x(t)\cos(2\pi f_c t)$

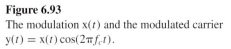

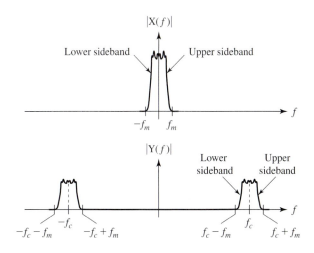

**Figure 6.93**
The modulation x($t$) and the modulated carrier
y($t$) = x($t$) cos(2$\pi f_c t$).

**Figure 6.94**
The modulation and the modulated carrier in the frequency
domain.

So now it can be seen that this kind of modulation has the effect of simply shifting the spectrum of the modulating signal up and down by the carrier frequency $f_c$ in the frequency domain (Figure 6.94).

So something which looks complicated in the time domain looks quite simple in the frequency domain. This is one of the advantages of frequency-domain analysis. This kind of amplitude modulation is called *double-sideband suppressed-carrier* (DSBSC) modulation, and it is the simplest to describe mathematically. The name comes from the fact that the two sidebands above and below zero frequency in the spectrum of x($t$) are translated into the two sidebands above and below $f_c$ and there is no impulse at the carrier frequency in the spectrum of the modulated signal.

DSBSC modulation is not used much in practice. However, an understanding of DSBSC modulation goes a long way toward understanding the more commonly used forms of modulation, so this is a good place to start. We have now accomplished one goal. The spectrum of the original signal which started out in a range of low frequencies has been shifted to a new range which can be located anywhere we desire by choosing the carrier frequency appropriately. The original signal resides in a bandwidth centered at zero, and the conventional nomenclature is that the original signal is at baseband. After the modulation, the signal information is in a different frequency band.

The solution to the problem of everyone talking at once in the same frequency range is to have everyone use a different frequency range by using a different carrier frequency. Take the case of amplitude modulation (AM) broadcast radio. There are many transmitting stations in any given geographic region simultaneously broadcasting. Each station is assigned a frequency band in which to broadcast. These frequency bands are 10 kHz wide. So a radio station modulates a carrier with its program source signal (the baseband signal). The carrier is at the center of its assigned frequency band. The modulated carrier then drives the transmitter. If the baseband signal has a bandwidth of less than 5 kHz, the station's broadcast signal will lie completely within its assigned frequency band. A receiver has to choose one station to listen to and reject the others. Its antenna receives energy from all stations and converts them all into a voltage at its terminals. Therefore, the receiver has to somehow select one frequency band to listen to and reject all others.

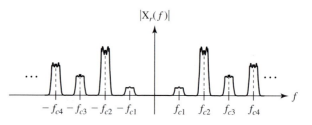

**Figure 6.95**
Spectrum of signal received by receiver antenna.

There is more than one way to select a single station to receive. But the most common way is to use the modulation idea again, but this time the operation is called *demodulation*. Let's suppose that the signal received by the antenna $x_r(t)$ is the sum of signals from several radio stations in the area and that the spectrum of the antenna signal is as illustrated in Figure 6.95. Suppose the station we want to hear is the one centered at $f_{c3}$. We multiply the received antenna signal by the sinusoid at that frequency creating a demodulated signal $y_r(t)$.

$$y_r(t) = x_r(t)\cos(2\pi f_c t) = A[x_1(t)\cos(2\pi f_{c1}t) + x_2(t)\cos(2\pi f_{c2}t)$$
$$+ \cdots + x_N(t)\cos(2\pi f_{cN}t)]\cos(2\pi f_{c3}t) \quad \textbf{(6.138)}$$

or

$$y_r(t) = A\sum_{k=1}^{N} x_k(t)\cos(2\pi f_{ck}t)\cos(2\pi f_{c3}t). \quad \textbf{(6.139)}$$

In the frequency domain,

$$Y_r(f) = A\sum_{k=1}^{N} X_k(f) * \frac{1}{2}[\delta(f - f_{ck}) + \delta(f + f_{ck})] * \frac{1}{2}[\delta(f - f_{c3}) + \delta(f + f_{c3})]$$
$$\textbf{(6.140)}$$

or

$$Y_r(f) = \frac{A}{4}\sum_{k=1}^{N} X_k(f) * [\delta(f - f_{c3} - f_{ck}) + \delta(f + f_{c3} - f_{ck})$$
$$+ \delta(f - f_{c3} + f_{ck}) + \delta(f + f_{c3} + f_{ck})]. \quad \textbf{(6.141)}$$

or

$$Y_r(f) = \frac{A}{4}\sum_{k=1}^{N} [X_k(f - f_{c3} - f_{ck}) + X_k(f + f_{c3} - f_{ck})$$
$$+ X_k(f - f_{c3} + f_{ck}) + X_k(f + f_{c3} + f_{ck})]. \quad \textbf{(6.142)}$$

This result looks complicated, but it is really not. Again we are just shifting the incoming signal up and down in frequency space and adding as illustrated in Figure 6.96.

Notice that the information spectrum that was centered at $f_{c3}$ has moved up and down and is now centered at zero (and also at $\pm 2f_{c3}$). We can now recover the original signal that was modulated up by the transmitter to $f_{c3}$ by applying a lowpass filter to this signal which passes only the signal power contained in the bandwidth of the desired information which is now centered at zero. This is not exactly how a typical AM receiver works, but many of the same processes are used in a typical AM receiver, and this technique does work.

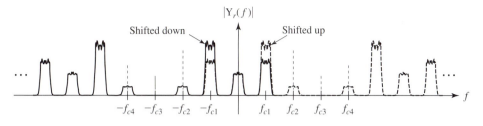

**Figure 6.96**
Receiver signal after demodulation.

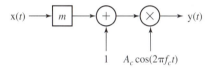

**Figure 6.97**
A double-sideband transmitted-carrier
modulator.

One problem with this technique is that the sinusoid at a frequency of $f_{c3}$ which is used in demodulation, the so-called local oscillator in the receiver, must not only be at exactly the right frequency $f_{c3}$ but must also be in phase with the carrier as received (or at least close) for best results. If the frequency of the local oscillator drifts just the slightest bit, the receiver will not work right. An annoying tone called the beat frequency will be heard as the local oscillator drifts off the exact frequency. The *beat frequency* is the difference between the carrier frequency and the local oscillator frequency. As a result, for this technique to work, the local oscillator's frequency and phase must be locked to the carrier phase. This is most commonly done with a device called a *phase-locked loop*. This type of demodulation is called *synchronous* demodulation because of the requirement that the carrier and the local oscillator be in phase (synchronized).

We use the term *tuning* a radio receiver to pick up the desired station. When we tune to a station, we are simply changing the frequency of the local oscillator in the receiver to cause a different station's signal to appear centered at zero (at baseband). There are simpler and more economical ways of doing the demodulation that are used in most standard AM receivers. In AM radio, the modulation is double-sideband transmitted-carrier modulation. But variants of DSBSC modulation are actually used in many types of communication systems which use modulation.

**Double-Sideband Transmitted-Carrier Modulation**    As previously mentioned, DSBSC modulation is not widely used. A modulation technique that is widely used is double-sideband transmitted-carrier (DSBTC) modulation. This is the technique used by commercial AM radio transmitters and by most international shortwave transmitters. It is very similar to DSBSC, the only difference being the addition of a constant to the signal $x(t)$ before modulation (Figure 6.97).

To simplify the analysis, assume that the signal $x(t)$ is normalized so that its maximum negative excursion is $-1$ (in some appropriate system of units). Then in this implementation $m$ is called the *modulation index*. (For most practical modulating signals, if the maximum negative excursion is $-1$, the maximum positive excursion is

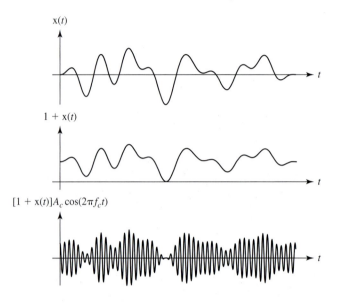

x(t)

1 + x(t)

$[1 + x(t)]A_c \cos(2\pi f_c t)$

**Figure 6.98**
DSBTC modulation and modulated carrier with $m = 1$.

approximately $+1$.) The response of the modulator is

$$y(t) = [1 + m\mathrm{x}(t)]A_c \cos(2\pi f_c t) \tag{6.143}$$

(Figure 6.98). Fourier-transforming (6.143),

$$Y(f) = [\delta(f) + m\mathrm{X}(f)] * \frac{A_c}{2}[\delta(f - f_c) + \delta(f + f_c)] \tag{6.144}$$

or

$$Y(f) = \frac{A_c}{2}\{[\delta(f - f_c) + \delta(f + f_c)] + m[\mathrm{X}(f - f_c) + \mathrm{X}(f + f_c)]\} \tag{6.145}$$

(Figure 6.99).

Looking at the spectrum we can see where the name transmitted carrier came from. There is an impulse at the carrier frequency which was not present in DSBSC modulation. It is natural to wonder why this modulation technique is so widely used, given that it requires a slightly more complicated system to implement. The reason is that, even though DSBTC modulation is a little more complicated than DSBSC modulation, DSBTC *demodulation* is much simpler than DSBSC demodulation. For each commercial AM radio station there is one transmitter which modulates the carrier with the baseband signal and thousands or even millions of receivers which demodulate the modulated carrier signal to re-create the baseband signal. DSBTC demodulation is very simple using a circuit called an *envelope* detector. Its operation is best understood in the time domain. In DSBTC modulation, the modulated carrier traces out the shape of the baseband signal with the positive peaks of the carrier oscillation (Figure 6.100).

The envelope detector is a circuit which follows the peaks of the modulated carrier, thereby approximately reproducing the baseband signal (Figure 6.101). The reproduction of the baseband signal represented in Figure 6.101 is not very good, but it does illustrate the concept of the operation of an envelope detector. In actual practice the carrier frequency would be much higher than represented in this figure and the

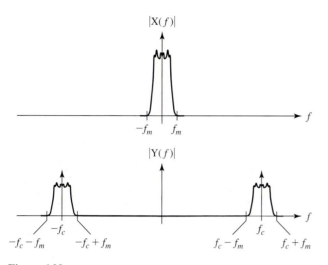

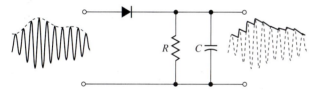

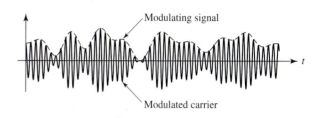

**Figure 6.99**
Spectra of baseband signal and DSBTC signal.

**Figure 6.100**
Relation between baseband signal and modulated carrier.

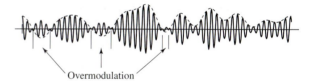

**Figure 6.101**
Envelope detector circuit.

**Figure 6.102**
Overmodulation.

reproduction of the baseband signal would be much better. The explanation of the operation of the envelope detector was done in the time domain. That is because the envelope detector is a nonlinear system, and therefore linear system theory does not apply. No local oscillator or synchronization is required for envelope detection, so this demodulation technique is called *asynchronous* demodulation.

A DSBTC signal can also be demodulated by the same demodulation technique used for the DSBSC signal in the previous section but that requires a local oscillator in the receiver generating a sinusoid in phase with the received carrier. The envelope detector is much simpler and less expensive.

In Figures 6.98 to 6.100 the modulation index was $m = 1$. If $m > 1$, $1 + m\mathrm{x}(t)$ can go negative, overmodulation occurs, and the envelope detector cannot recover the original baseband signal without some distortion (Figure 6.102).

**Single-Sideband Modulation and Demodulation**   The amplitude spectrum $\mathrm{X}(f)$ of any real signal $\mathrm{x}(t)$ has the quality that

$$\mathrm{X}(f) = \mathrm{X}^*(-f). \tag{6.146}$$

Therefore, the information in $\mathrm{X}(f)$, for $f \geq 0$ only, is sufficient to reconstruct the signal exactly. That fact underlies the concept of single-sideband suppressed-carrier (SSBSC) modulation. In DSBSC modulation, the amplitude spectrum centered at the

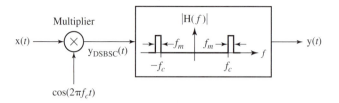

**Figure 6.103**
Single-sideband suppressed-carrier modulator.

carrier frequency (and at the negative of the carrier frequency) has information from $X(f)$ over the frequency range $-f_m < f < f_m$. But only half of that amplitude spectrum needs to be transmitted if the receiver is designed correctly. The advantage of transmitting only half the amplitude spectrum is that only half as much bandwidth is needed as in DSBSC modulation.

An SSBSC modulator is almost the same as a DSBSC modulator. The difference is a filter which removes either the upper or lower sideband before transmitting (Figure 6.103). The response from the multiplier is the same as it was in the DSBSC case earlier,

$$y_{\text{DSBSC}}(t) = x(t) \cos(2\pi f_c t). \qquad \textbf{(6.147)}$$

In the frequency domain the amplitude spectrum of the response of the multiplier is

$$Y_{\text{DSBSC}}(f) = \frac{1}{2}[X(f - f_c) + X(f + f_c)]. \qquad \textbf{(6.148)}$$

The filter in Figure 6.103 removes the lower sideband and leaves the upper sideband. The amplitude spectrum which results is

$$Y(f) = \frac{1}{2}[X(f - f_c) + X(f + f_c)]H(f) \qquad \textbf{(6.149)}$$

(Figure 6.104).

The demodulation process for SSBSC is the same as the first technique introduced for DSBSC, multiplication of the received signal by a local oscillator in phase with the carrier (Figure 6.105). If this signal is now lowpass filtered, the original spectrum is recovered. The original signal is completely recovered because all the information is in a single sideband. This type of modulation is much more easily understood using frequency-domain analysis than using time-domain analysis.

**Quadrature Carrier Modulation**    It is possible to transmit two signals simultaneously in the same bandwidth and still separate them with a receiver. The two signals modulate two carriers which are at the same frequency but are 90° out of phase making them orthogonal (Figure 6.106). The name *quadrature carrier* modulator comes from the fact that the sine and cosine carriers are *in quadrature* which means 90° out of phase, or orthogonal.

The modulator response signal is $y(t) = x_1(t) \sin(2\pi f_c t) + x_2(t) \cos(2\pi f_c t)$. Its CTFT is

$$Y(f) = X_1(f) * \frac{j}{2}[\delta(f + f_c) - \delta(f - f_c)] + X_2(f) * \frac{1}{2}[\delta(f - f_c) + \delta(f + f_c)]$$

$$\textbf{(6.150)}$$

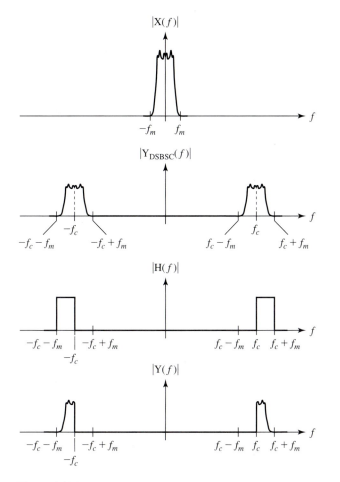

**Figure 6.104**
Operation of an SSBSC modulator.

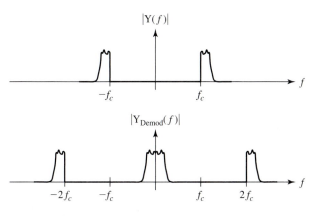

**Figure 6.105**
SSBSC demodulation.

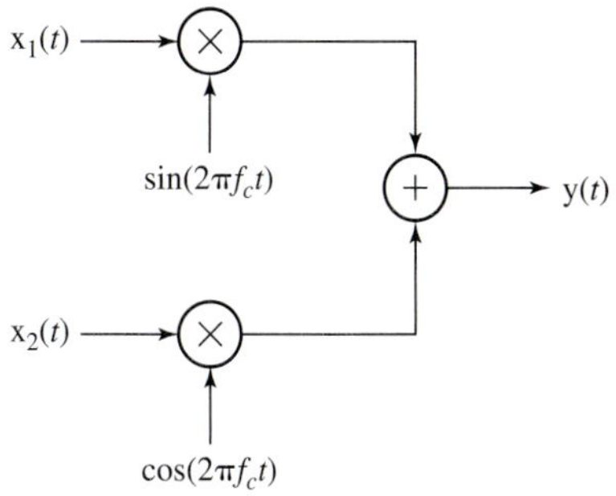

**Figure 6.106**
Quadrature carrier modulator.

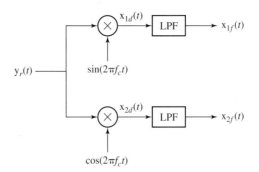

**Figure 6.107**
Quadrature amplitude demodulator.

or

$$Y(f) = \frac{j}{2}[X_1(f + f_c) - X_1(f - f_c)] + \frac{1}{2}[X_2(f - f_c) + X_2(f + f_c)]. \qquad \textbf{(6.151)}$$

The demodulation process multiplies the received signal $y(t)$ by local oscillators which are in phase with the sine and cosine carriers (Figure 6.107).

The two responses from the demodulator are

$$x_{1d}(t) = y(t)\sin(2\pi f_c t) = x_1(t)\sin^2(2\pi f_c t) + x_2(t)\sin(2\pi f_c t)\cos(2\pi f_c t) \qquad \textbf{(6.152)}$$

and

$$x_{2d}(t) = y(t)\cos(2\pi f_c t) = x_1(t)\sin(2\pi f_c t)\cos(2\pi f_c t) + x_2(t)\cos^2(2\pi f_c t) \qquad \textbf{(6.153)}$$

The CTFTs of these signals are

$$X_{1d}(f) = \left[\frac{j}{2}[X_1(f + f_c) - X_1(f - f_c)] + \frac{1}{2}[X_2(f - f_c) + X_2(f + f_c)]\right]$$
$$* \frac{j}{2}[\delta(f + f_c) - \delta(f - f_c)] \qquad \textbf{(6.154)}$$

and

$$X_{2d}(f) = \left[\frac{j}{2}[X_1(f + f_c) - X_1(f - f_c)] + \frac{1}{2}[X_2(f - f_c) + X_2(f + f_c)]\right]$$
$$* \frac{1}{2}[\delta(f - f_c) + \delta(f + f_c)] \qquad \textbf{(6.155)}$$

Carrying out the convolutions and simplifying,

$$X_{1d}(f) = \frac{1}{2}X_1(f) - \frac{1}{4}X_1(f + 2f_c) + \frac{j}{4}X_2(f + 2f_c)$$
$$- \frac{1}{4}X_1(f - 2f_c) - \frac{j}{4}X_2(f - 2f_c) \qquad \textbf{(6.156)}$$

and

$$X_{2d}(f) = \frac{1}{2}X_2(f) + \frac{j}{4}X_1(f + 2f_c) - \frac{j}{4}X_1(f - 2f_c)$$
$$+ \frac{1}{4}X_2(f - 2f_c) + \frac{1}{4}X_2(f + 2f_c). \qquad \textbf{(6.157)}$$

The upper response signal $X_{1r}(f)$ consists of the original signal $X_1(f)$ plus some other signals which are centered at $\pm 2 f_c$. Therefore, the lowpass filters recover the original signals.

This is a synchronous demodulation method and depends very much on having two local oscillators at exactly the same frequency as the carrier and with exactly the right phases. If the phases are wrong, there will be crosstalk between the two response signals. *Crosstalk* means that some of the power of one signal will appear in the other signal, and vice versa. Therefore, although two signals are squeezed into the same bandwidth and can be theoretically separated because their carriers are in quadrature, the demodulation process requires high precision and is much more difficult to implement than an envelope detector.

## PHASE AND GROUP DELAY

Now that we have explored the basic modulation methods, we are ready to consider another important phenomenon, group delay. We have seen in the analysis of ideal filters and in the time-shifting property of the Fourier transform that a simple delay of a time-domain signal corresponds to a linear variation of phase with frequency in the frequency domain,

$$\text{x}(t - t_0) \xleftrightarrow{\ \mathcal{F}\ } \text{X}(j\omega)e^{-j\omega t_0} \qquad \text{or} \qquad \text{x}[n - n_0] \xleftrightarrow{\ \mathcal{F}\ } \text{X}(j\Omega)e^{-j\Omega n_0} \qquad \textbf{(6.158)}$$

(This analysis will be done using radian frequency because the notation is somewhat more compact.) But most system transfer functions have a nonlinear dependence of phase on frequency. An important consideration in system design is how to interpret the significance of a nonlinear phase variation with frequency.

Fourier analysis is the analysis of signals viewing them as linear combinations of complex sinusoids. The phase of each excitation complex sinusoid at any frequency $\omega$ is changed to the phase of the response complex sinusoid at that same frequency according to the value of the transfer function at that frequency,

$$\text{Y}(j\omega) = \text{X}(j\omega)\text{H}(j\omega). \qquad \textbf{(6.159)}$$

So each complex sinusoid is delayed in time by an amount corresponding to the phase shift of $\text{H}(j\omega)$. This seems simple enough until we consider that if the phase shifts a half cycle $\pi$ rad, it is not clear by simply observing the complex sinusoid whether the shift is $\pi$ or $-\pi$ rad or, for that matter, $\pi \pm 2m\pi$ rad, where $m$ is an integer. Along with the phase ambiguity is the corresponding time ambiguity in the time domain. Of more significance in most practical system design is how the shape of an arbitrary signal which is a linear combination of complex sinusoids is affected by the variation of phase shift with frequency.

To illustrate a nonobvious effect of nonlinear phase delay on a signal, suppose the excitation signal x(t) is a sinusoidal carrier at a frequency $\omega_c$, which is DSBSC modulated by a modulation sinusoid at a frequency $\omega_m$ and that $\omega_m \ll \omega_c$. Then the excitation can be expressed as

$$\text{x}(t) = A\cos(\omega_m t)\cos(\omega_c t). \qquad \textbf{(6.160)}$$

The CTFT of this signal is

$$\text{X}(j\omega) = \frac{A\pi}{2}[\delta(\omega - \omega_c - \omega_m) + \delta(\omega - \omega_c + \omega_m)$$
$$+ \, \delta(\omega + \omega_c - \omega_m) + \delta(\omega + \omega_c + \omega_m)]. \qquad \textbf{(6.161)}$$

Suppose the system excited by this signal has a transfer function whose magnitude is a constant, one, over the frequency range $\omega_c - \omega_m < |\omega| < \omega_c + \omega_m$ and whose phase is given by $\phi(\omega)$ and that, as is always true for a real system $\phi(\omega) = -\phi(-\omega)$. The response of the system is

$$Y(j\omega) = \frac{A\pi}{2}[\delta(\omega - \omega_c - \omega_m) + \delta(\omega - \omega_c + \omega_m)$$
$$+ \delta(\omega + \omega_c - \omega_m) + \delta(\omega + \omega_c + \omega_m)]e^{j\phi(\omega)} \qquad \textbf{(6.162)}$$

or, using the equivalence property of the impulse,

$$Y(j\omega) = \frac{A\pi}{2}\big[\delta(\omega - \omega_c - \omega_m)e^{j\phi(\omega_c+\omega_m)} + \delta(\omega - \omega_c + \omega_m)e^{j\phi(\omega_c-\omega_m)}$$
$$+ \delta(\omega + \omega_c - \omega_m)e^{j\phi(-\omega_c+\omega_m)} + \delta(\omega + \omega_c + \omega_m)e^{j\phi(-\omega_c-\omega_m)}\big].$$
$$\textbf{(6.163)}$$

This expression can be written equivalently as

$$Y(j\omega) = \frac{A\pi}{2}\big[\delta(\omega - \omega_c - \omega_m)e^{j\omega(\phi(\omega_c+\omega_m)/(\omega_c+\omega_m))}$$
$$+ \delta(\omega - \omega_c + \omega_m)e^{j\omega(\phi(\omega_c-\omega_m)/(\omega_c-\omega_m))}$$
$$+ \delta(\omega + \omega_c - \omega_m)e^{-j\omega(\phi(-\omega_c+\omega_m)/(\omega_c-\omega_m))}$$
$$+ \delta(\omega + \omega_c + \omega_m)e^{-j\omega(\phi(-\omega_c-\omega_m)/(\omega_c+\omega_m))}\big] \qquad \textbf{(6.164)}$$

or, using $\phi(\omega) = -\phi(-\omega)$,

$$Y(j\omega) = \frac{A\pi}{2}\big[\delta(\omega - \omega_c - \omega_m)e^{j\omega(\phi(\omega_c+\omega_m)/(\omega_c+\omega_m))}$$
$$+ \delta(\omega + \omega_c + \omega_m)e^{j\omega(\phi(\omega_c+\omega_m)/(\omega_c+\omega_m))}$$
$$+ \delta(\omega - \omega_c + \omega_m)e^{j\omega(\phi(\omega_c-\omega_m)/(\omega_c-\omega_m))}$$
$$+ \delta(\omega + \omega_c - \omega_m)e^{j\omega(\phi(\omega_c-\omega_m)/(\omega_c-\omega_m))}\big] \qquad \textbf{(6.165)}$$

or

$$Y(j\omega) = \frac{A\pi}{2}\big[(\delta(\omega - \omega_c - \omega_m) + \delta(\omega + \omega_c + \omega_m))e^{j\omega(\phi(\omega_c+\omega_m)/(\omega_c+\omega_m))}$$
$$+ (\delta(\omega - \omega_c + \omega_m) + \delta(\omega + \omega_c - \omega_m))e^{j\omega(\phi(\omega_c-\omega_m)/(\omega_c-\omega_m))}\big].$$
$$\textbf{(6.166)}$$

The inverse CTFT of (6.166) is

$$y(t) = \frac{A}{2}\left[\cos\left((\omega_c + \omega_m)\left[t + \frac{\phi(\omega_c + \omega_m)}{\omega_c + \omega_m}\right]\right)\right.$$
$$\left. + \cos\left((\omega_c - \omega_m)\left[t + \frac{\phi(\omega_c - \omega_m)}{\omega_c - \omega_m}\right]\right)\right] \qquad \textbf{(6.167)}$$

or

$$y(t) = \frac{A}{2}[\cos((\omega_c + \omega_m)t + \phi(\omega_c + \omega_m)) + \cos((\omega_c - \omega_m)t + \phi(\omega_c - \omega_m))].$$
$$\textbf{(6.168)}$$

Using the trigonometric identity

$$\cos(x)\cos(y) = \frac{1}{2}[\cos(x + y) + \cos(x - y)], \qquad \textbf{(6.169)}$$

we can express (6.168) as

$$y(t) = A \cos\left(\omega_c t + \frac{\phi(\omega_c + \omega_m) + \phi(\omega_c - \omega_m)}{2}\right)$$

$$\times \cos\left(\omega_m t + \frac{\phi(\omega_c + \omega_m) - \phi(\omega_c - \omega_m)}{2}\right) \tag{6.170}$$

or

$$y(t) = A \cos\left(\omega_c\left(t + \frac{\phi(\omega_c + \omega_m) + \phi(\omega_c - \omega_m)}{2\omega_c}\right)\right)$$

$$\times \cos\left(\omega_m\left(t + \frac{\phi(\omega_c + \omega_m) - \phi(\omega_c - \omega_m)}{2\omega_m}\right)\right). \tag{6.171}$$

This result clearly shows that the carrier is shifted in time by $[\phi(\omega_c + \omega_m) + \phi(\omega_c - \omega_m)]/2\omega_c$ seconds and that the modulation is shifted in time by $[\phi(\omega_c + \omega_m) - \phi(\omega_c - \omega_m)]/2\omega_m$ seconds. What these times actually are depends on the nature of the phase-shift function $\phi(\omega)$. Suppose first that $\phi(\omega) = -K\omega$, where $K$ is a positive constant. Then the time shift of the carrier is

$$\frac{K(\omega_c + \omega_m) + K(\omega_c - \omega_m)}{2\omega_c} = -K, \tag{6.172}$$

the time shift of the modulation is

$$\frac{\phi(\omega_c + \omega_m) - \phi(\omega_c - \omega_m)}{2\omega_m} = -K. \tag{6.173}$$

The two time shifts are exactly the same, as they should be for a system with a simple linear phase shift. [Notice that because of the plus sign in (6.171) in the arguments of the cosines that these shifts are delays in time.] Now suppose the phase shift is

$$\phi(\omega) = -\tan^{-1}\left(2\frac{\omega}{\omega_c}\right) \tag{6.174}$$

(a typical single-pole lowpass filter phase shift) and let $\omega_c = 10\omega_m$. Then the time delay of the carrier is $1.107/\omega_c$, and the time delay of the modulation is $0.4/\omega_c$. The two time delays differ by a factor of about 2.75. These effects are illustrated in Figures 6.108 and 6.109 for $\omega_c = 2\pi \times 1000$ and $\omega_m = 2\pi \times 20$. From the magnified view in Figure 6.109 it is apparent that the timing relationship between the carrier and modulation has changed due to the nonlinear phase shift of the system transfer function.

Notice that the expression for the modulation time shift $[\phi(\omega_c + \omega_m) - \phi(\omega_c - \omega_m)]/2\omega_m$ looks a lot like the definition of a derivative. In fact, in the limit as $\omega_m \to 0$, the modulation time shift is $[(d/df)(\phi(\omega))]_{\omega=\omega_c}$. In that same limit, the time shift of the carrier, $[\phi(\omega_c + \omega_m) + \phi(\omega_c - \omega_m)]/2\omega_c$, is $\phi(\omega_c)/\omega_c$. So the carrier time shift is proportional to the phase shift and, for narrowband modulation $\omega_m \ll \omega_c$, the modulation time shift is proportional to the derivative with respect to frequency of the phase shift. This holds approximately true for any modulation in a narrow band centered at the carrier frequency. Group delay is defined as

$$\tau(\omega) = -\frac{d}{d\omega}(\phi(\omega)). \tag{6.175}$$

(The negative sign is there because for a positive time *delay* the quantity, $(d/d\omega)(\phi(\omega))$ must be negative.) A plot of the phase shift $\phi(\omega) = -\tan^{-1}(2(\omega/\omega_c))$ shows graphically the difference between phase delay and group delay (Figure 6.110).

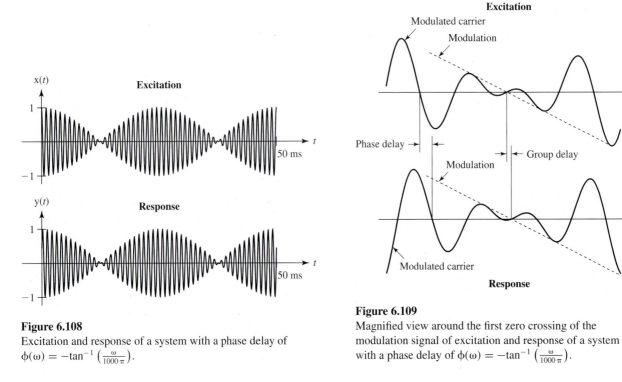

**Figure 6.108**
Excitation and response of a system with a phase delay of
$\phi(\omega) = -\tan^{-1}\left(\frac{\omega}{1000\,\pi}\right)$.

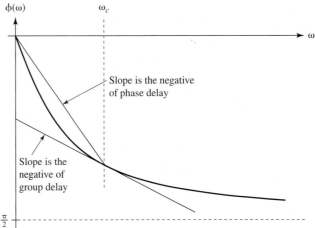

**Figure 6.109**
Magnified view around the first zero crossing of the
modulation signal of excitation and response of a system
with a phase delay of $\phi(\omega) = -\tan^{-1}\left(\frac{\omega}{1000\,\pi}\right)$.

**Figure 6.110**
Relation between phase shift, phase delay, and group delay.

### EXAMPLE 6.7

A system has a transfer function

$$H(j\omega) = \frac{(j\omega - z_1)(j\omega - z_1^*)}{(j\omega - p_1)(j\omega - p_1^*)} = \frac{(j\omega)^2 - j\omega(2z_{1r}) + |z_1|^2}{(j\omega)^2 + j\omega(2\zeta\,\omega_0) + \omega_0^2}, \qquad (6.176)$$

where $\omega_0^2 = |p_1|^2$ and $2\zeta\,\omega_0 = -p_1 - p_1^* = -2p_{1r}$. Find and plot its group delay versus frequency and its impulse response versus time for $z_1 = 1 + j10$ and $p_1 = -1 + j10$.

■ **Solution**

First examine the magnitude of the transfer function. The square of the magnitude is

$$|H(j\omega)|^2 = H(j\omega)H^*(j\omega) = \frac{(j\omega - z_1)(j\omega - z_1^*)}{(j\omega - p_1)(j\omega - p_1^*)} \frac{(-j\omega - z_1^*)(-j\omega - z_1)}{(-j\omega - p_1^*)(-j\omega - p_1)} \qquad \textbf{(6.177)}$$

$$|H(j\omega)|^2 = \frac{\left((\omega - z_{1i})^2 + z_{1r}^2\right)\left((\omega + z_{1i})^2 + z_{1r}^2\right)}{\left((\omega - p_{1i})^2 + p_{1r}^2\right)\left((\omega + p_{1i})^2 + p_{1r}^2\right)} \qquad \textbf{(6.178)}$$

where the subscripts $r$ and $i$ indicate real and imaginary parts, respectively. Since, in this case, $z_{1i} = p_{1i}$ and $z_{1r}^2 = p_{1r}^2$, the magnitude of this transfer function is a constant,

$$|H(j\omega)| = 1. \qquad \textbf{(6.179)}$$

This type of transfer function is called an *all-pass* function. Its magnitude is independent of frequency, but its phase is not. The phase is

$$\phi(\omega) = \angle(j\omega - z_1) + \angle(j\omega - z_1^*) - \angle(j\omega - p_1) - \angle(j\omega - p_1^*) \qquad \textbf{(6.180)}$$

which reduces to

$$\phi(\omega) = \tan^{-1}\left(\frac{\omega - z_{1i}}{-z_{1r}}\right) + \tan^{-1}\left(\frac{\omega + z_{1i}}{-z_{1r}}\right) - \tan^{-1}\left(\frac{\omega - p_{1i}}{-p_{1r}}\right) - \tan^{-1}\left(\frac{\omega + p_{1i}}{-p_{1r}}\right).$$
$$\textbf{(6.181)}$$

Using $\frac{d}{dz}(\tan^{-1}(z)) = \frac{1}{1+z^2}$ the group delay is

$$\tau(\omega) = \frac{1/z_{1r}}{1 + ((\omega - z_{1i})/-z_{1r})^2} + \frac{1/z_{1r}}{1 + ((\omega + z_{1i})/-z_{1r})^2}$$
$$- \frac{1/p_{1r}}{1 + ((\omega - p_{1i})/-p_{1r})^2} - \frac{1/p_{1r}}{1 + ((\omega + p_{1i})/-p_{1r})^2}. \qquad \textbf{(6.182)}$$

or, substituting parameter values,

$$\tau(\omega) = 2\left[\frac{1}{1 + (\omega - 10)^2} + \frac{1}{1 + (\omega + 10)^2}\right] \qquad \textbf{(6.183)}$$

(Figure 6.111).

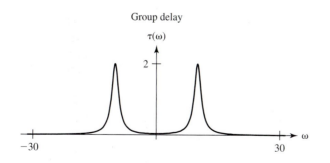

Group delay
$\tau(\omega)$

**Figure 6.111**
Group delay of the system with the transfer function
$H(j\omega) = \frac{(j\omega - 1 - j10)(j\omega - 1 + j10)}{(j\omega + 1 + j10)(j\omega + 1 - j10)}.$

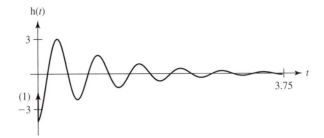

**Figure 6.112**

Impulse response of the system with the transfer function
$H(j\omega) = \frac{(j\omega-1-j10)(j\omega-1+j10)}{(j\omega+1+j10)(j\omega+1-j10)}$.

The impulse response is the inverse CTFT of $H(j\omega)$,

$$
h(t) = \frac{1}{\sqrt{1-\zeta^2}} \left\{ \frac{d}{dt}\left( e^{-\omega_0\zeta t} \cos\left[\omega_0\sqrt{1-\zeta^2}\,t + \tan^{-1}\left(\frac{\zeta}{\sqrt{1-\zeta^2}}\right)\right] u(t)\right) \right.
$$

$$
- (2z_{1r})e^{-\omega_0\zeta t}\cos\left[\omega_0\sqrt{1-\zeta^2}\,t + \tan^{-1}\left(\frac{\zeta}{\sqrt{1-\zeta^2}}\right)\right]u(t)
$$

$$
\left. + |z_1|^2 \frac{e^{-\omega_0\zeta t}\sin\left(\omega_0\sqrt{1-\zeta^2}\,t\right)}{\omega_0}u(t) \right\}. \qquad \textbf{(6.184)}
$$

or

$$
h(t) = \delta(t) + \left\{ |z_1|^2 \frac{e^{-\omega_0\zeta t}\sin\left(\omega_0\sqrt{1-\zeta^2}\,t\right)}{\omega_0\sqrt{1-\zeta^2}} \right.
$$

$$
- e^{-\omega_0\zeta t}\omega_0\sin\left[\omega_0\sqrt{1-\zeta^2}\,t + \tan^{-1}\left(\frac{\zeta}{\sqrt{1-\zeta^2}}\right)\right]
$$

$$
\left. - \frac{\omega_0\zeta + 2z_{1r}}{\sqrt{1-\zeta^2}}e^{-\omega_0\zeta t}\cos\left[\omega_0\sqrt{1-\zeta^2}\,t + \tan^{-1}\left(\frac{\zeta}{\sqrt{1-\zeta^2}}\right)\right] \right\} u(t) \quad \textbf{(6.185)}
$$

(Figure 6.112).

Notice that the ringing part of the impulse response has a characteristic ringing rate of about 1.6 Hz (about 10 rad/s) which is the frequency of the peak of the group delay. The impulse has frequency components uniformly distributed over all frequencies. The part of the excitation impulse near the frequency, $f = 1.6$ Hz, is delayed more than the rest of the excitation impulse and that is what shows up occurring at a later time. ∎

**Pulse-Amplitude Modulation**    Pulse-amplitude modulation is a technique used in various kinds of communication and control systems. It is also important because it forms a conceptual basis for the study of sampling in Chapter 7. It is similar to DSBSC modulation except that the carrier is not a sinusoid, but instead a periodic pulse train $p(t)$, of pulses of width $w$, fundamental period $T_s$, and height one (Figure 6.113).

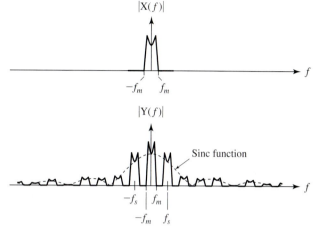

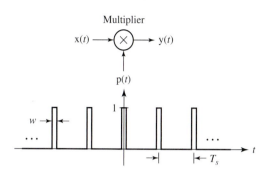

**Figure 6.113**
Pulse train.

**Figure 6.114**
Magnitude CTFT of excitation and response signals.

The pulse train can be mathematically described by

$$p(t) = \text{rect}\left(\frac{t}{w}\right) * \frac{1}{T_s} \text{comb}\left(\frac{t}{T_s}\right). \tag{6.186}$$

If the excitation of the pulse-amplitude modulator is $x(t)$, the response is

$$y(t) = x(t)p(t) = x(t)\left[\text{rect}\left(\frac{t}{w}\right) * \frac{1}{T_s} \text{comb}\left(\frac{t}{T_s}\right)\right]. \tag{6.187}$$

The CTFT of $y(t)$ is

$$Y(f) = X(f) * w \text{ sinc}(wf) \text{ comb}\left(\frac{f}{f_s}\right), \tag{6.188}$$

where $f_s = 1/T_s$ is the pulse repetition rate (pulse train fundamental frequency) and

$$Y(f) = X(f) * \left[wf_s \sum_{k=-\infty}^{\infty} \text{sinc}(wkf_s)\,\delta(f - kf_s)\right] \tag{6.189}$$

$$Y(f) = wf_s \sum_{k=-\infty}^{\infty} \text{sinc}(wkf_s)\,X(f - kf_s). \tag{6.190}$$

The CTFT $Y(f)$ of the response is a set of replicas of the CTFT of the excitation signal $x(t)$, repeated periodically at integer multiples of the pulse repetition rate $f_s$ and also multiplied by the value of a sinc function whose width is determined by the pulse width $w$ (Figure 6.114).

Replicas of the spectrum of the excitation signal occur multiple times in the spectrum of the response signal, each centered at an integer multiple of the pulse repetition rate and multiplied by a different constant. The excitation signal can be recovered from the response signal by a lowpass filter, if the bandwidth of the excitation signal is small enough that the replicas don't overlap. We will revisit this idea in considerably more detail in Chapter 7.

The excitation signal could also be recovered by a synchronous demodulation technique in which a replica centered at a nonzero multiple of the pulse repetition rate

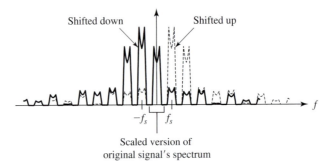

Scaled version of
original signal's spectrum

**Figure 6.115**
Synchronous demodulation of a PAM signal with a sinusoid
at a frequency $f_s$ equal to the pulse repetition rate.

is shifted to baseband by multiplying the pulse-amplitude modulation signal by a sinusoid at that same multiple of the pulse repetition rate (Figure 6.115).

One might well wonder why anyone would want to go to this trouble when the baseband replica of the excitation spectrum can be recovered by a simple lowpass filter. The answer is that in some systems the baseband replica may be corrupted by noise or an interference signal and the other replicas may be cleaner.

**Discrete-Time Modulation**  Modulation can also be used in DT systems in a manner similar to the way it is used in CT systems. The simplest form of DT modulation is DSBSC modulation in which we multiply a DT carrier signal c[n] by a DT modulation signal x[n]. Let the carrier be the DT sinusoid

$$c[n] = \cos(2\pi F_0 n). \tag{6.191}$$

Then the response of the DT modulator is

$$y[n] = x[n]c[n] = x[n]\cos(2\pi F_0 n) \tag{6.192}$$

(Figure 6.116).

The DT frequency-domain counterpart to multiplication in the DT domain is periodic convolution,

$$Y(F) = X(F) \circledast C(F) = X(F) \circledast \left\{ \frac{1}{2}[\delta(F - F_0) + \delta(F + F_0)] * \text{comb}(F) \right\} \tag{6.193}$$

or

$$Y(F) = \frac{1}{2}[X(F - F_0) + X(F + F_0)], \tag{6.194}$$

(Figure 6.117), which is very similar to the analogous result for DSBSC CT modulation,

$$Y(f) = \frac{1}{2}[X(f - f_0) + X(f + f_0)]. \tag{6.195}$$

If this type of modulation is to be used to accomplish frequency multiplexing, the sum of the DT bandwidths of all the signals must be less than one-half.

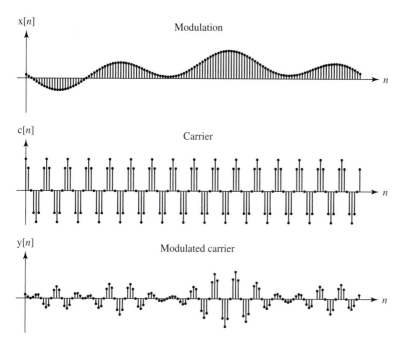

**Figure 6.116**
Modulation, carrier, and modulated carrier in a DT DSBSC system.

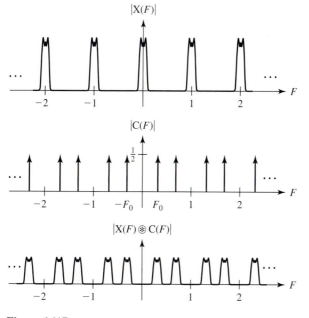

**Figure 6.117**
DTFTs of DT modulation, DT carrier, and DT modulated
carrier.

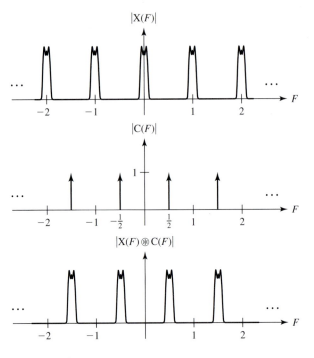

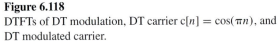

**Figure 6.118**
DTFTs of DT modulation, DT carrier c[n] = cos($\pi n$), and
DT modulated carrier.

One particularly simple and interesting type of discrete-time DSBSC modulation is to use a carrier c[n] = cos($\pi n$). This is a DT cosine formed by sampling a CT cosine at a sampling rate which is exactly twice its frequency. It is particularly simple because it is just the sequence, . . . 1, −1, 1, −1, 1, −1, . . . . The DTFTs that result when this carrier is used are illustrated in Figure 6.118.

This type of modulation inverts the frequency spectrum of a DT modulation. If it is initially a lowpass spectrum, it becomes highpass, and vice versa. This is a very easy type of modulation to implement because it consists of simply changing the sign of every other value of the DT modulation signal. The demodulation to recover the original signal is to do exactly the same process again, putting all the frequency components back in their original positions.

One interesting use of this type of modulation is to convert a lowpass DT filter into a highpass DT filter. If we modulate this type of carrier with a signal and then pass it through a lowpass DT filter, the frequencies that were originally low will be high and will not pass through and the frequencies that were originally high will be low and will pass through. Then we can demodulate the output of the filter by exactly the same type of modulation, converting the high frequencies (the original low frequencies) back to low frequencies. Using this technique we can use one type of DT filter for both lowpass and highpass filtering.

## 6.10 SPECTRAL ANALYSIS

In the previous sections we explored the operation of filters and suggested a system that could be used to measure the power spectrum of a signal. However, this system is impractical and another system called a *spectrum analyzer* is commonly used for this

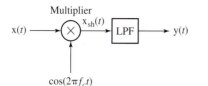

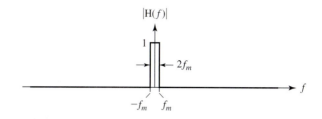

**Figure 6.119**
The essential components of a swept-frequency spectrum analyzer.

**Figure 6.120**
Idealized magnitude response of the lowpass filter.

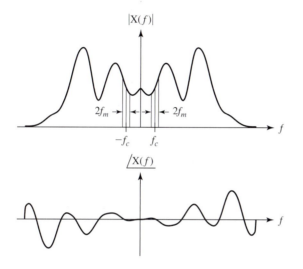

**Figure 6.121**
Spectrum of the excitation.

purpose. As an explanation of how a spectrum analyzer works and as another example of analyzing systems directly in the frequency domain, let us analyze the basic operation of a spectrum analyzer. A simplified block diagram of the heart of a typical swept-frequency spectrum analyzer is illustrated in Figure 6.119.

A swept-frequency spectrum analyzer multiplies an incoming signal by a sinusoid, modulation again. The product is then processed by the block called LPF which stands for lowpass filter. To keep the explanation simple for now, assume the lowpass filter is ideal as illustrated in Figure 6.120.

The operation of multiplying the excitation, $x(t)$, by a sinusoid is described in the time domain by

$$x_{sh}(t) = x(t)\cos(2\pi f_c t). \tag{6.196}$$

We can find the CTFT of both sides.

$$X_{sh}(f) = X(f) * \frac{1}{2}[\delta(f - f_c) + \delta(f + f_c)] \tag{6.197}$$

or

$$X_{sh}(f) = \frac{1}{2}[X(f - f_c) + X(f + f_c)]. \tag{6.198}$$

Suppose the CTFT of the excitation signal happened to have the shape illustrated Figure 6.121. The frequency of the sinusoid $\pm f_c$ is indicated on the graph of the

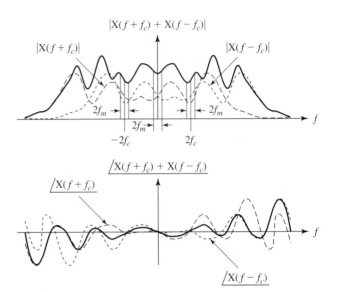

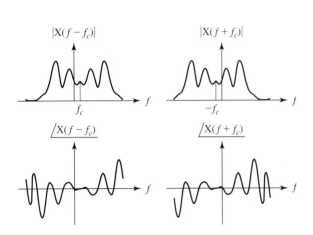

**Figure 6.122**
Spectrum of the excitation shifted both up and down by $f_c$.

**Figure 6.123**
Spectrum of the excitation after multiplication by a sinusoid at $f_c$.

magnitude spectrum of the excitation. Also indicated are two limits, above and below $\pm f_c$, at $\pm f_c \pm f_m$ where $f_m$ is the highest frequency the lowpass filter will pass. Notice that the magnitude spectrum is an *even* function of frequency and the phase spectrum is an *odd* function of frequency as proven earlier in Chapter 5. The two individual shifted spectra $X(f - f_c)$ and $X(f + f_c)$ would appear as illustrated in Figure 6.122.

The spectral regions defined by $f_c - f_m < |f| < f_c + f_m$ move both up and down in frequency. The region $f_c - f_m < f < f_c + f_m$ moves down to the region $-f_m < f < f_m$ and up to the region $2f_c - f_m < f < 2f_c + f_m$. The region $-f_c - f_m < f < -f_c + f_m$ moves up to the region $-f_m < f < f_m$ and down to the region $-2f_c - f_m < f < -2f_c + f_m$. The sum of the two shifted spectra is the spectrum illustrated in Figure 6.123. (Do not expect that the magnitude of the sum of the two shifted spectra will be the sum of the magnitudes of the two shifted spectra. The sum depends on the phase also.)

The lowpass filter removes all signal power except the signal power in the region $-f_m < f < f_m$. Remember that this signal power came from the regions defined by $f_c - f_m < |f| < f_c + f_m$ in the original excitation signal. Therefore, the response of the spectrum analyzer is a signal whose signal power is proportional to the signal power of the original signal in those frequency ranges. When we shift up and down by the same frequency $f_c$, the overlap near zero frequency is always of the original spectrum and its complex conjugate in a region for which $|f|$ is near $|f_c|$.

Now imagine that $f_c$ is changed to a new value. The amount of shift in the frequency domain would change, and the system response signal power would be proportional to the signal power of the original signal in a different spectral region centered at the new $\pm f_c$ with a width of $2f_m$. In a real spectrum analyzer there is a variable-frequency sinusoidal generator, called a sweep generator, which sweeps through a range of frequencies. The spectrum analyzer's response signal-power indicates the excitation signal-power in a small range of frequencies around the sweep frequency. The response signal-power is plotted on a screen as a function of the sweep frequency and the resulting display is called the excitation signal's *power spectrum*.

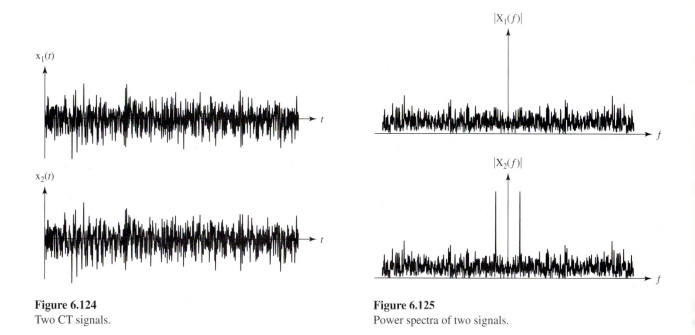

**Figure 6.124**
Two CT signals.

**Figure 6.125**
Power spectra of two signals.

(In Chapter 8 we will define *power spectral density* which will be closely related to the power spectrum.)

An important use of a spectrum analyzer is in spectral analysis of the content of signals. As an example consider the two signals $x_1(t)$ and $x_2(t)$ illustrated in Figure 6.124. They look very similar. Are they identical? From a direct visual inspection some small differences are apparent. But exactly what is the difference? In Figure 6.125 are plots of the power spectra of these two signals $X_1(f)$ and $X_2(f)$ on the same scale. Now one value of spectral analysis becomes clear. The plots of the power spectra of the signals make obvious the fact that the signals are definitely different and how they are different. The second signal has a strong sinusoidal component which shows up as two tall narrow spikes in the amplitude spectrum. The two signals $x_1(t)$ and $x_2(t)$ are random. Therefore, it would be impossible to write a mathematical expression to be transformed. But a spectrum analyzer can still display the power spectra.

The full importance of the Fourier transform can only be understood after some other concepts like impulse sampling, power spectral density, correlation, and spectral estimation, are introduced, which are derived, analyzed and implemented through the use of the Fourier transform. Also the Fourier transform is a natural stepping stone to other important transforms like the Laplace transform and the $z$ transform. All these topics will be explored in succeeding chapters.

## 6.11  SUMMARY OF IMPORTANT POINTS

1. The frequency response and impulse response of LTI systems are related through the Fourier transform.
2. The characterization of systems in the frequency domain allows for generalized design procedures for systems which process certain types of signals.
3. An ideal filter is distortionless within its passband.
4. Ideal filters cannot be built but can, in some important aspects, be arbitrarily closely approximated.

5. Filtering techniques can be applied to images just as they can to signals.
6. The Bode diagram technique can be used to do quick, approximate system analysis and design.
7. Discrete-time filters have several advantages over CT filters.
8. Communication systems which use frequency multiplexing are conveniently analyzed using Fourier methods.
9. Pulse-amplitude modulation creates multiple frequency-domain replicas of the signal that was modulated. This concept will be very important later in the study of sampling.
10. All the ideas that apply to CT filtering and modulation systems apply in a similar way to DT filtering and modulation systems.
11. Spectral analysis of signals can reveal information that is difficult to detect using time-domain methods.

## EXERCISES WITH ANSWERS

1. A system has an impulse response

$$h_1(t) = 3e^{-10t}u(t)$$

and another system has an impulse response

$$h_2(t) = \delta(t) - 3e^{-10t}u(t).$$

  *a.* Sketch the magnitude and phase of the transfer function of these two systems in a parallel connection.

  *b.* Sketch the magnitude and phase of the transfer function of these two systems in a cascade connection.

**Answers:**

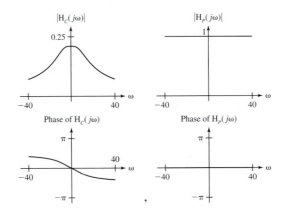

2. Below are some pairs of signals, $x(t)$ and $y(t)$. In each case decide whether or not $y(t)$ is a distorted version of $x(t)$.

  *a.*

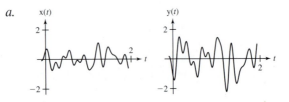

*b.*

*c.*

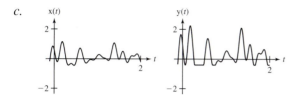

*d.*

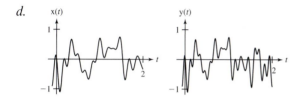

*e.*

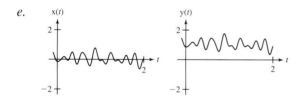

*f.*

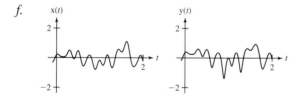

## Answers:

Two undistorted and the rest distorted

3. Classify each of these transfer functions as having a lowpass, highpass, bandpass, or bandstop frequency response.

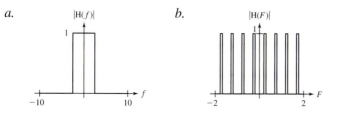

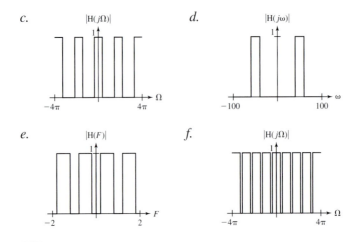

**Answers:**

Two lowpass, two bandpass, one highpass, one bandstop

4.  Classify each of these transfer functions as having a lowpass, highpass, bandpass, or bandstop frequency response.

   a.  $H(f) = 1 - \mathrm{rect}\left(\dfrac{|f| - 100}{10}\right)$

   b.  $H(F) = \mathrm{rect}(10F) * \mathrm{comb}(F)$

   c.  $H(j\Omega) = \left[\mathrm{rect}\left(20\pi\left(\Omega - \dfrac{\pi}{4}\right)\right) + \mathrm{rect}\left(20\pi\left(\Omega + \dfrac{\pi}{4}\right)\right)\right] * \mathrm{comb}\left(\dfrac{\Omega}{2\pi}\right)$

**Answers:**

Bandpass, lowpass, bandstop

5.  A system has an impulse response

$$h(t) = 10\,\mathrm{rect}\left(\frac{t - 0.01}{0.02}\right).$$

   What is its null bandwidth?

**Answer:**

50

6.  A system has an impulse response

$$h[n] = \left(\frac{7}{8}\right)^n u[n].$$

   What is its half-power DT-frequency bandwidth?

**Answer:**

0.1337 rad

7.  Determine whether or not the CT systems with these transfer functions are causal.

   a.  $H(f) = \mathrm{sinc}(f)$     b.  $H(f) = \mathrm{sinc}(f)e^{-j\pi f}$

   c.  $H(j\omega) = \mathrm{rect}(\omega)$     d.  $H(j\omega) = \mathrm{rect}(\omega)e^{-j\omega}$

   e.  $H(f) = A$     f.  $H(f) = Ae^{j2\pi f}$

**Answer:**

Two causal, four noncausal

8. Determine whether or not the DT systems with these transfer functions are causal.

   a. $H(F) = \dfrac{\sin(7\pi F)}{\sin(\pi F)}$

   b. $H(F) = \dfrac{\sin(7\pi F)}{\sin(\pi F)}e^{-j2\pi F}$

   c. $H(F) = \dfrac{\sin(3\pi F)}{\sin(\pi F)}e^{-j2\pi F}$

   d. $H(F) = \text{rect}(10F) * \text{comb}(F)$

**Answer:**

One causal, three noncausal

9. Find and sketch the frequency response of each of these circuits given the indicated excitation and response.

   a. Excitation $v_i(t)$, response $v_L(t)$

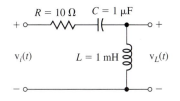

   b. Excitation $v_i(t)$, response $i_C(t)$

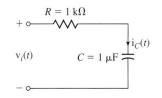

   c. Excitation $v_i(t)$, response $v_R(t)$

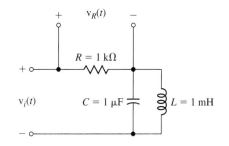

   d. Excitation $i_i(t)$, response $v_R(t)$

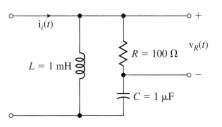

**Answers:**

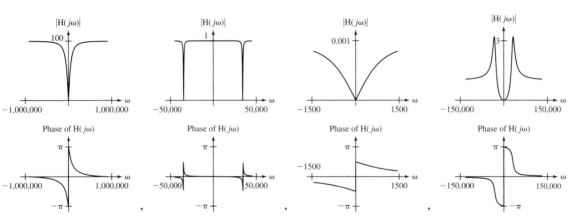

**10.** Classify each of these transfer functions as having a lowpass, highpass, bandpass, or bandstop frequency response.

*a.* $\quad H(f) = \dfrac{1}{1 + jf}$

*b.* $\quad H(f) = \dfrac{jf}{1 + jf}$

*c.* $\quad H(j\omega) = -\dfrac{j10\omega}{100 - \omega^2 + j10\omega}$

*d.* $\quad H(F) = \dfrac{\sin(3\pi F)}{\sin(\pi F)}$

*e.* $\quad H(j\Omega) = j[\sin(\Omega) + \sin(2\Omega)]$

**Answers:**

Two lowpass, two bandpass, one highpass

**11.** Plot the magnitude frequency responses, both on a linear-magnitude and on a log-magnitude scale, of the systems with these transfer functions, over the frequency range specified.

*a.* $\quad H(f) = \dfrac{20}{20 - 4\pi^2 f^2 + j42\pi f} \qquad -100 < f < 100$

*b.* $\quad H(j\omega) = \dfrac{2 \times 10^5}{(100 + j\omega)(1700 - \omega^2 + j20\omega)} \qquad -500 < \omega < 500$

**Answers:**

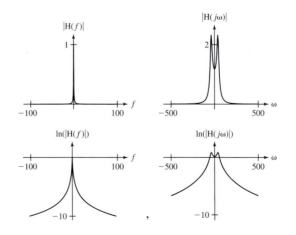

**12.** Draw asymptotic and exact magnitude and phase Bode diagrams for the frequency responses of the following circuits and systems.

*a.* An $RC$ lowpass filter with $R = 1\ M\Omega$ and $C = 0.1\ \mu F$.

*b.*

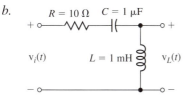

**Answers:**

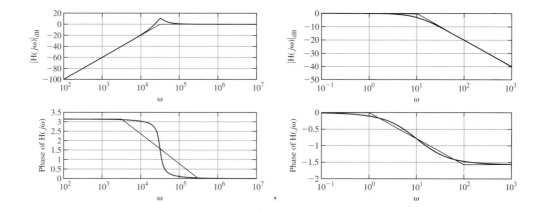

,

**13.** Find the transfer functions $H(f) = V_o(f)/V_i(f)$ of these active filters and identify them as lowpass, highpass, bandpass, or bandstop.

*a.*

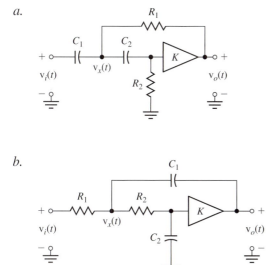

*b.*

**Answers:**
Highpass and lowpass

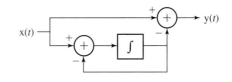

**Figure E14**

**14.** Show that the system in Figure E14 has a highpass frequency response.

**Answer:**

$$H(j\omega) = \frac{Y(j\omega)}{X(j\omega)} = \frac{j\omega}{j\omega + 1}$$

**15.** Draw the block diagram of a system with a bandpass frequency response using two integrators as functional blocks. Then find its transfer function and verify that it has a bandpass frequency response.

**Answer:**

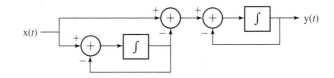

**16.** Find the transfer function $H(j\Omega) = Y(j\Omega)/X(j\Omega)$, and sketch the frequency response of each of these DT filters over the range $-4\pi < \Omega < 4\pi$.

*a.*

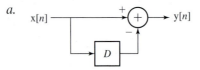

*b.*

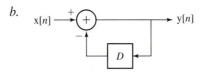

*c.*

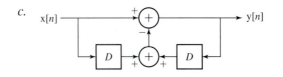

*d.*

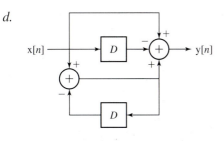

**Answers:**

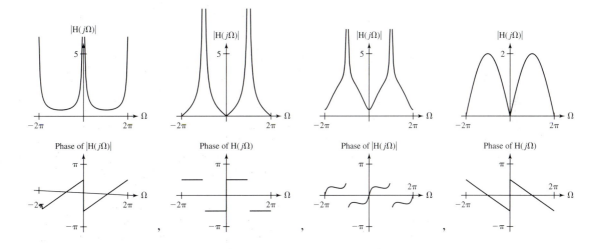

17. Find the minimum stopband attenuation of a moving-average filter with $N = 3$. Define the stopband as the frequency region $F_c < F < \frac{1}{2}$, where $F_c$ is the DT frequency of the first null in the frequency response.

**Answer:**

11.35 dB of attenuation

18. In the system in Figure E18, $x_t(t) = \text{sinc}(t)$, $f_c = 10$, and the cutoff frequency of the lowpass filter is 1 Hz. Plot the following signals and the magnitudes and phases of their CTFTs:

   *a.* $x_t(t)$

   *b.* $y_t(t)$

   *c.* $y_d(t)$

   *d.* $y_f(t)$

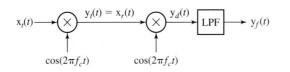

**Figure E18**

**Answers:**

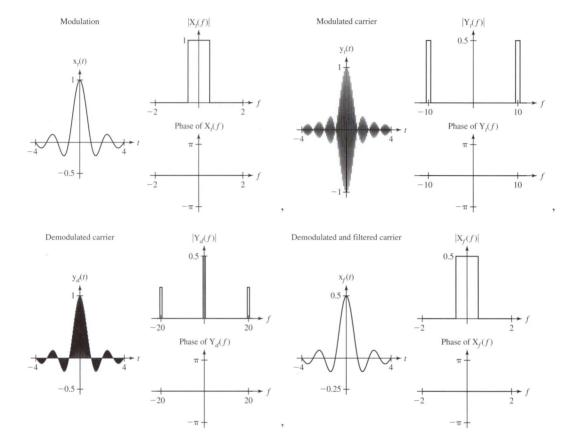

19.  In the system in Figure E19, $x_t(t) = \text{sinc}(10t) * \text{comb}(t)$, $m = 1$, $f_c = 100$, and the cutoff frequency of the lowpass filter is 10 Hz. Plot the following signals, and the magnitudes and phases of their CTFTs:

a.   $x_t(t)$

b.   $y_t(t)$

c.   $y_d(t)$

d.   $y_f(t)$

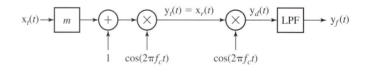

**Figure E19**

**Answers:**

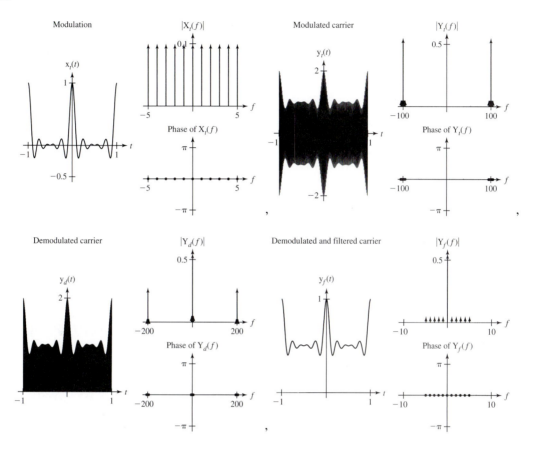

20. An $RC$ lowpass filter with a time constant of 16 ms is excited by a DSBSC signal

$$x(t) = \sin(2\pi t) \cos(20\pi t).$$

Find the phase and group delays at the carrier frequency.

**Answers:**

12.54 ms, 7.95 ms

21. A pulse train

$$p(t) = \text{rect}(100t) * 10 \, \text{comb}(10t)$$

is modulated by a signal

$$x(t) = \sin(4\pi t).$$

Plot

a.  The response of the modulator $y(t)$
b.  The CTFTs of the excitation and response

**Answers:**

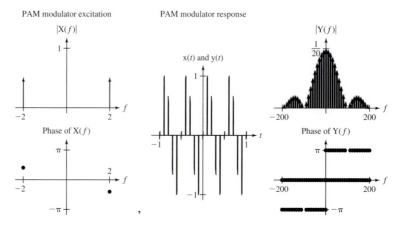

22.  In the system in Figure E22, let the excitation be

$$x(t) = \text{rect}(1000t) * 250 \text{ comb}(250t)$$

and let the filter be ideal, with unity passband gain. Plot the signal power of the response y(t) of this system versus the sweep frequency $f_c$ over the range $0 < f_c < 2000$ for the following LPF bandwidths:

*a.*   5 Hz      *b.*   50 Hz      *c.*   500 Hz

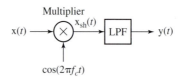

**Figure E22**

**Answer:**

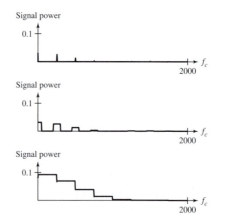

# EXERCISES WITHOUT ANSWERS

**23.** A signal $x(t)$ is described by

$$x(t) = 500 \text{ rect}(1000t) * \text{comb}(500t)$$

    *a.* If $x(t)$ is the excitation of an ideal lowpass filter with a cutoff frequency of 3 kHz, plot the excitation $x(t)$ and the response $y(t)$ on the same scale and compare.

    *b.* If $x(t)$ is the excitation of an ideal bandpass filter with a low cutoff frequency of 1 kHz and a high cutoff frequency of 5 kHz, plot the excitation $x(t)$ and the response $y(t)$ on the same scale and compare.

**24.** Determine whether or not the CT systems with these transfer functions are causal.

    *a.* $H(j\omega) = \dfrac{2}{j\omega}$

    *b.* $H(j\omega) = \dfrac{10}{6 + j4\omega}$

    *c.* $H(j\omega) = \dfrac{4}{25 - \omega^2 + j6\omega}$

    *d.* $H(j\omega) = \dfrac{4}{25 - \omega^2 + j6\omega} e^{j\omega}$

    *e.* $H(j\omega) = \dfrac{4}{25 - \omega^2 + j6\omega} e^{-j\omega}$

    *f.* $H(j\omega) = \dfrac{j\omega + 9}{45 - \omega^2 + j6\omega}$

    *g.* $H(j\omega) = \dfrac{49}{49 + \omega^2}$

**25.** Determine whether or not the DT systems with these transfer functions are causal.

    *a.* $H(F) = [\text{rect}(10F) * \text{comb}(F)]e^{-j20\pi F}$

    *b.* $H(F) = j\sin(2\pi F)$

    *c.* $H(F) = 1 - e^{-j4\pi F}$

    *d.* $H(j\Omega) = \dfrac{8e^{j\Omega}}{8 - 5e^{-j\Omega}}$

**26.** Find and sketch the frequency response of each of these circuits given the indicated excitation and response.

    *a.* Excitation $v_i(t)$, response $v_{C2}(t)$

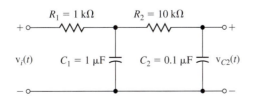

*b.* Excitation $v_i(t)$, response $i_{C1}(t)$

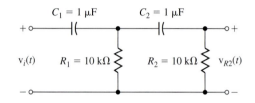

*c.* Excitation $v_i(t)$, response $v_{R2}(t)$

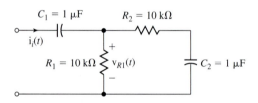

*d.* Excitation $i_i(t)$, response $v_{R1}(t)$

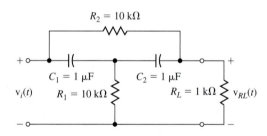

*e.* Excitation $v_i(t)$, response $v_{RL}(t)$

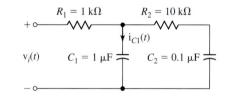

27. Find and sketch versus frequency, the magnitude and phase of the input impedance $Z_{in}(j\omega) = V_i(j\omega)/I_i(j\omega)$ and transfer function $H(j\omega) = V_o(j\omega)/V_i(j\omega)$ for each of these filters.

*a.*

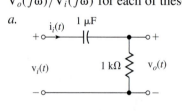

*b.*

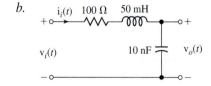

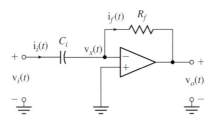

**Figure E30**

28. The signal $x(t)$ in Exercise 23 is the input voltage signal of an $RC$ lowpass filter with $R = 1\ k\Omega$ and $C = 0.3\ \mu F$. Sketch the input and output voltage signals versus time on the same scale.

29. Draw asymptotic and exact magnitude and phase Bode diagrams for the frequency responses of the following circuits and systems.

*a.*

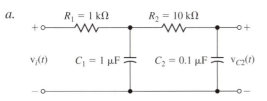

*b.*

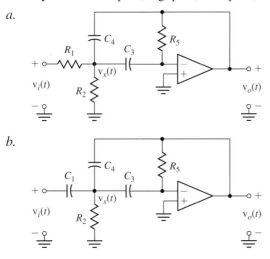

$X(j\omega) \longrightarrow \boxed{\dfrac{10}{j\omega + 10}} \longrightarrow \boxed{\dfrac{j\omega}{j\omega + 10}} \longrightarrow Y(j\omega)$

*c.* A system whose transfer function is $H(j\omega) = \dfrac{j20\omega}{10{,}000 - \omega^2 + j20\omega}$.

30. Find the transfer function for the circuit shown in Figure E30. What function does it perform?

31. Design an active highpass filter using an ideal operational amplifier, two resistors, and one capacitor, and derive its transfer function to verify that it is highpass.

32. Find the transfer functions $H(f) = V_o(f)/V_i(f)$ of these active filters and identify them as lowpass, highpass, bandpass, or bandstop.

*a.*

*b.*

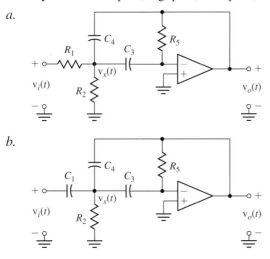

*c.*

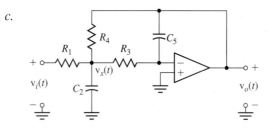

33. When music is recorded on analog magnetic tape and later played back, a high-frequency noise component, called tape *hiss*, is added to the music. For purposes of analysis assume that the spectrum of the music is flat at $-30$ dB across the audio spectrum from 20 Hz to 20 kHz. Also assume that the spectrum of the signal played back on the tape deck has an added component making the playback signal have a Bode diagram as illustrated in Figure E33. The extra high-frequency noise could be attenuated by a lowpass filter, but that would also attenuate the high-frequency components of the music, reducing its fidelity. One solution to the problem is to preemphasize the high-frequency part of the music during the recording process so that when the lowpass filter is applied to the playback the net effect on the music is zero but the hiss has been attenuated. Design an active filter which could be used during the recording process to do the preemphasis.

34. One problem with causal CT filters is that the response of the filter always lags the excitation. This problem cannot be eliminated if the filtering is done in real time, but if the signal is recorded for later off-line filtering, one simple way of eliminating the lag effect is to filter the signal, record the response, and then filter that recorded response with the same filter but playing the signal back through the system backwards. Suppose the filter is a single-pole filter with a transfer function of the form,

$$H(j\omega) = \frac{1}{1 + j(\omega/\omega_c)},$$

where $\omega_c$ is the cutoff frequency (half-power frequency) of the filter.
   a. What is the effective transfer function of the entire process of filtering the signal forward, and then backward?
   b. What is the effective impulse response?

35. Repeat Exercise 18 but with the second $\cos(2\pi f_c t)$ replaced by $\sin(2\pi f_c t)$.

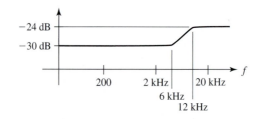

**Figure E33**
Bode diagram of playback signal.

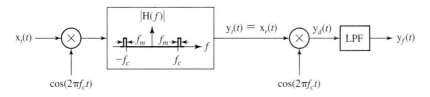

**Figure E36**

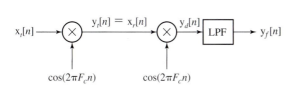

**Figure E39**

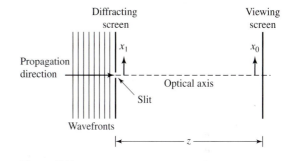

**Figure E41**
One-dimensional diffraction of light through a slit.

**36.** In the system in Figure E36, $x_t(t) = \text{sinc}(t)$, $f_c = 10$, and the cutoff frequency of the lowpass filter is 1 Hz. Plot the signals $x_t(t)$, $y_t(t)$, $y_d(t)$, and $y_f(t)$ and the magnitudes and phases of their CTFTs

**37.** A quadrature modulator modulates a sine carrier $\sin(20\pi t)$ with a signal $x_1(t) = \text{sinc}(t)$ and a cosine carrier $\cos(20\pi t)$ with a signal $x_2(t) = \text{rect}(t)$. The quadrature demodulator has a phase error making its local oscillators be $\sin(20\pi t - (\pi/6))$ and $\cos(20\pi t - (\pi/6))$. Plot the two demodulated and filtered signals $x_{1f}(t)$ and $x_{2f}(t)$.

**38.** A pulse train

$$p(t) = \frac{1}{w}\, \text{rect}\left(\frac{t}{w}\right) * 4\,\text{comb}(4t)$$

is modulated by a signal

$$x(t) = \text{sinc}(t).$$

Plot the response of the modulator $y(t)$ and the CTFTs of the excitation and response for

*a.*  $w = 10$ ms     *b.*  $w = 1$ ms

**39.** In the system in Figure E39, $x_t[n] = \text{sinc}(n/20)$, $F_c = \frac{1}{4}$, and the cutoff DT frequency of the lowpass filter is $\frac{1}{20}$. Plot the signals $x_t[n]$, $y_t[n]$, $y_d[n]$, and $y_f[n]$ and the magnitudes and phases of their DTFTs.

**40.** Repeat Exercise 22 but with an excitation

$$x(t) = \text{rect}(1000t) * 20\,\text{comb}(20t).$$

**41.** The diffraction of light can be approximately described through the use of the Fourier transform. Consider an opaque screen with a small slit being illuminated from the left by a normally incident uniform plane light wave (Figure E41). If $z \gg \pi x_1^2/\lambda$ is a good approximation for any $x_1$ in the slit, then the electric field strength of the light striking the viewing screen can be accurately described by

$$E_0(x_0) = K \frac{e^{j(2\pi z/\lambda)}}{j\lambda z} e^{j(\pi/\lambda z)x_0^2} \int_{-\infty}^{\infty} E_1(x_1)\, e^{-j(2\pi/\lambda z)x_0 x_1}\, dx_1$$

where $E_1$ = field strength at diffracting screen
$\quad\quad E_0$ = field strength at viewing screen
$\quad\quad K$ = constant of proportionality
$\quad\quad \lambda$ = wavelength of light

The integral is a Fourier transform with different notation. The field strength at the viewing screen can be written as

$$E_0(x_0) = K \frac{e^{j(2\pi z/\lambda)}}{j\lambda z} e^{j(\pi/\lambda z)x_0^2}\, \mathcal{F}[E_1(t)]_{f \to x_0/\lambda z}.$$

The intensity $I_0(x_0)$ of the light at the viewing screen is the square of the magnitude of the field strength

$$I(x_0) = |E_0(x_0)|^2.$$

a.  Plot the intensity of light at the viewing screen if the slit width is 1 mm, the wavelength of light is 500 nm, the distance $z$ is 100 m, the constant of proportionality is $10^{-3}$, and the electric field strength at the diffraction screen is 1 V/m.

b.  Now let the slit be replaced by two slits each 0.1 mm in width, separated by 1 mm (center-to-center) and centered on the optical axis. Plot the intensity of light at the viewing screen if the other parameters are the same as in part (a).

42.  In Figure E42a is a circuit diagram of a half-wave rectifier followed by a capacitor to smooth the response voltage. Model the diode as ideal, and let the input voltage signal be a cosine at 60 Hz with an amplitude of $120\sqrt{2}$ V. Let the $RC$ time constant be 0.1 s. Then the output voltage signal will look as

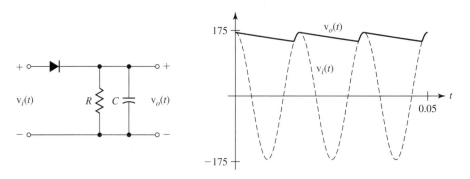

**Figure E42**
(a) Half-wave rectifier with a capacitive smoothing filter, and (b) excitation and response voltages.

illustrated in Figure E42b. Find and plot the magnitude of the CTFT of the output voltage signal.

43. Create a discrete-space image consisting of 96 by 96 pixels. Let the image be like a checkerboard consisting of 8 by 8 alternating black and white squares.

   *a.* Filter the image row by row and then column by column with a DT filter whose impulse response is

   $$h[n] = 0.2(0.8)^n u[n],$$

   and display the image on the screen using the `image` command in MATLAB.

   *b.* Filter the image row by row and then column by column with a DT filter whose impulse response is

   $$h[n] = \delta[n] - 0.2(0.8)^n u[n],$$

   and display the image on the screen using the `image` command in MATLAB.

44. In the system of Figure E44 let the CTFT of the excitation be $X(f) = \text{tri}(f/f_c)$. This system is sometimes called a scrambler because it moves the frequency components of a signal to new locations making it unintelligible.

   *a.* Using only an analog multiplier and an ideal filter, design a descrambler which would recover the original signal.

   *b.* Sketch the magnitude spectrum of each of the signals in the scrambler–descrambler system.

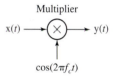

Multiplier

**Figure E44**
A scrambler.

45. Electronic amplifiers that handle very low frequency signals are difficult to design because thermal drifts of offset voltages cannot be distinguished from the signals. For this reason a popular technique for designing low-frequency amplifiers is the so-called chopper-stabilized amplifier (Figure E45).

   A chopper-stabilized amplifier "chops" the input voltage signal by switching it on and off periodically. This action is equivalent to a pulse-amplitude modulation in which the pulse train being modulated by the excitation is a 50 percent duty-cycle square wave which alternates between zero and one. Then the chopped signal is bandpass-filtered to remove any slow thermal drift signals from the first amplifier. Then the amplified signal is chopped again at exactly the same rate and in phase with the chopping signal

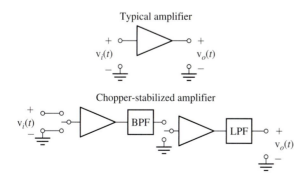

**Figure E45**
A chopper-stabilized amplifier.

used at the input of the first amplifier. Then this signal may be further amplified. The last step is to lowpass-filter the signal out of the last amplifier to recover an amplified version of the original signal. (This is a simplified model, but it illustrates the essential features of a chopper-stabilized amplifier.)

Let the following be the parameters of the chopper-stabilized amplifier:

| | |
|---|---|
| Chopping frequency | = 500 Hz |
| Gain of the first amplifier | = 100 V/V |
| Bandpass filter | = unity-gain, ideal, zero-phase; passband of $250 < |f| < 750$ |
| Gain of the second amplifier | = 10 V/V |
| Lowpass filter | = unity-gain, ideal, zero-phase; bandwidth of 100 Hz |

Let the input voltage signal have a 100-Hz bandwidth. What is the effective DC gain of this chopper-stabilized amplifier?

46. A common problem in over-the-air television signal transmission is *multipath* distortion of the received signal due to the transmitted signal bouncing off structures. Typically a strong main signal arrives at some time and a weaker ghost signal arrives later. So if the transmitted signal is $x_t(t)$, the received signal is

$$x_r(t) = K_m x_t(t - t_m) + K_g x_t(t - t_g),$$

where $K_m \gg K_g$ and $t_g > t_m$.

*a.* What is the transfer function of this communication channel?

*b.* What would be the transfer function of an equalization system that would compensate for the effects of multipath?

### CHAPTER 7

# Sampling and the Discrete Fourier Transform

## 7.1 INTRODUCTION AND GOALS

As indicated in Chapter 6, in the application of signal processing to real signals in real systems, we often do not have a mathematical description of the signals. We must measure and analyze them to discover their characteristics. If the signal is unknown, the process of analysis begins with the acquisition of the signal. *Acquisition* means measuring and recording the signal over a period of time. This could be done with a tape recorder or other recording device, but by far the most common technique of acquiring signals today is by sampling. As first introduced in Chapter 2, *sampling* a signal is the process of acquiring its values only at discrete points in time. The main reason we acquire signals in this way is that most signal processing and analysis today is done using digital computers. A digital computer requires that all information it processes be in the form of numbers. Therefore, the samples are acquired and stored as numbers. Since the memory and mass storage capacity of a computer are finite, it can only handle a finite number of numbers. Therefore, if a digital computer is to be used to analyze a signal, it can only be sampled for a finite time. The salient question addressed in this chapter is, To what extent do the samples accurately describe the signal from which they are taken? We will see that whether, and how much, information is lost by sampling depends on the way the samples are taken. We will find that under certain circumstances all, or practically all, the signal information can be stored in a finite number of samples.

The processing of DT (sampled) signals is becoming more important every day. Since the operations done on DT signals are performed by computers which operate on numbers stored as digits, the common term for the processing of DT signals is *digital signal processing* (DSP). Many filtering operations that were once done with analog filters now use digital filters which operate on samples from a signal, instead of the original CT signal. Modern cellular telephone systems use DSP to improve voice quality, separate channels, and switch users between cells. Long-distance telephone communication systems use DSP to efficiently use long trunk lines and microwave links. Television sets use DSP to improve picture quality. Robotic vision is based on signals from cameras which digitize (sample) an image and then analyze it with computation techniques to recognize features. Modern control systems in automobiles, manufacturing plants, and scientific instrumentation usually have embedded processors which analyze signals and make decisions using DSP.

1. To understand how signals are sampled

2. To determine how a CT signal must be sampled and to what extent the samples describe the signal

3. To learn how to reconstruct a CT signal from its samples

4. To apply to DT signals all the concepts of sampling developed for the sampling of CT signals

5. To learn how to use the discrete Fourier transform and see how it is related to other Fourier methods

6. To learn how the fast Fourier transform algorithm increases the speed of computation of the discrete Fourier transform

## 7.2 SAMPLING METHODS

Sampling of electrical signals, usually voltages, is most commonly done with two devices, the sample-and-hold (S/H) and the analog-to-digital converter (ADC). Sometimes these two devices are packaged together in one electronic module. The excitation of the S/H is the analog voltage at its input, and when the S/H is clocked, it reproduces that voltage at its output as its response and holds that voltage until it is clocked to acquire another voltage (Figure 7.1). In the figure the signal $c(t)$ is the clock signal. The acquisition of the input voltage signal of the S/H occurs during the *aperture time,* which is the width of a clock pulse. During the clock pulse the output voltage signal very

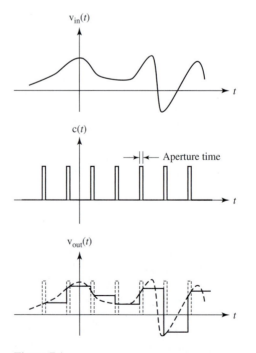

**Figure 7.1**
Operation of a sample-and-hold device.

quickly moves from its previous value to track the excitation. At the end of the clock pulse the output voltage signal is held at a fixed value until the next clock pulse occurs.

An ADC accepts an analog voltage or current excitation at its input and converts it into a set of binary bits (a *code*) as its response. The ADC response can be serial or parallel. If the ADC has a serial response, it produces on an output *pin,* a single response voltage or current that is a timed sequence of high and low voltages representing the ones and zeros of the set of binary bits. If the ADC has a parallel response, there is a response voltage or current for each bit and each bit appears simultaneously on a dedicated output pin of the ADC as a high or low voltage or current representing a one or a zero in the set of binary bits (Figure 7.2). Usually an ADC is preceded by a S/H to keep its excitation constant during the conversion time.

The excitation of the ADC is a CT signal, and the response is a DT signal. The response of the ADC is not only discrete-time but also quantized and encoded. The number of binary bits produced by the ADC is finite. Therefore, the number of unique bit patterns it can produce is also finite. If the number of bits the ADC produces is $n$, the number of unique bit patterns it can produce is $2^n$. *Quantization* is the effect of converting a continuum of (infinitely many) excitation values into a finite number of response values. Since the response has an error due to quantization, it is as though the signal has noise on it and this noise is called quantization noise. If the number of bits used to represent the response is large enough, quantization noise is often negligible in comparison with other noise sources. After quantization, the ADC encodes the signal also. *Encoding* is the conversion from an analog voltage to a binary bit pattern. So the excitation of an ADC is an analog (CT) voltage, and the response is a sequence of binary numbers or codes. The relation between the excitation and response of an ADC whose input signal voltage range is $-V_0 < v_{in}(t) < +V_0$ is illustrated in Figure 7.3 for a 3-bit ADC. (A 3-bit ADC is rarely, if ever, actually used, but it does illustrate the quantization effect nicely because the number of unique bit patterns is small and the quantization noise is large.) The effects of quantization are easy to see in a sinusoid quantized by 3 bits (Figure 7.4). When the signal is quantized to 8 bits, the quantization error is much smaller (Figure 7.5).

The opposite of analog-to-digital conversion is obviously digital-to-analog conversion, and the device that does that is called a digital-to-analog converter (DAC). A DAC accepts binary bit patterns as its excitation and produces an analog voltage as

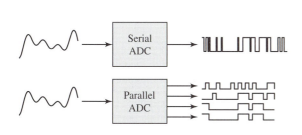

**Figure 7.2**
Serial and parallel ADC operation.

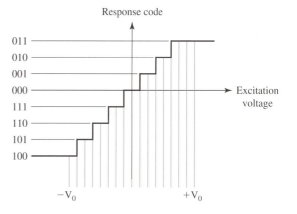

**Figure 7.3**
ADC excitation-response relationship.

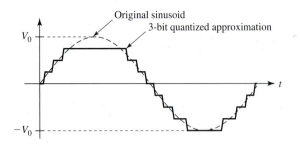

**Figure 7.4**
Sinusoid quantized to 3 bits.

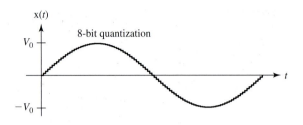

**Figure 7.5**
Sinusoid quantized to 8 bits.

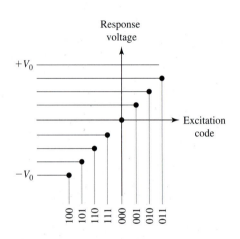

**Figure 7.6**
DAC excitation–response relationship.

its response. Since the number of unique bit patterns it can accept is finite, the DAC response signal is an analog voltage that is quantized. The relation between excitation and response for a DAC is shown in Figure 7.6.

In the material to follow, the effects of quantization will not be considered. The model for analyzing the effects of sampling will be that the sampler is ideal in the sense that the response signal's quantization noise is zero.

In Chapter 6 we introduced the idea of sampling a signal by multiplying a pulse train by the signal and called that type of modulation *pulse-amplitude modulation* (PAM). We will now apply that theory to the process of sampling a signal with a S/H. Let a sampled signal $x_p(t)$ be equal to the signal being sampled $x(t)$ during the aperture time of a S/H and zero otherwise. Let the aperture time of the S/H be $w$, and let the time between samples be $T_s$. Then, from Chapter 6,

$$p(t) = \text{rect}\left(\frac{t}{w}\right) * \frac{1}{T_s}\,\text{comb}\left(\frac{t}{T_s}\right), \tag{7.1}$$

$$x_p(t) = x(t)p(t) = x(t)\,\text{rect}\left(\frac{t}{w}\right) * \frac{1}{T_s}\,\text{comb}\left(\frac{t}{T_s}\right), \tag{7.2}$$

and

$$X_p(f) = wf_s \sum_{k=-\infty}^{\infty} \text{sinc}(wkf_s)\, X(f - kf_s). \tag{7.3}$$

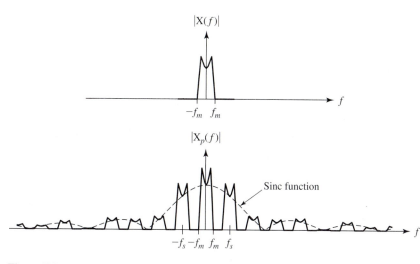

**Figure 7.7**
Magnitudes of the CTFTs of the original and sampled signals.

The CTFT of $x_p(t)$, $X_p(f)$, is a set of replicas of the CTFT of the original signal $x(t)$ repeated periodically at integer multiples of the sampling rate $f_s$ and also multiplied by the value of a sinc function whose width is determined by the aperture time $w$ of the S/H (Figure 7.7).

The shorter the aperture time of the S/H, the wider the sinc function is. An ideal S/H would have an aperture time of zero so as to acquire the signal instantaneously and allow very fast sampling. As the aperture time goes to zero, at a constant sampling rate, the CTFT of $x_p(t)$ goes to zero because the power of the PAM signal goes to zero. But if we now modify the sampling process to compensate for that effect by making the *area* of each sampling pulse one instead of the *height,* we get

$$p(t) = \frac{1}{w} \, \text{rect}\left(\frac{t}{w}\right) * \frac{1}{T_s} \, \text{comb}\left(\frac{t}{T_s}\right), \qquad (7.4)$$

and, finding the CTFT of $x_p(t)$,

$$X_p(f) = f_s \sum_{k=-\infty}^{\infty} \text{sinc}(wkf_s) \, X(f - kf_s). \qquad (7.5)$$

As the aperture time $w$ approaches zero, the sinc function becomes infinitely wide and we get

$$\lim_{w \to 0} X_p(f) = X_\delta(f) = f_s \sum_{k=-\infty}^{\infty} X(f - kf_s). \qquad (7.6)$$

Of course, in that same limit,

$$\lim_{w \to 0} \frac{1}{w} \, \text{rect}\left(\frac{t}{w}\right) = \delta(t), \qquad (7.7)$$

$$\lim_{w \to 0} p(t) = \frac{1}{T_s} \, \text{comb}\left(\frac{t}{T_s}\right) = f_s \, \text{comb}(f_s t), \qquad (7.8)$$

and

$$\lim_{w \to 0} x_p(t) = x_\delta(t) = x(t) f_s \, \text{comb}(f_s t). \qquad (7.9)$$

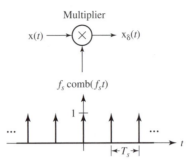

**Figure 7.8**
An impulse modulator producing
an impulse-sampled signal.

So p($t$) becomes a periodic sequence of unit impulses, spaced apart by $T_s$ in time. This ideal sampler limit represents what is called *impulse sampling* or sometimes *impulse modulation* (Figure 7.8). By using this model we can explore the relation between a signal and samples taken from it and discover how fast we must sample to preserve the information in the signal.

## 7.3  REPRESENTING A CONTINUOUS-TIME SIGNAL BY SAMPLES

### QUALITATIVE CONCEPTS

If we are to use samples from a CT signal, instead of the signal itself, the most important and basic question to answer is how to sample the signal so as to retain the information it carries. If the CT signal can be exactly reconstructed from the samples, then the samples contain all the information in the signal. We must decide how fast to sample the CT signal and how long to sample it. As an introduction to the questions involved in deciding how to sample a signal, consider the CT signal x($t$) (Figure 7.9a).

Suppose this signal is sampled at the sampling rate illustrated in Figure 7.9b. Most people would probably intuitively say that there are enough samples here to describe the signal adequately by drawing a smooth curve through the points. It seems that little if any information is lost in sampling because we could apparently reconstruct the signal from the samples. How about the sampling rate in Figure 7.9c? Is this sampling rate adequate? How about the rate in Figure 7.9d? Most people would probably agree that the sampling rate in Figure 7.9d is inadequate. The intuitive reason for saying that is that a naturally drawn smooth curve through the last sample set would not look very much like the original curve. Although the last sampling rate was inadequate for this signal, it might be just fine for another signal (Figure 7.10). It seems adequate for the signal of Figure 7.10 because it is much smoother and more slowly varying.

So there is some minimum rate at which samples can be taken while retaining the information in the signal and it depends on how fast the signal varies with time. That is, it depends on the *frequency content* of the signal. We have an intuitive feel about how fast is fast enough, but we would like to make the decision exact and concrete with some kind of mathematical justification for the choice. The question of how fast

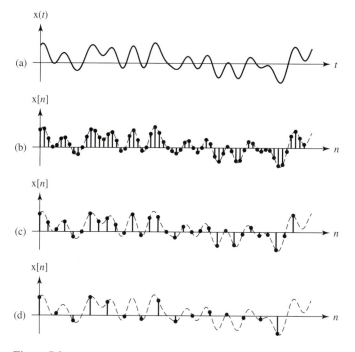

**Figure 7.9**
(a) A CT signal, and (b) to (d) DT signals formed by sampling
the CT signal at different rates.

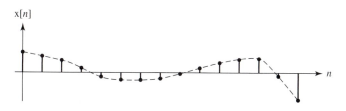

**Figure 7.10**
A DT signal formed by sampling a slowly varying signal.

samples have to be taken to describe a signal was answered definitively by Claude
Shannon with his, now famous, sampling theorem.

To show Shannon's result it is necessary to build a precise mathematical frame-
work of notation and technique, first to describe the sampling process and then to show
its capabilities and limitations. We have come to this point in signal analysis with very
good analysis techniques for CT and DT signals. It is now time to apply them to the
sampling process.

## SHANNON'S SAMPLING THEOREM

So far we have been considering CT and DT signals separately. It was shown in
Chapter 5 that if we sample a CT signal $x(t)$ to form a DT signal $x[n]$, that there is an
information equivalence between $x[n]$ and a CT signal $x_\delta(t)$ which consists only of

Claude Elwood Shannon

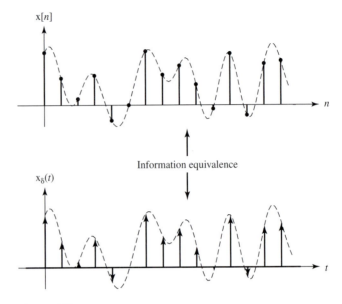

**Figure 7.11**
Information equivalence between a DT signal formed by sampling
a CT signal and an impulse signal formed by impulse-sampling a
CT signal.

impulses whose strengths are the same as the values of $x[n]$,

$$x_\delta(t) = \sum_{n=-\infty}^{\infty} x[n]\delta(t - nT_s) = x(t)\,f_s\,\text{comb}(f_s t), \tag{7.10}$$

where $f_s = 1/T_s$ is the sampling rate (Figure 7.11). This is seen in the relationship
between the DTFT of $x[n]$, $X(F)$, and the CTFT of $x_\delta(t)$,

$$X_\delta(f) = X_{\text{DTFT}}\left(\frac{f}{f_s}\right). \tag{7.11}$$

This information equivalence is important because if it can be shown that $x[n]$ not only
contains all the information in $x_\delta(t)$ but also contains all the information in $x(t)$, then
it follows that we could (at least in principle) reconstruct $x(t)$ from its samples.

In the following exploration of sampling we will use a CT signal as an example for
comparison of methods and concepts, a sinc function,

$$x(t) = A \operatorname{sinc}\left(\frac{t}{w}\right). \tag{7.12}$$

We begin by finding the CTFT of the signal,

$$X_{\text{CTFT}}(f) = Aw \operatorname{rect}(wf). \tag{7.13}$$

[In this development the CTFT of $x(t)$ will be denoted by $X_{\text{CTFT}}(f)$ and the DTFT of
$x[n]$ will be denoted by $X_{\text{DTFT}}(F)$ to avoid confusion between the two functions. Each
function is of a different continuous independent variable since both are used and the
transformation $f \to f_s F$ is used in relating them to each other. The CTFT of $x_\delta(t)$ will
simply be denoted by $X_\delta(f)$ since there is no DTFT with which it could be confused.]
The CT signal and the magnitude of its CTFT are illustrated in Figure 7.12. One reason
this signal was chosen as an example is that its CTFT is zero for frequencies
$|f| > 1/2w$. This makes it a band-limited signal.

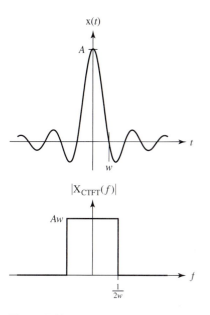

**Figure 7.12**
Example CT signal and the magnitude
of its CTFT.

Next we sample and impulse-sample $x(t)$ with a time between samples $T_s$ yielding the DT signal

$$x[n] = x(nT_s) = A \operatorname{sinc}\left(\frac{nT_s}{w}\right) \tag{7.14}$$

and the information-equivalent CT impulse signal

$$x_\delta(t) = A \operatorname{sinc}\left(\frac{t}{w}\right) f_s \operatorname{comb}(f_s t) = A \sum_{n=-\infty}^{\infty} \operatorname{sinc}\left(\frac{nT_s}{w}\right) \delta(t - nT_s). \tag{7.15}$$

Since $x[n]$ is, in general, an aperiodic DT signal, the appropriate Fourier method for analysis is the DTFT, and the DTFT is

$$X_{\text{DTFT}}(F) = Aw f_s \operatorname{rect}(Fw f_s) * \operatorname{comb}(F). \tag{7.16}$$

The DT signal and its DTFT are illustrated in Figure 7.13 for two different sampling rates.

When comparing the CTFT of the CT signal and the DTFT of the DT signal formed by sampling it, there are some obvious similarities. For this example signal, the CTFT is a rectangle function and the DTFT is a periodic repetition of rectangle functions. The CTFT is

$$X_{\text{CTFT}}(f) = Aw \operatorname{rect}(wf), \tag{7.17}$$

and the DTFT is

$$X_{\text{DTFT}}(F) = Aw f_s \operatorname{rect}(Fw f_s) * \operatorname{comb}(F) \tag{7.18}$$

or

$$X_{\text{DTFT}}(F) = Aw f_s \sum_{k=-\infty}^{\infty} \operatorname{rect}((F - k)w f_s). \tag{7.19}$$

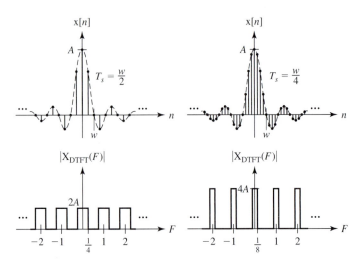

**Figure 7.13**
Example DT signal and the magnitude of its DTFT, for two different sampling rates.

If we take from the summation in (7.19) the $k = 0$ rectangle, $Awf_s \, \text{rect}(Fwf_s)$, and make the change of variable $F \to f/f_s$, we get the functional transformation

$$Awf_s \, \text{rect}(Fwf_s) \to Awf_s \, \text{rect}(wf). \tag{7.20}$$

If we then multiply this result by $T_s$, we get

$$T_s \, [Awf_s \, \text{rect}(Fwf_s)] = Aw \, \text{rect}(wf) = X_{\text{CTFT}}(f). \tag{7.21}$$

So, from this example at least, it seems that one way to recover the CT signal from the DT signal formed by sampling it is to follow these five steps:

1. Find the DTFT of the DT signal.
2. Isolate the $k = 0$ function from step 1.
3. Make the change of variable $F \to f/f_s$ in the result of step 2.
4. Multiply the result of step 3 by $T_s$.
5. Find the inverse CTFT of the result of step 4.

In the previous illustrations, the time between samples $T_s$ was always less than $w$. What happens if $T_s$ is greater than $w$? Then in the expression

$$X_{\text{DTFT}}(F) = Awf_s \sum_{k=-\infty}^{\infty} \text{rect}((F - k)wf_s) \tag{7.22}$$

the rectangle functions *overlap* in the DTFT summation and the shape of $X_{\text{CTFT}}(f)$ is no longer obvious when observing $X_{\text{DTFT}}(F)$ (Figure 7.14). When this happens, it is no longer possible, simply by looking at the DTFT, to extract the CTFT of the original CT signal and thereby reconstruct it.

At this point the information equivalence between $x[n]$ and $x_\delta(t)$ becomes very useful. Imagine that we form $x_\delta(t)$ by impulse-sampling $x(t)$ as indicated by

$$x_\delta(t) = \sum_{n=-\infty}^{\infty} x[n]\delta(t - nT_s) = x(t)f_s \, \text{comb}(f_s t). \tag{7.23}$$

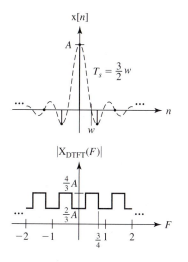

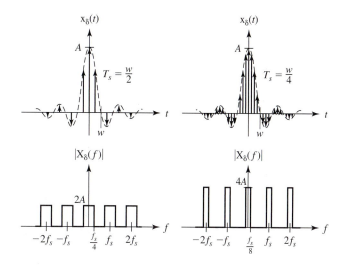

**Figure 7.14**
An undersampled signal and its DTFT.

**Figure 7.15**
Impulse-sampled example signal and the magnitude of its CTFT, for two different sampling rates.

Then, since $x_\delta(t)$ is a CT function, we can find its CTFT,

$$X_\delta(f) = X_{\text{CTFT}}(f) * \text{comb}(T_s f) = f_s \sum_{k=-\infty}^{\infty} X_{\text{CTFT}}(f - kf_s), \qquad (7.24)$$

where $f_s = 1/T_s$. For the example signal,

$$X_{\text{CTFT}}(f) = A w \, \text{rect}(wf). \qquad (7.25)$$

Therefore,

$$X_\delta(f) = f_s \sum_{k=-\infty}^{\infty} A w \, \text{rect}(w(f - kf_s)), \qquad (7.26)$$

and this is the same as

$$X_{\text{DTFT}}(F)|_{F \to f/f_s} = A w f_s \sum_{k=-\infty}^{\infty} \text{rect}\left(\left(\frac{f}{f_s} - k\right) w f_s\right)$$

$$= A w f_s \sum_{k=-\infty}^{\infty} \text{rect}((f - kf_s)w) \qquad (7.27)$$

(Figure 7.15).

If $0 < T_s < w$, in the frequency range $-(f_s/2) < f < f_s/2$, $X_{\text{CTFT}}(f)$ and $X_\delta(f)$ are identical except for a scaling factor $f_s$. Therefore, if $x_\delta(t)$ were filtered by an ideal lowpass filter whose corner frequency is somewhere between $1/2w$ and $f_s - (1/2w)$ and whose gain is $T_s$, the filter output would be exactly the same as the original signal $x(t)$, but only if $0 < T_s < w$ (Figure 7.16).

If $T_s \geq w$, the rectangles in

$$X_\delta(f) = f_s \sum_{k=-\infty}^{\infty} A w \, \text{rect}(w(f - kf_s)) \qquad (7.28)$$

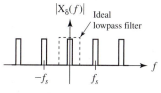

**Figure 7.16**
Recovery of the original CT signal using an ideal lowpass filter.

overlap and now we cannot recover the original signal by filtering with an ideal low-pass filter.

This analysis was for an example signal, a sinc function. We can now generalize the results. The sinc function is band-limited because above some maximum frequency its CTFT is zero. The reason we were able to recover the information in an impulse-sampled version of that signal was that when we impulse-sampled with a time between samples $0 < T_s < w$, the shape of the CTFT of the impulse-sampled signal and the CTFT of the original signal were identical in the frequency range $-(f_s/2) < f < f_s/2$. This occurred because the *replicas* of the original signal's CTFT which appear in the CTFT of the impulse-sampled signal did not overlap. These replicas are called *aliases*. If the original CT signal is not band-limited, the aliases will overlap and we cannot recover the original signal from the samples with an ideal lowpass filter. The requirement $0 < T_s < w$ is equivalent to the requirement $f_s > 1/w = 2f_m$ where $f_m$ is the highest frequency present in the original signal. Therefore, in order to be able to recover a CT signal from samples taken from it, the sampling rate must be more than twice the highest frequency present in the signal.

This description of sampling effects was couched in terms of impulse-sampling and the CTFT of the impulse-sampled signal. An analogous argument was made earlier in terms of sampling the CT signal to form a DT signal and then manipulating the DTFT of that signal. The two approaches to the analysis of sampling effects yield the same conclusion.

Suppose the magnitude of the CTFT, $|X(f)|$, of a band-limited CT signal $x(t)$ is as illustrated in Figure 7.17. Then impulse-sample $x(t)$ to form $x_\delta(t)$. How $|X_\delta(f)|$ will look depends on the relationship between $f_s$ and $f_m$. Let $f_s = 4f_m$. Then $|X_\delta(f)|$ will look as illustrated in Figure 7.18. These shifted versions of the original spectrum that show up at integer multiples of the sampling rate are called aliases because they look like the original spectrum but appear in a different place. (In the most common use of the word *alias,* criminals use aliases when they appear in a different place also.) Notice that in this case it would be easy (in principle) to recover the original signal from the impulse-sampled signal by simply lowpass-filtering the impulse-sampled signal with an ideal unity-gain lowpass filter whose cutoff frequency is somewhere between $f_m$ and $f_s - f_m$ and then dividing the result by $f_s$.

Now let $f_s = 2f_m$. The nonzero portions of the aliases now just touch (Figure 7.19), and an ideal lowpass filter could still recover the original signal from the DT signal if its

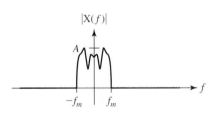

**Figure 7.17**
Magnitude of the amplitude spectrum of a band-limited signal.

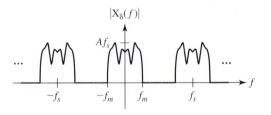

**Figure 7.18**
Magnitude of the amplitude spectrum of a strictly band-limited signal which has been impulse-sampled at four times its highest frequency.

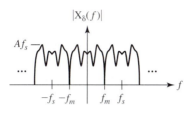

**Figure 7.19**
Magnitude of the amplitude spectrum of a strictly band-limited signal which has been impulse-sampled at twice its highest frequency.

cutoff frequency were set at exactly $f_m$ [and if there were no impulse in X($f$) at exactly $f_m$]. If the sampling rate were any lower than $2f_m$, the aliases would overlap and no filter could recover the original signal directly from the impulse-sampled signal. (In the jargon of sampling theory, if the aliases overlap, the impulse-sampled signal is said to be aliased. This can be prevented by prefiltering a signal with an analog anti-aliasing filter which restricts the signal bandwidth to less than half the sampling rate before the sampling occurs.)

Now the most common form of Shannon's sampling theorem can be stated.

> If a signal is sampled for all time at a rate more than twice the highest frequency at which its CTFT is nonzero, it can be exactly reconstructed from the samples.

The highest frequency present in a signal $f_m$ is called the *Nyquist frequency*. The minimum rate at which a signal can be sampled and still be reconstructed from its samples is called the *Nyquist rate,* and it is always $2f_m$. (Harry Nyquist of Bell Labs was a pioneer in signal and system analysis.) Rate and frequency both describe something which happens periodically. In this text, the word *frequency* will refer to the frequencies present in a signal, and the word *rate* will refer to the way a signal is sampled. A signal sampled at greater than its Nyquist rate is said to be *oversampled* and a signal sampled at less than its Nyquist rate is said to be *undersampled*.

Harry Nyquist,
2/7/1889–4/4/1976

## ALIASING

The phenomenon of aliasing is not an exotic mathematical concept that is outside the experience of ordinary people. Almost everyone has observed aliasing, but probably without knowing what to call it. A very common example of aliasing sometimes occurs while you are watching television. Suppose you are watching a Western movie on television and there is a picture of a horse-drawn wagon with spoked wheels. If the wheels on the wagon gradually rotate faster and faster, a point is reached at which the wheels appear to stop rotating forward and begin to appear to rotate backward even though the wagon or buggy is obviously moving forward. If the speed of rotation were increased further, the wheels would eventually appear to stop and then rotate forward again. This is an example of the phenomenon of aliasing.

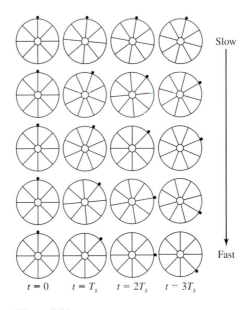

**Figure 7.20**
Wagon wheel angular positions at four
sampling times.

Although it is not apparent to the human eye, the image on a television screen is
flashed upon the screen 30 times per second (in the United States). That is, the image
is *sampled* at a rate of 30 Hz. Figure 7.20 shows the positions of a spoked wheel at four
sampling instants for several different rotational velocities, starting with a lower rotational velocity at the top and progressing toward a higher rotational velocity at the bottom. (A small index dot has been added to the wheel to help you see the actual rotation
of the wheel, as opposed to the apparent rotation.)

This wheel has eight spokes, so upon rotation by one-eighth of a complete revolution the wheel looks exactly the same as it did in its initial position. Therefore, the
*image* of the wheel has an angular period of $\pi/4$ rad, or $45°$, the angular spacing between spokes. If the rotational velocity of the wheel is $f_0$ revolutions per second (Hz)
the image fundamental frequency is $8 f_0$ Hz. The image repeats exactly eight times in
one complete wheel rotation. On the top row the wheel is rotating slowly, and in the
second, third, and fourth images in the top row the spokes have rotated by $5°$, $10°$,
and $15°$ clockwise. The eye and brain of the observer interpret the succession of images to mean that the wheel is rotating clockwise because of the progression of angles at the sampling instants. In this case the wheel appears to be (and is) rotating at
an image rotational frequency of $-(5/T_s)$ degrees/s. In the second row, the angles of
rotation are $0°$, $20°$, $40°$, and $60°$ clockwise. The wheel still (correctly) appears to be
rotating clockwise but now at a rotational frequency of $-(20/T_s)$ degrees/s. In the
third row, the wheel rotates clockwise by $22.5°$ between samples. Now the ambiguity caused by sampling begins. If the index dot were not there, it would be impossible to determine whether the wheel is rotating at a rotational frequency of
$-(22.5°/T_s)$ or $+(22.5°/T_s)$ because the image samples are identical for those two
cases. It is impossible, by simply looking at the sample images, to determine whether

the rotation is clockwise or counterclockwise. In the fourth row the wheel is rotating $40°$ clockwise between samples. Now (ignoring the index dot) the wheel definitely appears to be rotating at $+(5/T_s)$ degrees/s instead of at the actual rotational frequency of $-(40/T_s)$ degrees/s. The perception of the human brain would be that the wheel is rotating $5°$ counterclockwise between samples instead of $40°$ clockwise. In the bottom row the wheel rotates clockwise $45°$ between samples. Now the wheel appears to be standing still even though it is rotating clockwise. Its angular velocity seems to be zero because it is being sampled at a rate exactly equal to the image fundamental frequency.

**EXAMPLE 7.1**

Find the Nyquist frequency and Nyquist rate for each of the following signals.

a.  $x(t) = 25 \cos(500\pi t)$

b.  $x(t) = 15 \text{ rect}\left(\dfrac{t}{2}\right)$

c.  $x(t) = 10 \text{ sinc}(5t)$

d.  $x(t) = 2 \text{ sinc}(5000t) \sin(500{,}000\pi t)$

■ **Solution**

a.
$$X(f) = \frac{25}{2}[\delta(f - 250) + \delta(f + 250)] \tag{7.29}$$

The highest frequency (and the only frequency) present in this signal is 250 Hz. The Nyquist frequency is 250 Hz and the Nyquist rate is 500 Hz.

b.
$$X(f) = 30 \text{ sinc}(2f) \tag{7.30}$$

Since the sinc function never goes to zero and stays there at a finite frequency, the highest frequency in the signal is infinite and the Nyquist frequency and rate are also infinite. The rectangle function is not band-limited.

c.
$$X(f) = 2 \text{ rect}\left(\frac{f}{5}\right) \tag{7.31}$$

The highest frequency present in $x(t)$ is the value of $f$ at which the rect function has its discontinuous transition from one to zero, $f = 2.5$ Hz. Therefore, the Nyquist frequency is 2.5 Hz and the Nyquist rate is 5 Hz.

d.
$$X(f) = \frac{1}{2500} \text{ rect}\left(\frac{f}{5000}\right) * \frac{j}{2}[\delta(f + 250 \text{ kHz}) - \delta(f - 250 \text{ kHz})] \tag{7.32}$$

$$X(f) = \frac{j}{5000}\left[\text{rect}\left(\frac{f + 250 \text{ kHz}}{5000}\right) - \text{rect}\left(\frac{f - 250 \text{ kHz}}{5000}\right)\right] \tag{7.33}$$

The highest frequency in $x(t)$ occurs at

$$f = 252.5 \text{ kHz.} \tag{7.34}$$

Therefore the Nyquist frequency is 252.5 kHz and the Nyquist rate is 505 kHz.

**EXAMPLE 7.2**

Suppose a signal which is to be acquired by a data-acquisition system is known to have an amplitude spectrum that is flat out to 100 kHz and drops suddenly there to zero. Suppose further that the fastest rate at which our data-acquisition system can sample the signal is 60 kHz. Design an anti-aliasing $RC$ lowpass filter which will reduce the signal's amplitude spectrum at 30 kHz to less than 1 percent of its value at very low frequencies so that aliasing will be minimized.

#### ■ Solution

The transfer function of a unity-gain $RC$ lowpass filter is given by

$$\mathrm{H}(f) = \frac{1}{j2\pi f RC + 1}. \tag{7.35}$$

The squared magnitude of the transfer function is given by

$$|\mathrm{H}(f)|^2 = \frac{1}{(2\pi f RC)^2 + 1}. \tag{7.36}$$

Set the $RC$ time constant so that at 30 kHz, the squared magnitude of $\mathrm{H}(f)$ is $(0.01)^2$. That is,

$$|\mathrm{H}(30{,}000)|^2 = \frac{1}{(2\pi \times 30{,}000 \times RC)^2 + 1} = (0.01)^2. \tag{7.37}$$

Solving for $RC$,

$$RC = 0.0005305. \tag{7.38}$$

The corner frequency ($-3$ dB frequency) of this $RC$ lowpass filter is 300 Hz which is 100 times lower than the Nyquist frequency of 30 kHz (Figure 7.21). It must be set this low to meet the specification using a single-pole filter because its transfer function rolls off so slowly with frequency. For this reason most real anti-aliasing filters are designed with much faster roll-offs.

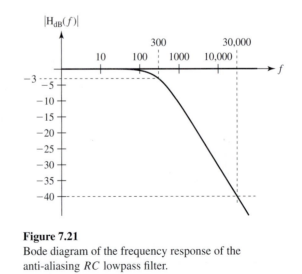

**Figure 7.21**
Bode diagram of the frequency response of the anti-aliasing $RC$ lowpass filter.

## TIME-LIMITED AND BAND-LIMITED SIGNALS

Recall that the original mathematical statement of the way a signal is impulse-sampled is

$$x_\delta(t) = \sum_{n=-\infty}^{\infty} x(nT_s)\delta(t - nT_s). \tag{7.39}$$

Since the summation is from $n = -\infty$ to $+\infty$, in general, infinitely many samples are needed to describe $x(t)$ exactly. Shannon's sampling theorem is predicated on sampling this way. So, even though the minimum sampling rate has been found, and may be finite, one must (in general) still take infinitely many samples to *exactly* reconstruct the original signal from its samples, even if it is band-limited and we sample at a rate more than twice its highest frequency. We will come back to the problem of needing infinitely many samples soon.

It is tempting to think that if a signal is *time-limited* (having nonzero values only over a finite period of time), one could then sample only over that time, knowing all the other samples are zero and have all the information in the signal. The problem with that idea is that no time-limited signal can also be band-limited, and, therefore, no finite sampling rate is adequate.

The fact that a signal cannot be simultaneously time-limited and band-limited is a fundamental law of Fourier analysis. The validity of this law can be demonstrated by the following argument. Let a signal $x(t)$ have no nonzero values outside the time range $t_1 < t < t_2$. Let its CTFT be $X(f)$. Let us hypothesize for now that $x(t)$ is also band-limited, that is, that the magnitude of $X(f)$ is zero for frequencies $f$ greater in magnitude than $f_m$ where $f_m$ is finite. If $x(t)$ is time-limited to the time range $t_1 < t < t_2$, then it can be multiplied by a rectangle function whose nonzero portion covers this same time range without changing the signal. That is

$$x(t) = x(t) \operatorname{rect}\left(\frac{t - t_0}{\Delta t}\right) \tag{7.40}$$

where $t_0 = (t_1 + t_2)/2$ and $\Delta t = t_2 - t_1$ (Figure 7.22).

Finding the CTFT of both sides of (7.40),

$$X(f) = X(f) * \Delta t \operatorname{sinc}(\Delta t f)e^{-j2\pi f t_0}. \tag{7.41}$$

This last equation says that $X(f)$ is unaffected by being convolved with a sinc function. Since $\operatorname{sinc}(\Delta t f)$ has an infinite extent, if it is convolved with an $X(f)$ which has a finite extent, as hypothesized, the convolution of the two will have an infinite extent. Therefore, the last equation cannot be satisfied by any $X(f)$ which has a finite extent.

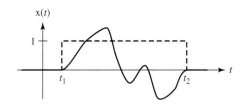

**Figure 7.22**
A time-limited function and a rectangle
limited to the same time.

This violates the original hypothesis, thereby proving that if a signal is time-limited it cannot be band-limited. The converse, a band-limited signal cannot be time-limited, can be proven by a similar argument. A signal can be simultaneously *unlimited* in both time and frequency, but it cannot be simultaneously *limited* in both time and frequency.

## SAMPLING BANDPASS SIGNALS

Shannon's sampling theorem, as previously stated, was based on a simple idea: If we sample fast enough, the aliases do not overlap and the original signal can be recovered. We found that if we sample faster than twice the highest frequency in the signal that we can recover the signal from the samples. That is true for all signals, but for some signals the minimum sampling rate can be reduced.

In making the argument that we must sample at a rate greater than twice the highest frequency in the signal, we were implicitly assuming that if we sampled any slower than that the aliases would overlap. In the spectra used earlier to illustrate the ideas, the aliases would overlap. But that is not true of all signals. For example, let a CT signal have a bandpass spectrum that is nonzero only for $f_1 < |f| < f_2$. Then the bandwidth of this signal is $f_2 - f_1$ (Figure 7.23). If we sample this signal with $f_s < 2f_2$, we could get the aliases illustrated in Figure 7.24. These aliases do not overlap. Therefore, it must be possible, with the right kind of signal processing, to recover the signal from the samples. In this case the right kind of signal processing would be to filter the impulse-sampled signal with an ideal bandpass filter which just covers the range $f_1 < |f| < f_2$. In the previous statement of the sampling theorem we had to know the highest frequency in the signal to know how fast to sample and how to filter the impulse-sampled signal to recover the original signal. In this more general statement of the sampling theorem we must know the band of frequencies the signal occupies and use an ideal filter which covers that band to recover the signal.

The choice of sampling rate in Figure 7.24 was fortuitous. We could have chosen a different rate for which the aliases would overlap. This could have happened even with a somewhat higher sampling rate. If we sample at a rate higher than twice the highest frequency, then there is no way the aliases can overlap. The general formula for the minimum possible sampling rate above which a bandpass signal can be recovered from the samples can be shown to be

$$f_s > \frac{2f_2}{\text{Largest integer not exceeding } f_2/(f_2 - f_1)} \tag{7.42}$$

Notice that if $f_1 = 0$, this reduces to Shannon's sampling theorem as previously stated. In the special case in which $f_2 = m(f_2 - f_1)$, where $m$ is an integer, the formula

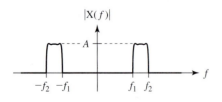

**Figure 7.23**
A bandpass signal spectrum.

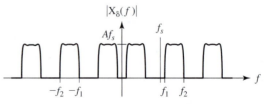

**Figure 7.24**
The spectrum of an impulse-sampled bandpass signal.

becomes

$$f_s > \frac{2f_2}{m} = 2(f_2 - f_1) \tag{7.43}$$

which says that the absolute minimum sampling rate in the most favorable situation is twice the bandwidth of the signal, not the highest frequency. But be careful. Some sampling rates which are higher but still less than twice the highest frequency will cause aliases to overlap.

In most real engineering design situations, choosing the sampling rate to be more than twice the highest frequency in the signal is the practical solution and, as we will soon see, that rate is usually considerably above the Nyquist rate in order to simplify signal reconstruction.

## INTERPOLATION

Exactly how could a signal be reconstructed from its samples, assuming it had been properly sampled? The description of the process of reconstruction in the frequency domain was to filter the impulse-sampled signal with an ideal lowpass filter which cuts off above $f_m$ and below $f_s - f_m$ and has a gain of $T_s$ (Figure 7.25).

Let the cutoff frequency of the filter be $f_c$. Then

$$X(f) = T_s \, \text{rect}\left(\frac{f}{2f_c}\right) X_\delta(f) \qquad f_m < f_c < (f_s - f_m). \tag{7.44}$$

What is the equivalent operation in the time domain? Taking the inverse transform,

$$x(t) = 2f_c T_s \, \text{sinc}(2f_c t) * x_\delta(t) = 2\frac{f_c}{f_s} \, \text{sinc}(2f_c t) * x_\delta(t) \tag{7.45}$$

and since

$$x_\delta(t) = \sum_{n=-\infty}^{\infty} x(nT_s)\delta(t - nT_s), \tag{7.46}$$

we can say that

$$x(t) = 2\frac{f_c}{f_s} \sum_{n=-\infty}^{\infty} x(nT_s) \, \text{sinc}(2f_c(t - nT_s)). \tag{7.47}$$

The reconstruction process consists of replacing each sample by a sinc function, centered at the time of the sample and scaled by the sample value times $2(f_c/f_s)$ and then

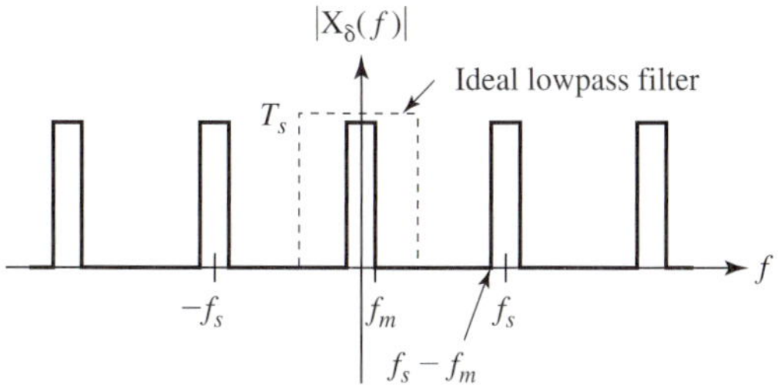

**Figure 7.25**
Rejecting aliases with an ideal lowpass filter.

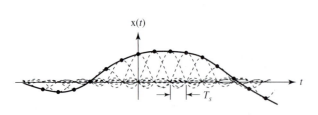

x(t)

**Figure 7.26**
Interpolation process for a signal sampled at its Nyquist rate.

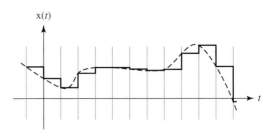

x(t)

**Figure 7.27**
Zero-order hold signal reconstruction.

adding all the functions so created. The process of filling in a signal's values between samples is called *interpolation*.

Suppose the signal is sampled at exactly the Nyquist rate $f_s = 2 f_m$. Now the requirement $f_m < f_c < f_s - f_m$ cannot be met since $f_m = f_s - f_m$. In this case, we must allow the filter cutoff frequency and the maximum frequency in the signal to be the same. This will work as long as the signal's spectrum does not have an impulse at $f_m$. (If there is an impulse at $f_m$, it will be aliased in the sampling process.) Then the interpolation process is described by the simpler expression

$$x(t) = \sum_{n=-\infty}^{\infty} x(nT_s) \operatorname{sinc}\left(\frac{t - nT_s}{T_s}\right). \tag{7.48}$$

Now interpolation consists simply of multiplying each sinc function by its corresponding sample value and then adding all the scaled and shifted sinc functions as illustrated in Figure 7.26.

This interpolation method reconstructs the signal exactly, but it is based on an assumption which is never justified in practice, the availability of infinitely many samples. The interpolated value at any point is the sum of contributions from infinitely many weighted sinc functions. But since, as a practical matter, we cannot acquire infinitely many samples, we must approximately reconstruct the signal using a finite number of samples. There are many techniques that can be used, and the selection of the one to be used in any given situation depends on what accuracy of reconstruction is required and how oversampled the signal is.

Probably the simplest approximate reconstruction idea is to let the reconstruction always be the value of the most recent sample (Figure 7.27). This is a simple technique because the samples, in the form of numerical codes, can be the excitation of a DAC which is clocked to produce a new response signal with every clock pulse. The signal produced by this technique has a stair-step shape which follows (and lags) the original signal. This type of signal reconstruction can be modeled (except for quantization effects) by passing the impulse-sampled signal through a system called a *zero-order hold* whose impulse response is

$$h(t) = \begin{cases} 1 & 0 < t < T_s \\ 0 & \text{otherwise} \end{cases} = \operatorname{rect}\left(\frac{t - (T_s/2)}{T_s}\right) \tag{7.49}$$

(Figure 7.28). We can compare this reconstruction system to the ideal lowpass reconstruction filter by looking at the transfer function of the zero-order hold,

$$H(f) = T_s \operatorname{sinc}(T_s f) e^{-j\pi f T_s} \tag{7.50}$$

(Figure 7.29).

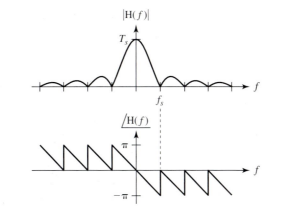

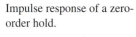

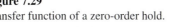

**Figure 7.28**
Impulse response of a zero-order hold.

**Figure 7.29**
Transfer function of a zero-order hold.

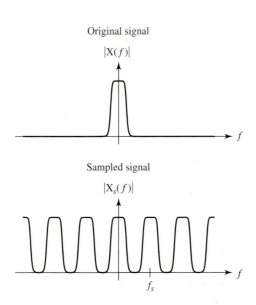

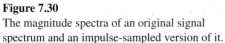

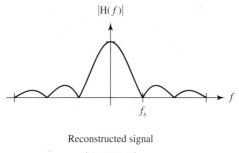

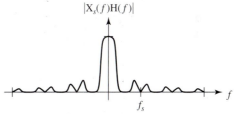

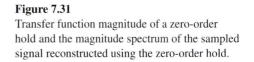

**Figure 7.30**
The magnitude spectra of an original signal spectrum and an impulse-sampled version of it.

**Figure 7.31**
Transfer function magnitude of a zero-order hold and the magnitude spectrum of the sampled signal reconstructed using the zero-order hold.

An ideal reconstruction filter would include the signal bandwidth with no distortion and exclude all the aliases. The zero-order hold does not have an absolute bandwidth like the ideal reconstruction filter because its transfer function magnitude is not zero for all frequencies beyond some finite frequency. Instead its transfer function has a null at the center of each alias and generally diminishes with frequency. Figures 7.30 through 7.32 illustrate an original signal spectrum, the spectrum of that signal after it has been impulse-sampled, and the effects of the zero-order hold in reconstructing the original signal from the samples.

The zero-order hold reduces the effect of aliases, but it does not completely eliminate them. Also, the zero-order hold does not have a perfectly flat top at low frequencies

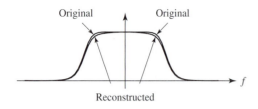

**Figure 7.32**
A comparison between the original signal and the reconstructed signal in the bandwidth of the original signal showing the effect of the rounded top of the zero-order hold transfer function.

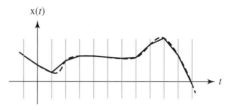

**Figure 7.33**
Signal reconstruction by straight-line interpolation.

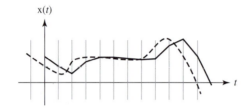

**Figure 7.34**
Straight-line signal reconstruction delayed by one sample time.

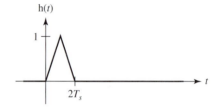

**Figure 7.35**
Impulse response of a first-order hold.

like the ideal reconstruction filter, so it introduces some distortion. One popular way of further reducing the effects of the aliases is to follow the zero-order hold with a practical lowpass filter which smooths out the steps caused by the zero-order hold. The zero-order hold, (7.49), inevitably causes a delay relative to the original signal because it is causal.

Another natural reconstruction idea is to interpolate between samples with straight lines (Figure 7.33). This is obviously a better approximation to the original signal, but it is a little harder to implement. As drawn in Figure 7.33, the value of the interpolated signal at any time depends on the value of the previous sample and the value of the next sample. This cannot be done in real time because the value of the next sample is not known. But if we are willing to delay the reconstructed signal by one sample time $T_s$, we can make the reconstruction process occur in real time and the reconstructed signal would appear as in Figure 7.34.

This interpolation can be accomplished by following the zero-order hold, (7.49), by an identical zero-order hold. This means that the impulse response of such a signal-reconstruction filter would be the convolution of the zero-order hold impulse response with itself,

$$\mathrm{h}(t) = \mathrm{rect}\left(\frac{t - (T_s/2)}{T_s}\right) * \mathrm{rect}\left(\frac{t - (T_s/2)}{T_s}\right) = \mathrm{tri}\left(\frac{t - T_s}{T_s}\right) \qquad \textbf{(7.51)}$$

(Figure 7.35). This type of filter is called a *first-order* hold. Its transfer function is

$$\mathrm{H}(f) = T_s\,\mathrm{sinc}^2(T_s f)e^{-j2\pi f T_s}. \qquad \textbf{(7.52)}$$

This transfer function is similar to that of a zero-order hold but attenuates aliases more because its magnitude diminishes faster with increasing frequency.

If two zero-order holds are better than one, are three better than two? The answer is generally yes if we are only considering the smoothness of the reconstruction and not any other criteria like system complexity or cost or delay. Any *n*th-order hold convolved with a zero-order hold creates an $(n + 1)$th-order hold which smooths the signal more but at the same time delays the reconstructed signal more. The acceptance of delay in signal reconstruction to get a smoother reconstruction is an inherent design trade-off and stems from the same concept that applies to the design of almost-ideal filters, in which the closer we approach the ideal filter, the longer we must wait for the response.

One very familiar example of the use of sampling and signal reconstruction is the playback of an audio compact disk (CD). A CD stores samples of a musical signal which have been taken at a rate of 44.1 kHz. Half of that sampling rate is 22.05 kHz. The frequency response of the human ear is conventionally taken to span from about 20 Hz to 20 kHz with some variability in that range. So the sampling rate is a little more than twice the highest frequency the human ear can detect.

## SAMPLING A SINUSOID

The whole point of Fourier analysis is that any signal can be decomposed into sinusoids (real or complex). Therefore, let's explore sampling by looking at some real sinusoids sampled above, below, and at the Nyquist rate. In each example a sample occurs at time $t = 0$. This sets a definite phase relationship between an exactly described mathematical signal and the way it is sampled. (This is arbitrary, but there must always be a sampling-time reference, and when we get to sampling for finite times, the first sample will always be at time $t = 0$, unless otherwise stated.)

*Case 1*  A cosine sampled at a rate which is four times its frequency or at twice its Nyquist rate (Figure 7.36).  It is clear here that the sample values and the knowledge that the signal is sampled fast enough are adequate to uniquely describe this sinusoid. No other sinusoid of this, or any other frequency, below half the sampling rate could pass exactly through all the samples in the full time range $-\infty < n < +\infty$. In fact no other signal of any kind which is band-limited to below half the sampling rate could pass exactly through all the samples.

*Case 2*  A cosine sampled at twice its frequency or at its Nyquist rate (Figure 7.37). Is this sampling adequate to uniquely determine the signal? No. Consider the sinusoidal signal in Figure 7.38 which is of the same frequency and passes exactly through the same samples. This is a special case which illustrates the subtlety mentioned earlier in the sampling theorem. To be sure of exactly reconstructing any general signal from its samples, the sampling rate must be more than the Nyquist rate instead of at least the Nyquist rate. In earlier examples, it did not matter because the signal power

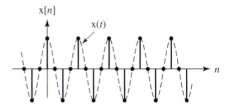

**Figure 7.36**
Cosine sampled at twice its Nyquist rate.

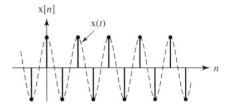

**Figure 7.37**
Cosine sampled at its Nyquist rate.

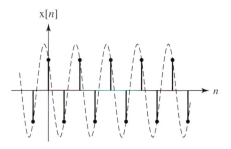

**Figure 7.38**
Sinusoid with the same samples as a cosine
sampled at its Nyquist rate.

at exactly the Nyquist frequency was zero (no impulse in the amplitude spectrum
there). If there is a sinusoid in a signal, exactly at its band limit, the sampling must exceed the Nyquist rate for exact reconstruction, in general. Notice that there is no ambiguity about the frequency of the signal. But there is ambiguity about the amplitude and
phase as illustrated in the figures. If the sinc function interpolation procedure derived
earlier were applied to the samples in Figure 7.38, the cosine in Figure 7.37 which was
sampled at its peaks would result.

Any sinusoid at some frequency can be expressed as the sum of an unshifted cosine of some amplitude at the same frequency and an unshifted sine of some amplitude
at the same frequency. The amplitudes of the unshifted sine and cosine depend on the
phase of the original sinusoid.

$$A \cos(2\pi f_0 t + \theta) = A \cos(2\pi f_0 t) \cos(\theta) - A \sin(2\pi f_0 t) \sin(\theta) \qquad \textbf{(7.53)}$$

$$A \cos(2\pi f_0 t + \theta) = \underbrace{A \cos(\theta)}_{A_c} \cos(2\pi f_0 t) + \underbrace{[-A \sin(\theta)]}_{A_s} \sin(2\pi f_0 t) \qquad \textbf{(7.54)}$$

$$A \cos(2\pi f_0 t + \theta) = A_c \cos(2\pi f_0 t) + A_s \sin(2\pi f_0 t) \qquad \textbf{(7.55)}$$

When a sinusoid is sampled at exactly the Nyquist rate, the sinc function interpolation
always yields the cosine part and drops the sine part, an effect of aliasing. The cosine
part of a general sinusoid is often called the *in-phase* part, and the sine part is often
called the *quadrature* part. The dropping of the quadrature part of a sinusoid can easily
be seen in the time domain by sampling an unshifted sine function at exactly the
Nyquist rate. All the samples are zero (Figure 7.39).

If we were to add a sine function of any amplitude at exactly this frequency (half
the sampling rate) to any signal and then sample the new signal, the samples would be
the same as if the sine function were not there because its value is exactly zero at each
sample time (Figure 7.40). Therefore the quadrature, or sine, part of a signal which is
at exactly half the sampling rate does not show up when the signal is sampled.

*Case 3* A sinusoid sampled at slightly more than the Nyquist rate (Figure 7.41).
Now, because the sampling rate is higher than the Nyquist rate, the samples do not all
occur at zero crossings and there is enough information in the samples to reconstruct
the signal. There is only one sinusoid whose frequency is less than half the sampling
rate; of a unique amplitude, phase, and frequency; and which passes exactly through
all these samples.

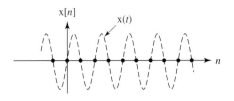

**Figure 7.39**
Sine sampled at its Nyquist rate.

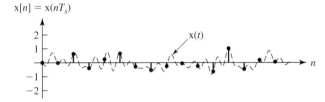

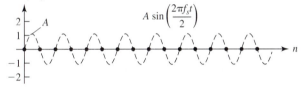

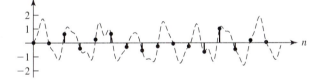

**Figure 7.40**
Effect on samples of adding a sine at half the sampling rate.

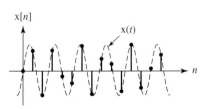

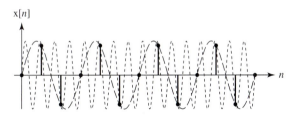

**Figure 7.41**
Sine sampled at slightly above its
Nyquist rate.

**Figure 7.42**
Two sinusoids of different frequencies which have
the same sample values.

***Case 4*** Two sinusoids of different frequencies sampled at the same rate with the same sample values (Figure 7.42). In this case, the lower-frequency sinusoid is oversampled and the higher-frequency sinusoid is undersampled. This illustrates the ambiguity caused by undersampling. If we only had access to the samples from the higher-frequency sinusoid, we would most likely interpret them as having come from the lower-frequency sinusoid.

Recall that the spectrum of a sampled signal is the spectrum of the original signal, except multiplied by the sampling rate and repeated at integer multiples of the sampling rate. That being the case, if a sinusoid

$$x_1(t) = A \cos(2\pi f_0 t + \theta) \tag{7.56}$$

is sampled at a rate $f_s$, the samples will be the same as the samples from another sinusoid

$$x_2(t) = A \cos(2\pi (f_0 + kf_s)t + \theta). \tag{7.57}$$

where $k$ is any integer (including negative integers). This is easily shown by expanding the argument of $x_2(t)$,

$$x_2(t) = A \cos(2\pi f_0 t + 2\pi (kf_s)t + \theta). \tag{7.58}$$

The samples occur at times $nT_s$, where $n$ is an integer. Therefore, the $n$th sample values of the two sinusoids are

$$x_1(nT_s) = A \cos(2\pi f_0 n T_s + \theta) \qquad \text{and}$$

$$x_2(nT_s) = A \cos(2\pi f_0 n T_s + 2\pi (kf_s)nT_s + \theta) \tag{7.59}$$

and, since $f_s T_s = 1$, the second equation simplifies to

$$x_2(nT_s) = A \cos(2\pi f_0 n T_s + 2k\pi n + \theta). \tag{7.60}$$

Then, since $kn$ is the product of integers and, therefore, also an integer, and since adding an integer multiple of $2\pi$ to the argument of a sinusoid does not change its value,

$$x_2(nT_s) = A \cos(2\pi f_0 n T_s + 2k\pi n + \theta) = A \cos(2\pi f_0 n T_s + \theta) = x_1(nT_s). \tag{7.61}$$

## 7.4 SAMPLING DISCRETE-TIME SIGNALS

In Sections 7.1 to 7.3 all the signals that were sampled were CT signals. DT signals can also be sampled. Just as in sampling CT signals, the main concern in sampling DT signals is whether the information in the signal is preserved by the sampling process. There are two complementary processes used in DT signal processing to change the sampling rate of a signal, decimation and interpolation. *Decimation* is a process of reducing the number of samples, and *interpolation* is a process of increasing the number of samples. We will consider decimation first.

We impulse sampled a CT signal by multiplying it by a CT impulse train, a CT comb function. Analogously, we can sample a DT signal by multiplying it by a DT impulse train, a DT comb function. Let the DT signal to be sampled be $x[n]$. Then the sampled signal would be

$$x_s[n] = x[n] \, \text{comb}_{N_s}[n], \tag{7.62}$$

where $N_s$ is the discrete time between samples and the DT sampling rate is $F_s = 1/N_s$ (Figure 7.43). The DTFT of the sampled signal is

$$X_s(F) = X(F) \circledast \text{comb}(N_s F) = X(F) \circledast \text{comb}\left(\frac{F}{F_s}\right) \tag{7.63}$$

(Figure 7.44).

The similarity of DT sampling to CT sampling is obvious. In both cases, if the aliases do not overlap, the original signal can be recovered from the samples and there

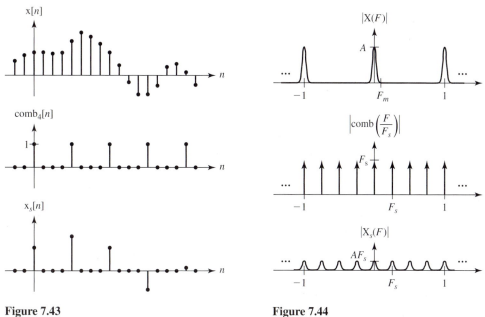

**Figure 7.43**
An example of DT sampling.

**Figure 7.44**
DTFT of a DT signal and a sampled version of it.

is a minimum sampling rate for recovery of the signals. The sampling rate must satisfy the inequality $F_s > 2F_m$, where $F_m$ is the DT frequency above which the DTFT of the original DT signal is zero (in the base fundamental period, $|F| < \frac{1}{2}$). That is, for $F_m < |F| < 1 - F_m$ the DTFT of the original signal is zero. A DT signal which satisfies this requirement is band-limited in the DT sense.

Just as with CT sampling, if a signal is properly sampled, we can reconstruct it from the samples using interpolation. The process of recovering the original signal is described in the DT frequency domain as a lowpass filtering operation,

$$X(F) = X_s(F) \left[ \frac{1}{F_s} \text{ rect} \left( \frac{F}{2F_c} \right) * \text{comb}(F) \right], \tag{7.64}$$

where $F_c$ is the cutoff DT frequency of the ideal lowpass DT filter. The equivalent operation in the DT domain is a DT convolution,

$$x[n] = x_s[n] * \frac{2F_c}{F_s} \text{ sinc}(2F_c n). \tag{7.65}$$

In the practical application of sampling DT signals, it does not make much sense to retain all those zero values between the sampling points because we already know they are zero. Therefore, it is common to create a new signal $x_d[n]$ which has only the nonzero values of the DT signal $x_s[n]$ at integer multiples of the sampling interval $N_s$. The process of forming this new signal is called *decimation*. Decimation was briefly discussed in Chapter 2. The relations between the signals are given by

$$x_d[n] = x_s[N_s n] = x[N_s n]. \tag{7.66}$$

This operation is DT time-scaling which, for $N_s > 1$, causes DT time compression and the corresponding effect in the DT frequency domain is DT frequency expansion. The

DTFT of $x_d[n]$ is

$$X_d(F) = \sum_{n=-\infty}^{\infty} x_d[n] e^{-j2\pi Fn} = \sum_{n=-\infty}^{\infty} x_s[N_s n] e^{-j2\pi Fn} \tag{7.67}$$

We can make a change of variable $m = N_s n$, yielding

$$X_d(F) = \sum_{\substack{m=-\infty \\ m=\text{integer} \\ \text{multiple of } N_s}}^{\infty} x_s[m] e^{-j2\pi F(m/N_s)} \tag{7.68}$$

Now, taking advantage of the fact that all the extra values of $x_s[n]$ between the allowed values, $m = $ integer multiple of $N_s$, are zero, we can include the zeros in the summation, yielding

$$X_d(F) - \sum_{m=-\infty}^{\infty} x_s[m] e^{-j2\pi(F/N_s)m} = X_s\left(\frac{F}{N_s}\right). \tag{7.69}$$

So the DTFT of the decimated signal is a DT frequency-scaled version of the DTFT of the sampled signal (Figure 7.45). Observe carefully that the DTFT of the decimated signal is *not* a DT frequency-scaled version of the DTFT of the original signal, but rather a DT frequency-scaled version of the DT-sampled original signal,

$$X_d(F) = X_s\left(\frac{F}{N_s}\right) \neq X\left(\frac{F}{N_s}\right). \tag{7.70}$$

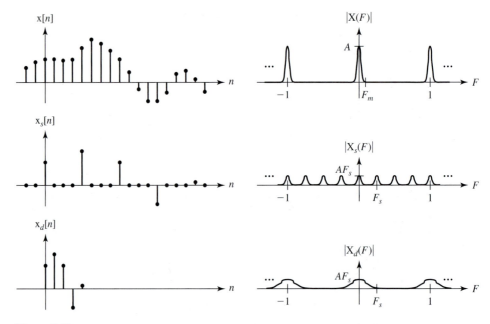

**Figure 7.45**
Comparison of the DT domain and DT-frequency domain effects of sampling and decimation.

The term *downsampling* is sometimes used instead of decimation. This term comes from the idea that the DT signal was produced by sampling a CT signal. If the CT signal was oversampled by some factor, then the DT signal can be decimated by the same factor without losing information about the original CT signal, thus reducing the effective sampling rate or downsampling.

The opposite of decimation is interpolation or *upsampling*. The process is simply the reverse of decimation. First extra zeros are placed between samples, and then the signal so created is filtered by an ideal DT lowpass filter. Let the original DT signal be $x[n]$ and let the signal created by adding $N_s - 1$ zeros between samples be $x_s[n]$. Then

$$x_s[n] = \begin{cases} x\left[\dfrac{n}{N_s}\right] & \dfrac{n}{N_s} \text{ is an integer} \\ 0 & \text{otherwise} \end{cases}.$$

This DT expansion of $x[n]$ to form $x_s[n]$ is the exact opposite of the DT compression of $x_s[n]$ to form $x_d[n]$ in decimation, so we should expect the effect in the DT frequency domain to be the opposite also, a DT expansion by a factor of $N_s$ creates a DT frequency compression by the same factor

$$X_s(F) = X(N_s F) \tag{7.71}$$

(Figure 7.46).

The signal $x_s[n]$ can be lowpass-filtered to interpolate between the nonzero values. If we use an ideal unity-gain lowpass filter with a transfer function

$$H(F) = \text{rect}(N_s F) * \text{comb}(F), \tag{7.72}$$

we get an interpolated signal,

$$X_i(F) = X_s(F)[\text{rect}(N_s F) * \text{comb}(F)], \tag{7.73}$$

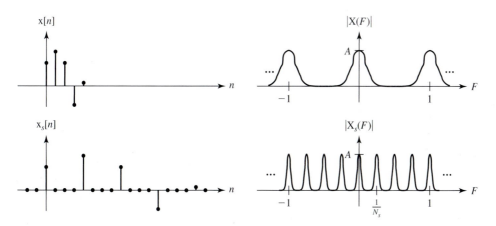

**Figure 7.46**
Effects, in both the DT and DT-frequency domains, of inserting $N_s - 1$ zeros between samples.

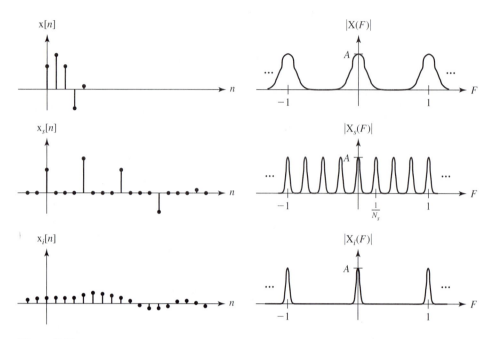

**Figure 7.47**
Comparison of the DT domain and DT-frequency domain effects of expansion and interpolation.

and the equivalent in the DT domain is

$$x_i[n] = x_s[n] * \frac{1}{N_s} \, \text{sinc}\left(\frac{n}{N_s}\right) \tag{7.74}$$

(Figure 7.47). Notice that the interpolation using the unity-gain ideal lowpass filter introduced a gain factor of $1/N_s$, reducing the amplitude of the interpolated signal $x_i[n]$ relative to the original signal $x[n]$. This can be compensated for by using an ideal lowpass filter with a gain of $N_s$,

$$H(F) = N_s \, \text{rect}(N_s F) * \text{comb}(F), \tag{7.75}$$

instead of unity gain.

## EXAMPLE 7.3

Sample the DSBSC signal

$$x(t) = 5 \sin(2000\pi t) \cos(20{,}000\pi t) \tag{7.76}$$

at 80 kHz over one fundamental period to form a DT signal $x[n]$, take every fourth sample of $x[n]$ to form $x_s[n]$, and decimate $x_s[n]$ to form $x_d[n]$. Then upsample $x_d[n]$ by a factor of eight to form $x_i[n]$.

### ■ Solution
See Figures 7.48 and 7.49.

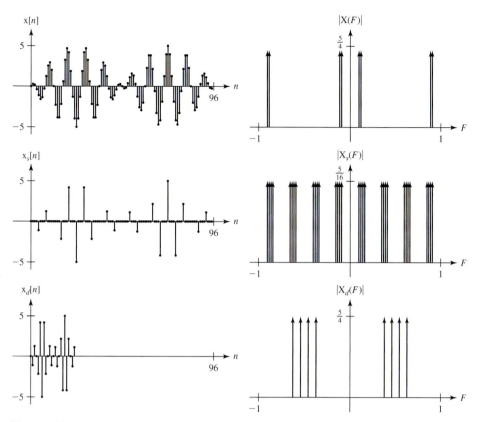

**Figure 7.48**
Original, sampled, and decimated DT signals and their DTFTs.

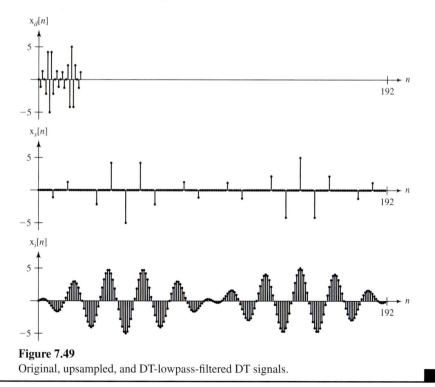

**Figure 7.49**
Original, upsampled, and DT-lowpass-filtered DT signals.

## 7.5  BAND-LIMITED PERIODIC SIGNALS

In Section 7.3 we saw what the requirements were for adequately sampling a signal. We also learned that, in general, for perfect reconstruction of the signal, infinitely many samples are required. Since a computer has a finite storage capability, it is important to explore methods of DT signal analysis using a finite number of samples.

One type of signal which can be described by a finite number of samples is a band-limited, periodic signal. Knowledge of what happens in one period is sufficient to describe all periods, and one period is finite in length (Figure 7.50). Therefore, a finite number of samples over exactly one fundamental period of a band-limited, periodic signal taken at a rate above the Nyquist rate is a complete description of the signal.

Let the DT signal formed by sampling a band-limited, periodic signal $x(t)$ above its Nyquist rate be the periodic DT signal $x[n]$, and let an impulse-sampled version of $x(t)$ sampled at the same rate be $x_\delta(t)$ (Figure 7.51). Only one period of samples is shown in Figure 7.51 to emphasize that one period of samples is enough to completely describe the band-limited periodic signal. Using the Fourier relationships derived in Chapter 5, we can find the appropriate Fourier transforms of these signals (Figure 7.52).

The CTFT of $x(t)$ consists only of impulses because it is periodic, and it consists of a finite number of impulses because it is band-limited. So a finite number of numbers completely characterizes the signal in both the time and frequency domains. If we

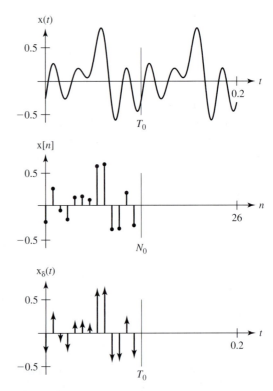

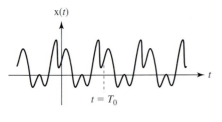

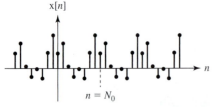

**Figure 7.50**

A band-limited periodic CT signal and a DT signal formed by sampling it 8 times per fundamental period.

**Figure 7.51**

A band-limited periodic CT signal, and a DT signal and a CT impulse signal created by sampling it above its Nyquist rate.

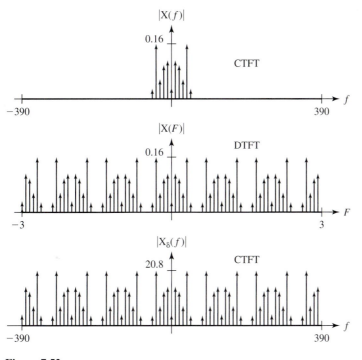

**Figure 7.52**
Magnitudes of the Fourier transforms of the three time-domain signals of Figure 7.51.

multiply the impulse strengths in $X(f)$ by the sampling rate $f_s$, we get the impulse strengths in the same frequency range of $X_\delta(f)$.

**EXAMPLE 7.4**

Find the CTFS harmonic function for the signal $x(t) = 4 + 2\cos(20\pi t) - 3\sin(40\pi t)$ by sampling above the Nyquist rate over exactly one fundamental period and finding the DTFS harmonic function of the samples.

**■ Solution**

There are exactly three frequencies present in the signal, 0, 10, and 20 Hz. Therefore, the highest frequency present in the signal is 20 Hz and the Nyquist rate is 40 Hz. The fundamental frequency is the greatest common divisor of 10 and 20 Hz, which is 10 Hz. So we must sample for $\frac{1}{10}$ s. If we were to sample at the Nyquist rate for exactly one fundamental period, we would get four samples. If we are to sample for *exactly* one fundamental period above the Nyquist rate, we must take five or more samples in one fundamental period. To keep the calculations simple we will sample eight times in one fundamental period. That is a sampling rate of 80 Hz. Then, beginning the sampling at time $t = 0$, the samples are

$$\{x[0], x[1], \ldots, x[7]\} = \left\{6, 1 + \sqrt{2}, 4, 7 - \sqrt{2}, 2, 1 - \sqrt{2}, 4, 7 + \sqrt{2}\right\}. \tag{7.77}$$

Using the formula for finding the DTFS harmonic function of a DT function,

$$X_{\text{DTFS}}[k] = \frac{1}{N_0} \sum_{n=\langle N_0 \rangle} x[n] e^{-j2\pi(kn/N_0)} \tag{7.78}$$

we get

$$\{X_{DTFS}[0], X_{DTFS}[1], \ldots, X_{DTFS}[7]\} = \left\{4, 1, \ j\frac{3}{2}, 0, 0, 0, -j\frac{3}{2}, 1\right\}. \tag{7.79}$$

This is one fundamental period of the DTFS harmonic function $X_{DTFS}[k]$ of the DT function $x[n]$. Finding the CTFS harmonic function of $x(t) = 4 + 2\cos(20\pi t) - 3\sin(40\pi t)$ directly using

$$X_{CTFS}[k] = \frac{1}{T_0} \int_{T_0} x(t) e^{-j2\pi(kf_0)t} \, dt \tag{7.80}$$

we get

$$\{X_{CTFS}[-4], X_{CTFS}[-3], \ldots, X_{CTFS}[4]\} = \left\{0, 0, -j\frac{3}{2}, 1, 4, 1, j\frac{3}{2}, 0, 0\right\}. \tag{7.81}$$

In the two results, the values, $\{X[0], X[1], X[2], X[3], X[4]\}$ are the same, and, using the fact that $X_{DTFS}[k]$ is periodic with fundamental period 8, $\{X[-4], X[-3], X[-2], X[-1]\}$ are the same also.

Now let's violate the sampling theorem by sampling at the Nyquist rate. In this case there are four samples,

$$\{x[0], x[1], x[2], x[3]\} = \{6, 4, 2, 4\} \tag{7.82}$$

and the DTFS harmonic function is

$$\{X[0], X[1], X[2], X[3]\} = \{4, 1, 0, 1\}. \tag{7.83}$$

The CTFS harmonic function is

$$\{X[-2], X[-1], \ldots, X[2]\} = \left\{-j\frac{3}{2}, 1, 4, 1, j\frac{3}{2}\right\}. \tag{7.84}$$

The $j\frac{3}{2}$'s are missing from the DTFS harmonic function. These are the coefficients of the sine function at 40 Hz. This is a demonstration that when we sample a sine function at exactly the Nyquist rate, we don't see it in the samples because we sample it exactly at its zero crossings. ∎

A thoughtful reader may have noticed that the description of a signal based on samples in the time domain from one fundamental period consists of a finite set of numbers $x[n]$, $n_0 \leq n < n_0 + N_0$, which contains $N_0$ independent real numbers and the corresponding DTFS harmonic function description of the signal in the frequency domain consists of the finite set of numbers $X[k]$, $k_0 \leq k < k_0 + N_0$, which contains $N_0$ complex numbers and, therefore, $2N_0$ real numbers (two real numbers for each complex number, the real and imaginary parts). So it would seem that the description in the time domain is more efficient than in the frequency domain since it is accomplished with fewer real numbers. But how can this be when the set $X[k]$, $k_0 \leq k < k_0 + N_0$, is calculated directly from the set $x[n]$, $n_0 \leq n < n_0 + N_0$, with no extra information? A closer examination of the relationship between the two sets of numbers will reveal that this apparent difference is an illusion.

Consider first the coefficient $X[0]$. It can be computed by the DTFS harmonic function formula as

$$X[0] = \frac{1}{N_0} \sum_{n=\langle N_0 \rangle} x[n]. \tag{7.85}$$

Since all the x[$n$]'s are real, X[0] must also be real because it is simply the average of all the x[$n$]'s. There are two cases to consider next, $N_0$ even and $N_0$ odd.

***Case 1*** $N_0$ even.   For simplicity, and without loss of generality, in

$$X[k] = \frac{1}{N_0} \sum_{n=\langle N_0 \rangle} x[n] e^{-j\pi(kn/N_0)} = \frac{1}{N_0} \sum_{n=k_0}^{k_0+N_0-1} x[n] e^{-j\pi(kn/N_0)} \quad \textbf{(7.86)}$$

let $k_0 = -(N_0/2)$. Then

$$X[k_0] = X\left[-\frac{N_0}{2}\right] = \frac{1}{N_0} \sum_{n=\langle N_0 \rangle} x[n] e^{-j\pi n} = \frac{1}{N_0} \sum_{n=\langle N_0 \rangle} x[n](-1)^n \quad \textbf{(7.87)}$$

and X[$k_0$] is guaranteed to be real. All the DTFS harmonic function values in one period, other than X[0] and X[$-(N_0/2)$], occur in pairs X[$k$] and X[$-k$]. Next recall that for any real x[$n$], X[$k$] = X*[$-k$]. That is, once we know X[$k$] we also know X[$-k$]. So, even though each X[$k$] contains two real numbers, and each X[$-k$] does also, X[$-k$] does not add any information since we already know that X[$k$] = X*[$-k$]. That is, X[$-k$] is not *independent* of X[$k$]. So now we have, as independent numbers, X[0], X[$-(N_0/2)$], and X[$k$] for positive $k$. All the X[$k$]'s from $k = 1$ to $(N_0/2) - 1$ yield a total of $2((N_0/2) - 1) = N_0 - 2$ independent real numbers. Add the two guaranteed-real coefficients X[0] and X[$-(N_0/2)$], and we finally have a total of $N_0$ independent real numbers in the frequency-domain description of this signal.

***Case 2*** $N_0$ odd.   For simplicity, and without loss of generality, let $k_0 = -((N_0 - 1)/2)$. In this case, we simply have X[0] plus $(N_0 - 1)/2$ complex conjugate pairs X[$k$] and X[$-k$]. We have already seen that X[$k$] = X*[$-k$]. So we have the real number X[0] and two independent real numbers per complex conjugate pair or $N_0 - 1$ independent real numbers for a total of $N_0$ independent real numbers.

The information content in the form of independent real numbers is conserved in the process of converting from the time to the frequency domain.

## 7.6 THE DISCRETE FOURIER TRANSFORM AND ITS RELATION TO OTHER FOURIER METHODS

The most commonly used Fourier analysis technique in the world is the so-called fast Fourier transform which is an efficient algorithm for computing the discrete Fourier transform (DFT). The DFT is almost identical to the DTFS. The only real differences are a scaling factor and an assumption that the first sample of the CT signal occurs at time $t = 0$. The DTFS of a set of samples x[$n$] = x($nT_s$), $0 \le n < N_F$, from a CT signal x($t$) is defined by the transform pair

$$x[n] = \sum_{k=0}^{N_F-1} X[k] e^{j2\pi(nk/N_F)} \xleftrightarrow{\;\mathcal{FS}\;} X[k] = \frac{1}{N_F} \sum_{n=0}^{N_F-1} x[n] e^{-j2\pi(nk/N_F)} \quad \textbf{(7.88)}$$

where x[$n$] = x($nT_s$). The DTFS harmonic function X[$k$] is periodic with fundamental period $N_F$, and, in general, the representation x[$n$] = $\sum_{k=0}^{N_F-1} X[k] e^{j2\pi(nk/N_F)}$ is only valid for $0 \le n < N_F$. If x[$n$] is periodic with fundamental period $N_0$ and $N_F = N_0$, then the representation x[$n$] = $\sum_{k=0}^{N_F-1} X[k] e^{j2\pi(nk/N_F)}$ is valid for any $n$.

The DFT of that same set of samples is defined by the transform pair

$$x[n] = \frac{1}{N_F} \sum_{k=0}^{N_F-1} X[k] e^{j2\pi(nk/N_F)} \xleftrightarrow{\;\mathcal{DFT}\;} X[k] = \sum_{n=0}^{N_F-1} x[n] e^{-j2\pi(nk/N_F)}. \quad \textbf{(7.89)}$$

So the relation between the DTFS harmonic function and the DFT is

$$\mathrm{X}_{\mathrm{DFT}}[k] = N_F \mathrm{X}_{\mathrm{DTFS}}[k]. \tag{7.90}$$

One of the most important practical applications of the DFT is its use as an approximation to the CTFT. In the development of the relationship between the CTFT and the DFT which follows, all the processing steps from the original CT function to the DFT will be illustrated by an example signal.

Let a CT signal $\mathrm{x}(t)$ be sampled and let the total number of samples taken be

$$N_F = T_F f_s, \tag{7.91}$$

where $T_F$ is the total sampling time and $f_s$ is the sampling frequency. Then the time between samples is $T_s$ where

$$T_s = \frac{1}{f_s}. \tag{7.92}$$

The example original signal in both the time and frequency domains is shown in Figure 7.53.

The first processing step in converting from the CTFT to the DFT is to sample the CT signal $\mathrm{x}(t)$ to form a DT signal $\mathrm{x}_s[n]$,

$$\mathrm{x}_s[n] = \mathrm{x}(nT_s). \tag{7.93}$$

The frequency-domain counterpart of the DT function is its DTFT. Using the relationships among Fourier methods derived in Chapter 5, we can write the DTFT of $\mathrm{x}_s[n]$, $\mathrm{X}_s(F)$, in terms of the CTFT of $\mathrm{x}(t)$, $\mathrm{X}(f)$. It is

$$\mathrm{X}_s(F) = f_s \mathrm{X}(f_s F) * \mathrm{comb}(F) = f_s \sum_{n=-\infty}^{\infty} \mathrm{X}(f_s(F - n)), \tag{7.94}$$

a frequency-scaled and periodically repeated version of $\mathrm{X}(f)$ (Figure 7.54).

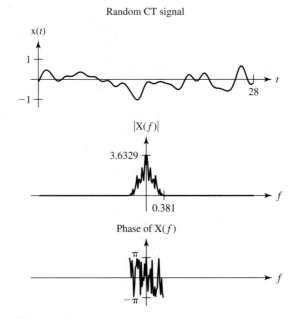

**Figure 7.53**
An original CT signal and its CTFT.

DT signal formed by sampling the CT signal

Windowed DT signal

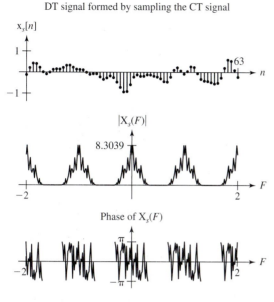

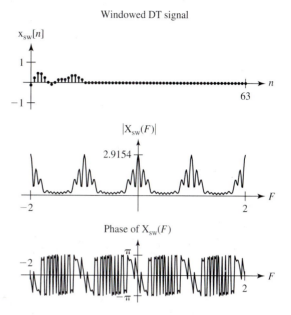

**Figure 7.54**
Original signal, time sampled to form a DT signal,
and the DTFT of the DT signal.

**Figure 7.55**
Original signal, time-sampled and windowed to form
a DT signal, and the DTFT of that DT signal.

Next, we must limit the number of samples to those occurring in the total DT sampling time $N_F$. Let the time of the first sample be $n = 0$. (This is the default assumption in the DFT. Other time references could be used, but the effect of a different time reference is simply a phase shift which varies linearly with frequency.) This can be accomplished by multiplying $x_s[n]$ by a window function,

$$w[n] = \begin{cases} 1 & 0 \le n < N_F \\ 0 & \text{otherwise} \end{cases} \qquad (7.95)$$

as illustrated in Figure 7.55. This window function has exactly $N_F$ nonzero values, the first one being at discrete time $n = 0$. Call the sampled-and-windowed DT signal $x_{sw}[n]$. Then

$$x_{sw}[n] = w[n]x_s[n] = \begin{cases} x_s[n] & 0 \le n < N_F \\ 0 & \text{otherwise} \end{cases} . \qquad (7.96)$$

The process of limiting a signal to the finite range $N_F$ in discrete time is called *windowing,* because we are considering only that part of the sampled signal which can be seen through a DT window of finite length. The window function need not be a rectangle. Other window shapes are often used in practice to minimize an effect called *leakage* (described later) in the frequency domain. The DTFT of $x_{sw}[n]$ is the periodic convolution of the DTFT of the DT signal $x[n]$ and the DTFT of the window function $w[n]$,

$$X_{sw}(F) = W(F) \circledast X_s(F). \qquad (7.97)$$

The DTFT of the window function is

$$W(F) = \sum_{n=-\infty}^{\infty} w[n]e^{-j2\pi Fn} = \sum_{n=0}^{N_F-1} e^{-j2\pi Fn} \qquad (7.98)$$

or

$$W(F) = \frac{1 - e^{-j2\pi F N_F}}{1 - e^{-j2\pi F}} = \frac{e^{-j\pi F N_F}}{e^{-j\pi F}} \frac{e^{j\pi F N_F} - e^{-j\pi F N_F}}{e^{j\pi F} - e^{-j\pi F}} = e^{-j\pi F(N_F - 1)} \frac{\sin(\pi F N_F)}{\sin(\pi F)}$$

(7.99)

or, expressed as a Dirichlet function,

$$W(F) = e^{-j\pi F(N_F - 1)} N_F \, \mathrm{drcl}(F, N_F).$$

(7.100)

Then

$$X_{sw}(F) = e^{-j\pi F(N_F - 1)} N_F \, \mathrm{drcl}(F, N_F) \circledast f_s \sum_{n=-\infty}^{\infty} X(f_s(F - n))$$

(7.101)

or, using the fact that periodic convolution with a periodic signal is equivalent to aperiodic convolution with one fundamental period of the periodic signal,

$$X_{sw}(F) = f_s[e^{-j\pi F(N_F - 1)} N_F \, \mathrm{drcl}(F, N_F)] * X(f_s F).$$

(7.102)

So the effect in the DT-frequency domain of windowing in discrete time is that the Fourier transform of the time-sampled signal has been periodically convolved with

$$W(F) = e^{-j\pi F(N_F - 1)} N_F \, \mathrm{drcl}(F, N_F)$$

(7.103)

(Figure 7.56).

The convolution will tend to spread $X_s(F)$ in the DT frequency domain, which causes the power of $X_s(F)$ at any frequency to leak over into adjacent frequencies in

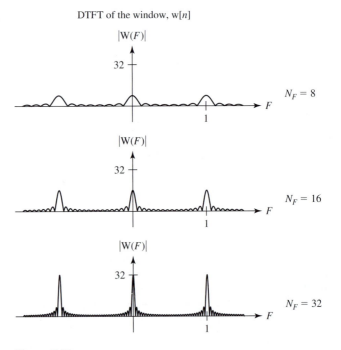

**Figure 7.56**

Magnitude of the DTFT of the rectangular window function,
$$w[n] = \begin{cases} 1, & 0 \le n < N_F \\ 0, & \text{otherwise} \end{cases}, \text{ for three different window widths.}$$

$X_{sw}(F)$. This is where the term *leakage* comes from. The use of a different window function whose DTFT is more confined in the DT frequency domain minimizes (but can never completely eliminate) leakage. As can be seen in Figure 7.56, as the number of samples $N_F$ increases, the width of the main lobe of each fundamental period of this function decreases, reducing leakage. So another way to reduce leakage is to use a larger set of samples.

At this point in the process we have a finite sequence of numbers from the DT signal, but the DTFT of the windowed signal is a periodic function in continuous DT frequency $F$ and is, therefore, not appropriate for computer storage and manipulation. The fact that the DT domain function has become time-limited by the windowing process and the fact that the DT frequency-domain function is periodic allow us to sample in the DT frequency domain over one fundamental period to completely describe the DT frequency-domain function. It is natural at this point to wonder how a frequency-domain function must be sampled to be able to reconstruct it from its samples. The answer is almost identical to the answer for sampling time-domain signals except that *time* and *frequency* have exchanged roles. The only real difference is that the functions we deal with in the frequency domain are a little more general because they are usually complex instead of purely real like the usual time-domain signals. The relations between the time and frequency domains are almost identical because of the duality of the forward and inverse Fourier transforms.

In Chapter 5 we found that sampling in the DT frequency domain corresponds to periodic repetition in the DT domain through the relationship,

$$X_p[k] = \frac{1}{N_F} X\left(\frac{k}{N_F}\right), \tag{7.104}$$

where $x_p[n]$ is a periodic DT-domain function formed by periodically repeating an aperiodic DT-domain function $x[n]$, $X_p[k]$ is the DTFS harmonic function of $x_p[n]$, and $N_F$ is the fundamental period of the periodic repetition (Figure 7.57).

Therefore, if we form a periodic repetition of $x_{sw}[n]$,

$$x_{sws}[n] = \sum_{m=-\infty}^{\infty} x_{sw}[n - m N_F]. \tag{7.105}$$

with fundamental period $N_F$, its DTFS harmonic function is

$$X_{sws}[k] = \frac{1}{N_F} X_{sw}\left(\frac{k}{N_F}\right) \qquad k \text{ is an integer} \tag{7.106}$$

or, from (7.102),

$$X_{sws}[k] = \frac{f_s}{N_F}\left[e^{-j\pi F(N_F-1)} N_F \operatorname{drcl}(F, N_F) * X(f_s F)\right]_{F \to k/N_F}. \tag{7.107}$$

The effect of the last operation, sampling in the DT frequency domain, is sometimes called *picket fencing*. The effect, in the DT domain, of sampling in the DT frequency domain is to periodically repeat the windowed DT function, with a fundamental period of $N_F$ (Figure 7.58). Since its nonzero length is exactly $N_F$, this is a periodic repetition of $x_{sw}[n]$ with a fundamental period equal to its length, so the multiple replicas of $x_{sw}[n]$ do not overlap but instead just touch. Therefore, $x_{sw}[n]$ can be recovered from $x_{sws}[n]$ by simply isolating one fundamental period of $x_{sws}[n]$ in the DT range $0 \le n < N_F$.

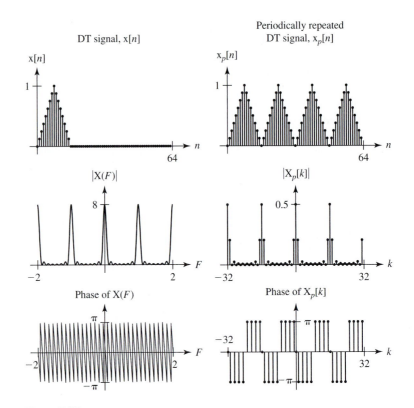

**Figure 7.57**
The equivalence of sampling in the frequency domain and periodic repetition in the time domain.

The result,

$$X_{sws}[k] = \frac{f_s}{N_F}\left[ e^{-j\pi F(N_F - 1)} N_F \, \mathrm{drcl}(F, N_F) * X(f_s F)\right]_{F \to k/N_F} \qquad (7.108)$$

is the DTFS harmonic function of a periodic extension of the DT signal formed by sampling the original CT signal over a finite time. Since the DFT is the same as the DTFS harmonic function except for a scale factor of $N_F$, the equivalent expression in terms of the DFT is

$$X_{sws,\mathrm{DFT}}[k] = N_F X_{sws,\mathrm{DTFS}}[k] = f_s\left[ e^{-j\pi F(N_F - 1)} N_F \, \mathrm{drcl}(F, N_F) * X(f_s F)\right]_{F \to k/N_F}. \qquad (7.109)$$

The DFT of samples from a CT signal can be used to approximate the CTFT of the signal. The CTFT of a signal $x(t)$ is

$$X(f) = \int_{-\infty}^{\infty} x(t) e^{-j2\pi ft}\, dt. \qquad (7.110)$$

When we apply this to signals that are causal, we get

$$X(f) = \int_{0}^{\infty} x(t) e^{-j2\pi ft}\, dt. \qquad (7.111)$$

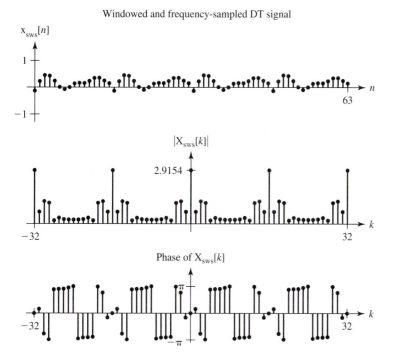

**Figure 7.58**
Original signal, time-sampled, windowed, and periodically repeated, to form
a periodic DT signal and the DTFS harmonic function of that signal.

We can write this integral in the form

$$X(f) = \sum_{n=0}^{\infty} \int_{nT_s}^{(n+1)T_s} x(t)e^{-j2\pi ft}\,dt. \tag{7.112}$$

If $T_s$ is small enough, the variation of $x(t)$ in the time interval $nT_s < t < (n+1)T_s$ is
small and the CTFT can be approximated by

$$X(f) \cong \sum_{n=0}^{\infty} x(nT_s) \int_{nT_s}^{(n+1)T_s} e^{-j2\pi ft}\,dt \tag{7.113}$$

or

$$X(f) \cong \sum_{n=0}^{\infty} x(nT_s) \left[\frac{e^{-j2\pi ft}}{-j2\pi f}\right]_{nT_s}^{(n+1)T_s} = \sum_{n=0}^{\infty} x(nT_s)\frac{e^{-j2\pi fnT_s} - e^{-j2\pi f(n+1)T_s}}{j2\pi f} \tag{7.114}$$

or

$$X(f) \cong \frac{1 - e^{-j2\pi fT_s}}{j2\pi f} \sum_{n=0}^{\infty} x(nT_s)e^{-j2\pi fnT_s} = T_s e^{-j\pi fT_s}\,\mathrm{sinc}(T_s f) \sum_{n=0}^{\infty} x(nT_s)e^{-j2\pi fnT_s}$$

$$\tag{7.115}$$

(Figure 7.59).

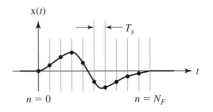

**Figure 7.59**
A CT signal and multiple intervals
on which the CTFT integral can be
evaluated.

If the signal $x(t)$ is an energy signal, then beyond some time its size must become negligible and we can replace the infinite range of $n$ in the summation with a finite range $0 \leq n < N_F$, yielding

$$X(f) \cong T_s e^{-j\pi f T_s} \operatorname{sinc}(T_s f) \sum_{n=0}^{N_F-1} x(nT_s) e^{-j2\pi f nT_s}. \tag{7.116}$$

Now if we compute the CTFT only at integer multiples of $f_s/N_F$,

$$X\left(k\frac{f_s}{N_F}\right) \cong T_s e^{-j\pi k \frac{f_s}{N_F} T_s} \operatorname{sinc}\left(T_s k \frac{f_s}{N_F}\right) \sum_{n=0}^{N_F-1} x(nT_s) e^{-j2\pi k(f_s/N_F)nT_s} \tag{7.117}$$

or

$$X\left(k\frac{f_s}{N_F}\right) \cong T_s e^{-j(\pi k/N_F)} \operatorname{sinc}\left(\frac{k}{N_F}\right) \sum_{n=0}^{N_F-1} x(nT_s) e^{-j2\pi(kn/N_F)}. \tag{7.118}$$

The summation in (7.118) is the DFT of $x[n] = x(nT_s)$. Therefore,

$$X(kf_F) \cong T_s e^{-j(\pi k/N_F)} \operatorname{sinc}\left(\frac{k}{N_F}\right) X_{\text{DFT}}[k]. \tag{7.119}$$

For those harmonic numbers $k$, for which $k \ll N_F$,

$$X(kf_F) \cong T_s X_{\text{DFT}}[k]. \tag{7.120}$$

So if we oversample by a large factor and sample a large number of times, the approximation in (7.120) becomes accurate for frequencies well below half the sampling rate.

We will now look at a special case of application of the DFT. Suppose that the original signal $x(t)$ is band-limited with maximum frequency $f_m$ and periodic with fundamental period $T_0$ and that we sample it $N_F$ times at a rate above its Nyquist rate over exactly one fundamental period (Figure 7.60). If the signal were sampled at exactly the Nyquist rate over one fundamental period, the number of samples would be the even integer $N_0 = 2f_m/f_0$ (because the highest frequency $f_m$ in a periodic signal must be an integer multiple of the fundamental frequency $f_0$). Therefore, the number of samples must be an integer $N_F > N_0$ and $f_s = N_F f_0$. Since the signal is periodic, it has a CTFS representation

$$x(t) = \sum_{q=-\infty}^{\infty} X_{\text{CTFS}}[q] e^{j2\pi(qf_F)t}, \tag{7.121}$$

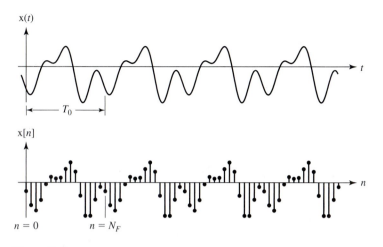

**Figure 7.60**
A band-limited periodic signal and a DT signal formed by sampling
it above its Nyquist rate.

and because it is band-limited,

$$X_{CTFS}[q] = 0 \qquad |q| > \frac{f_m}{f_0} = \frac{N_0}{2} \tag{7.122}$$

and, therefore,

$$x(t) = \sum_{q=-(N_0/2)}^{N_0/2} X_{CTFS}[q] e^{j2\pi(qf_F)t}. \tag{7.123}$$

Relating the CTFT to the CTFS harmonic function,

$$X(f) = \sum_{q=-(N_0/2)}^{N_0/2} X_{CTFS}[q]\delta(f - qf_0) \tag{7.124}$$

(Figure 7.61). (The index $q$ is used here instead of $k$ because $k$ will be used as the independent variable in the DTFS harmonic function of the time-sampled, windowed, and DT frequency-sampled signal $X_{sws}[k]$.)

The DT frequency-domain form of $X_{sws}[k]$ is

$$X_{DFT}[k] = f_s \left[ e^{-j\pi F(N_F-1)} N_F \, \text{drcl}(F, N_F) * X(f_s F) \right]_{F \to k/N_F}, \tag{7.125}$$

where it is to be understood that this $X_{DFT}[k]$ is the DFT of the samples, not the DTFS harmonic function of the samples. Substituting in $X(f) = \sum_{q=-(N_0/2)}^{N_0/2} X_{CTFS}[q]\,\delta(f - qf_0)$ and making the change of variable $f \to f_s F$ as indicated in (7.125),

$$X_{DFT}[k]$$

$$= f_s \left[ e^{-j\pi F(N_F-1)} N_F \, \text{drcl}(F, N_F) * \sum_{q=-(N_0/2)}^{N_0/2} X_{CTFS}[q]\delta(f_s F - qf_0) \right]_{F \to k/N_F}. \tag{7.126}$$

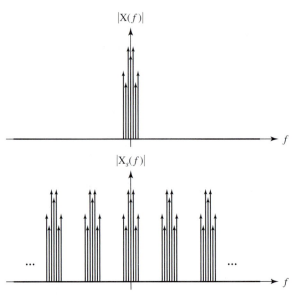

**Figure 7.61**
CTFTs of the original signal and the impulse-sampled signal.

Rearranging and using the scaling property of the impulse,

$$
\begin{aligned}
&X_{\text{DFT}}[k] \\
&= \left[ \sum_{q=-(N_0/2)}^{N_0/2} X_{\text{CTFS}}[q] \left\{ e^{-j\pi F(N_F-1)} N_F \, \text{drcl}(F, N_F) * \delta\left(F - q\frac{f_0}{f_s}\right) \right\} \right]_{F \to k/N_F}.
\end{aligned}
$$

$$(7.127)$$

Using $N_F = f_s/f_0$ and performing the indicated convolution,

$$
\begin{aligned}
&X_{\text{DFT}}[k] \\
&= \left[ \sum_{q=-(N_0/2)}^{N_0/2} X_{\text{CTFS}}[q] \left\{ e^{-j\pi(F-(q/N_F))(N_F-1)} N_F \, \text{drcl}\left(F - \frac{q}{N_F}, N_F\right) \right\} \right]_{F \to k/N_F}.
\end{aligned}
$$

$$(7.128)$$

Making the change of variable $F \to k/N_F$ as indicated in (7.128),

$$
X_{\text{DFT}}[k] = \left[ \sum_{q=-(N_0/2)}^{N_0/2} X_{\text{CTFS}}[q] \left\{ e^{-j\pi((k-q)/N_F)(N_F-1)} N_F \, \text{drcl}\left(\frac{k-q}{N_F}, N_F\right) \right\} \right].
$$

$$(7.129)$$

The Dirichlet function $\text{drcl}(t, N)$ is zero when $t$ is an integer multiple of $1/N$, unless that $t$ is an integer. When $t$ is an integer, the Dirichlet function is either $+1$ or $-1$. In (7.129) since $k$ and $q$ are both integers, $k - q$ is also an integer. Therefore, the Dirichlet function $\text{drcl}((k-q)/N_F, N_F)$ is zero for every value of $(k-q)/N_F$ except when it is an integer. When $(k-q)/N_F$ is an integer, the value of $e^{-j\pi((k-q)/N_F)(N_F-1)} \times N_F \, \text{drcl}((k-q)/N_F, N_F)$ is $N_F$.

    Since the $q$ summation in (7.129) is over the range $-(N_0/2) \le q \le N_0/2$ for values of $k$ in the range $-(N_0/2) \le k \le N_0/2$, $(k-q)/N_F = m$ can only be satisfied for

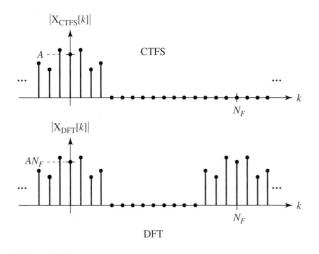

**Figure 7.62**
Relationship between the CTFS harmonic function of a
band-limited periodic signal and the DFT of samples from
one fundamental period of that signal.

$m = 0$ which means $k = q$. Therefore

$$X_{\text{DFT}}[k] = N_F X_{\text{CTFS}}[k] \qquad -(N_0/2) \leq k \leq N_0/2. \qquad \textbf{(7.130)}$$

For other values of $k$ the relation is the same except that an integer multiple of $N_F$ must be added to $q$. That is, $X_{\text{DFT}}[k]$ is a periodic repetition of $N_F X_{\text{CTFS}}[k]$ with fundamental period $N_F$, and

$$X_{\text{DFT}}[k] = N_F X_{\text{CTFS}}[k] * \text{comb}_{N_F}[k] \qquad \textbf{(7.131)}$$

(Figure 7.62).

In words, if a signal $x(t)$ is periodic and band-limited and sampled $N_F$ times at a rate greater than its Nyquist rate, over exactly one fundamental period, the DFT of that set of samples is $N_F$ times a periodic repetition of the CTFS harmonic function $X_{\text{CTFS}}[k]$ of the original signal $x(t)$ with fundamental period $N_F$. So in the special case of band-limited, periodic signals sampled above the Nyquist rate over exactly one fundamental period, the DFT of the samples can be converted exactly into the CTFS (and, therefore, the CTFT) of the original signal.

Below is a MATLAB program to compute the CTFS of a signal based on samples from it under the assumption that exactly one fundamental period is sampled an integer number of times at a rate above the Nyquist rate.

```
%    Function to compute an approximation to the continuous-time
%    Fourier series (CTFS) X[k] of a signal x(t) based on an input
%    set of data x(n*Ts), n = 0 to NF - 1, at the harmonic numbers k,
%    where NF is the total number of samples. The vector returned is a
%    vector of X[k]'s at the k's input to the function. If an input
%    k vector is not provided, the returned k's will be in the range,
%    -NF/2 ≤ k ≤ NF/2.
%
%    The computation is done based on the assumption that the set of
%    data x(n*Ts) is from one period of the periodic signal x(t) and
%    that the sampling is done in accordance with the sampling theorem.
```

```
%       Based on those assumptions any CTFS components at harmonic numbers
%       at or above NF/2 will have zero amplitude and that value will be
%       returned for any k at or above NF/2 in absolute value.
%
%       t and x must be column vectors of real numbers and must have the
%       same length.
%
%       [X,k] = CTFS(x,t,k)
function [X,k] = CTFS(x,t,k)

    NF = length(x) ;

    %       Compute the sampling interval, sampling frequency, and
    %       fundamental frequency of x(t).

    Ts = t(2) - t(1) ; fs = 1/Ts ; fF = fs/NF ;

    %       If the k vector is not input, generate one to cover -NF/2 to
    %       NF/2.

    if nargin < 3, k = [-NF/2:NF/2]' ; end

    %       Compute one period of the CTFS.

    Xper = DTFS(x) ; kvec = [0:NF-1]' ;

    %       If the first sample is not at time t = 0, phase shift the
    %       CTFS accordingly.

    Xper = Xper.*exp(-j*2*pi*(kvec*fF)*t(1)) ;

    %       Compute the CTFS at the input-k vector values assuming here
    %       that they repeat periodically like the DTFS.

    X = Xper(mod(k,NF)+1) ;

    %       Set the values of the CTFS harmonic amplitudes to zero for
    %       k's outside the range -NF/2 to NF/2.

    X(find(abs(k)>=NF/2)) = 0 ;
```

## 7.7  EXAMPLES OF THE USE OF THE DISCRETE FOURIER TRANSFORM

The following examples will illustrate some of the features and limitations of the DFT as a Fourier analysis tool.

### EXAMPLE 7.5

The band-limited, periodic signal

$$x(t) = 1 + \cos(8\pi t) + \sin(4\pi t) \tag{7.132}$$

is sampled at the Nyquist rate (Figure 7.63). Find the sample values over one fundamental period, and find the DFT of the sample values. Find the CTFS harmonic function of the signal.

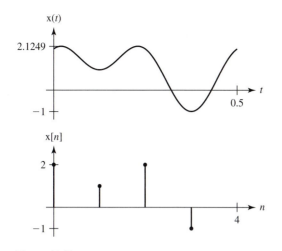

**Figure 7.63**
A CT signal and a DT signal formed by sampling it
at its Nyquist rate over one fundamental period.

■ **Solution**

The highest frequency present in the signal is 4 Hz. Therefore, the samples must be taken at
8 Hz. The fundamental period of the signal is 0.5 s. Therefore, four samples are required.
Assuming that the first sample is taken at time $t = 0$, the samples are

$$\{x[0], x[1], x[2], x[3]\} = \{2, 1, 2, -1\}. \tag{7.133}$$

From the DFT definition,

$$X_{DFT}[k] = \sum_{n=0}^{N_0-1} x[n]e^{-j2\pi(nk/N)}, \tag{7.134}$$

$$X_{DFT}[0] = \sum_{n=0}^{3} x[n] = 4,$$

$$X_{DFT}[1] = \sum_{n=0}^{3} x[n]e^{-j(\pi n/2)} = 2 - j - 2 - j = -j2, \tag{7.135}$$

$$X_{DFT}[2] = \sum_{n=0}^{3} x[n]e^{-j\pi n} = 2 - 1 + 2 + 1 = 4, \tag{7.136}$$

$$X_{DFT}[3] = \sum_{n=0}^{3} x[n]e^{-j(3\pi n/2)} = 2 + j - 2 + j = j2. \tag{7.137}$$

Therefore, the DFT is

$$\{X[0], X[1], X[2], X[3]\}_{DFT} = \{4, -j2, 4, j2\}. \tag{7.138}$$

The CTFT of the original signal is

$$X(f) = \delta(f) + \frac{1}{2}[\delta(f-4) + \delta(f+4)] + \frac{j}{2}[\delta(f+2) - \delta(f-2)] \tag{7.139}$$

or, ordering the impulses with increasing frequency,

$$X(f) = \frac{1}{2}\delta(f+4) + \frac{j}{2}\delta(f+2) + \delta(f) - \frac{j}{2}\delta(f-2) + \frac{1}{2}\delta(f-4) \quad \textbf{(7.140)}$$

which is of the form,

$$X(f) = \sum_{k=-(N_0/2)}^{N_0/2} X_{\text{CTFS}}[k]\delta(f - kf_0), \quad \textbf{(7.141)}$$

where $X_{\text{CTFS}}[k]$ is the CTFS harmonic function $f_0 = 1/T_0$ and $T_0$ is the fundamental period of the signal. So the CTFS harmonic function of the band-limited, periodic signal from which samples (over one fundamental period) were taken is

$$\{X[-2], X[-1], X[0], X[1], X[2]\}_{\text{CTFS}} = \left\{\frac{1}{2}, +\frac{j}{2}, 1, -\frac{j}{2}, \frac{1}{2}\right\}. \quad \textbf{(7.142)}$$

If we divide the DFT results by the number of points, 4, we get

$$\frac{1}{4}\{X[0], X[1], X[2], X[3]\}_{\text{DFT}} = \left\{1, -\frac{j}{2}, 1, +\frac{j}{2}\right\}. \quad \textbf{(7.143)}$$

Using the periodicity of the DFT, we see that we get the correct values for $X[-1]$, $X[0]$, and $X[1]$ but not for $X[2]$ and $X[-2]$. They are wrong by a factor of two because of aliasing. We did not sample *above* the Nyquist rate; we sampled *at* the Nyquist rate. ∎

In Example 7.5 the signal was sampled at exactly the Nyquist rate for exactly one fundamental period. What would happen if we sampled at twice the Nyquist rate for exactly one fundamental period or at the Nyquist rate for exactly two fundamental periods?

### EXAMPLE 7.6

The band-limited, periodic signal

$$x(t) = 1 + \cos(8\pi t) + \sin(4\pi t) \quad \textbf{(7.144)}$$

is sampled at *twice* the Nyquist rate (Figure 7.64). Find the sample values over one fundamental period and find the DFT of the sample values. Also find the CTFS harmonic function of the signal.

■ **Solution**

The highest frequency present in the signal is 4 Hz. Therefore, the samples must be taken at 16 Hz. The fundamental period of the signal is 0.5 s. Therefore eight samples are required. Assuming that the first sample is taken at time $t = 0$, the samples are

$$\{x[0], \ldots, x[7]\} = \left\{2, 1 + \frac{1}{\sqrt{2}}, 1, 1 + \frac{1}{\sqrt{2}}, 2, 1 - \frac{1}{\sqrt{2}}, -1, 1 - \frac{1}{\sqrt{2}}\right\} \quad \textbf{(7.145)}$$

and the DFT of those samples is

$$\{X[0], \ldots, X[7]\}_{\text{DFT}} = \{8, -j4, 4, 0, 0, 0, 4, j4\}. \quad \textbf{(7.146)}$$

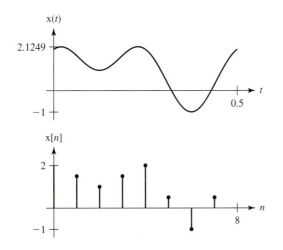

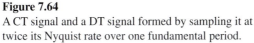

**Figure 7.64**
A CT signal and a DT signal formed by sampling it at
twice its Nyquist rate over one fundamental period.

The CTFS harmonic function of the original signal is the same as before,

$$\{X[-2], X[-1], X[0], X[1], X[2]\}_{CTFS} = \left\{\frac{1}{2}, +\frac{j}{2}, 1, -\frac{j}{2}, \frac{1}{2}\right\}, \qquad \textbf{(7.147)}$$

and dividing the DFT result by 8,

$$\frac{1}{8}\{X[0], \ldots, X[7]\}_{DFT} = \left\{1, -\frac{j}{2}, \frac{1}{2}, 0, 0, 0, \frac{1}{2}, +\frac{j}{2}\right\}. \qquad \textbf{(7.148)}$$

Using the periodicity of the DFT we see that these results agree. In this case, we sampled twice
as fast as in Example 7.5. What we got for our trouble was information about higher frequencies that might have been present in the signal and no aliasing because we sampled above the
Nyquist rate. Of course, since we used the same signal, there were not any higher frequencies
present and the extra X[$k$]'s, $\{X[3], X[4], X[5]\}_{DFT}$, were all zero. ∎

**EXAMPLE 7.7**

The band-limited, periodic signal

$$x(t) = 1 + \cos(8\pi t) + \sin(4\pi t) \qquad \textbf{(7.149)}$$

is sampled at the Nyquist rate (Figure 7.65). Find the sample values over two fundamental periods, find the DFT of the sample values. Also find the CTFS harmonic function of the signal.

**■ Solution**
The highest frequency present in the signal is 4 Hz. Therefore the samples must be taken at
8 Hz. The fundamental period of the signal is 0.5 s. Therefore, eight samples are required.
Assuming that the first sample is taken at time $t = 0$, the samples are

$$\{x[0], \ldots, x[7]\} = \{2, 1, 2, -1, 2, 1, 2, -1\} \qquad \textbf{(7.150)}$$

and the DFT of those samples is

$$\{X[0], \ldots, X[7]\}_{DFT} = \{8, 0, -j4, 0, 8, 0, j4, 0\}. \qquad \textbf{(7.151)}$$

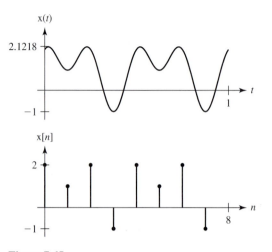

**Figure 7.65**
A CT signal and a DT signal formed by sampling it
at its Nyquist rate over two fundamental periods.

The CTFS harmonic function of the original signal is still the same,

$$\{X[-2], X[-1], X[0], X[1], X[2]\}_{\text{CTFS}} = \left\{\frac{1}{2}, +\frac{j}{2}, 1, -\frac{j}{2}, \frac{1}{2}\right\}. \qquad (7.152)$$

Comparing the CTFS harmonic function and the DFT,

$$\frac{1}{8}\{X[0], \ldots, X[7]\}_{\text{DFT}} = \left\{1, 0, -\frac{j}{2}, 0, 1, 0, +\frac{j}{2}, 0\right\}. \qquad (7.153)$$

The fundamental of the CTFS corresponds to the second harmonic of the DFT because we sampled over two fundamental periods. Therefore, the results correspond correctly, again except for the highest harmonic which is wrong because of aliasing. As in Example 7.5 we sampled at the Nyquist rate instead of above it. As in Example 7.6 we get extra information about the signal. By sampling twice as long, we could recognize frequencies twice as low (fundamental periods twice as long) that might have been present in the signal. That made the lowest nonzero frequency in the DFT half what it was before. Also, since $X_{\text{DFT}}[k]$ occurs at integer multiples of the lowest nonzero frequency, the whole frequency-domain plot has twice the resolution it had in Examples 7.5 and 7.6. The sampling rate is the same as Example 7.5; therefore, the highest frequency that can be found is the same as in Example 7.5 and half that in Example 7.6. ∎

**EXAMPLE 7.8**

Sample the signal

$$x(t) = 5\sin(\pi t)\,\text{rect}\left(\frac{t-2}{4}\right) \qquad (7.154)$$

beginning at time $t = 0$,

a.   16 times at 4 Hz
b.   32 times at 4 Hz

c.   64 times at 4 Hz

d.   32 times at 8 Hz

e.   64 times at 8 Hz

In each case find the DFT of the samples and plot comparisons of the signal and its samples in the time domain and comparisons of the magnitude of the CTFT of the signal and the magnitude of the product of the DFT of the samples and the sampling interval $T_s$.

■ **Solution**

The CTFT of $x(t)$ is

$$X(f) = j10\left[ \operatorname{sinc}\left(4\left(f + \frac{1}{2}\right)\right) e^{-j4\pi(f+(1/2))} - \operatorname{sinc}\left(4\left(f - \frac{1}{2}\right)\right) e^{-j4\pi(f-(1/2))} \right]$$

**(7.155)**

a.   The signal sampled 16 times at 4 Hz is shown in Figure 7.66. The DFT repeats periodically with fundamental period $N_F = 16$ or, in terms of frequency, with fundamental period $f_s = N_F f_F$ Hz, but in the frequency range $-(f_s/2) < f < f_s/2$ the DFT (multiplied by the sampling interval $T_s$) seems to approximate samples of the CTFT at integer multiples of the fundamental frequency $f_F = f_s/N_F$ of the DFT. The resolution of the DFT is not very good. Since all the samples except two in the frequency range $-(f_s/2) < f < f_s/2$ occur at zeros of the CTFT, if we just looked at the DFT result without knowing the CTFT, we would conclude that the CTFT had two impulses at equal positive and negative frequencies and that, therefore, the original signal was a sinusoid. Remember that the DFT applies exactly to periodic signals and the set of samples used here is from exactly two fundamental periods of a sinusoid. In the absence of other information, the logical conclusion from the samples is that the sample pattern repeats periodically and that the signal is therefore a sinusoid, instead of the actual signal which is a time-limited version of a sinusoid. Taking more samples would alleviate this problem.

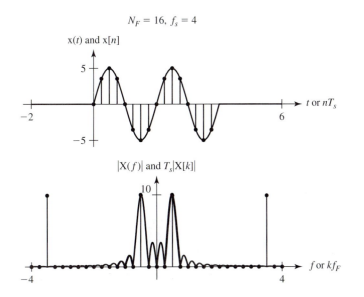

**Figure 7.66**
Signal sampled 16 times at 4 Hz.

b. The signal sampled 32 times at 4 Hz is shown in Figure 7.67. Here twice as many samples were taken as in part (a). The extra samples were all zero. This kind of extension of the sampling of a signal with extra zeros is called *zero padding*. The inclusion of the extra zeros doubles the total sampling time and also doubles the resolution of the DFT. Now we have DFT values that fall between zero crossings of the CTFT, and we can begin to see, by observing the DFT only, that the original signal is not simply a sinusoid. The agreement between the DFT and the CTFT seems very good at low frequencies, but notice that at frequencies close to half the sampling rate, the agreement between the DFT and CTFT is not so good. This difference is easier to see on a logarithmic magnitude plot (Figure 7.68). The difference is caused by aliasing. The original signal is not band-limited so the aliases overlap and, in this case, that causes a significant error near half the sampling rate.

c. The signal sampled 64 times at 4 Hz is shown in Figure 7.69. Here the number of samples was doubled again. This again doubles the resolution of the DFT but does not really help the aliasing problem. A higher sampling rate would reduce errors due to aliasing.

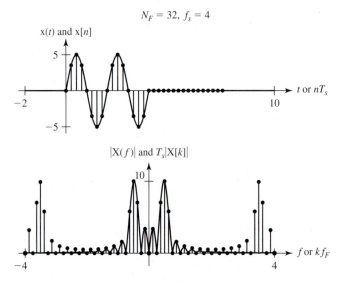

**Figure 7.67**
Signal sampled 32 times at 4 Hz.

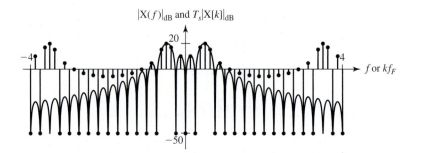

**Figure 7.68**
Logarithmic magnitude plot; signal sampled 32 times at 4 Hz.

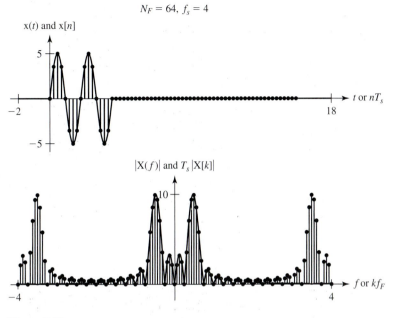

**Figure 7.69**
Signal sampled 64 times at 4 Hz.

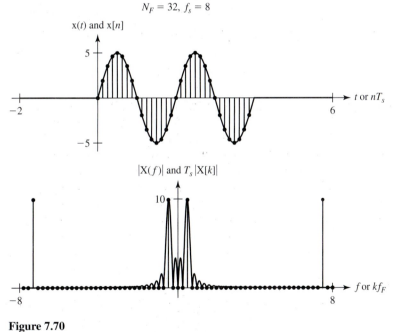

**Figure 7.70**
Signal sampled 32 times at 8 Hz.

d.  The signal sampled 32 times at 8 Hz is shown in Figure 7.70. Here the sampling rate is doubled and the number of samples is the same as in part (b). Again, as in part (a) the DFT seems to be indicating that the signal from which the samples was taken was a pure sinusoid because exactly two cycles of a sinusoid were sampled. If we now increase the number of samples at this sampling rate, we will get better frequency-domain resolution and have a reduced aliasing error.

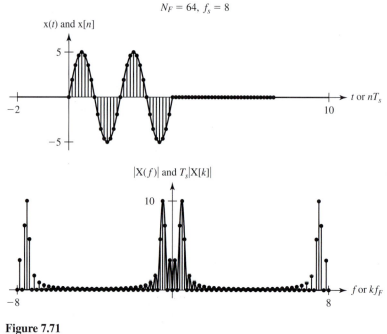

**Figure 7.71**
Signal sampled 64 times at 8 Hz.

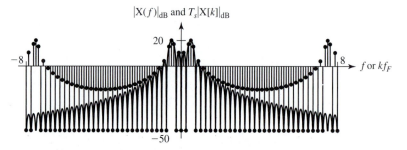

**Figure 7.72**
Logarithmic magnitude plot; signal sampled 64 times at 8 Hz.

e.   The signal sampled 64 times at 8 Hz is shown in Figure 7.71. Here we have sampled
     64 times at 8 Hz. Aliasing error is reduced and the frequency-domain resolution is good
     enough to see that the signal is not simply a sinusoid (Figure 7.72).

     This example reinforces the general principle stated earlier that sampling longer improves
frequency-domain resolution and sampling at a higher rate reduces errors due to aliasing. So a
good general rule in using the DFT to approximate the CTFT is to sample as fast as possible for
as long as possible. In the theoretical limit in which we sample infinitely fast for an infinite time,
all the information in the CTFT is preserved in the DFT. The DFT approaches the CTFT in that
limit. Of course, in any practical situation there are limits imposed by real samplers. Real samplers can only sample at a finite rate, and real computer memories can only store a finite number of data values.

Examples 7.5 to 7.8 analyzed samples from known mathematical functions to demonstrate some features of the DFT. Example 7.9 is more realistic in that the signal is not a known mathematical function.

### EXAMPLE 7.9

Suppose 16 samples are taken from a signal at 1-ms intervals and that the samples are the ones graphed in Figure 7.73 (with the usual assumption that the first sample occurs at time $t = 0$). The reason for taking the samples is to gain information about the signal that was sampled. What do we know so far? We know the value of the signal at 16 points. If we are to draw any more conclusions than that, we must have some other information or make some assumptions.

**■ Solution**

What happened before the first sample and after the last sample? What would be reasonable to assume? We could assume that the signal varies in a similar manner outside this range of samples. That similar variation could take on many different forms. So this assumption is not mathematically precise. One possible form might be the signal in Figure 7.74a. We could assume that the signal is zero outside this range of samples (Figure 7.74b). But, if it is, we know that we cannot sample it adequately because a signal that is time-limited is not band-limited. The usual assumption is that the set of samples we took is reasonably representative of the total signal. (If that is not true, the analysis won't mean much.) That is, we assume that the signal outside this time range is similar to the signal inside this time range. To make that assumption precise, we assume that the signal before and after the samples is as similar to the signal during the sampling as possible. We assume that if we sampled some more, we would simply repeat the set of samples we got, over and over again (Figure 7.74c). That is very probably not exactly true. But what would be a better assumption? If the sample set we took is typical, then the assumption that the signal just keeps doing the same thing again and again is the best one we can make. Using that assumption we can say that the samples we took are from one fundamental period of a periodic signal. We assume that if we had kept sampling, we would simply have repeated the samples again and again.

The next logical question is, What happened between the samples? Again we don't really know. Below are some illustrations of what the signal that was sampled could have looked like (Figure 7.75). In each of the three signals in the figure, the sample values are exactly the same but the signals are different. Unless we know something else about the signal that was sampled, any of these signals could theoretically be the actual signal sampled. But if the signal was properly sampled according to Shannon's sampling theorem (at a rate at more than twice its maximum frequency), only one of these candidate signals could be the one sampled, the last one in Figure 7.75. So now we have narrowed down the possible signals from which the samples could

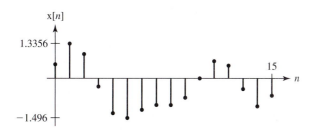

**Figure 7.73**
A DT signal formed by sampling an unknown CT signal for a finite time.

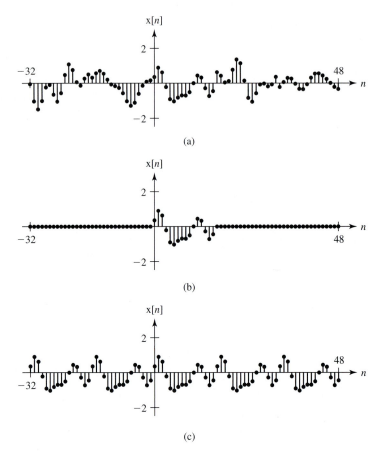

**Figure 7.74**
Three possible extensions of the original samples.

have come to only one, a band-limited, periodic signal which passes through the points. We could now take the original set of samples and from it make the best estimate (based on our assumptions) of the CT signal it came from. That is exactly how the CT signal in Figure 7.75c was created.

Instead of trying to reconstruct the original signal from its samples, it is more common in signal analysis to use the DFT to look at the frequency content of signals. We know how to find the CTFS harmonic function by using the DFT. What is the relation between the CTFS harmonic function and the CTFT of the original signal? It was shown previously that

$$X(f) = \sum_{k=-(N_0/2)}^{N_0/2} X_{\text{CTFS}}[k]\delta(f - kf_0). \tag{7.156}$$

That is, the CTFT for the assumed band-limited, periodic signal is a finite set of impulses spaced apart by the fundamental frequency $f_0$. Using the relation between the CTFS harmonic function and the DFT derived above for band-limited, periodic signals,

$$X(f) = \frac{1}{N_F} \sum_{k=-(N_F/2)}^{N_F/2} X_{\text{DFT}}[k]\delta(f - kf_0) \qquad -\frac{N_F}{2} \leq k \leq \frac{N_F}{2} \tag{7.157}$$

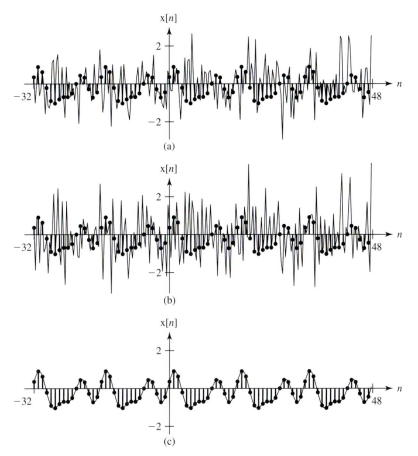

**Figure 7.75**
Three signals, all of which have the original sample values.

or

$$
\begin{aligned}
X(f) = \frac{1}{N_F} \Bigg\{ &\underbrace{X_{\text{DFT}}\left[-\frac{N_F}{2}\right]}_{=0} \delta\left(f - \left(-\frac{N_F}{2}\right)f_0\right) \\
&+ X_{\text{DFT}}\left[-\frac{N_F}{2}+1\right]\delta\left(f - \left(-\frac{N_F}{2}+1\right)f_0\right) + \cdots \\
&+ X_{\text{DFT}}[0]\delta(f) + \cdots + X_{\text{DFT}}\left[\frac{N_F}{2}-1\right]\delta\left(f - \left(\frac{N_F}{2}-1\right)f_0\right) \\
&+ \underbrace{X_{\text{DFT}}\left[\frac{N_F}{2}\right]}_{=0}\delta\left(f - \frac{N_F}{2}f_0\right) \Bigg\}.
\end{aligned}
$$

$$(7.158)$$

(Notice that the CTFS harmonic function components at harmonic numbers $-(N_F/2)$ and $N_F/2$ are always zero if the signal is properly sampled at more than twice the Nyquist frequency because then there is no signal power at half the sampling rate. As the sampling rate is increased, more and more of the components near half the sampling rate will also be zero.) ∎

**EXAMPLE 7.10**

Sample a sinusoidal function and find the DFT of the samples and the CTFS harmonic function of the periodic repetition.

■ **Solution**

This problem description, like many real engineering problems, is vague and ill defined. We must make some reasonable choices for sampling rates and times so that the results will be useful. Let the CT signal be a unit-amplitude cosine and let the fundamental period be 10 ms and the total sampling time 20 ms and take 32 samples in that time. The cosine is described by

$$x(t) = \cos(200\pi t) \tag{7.159}$$

and its CTFT is

$$X(f) = \frac{1}{2}[\delta(f - 100) + \delta(f + 100)]. \tag{7.160}$$

Since the cosine's frequency is 100 Hz and the sampling rate is 1.6 kHz, the signal will definitely be oversampled and no aliasing will occur. The results are illustrated in Figure 7.76.

In this case the signal is band-limited and periodic and the sampling is done over an integer number of fundamental periods. Therefore, one should expect an exact correspondence between the CTFT of the CT signal and DFT of the samples. The CTFT of the original sinusoid has two impulses, one at $+f_0$ and the other at $-f_0$, where $f_0$ is the cosine's frequency. For a unit-amplitude sinusoid like this one the strengths of the impulses should each be $\frac{1}{2}$. The cosine's frequency is 100 Hz. The frequency-domain resolution of the DFT is the reciprocal of the

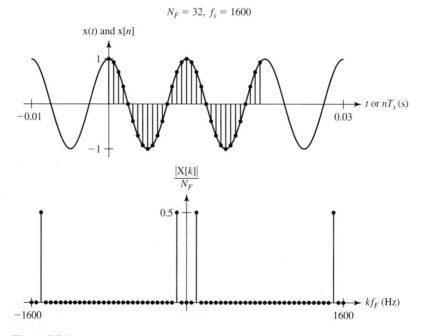

**Figure 7.76**
A cosine sampled over two fundamental periods and the magnitude of its DFT, divided by the number of samples $N_F$.

total sampling time, or 50 Hz. Therefore, the DFT should have nonzero values only at the second harmonic of 50 Hz, which it does. When the DFT result is divided by the number of samples $N_F$, the discrete-harmonic-number impulses in the DFT have the same strength as the continuous-frequency impulses in the CTFT of the CT sinusoid.

For the nonperiodic, energy signal of Example 7.8, the DFT was scaled by multiplying by the sampling interval $T_s$ and the DFT of the samples approximated samples of the CTFT of the CT signal that was sampled. For this periodic signal, the scaling of the DFT was done by dividing it by the number of samples $N_F$. Why are these factors different? First, realize that since the CTFT of a periodic signal consists only of impulses, it cannot be sampled in any meaningful sense. So the DFT of a periodic signal must be scaled to yield the strengths of the impulses, not their amplitudes, which are undefined. In the case of nonperiodic energy signals, the CTFT is a continuous-frequency function with no impulses. In this case a correspondence must be made between the strengths of the DFT impulses and the samples of the CTFT. One way to see the correspondence is to realize that the CTFT is a *spectral density* function and, therefore, has the units of the signal transformed divided by frequency. For example, if the CT signal has units of volts, its CTFT has units of volts per hertz. The DFT is computed by forming various linear combinations of samples of the CT function; therefore, its units would be the same as the signal units, in this case, just volts. To convert that to an approximation of the CTFT we must divide by some frequency to make the units right. But what frequency? If we equate the amplitude in each resolution range of the DFT to the amplitude spectral density of the CTFT, the appropriate division factor is the resolution bandwidth of the DFT which is $f_s/N_F$. So if we take the dividing factor for periodic functions $N_F$ and multiply it by $f_s/N_F$ to form a new dividing factor for nonperiodic energy signals $f_s$, the effect is the same as multiplying by the sampling interval $T_s$ because $f_s = 1/T_s$.

---

**EXAMPLE 7.11**

---

Sample a sinusoid over a noninteger number of fundamental periods and observe the effect on the DFT.

### ■ Solution

Let the sinusoid be a cosine whose fundamental period is $66\frac{2}{3}$ ms and sample it 32 times in 100 ms. The results are illustrated in Figure 7.77.

The original CT cosine has a CTFT with exactly two impulses at $+15$ and $-15$ Hz. But the DFT has nonzero components at every harmonic of its fundamental frequency which is 10 Hz because the total sampling time is 100 ms. Since 15 Hz is not an integer multiple of 10 Hz, there is no resolved frequency component in the DFT at exactly the cosine's frequency. But the two strongest components are at 10 and 20 Hz, which bracket the actual cosine frequency of 15 Hz. Therefore, one could say that the DFT is attempting to report the nature of the signal from which the samples came the best it can given the poor sampling choice. This spreading of the signal's power from the exact location into adjacent locations is an example of leakage. That is, the power at 15 Hz has leaked into components at 10, 20, 30 Hz, etc. because the original signal was not sampled over an integer number of fundamental periods. This problem could be solved by sampling over an integer number of fundamental periods. But it could also be greatly reduced by sampling for a much longer time, even if that time is not an integer multiple of the cosine's fundamental period, because with a longer sampling time, the frequency-domain resolution gets better and the bulk of the signal's power can be placed much closer to the actual frequency of 15 Hz. Figure 7.78 shows the result of sampling over six and one-half fundamental periods with all other parameters unchanged. Now, even though there is still not a resolved component at the

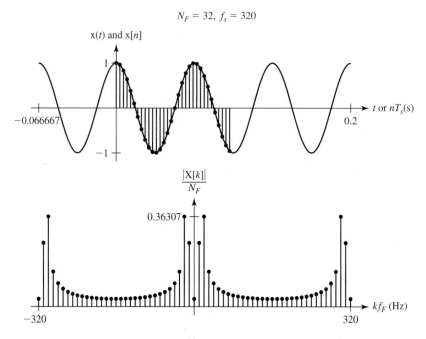

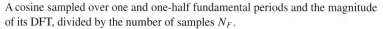

**Figure 7.77**

A cosine sampled over one and one-half fundamental periods and the magnitude of its DFT, divided by the number of samples $N_F$.

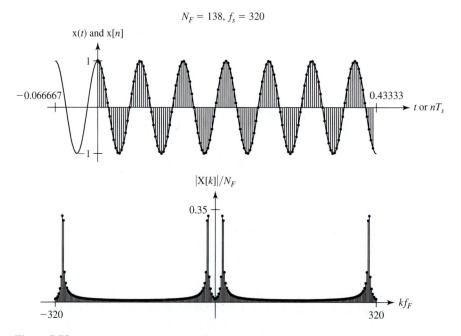

**Figure 7.78**

A cosine sampled over six and one-half fundamental periods and the magnitude of its DFT, divided by the number of samples $N_F$.

frequency of the CT signal, 15 Hz, because of the greater number of points and the consequent higher resolution of the DFT, there are components much closer to 15 Hz than in the previous case and the leakage is spread less widely. ∎

## 7.8 THE FAST FOURIER TRANSFORM

The forward DFT is defined by

$$X[k] = \sum_{n=0}^{N_F-1} x[n]e^{-j2\pi(nk/N_F)}. \tag{7.161}$$

A straightforward way of computing the DFT would be by the following algorithm (written in MATLAB) which directly implements the operations indicated in (7.161).

```
.
.
.
%(Acquire the input data in an array, "x", with "NF" elements.)
.
.
.
%
%      Initialize the DFT array to a column vector of 0s.
%
X=zeros(NF,1) ;
%
%      Compute the Xn's in a nested, double "for" loop.
%
for n=0:NF-1
      for k=0:NF-1
            X(n+1)=X(n+1)+x(k+1)*exp(-j*2*pi*n*k/NF) ;
      end
end
.
.
.
```

(By the way, one should never actually write this program in MATLAB because the DFT is already built into MATLAB as an intrinsic function called `fft`.)

The computation of a DFT using this algorithm requires $N^2$ complex multiply-add operations. Therefore, the number of computations increases as the square of the number of elements in the input vector which is being transformed. In 1965 Cooley and Tukey popularized an algorithm which is much more efficient in computing time for large input arrays whose length is an integer power of 2. This algorithm for computing the DFT is the so-called fast Fourier transform (FFT).

The operation of the FFT algorithm can be illustrated by an example, computing the DFT of a set of four data samples using the algorithm. Let the set of samples from a signal be designated as the DT signal $x_0[n]$ so that the set of input data for the algorithm is $\{x_0[0], x_0[1], x_0[2], x_0[3]\}$. The DFT formula computes the DFT according to

$$X[k] = \sum_{n=0}^{N_F-1} \text{x}[n]e^{-j2\pi(kn/N_F)}. \tag{7.162}$$

It is convenient to use the notation

$$W = e^{-j(2\pi/N_F)}, \tag{7.163}$$

For this case of four data points, we can then write the DFT in a matrix form as

$$\begin{bmatrix} X[0] \\ X[1] \\ X[2] \\ X[3] \end{bmatrix} = \begin{bmatrix} W^0 & W^0 & W^0 & W^0 \\ W^0 & W^1 & W^2 & W^3 \\ W^0 & W^2 & W^4 & W^6 \\ W^0 & W^3 & W^6 & W^9 \end{bmatrix} \begin{bmatrix} \text{x}_0[0] \\ \text{x}_0[1] \\ \text{x}_0[2] \\ \text{x}_0[3] \end{bmatrix}. \tag{7.164}$$

Performing the usual direct matrix multiplication would require $N^2$ complex multiplications and $N(N-1)$ complex additions. We can rewrite (7.164) in the form,

$$\begin{bmatrix} X[0] \\ X[1] \\ X[2] \\ X[3] \end{bmatrix} = \begin{bmatrix} 1 & 1 & 1 & 1 \\ 1 & W^1 & W^2 & W^3 \\ 1 & W^2 & W^0 & W^2 \\ 1 & W^3 & W^2 & W^1 \end{bmatrix} \begin{bmatrix} \text{x}_0[0] \\ \text{x}_0[1] \\ \text{x}_0[2] \\ \text{x}_0[3] \end{bmatrix} \tag{7.165}$$

because $W^n = W^{n+mN_F}$, where $m$ is an integer. The next step is not obvious. It is possible to factor the matrix into the product of two matrices,

$$\begin{bmatrix} X[0] \\ X[2] \\ X[1] \\ X[3] \end{bmatrix} = \begin{bmatrix} 1 & W^0 & 0 & 0 \\ 1 & W^2 & 0 & 0 \\ 0 & 0 & 1 & W^1 \\ 0 & 0 & 1 & W^3 \end{bmatrix} \begin{bmatrix} 1 & 0 & W^0 & 0 \\ 0 & 1 & 0 & W^0 \\ 1 & 0 & W^2 & 0 \\ 0 & 1 & 0 & W^2 \end{bmatrix} \begin{bmatrix} \text{x}_0[0] \\ \text{x}_0[1] \\ \text{x}_0[2] \\ \text{x}_0[3] \end{bmatrix}. \tag{7.166}$$

The proof of this factorization will not be presented here, but it can be found in Brigham (1974). Notice that the order of the DFT result has been changed. The "1" and "2" elements have exchanged positions in the left-side vector. It is sufficient for our purposes here to verify that this factorization is indeed correct by multiplying out the matrices. The reader is invited to do so. When the matrices are multiplied, the "1" and "2" rows will also be exchanged making the matrix equation equivalent to the original version in (7.165).

Now calculate the number of multiplications and additions required. First identify the result of multiplying the second square matrix by the input data set as

$$\begin{bmatrix} \text{x}_1[0] \\ \text{x}_1[1] \\ \text{x}_1[2] \\ \text{x}_1[3] \end{bmatrix} = \begin{bmatrix} 1 & 0 & W^0 & 0 \\ 0 & 1 & 0 & W^0 \\ 1 & 0 & W^2 & 0 \\ 0 & 1 & 0 & W^2 \end{bmatrix} \begin{bmatrix} \text{x}_0[0] \\ \text{x}_0[1] \\ \text{x}_0[2] \\ \text{x}_0[3] \end{bmatrix}. \tag{7.167}$$

The first element is

$$\text{x}_1[0] = \text{x}_0[0] + W^0\text{x}_0[2]. \tag{7.168}$$

This computation requires one multiplication and one addition. (Although $W^0$ is one, we will leave this as a multiplication to lead to a general conclusion.) Similarly $\text{x}_1[1]$

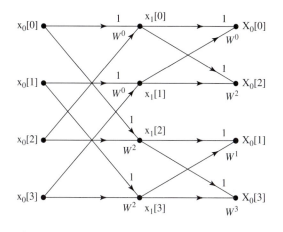

**Figure 7.79**
Signal flow graph for a four-point FFT.

requires one multiplication and one addition. But $x_1[2]$ requires only one addition because $W^0 = -W^2$ and the product $W^0 x_0[2]$ has already been computed in the computation of the first element and can, therefore, just be stored until needed and then subtracted instead of added. Similarly, $x_1[3]$ requires only one more addition. So far we have two multiplications and four additions. By invoking similar symmetry conditions on the second matrix multiplication we find that two more multiplications and four more additions are required. So, overall, four multiplications and eight additions are required. Compare that with the 16 multiplications and 12 additions required in the direct DFT computation in (7.164). Since, computationally, multiplications generally require much more computation time than additions, the FFT algorithm for four points is approximately four times faster than the direct DFT. The vector which results from this type of computation is *scrambled* relative to the original vector, but the unscrambling operation is computationally quite fast, so it does not materially affect the speed ratio.

It is instructive to view the FFT calculation process in *signal-flow* graph form. The four-point FFT algorithm is shown in Figure 7.79. This signal-flow graph illustrates how the computations using matrix factorization are done for a four-point FFT. Figure 7.80 is the signal-flow graph for a 16-point FFT.

By counting the number of multiplications for each data vector length which is an integer power of 2, we can inductively find a formula for the total number of multiplications required and compare it to the number required for the direct DFT. The number of multiplications for an FFT of length $N = 2^p$, where $p$ is an integer, is $Np/2$. Therefore, the speed ratio for the FFT as opposed to the direct DFT is approximately

$$\frac{N^2}{Np/2} = \frac{2N}{p} \qquad (7.169)$$

as tabulated in Table 7.1.

These speed improvement factors do not apply if $p$ is not an integer. For this reason, practically all actual DFT analysis is done with the FFT using data vector lengths which are an integer power of 2. (In MATLAB if the input vector is an integer power of 2 in length, the algorithm used in the MATLAB function `fft` is the FFT algorithm just discussed. If it is not an integer power of 2 in length, the DFT is still computed, but the speed suffers because a less-efficient algorithm must be used.)

James W. Cooley

John Wilder Tukey

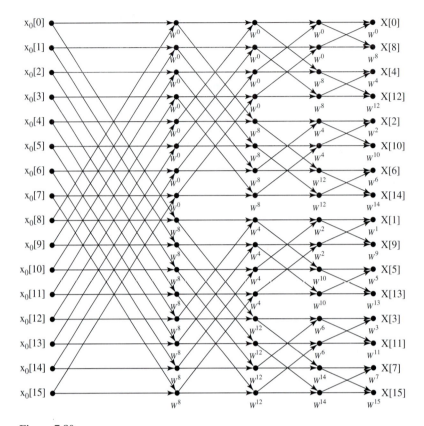

**Figure 7.80**
Signal flow graph for a 16-point FFT.

**Table 7.1**  Speed ratio of FFT to direct DFT versus number of points

| $p$ | $N$ | Speed Ratio FFT/DFT |
|---|---|---|
| 2 | 4 | 4.00 |
| 3 | 8 | 5.33 |
| 4 | 16 | 8.00 |
| 5 | 32 | 12.80 |
| 6 | 64 | 21.33 |
| 7 | 128 | 36.57 |
| 8 | 256 | 64.00 |
| 9 | 512 | 113.78 |
| 10 | 1,024 | 204.80 |
| 11 | 2,048 | 372.36 |
| 12 | 4,096 | 682.67 |
| 13 | 8,192 | 1,260.31 |
| 14 | 16,384 | 2,340.57 |
| 15 | 32,768 | 4,369.07 |
| 16 | 65,536 | 8,192.00 |

## 7.9 SUMMARY OF IMPORTANT POINTS

1. A sampled signal has a Fourier spectrum that is a periodically repeated version of the spectrum of the signal sampled. Each repetition is called an alias.
2. If the aliases in the spectrum of the sampled signal do not overlap, the original signal can be recovered from the samples.
3. If the signal is sampled at a rate more than twice its highest frequency, the aliases will not overlap.
4. A signal cannot be simultaneously time-limited and band-limited.
5. In the case of bandpass signals the absolute minimum sampling rate needed to recover the original signal is twice the bandwidth.
6. The ideal interpolating function is the sinc function, but since it is noncausal, other methods must be used in practice.
7. Discrete-time signals can be sampled in much the same way that continuous-time signals are sampled, and the consequences are analogous.
8. A band-limited periodic signal can be completely described by a finite set of numbers.
9. The discrete Fourier transform (DFT) is almost exactly the same as the DTFS, the only real difference being a scale factor.
10. The CTFT of a CT signal and the DFT of samples from it are related through the operations, sampling in time, windowing, and sampling in frequency.
11. The DFT can be used to approximate the CTFT or the CTFS, and as the sampling rate and/or number of samples are increased, the approximation gets better.
12. The fast Fourier transform (FFT) is a very efficient algorithm for computing the DFT which takes advantage of symmetries that occur when the number of points is an integer power of 2.

## EXERCISES WITH ANSWERS

1. Sample the signal

$$x(t) = 10 \operatorname{sinc}(500t)$$

by multiplying it by the pulse train

$$p(t) = \operatorname{rect}(10^4 t) * 1000 \operatorname{comb}(1000t)$$

to form the signal $x_p(t)$. Sketch the magnitude of the CTFT, $X_p(f)$, of $x_p(t)$.

**Answer:**

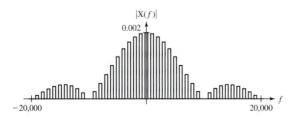

2.   Let

$$x(t) = 10 \, \text{sinc}(500t)$$

as in Exercise 1 and form a signal,

$$x_p(t) = [1000x(t) \, \text{comb}(1000t)] * \text{rect}(10^4 t).$$

Sketch the magnitude of the CTFT, $X_p(f)$, of $x_p(t)$ and compare it to the result of Exercise 1.

3.   *a.*   Given a CT signal

$$x(t) = \text{tri}(100t),$$

form a DT signal $x[n]$ by sampling $x(t)$ at a rate $f_s = 800$ and form an information-equivalent CT impulse signal $x_\delta(t)$ by multiplying $x(t)$ by a periodic sequence of unit impulses whose fundamental frequency is the same, $f_0 = f_s = 800$. Sketch the magnitude of the DTFT of $x[n]$ and the CTFT of $x_\delta(t)$.

*b.*   Change the sampling rate to $f_s = 5000$ and repeat part (a).

**Answers:**

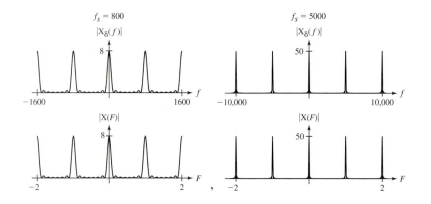

4.   *a.*   Given a band-limited CT signal

$$x(t) = \text{sinc}\left(\frac{t}{4}\right)\cos(2\pi t),$$

form a DT signal $x[n]$ by sampling $x(t)$ at a rate $f_s = 4$ and form an information-equivalent CT impulse signal $x_\delta(t)$ by multiplying $x(t)$ by a periodic sequence of unit impulses whose fundamental frequency is the same, $f_0 = f_s = 4$. Sketch the magnitude of the DTFT of $x[n]$ and the CTFT of $x_\delta(t)$.

*b.*   Change the sampling rate to $f_s = 2$ and repeat part (a).

**Answers:**

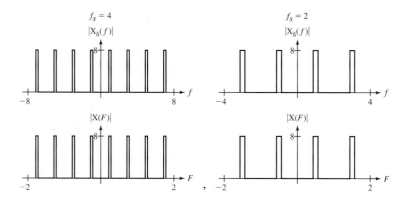

5. Find the Nyquist rates for these signals.
    a. $x(t) = \text{sinc}(20t)$
    b. $x(t) = 4 \text{ sinc}^2(100t)$
    c. $x(t) = 8 \sin(50\pi t)$
    d. $x(t) = 4 \sin(30\pi t) + 3 \cos(70\pi t)$
    e. $x(t) = \text{rect}(300t)$
    f. $x(t) = -10 \sin(40\pi t) \cos(300\pi t)$

**Answers:**

200,   340,   70,   50,   infinite,   20

6. Sketch these time-limited signals, and find and sketch the magnitude of their CTFTs and confirm that they are not band-limited.

    a. $x(t) = 5 \text{ rect}\left(\dfrac{t}{100}\right)$

    b. $x(t) = 10 \text{ tri}(5t)$

    c. $x(t) = \text{rect}(t)[1 + \cos(2\pi t)]$

    d. $x(t) = \text{rect}(t)[1 + \cos(2\pi t)] \cos(16\pi t)$

**Answers:**

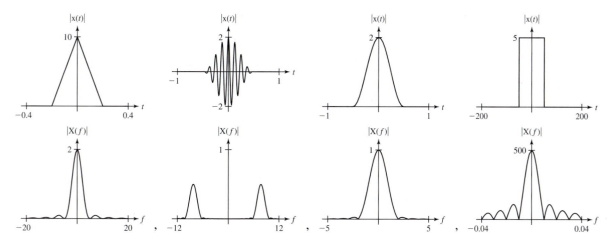

7.  Sketch the magnitudes of these band-limited-signal CTFTs, and find and sketch their inverse CTFTs and confirm that they are not time-limited.

  a.  $X(f) = \text{rect}(f)e^{-j4\pi f}$
  b.  $X(f) = \text{tri}(100f)e^{j\pi f}$
  c.  $X(f) = \delta(f - 4) + \delta(f + 4)$
  d.  $X(f) = j[\delta(f + 4) - \delta(f - 4)] * \text{rect}(8f)$

**Answers:**

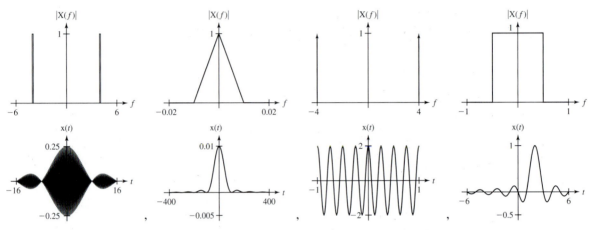

8.  Sample the CT signal

$$x(t) = \sin(2\pi t)$$

at a sampling rate $f_s$. Then, using MATLAB, plot the interpolation between samples in the time range $-1 < t < 1$ using the approximation

$$x(t) \cong 2\frac{f_c}{f_s}\sum_{n=-N}^{N} x(nT_s)\,\text{sinc}(2f_c(t - nT_s))$$

with these combinations of $f_s$, $f_c$, and $N$.

  a.  $f_s = 4$, $f_c = 2$, $N = 1$     b.  $f_s = 4$, $f_c = 2$, $N = 2$
  c.  $f_s = 8$, $f_c = 4$, $N = 4$     d.  $f_s = 8$, $f_c = 2$, $N = 4$
  e.  $f_s = 16$, $f_c = 8$, $N = 8$    f.  $f_s = 16$, $f_c = 8$, $N = 16$

**Answers:**

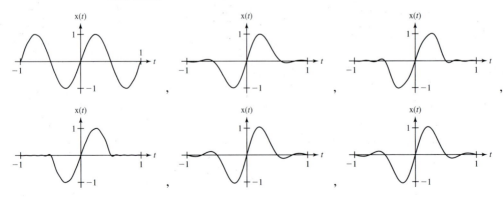

9. For each signal and specified sampling rate, plot the original signal and an interpolation between samples of the signal using a zero-order hold, over the time range $-1 < t < 1$. (The MATLAB function `stairs` could be useful here.)

   a. $x(t) = \sin(2\pi t)$,  $f_s = 8$
   b. $x(t) = \sin(2\pi t)$,  $f_s = 32$
   c. $x(t) = \text{rect}(t)$,  $f_s = 8$
   d. $x(t) = \text{tri}(t)$,  $f_s = 8$

**Answers:**

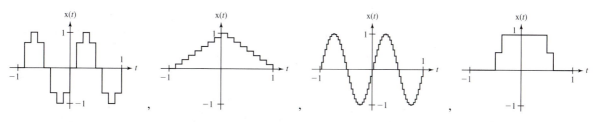

10. For each signal in Exercise 9, lowpass-filter the signal interpolated with zero-order hold with a single-pole lowpass filter whose $-3$-dB frequency is one-fourth of the sampling rate.

**Answers:**

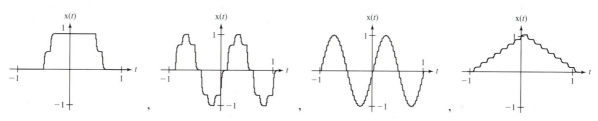

11. Repeat Exercise 9 except use a first-order hold instead of a zero-order hold.

**Answers:**

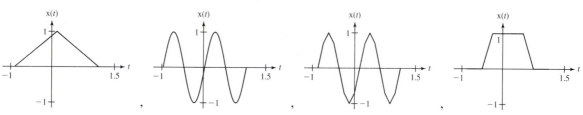

12. Sample the two signals

$$x_1(t) = e^{-t^2} \quad \text{and} \quad x_2(t) = e^{-t^2} + \sin(8\pi t)$$

in the time interval $-3 < t < 3$ at 8 Hz and demonstrate that the sample values are the same.

13. For each pair of signals below, sample at the specified rate and find the DTFT of the sampled signals. In each case, explain, by examining the DTFTs of the two signals, why the samples are the same.

   a. $x(t) = 4\cos(16\pi t)$  and  $x(t) = 4\cos(76\pi t)$,  $f_s = 30$
   b. $x(t) = 6\,\text{sinc}(8t)$  and  $x(t) = 6\,\text{sinc}(8t)\cos(400\pi t)$,  $f_s = 100$
   c. $x(t) = 9\cos(14\pi t)$  and  $x(t) = 9\cos(98\pi t)$,  $f_s = 56$

**Answers:**

$$75 \, \text{rect} \left( \frac{25}{2} F \right) * \text{comb}(F), \quad 2 \left[ \text{comb} \left( F - \frac{8}{30} \right) + \text{comb} \left( F + \frac{8}{30} \right) \right],$$

$$\frac{9}{2} \left[ \text{comb} \left( F - \frac{1}{8} \right) + \text{comb} \left( F + \frac{1}{8} \right) \right]$$

14.  For each sinusoid, find the two other sinusoids whose frequencies are nearest the frequency of the given sinusoid and which, when sampled at the specified rate, have exactly the same samples.

   a.   $x(t) = 4 \cos(8\pi t), \quad f_s = 20$       b.   $x(t) = 4 \sin(8\pi t), \quad f_s = 20$
   c.   $x(t) = 2 \sin(-20\pi t), \quad f_s = 50$       d.   $x(t) = 2 \cos(-20\pi t), \quad f_s = 50$
   e.   $x(t) = 5 \cos \left( 30\pi t + \frac{\pi}{4} \right), \quad f_s = 50$

**Answers:**

$$-2 \sin(-80\pi t) \quad \text{and} \quad 2 \sin(-120\pi t),$$

$$5 \cos \left( 130\pi t + \frac{\pi}{4} \right) \quad \text{and} \quad 5 \cos \left( -70\pi t + \frac{\pi}{4} \right),$$

$$4 \sin(48\pi t) \quad \text{and} \quad -4 \sin(32\pi t), \quad 2 \cos(80\pi t) \quad \text{and} \quad 2 \cos(-120\pi t),$$

$$4 \cos(48\pi t) \quad \text{and} \quad 4 \cos(32\pi t)$$

15.  For each DT signal, plot the original signal and the sampled signal for the specified sampling interval.

   a.   $x[n] = \sin \left( \frac{2\pi n}{24} \right), \quad N_s = 4$

   b.   $x[n] = \text{rect}_9[n], \quad N_s = 2$

   c.   $x[n] = \cos \left( \frac{2\pi n}{48} \right) \cos \left( \frac{2\pi n}{8} \right), \quad N_s = 2$

   d.   $x[n] = \left( \frac{9}{10} \right)^n u[n], \quad N_s = 6$

**Answers:**

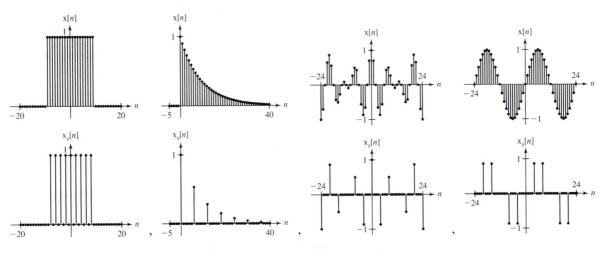

16.  For each signal in Exercise 15, sketch the magnitude of the DTFT of the original signal and the sampled signal.

**Answers:**

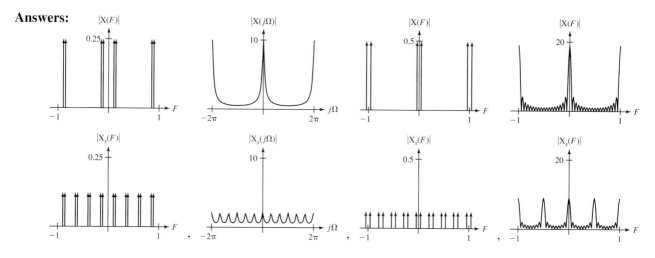

**17.** For each DT signal, plot the original signal and the decimated signal for the specified sampling interval. Also plot the magnitudes of the DTFT's of both signals.

a. $\text{x}[n] = \text{tri}\left(\dfrac{n}{10}\right),\ N_s = 2$    b. $\text{x}[n] = (0.95)^n \sin\left(\dfrac{2\pi n}{10}\right) \text{u}[n],\ N_s = 2$

c. $\text{x}[n] = \cos\left(\dfrac{2\pi n}{8}\right),\ N_s = 7$

**Answers:**

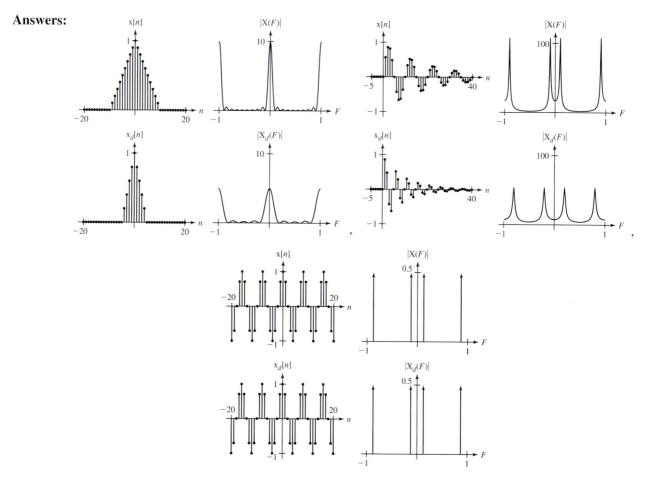

18. For each signal in Exercise 17, insert the specified number of zeros between samples, lowpass DT filter the signals with the specified cutoff frequency, and plot the resulting signal and the magnitude of its DTFT.

    a.   Insert 1 zero between points. Cutoff frequency is $F_c = 0.1$.

    b.   Insert 4 zeros between points. Cutoff frequency is $F_c = 0.2$.

    c.   Insert 4 zeros between points. Cutoff frequency is $F_c = 0.02$.

**Answers:**

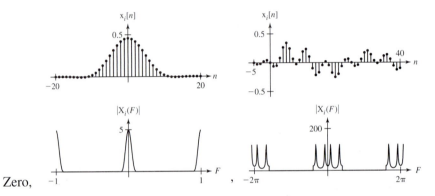

Zero,

19. Sample the following CT signals x(t) to form DT signals x[n]. Sample at the Nyquist rate and then at the next higher rate for which the number of samples per cycle is an integer. Plot the CT and DT signals and the magnitudes of the CTFTs of the CT signals and the DTFTs of the DT signals.

    a.   $x(t) = 2\sin(30\pi t) + 5\cos(18\pi t)$

    b.   $x(t) = 6\sin(6\pi t)\cos(24\pi t)$

**Answers:**

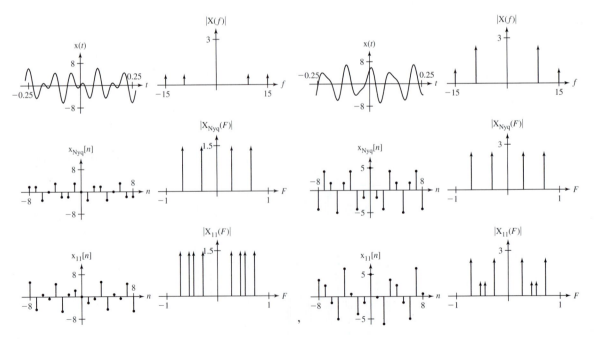

**20.** For each of these signals find the DTFS harmonic function over one fundamental period and show that $X[N_0/2]$ is real.

    *a.*   $x[n] = \text{rect}_2[n] * \text{comb}_{12}[n]$

    *b.*   $x[n] = \text{rect}_2[n+1] * \text{comb}_{12}[n]$

    *c.*   $x[n] = \cos\left(\dfrac{14\pi n}{16}\right) \cos\left(\dfrac{2\pi n}{16}\right)$

    *d.*   $x[n] = \cos\left(\dfrac{12\pi n}{14}\right) \cos\left(\dfrac{2\pi (n-3)}{14}\right)$

**Answers:**

$$\frac{1}{4}(\text{comb}_{16}[k-8] + \text{comb}_{16}[k-6] + \text{comb}_{16}[k+6] + \text{comb}_{16}[k+8]),$$

$$\frac{1}{12}\frac{\sin(5(k\pi/12))}{\sin(k\pi/12)}e^{j(\pi k/6)},$$

$$\frac{1}{4}(\text{comb}_{14}[k-7] + \text{comb}_{14}[k-5] + \text{comb}_{14}[k+5] + \text{comb}_{14}[k+7])e^{j(3\pi k/7)},$$

$$\frac{1}{12}\frac{\sin(5(k\pi/12))}{\sin(k\pi/12)}$$

**21.** Start with a signal

$$x(t) = 8\cos(30\pi t)$$

and sample, window, and periodically repeat it using a sampling rate of $f_s = 60$ and a window width of $N_F = 32$. For each signal in the process, plot the signal and its transform, either CTFT or DTFT.

**Answers:**

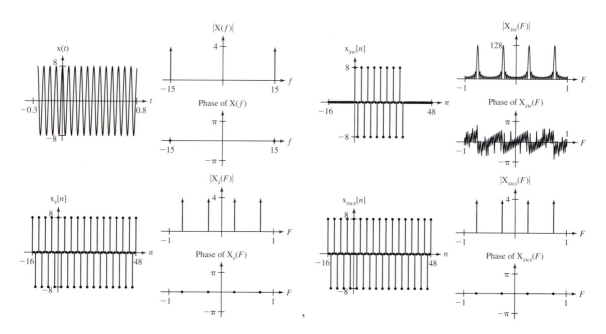

,

22. Sometimes window shapes other than a rectangle are used. Using MATLAB, find and plot the magnitudes of the DFTs of these window functions, with $N = 32$.

    *a.* von Hann or Hanning

    $$w[n] = \frac{1}{2}\left[1 - \cos\left(\frac{2\pi n}{N-1}\right)\right] \qquad 0 \le n < N$$

    *b.* Bartlett

    $$w[n] = \begin{cases} \dfrac{2n}{N-1} & 0 \le n \le \dfrac{N-1}{2} \\ 2 - \dfrac{2n}{N-1} & \dfrac{N-1}{2} \le n < N \end{cases}$$

    *c.* Hamming

    $$w[n] = 0.54 - 0.46\cos\left(\frac{2\pi n}{N-1}\right) \qquad 0 \le n < N$$

    *d.* Blackman

    $$w[n] = 0.42 - 0.5\cos\left(\frac{2\pi n}{N-1}\right) + 0.08\cos\left(\frac{4\pi n}{N-1}\right) \qquad 0 \le n < N$$

    **Answers:**

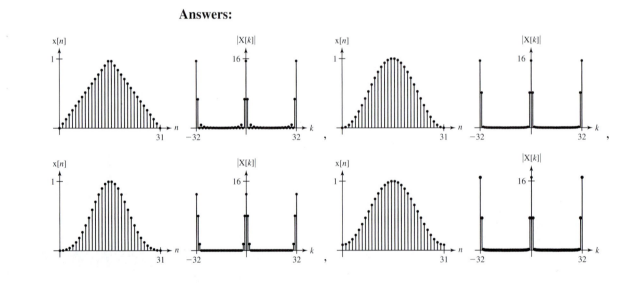

23. Sample the following signals at the specified rates for the specified times and plot the magnitudes of the DFTs versus harmonic number in the range $-(N_F/2) < k < (N_F/2) - 1$.

    *a.* $x(t) = \cos(2\pi t)$,   $f_s = 2$, $N_F = 16$
    *b.* $x(t) = \cos(2\pi t)$,   $f_s = 8$, $N_F = 16$
    *c.* $x(t) = \cos(2\pi t)$,   $f_s = 16$, $N_F = 256$
    *d.* $x(t) = \cos(3\pi t)$,   $f_s = 2$, $N_F = 16$
    *e.* $x(t) = \cos(3\pi t)$,   $f_s = 8$, $N_F = 16$
    *f.* $x(t) = \cos(3\pi t)$,   $f_s = 16$, $N_F = 256$

**Answers:**

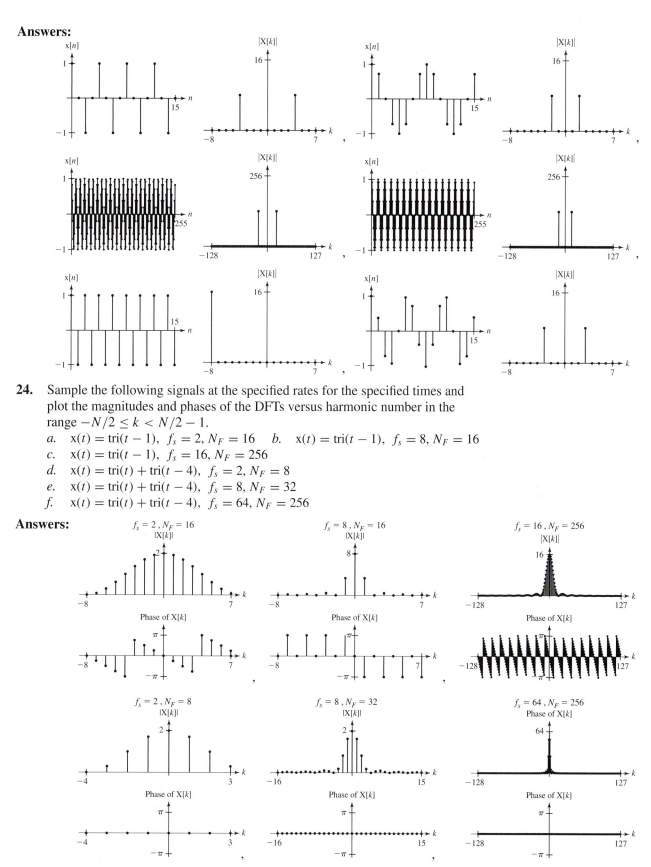

24. Sample the following signals at the specified rates for the specified times and plot the magnitudes and phases of the DFTs versus harmonic number in the range $-N/2 \le k < N/2 - 1$.

    a. $x(t) = \text{tri}(t - 1)$, $f_s = 2$, $N_F = 16$   b. $x(t) = \text{tri}(t - 1)$, $f_s = 8$, $N_F = 16$

    c. $x(t) = \text{tri}(t - 1)$, $f_s = 16$, $N_F = 256$

    d. $x(t) = \text{tri}(t) + \text{tri}(t - 4)$, $f_s = 2$, $N_F = 8$

    e. $x(t) = \text{tri}(t) + \text{tri}(t - 4)$, $f_s = 8$, $N_F = 32$

    f. $x(t) = \text{tri}(t) + \text{tri}(t - 4)$, $f_s = 64$, $N_F = 256$

**Answers:**

**25.** Sample each CT signal, x($t$), $N_F$ times at the rate $f_s$, creating the DT signal x[$n$]. Plot x($t$) versus $t$ and x[$n$] versus $nT_s$ over the time range $0 < t < N_F T_s$. Find the DFT X[$k$] of the $N_F$ samples. Then plot the magnitude and phase of X($f$) versus $f$ and $T_s$X[$k$] versus $k\,\Delta f$ over the frequency range $-(f_s/2) < f < f_s/2$, where $\Delta f = f_s/N_F$. Plot $T_s$X[$k$] as a continuous function of $k\Delta f$ using the MATLAB `plot` command.

a.   $x(t) = 5\,\text{rect}(2(t-2))$,   $f_s = 16,\ N_F = 64$

b.   $x(t) = 3\,\text{sinc}\left(\dfrac{t-20}{5}\right)$,   $f_s = 1,\ N_F = 40$

c.   $x(t) = 2\,\text{rect}(t-2)\,\sin(8\pi t)$,   $f_s = 32,\ N_F = 128$

d.   $x(t) = 10\left[\text{tri}\left(\dfrac{t-2}{2}\right) - \text{tri}\left(\dfrac{t-6}{2}\right)\right]$,   $f_s = 8,\ N_F = 64$

e.   $x(t) = 5\cos(2\pi t)\cos(16\pi t)$,   $f_s = 64,\ N_F = 128$

**Answers:**

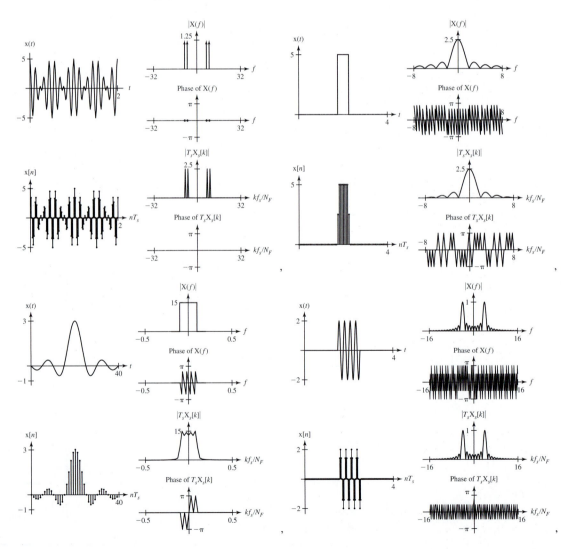

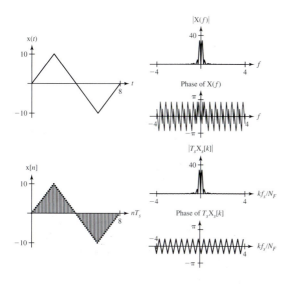

**26.** Sample each CT signal, $x(t)$, $N_F$ times at the rate $f_s$ creating the DT signal $x[n]$. Plot $x(t)$ versus $t$ and $x[n]$ versus $nT_s$ over the time range $0 < t < N_F T_s$. Find the DFT $X[k]$ of the $N_F$ samples. Then plot the magnitude and phase of $X(f)$ versus $f$ and $X[k]/N_F$ versus $k\,\Delta f$ over the frequency range $-(f_s/2) < f < f_s/2$, where $\Delta f = f_s/N_F$. Plot $X[k]/N_F$ as an impulse function of $k\Delta f$ using the MATLAB `stem` command to represent the impulses.

*a.* $x(t) = 4\cos(200\pi t), \quad f_s = 800, \ N_F = 32$

*b.* $x(t) = 6\,\text{rect}(2t) * \text{comb}(t), \quad f_s = 16, \ N_F = 128$

*c.* $x(t) = 6\,\text{sinc}(4t) * \text{comb}(t), \quad f_s = 16, \ N_F = 128$

*d.* $x(t) = 5\cos(2\pi t)\cos(16\pi t), \quad f_s = 64, \ N_F = 128$

**Answers:**

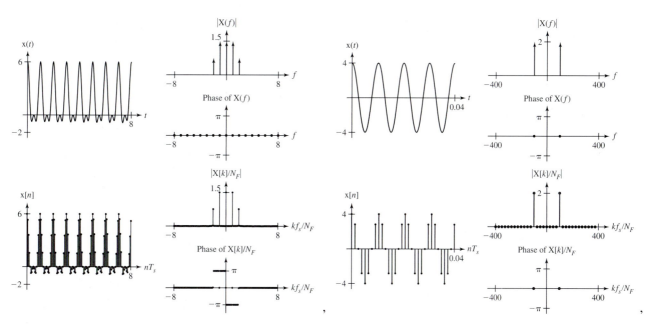

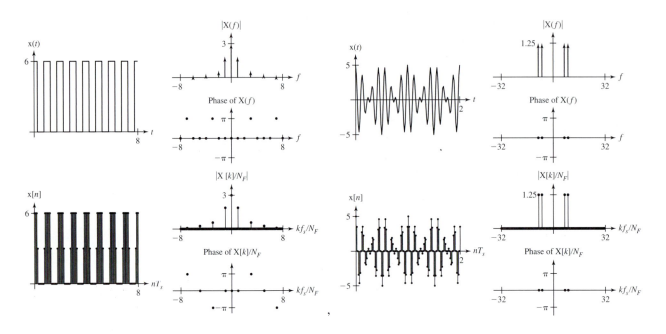

## EXERCISES WITHOUT ANSWERS

**27.** Using MATLAB (or an equivalent mathematical computer tool) plot the signal
$$x(t) = 3\cos(20\pi t) - 2\sin(30\pi t)$$
over a time range of $0 < t < 400$ ms. Also plot the DT signal formed by sampling this function at the following sampling intervals:

a.  $T_s = \frac{1}{120}$ s      b.  $T_s = \frac{1}{60}$ s

c.  $T_s = \frac{1}{30}$ s      d.  $T_s = \frac{1}{15}$ s

Based on what you observe, what can you say about how fast this signal should be sampled so that it could be reconstructed from the samples?

**28.** A signal $x(t) = 20\cos(1000\pi t)$ is impulse-sampled at a sampling rate of 2 kHz. Plot two fundamental periods of the impulse-sampled signal $x_\delta(t)$. (Let the one sample be at time $t = 0$.) Then plot four fundamental periods, centered at 0 Hz, of the CTFT $X_\delta(f)$ of the impulse-sampled signal $x_\delta(t)$. Change the sampling rate to 500 Hz and repeat.

**29.** A signal $x(t) = 10\,\text{rect}(t/4)$ is impulse-sampled at a sampling rate of 2 Hz. Plot the impulse-sampled signal $x_\delta(t)$ on the interval $-4 < t < 4$. Then plot three fundamental periods, centered at $f = 0$ of the CTFT $X_\delta(f)$ of the impulse-sampled signal $x_\delta(t)$. Change the sampling rate to $\frac{1}{2}$ Hz and repeat.

**30.** A signal $x(t) = 4\,\text{sinc}(10t)$ is impulse-sampled at a sampling rate of 20 Hz. Plot the impulse-sampled signal $x_\delta(t)$ on the interval $-0.5 < t < 0.5$. Then plot three fundamental periods, centered at $f = 0$, of the CTFT $X_\delta(f)$ of the impulse-sampled signal $x_\delta(t)$. Change the sampling rate to 4 Hz and repeat.

**31.** A DT signal $x[n]$ is formed by sampling a CT signal $x(t) = 20\cos(8\pi t)$ at a sampling rate of 20 Hz. Plot $x[n]$ over 10 fundamental periods versus discrete time. Then do the same for sampling frequencies of 8 and 6 Hz.

**32.** A DT signal $x[n]$ is formed by sampling a CT signal $x(t) = -4\sin(200\pi t)$ at a sampling rate of 400 Hz. Plot $x[n]$ over 10 fundamental periods versus discrete time. Then do the same for sampling frequencies of 200 and 60 Hz.

**33.** Find the Nyquist rates for these signals.

   *a.*   $x(t) = 15 \operatorname{rect}(300t) \cos(10^4 \pi t)$     *b.*   $x(t) = 7 \operatorname{sinc}(40t) \cos(150\pi t)$

   *c.*   $x(t) = 15[\operatorname{rect}(500t) * 100 \operatorname{comb}(100t)] \cos(10^4 \pi t)$

   *d.*   $x(t) = 4[\operatorname{sinc}(500t) * \operatorname{comb}(200t)]$

   *e.*   $x(t) = -2[\operatorname{sinc}(500t) * \operatorname{comb}(200t)] \cos(10^4 \pi t)$

**34.** On one graph, plot the DT signal formed by sampling the following three CT functions at a sampling rate of 30 Hz.

   *a.*   $x_1(t) = 4 \sin(20\pi t)$     *b.*   $x_2(t) = 4 \sin(80\pi t)$

   *c.*   $x_2(t) = -4 \sin(40\pi t)$

**35.** Plot the DT signal $x[n]$ formed by sampling the CT signal

$$x(t) = 10 \sin(8\pi t)$$

at twice the Nyquist rate and $x(t)$ itself. Then on the same graph plot at least two other CT sinusoids which would yield exactly the same samples if sampled at the same times.

**36.** Plot the magnitude of the CTFT of

$$x(t) = 25 \operatorname{sinc}^2\left(\frac{t}{6}\right).$$

Infinitely many samples would be required to exactly reconstruct $x(t)$ from its samples. If a practical compromise were made in which sampling was performed over the minimum possible time which could contain 99 percent of the energy of this waveform, how many samples would be required?

**37.** Plot the magnitude of the CTFT of

$$x(t) = 8 \operatorname{rect}(3t).$$

This signal is not band-limited, so it cannot be sampled adequately to exactly reconstruct the signal from the samples. As a practical compromise, assume that a bandwidth which contains 99 percent of the energy of $x(t)$ is great enough to practically reconstruct $x(t)$ from its samples. What is the minimum required sampling rate in this case?

**38.** A signal $x(t)$ is periodic, and one fundamental period of the signal is described by

$$x(t) = \begin{cases} 3t & 0 < t < 5.5 \\ 0 & 5.5 < t < 8 \end{cases}.$$

Find the samples of this signal over one fundamental period sampled at a rate of 1 Hz (beginning at time $t = 0$). Then plot, on the same scale, two fundamental periods of the original signal and two fundamental periods of a periodic signal which is band-limited to 0.5 Hz or less that would have these same samples.

**39.** How many sample values are required to yield enough information to exactly describe these band-limited periodic signals?

   *a.*   $x(t) = 8 + 3 \cos(8\pi t) + 9 \sin(4\pi t)$

   *b.*   $x(t) = 8 + 3 \cos(7\pi t) + 9 \sin(4\pi t)$

**40.** Sample the CT signal

$$x(t) = 15\left[\operatorname{sinc}(5t) * \frac{1}{2} \operatorname{comb}\left(\frac{t}{2}\right)\right] \sin(32\pi t)$$

to form the DT signal $x[n]$. Sample at the Nyquist rate and then at the next higher rate for which the number of samples per cycle is an integer. Plot the

CT and DT signals and the magnitude of the CTFT of the CT signal and the DTFT of the DT signal.

41. Without using a computer, find the forward DFT of the following sequence of data and then find the inverse DFT of that sequence and verify that you get back the original sequence

$$\{3, 4, 1, -2\}.$$

42. Redo Example 7.5 except use

$$x(t) = 1 + \sin(8\pi t) + \cos(4\pi t)$$

as the signal being sampled. Explain any apparent discrepancies that arise.

43. Sample the band-limited periodic signal $x(t) = 15\cos(300\pi t) + 40\sin(200\pi t)$ at exactly its Nyquist rate over exactly one fundamental period of $x(t)$. Find the DFT of those samples. From the DFT find the CTFS harmonic function. Plot the CTFS representation of the signal that results and compare it with $x(t)$. Explain any differences. Repeat for a sampling rate of twice the Nyquist rate.

44. Sample the band-limited periodic signal $x(t) = 8\cos(50\pi t) - 12\sin(80\pi t)$ at exactly its Nyquist rate over exactly one fundamental period of $x(t)$. Find the DFT of those samples. From the DFT find the CTFS harmonic function. Plot the CTFS representation of the signal that results and compare it with $x(t)$. Explain any differences. Repeat for a sampling rate of twice the Nyquist rate.

45. Using MATLAB,
    a. Generate a pseudo-random sequence of 256 data points in a vector x, using the `randn` function which is built into MATLAB.
    b. Find the DFT of that sequence of data and put it in a vector X.
    c. Set a vector Xlpf equal to X.
    d. Change all the values in Xlpf to zero except the first and last eight points.
    e. Take the real part of the inverse DFT of Xlpf and put it in a vector xlpf.
    f. Generate a set of 256 sample times t, which begin with 0 and are uniformly separated by 1.
    g. Plot x and xlpf versus t on the same scale and compare.
    What kind of effect does this operation have on a set of data? Why is the output array called xlpf?

46. Sample the signal $x(t) = \text{rect}(t)$ at three different frequencies, 8, 16, and 32 Hz for 2 s. Plot the magnitude of the DFT in each case. Which of these sampling frequencies yields a magnitude plot that looks most like the magnitude of the CTFT of $x(t)$?

47. Sample the signal, $x(t) = \text{rect}(t)$, at 8 Hz for three different total times, 2, 4, and 8 s. Plot the magnitude of the DFT in each case. Which of these total sampling times yields a magnitude plot that looks most like the magnitude of the CTFT of $x(t)$?

48. Sample the signal $x(t) = \cos(\pi t)$ at three different frequencies, 2, 4, and 8 Hz for 5 s. Plot the magnitude of the DFT in each case. Which of these sampling frequencies yields a magnitude plot that looks most like the magnitude of the CTFT of $x(t)$?

49. Sample the signal $x(t) = \cos(\pi t)$, at 8 Hz for three different total times, 5, 9, and 13 s. Plot the magnitude of the DFT in each case. Which of these total sampling times yields a magnitude plot that looks most like the magnitude of the CTFT of $x(t)$?

# Correlation, Energy Spectral Density, and Power Spectral Density

## 8.1 INTRODUCTION AND GOALS

In signal and system analysis the characteristics of individual signals are, of course, important, but often the relationships between signals are just as important. Relationships between signals often indicate whether the physical phenomena which caused the signals are related or whether one signal is a modified version of the other. The relationship between two signals in a system can be used to measure system characteristics. For example, in a flowing liquid stream a heater can be placed upstream of a temperature sensor (Figure 8.1). Then the heater power can be modulated with a known signal shape. By knowing the spacing $d$ between the heater and the temperature sensor and by observing the signal coming from the downstream temperature sensor and watching for that same shape (or a similar shape) to arrive at a later time, the flow rate can be determined from the spacing $d$ and the time delay between the signals. This is just a very simple system whose excitation–response relationship is like a filter with a time delay and some frequency-dependent attenuation, and the relationship between the two signals indicates what the attenuation and time delay are. The relationship between signals often indicates whether one depends on the other, both depend on some common phenomenon, or they are independent.

In this chapter we will explore the mathematical techniques of comparing two signals. These comparison methods can be applied to all kinds of signals, continuous-time and discrete-time, deterministic and random. An exploration of the properties of random signals is beyond the scope of this text, but the basic ideas of how to compare signals will be presented here with examples using both random and nonrandom signals. However, the exercises use only nonrandom signals.

### CHAPTER GOALS

1. To develop an understanding of how to mathematically define the similarity between two signals in the time domain

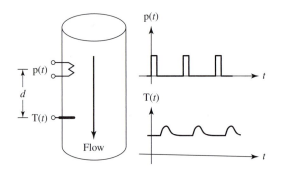

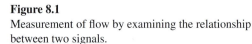

**Figure 8.1**
Measurement of flow by examining the relationship
between two signals.

2. To develop an understanding of how to mathematically define the similarity
   between two signals in the frequency domain

3. To relate the time-domain and frequency-domain methods to each other through
   the Fourier transform

## 8.2  CORRELATION AND THE CORRELOGRAM

How do we determine whether two signals are correlated? The natural answer is to
simply look at them and try to detect any similarity between them. Human beings are
very good at seeing similarities between images, especially faces. It is an important
evolutionary survival skill. We can each recognize a large number of people as distinct
individuals. **We** can also read **text** that is *handwritten* or printed in different fonts, in
UPPERCASE or lowercase. But we need a mathematical method to precisely and
quantitatively indicate the correlation between signals.

Figures 8.2 to 8.5 illustrate pairs of signals. Each pair of signals is plotted versus
time, and then the two signals are plotted against each other. This third plot is called a
*correlogram,* and it can help in determining whether or not two signals are correlated.

It might not have been obvious at first glance that the two DT signals in Figure 8.2
are very similar, but the correlogram illustrates this relationship very clearly. When the
second signal is plotted against the first, the correlogram follows a straight line through
the origin with a negative slope. The correlogram is simply indicating that when $x_1[n]$
goes positive from zero, $x_2[n]$ always goes negative from zero by a proportional
amount, and vice versa. In this example, the slope of the correlogram line is $-1$. That
means that when $x_1[n]$ deviates positively from zero, $x_2[n]$ deviates negatively from
zero by the same amount. This indicates that there is a simple mathematical relation
between the two signals,

$$x_2[n] = -x_1[n]. \tag{8.1}$$

If a correlogram tends to form a straight line, or close to a straight line, the two signals
used to form the correlogram are said to be highly *correlated*. The closer the correlogram is to a line, the more correlated the signals are. If the line has a positive slope, the
signals are positively correlated, and if the line has a negative slope, the signals are
negatively correlated.

The two CT signals in Figure 8.3 have similar characteristics. That is, they deviate
about the same amount from zero, their average values both appear to be about zero, and

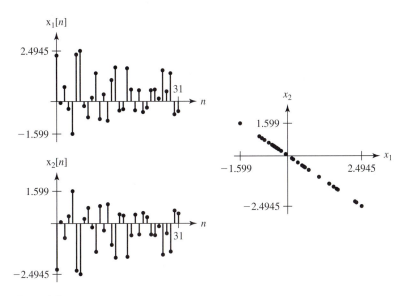

**Figure 8.2**
A pair of DT signals and their correlogram.

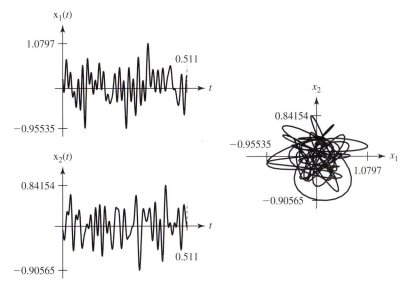

**Figure 8.3**
A pair of CT signals and their correlogram.

they tend to vary as a function of time at the same general rate. But are they correlated? There is no obvious similarity from just examining them, and the correlogram confirms that there is no general trend of one signal varying in the same direction as the other or in the opposite direction. Since there is no apparent linearity in the correlogram, we would conclude, based on this evidence, that these two signals are uncorrelated.

As in Figure 8.3, the two DT signals in Figure 8.4 also have similar characteristics, but they are certainly not identical. The correlogram confirms that there is some similarity because the points plotted stay fairly close to a straight line with a positive slope. That is, there are more points in the first and third quadrants than in the second and

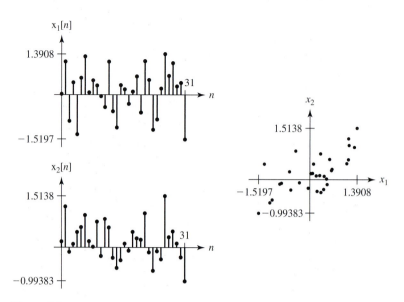

**Figure 8.4**
A pair of DT signals and their correlogram.

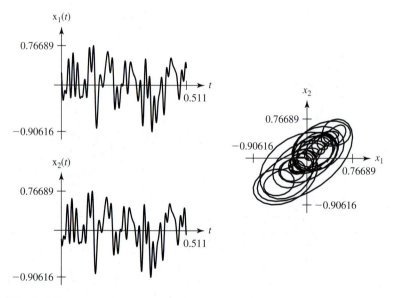

**Figure 8.5**
A pair of CT signals and their correlogram.

fourth quadrants. The correlogram is telling us that these two signals are not completely correlated, but they are not completely uncorrelated either. There is some relationship between them, but it is not a simple proportionality as it was for the signals in Figure 8.2. A typical situation which would cause this kind of relationship would be that $x_2[n]$ is some constant $K$ times $x_1[n]$ plus a third signal, usually random noise $n[n]$. The relationship could be described mathematically by

$$x_2[n] = K x_1[n] + n[n]. \tag{8.2}$$

Figure 8.5 is different because even though we can see by looking at the two CT signals that their shapes are very similar, the correlogram tells us that they are not

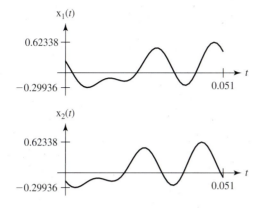

**Figure 8.6**
A magnified view of the two signals in Figure 8.5.

exactly proportional to each other because it is not a straight line (although it tends to lie in the first and third quadrants more than in the second and fourth). But it does have an interesting shape: pseudo-elliptical shapes centered on a positive-slope line. What is this shape telling us? If you look closely at the two time plots, you will notice a very small time shift between them (Figure 8.6). The second signal is a time-shifted version of the first. In this case, the second signal moves in the same direction as the first but earlier in time. So there is a relation between them, but with time shift. That kind of relationship can be described mathematically by

$$x_2(t) = K x_1(t - \tau), \tag{8.3}$$

where, in this case, $K = 1$ and $\tau < 0$. If we were to shift the second signal a little later in time, we would get a straight-line correlogram with a positive slope indicating a strong positive correlation.

One way to see why the correlogram has this distinctive shape when there is a time delay between the signals is to plot a correlogram for two very simple DT signals, two sinusoids of the same frequency with a 45° phase shift (a one-eighth fundamental period time delay) between them (Figure 8.7). If the phase shift is changed to 90°, we get a correlogram like Figure 8.8.

Two terms commonly used in descriptions of relations between signals are *correlation* and *independence*. We have already, at least qualitatively, defined correlation. Positive correlation simply means the tendency of two signals to move in the same direction at the same time, and negative correlation means the tendency of two signals to move in opposite directions at the same time. The commonly accepted definition of independence is that if two signals are independent that means that there is no commonality between them. That is, there is no mathematical relationship between the generation of one and the generation of the other.

Since independence and correlation seem to be opposite concepts, it is tempting at this point to think that if two signals are not independent they are correlated, but that is not generally true. This last correlogram (Figure 8.8) is a good illustration of the difference between correlation and dependence. The two CT signals are obviously not independent since they are sinusoids of the same frequency and there is a simple mathematical relationship between them. Knowing one of them and the phase difference would allow calculation of the other. But they are *uncorrelated*. This is indicated by the lack of any discernible linearity in the correlogram. As we will soon see, this can be shown mathematically from the definition of correlation.

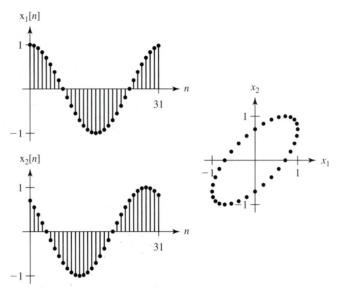

**Figure 8.7**
A correlogram for two DT sinusoids with a 45° phase difference.

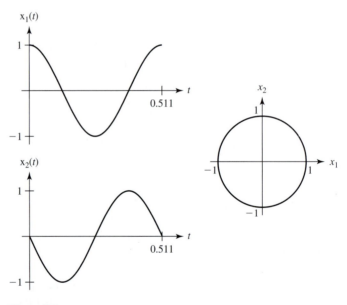

**Figure 8.8**
A correlogram for two CT sinusoids with a 90° phase difference.

Figure 8.9 presents another interesting type of correlogram. The two DT signals look quite different, and the correlogram certainly does not tend to form a straight line, yet, looking at the correlogram, the feeling that there is some mathematical relationship between the two signals is irresistible. Even though the two signals are not linearly related it is apparent from the correlogram that they are *nonlinearly* related. According to the usual definition of correlation, these signals are not highly correlated, but they are obviously closely *related* because the correlogram, although not linear, forms a

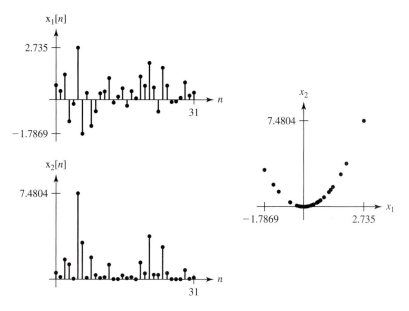

**Figure 8.9**
A pair of DT signals and their correlogram.

very definite single smooth curve. In this case, the actual mathematical relationship between the two signals is $x_2[n] = x_1^2[n]$. Correlation is ordinarily defined based on a linear relationship between signals. In this case, we could show that by plotting the square of $x_1[n]$ versus $x_2[n]$. Then we would get a straight line, and we would say that $x_1^2[n]$ and $x_2[n]$ are highly correlated.

Plotting correlograms in MATLAB is quite simple. For DT signals we plot one signal against the other, plotting only dots. For example,

```
        .
        .
        .
%       Assign values of one DT signal to x1 and the other DT signal to x2.
        .
plot(x1,x2,'k.') ;
        .
        .
```

For CT signals we must first sample the signals well above the higher of the two Nyquist rates, and then plot one signal against the other, plotting lines between points. For example,

```
        .
        .
        .
%       Assign samples of one CT signal to x1 and samples from the other
%       CT signal to x2.
        .
plot(x1,x2,'k') ;
        .
        .
```

## 8.3 THE CORRELATION FUNCTION

### CONCEPTUAL BASIS

The correlogram is useful as a visualization tool, but it would be nice to have a precise mathematical way of expressing the relationship between two signals. *Correlation* is the mathematical technique which indicates whether two signals are related and in a precise quantitative way how much they are related.

The mathematical calculation of correlation is based on an analysis of whether two signals tend to move together. That is, if one signal moves in a positive direction and the other signal also moves in a positive direction at the same time, they are correlated, at least for that time. The same is true if they both move in a negative direction together. If, over a long period of time, the signals tend to move in the same direction at the same time, they are said to be *positively* correlated. If, over a long period of time, two signals tend to move in opposite directions at the same time, they are also correlated, but in a *negative* sense. If, over a long period of time, the two signals tend to move in the same direction about half the time and in opposite directions the other half of the time, they are said to be uncorrelated. (This is true of the two sinusoids previously mentioned which were 90° out of phase.) These statements are not mathematically precise, but they describe in a conceptual way how correlation is calculated.

The mathematical definition of correlation must somehow embody these ideas about how signals move relative to each other. This is done by looking at the average value of the product of the functions. Consider first two signals each of which has an average value of zero (Figures 8.10 and 8.11). If they tend to move together in the same direction, their product tends to be positive. If they are both positive, the product is positive, and if they are both negative, their product is still positive. Similarly, if they

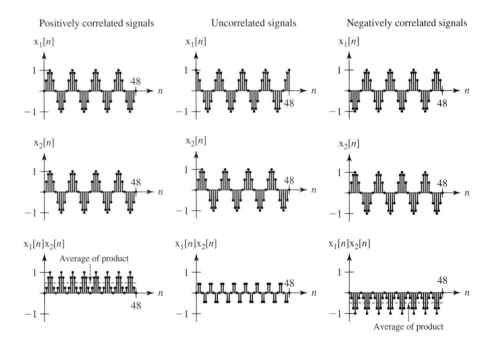

**Figure 8.10**
Correlation of DT sinusoids with zero average value.

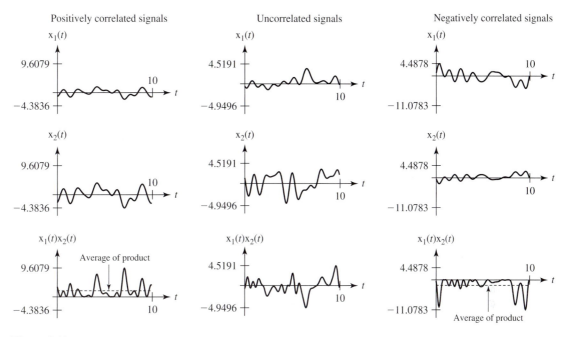

**Figure 8.11**
Correlation of CT random signals with zero average value.

move in opposite directions most of the time, their product will tend to be negative most of the time. Therefore, the average of their product over a long period of time is a good measure of how correlated they are and in which sense.

If the average values of the signals are both *nonzero,* then a bias will be added to the product, but the *variation around that bias* will still indicate whether their variations are moving in the same or opposite directions (Figures 8.12 and 8.13). If the average of the product of the signals is greater than the product of the average values of the two individual signals, the signals are positively correlated. If the average of the product is less than the product of the averages, the signals are negatively correlated. If the average of the product equals the product of the averages, the signals are uncorrelated. A careful look at the uncorrelated case in Figure 8.13 would reveal that the average of the product and the product of the averages are not *exactly* the same, although they are very close. This occurs because the average is taken over a short time. As the time over which the average is taken is increased, these two values approach the same limit.

## ENERGY SIGNALS

The mathematical definition of correlation depends on the type of signals being analyzed. There are two commonly accepted definitions, one for energy signals and one for power signals. For two CT energy signals $x(t)$ and $y(t)$, correlation is defined by $\int_{-\infty}^{\infty} x(t)y^*(t)\, dt$. For two DT energy signals $x[n]$ and $y[n]$, correlation is defined by $\sum_{n=-\infty}^{\infty} x[n]y^*[n]$. For the common case in which both signals are real, these definitions simplify to $\int_{-\infty}^{\infty} x(t)y(t)\, dt$ and $\sum_{n=-\infty}^{\infty} x[n]y[n]$.

It is much more common in signal and system analysis to refer to the *correlation function* instead of just the correlation. The correlation function is a mathematical expression of how correlated two signals are as a function of how much one of them is shifted. The correlation between two functions is a single number. The correlation

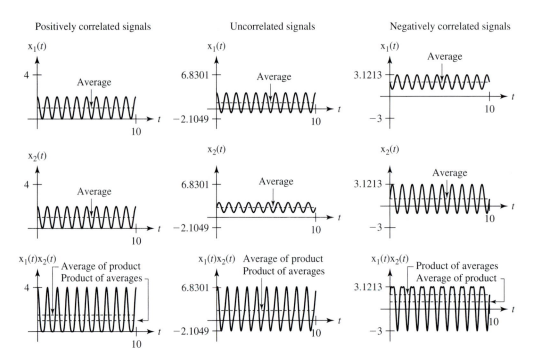

**Figure 8.12**
Correlation of CT sinusoids with nonzero average value.

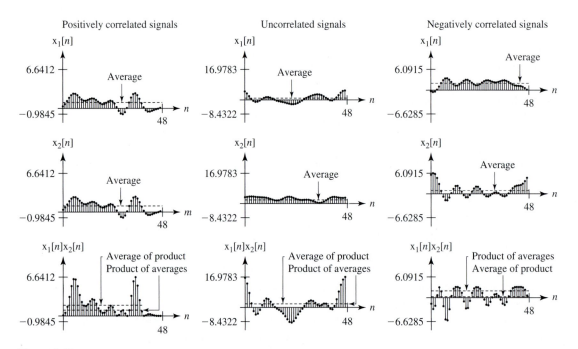

**Figure 8.13**
Correlation of DT random signals with nonzero average value.

function between two functions is itself a function, a function of the shift amount. The mathematical definition of the correlation function $R_{xy}$ between two CT energy signals $x(t)$ and $y(t)$ is

$$R_{xy}(\tau) = \int_{-\infty}^{\infty} x(t)y^*(t+\tau)\,dt = \int_{-\infty}^{\infty} x(t-\tau)y^*(t)\,dt \qquad (8.4)$$

or, if both signals $x(t)$ and $y(t)$ are real,

$$R_{xy}(\tau) = \int_{-\infty}^{\infty} x(t)y(t+\tau)\,dt = \int_{-\infty}^{\infty} x(t-\tau)y(t)\,dt. \qquad (8.5)$$

For DT energy signals,

$$R_{xy}[m] = \sum_{n=-\infty}^{\infty} x[n]y^*[n+m] = \sum_{n=-\infty}^{\infty} x[n-m]y^*[n], \qquad (8.6)$$

or, if both signals $x[n]$ and $y[n]$, are real,

$$R_{xy}[m] = \sum_{n=-\infty}^{\infty} x[n]y[n+m] = \sum_{n=-\infty}^{\infty} x[n-m]y[n]. \qquad (8.7)$$

---

Different authors use different definitions of correlation. The differences occur in the specification of which signal is to be shifted, the direction in which it is to be shifted, and the symbol to be used for the shift variable. Typical definitions for CT signals are

$$R_{xy}(\tau) = \int_{-\infty}^{\infty} x(t+\tau)y(t)\,dt, \qquad R_{xy}(\tau) = \int_{-\infty}^{\infty} x(t)y(t-\tau)\,dt, \qquad (8.8)$$

$$R_{xy}(t) = \int_{-\infty}^{\infty} x(t+\tau)y(\tau)\,d\tau, \qquad R_{xy}(t) = \int_{-\infty}^{\infty} x(\tau)y(\tau-t)\,d\tau. \qquad (8.9)$$

It would be nice, of course, if everyone could agree on a common definition. But all that is really important is that a definition is established and used consistently in any text. The fundamental characteristics of correlation and the implications for signal and system analysis are the same regardless of which definition is used.

---

Notice the similarity between the *correlation* function for two energy signals and the *convolution* of two signals presented earlier. The convolution of two signals x and y is

$$x(t) * y(t) = \int_{-\infty}^{\infty} x(t-\tau)y(\tau)\,d\tau \qquad \text{or} \qquad x[n] * y[n] = \sum_{m=-\infty}^{\infty} x[n-m]y[m].$$

$$(8.10)$$

The only difference is that in convolution one of the signals is flipped in time (time-inverted) before the shifting process occurs, and in correlation the flipping process is omitted. Therefore, for energy signals, there is a simple mathematical relationship between correlation and convolution,

$$R_{xy}(\tau) = x(-\tau) * y(\tau) \qquad \text{or} \qquad R_{xy}[m] = x[-m] * y[m]. \qquad \textbf{(8.11)}$$

Since there is such a close relationship between convolution and correlation for energy signals, we can use the multiplication–convolution duality of the Fourier transform to help in calculating correlations like we did earlier for convolutions. Convolution in the time domain corresponds to multiplication in the frequency domain. Therefore, using

$$x(-t) \xleftrightarrow{\;\mathcal{F}\;} X^*(f) \qquad \text{and} \qquad x[-n] \xleftrightarrow{\;\mathcal{F}\;} X^*(F), \qquad \textbf{(8.12)}$$

the correlation function for energy signals can be expressed as

$$R_{xy}(\tau) \xleftrightarrow{\;\mathcal{F}\;} X^*(f)Y(f) \qquad \textbf{(8.13)}$$

or

$$R_{xy}[m] \xleftrightarrow{\;\mathcal{F}\;} X^*(F)Y(F). \qquad \textbf{(8.14)}$$

## POWER SIGNALS

The correlation function between two CT power signals $x(t)$ and $y(t)$ is mathematically defined by

$$R_{xy}(\tau) = \lim_{T \to \infty} \frac{1}{T} \int_T x(t)y^*(t + \tau)\, dt = \lim_{T \to \infty} \frac{1}{T} \int_T x(t - \tau)y^*(t)\, dt. \qquad \textbf{(8.15)}$$

If $x(t)$ and $y(t)$ are both real,

$$R_{xy}(\tau) = \lim_{T \to \infty} \frac{1}{T} \int_T x(t)y(t + \tau)\, dt = \lim_{T \to \infty} \frac{1}{T} \int_T x(t - \tau)y(t)\, dt. \qquad \textbf{(8.16)}$$

The correlation function between two DT power signals $x[n]$ and $y[n]$ is mathematically defined by

$$R_{xy}[m] = \lim_{N \to \infty} \frac{1}{N} \sum_{n=\langle N \rangle} x[n]y^*[n + m] = \lim_{N \to \infty} \frac{1}{N} \sum_{n=\langle N \rangle} x[n - m]y^*[n]. \qquad \textbf{(8.17)}$$

If $x[n]$ and $y[n]$ are both real,

$$R_{xy}[m] = \lim_{N \to \infty} \frac{1}{N} \sum_{n=\langle N \rangle} x[n]y[n + m] = \lim_{N \to \infty} \frac{1}{N} \sum_{n=\langle N \rangle} x[n - m]y[n]. \qquad \textbf{(8.18)}$$

An important special case of correlation of power signals is the correlation between two periodic signals whose fundamental periods are such that the product of the two signals is also periodic. This will happen any time the fundamental periods of the two periodic signals have a finite *least common multiple* (LCM). (Recall that the LCM of two numbers is the smallest number which, when divided by each of the two numbers, yields an integer. For example, the LCM of 3 and 4 is 12, the LCM of 10 and 12 is 60, and the LCM of 6 and 9 is 18.)

For two periodic functions whose product has a period $T$ or $N$, the general form of the correlation function (for real power functions)

$$R_{xy}(\tau) = \lim_{T \to \infty} \frac{1}{T} \int_T x(t)y(t + \tau)\, dt \qquad \text{or} \qquad R_{xy}[m] = \lim_{N \to \infty} \frac{1}{N} \sum_{n=\langle N \rangle} x[n]y[n + m]$$

$$\textbf{(8.19)}$$

can be replaced by

$$R_{xy}(\tau) = \frac{1}{T} \int_T x(t)y(t + \tau)\, dt \quad \text{or} \quad R_{xy}[m] = \frac{1}{N} \sum_{n=\langle N \rangle} x[n]y[n + m] \quad \textbf{(8.20)}$$

because the integral over one period of the product, divided by the period (which is the average of the integrand over one period) is the same as the average over any integer number of periods, including infinitely many periods. The right-hand sides of the two equations in (8.20) are very similar to periodic convolutions. In fact we can express the correlation between two periodic signals over any period they have in common as a periodic convolution,

$$R_{xy}(\tau) = \frac{x(-\tau) \circledast y(\tau)}{T} \quad \text{and} \quad R_{xy}[m] = \frac{x[-m] \circledast y[m]}{N}, \quad \textbf{(8.21)}$$

over that common period or, using the CTFS or DTFS and their multiplication–convolution duality property,

$$x(t) \circledast y(t) \xleftrightarrow{\quad \mathcal{FS} \quad} T_0 X[k]Y[k] \quad \textbf{(8.22)}$$

or

$$x[n] \circledast y[n] \xleftrightarrow{\quad \mathcal{FS} \quad} N_0 Y[k]X[k], \quad \textbf{(8.23)}$$

$$R_{xy}(\tau) \xleftrightarrow{\quad \mathcal{FS} \quad} X^*[k]Y[k] \quad \text{and} \quad R_{xy}[m] \xleftrightarrow{\quad \mathcal{FS} \quad} X^*[k]Y[k], \quad \textbf{(8.24)}$$

where, in each case, the Fourier-series representation is taken over a time $T$ or $N$, which is any period which is common to both functions.

The reason we have two definitions of the correlation function is that if we applied the definition for energy signals,

$$R_{xy}(\tau) = \int_{-\infty}^{\infty} x(t)y(t + \tau)\, dt \quad \text{or} \quad R_{xy}[m] = \sum_{n=-\infty}^{\infty} x[n]y[n + m], \quad \textbf{(8.25)}$$

to a power signal, the result would be infinite and if we applied the definition for power signals,

$$R_{xy}(\tau) = \lim_{T \to \infty} \frac{1}{T} \int_T x(t)y(t + \tau)\, dt \quad \text{or} \quad R_{xy}[m] = \lim_{N \to \infty} \frac{1}{N} \sum_{n=\langle N \rangle} x[n]y[n + m], \quad \textbf{(8.26)}$$

to an energy signal the result would be zero. It is natural to ask at this point what to use if one signal is an energy signal and the other signal is a power signal. The answer is to use the energy signal definition,

$$R_{xy}(\tau) = \int_{-\infty}^{\infty} x(t)y(t + \tau)\, dt \quad \text{or} \quad R_{xy}[m] = \sum_{n=-\infty}^{\infty} x[m]y[n + m]. \quad \textbf{(8.27)}$$

The finite energy of the energy signal will keep the integral of the product from being infinite.

As stated previously, the correlation function is more general than just the correlation because it is a function of how much the second function is shifted. Some

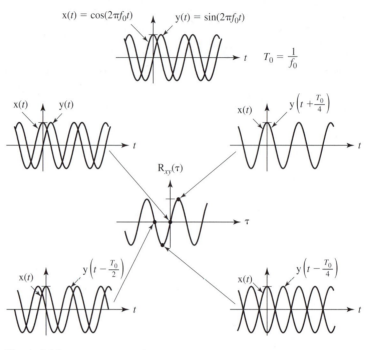

**Figure 8.14**
Graphical illustration of the correlation between a cosine and a sine at different shifts.

functions are uncorrelated at one shift and highly correlated at another shift, for example, a CT sine and a cosine of the same frequency. If neither one is shifted, they are uncorrelated. If one is shifted by 90°, they are highly correlated, either positively or negatively (Figure 8.14).

## EXAMPLE 8.1

Find the correlation function for the energy signals in Figure 8.15.

### ■ Solution

*Method 1:*

$$x_1(t) = 4 \operatorname{rect}\left(\frac{t}{4}\right) \xleftrightarrow{\mathcal{F}} X_1(f) = 16 \operatorname{sinc}(4f) \tag{8.28}$$

$$x_2(t) = \operatorname{rect}\left(\frac{t+1}{2}\right) - \operatorname{rect}\left(\frac{t-1}{2}\right) \xleftrightarrow{\mathcal{F}} X_2(f) = 2 \operatorname{sinc}(2f)(e^{j2\pi f} - e^{-j2\pi f}) \tag{8.29}$$

$$R_{12}(\tau) \xleftrightarrow{\mathcal{F}} X_1^*(f)X_2(f) = 32 \operatorname{sinc}(4f) \operatorname{sinc}(2f)(e^{j2\pi f} - e^{-j2\pi f}) \tag{8.30}$$

Using

$$\frac{a+b}{2} \operatorname{tri}\left(\frac{2t}{a+b}\right) - \frac{a-b}{2} \operatorname{tri}\left(\frac{2t}{a-b}\right) \xleftrightarrow{\mathcal{F}} |ab| \operatorname{sinc}(af) \operatorname{sinc}(bf) \qquad a > b > 0 \tag{8.31}$$

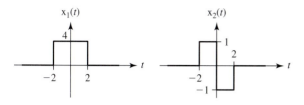

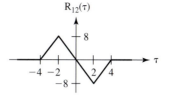

**Figure 8.15**
Two energy signals.

**Figure 8.16**
Correlation function.

from Appendix E,

$$12 \text{ tri}\left(\frac{2t}{3}\right) - 4 \text{ tri}(2t) \xleftrightarrow{\mathcal{F}} 32 \text{ sinc}(4f) \text{ sinc}(2f). \tag{8.32}$$

Then using the time-shifting property of the CTFT,

$$4\left\{3 \text{ tri}\left(\frac{2(t+1)}{3}\right) - \text{tri}\,(2(t+1)) - 3 \text{ tri}\left(\frac{2(t-1)}{3}\right) + \text{tri}(2(t-1))\right\} \xleftrightarrow{\mathcal{F}}$$
$$32 \text{ sinc}(4f) \text{ sinc}(2f)(e^{j2\pi f} - e^{-j2\pi f}). \tag{8.33}$$

Therefore,

$$R_{12}(\tau) = 4\left\{3 \text{ tri}\left(\frac{2(\tau+1)}{3}\right) - \text{tri}(2(\tau+1)) - 3 \text{ tri}\left(\frac{2(\tau-1)}{3}\right) + \text{tri}(2(\tau-1))\right\}. \tag{8.34}$$

The function $3 \text{ tri}(2(t+1)/3) - \text{tri}(2(t+1))$ is a trapezoid of height, 2, whose lower base extends from $-4$ to $2$ and whose upper base extends from $-2$ to $0$. It is, therefore, centered at $-1$. The function $3 \text{ tri}(2(t-1)/3) - \text{tri}(2(t-1))$ is identical except shifted to the right by 2 to be centered at $+1$. When we subtract the second function from the first and multiply by 4, we get the function in Figure 8.16.

*Method 2:* The definition of the correlation function for energy signals is

$$R_{xy}(\tau) = \int_{-\infty}^{\infty} x(t)y(t+\tau)\,dt \tag{8.35}$$

or, in this case,

$$R_{12}(\tau) = \int_{-\infty}^{\infty} x_1(t)x_2(t+\tau)\,dt. \tag{8.36}$$

The integral depends on the shift amount $\tau$ as illustrated in Figure 8.17.

*Case 1.* $\tau > 4$. In this case the signals do not overlap and the correlation function is zero.

*Case 2.* $2 < \tau < 4$. For this case the correlation function is

$$R_{12}(\tau) = \int_{-2}^{-\tau+2} 4 \times (-1)\,dt = 4(\tau - 4). \tag{8.37}$$

*Case 3.* $0 < \tau < 2$. For this case the correlation function is

$$R_{12}(\tau) = \int_{-2}^{-\tau} 4 \times (+1)\,dt + \int_{-\tau}^{-\tau+2} 4 \times (-1)\,dt = 4\left(\int_{-2}^{-\tau} dt - \int_{-\tau}^{-\tau+2} dt\right) = -4\tau. \tag{8.38}$$

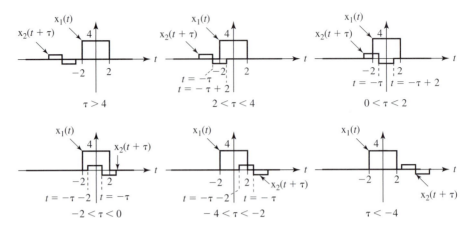

**Figure 8.17**
Six cases of the shift amount $\tau$.

*Case 4.* $-2 < \tau < 0$.    Because of the even symmetry of the first signal and the odd symmetry of the second signal, the result is the same as case 3, $R_{12}(\tau) = -4\tau$.

*Case 5.* $-4 < \tau < 2$.    Again, from symmetry considerations,

$$R_{12}(\tau) = 4(\tau + 4).    \tag{8.39}$$

*Case 6.* $\tau < -4$.    In this case the signals do not overlap and the correlation function is again zero. When this result is graphed, it looks exactly like the previous result in Figure 8.16.

---

## EXAMPLE 8.2

Find the correlation between the DT power signals,

$$x[n] = 5 \cos\left(\frac{2\pi n}{5}\right)    \tag{8.40}$$

and

$$y[n] = 2 \cos\left(\frac{2\pi n}{7}\right).    \tag{8.41}$$

### ■ Solution

*Method 1:*    Use the relation

$$R_{xy}[m] \overset{\mathcal{FS}}{\longleftrightarrow} X^*[k]Y[k].    \tag{8.42}$$

Before we can use this result we must find a common period for the two signals. The two individual periods are 5 and 7. The least common multiple of those two periods is 35. The two DTFS harmonic functions are

$$X[k] = \frac{5}{2}(\text{comb}_{35}[k-7] + \text{comb}_{35}[k+7])    \tag{8.43}$$

and

$$Y[k] = (\text{comb}_{35}[k-5] + \text{comb}_{35}[k+5]).    \tag{8.44}$$

Therefore,

$$R_{xy}[m] \xleftrightarrow{\ \mathcal{FS}\ } \frac{5}{2}(\text{comb}_{35}[k-5] + \text{comb}_{35}[k+5])(\text{comb}_{35}[k-7] + \text{comb}_{35}[k+7]). \tag{8.45}$$

This is the product of two periodic sequences of DT impulses. Therefore, the product is zero except where $X[k]$ and $Y[k]$ both have a non-zero impulse which occurs at the same value of $k$. The non-zero impulses in $X[k]$ and $Y[k]$ *never* occur at the same value of $k$. Therefore, the correlation is zero,

$$R_{xy}[m] = 0. \tag{8.46}$$

*Method 2:*   The general expression for the correlation function for DT power signals is

$$R_{xy}[m] = \lim_{N \to \infty} \frac{1}{N} \sum_{n=\langle N \rangle} x[n]y[n+m]. \tag{8.47}$$

Applying that to $x[n]$ and $y[n]$ we get

$$R_{xy}[m] = \lim_{N \to \infty} \frac{1}{N} \sum_{n=\langle N \rangle} 5 \cos\left(\frac{2\pi n}{5}\right) 2 \cos\left(\frac{2\pi (n+m)}{7}\right). \tag{8.48}$$

Using

$$\cos(x)\cos(y) = \frac{1}{2}[\cos(x-y) + \cos(x+y)] \tag{8.49}$$

we get

$$R_{xy}[m] = \lim_{N \to \infty} \frac{5}{N} \sum_{n=\langle N \rangle} \left[ \cos\left(\frac{2\pi n}{5} - \frac{2\pi (n+m)}{7}\right) + \cos\left(\frac{2\pi n}{5} + \frac{2\pi (n+m)}{7}\right) \right] \tag{8.50}$$

$$R_{xy}[m] = \lim_{N \to \infty} \frac{5}{N} \sum_{n=\langle N \rangle} \left[ \cos\left(\frac{4\pi n}{35} - \frac{2\pi m}{7}\right) + \cos\left(\frac{24\pi n}{35} + \frac{2\pi m}{7}\right) \right]. \tag{8.51}$$

Then, using

$$\cos(x+y) = \cos(x)\cos(y) - \sin(x)\sin(y) \tag{8.52}$$

$$R_{xy}[m] = \lim_{N \to \infty} \frac{5}{N} \left[ \cos\left(-\frac{2\pi m}{7}\right) \sum_{n=\langle N \rangle} \cos\left(\frac{4\pi n}{35}\right) - \sin\left(-\frac{2\pi m}{7}\right) \sum_{n=\langle N \rangle} \sin\left(\frac{4\pi n}{35}\right) \right.$$

$$\left. + \cos\left(\frac{2\pi m}{7}\right) \sum_{n=\langle N \rangle} \cos\left(\frac{24\pi n}{35}\right) - \sin\left(\frac{2\pi m}{7}\right) \sum_{n=\langle N \rangle} \sin\left(\frac{24\pi n}{35}\right) \right]. \tag{8.53}$$

Since the starting point of the summation is arbitrary, let it be $n = 0$ in each case. Then

$$R_{xy}[m] = \lim_{N \to \infty} \frac{5}{N} \left[ \cos\left(\frac{2\pi m}{7}\right) \sum_{n=0}^{N-1} \cos\left(\frac{4\pi n}{35}\right) + \sin\left(\frac{2\pi m}{7}\right) \sum_{n=0}^{N-1} \sin\left(\frac{4\pi n}{35}\right) \right.$$

$$\left. + \cos\left(\frac{2\pi m}{7}\right) \sum_{n=0}^{N-1} \cos\left(\frac{24\pi n}{35}\right) - \sin\left(\frac{2\pi m}{7}\right) \sum_{n=0}^{N-1} \sin\left(\frac{24\pi n}{35}\right) \right]. \tag{8.54}$$

Using the exponential definitions of the sine and cosine,

$$R_{xy}[m] = \lim_{N \to \infty} \frac{5}{2N} \left[ \cos\left(\frac{2\pi m}{7}\right) \sum_{n=0}^{N-1} \left(e^{j(4\pi n/35)} + e^{-j(4\pi n/35)}\right) \right.$$

$$-j\sin\left(\frac{2\pi m}{7}\right) \sum_{n=0}^{N-1} \left(e^{j(4\pi n/35)} - e^{-j(4\pi n/35)}\right)$$

$$+ \cos\left(\frac{2\pi m}{7}\right) \sum_{n=0}^{N-1} \left(e^{j(24\pi n/35)} + e^{-j(24\pi n/35)}\right)$$

$$\left. + j\sin\left(\frac{2\pi m}{7}\right) \sum_{n=0}^{N-1} \left(e^{j(24\pi n/35)} - e^{-j(24\pi n/35)}\right) \right].$$

(8.55)

Then using

$$\sum_{n=0}^{N-1} r^n = \begin{cases} N & r = 1 \\ \dfrac{1 - r^N}{1 - r} & \text{otherwise} \end{cases}$$

(8.56)

we get

$$R_{xy}[m] = \frac{5}{2} \left[ \cos\left(\frac{2\pi m}{7}\right) \lim_{N \to \infty} \frac{1}{N} \left(\frac{1 - e^{j(4\pi N/35)}}{1 - e^{j(4\pi/35)}} + \frac{1 - e^{-j(4\pi N/35)}}{1 - e^{-j(4\pi/35)}}\right) \right.$$

$$-j\sin\left(\frac{2\pi m}{7}\right) \lim_{N \to \infty} \frac{1}{N} \left(\frac{1 - e^{j(4\pi N/35)}}{1 - e^{j(4\pi/35)}} - \frac{1 - e^{-j(4\pi N/35)}}{1 - e^{-j(4\pi/35)}}\right)$$

$$+ \cos\left(\frac{2\pi m}{7}\right) \lim_{N \to \infty} \frac{1}{N} \left(\frac{1 - e^{j(24\pi N/35)}}{1 - e^{j(4\pi/35)}} + \frac{1 - e^{-j(24\pi N/35)}}{1 - e^{-j(4\pi/35)}}\right)$$

$$\left. + j\sin\left(\frac{2\pi m}{7}\right) \lim_{N \to \infty} \frac{1}{N} \left(\frac{1 - e^{j(24\pi N/35)}}{1 - e^{j(4\pi/35)}} - \frac{1 - e^{-j(24\pi N/35)}}{1 - e^{-j(4\pi/35)}}\right) \right].$$

(8.57)

Now examine one of the fractional terms in parentheses in this expression, $(1 - e^{j(4\pi N/35)})/(1 - e^{j(4\pi/35)})$. The numerator can never exceed two in magnitude, no matter what value $N$ has and the denominator is a finite constant. Therefore, as $N$ approaches infinity this fraction is bounded. The same can be said of all the other fractions of the same form. The factor $1/N$ multiplying each fraction makes the correlation approach zero as $N$ approaches infinity. Therefore,

$$R_{xy}[m] = 0.$$

(8.58)

These two DT power signals are completely uncorrelated. The lack of correlation is a consequence of the fact that the two signals have different frequencies and the correlation of a power signal is computed over all discrete time $n$.

■

The result of Example 8.2 leads to an important general conclusion. The correlation between two sinusoids of different frequencies is zero. Let $x_1(t) = A_1 \cos(2\pi f_{01} t + \theta_1)$ and $x_2(t) = A_2 \cos(2\pi f_{02} t + \theta_2)$. Their CTFS harmonic functions consist of impulses at different locations and the product is zero. Therefore, the correlation function is also zero. We can also show that the correlation is zero directly from the definition. The correlation is

$$R_{12}(\tau) = \lim_{T \to \infty} \frac{1}{T} \int_{-(T/2)}^{T/2} A_1 \cos(2\pi f_{01}t + \theta_1) A_2 \cos(2\pi f_{02}(t + \tau) + \theta_2) \, dt. \quad \textbf{(8.59)}$$

We can use the trigonometric identity,

$$\cos(x)\cos(y) = \frac{1}{2}[\cos(x - y) + \cos(x + y)], \quad \textbf{(8.60)}$$

to write

$$R_{12}(\tau) = \lim_{T \to \infty} \frac{A_1 A_2}{2T} \int_{-(T/2)}^{T/2} [\cos(2\pi(f_{01} - f_{02})t - 2\pi f_{02}\tau + \theta_1 - \theta_2)$$
$$+ \cos(2\pi(f_{01} + f_{02})t + 2\pi f_{02}\tau + \theta_1 + \theta_2)] \, dt.$$
$$\textbf{(8.61)}$$

If $f_{01} \neq f_{02}$, then

$$R_{12}(\tau) = \lim_{T \to \infty} \frac{A_1 A_2}{2T} \left[ \underbrace{\frac{\sin(2\pi(f_{01} - f_{02})t - 2\pi f_{02}\tau + \theta_1 - \theta_2)}{2\pi(f_{01} - f_{02})}}_{\text{Bounded}} \right.$$

$$\left. + \underbrace{\frac{\sin(2\pi(f_{01} + f_{02})t + 2\pi f_{02}\tau + \theta_1 + \theta_2)}{2\pi(f_{01} + f_{02})}}_{\text{Bounded}} \right]_{-(T/2)}^{T/2} .$$
$$\textbf{(8.62)}$$

In the limit as $T$ approaches infinity, the division by $T$ of a bounded quantity is zero. Therefore, if $f_{01} \neq f_{02}$, $R_{12}(\tau) = 0$.

## 8.4 AUTOCORRELATION

### RELATION TO SIGNAL ENERGY AND SIGNAL POWER

A very important special case of the correlation function is the correlation of a function with itself. This type of correlation function is called the *autocorrelation* function. If $x(t)$ is an energy signal, its autocorrelation is

$$R_{xx}(\tau) = \int_{-\infty}^{\infty} x(t)x(t + \tau) \, dt \qquad \text{or} \qquad R_{xx}[m] = \sum_{n=-\infty}^{\infty} x[n]x[n + m]. \quad \textbf{(8.63)}$$

At a shift of zero that becomes

$$R_{xx}(0) = \int_{-\infty}^{\infty} x^2(t) \, dt \qquad \text{or} \qquad R_{xx}[0] = \sum_{n=-\infty}^{\infty} x^2[n], \quad \textbf{(8.64)}$$

which is the *total signal energy* of the signal.

If $x(t)$ or $x[n]$ is a power signal, the autocorrelation at zero shift is

$$R_{xx}(0) = \lim_{T \to \infty} \frac{1}{T} \int_T x^2(t)\, dt \qquad R_{xx}[0] = \lim_{N \to \infty} \frac{1}{N} \sum_{n=\langle N \rangle} x^2[n], \qquad \textbf{(8.65)}$$

which is the *average signal power* of the signal.

## PROPERTIES OF AUTOCORRELATION

The autocorrelation depends on the choice of the shift amount, so we cannot say what the autocorrelation function looks like until we know what the function is. But we *can* say that the value of the autocorrelation can never be bigger than it is at zero shift. That is,

$$R_{xx}(0) \geq R_{xx}(\tau) \qquad \text{or} \qquad R_{xx}[0] \geq R_{xx}[m] \qquad \textbf{(8.66)}$$

because at a zero shift, the correlation with itself is obviously as large as it can get since the shifted and unshifted versions coincide. Also,

$$R_{xx}(-\tau) = \int_{-\infty}^{\infty} x(t)x(t-\tau)\, dt \qquad \text{or} \qquad R_{xx}(-\tau) = \lim_{T \to \infty} \frac{1}{T} \int_{-(T/2)}^{T/2} x(t)x(t-\tau)\, dt.$$
$$\textbf{(8.67)}$$

Then if we make the change of variable

$$t' = t - \tau \qquad \text{and} \qquad dt' = dt, \qquad \textbf{(8.68)}$$

we can show that

$$R_{xx}(\tau) = R_{xx}(-\tau). \qquad \textbf{(8.69)}$$

It can be shown by a similar technique that

$$R_{xx}[m] = R_{xx}[-m] \qquad \textbf{(8.70)}$$

or, in words, all autocorrelation functions (but not all correlation functions) are even functions.

Another characteristic of the autocorrelation function is that a time shift of a signal does not change its autocorrelation. Let $R_{xx}[m]$ be the autocorrelation function of a DT energy signal $x[n]$. Then

$$R_{xx}[m] = \sum_{n=-\infty}^{\infty} x[n]x[n+m]. \qquad \textbf{(8.71)}$$

Now let $y[n] = x[n - n_0]$. Then

$$R_{yy}[m] = \sum_{n=-\infty}^{\infty} y[n]y[n+m] = \sum_{n=-\infty}^{\infty} x[n-n_0]x[n-n_0+m]. \qquad \textbf{(8.72)}$$

We can make a change of variable $q = n - n_0$. Then

$$R_{yy}[m] = \sum_{q=-\infty}^{\infty} x[q]x[q+m] = R_{xx}[m], \qquad \textbf{(8.73)}$$

proving that the autocorrelation functions of $x[n]$ and $y[n]$ are the same, regardless of what the value of $n_0$ is. The same rule holds for CT energy signals and CT and DT power signals.

The autocorrelation of a sum of sinusoids of different frequencies is the sum of the autocorrelations of the individual sinusoids. To demonstrate this idea let a CT power signal $x(t)$ be a sum of two sinusoids $x_1(t)$ and $x_2(t)$, where

$$x_1(t) = A_1 \cos(2\pi f_{01} t + \theta_1) \qquad \text{and}$$

$$x_2(t) = A_2 \cos(2\pi f_{02} t + \theta_2) \qquad f_{01} \neq f_{02}. \tag{8.74}$$

The autocorrelation of this signal is

$$R_x(\tau) = \lim_{T \to \infty} \frac{1}{T} \int_{-(T/2)}^{T/2} x(t) x(t + \tau) \, d\tau \tag{8.75}$$

$$R_x(\tau) = \lim_{T \to \infty} \frac{1}{T} \int_{-(T/2)}^{T/2} [x_1(t) x_1(t + \tau) + x_1(t) x_2(t + \tau)$$

$$+ \, x_2(t) x_1(t + \tau) + x_2(t) x_2(t + \tau)] \, d\tau$$

$$R_x(\tau) = \left\{ \underbrace{\lim_{T \to \infty} \frac{1}{T} \int_{-(T/2)}^{T/2} x_1(t) x_1(t + \tau) \, d\tau}_{= R_1(\tau)} + \underbrace{\lim_{T \to \infty} \frac{1}{T} \int_{-(T/2)}^{T/2} x_1(t) x_2(t + \tau) \, d\tau}_{= R_{12}(\tau)} \right.$$

$$\left. + \underbrace{\lim_{T \to \infty} \frac{1}{T} \int_{-(T/2)}^{T/2} x_2(t) x_1(t + \tau) \, d\tau}_{= R_{21}(\tau)} + \underbrace{\lim_{T \to \infty} \frac{1}{T} \int_{-(T/2)}^{T/2} x_2(t) x_2(t + \tau) \, d\tau}_{= R_2(\tau)} \right\}.$$

The correlations $R_{12}(\tau)$ and $R_{21}(\tau)$ are both zero because they are the correlations between sinusoids of different frequencies. Therefore,

$$R_x(\tau) = R_1(\tau) + R_2(\tau). \tag{8.76}$$

## AUTOCORRELATION EXAMPLES

Figures 8.18 and 8.19 show some graphical examples of some energy signals and their autocorrelation functions. Figure 8.18 is an illustration of the autocorrelation functions for three random DT energy signals. Since they are random, these signals are all different, but they do have some similar properties. One of the similarities is seen in their autocorrelation functions. All three autocorrelation functions have a sharp peak at $m = 0$ and then very quickly go to a small random fluctuation around zero for even very small nonzero values of the shift $m$. The autocorrelation function is describing an important feature of these signals. Each of them changes very rapidly with time to new values that have practically no correlation with past or future values, even at a very short time in the past or future.

  Figure 8.19 is an illustration of the autocorrelation functions for two CT sinusoidal bursts. These waveforms are typical of communication signals which encode binary

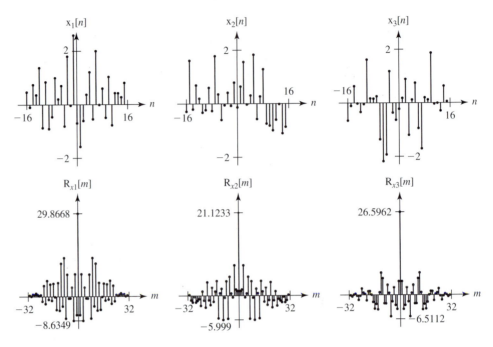

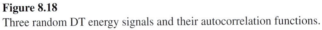

**Figure 8.18**
Three random DT energy signals and their autocorrelation functions.

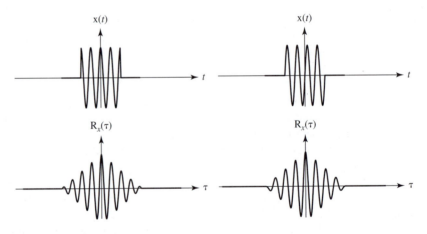

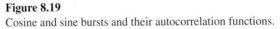

**Figure 8.19**
Cosine and sine bursts and their autocorrelation functions.

data for transmission. Notice that, even though one is a cosine burst and the other is a sine burst, their autocorrelation functions are almost identical. Also notice that, even though the sine function is odd, its autocorrelation function is even. The autocorrelation function is even because it indicates how a function relates to itself when shifted, not how the function itself varies with time. For these two signals, the relation of each to a shifted version of itself is almost exactly the same. (When we get to power signals, we will see that a cosine and sine of the same frequency and amplitude have exactly

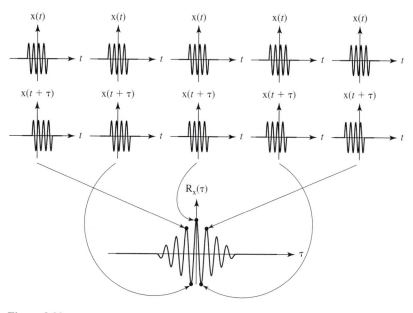

**Figure 8.20**
Relation of shift amount $\tau$ to autocorrelation.

the same autocorrelation function.) The fact that these two autocorrelation functions are so close comes from the same basic reason that when a function is time-shifted its autocorrelation function does not change. In the case of the two sinusoidal bursts, the sine burst is not simply a time-shifted version of the cosine burst, but it *almost* is. That is why the two autocorrelation functions are almost the same.

Try to visualize the shifting process inherent in the autocorrelation (Figure 8.20). At zero shift either sinusoidal burst and its shifted version coincide and the area under the product is a maximum. That is why the autocorrelation functions have a maximum value at $\tau = 0$. Then as we shift one version of the signal, at a shift of half a fundamental period of the underlying sinusoid, the positive and negative peaks line up and we get a large negative area under the product. Then as we shift another half fundamental period, the positive peaks line up again but now, because of the shift, the peaks on opposite ends of the two versions are multiplied by zero. Therefore, although the area under the product reaches a positive peak, it is a smaller positive peak than the one for zero shift. As the shift proceeds further, the peaks, both positive and negative, gradually fall to zero because of the diminishing overlap between the nonzero portions of the signals.

Correlation is the basis of a widely used technique in communication systems called *matched* filtering. In digital communication systems the only important thing is that the 1s and 0s in the data stream be distinguishable from each other so the receiver can reproduce the bit pattern that was transmitted. A 1 is sent as a signal of some shape, and a 0 is sent as a signal of some different shape, ideally a *very* different shape. The 1s and 0s could be sent as different voltage-level pulses, or as sinusoidal bursts with different phases or frequencies, or in a variety of other ways. The object of the receiver is to recognize the bits. The designer of the communication system knows the shapes of the signals representing the bits, so the receiver is designed to optimally detect those shapes in the presence of noise which is always present in any system at some level.

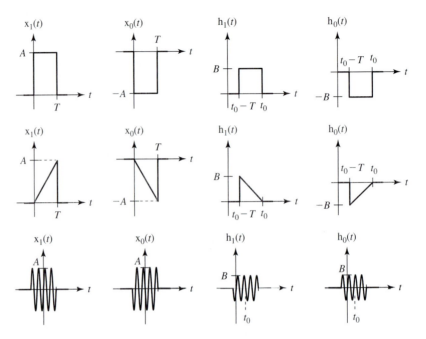

**Figure 8.21**
Some signals representing 1s and 0s and the impulse responses of matched filters
designed to optimally detect them in the presence of noise.

It has been shown that in the presence of the most common type of random noise,
the best way to detect a signal of a certain shape is to use a filter which is matched to
that shape. Let the signal representing a 1 be $x_1(t)$ and let the signal representing a 0
be $x_0(t)$. A matched filter is an LTI system whose impulse response $h(t)$ is a scaled, and
perhaps shifted, version of the time inverse of the signal to be detected. An example of
typical shapes for 1s and 0s and the corresponding matched filter impulse responses
are illustrated in Figure 8.21.

Suppose we are designing the part of the system which detects 1s. Let a transmitted 1 be $x_{1T}(t) = x_1(t)$ and let a received 1 be $x_{1R}(t) = Ax_1(t - t_D)$, where $A$ is some
constant representing the attenuation in transmission and $t_D$ is a constant representing
the propagation delay in transmission. The impulse response of that system would be
$h_1(t) = Bx_1(-t + t_0)$, where $B$ is an arbitrary constant. The response $y_1(t)$ of the
system is the convolution of the excitation (the received signal) with the impulse
response,

$$y_1(t) = x_{1R}(t) * h_1(t) = Ax_1(t - t_D) * Bx_1(-t + t_0), \qquad \textbf{(8.77)}$$

$$y_1(t) = AB \int_{-\infty}^{\infty} x_1(\tau - t_D)x_1(-(t - \tau) + t_0)\, d\tau,$$

or

$$y_1(t) = AB \int_{-\infty}^{\infty} x_1(\tau - t_D)x_1(\tau - (t - t_0))\, d\tau.$$

Making the change of variable $\tau - t_D = \lambda$,

$$y_1(t) = AB \int_{-\infty}^{\infty} x_1(\lambda) x_1(\lambda - (t - t_D - t_0)) \, d\lambda. \qquad \textbf{(8.78)}$$

Applying the definition of autocorrelation and the fact that it is an even function,

$$R_x(\tau) = \int_{-\infty}^{\infty} x(t) x(t + \tau) \, d\tau = \int_{-\infty}^{\infty} x(t) x(t - \tau) \, d\tau, \qquad \textbf{(8.79)}$$

for CT energy signals, we get

$$y_1(t) = AB \times R_{x_1}(t - t_D - t_0). \qquad \textbf{(8.80)}$$

Figure 8.22 is an illustration of a signal, without noise and with noise, and the response of a matched filter for each case. The matched filter response for a 1 is a scaled version of the autocorrelation function of the signal representing a 1, delayed in time by the propagation delay $t_0$. For this reason another common name for the matched filter is the *correlation* filter. An autocorrelation is a maximum when its argument is zero, so the response of the matched filter is a maximum when $t = t_D + t_0$, and if a 1 is present in the signal, the matched filter will be a maximum at that time. If the signal representing a 0 is the negative of the signal representing a 1, the same filter can be used to detect both. If, at the end of a bit time, the signal from the matched filter is positive, then the bit is probably a 1, and if it is negative, the bit is probably a 0.

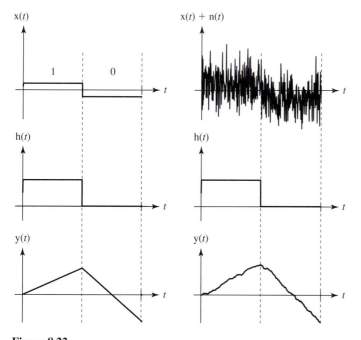

**Figure 8.22**
A 1 followed by a 0, a filter impulse response matched to the 1, and
the response of the filter with and without noise.

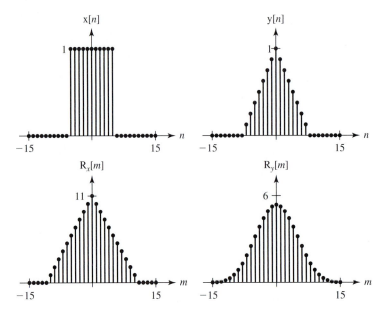

**Figure 8.23**
A rectangle, triangle, and their autocorrelation functions.

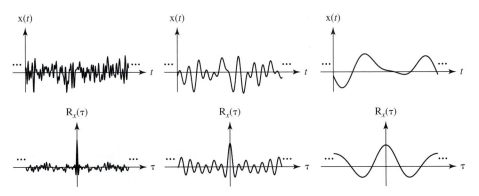

**Figure 8.24**
Three different random power signals and their autocorrelation functions.

Figure 8.23 is an illustration of the autocorrelation functions for two familiar signal shapes, a DT rectangle and a DT triangle. Figure 8.24 illustrates the autocorrelation functions for three random CT power signals. Notice how the rate of variation of the autocorrelation functions with time indicates generally how fast the signals themselves change with time. In other words, the autocorrelation function tells us something about the frequency content of the signal. We will soon make that relationship concrete when we define energy spectral density and power spectral density.

Figure 8.25 illustrates the autocorrelation functions for a cosine and a sine. As indicated earlier, since a sine of the same frequency and amplitude as a cosine is just a time-shifted cosine, the two autocorrelation functions must be the same.

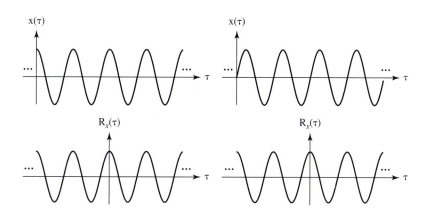

**Figure 8.25**
A cosine and a sine of the same amplitude and frequency and their identical autocorrelation functions.

Find the autocorrelation of the CT power signal in Figure 8.26.

■ **Solution**

This is a power signal, so the autocorrelation is

$$R_x(\tau) \xleftrightarrow{\mathcal{FS}} X^*[k]X[k] = |X[k]|^2. \tag{8.81}$$

The signal is described by

$$x(t) = A \, \text{rect}\left(\frac{2t}{T_0}\right) * \frac{1}{T_0} \, \text{comb}\left(\frac{t}{T_0}\right), \tag{8.82}$$

and its CTFS harmonic function is

$$X[k] = \frac{A}{2} \, \text{sinc}\left(\frac{k}{2}\right). \tag{8.83}$$

Therefore, the autocorrelation is

$$R_x(\tau) \xleftrightarrow{\mathcal{FS}} \left| \frac{A}{2} \, \text{sinc}\left(\frac{k}{2}\right) \right|^2 = \left(\frac{A}{2}\right)^2 \text{sinc}^2\left(\frac{k}{2}\right) \tag{8.84}$$

and, using

$$\text{tri}\left(\frac{t}{w}\right) * \frac{1}{T_0} \, \text{comb}\left(\frac{t}{T_0}\right) \xleftrightarrow{\mathcal{FS}} \frac{w}{T_0} \, \text{sinc}^2\left(\frac{w}{T_0}k\right), \tag{8.85}$$

$$R_x(\tau) = \frac{A^2}{2} \, \text{tri}\left(\frac{2\tau}{T_0}\right) * \frac{1}{T_0} \, \text{comb}\left(\frac{\tau}{T_0}\right) \tag{8.86}$$

(Figure 8.27).

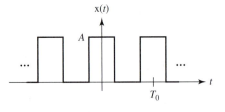

**Figure 8.26**
A square-wave signal.

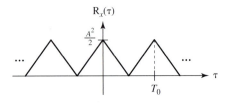

**Figure 8.27**
Autocorrelation of the square-wave signal.

## EXAMPLE 8.4

Find the autocorrelation of the DT energy signal

$$x[n] = \cos(\pi n) \, \text{sinc}\left(\frac{n}{2}\right) \tag{8.87}$$

(Figure 8.28).

### ■ Solution

We can use

$$R_x[m] \xleftrightarrow{\mathcal{F}} X^*(F)X(F) \tag{8.88}$$

to help find this autocorrelation. The DTFT of $x[n]$ is

$$X(F) = \frac{1}{2}\left[\text{comb}\left(F - \frac{1}{2}\right) + \text{comb}\left(F + \frac{1}{2}\right)\right] \circledast (2\,\text{rect}(2F) * \text{comb}(F)) \tag{8.89}$$

or

$$X(F) = \left[\delta\left(F - \frac{1}{2}\right) + \delta\left(F + \frac{1}{2}\right)\right] * (\text{rect}(2F) * \text{comb}(F))$$

or

$$X(F) = \text{rect}\left(2\left(F - \frac{1}{2}\right)\right) * \text{comb}(F) + \text{rect}\left(2\left(F + \frac{1}{2}\right)\right) * \text{comb}(F).$$

This is the sum of two periodic rectangular functions, which, because of the two DT frequency shifts $F - \frac{1}{2}$ and $F + \frac{1}{2}$ exactly coincide. So the sum is just twice either of the two periodic rectangular functions (Figure 8.29).

Since $X(F)$ is purely real, $X(F) = X^*(F)$. Although it is possible to find the inverse DTFT of (8.89) analytically, it is much easier to simply look at Figure 8.29 and write a simpler expression for $X(F)$ before doing the inverse DTFT of $X^*(F)X(F)$. $X(F)$ is just a periodically repeated rectangle with a height of 2, width of $\frac{1}{2}$, and fundamental period of 1.

$$X(F) = 2\,\text{rect}\left(2\left(F - \frac{1}{2}\right)\right) * \text{comb}(F). \tag{8.90}$$

Therefore,

$$R_x[m] \xleftrightarrow{\mathcal{F}} \left(2\,\text{rect}\left(2\left(F - \frac{1}{2}\right)\right) * \text{comb}(F)\right)^2. \tag{8.91}$$

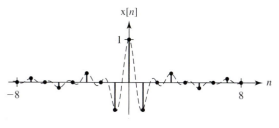

**Figure 8.28**
A DT signal.

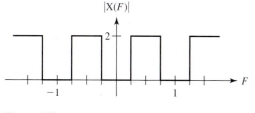

**Figure 8.29**
Magnitude of the DTFT of $x[n] = \cos(\pi n) \operatorname{sinc}\left(\dfrac{n}{2}\right)$.

Since the rectangle function squared is equal to itself, [†]

$$\operatorname{rect}^2(F) = \operatorname{rect}(F), \qquad (8.92)$$

the inverse Fourier transform of the rectangle function squared convolved with $\operatorname{comb}(F)$ is the same as the inverse Fourier transform of the rectangle function convolved with $\operatorname{comb}(F)$,

$$\operatorname{sinc}\left(\frac{n}{w}\right) \xleftrightarrow{\;\mathcal{F}\;} w \operatorname{rect}^2(wF) * \operatorname{comb}(F), \qquad (8.93)$$

and, using the frequency-shifting property of the DTFT,

$$e^{j2\pi F_0 n} x[n] \xleftrightarrow{\;\mathcal{F}\;} X(F - F_0), \qquad (8.94)$$

we get

$$R_x[m] = 2\operatorname{sinc}\left(\frac{n}{2}\right)e^{j\pi n} = 2\operatorname{sinc}\left(\frac{n}{2}\right)\left(\cos(\pi n) + j\underbrace{\sin(\pi n)}_{=0}\right) \qquad (8.95)$$

or

$$R_x[m] = 2\cos(\pi n)\operatorname{sinc}\left(\frac{n}{2}\right). \qquad (8.96)$$

So we get the completely counterintuitive result that the autocorrelation function for $\cos(\pi n)$ $\operatorname{sinc}(n/2)$ is $2\cos(\pi n)\operatorname{sinc}(n/2)$. Except for a factor of two, it is its own autocorrelation! ■

The most important use of autocorrelation is in the analysis of the effect of LTI systems on random signals. Consider the following qualitative argument to see how autocorrelation describes a random signal. Let a signal $x(t)$ be a linear combination of sinusoids of different frequencies,

$$x(t) = \sum_{k=1}^{N} A_k \cos(2\pi f_{0k} t + \theta_k). \qquad (8.97)$$

Then, since the sinusoids are all of different frequencies,

$$R_x(\tau) = \sum_{k=1}^{N} R_k(\tau), \qquad (8.98)$$

---

[†]The square of the rectangle function is not exactly equal to itself because the value of $\frac{1}{2}$ at its discontinuity becomes a value of $\frac{1}{4}$ when squared. But this difference has no practical consequences. The transforms of rect and $\operatorname{rect}^2$ are identical.

where $R_k(\tau)$ is the autocorrelation of $A_k \cos(2\pi f_{0k}t + \theta_k)$. Also $R_k(\tau)$ is independent of the choice of $\theta_k$. Now imagine that we were to form several versions of $x(t)$ by using randomly chosen phase shifts $\theta_k$, but the same amplitudes and frequencies.

In each group of four signals in Figures 8.30 and 8.31 the signals are all different but they have identical autocorrelation functions. Looking at the signals in each group of four it is apparent that they are similar but not exactly the same. Their common

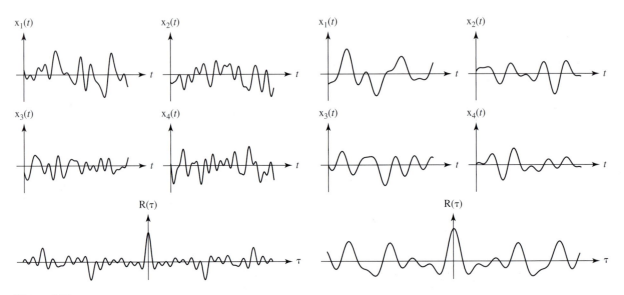

**Figure 8.30**
Two illustrations of groups of four random signals with identical autocorrelation functions.

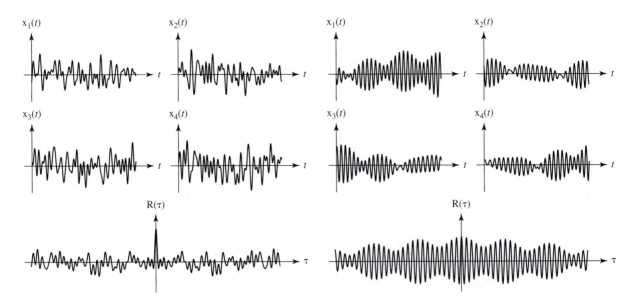

**Figure 8.31**
Two more illustrations of groups of four random signals with identical autocorrelation functions.

characteristics (the amplitudes and frequencies of the sinusoids which form them) are described by the autocorrelation function. The autocorrelation function describes a signal generally, but not exactly. It is the best description of a random signal, short of an exact description which, as a practical matter, is often either not available or not needed.

## 8.5 CROSS CORRELATION

### PROPERTIES OF CROSS CORRELATION

A common term for the correlation function between two different signals is *cross correlation* to distinguish it from autocorrelation. Autocorrelation is simply a special case of the cross-correlation function. Cross correlation is more general than autocorrelation, so the properties are not as numerous, but there is one property that is sometimes useful,

$$R_{xy}(\tau) = R_{yx}(-\tau) \qquad \text{or} \qquad R_{xy}[m] = R_{yx}[-m]. \qquad \textbf{(8.99)}$$

Notice that when $y(t) = x(t)$ or $y[n] = x[n]$ this property reduces to the property of autocorrelation functions that they are even functions of shift.

### EXAMPLES OF CROSS CORRELATION

As an illustration of cross correlation, suppose that the two CT power signals $x(t)$ and $y(t)$ are the ones illustrated in Figure 8.32. (These illustrations show the signals for a finite time. They are assumed to be power signals which are similar for other time ranges.) Their cross-correlation function is illustrated in Figure 8.33. It might not have been obvious at first glance that the two waveforms were highly correlated, but one look at the largest peak in the cross-correlation function indicates that they are. The peak occurs at a shift which is exactly the amount of shift between $x(t)$ and $y(t)$ at which they line up. That is, if the $y(t)$ signal is shifted to the left by that amount, all the peaks of $x(t)$ and $y(t)$ coincide in time and there is maximum similarity or correlation

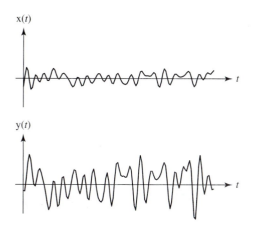

**Figure 8.32**
Two signals from which to find a cross correlation.

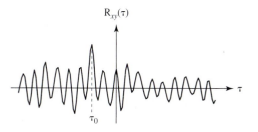

**Figure 8.33**
Cross correlation of two signals.

between the two waveforms. Figure 8.34 is a plot of the two signals with y($t$) shifted to emphasize the point.

In signal analysis it is sometimes very important whether two signals are correlated or not. When two signals are added, the signal power in the sum depends strongly on whether the two signals are correlated. Consider three DT signals x[$n$], y[$n$], and z[$n$] (Figure 8.35). All are sinusoidal with the same amplitude and frequency. Now consider the signals x[$n$] + y[$n$] and x[$n$] + z[$n$] (Figure 8.36). The signal x[$n$] + y[$n$] definitely has a greater amplitude than the signal x[$n$] + z[$n$] and, therefore, has a greater average signal power. This occurs basically because x[$n$] and y[$n$] are positively correlated (they are equal) and x[$n$] and z[$n$] are uncorrelated (they are 90° out of phase).

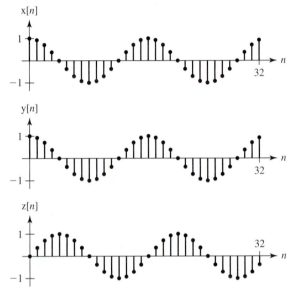

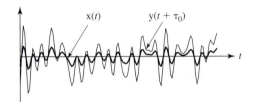

**Figure 8.34**
Original functions with y($t$) shifted to show correlation.

**Figure 8.35**
Three DT sinusoidal signals.

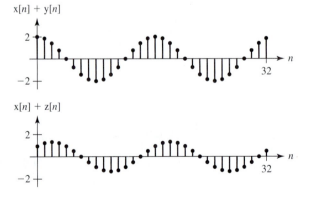

**Figure 8.36**
Sums of DT sinusoidal signals.

Next consider three CT random signals $x(t)$, $y(t)$, and $z(t)$ (Figure 8.37) plotted on the same scale, all of which have an average value of zero and exactly the same signal power. Plots of $x(t) + y(t)$ and $x(t) + z(t)$ on the same scale are shown in Figure 8.38. It should be apparent that there is a qualitative difference between $x(t) + y(t)$ and $x(t) + z(t)$. That is, $x(t) + z(t)$ generally deviates farther from zero than $x(t) + y(t)$ does.

The power in a signal is proportional to its square. When $x(t) + y(t)$ and $x(t) + z(t)$ are squared, the difference becomes more apparent. Again they are plotted on exactly the same scale (Figure 8.39).

The average power of a signal is proportional to the mean of its square. The mean of the square of $x(t) + z(t)$ is bigger than the mean of the square of $x(t) + y(t)$ by a factor of two. A plot of the cross-correlation functions between the signals, on exactly the same scale, reveals why (Figure 8.40).

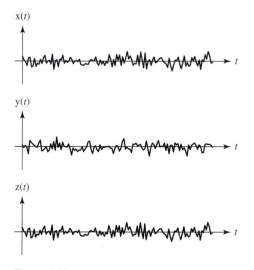

**Figure 8.37**
Three random signals.

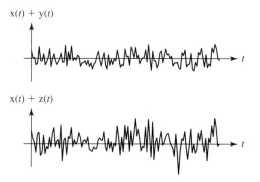

**Figure 8.38**
Sums of random signals.

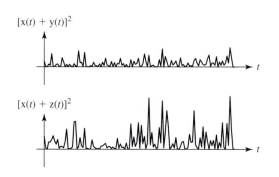

**Figure 8.39**
Squares of sums of random signals.

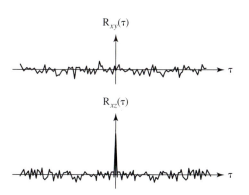

**Figure 8.40**
Cross correlations between signals.

It is now apparent that $x(t)$ and $z(t)$ are highly correlated at zero shift. (They are in fact identical.) But $x(t)$ and $y(t)$ are not well correlated at all, at any shift. When $x(t)$ and $z(t)$ are added, the result is simply $2x(t)$. Everywhere $x(t)$ is positive so is $z(t)$, and everywhere $x(t)$ is negative so is $z(t)$. The squaring operation makes both positive and negative sums positive.

What would happen to the average power of $x(t) + z(t)$ if $x(t)$ were equal to the negative of $z(t)$? Obviously then $x(t) + z(t)$ would be zero everywhere and the average power would also be zero. The cross correlation between $x(t)$ and $z(t)$ would then have the form illustrated in Figure 8.41.

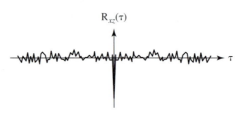

**Figure 8.41**
Cross correlation between two signals with
perfect negative correlation at zero shift.

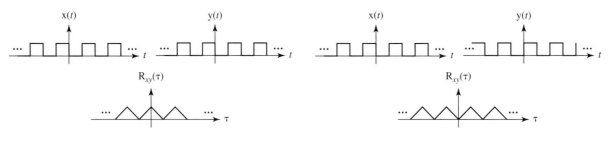

**Figure 8.42**
Cross correlations between periodic, nonsinusoidal CT signals.

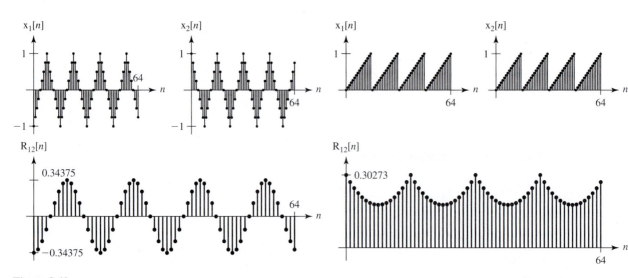

**Figure 8.43**
Cross correlations between periodic, nonsinusoidal DT signals.

What if $z(t)$ were time-shifted a little before being added to $x(t)$? Notice that the correlation between $x(t)$ and $z(t)$ is very high at zero shift but quickly goes to a very low value for even a small shift. When $z(t)$ is shifted even a small amount, the power in $x(t) + z(t)$ immediately goes to the same as in $x(t) + y(t)$, because the correlation goes to approximately zero.

In Figures 8.42 and 8.43 are some examples of pairs of signals and their cross correlations.

## 8.6 CORRELATION AND THE FOURIER SERIES

Recall the formulas for the trigonometric CTFS harmonic function of a periodic signal over exactly one fundamental period,

$$X_c[k] = \frac{2}{T_0} \int_{T_0} x(t) \cos(2\pi(kf_0)t)\,dt \qquad k = 1, 2, 3, \ldots, \qquad \textbf{(8.100)}$$

and

$$X_s[k] = \frac{2}{T_0} \int_{T_0} x(t) \sin(2\pi(kf_0)t)\,dt \qquad k = 1, 2, 3, \ldots. \qquad \textbf{(8.101)}$$

Each value of $X_c[k]$ or $X_s[k]$ is simply twice the cross correlation, at zero shift, between the function $x(t)$ and sines and cosines of different fundamental periods. That is,

$$X_c[k] = 2R_{xc}(0) \qquad X_s[k] = 2R_{xs}(0) \qquad \textbf{(8.102)}$$

where

$$c(t) = \cos(2\pi(kf_0)t) \qquad \text{and} \qquad s(t) = \sin(2\pi(kf_0)t). \qquad \textbf{(8.103)}$$

Similarly,

$$X[k] = \frac{1}{T_0} \int_{T_0} x(t)e^{-j2\pi(kf_0)t}\,dt = R_{xz}(0), \qquad \textbf{(8.104)}$$

where

$$z(t) = e^{+j2\pi(kf_0)t}. \qquad \textbf{(8.105)}$$

(Notice that in the equation for $X[k]$, the general form of cross correlation for complex functions must be used. That is what makes the sign in the exponent of $e$ in the last equation positive instead of negative.)

Now the representation of a signal by a Fourier series, which is a process of decomposing a signal into a linear combination of sinusoidal functions, can be seen as a process of correlating the signal with the sinusoids to find out whether any particular sinusoid or complex exponential is present in the signal and, if so, how much of it is there.

## 8.7 ENERGY SPECTRAL DENSITY (ESD)

In the remaining sections of this chapter there will be a discussion of energy spectral density (ESD) and then power spectral density (PSD) and their relation to autocorrelation. During the development of these concepts it is natural to wonder why autocorrelation, ESD, and PSD are necessary and useful since what happens to a signal seems to be completely and directly determined by the use of linear system concepts and the Fourier transform without appealing to the concepts of autocorrelation, ESD, and PSD.

But this is only true if one has an exact description of the signal. As mentioned earlier, although most real signals in real systems do not have an exact description, the autocorrelation and power spectral density can still be found (or at least estimated). This type of signal is called a *random* signal. The best way to analyze random signals as they progress through systems is through their autocorrelation, ESD, and/or PSD. Since random variables are not covered in this text, we will apply the ideas of autocorrelation, ESD, and PSD to deterministic signals in this chapter to illustrate the principles involved.

## DEFINITION AND DERIVATION OF ENERGY SPECTRAL DENSITY

Parseval's theorem relates the total signal energy in a signal $x(t)$ or $x[n]$ to its Fourier transform $X(f)$ or $X(F)$ through

$$E_x = \int_{-\infty}^{\infty} |x(t)|^2 \, dt = \int_{-\infty}^{\infty} |X(f)|^2 \, df \quad \text{or} \quad E_x = \sum_{n=-\infty}^{\infty} |x[n]|^2 = \int_{1} |X(F)|^2 \, dF$$

(8.106)

The quantity $|X(f)|^2$ or $|X(F)|^2$ is called the *energy spectral density* and is given the symbol $\Psi$. That is,

$$\Psi_x(f) = |X(f)|^2 \quad \text{or} \quad \Psi_x(F) = |X(F)|^2.$$

(8.107)

It is called energy spectral density because it describes mathematically the variation of the signal energy with frequency. If $x(t)$ or $x[n]$ is a real function, $\Psi_x(f)$ or $\Psi_x(F)$ is even, nonnegative, and real. Therefore, the signal energy can be written as

$$E_x = 2 \int_{0}^{\infty} \Psi_x(f) \, df \quad \text{or} \quad E_x = 2 \int_{0}^{1/2} \Psi_x(F) \, dF.$$

(8.108)

## EFFECTS OF SYSTEMS ON ESD

The usefulness of the concept of energy spectral density can be seen if we analyze the effect of bandpass-filtering an excitation CT signal $x(t)$ to create a response signal $y(t)$. If the filter is ideal, with unity gain and linear phase in its passband, that part of the signal within the passband will be unaffected (except possibly for a time shift) and that part of the signal outside the passband will be eliminated. Signal energy of a CT signal is found by integrating ESD over all frequencies. If the signal has no ESD over some range of frequencies, the range of the integral need only cover the range over which the signal is nonzero. Then the total signal energy of $y(t)$ can be found by integrating its ESD,

$$E_y = 2 \int_{0}^{\infty} \Psi_y(f) \, df = 2 \int_{0}^{\infty} |Y(f)|^2 \, df = 2 \int_{0}^{\infty} |H(f)X(f)|^2 \, df$$

(8.109)

$$E_y = 2 \int_{0}^{\infty} |H(f)|^2 \Psi_x(f) \, df = 2 \int_{f_L}^{f_H} \Psi_x(f) \, df.$$

(8.110)

This integral can also be thought of as that part of the signal energy of $x(t)$ which lies within the filter's passband. In general, the ESD of the response of a linear CT system is related to the ESD of the excitation by

$$\Psi_y(f) = |H(f)|^2 \, \Psi_x(f) = H(f)H^*(f)\Psi_x(f) \qquad \textbf{(8.111)}$$

and the ESD of the response of a linear DT system is related to the ESD of the excitation by

$$\Psi_y(F) = |H(F)|^2 \, \Psi_x(F) = H(F)H^*(F)\Psi_x(F). \qquad \textbf{(8.112)}$$

The units of ESD depend on the units of the signal to which it applies and whether the signal is continuous-time or discrete-time. For example, if the signal unit is the volt (V) and it is a CT signal, its Fourier transform has units of V/Hz and its ESD has units of $(V/Hz)^2$ or $(V \text{-} s)^2$. These units can be rearranged into a more meaningful form as $V^2 \cdot s/Hz$, which expresses the ESD as a signal energy in $V^2 \cdot s$ per unit frequency in Hz. For DT signals, the unit of ESD is simply the square of the signal unit, whatever that may be.

## THE ESD CONCEPT

The ESD of a signal is a description of the distribution of the signal energy of the signal versus frequency. In the signal-processing discipline there are two ways of conceiving ESD, double-sided and single-sided. Mathematically, double-sided ESDs are more convenient in complicated system analysis, but, because of the conceptual difficulty of imagining a negative frequency, ESDs are often discussed as though all the signal energy resides in positive frequency space. Since the double-sided ESD (the one previously derived) is an even function, the relation between double- and single-sided ESDs is simple. The single-sided ESD of a signal is twice the double-sided ESD of the same signal for positive frequencies and zero for negative frequencies. Defined this way the total energy in a signal is the integral over all frequency space of either ESD.

The name, energy spectral density, comes from the fact that ESD is a mathematical functional description of how the signal energy of a signal is distributed in frequency. Figure 8.44 is a conceptual block diagram of how the ESD of a CT signal could be measured using an array of filters, squarers, integrators, and dividers to estimate the ESD versus frequency. It is rarely, if ever, actually measured this way, but the diagram does aid understanding of what ESD really is.

## RELATION OF ESD TO AUTOCORRELATION

For energy signals, the time-domain counterpart to ESD is autocorrelation. The autocorrelation of an energy signal $x(t)$ or $x[n]$ is the inverse Fourier transform of its ESD,

$$R_x(t) \overset{\mathcal{F}}{\longleftrightarrow} \Psi_x(f) \qquad \text{or} \qquad R_x[n] \overset{\mathcal{F}}{\longleftrightarrow} \Psi_x(F). \qquad \textbf{(8.113)}$$

This can be proven by the following logic. From the definition of ESD,

$$\Psi_x(f) = |X(f)|^2 \qquad \text{or} \qquad \Psi_x(F) = |X(F)|^2, \qquad \textbf{(8.114)}$$

we can write

$$R_x(t) \overset{\mathcal{F}}{\longleftrightarrow} X^*(f)X(f) \qquad \text{or} \qquad R_x[n] \overset{\mathcal{F}}{\longleftrightarrow} X^*(F)X(F). \qquad \textbf{(8.115)}$$

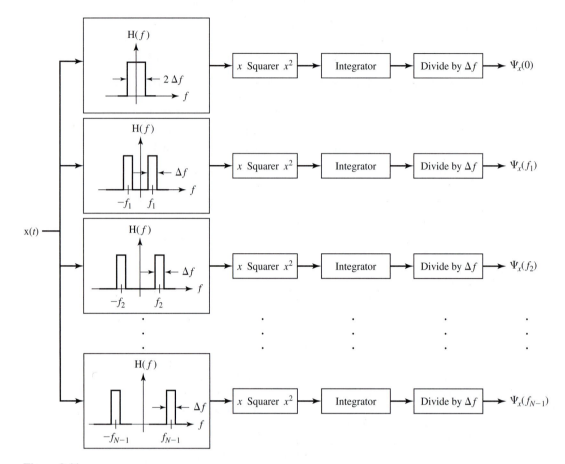

**Figure 8.44**
Conceptual block diagram illustrating the concept of energy spectral density for a CT signal.

Translating multiplication in the frequency domain into convolution in the time domain, and using the Fourier transform property relating to complex conjugates,

$$R_x(t) = x(-t) * x(t) = \int_{-\infty}^{\infty} x(-\tau)x(t - \tau)\, d\tau \qquad (8.116)$$

or

$$R_x[n] = x[-n] * x[n] = \sum_{m=-\infty}^{\infty} x[-m]x[n - m], \qquad (8.117)$$

and these can be simplified to

$$R_x(t) = \int_{-\infty}^{\infty} x(\tau)x(\tau + t)\, d\tau \qquad (8.118)$$

or

$$R_x[n] = \sum_{m=-\infty}^{\infty} x[m]x[m + n], \qquad (8.119)$$

which are exactly the definitions of autocorrelation for CT and DT energy signals. (The two symbols $t$ and $\tau$, or $n$ and $m$, have exchanged places, but that does not invalidate the result.)

## 8.8 POWER SPECTRAL DENSITY (PSD)

### DEFINITION AND DERIVATION OF POWER SPECTRAL DENSITY

PSD has the same relation to power signals as ESD has to energy signals. Many signals in systems are thought of and analyzed as if they were power signals even though they are not, since no real signal can endure an infinite time. But they are often steady-state signals which have been active a long time and are expected to continue for a long time.

Since the total signal energy of a power signal cannot be found, let us first find the ESD of *a truncated version* of a CT signal $x_T(t)$.

$$x_T(t) = \begin{cases} x(t) & |t| < \dfrac{T}{2} \\ 0 & \text{otherwise} \end{cases} = \text{rect}\left(\frac{t}{T}\right)x(t), \tag{8.120}$$

and

$$\Psi_{x_T}(f) = |X_T(f)|^2, \tag{8.121}$$

where

$$X_T(f) = \int_{-\infty}^{\infty} x_T(t)e^{-j2\pi ft}\,dt = \int_{-(T/2)}^{T/2} x(t)e^{-j2\pi ft}\,dt. \tag{8.122}$$

The average signal power of the signal $x_T(t)$ in this time interval is the signal energy of the signal in this time interval divided by the length of the time interval. Therefore, it is analogous and logical to define the PSD of the truncated signal as its ESD divided by the time,

$$G_{x_T}(f) = \frac{\Psi_{x_T}(f)}{T} = \frac{1}{T}|X_T(f)|^2. \tag{8.123}$$

As the time interval $T$ becomes larger, the PSD of this truncated signal approaches that of the original signal. Therefore,

$$G_x(f) = \lim_{T\to\infty} G_{x_T}(f) = \lim_{T\to\infty} \frac{1}{T}|X_T(f)|^2. \tag{8.124}$$

In a manner analogous to that derived for ESD, the power of a finite-signal-power signal in a bandwidth $f_L$ to $f_H$ is given by

$$\text{Power} = 2\int_{f_L}^{f_H} G(f)\,df. \tag{8.125}$$

The equivalent result for the PSD of a DT signal is

$$G_x(F) = \lim_{N\to\infty} G_{x_N}(F) = \lim_{N\to\infty} \frac{1}{N}|X_N(F)|^2. \tag{8.126}$$

### EFFECTS OF SYSTEMS ON PSD

The relationship between the PSD of a excitation and the PSD of the response of a linear system is similar to the relation between the ESD of an excitation and the ESD of a response of a linear system. The PSD of the response of a linear system is related to the PSD of the excitation by

$$G_y(f) = |H(f)|^2 G_x(f) = H(f)H^*(f)G_x(f) \tag{8.127}$$

or

$$G_y(F) = |H(F)|^2 G_x(F) = H(F)H^*(F)G_x(F). \qquad \textbf{(8.128)}$$

This is a very important result and is the starting point for most analyses of how noise propagates through an LTI system.

The units of PSD again depend on the units of the underlying signal to which it applies and whether it is CT or DT. If the signal unit of a CT signal is the amp (A), the units of PSD are $A^2$/Hz. If the signal unit is the volt, the units of PSD are $V^2$/Hz. Since signal power is the integral of PSD over a range of frequency the Hz is integrated out. Therefore, the signal power of a current signal has units of $A^2$ and the signal power of a voltage signal has units of $V^2$. For DT signals, the unit is simply the square of the signal unit. For convenience, in many analyses in which the signal units are consistent throughout a system, analysis is done without using units. But in any analysis in which the final result must be related back to a physical quantity, the units must ultimately be considered and shown to be consistent.

## THE PSD CONCEPT

One way to visualize the concept of PSD is to imagine that a CT signal is processed by a system as illustrated in Figure 8.45. The signal power of the CT signal x(t) is first split into small frequency ranges by ideal bandpass filters, each with bandwidth

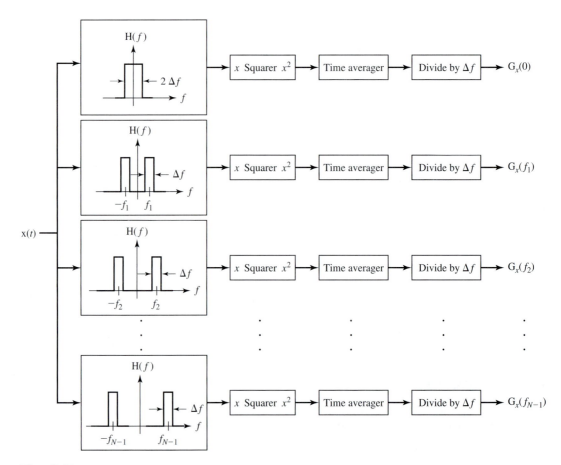

**Figue 8.45**
Block diagram illustrating the concept of power spectral density.

$\Delta f$. Each signal so formed is then squared (to form *instantaneous* signal power) and time-averaged (to form *average* signal power), then divided by $\Delta f$ to form time-average signal power per unit frequency. Then the outputs $G_x(f_k)$ are estimates of the PSD at discrete frequencies. If $N$ goes to infinity, the outputs $G_x(f_k)$ cover all frequency space. The exact single-sided PSD of $x(t)$ is simply the limit of this process as $\Delta f$ approaches zero and $N$ approaches infinity so that the coverage is uniform and continuous over all frequency space.

As an example of what the signals in the system of Figure 8.45 might look like, let the input signal be $x(t)$ and let $N = 4$. Some of the signals are illustrated in Figure 8.46.

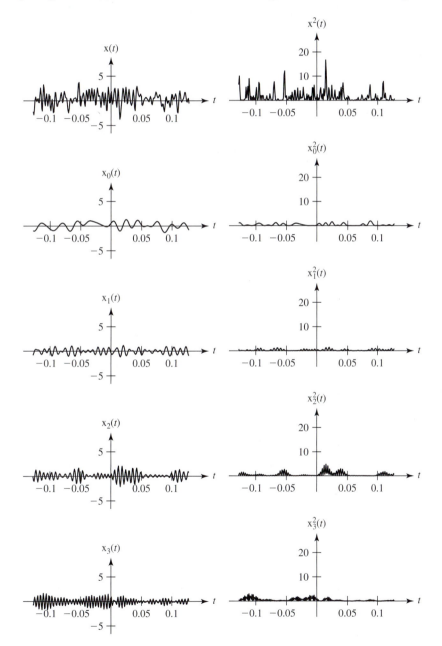

**Figure 8.46**
Typical signals in the conceptual system of Figure 8.45 with $N = 4$.

## RELATION OF PSD TO AUTOCORRELATION

For power signals, the time-domain counterpart to PSD is autocorrelation. The autocorrelation of a power signal is the inverse Fourier transform of the PSD. That is,

$$R(t) \xleftrightarrow{\mathcal{F}} G(f) \quad \text{or} \quad R[n] \xleftrightarrow{\mathcal{F}} G(F).\qquad(8.129)$$

The proof is similar to the one presented earlier for energy signals.

---

**EXAMPLE 8.5**

Find the PSD of the power signal in Figure 8.47.

### ■ Solution

We have already found the autocorrelation function (Figure 8.48) for this signal in Example 8.3. This function can be compactly described by

$$R_x(t) = \frac{A^2}{2} \operatorname{tri}\left(\frac{2t}{T_0}\right) * \frac{1}{T_0} \operatorname{comb}\left(\frac{t}{T_0}\right).\qquad(8.130)$$

Now the PSD can be found by Fourier-transforming the autocorrelation,

$$G_x(f) = \frac{A^2 T_0}{4} \operatorname{sinc}^2\left(\frac{T_0 f}{2}\right) \operatorname{comb}(T_0 f)\qquad(8.131)$$

or

$$G_x(f) = \frac{A^2}{4} \sum_{n=-\infty}^{\infty} \operatorname{sinc}^2\left(\frac{n}{2}\right)\delta\left(f - \frac{n}{T_0}\right) = \frac{A^2}{4} \sum_{n=-\infty}^{\infty} \operatorname{sinc}^2\left(\frac{n}{2}\right)\delta(f - nf_0)\qquad(8.132)$$

(Figure 8.49). The PSD indicates that the signal has significant power at frequencies of zero and the fundamental frequency of the signal $f_0 = 1/T_0$.

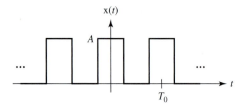

**Figure 8.47**
A square-wave signal.

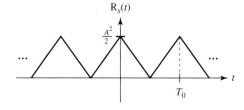

**Figure 8.48**
Autocorrelation of the square-wave signal.

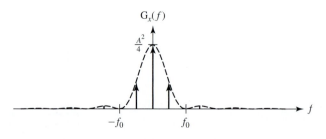

**Figure 8.49**
PSD of the square-wave signal.

Suppose we decide to take the other approach to finding the PSD, the direct approach using the definition

$$G_x(f) = \lim_{T \to \infty} \frac{1}{T} |X_T(f)|^2. \tag{8.133}$$

The time signal $x(t)$ is

$$x(t) = A \, \text{rect}\left(\frac{2t}{T_0}\right) * \frac{1}{T_0} \, \text{comb}\left(\frac{t}{T_0}\right). \tag{8.134}$$

The truncated time signal is

$$x_T(t) = A \left\{ \text{rect}\left(\frac{2t}{T_0}\right) * \frac{1}{T_0} \, \text{comb}\left(\frac{t}{T_0}\right) \right\} \text{rect}\left(\frac{t}{T}\right). \tag{8.135}$$

Its Fourier transform $X_T(f)$ is

$$X_T(f) = A \left[ \frac{T_0}{2} \, \text{sinc}\left(\frac{T_0 f}{2}\right) \text{comb}(T_0 f) \right] * T \, \text{sinc}(Tf). \tag{8.136}$$

Then

$$G_x(f) = \lim_{T \to \infty} \frac{1}{T} \left| A \left[ \frac{T_0}{2} \, \text{sinc}\left(\frac{T_0 f}{2}\right) \text{comb}(T_0 f) \right] * T \, \text{sinc}(Tf) \right|^2 \tag{8.137}$$

$$G_x(f) = \lim_{T \to \infty} \frac{1}{T} \left| A \left[ \sum_{n=-\infty}^{\infty} \frac{1}{2} \, \text{sinc}\left(\frac{n}{2}\right) \delta(f - nf_0) \right] * T \, \text{sinc}(Tf) \right|^2 \tag{8.138}$$

$$G_x(f) = \lim_{T \to \infty} \frac{1}{T} \left| \frac{A}{2} \sum_{n=-\infty}^{\infty} \text{sinc}\left(\frac{n}{2}\right) T \, \text{sinc}[T(f - nf_0)] \right|^2. \tag{8.139}$$

As $T$ gets larger, the sinc functions in the summation become thinner and overlap less and in the limit do not overlap at all. In that limit, the square of the summation equals the summation of the squares of the individual functions since they do not overlap. Therefore,

$$G_x(f) = \lim_{T \to \infty} \frac{1}{T} \sum_{n=-\infty}^{\infty} \frac{A^2}{4} \, \text{sinc}^2\left(\frac{n}{2}\right) T^2 \, \text{sinc}^2[T(f - nf_0)]. \tag{8.140}$$

Taking the limit inside,

$$G_x(f) = \sum_{n=-\infty}^{\infty} \frac{A^2}{4} \, \text{sinc}^2\left(\frac{n}{2}\right) \lim_{T \to \infty} \{T \, \text{sinc}^2[T(f - nf_0)]\}. \tag{8.141}$$

To properly interpret the limiting process on the $\text{sinc}^2$ function consider the following:

$$\mathcal{F}^1[a \, \text{sinc}^2(af)] = \text{tri}\left(\frac{t}{a}\right). \tag{8.142}$$

Then, using the fact that the area under a frequency-domain function is its inverse transform evaluated at $t = 0$, the area under the $\text{sinc}^2$ function is one. If $a$ is allowed to approach infinity, the area under the $\text{sinc}^2$ function stays constant at one because the triangle function is just getting wider and its value at $t = 0$ stays the same. At the same time the width of the $\text{sinc}^2$ function is decreasing toward zero. A function whose area is constant while its width approaches zero is an impulse (in the limit). Therefore,

$$G_x(f) = \frac{A^2}{4} \sum_{n=-\infty}^{\infty} \text{sinc}^2\left(\frac{n}{2}\right) \delta(f - nf_0), \tag{8.143}$$

which agrees with the previous result after considerably more mathematical and conceptual effort. ∎

## EXAMPLE 8.6

Find the PSD of the DT signal

$$x[n] = \text{comb}_{N_0}[n]. \tag{8.144}$$

### ■ Solution

First find the autocorrelation function of this periodic signal using

$$\text{comb}_{N_0}[n] \xleftrightarrow{\ \mathcal{FS}\ } \frac{1}{N_0}, \tag{8.145}$$

$$R_X[m] \xleftrightarrow{\ \mathcal{FS}\ } X^*[k]X[k] = \frac{1}{N_0^2}. \tag{8.146}$$

Then, again using (8.145),

$$R_x[m] = \frac{1}{N_0} \text{comb}_{N_0}[m]. \tag{8.147}$$

The PSD is the DTFT of the autocorrelation function which is

$$G_x(F) = \frac{1}{N_0} \text{comb}(N_0 F). \tag{8.148}$$

We can check the reasonableness of this result by finding the average signal power from the PSD. It is

$$P_x = \int_1 G_x(F)\,dF = \frac{1}{N_0} \int_1 \text{comb}(N_0 F)\,dF \tag{8.149}$$

$$P_x = \frac{1}{N_0} \int_1 \sum_{n=-\infty}^{\infty} \delta(N_0 F - n)\,dF = \frac{1}{N_0^2} \int_1 \sum_{n=-\infty}^{\infty} \delta\left(F - \frac{n}{N_0}\right) dF. \tag{8.150}$$

We can choose any DT interval of length one for the integration. No matter which one we choose there are exactly $N_0$ impulses in it, so the integral evaluates to $N_0$. Then the average power is

$$P_x = \frac{1}{N_0}. \tag{8.151}$$

This result implies that as the fundamental period increases, the average power decreases. Since each impulse in the comb function has the same energy, this is reasonable. When the impulses occur less frequently, the average power decreases. Also this agrees with the autocorrelation function evaluated at $m = 0$. Therefore, the answer seems reasonable. ∎

## 8.9 SUMMARY OF IMPORTANT POINTS

1. Relationships between signals are often as important as the signals themselves.
2. The correlogram is a good way to illustrate whether and how much two signals are correlated.
3. Correlation and independence are not exactly opposite concepts although for many signals they seem to be.
4. The correlation function indicates how correlated two signals are as a function of how much one of them is shifted in time.
5. There are two definitions of the correlation function, one for energy signals and one for power signals.
6. Correlation and convolution are closely related mathematical processes.
7. The correlation of a signal with a shifted version of itself is called autocorrelation.
8. Autocorrelation is closely related to signal energy or power and contains important information about how rapidly a signal varies in time.
9. The Fourier series harmonic function can be viewed as the correlation of a signal with a succession of sinusoids, complex or real.
10. Energy spectral density and power spectral density are the frequency-domain counterparts of autocorrelation, related through the Fourier transform.
11. Energy spectral density and power spectral density indicate how the energy or power of a signal varies with frequency.

## EXERCISES WITH ANSWERS

1. Plot correlograms of the following pairs of CT and DT signals.

   a. $x_1(t) = \cos(2\pi t)$, $x_2(t) = 2\cos(4\pi t)$

   b. $x_1[n] = \sin\left(\dfrac{2\pi n}{16}\right)$, $x_2[n] = 2\cos\left(\dfrac{2\pi n}{8}\right)$

   c. $x_1(t) = e^{-t}u(t)$, $x_2(t) = e^{-2t}u(t)$

   d. $x_1[n] = e^{-(n/10)}\cos\left(\dfrac{2\pi n}{10}\right)u[n]$, $x_2[n] = e^{-(n/10)}\sin\left(\dfrac{2\pi n}{10}\right)u[n]$

**Answers:**

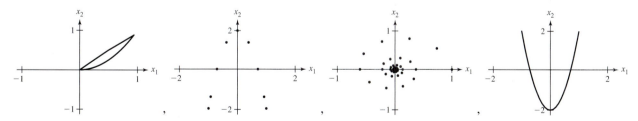

2. Plot correlograms of the following pairs of CT and DT signals.

   a. $x_1(t) = \cos(2\pi t)$, $x_2(t) = \cos^2(2\pi t)$

   b. $x_1[n] = n$, $x_2[n] = n^3$, $\quad -10 < n < 10$

   c. $x_1(t) = t$, $x_2(t) = 2 - t^2$, $\quad -4 < t < 4$

**Answers:**

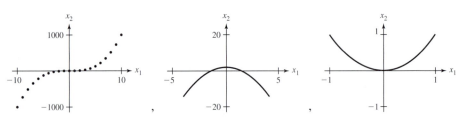

3. In MATLAB generate two vectors x1 and x2 representing DT signals using the following code fragment,

```
x1 = randn(100,1) ; x2 = randn(100,1) ; x3 = randn(100,1) ;
```

Then plot correlograms of the following pairs of DT signals.

*a.* x1 and x2

*b.* x1 and x1+x2

*c.* x1+x2 and x1+x3

*d.* x1+x2/10 and -x1+x3/10

**Answers:**

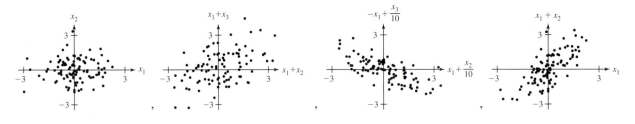

4. Plot the correlation function for each of the following pairs of energy signals.

*a.* $x_1(t) = 4\,\text{rect}(t)$, $x_2(t) = -3\,\text{rect}(2t)$

*b.* $x_1[n] = 2\,\text{rect}_3[n]$, $x_2[n] = 5\,\text{rect}_8[n]$

*c.* $x_1(t) = 4e^{-t}\,u(t)$, $x_2(t) = 4e^{-t}\,u(t)$

*d.* $x_1[n] = 2e^{-(n/16)} \sin\left(\dfrac{2\pi n}{8}\right) u[n]$,

$$x_2[n] = -3e^{-(n/16)} \sin\left(\dfrac{2\pi n}{8} - \dfrac{\pi}{4}\right) u[n]$$

**Answers:**

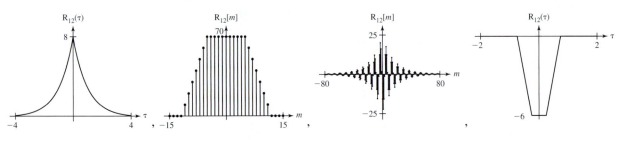

**5.** Plot the correlation function for each of the following pairs of power signals.

    *a.*   $x_1(t) = 6\sin(12\pi t)$, $x_2(t) = 5\cos(12\pi t)$

    *b.*   $x_1[n] = 6\sin\left(\dfrac{2\pi n}{12}\right)$, $x_2[n] = 5\sin\left(\dfrac{2\pi n}{12}\right)$

    *c.*   $x_1(t) = 6\sin(12\pi t)$, $x_2(t) = 5\sin\left(12\pi t - \dfrac{\pi}{4}\right)$

**Answers:**

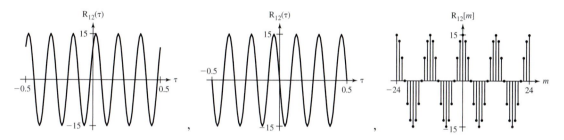

**6.** Find the autocorrelations of the following CT and DT energy and power signals and show that, at zero shift, the value of the autocorrelation is the signal energy or power and that all the properties of autocorrelation functions are satisfied.

    *a.*   $x(t) = e^{-3t}\,u(t)$

    *b.*   $x[n] = \text{rect}_5[n-5]$

    *c.*   $x(t) = \text{rect}\left(2\left(t - \dfrac{1}{4}\right)\right) - \text{rect}\left(2\left(t - \dfrac{3}{4}\right)\right)$

**Answers:**

$$\frac{1}{6}e^{-3|t|}, \quad 11\,\text{tri}\left(\frac{m}{11}\right), \quad \text{tri}(2t) - \frac{1}{2}\left[\text{tri}\left(2\left(t - \frac{1}{2}\right)\right) + \text{tri}\left(2\left(t + \frac{1}{2}\right)\right)\right]$$

**7.** Find the autocorrelation functions of the following power signals.

    *a.*   $x(t) = 5\sin(24\pi t) - 2\cos(18\pi t)$

    *b.*   $x[n] = -4\sin\left(\dfrac{2\pi n}{36}\right) - 2\cos\left(\dfrac{2\pi n}{40}\right)$

**Answers:**

$$\frac{25}{2}\cos(24\pi t) + 2\cos(18\pi t), \quad 8\cos\left(\frac{2\pi n}{36}\right) + 2\cos\left(\frac{2\pi n}{40}\right)$$

**8.** A signal is sent from a transmitter to a receiver and is corrupted by noise along the way. The signal shape is of the functional form

$$x(t) = A\sin(2\pi f_0 t)\,\text{rect}\left(\frac{f_0}{4}\left(t - \frac{1}{2f_0}\right)\right).$$

What is the transfer function of a matched filter for this signal?

**Answer:**

$$H(f) = j\frac{2K}{f_0}e^{-j2\pi f(t_0-(1/2f_0))}\left[\text{sinc}\left(\frac{4(f+f_0)}{f_0}\right) - \text{sinc}\left(\frac{4(f-f_0)}{f_0}\right)\right]$$

9.  Find the signal power of the following sums or differences of signals and compare it to the power in the individual signals. How does the comparison relate to the correlation between the two signals that are summed or differenced?

    *a.*  $x(t) = \sin(2\pi t) + \cos(2\pi t)$

    *b.*  $x(t) = \sin(2\pi t) + \cos\left(2\pi t - \frac{\pi}{4}\right)$

    *c.*  $x[n] = \text{rect}_2[n] * \text{comb}_{10}[n] - \text{tri}\left(\frac{n}{2}\right) * \text{comb}_{10}[n]$

    *d.*  $x[n] = \text{rect}_2[n] * \text{comb}_{10}[n] + \text{tri}\left(\frac{n-5}{2}\right) * \text{comb}_{10}[n]$

**Answers:**

$0.65 = 0.5 + 0.15, \quad 1 = 0.5 + 0.5, \quad 0.25 < 0.5 + 0.15, \quad 1.707 > 0.5 + 0.5$

10.  Find the cross-correlation functions of the following pairs of periodic signals.

    *a.*  $x_1(t) = \text{rect}\left(\frac{t}{6}\right) * \text{comb}\left(\frac{t}{24}\right),$

        $x_2(t) = \text{rect}\left(\frac{t-3}{6}\right) * \text{comb}\left(\frac{t}{24}\right)$

    *b.*  $x_1[n] = \sin^2\left(\frac{2\pi n}{8}\right), x_2[n] = \sin^2\left(\frac{2\pi n}{10}\right)$

    *c.*  $x_1(t) = e^{-j10\pi t}, x_2(t) = \cos(10\pi t)$

**Answers:**

$$\frac{1}{2}e^{-j10\pi\tau}, \quad \frac{1}{4}, \quad 6\,\text{tri}\left(\frac{\tau-3}{6}\right) * \text{comb}\left(\frac{\tau}{24}\right)$$

11.  Find the ESDs of the following energy signals.

    *a.*  $x[n] = A\delta[n - n_0]$

    *b.*  $x(t) = e^{-100t}\,u(t)$

    *c.*  $x[n] = 10\left(\frac{7}{8}\right)^n \sin\left(\frac{2\pi n}{12}\right)u[n]$

    *d.*  $x(t) = A\,\text{tri}\left(\frac{t-t_0}{w}\right)$

**Answers:**

$$A^2, \quad (Aw)^2 \, \text{sinc}^4(wf), \quad \frac{1}{10^4 + \omega^2},$$

$$100 \frac{(0.458)^2}{[1 - 1.515 \cos{(\Omega)} + 0.7656 \cos{(2\Omega)}]^2 + [1.515 \sin{(\Omega)} - 0.7656 \sin{(2\Omega)}]^2}$$

**12.** Find the ESD of the response $y(t)$ or $y[n]$ of each system with impulse response $h(t)$ or $h[n]$ to the excitation $x(t)$ or $x[n]$.

  a.  $x[n] = \delta[n]$, $h[n] = \left(-\dfrac{9}{10}\right)^n u[n]$

  b.  $x(t) = e^{-100t} u(t)$, $h(t) = e^{-100t} u(t)$

  c.  $x[n] = \text{rect}_3[n]$, $h[n] = \text{rect}_2[n - 2]$

  d.  $x(t) = 4e^{-t} \cos(2\pi t) \, u(t)$,

  $h(t) = \text{rect}\left(t - \dfrac{1}{2}\right)$

**Answers:**

$$\frac{1 + \omega^2}{[(2\pi)^2 + 1 - \omega^2]^2 + 4\omega^2} \, \text{sinc}^2\left(\frac{\omega}{2\pi}\right), \quad \frac{\sin^2{(7\pi F)}}{\sin^2{(\pi F)}} \frac{\sin^2{(5\pi F)}}{\sin^2{(\pi F)}},$$

$$\left(\frac{1}{10^4 + \omega^2}\right)^2, \quad \frac{100}{181 + 180 \cos{(\Omega)}}$$

**13.** Find the PSDs of these signals.

  a.  $x(t) = A \cos(2\pi f_0 t + \theta)$

  b.  $x(t) = 3 \, \text{rect}(100t) * \text{comb}(25t)$

  c.  $x[n] = 8 \sin\left(\dfrac{2\pi n}{12}\right)$

  d.  $x[n] = 3 \, \text{rect}_4[n] * \text{comb}_{20}[n]$

**Answers:**

$$3.6 \times 10^{-5} \, \text{sinc}^2\left(\frac{f}{100}\right) \text{comb}\left(\frac{f}{25}\right), \quad \frac{9}{20} \frac{\sin^2{(9\pi F)}}{\sin^2{(\pi F)}} \text{comb}(20F),$$

$$16\left[\text{comb}\left(F - \frac{1}{12}\right) + \text{comb}\left(F + \frac{1}{12}\right)\right], \quad \frac{A^2}{4}[\delta(f - f_0) + \delta(f + f_0)]$$

**14.** Find the PSD of the response $y(t)$ or $y[n]$ of each system with impulse response $h(t)$ or $h[n]$ to the excitation $x(t)$ or $x[n]$.

  a.  $x(t) = 4 \cos\left(32\pi t - \dfrac{\pi}{4}\right)$, $h(t) = e^{-(t/10)} u(t)$

    b.　$x(t) = 4 \operatorname{comb}(2t)$, $h(t) = \operatorname{rect}(t - 1)$

    c.　$x[n] = 2 \operatorname{comb}_8[n]$, $h[n] = \left(\dfrac{11}{12}\right)^n u[n - 1]$

    d.　$x[n] = (-0.9)^n u[n]$, $h[n] = (0.5)^n u[n]$

**Answers:**

$$0, \quad 400\frac{\delta(f - 16) + \delta(f + 16)}{1 + (320\pi)^2},$$

$$8 \operatorname{comb}\left(\frac{f}{2}\right) \operatorname{sinc}^2(f), \quad \frac{0.42 \operatorname{comb}(8F)}{1 - 1.8333 \cos(2\pi F) + 0.8403}$$

## EXERCISES WITHOUT ANSWERS

15.　Plot correlograms of the following pairs of CT and DT signals.

    a.　$x_1(t) = \left[\operatorname{tri}\left(4\left(t - \dfrac{1}{4}\right)\right) - \operatorname{tri}\left(4\left(t - \dfrac{3}{4}\right)\right)\right] * \operatorname{comb}(t)$,

       $x_2(t) = \sin(2\pi t)$

    b.　$x_1[n] = \left[\operatorname{tri}\left(\dfrac{n - 8}{8}\right) - \operatorname{tri}\left(\dfrac{n - 24}{8}\right)\right] * \operatorname{comb}_{32}[n]$,

       $x_2[n] = \cos\left(\dfrac{2\pi n}{32}\right)$

    c.　$x_1(t) = \left[\operatorname{tri}\left(4\left(t - \dfrac{1}{4}\right)\right) - \operatorname{tri}\left(4\left(t - \dfrac{3}{4}\right)\right)\right] * \operatorname{comb}(t)$,

       $x_2(t) = \left[\operatorname{tri}(4t) - \operatorname{tri}\left(4\left(t - \dfrac{1}{2}\right)\right)\right] * \operatorname{comb}(t)$

    d.　$x_1[n] = \left[\operatorname{rect}\left(\dfrac{n - 8}{16}\right) - \operatorname{rect}\left(\dfrac{n - 24}{16}\right)\right] * \operatorname{comb}_{32}[n]$,

       $x_2[n] = \sin\left(\dfrac{2\pi n}{32}\right)$

16.　Plot a correlogram for the following sets of samples from two signals x and y. In each case, from the nature of the correlogram determine what relationship, if any, exists between the two sets of data.

    a.　$x = \{6, 5, 8, -2, 3, -10, 9, -2, -4, 3, -2, 6, 0, -5, -7, 1, 9, 9, 4, -6\}$;
       $y = \{-1, -10, -4, 4, 5, -2, -3, -5, -9, 2, 6, -5, -1, -10, -9, 0, 4,$
          $-10, 9, -1\}$

    b.　$x = \{4, 6, 0, 0, 5, -6, 8, -9, 0, 8, 7, 2, -5, -3, -4, -4, 8, 0, 4, 7\}$;
       $y = \{-11, -13, 3, -1, -8, 10, -16, 16, 1, -17, -14, -3, 9, 7, 12, 9,$
          $-17, 1, -8, -17\}$

c.  x = {0, 6, 11, 16, 19, 20, 19, 16, 11, 6, −0, −7, −12, −17, −20, −20,
    −20, −17, −12, −7};
    y = {19, 15, 10, 8, 3, −9, −12, −19, −19, −25, −19, −17, −12, −5, −1,
    5, 8, 12, 17, 20}

17. Plot the correlation function for each of the following pairs of energy signals.

   a.  $x_1(t) = \text{rect}(t)\sin(10\pi t)$,
       $x_2(t) = \text{rect}(t)\cos(10\pi t)$

   b.  $x_1[n] = \delta[n-1] - \delta[n+1]$,
       $x_2[n] = -\delta[n-1] + \delta[n+1]$

   c.  $x_1(t) = e^{-t^2}$,  $x_2(t) = e^{-2t^2}$

18. Plot the correlation function for each of the following pairs of power signals.

   a.  $x_1[n] = -3\sin\left(\dfrac{2\pi n}{20}\right)$,

       $x_2[n] = 8\sin\left(\dfrac{2\pi n}{10}\right)$

   b.  $x_1(t) = \text{rect}(4t) * \text{comb}(t)$,
       $x_2(t) = \text{rect}(4t) * \text{comb}(t)$

   c.  $x_1(t) = 4\,\text{rect}(t) * \dfrac{1}{2}\text{comb}\left(\dfrac{t}{2}\right) - 2$,

       $x_2(t) = 4\,\text{rect}(t-1) * \dfrac{1}{2}\text{comb}\left(\dfrac{t}{2}\right) - 2$

19. Find the autocorrelations of the following CT and DT energy and power signals and show that, at zero shift, the value of the autocorrelation is the signal energy or power and that all the properties of autocorrelation functions are satisfied.

   a.  $x[n] = \delta[n] + \delta[n-1] + \delta[n-2] + \delta[n-3]$
   b.  $x(t) = A\cos(2\pi f_0 t + \theta)$
   c.  $x[n] = \text{comb}_{12}[n]$

20. Find and sketch the autocorrelation function of

$$x(t) = 10\,\text{rect}(2t) * \dfrac{1}{4}\text{comb}\left(\dfrac{t}{4}\right).$$

   Check to be sure that its value at zero shift is the same as the average signal power of $x(t)$.

21. Find all cross-correlation and autocorrelation functions for these three signals:

$$x_1(t) = \cos(2\pi t) \qquad x_2(t) = \sin(2\pi t) \qquad x_3(t) = \cos(4\pi t)$$

   Check your autocorrelation answers by finding the average power of each signal.

22. Find and sketch the cross correlation between a unit-amplitude, 1-Hz cosine and a 50 percent duty-cycle square wave which has a peak-to-peak amplitude of two, a fundamental period of one, an average value of zero and is an even function.

23. Find and sketch the ESD of each of these signals:

    a.  $x(t) = A \operatorname{rect}\left(\dfrac{t}{w}\right)$

    b.  $x(t) = A \operatorname{rect}\left(\dfrac{t+1}{w}\right)$

    c.  $x(t) = A \operatorname{sinc}\left(\dfrac{t}{w}\right)$

    d.  $x(t) = \dfrac{1}{\sqrt{2\pi}} e^{-(t^2/2)}$

24. Find the PSDs of

    a.  $x(t) = A$

    b.  $x(t) = A \cos(2\pi f_0 t)$

    c.  $x(t) = A \sin(2\pi f_0 t)$

25. Which of the following functions could not be the autocorrelation function of a real signal and why?

    a.  $R(\tau) = \operatorname{tri}(\tau)$

    b.  $R(\tau) = A \sin(2\pi f_0 \tau)$

    c.  $R(\tau) = \operatorname{rect}(\tau)$

    d.  $R(\tau) = A \operatorname{sinc}(B\tau)$

CHAPTER 9

# The Laplace Transform

## 9.1 INTRODUCTION AND GOALS

The CTFT is a powerful tool for CT signal and system analysis, but it has its limitations. There are some useful signals which do not have a CTFT, even in the generalized sense which allows for impulses in the CTFT of a signal. The CTFT expresses signals as linear combinations of complex sinusoids. The Laplace transform expresses signals as linear combinations of complex exponentials, which are the eigenfunctions of the differential equations which describe continuous-time LTI systems. Complex sinusoids are a special case of complex exponentials. Therefore, the Laplace transform is more general than the CTFT. The Laplace transform can describe functions that the CTFT cannot describe. The impulse responses of LTI systems completely characterize them. Because the Laplace transform describes the impulse responses of LTI systems as linear combinations of complex exponentials, the eigenfunctions of LTI systems, it directly encapsulates the characteristics of a system in a powerful way. Many system analysis and design techniques are based on the Laplace transform.

### CHAPTER GOALS

1. To develop a new transform method, the Laplace transform, which is applicable to more signals and systems than the Fourier transform

2. To define the range of signals to which the Laplace transform applies

3. To show the relationship between the Laplace and Fourier transforms

4. To show the relationship between the Laplace transform of the impulse response of an LTI system and the eigenfunctions of that system

5. To derive and illustrate the properties of the Laplace transform, especially those which do not have a direct counterpart in the Fourier transform

6. To show how the Laplace transform can be used to solve differential equations with initial conditions

## 9.2 DEVELOPMENT OF THE LAPLACE TRANSFORM

### DERIVATION AND DEFINITION

When we extended the Fourier series to the Fourier transform, we let the fundamental period of a periodic signal increase to infinity making the discrete frequencies $kf_0$ in the CTFS merge into the continuum of frequencies $f$ in the CTFT. This led to the two alternate definitions of the Fourier transform,

$$X(j\omega) = \int_{-\infty}^{\infty} x(t)e^{-j\omega t}\, dt \qquad x(t) = \frac{1}{2\pi} \int_{-\infty}^{\infty} X(j\omega)e^{+j\omega t}\, d\omega \tag{9.1}$$

and

$$X(f) = \int_{-\infty}^{\infty} x(t)e^{-j2\pi ft}\, dt \qquad x(t) = \int_{-\infty}^{\infty} X(f)e^{+j2\pi ft}\, df. \tag{9.2}$$

There are two common approaches to introducing the Laplace transform. One approach is to conceive the Laplace transform as a generalization of the Fourier transform by expressing functions as linear combinations of complex exponentials instead of as linear combinations of the more restricted class of functions, complex sinusoids, used in the Fourier transform. The other approach is to exploit the unique nature of the complex exponential as the eigenfunction of the differential equations which describe linear systems and to realize that an LTI system excited by a complex exponential responds with another complex exponential. The relation between the excitation and response complex exponentials of an LTI system is the Laplace transform. Therefore, the Laplace transform is a powerful way of characterizing a system. We will consider both approaches.

The Fourier transform expresses a time-domain signal as a linear combination of complex sinusoids of the form $e^{j\omega t}$ or $e^{j2\pi ft}$. The form can be generalized by changing the complex sinusoids to complex exponentials of the form $e^{st}$, where the variable $s$ can take on general complex values as opposed to $j\omega$ or $j2\pi f$ which take on only imaginary values (because $\omega$ and $f$ are *real* variables associated with the *real* physical concept of frequency). If we simply generalize the forward Fourier transform by replacing complex sinusoids with complex exponentials, we get the transform,

$$\mathcal{L}(x(t)) = X(s) = \int_{-\infty}^{\infty} x(t)e^{-st}\, dt, \tag{9.3}$$

which defines a forward Laplace transform, where the notation, $\mathcal{L}(\ )$, means "Laplace transform of."

Since we allow $s$ to have values anywhere in the complex plane, it has a real part and an imaginary part. Let $s$ be expressed as

$$s = \sigma + j\omega. \tag{9.4}$$

Then, for the special case in which the real part of $s$, $\sigma$, is zero and the Fourier transform of the function $x(t)$ exists in the strict sense, the forward Laplace transform is equivalent to a forward Fourier transform. Using $s = \sigma + j\omega$ in the forward Laplace transform we get

$$X(s) = \int_{-\infty}^{\infty} x(t)e^{-(\sigma+j\omega)t}\, dt = \int_{-\infty}^{\infty} [x(t)e^{-\sigma t}]e^{-j\omega t}\, dt = \mathcal{F}(x(t)e^{-\sigma t}). \tag{9.5}$$

Pierre-Simon Laplace,
3/23/1749–3/2/1827

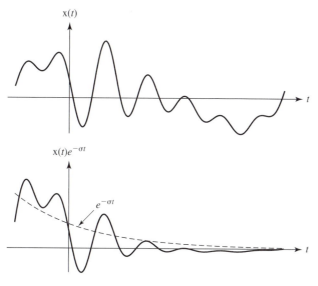

**Figure 9.1**
The effect of the decaying-exponential convergence factor on
the original function.

So one way of conceptualizing the Laplace transform is to realize that it is equivalent
to a Fourier transform of the product of the function $x(t)$ and a real exponential convergence factor of the form $e^{-\sigma t}$ as illustrated in Figure 9.1.

It is natural to wonder what has been gained by introducing the extra factor $e^{-\sigma t}$
into the transformation process. This factor allows us, in some cases, to find transforms
of functions for which the Fourier transform cannot be found. As mentioned in Chapter 5, the Fourier transforms of some functions do not (strictly speaking) exist. For
example, the function

$$g(t) = A\mathrm{u}(t) \tag{9.6}$$

would have the Fourier transform

$$G(j\omega) = \int_{-\infty}^{\infty} A\mathrm{u}(t)e^{-j\omega t}\,dt = A\int_{0}^{\infty} e^{-j\omega t}\,dt \qquad \text{or}$$

$$G(f) = \int_{-\infty}^{\infty} A\mathrm{u}(t)e^{-j2\pi ft}\,dt = A\int_{0}^{\infty} e^{-j2\pi ft}\,dt. \tag{9.7}$$

These integrals do not converge. The technique used to make the Fourier transform
converge was to multiply the signal by a convergence factor $e^{-\sigma|t|}$, where $\sigma$ is a posi-
tive real constant. Then the Fourier transform of the modified signal can be found and
the limit taken as $\sigma$ approaches zero. The Fourier transform found by this technique
was called a generalized Fourier transform in which the impulse was allowed as a part
of the transform. Notice that, for time $t > 0$, this convergence factor is the same in the
Laplace transform and the generalized Fourier transform, but in the Laplace transform
the limit as $\sigma$ approaches zero is not taken. As we will soon see there are other useful
functions which do not even have a generalized Fourier transform.

Now, to formally derive the forward and inverse Laplace transforms from the Fourier transform, we take the Fourier transform of

$$g_\sigma(t) = g(t)e^{-\sigma t} \tag{9.8}$$

instead of the original function $g(t)$. That integral would then be

$$\mathcal{F}(g_\sigma(t)) = G_\sigma(j\omega) = \int_{-\infty}^{\infty} g_\sigma(t)e^{-j\omega t}\,dt = \int_{-\infty}^{\infty} g(t)e^{-(\sigma+j\omega)t}\,dt. \tag{9.9}$$

This integral may or may not converge, depending on the nature of the function $g(t)$ and the choice of the value of $\sigma$. We will soon explore the conditions under which the integral converges. Using the notation $s = \sigma + j\omega$,

$$\mathcal{F}(g_\sigma(t)) = \mathcal{L}(g(t)) = G_{\mathcal{L}}(s) = \int_{-\infty}^{\infty} g(t)e^{-st}\,dt. \tag{9.10}$$

This is the Laplace transform of $g(t)$ *if* the integral converges.

The inverse Fourier transform would be

$$\mathcal{F}^{-1}(G_\sigma(j\omega)) = g_\sigma(t) = \frac{1}{2\pi}\int_{-\infty}^{\infty} G_\sigma(j\omega)e^{+j\omega t}\,d\omega = \frac{1}{2\pi}\int_{-\infty}^{\infty} G_{\mathcal{L}}(s)e^{+j\omega t}\,d\omega. \tag{9.11}$$

Using

$$s = \sigma + j\omega \qquad \text{and} \qquad ds = j\,d\omega \tag{9.12}$$

we get

$$g_\sigma(t) = \frac{1}{j2\pi}\int_{\sigma-j\infty}^{\sigma+j\infty} G_{\mathcal{L}}(s)e^{+(s-\sigma)t}\,ds = \frac{e^{-\sigma t}}{j2\pi}\int_{\sigma-j\infty}^{\sigma+j\infty} G_{\mathcal{L}}(s)e^{+st}\,ds \tag{9.13}$$

or, dividing both sides by $e^{-\sigma t}$,

$$g(t) = \frac{1}{j2\pi}\int_{\sigma-j\infty}^{\sigma+j\infty} G_{\mathcal{L}}(s)e^{+st}\,ds, \tag{9.14}$$

which defines an inverse Laplace transform. When we are dealing only with Laplace transforms, the $\mathcal{L}$ subscript will not be needed to avoid confusion with Fourier transforms and the inverse transform can be written as

$$g(t) = \frac{1}{j2\pi}\int_{\sigma-j\infty}^{\sigma+j\infty} G(s)e^{+st}\,ds. \tag{9.15}$$

The result (9.15) shows that a function can be expressed as a linear combination of complex exponentials. This is a generalization of the fact that a function can be expressed as a linear combination of complex sinusoids.

The other approach to developing and understanding the Laplace transform is to consider the response of an LTI system excited by a complex exponential of the form $x(t) = Ae^{st}$, where $s$ can be any complex number. The response is the convolution of the excitation with the impulse response,

$$y(t) = h(t) * x(t) = \int_{-\infty}^{\infty} h(\tau)x(t - \tau)\, d\tau = \int_{-\infty}^{\infty} h(\tau)Ae^{s(t-\tau)}\, d\tau$$

$$= \underbrace{Ae^{st}}_{x(t)} \underbrace{\int_{-\infty}^{\infty} h(\tau)e^{-s\tau}\, d\tau}_{\substack{\text{Laplace transform} \\ \text{of } h(t)}} . \tag{9.16}$$

This result shows that the response of an LTI system to a complex-exponential excitation of the form $Ae^{st}$ is a complex exponential of the same form, except multiplied by $H(s)$, the Laplace transform of the impulse response of the system. We can express most useful signals as a linear combination of complex exponentials. Therefore, the response to an excitation can be found by multiplying the Laplace transform of the excitation (which expresses the excitation as a linear combination of complex exponentials) by the Laplace transform of the impulse response. This is directly analogous to the corresponding Fourier transform result that an excitation to an LTI system of the form $Ae^{j\omega t}$ produces a response $Ae^{j\omega t}H(j\omega)$, where $H(j\omega)$ is the Fourier transform of the impulse response of that LTI system. The function $Ae^{st}$ is more general than the function $Ae^{j\omega t}$; therefore, the Laplace transform is more general than the Fourier transform. The Fourier transform is really just a special case of the Laplace transform defined by (9.3), with some notation changes.

## REGION OF CONVERGENCE

As mentioned earlier, there are useful functions which do not have even a generalized Fourier transform, for example, the causal function

$$g_1(t) = Ae^{\alpha t}u(t) \qquad \alpha > 0 \tag{9.17}$$

(Figure 9.2). This is a function that increases without bound as $t$ increases. Even though this function does not have a Fourier transform, it *does* have a Laplace transform. The Laplace transform is

$$G_1(s) = \int_{-\infty}^{\infty} Ae^{\alpha t}u(t)e^{-st}\, dt = A\int_{0}^{\infty} e^{-(s-\alpha)t}\, dt = A\int_{0}^{\infty} e^{(\alpha-\sigma)t}e^{-j\omega t}\, dt. \tag{9.18}$$

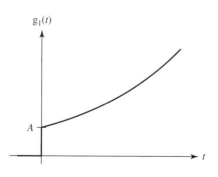

**Figure 9.2**
A causal function that is not Fourier-transformable.

Does this integral converge? It converges if $\alpha - \sigma$ is negative, that is, if $\sigma > \alpha$. If $\sigma > \alpha$, the function $e^{(\alpha - \sigma)t}$ approaches zero as $t$ approaches positive infinity. The specification $\sigma > \alpha$ defines what is called the *region of convergence* (ROC). The Laplace transform exists for those values of $s$ in the complex plane for which $\sigma > \alpha$. In other words, if the real part of $s$ is large enough, even those functions which increase exponentially with time and are, therefore, unbounded, have a Laplace transform. Completing the integral in (9.18),

$$G_1(s) = \frac{A}{s - \alpha} \qquad \sigma = \mathrm{Re}(s) > \alpha. \tag{9.19}$$

This transform result $G_1(s)$ goes to infinity at a finite value of $s$, $s = \alpha$. This point in the complex $s$ plane is called a *pole* of $G_1(s)$. Points in the complex $s$ plane at which the transform goes to zero are called *zeros* of $G_1(s)$. In this case there are no finite zeros of $G_1(s)$. It is often informative to plot the locations of the finite poles and zeros of a Laplace-domain function in the complex $s$ plane. The constellation of poles and zeros conveys a lot (but not all) about the nature of the function at a glance. The pole–zero plot for $G_1(s)$ is illustrated in Figure 9.3 along with the region of convergence in the $s$ plane. The region of convergence is that part of the $s$ plane for which the real part of $s$ is greater than $\alpha$.

Now consider another function, the anticausal function $g_2(t) = Ae^{-\alpha t}u(-t) = g_1(-t)$, $\alpha > 0$ (Figure 9.4). The Laplace-transform integral is

$$G_2(s) = \int_{-\infty}^{\infty} Ae^{-\alpha t}u(-t)e^{-st}\,dt = \int_{-\infty}^{0} Ae^{-(s+\alpha)t}\,dt. \tag{9.20}$$

The integral converges if $\sigma < -\alpha$, and the transform is

$$G_2(s) = -\frac{A}{s + \alpha} = G_1(-s) \qquad \sigma < -\alpha. \tag{9.21}$$

The pole–zero plot and region of convergence for this function are illustrated in Figure 9.5.

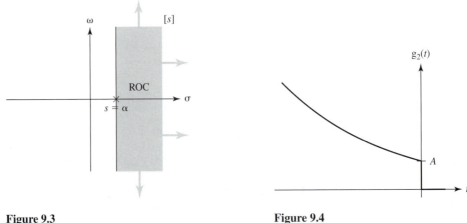

**Figure 9.3**
Pole–zero plot and region of
convergence for $G_1(s)$.

**Figure 9.4**
An anticausal function that is not Fourier-
transformable.

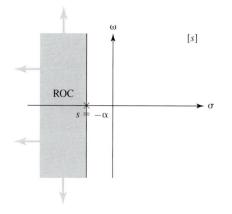

**Figure 9.5**
Pole–zero plot and region of convergence
for $G_2(s)$.

Now let the function to be transformed be $g(t) = Ae^{\alpha t}$. The transform integral
becomes

$$G(s) = \int_{-\infty}^{\infty} Ae^{\alpha t} e^{-st} \, dt = A \int_{-\infty}^{\infty} e^{\alpha t} e^{-\sigma t} e^{-j\omega t} \, dt = A \int_{-\infty}^{\infty} e^{(\alpha-\sigma)t} e^{-j\omega t} \, dt. \qquad \textbf{(9.22)}$$

This integral does not converge. No matter what value we choose for $\sigma$, we cannot
evaluate the integral at one of its limits, either lower or upper.

**EXAMPLE 9.1**

Find the Laplace transform of

$$x(t) = e^{-t}u(t) + e^{-2t}u(t). \qquad \textbf{(9.23)}$$

■ **Solution**

Using the definition

$$x(t) \xleftrightarrow{\mathcal{L}} \int_{-\infty}^{\infty} [e^{-t}u(t) + e^{-2t}u(t)]e^{-st} \, dt = \int_{0}^{\infty} \left[ e^{-(s+1)t} + e^{-(s+2)t} \right] dt \qquad \sigma > -1$$

$$\textbf{(9.24)}$$

or

$$x(t) \xleftrightarrow{\mathcal{L}} \frac{1}{s+1} + \frac{1}{s+2} \qquad \sigma > -1. \qquad \textbf{(9.25)}$$

The ROC is $\sigma > -1$. If we had found the Laplace transforms of $e^{-t}u(t)$ and $e^{-2t}u(t)$ separately,
we would have found the two ROCs $\sigma > -1$ and $\sigma > -2$, respectively. So the overall ROC is
the region which is common to both ROCs, $ROC = ROC_1 \cap ROC_2$. ■

To illustrate the importance of specifying, not only the algebraic form of the
Laplace transform, but also its ROC, consider the Laplace transforms

$$e^{-\alpha t}u(t) \xleftrightarrow{\mathcal{L}} \frac{1}{s+\alpha} \qquad \sigma > -\alpha \qquad \textbf{(9.26)}$$

and

$$-e^{-\alpha t}u(-t) \overset{\mathcal{L}}{\longleftrightarrow} \frac{1}{s+\alpha} \qquad \sigma < -\alpha. \qquad \textbf{(9.27)}$$

The algebraic expression for the Laplace transform is the same in each case, but the ROCs are totally different; in fact, they are mutually exclusive. That means, for example, that the Laplace transform of the sum of these two functions cannot be found because we cannot find a region in the $s$ plane which is common to the ROCs of both $e^{-\alpha t}u(t)$ and $-e^{-\alpha t}u(-t)$.

**EXAMPLE 9.2**

Find the Laplace transform of

$$x(t) = e^{-t}u(t) + e^{2t}u(-t). \qquad \textbf{(9.28)}$$

■ **Solution**

The Laplace transform of this sum is the sum of the Laplace transforms of the individual terms $e^{-t}u(t)$ and $e^{2t}u(-t)$, and the ROC of the sum is the region in the $s$ plane which is common to the two ROCs.

$$e^{-t}u(t) \overset{\mathcal{L}}{\longleftrightarrow} \frac{1}{s+1} \qquad \sigma > -1 \qquad \textbf{(9.29)}$$

and

$$e^{2t}u(-t) \overset{\mathcal{L}}{\longleftrightarrow} -\frac{1}{s-2} \qquad \sigma < 2. \qquad \textbf{(9.30)}$$

In this case, there is a region in the $s$ plane which is common to both ROCs, $-1 < \sigma < 2$, so

$$e^{-t}u(t) + e^{2t}u(-t) \overset{\mathcal{L}}{\longleftrightarrow} \frac{1}{s+1} - \frac{1}{s-2} \qquad -1 < \sigma < 2 \qquad \textbf{(9.31)}$$

(Figure 9.6).

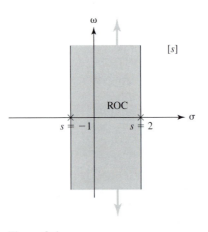

**Figure 9.6**
ROC for the Laplace transform of
$x(t) = e^{-t}u(t) + e^{2t}u(-t)$.

Notice that, in Example 9.2, the ROC contains the $\omega$ axis ($\sigma = 0$). That means that the integral $\int_{-\infty}^{\infty} [e^{-t}u(t) + e^{2t}u(-t)]e^{-j\omega t}\,dt$ converges. Since this is exactly the Fourier transform of $e^{-t}u(t) + e^{2t}u(-t)$, that implies that its Fourier transform exists. We can say generally that if the region of convergence of the Laplace transform contains the $\omega$ axis, the Fourier transform exists.

If a function $x(t)$ is time-limited to $t_1 < t < t_2$ and bounded, its Laplace transform is

$$X(s) = \int_{-\infty}^{\infty} x(t)e^{-st}\,dt = \int_{t_1}^{t_2} x(t)e^{-st}\,dt. \qquad (9.32)$$

This integral converges for any finite value of $s$. Therefore, for functions of this type the region of convergence is the entire $s$ plane.

## THE UNILATERAL LAPLACE TRANSFORM

In the exploration of the Laplace transform so far it is apparent that if we consider the full range of possible signals to transform, sometimes a region of convergence can be found and sometimes it cannot be found. If we leave out some pathological functions like $t^t$ or $e^{t^2}$ which grow faster than an exponential (and have no known engineering usefulness), and restrict ourselves to functions which are zero before or after time $t = 0$, the Laplace transform and its ROC become considerably simpler. The quality that made the functions $g_1(t) = Ae^{\alpha t}u(t), \alpha > 0$, and $g_2(t) = Ae^{-\alpha t}u(-t), \alpha > 0$, Laplace-transformable was that each of them was restricted by the unit step function to be zero over a semi-infinite range of time. The function $g_1(t) = Ae^{\alpha t}u(t), \alpha > 0$, is called a *causal* function because it is zero before time $t = 0$. The function $g_2(t) = Ae^{-\alpha t}u(-t), \alpha > 0$, is called an *anticausal* function because it is zero after time $t = 0$. Now we can generalize what we have seen so far. Any function which is defined as zero either before or after some finite time $t = t_0$, and whose variation with time over the rest of time is no faster than an exponential function, has a Laplace transform, and the region of convergence of the transform always exists and is determined by the functional behavior.

Even a function as benign as $g(t) = A$, which is bounded for all $t$, causes problems, because a single convergence factor which makes the Laplace transform converge for all time cannot be found. But the function $g(t) = Au(t)$ *is* Laplace-transformable. The presence of the unit step allows the choice of a convergence factor for positive time which makes the Laplace transform integral converge. For this reason, a modification of the Laplace transform which avoids many convergence issues is conventionally used. (We shall soon see that there are also other reasons to use the modified form.) Let us now redefine the Laplace transform as

$$G(s) = \int_{0^-}^{\infty} g(t)e^{-st}\,dt. \qquad (9.33)$$

Only the lower limit of integration has changed. With this new definition any function which grows no faster than an exponential in positive time has a Laplace transform.

The Laplace transform defined by

$$G(s) = \int_{-\infty}^{\infty} g(t)e^{-st}\,dt \qquad (9.34)$$

is conventionally called the *two-sided* or *bilateral* Laplace transform. The Laplace transform defined by

$$G(s) = \int_{0^-}^{\infty} g(t)e^{-st}\,dt \tag{9.35}$$

is conventionally called the *one-sided* or *unilateral* Laplace transform. The unilateral Laplace transform is restrictive in the sense that it excludes the negative-time behavior of functions which are nonzero for negative time. But since, in the analysis of any real system, a time origin can be chosen to make all signals zero before that time, this is not really a practical problem and actually has some advantages. Since the lower limit of integration is $t = 0^-$, any functional behavior of $g(t)$ before time $t = 0$ is irrelevant to the transform. This means that any other function which has the same behavior at or after time $t = 0$ will have the same transform. Therefore, for the transform to be unique to one time-domain function, it should only be applied to functions which are zero before time $t = 0$.

> Even for times $t > 0$, the transform is not actually unique to a single time-domain function. As mentioned in Chapter 2 in the discussion of the definition of the unit step function, all the definitions have exactly the same transform and yet their values are different at the discontinuity time $t = 0$. This is a mathematical point without any real engineering impact. If two functions differ in value at isolated points, their effect as excitations on any real system will be identical because there is no signal energy in a signal at an isolated point (unless there is an impulse at the isolated point) and real systems respond to the energy of excitation signals.

The inverse unilateral Laplace transform is exactly the same as derived for the bilateral Laplace transform,

$$g(t) = \frac{1}{j2\pi} \int_{\sigma-j\infty}^{\sigma+j\infty} G(s)e^{+st}\,ds. \tag{9.36}$$

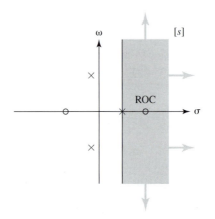

**Figure 9.7**
ROC for a unilateral Laplace transform.

It is common to see the Laplace-transform pair written as

$$\mathcal{L}(\mathrm{g}(t)) = \mathrm{G}(s) = \int_{0^-}^{\infty} \mathrm{g}(t)e^{-st}\,dt \qquad \mathcal{L}^{-1}(\mathrm{G}(s)) = \mathrm{g}(t) = \frac{1}{j2\pi} \int_{\sigma - j\infty}^{\sigma + j\infty} \mathrm{G}(s)e^{+st}\,ds.$$

$$\textbf{(9.37)}$$

The unilateral Laplace transform has a simple ROC. It is always the region of the $s$ plane for which $\sigma$ lies to the right of all the poles of the transform (Figure 9.7).

It is conventional terminology to refer to the $s$ domain as the *complex frequency domain* since $s$ is a complex variable and can range over the entire complex plane and its units are radians per second. The variable $\omega$ or $f$ is a real variable and the domain into which a time-domain function is Fourier-transformed is sometimes called the *real frequency* domain. From this point on, the unilateral Laplace transform will be referred to as simply *the* Laplace transform and the bilateral Laplace transform will be designated specifically.

EXAMPLE 9.3

Find the Laplace transform of $e^{-\alpha t}\mathrm{u}(t)$.

### ■ Solution

$$e^{-\alpha t}\mathrm{u}(t) \overset{\mathcal{L}}{\longleftrightarrow} \int_{0^-}^{\infty} e^{-\alpha t}\mathrm{u}(t)e^{-st}\,dt = \int_{0^-}^{\infty} e^{-(s+\alpha)t}\,dt. \qquad \textbf{(9.38)}$$

This integral converges for any value of $s$ whose real part $\sigma$ is greater than $-\alpha$. Therefore,

$$e^{-\alpha t}\mathrm{u}(t) \overset{\mathcal{L}}{\longleftrightarrow} \left[\frac{e^{-(s+\alpha)t}}{-(s+\alpha)}\right]_{0^-}^{\infty} = \frac{1}{s+\alpha} \qquad \sigma > -\alpha \qquad \textbf{(9.39)}$$

(Figure 9.8).

For the special case of $\alpha = 0$, $e^{-\alpha t}\mathrm{u}(t)$ becomes simply $\mathrm{u}(t)$ and

$$\mathrm{u}(t) \overset{\mathcal{L}}{\longleftrightarrow} \frac{1}{s} \qquad \sigma > 0 \qquad \textbf{(9.40)}$$

(Figure 9.9).

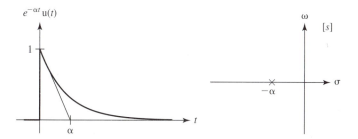

**Figure 9.8**
A decaying exponential and the pole–zero plot of its Laplace transform.

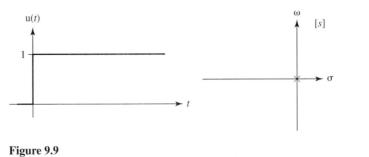

**Figure 9.9**
Unit step and the pole–zero plot of its Laplace transform.

**EXAMPLE 9.4**

Find the Laplace transforms of the damped sinusoids $e^{-\alpha t} \cos(\omega_0 t)\, u(t)$ and $e^{-\alpha t} \sin(\omega_0 t)\, u(t)$.

■ **Solution**

$$e^{-\alpha t} \cos(\omega_0 t)\, u(t) \xleftrightarrow{\mathcal{L}} \int_{0^-}^{\infty} e^{-\alpha t} \cos(\omega_0 t)\, u(t) e^{-st}\, dt = \int_{0^-}^{\infty} e^{-\alpha t} \frac{e^{j\omega_0 t} + e^{-j\omega_0 t}}{2} e^{-st}\, dt \qquad (9.41)$$

$$e^{-\alpha t} \cos(\omega_0 t)\, u(t) \xleftrightarrow{\mathcal{L}} \frac{1}{2} \int_{0^-}^{\infty} \left( e^{(j\omega_0 - s - \alpha)t} + e^{-(j\omega_0 + s + \alpha)t} \right) dt \qquad (9.42)$$

$$e^{-\alpha t} \cos(\omega_0 t)\, u(t) \xleftrightarrow{\mathcal{L}} \frac{1}{2} \left[ \frac{e^{(j\omega_0 - s - \alpha)t}}{j\omega_0 - (s + \alpha)} + \frac{e^{-(j\omega_0 + s + \alpha)t}}{-j\omega_0 - (s + \alpha)} \right]_{0^-}^{\infty} \qquad (9.43)$$

$$e^{-\alpha t} \cos(\omega_0 t)\, u(t) \xleftrightarrow{\mathcal{L}} \frac{s + \alpha}{(s + \alpha)^2 + \omega_0^2} \qquad \sigma > -\alpha \qquad (9.44)$$

$$e^{-\alpha t} \sin(\omega_0 t)\, u(t) \xleftrightarrow{\mathcal{L}} \int_{0^-}^{\infty} e^{-\alpha t} \sin(\omega_0 t)\, u(t) e^{-st}\, dt = \int_{0^-}^{\infty} e^{-\alpha t} \frac{e^{j\omega_0 t} - e^{-j\omega_0 t}}{j2} e^{-st}\, dt \qquad (9.45)$$

$$e^{-\alpha t} \sin(\omega_0 t)\, u(t) \xleftrightarrow{\mathcal{L}} \frac{1}{j2} \int_{0^-}^{\infty} \left( e^{(j\omega_0 - s - \alpha)t} - e^{-(j\omega_0 + s + \alpha)t} \right) dt \qquad (9.46)$$

$$e^{-\alpha t} \sin(\omega_0 t)\, u(t) \xleftrightarrow{\mathcal{L}} \frac{1}{j2} \left[ \frac{e^{(j\omega_0 - s - \alpha)t}}{j\omega_0 - (s + \alpha)} - \frac{e^{-(j\omega_0 + s + \alpha)t}}{-j\omega_0 - (s + \alpha)} \right]_{0^-}^{\infty} \qquad (9.47)$$

$$e^{-\alpha t} \sin(\omega_0 t)\, u(t) \xleftrightarrow{\mathcal{L}} \frac{\omega_0}{(s + \alpha)^2 + \omega_0^2} \qquad \sigma > -\alpha \qquad (9.48)$$

Using the results of Example 9.4 it follows that undamped sinusoids ($\alpha = 0$) have the Laplace transforms

$$\cos(\omega_0 t)\, u(t) \xleftrightarrow{\mathcal{L}} \frac{s}{s^2 + \omega_0^2} \qquad \sigma > 0 \qquad (9.49)$$

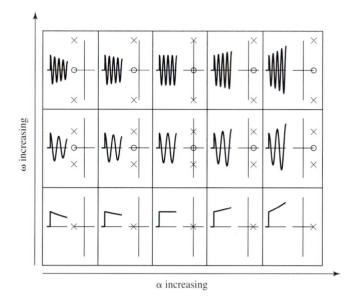

**Figure 9.10**
Illustration, in both the time and frequency domains, of the effects of
the decay rate parameter $\alpha$ and the radian frequency $\omega_0$.

and

$$\sin(\omega_0 t)\, u(t) \overset{\mathcal{L}}{\longleftrightarrow} \frac{\omega_0}{s^2 + \omega_0^2} \qquad \sigma > 0, \tag{9.50}$$

and that a decaying exponential ($\omega_0 = 0$) has the Laplace transform

$$e^{-\alpha t} u(t) \overset{\mathcal{L}}{\longleftrightarrow} \frac{1}{s + \alpha} \qquad \sigma > -\alpha \tag{9.51}$$

as we saw in Example 9.3. Consideration of the time-domain signal $e^{\alpha t} \cos(\omega_0 t)\, u(t)$
and its Laplace transform, $(s - \alpha)/[(s - \alpha)^2 + \omega_0^2]$, leads to the diagram in Figure 9.10 which relates exponential growth rate $\alpha$ and undamped radian frequency $\omega_0$
to pole and zero locations.

**EXAMPLE 9.5**

Find the Laplace transform of $\delta(t)$.

**■ Solution**
The Laplace-transform integral can be evaluated using the sampling property of the impulse.

$$\delta(t) \overset{\mathcal{L}}{\longleftrightarrow} \int_{0^-}^{\infty} \delta(t) e^{-st}\, dt = [e^{-st}]_{t=0} = 1. \tag{9.52}$$

The ROC is the entire $s$ plane. ■

## 9.3 PROPERTIES OF THE LAPLACE TRANSFORM

The Laplace transform has properties similar to the properties of the CTFT. If

$$\mathcal{L}(g(t)) = G(s) \qquad \text{and} \qquad \mathcal{L}(h(t)) = H(s) \tag{9.53}$$

and $g(t) = 0$ for $t < 0$ and $h(t) = 0$ for $t < 0$, then the following properties can be
shown to apply.

## LINEARITY

The linearity property is exactly the same for the Laplace and CTFT transforms and is proven in the same way.

$$\alpha g(t) + \beta h(t) \xleftrightarrow{\mathcal{L}} \alpha G(s) + \beta H(s) \qquad (9.54)$$

## TIME SHIFTING

Let $t_0$ be a positive real constant. Then

$$g(t - t_0) \xleftrightarrow{\mathcal{L}} \int_{0^-}^{\infty} g(t - t_0)e^{-st}\, dt. \qquad (9.55)$$

Make the change of variable,

$$\lambda = t - t_0 \qquad \text{and} \qquad d\lambda = dt \qquad (9.56)$$

and (9.55) becomes

$$g(t - t_0) \xleftrightarrow{\mathcal{L}} \int_{-t_0^-}^{\infty} g(\lambda)e^{-s(\lambda + t_0)}\, d\lambda. \qquad (9.57)$$

If $g(t) = 0$ for $t < 0$,

$$g(t - t_0) \xleftrightarrow{\mathcal{L}} e^{-st_0} \int_{0^-}^{\infty} g(\lambda)e^{-s\lambda}\, d\lambda = e^{-st_0}G(s). \qquad (9.58)$$

The time-shifting property of the Laplace transform is

$$g(t - t_0) \xleftrightarrow{\mathcal{L}} G(s)e^{-st_0} \qquad t_0 > 0. \qquad (9.59)$$

This property is only valid for time shifts to the right (time delays) because only for delayed signals is the entire nonzero part of the signal still included in the integral from $0^-$ to infinity. If a signal were shifted to the left (advanced in time), some of it might occur before time $t = 0$ and not be included within the limits of the Laplace transform integral. That would destroy the unique relation between the transform of the signal and the transform of its shifted version, making it impossible to relate them in any general way (Figure 9.11).

## COMPLEX-FREQUENCY SHIFTING

Let $s_0$ be a constant. Then

$$e^{s_0 t}g(t) \xleftrightarrow{\mathcal{L}} \int_{0^-}^{\infty} e^{s_0 t}g(t)e^{-st}\, dt \qquad (9.60)$$

$$e^{s_0 t}g(t) \xleftrightarrow{\mathcal{L}} \int_{0^-}^{\infty} g(t)e^{-(s - s_0)t}\, dt = G(s - s_0). \qquad (9.61)$$

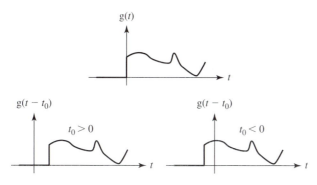

**Figure 9.11**
Shifts of a causal function.

The complex-frequency-shifting property of the Laplace transform is

$$e^{s_0 t} g(t) \xleftrightarrow{\mathcal{L}} G(s - s_0). \qquad (9.62)$$

## TIME SCALING

Let $a$ be any positive real constant. Then the Laplace transform of $g(at)$ is

$$g(at) \xleftrightarrow{\mathcal{L}} \int_{0^-}^{\infty} g(at) e^{-st} \, dt. \qquad (9.63)$$

Let $\lambda = at$ and $d\lambda = a \, dt$. Then

$$g(at) \xleftrightarrow{\mathcal{L}} \int_{0^-}^{\infty} g(\lambda) e^{-s(\lambda/a)} \frac{d\lambda}{a} = \frac{1}{a} \int_{0^-}^{\infty} g(\lambda) e^{-(s/a)\lambda} \, d\lambda = \frac{1}{a} G\left(\frac{s}{a}\right) \qquad a > 0$$

$$g(at) \xleftrightarrow{\mathcal{L}} \frac{1}{a} G\left(\frac{s}{a}\right) \qquad a > 0.$$

The constant $a$ cannot be negative because that would turn a causal signal into a non-causal signal and the unilateral Laplace transform is only valid for causal signals. Just as we found with the Fourier transform, a compression of the time signal corresponds to an expansion of its Laplace transform, and vice versa.

**EXAMPLE 9.6**

Find the Laplace transforms of

$$x(t) = u(t) - u(t - a) \qquad (9.64)$$

and

$$x(2t) = u(2t) - u(2t - a). \qquad (9.65)$$

### ■ Solution

We have already found the Laplace transform of u($t$), $1/s$. Using the linearity and time-shifting properties,

$$\text{u}(t) - \text{u}(t - a) \overset{\mathcal{L}}{\longleftrightarrow} \frac{1 - e^{-as}}{s}. \qquad (9.66)$$

Now, using the time-scaling property,

$$\text{u}(2t) - \text{u}(2t - a) \overset{\mathcal{L}}{\longleftrightarrow} \frac{1}{2}\frac{1 - e^{-a(s/2)}}{s/2} = \frac{1 - e^{-(as/2)}}{s}. \qquad (9.67)$$

■

## FREQUENCY SCALING

Let $a$ be any positive real constant. Then, using the time-scaling property of the Laplace transform,

$$\text{g}(at) \overset{\mathcal{L}}{\longleftrightarrow} \frac{1}{a}\text{G}\left(\frac{s}{a}\right) \qquad a > 0. \qquad (9.68)$$

Let $b = 1/a$. Then

$$\text{g}\left(\frac{t}{b}\right) \overset{\mathcal{L}}{\longleftrightarrow} b\text{G}(bs) \qquad b > 0 \qquad (9.69)$$

or

$$\frac{1}{b}\text{g}\left(\frac{t}{b}\right) \overset{\mathcal{L}}{\longleftrightarrow} \text{G}(bs) \qquad b > 0 \qquad (9.70)$$

and the frequency-scaling property of the Laplace transform is

$$\boxed{\frac{1}{a}\text{g}\left(\frac{t}{a}\right) \overset{\mathcal{L}}{\longleftrightarrow} \text{G}(as) \qquad a > 0.} \qquad (9.71)$$

## TIME DIFFERENTIATION ONCE

From the definition of the Laplace transform,

$$\text{G}(s) = \int_{0^-}^{\infty} \text{g}(t)e^{-st}\,dt. \qquad (9.72)$$

Evaluate the integral by parts using

$$\int u\,dv = uv - \int v\,du \qquad (9.73)$$

and let

$$u = \text{g}(t) \qquad \text{and} \qquad dv = e^{-st}\,dt. \qquad (9.74)$$

Then

$$du = \frac{d}{dt}(\text{g}(t))\,dt \qquad \text{and} \qquad v = -\frac{1}{s}e^{-st} \qquad (9.75)$$

and

$$\int_{0^-}^{\infty} g(t)e^{-st}\,dt = g(t)\left(-\frac{1}{s}\right)e^{-st}\bigg|_{0^-}^{\infty} + \frac{1}{s}\int_{0^-}^{\infty}\frac{d}{dt}(g(t))e^{-st}\,dt \qquad \textbf{(9.76)}$$

$$G(s) = \frac{1}{s}g(0^-) + \frac{1}{s}\int_{0^-}^{\infty}\frac{d}{dt}(g(t))e^{-st}\,dt \qquad \textbf{(9.77)}$$

[where it is understood that $\text{Re}(s) = \sigma$ is chosen to make $G(s)$ exist]. Then

$$\mathcal{L}\left\{\frac{d}{dt}(g(t))\right\} = \int_{0^-}^{\infty}\frac{d}{dt}(g(t))e^{-st}\,dt = sG(s) - g(0^-) \qquad \textbf{(9.78)}$$

and the time-differentiation-once property of the Laplace transform is

$$\boxed{\frac{d}{dt}(g(t)) \xleftrightarrow{\mathcal{L}} sG(s) - g(0^-).} \qquad \textbf{(9.79)}$$

This is one of the most important properties of the (unilateral) Laplace transform. It has no counterpart in the Fourier transform because the Laplace transform has a starting point in time and the Fourier transform does not. This is the property that makes the solution of transient problems easier using the Laplace transform than the Fourier transform. When using the differentiation property in solving differential equations, the initial condition $g(0^-)$ is automatically called for in the proper form as an inherent part of the transform process.

## TIME DIFFERENTIATION TWICE

This property can be proven using the property of time differentiation once and applying it to a time derivative to form a second derivative. The second time derivative of a function $g(t)$ is

$$\frac{d^2}{dt^2}(g(t)) = \frac{d}{dt}\left(\frac{d}{dt}(g(t))\right). \qquad \textbf{(9.80)}$$

Therefore, using

$$\frac{d}{dt}(g(t)) \xleftrightarrow{\mathcal{L}} sG(s) - g(0^-) \qquad \textbf{(9.81)}$$

we get

$$\mathcal{L}\left\{\frac{d^2}{dt^2}(g(t))\right\} = s\mathcal{L}\left\{\frac{d}{dt}(g(t))\right\} - \frac{d}{dt}(g(t))\bigg|_{t=0^-} \qquad \textbf{(9.82)}$$

$$\mathcal{L}\left\{\frac{d^2}{dt^2}(g(t))\right\} = s\{sG(s) - g(0^-)\} - \frac{d}{dt}(g(t))\bigg|_{t=0^-}$$

$$= s^2G(s) - sg(0^-) - \frac{d}{dt}(g(t))\bigg|_{t=0^-} \qquad \textbf{(9.83)}$$

The time-differentiation-twice property of the Laplace transform is

$$
\frac{d^2}{dt^2}(g(t)) \overset{\mathcal{L}}{\longleftrightarrow} s^2 G(s) - s g(0^-) - \frac{d}{dt}(g(t))_{t=0^-}. \tag{9.84}
$$

Like the property of time differentiation once, this property is also important in the solution of differential equations, again because it automatically handles the initial conditions in such a systematic way. It can be extended to any number of derivatives, although, in practice, the first and second derivatives are the ones most often needed.

## COMPLEX-FREQUENCY DIFFERENTIATION

From the definition of the Laplace transform,

$$
G(s) = \int_{0^-}^{\infty} g(t) e^{-st} \, dt. \tag{9.85}
$$

Differentiating with respect to $s$,

$$
\frac{d}{ds}(G(s)) = \frac{d}{ds} \int_{0^-}^{\infty} g(t) e^{-st} \, dt = \int_{0^-}^{\infty} \frac{d}{ds}(g(t) e^{-st}) \, dt
$$

$$
= \int_{0^-}^{\infty} -t g(t) e^{-st} \, dt = \mathcal{L}(-t g(t)) \tag{9.86}
$$

$$
-t g(t) \overset{\mathcal{L}}{\longleftrightarrow} \frac{d}{ds}(G(s)). \tag{9.87}
$$

## MULTIPLICATION–CONVOLUTION DUALITY

The time-domain convolution of $g(t)$ with $h(t)$ is

$$
g(t) * h(t) = \int_{-\infty}^{\infty} g(\tau) h(t - \tau) \, d\tau. \tag{9.88}
$$

Since $g(t)$ is zero for time $t < 0$,

$$
g(t) * h(t) = \int_{0^-}^{\infty} g(\tau) h(t - \tau) \, d\tau. \tag{9.89}
$$

From the definition of the Laplace transform,

$$
\mathcal{L}[g(t) * h(t)] = \int_{0^-}^{\infty} \left[ \int_{0^-}^{\infty} g(\tau) h(t - \tau) \, d\tau \right] e^{-st} \, dt \tag{9.90}
$$

$$
\mathcal{L}[g(t) * h(t)] = \int_{0^-}^{\infty} g(\tau) \left[ \int_{0^-}^{\infty} e^{-st} h(t - \tau) \, dt \right] d\tau.
$$

Since h($t$) is zero for time $t < 0$,

$$\mathcal{L}[g(t) * h(t)] = \int_{0^-}^{\infty} g(\tau) \left[ \int_{\tau^-}^{\infty} e^{-st} h(t - \tau) \, dt \right] d\tau. \qquad (9.91)$$

Let $\lambda = t - \tau$ and $d\lambda = dt$. Then

$$\mathcal{L}[g(t) * h(t)] = \int_{0^-}^{\infty} g(\tau) \left[ \int_{0^-}^{\infty} e^{-s(\lambda+\tau)} h(\lambda) \, d\lambda \right] d\tau \qquad (9.92)$$

$$\mathcal{L}[g(t) * h(t)] = \int_{0^-}^{\infty} e^{-s\tau} g(\tau) \left[ \underbrace{\int_{0^-}^{\infty} e^{-s\lambda} h(\lambda) \, d\lambda}_{H(s)} \right] d\tau \qquad (9.93)$$

$$\mathcal{L}[g(t) * h(t)] = H(s) \int_{0^-}^{\infty} e^{-s\tau} g(\tau) \, d\tau = G(s)H(s).$$

The time-domain convolution property of the Laplace transform is

$$\boxed{g(t) * h(t) \overset{\mathcal{L}}{\longleftrightarrow} G(s)H(s).} \qquad (9.94)$$

The Laplace transform of a product of time-domain functions is

$$\mathcal{L}[g(t)h(t)] = \int_{0^-}^{\infty} g(t)h(t)e^{-st} \, dt \qquad (9.95)$$

$$\mathcal{L}[g(t)h(t)] = \int_{0^-}^{\infty} \left[ \frac{1}{j2\pi} \int_{\sigma-j\infty}^{\sigma+j\infty} G(w)e^{wt} \, dw \right] h(t)e^{-st} \, dt,$$

where $\sigma$ is chosen to make $G(s)$ and $H(s)$ exist. Doing the $t$ integration first,

$$\mathcal{L}[g(t)h(t)] = \frac{1}{j2\pi} \int_{\sigma-j\infty}^{\sigma+j\infty} G(w) \left[ \int_{0^-}^{\infty} h(t)e^{-(s-w)t} \, dt \right] dw. \qquad (9.96)$$

If H($s$) exists, then

$$\int_{0^-}^{\infty} h(t)e^{-(s-w)t} \, dt = H(s - w) \qquad (9.97)$$

and

$$\mathcal{L}[g(t)h(t)] = \frac{1}{j2\pi} \int_{\sigma-j\infty}^{\sigma+j\infty} G(w)H(s - w) \, dw. \qquad (9.98)$$

Therefore,

$$\boxed{g(t)h(t) \overset{\mathcal{L}}{\longleftrightarrow} \frac{1}{j2\pi} \int_{\sigma-j\infty}^{\sigma+j\infty} G(w)H(s - w) \, dw.} \qquad (9.99)$$

The integral in (9.99) is almost an aperiodic convolution in the sense previously defined in Chapter 3 but not exactly. This is a contour integral in the complex plane and is beyond the scope of this text.

The multiplication–convolution duality property is important because it is the basis of the idea of the transfer function just as it was with the Fourier transform. The basic system operation of convolving the excitation with the impulse response in the time domain to get the time-domain response

$$y(t) = x(t) * h(t) \tag{9.100}$$

is converted to the multiplication of the excitation by the transfer function in the frequency domain to get the frequency-domain response

$$Y(s) = X(s)H(s). \tag{9.101}$$

## INTEGRATION

The integration property is easy to prove, using the convolution property just proven in the section on multiplication–convolution duality and the fact that

$$g(t) * u(t) = \int_{-\infty}^{\infty} g(\tau)u(t - \tau)\, d\tau = \int_{0^-}^{t} g(\tau)\, d\tau \tag{9.102}$$

$$g(t) * u(t) \xleftrightarrow{\mathcal{L}} G(s)U(s) = \frac{1}{s}G(s). \tag{9.103}$$

Therefore,

$$\int_{0^-}^{t} g(\tau)\, d\tau \xleftrightarrow{\mathcal{L}} \frac{1}{s}G(s). \tag{9.104}$$

## INITIAL VALUE THEOREM

Using the time-differentiation-once property of the Laplace transform,

$$\mathcal{L}\left\{\frac{d}{dt}(g(t))\right\} = \int_{0^-}^{\infty} \frac{d}{dt}(g(t))e^{-st}\, dt = sG(s) - g(0^-). \tag{9.105}$$

Let $s \to \infty$; then

$$\lim_{s \to \infty} \int_{0^-}^{\infty} \frac{d}{dt}(g(t))e^{-st}\, dt = \lim_{s \to \infty} [sG(s) - g(0^-)] \tag{9.106}$$

$$\int_{0^-}^{\infty} \lim_{s \to \infty} \left\{\frac{d}{dt}(g(t))e^{-st}\right\} dt = \lim_{s \to \infty} [sG(s) - g(0^-)]. \tag{9.107}$$

***Case 1*** g(t) is continuous at $t = 0$. If the Laplace transform of g(t), G(s), exists for $\mathrm{Re}(s) = \sigma > \sigma_0$, the quantity $(d/dt)(g(t))e^{-st}$ approaches zero as $s$ approaches infinity and

$$0 = \lim_{s \to \infty} [sG(s) - g(0^-)] \tag{9.108}$$

$$g(0^-) = \lim_{s \to \infty} sG(s) \tag{9.109}$$

and, since $g(t)$ is continuous at $t = 0$, $g(0^-) = g(0^+)$ and

$$g(0^+) = \lim_{s \to \infty} sG(s). \tag{9.110}$$

***Case 2*** $g(t)$ is discontinuous at $t = 0$.    In this case, the discontinuity of $g(t)$ at $t = 0$ means that the derivative of $g(t)$ has an impulse at $t = 0$ and the strength of the impulse is $g(0^+) - g(0^-)$. Now the integral $\lim_{s \to \infty} \int_{0^-}^{\infty} (d/dt)(g(t))e^{-st}\, dt$ becomes

$$\lim_{s \to \infty} \int_{0^-}^{\infty} \frac{d}{dt}(g(t))e^{-st}\, dt = \lim_{s \to \infty} \int_{0^-}^{0^+} [g(0^+) - g(0^-)]\delta(t)e^{-st}\, dt + \underbrace{\lim_{s \to \infty} \int_{0^+}^{\infty} \frac{d}{dt}(g(t))e^{-st}\, dt}_{=0}$$

$$\tag{9.111}$$

and, using the sampling property of the impulse in the first integral of (9.111),

$$\lim_{s \to \infty} \int_{0^-}^{\infty} \frac{d}{dt}(g(t))e^{-st}\, dt = \lim_{s \to \infty} [g(0^+) - g(0^-)] = g(0^+) - g(0^-). \tag{9.112}$$

Therefore,

$$g(0^+) - g(0^-) = \lim_{s \to \infty} [sG(s) - g(0^-)] = \lim_{s \to \infty} sG(s) - g(0^-) \tag{9.113}$$

or

$$\boxed{g(0^+) = \lim_{s \to \infty} sG(s),} \tag{9.114}$$

and the result is the same as in case 1.

## FINAL VALUE THEOREM

From the time-differentiation-once property of the Laplace transform,

$$\lim_{s \to 0} \int_{0^-}^{\infty} \frac{d}{dt}(g(t))e^{-st}\, dt = \lim_{s \to 0} [sG(s) - g(0^-)] \tag{9.115}$$

$$\int_{0^-}^{\infty} \lim_{s \to 0} \left\{ \frac{d}{dt}(g(t))e^{-st} \right\} dt = \lim_{s \to 0} [sG(s) - g(0^-)] \tag{9.116}$$

$$\int_{0^-}^{\infty} \frac{d}{dt}(g(t))\, dt = \lim_{s \to 0} [sG(s) - g(0^-)] \tag{9.117}$$

$$\lim_{t \to \infty} [g(t) - g(0^-)] = \lim_{s \to 0} [sG(s) - g(0^-)]. \tag{9.118}$$

Then, if $\lim_{t \to \infty} g(t)$ exists, the final value theorem of the Laplace transform is

$$\boxed{\lim_{t \to \infty} g(t) = \lim_{s \to 0} s G(s).} \tag{9.119}$$

It must be stressed that this property only applies if $\lim_{t \to \infty} g(t)$ exists. It is possible that $\lim_{s \to 0} s G(s)$ exists but $\lim_{t \to \infty} g(t)$ does not. For example, suppose

$$X(s) = \frac{s}{s^2 + \omega_0^2}. \tag{9.120}$$

Then

$$\lim_{s \to 0} s X(s) = \lim_{s \to 0} \frac{s^2}{s^2 + \omega_0^2} = 0. \tag{9.121}$$

But the inverse Laplace transform of $X(s)$ is

$$x(t) = \cos(\omega_0 t) \tag{9.122}$$

and $\lim_{t \to \infty} x(t)$ does not exist. Therefore, the conclusion from (9.121) that the final value of $x(t)$ is zero, is wrong.

## EXAMPLE 9.7

Find the final value of the response $y(t)$ of a system whose transfer function is

$$H(s) = \frac{s+3}{s^2 + 4s + 5} \tag{9.123}$$

when the system is excited by a unit step and when it is excited by a unit impulse.

### ■ Solution

If the system is excited by a unit step, the Laplace transform of the response is

$$H_{-1}(s) = \frac{1}{s} \frac{s+3}{s^2 + 4s + 5} \tag{9.124}$$

and the final value of $h_{-1}(t)$ is then

$$\lim_{t \to \infty} h_{-1}(t) = \lim_{s \to 0} s H_{-1}(s) = \lim_{s \to 0} s \frac{1}{s} \frac{s+3}{s^2 + 4s + 5} = \frac{3}{5}. \tag{9.125}$$

If the system is excited by a unit impulse, the Laplace transform of the response is

$$H(s) = \frac{s+3}{s^2 + 4s + 5} \tag{9.126}$$

and the final value of $h(t)$ is then

$$\lim_{t \to \infty} h(t) = \lim_{s \to 0} s H(s) = \lim_{s \to 0} s \frac{s+3}{s^2 + 4s + 5} = 0. \tag{9.127}$$

## SUMMARY OF PROPERTIES OF THE UNILATERAL LAPLACE TRANSFORM

Linearity
$$\alpha g(t) + \beta h(t) \overset{\mathcal{L}}{\longleftrightarrow} \alpha G(s) + \beta H(s)$$

Time shifting
$$g(t - t_0) \overset{\mathcal{L}}{\longleftrightarrow} G(s)e^{-st_0} \qquad t_0 > 0$$

Complex-frequency shifting
$$e^{at}g(t) \overset{\mathcal{L}}{\longleftrightarrow} G(s - a)$$

Time scaling
$$g(at) \longleftrightarrow \frac{1}{a}G\left(\frac{s}{a}\right) \qquad a > 0$$

Frequency scaling
$$\frac{1}{a}g\left(\frac{t}{a}\right) \overset{\mathcal{L}}{\longleftrightarrow} G(as) \qquad a > 0$$

Time differentiation once
$$\frac{d}{dt}(g(t)) \overset{\mathcal{L}}{\longleftrightarrow} sG(s) - g(0^-)$$

Time differentiation twice
$$\frac{d^2}{dt^2}(g(t)) \overset{\mathcal{L}}{\longleftrightarrow} s^2 G(s) - sg(0^-) - \frac{d}{dt}(g(t))_{t=0^-}$$

Complex frequency differentiation
$$-tg(t) \overset{\mathcal{L}}{\longleftrightarrow} \frac{d}{ds}(G(s))$$

Multiplication–convolution duality
$$g(t) * h(t) \overset{\mathcal{L}}{\longleftrightarrow} G(s)H(s)$$

$$g(t)h(t) \overset{\mathcal{L}}{\longleftrightarrow} \frac{1}{j2\pi}\int_{\sigma-j\infty}^{\sigma+j\infty} G(w)H(s-w)\,dw$$

Integration
$$\int_{0^-}^{t} g(\tau)\,d\tau \overset{\mathcal{L}}{\longleftrightarrow} \frac{G(s)}{s}$$

Initial value theorem
$$g(0^+) = \lim_{s\to\infty} sG(s)$$

Final value theorem
$$\lim_{t\to\infty} g(t) = \lim_{s\to 0} sG(s) \qquad \text{if } \lim_{t\to\infty} g(t) \text{ exists}$$

These properties, along with the table of common Laplace transforms in Appendix F, can be used to solve a wide variety of practical engineering problems.

# 9.4 THE INVERSE LAPLACE TRANSFORM USING PARTIAL-FRACTION EXPANSION

The table of Laplace transforms in Appendix F was developed using the integral definitions of the forward and inverse Laplace transforms. In engineering practice it is rare to use the integral definitions to find forward or inverse transforms. It is much more common to use the tables and properties to find transforms because nearly every real engineering problem involves linear combinations of functions which appear in standard tables.

A very common type of problem in signal and system analysis using Laplace methods is to find the inverse transform of an *s*-domain function in the form of a ratio of polynomials in $s$,

$$G(s) = \frac{b_N s^N + b_{N-1} s^{N-1} + \cdots + b_1 s + b_0}{s^D + a_{D-1} s^{D-1} + \cdots + a_1 s + a_0} \tag{9.128}$$

where the numerator and denominator coefficients $b$ and $a$, respectively, are constants. Since the order of the numerator and denominator are arbitrary, this function does not appear in standard tables of Laplace transforms. But, under certain very common conditions, using a technique called *partial-fraction expansion,* it can be expressed as a sum of functions which *do* appear in standard tables of Laplace transforms.

It is always possible (in principle at least) to factor the denominator polynomial and to put the function into the form

$$G(s) = \frac{b_N s^N + b_{N-1} s^{N-1} + \cdots + b_1 s + b_0}{(s - p_1)(s - p_2) \cdots (s - p_D)} \tag{9.129}$$

where the $p$'s are the poles of $G(s)$. Let's assume, for now, the simplest case, that there are no repeated poles and that $D > N$, making the fraction proper in $s$. Once the poles have been identified we should be able to write the function in the partial-fraction form,

$$G(s) = \frac{K_1}{s - p_1} + \frac{K_2}{s - p_2} + \cdots + \frac{K_D}{s - p_D}, \tag{9.130}$$

*if* we can find the correct values of the $K$'s. For this form of the function to be correct the equation

$$\frac{b_N s^N + b_{N-1} s^{N-1} + \cdots + b_1 s + b_0}{(s - p_1)(s - p_2) \cdots (s - p_D)} = \frac{K_1}{s - p_1} + \frac{K_2}{s - p_2} + \cdots + \frac{K_D}{s - p_D} \tag{9.131}$$

must be satisfied for any arbitrary value of $s$. This equation can be solved by putting the right side into the form of a single fraction with a common denominator which is the same as the left-side denominator, and then setting the coefficients of each power of $s$ in the numerators equal and solving those $D$ equations for the $D$ number of $K$'s. But there is an easier way. Multiply both sides of (9.131) by $s - p_1$.

$$(s - p_1) \frac{b_N s^N + b_{N-1} s^{N-1} + \cdots + b_1 s + b_0}{(s - p_1)(s - p_2) \cdots (s - p_D)}$$

$$= \left[ (s - p_1) \frac{K_1}{s - p_1} + (s - p_1) \frac{K_2}{s - p_2} + \cdots + (s - p_1) \frac{K_D}{s - p_D} \right] \tag{9.132}$$

or

$$\frac{b_N s^N + b_{N-1} s^{N-1} + \cdots + b_1 s + b_0}{(s - p_2) \cdots (s - p_D)}$$

$$= K_1 + (s - p_1) \frac{K_2}{s - p_2} + \cdots + (s - p_1) \frac{K_D}{s - p_D}. \tag{9.133}$$

Since (9.133) must be satisfied for any arbitrary value of $s$, let $s = p_1$. All the factors $(s - p_1)$ on the right side become zero, and (9.133) becomes

$$K_1 = \frac{b_N p_1^N + b_N p_1^{N-1} + \cdots + b_1 p_1 + b_0}{(p_1 - p_2) \cdots (p_1 - p_D)} \tag{9.134}$$

and we immediately have the value of $K_1$. We can use the same technique to find all the other $K$'s. Then, using the Laplace transform pair

$$e^{-at}u(t) \overset{\mathcal{L}}{\longleftrightarrow} \frac{1}{s+a} \tag{9.135}$$

we can find the inverse Laplace transform as

$$g(t) = (K_1 e^{p_1 t} + K_2 e^{p_2 t} + \cdots + K_D e^{p_D t})u(t). \tag{9.136}$$

The most common situation in practice is that there are no repeated poles, but let's see what happens if we have two poles which are identical,

$$G(s) = \frac{b_N s^N + b_{N-1} s^{N-1} + \cdots + b_1 s + b_0}{(s - p_1)^2 (s - p_3) \cdots (s - p_D)}. \tag{9.137}$$

If we try the same technique to find the partial-fraction form, we get

$$G(s) = \frac{K_{11}}{s - p_1} + \frac{K_{12}}{s - p_1} + \frac{K_3}{s - p_3} + \cdots + \frac{K_D}{s - p_D}. \tag{9.138}$$

But this can be written as

$$G(s) = \frac{K_{11} + K_{12}}{s - p_1} + \frac{K_3}{s - p_3} + \cdots + \frac{K_D}{s - p_D} = \frac{K_1}{s - p_1} + \frac{K_3}{s - p_3} + \cdots + \frac{K_D}{s - p_D}. \tag{9.139}$$

We see then, that the sum of two arbitrary constants $K_{11} + K_{12}$ is really only a single arbitrary constant; there are really only a $D - 1$ number of $K$'s, not a $D$ number of $K$'s; and when we form the common denominator, it is not the same as the denominator of the original function. We could change the form of the partial-fraction expansion to

$$G(s) = \frac{K_1}{(s - p_1)^2} + \frac{K_3}{s - p_3} + \cdots + \frac{K_D}{s - p_D}. \tag{9.140}$$

Then, if we tried to solve the equation by finding a common denominator and equating equal powers of $s$, we would find that we have $D$ equations in $D - 1$ unknowns and there is no unique solution. The solution to this problem is to find a partial-fraction expansion in the form

$$G(s) = \frac{K_{12}}{(s - p_1)^2} + \frac{K_{11}}{s - p_1} + \frac{K_3}{s - p_3} + \cdots + \frac{K_D}{s - p_D}. \tag{9.141}$$

We can find $K_{12}$ by multiplying both sides of

$$\frac{b_N s^N + b_{N-1} s^{N-1} + \cdots + b_1 s + b_0}{(s - p_1)^2 (s - p_3) \cdots (s - p_D)}$$

$$= \frac{K_{12}}{(s - p_1)^2} + \frac{K_{11}}{s - p_1} + \frac{K_3}{s - p_3} + \cdots + \frac{K_D}{s - p_D} \tag{9.142}$$

by $(s - p_1)^2$ yielding

$$\frac{b_N s^N + b_{N-1} s^{N-1} + \cdots + b_1 s + b_0}{(s - p_3) \cdots (s - p_D)}$$

$$= \left[ K_{12} + (s - p_1) K_{11} + (s - p_1)^2 \frac{K_3}{s - p_3} + \cdots + (s - p_1)^2 \frac{K_D}{s - p_D} \right] \tag{9.143}$$

and then letting $s = p_1$, yielding

$$K_{12} = \frac{b_N p_1^N + b_{N-1} p_1^{N-1} + \cdots + b_1 p_1 + b_0}{(p_1 - p_3) \cdots (p_1 - p_D)}.$$

But when we try to find $K_{11}$ by the usual technique, we encounter another problem,

$$(s - p_1) \frac{b_N s^N + b_{N-1} s^{N-1} + \cdots + b_1 s + b_0}{(s - p_1)^2 (s - p_3) \cdots (s - p_D)}$$

$$= \left[ (s - p_1) \frac{K_{12}}{(s - p_1)^2} + (s - p_1) \frac{K_{11}}{s - p_1} + (s - p_1) \frac{K_3}{s - p_3} + \cdots + (s - p_1) \frac{K_D}{s - p_D} \right]$$

(9.144)

or

$$\frac{b_N s^N + b_{N-1} s^{N-1} + \cdots + b_1 s + b_0}{(s - p_1)(s - p_3) \cdots (s - p_D)} = \frac{K_{12}}{s - p_1} + K_{11}. \qquad (9.145)$$

Now if we set $s = p_1$ we get division by zero on both sides of the equation and we cannot solve it for $K_{11}$. But we can avoid this problem by multiplying through by $(s - p_1)^2$ yielding

$$\frac{b_N s^N + b_{N-1} s^{N-1} + \cdots + b_1 s + b_0}{(s - p_3) \cdots (s - p_D)}$$

$$= \left[ K_{12} + (s - p_1) K_{11} + (s - p_1)^2 \frac{K_3}{s - p_3} + \cdots + (s - p_1)^2 \frac{K_D}{s - p_D} \right] \qquad (9.146)$$

as in (9.143), and then differentiating with respect to $s$, yielding

$$\frac{d}{ds} \left[ \frac{b_N s^N + b_{N-1} s^{N-1} + \cdots + b_1 s + b_0}{(s - p_3) \cdots (s - p_D)} \right]$$

$$= \left[ K_{11} + \frac{(s - p_3)2(s - p_1) - (s - p_1)^2}{(s - p_3)^2} K_3 + \cdots \right.$$

$$\left. + \frac{(s - p_q)2(s - p_1) - (s - p_1)^2}{(s - p_D)^2} K_D \right]. \qquad (9.147)$$

Then setting $s = p_1$ and solving for $K_{11}$,

$$K_{11} = \frac{d}{ds} \left[ \frac{b_N s^N + b_{N-1} s^{N-1} + \cdots + b_1 s + b_0}{(s - p_3) \cdots (s - p_D)} \right]_{s \to p_1} = \frac{d}{ds} [(s - p_1)^2 G(s)]_{s \to p_1}. \qquad (9.148)$$

If there were a higher-order repeated root such as a triple, quadruple, etc., we could find the coefficients by extending this differentiation idea to multiple derivatives. In general, if H($s$) is of the form

$$H(s) = \frac{b_N s^N + b_{N-1} s^{N-1} + \cdots + b_1 s + b_0}{(s - p_1)(s - p_2) \cdots (s - p_{D-1})(s - p_D)^m} \qquad (9.149)$$

with $D - 1$ distinct poles and a repeated $D$th pole of order $m$, it can be written as

$$\mathrm{H}(s) = \frac{K_1}{s - p_1} + \frac{K_2}{s - p_2} + \cdots + \frac{K_{D-1}}{s - p_{D-1}} + \frac{K_{D,m}}{(s - p_D)^m}$$
$$+ \frac{K_{D,m-1}}{(s - p_D)^{m-1}} + \cdots + \frac{K_{D,1}}{s - p_D}, \tag{9.150}$$

where the $K$'s for the unrepeated poles are found as before and where the $K$ for repeated pole $q$ of order $m$ is

$$K_{qk} = \frac{1}{(m - k)!} \frac{d^{m-k}}{ds^{m-k}} [(s - p_q)^m \mathrm{H}(s)]_{s \to p_q} \qquad k = 1, 2, \ldots, m \tag{9.151}$$

and it is understood that $0! = 1$.

Let's now examine the effect of a violation of one of the assumptions in the original explanation of the partial-fraction expansion method, the assumption that $G(s)$ is a proper fraction in $s$. If $N \geq D$, we cannot expand in partial fractions because the partial-fraction expression is in the form

$$G(s) = \frac{K_1}{s - p_1} + \frac{K_2}{s - p_2} + \cdots + \frac{K_D}{s - p_D}, \tag{9.152}$$

and, if we were to combine these fractions by finding a common denominator, the resulting numerator could not have a power of $s$ greater than $D - 1$. Therefore, any ratio of polynomials in $s$ that is to be expanded in partial fractions must be proper in $s$. This is not really much of a restriction because, if the fraction is improper in $s$, we can always synthetically divide the numerator by the denominator until we have a remainder that is of a lower order than the denominator. Then we will have an expression consisting of the sum of terms with nonnegative integer powers of $s$ plus a proper fraction in $s$. The terms with nonnegative powers of $s$ have inverse Laplace transforms which are impulses and higher-order singularities (see Example 9.9).

Now that we have seen how to find an inverse transform using partial-fraction expansion we can show under what conditions a function of the form

$$G(s) = \frac{b_N s^N + b_{N-1} s^{N-1} + \cdots + b_1 s + b_0}{s^D + a_{D-1} s^{D-1} + \cdots + a_1 s + a_0} \tag{9.153}$$

has an inverse transform for which the final value theorem applies. First, if the fraction is improper in $s$, then the numerator should be synthetically divided by the denominator until a proper fraction is formed. Then the denominator is factored and, if the poles are distinct, the function can be expressed in the partial-fraction form

$$G(s) = \frac{K_1}{s - p_1} + \frac{K_2}{s - p_2} + \cdots + \frac{K_D}{s - p_D}. \tag{9.154}$$

The form of the corresponding time-domain function is

$$g(t) = K_1 e^{p_1 t} + K_2 e^{p_2 t} + \cdots + K_D e^{p_D t}. \tag{9.155}$$

If the poles are all in the open left half-plane, all the terms in (9.155) approach zero as time approaches infinity, $\lim_{t \to \infty} g(t)$ is zero, and the final value theorem applies. If exactly one of the poles is at zero, then one of the terms in $g(t)$ is a constant and $\lim_{t \to \infty} g(t)$ still exists, but is not zero, and the final value theorem still applies. Let the single pole at zero be $p_1$. Then

$$g(t) = K_1 + K_2 e^{p_2 t} + \cdots + K_D e^{p_D t} \tag{9.156}$$

and

$$\lim_{t \to \infty} g(t) = \lim_{t \to \infty} (K_1 + K_2 e^{p_2 t} + \cdots + K_D e^{p_D t}) = K_1. \tag{9.157}$$

The corresponding calculation in the frequency domain is

$$\lim_{s \to 0} sG(s) = \lim_{s \to 0} s \left[ \frac{K_1}{s} + \frac{K_2}{s - p_2} + \cdots + \frac{K_D}{s - p_D} \right] = K_1. \tag{9.158}$$

If there is a pole on the $\omega$ axis other than at zero, there is at least one complex-conjugate pair of poles on the $\omega$ axis, $g(t)$ contains an undamped sinusoid, and $\lim_{t \to \infty} g(t)$ does not exist.

If there are any repeated poles on the $\omega$ axis, even at zero, $\lim_{t \to \infty} g(t)$ does not exist because the repeated pole introduces a time-domain function of the form $Kt$ or $Kt \cos(\omega_0 t + \theta)$, each of which grows with time. So we can summarize by saying that if there are any poles in the right half-plane or if there is more than one pole on the $\omega$ axis, the final-value theorem does not apply.

MATLAB has a function `residue` for finding residues, which can be used in finding partial-fraction expansions. The syntax is

$$[\text{r}, \text{p}, \text{k}] \; = \; \text{residue}(\text{b}, \text{a})$$

where   b = vector of coefficients of descending powers of $s$ in numerator of expression

       a = vector of coefficients of descending powers of $s$ in denominator of expression

       r = vector of residues

       p = vector of pole locations

       k = vector of so-called direct terms which result when degree of numerator is equal to or greater than degree of denominator.

The vectors a and b must always include all powers of $s$ down through zero. The *residues* are the numerators in the partial-fraction expansion.

For example, suppose we want to expand the expression

$$H(s) = \frac{s^2 + 3s + 1}{s^4 + 5s^3 + 2s^2 + 7s + 3} \tag{9.159}$$

in partial fractions. In MATLAB,

```
»b = [1 3 1] ; a = [1 5 2 7 3] ;
»[r,p,k] = residue(b,a) ;
»r

r =

   -0.0856
    0.0496 - 0.2369i
    0.0496 + 0.2369i
   -0.0135
```

```
»p

p =

  -4.8587
   0.1441 + 1.1902i
   0.1441 - 1.1902i
  -0.4295

»k

k =

     []

»
```

So there are four poles, at $-4.8587, 0.1441 + j1.1902, 0.1441 - j1.1902$, and $-0.4295$, and the residues at those poles are $-0.0856, 0.0496 - j0.2369, 0.0496 + j0.2369$, and $-0.0135$, respectively. There are no direct terms because H($s$) is a proper fraction in $s$. Now we can write H($s$) as

$$H(s) = \frac{0.0496 - j0.2369}{s - 0.1441 - j1.1902} + \frac{0.0496 + j0.2369}{s - 0.1441 + j1.1902}$$

$$- \frac{0.0856}{s + 0.48587} - \frac{0.0135}{s + 0.4295} \qquad \textbf{(9.160)}$$

or, combining the two terms with complex poles and residues into one term,

$$H(s) = \frac{0.0991s + 0.5495}{s^2 - 0.2883s + 1.437} - \frac{0.0856}{s + 0.48587} - \frac{0.0135}{s + 0.4295}. \qquad \textbf{(9.161)}$$

**EXAMPLE 9.8**

Find the inverse Laplace transform of G($s$) = $\dfrac{10s}{(s + 1)(s + 3)}$.

■ **Solution**

We can expand this expression in partial fractions yielding

$$G(s) = -\frac{5}{s + 1} + \frac{15}{s + 3}. \qquad \textbf{(9.162)}$$

Then, using

$$e^{-at}u(t) \xleftrightarrow{\;\mathcal{L}\;} \frac{1}{s + a}, \qquad \textbf{(9.163)}$$

from the Laplace transform table in Appendix F, we get

$$g(t) = 5(3e^{-3t} - e^{-t})u(t). \qquad \textbf{(9.164)}$$

**EXAMPLE 9.9**

Find the inverse Laplace transform of $G(s) = \dfrac{10s^2}{(s+1)(s+3)}e^{-s}$.

### ■ Solution

The coefficient of $e^{-s}$ is an improper fraction in $s$. Synthetically dividing the numerator by the denominator we get

$$
s^2 + 4s + 3 \overline{)\begin{array}{l} 10 \\ 10s^2 \end{array}}
\qquad \text{or} \qquad
\frac{10s^2}{(s+1)(s+3)} = 10 - \frac{40s + 30}{s^2 + 4s + 3}.
\qquad \textbf{(9.165)}
$$

$$
\underline{10s^2 + 40s + 30}
$$
$$
-40s - 30
$$

Therefore,

$$
G(s) = e^{-s}\left[ 10 - \frac{40s + 30}{(s+1)(s+3)} \right].
\qquad \textbf{(9.166)}
$$

Expanding the (proper) fraction in $s$ in partial fractions,

$$
G(s) = \left[ 10 - 5\left( \frac{9}{s+3} - \frac{1}{s+1} \right) \right] e^{-s}.
\qquad \textbf{(9.167)}
$$

Then, using

$$
e^{-at}u(t) \xleftrightarrow{\;\mathcal{L}\;} \frac{1}{s+a}
\qquad \text{and} \qquad
\delta(t) \xleftrightarrow{\;\mathcal{L}\;} 1
\qquad \textbf{(9.168)}
$$

and the time-shifting property of the Laplace transform, we get

$$
g(t) = 10\delta(t-1) - 5\big(9e^{-3(t-1)} - e^{-(t-1)}\big)u(t-1)
\qquad \textbf{(9.169)}
$$

(Figure 9.12).

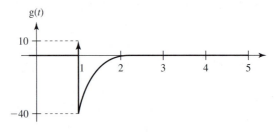

**Figure 9.12**
Inverse Laplace transform of
$G(s) = \dfrac{10s^2}{(s+1)(s+3)}e^{-s}$.

**EXAMPLE 9.10**

Find the inverse Laplace transform of $G(s) = \dfrac{s}{(s+3)(s^2+4s+5)}$.

**■ Solution**

If we take the usual route of finding a partial-fraction expansion, we must first factor the denominator,

$$G(s) = \frac{s}{(s+3)(s+2+j)(s+2-j)}. \tag{9.170}$$

Then, expanding in partial fractions,

$$G(s) = -\frac{\frac{3}{2}}{s+3} + \frac{(3-j)/4}{s+2+j} + \frac{(3+j)/4}{s+2-j}. \tag{9.171}$$

With complex roots like this we have a choice. We can (1) continue as though they were real roots, find a time-domain expression, and then simplify it or (2) combine the last two fractions into one fraction with all real coefficients and find its inverse Laplace transform by looking up that form in a table.

*Method 1:*

$$g(t) = \left(-\frac{3}{2}e^{-3t} + \frac{3-j}{4}e^{-(2+j)t} + \frac{3+j}{4}e^{-(2-j)t}\right)u(t). \tag{9.172}$$

This is a correct expression for $g(t)$, but it is not in the most convenient form. We can manipulate it into an expression containing only real-valued functions.

$$g(t) = \left(-\frac{3}{2}e^{-3t} + \frac{3e^{-(2+j)t} + 3e^{-(2-j)t} - je^{-(2+j)t} + je^{-(2-j)t}}{4}\right)u(t) \tag{9.173}$$

$$g(t) = \left(-\frac{3}{2}e^{-3t} + e^{-2t}\frac{3(e^{-jt} + e^{jt}) - j(e^{-jt} - e^{jt})}{4}\right)u(t)$$

$$g(t) = \frac{3}{2}\left\{e^{-2t}\left[\cos(t) - \frac{1}{3}\sin(t)\right] - e^{-3t}\right\}u(t).$$

*Method 2:*

$$G(s) = \frac{-\frac{3}{2}}{s+3} + \frac{1}{4}\frac{(3-j)(s+2-j) + (3+j)(s+2+j)}{s^2+4s+5} \tag{9.174}$$

$$G(s) = \frac{-\frac{3}{2}}{s+3} + \frac{1}{4}\frac{6s+10}{s^2+4s+5} = \frac{-\frac{3}{2}}{s+3} + \frac{6}{4}\frac{s+\frac{5}{3}}{(s+2)^2+1}$$

$$G(s) = \frac{-\frac{3}{2}}{s+3} + \frac{3}{2}\left[\frac{s+2}{(s+2)^2+1} - \frac{\frac{1}{3}}{(s+2)^2+1}\right].$$

Then using

$$e^{-\alpha t}\cos(\beta t)\,u(t) \overset{\mathcal{L}}{\longleftrightarrow} \frac{s+\alpha}{(s+\alpha)^2+\beta^2} \qquad \sigma > -\alpha \tag{9.175}$$

and

$$e^{-\alpha t} \sin(\beta t)\, u(t) \overset{\mathcal{L}}{\longleftrightarrow} \frac{\omega_0}{(s+\alpha)^2 + \beta^2} \qquad \sigma > -\alpha \tag{9.176}$$

$$g(t) = \frac{3}{2}\left\{ e^{-2t}\left[\cos(t) - \frac{1}{3}\sin(t)\right] - e^{-3t}\right\} u(t). \tag{9.177}$$

Realizing that there are two complex roots, another approach is to find the partial-fraction expansion in the form

$$G(s) = \frac{A}{s+3} + \frac{Bs + C}{s^2 + 4s + 5}. \tag{9.178}$$

$A$ is found exactly as before to be $-\frac{3}{2}$. Since (9.178) must be satisfied for any arbitrary value of $s$ and

$$G(s) = \frac{s}{(s+3)(s^2 + 4s + 5)}, \tag{9.179}$$

we can write

$$\left[\frac{s}{(s+3)(s^2 + 4s + 5)}\right]_{s=0} = \left[\frac{-\frac{3}{2}}{s+3} + \frac{Bs + C}{s^2 + 4s + 5}\right]_{s=0} \tag{9.180}$$

or

$$0 = -\frac{1}{2} + \frac{C}{5} \Rightarrow C = \frac{5}{2}. \tag{9.181}$$

Then

$$\frac{s}{(s+3)(s^2 + 4s + 5)} = \frac{-\frac{3}{2}}{s+3} + \frac{Bs + \frac{5}{2}}{s^2 + 4s + 5} \tag{9.182}$$

and we can find $B$ by letting $s$ be any convenient number, for example, one. Then

$$\frac{1}{40} = -\frac{3}{8} + \frac{B + \frac{5}{2}}{10} \Rightarrow B = \frac{3}{2} \tag{9.183}$$

and

$$G(s) = \frac{-\frac{3}{2}}{s+3} + \frac{3}{2}\frac{s + \frac{5}{3}}{s^2 + 4s + 5}. \tag{9.184}$$

This result is identical to (9.174), and the rest of the solution is, therefore, the same. ∎

## EXAMPLE 9.11

Find the inverse Laplace transform of

$$G(s) = \frac{s+5}{s^2(s+2)}. \tag{9.185}$$

■ **Solution**
This function has a repeated pole at zero. Therefore, the form of the partial fraction expansion must be

$$G(s) = \frac{K_{11}}{s^2} + \frac{K_{12}}{s} + \frac{K_3}{s+2}. \tag{9.186}$$

We find $K_{11}$ by multiplying $G(s)$ by $s^2$, and setting $s$ to zero in the remaining expression, yielding

$$K_{11} = [s^2 G(s)]_{s \to 0} = \frac{5}{2}. \qquad (9.187)$$

We find $K_{12}$ by multiplying $G(s)$ by $s^2$, differentiating with respect to $s$, and setting $s$ to zero in the remaining expression, yielding

$$K_{11} = \frac{d}{ds}[s^2 G(s)]_{s \to 0} = \frac{d}{ds}\left[\frac{s+5}{s+2}\right]_{s \to 0} = \left[\frac{(s+2) - (s+5)}{(s+2)^2}\right]_{s \to 0} = -\frac{3}{4}. \qquad (9.188)$$

We find $K_3$ by the usual method to be $\frac{3}{4}$. So

$$G(s) = \frac{5}{2s^2} - \frac{3}{4s} + \frac{3}{4(s+2)} \qquad (9.189)$$

and the inverse transform is

$$g(t) = \left(\frac{5}{2}t - \frac{3}{4} + \frac{3}{4}e^{-2t}\right)u(t) = \frac{10t - 3(1 - e^{-2t})}{4}u(t). \qquad (9.190)$$

## 9.5 LAPLACE TRANSFORM–FOURIER TRANSFORM EQUIVALENCE

The Laplace transform is really just a generalization of the CTFT, analyzing functions as linear combinations of general complex exponentials instead of linear combinations of a special case of complex exponentials, complex sinusoids. For many common functions, the Laplace and Fourier transforms are very simply related. For any function $g(t)$ which is zero before time $t = 0$ and whose Laplace-transform ROC includes the $\omega$ axis, the CTFT $G_{\mathcal{F}}(j\omega)$ or $G_{\mathcal{F}}(f)$ can be found from the Laplace transform $G_{\mathcal{L}}(s)$ by the functional transformation

$$G_{\mathcal{F}}(j\omega) = G_{\mathcal{L}}(s)|_{s \to j\omega} \qquad \text{or} \qquad G_{\mathcal{F}}(f) = G_{\mathcal{L}}(s)|_{s \to j2\pi f}. \qquad (9.191)$$

Notice that, because of the notation used for the $\omega$ form of the CTFT, the functions $G_{\mathcal{F}}(\ ) = G_{\mathcal{L}}(\ )$ are mathematically the same function and converting back and forth between $\omega$-form CTFTs and the Laplace transform is simply a process of exchanging the functional arguments $s$ and $j\omega$. So we don't need the $\mathcal{F}$ and $\mathcal{L}$ subscripts and can simply write

$$G(j\omega) = G(s)|_{s \to j\omega}. \qquad (9.192)$$

That is the principal reason that the $\omega$ form of the CTFT of a function $x(t)$ was written with the functional notation $X(j\omega)$ instead of $X(\omega)$.

## 9.6 SOLUTION OF DIFFERENTIAL EQUATIONS WITH INITIAL CONDITIONS

The power of the Laplace transform lies in its use in the analysis of linear system dynamics. This comes about because linear systems are described by linear differential equations and, after Laplace transformation, differentiation is represented by simple multiplication by $s$. Therefore, the solution of the differential equation is transformed into the solution of an algebraic equation. This could all be said of the Fourier transform also, but the unilateral Laplace transform is especially convenient for transient analysis of systems whose excitation begins at an initial time which can be identified as $t = 0$ and of unstable systems or systems driven by forcing functions which are unbounded as time increases.

## EXAMPLE 9.12

Solve the differential equation

$$\frac{d^2}{dt^2}(x(t)) + 7\frac{d}{dt}(x(t)) + 12x(t) = 0 \tag{9.193}$$

for times $t > 0$, subject to the initial conditions

$$x(0^-) = 2 \quad \text{and} \quad \frac{d}{dt}(x(t))_{t=0^-} = -4. \tag{9.194}$$

### ■ Solution

First Laplace-transform both sides of the equation,

$$s^2 X(s) - sx(0^-) - \frac{d}{dt}(x(t))_{t=0^-} + 7[sX(s) - x(0^-)] + 12X(s) = 0. \tag{9.195}$$

Then solve for $X(s)$,

$$X(s) = \frac{sx(0^-) + 7x(0^-) + \frac{d}{dt}(x(t))_{t=0^-}}{s^2 + 7s + 12} \tag{9.196}$$

or

$$X(s) = \frac{2s + 10}{s^2 + 7s + 12}.$$

Expanding $X(s)$ in partial fractions,

$$X(s) = \frac{4}{s+3} - \frac{2}{s+4}. \tag{9.197}$$

From the Laplace transform table in Appendix F,

$$e^{-\alpha t}u(t) \overset{\mathcal{L}}{\longleftrightarrow} \frac{1}{s+\alpha}. \tag{9.198}$$

Therefore, inverse Laplace transforming,

$$x(t) = (4e^{-3t} - 2e^{-4t})u(t). \tag{9.199}$$

Substituting this result into the original differential equation for times $t \geq 0$,

$$\frac{d^2}{dt^2}[4e^{-3t} - 2e^{-4t}] + 7\frac{d}{dt}[4e^{-3t} - 2e^{-4t}] + 12[4e^{-3t} - 2e^{-4t}] = 0 \tag{9.200}$$

$$36e^{-3t} - 32e^{-4t} - 84e^{-3t} + 56e^{-4t} + 48e^{-3t} - 24e^{-4t} = 0 \tag{9.201}$$

$$0 = 0, \tag{9.202}$$

which proves that the $x(t)$ found actually solves the differential equation. Also

$$x(0^-) = 4 - 2 = 2 \quad \text{and} \quad \frac{d}{dt}[x(t)]_{t=0^-} = -12 + 8 = -4, \tag{9.203}$$

which verifies that the solution also satisfies the stated initial conditions.

EXAMPLE 9.13

Let the lowpass filter in Figure 9.13 be excited by a unit voltage impulse at time $t = \tau, \tau > 0$. Find the response $v_{out}(t)$.

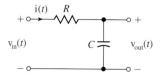

**■ Solution**

The differential equation describing this circuit for the general case in which the initial capacitor voltage $v_{out}(0^-)$ may not be zero, is

**Figure 9.13**
An $RC$ lowpass filter.

$$C v'_{out}(t) = \frac{v_{in}(t) - v_{out}(t)}{R} \tag{9.204}$$

Laplace transforming,

$$C[s V_{out}(s) - v_{out}(0^-)] = \frac{V_{in}(s) - V_{out}(s)}{R}. \tag{9.205}$$

For impulse excitation,

$$C[s V_{out}(s) - v_{out}(0^-)] = \frac{e^{-\tau s} - V_{out}(s)}{R}. \tag{9.206}$$

Rearranging and solving for $V_{out}(s)$,

$$V_{out}(s) = \frac{e^{-\tau s} + RC v_{out}(0^-)}{sRC + 1}. \tag{9.207}$$

Inverse Laplace transforming,

$$v_{out}(t) = \frac{e^{-(t-\tau)/RC}}{RC} u(t - \tau) + v_{out}(0^-) e^{-(t/RC)} u(t). \tag{9.208}$$

The first term is the response to the impulse excitation, and the second term is the decay of the initial capacitor voltage. Applying the initial value theorem

$$g(0^+) = \lim_{s \to \infty} s G(s) \tag{9.209}$$

to the $s$-domain expression for the output voltage, we get

$$v_{out}(0^+) = \lim_{s \to \infty} s \left\{ \frac{1}{RC} \frac{e^{-s\tau}}{s + (1/RC)} + \frac{v_{out}(0^-)}{s + (1/RC)} \right\} = v_{out}(0^-) \qquad \tau > 0. \qquad \text{Check}. \tag{9.210}$$

Notice what happens if we let $\tau$ be zero. In that case,

$$v_{out}(0^+) = \lim_{s \to \infty} s \left\{ \frac{1}{RC} \frac{1}{s + (1/RC)} + \frac{v_{out}(0^-)}{s + (1/RC)} \right\} = \frac{1}{RC} + v_{out}(0^-). \tag{9.211}$$

This simply indicates (correctly) that if the impulse occurs at time $t = 0$, the capacitor voltage at $t = 0^+$ changes from $v_{out}(0^+) = v_{out}(0^-)$ to $v_{out}(0^+) = (1/RC) + v_{out}(0^-)$ because of the charge that is dumped onto it by the impulse.

## 9.7  THE BILATERAL LAPLACE TRANSFORM

We began the chapter defining the forward Laplace transform by the integral

$$\mathcal{L}(x(t)) = X(s) = \int_{-\infty}^{\infty} x(t) e^{-st} \, dt. \tag{9.212}$$

Later, after examining the ROC for various types of signals, we found it convenient to restrict this definition to the unilateral Laplace transform. Although most practical analysis of systems is done using the unilateral form, the bilateral form is more general and has some usefulness in analyzing noncausal systems and/or systems with noncausal excitations. Also the bilateral Laplace transform can be considered the mother of all transforms because the unilateral Laplace transform, the Fourier transform, and the $z$ transform (to be introduced in Chapter 11) are all, in a very real sense, just special cases of the bilateral Laplace transform, some with notation changes. Now that we are familiar with the unilateral form, we can extend to the bilateral form by showing that the unilateral transform pairs can be used to find bilateral transform pairs.

## CALCULATION USING THE UNILATERAL LAPLACE TRANSFORM

Any signal can be expressed as the sum of three parts, the anticausal part occurring before time $t = 0$, the part occurring at time $t = 0$, and the causal part occurring after time $t = 0$,

$$x(t) = x_{ac}(t) + x_0(t) + x_c(t) \tag{9.213}$$

where

$$x_{ac}(t) = \begin{cases} x(t) & t < 0 \\ 0 & \text{otherwise} \end{cases} \tag{9.214}$$

$$x_0(t) = \begin{cases} x(t) & t = 0 \\ 0 & \text{otherwise} \end{cases} \tag{9.215}$$

$$x_c(t) = \begin{cases} x(t) & t > 0 \\ 0 & \text{otherwise} \end{cases} \tag{9.216}$$

(Figure 9.14). If the signal does not have an impulse at time $t = 0$ (like the first signal in Figure 9.14), the part of the signal occurring at time $t = 0$ has no effect on the Laplace transform and can be neglected because it has no signal energy. If the signal has an impulse at time $t = 0$, its effect can be considered separately and added to the transforms of the other two parts. The bilateral Laplace transform of $x(t)$ is

$$X(s) = \int_{-\infty}^{\infty} x(t)e^{-st}\, dt = \int_{-\infty}^{0^-} x(t)e^{-st}\, dt + \int_{0^-}^{0^+} x(t)e^{-st}\, dt + \int_{0^+}^{\infty} x(t)e^{-st}\, dt \tag{9.217}$$

or

$$X(s) = X_{ac}(s) + X_0(s) + X_c(s), \tag{9.218}$$

where

$$X_{ac}(s) = \int_{-\infty}^{0^-} x(t)e^{-st}\, dt \qquad X_0(s) = \int_{0^-}^{0^+} x(t)e^{-st}\, dt \qquad X_c(s) = \int_{0^+}^{\infty} x(t)e^{-st}\, dt. \tag{9.219}$$

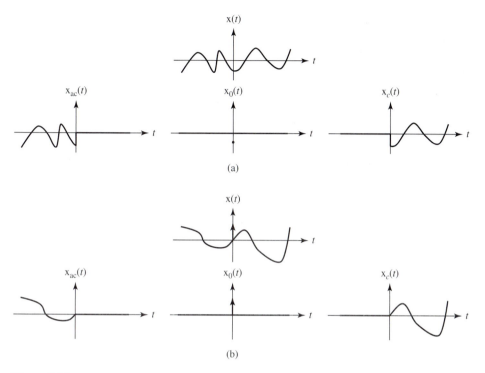

(a)

(b)

**Figure 9.14**
Two signals and their three parts.

Making the change of variable $t \to -t \Rightarrow dt \to -dt$ in the anticausal transform, we get

$$X_{ac}(s) = -\int_{\infty}^{0^+} x(-t)e^{st}\, dt = \int_{0^+}^{\infty} x(-t)e^{st}\, dt. \tag{9.220}$$

If (9.220) defines $X_{ac}(s)$, then $X_{ac}(-s)$ can be found by negating $s$ everywhere it occurs, yielding

$$X_{ac}(-s) = \int_{0^+}^{\infty} x(-t)e^{-st}\, dt = \int_{0^+}^{\infty} x_{ac}(-t)e^{-st}\, dt, \tag{9.221}$$

which is the unilateral Laplace transform of the time inverse of the anticausal part of the signal (which is causal).

The procedure for finding a bilateral Laplace transform using unilateral Laplace transforms is

1. Find the unilateral Laplace transform $X_c(s)$ of the causal signal $x_c(t)$ along with its ROC, the region to the right of its rightmost pole.
2. Find the unilateral Laplace transform $X_{ac}(-s)$ of the causal signal $x_{ac}(-t)$ along with its ROC, the region to the right of its rightmost pole.
3. Make the change of variable $s \to -s$ in $X_{ac}(-s)$ and in its ROC, yielding $X_{ac}(s)$, along with its ROC, the region to the left of its leftmost pole.

4.  If there is an impulse at time $t = 0$, find its Laplace transform as $X_0(s)$ along with its ROC, the entire $s$ plane. Otherwise $X_0(s) = 0$.

5.  Add $X_c(s)$, $X_0(s)$, and $X_{ac}(s)$ to form $X(s)$. The ROC of $X(s)$ is the region of the $s$ plane which is common to the ROCs of $X_c(s)$ and $X_{ac}(s)$. If such a region does not exist, the bilateral Laplace transform of $x(t)$ does not exist either.

## PROPERTIES

Some of the properties of the bilateral Laplace transform are not the same as the corresponding properties of the unilateral Laplace transform. The properties are summarized below without proof. The proofs are similar to the corresponding property proofs for the unilateral Laplace transform. One significant difference is that the region of convergence must be more carefully taken into account when applying the properties of the bilateral transform. Let $G(s) = \mathcal{L}(g(t))$ and $H(s) = \mathcal{L}(h(t))$ and let the ROC of G be $R_G$ and let the ROC of H be $R_H$.

Linearity
$$\alpha g(t) + \beta h(t) \xleftrightarrow{\mathcal{L}} \alpha G(s) + \beta H(s)$$
$$\text{ROC} = R_G \cap R_H$$

Time shifting
$$g(t - t_0) \xleftrightarrow{\mathcal{L}} G(s)e^{-st_0} \qquad \text{ROC} = R_G$$

Time scaling
$$g(at) \xleftrightarrow{\mathcal{L}} \frac{1}{|a|}G\left(\frac{s}{a}\right) \qquad \text{ROC} = \frac{R_G}{a}$$

Complex-frequency shifting
$$e^{s_0 t}g(t) \xleftrightarrow{\mathcal{L}} G(s - s_0) \qquad \begin{array}{l} \text{ROC} = R_G \\ \text{shifted to the right} \\ \text{by } s_0 \end{array}$$

Frequency scaling
$$\frac{1}{|a|}g\left(\frac{t}{a}\right) \xleftrightarrow{\mathcal{L}} G(as) \qquad \text{ROC} = aR_G$$

Time differentiation
$$\frac{d}{dt}(g(t)) \xleftrightarrow{\mathcal{L}} sG(s) \qquad \text{ROC} = R_G \text{ at least}$$

Complex-frequency differentiation
$$-tg(t) \xleftrightarrow{\mathcal{L}} \frac{d}{ds}(G(s)) \qquad \text{ROC} = R_G$$

Convolution
$$g(t) * h(t) \xleftrightarrow{\mathcal{L}} G(s)H(s)$$
$$\text{ROC} = R_G \cap R_H \text{ at least}$$

Integration
$$\int_{-\infty}^{t} g(\tau)\,d\tau \xleftrightarrow{\mathcal{L}} \frac{G(s)}{s}$$
$$\text{ROC} = R_G \cap \{\text{Re}(s) > 0\} \text{ at least}$$

---

**EXAMPLE 9.14**

Find the bilateral Laplace transform of

$$x(t) = e^{-at}\cos(\omega_{01}t)\,u(t) + e^{bt}\cos(\omega_{02}t)\,u(-t). \qquad \textbf{(9.222)}$$

■ **Solution**

This signal is already written as the sum of a causal and an anticausal signal, and there is no impulse at time $t = 0$. First we find the unilateral transform of the causal part

$$x_c(t) = e^{-at}\cos(\omega_{01}t)\,u(t) \qquad \textbf{(9.223)}$$

starting with the table entry from Appendix F,

$$e^{-\alpha t} \cos(\omega_0 t)\, u(t) \xleftrightarrow{\;\mathcal{L}\;} \frac{s + \alpha}{(s + \alpha)^2 + \omega_0^2} \qquad \sigma > -\alpha. \qquad (9.224)$$

Then

$$x_c(t) = e^{-at} \cos(\omega_{01} t)\, u(t) \xleftrightarrow{\;\mathcal{L}\;} X_c(s) = \frac{s + a}{(s + a)^2 + \omega_{01}^2} \qquad \sigma > -a. \qquad (9.225)$$

Next we find the unilateral transform of the time inverse of the anticausal signal,

$$x_{ac}(-t) = e^{-bt} \cos(-\omega_{02} t)\, u(t). \qquad (9.226)$$

From (9.224)

$$e^{-bt} \cos(\omega_{02} t)\, u(t) \xleftrightarrow{\;\mathcal{L}\;} \frac{s + b}{(s + b)^2 + \omega_{02}^2} \qquad \sigma > -b, \qquad (9.227)$$

and, because the cosine is an even function,

$$x_{ac}(-t) = e^{-bt} \cos(-\omega_{02} t)\, u(t) \xleftrightarrow{\;\mathcal{L}\;} X_{ac}(-s) = \frac{s + b}{(s + b)^2 + \omega_{02}^2} \qquad \sigma > -b. \qquad (9.228)$$

Therefore,

$$X(s) = \frac{s + a}{(s + a)^2 + \omega_{01}^2} + \frac{-s + b}{(-s + b)^2 + \omega_{02}^2} \qquad \sigma > -a \quad \text{and} \quad -\sigma > -b \qquad (9.229)$$

or

$$X(s) = \frac{s + a}{(s + a)^2 + \omega_{01}^2} - \frac{s - b}{(s - b)^2 + \omega_{02}^2} \qquad -a < \sigma < b. \qquad (9.230)$$

If $b > -a$, then the bilateral Laplace transform of $x(t)$ exists. Otherwise it does not. The condition $b > -a$ can be satisfied in infinitely many ways, some of which are illustrated in Figure 9.15. If $b$ and $a$ are both positive (as in Figure 9.15a), then the ROC contains the $\omega$ axis and $x(t)$ is also Fourier-transformable.

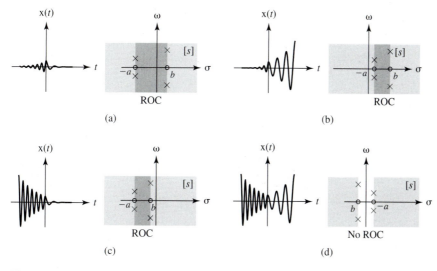

**Figure 9.15**
Four noncausal signals with their pole–zero plots and ROCs.

**EXAMPLE 9.15**

The noncausal signal

$$x(t) = e^{-3t}u(t) + e^{-t}u(-t) \tag{9.231}$$

is the excitation of a noncausal highpass filter whose impulse response is

$$h(t) = \delta(t) - e^{-2|t|} \tag{9.232}$$

(Figure 9.16). Find the response of the system $y(t)$.

■ **Solution**

The bilateral Laplace transform of the noncausal excitation $x(t)$ is

$$X(s) = \frac{1}{s+3} - \frac{1}{s+1} = -\frac{2}{(s+3)(s+1)} \qquad -3 < \sigma < -1. \tag{9.233}$$

The impulse response is also noncausal, so its bilateral Laplace transform is found by the same general method as that of a noncausal signal,

$$H(s) = 1 - \frac{1}{s+2} + \frac{1}{s-2} = \frac{s^2}{(s+2)(s-2)} \qquad -2 < \sigma < 2. \tag{9.234}$$

The bilateral Laplace transform of the response is the product of the bilateral Laplace transforms of the excitation and the impulse response, and its ROC is the region of the $s$ plane common to both ROCs,

$$Y(s) = -\frac{2s^2}{(s+2)(s-2)(s+3)(s+1)} \qquad -2 < \sigma < -1, \tag{9.235}$$

or, expanding in partial fractions,

$$Y(s) = -\left[ \frac{2}{s+2} + \frac{\frac{2}{15}}{s-2} - \frac{\frac{9}{5}}{s+3} - \frac{\frac{1}{3}}{s+1} \right] \qquad -2 < \sigma < -1. \tag{9.236}$$

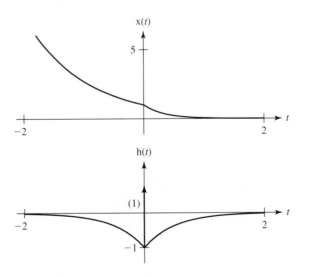

**Figure 9.16**
A noncausal excitation and a noncausal impulse response.

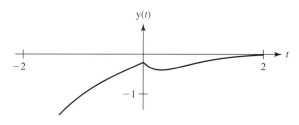

**Figure 9.17**
Response of the noncausal system to the noncausal
excitation.

There are two poles $s = 2$ and $s = -1$ to the right of the ROC and two poles $s = -3$ and $s = -2$ to the left of the ROC. The poles to the right determine the anticausal response, and the poles to the left determine the causal response. The inverse bilateral transform is found by reversing the process of finding the forward bilateral transform. Inverse-transform the causal part just like a unilateral inverse Laplace transform. Change the sign of $s$ in the anti-causal part, find the unilateral inverse Laplace transform, and then make the transformation $t \to -t$.

$$y(t) = - \left[ 2e^{-2t}u(t) - \frac{9}{5}e^{-3t}u(t) - \frac{2}{15}e^{2t}u(-t) + \frac{1}{3}e^{-t}u(-t) \right] \qquad \textbf{(9.237)}$$

or

$$y(t) = \frac{(27e^{-3t} - 30e^{-2t})u(t) + (2e^{2t} - 5e^{-t})u(-t)}{15} \qquad \textbf{(9.238)}$$

(Figure 9.17). ∎

## 9.8 SUMMARY OF IMPORTANT POINTS

1. Laplace transforms can be found for some functions for which the Fourier transform, even in its generalized form, does not exist.
2. Laplace transforms represent functions as combinations of complex exponentials, the eigenfunctions of LTI systems, instead of as combinations of complex sinusoids.
3. A Laplace transform is defined only in its region of convergence in the $s$ plane.
4. Restriction of the Laplace transform to the unilateral form simplifies consideration of the region of convergence and has some advantages in practical applications of Laplace transforms.
5. In most practical situations the inverse Laplace transform is found using the technique of partial-fraction expansion.
6. If the region of convergence of the Laplace transform of a function contains the $\omega$ axis, the function also has a Fourier transform.
7. The unilateral Laplace transform is very convenient in the solution of differential equations with initial conditions.
8. The bilateral Laplace transform can be found using unilateral Laplace transform tables and can be used to analyze noncausal signals and/or systems.

## EXERCISES WITH ANSWERS

1. Sketch the pole–zero plot and region of convergence (if it exists) for these signals.

    a. $x(t) = e^{-8t}u(t)$

    b. $x(t) = e^{3t}\cos(20\pi t)\, u(-t)$

    c. $x(t) = e^{2t}u(-t) - e^{-5t}u(t)$

**Answers:**

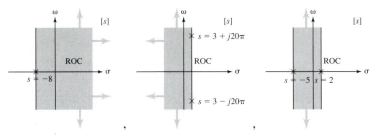

2. Starting with the definition of the Laplace transform

$$\mathcal{L}(g(t)) = G(s) = \int_{0^-}^{\infty} g(t)e^{-st}\, dt,$$

    find the Laplace transforms of these signals.

    a. $x(t) = e^t u(t)$        b. $x(t) = e^{2t}\cos(200\pi t)\, u(t)$

    c. $x(t) = \text{ramp}(t)$        d. $x(t) = te^t u(t)$

**Answers:**

$$\frac{1}{s-1}, \text{Re}(s) = \sigma > 1; \quad \frac{1}{s^2}, \text{Re}(s) = \sigma > 0;$$

$$\frac{s-2}{(s-2)^2 + (200\pi)^2}, \text{Re}(s) = \sigma > 2; \quad \frac{1}{(s-1)^2}, \text{Re}(s) = \sigma > 1$$

3. Using the time-shifting property, find the Laplace transform of these signals.

    a. $x(t) = u(t) - u(t-1)$        b. $x(t) = 3e^{-3(t-2)}u(t-2)$

    c. $x(t) = 3e^{-3t}u(t-2)$        d. $x(t) = 5\sin(\pi(t-1))u(t-1)$

**Answers:**

$$\frac{3e^{-2s-6}}{s+3}, \quad \frac{1-e^{-s}}{s}, \quad \frac{5\pi e^{-s}}{s^2+\pi^2}, \quad \frac{3e^{-2s}}{s+3}$$

4. Using the complex-frequency-shifting property, find and sketch the inverse Laplace transform of

$$X(s) = \frac{1}{(s+j4)+3} + \frac{1}{(s-j4)+3}.$$

**Answer:**

**5.** Using the time-scaling property, find the Laplace transforms of these signals.

    *a.*   $x(t) = \delta(4t)$        *b.*   $x(t) = u(4t)$

**Answers:**

$\dfrac{1}{s}$, $\text{Re}(s) > 0$;    $\dfrac{1}{4}$, all $s$

**6.** Using the time-differentiation property, find the Laplace transforms of these signals.

    *a.*   $x(t) = \dfrac{d}{dt}(u(t))$             *b.*   $x(t) = \dfrac{d}{dt}(e^{-10t}u(t))$

    *c.*   $x(t) = \dfrac{d}{dt}(4\sin(10\pi t)\,u(t))$      *d.*   $x(t) = \dfrac{d}{dt}(10\cos(15\pi t)\,u(t))$

**Answers:**

$\dfrac{40\pi s}{s^2 + (10\pi)^2}$, $\text{Re}(s) > 0$;    $\dfrac{10s^2}{s^2 + (15\pi)^2}$, $\text{Re}(s) > 0$;    1, All $(s)$

$\dfrac{s}{s + 10}$, $\text{Re}(s) > -10$

**7.** Using multiplication–convolution duality, find the Laplace transforms of these signals and sketch the signals versus time.

    *a.*   $x(t) = e^{-t}u(t) * u(t)$

    *b.*   $x(t) = e^{-t}\sin(20\pi t)\,u(t) * u(t)$

    *c.*   $x(t) = 8\cos\left(\dfrac{\pi t}{2}\right)u(t) * [u(t) - u(t - 1)]$

    *d.*   $x(t) = 8\cos(2\pi t)\,u(t) * [u(t) - u(t - 1)]$

**Answers:**

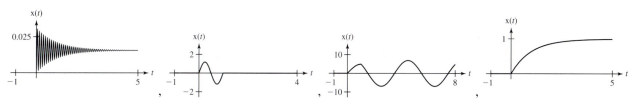

**8.** Using the initial and final value theorems, find the initial and final values (if possible) of the signals with the following Laplace transforms.

    *a.*   $X(s) = \dfrac{10}{s + 8}$            *b.*   $X(s) = \dfrac{s + 3}{(s + 3)^2 + 4}$

    *c.*   $X(s) = \dfrac{s}{s^2 + 4}$             *d.*   $X(s) = \dfrac{10s}{s^2 + 10s + 300}$

    *e.*   $X(s) = \dfrac{8}{s(s + 20)}$          *f.*   $X(s) = \dfrac{8}{s^2(s + 20)}$

**Answers:**

10, Does not apply,   0,   1,   0,   0,   does not apply,   $\frac{2}{5}$,   1,   10,   0,   0

9. Find the inverse Laplace transforms of these functions.

a. $X(s) = \dfrac{24}{s(s + 8)}$

b. $X(s) = \dfrac{20}{s^2 + 4s + 3}$

c. $X(s) = \dfrac{5}{s^2 + 6s + 73}$

d. $X(s) = \dfrac{10}{s(s^2 + 6s + 73)}$

e. $X(s) = \dfrac{4}{s^2(s^2 + 6s + 73)}$

f. $X(s) = \dfrac{2s}{s^2 + 2s + 13}$

g. $X(s) = \dfrac{s}{s + 3}$

h. $X(s) = \dfrac{s}{s^2 + 4s + 4}$

i. $X(s) = \dfrac{s^2}{s^2 - 4s + 4}$

j. $X(s) = \dfrac{10s}{s^4 + 4s^2 + 4}$

**Answers:**

$$2\left\{ e^{-t}\left[ \cos\left(\sqrt{12}t\right) - \frac{\sin\left(\sqrt{12}t\right)}{\sqrt{12}} \right] \right\} u(t), \quad 10(e^{-t} - e^{-3t})\, u(t),$$

$$e^{-2t}(1 - 2t)\, u(t), \quad \frac{10}{73}\left[ 1 - \sqrt{\frac{73}{64}}\, e^{-3t} \cos(8t - 0.3588) \right] u(t),$$

$$\delta(t) + 4e^{2t}(t + 1)\, u(t), \quad \frac{1}{(73)^2}\left[ 292t - 24 + 24e^{-3t}\left( \cos(8t) - \frac{55}{48}\sin(8t) \right) \right] u(t),$$

$$\frac{5}{8}e^{-3t}\sin(8t)\, u(t), \quad \delta(t) - 3e^{-3t}u(t), \quad 3(1 - e^{-8t})\, u(t), \quad \frac{5\sqrt{2}}{2}t\sin(\sqrt{2}t)\, u(t)$$

10. Using a table of Laplace transforms, find the CTFTs of these signals.

a. $x(t) = 10e^{-100t}u(t)$

b. $x(t) = 3e^{-50t}\cos(100\pi t)\, u(t)$

**Answers:**

$$3\frac{j\omega + 50}{(j\omega + 50)^2 + (100\pi)^2}, \quad \frac{10}{j\omega + 100}$$

11. Using the Laplace transform, solve these differential equations for $t \geq 0$.

a. $x'(t) + 10x(t) = u(t), \quad x(0^-) = 1$

b. $x''(t) - 2x'(t) + 4x(t) = u(t), \quad x(0^-) = 0, \quad \left[\dfrac{d}{dt}x(t)\right]_{t=0^-} = 4$

c. $x'(t) + 2x(t) = \sin(2\pi t)\, u(t), \quad x(0^-) = -4$

**Answers:**

$$\frac{1}{4}\left( 1 - e^{t}\cos\left(\sqrt{3}t\right) + \frac{17}{\sqrt{3}}e^{t}\sin\left(\sqrt{3}t\right) \right) u(t), \quad \frac{1 + 9e^{-10t}}{10}u(t),$$

$$x(t) = \left[ \frac{2\pi e^{-2t} - 2\pi\cos(2\pi t) + 2\sin(2\pi t)}{4 + (2\pi)^2} - 4e^{-2t} \right] u(t)$$

12. Using the Laplace transform, find and sketch the time-domain response $y(t)$ of the systems with these transfer functions to the sinusoidal excitation $x(t) = A\cos(10\pi t)\, u(t)$.

a. $H(s) = \dfrac{1}{s + 1}$

b. $H(s) = \dfrac{s - 2}{(s - 2)^2 + 16}$

**Answers:**

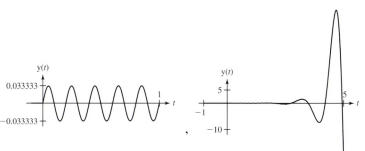

**13.** Write the differential equations describing these systems, and find and sketch the indicated responses.

   a.    $x(t) = u(t)$, $y(t)$ is the response, $y(0^-) = 0$

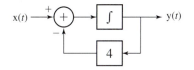

   b.    $v(0^-) = 10$, $v(t)$ is the response

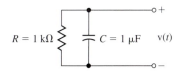

**Answers:**

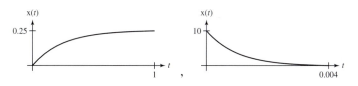

**14.** Find the three parts $x_{ac}(t)$, $x_0(t)$, and $x_c(t)$ of the following signals.

   a.    $x(t) = e^{-10t} u(t) - e^{2t} u(-t)$          b.    $x(t) = K$

   c.    $x(t) = u(t)$                              d.    $x(t) = \dfrac{d}{dt}(u(t))$

**Answers:**
$x_{ac}(t) = 0$, $x_0(t) = 0$, $x_c(t) = u(t)$;    $x_{ac}(t) = -e^{2t} u(-t)$, $x_0(t) = 0$, $x_c(t) = e^{-10t} u(t)$;
$x_{ac}(t) = 0$, $x_0(t) = \delta(t)$, $x_c(t) = 0$;    $x_{ac}(t) = K u(-t)$, $x_0(t) = 0$, $x_c(t) = K u(t)$

**15.** Find the bilateral Laplace transforms of these signals.

   a.    $x(t) = 3e^{-7t} u(t) - 12e^{4t} u(-t)$          b.    $x(t) = 50e^{-10|t|}$

**Answers:**
$3\dfrac{5s + 24}{s^2 + 3s - 28}$, $-7 < \text{Re}(s) < 4$;    $-\dfrac{1000}{s^2 - 100}$, $-10 < \text{Re}(s) < 10$

**16.** Find the responses y(t) of these systems h(t) to the corresponding excitations x(t).

    *a.*   $h(t) = e^{-5t}u(t)$,   $x(t) = 3e^{-7t}u(t) - 12e^{4t}u(-t)$

    *b.*   $h(t) = \text{tri}(t)$,   $x(t) = e^{-t}u(t)$

    *c.*   $h(t) = e^{-10t}u(t)$,   $x(t) = 50e^{-10|t|}$

**Answers:**

$$\left\{ \begin{array}{l} \left[\text{ramp}(t+1) - 1 + e^{-(t+1)}\right]u(t+1) \\ -2\left[\text{ramp}(t) - 1 + e^{-t}\right]u(t) \\ + \left[\text{ramp}(t-1) - 1 + e^{-(t-1)}\right]u(t-1) \end{array} \right\} \quad \frac{1}{6}e^{-5t} - \frac{3}{2}e^{-7t}u(t) - \frac{4}{3}e^{4t}u(-t),$$

$$50\left\{ te^{-10t}u(t) + \frac{1}{20}[e^{t}u(-t) + e^{-t}u(t)] \right\}$$

## EXERCISES WITHOUT ANSWERS

**17.** Sketch the pole–zero plot and region of convergence (if it exists) for these signals.

    *a.*   $x(t) = e^{-t}u(-t) - e^{-4t}u(t)$      *b.*   $x(t) = e^{-2t}u(-t) - e^{t}u(t)$

**18.** Using the integral definition find the unilateral Laplace transform of these time functions.

    *a.*   $g(t) = e^{-at}u(t)$

    *b.*   $g(t) = e^{-a(t-\tau)}u(t-\tau), \quad \tau > 0$

    *c.*   $g(t) = e^{-a(t+\tau)}u(t+\tau), \quad \tau > 0$

    *d.*   $g(t) = \sin(\omega_0 t)\, u(t)$

    *e.*   $g(t) = \text{rect}(t)$

    *f.*   $g(t) = \text{rect}\left(t - \frac{1}{2}\right)$

**19.** Using MATLAB (or any other appropriate computer mathematics tool) do the inversion integral of

$$G(s) = \frac{1}{s + 10}$$

numerically. That is, approximate the inversion integral with a summation of the form

$$g(t) \doteq \mathcal{L}^{1}(G(s)) = \frac{1}{j2\pi} \sum_{n=-N}^{N} \frac{e^{s_n t}}{s_n + 10} \Delta s_n$$

$$= \frac{1}{j2\pi} \sum_{n=-N}^{N} \frac{e^{(\sigma + jn\Delta\omega)t}}{\sigma + jn\Delta\omega + 10} j\Delta\omega, \, \sigma > 0.$$

Choose the combination of large $N$ and small $\Delta\omega$ so that the summation will range over a contour from well below to well above the real axis. Plot $g(t)$ versus $t$ by computing the value of $g(t)$ at every value of $t$ from the given summation approximation to the inversion integral. Compare to the analytical result. Try at least three different values of $\sigma$ to see the effect on the result. (Ideally there is no effect of changing $\sigma$ as long as it is greater than $-10$, but actually, in this numerical approximation, there will be some small effects.)

**20.** Using a table of unilateral Laplace transforms and the properties, find the unilateral Laplace transforms of the following functions.

    *a.*    $g(t) = 5\sin(2\pi(t-1))\,u(t-1)$

    *b.*    $g(t) = 5\sin(2\pi t)\,u(t-1)$

    *c.*    $g(t) = 2\cos(10\pi t)\cos(100\pi t)\,u(t)$

    *d.*    $g(t) = \dfrac{d}{dt}(u(t-2))$         *e.*    $g(t) = \displaystyle\int_{0^+}^{t} u(\tau)\,d\tau$

    *f.*    $g(t) = \dfrac{d}{dt}(5e^{-(t-\tau)/2}u(t-\tau)),\ \tau > 0$

    *g.*    $g(t) = 2e^{-5t}\cos(10\pi t)\,u(t)$

    *h.*    $x(t) = 5\sin(\pi t - \frac{\pi}{8})\,u(t)$

**21.** Given

$$g(t) \overset{\mathcal{L}}{\longleftrightarrow} \frac{s+1}{s(s+4)}$$

find the Laplace transforms of

    *a.*    $g(2t)$       *b.*    $\dfrac{d}{dt}(g(t))$

    *c.*    $g(t-4)$     *d.*    $g(t)*g(t)$

**22.** Find the time-domain functions which are the inverse Laplace transforms of these functions. Then, using the initial and final value theorems verify that they agree with the time-domain functions.

    *a.*    $G(s) = \dfrac{4s}{(s+3)(s+8)}$     *b.*    $G(s) = \dfrac{4}{(s+3)(s+8)}$

    *c.*    $G(s) = \dfrac{s}{s^2+2s+2}$     *d.*    $G(s) = \dfrac{e^{-2s}}{s^2+2s+2}$

**23.** Given

$$e^{-4t}u(t) \overset{\mathcal{L}}{\longleftrightarrow} G(s)$$

find the inverse Laplace transforms of

    *a.*    $G\!\left(\dfrac{s}{3}\right)$     *b.*    $G(s-2)+G(s+2)$     *c.*    $\dfrac{G(s)}{s}$

**24.** The CTFT of

$$x(t) = e^{-|t|}$$

exists, but the (unilateral) Laplace transform does not. Explain why.

**25.** Compare the CTFT and the Laplace transform of a unit step. Explain why the CTFT cannot be found from the Laplace transform.

**26.** Show that the following common Laplace transform pairs can be derived using only the impulse transformation $\delta(t) \overset{\mathcal{L}}{\longleftrightarrow} 1$ and the properties of the Laplace transform.

    *a.*    $u(t) \overset{\mathcal{L}}{\longleftrightarrow} \dfrac{1}{s}$     *b.*    $e^{-\alpha t}u(t) \overset{\mathcal{L}}{\longleftrightarrow} \dfrac{1}{s+\alpha}$

    *c.*    $\cos(\omega_0 t)\,u(t) \overset{\mathcal{L}}{\longleftrightarrow} \dfrac{s}{s+\omega_0^2}$

27.  Given an LTI system transfer function $H(s)$, find the time-domain response to the corresponding excitation $x(t)$.

    *a.*  $H(s) = \dfrac{1}{s+1}$, $x(t) = \sin(2\pi t)\,u(t)$     *b.*  $H(s) = \dfrac{3}{s+2}$, $x(t) = u(t)$

    *c.*  $H(s) = \dfrac{3s}{s+2}$, $x(t) = u(t)$

    *d.*  $H(s) = \dfrac{5s}{s^2+2s+2}$, $x(t) = u(t)$

    *e.*  $H(s) = \dfrac{5s}{s^2+2s+2}$, $x(t) = \sin(2\pi t)\,u(t)$

28.  Write the differential equations describing these systems, and find and sketch the indicated responses.

    *a.*  $x(t) = u(t)$, $y(t)$ is the response, $y(0^-) = -5$, $\left[\dfrac{d}{dt}(y(t))\right]_{t=0^-} = 10$

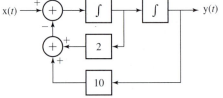

    *b.*  $i_s(t) = u(t)$, $v(t)$ is the response, no initial energy storage

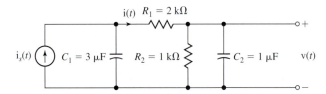

    *c.*  $i_s(t) = \cos(2000\pi t)\,u(t)$, $v(t)$ is the response, no initial energy storage

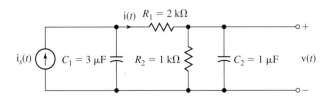

29.  Find the three parts $x_{ac}(t)$, $x_0(t)$, and $x_e(t)$ of the following signals.

    *a.*  $x(t) = t$                *b.*  $x(t) = \sin(\omega t)$

    *c.*  $x(t) = \dfrac{d}{dt}(\mathrm{sgn}(t))$

30.  Find the bilateral Laplace transforms of these signals.

    *a.*  $x(t) = \mathrm{rect}(t)$     *b.*  $x(t) = \mathrm{rect}(t)\sin(20\pi t)$

    *c.*  $x(t) = [e^{-2t}u(t) - e^{2t}u(-t)]\sin(2\pi t)$

### CHAPTER 10

# Laplace Transform Analysis of Signals and Systems

## 10.1 INTRODUCTION AND GOALS

In this chapter we will explore various applications of the Laplace transform to system analysis. The Laplace transform is a very powerful tool for analysis and design of systems. It enables an engineer not only to find the total response to an arbitrary excitation but to generalize from the system transfer function to its stability and its response to various types of signals. After we have become more familiar with Laplace analysis methods, we will expand them to more complicated systems with multiple inputs and outputs.

### CHAPTER GOALS

1. To illustrate the application of the Laplace transform and analysis techniques based on the Laplace transform to the design and analysis of systems through examples

2. To assess the stability of a system directly from its transfer function

3. To see how system responses to standard signals reveal system characteristics

4. To develop systematic methods of analysis for multiple-input, multiple-output systems using the Laplace transform

## 10.2 TRANSFER FUNCTIONS FROM CIRCUITS AND SYSTEM DIAGRAMS

Much signal and system analysis is done by engineers without ever referring directly to a time-domain quantity. Transfer functions in the $s$ domain are written directly from system diagrams. Much system design is done using only the frequency-domain concepts, frequency response, and bandwidth. The analysis of CT filters is an example of frequency-domain signal and system analysis.

For electrical engineers the most common system analysis is circuit analysis. Circuit analysis can be done in the time domain, but it is usually done in the frequency domain because of the power of linear algebra in expressing system interrelationships in terms of algebraic (instead of differential) equations. Circuits are interconnections of circuit elements such as resistors, capacitors, inductors, transistors, diodes, transformers, voltage sources, and current sources. To the extent that these elements can be characterized by linear frequency-domain relationships, the circuit can be analyzed by frequency-domain techniques. Nonlinear elements such as transistors, diodes, and transformers can often be modeled approximately over small-signal ranges as linear devices. These models consist of linear resistors, capacitors, and inductors plus dependent voltage and current sources, all of which can be characterized by LTI system transfer functions.

As an example of frequency-domain circuit analysis using Laplace methods consider the circuit of Figure 10.1. This figure illustrates the circuit in the time domain. This circuit can be described by two coupled integrodifferential equations,

$$R_1 i_1(t) + L\left[\frac{d}{dt}(i_1(t)) - \frac{d}{dt}(i_2(t))\right] = v_g(t) \tag{10.1}$$

$$L\left[\frac{d}{dt}(i_2(t)) - \frac{d}{dt}(i_1(t))\right] + \frac{1}{C}\int_{0^-}^{t} i_2(\lambda)\,d\lambda + v_c(0^-) + R_2 i_2(t) = 0. \tag{10.2}$$

If we Laplace-transform both equations, we get

$$R_1 I_1(s) + L[s I_1(s) - i_1(0^+) - s I_2(s) + i_2(0^+)] = V_g(s) \tag{10.3}$$

$$L[s I_2(s) - i_2(0^+) - s I_1(s) + i_1(0^+)] + \frac{1}{sC} I_2(s) + \frac{v_c(0^-)}{s} + R_2 I_2(s) = 0. \tag{10.4}$$

If there is initially no energy stored in the circuit, these equations simplify to

$$R_1 I_1(s) + L[s I_1(s) - s I_2(s)] = V_g(s) \tag{10.5}$$

$$L[s I_2(s) - s I_1(s)] + \frac{1}{sC} I_2(s) + R_2 I_2(s) = 0. \tag{10.6}$$

It is common to rewrite the equations in the form,

$$R_1 I_1(s) + s L I_1(s) - s L I_2(s) = V_g(s) \tag{10.7}$$

$$s L I_2(s) - s L I_1(s) + \frac{1}{sC} I_2(s) + R_2 I_2(s) = 0 \tag{10.8}$$

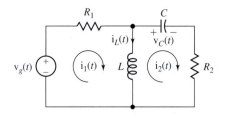

**Figure 10.1**
Time-domain circuit diagram of an *RLC* circuit.

or

$$Z_{R_1}(s)I_1(s) + Z_L(s)I_1(s) - Z_L(s)I_2(s) = V_g(s) \qquad \textbf{(10.9)}$$

$$Z_L(s)I_2(s) - Z_L(s)I_1(s) + Z_C(s)I_2(s) + Z_{R_2}(s)I_2(s) = 0, \qquad \textbf{(10.10)}$$

where

$$Z_{R_1}(s) = R_1 \qquad Z_{R_2}(s) = R_2 \qquad Z_L(s) = sL \qquad Z_C(s) = \frac{1}{sC}. \qquad \textbf{(10.11)}$$

The equations are written this way to emphasize the impedance concept of frequency-domain circuit analysis. The coefficients $sL$ and $1/sC$ are the impedances of the inductor and capacitor, respectively. Impedance is a generalization of the concept of resistance. Using this concept, frequency-domain equations can be written directly from circuit diagrams using relations similar to Ohm's law for resistors,

$$V_R(s) = Z_R I(s) = RI(s) \quad V_L(s) = Z_L I(s) = sLI(s) \quad V_C(s) = Z_C I(s) = \frac{1}{sC}I(s).$$
$$\textbf{(10.12)}$$

Now the circuit of Figure 10.1 can be conceived in the frequency domain as the circuit of Figure 10.2. The circuit equations can be written directly from Figure 10.2 as two mesh equations in the complex-frequency domain without ever writing the time-domain equations.

$$R_1 I_1(s) + sLI_1(s) - sLI_2(s) = V_g(s) \qquad \textbf{(10.13)}$$

$$sLI_2(s) - sLI_1(s) + \frac{1}{sC}I_2(s) + R_2 I_2(s) = 0 \qquad \textbf{(10.14)}$$

These circuit equations can be interpreted in a system sense as integration, differentiation, and/or multiplication by a constant and summation of signals, in this case, $I_1(s)$ and $I_2(s)$.

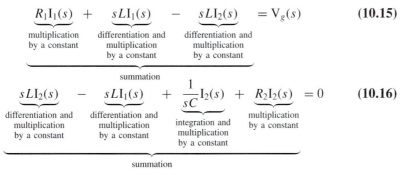

$$\underbrace{R_1 I_1(s)}_{\substack{\text{multiplication} \\ \text{by a constant}}} + \underbrace{sLI_1(s)}_{\substack{\text{differentiation and} \\ \text{multiplication} \\ \text{by a constant}}} - \underbrace{sLI_2(s)}_{\substack{\text{differentiation and} \\ \text{multiplication} \\ \text{by a constant}}} = V_g(s) \qquad \textbf{(10.15)}$$

$$\underbrace{\hspace{2cm}}_{\text{summation}}$$

$$\underbrace{sLI_2(s)}_{\substack{\text{differentiation and} \\ \text{multiplication} \\ \text{by a constant}}} - \underbrace{sLI_1(s)}_{\substack{\text{differentiation and} \\ \text{multiplication} \\ \text{by a constant}}} + \underbrace{\frac{1}{sC}I_2(s)}_{\substack{\text{integration and} \\ \text{multiplication} \\ \text{by a constant}}} + \underbrace{R_2 I_2(s)}_{\substack{\text{multiplication} \\ \text{by a constant}}} = 0 \qquad \textbf{(10.16)}$$

$$\underbrace{\hspace{2cm}}_{\text{summation}}$$

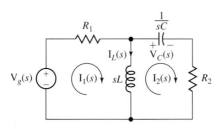

**Figure 10.2**
Frequency-domain circuit diagram of an
$RLC$ circuit.

A block diagram could be drawn for this system using integrators, gain blocks, and summers. (We will explore some techniques for doing just that in the section on state-space methods presented later.)

Other kinds of continuous-time systems can also be modeled by interconnections of integrators, gain blocks, and summers. These elements may represent various physical systems which have the same mathematical relationships between an excitation and a response. As a very simple example, suppose a mass $m$ is acted upon by a force (an excitation) $f(t)$. It responds by moving. The response could be the position $p(t)$ of the mass in some appropriate coordinate system. According to classical Newtonian mechanics, the acceleration of a body in any coordinate direction is proportional to the force applied to the body in that direction divided by the mass of the body,

$$\frac{d^2}{dt^2}(p(t)) = \frac{f(t)}{m}. \tag{10.17}$$

This can be directly stated in the Laplace domain as (assuming the initial position and velocity are zero)

$$s^2 P(s) = \frac{F(s)}{m}. \tag{10.18}$$

So this very simple system could be modeled by a multiplication by a constant and two integrators (Figure 10.3).

We can also represent as block diagrams more complicated systems as in Figure 10.4. The positions $x_1$ and $x_2$ are the distances from the rest positions of masses $m_1$ and $m_2$, respectively. The sum of forces on mass $m_1$ is

$$f(t) - K_d x_1'(t) - K_{s1}[x_1(t) - x_2(t)] = m_1 x_1''(t). \tag{10.19}$$

The sum of forces on mass $m_2$ is

$$K_{s1}[x_1(t) - x_2(t)] - K_{s2}x_2(t) = m_2 x_2''(t). \tag{10.20}$$

Laplace-transforming both equations,

$$\begin{aligned} F(s) - K_d s X_1(s) - K_{s1}[X_1(s) - X_2(s)] &= m_1 s^2 X_1(s) \\ K_{s1}[X_1(s) - X_2(s)] - K_{s2}X_2(s) &= m_2 s^2 X_2(s) \end{aligned}. \tag{10.21}$$

We can model the mechanical system with a block diagram (Figure 10.5).

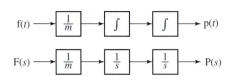

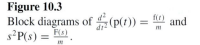

**Figure 10.3**
Block diagrams of $\frac{d^2}{dt^2}(p(t)) = \frac{f(t)}{m}$ and $s^2 P(s) = \frac{F(s)}{m}$.

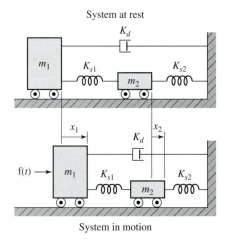

**Figure 10.4**
A mechanical system.

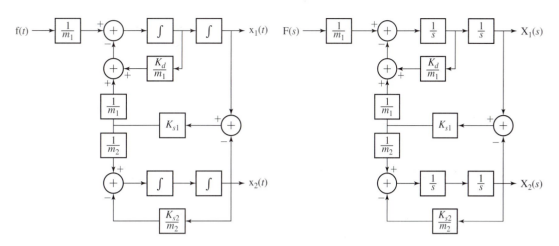

**Figure 10.5**
Time-domain and frequency-domain block diagrams of the mechanical system of Figure 10.4.

## 10.3  SYSTEM STABILITY

A very important consideration in system analysis is system stability. As was shown in Chapter 3, a CT system is stable if its impulse response is absolutely integrable. The impulse response of a causal system is absolutely integrable if it decays exponentially as time increases. The Laplace transform of the impulse response is the transfer function. For systems which can be described by differential equations of the form

$$\sum_{k=0}^{D} a_k \frac{d^k}{dt^k}(y(t)) = \sum_{k=0}^{N} b_k \frac{d^k}{dt^k}(x(t)), \qquad (10.22)$$

where $a_D = 1$, without loss of generality, the transfer function is of the form

$$H(s) = \frac{Y(s)}{X(s)} = \frac{\displaystyle\sum_{k=0}^{N} b_k s^k}{\displaystyle\sum_{k=0}^{D} a_k s^k} = \frac{b_N s^N + b_{N-1} s^{N-1} + \cdots + b_1 s + b_0}{s^D + a_{D-1} s^{D-1} + \cdots + a_1 s + a_0}. \qquad (10.23)$$

The denominator can always be factored (in principle at least), so the transfer function can also be written in the form,

$$H(s) = \frac{Y(s)}{X(s)} = \frac{b_N s^N + b_{N-1} s^{N-1} + \cdots + b_1 s + b_0}{(s - p_1)(s - p_2) \cdots (s - p_D)}. \qquad (10.24)$$

If there are any pole–zero pairs which lie at exactly the same location in the $s$ plane, they cancel in the transfer function and should be removed before the transfer function is examined for stability. If $N < D$ and none of the poles is repeated, then the transfer function can be expressed in partial-fraction form as

$$H(s) = \frac{K_1}{s - p_1} + \frac{K_2}{s - p_2} + \cdots + \frac{K_D}{s - p_D}, \qquad (10.25)$$

and the impulse response is then of the form

$$h(t) = K_1 e^{p_1 t} + K_2 e^{p_2 t} + \cdots + K_D e^{p_D t}, \qquad (10.26)$$

**Table 10.1**  Conditions for system stability, marginal stability, and instability

| Stability | Marginal Stability | Instability |
|---|---|---|
| All poles in the open LHP | One or more simple poles on the ω axis but no repeated poles on the ω axis and no poles in the open RHP | One or more poles in the open RHP or one or more multiple poles on the ω axis |

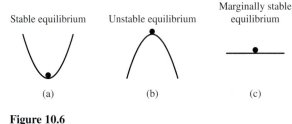

Figure 10.6
Illustrations of three types of stability.

where the $p$'s are the poles of the transfer function. For h($t$) to be absolutely integrable, the real part of each of the $p$'s must be negative; therefore, all the poles of the transfer function must lie in the open left half-plane (LHP). The term *open left half-plane* means the left half-plane not including the ω axis. If there are simple poles on the ω axis and no poles are in the right half-plane (RHP), the system is called *marginally stable* because, even though the impulse response does not decay with time, it does not grow either. Marginal stability is a special case of instability. If there are any multiple poles on the ω axis or any poles in the right half-plane, the system is unstable. These conditions are summarized in Table 10.1.

An analogy that is sometimes helpful in remembering the different descriptions of system stability or instability is to consider a sphere placed on different kinds of surfaces (Figure 10.6). If we excite the system in Figure 10.6a by applying an impulse of horizontal force to the sphere, it responds by moving and then rolling back and forth. If there is even the slightest bit of rolling friction (or any other loss mechanism like air resistance), the sphere eventually returns to its initial equilibrium position. This is an example of a stable system. If there is no friction (or any other loss mechanism), the sphere will oscillate back and forth forever but will remain confined near the relative low point of the surface. Its response does not grow with time, but it does not decay either. In this case the system is marginally stable.

If we excite the sphere in Figure 10.6b even the slightest bit, the sphere rolls down the hill and never returns. If the hill is infinitely high, the sphere's speed will approach infinity, an unbounded response to a bounded excitation. This is an unstable system.

In Figure 10.6c if we excite the sphere with an impulse of horizontal force, it responds by rolling. If there is any loss mechanism, the sphere eventually comes to rest but not at its original point. This is a bounded response to a bounded excitation, and the system is stable. If there is no loss mechanism, the sphere will roll forever. This is marginal stability again.

# 10.4 PARALLEL, CASCADE, AND FEEDBACK CONNECTIONS

Earlier we found the impulse and frequency responses of cascade and parallel connections of systems. The results for these types of systems are the same when the transfer functions are expressed in terms of Laplace transforms as when expressed in terms of Fourier transforms (Figures 10.7 and 10.8).

Another type of connection is very important in system analysis, the feedback connection (Figure 10.9). The transfer function $H_1(s)$ is the *forward* path, and the transfer function $H_2(s)$, is the *feedback* path. In the control-system literature it is common to call the forward path transfer function $H_1(s)$ the *plant* because it is usually an established system designed to produce something and the feedback path transfer function $H_2(s)$ the *sensor* because it is usually a system added to the plant to help control it or stabilize it by sensing the plant response and feeding it back to the summing point at the plant input. The input signal of the forward path (plant) is called the *error* signal and is given by

$$E(s) = X(s) - H_2(s)Y(s). \tag{10.27}$$

The output signal from $H_1(s)$,

$$Y(s) = H_1(s)E(s), \tag{10.28}$$

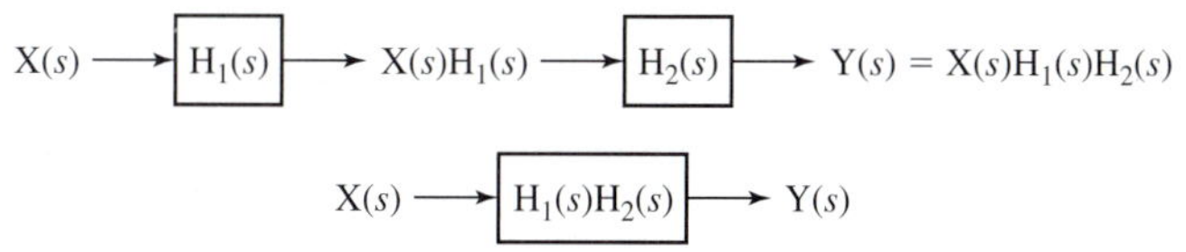

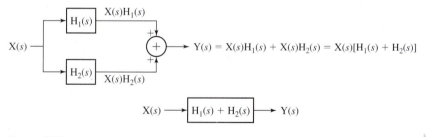

**Figure 10.7**
Cascade connection of systems.

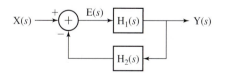

**Figure 10.8**
Parallel connection of systems.

**Figure 10.9**
Feedback connection of systems.

is the input signal for the feedback path $H_2(s)$. Combining equations and solving for the overall transfer function,

$$H(s) = \frac{Y(s)}{X(s)} = \frac{H_1(s)}{1 + H_1(s)H_2(s)}. \qquad (10.29)$$

In the block diagram illustrating feedback in Figure 10.9 the feedback signal is subtracted from the input signal. This is a very common convention in feedback system analysis and stems from the history of feedback used as negative feedback to stabilize a system. It is common in system analysis to give the product of the forward and feedback path transfer functions a special name, *loop transfer function,*

$$T(s) = H_1(s)H_2(s), \qquad (10.30)$$

because it shows up so much in feedback system analysis. In electronic feedback amplifier design this is sometimes called the *loop transmission*. It is given this name, or the name *loop transfer function,* because it represents what happens to a signal as it goes from any point in the loop, around the loop exactly one time, and back to the starting point (except for the effect of the minus sign on the summer). So the gain of the feedback system is the forward-path gain $H_1(s)$ divided by one plus the loop transfer function,

$$H(s) = \frac{H_1(s)}{1 + T(s)}. \qquad (10.31)$$

Notice that when $H_2(s)$ goes to zero (meaning there is no feedback), $T(s)$ does also and the system gain $H(s)$ becomes the same as the forward-path gain $H_1(s)$.

Feeding the forward-path output signal back to alter its own input signal is often called *closing the loop* for obvious reasons. If there is no feedback path, the system is said to be operating *open loop*. Politicians, business executives, and other movers and shakers in our society want to be "in the loop." This terminology probably came from feedback loop concepts because if one is in the loop, he or she has the chance of affecting the system performance and, therefore, has power in the political, economic, or social system in which he or she operates.

The MATLAB control system toolbox contains many helpful commands for the analysis of systems. They are based on the idea of a *system object*, a special type of variable in MATLAB for the description of systems. One way of creating a system description in MATLAB is through the use of the `tf` (transfer function) command. The syntax for creating a system object with `tf` is

```
sys = tf(num,den) .
```

This command creates a system object `sys` from two vectors `num` and `den`. The two vectors are the coefficients of $s$, in descending order, in the numerator and denominator of a transfer function. For example, let the transfer function be

$$H_1(s) = \frac{s^2 + 4}{s^5 + 4s^4 + 7s^3 + 15s^2 + 31s + 75}. \qquad (10.32)$$

In MATLAB we can form $H_1(s)$ with

```
»num = [1 0 4] ;
»den = [1 4 7 15 31 75] ;
»H1 = tf(num,den) ;
»H1
```

```
Transfer function:
                    s^2 + 4
-------------------------------------------------------------
s^5 + 4 s^4 + 7 s^3 + 15 s^2 + 31 s + 75
```

Alternatively we can form a system description by specifying the zeros, poles, and frequency-independent gain of a system using the `zpk` command. The syntax is

```
sys = zpk(z,p,k),
```

where $z$ = vector of zeros of the system
$\quad\quad p$ = vector of poles of the system
$\quad\quad k$ = frequency-independent gain

For example, suppose we know that a system has a transfer function

$$H_2(s) = 20\frac{s+4}{(s+3)(s+10)}. \tag{10.33}$$

We can form the system description with

```
»z = [-4] ;
»p = [-3 -10] ;
»k = 20 ;
»H2 = zpk(z,p,k) ;
»H2

Zero/pole/gain:
   20 (s+4)
---------------------------
(s+3) (s+10)
```

We can convert one type of system description to the other type.

```
»tf(H2)

Transfer function:
   20 s + 80
-------------------------------------
s^2 + 13 s + 30

»zpk(H1)

Zero/pole/gain:
                         (s^2 + 4)
-------------------------------------------------------------------------
(s+3.081) (s^2 + 2.901s + 5.45) (s^2 - 1.982s + 4.467)
```

We can get information about systems from their descriptions using the two commands `tfdata` and `zpkdata`. For example,

```
»[num,den] = tfdata(H2,'v') ;
»num

num =

     0   20   80

»den
```

```
den =

    1    13    30
```

or

```
»[z,p,k] = zpkdata(H1,'v') ;
»z

z =

     0 + 2.0000i
     0 - 2.0000i

»p

p =

   -3.0807
   -1.4505 + 1.8291i
   -1.4505 - 1.8291i
    0.9909 + 1.8669i
    0.9909 - 1.8669i

»k

k =

     1
```

The 'v' argument in these commands indicates that the answers should be returned in vector form. This last result indicates that the transfer function $H_1(s)$ has zeros at $\pm j2$ and poles at $-3.0807$, $-1.4505 \pm j1.829$, and $0.9909 \pm j1.8669$ (and is, therefore, unstable).

The real power of the control system toolbox comes in interconnecting systems. Suppose we want the overall transfer function $H(s) = H_1(s)H_2(s)$ of these two systems in a cascade connection. In MATLAB,

```
»Hc = H1*H2 ;
»Hc

Zero/pole/gain:
                        20 (s+4) (s^2 + 4)
-----------------------------------------------------------------------
(s+3.081) (s+3) (s+10) (s^2 + 2.901s + 5.45) (s^2 - 1.982s + 4.467)

»tf(Hc)

Transfer function:
              20 s^3 + 80 s^2 + 80 s + 320
-----------------------------------------------------------------------
s^7 + 17 s^6 + 89 s^5 + 226 s^4 + 436 s^3 + 928 s^2 + 1905 s + 2250
```

If we want to know what the transfer function of these two systems in parallel would be,

```
»Hp = H1 + H2 ;
»Hp
```

```
Zero/pole/gain:
20 (s+4.023) (s+3.077) (s^2 + 2.881s + 5.486) (s^2 - 1.982s + 4.505)
---------------------------------------------------------------------------
(s+3.081) (s+3) (s+10) (s^2 + 2.901s + 5.45) (s^2 - 1.982s + 4.467)

»tf(Hp)

Transfer function:
 20 s^6 + 160 s^5 + 461 s^4 + 873 s^3 + 1854 s^2 + 4032 s + 6120
---------------------------------------------------------------------------
s^7 + 17 s^6 + 89 s^5 + 226 s^4 + 436 s^3 + 928 s^2 + 1905 s + 2250
```

Once we have a system described, we can graph its step response with `step`, its impulse response with `impulse`, and a Bode diagram of its frequency response with `bode`. We can also plot its pole–zero diagram using the MATLAB command `pzmap`. MATLAB has a function called `freqresp` which does frequency-response plots. The syntax is

`H = freqresp(sys,w)`,

where `sys` = MATLAB system description

   `w` = vector of radian frequencies ($\omega$)

   `H` = frequency response of system at those radian frequencies

There are many other useful commands in the control system toolbox which can be examined by typing `help control`.

## 10.5 ANALYSIS OF FEEDBACK SYSTEMS

### BENEFICIAL EFFECTS OF FEEDBACK

Feedback is used for many different purposes. One interesting effect of feedback can be seen in a system like Figure 10.10. In this feedback system the forward-path gain is simply a frequency-independent gain $K$. The overall transfer function is then

$$H(s) = \frac{K}{1 + K\,H_2(s)}. \qquad (10.34)$$

If $K$ is large enough, then, at least for some values of $s$, $K\,H_2(s) \gg 1$ and $H(s) \cong 1/H_2(s)$. In words, if $K$ is large enough, the overall transfer function of the feedback system performs the approximate inverse of the operation of the feedback path. That means that if we were to cascade-connect a system with transfer function $H_2(s)$ to this feedback system, the overall system transfer function would be approximately one (Figure 10.11).

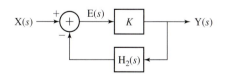

**Figure 10.10**
A feedback system.

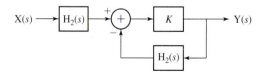

**Figure 10.11**
A system cascaded with another system designed to be its approximate inverse.

It is natural to wonder at this point what has been accomplished because the system of Figure 10.11 seems to have no net effect. There are real situations in which a signal has been changed by some kind of unavoidable system effect and it is desired to restore the original signal. This is very common in communication systems in which a signal has been sent over a channel which ideally would not change the signal but actually does for reasons beyond the control of the designer. An equalization filter can be used to restore the original signal. It is designed to, as nearly as possible, have the inverse of the effect of the channel on the signal. Some systems designed to measure physical phenomena use sensors which have inherently lowpass transfer functions, usually because of some unavoidable mechanical or thermal inertia. The measurement system can be made to respond more quickly by cascading the sensor with an electronic signal-processing system whose transfer function is the approximate inverse of the sensor's transfer function.

Another important use of feedback is to reduce the sensitivity of a system to parameter changes. A very common example of this benefit is in the use of feedback in an operational amplifier configured as in Figure 10.12.

A typical approximate expression for the gain of an operational amplifier with the noninverting input grounded [$H_1(s)$ in the feedback block diagram] is

$$H_1(s) = \frac{V_o(s)}{V_e(s)} = -\frac{A_0}{1 - (s/p)}, \tag{10.35}$$

where $A_0$ is the magnitude of the operational amplifier voltage gain at low frequencies and $p$ is a single pole on the negative real axis of the $s$ plane. The overall transfer function can be found using standard circuit analysis techniques. But it can also be found by using feedback concepts. The error voltage $V_e(s)$ is a function of $V_i(s)$ and $V_o(s)$. Since the input impedance of the operational amplifier is typically very large compared with the two external impedances $Z_i(s)$ and $Z_f(s)$, the error voltage is

$$V_e(s) = V_o(s) + [V_i(s) - V_o(s)] \frac{Z_f(s)}{Z_i(s) + Z_f(s)} \tag{10.36}$$

or

$$V_e(s) = V_o(s) \frac{Z_i(s)}{Z_i(s) + Z_f(s)} + V_i(s) \frac{Z_f(s)}{Z_i(s) + Z_f(s)}.$$

So we can model the system using the block diagram in Figure 10.13.

According to the general feedback-system transfer function derived in (10.29),

$$H(s) = \frac{Y(s)}{X(s)} = \frac{H_1(s)}{1 + H_1(s)H_2(s)}, \tag{10.37}$$

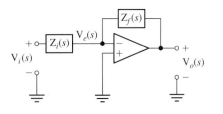

**Figure 10.12**
An inverting voltage amplifier using an operational amplifier with feedback.

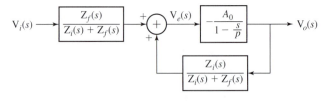

**Figure 10.13**
Block diagram of an inverting voltage amplifier using feedback on an operational amplifier.

the amplifier transfer function should be

$$\frac{V_o(s)}{V_i(s)(Z_f(s)/(Z_i(s)+Z_f(s)))} = \frac{-(A_0/(1-(s/p))}{1+[-(A_0/(1-(s/p)))][-(Z_i(s)/(Z_i(s)+Z_f(s)))]}.$$

**(10.38)**

[Note that the sign of the feedback transfer function was reversed because in Figure 10.13 the feedback polarity was positive which is the opposite of the feedback polarity assumed in the derivation of (10.29) and (10.37)]. Simplifying, and forming the ratio of $V_o(s)$ to $V_i(s)$ as the desired overall transfer function,

$$\frac{V_o(s)}{V_i(s)} = \frac{-A_0 Z_f(s)}{(1-(s/p)+A_0)Z_i(s)+(1-(s/p))Z_f(s)}.$$

**(10.39)**

If the low-frequency gain magnitude $A_0$ is very large (which it usually is), then we can approximate this transfer function at low frequencies as

$$\frac{V_o(s)}{V_i(s)} \cong -\frac{Z_f(s)}{Z_i(s)}.$$

**(10.40)**

This is the well-known ideal operational amplifier formula for the gain of an inverting voltage amplifier. In this case being *large* means that $A_0$ is large enough that the denominator of the transfer function is approximately $A_0 Z_i(s)$, which means that

$$|A_0| \gg \left|1-\frac{s}{p}\right| \quad \text{and} \quad |A_0| \gg \left|1-\frac{s}{p}\right|\left|\frac{Z_f(s)}{Z_i(s)}\right|.$$

**(10.41)**

The exact value of $A_0$ is not important as long as it is very large; this fact represents the reduction in the system's sensitivity to changes in parameter values in (at least some of) its components.

To illustrate the effects of feedback on amplifier performance let

$$A_0 = 10^7 \quad \text{and} \quad p = -100.$$

**(10.42)**

Also, let $Z_f(s)$ be a resistor of $10 \text{ k}\Omega$ and let $Z_i(s)$ be a resistor of $1 \text{ k}\Omega$. Ideally this is an inverting voltage amplifier. Then the overall system transfer function is

$$\frac{V_o(s)}{V_i(s)} = \frac{-10^8}{11(1+(s/100))+10^7}.$$

**(10.43)**

The numerical value of the transfer function at a real radian frequency of $\omega = 100$ (a cyclic frequency of $f = 100/2\pi \cong 15.9 \text{ Hz}$) is

$$\frac{V_o(s)}{V_i(s)} = \frac{-10^8}{11+j11+10^7} = -9.999989 + j0.000011.$$

**(10.44)**

Now let the operational amplifier's low-frequency gain be reduced by a factor of 10 to $A_0 = 10^6$. When we recalculate the transfer function at 15.9 Hz, we get

$$\frac{V_o(s)}{V_i(s)} = \frac{-10^7}{11+j11+10^6} = -9.99989 + j0.00011,$$

**(10.45)**

a change of approximately 0.001 percent in the magnitude of the transfer function. So a change in the forward-path transfer function of a factor of 10 produced a change in the overall system transfer function magnitude of about 0.001 percent. The feedback connection made the overall transfer function very insensitive to changes in the operational

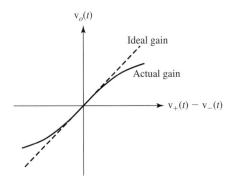

**Figure 10.14**
Linear and nonlinear operational amplifier gain.

amplifier gain, even very large changes. In amplifier design this is a very nice result because resistors, and especially resistor ratios, can be made very insensitive to environmental factors and can hold the system transfer function almost constant, even if components in the operational amplifier change by large percentages from their nominal values.

Another consequence of the relative insensitivity of the system transfer function to the gain $A_0$ of the operational amplifier is that if $A_0$ is a function of signal level, making the operational amplifier gain nonlinear, as long as $A_0$ is large, the system transfer function is still very accurate (Figure 10.14).

Another beneficial effect of feedback can be seen by calculating the bandwidth of the operational amplifier itself and comparing that to the bandwidth of the inverting amplifier with feedback. The corner frequency of the operational amplifier itself in this example is 15.9 Hz. The corner frequency of the inverting amplifier with feedback is the frequency at which the real and imaginary parts of the denominator of the overall transfer function are equal in magnitude which occurs at a real cyclic frequency of $f \cong 14.5\,\mathrm{MHz}$. This is an increase in bandwidth by a factor of approximately 910,000. It is hard to overstate the importance of feedback principles in improving the performance of systems in many ways.

The transfer function of the operational amplifier is a very large number at low frequencies. So the operational amplifier has a large voltage gain at low frequencies. The voltage gain of the feedback amplifier is typically much smaller. So, in using feedback, we have lost voltage gain but obtained gain stability and bandwidth (among other things). In effect, we have traded gain for improvements in other amplifier characteristics.

Feedback can be used to stabilize an otherwise unstable system. The F-117 Stealth Fighter is inherently aerodynamically unstable. It can only fly under a pilot's control with the help of a computer-controlled feedback system which senses the aircraft's position, speed, and attitude and constantly compensates when it starts to go unstable. A very simple example of stabilization of an unstable system using feedback would be a system whose forward-path transfer function is

$$\mathrm{H}_1(s) = \frac{1}{s - p} \qquad p > 0. \tag{10.46}$$

Obviously, with a pole in the right half-plane this system is unstable. If we use a feedback-path transfer function which is a frequency-independent gain $K$, we get the overall system transfer function

$$H(s) = \frac{1/(s - p)}{1 + (K/(s - p))} = \frac{1}{s - p + K}. \qquad \textbf{(10.47)}$$

For any value of $K$ satisfying $K > p$, the system is stable.

## INSTABILITY CAUSED BY FEEDBACK

Although feedback can have many very beneficial effects, there is another effect of feedback in systems that is also very important and can be a problem rather than an advantage. The addition of feedback to a stable system can cause it to become unstable. The overall feedback system gain is

$$H(s) = \frac{Y(s)}{X(s)} = \frac{H_1(s)}{1 + H_1(s)H_2(s)}. \qquad \textbf{(10.48)}$$

Even though all the poles of $H_1(s)$ and $H_2(s)$ may lie in the open left-half plane, the poles of $H(s)$ may not.

Almost everyone has experienced a system made unstable by feedback. Often when large crowds gather to hear someone speak, a public-address (PA) system is used. The speaker speaks into a microphone and her voice is amplified and fed to one or more speakers so everyone in the audience can hear her voice. Of course, the sound emanating from the speakers is also detected and amplified by the microphone and amplifier. This is an example of feedback because the output signal of the PA system (the sound from the speakers) is fed back to the input of the system as an input signal (sound into the microphone). Anyone who has ever heard it will never forget the sound of the PA system when it goes unstable, usually a very loud tone. And we probably know the usual solution, turn down the amplifier gain. This tone can occur even when no one is speaking into the microphone. Why does the system go unstable with no apparent input signal and why does turning down the amplifier gain not just reduce the volume of the tone, but eliminate it entirely?

Albert Einstein was famous for the *Gedankenversuch* (thought experiment). We can understand the feedback phenomenon through a thought experiment. Imagine that we have a microphone, amplifier, and speaker in the middle of a desert with no one around and no wind or other acoustic disturbance and that the amplifier gain is initially turned down to zero. If we tap on the microphone, we hear only the direct sound of tapping and nothing from the speakers because the amplifier gain is zero. Then we turn the amplifier gain up a little. Now when we tap on the microphone we hear the tap directly but also some sound from the speakers, slightly delayed because of the distance the sound has to travel from the speakers to our ears. As we turn the gain up more and more, the tapping sound from the speakers rises in volume (Figure 10.15). [In the figure, $p(t)$ is acoustic pressure as a function of time.] As we increase the round-trip gain, when we tap on the microphone, we gradually notice a change, not just in the volume, but also in the nature of the sound from the speakers. (*Round-trip gain* is the magnitude of the ratio of a signal at some point in a feedback system to the signal at the same point on the previous round-trip through the system.) We hear not only the tap, but we hear what is commonly called *reverberation,* multiple echoes of the tap. These multiple echoes are caused by the sound of the tap coming from the speaker to the microphone, being amplified and going to the speaker again, and returning to the microphone again multiple times. As the gain is increased, this phenomenon becomes more

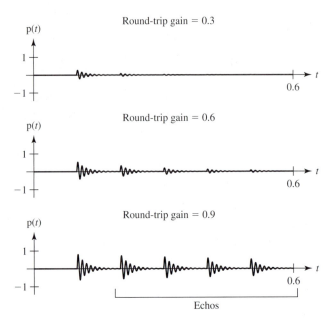

**Figure 10.15**
Sound from tapping on the microphone of a PA system for three
different system round-trip gains.

obvious and at some gain level a loud tone begins and continues, without any tapping
or any other acoustic input to the microphone, until we turn the gain back down. Why?

At some level of gain, any signal from the microphone, no matter how weak, is
amplified, fed to the speaker, returns to the microphone, and causes a new signal in the
microphone which is the same strength as the original signal. At this gain the signal
never dies; it just keeps on circulating. If the gain is made slightly higher, the signal
grows every time it makes the round trip from microphone to speaker and back. If the
PA system were truly linear, that signal would increase without bound. But the PA system is not truly linear and at some volume level the amplifier is driving the speaker as
hard as it can and the sound level does not increase any more.

It is natural to wonder how this process begins without any acoustic input to the
microphone. First, as a practical matter, it is impossible to arrange to have absolutely
no sound strike the microphone. Second, even if that were possible, the amplifier has
inherent random noise processes which cause an acoustic signal from the speaker, and
that is enough to start the feedback process.

Now carry the experiment a little further. With the amplifier gain high enough to
cause the tone, we move the speaker farther from the microphone. As we move the
speaker away, the pitch of the loud tone changes and at some distance the tone stops.
The pitch changes because the frequency of the tone depends on the time sound takes
to propagate from the speaker to the microphone. The loud tone stops at some distance
because the sound intensity from the speaker is reduced as it is moved farther away and
the return signal due to feedback is less than the original signal and the signal dies
away instead of increasing in power.

Now we will mathematically model the PA system with the tools we have
learned and see exactly how feedback instability occurs (Figure 10.16). To keep the
model simple, yet illustrative, we will let the transfer functions of the microphone,

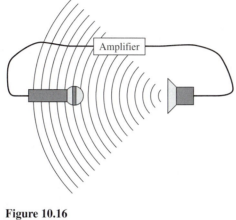

**Figure 10.16**
A PA system.

amplifier, and speaker be the constants $K_m$, $K_A$, and $K_s$. Then we model the propagation of sound from the speaker to the microphone as a simple delay with a gain that is inversely proportional to the square of the distance $d$ from the speaker to the microphone

$$s_m(t) = K \frac{s_s(t - (d/v))}{d^2}, \qquad (10.49)$$

where $s_s(t)$ = sound from speaker
   $s_m(t)$ = sound arriving at microphone
      $v$ = speed of sound in air
      $K$ = a constant

Laplace-transforming both sides of (10.49),

$$S_m(s) = \frac{K}{d^2} S_s(s) e^{-(d/v)s}. \qquad (10.50)$$

Then we can model the PA system as a feedback system with a forward-path transfer function

$$H_1(s) = K_m K_A K_s \qquad (10.51)$$

and a feedback-path transfer function

$$H_2(s) = \frac{K}{d^2} e^{-(d/v)s} \qquad (10.52)$$

(Figure 10.17).

The overall transfer function is then

$$H(s) = \frac{K_m K_A K_s}{1 - (K_m K_A K_s K/d^2) e^{-(d/v)s}}. \qquad (10.53)$$

[The sign in the denominator is minus instead of plus because the feedback polarity is the opposite of the polarity assumed in the general feedback system transfer function result (10.29)]. The poles $p$ of this system transfer function lie at the zeros of

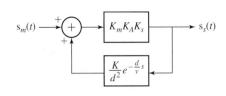

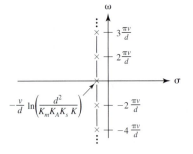

**Figure 10.17**
Block diagram of a PA system.

**Figure 10.18**
Pole–zero diagram of the PA system.

$1 - \left(K_m K_A K_s K / d^2\right) e^{-(d/v)p}$. Solving,

$$1 - \frac{K_m K_A K_s K}{d^2} e^{-(d/v)p} = 0 \tag{10.54}$$

or

$$e^{-(d/v)p} = \frac{d^2}{K_m K_A K_s K} \tag{10.55}$$

or

$$-\frac{d}{v} p = \log\left(\frac{d^2}{K_m K_A K_s K}\right) = \ln\left(\frac{d^2}{K_m K_A K_s K}\right) + j2n\pi \qquad n \text{ is an integer} \tag{10.56}$$

or

$$p = -\frac{v}{d}\left[\ln\left(\frac{d^2}{K_m K_A K_s K}\right) + j2n\pi\right] \tag{10.57}$$

(Figure 10.18).

This is a little different from the usual systems we have been analyzing because this system has infinitely many poles, one for each integer $n$. But that is not a problem in this analysis because we are only trying to establish the conditions under which the system is stable. As we have already seen, stability requires that all poles lie in the open left half-plane. That means, in this case, that

$$-\frac{v}{d}\ln\left(\frac{d^2}{K_m K_A K_s K}\right) < 0 \tag{10.58}$$

or

$$\ln\left(\frac{d^2}{K_m K_A K_s K}\right) > 0 \tag{10.59}$$

or

$$\frac{K_m K_A K_s K}{d^2} < 1. \tag{10.60}$$

In words, the product of all the transfer function magnitudes around the feedback loop must be less than one. This makes common sense because if the product of all the transfer function magnitudes around the loop exceeds one, that means that when a signal makes a complete round-trip through the feedback loop, it is bigger when it comes back than when it left and that causes it to grow without bound. So when we turn down the amplifier gain to stop the loud tone caused by feedback, we are satisfying (10.60).

Suppose we increase the loop gain $K_m K_A K_s K / d^2$ by turning up the amplifier gain $K_A$. The poles move to the right, parallel to the $\sigma$ axis, and at some gain value they reach the $\omega$ axis. Now suppose instead we increase the loop gain by moving the microphone and speaker closer together. This moves the poles to the right but also away from the $\sigma$ axis so that when we reach marginal stability the poles are all at higher radian frequencies.

When the system is exactly at marginal stability, the frequencies of oscillation are determined by the locations of the poles on the $\omega$ axis. Those locations are

$$\omega = -\frac{v}{d} 2n\pi. \qquad (10.61)$$

This indicates that the frequencies of oscillation are determined by the speed of sound in air and the distance between the speaker and microphone. Let $n$ be one. Then (10.61) is simply saying that the system oscillates in such a way that the propagation time from speaker to microphone is exactly one fundamental period of that frequency. If that is true, when the speaker sound reaches the microphone, it arrives exactly in phase (actually $2\pi$ rad out of phase, which is equivalent to being in phase) with a sound of that frequency and reinforces it. For higher integer values of $n$ we get frequencies which arrive with higher multiples of $2\pi$ rad of phase shift and, therefore, also reinforce. So the pole–zero diagram is telling us that the system will oscillate if the amplifier gain magnitude is large enough and it will oscillate at frequencies for which the feedback signal is in phase with the original signal.

The system which obeys this simple model can oscillate at multiple frequencies simultaneously. In reality that is unlikely. A real PA system microphone, amplifier, and speaker would have transfer functions which are functions of frequency and would, therefore, change the pole locations so that only one pair of poles would lie on the $\omega$ axis at marginal stability. If the gain is turned up above the gain for marginal stability, the system is driven into a nonlinear mode of operation and linear system analysis methods fail to predict exactly how it will oscillate. But linear system methods do predict accurately that it *will* oscillate and that is very important.

## STABLE OSCILLATION USING FEEDBACK

The oscillation of the PA system in the last section was an undesirable system response. But some systems are designed to oscillate. Examples are laboratory function generators, computer clocks, local oscillators in radio receivers, quartz crystals in wristwatches, and a pendulum on a grandfather clock. Some systems are designed to oscillate in a nonlinear mode in which they simply alternate between two or more unstable states and their response signals are not necessarily sinusoidal. Free-running computer clocks are a good example of this. But some systems are designed to operate as an LTI system in a marginally stable mode with a true sinusoidal oscillation. Since marginal stability requires that the system have poles on the $\omega$ axis of the $s$ plane, this mode of operation is very exacting. The slightest movement of the system poles due to any parameter variation will cause the oscillation either to grow or decay with time. So systems which operate in this mode must have some mechanism for keeping the poles on the $\omega$ axis.

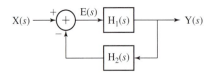

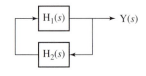

**Figure 10.19**
Prototype feedback system.

**Figure 10.20**
Oscillator feedback system.

The prototype feedback diagram (Figure 10.19) has an excitation and a response. A system designed to oscillate does not have an (apparent) excitation; that is, $X(s) = 0$ (Figure 10.20). (Notice that in this diagram the output signal of the feedback path is the input signal of the forward path with no external input signal added and no sign change.) How can we have a response if we have no excitation? The short answer is that we cannot. However, it is important to realize that every system is constantly being excited whether we intend it or not. Every system has random noise processes which cause signal fluctuations. The system responds to these noise fluctuations just as it would to an intentional excitation.

The key to having a stable oscillation is having a transfer function with poles on the $\omega$ axis of the form

$$H(s) = \frac{A}{s + \omega_0^2}. \qquad (10.62)$$

Then the system gain at the real radian frequency $\omega_0$ ($s = \pm j\omega_0$) is infinite, implying that the response is infinitely greater than the excitation. That could mean either that a finite excitation produces an infinite response or that a zero excitation produces a finite response. In either case the ratio of response to excitation is infinite. Therefore, a system with poles on the $\omega$ axis can produce a stable nonzero response with no excitation.

One very interesting and important example of a system designed to oscillate in a marginally stable mode is a laser. The acronym LASER stands for *light amplification by stimulated emission of radiation*. A laser is not actually a light amplifier (although, internally, light amplification does occur), it is a light oscillator, but the acronym for light oscillation by stimulated emission of radiation, LOSER, described itself and did not catch on.

Even though the laser is an oscillator, light amplification is an inherent process in its operation. A laser is filled with a medium which has been pumped by an external power source in such a way that light of the right wavelength propagating through the pumped medium experiences an increase in power as it propagates (Figure 10.21). The device illustrated in this figure is a one-pass, traveling-wave light amplifier, not a laser. The oscillation of light in a laser is caused by introducing into the one-pass traveling-wave light amplifier, mirrors on each end which reflect some or all of the light striking them. At each mirror some or all of the light is fed back into the pumped laser medium for further amplification (Figure 10.22).

It would be possible, in principle, to introduce light at one end of this device through a partial mirror and amplify it. Such a device is called a *regenerative traveling-wave light amplifier*. But it is much more common to make the mirror at one end as reflective as possible, essentially reflecting all the light that strikes it, and to make the mirror at the other end a partial mirror, reflecting some of the light that strikes it and transmitting the rest.

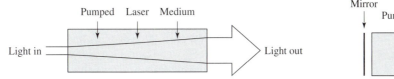

Figure 10.21
A one-pass traveling-wave light amplifier.

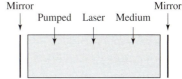

Figure 10.22
A laser.

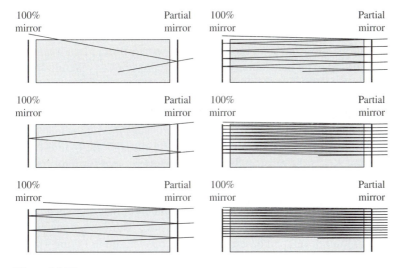

**Figure 10.23**
Multiple light reflections at different initial angles.

A laser operates without any external source of light. The light that it emits begins in the pumped laser medium itself. A phenomenon called *spontaneous emission* causes light to be generated at random times and in random directions in the pumped medium. Any such light which happens to propagate directly toward a mirror gets amplified on its way to the mirror, and then reflected and further amplified as it bounces between the mirrors. The closer the propagation is to normal to the mirrors, the longer the beam bounces and the more it is amplified by the multiple passes through the laser medium. In steady-state operation the light which is normal to the mirrors has the highest power of all the light propagation inside the laser cavity because it has the greatest gain advantage. One mirror is always a partial mirror so some light transmits at each bounce off that mirror. This light constitutes the output light beam of the laser (Figure 10.23).

In order for light oscillation to be sustained, the loop transfer function of the system must be the real number −1 under the assumed negative feedback sign on the prototype feedback system of Figure 10.19 or it must be the real number +1 under the assumption of the oscillator system of Figure 10.20. Under either assumption, for stable oscillation, the light, as it travels from a starting point to one mirror, back to the other mirror, and then back to the starting point, must experience an overall gain magnitude of one and a phase shift of an integer multiple of $2\pi$ rad. This simply means that the wavelength of the light must be such that it fits into the laser cavity with exactly an integer number of waves in one round-trip path.

It is important to realize here that the wavelength of light in lasers is typically somewhere in the range from 100 nm to many microns (ultraviolet to far infrared) and

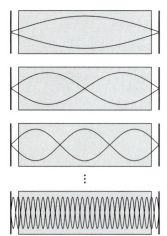

**Figure 10.24**
Illustrations of wavelengths which fit into the laser cavity an integer number of times.

lengths of laser cavities are typically in the range of a few centimeters to more than a meter in some cases. Therefore, as light propagates between the mirrors it may experience more than a million radians of phase shift and, even in the shortest cavities, the phase shift is a large multiple of $2\pi$ rad. So in a laser the exact wavelength of oscillation is determined by which optical wavelength fits into the round-trip path with exactly an integer number of waves. There are infinitely many wavelengths which satisfy this criterion, the wave which fits into the round trip exactly once plus all its harmonics (Figure 10.24). Although all these wavelengths of light could theoretically oscillate, there are other mechanisms (atomic or molecular resonances, wavelength-selective mirrors, etc.) which limit the actual oscillation to a small number of these wavelengths which experience enough gain to oscillate.

A laser can be modeled by a block diagram with a forward path and a feedback path (Figure 10.25). The constants $K_F$ and $K_R$ represent the magnitude of the gain experienced by the electric field of the light as it propagates from one mirror to the other along the forward and reverse paths, respectively. The factors $e^{-(L/c)s}$ account for the phase shift due to propagation time, where $L$ is the distance between the mirrors and $c$ is the speed of light in the laser cavity. The constant $K_{to}$ is the electric field transmission coefficient for light exiting the laser cavity through the output partial mirror and the constant $K_{ro}$ is the electric field reflection coefficient for light reflected at the output partial mirror back into the laser cavity. The constant $K_r$ is the electric field reflection coefficient for light reflected at the 100 percent mirror back into the laser cavity. $K_{to}$, $K_{ro}$, and $K_r$ are, in general, complex, indicating that there is a phase shift of the electric field during reflection and transmission. The loop transfer function is (using the definition developed based on the sign convention in Figure 10.19)

$$T(s) = -K_F K_{ro} K_R K_r e^{-(2L/c)s}. \tag{10.63}$$

Its value is $-1$ when

$$|K_F K_{ro} K_R K_r| = 1 \tag{10.64}$$

and

$$e^{-(2L/c)s} = 1 \tag{10.65}$$

or, equivalently,

$$s = -j2\pi n \left( \frac{c}{2L} \right) \qquad n \text{ is any integer,} \tag{10.66}$$

where the quantity $c/2L$ is the round-trip travel time for the propagating light wave. These are values of $s$ on the $\omega$ axis at harmonics of a fundamental radian frequency $2\pi(c/2L)$. Since this is the fundamental frequency, it is also the spacing between frequencies which is conventionally called the *axial mode spacing* $\Delta\omega_{ax}$.

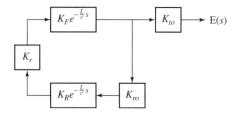

**Figure 10.25**
Laser block diagram.

When a laser is first turned on, the medium is pumped and a light beam starts through spontaneous emission. It grows in intensity because, at first, the magnitude of the round-trip gain is greater than one ($|K_F K_{ro} K_R K_r| > 1$). But, as it grows, it extracts energy from the pumped medium and that reduces the gains $K_F$ and $K_R$. An equilibrium is reached when the beam strength is exactly the right magnitude to keep the round-trip gain magnitude $|K_F K_{ro} K_R K_r|$ at exactly one. The pumping and light-amplification mechanisms in a laser together form a self-limiting process which stabilizes at a round-trip gain magnitude of one. So, as long as there is enough pumping power and the mirrors are reflective enough to achieve a round-trip gain magnitude of one at some very low output power, the laser will oscillate stably.

If the pumping power is increased, the output power will increase to extract more power from the pumped medium and reduce the round-trip gain magnitude back to one. If the pumping power is reduced, the output power will decrease extracting less power from the pumped medium and will increase the round-trip gain back to one. But if the pumping power is reduced too much, because of other loss mechanisms present in the cavity, it will not inject enough power into the pumped medium to sustain oscillation, even with a zero-output-power light beam and the laser will stop oscillating. This is an example of a system which is self-limiting in a way that settles into a stable sinusoidal oscillation without any system nonlinearity.

## THE ROUTH–HURWITZ STABILITY TEST

We have already seen that the requirement for system stability is that the poles all lie in the open left half-plane. We can find the pole locations by factoring the denominator of the transfer function, and any denominator can be factored, at least numerically, by using a mathematical tool like MATLAB. But factoring is not necessary to determine stability. There is an analysis technique called the *Routh–Hurwitz test* which can determine whether or not the system is stable without factoring the denominator. Also, it lends some insight into system characteristics and is useful in conjunction with other techniques like the root-locus method to be explored in the next section. The derivation of the rules used in the Routh–Hurwitz test are beyond the scope of this text, but we can explore their use and gain some intuitive understanding of their validity.

Let the form of a transfer function be

$$H(s) = \frac{N(s)}{D(s)} \tag{10.67}$$

and let the denominator $D(s)$ be of the form

$$D(s) = a_D s^D + a_{D-1} s^{D-1} + \cdots + a_1 s + a_0, \tag{10.68}$$

where $a_D$ is nonzero. The first step in the Routh–Hurwitz test is to construct the Routh array (Figure 10.26).

The *Routh array* is an array of numbers which has $D + 1$ rows and $(D/2) + 1$ columns for $D$ even and $(D + 1)/2$ columns for $D$ odd. The first two rows contain the coefficients of the denominator polynomial. The entries in the following row are found by the formulas

$$b_{D-2} = -\frac{\begin{vmatrix} a_D & a_{D-2} \\ a_{D-1} & a_{D-3} \end{vmatrix}}{a_{D-1}}, \qquad b_{D-4} = -\frac{\begin{vmatrix} a_D & a_{D-4} \\ a_{D-1} & a_{D-5} \end{vmatrix}}{a_{D-1}}, \tag{10.69}$$

etc. The entries on succeeding rows are computed by the same process based on previous row entries. If an entry evaluates to zero, replace the zero by an $\varepsilon$ (an arbitrarily

| $D$ | $a_D$ | $a_{D-2}$ | $a_{D-4}$ | $\cdots$ | $a_0$ | | $D$ | $a_D$ | $a_{D-2}$ | $a_{D-4}$ | $\cdots$ | $a_1$ |
|---|---|---|---|---|---|---|---|---|---|---|---|---|
| $D-1$ | $a_{D-1}$ | $a_{D-3}$ | $a_{D-5}$ | $\cdots$ | $0$ | | $D-1$ | $a_{D-1}$ | $a_{D-3}$ | $a_{D-5}$ | $\cdots$ | $a_0$ |
| $D-2$ | $b_{D-2}$ | $b_{D-4}$ | $b_{D-6}$ | $\cdots$ | $0$ | | $D-2$ | $b_{D-2}$ | $b_{D-4}$ | $b_{D-6}$ | $\cdots$ | $0$ |
| $D-3$ | $c_{D-3}$ | $c_{D-5}$ | $c_{D-7}$ | $\cdots$ | $0$ | | $D-3$ | $c_{D-3}$ | $c_{D-5}$ | $c_{D-7}$ | $\cdots$ | $0$ |
| $\vdots$ | $\vdots$ | $\vdots$ | $\vdots$ | $\vdots$ | $\vdots$ | | $\vdots$ | $\vdots$ | $\vdots$ | $\vdots$ | $\vdots$ | $\vdots$ |
| $2$ | $d_2$ | $d_0$ | $0$ | $0$ | $0$ | | $2$ | $d_2$ | $d_0$ | $0$ | $0$ | $0$ |
| $1$ | $e_1$ | $0$ | $0$ | $0$ | $0$ | | $1$ | $e_1$ | $0$ | $0$ | $0$ | $0$ |
| $0$ | $f_0$ | $0$ | $0$ | $0$ | $0$ | | $0$ | $f_0$ | $0$ | $0$ | $0$ | $0$ |
| | | $D$ even | | | | | | | $D$ odd | | | |

**Figure 10.26**
The Routh array.

small real number, either positive or negative) and continue. In the process of calculating succeeding row entries, drop any higher powers of $\varepsilon$ to simplify the computations. The process continues until we reach row zero. If there are any zeros or sign changes in the $a_D$ column, the system is unstable. The number of sign changes in the $a_D$ column is the number of poles in the right half-plane. If an $\varepsilon$ is used, it is considered either positive or negative. The number of sign changes will be the same either way. If there are no zeros or sign changes in the first column, the system is stable. If a row which has entries that are all zeros occurs before the row indexed by zero, the system has at least two poles of equal order that lie at locations in the complex plane that are radially opposite each other and equidistant from the origin. That means that either there is a pole in the right half-plane or two poles on the $\omega$ axis. In such a case the system cannot be strictly stable, but it can be marginally stable.

## EXAMPLE 10.1

Using the Routh–Hurwitz stability test determine whether the systems whose transfer functions are

$$H_1(s) = \frac{2s^2 + 4s - 3}{s^4 + 2s^3 + 8s^2 + 3s + 4} \tag{10.70}$$

and

$$H_2(s) = \frac{8s^2 + s + 10}{6s^4 + s^3 + 2s^2 + 4s + 1} \tag{10.71}$$

are stable.

### ■ Solution
The Routh array for $H_1(s)$ is

| | | |
|---|---|---|
| $1$ | $8$ | $4$ |
| $2$ | $3$ | $0$ |
| $\frac{13}{2}$ | $4$ | $0$ |
| $\frac{23}{13}$ | $0$ | $0$ |
| $4$ | $0$ | $0$ |

and the system is stable. This can be confirmed by factoring the denominator to find the poles. The poles are

$$-0.8547 + j2.4890 \qquad -0.8547 - j2.4890 \qquad -0.1453 + j0.7460 \qquad -0.1453 - j0.7460$$

which all lie in the open left half-plane.

The Routh array for $H_2(s)$ is

$$
\begin{array}{ccc}
6 & 2 & 1 \\
1 & 4 & 0 \\
-22 & 1 & 0 \\
\frac{89}{22} & 0 & 0 \\
1 & 0 & 0
\end{array}
$$

and the system is unstable with two poles in the right half-plane as indicated by the two sign changes in the first column. The poles in this case are

$$0.3865 + j0.8474 \qquad 0.3865 - j0.8474 \qquad -0.6390 \qquad -0.3007$$

and two lie in the right half-plane as indicated by the Routh array. ∎

**EXAMPLE 10.2**

Use the Routh–Hurwitz stability test to determine criteria for stability of a general second-order system whose transfer function is of the form

$$H(s) = \frac{N(s)}{s^2 + a_1 s + a_0}. \qquad (10.72)$$

**■ Solution**

The Routh array is

$$
\begin{array}{cc}
1 & a_0 \\
a_1 & 0 \\
a_0 & 0
\end{array}. \qquad (10.73)
$$

This result simply indicates that if $a_1$ and $a_0$ are both positive, there are no poles in the right half-plane and the system is stable. ∎

## THE ROOT-LOCUS METHOD

A very common situation in feedback system analysis is a system of the form illustrated in Figure 10.27. There is an adjustable gain parameter $K$ (which is conventionally taken to be nonnegative) and the choice of its value has a strong effect on the system's dynamics. The overall system transfer function is

$$H(s) = \frac{K H_1(s)}{1 + K H_1(s) H_2(s)} \qquad (10.74)$$

and the loop transfer function is

$$T(s) = K H_1(s) H_2(s). \qquad (10.75)$$

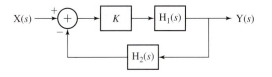

**Figure 10.27**
A common type of feedback system.

The poles of H($s$) are the zeros of $1 + T(s)$. The loop transfer function can be written in the form of $K$ times a numerator divided by a denominator,

$$T(s) = K\frac{P(s)}{Q(s)}, \tag{10.76}$$

so the poles of H($s$) occur where

$$1 + K\frac{P(s)}{Q(s)} = 0, \tag{10.77}$$

which can be expressed in the two alternative forms

$$Q(s) + KP(s) = 0 \tag{10.78}$$

and

$$\frac{Q(s)}{K} + P(s) = 0. \tag{10.79}$$

From (10.76), we see that if T($s$) is proper [Q($s$) is of higher order than P($s$)], the finite zeros of Q($s$) constitute all the poles of T($s$) and the zeros of P($s$) are all finite zeros of T($s$), but, because the order of P($s$) is less than the order of Q($s$), there are also one or more zeros of T($s$) at infinity.

The full range of possible adjustment of $K$ is from zero to infinity. First let $K$ approach zero. In that limit, from (10.78), the zeros of $1 + T(s)$, which are the poles of H($s$), are the zeros of

$$Q(s) = 0, \tag{10.80}$$

and the poles of H($s$) are, therefore, the poles of T($s$) because $T(s) = K(P(s)/Q(s))$. Now consider the opposite case, where $K$ approaches infinity. In that limit, from (10.79), the zeros of $1 + T(s)$ are the zeros of

$$P(s) = 0 \tag{10.81}$$

and the poles of H($s$) are the zeros of T($s$) (including any zeros at infinity). So the loop transfer function poles and zeros are very important in the analysis of the closed-loop system.

As the gain factor $K$ moves from zero to infinity, the poles of the closed-loop system move from the poles of the loop transfer function to the zeros of the loop transfer function (some of which may be at infinity). A *root-locus* plot is a plot of the locations of the closed-loop poles as the gain factor $K$ is varied from zero to infinity. The name root locus comes from the location (locus) of a root of $1 + T(s)$ as the gain factor $K$ is varied.

We will first examine two simple examples of the root-locus method and then establish some general rules for finding the root locus of any system. Consider first a system whose forward-path gain is

$$H_1(s) = \frac{K}{(s+1)(s+2)} \tag{10.82}$$

and whose feedback-path gain is

$$H_2(s) = 1. \tag{10.83}$$

Then

$$T(s) = \frac{K}{(s+1)(s+2)} \tag{10.84}$$

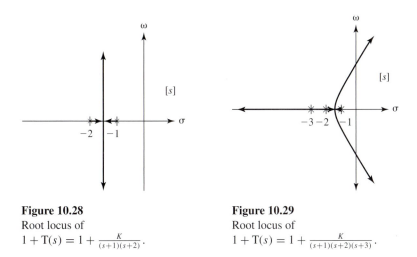

**Figure 10.28**
Root locus of
$1 + T(s) = 1 + \frac{K}{(s+1)(s+2)}$.

**Figure 10.29**
Root locus of
$1 + T(s) = 1 + \frac{K}{(s+1)(s+2)(s+3)}$.

and the root-locus plot begins at $s = -1$ and $s = -2$, the poles of T($s$). All the zeros of T($s$) are at infinity and those are the zeros that the root locus approaches as the gain factor $K$ is increased (Figure 10.28).

The roots of $1 + $ T($s$) are the roots of

$$(s + 1)(s + 2) + K = s^2 + 3s + 2 + K = 0 \qquad (10.85)$$

and, using the quadratic formula, the roots are at $(-3 \pm \sqrt{1 - 4K})/2$. For $K = 0$ we get roots at $s = -1$ and $s = -2$, the poles of T($s$). For $K = \frac{1}{4}$ we get a double root at $-\frac{3}{2}$. For $K > \frac{1}{4}$ we get two complex-conjugate roots whose imaginary parts go to plus and minus infinity as $K$ increases but whose real parts stay at $-\frac{3}{2}$. Since this root locus extends to infinity in the imaginary dimension with a real part that always places the roots in the left half-plane, this system is stable for any value of $K$.

Now add one pole to the forward-path transfer function making it

$$H_1(s) = \frac{K}{(s + 1)(s + 2)(s + 3)}. \qquad (10.86)$$

The new root locus is the locus of solutions to the equation

$$s^3 + 6s^2 + 11s + 6 + K = 0 \qquad (10.87)$$

(Figure 10.29). At or above the value of $K$ for which two branches of the root locus cross the $\omega$ axis, this system is unstable. So, in this case, a system which is open-loop stable can be made unstable by using feedback. We can find the value of $K$ at which the poles cross into the right half-plane using the Routh–Hurwitz test. The system transfer function is

$$H(s) = \frac{K}{s^3 + 6s^2 + 11s + 6 + K}, \qquad (10.88)$$

and the Routh array is illustrated in Figure 10.30. So the critical value of the gain parameter $K$ is 60. If $K = 60$, the gain expression is

$$H(s) = \frac{K}{s^3 + 6s^2 + 11s + 66} = \frac{K}{(s + 6)(s + j\sqrt{11})(s - j\sqrt{11})} \qquad (10.89)$$

and we see that there are two poles $s = \pm j\sqrt{11}$ on the $\omega$ axis of the $s$ plane.

| $D$ | 1 | 11 |
|---|---|---|
| $D - 1$ | 6 | $6 + K$ |
| $D - 2$ | $\frac{60 - K}{6}$ | 0 |
| $D - 3$ | $6 + K$ | 0 |
| $D - 4$ | 0 | 0 |

**Figure 10.30**
The Routh array for
$H(s) = \frac{K}{s^3 + 6s^2 + 11s + 6 + K}$.

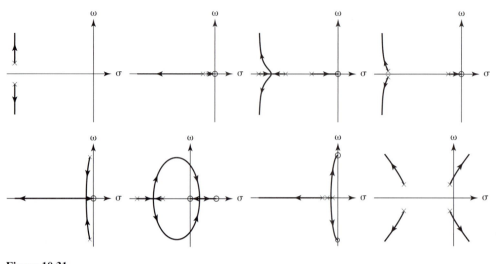

**Figure 10.31**
Example root-locus plots.

Figure 10.31 illustrates some root-locus plots for different numbers and different locations of the poles and zeros of $1 + T(s)$. The rules for root-locus plotting are

1. Each root-locus branch begins on a pole of $T(s)$ and terminates on a zero of $T(s)$.
2. Any portion of the real axis for which the sum of the number of real poles and/or real zeros lying to its right on the real axis is odd is a part of the root locus.
3. The root locus is symmetrical about the real axis.
4. If the number of poles of $T(s)$ exceeds the number of zeros of $T(s)$ by an integer $m$, then $m$ branches of the root locus terminate on zeros of $T(s)$ which lie at infinity. Each of these branches approaches a straight-line asymptote, and the angles of these asymptotes are at the angles $k\pi/m$, $k = 1, 3, 5, \ldots$, with respect to the positive real axis. These asymptotes intersect on the real axis at the location

$$\sigma = \frac{1}{m}\left(\sum \text{finite poles} - \sum \text{finite zeros}\right). \qquad \textbf{(10.90)}$$

The MATLAB control toolbox has a command for plotting the root locus of a system transfer function. The syntax is `rlocus(sys)`, where `sys` is a MATLAB system-description object.

### GAIN-MARGIN AND PHASE-MARGIN ANALYSIS OF SYSTEM STABILITY

In practical feedback system design, because of the uncertainty in our knowledge of some parameters, we typically design a system with a margin of error so that if our estimates of some parameters are somewhat inaccurate we still can have a reasonable assurance that the system is stable. Gain-margin analysis and phase-margin analysis are two methods of examining how much margin of error we have. These analyses are done by inspection of Bode diagrams of the magnitude and phase of the loop transfer function of a feedback system.

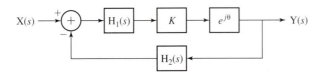

**Figure 10.32**
A feedback system with an additional gain block and
an additional phase block.

One way to appreciate the results of gain-margin or phase-margin analysis is to imagine that the typical feedback system with a forward-path gain of $H_1(s)$ and a feedback-path gain of $H_2(s)$ has two additional blocks in the loop, a gain block and a phase block (Figure 10.32). The gain block represents the factor by which the gain of the system might be in error, and the phase block represents the angle by which the phase of the system might be in error. It is usually not difficult to arrange to insert into a system an arbitrary frequency-independent gain to compensate for the gain error represented by the $K$ block in Figure 10.32. But it is an altogether different proposition to try to insert into a system a frequency-independent phase shift to compensate for any phase errors. Typically the phase shift of any component of a system is a function of frequency, not a fixed phase independent of frequency.

We know from our exploration of the root-locus method that if a system is to be unstable, the zeros of $1 + T(s)$ in the root-locus plot must cross the $\omega$ axis of the $s$ plane. Another way of saying the same thing is that system instability will occur if, for any real value of $\omega$,

$$T(j\omega) = -1. \tag{10.91}$$

The number $-1$ has a magnitude of one and a phase of $-\pi$ rad. Therefore, if, for any real frequency $\omega$, the magnitude of the loop transfer function is one and the phase is $-\pi$, the system is unstable because it can oscillate at that frequency.

Suppose a feedback system has the same three-pole loop transfer function

$$T(s) = \frac{K}{s^3 + 6s^2 + 11s + 6} \tag{10.92}$$

used in the discussion of root-locus methods but with a specific gain parameter value of $K = 10$. Its Bode diagram is illustrated in Figure 10.33. The gain margin is the factor by which the gain would have to be multiplied to cause the magnitude of the loop transfer function to be one at the frequency at which the phase is $-\pi$. Reading from the figure, the gain margin in this example is about 15.6 dB, which is equivalent to a factor of about six. (We know from the root-locus example that the factor is exactly six.) The phase margin is the difference between the phase where the loop transfer magnitude is one and a phase of $-\pi$. It is, therefore, positive for stable systems and negative for unstable systems. The phase margin in this example is about $+1.5$ rad, or about $+86°$.

## STEADY-STATE TRACKING ERRORS
## IN UNITY-GAIN FEEDBACK SYSTEMS

One very common type of feedback system is one in which the purpose of the system is to make the output signal track the input signal using unity-gain feedback $[H_2(s) = 1]$ (Figure 10.34). This type of system is called *unity gain* because the output

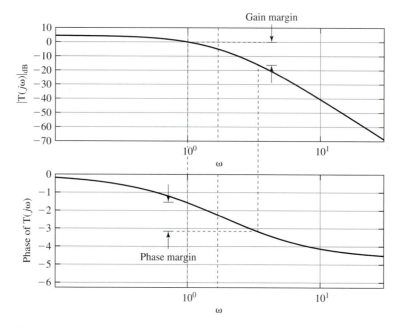

**Figure 10.33**
Gain and phase margins for $T(s) = \frac{10}{s^3+6s^2+11s+6}$.

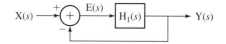

**Figure 10.34**
A unity-gain feedback system.

signal is always compared directly with the input signal and, if there is any difference (error signal), it will be amplified by the forward-path gain of the system in an attempt to correct the output signal. If the forward-path gain of the system is large, that forces the error signal to be small, making the output and input signals closer together. Whether or not the error signal can be forced to zero depends on the forward-path transfer function $H_1(s)$ and the type of excitation.

It is natural to wonder at this point what the purpose is of a system whose output signal equals its input signal. What have we gained? If the system is an electronic amplifier and the signals are voltages, we have a voltage gain of one, but the input impedance could be very high and the response voltage could drive a very low impedance so that the actual power, in watts, delivered by the output signal is much greater than the actual power supplied by the input signal. In other systems the input signal could be a voltage set by a low-power amplifier or a potentiometer and the output signal could be the position of some large mechanical device like a crane, an artillery piece, or an astronomical telescope. In this case the feedback-path transfer function can have a magnitude of one, but that could mean 1 V for a 1-m position or some other combination of units. It is the dissimilarity of the units that allows for a real power gain.

Now we will mathematically determine the nature of the steady-state error. The term *steady state* means mathematically the behavior as time approaches infinity. The error signal is

$$E(s) = X(s) - Y(s) = X(s) - H_1(s)E(s). \qquad \textbf{(10.93)}$$

Solving for E($s$),

$$E(s) = \frac{X(s)}{1 + H_1(s)}. \qquad \textbf{(10.94)}$$

We can find the steady-state value of the error signal using the final value theorem,

$$\lim_{t \to \infty} e(t) = \lim_{s \to 0} sE(s) = \lim_{s \to 0} s \frac{X(s)}{1 + H_1(s)}. \qquad \textbf{(10.95)}$$

If the input signal is a step of the form x($t$) = Au($t$), then X($s$) = $\frac{A}{s}$ and

$$\lim_{t \to \infty} e(t) = \lim_{s \to 0} \frac{A}{1 + H_1(s)} \qquad \textbf{(10.96)}$$

and there is zero steady-state error if $\lim_{s \to 0} \frac{1}{1+H_1(s)}$ is zero. If $H_1(s)$ is in the familiar form of a ratio of polynomials in $s$,

$$H_1(s) = \frac{b_N s^N + b_{N-1}s^{N-1} + \cdots + b_2 s^2 + b_1 s + b_0}{a_D s^D + a_{D-1}s^{D-1} + \cdots + a_2 s^2 + a_1 s + a_0}, \qquad \textbf{(10.97)}$$

then

$$\lim_{t \to \infty} e(t) = \lim_{s \to 0} \frac{1}{1 + \dfrac{b_N s^N + b_{N-1}s^{N-1} + \cdots + b_2 s^2 + b_1 s + b_0}{a_D s^D + a_{D-1}s^{D-1} + \cdots + a_2 s^2 + a_1 s + a_0}} = \frac{a_0}{a_0 + b_0} \qquad \textbf{(10.98)}$$

and, if $a_0 = 0$ and $b_0 \neq 0$, the steady-state error is zero. If $a_0 = 0$, then $H_1(s)$ can be expressed in the form

$$H_1(s) = \frac{b_N s^N + b_{N-1}s^{N-1} + \cdots + b_2 s^2 + b_1 s + b_0}{s(a_D s^{D-1} + a_{D-1}s^{D-2} + \cdots + a_2 s + a_1)} \qquad \textbf{(10.99)}$$

and it is immediately apparent that $H_1(s)$ has a pole at zero. So we can summarize by saying that if a stable unity-gain feedback system has a forward-path transfer function with a pole at zero, the steady-state error for a step excitation is zero. If there is no pole at zero, the steady-state error is $a_0/(a_0 + b_0)$ and the larger $b_0$ is in comparison with $a_0$, the smaller the steady-state error. This makes sense from another point of view because if the forward-path gain is of the form (10.97), the closed-loop, low-frequency gain is $b_0/(a_0 + b_0)$ which approaches one for $b_0 \gg a_0$, indicating that the input and output signals approach the same value.

A unity-gain feedback system with a forward-path transfer function $H_1(s)$ that has no poles at zero is called a *type 0* system. If it has one pole at zero, the system is a type 1 system. In general, any unity-gain feedback system is a type $n$ system, where $n$ is the number of poles at zero in $H_1(s)$. So, summarizing using the new terminology,

1.  A stable type 0 system has a finite steady-state error for step excitation.
2.  A stable type $n$ system, $n \geq 1$, has a zero steady-state error for step excitation.

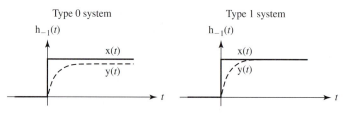

**Figure 10.35**
Type 0 and type 1 system responses to a step.

Figure 10.35 illustrates typical steady-state responses to step excitation for type 0 and type 1 systems.

Now we will consider a ramp excitation $x(t) = At u(t)$ whose Laplace transform is $X(s) = A/s^2$. The steady-state error is

$$\lim_{t\to\infty} e(t) = \lim_{s\to 0} \frac{A}{s\,[1 + H_1(s)]}. \qquad (10.100)$$

Again, if $H_1(s)$ is a ratio of polynomials in $s$,

$$\lim_{t\to\infty} e(t) = \lim_{s\to 0} \frac{1}{s} \frac{1}{1 + \dfrac{b_N s^N + b_{N-1} s^{N-1} + \cdots + b_2 s^2 + b_1 s + b_0}{a_D s^D + a_{D-1} s^{D-1} + \cdots + a_2 s^2 + a_1 s + a_0}} \qquad (10.101)$$

or

$$\lim_{t\to\infty} e(t) = \lim_{s\to 0} \frac{a_D s^D + a_{D-1} s^{D-1} + \cdots + a_2 s^2 + a_1 s + a_0}{s[a_D s^D + a_{D-1} s^{D-1} + \cdots + a_2 s^2 + a_1 s + a_0}$$
$$+ b_N s^N + b_{N-1} s^{N-1} + \cdots + b_2 s^2 + b_1 s + b_0].$$

This limit depends on the values of the $a$'s and $b$'s. If $a_0 \neq 0$, the steady-state error is infinite. If $a_0 = 0$ and $b_0 \neq 0$, the limit is $a_1/b_0$ indicating that the steady-state error is a nonzero constant. If $a_0 = 0$, $a_1 = 0$, and $b_0 \neq 0$, the steady-state error is zero. The condition $a_0 = 0$ and $a_1 = 0$ means there is a double pole at zero in the forward-path transfer function. So for a type 2 system, the steady-state error under ramp excitation is zero. Summarizing,

1. A stable type 0 system has an infinite steady-state error for ramp excitation.
2. A stable type 1 system has a finite steady-state error for ramp excitation.
3. A stable type $n$ system, $n \geq 2$, has a zero steady-state error for ramp excitation.

Figure 10.36 illustrates typical steady-state responses to ramp excitation for stable type 0, type 1, and type 2 systems.

These results can be extrapolated to higher-order excitations, $[At^2 u(t), At^3 u(t),$ etc.]. When the highest power of $s$ in the denominator of the transform of the excitation is the same as, or lower than, the type number (0, 1, 2, etc.) of the system, the steady-state error is zero. This result was illustrated with forward-path transfer functions in the form of a ratio of polynomials, but the result can be shown to be true for any form of transfer function based only on the number of poles at zero.

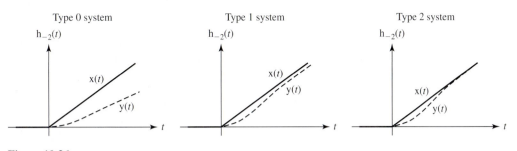

**Figure 10.36**
Type 0, 1, and 2 system responses to a ramp.

# 10.6 BLOCK-DIAGRAM REDUCTION AND MASON'S THEOREM

Some system block diagrams are large and complicated, with many components and interconnections. It is often desirable to find a mathematical relationship between an excitation and a response from the block diagram. One way to do that is to write all the equations relating component excitations and responses and then solve them for the ratio of the overall system response to its excitation. But there are two other ways which are very useful in some situations and lend insight into system operation, block-diagram reduction and Mason's theorem.

We have already seen examples of block-diagram reduction when we found the equivalent transfer function for two systems connected in cascade, parallel, or feedback configurations. There are three other useful operations which help reduce block diagrams, moving a pickoff point, moving a summer, and combining summers. Figure 10.37 illustrates how to move a pickoff point without changing any signals, Figure 10.38 illustrates how to move a summer without changing any signals, and Figure 10.39 illustrates how to combine two summers.

As an example of the use of block-diagram reduction consider the system in Figure 10.40. First we move the leftmost pickoff point to the right past the 10 (Figure 10.41). Then we move the first summer to the right past the $1/s$ block (Figure 10.42). Now we can combine the two summers into one summer (Figure 10.43). We can combine the two parallel blocks $1/s$ and $1/10(s+3)$ into one block

$$\frac{1}{s} + \frac{1}{10(s+3)} = \frac{11s+30}{10s(s+3)}$$

(Figure 10.44). Then we can combine the two cascade connected blocks 10 and $(11s+30)/10s(s+3)$ into one block (Figure 10.45). We can now reduce the feedback loop, using the general relationship derived in the Section 10.4

$$H(s) = \frac{H_1(s)}{1 + H_1(s)H_2(s)}, \tag{10.102}$$

where, in this case, $H_1(s) = 1$ and $H_2(s) = 3/s(s+8)$. The equivalent transfer function for the feedback loop is then

$$H(s) = \frac{s(s+8)}{s^2 + 8s + 3} \tag{10.103}$$

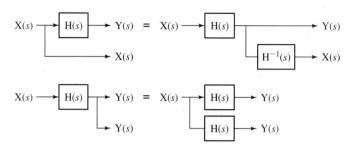

**Figure 10.37**
Moving a pickoff point.

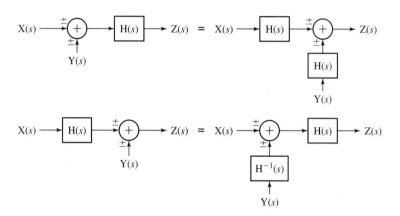

**Figure 10.38**
Moving a summer.

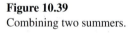

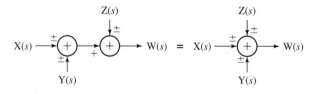

**Figure 10.39**
Combining two summers.

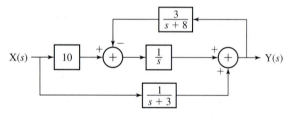

**Figure 10.40**
A system to be reduced by block-diagram reduction techniques.

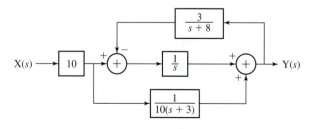

**Figure 10.41**
First block-diagram reduction step.

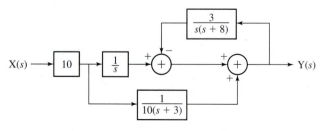

**Figure 10.42**
Second block-diagram reduction step.

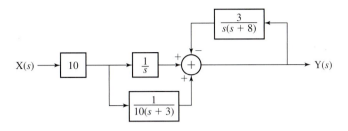

**Figure 10.43**
Third block-diagram reduction step.

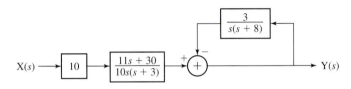

**Figure 10.44**
Fourth block-diagram reduction step.

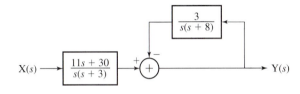

**Figure 10.45**
Fifth block-diagram reduction step.

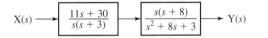

**Figure 10.46**
Sixth block-diagram reduction step.

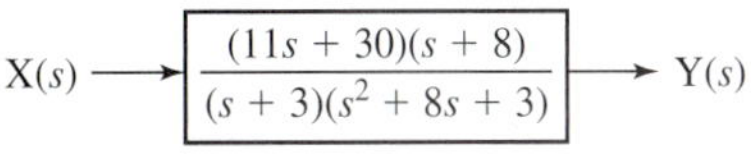

**Figure 10.47**
Seventh block-diagram reduction step.

(Figure 10.46). Finally, we can combine these two cascaded systems into one overall transfer function (Figure 10.47).

An alternative way to find the overall gain of a system is through Mason's theorem. This theorem uses the transfer functions of all paths from the system input to the system output and the loop transfer functions of all feedback loops in the system. Let the number of paths from input to output be $N_p$, and let the number of feedback loops be $N_L$. Let $P_i(s)$ be the transfer function of the $i$th path from input to output, and let $T_i(s)$ be the loop transfer function of the $i$th feedback loop. [The loop transfer function is defined as in (10.30) under the assumption of negative feedback polarity. If the polarity is positive, the sign of the loop transmission is changed.] Define a determinant $\Delta(s)$ by

$$\Delta(s) = 1 + \sum_{i=1}^{N_L} T_i(s) + \sum_{\substack{i\text{th loop and} \\ j\text{th loop not} \\ \text{sharing a signal}}} T_i(s)T_j(s) + \sum_{\substack{i\text{th, }j\text{th, }k\text{th} \\ \text{loops not} \\ \text{sharing a signal}}} T_i(s)T_j(s)T_k(s) + \cdots .$$

$$(10.104)$$

Mason's theorem states that the overall system transfer function is

$$H(s) = \frac{\displaystyle\sum_{i=1}^{N_p} P_i(s)\,\Delta_i(s)}{\Delta(s)}$$

$$(10.105)$$

where $\Delta_i(s)$ is the same as $\Delta(s)$ except that all feedback loops which share a signal with the $i$th path, $P_i(s)$, are excluded.

We can apply Mason's theorem to the system of Figure 10.40. There are two paths from input to output with transfer functions

$$P_1(s) = \frac{10}{s} \quad \text{and} \quad P_2(s) = \frac{1}{s+3}, \tag{10.106}$$

and there is one feedback loop. Therefore, $N_p = 2$ and $N_L = 1$. Then

$$\Delta(s) = 1 + \frac{1}{s}\frac{3}{s+8} = 1 + \frac{3}{s(s+8)}. \tag{10.107}$$

Since the feedback loop shares a signal with both paths, $\Delta_1(s) = \Delta_2(s) = 1$ and

$$H(s) = \frac{\sum\limits_{i=1}^{N_p} P_i(s)\,\Delta_i(s)}{\Delta(s)} = \frac{10/s + (1/(s+3))}{1 + (3/(s(s+8)))} \tag{10.108}$$

or, after simplification,

$$H(s) = \frac{(11s+30)/s(s+3)}{(s^2+8s+3)/s(s+8)} = \frac{(s+8)(11s+30)}{(s+3)(s^2+8s+3)} \tag{10.109}$$

which is the same as the result from the block-diagram reduction process.

Block-diagram reduction can also be done using MATLAB's control toolbox. When two systems are connected in cascade, their transfer functions multiply and that is done with the overloaded operator $*$ or the `series` command. When two systems are connected in parallel, their transfer functions add and that is done with the overloaded operator $+$ or the `parallel` command. When two systems are connected in a feedback arrangement (with the assumption of negative feedback used in this text), their transfer functions can be combined with the `feedback` command. The syntax of the `feedback` command is

```
sys = feedback(sys1,sys2)
```

where `sys1` is the system description of the forward path and `sys2` is the system description of the feedback path. For example,

```
»H1 = tf([1 0],[1 3 2]) ;
»H2 = tf(1,[1 0]) ;
»H = feedback(H1,H2) ;
»H1

Transfer function:
        s
---------------
s^2 + 3 s + 2

»H2

Transfer function:
1
-
s

»H
```

Transfer function:
```
        s^2
---------------------------------------
s^3 + 3 s^2 + 3 s
```

Notice that the last system description, although correct, is not in the best form because the numerator and denominator could both be divided by $s$ to simplify the expression. The command `minreal` (minimum realization) in the MATLAB control toolbox accomplishes this.

»minreal(H)

Transfer function:
```
       s
-------------------------------
s^2 + 3 s + 3
```

**EXAMPLE 10.3**

Find the transfer function of the system in Figure 10.48 by block-diagram reduction and by Mason's theorem.

■ **Solution**

*Block-diagram reduction:*    We can move the pickoff point for the $s/(s + 1)$ block to the right, past the second $1/s$ block (Figure 10.49). We then can reduce the inner feedback loop involving $6/(s + 2)$ to a single block (Figure 10.50). We can combine some cascaded blocks

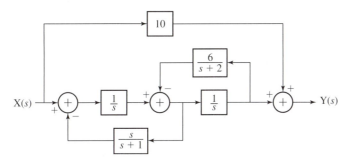

**Figure 10.48**
A system.

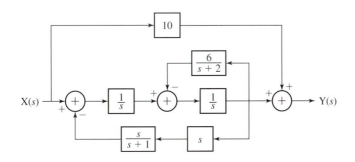

**Figure 10.49**
First block-diagram reduction step.

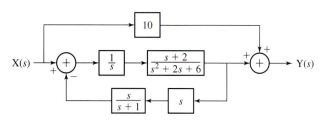

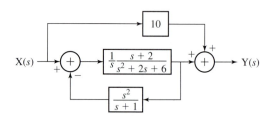

**Figure 10.50**
Second block-diagram reduction step.

**Figure 10.51**
Third block-diagram reduction step.

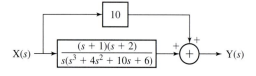

**Figure 10.52**
Fourth block-diagram reduction step.

**Figure 10.53**
Last block-diagram reduction step.

(Figure 10.51) and then reduce the remaining feedback loop to a single block (Figure 10.52). Finally, we combine the two parallel blocks into a single block (Figure 10.53).

*Mason's theorem:* There are two paths from input to output,

$$P_1(s) = 10 \quad \text{and} \quad P_2(s) = \frac{1}{s^2}. \tag{10.110}$$

There are two feedback loops with loop transfer functions

$$T_1(s) = \frac{1}{s+1} \quad \text{and} \quad T_2(s) = \frac{6}{s(s+2)}. \tag{10.111}$$

These two feedback loops share a common signal. Then, according to Mason's theorem,

$$\Delta(s) = 1 + \frac{1}{s+1} + \frac{6}{s(s+2)} \tag{10.112}$$

and

$$\Delta_1(s) = \Delta(s) \quad \text{and} \quad \Delta_2(s) = 1. \tag{10.113}$$

Then the transfer function is

$$H(s) = \frac{10[1 + (1/(s+1)) + (6/s(s+2))] + (1/s^2)}{1 + (1/(s+1)) + (6/s(s+2))} = 10 + \frac{1/s^2}{1 + (1/(s+1)) + (6/s(s+2))} \tag{10.114}$$

or, after simplification,

$$H(s) = 10 + \frac{(s+1)(s+2)}{s(s^3 + 4s^2 + 10s + 6)}. \tag{10.115}$$

## 10.7 SYSTEM RESPONSES TO STANDARD SIGNALS

We have seen in previous signal and system analysis that an LTI system is completely characterized by its impulse response. Although that is certainly true, it is useful for pedagogical purposes to analyze the response to some other standard signals, most importantly a unit step and a suddenly-applied sinusoid.

## UNIT STEP RESPONSE

Let the transfer function of an LTI system be of the form

$$H(s) = \frac{N(s)}{D(s)}, \tag{10.116}$$

where $N(s)$ is of a lower degree in $s$ than $D(s)$. Then the Laplace transform of the response $Y(s)$ to an excitation whose Laplace transform is $X(s)$ is

$$Y(s) = \frac{N(s)}{D(s)}X(s). \tag{10.117}$$

Let the excitation be a unit step. Then the Laplace transform of the response is

$$Y(s) = H_{-1}(s) = \frac{N(s)}{sD(s)}. \tag{10.118}$$

Using the partial-fraction expansion technique, this can be separated into two terms

$$Y(s) = \frac{N_1(s)}{D(s)} + \frac{K}{s}, \tag{10.119}$$

where $K = H(0)$. If the system is stable, the roots of $D(s)$ are all in the open left half-plane and the inverse Laplace transform of $N_1(s)/D(s)$ is called the *transient response* because it decays to zero as time $t$ approaches infinity. So the steady-state response of the system to a unit step excitation is the inverse Laplace transform of $H(0)/s$ which is $H(0)u(t)$. The expression

$$Y(s) = \frac{N_1(s)}{D(s)} + \frac{K}{s} \tag{10.120}$$

has two terms. The first term has poles which are identical to the system poles, and the second term has a pole at the same location as the Laplace transform of the unit step excitation.

This result can be generalized to an arbitrary excitation. If the Laplace transform of the excitation is

$$X(s) = \frac{N_x(s)}{D_x(s)}, \tag{10.121}$$

then the Laplace transform of the system response is

$$Y(s) = \frac{N(s)}{D(s)}X(s) = \frac{N(s)}{D(s)}\frac{N_x(s)}{D_x(s)} = \underbrace{\frac{N_1(s)}{D(s)}}_{\substack{\text{same poles} \\ \text{as system}}} + \underbrace{\frac{N_{x1}(s)}{D_x(s)}}_{\substack{\text{same poles} \\ \text{as excitation}}} \tag{10.122}$$

Now let's examine the unit-step response of some simple systems. The simplest system is a first-order system whose transfer function is of the form

$$H(s) = \frac{A}{1 - (s/p)}, \tag{10.123}$$

where $A$ is the low-frequency gain of the system and $p$ is the pole location in the $s$ plane. The Laplace transform of the step response is

$$Y(s) = H_{-1}(s) = \frac{A}{(1 - (s/p))s} = \frac{A/p}{1 - (s/p)} + \frac{A}{s} = \frac{A}{s} - \frac{A}{s - p}. \tag{10.124}$$

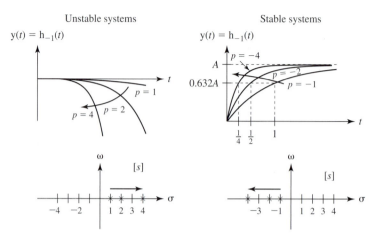

**Figure 10.54**
Responses of a first-order system to a unit step excitation and the
corresponding pole–zero diagrams.

Inverse Laplace-transforming,

$$y(t) = A(1 - e^{pt})u(t). \tag{10.125}$$

If $p$ is positive, the system is unstable and the magnitude of the response to a unit step
increases exponentially with time (Figure 10.54).

The speed of the exponential increase depends on the magnitude of $p$, being
greater for a larger magnitude of $p$. If $p$ is negative, the system is stable and the response approaches a constant $A$ with time. The speed of the approach to $A$ depends on
the magnitude of $p$, being greater for a larger magnitude of $p$. The negative reciprocal
of $p$ is called the *time constant* $\tau$ of the system,

$$\tau = -\frac{1}{p}, \tag{10.126}$$

and, for a stable system, the response to a unit step moves 63.2 percent of the distance
to the final value in a time equal to one time constant.

Now consider a second-order system whose transfer function is of the form

$$H(s) = \frac{A\omega_0^2}{s^2 + 2\zeta\omega_0 s + \omega_0^2} \qquad \omega_0 > 0. \tag{10.127}$$

This form of a second-order system transfer function has three parameters, the low-
frequency gain $A$, the damping factor $\zeta$, and the undamped resonant radian frequency
$\omega_0$. The form of the unit step response depends on these parameter values. The system
unit-step response is

$$Y(s) = H_{-1}(s) = \frac{A\omega_0^2}{s(s^2 + 2\zeta\omega_0 s + \omega_0^2)}$$

$$= \frac{A\omega_0^2}{s[s + \omega_0(\zeta + \sqrt{\zeta^2 - 1})][s + \omega_0(\zeta - \sqrt{\zeta^2 - 1})]}. \tag{10.128}$$

This can be expanded in partial fractions (if $\zeta \neq \pm 1$) as

$$Y(s) = A\left[\frac{1}{s} + \frac{1/2\left(\zeta^2 - 1 + \zeta\sqrt{\zeta^2 - 1}\right)}{s + \omega_0\left(\zeta + \sqrt{\zeta^2 - 1}\right)} + \frac{1/2\left(\zeta^2 - 1 - \zeta\sqrt{\zeta^2 - 1}\right)}{s + \omega_0\left(\zeta - \sqrt{\zeta^2 - 1}\right)}\right].$$

**(10.129)**

The time-domain response is then

$$y(t) = A\left[\frac{e^{-\omega_0\left(\zeta + \sqrt{\zeta^2 - 1}\right)t}}{2\left(\zeta^2 - 1 + \zeta\sqrt{\zeta^2 - 1}\right)} + \frac{e^{-\omega_0\left(\zeta - \sqrt{\zeta^2 - 1}\right)t}}{2\left(\zeta^2 - 1 - \zeta\sqrt{\zeta^2 - 1}\right)} + 1\right]u(t).$$

**(10.130)**

For the special case of $\zeta = \pm 1$ the system unit step response is

$$Y(s) = H_{-1}(s) = \frac{A\omega_0^2}{(s \pm \omega_0)^2 s},$$

**(10.131)**

the two poles are identical, the partial fraction expansion is

$$Y(s) = A\left[\frac{1}{s} - \frac{\pm\omega_0}{(s \pm \omega_0)^2} - \frac{1}{s \pm \omega_0}\right],$$

**(10.132)**

and the time-domain response is

$$y(t) = A[1 - (1 \pm \omega_0 t)e^{\mp\omega_0 t}]u(t) = Au(t)\begin{cases} 1 - (1 + \omega_0 t)e^{-\omega_0 t} & \zeta = 1 \\ 1 - (1 - \omega_0 t)e^{+\omega_0 t} & \zeta = -1 \end{cases}.$$

**(10.133)**

It is difficult, just by examining the mathematical functional form of the unit-step response, to immediately determine what the unit-step function response will look like for an arbitrary choice of parameters. To explore the effect of the parameters let's first set $A$ and $\omega_0$ constant and examine the effect of the damping factor $\zeta$. Let $A = 1$ and let $\omega_0 = 1$. Then the unit-step response and the corresponding pole–zero diagrams are as illustrated in Figure 10.55 for six choices of $\zeta$.

We can see why these different types of behavior occur if we examine the unit-step response

$$y(t) = h_{-1}(t) = A\left[\frac{e^{-\omega_0\left(\zeta + \sqrt{\zeta^2 - 1}\right)t}}{2\left(\zeta^2 - 1 + \zeta\sqrt{\zeta^2 - 1}\right)} + \frac{e^{-\omega_0\left(\zeta - \sqrt{\zeta^2 - 1}\right)t}}{2\left(\zeta^2 - 1 - \zeta\sqrt{\zeta^2 - 1}\right)} + 1\right]u(t),$$

**(10.134)**

in particular the exponents of $e$, $-\omega_0(\zeta \pm \sqrt{\zeta^2 - 1})t$. The signs of the real parts of these exponents determine whether the response grows or decays with time $t > 0$. For times, $t < 0$ the response is zero because of the unit step $u(t)$.

***Case 1*** $\zeta < 0$. If $\zeta < 0$, then the exponent of $e$ in both terms in (10.134) has a positive real part for positive time and the step response therefore grows with time and the system is unstable. The exact form of the unit-step response depends on the value of $\zeta$. It is a simple increasing exponential for $\zeta < -1$ and an exponentially growing sinusoid for $-1 < \zeta < 0$. But either way the system is unstable.

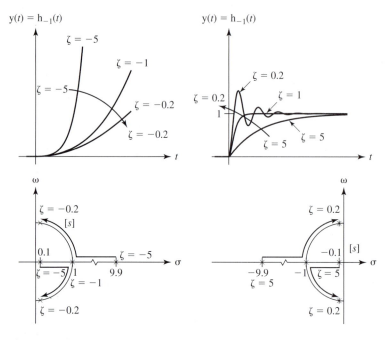

**Figure 10.55**
Second-order system responses to a unit step and the corresponding pole–zero diagrams.

**Case 2** $\zeta > 0$.   If $\zeta > 0$, then the exponent of $e$ in both terms in (10.134) has a negative real part for positive time and the step response, therefore, decays with time and the system is stable.

> *Case 2a.*  $\zeta > 1$.   If $\zeta > 1$, then $\zeta^2 - 1 > 0$, and the coefficients of $t$ in (10.134), $-\omega_0(\zeta \pm \sqrt{\zeta^2 - 1})t$, are both negative real numbers and the unit-step response is in the form of a constant plus the sum of two decaying exponentials. This case, $\zeta > 1$, is called the *overdamped* case.
>
> *Case 2b.*   $0 < \zeta < 1$.   If $0 < \zeta < 1$, then $\zeta^2 - 1 < 0$, and the coefficients of $t$ in (10.134), $-\omega_0(\zeta \pm \sqrt{\zeta^2 - 1})t$, are both complex numbers in a complex-conjugate pair with negative real parts, and the unit-step response is in the form of a constant plus the sum of two sinusoids multiplied by a decaying exponential. Even though the response "rings" or overshoots, it still settles to a constant value and is, therefore, the response of a stable system. This case, $0 < \zeta < 1$, is called the *underdamped* case.
>
> *Case 2c.*  $\zeta = 1$.   The dividing line between the overdamped and underdamped cases is the case $\zeta = 1$. This condition is called *critical damping*.

Now let's examine the effect of changing $\omega_0$ while holding the other parameters constant. Let $A = 1$ and $\zeta = 0.5$. The step response is illustrated in Figure 10.56 for three values of $\omega_0$. Since $\omega_0$ is the undamped resonant radian frequency, it is logical that it would affect the ringing rate of the step response. The response of any system to a step excitation can be found using the MATLAB control toolbox command `step`.

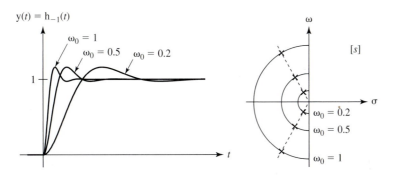

**Figure 10.56**
Second-order system response for three different values of $\omega_0$ and the corresponding pole–zero plots.

## RESPONSE TO A SUDDENLY-APPLIED SINUSOID

Now let's examine the response of a system to another standard type of excitation, a suddenly-applied sinusoid. Again let the system transfer function be of the form

$$H(s) = \frac{N(s)}{D(s)}. \tag{10.135}$$

Then the response to a suddenly-applied unit amplitude cosine $\cos(\omega_0 t)\, u(t)$ would be

$$Y(s) = \frac{N(s)}{D(s)} \frac{s}{s^2 + \omega_0^2}. \tag{10.136}$$

This can be separated into partial fractions in the form

$$Y(s) = \frac{N_1(s)}{D(s)} + \frac{1}{2}\frac{H(-j\omega_0)}{s + j\omega_0} + \frac{1}{2}\frac{H(j\omega_0)}{s - j\omega_0} = \frac{N_1(s)}{D(s)} + \frac{1}{2}\frac{H^*(j\omega_0)}{s + j\omega_0} + \frac{1}{2}\frac{H(j\omega_0)}{s - j\omega_0} \tag{10.137}$$

or

$$Y(s) = \frac{N_1(s)}{D(s)} + \frac{1}{2}\frac{H^*(j\omega_0)(s - j\omega_0) + H(j\omega_0)(s + j\omega_0)}{s^2 + \omega_0^2}$$

$$Y(s) = \frac{N_1(s)}{D(s)} + \frac{1}{2}\left\{ \frac{s}{s^2 + \omega_0^2}[H(j\omega_0) + H^*(j\omega_0)] + \frac{j\omega_0}{s^2 + \omega_0^2}[H(j\omega_0) - H^*(j\omega_0)] \right\} \tag{10.138}$$

or

$$Y(s) = \frac{N_1(s)}{D(s)} + \mathrm{Re}(H(j\omega_0))\frac{s}{s^2 + \omega_0^2} - \mathrm{Im}(H(j\omega_0))\frac{\omega_0}{s^2 + \omega_0^2}$$

The inverse Laplace transform of the term $\mathrm{Re}(H(j\omega_0))s/(s^2 + \omega_0^2)$, is a cosine at $\omega_0$ with an amplitude of $\mathrm{Re}(H(j\omega_0))$, and the inverse Laplace transform of the term,

$Im(H(j\omega_0))\omega_0/(s^2 + \omega_0^2)$, is a sine at $\omega_0$ with an amplitude of $Im(H(j\omega_0))$. That is,

$$y(t) = \mathcal{L}^{-1}\left(\frac{N_1(s)}{D(s)}\right) + [Re(H(j\omega_0))\cos(\omega_0 t) - Im(H(j\omega_0))\sin(\omega_0 t)]\,u(t)$$

(10.139)

or, using $Re(A)\cos(\omega_0 t) - Im(A)\sin(\omega_0 t) = |A|\cos(\omega_0 t + \angle A)$,

$$y(t) = \mathcal{L}^{-1}\left(\frac{N_1(s)}{D(s)}\right) + |H(j\omega_0)|\cos(\omega_0 t + \angle H(j\omega_0))\,u(t). \qquad \textbf{(10.140)}$$

If the system is stable, the roots of $D(s)$ are all in the open left half-plane and the inverse Laplace transform of $N_1(s)/D(s)$, the transient response, decays to zero as time $t$ approaches infinity. Therefore, the steady-state response which persists after the transient response has died away is a sinusoid of the same frequency as the excitation and with an amplitude and phase determined by the transfer function evaluated at $s = j\omega_0$. The steady-state response is exactly the same as the response obtained by using Fourier methods because the Fourier methods assume that the excitation is a true sinusoid, not a *suddenly-applied* sinusoid, and, therefore, there is no transient response in the solution.

## EXAMPLE 10.4

Find the total response of a system characterized by the transfer function

$$H(s) = \frac{10}{s + 10} \qquad \textbf{(10.141)}$$

to a suddenly-applied unit-amplitude cosine at a frequency of 2 Hz.

### ■ Solution

The radian frequency $\omega_0$ of the excitation is $4\pi$. Therefore, the Laplace transform of the response is

$$Y(s) = \frac{10}{s + 10}\frac{s}{s^2 + (4\pi)^2}$$

(10.142)

$$Y(s) = \frac{-0.388}{s + 10} + Re(H(j4\pi))\frac{s}{s^2 + (4\pi)^2} - Im(H(j4\pi))\frac{\omega_0}{s^2 + (4\pi)^2}$$

and the time-domain response is

$$y(t) = \mathcal{L}^{-1}\left(\frac{-0.388}{s + 10}\right) + |H(j4\pi)|\cos(4\pi t + \angle H(j4\pi))\,u(t) \qquad \textbf{(10.143)}$$

or

$$y(t) = \left[-0.388e^{-10t} + \left|\frac{10}{j4\pi + 10}\right|\cos(4\pi t - \angle(j4\pi + 10))\right]u(t)$$

or

$$y(t) = [-0.388e^{-10t} + 0.623\cos(4\pi t - 0.899)]u(t).$$

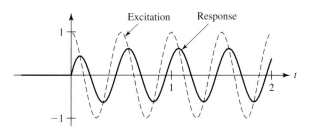

**Figure 10.57**
Excitation and response of a first-order system excited
by a suddenly applied cosine.

This response is illustrated in Figure 10.57. Looking at the graph we see that the response appears to reach steady state in less than 1s. This is reasonable given that the transient response has a time constant of one-tenth of a second. After the response reaches steady state, its amplitude is about 62 percent of the excitation amplitude and its phase is shifted so that it lags the excitation by about a $52°$ phase shift which is equivalent to about a 72-ms time delay.

If we solve for the response of the system using Fourier methods, we write the transfer function as

$$H(j\omega) = \frac{10}{j\omega + 10}.$$                     **(10.144)**

If we make the excitation of the system a cosine (not a suddenly-applied cosine), it is

$$x(t) = \cos(4\pi t)$$                     **(10.145)**

and its CTFT is

$$X(j\omega) = \pi\left[\delta(\omega - 4\pi) + \delta(\omega + 4\pi)\right].$$                     **(10.146)**

Then the system response is

$$Y(j\omega) = \pi\left[\delta(\omega - 4\pi) + \delta(\omega + 4\pi)\right]\frac{10}{j\omega + 10} = 10\pi\left[\frac{\delta(\omega - 4\pi)}{j4\pi + 10} + \frac{\delta(\omega + 4\pi)}{-j4\pi + 10}\right]$$

**(10.147)**

or

$$Y(j\omega) = 10\pi\frac{10\left[\delta(\omega - 4\pi) + \delta(\omega + 4\pi)\right] + j4\pi\left[\delta(\omega + 4\pi) - \delta(\omega - 4\pi)\right]}{16\pi^2 + 100}.$$

Inverse Fourier transforming,

$$y(t) = 0.388\cos(4\pi t) + 0.487\sin(4\pi t)$$                     **(10.148)**

or, using

$$\text{Re}(A)\cos(\omega_0 t) - \text{Im}(A)\sin(\omega_0 t) = |A|\cos(\omega_0 t + \angle A),$$                     **(10.149)**

$$y(t) = 0.623\cos(4\pi t - 0.899).$$                     **(10.150)**

This is exactly the same as the steady-state response part of the previous solution, which was found using Laplace transforms, after the transient response has died away.

## 10.8 POLE–ZERO DIAGRAMS AND GRAPHICAL CALCULATION OF FREQUENCY RESPONSE

Let $g(t)$ be a time-domain function whose Laplace transform has all its poles in the open left half-plane. Let the Laplace transform of $g(t)$ be $G(s)$. Then the Fourier transform of $g(t)$ is $G(j\omega)$. The Laplace transform of the impulse response $h(t)$ of an LTI system is the complex-frequency transfer function $H(s)$ and the Fourier transform is the real-frequency transfer function $H(j\omega)$. The variation of $H(j\omega)$ with radian frequency $\omega$ is also called the frequency response of the system. Therefore, the frequency response of a stable system can be obtained directly from the Laplace-domain transfer function by letting $s$ be $j\omega$. If the cyclic frequency form is preferred, it is simply

$$H_f(f) = H(j2\pi f). \qquad (10.151)$$

[The $f$ subscript is there because $H_f$ and $H$ are two different functions, that is, $H_f(f) \neq H(f)$].

In practice, the most common kind of transfer function is one which can be expressed as a ratio of polynomials in $s$,

$$H(s) = \frac{N(s)}{D(s)}. \qquad (10.152)$$

This type of transfer function can always be factored into the form

$$H(s) = A\frac{(s - z_1)(s - z_2)\cdots(s - z_N)}{(s - p_1)(s - p_2)\cdots(s - p_D)}. \qquad (10.153)$$

To graph the frequency response, let $s$ be restricted to $j\omega$, where $\omega$ is real. This can be conceived graphically by imagining that $s$ ranges only along the imaginary axis of the $s$ plane. Then the frequency response of the system is

$$H(j\omega) = A\frac{(j\omega - z_1)(j\omega - z_2)\cdots(j\omega - z_N)}{(j\omega - p_1)(j\omega - p_2)\cdots(j\omega - p_D)}. \qquad (10.154)$$

To illustrate a graphical interpretation of this result with an example let the transfer function be

$$H(s) = \frac{3s}{s + 3}. \qquad (10.155)$$

This transfer function has a zero at $s = 0$ and a pole at $s = -3$ (Figure 10.58). Converting the transfer function to a frequency response,

$$H(j\omega) = 3\frac{j\omega}{j\omega + 3}. \qquad (10.156)$$

The frequency response is three times the ratio of $j\omega$ to $j\omega + 3$. The numerator and denominator can be conceived as vectors in the $s$ plane as illustrated in Figure 10.59 for an arbitrary choice of $\omega$.

As the frequency $\omega$ is changed, the vectors change also. The magnitude of the frequency response at any particular frequency is three times the numerator vector magnitude divided by the denominator vector magnitude,

$$|H(j\omega)| = 3\frac{|j\omega|}{|j\omega + 3|}. \qquad (10.157)$$

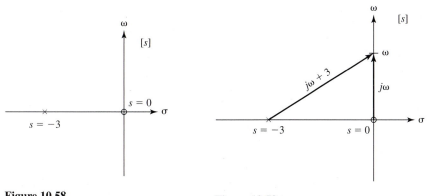

**Figure 10.58**
Pole–zero plot for H$(s) = \frac{3s}{s+3}$.

**Figure 10.59**
Diagram showing the vectors $j\omega$ and $j\omega + 3$.

The phase of the frequency response at any particular frequency is the phase of the constant $+3$ (which is obviously zero), plus the phase of the numerator $j\omega$ (a constant $\pi/2$ rad for positive frequencies and a constant $-(\pi/2)$ rad for negative frequencies) minus the phase of the denominator $j\omega + 3$,

$$\angle H(j\omega) = \underbrace{\angle 3}_{=0} + \angle j\omega - \angle(j\omega + 3). \tag{10.158}$$

At frequencies approaching zero from the positive side, the numerator vector length approaches zero and the denominator vector length approaches a minimum value of three, making the overall frequency-response magnitude approach zero. In that same limit, the phase of $j\omega$ is $\pi/2$ rad and the phase of $j\omega + 3$ approaches zero, so the overall frequency-response phase approaches $\pi/2$ rad,

$$\lim_{\omega \to 0^+} |H(j\omega)| = \lim_{\omega \to 0^+} 3\frac{|j\omega|}{|j\omega + 3|} = 0 \tag{10.159}$$

and

$$\lim_{\omega \to 0^+} \angle H(j\omega) = \lim_{\omega \to 0^+} \angle j\omega - \lim_{\omega \to 0^+} \angle(j\omega + 3) = \frac{\pi}{2} - 0 = \frac{\pi}{2}. \tag{10.160}$$

At frequencies approaching zero from the negative side, the numerator vector length approaches zero and the denominator vector length approaches a minimum value of three, making the overall frequency-response magnitude approach zero, as before. In that same limit, the phase of $j\omega$ is $-(\pi/2)$ rad and the phase of $j\omega + 3$ approaches zero, so the overall frequency-response phase approaches $-(\pi/2)$ rad,

$$\lim_{\omega \to 0^-} |H(j\omega)| = \lim_{\omega \to 0^-} 3\frac{|j\omega|}{|j\omega + 3|} = 0 \tag{10.161}$$

and

$$\lim_{\omega \to 0^-} \angle H(j\omega) = \lim_{\omega \to 0^-} \angle j\omega - \lim_{\omega \to 0^-} \angle(j\omega + 3) = -\frac{\pi}{2} - 0 = -\frac{\pi}{2}. \tag{10.162}$$

At frequencies approaching positive infinity, the two vector lengths approach the same value and the overall frequency-response magnitude approaches three. In that same limit, the phase of $j\omega$ is $\pi/2$ rad and the phase of $j\omega + 3$ approaches $\pi/2$ rad, so

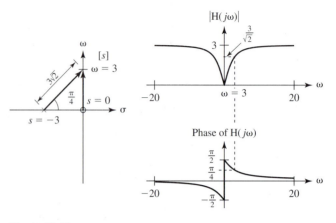

**Figure 10.60**
Magnitude and phase frequency response of a system whose
transfer function is $H(s) = \frac{3s}{s+3}$.

the overall frequency-response phase approaches zero,

$$\lim_{\omega \to +\infty} |H(j\omega)| = \lim_{\omega \to +\infty} 3\frac{|j\omega|}{|j\omega + 3|} = 3 \qquad (10.163)$$

and

$$\lim_{\omega \to +\infty} \angle H(j\omega) = \lim_{\omega \to +\infty} \angle j\omega - \lim_{\omega \to +\infty} \angle(j\omega + 3) = \frac{\pi}{2} - \frac{\pi}{2} = 0. \qquad (10.164)$$

At frequencies approaching negative infinity, the two vector lengths approach the same value and the overall frequency-response magnitude approaches three, as before. In that same limit, the phase of $j\omega$ is $-(\pi/2)$ rad and the phase of $j\omega + 3$ approaches $-(\pi/2)$ rad, so the overall frequency-response phase approaches zero,

$$\lim_{\omega \to -\infty} |H(j\omega)| = \lim_{\omega \to -\infty} 3\frac{|j\omega|}{|j\omega + 3|} = 3 \qquad (10.165)$$

and

$$\lim_{\omega \to -\infty} \angle H(j\omega) = \lim_{\omega \to -\infty} \angle j\omega - \lim_{\omega \to -\infty} \angle(j\omega + 3) = -\frac{\pi}{2} - \left(-\frac{\pi}{2}\right) = 0. \qquad (10.166)$$

These attributes of the frequency response inferred from the pole–zero plot are borne out by a graph of the magnitude and phase frequency response (Figure 10.60).

## EXAMPLE 10.5

Find the magnitude and phase frequency response of a system whose transfer function is

$$H(s) = \frac{s^2 + 2s + 17}{s^2 + 4s + 104}. \qquad (10.167)$$

■ **Solution**
This can be factored into

$$H(s) = \frac{(s + 1 - j4)(s + 1 + j4)}{(s + 2 - j10)(s + 2 + j10)}. \qquad (10.168)$$

So the poles and zeros of this transfer function are

$$z_1 = -1 + j4 \qquad z_2 = -1 - j4 \tag{10.169}$$

and

$$p_1 = -2 + j10 \qquad p_2 = -2 - j10 \tag{10.170}$$

as illustrated in Figure 10.61. Converting the transfer function to a frequency response,

$$H(j\omega) = \frac{(j\omega + 1 - j4)(j\omega + 1 + j4)}{(j\omega + 2 - j10)(j\omega + 2 + j10)}. \tag{10.171}$$

The magnitude of the frequency response at any particular frequency is the product of the numerator vector magnitudes divided by the product of the denominator vector magnitudes,

$$|H(j\omega)| = \frac{|j\omega + 1 - j4| \; |j\omega + 1 + j4|}{|j\omega + 2 - j10| \; |j\omega + 2 + j10|}. \tag{10.172}$$

The phase of the frequency response at any particular frequency is the sum of the numerator vector angles minus the sum of the denominator vector angles,

$$\angle H(j\omega) = \angle(j\omega + 1 - j4) + \angle(j\omega + 1 + j4) - [\angle(j\omega + 2 - j10) + \angle(j\omega + 2 + j10)]. \tag{10.173}$$

This transfer function has no poles or zeros on the $\omega$ axis. Therefore, its frequency response is neither zero nor infinite at any real frequency. But the poles and zeros are near the $\omega$ axis and, because of that proximity, will strongly influence the frequency response for real frequencies near those poles and zeros. For a real frequency $\omega$ near pole $p_1$, the denominator factor $j\omega + 2 - j10$ becomes very small and that makes the overall frequency-response magnitude become very large. Conversely, for a real frequency $\omega$, near zero $z_1$, the numerator factor $j\omega + 1 - j4$ becomes very small and that makes the overall frequency-response magnitude become very small. So, not only does the frequency-response magnitude go to zero at zeros and to infinity at poles, it becomes small near zeros and it becomes large near poles.

The frequency-response magnitude and phase are illustrated in Figure 10.62. Frequency response can be plotted using the MATLAB control toolbox command `bode`, and pole–zero diagrams can be plotted using the MATLAB control toolbox command `pzmap`.

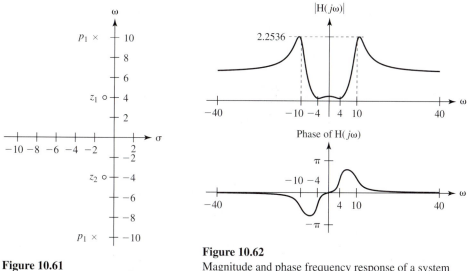

**Figure 10.61**
Pole–zero plot of $H(s) = \frac{s^2 + 2s + 17}{s^2 + 4s + 104}$.

**Figure 10.62**
Magnitude and phase frequency response of a system whose transfer function is $H(s) = \frac{s^2 + 2s + 17}{s^2 + 4s + 104}$.

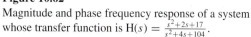

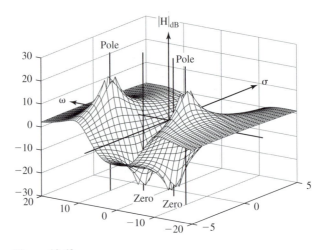

**Figure 10.63**
Surface plot of the magnitude of $H(s) = \frac{s^2+2s+17}{s^2+4s+104}$ in decibels.

By using this graphical concept to interpret pole–zero plots one can, with practice, perceive approximately how the frequency response looks. There is one aspect of the transfer function which is not evident in the pole–zero plot. The frequency-independent gain $A$ has no effect on the pole–zero plot and, therefore, cannot be determined by observing it. But all the dynamic behavior of the system is determinable from the pole–zero plot, to within a gain constant.

Another way of seeing the relation between pole-and-zero locations and frequency response is to plot the magnitude of the transfer function as a surface above the complex $s$ plane. For example, the transfer function

$$H(s) = \frac{s^2 + 2s + 17}{s^2 + 4s + 104} \tag{10.174}$$

in Example 10.5 would have the plot in Figure 10.63. The poles and zeros and their influence on frequency-response magnitude are clearly seen in this figure. (The plots are incomplete near the poles and zeros because the magnitude of the transfer function in decibels approaches plus or minus infinity at those locations.)

## 10.9 BUTTERWORTH FILTERS

In Chapter 8 we explored the frequency response of ideal and practical filters of several kinds. A very popular type of filter is the Butterworth filter. An $n$th-order lowpass Butterworth filter has a transfer function whose squared magnitude is of the form

$$|H(j\omega)|^2 = \frac{1}{1 + (\omega/\omega_c)^{2n}}. \tag{10.175}$$

The lowpass Butterworth filter is designed to be maximally flat for frequencies in its passband, $\omega < \omega_c$, meaning its variation with frequency in the passband is monotonic and approaches a zero derivative as the frequency approaches zero. Figure 10.64 illustrates the frequency response of a lowpass Butterworth filter with a corner frequency of $\omega_c = 1$ for four different orders $n$.

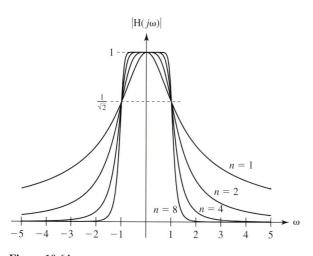

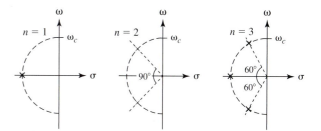

**Figure 10.64**

Lowpass Butterworth filter magnitude frequency responses for a corner frequency $\omega_c = 1$ and four different orders.

**Figure 10.65**

Lowpass Butterworth filter pole locations.

The lowpass Butterworth filter is interesting in the study of Laplace transforms because its poles lie on a semicircle in the open left-half plane whose radius is $\omega_c$, as illustrated in Figure 10.65. The number of poles is $n$ and the angular spacing between poles is always $\pi/n$. If $n$ is odd, there is a pole on the negative real axis and all the other poles occur in complex-conjugate pairs. If $n$ is even, all the poles occur in complex-conjugate pairs. Using these properties, the transfer function of a Butterworth filter can always be found.

The MATLAB signal toolbox has functions for designing CT Butterworth filters. The MATLAB function call

```
[z,p,k] = buttap(N) ;
```

returns the finite zeros in the vector $z$, the finite poles in the vector $p$, and the gain in the scalar $k$ for an Nth-order unity-gain Butterworth lowpass filter with a corner frequency $\omega_c = 1$. (Of course, as we have already seen, there are no finite zeros in a Butterworth filter transfer function, so $z$ is always an empty vector and, since the filter is unity-gain, $k$ is always one. The zeros and gain are included in the returned data because this form of returned data is used for more than just Butterworth filters. For other types of filters, there may be finite zeros and the gain may not be one.)

It is natural to wonder at this point how to use the information returned by MATLAB to design a filter whose corner frequency is not $\omega_c = 1$ or how to design a bandpass, highpass, or bandstop Butterworth filter. Once a lowpass Butterworth filter of a given order with a corner frequency $\omega_c = 1$ has been designed, the conversion of that filter to another form is simply a matter of a transformation of the frequency variable, which is the subject of Section 10.10.

## 10.10 FREQUENCY TRANSFORMATIONS

A very common and useful design technique is to design a transfer function on a normalized basis and then denormalize it to meet specific requirements. This is very commonly done because the design of normalized filters is numerically simpler than general filter design and the denormalization is a straightforward process once the normalized design is complete. The Butterworth filters of Section 10.9 are a good example of this type of design. MATLAB (and many books on filter design) allows the designer to quickly and easily design an $n$th-order lowpass Butterworth filter with unity gain and a corner frequency $\omega_c = 1$. Denormalizing the gain to a nonunity gain is trivial since it simply involves changing the gain coefficient. Changing the corner frequency or the filter type is somewhat more involved.

To change from a unity radian corner frequency $\omega_c = 1$ to a general corner frequency $\omega_c \neq 1$, simply make the independent-variable transformation $s \rightarrow s/\omega_c$. For example, a first-order unity-gain normalized Butterworth filter has a transfer function

$$H(s) = \frac{1}{1+s}. \qquad \textbf{(10.176)}$$

If we want to move the corner frequency to $\omega_c = 10$, the new transfer function is

$$H_{10}(s) = H(s)|_{s \rightarrow s/10} = \frac{1}{1+(s/10)} = \frac{10}{s+10}. \qquad \textbf{(10.177)}$$

This is the transfer function of a unity-gain lowpass filter with a corner radian frequency $\omega_c = 10$.

The real power of the transformation process is seen in converting a lowpass filter to a highpass filter. If we make the transformation $s \rightarrow 1/s$, then

$$H_{HP}(s) = H(s)|_{s \rightarrow 1/s} = \frac{1}{1+(1/s)} = \frac{s}{s+1} \qquad \textbf{(10.178)}$$

where $H_{HP}(s)$ is the transfer function of a first-order unity-gain highpass Butterworth filter with a corner frequency $\omega_c = 1$. We can also do both transformations simultaneously by making the transformation $s \rightarrow \omega_c/s$.

We can also transform a lowpass filter into a bandpass filter by making the transformation

$$s \rightarrow \frac{s^2 + \omega_L \omega_H}{s(\omega_H - \omega_L)}, \qquad \textbf{(10.179)}$$

where $\omega_L$ is the lower positive corner frequency of the bandpass filter and $\omega_H$ is the higher positive corner frequency. For example, let's design a first-order unity-gain bandpass filter with a passband from $\omega = 100$ to $200$ (Figure 10.66).

$$H_{BP}(s) = H(s)|_{s \rightarrow (s^2 + \omega_L \omega_H)/s(\omega_H - \omega_L)} = \frac{1}{(s^2 + \omega_L \omega_H)/s(\omega_H - \omega_L) + 1} \qquad \textbf{(10.180)}$$

Simplifying and inserting numerical values,

$$H_{BP}(s) = \frac{100s}{s^2 + 100s + 20,000} = \frac{100s}{(s+50+j132.2)(s+50-j132.2)}. \qquad \textbf{(10.181)}$$

Finally, we can transform a lowpass filter into a bandstop filter with the transformation

$$s \rightarrow \frac{s(\omega_H - \omega_L)}{s^2 + \omega_L \omega_H}. \qquad \textbf{(10.182)}$$

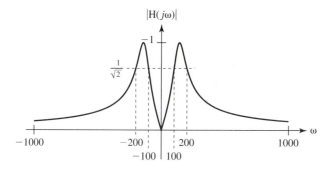

**Figure 10.66**
Magnitude frequency response of a unity-gain, first-order
bandpass Butterworth filter.

MATLAB has commands for the frequency transformation of normalized filters.
They are

| | |
|---|---|
| lp2bp | Lowpass to bandpass analog filter transformation. |
| lp2bs | Lowpass to bandstop analog filter transformation. |
| lp2hp | Lowpass to highpass analog filter transformation. |
| lp2lp | Lowpass to lowpass analog filter transformation. |

The syntax for lp2bp is

```
[numt,dent] = lp2bp(num,den,w0,bw)
```

where num, den = vectors of coefficients of $s$ in the numerator and denominator,
respectively, of the normalized lowpass filter transfer function

w0 = radian-frequency center frequency of bandpass filter

bw = radian-frequency bandwidth of bandpass filter

numt, dent = vectors of coefficients of $s$ in numerator and denominator,
respectively, of bandpass filter transfer function

The syntax of each of the other commands is similar.

As an example, we can design a normalized lowpass Butterworth filter with
buttap.

```
»[z,p,k] = buttap(3) ;
»z

z =

     []

»p

p =

  -0.5000 + 0.8660i
  -1.0000
  -0.5000 - 0.8660i

»k

k =

     1
```

This result indicates that a third-order normalized lowpass Butterworth filter has the transfer function

$$H_{LP}(s) = \frac{1}{(s + 1)(s + 0.5 + j0.866)(s + 0.5 - j0.866)}. \tag{10.183}$$

We can convert this to a ratio of polynomials using MATLAB system-object commands.

```
»[num,den] = tfdata(zpk(z,p,k),'v') ;
»num

num =

    0    0    0    1

»den

den =

  1.0000        2.0000 + 0.0000i  2.0000 + 0.0000i  1.0000 + 0.0000i
```

This result indicates that the normalized lowpass transfer function can also be written as

$$H_{LP}(s) = \frac{1}{s^3 + 2s^2 + 2s + 1}. \tag{10.184}$$

Using this result we can transform this normalized lowpass filter into a denormalized bandpass filter.

```
»[numt,dent] = lp2bp(num,den,8,2) ;
»numt

numt =

  Columns 1 through 4

         0                0.0000 - 0.0000i  0.0000 - 0.0000i  8.0000 -0.0000i

  Columns 5 through 7

    0.0000 - 0.0000i  0.0000 - 0.0000i  0.0000 - 0.0000i

»dent

dent =

    1.0e+05 *

  Columns 1 through 4

    0.0000                0.0000 + 0.0000i  0.0020 + 0.0000i  0.0052 + 0.0000i

  Columns 5 through 7

    0.1280 + 0.0000i   0.1638 + 0.0000i   2.6214 - 0.0000i

»bpf = tf(numt,dent) ;
»bpf
```

```
Transfer function:

1.542e-14 s^5 + 2.32e-13 s^4 + 8 s^3 + 3.644e-11 s^2 + 9.789e-11 s

                                                                  +

9.952e-10

-------------------------------------------------------------------
---
  s^6 + 4 s^5 + 200 s^4 + 520 s^3 + 1.28e04 s^2 + 1.638e04 s + 2.621e05
```

»

This result indicates that the bandpass-filter transfer function can be written as

$$H_{\text{BP}}(s) = \frac{8s^3}{s^6 + 4s^5 + 200s^4 + 520s^3 + 12{,}800s^2 + 16{,}380s + 26{,}2100}. \quad \textbf{(10.185)}$$

(The extremely small nonzero coefficients in the numerator of the transfer function reported by MATLAB are the result of round-off errors in the MATLAB calculations and have been neglected. Notice they did not appear in `numt`.)

## 10.11 ANALOG FILTER DESIGN WITH MATLAB

We have just seen how the MATLAB command `buttap` can be used to design a normalized Butterworth filter and how to denormalize it to other Butterworth filters. There are several other MATLAB commands which are useful in analog filter design. There are four other "... ap" commands, `cheb1ap`, `cheb2ap`, `ellipap`, and `besselap`, which design normalized analog filters of optimal types other than the Butterworth filter. The other optimal analog filter types are the Chebyshev (sometimes spelled Tchebysheff or Tchebischeff) filter, the elliptical filter, and the Bessel filter. Each of these filter types optimizes the performance of the filter according to a different criterion.

The Chebyshev filter is similar to the Butterworth filter, but it has an extra degree of design freedom. The Butterworth filter is called *maximally flat* because it is monotonic in the passband and stopband and approaches a flat response in the passband as the order is increased. There are two types of Chebyshev filters. The type-one Chebyshev filter has a frequency response that is not monotonic in the passband but is monotonic in the stopband. Its frequency response ripples in the passband. The presence of ripple in the passband is usually not in itself desirable, but it allows the transition from the passband to the stopband to be faster than that of a Butterworth filter of the same order. In other words, we trade passband flatness for a narrower transition band. The more ripple we allow in the passband, the narrower the transition band can be. The type-two Chebyshev filter is just the opposite. It has a monotonic passband and ripple in the stopband and, for the same filter order, also allows for a narrower transition band than a Butterworth filter.

The elliptical filter has ripple in both the passband and stopband and, for the same filter order, it has an even narrower transition band than either of the two types of Chebyshev filters. The Bessel filter is optimized on a different basis. It is optimized for linearity of the phase in the passband rather than for a flat magnitude response in the passband and/or stopband or a narrow transition band.

The syntax for each of these normalized analog filter designs is given here.

```
[z,p,k] = cheb1ap(N,Rp) ;
[z,p,k] = cheb2ap(N,Rs) ;
```

```
[z,p,k] = ellipap(N,Rp,Rs) ;
[z,p,k] = besselap(N) ;
```

where N = order of filter

      Rp = allowable ripple in passband, dB

      Rs = allowable ripple in stopband, dB

Once a filter has been designed, its frequency response can be found using either bode, which was introduced earlier, or freqs. The function freqs has the syntax

```
H = freqs(num,den,w) ;
```

where H is a vector of responses at the real radian-frequency points in the vector w, and num and den are vectors containing the coefficients of *s* in the numerator and denominator, respectively, of the filter transfer function.

## EXAMPLE 10.6

Using MATLAB, design a normalized fourth-order lowpass Butterworth filter, transform it into a denormalized bandstop filter with corner frequencies of 55 and 65 Hz, and then compare its frequency response with a type-one Chebyshev bandstop filter of the same order and corner frequencies and an allowable ripple in the passband of 0.3 dB.

### ■ Solution

```
%   Butterworth design

%   Design a normalized fourth-order Butterworth lowpass filter
%   and put the zeros, poles, and gain in zb, pb and kb.

[zb,pb,kb] = buttap(4) ;

%   Use MATLAB system tools to obtain the numerator and
%   denominator coefficient vectors, numb and denb.

[numb,denb] = tfdata(zpk(zb,pb,kb),'v') ;

%   Set the cyclic center frequency and bandwidth and then set
%   the corresponding radian center frequency and bandwidth.

f0 = 60 ; fbw = 10 ; w0 = 2*pi*f0 ; wbw = 2*pi*fbw ;

%   Denormalize the lowpass Butterworth to a bandstop Butterworth
%   with a stopband between 55 and 65 Hz.

[numbsb,denbsb] = lp2bs(numb,denb,w0,wbw) ;

%   Create a vector of cyclic frequencies to use in plotting the
%   frequency response of the filter. Then create a corresponding
%   radian-frequency vector and compute the frequency response.

wbsb = 2*pi*[40:0.2:80]' ; Hbsb = freqs(numbsb,denbsb,wbsb) ;

%   Chebyshev design

%   Design a normalized fourth-order type-one Chebyshev lowpass filter
%   and put the zeros, poles, and gain in zc, pc and kc.

[zc,pc,kc] = cheb1ap(4,0.3) ; wc = wb ;
```

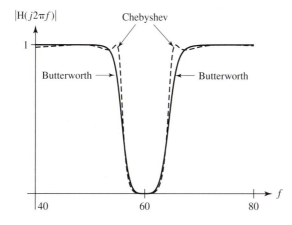

**Figure 10.67**
Comparison of the Butterworth and Chebyshev magnitude
frequency responses.

```
%   Use MATLAB system tools to obtain the numerator and
%   denominator coefficient vectors, numc and denc.

[numc,denc] = tfdata(zpk(zc,pc,kc),'v') ;

%   Denormalize the lowpass Chebyshev to a bandstop Chebyshev
%   with a stopband between 55 and 65 Hz.

[numbsc,denbsc] = lp2bs(numc,denc,w0,wbw) ;

%   Use the same radian-frequency vector used in the Butterworth
%   design and compute the frequency response of the Chebyshev
%   bandstop filter.

wbsc = wbsb ; Hbsc = freqs(numbsc,denbsc,wbsc) ;
```

The magnitude frequency responses are compared in Figure 10.67. Notice that the Butter-worth filter is monotonic in the passbands while the Chebyshev filter is not, but that the Cheby-shev filter has a steeper slope in the transition between passband and stop bands and slightly better stopband attenuation.  ◼

## 10.12  STANDARD REALIZATIONS OF SYSTEMS

The process of system design, as opposed to system analysis, is to develop a desired transfer function for a class of excitations which yields a desired response or re-sponses. Once we have found the desired transfer function the next logical step is to actually build, or perhaps simulate, the system. The first step in building or simulating a system is to form a block diagram which describes the interaction among all the sig-nals in the system. This step is called *realization* arising from the concept of making a real system instead of just a set of equations which describe its behavior. There are sev-eral standard types of system realizations. We will explore three here.

The first standard system realization is commonly called the *canonical* or *direct* form. It can be realized directly from the general form of a transfer function as the ratio of two polynomials,

$$H(s) = \frac{Y(s)}{X(s)} = \frac{\sum_{k=0}^{N} b_k s^k}{\sum_{k=0}^{N} a_k s^k} = \frac{b_N s^N + b_{N-1} s^{N-1} + \cdots + b_1 s + b_0}{s^N + a_{N-1} s^{N-1} + \cdots + a_1 s + a_0} \qquad a_N = 1$$

(10.186)

for a system described by an $N$th-order differential equation. Here the nominal orders of the numerator and denominator are both assumed to be $N$. (If the numerator order is actually less than $N$, then some of the higher-order $b$ coefficients will be zero.) The transfer function can be thought of as the product of two transfer functions,

$$H_1(s) = \frac{Y_1(s)}{X(s)} = \frac{1}{s^N + a_{N-1} s^{N-1} + \cdots + a_1 s + a_0} \qquad (10.187)$$

and

$$H_2(s) = \frac{Y(s)}{Y_1(s)} = b_N s^N + b_{N-1} s^{N-1} + \cdots + b_1 s + b_0 \qquad (10.188)$$

(Figure 10.68), where the response of the first system $Y_1(s)$ is the excitation of the second system.

We can draw a block diagram of $H_1(s)$ by rewriting (10.187) as

$$X(s) = [s^N + a_{N-1} s^{N-1} + \cdots + a_1 s + a_0] Y_1(s) \qquad (10.189)$$

or

$$X(s) = s^N Y_1(s) + a_{N-1} s^{N-1} Y_1(s) + \cdots + a_1 s Y_1(s) + a_0 Y_1(s) \qquad (10.190)$$

or

$$s^N Y_1(s) = X(s) - [a_{N-1} s^{N-1} Y_1(s) + \cdots + a_1 s Y_1(s) + a_0 Y_1(s)] \qquad (10.191)$$

(Figure 10.69). Now we can immediately synthesize the overall response $Y(s)$ as a linear combination of the various powers of $s$ multiplying $Y_1(s)$ (Figure 10.70).

The second standard system realization is the *cascade* form. The numerator and denominator of the general transfer function form

$$H(s) = \frac{Y(s)}{X(s)} = \frac{\sum_{k=0}^{N} b_k s^k}{\sum_{k=0}^{D} a_k s^k} = \frac{b_N s^N + b_{N-1} s^{N-1} + \cdots + b_1 s + b_0}{s^D + a_{D-1} s^{D-1} + \cdots + a_1 s + a_0} \qquad a_D = 1,$$

(10.192)

where $N \leq D$, can be factored yielding a transfer function expression of the form

$$H(s) = A \frac{s - z_1}{s - p_1} \frac{s - z_2}{s - p_2} \cdots \frac{s - z_N}{s - p_N} \frac{1}{s - p_{N+1}} \frac{1}{s - p_{N+2}} \cdots \frac{1}{s - p_D}. \qquad (10.193)$$

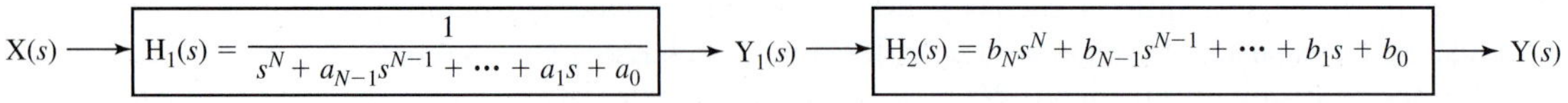

$$X(s) \longrightarrow \boxed{H_1(s) = \frac{1}{s^N + a_{N-1} s^{N-1} + \cdots + a_1 s + a_0}} \longrightarrow Y_1(s) \longrightarrow \boxed{H_2(s) = b_N s^N + b_{N-1} s^{N-1} + \cdots + b_1 s + b_0} \longrightarrow Y(s)$$

**Figure 10.68**
A system conceived as two cascaded systems.

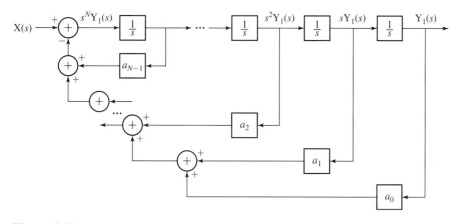

**Figure 10.69**
Realization of $H_1(s)$.

Any of the component fractions $Y_k(s)/X_k(s) = (s - z_k)/(s - p_k)$ or $Y_k(s)/X_k(s) = 1/(s - p_k)$ represents a subsystem that can be realized by writing the relationship as

$$H_k(s) = \underbrace{\frac{1}{s - p_k}}_{H_{k1}(s)} \underbrace{(s - z_k)}_{H_{k2}(s)} \quad \text{or} \quad H_k(s) = \frac{1}{s - p_k} \qquad \textbf{(10.194)}$$

and realizing it as a canonical system (Figure 10.71). Then the entire original system can be realized in cascade form (Figure 10.72).

A problem sometimes arises with this type of cascade realization. Sometimes the first-order subsystems have complex poles. This necessitates multiplication by complex numbers and that often cannot be done in a system simulation. In such cases, two subsystems with two complex conjugate poles should be combined into one second-order subsystem of the form

$$H_k(s) = \frac{s + b_0}{s^2 + a_1 s + a_0}, \qquad \textbf{(10.195)}$$

which can always be realized with real coefficients (Figure 10.73).

The last standard realization of a system is the *parallel* realization. This can be accomplished by expanding the standard transfer function form (10.195) in partial fractions of the form

$$H(s) = \frac{K_1}{s - p_1} + \frac{K_2}{s - p_2} + \cdots + \frac{K_D}{s - p_D} \qquad \textbf{(10.196)}$$

(Figure 10.74).

When systems are simulated by computational methods, the form of the system realization has an effect on the precision, and sometimes the stability, of the realization. Generally speaking, the cascade and parallel realizations are less sensitive to round-off errors in the computations done in the simulations than is the canonical realization. This is basically because the computations in the cascade and parallel realizations are more localized so there is less likelihood of a numerical error in one location propagating into errors in multiple other locations.

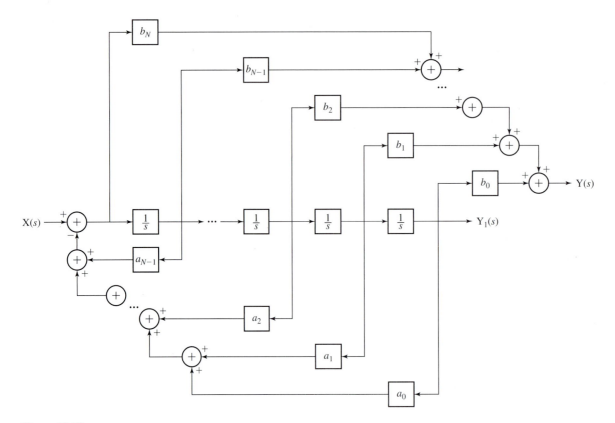

**Figure 10.70**
Overall canonical system realization.

$$H_k(s) = \frac{s - z_k}{s - p_k} \qquad\qquad H_k(s) = \frac{1}{s - p_k}$$

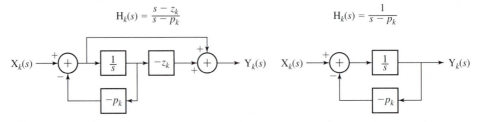

**Figure 10.71**
Canonical realization of a single subsystem in the cascade realization.

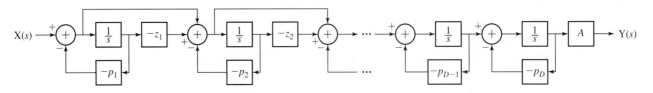

**Figure 10.72**
Overall cascade system realization.

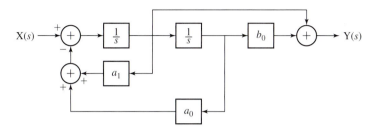

**Figure 10.73**
A standard-form second-order subsystem.

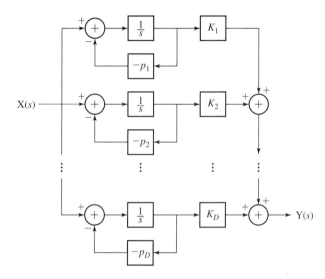

**Figure 10.74**
Overall parallel system realization.

## 10.13  STATE-SPACE SIGNAL AND SYSTEM ANALYSIS

Most of the analysis so far has been of relatively simple systems, with one input and one output. That is as it should be because understanding of signal and system analysis must build from simple concepts to more complicated ones. We now have the tools necessary to tackle larger CT systems. (After we explore the $z$ transform in Chapter 11, we will have the tools to tackle larger DT systems.) The analysis of a large system can quickly become very tedious and error-prone because of the size of the system of equations needed to describe it and the number of algebraic manipulations required to find a solution to those equations. Therefore, it is necessary to develop some systematic procedures to enable us to grapple with large systems and find solutions without errors and without spending inordinate amounts of time. A very popular method of analyzing large systems is through *state-variable* analysis. A set of state variables is a set of signals in a system which, together with the system excitation completely determines the state of the system at any future time. Take the case of the $RC$ lowpass filter. We needed to know the initial capacitor voltage in order to solve for the arbitrary constant and get an exact solution for the future response voltage. In the $RLC$ circuit we needed both the initial capacitor voltage and the initial inductor current. The capacitor voltage and inductor current are simple examples of state variables. Their values completely define the state (or condition) of the system at any time. Once we know them, and the system dynamics, and the excitations we can calculate anything else we want to know about the system at any future time.

Every system has an order. The *order* of a system is the same as the number of state variables necessary to uniquely establish its state. If the system is described by one differential or difference equation, the order of the system is the same as the order of the equation. If the system is described by multiple independent equations, its order is the sum of the orders of the equations. The number of state variables required by a system sets the size of the state vector and, therefore, the number of dimensions in the state space which is just a specific example of a vector space. Then the state of the system can be conceptualized as a position in the state space. Common terminology is that as the system responds to its excitations, the state of the system follows a *trajectory* through that *space*.

A system's state variables are not unique. One person might choose one set and another person might choose another set and both sets could be correct and complete. However, in many cases there may be one set of state variables that is more convenient than any another for some analysis purpose.

State-variable analysis has the following desirable characteristics:

1. It reduces the probability of analysis errors by making the process systematic.
2. It describes all the important system signals, both internal and external.
3. It lends insight into system dynamics and that can help in system design optimization.
4. It can be formulated using matrix methods and, when that is done, the state of the system and the responses of the system can be described by two matrix equations.
5. When state-variable analysis techniques are combined with transform techniques, they are even more powerful in the analysis of complicated systems.

To introduce state-space analysis techniques we will begin by applying them to a very simple system, a parallel $RLC$ circuit (Figure 10.75). Let the excitation be designated as the current at the input port $i_{\text{in}}(t)$ and let the responses be designated as the voltage at the output port $v_{\text{out}}(t)$ and the current through the resistor $i_R(t)$. Summing the currents leaving and entering the top node, we get

$$G v_{\text{out}}(t) + \frac{1}{L} \int_{-\infty}^{t} v_{\text{out}}(\lambda)\, d\lambda + C v'_{\text{out}}(t) = i_{\text{in}}(t) \qquad \textbf{(10.197)}$$

where $G = 1/R$. This is an integrodifferential equation. We could differentiate it with respect to time and form a second-order differential equation. Therefore, this is a second-order system.

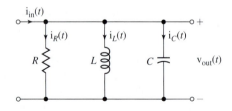

**Figure 10.75**
A parallel $RLC$ circuit.

Instead of immediately trying to solve the system equation in its present form we will reformulate the information it contains. We identify the capacitor voltage $v_C(t)$ and the inductor current $i_L(t)$ as state variables. The standard state-variable description of a system has two sets of equations, the system equations and the output equations. The system equations are written in a standard form. Each one has the derivative of a state variable on the left side and some linear combination of the state variables and excitations on the right side. Using Ohm's law, Kirchhoff's laws, and the defining equations for inductors and capacitors we can write the system equations

$$i_L'(t) = \frac{1}{L} v_C(t) \qquad \textbf{(10.198)}$$

and

$$v_C'(t) = -\frac{1}{C} i_L(t) - \frac{G}{C} v_C(t) + \frac{1}{C} i_{\text{in}}(t). \qquad \textbf{(10.199)}$$

The output equations express the responses as linear combinations of the state variables. In this case they would be

$$v_{\text{out}}(t) = v_C(t) \qquad \textbf{(10.200)}$$

and

$$i_R(t) = G v_C(t). \qquad \textbf{(10.201)}$$

The system equations can be reformulated in a standard matrix form as

$$\begin{bmatrix} i_L'(t) \\ v_C'(t) \end{bmatrix} = \begin{bmatrix} 0 & 1/L \\ -(1/C) & -(G/C) \end{bmatrix} \begin{bmatrix} i_L(t) \\ v_C(t) \end{bmatrix} + \begin{bmatrix} 0 \\ 1/C \end{bmatrix} [i_{\text{in}}(t)], \qquad \textbf{(10.202)}$$

and the output equations can be written in a standard matrix form as

$$\begin{bmatrix} v_{\text{out}}(t) \\ i_R(t) \end{bmatrix} = \begin{bmatrix} 0 & 1 \\ 0 & G \end{bmatrix} \begin{bmatrix} i_L(t) \\ v_C(t) \end{bmatrix} + \begin{bmatrix} 0 \\ 0 \end{bmatrix} [i_{\text{in}}(t)]. \qquad \textbf{(10.203)}$$

The state variables seem to be a lot like responses. The distinction between state variables and responses comes only in the way they are used. The state variables are a set of system signals which completely describe the state of the system. The responses of a system are the signals we designate arbitrarily as responses for whatever system design purpose we may have in any particular system analysis. A state variable can also be a response. But even if a state variable and a response are the same in the analysis of a particular system, in the standard state-space equation forms we give them separate names, just to be systematic. That may seem to be a waste of time, but in large system analysis it is actually a good idea and can prevent analysis errors.

The state-variable formulation of system equations makes the process of drawing a system block-diagram realization very easy and systematic. For this example the system block diagram can be drawn directly from the system equations as illustrated in Figure 10.76.

We will refer to the vector of state variables as the vector $\mathbf{q}(t)$, the vector of excitations as the vector $\mathbf{x}(t)$, and the vector of responses as the vector $\mathbf{y}(t)$. The matrix which multiplies $\mathbf{q}(t)$ in the system equation (10.203) is conventionally called $\mathbf{A}$, and the matrix which multiplies $\mathbf{x}(t)$ in the system equation is conventionally called $\mathbf{B}$. The

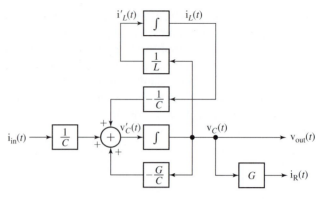

**Figure 10.76**
A state-variable system block diagram of the parallel $RLC$ circuit.

matrix which multiplies $\mathbf{q}(t)$ in the output equation (10.204) is conventionally called **C**, and the matrix which multiplies the $\mathbf{x}(t)$ in the output equation is conventionally called **D**. Using this notation we can write the matrix system equation as

$$\mathbf{q}'(t) = \mathbf{A}\mathbf{q}(t) + \mathbf{B}\mathbf{x}(t) \qquad (10.204)$$

where, in this case,

$$\mathbf{q}(t) = \begin{bmatrix} i_L(t) \\ v_C(t) \end{bmatrix}$$

$$\mathbf{A} = \begin{bmatrix} 0 & 1/L \\ -(1/C) & -(G/C) \end{bmatrix}$$

$$\mathbf{B} = \begin{bmatrix} 0 \\ 1/C \end{bmatrix}$$

$$\mathbf{x}(t) = [i_{in}(t)]$$

and we can write the equation for the response as

$$\mathbf{y}(t) = \mathbf{C}\mathbf{q}(t) + \mathbf{D}\mathbf{x}(t) \qquad (10.205)$$

where, in this case,

$$\mathbf{y}(t) = \begin{bmatrix} v_{out}(t) \\ i_R(t) \end{bmatrix} = \text{vector of responses}$$

$$\mathbf{C} = \begin{bmatrix} 0 & 1 \\ 0 & G \end{bmatrix}$$

$$\mathbf{D} = \begin{bmatrix} 0 \\ 0 \end{bmatrix}$$

(The equation for the response is conventionally called the *output* equation.) No matter how complicated the system may be, with the proper assignment of state-variable vectors and matrices, the system and output equations of LTI systems can always be written as these two matrix equations. In this relatively simple example the power of this formulation may not be evident because the solution of a system this simple is not difficult using classical techniques. But when the system gets larger, this systematic technique compares very favorably with less systematic techniques.

> Some authors use the symbol $\mathbf{x}$ to represent the vector of state variables instead of the symbol $\mathbf{q}$. This could be confusing because, in all previous material, we (and most authors) have consistently used the symbol $x(t)$ to represent an excitation. Some authors use the symbol $\mathbf{u}$ for the vector of excitations instead of the symbol $\mathbf{x}$. Also, in previous material we (and most other authors) use the symbol $u(t)$ to represent the unit step function. So, even though $\mathbf{u}$ is boldface and $u(t)$ is not, it should be less confusing to use the symbol $\mathbf{x}$ as the excitation rather than the symbol $\mathbf{u}$, especially since the symbol $x(t)$ has been used up until now to represent an excitation in a single-input system.

So far we have only described the system but have not solved the equations. One of the really powerful aspects of state-space formulation of system analysis is the straightforward and systematic way the equations can be solved. The state equations are

$$\begin{aligned}\mathbf{q}'(t) &= \mathbf{A}\mathbf{q}(t) + \mathbf{B}\mathbf{x}(t) \\ \mathbf{y}(t) &= \mathbf{C}\mathbf{q}(t) + \mathbf{D}\mathbf{x}(t)\end{aligned} \qquad (10.206)$$

Obviously if we can find the solution vector $\mathbf{q}(t)$ to the system equation, we can immediately calculate the response vector $\mathbf{y}(t)$ because the excitation vector $\mathbf{x}(t)$ is known. So the solution process is to first find the solution of the system equation.

We could find a time-domain solution directly from these matrix equations, but it is easier to use the Laplace transform to help find the solution. Laplace-transforming the system equation we get

$$s\mathbf{Q}(s) - \mathbf{q}(0^-) = \mathbf{A}\mathbf{Q}(s) + \mathbf{B}\mathbf{X}(s) \qquad (10.207)$$

or

$$[s\mathbf{I} - \mathbf{A}]\mathbf{Q}(s) = \mathbf{B}\mathbf{X}(s) + \mathbf{q}(0^-). \qquad (10.208)$$

We can solve this equation for $\mathbf{Q}(s)$ by multiplying both sides by $[s\mathbf{I} - \mathbf{A}]^{-1}$ yielding

$$\mathbf{Q}(s) = [s\mathbf{I} - \mathbf{A}]^{-1}[\mathbf{B}\mathbf{X}(s) + \mathbf{q}(0^-)]. \qquad (10.209)$$

The matrix $[s\mathbf{I} - \mathbf{A}]^{-1}$ is conventionally designated by the symbol $\mathbf{\Phi}(s)$. Using that notation (10.209) becomes

$$\mathbf{Q}(s) = \mathbf{\Phi}(s)[\mathbf{B}\mathbf{X}(s) + \mathbf{q}(0^-)] = \underbrace{\mathbf{\Phi}(s)\mathbf{B}\mathbf{X}(s)}_{\substack{\text{zero-state}\\\text{response}}} + \underbrace{\mathbf{\Phi}(s)\mathbf{q}(0^-)}_{\substack{\text{zero-input}\\\text{response}}} \qquad (10.210)$$

and the state vector is seen as consisting of two parts, a zero-state response and a zero-input response. We can now find the time-domain solution by inverse Laplace-transforming (10.210),

$$\mathbf{q}(t) = \underbrace{\phi(t) * \mathbf{B}\mathbf{x}(t)}_{\substack{\text{zero-state}\\\text{response}}} + \underbrace{\phi(t)\mathbf{q}(0^-)}_{\substack{\text{zero-input}\\\text{response}}} \qquad (10.211)$$

where $\phi(t) \overset{\mathcal{L}}{\longleftrightarrow} \mathbf{\Phi}(s)$ and $\phi(t)$ is called the state transition matrix. The name *state transition matrix* comes from the fact that once the initial state and excitations are known, $\phi(t)$ is what allows us to calculate the state at any future time. In other words, $\phi(t)$ lets us calculate the way the system makes a transition from one state to another.

We will now apply this method to the example. The matrices in the state equation are

$$\mathbf{q}(t) = \begin{bmatrix} i_L(t) \\ v_C(t) \end{bmatrix} \quad \mathbf{A} = \begin{bmatrix} 0 & 1/L \\ -(1/C) & -(G/C) \end{bmatrix} \quad \mathbf{B} = \begin{bmatrix} 0 \\ 1/C \end{bmatrix} \quad \mathbf{x}(t) = [i_{\text{in}}(t)].$$

(10.212)

To make the problem concrete let the excitation current be

$$i(t) = A u(t),$$

(10.213)

let the initial conditions be

$$\mathbf{q}(0^-) = \begin{bmatrix} i_L(0^-) \\ v_C(0^-) \end{bmatrix} = \begin{bmatrix} 0 \\ 1 \end{bmatrix},$$

(10.214)

and let the component values be $R = \frac{1}{3}$, $C = 1$, and $L = 1$. Then

$$\mathbf{\Phi}(s) = (s\mathbf{I} - \mathbf{A})^{-1} = \begin{bmatrix} s & -(1/L) \\ 1/C & s + (G/C) \end{bmatrix}^{-1}$$

$$= \frac{\begin{bmatrix} s + (G/C) & -(1/C) \\ 1/L & s \end{bmatrix}^T}{s^2 + (G/C)s + (1/LC)} = \frac{\begin{bmatrix} s + (G/C) & 1/L \\ -(1/C) & s \end{bmatrix}}{s^2 + (G/C)s + (1/LC)}$$

(10.215)

and the solution for the state variables in the Laplace domain is

$$\mathbf{Q}(s) = \mathbf{\Phi}(s)[\mathbf{B}\mathbf{X}(s) + \mathbf{q}(0^-)]$$

$$= \frac{\begin{bmatrix} s + (G/C) & 1/L \\ -(1/C) & s \end{bmatrix}}{s^2 + (G/C)s + (1/LC)} \begin{bmatrix} 0 \\ 1/C \end{bmatrix} \begin{bmatrix} 1 \\ s \end{bmatrix}$$

$$+ \frac{\begin{bmatrix} s + (G/C) & 1/L \\ -(1/C) & s \end{bmatrix}}{s^2 + (G/C)s + (1/LC)} \begin{bmatrix} 0 \\ 1 \end{bmatrix}$$

(10.216)

or

$$\mathbf{Q}(s) = \frac{\begin{bmatrix} 1/sLC \\ 1/C \end{bmatrix} + \begin{bmatrix} 1/L \\ s \end{bmatrix}}{s^2 + (G/C)s + (1/LC)}$$

$$= \begin{bmatrix} \dfrac{1}{sLC(s^2 + (G/C)s + (1/LC))} + \dfrac{1}{L(s^2 + (G/C)s + (1/LC))} \\ \dfrac{1}{C(s^2 + (G/C)s + (1/LC))} + \dfrac{s}{s^2 + (G/C)s + (1/LC)} \end{bmatrix}.$$

(10.217)

Substituting in numerical component values we get

$$\mathbf{Q}(s) = \begin{bmatrix} \dfrac{1}{s(s^2 + 3s + 1)} + \dfrac{1}{s^2 + 3s + 1} \\ \dfrac{1}{s^2 + 3s + 1} + \dfrac{s}{s^2 + 3s + 1} \end{bmatrix}$$

(10.218)

or in partial-fraction form,

$$\mathbf{Q}(s) = \begin{bmatrix} \dfrac{1}{s} + \dfrac{0.17}{s+2.62} - \dfrac{1.17}{s+0.382} - \dfrac{0.447}{s+2.62} + \dfrac{0.447}{s+0.382} \\[2mm] -\dfrac{0.447}{s+2.62} + \dfrac{0.447}{s+0.382} + \dfrac{1.17}{s+2.62} - \dfrac{0.17}{s+0.382} \end{bmatrix} \quad \textbf{(10.219)}$$

$$\mathbf{Q}(s) = \begin{bmatrix} \dfrac{1}{s} - \dfrac{0.277}{s+2.62} - \dfrac{0.723}{s+0.382} \\[2mm] \dfrac{0.723}{s+2.62} + \dfrac{0.277}{s+0.382} \end{bmatrix} \quad \textbf{(10.220)}$$

Inverse Laplace-transforming,

$$\mathbf{q}(t) = \begin{bmatrix} 1 - 0.277e^{-2.62t} - 0.723e^{-0.382t} \\ 0.723e^{-0.382t} + 0.277e^{-2.62t} \end{bmatrix} u(t). \quad \textbf{(10.221)}$$

Now we can find the responses immediately using the ouput equation $\mathbf{y}(t) = \mathbf{Cq}(t) + \mathbf{Dx}(t)$,

$$\mathbf{y}(t) = \begin{bmatrix} 0 & 1 \\ 0 & G \end{bmatrix} \mathbf{q} + \begin{bmatrix} 0 \\ 0 \end{bmatrix} \mathbf{x} = \begin{bmatrix} 0 & 1 \\ 0 & 3 \end{bmatrix} \begin{bmatrix} 1 - 0.277e^{-2.62t} - 0.723e^{-0.382t} \\ 0.723e^{-0.382t} + 0.277e^{-2.62t} \end{bmatrix} u(t) \quad \textbf{(10.222)}$$

or

$$\mathbf{y}(t) = \begin{bmatrix} 0.723e^{-0.382t} + 0.277e^{-2.62t} \\ 2.169e^{-0.382t} + 0.831e^{-2.62t} \end{bmatrix} u(t). \quad \textbf{(10.223)}$$

We can use the state-space analysis technique to find the matrix transfer function of the system. The transfer function is defined only for the zero-state response. Starting with

$$s\mathbf{Q}(s) - \mathbf{q}(0^-) = \mathbf{AQ}(s) + \mathbf{BX}(s), \quad \textbf{(10.224)}$$

and requiring that the initial state $\mathbf{q}(0^-)$ be zero, we can solve for $\mathbf{Q}(s)$ as

$$\mathbf{Q}(s) = [s\mathbf{I} - \mathbf{A}]^{-1} \mathbf{BX}(s) = \mathbf{\Phi}(s)\mathbf{BX}(s). \quad \textbf{(10.225)}$$

Then the response $\mathbf{Y}(s)$ is

$$\mathbf{Y}(s) = \mathbf{CQ}(s) + \mathbf{DX}(s) = \mathbf{C\Phi}(s)\mathbf{BX}(s) + \mathbf{DX}(s) = [\mathbf{C\Phi}(s)\mathbf{B} + \mathbf{D}]\,\mathbf{X}(s). \quad \textbf{(10.226)}$$

Therefore, since the system response is the product of the system transfer function and the system excitation, the matrix transfer function is

$$\mathbf{H}(s) = \mathbf{C\Phi}(s)\mathbf{B} + \mathbf{D}. \quad \textbf{(10.227)}$$

This transfer function relates all the excitations of the system to all the responses of the system through

$$\mathbf{Y}(s) = \mathbf{H}(s)\mathbf{X}(s). \quad \textbf{(10.228)}$$

Since $\mathbf{\Phi}(s) = [s\mathbf{I} - \mathbf{A}]^{-1}$,

$$\mathbf{H}(s) = \mathbf{C}\,[s\mathbf{I} - \mathbf{A}]^{-1}\,\mathbf{B} + \mathbf{D}. \quad \textbf{(10.229)}$$

Examine $[s\mathbf{I} - \mathbf{A}]^{-1}$. Since it is the inverse of $[s\mathbf{I} - \mathbf{A}]$, it is the adjoint of $[s\mathbf{I} - \mathbf{A}]$, divided by the determinant $|s\mathbf{I} - \mathbf{A}|$. So every element in $[s\mathbf{I} - \mathbf{A}]^{-1}$ has a denominator which is $|s\mathbf{I} - \mathbf{A}|$ (unless some factors in the transpose of the matrix of cofactors of

$s\mathbf{I} - \mathbf{A}$ cancel some factors in $|s\mathbf{I} - \mathbf{A}|$). Premultiplying by $\mathbf{C}$ and postmultiplying by $\mathbf{B}$ does not change that fact because $\mathbf{C}$ and $\mathbf{B}$ are matrices of constants. The addition of the $\mathbf{D}$ matrix does not change the denominators of the elements of $\mathbf{H}(s)$ either because it is also a matrix of constants. Therefore, the denominator of every element of $\mathbf{H}(s)$ is $|s\mathbf{I} - \mathbf{A}|$ (unless some pole–zero cancellation occurred). All elements of $\mathbf{H}(s)$, and therefore all transfer functions from all excitations to all responses, have the same poles. This leads to an important idea. Even though transfer function is defined as the ratio of a response to an excitation, the poles of any system transfer function are determined by the system itself, not by the excitations or the responses. Those poles are the zeros of $|s\mathbf{I} - \mathbf{A}|$ (except for any pole–zero cancellation) and the zeros of $|s\mathbf{I} - \mathbf{A}|$ are the eigenvalues of $\mathbf{A}$.

The previous example problem could have been solved using a different set of state variables. For example, the resistor current $i_R(t)$ and the inductor current $i_L(t)$ could have been chosen as the state variables. Then the system equation would be

$$\begin{bmatrix} i'_R(t) \\ i'_L(t) \end{bmatrix} = \begin{bmatrix} -(G/C) & -(G/C) \\ 1/LG & 0 \end{bmatrix} \begin{bmatrix} i_R(t) \\ i_L(t) \end{bmatrix} + \begin{bmatrix} G/C \\ 0 \end{bmatrix} [i_{in}(t)] \qquad \textbf{(10.230)}$$

and the output equation would be

$$\begin{bmatrix} v_{out}(t) \\ i_R(t) \end{bmatrix} = \begin{bmatrix} 1/G & 0 \\ 1 & 0 \end{bmatrix} \begin{bmatrix} i_R(t) \\ i_L(t) \end{bmatrix} + \begin{bmatrix} 0 \\ 0 \end{bmatrix} [i_{in}(t)]. \qquad \textbf{(10.231)}$$

In solving for the state variables we find

$$\mathbf{\Phi}(s) = [s\mathbf{I} - \mathbf{A}]^{-1} = \begin{bmatrix} s + (G/C) & G/C \\ -(1/LG) & s \end{bmatrix}^{-1} = \frac{\begin{bmatrix} s & -(G/C) \\ 1/LG & s + (G/C) \end{bmatrix}}{s^2 + (G/C)s + (1/LC)}. \qquad \textbf{(10.232)}$$

It is important here to note that the determinant $|s\mathbf{I} - \mathbf{A}|$ is exactly the same as it was for the first set of state variables. That can be shown to be generally true. That is, the determinant $|s\mathbf{I} - \mathbf{A}|$ is independent of the choice of state variables. The matrix $\mathbf{A}$ changes, but the determinant $|s\mathbf{I} - \mathbf{A}|$ does not. Therefore, the determinant $|s\mathbf{I} - \mathbf{A}|$ is saying something fundamental about the system itself, not any particular choice of the way we analyze the system. Recall that in solving systems of differential equations, the determinant $|\lambda\mathbf{I} - \mathbf{A}|$ was called the *characteristic equation*. It was given that name because it characterizes the system of differential equations and it is independent of the method chosen to solve the equations. State equations are systems of differential equations which describe systems. Therefore, the invariance of $|s\mathbf{I} - \mathbf{A}|$ in the solution of state equations should be expected from the invariance of $|\lambda\mathbf{I} - \mathbf{A}|$ in the solution of systems of differential equations.

Any set of state variables can be transformed to another set through a linear transformation. Suppose we are using a state-variable vector $\mathbf{q}_1(t)$ and we decide to use another state-variable vector $\mathbf{q}_2(t)$, which is related to $\mathbf{q}_1(t)$ by

$$\mathbf{q}_2(t) = \mathbf{T}\mathbf{q}_1(t), \qquad \textbf{(10.233)}$$

where $\mathbf{T}$ is the transformation matrix relating the two state-variable vectors. Then

$$\mathbf{q}'_2(t) = \mathbf{T}\mathbf{q}'_1(t) = \mathbf{T}(\mathbf{A}_1\mathbf{q}_1(t) + \mathbf{B}_1\mathbf{x}(t)) = \mathbf{T}\mathbf{A}_1\mathbf{q}_1(t) + \mathbf{T}\mathbf{B}_1\mathbf{x}(t). \qquad \textbf{(10.234)}$$

From (10.234), $\mathbf{q}_1(t) = \mathbf{T}^{-1}\mathbf{q}_2(t)$; therefore,

$$\mathbf{q}'_2(t) = \mathbf{T}\mathbf{A}_1\mathbf{T}^{-1}\mathbf{q}_2(t) + \mathbf{T}\mathbf{B}_1\mathbf{x}(t) = \mathbf{A}_2\mathbf{q}_2(t) + \mathbf{B}_2\mathbf{x}(t), \qquad \textbf{(10.235)}$$

where $\mathbf{A}_2 = \mathbf{TA}_1\mathbf{T}^{-1}$ and $\mathbf{B}_2 = \mathbf{TB}_1$. In the output equation, we get

$$y(t) = \mathbf{C}_1\mathbf{q}_1(t) + \mathbf{D}_1\mathbf{x}(t) = \mathbf{C}_1\mathbf{T}^{-1}\mathbf{q}_2(t) + \mathbf{D}_1\mathbf{x}(t) = \mathbf{C}_2\mathbf{q}_2(t) + \mathbf{D}_2\mathbf{x}(t), \quad \textbf{(10.236)}$$

where $\mathbf{C}_2 = \mathbf{C}_1\mathbf{T}^{-1}$ and $\mathbf{D}_2 = \mathbf{D}_1$. The eigenvalues of $\mathbf{A}_1$ are determined by the system. When we choose a different set of state variables by transforming one set to another through the transformation matrix $\mathbf{T}$, we are not changing the system, only the way we analyze it. Therefore, the eigenvalues of $\mathbf{A}_1$ and $\mathbf{A}_2 = \mathbf{TA}_1\mathbf{T}^{-1}$ should be the same. That can be proven by the following argument. Consider the product,

$$\mathbf{T}\left[s\mathbf{I} - \mathbf{A}_1\right]\mathbf{T}^{-1} = s\underbrace{\mathbf{TIT}^{-1}}_{\mathbf{I}} - \underbrace{\mathbf{TA}_1\mathbf{T}^{-1}}_{\mathbf{A}_2} = s\mathbf{I} - \mathbf{A}_2. \quad \textbf{(10.237)}$$

Taking the determinant of both sides of (10.237),

$$|\mathbf{T}[s\mathbf{I} - \mathbf{A}_1]\mathbf{T}^{-1}| = |s\mathbf{I} - \mathbf{A}_2|. \quad \textbf{(10.238)}$$

Now we can use two determinant properties from Appendix J. The determinant of a product of two matrices is the product of their determinants, and the determinant of the inverse of a matrix is the reciprocal of the determinant of the matrix. Applying those properties to (10.238) we get

$$|\mathbf{T}||[s\mathbf{I} - \mathbf{A}_1]||\mathbf{T}^{-1}| = |s\mathbf{I} - \mathbf{A}_2|. \quad \textbf{(10.239)}$$

Determinants are scalars; therefore multiplication of determinants is commutative and associative and

$$\underbrace{|\mathbf{T}||\mathbf{T}^{-1}|}_{1} \; |s\mathbf{I} - \mathbf{A}_1| = |s\mathbf{I} - \mathbf{A}_2| \quad \textbf{(10.240)}$$

and, finally,

$$|s\mathbf{I} - \mathbf{A}_1| = |s\mathbf{I} - \mathbf{A}_2|. \quad \textbf{(10.241)}$$

Since the determinants are the same, their roots are also the same, proving that the eigenvalues of a system are invariant to the choices of state variables and responses.

If the eigenvalues of a system are distinct, it is possible to choose state variables in such a way that the system matrix $\mathbf{A}$ is diagonal. If $\mathbf{A}$ is diagonal, then it is of the form

$$\mathbf{A} = \begin{bmatrix} a_{11} & 0 & \cdots & 0 \\ 0 & a_{22} & \cdots & 0 \\ \vdots & \vdots & \ddots & \vdots \\ 0 & 0 & \cdots & a_{NN} \end{bmatrix}, \quad \textbf{(10.242)}$$

where $N$ is the order of the system. Then the determinant $|s\mathbf{I} - \mathbf{A}|$ is

$$|s\mathbf{I} - \mathbf{A}| = (s - a_{11})(s - a_{22}) \cdots (s - a_{NN}). \quad \textbf{(10.243)}$$

Since this is in factored form, the roots are exactly $a_{11}, a_{22}, \dots, a_{NN}$. Therefore, if the system matrix $\mathbf{A}$ is diagonal, the elements on the diagonal are the eigenvalues of the system and the matrix can be expressed in the form

$$\mathbf{A} = \Lambda = \begin{bmatrix} \lambda_1 & 0 & \cdots & 0 \\ 0 & \lambda_2 & \cdots & 0 \\ \vdots & \vdots & \ddots & \vdots \\ 0 & 0 & \cdots & \lambda_N \end{bmatrix} \quad \textbf{(10.244)}$$

(where $\Lambda$ is a capitalized $\lambda$). Now suppose we have a system matrix $\mathbf{A}$ which is not diagonal and we want to find a transformation $\mathbf{T}$ that makes it diagonal. Then

$$\Lambda = \mathbf{TAT}^{-1}. \tag{10.245}$$

Postmultiplying both sides by $\mathbf{T}$,

$$\Lambda\mathbf{T} = \mathbf{TA}. \tag{10.246}$$

Since $\Lambda$ and $\mathbf{A}$ are known, this equation can be solved for $\mathbf{T}$. Notice that if were to find a solution $\mathbf{T}$ of (10.246) and multiply that $\mathbf{T}$ by a scalar $K$ to create another transformation matrix $\mathbf{T}_2 = K\mathbf{T}$, we could say

$$\Lambda\mathbf{T}_2 = \Lambda K\mathbf{T} = K\Lambda\mathbf{T} \tag{10.247}$$

and then, using (10.247), we get

$$\Lambda\mathbf{T}_2 = K\mathbf{TA} = \mathbf{T}_2\mathbf{A} \tag{10.248}$$

or simply

$$\Lambda\mathbf{T}_2 = \mathbf{T}_2\mathbf{A} \tag{10.249}$$

which, except for the name of the transformation matrix, is the same as (10.246) proving that the solution $\mathbf{T}$ is not unique.

Once we have found a transformation that diagonalizes the system matrix, we then have a system of equations of the form

$$\begin{bmatrix} q_1'(t) \\ q_2'(t) \\ \vdots \\ q_N'(t) \end{bmatrix} = \begin{bmatrix} \lambda_1 & 0 & \cdots & 0 \\ 0 & \lambda_2 & \cdots & 0 \\ \vdots & \vdots & \ddots & \vdots \\ 0 & 0 & \cdots & \lambda_N \end{bmatrix} \begin{bmatrix} q_1(t) \\ q_2(t) \\ \vdots \\ q_N(t) \end{bmatrix} + \mathbf{Bx}(t) \tag{10.250}$$

Since $\mathbf{B}$ and $\mathbf{x}(t)$ are known, this matrix equation is equivalent to a set of $N$ uncoupled differential equations in $N$ unknowns, $q_1, q_2, \ldots, q_N$. Each equation can be solved without reference to the others. So the diagonalization of the system matrix converts the solution of $N$ coupled first-order simultaneous differential equations into $N$ solutions of one first-order differential equation each.

The MATLAB system-object concept includes CT state-space models of systems. The fundamental function is `ss` and its syntax is

```
sys = ss(A,B,C,D) ;
```

where A, B, C, and D are the state-space-representation matrices of the same name. The function `ssdata` extracts state-space matrices from a system description in a manner analogous to `zpkdata` and `tfdata`. The function `ss2ss` transforms a state-space model to another state-space model. The syntax is

```
sys = ss2ss(sys,T) ;
```

where T is the transformation matrix.

## 10.14  SUMMARY OF IMPORTANT POINTS

1. A stable system has a transfer function with all its poles in the open left half of the $s$ plane.
2. Feedback techniques are often very important in improving the performance of systems.

**3.** Feedback can stabilize an unstable system, but it can also destabilize a stable system.

**4.** Feedback can be used to create a marginally stable, oscillating system.

**5.** The Routh–Hurwitz, root-locus, and gain- and phase-margin analysis techniques are valuable tools for assessing system stability and performance.

**6.** Different types of unity-gain feedback systems have different tracking errors in response to standard signals.

**7.** Block diagrams can be directly reduced graphically, by using Mason's theorem or by using MATLAB.

**8.** The response of systems to standard signals like the unit step and sinusoid are useful in revealing system characteristics.

**9.** The frequency response of a system can be deduced from the pole–zero diagram of the system.

**10.** There are several standard methods of realization of systems from transfer functions.

**11.** In the systematic analysis of multiple-input, multiple-output systems, state-space analysis techniques are very useful and significantly reduce the probability of analysis error and promote understanding of system dynamics.

## EXERCISES WITH ANSWERS

**1.** For each circuit write the transfer function between the indicated excitation and indicated response. Express each transfer function in the standard form

$$H(s) = A \frac{s^N + b_{N-1}s^{N-1} + \cdots + b_2 s^2 + b_1 s + b_0}{s^D + a_{D-1}s^{D-1} + \cdots + a_2 s^2 + a_1 s + a_0}.$$

   *a.*  Excitation: $v_s(t)$        Response: $v_o(t)$

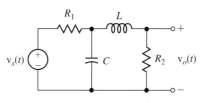

   *b.*  Excitation: $i_s(t)$        Response: $v_o(t)$

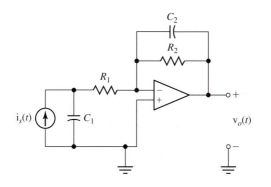

c.   Excitation: $v_s(t)$        Response: $i_1(t)$

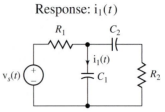

**Answers:**

$$\frac{1}{R_1}\frac{s^2 + s\dfrac{1}{R_2C_2}}{s^2 + s\left(\dfrac{1}{R_2C_2} + \dfrac{1}{R_2C_1} + \dfrac{1}{R_1C_1}\right) + \dfrac{1}{R_1R_2C_1C_2}},$$

$$\frac{R_2}{R_1LC}\frac{1}{s^2 + s\left(\dfrac{1}{R_1C} + \dfrac{R_2}{L}\right) + \dfrac{R_2 + R_1}{R_1LC}},$$

$$-\frac{1}{R_1C_1C_2}\frac{1}{s^2 + s\left(\dfrac{1}{R_2C_2} + \dfrac{1}{R_1C_1}\right) + \dfrac{1}{R_1R_2C_1C_2}}$$

**2.**   For each block diagram write the transfer function between the excitation $x(t)$ and the response $y(t)$.

a.

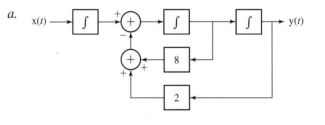

b.

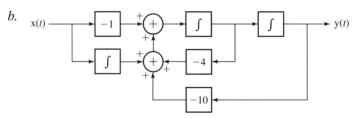

**Answers:**

$$\frac{1}{s^3 + 8s^2 + 2s}, \quad -\frac{s - 1}{s^3 + 4s^2 + 10s}$$

**3.**   Evaluate the stability of the systems with each of these transfer functions.

a.   $H(s) = -\dfrac{100}{s + 200}$          b.   $H(s) = \dfrac{80}{s - 4}$

c.   $H(s) = \dfrac{6}{s(s + 1)}$          d.   $H(s) = -\dfrac{15s}{s^2 + 4s + 4}$

e. $H(s) = 3\dfrac{s - 10}{s^2 + 4s + 29}$  f. $H(s) = 3\dfrac{s^2 + 4}{s^2 - 4s + 29}$

g. $H(s) = \dfrac{1}{s^2 + 64}$  h. $H(s) = \dfrac{10}{s^3 + 4s^2 + 29s}$

**Answers:**

Three are stable, four are unstable including two that are marginally stable

4. Find the overall transfer functions of these systems in the form of a single ratio of polynomials in $s$.

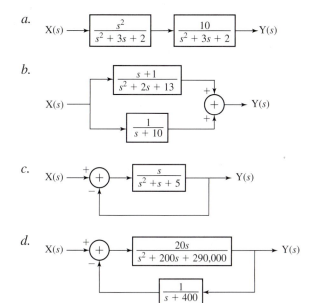

a.
b.
c.
d.

**Answers:**

$$20\dfrac{s^2 + 400s}{s^3 + 600s^2 + 370{,}020s + 1.16 \times 10^8}, \quad 2\dfrac{s^2 + \frac{13}{2}s + \frac{23}{2}}{s^3 + 12s^2 + 33s + 130},$$

$$10\dfrac{s^2}{s^4 + 6s^3 + 13s^2 + 12s + 4}, \quad \dfrac{s}{s^2 + 2s + 5}$$

5. In the feedback system given in Figure E5, find the overall system transfer function for these values of forward-path gain $K$.

  a. $K = 10^6$      b. $K = 10^5$
  c. $K = 10$        d. $K = 1$
  e. $K = -1$        f. $K = -10$

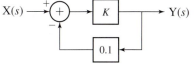

**Figure E5**

**Answers:**

5,   $-1.111$,   $-\infty$,   $0.909$,   $10$,   $10$

6. In the feedback system given in Figure E6, plot the response of the system to a unit step, for the time interval $0 < t < 10$, and then write the expression for the overall system transfer function and draw a pole–zero diagram, for these values of $K$.

   a.   $K = 20$       b.   $K = 10$

   c.   $K = 1$        d.   $K = -1$

   e.   $K = -10$      f.   $K = -20$

**Answers:**

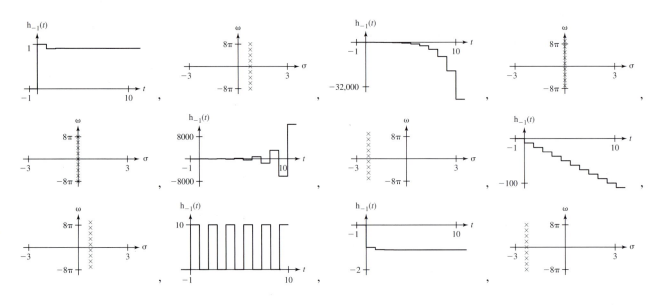

7. For what range of values of $K$ is the system given in Figure E7 stable? Plot the step responses for $K = 0$, $K = 4$, and $K = 8$.

**Answer:**

$K > 4$,

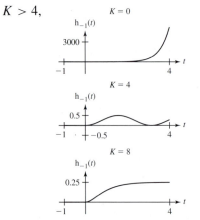

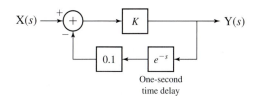

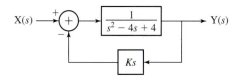

**Figure E6**                    **Figure E7**

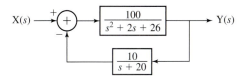

**Figure E8**

8.  Plot the impulse response and the pole–zero diagram for the forward-path and the overall system given in Figure E8.

**Answers:**

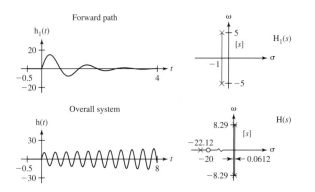

9.  Using the Routh–Hurwitz method, evaluate the stability of the system whose transfer function is

$$H(s) = \frac{s^3 + 3s + 10}{s^5 + 2s^4 + 10s^3 + 4s^2 + 8s + 20}.$$

**Answer:**
Unstable

10. Using the Routh–Hurwitz stability test, evaluate the stability of the system whose transfer function is of the general form

$$H(s) = \frac{N(s)}{s^3 + a_2 s^2 + a_1 s + a_0}.$$

What are the relations among $a_2$, $a_1$, and $a_0$ that ensure stability?

**Answer:**

$a_2 > 0, a_1a_2 > a_0, a_0 > 0$.

11. Plot the root locus for each of the systems which have these loop transfer functions, and identify the transfer functions that are stable for all positive real values of $K$.

    *a.*  $T(s) = \dfrac{K}{(s+3)(s+8)}$

    *b.*  $T(s) = \dfrac{Ks}{(s+3)(s+8)}$

    *c.*  $T(s) = \dfrac{Ks^2}{(s+3)(s+8)}$

    *d.*  $T(s) = \dfrac{K}{(s+1)(s^2+4s+8)}$

**Answers:**

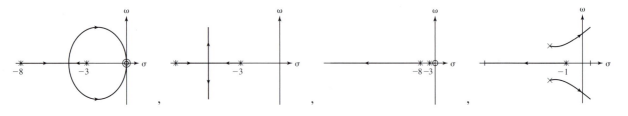

12. Use the block diagram of an inverting amplifier using an operational amplifier given in Figure E12, with $A_0 = 10^4$, $p = -2000\pi$, $Z_f = 10$ k$\Omega$, and $Z_i = 1$ k$\Omega$, to find the gain and phase margins of the amplifier.

**Answer:**

$90°$, infinity

13. Plot the unit-step and ramp responses of unity-gain feedback systems with these forward-path transfer functions.

    *a.*  $H_1(s) = \dfrac{100}{s+10}$          *b.*  $H_1(s) = \dfrac{100}{s(s+10)}$

    *c.*  $H_1(s) = \dfrac{100}{s^2(s+10)}$       *d.*  $H_1(s) = \dfrac{20}{(s+2)(s+6)}$

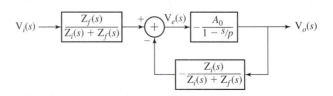

**Figure E12**

**Answers:**

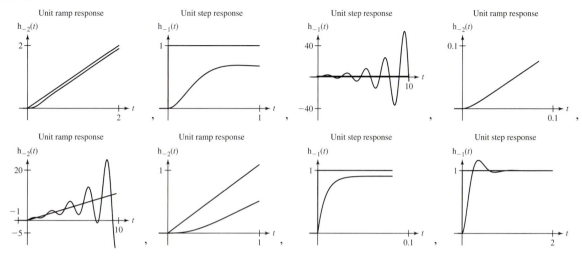

Unit ramp response $h_{-2}(t)$ · Unit step response $h_{-1}(t)$ · Unit step response $h_{-1}(t)$ · Unit ramp response $h_{-2}(t)$

Unit ramp response $h_{-2}(t)$ · Unit ramp response $h_{-2}(t)$ · Unit step response $h_{-1}(t)$ · Unit step response $h_{-1}(t)$

**14.** Reduce these block diagrams to a single block by block-diagram reduction. Check the answer using Mason's theorem.

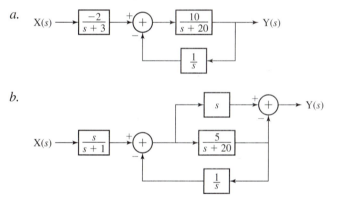

*a.*

*b.*

**Answers:**

$$X(s) \longrightarrow \boxed{\dfrac{-20s}{(s+3)\,(s^2+20s+10)}} \longrightarrow Y(s) \quad , \quad X(s) \longrightarrow \boxed{\dfrac{s^2(s^2+20s-5)}{(s+1)\,(s^2+20s+5)}} \longrightarrow Y(s)$$

**15.** Find the responses of the systems with these transfer functions to a unit step and a suddenly applied unit-amplitude 1-Hz cosine. Also find the responses to a true unit-amplitude 1-Hz cosine (not suddenly applied) using the CTFT and compare to the steady-state part of the total solution found using the Laplace transform.

*a.* $H(s) = \dfrac{1}{s}$

*b.* $H(s) = \dfrac{s}{s+1}$

*c.* $H(s) = \dfrac{s}{s^2+2s+40}$

*d.* $H(s) = \dfrac{s^2+2s+40}{s^2}$

**Answers:**

(Step responses) $[1 + 2t + 20t^2]u(t), \quad \text{ramp}(t), \quad \dfrac{e^{-t}\sin\left(\sqrt{39}\,t\right)}{\sqrt{39}}\,u(t), \quad e^{-t}u(t)$

**16.** For each pole–zero diagram sketch the approximate frequency-response magnitude.

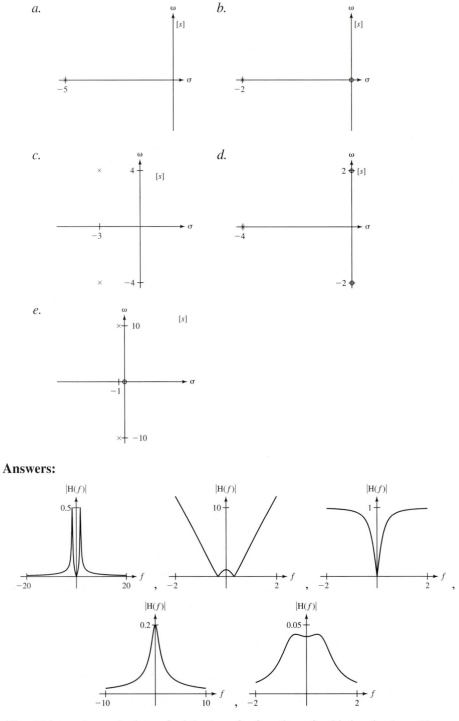

*a.*

*b.*

*c.*

*d.*

*e.*

**Answers:**

**17.** Using only a calculator, find the transfer function of a third-order ($n = 3$) lowpass Butterworth filter with corner frequency $\omega_c = 1$ and unity gain at zero frequency.

**Answer:**

$$\frac{1}{s^3 + 2s^2 + 2s + 1}$$

18. Using MATLAB, find the transfer function of an eighth-order lowpass Butterworth filter with corner frequency $\omega_c = 1$ and unity gain at zero frequency.

**Answer:**

$$\frac{1}{s^8 + 5.126s^7 + 13.1371s^6 + 21.8462s^5 + 25.6884s^4 + 21.8462s^3 + 13.1371s^2 + 5.126s + 1}$$

19. Find the transfer functions of these Butterworth filters.
    a. Second-order highpass with a corner frequency of 20 kHz and a passband gain of 5.
    b. Third-order bandpass with a center frequency of 5 kHz, a $-3$-dB bandwidth of 500 Hz, and a passband gain of 1.
    c. Fourth-order bandstop with a center frequency of 10 MHz, a $-3$-dB bandwidth of 50 kHz, and a passband gain of 1.

**Answers:**

$$\frac{3.1 \times 10^{10}s^3}{s^6 + 6283s^5 + 2.97 \times 10^9 s^4 + 1.24 \times 10^{13}s^3 + 2.93 \times 10^{18}s^2 + 6.09 \times 10^{21}s + 9.542 \times 10^{26}},$$

$$\frac{s^8 + 1.57 \times 10^{16}s^6 + 9.243 \times 10^{31}s^4 + 2.418 \times 10^{47}s^2 + 2.373 \times 10^{62}}{\left[ s^8 + 8.205 \times 10^5 s^7 + 1.57 \times 10^{16}s^6 + 9.665 \times 10^{21}s^5 + 9.24 \times 10^{31}s^4 + 3.729 \times 10^{37}s^3 + 2.419 \times 10^{47}s^2 + 5.256 \times 10^{52}s + 2.373 \times 10^{62} \right]},$$

$$\frac{5s^2}{s^2 + 1.777 \times 10^5 s + 1.579 \times 10^{10}}$$

20. Draw canonical system diagrams of the systems with these transfer functions.
    a. $H(s) = \dfrac{1}{s+1}$    b. $H(s) = 4\dfrac{s+3}{s+10}$

**Answers:**

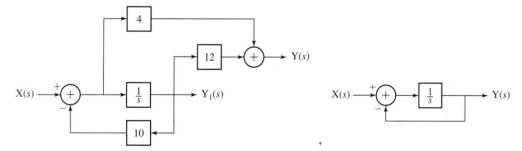

21. Draw cascade system diagrams of the systems with these transfer functions.
    a. $H(s) = \dfrac{s}{s+1}$    b. $H(s) = \dfrac{s+4}{(s+2)(s+12)}$
    c. $H(s) = \dfrac{20}{s(s^2 + 5s + 10)}$

**Answers:**

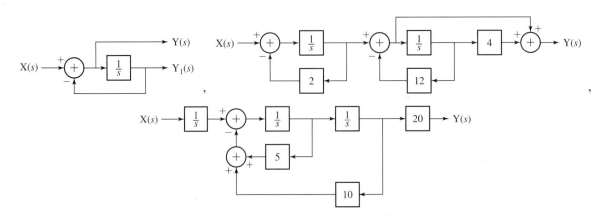

,

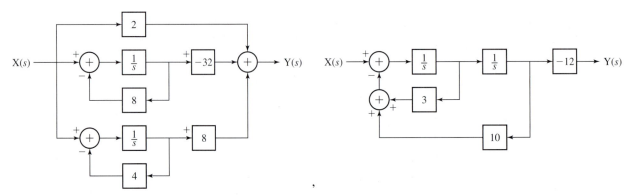

22.   Draw parallel system diagrams of the systems with these transfer functions.

a.   $H(s) = \dfrac{-12}{s^2 + 3s + 10}$

b.   $H(s) = \dfrac{2s^2}{s^2 + 12s + 32}$

**Answers:**

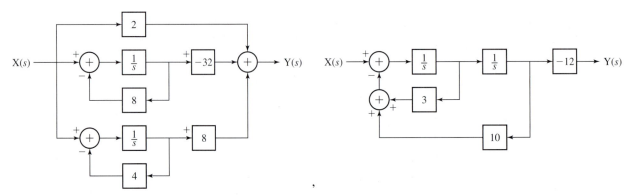

,

23.   Write state equations and output equations for the circuit of Figure E23 with the inductor current $i_L(t)$ and capacitor voltage $v_C(t)$ as the state variables, the voltage at the input $v_i(t)$ as the excitation, and the voltage at the output $v_L(t)$ as the response.

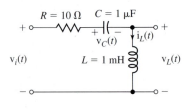

**Figure E23**
An $RLC$ circuit.

**Answer:**

$$\begin{bmatrix} v_C'(t) \\ i_L'(t) \end{bmatrix} = \begin{bmatrix} 0 & 1/C \\ -(1/L) & -(R/L) \end{bmatrix} \begin{bmatrix} v_C(t) \\ i_L(t) \end{bmatrix} + \begin{bmatrix} 0 \\ 1/L \end{bmatrix} v_i(t),$$

$$v_L(t) = \begin{bmatrix} -1 & -R \end{bmatrix} \begin{bmatrix} v_C(t) \\ i_L(t) \end{bmatrix} + v_i(t)$$

24. Write state equations and output equations for the circuit of Figure E24 with the inductor current $i_L(t)$ and capacitor voltage $v_C(t)$ as the state variables, the current at the input $i_i(t)$ as the excitation, and the voltage at the output $v_R(t)$ as the response.

**Answer:**

$$\begin{bmatrix} v_C'(t) \\ i_L'(t) \end{bmatrix} = \begin{bmatrix} 0 & -\frac{1}{C} \\ 1 & -\frac{R}{L} \end{bmatrix} \begin{bmatrix} v_C(t) \\ i_L(t) \end{bmatrix} + \begin{bmatrix} \frac{1}{C} \\ \frac{R}{L} \end{bmatrix} i_i(t),$$

$$v_R(t) = \begin{bmatrix} 0 & -R \end{bmatrix} \begin{bmatrix} v_C(t) \\ i_L(t) \end{bmatrix} + R i_i(t)$$

25. From the system transfer function

$$H(s) = \frac{s(s+3)}{s^2 + 2s + 9},$$

write a set of state equations and output equations using a minimum number of states.

**Answer:**

$$\begin{bmatrix} sQ_1(s) \\ sQ_2(s) \end{bmatrix} = \begin{bmatrix} 0 & 1 \\ -9 & -2 \end{bmatrix} \begin{bmatrix} Q_1(s) \\ Q_2(s) \end{bmatrix} + \begin{bmatrix} 0 \\ s^2 + 3s \end{bmatrix} X(s), \quad Y(s) = \begin{bmatrix} 1 & 0 \end{bmatrix} \begin{bmatrix} Q_1(s) \\ Q_2(s) \end{bmatrix}$$

26. Write state equations and output equations for the system whose block diagram is in Figure E26 using the responses of the integrators as the state variables.

**Answer:**

$$\begin{bmatrix} q_1'(t) \\ q_2'(t) \\ q_3'(t) \end{bmatrix} = \begin{bmatrix} 0 & 1 & 0 \\ -2 & -8 & 1 \\ 0 & 0 & 0 \end{bmatrix} \begin{bmatrix} q_1(t) \\ q_2(t) \\ q_3(t) \end{bmatrix} + \begin{bmatrix} 0 \\ 0 \\ 1 \end{bmatrix} x(t), \quad y(t) = \begin{bmatrix} 1 & 0 & 0 \end{bmatrix} \begin{bmatrix} q_1(t) \\ q_2(t) \\ q_3(t) \end{bmatrix}$$

27. A system is excited by the signal $x(t) = 3u(t)$ and the response is $y(t) = 0.961 e^{-1.5t} \sin(3.122t) u(t)$. Write a set of state equations and output equations using a minimum number of states.

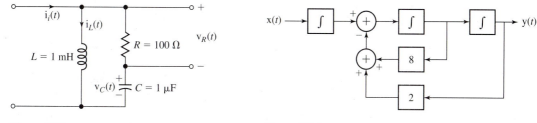

**Figure E24**
An $RLC$ circuit.

**Figure E26**
A system.

**Answer:**

$$\begin{bmatrix} sQ_1(s) \\ sQ_2(s) \end{bmatrix} = \begin{bmatrix} 0 & 1 \\ -12 & -3 \end{bmatrix} \begin{bmatrix} Q_1(s) \\ Q_2(s) \end{bmatrix} + \begin{bmatrix} 0 \\ s \end{bmatrix} X(s), \quad Y(s) = \begin{bmatrix} 1 & 0 \end{bmatrix} \begin{bmatrix} Q_1(s) \\ Q_2(s) \end{bmatrix}$$

**28.** A system is described by the differential equation

$$y''(t) + 4y'(t) + 7y(t) = 10\cos(200\pi t)\,u(t).$$

Write a set of state equations and output equations for this system.

**Answer:**

$$\begin{bmatrix} q_1'(t) \\ q_2'(t) \end{bmatrix} = \begin{bmatrix} 0 & 1 \\ -7 & -4 \end{bmatrix} \begin{bmatrix} q_1(t) \\ q_2(t) \end{bmatrix} + \begin{bmatrix} 0 \\ 1 \end{bmatrix} 10\cos(200\pi t)\,u(t),$$

$$y(t) = \begin{bmatrix} 1 & 0 \end{bmatrix} \begin{bmatrix} q_1(t) \\ q_2(t) \end{bmatrix}$$

**29.** A system is described by the state equations and output equations

$$\begin{bmatrix} q_1'(t) \\ q_2'(t) \end{bmatrix} = \begin{bmatrix} -2 & 1 \\ 3 & 0 \end{bmatrix} \begin{bmatrix} q_1(t) \\ q_1(t) \end{bmatrix} + \begin{bmatrix} 1 & 2 \\ -2 & 0 \end{bmatrix} \begin{bmatrix} x_1(t) \\ x_2(t) \end{bmatrix}$$

and

$$\begin{bmatrix} y_1(t) \\ y_2(t) \end{bmatrix} = \begin{bmatrix} 3 & 5 \\ -2 & 4 \end{bmatrix} \begin{bmatrix} q_1(t) \\ q_2(t) \end{bmatrix},$$

with excitation $\begin{bmatrix} x_1(t) \\ x_2(t) \end{bmatrix} = \begin{bmatrix} -\delta(t) \\ u(t) \end{bmatrix}$ and initial conditions, $\begin{bmatrix} q_1(0^-) \\ q_2(0^-) \end{bmatrix} = \begin{bmatrix} 0 \\ 3 \end{bmatrix}$.

Find the system response vector $\begin{bmatrix} y_1(t) \\ y_2(t) \end{bmatrix}$.

**Answer:**

$$\begin{bmatrix} 5e^{-3t} + 27e^t - 10 \\ 15e^{-3t} + 15e^t - 8 \end{bmatrix} u(t)$$

**30.** A system is described by the vector state equation and output equation

$$q'(t) = \mathbf{A}q(t) + \mathbf{B}x(t)$$

and

$$y(t) = \mathbf{C}q(t) + \mathbf{D}x(t),$$

where $\mathbf{A} = \begin{bmatrix} -1 & -3 \\ 2 & -7 \end{bmatrix}$, $\mathbf{B} = \begin{bmatrix} 1 & 0 \\ 0 & 1 \end{bmatrix}$, $\mathbf{C} = \begin{bmatrix} 2 & -3 \\ 0 & 4 \end{bmatrix}$, and $\mathbf{D} = \begin{bmatrix} 1 & 0 \\ 0 & 0 \end{bmatrix}$,

Define two new states, in terms of the old states, for which the $\mathbf{A}$ matrix is diagonal and rewrite the state equations.

**Answers:**

$$q_2'(t) = \begin{bmatrix} 0.8446 & -0.5354 \\ -0.3893 & 0.9211 \end{bmatrix} q_1'(t),$$

$$q_2'(t) = \begin{bmatrix} -2.2679 & 0 \\ 0 & -5.7321 \end{bmatrix} q_2(t) + \begin{bmatrix} 0.8446 & -0.5354 \\ -0.3893 & 0.9211 \end{bmatrix} x(t)$$

**31.** For the original state equations and output equations of Exercise 30 write a differential-equation description of the system.

**Answer:**

$$y_1'(t) = -4y_1(t) + \frac{3}{4}y_2(t) + 6x_1(t) - 3x_2(t) + x_1'(t),$$

$$y_2'(t) = 4y_1(t) - 4y_2(t) - 4x_1(t) - 4x_2(t)$$

## EXERCISES WITHOUT ANSWERS

**32.** Find the $s$-domain transfer functions for the given circuits and then draw block diagrams for them as systems with $V_i(s)$ as the excitation and $V_o(s)$ as the response.

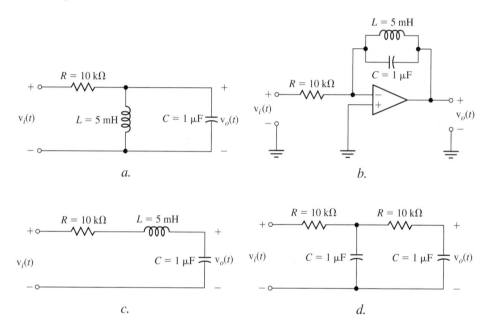

**33.** Determine whether the systems with these transfer functions are stable, marginally stable, or unstable.

a. $H(s) = \dfrac{s(s + 2)}{s^2 + 8}$  b. $H(s) = \dfrac{s(s - 2)}{s^2 + 8}$

c. $H(s) = \dfrac{s^2}{s^2 + 4s + 8}$  d. $H(s) = \dfrac{s^2}{s^2 - 4s + 8}$

e. $H(s) = \dfrac{s}{s^3 + 4s^2 + 8s}$

**34.** Find the expression for the overall system transfer function of the system given in Figure E34. Determine the values of $K$ for which the system is stable for each of the following situations:

a. $\beta = 1$  b. $\beta = -1$

c. $\beta = 10$

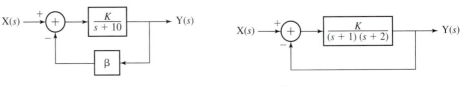

**Figure E34**　　　　　　　　　　　　　　　**Figure E35**

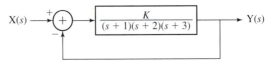

**Figure E36**

**35.** Find the expression for the overall system transfer function of the system given in Figure E35. For what positive values of $K$ is the system stable?

**36.** Find the expression for the overall system transfer function of the system given in Figure E36. Using MATLAB plot the paths of the poles of the overall system transfer function as a function of $K$. For what positive values of $K$ is the system stable?

**37.** Thermocouples are used to measure temperature in many industrial processes. A thermocouple is usually mechanically mounted inside a thermowell, a metal sheath which protects it from damage by vibration, bending stress, or other forces. One effect of the thermowell is that its thermal mass slows the effective time response of the thermocouple-thermowell combination compared with the inherent time response of the thermocouple alone. Let the actual temperature on the outer surface of the thermowell in kelvins be $T_s(t)$ and let the voltage developed by the thermocouple in response to temperature be $v_t(t)$. The response of the thermocouple to a 1 $K$ step change in the thermowell outer-surface temperature from $T_1$ to $T_1 + 1$ is

$$v_t(t) = K\left[T_1 + \left(1 - e^{-(t/0.2)}\right) u(t)\right],$$

where $K$ is the thermocouple temperature-to-voltage conversion constant.

*a.* Let the conversion constant be $K = 40$ μV/K. Design an active filter which processes the thermocouple voltage and compensates for its time lag making the overall system have a response to a 1 K step thermowell-surface temperature change that is itself a voltage step of 1 mV.

*b.* Suppose that the thermocouple also is subject to electromagnetic interference (EMI) from nearby high-power electrical equipment. Let the EMI be modeled as a sinusoid with an amplitude of 20 μV at the thermocouple terminals. Calculate the response of the thermocouple-filter combination to EMI frequencies of 1, 10, and 60 Hz. How big is the apparent temperature fluctuation caused by the EMI in each case?

**38.** A laser operates on the fundamental principle that a pumped medium amplifies a traveling-light beam as it propagates through the medium. Without mirrors, a laser becomes a single-pass traveling-wave amplifier (Figure E38a). This is a system without feedback. If we now place mirrors at each end of the pumped medium, we introduce feedback into the system. When the gain of the medium

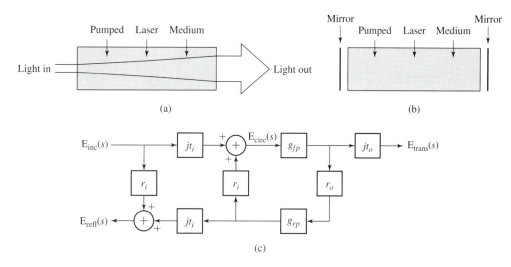

**Figure E38**
(a) A one-pass traveling-wave light amplifier, (b) a regenerative traveling-wave amplifier, and
(c) a block diagram of an RTWA.

becomes large enough, the system oscillates creating a coherent output light
beam. That is laser operation. If the gain of the medium is less than that
required to sustain oscillation, the system is known as a regenerative
traveling-wave amplifier (RTWA).

Let the electric field of a light beam incident on the RTWA from the left be
the excitation of the system $E_{inc}(s)$, and let the electric fields of the reflected
light $E_{refl}(s)$ and the transmitted light $E_{trans}(s)$ be the responses of the system
(Figure E38c).

Let the system parameters be as follows:

Electric field reflectivity of the input mirror $r_i = 0.99$
Electric field transmissivity of the input mirror $t_i = \sqrt{1 - r_i^2}$
Electric field reflectivity of the output mirror $r_o = 0.98$
Electric field transmissivity of the output mirror $t_o = \sqrt{1 - r_o^2}$
Forward and reverse path electric field gains $g_{fp}(s) = g_{rp}(s) = 1.01e^{-10^{-9}s}$

Find an expression for the frequency response $E_{trans}(f)/E_{inc}(f)$ of this
optical amplifier, and plot its magnitude over the frequency range
$3 \times 10^{14} \pm 5 \times 10^8$ Hz.

**39.** A classical example of the use of feedback is the phase-locked loop used to
demodulate frequency-modulated signals (Figure E39). The input signal $x(t)$ is
a frequency-modulated sinusoid. The phase detector detects the phase difference
between the input signal and the signal produced by the voltage-controlled
oscillator. The response of the phase detector is a voltage signal proportional to
phase difference. The loop filter then filters that voltage signal. Then the loop
filter response controls the frequency of the voltage-controlled oscillator. When
the input signal is at a constant frequency and the loop is locked, the phase
difference between the two phase-detector input signals is zero. (In an actual
phase detector the phase difference is 90° at lock. But that is not significant in
this analysis since that only causes a 90° phase shift and has no impact on system

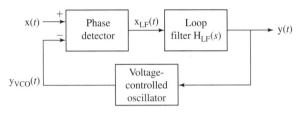

**Figure E39**
A phase-locked loop.

performance or stability.) As the frequency of the input signal $x(t)$ varies, the loop detects the accompanying phase variation and tracks it. The overall output signal $y(t)$ is a signal proportional to the frequency of the input signal.

The actual excitation, in a system sense, of this system is not $x(t)$, but rather the phase of $x(t)$, $\phi_x(t)$, because the phase detector detects differences in phase, not voltage. Let the frequency of $x(t)$ be $f_x(t)$. The relation between phase and frequency can be seen by examining a sinusoid. Let $x(t) = A\cos(2\pi f_0 t)$. The phase of this cosine is $2\pi f_0 t$ and, for a simple sinusoid ($f_0$ constant), it increases linearly with time. The frequency is $f_0$, the derivative of the phase. Therefore, the relation between phase and frequency for a frequency-modulated signal is

$$f_x(t) = \frac{1}{2\pi}\frac{d}{dt}(\phi_x(t)).$$

Let the frequency of the input signal be 100 MHz. Let the transfer function of the voltage-controlled oscillator be $10^8$ Hz/V. Let the transfer function of the loop filter be

$$H_{LF}(s) = \frac{1}{s + 1.2 \times 10^5}.$$

Let the transfer function of the phase detector be 1 V/rad. If the frequency of the excitation signal suddenly changes to 100.001 MHz, plot the change in the output signal $\Delta y(t)$.

**40.** Plot the root locus for each of the systems which have these loop transfer functions and identify the transfer functions that are stable for all positive real values of $K$.

a. $T(s) = \dfrac{K(s+10)}{(s+1)(s^2+4s+8)}$

b. $T(s) = \dfrac{K(s^2+10)}{(s+1)(s^2+4s+8)}$

c. $T(s) = \dfrac{K}{s^3+37s^2+332s+800}$

d. $T(s) = \dfrac{K(s-4)}{s+4}$

e. $T(s) = \dfrac{K(s-4)}{(s+4)^2}$

f. $T(s) = \dfrac{K(s+6)}{(s+5)(s+9)(s^2+4s+12)}$

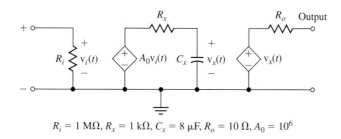

$R_i = 1\ \text{M}\Omega,\ R_x = 1\ \text{k}\Omega,\ C_x = 8\ \mu\text{F},\ R_o = 10\ \Omega,\ A_0 = 10^6$

**Figure E41**
Simple model of an operational amplifier.

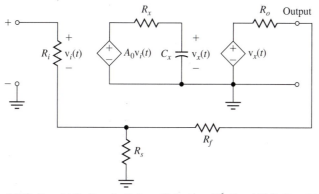

$R_i = 1\ \text{M}\Omega,\ R_x = 1\ \text{k}\Omega,\ C_x = 8\ \mu\text{F},\ R_o = 10\ \Omega,\ A_0 = 10^6,\ R_f = 10\ \text{k}\Omega,\ R_s = 5\ \text{k}\Omega$

**Figure E42**
An operational amplifier connected as a noninverting amplifier.

**41.** The circuit in Figure E41 is a simple approximate model of an operational amplifier with the inverting input grounded.

    *a.* Define the excitation of the circuit as the current of a current source applied to the noninverting input, and define the response as the voltage developed between the noninverting input and ground. Find the transfer function and graph its frequency response. This transfer function is the input impedance.

    *b.* Define the excitation of the circuit as the current of a current source applied to the output, and define the response as the voltage developed between the output and ground with the noninverting input grounded. Find the transfer function and graph its frequency response. This transfer function is the output impedance.

    *c.* Define the excitation of the circuit as the voltage of a voltage source applied to the noninverting input, and define the response as the voltage developed between the output and ground. Find the transfer function and graph its frequency response. This transfer function is the voltage gain.

**42.** Change the circuit of Figure E41 to the circuit of Figure E42. This is a feedback circuit which establishes a positive closed-loop voltage gain of the overall amplifier. Repeat steps (a), (b), and (c) of Exercise 41 for the feedback circuit and compare the results. What are the important effects of feedback for this circuit?

**43.** Plot the unit step and ramp responses of unity-gain feedback systems with these forward-path transfer functions.

a. $H_1(s) = \dfrac{20}{s(s+2)(s+6)}$

b. $H_1(s) = \dfrac{20}{s^2(s+2)(s+6)}$

c. $H_1(s) = \dfrac{100}{s^2 + 10s + 34}$

d. $H_1(s) = \dfrac{100}{s(s^2 + 10s + 34)}$

e. $H_1(s) = \dfrac{100}{s^2(s^2 + 10s + 34)}$

**44.** Draw pole–zero diagrams of these transfer functions.

a. $H(s) = \dfrac{(s+3)(s-1)}{s(s+2)(s+6)}$

b. $H(s) = \dfrac{s}{s^2 + s + 1}$

c. $H(s) = \dfrac{s(s+10)}{s^2 + 11s + 10}$

d. $H(s) = \dfrac{1}{(s+1)(s^2 + 1.618s + 1)(s^2 + 0.618s + 1)}$

**45.** A second-order system is excited by a unit step and the response is as illustrated in Figure E45. Write an expression for the transfer function of the system.

**46.** For each of the pole–zero plots determine whether the frequency response is that of a practical lowpass, bandpass, highpass, or bandstop filter.

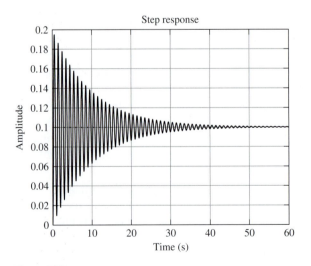

**Figure E45**
Step response of a second-order system.

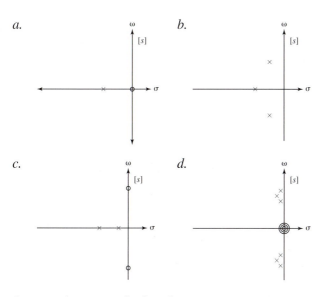

*a.*   *b.*

*c.*   *d.*

**47.**   A system has a transfer function

$$H(s) = \frac{A}{s^2 + 2\zeta\,\omega_0 s + \omega_0^2}.$$

*a.*   Let $\omega_0 = 1$. Then let $\zeta$ vary continuously from 0.1 to 10 and plot in the $s$ plane the paths that the two poles take while $\zeta$ is varying between those limits.

*b.*   Find the real-valued functional form of the impulse response for the case $\omega_0 = 1$ and $\zeta = 0.5$.

*c.*   Sketch the phase frequency response for the case $\omega_0 = 1$ and $\zeta = 0.1$.

*d.*   Find the $-3$-dB bandwidth for the case $\omega_0 = 1$ and $\zeta = 0.1$.

*e.*   The $Q$ of a system is a measure of how sharp its frequency response is near a resonance. It is defined as

$$Q = \frac{1}{2\zeta}.$$

For very high $Q$ systems what is the relationship between $Q$, $\omega_0$, and the $-3$-dB bandwidth?

**48.**   Draw canonical system diagrams of the systems with these transfer functions.

*a.*   $H(s) = 10\dfrac{s^2 + 8}{s^3 + 3s^2 + 7s + 22}$

*b.*   $H(s) = 10\dfrac{s + 20}{(s + 4)(s + 8)(s + 14)}$

**49.**   Draw cascade system diagrams of the systems with these transfer functions.

*a.*   $H(s) = -50\dfrac{s^2}{s^3 + 8s^2 + 13s + 40}$

*b.*   $H(s) = \dfrac{s^3}{s^3 + 18s^2 + 92s + 120}$

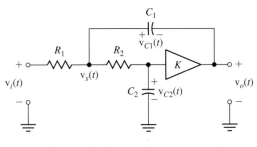

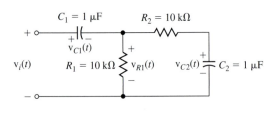

$R_1 = 6.8 \text{ k}\Omega$, $R_2 = 12 \text{ k}\Omega$, $C_1 = 6.8 \text{ nF}$, $C_2 = 6.8 \text{ nF}$, $K = 3$

**Figure E51**
A second-order $RC$ circuit.

**Figure E52**
A constant-$K$ lowpass filter.

50. Draw parallel system diagrams of the systems with these transfer functions.

    a.   $H(s) = 10\dfrac{s^3}{s^3 + 4s^2 + 9s + 3}$

    b.   $H(s) = \dfrac{5}{6s^3 + 77s^2 + 228s + 189}$

51. Write state equations and output equations for the circuit of Figure E51 with the two capacitor voltages $v_{C1}(t)$ and $v_{C2}(t)$ as the state variables, the voltage at the input $v_i(t)$ as the excitation, and the voltage $v_{R1}(t)$ as the response. Then, assuming the capacitors are initially uncharged, find the unit step response of the circuit.

52. Write state equations and output equations for the circuit of Figure E52 with the two capacitor voltages $v_{C1}(t)$ and $v_{C2}(t)$ as the state variables, the voltage at the input $v_i(t)$ as the excitation, and the voltage at the output $v_o(t)$ as the response. Then, find and plot the response voltage for a unit step excitation assuming that the initial conditions are

$$\begin{bmatrix} v_{C1}(0) \\ v_{C2}(0) \end{bmatrix} = \begin{bmatrix} 2 \\ -1 \end{bmatrix}.$$

# CHAPTER 11

# The $z$ Transform

## 11.1 INTRODUCTION AND GOALS

This chapter follows a path similar to that of Chapter 9 on the Laplace transform, except it applies to DT signals and systems instead of CT signals and systems. The $z$ transform is to the DTFT what the Laplace transform is to the CTFT. It increases the range of application of techniques for the DT frequency domain to include signals that do not have a DTFT. Also, like the Laplace transform for CT signals and systems, the $z$ transform gives more insight into system dynamics and stability.

### CHAPTER GOALS

1. To develop the $z$ transform as a more general analysis technique for DT systems than the DTFT

2. To see how the $z$ transform can be developed as a generalization of the DTFT

3. To see the $z$ transform as a result of the convolution process when a DT system is excited by its eigenfunction

4. To derive properties of the $z$ transform that are useful in finding the forward and inverse $z$ transforms of practical DT signals

5. To solve difference equations with initial conditions using the $z$ transform

6. To appreciate the relationship between the $z$ and Laplace transforms

## 11.2 DEVELOPMENT OF THE $z$ TRANSFORM

The Laplace transform is a generalization of the CTFT which allows for consideration of signals and impulse responses which do not have a CTFT. In Chapter 9 we saw how this generalization allows for analysis of signals and systems which cannot be analyzed with the Fourier transform and also how it gives insight into system performance through analysis of the locations of the poles and zeros of the transfer function in the $s$ plane. The $z$ transform is a generalization of the DTFT with similar advantages. The $z$ transform is to DT signal and system analysis what the Laplace transform is to CT signal and system analysis.

## DERIVATION AND DEFINITION

There are two approaches to deriving the $z$ transform which are analogous to the two approaches taken to derive the Laplace transform, generalizing the DTFT and exploiting the unique properties of complex exponentials as the eigenfunctions of LTI systems.

The DTFT is defined by

$$x[n] = \frac{1}{2\pi} \int_{2\pi} X(j\Omega) e^{j\Omega n}\, d\Omega \xleftrightarrow{\mathcal{F}} X(j\Omega) = \sum_{n=-\infty}^{\infty} x[n] e^{-j\Omega n} \qquad \textbf{(11.1)}$$

or

$$x[n] = \int_{1} X(F) e^{j2\pi F n}\, dF \xleftrightarrow{\mathcal{F}} X(F) = \sum_{n=-\infty}^{\infty} x[n] e^{-j2\pi F n}. \qquad \textbf{(11.2)}$$

The Laplace transform generalizes the Fourier transform by changing complex sinusoids of the form $e^{j\omega t}$ to complex exponentials of the form $e^{st}$, where $s = \sigma + j\omega$, and the extra degree of freedom is introduced by the new variable $\sigma$, the real part of $s$. If we followed an analogous path for DT signals, we would generalize DT complex sinusoids of the form $e^{j\Omega n}$ to DT complex exponentials of the form $e^{Sn}$, where $S = \Sigma + j\Omega$. In this development, we are using the already established idea of using lowercase letters $\omega$, $f$, $\sigma$, and $s$ to indicate CT real and complex frequencies and the corresponding uppercase letters $\Omega$, $F$, $\Sigma$, and $S$ to indicate DT real and complex frequencies. Strictly following the analogy to the Laplace transform, the new DT transform would be

$$X(S) = \sum_{n=-\infty}^{\infty} x[n] e^{-Sn} = \sum_{n=-\infty}^{\infty} x[n] e^{-(\Sigma + j\Omega)n} = \sum_{n=-\infty}^{\infty} (x[n] e^{-n\Sigma}) e^{-j\Omega n} \qquad \textbf{(11.3)}$$

and we could conceive of this new kind of transform as the DTFT of a modified version of the signal, the modification being the multiplication by a DT convergence factor $e^{-\Sigma n}$ (Figure 11.1).

Although this notation is a logical extension of the previous transform notation, it is not what is conventionally used for this new DT transform. Instead the new transform conventionally uses $z^n$ instead of $e^{Sn}$. This is consistent with our previous use of the notation $\alpha^n$ for a DT complex exponential instead of the equivalent notation $e^{\beta n}$,

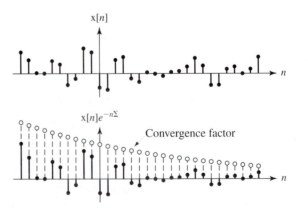

**Figure 11.1**
An original signal and that same signal multiplied by a convergence factor.

where $\alpha = e^{\beta}$ and both $\alpha$ and $\beta$ can, in general, be complex. The use of $z$ instead of $e^{S}$ simplifies the notation and is universally used in the signal and system discipline.

The other approach to deriving the $z$ transform is to realize that when we excite a discrete-time LTI system with a complex exponential of the form $x[n] = Az^{n}$, the response can be found by convolution to be

$$y[n] = x[n] * h[n] = Az^{n} * h[n] = \sum_{m=-\infty}^{\infty} h[m]Az^{(n-m)} = \underbrace{Az^{n}}_{x[n]} \underbrace{\sum_{m=-\infty}^{\infty} h[m]z^{-m}}_{z \text{ transform of } h[n]}.$$

**(11.4)**

Since any DT signal with engineering usefulness can be expressed as a linear combination of DT complex exponentials, the response to any excitation can be found by multiplying the $z$ transform of the excitation (which expresses the excitation as a linear combination of complex exponentials) by the $z$ transform of the impulse response. The DTFT is a special case of the $z$ transform with some notation changes.

Now we define the $z$ transform by

$$X(z) = \sum_{n=-\infty}^{\infty} x[n]z^{-n},$$

**(11.5)**

where $z$ can range anywhere in the complex plane. This contrasts with $e^{j\Omega}$ which can only be on the unit circle because radian frequency $\Omega$ is restricted to real values and is identified with the real physical concept of (radian) frequency. Equation (11.5) defines the forward bilateral $z$ transform. $z$ transformation can also be indicated by the notation

$$\mathcal{Z}(x[n]) = X(z)$$

**(11.6)**

or

$$x[n] \xleftrightarrow{\ \mathcal{Z}\ } X(z).$$

**(11.7)**

The DTFTs of some commonly used functions do not exist in the strict sense. For example, the unit sequence u[n] does not have a DTFT because the DTFT would be

$$X(j\Omega) = \sum_{n=-\infty}^{\infty} u[n]e^{-j\Omega n} = \sum_{n=0}^{\infty} e^{-j\Omega n}$$

**(11.8)**

and the summation does not converge. But, even though the DTFT does not exist, the $z$ transform

$$X(z) = \sum_{n=-\infty}^{\infty} u[n]z^{-n} = \sum_{n=0}^{\infty} z^{-n}$$

**(11.9)**

does exist for values of $z$ whose magnitudes are greater than one. The requirement that the magnitude of $z$ be greater than one for convergence defines a region of convergence (ROC) of the $z$ transform in the $z$ plane, the open exterior of the unit circle. The $z$-transform summation

$$X(z) = \sum_{n=0}^{\infty} z^{-n} \qquad |z| > 1$$

**(11.10)**

can be written in closed form as

$$X(z) = \frac{z}{z-1} = \frac{1}{1-z^{-1}} \qquad |z| > 1.$$

**(11.11)**

The two forms

$$X(z) = \frac{z}{z-1} \tag{11.12}$$

and

$$X(z) = \frac{1}{1 - z^{-1}} \tag{11.13}$$

are equal, but one or the other may be preferred in certain situations. For example, it is immediately obvious from (11.12) that this $z$ transform has a zero at $z = 0$ and a pole at $z = 1$. Although the pole and zero locations can be found by examining the second form, (11.13), they are not as immediately obvious. For reasons that we will soon see, the second form, (11.13), is often preferred in situations in which a DT system is being synthesized from a $z$-domain transfer function.

Earlier we found that the response y[$n$] of a discrete-time LTI system to a DT excitation x[$n$] is the convolution of the excitation with the system's impulse response h[$n$],

$$y[n] = \sum_{m=-\infty}^{\infty} x[m]h[n-m] = \sum_{m=-\infty}^{\infty} h[m]x[n-m], \tag{11.14}$$

and, for an excitation in the form of a DT complex exponential

$$x[n] = Az^n, \tag{11.15}$$

the response is then

$$y[n] = \sum_{m=-\infty}^{\infty} h[m]Az^{n-m} = Az^n \underbrace{\sum_{m=-\infty}^{\infty} h[m]z^{-m}}_{=\mathcal{Z}(h[n])} = Az^n H(z). \tag{11.16}$$

This shows that the response to a DT complex exponential is another DT complex exponential of the same form but with a different multiplier which is the $z$ transform of the impulse response. Since any DT signal with engineering usefulness can be expressed as a linear combination of DT complex exponentials, the $z$ transform of the response to any excitation can be found by multiplying the $z$ transform of the excitation by the $z$ transform of the impulse response,

$$Y(z) = X(z)H(z). \tag{11.17}$$

This is directly analogous to the corresponding CTFT and Laplace-transform results for CT systems excited by CT complex sinusoids or exponentials,

$$Y(j\omega) = X(j\omega)H(j\omega) \quad \text{and} \quad Y(s) = X(s)H(s), \tag{11.18}$$

and to the DTFT result for DT systems excited by DT complex sinusoids,

$$Y(j\Omega) = X(j\Omega)H(j\Omega). \tag{11.19}$$

## REGION OF CONVERGENCE

In a manner analogous to finding the Laplace transform of $Ae^{\alpha t}u(t)$, $\alpha > 0$, we can find the $z$ transform of the causal DT function $A\alpha^n u[n]$, $|\alpha| > 0$ (which does not have a DTFT), as

$$X(z) = A \sum_{n=-\infty}^{\infty} \alpha^n u[n]z^{-n} = A \sum_{n=0}^{\infty} \alpha^n z^{-n} = A \sum_{n=0}^{\infty} \left(\frac{\alpha}{z}\right)^n \tag{11.20}$$

(Figure 11.2). The summation in (11.20) converges if $|z| > |\alpha|$. This defines the region of convergence as the open exterior of a circle of radius $|\alpha|$ in the $z$ plane (Figure 11.3). The $z$-transform summation

$$X(z) = A \sum_{n=0}^{\infty} \alpha^n z^{-n} = A \sum_{n=0}^{\infty} \left(\frac{z}{\alpha}\right)^{-n} \qquad |z| > |\alpha| \qquad \textbf{(11.21)}$$

can be written in closed form as

$$X(z) = A \frac{z/\alpha}{(z/\alpha) - 1} = A \frac{z}{z - \alpha} = \frac{A}{1 - \alpha z^{-1}} \qquad |z| > |\alpha|. \qquad \textbf{(11.22)}$$

The function, $A\alpha^{-n}u[-n]$, $|\alpha| > 0$, is an anticausal DT function which does not have a DTFT (Figure 11.4). Its $z$ transform is

$$X(z) = A \sum_{n=-\infty}^{\infty} \alpha^{-n}u[-n]z^{-n} = A \sum_{n=-\infty}^{0} \alpha^{-n}z^{-n} = A \sum_{n=0}^{\infty} (\alpha z)^n, \qquad \textbf{(11.23)}$$

and the transform exists if $|\alpha z| < 1$ or $|z| < 1/|\alpha|$. Therefore, the region of convergence is the open interior of a circle of radius $1/|\alpha|$ in the $z$ plane (Figure 11.5). The

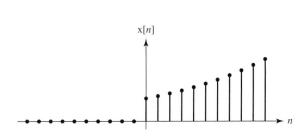

**Figure 11.2**
A growing causal DT exponential signal.

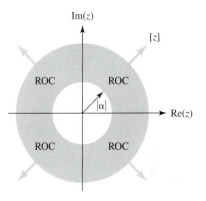

**Figure 11.3**
Region of convergence of the
$z$ transform of $A\alpha^n u[n]$, $|\alpha| > 0$.

**Figure 11.4**
A growing anticausal DT exponential signal.

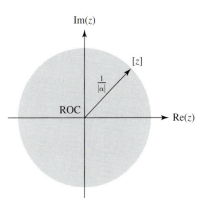

**Figure 11.5**
Region of convergence of the
$z$ transform of $A\alpha^{-n}u[-n]$, $|\alpha| > 0$.

*z*-transform summation

$$X(z) = A \sum_{n=-\infty}^{0} \alpha^{-n} z^{-n} = A \sum_{n=0}^{\infty} \alpha^n z^n \qquad |z| < \frac{1}{|\alpha|} \tag{11.24}$$

can be written in closed form as

$$X(z) = \frac{A}{1 - \alpha z} = \frac{A z^{-1}}{z^{-1} - \alpha} \qquad |z| < \frac{1}{|\alpha|}. \tag{11.25}$$

Just as we found in Chapter 9 that the bilateral Laplace transform of a constant did not exist because no single convergence factor could be found that made the transform integral converge, the bilateral *z* transform of a constant cannot be found either.

## EXAMPLE 11.1

Find the *z* transform of

$$x[n] = \left[ 3 \left( \frac{4}{5} \right)^n - \left( \frac{2}{3} \right)^{2n} \right] u[n]. \tag{11.26}$$

### ■ Solution

Using the definition,

$$X(z) = \sum_{n=-\infty}^{\infty} \left[ 3 \left( \frac{4}{5} \right)^n - \left( \frac{2}{3} \right)^{2n} \right] u[n] z^{-n} = \sum_{n=0}^{\infty} \left[ 3 \left( \frac{5}{4} \right)^{-n} - \left( \frac{3}{2} \right)^{-2n} \right] z^{-n}$$

$$X(z) = 3 \underbrace{\sum_{n=0}^{\infty} \left( \frac{5z}{4} \right)^{-n}}_{\text{ROC: } |z| > \frac{4}{5}} - \underbrace{\sum_{n=0}^{\infty} \left( \frac{9z}{4} \right)^{-n}}_{\text{ROC: } |z| > \frac{4}{9}} \tag{11.27}$$

$$X(z) = \frac{15z}{5z - 4} - \frac{9z}{9z - 4} \qquad |z| > \frac{4}{5}.$$

The region of convergence is illustrated in Figure 11.6.

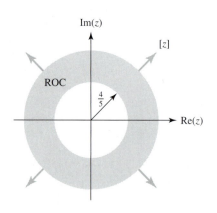

**Figure 11.6**
ROC of convergence of the *z* transform
of $x[n] = \left[ 3 \left( \frac{4}{5} \right)^n - \left( \frac{2}{3} \right)^{2n} \right] u[n]$.

EXAMPLE **11.2**

Find the *z* transform of

$$x[n] = 2^n u[n] + 3^n u[-n]. \tag{11.28}$$

■ **Solution**

Applying the definition,

$$X(z) = \sum_{n=-\infty}^{\infty} (2^n u[n] + 3^n u[-n]) z^{-n} = \sum_{n=0}^{\infty} 2^n z^{-n} + \sum_{n=-\infty}^{0} 3^n z^{-n} \tag{11.29}$$

$$X(z) = \sum_{n=0}^{\infty} \left(\frac{2}{z}\right)^n + \sum_{n=-\infty}^{0} \left(\frac{3}{z}\right)^n = \underbrace{\sum_{n=0}^{\infty} \left(\frac{2}{z}\right)^n}_{\text{ROC: } |z|>2} + \underbrace{\sum_{n=0}^{\infty} \left(\frac{z}{3}\right)^n}_{\text{ROC: } |z|<3} \tag{11.30}$$

$$X(z) = \frac{z}{z-2} - \frac{3}{z-3} \qquad 2 < |z| < 3.$$

(Figure 11.7)

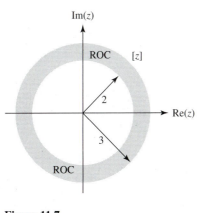

**Figure 11.7**
ROC for *z* transform of
$x[n] = 2^n u[n] + 3^n u[-n]$.

## THE UNILATERAL *z* TRANSFORM

The unilateral Laplace transform proved convenient for CT functions, and the unilateral *z* transform is convenient for DT functions for the same reasons. We can define a unilateral *z* transform that is only valid for DT functions which are zero before discrete time $n = 0$ and avoid, in most practical problems, any involved consideration of the region of convergence. Then, as we did with the Laplace transform, we will later show that any bilateral *z* transform can be found by using unilateral *z*-transform tables.

The unilateral *z* transform is defined by

$$X(z) = \sum_{n=0}^{\infty} x[n] z^{-n}. \tag{11.31}$$

From this point on the unilateral *z* transform will be referred to simply as *the z* transform and the bilateral *z* transform will be explicitly indicated.

**EXAMPLE 11.3**

Find the *z* transform of

$$x[n] = \sin(\Omega_0 n)\, u[n]. \tag{11.32}$$

**■ Solution**

We can write the sine function in terms of complex exponentials

$$\sin(\Omega_0 n)\, u[n] = \frac{e^{j\Omega_0} - e^{-j\Omega_0}}{j2} u[n]. \tag{11.33}$$

Then, using the transform

$$\mathcal{Z}(\alpha^n u[n]) = \frac{z}{z - \alpha} \tag{11.34}$$

found in (11.22), we can write

$$\mathcal{Z}(\sin(\Omega_0 n)\, u[n]) = \frac{1}{j2}\left[\frac{z}{z - e^{j\Omega_0}} - \frac{z}{z - e^{-j\Omega_0}}\right] = \frac{1}{j2}\frac{z(z - e^{-j\Omega_0}) - z(z - e^{j\Omega_0})}{(z - e^{j\Omega_0})(z - e^{-j\Omega_0})} \tag{11.35}$$

or, simplifying,

$$\mathcal{Z}(x[n]) = \frac{z\sin(\Omega_0)}{z^2 - 2z\cos(\Omega_0) + 1} = \frac{\sin(\Omega_0)z^{-1}}{1 - 2\cos(\Omega_0)z^{-1} + z^{-2}}. \tag{11.36}$$

A useful table of *z* transforms is given in Appendix G.

## 11.3  PROPERTIES OF THE *z* TRANSFORM

Given the *z*-transform pairs,

$$g[n] \stackrel{\mathcal{Z}}{\longleftrightarrow} G(z) \qquad g[n] = 0 \quad n < 0 \tag{11.37}$$

and

$$h[n] \stackrel{\mathcal{Z}}{\longleftrightarrow} H(z) \qquad h[n] = 0 \quad n < 0, \tag{11.38}$$

we can prove the properties of the *z* transform in the following sections.

### LINEARITY

The linearity property is exactly the same for the *z* transform as for all the other transforms and the proof is similar.

$$\boxed{\alpha g[n] + \beta h[n] \stackrel{\mathcal{Z}}{\longleftrightarrow} \alpha G(z) + \beta H(z)} \tag{11.39}$$

This simply shows that, like all other transform methods, the *z* transform is linear and superposition applies.

## TIME SHIFTING

There are two different cases to consider, negative and positive shifts in discrete time (Figure 11.8).

***Case 1*** Positive shifts in discrete time.   The assumption in (11.37) is that the signal is causal. Therefore, positive shifts in discrete time simply shift in leading zeros.

$$\mathcal{Z}(g[n - n_0]) = \sum_{n=0}^{\infty} g[n - n_0]z^{-n} = \sum_{n=n_0}^{\infty} g[n - n_0]z^{-n} \qquad n_0 \geq 0 \qquad \textbf{(11.40)}$$

Let $m = n - n_0$. Then

$$\mathcal{Z}(g[n - n_0]) = \sum_{m=0}^{\infty} g[m]z^{-(m+n_0)} = z^{-n_0} \sum_{m=0}^{\infty} g[m]z^{-m} = z^{-n_0}G(z) \qquad \textbf{(11.41)}$$

$$\boxed{g[n - n_0] \overset{\mathcal{Z}}{\longleftrightarrow} z^{-n_0}G(z) \qquad n_0 \geq 0}. \qquad \textbf{(11.42)}$$

This property only applies to causal signals. Otherwise a positive shift could shift in new nonzero signal values and the relationship between the transforms of the original and shifted signals would not be unique.

***Case 2*** Negative shifts in discrete time.   In this case, using the unilateral $z$ transform, we are in general cutting off part of the signal by shifting it to the left because the transform summation begins at $n = 0$. Therefore, if the property is to apply generally, we must find a way to restore the information that was cut off. Otherwise the transform of the unshifted signal and the shifted signal cannot be uniquely related.

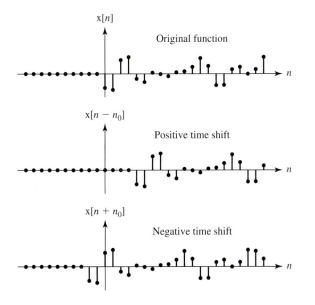

**Figure 11.8**
Shifts in discrete time.

Start with the definition of the $z$ transform

$$\mathcal{Z}(g[n+1]) = \sum_{n=0}^{\infty} g[n+1]z^{-n} = z \sum_{n=0}^{\infty} g[n+1]z^{-(n+1)}. \tag{11.43}$$

In (11.43) we have a summation which does not include the effect of g[0] in the original function. Let $m = n + 1$, then

$$\mathcal{Z}(g[n+1]) = z \sum_{m=1}^{\infty} g[m]z^{-m} = z \left( \sum_{m=0}^{\infty} g[m]z^{-m} - g[0] \right) = z(G(z) - g[0]). \tag{11.44}$$

By this process we now are including the effect of g[0], and the transform of the original signal and the shifted signal are uniquely related. We can extend this method to greater shifts, for example, a negative shift of two in discrete time,

$$\mathcal{Z}(g[n+2]) = \sum_{n=0}^{\infty} g[n+2]z^{-n} = z^2 \sum_{n=0}^{\infty} g[n+2]z^{-(n+2)}. \tag{11.45}$$

Let $m = n + 2$; then

$$\mathcal{Z}(g[n+2]) = z^2 \sum_{m=2}^{\infty} g[m]z^{-m} = z^2 \left( \sum_{m=0}^{\infty} g[m]z^{-m} - g[0] - z^{-1}g[1] \right) \tag{11.46}$$

$$\mathcal{Z}(g[n+2]) = z^2(G(z) - g[0] - z^{-1}g[1]).$$

Then, by induction, for greater shifts,

$$\boxed{g[n+n_0] \overset{\mathcal{Z}}{\longleftrightarrow} z^{n_0} \left( G(z) - \sum_{m=0}^{n_0-1} g[m]z^{-m} \right) \qquad n_0 > 0} . \tag{11.47}$$

The time-shifting property is very important in converting $z$-domain transfer-function expressions into actual DT systems and, other than the linearity property, is probably the most often used property of the $z$ transform.

## EXAMPLE 11.4

A DT system has a transfer function

$$H(z) = \frac{Y(z)}{X(z)} = \frac{z - \frac{1}{2}}{z^2 - z + \frac{2}{9}}. \tag{11.48}$$

Draw a system block diagram using delay and gain blocks.

### ■ Solution

We can rewrite the equation in the form,

$$\frac{Y(z)}{X(z)} = z^{-1} \frac{1 - \frac{1}{2}z^{-1}}{1 - z^{-1} + \frac{2}{9}z^{-2}}. \tag{11.49}$$

We can rearrange (11.49) into

$$Y(z)\left( 1 - z^{-1} + \frac{2}{9}z^{-2} \right) = \left( z^{-1} - \frac{1}{2}z^{-2} \right) X(z) \tag{11.50}$$

or

$$Y(z) - z^{-1}Y(z) + \frac{2}{9}z^{-2}Y(z) = z^{-1}X(z) - \frac{1}{2}z^{-2}X(z). \tag{11.51}$$

Now, using the time-shifting property, if

$$x[n] \xleftrightarrow{\ z\ } X(z) \qquad \text{and} \qquad y[n] \xleftrightarrow{\ z\ } Y(z), \tag{11.52}$$

then the inverse $z$ transform of (11.51) is

$$y[n] - y[n-1] + \frac{2}{9}y[n-2] = x[n-1] - \frac{1}{2}x[n-2] \tag{11.53}$$

or

$$y[n] = x[n-1] - \frac{1}{2}x[n-2] + y[n-1] - \frac{2}{9}y[n-2]. \tag{11.54}$$

Equation (11.54) is called a *recursion* relationship between the excitation x[n] and the response y[n] expressing the present value of the response (at discrete time n) as a linear combination of the present and past values of both the excitation and the response (at discrete times n, n − 1, n − 2, . . .). From it we can directly synthesize a block diagram of a system with the transfer function (11.48) (Figure 11.9). This is not the only way of drawing a block diagram to represent this system. We can rewrite the transfer function as

$$H(z) = \frac{Y(z)}{X(z)} = \frac{z - \frac{1}{2}}{\left(z - \frac{1}{3}\right)\left(z - \frac{2}{3}\right)} = \frac{z - \frac{1}{2}}{z - \frac{1}{3}} \times \frac{1}{z - \frac{2}{3}}. \tag{11.55}$$

If we let

$$Y_1(z) = \frac{z - \frac{1}{2}}{z - \frac{1}{3}}X(z) \qquad \text{and} \qquad Y(z) = \frac{1}{z - \frac{2}{3}}Y_1(z), \tag{11.56}$$

then

$$H_1(z) = \frac{Y_1(z)}{X(z)} = \frac{z - \frac{1}{2}}{z - \frac{1}{3}}, \quad H_2(z) = \frac{Y(z)}{Y_1(z)} = \frac{1}{z - \frac{2}{3}}, \quad \text{and} \quad H(z) = H_1(z)H_2(z), \tag{11.57}$$

and we can draw the system block diagram as the cascade connection of two simpler systems (Figure 11.10). There are also other ways of drawing block diagrams which we will explore in Chapter 12.

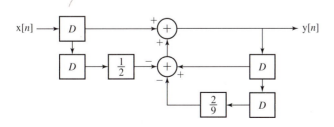

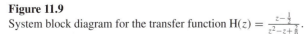

**Figure 11.9**
System block diagram for the transfer function $H(z) = \frac{z - \frac{1}{2}}{z^2 - z + \frac{2}{9}}$.

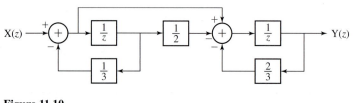

**Figure 11.10**
Alternate system block diagram for the transfer function $H(z) = \frac{z - \frac{1}{2}}{z^2 - z + \frac{2}{9}}$.

## CHANGE OF SCALE

If we compress or expand the $z$ transform of a signal in the $z$ domain, the equivalent effect in the DT domain is a multiplication by a complex exponential.

$$\mathcal{Z}(\alpha^n g[n]) = \sum_{n=0}^{\infty} \alpha^n g[n] z^{-n} = \sum_{n=0}^{\infty} g[n] \left( \frac{z}{\alpha} \right)^{-n} = G\left( \frac{z}{\alpha} \right) \qquad (11.58)$$

$$\boxed{\alpha^n g[n] \overset{\mathcal{Z}}{\longleftrightarrow} G\left( \frac{z}{\alpha} \right)}. \qquad (11.59)$$

A special case of this property is of particular interest. Let the constant $\alpha$ be $e^{j\Omega_0}$, where $\Omega_0$ is real. Then

$$e^{j\Omega_0 n} g[n] \overset{\mathcal{Z}}{\longleftrightarrow} G(z e^{-j\Omega_0}). \qquad (11.60)$$

Every value of $z$ is changed to $z e^{-j\Omega_0}$. This accomplishes a counterclockwise rotation of the transform $G(z)$ in the $z$ plane by the angle $\Omega_0$ because $e^{-j\Omega_0}$ has a magnitude of one and a phase of $-\Omega_0$. So any particular functional value $G(z_0)$ of the transform of $g[n]$ is converted to the value $G(z_0 e^{-j\Omega_0})$ of the transform of $e^{j\Omega_0 n} g[n]$ (Figure 11.11).

A multiplication by a complex sinusoid of the form $e^{j\Omega_0 n}$ in the DT domain corresponds to a rotation of its $z$ transform. Referring back for a moment to the idea that $z = e^{\Sigma + j\Omega}$ is a generalization of $e^{j\Omega}$, if we restrict the $z$ transform to real DT frequencies such that $\Sigma = 0$, we have $z = e^{j\Omega}$ and a multiplication by $e^{-j\Omega_0}$ yields $z e^{-j\Omega_0} = e^{j(\Omega - \Omega_0)}$, which can be interpreted as a *shift* in the $\Omega$ domain by an amount $\Omega_0$. So this property is analogous to the frequency-shifting property of the DTFT and will explain

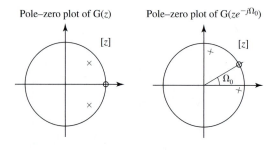

**Figure 11.11**
Illustration of the frequency-scaling property of the $z$ transform for the special case of a scaling by $e^{j\Omega_0}$.

similar effects if we do DT modulation of a DT signal. That is, a *rotation* in the $z$ domain is analogous to a *shift* in the DT frequency domain because of the relationship $z = e^{j\Omega}$.

**EXAMPLE 11.5**

Find the $z$ transforms of

$$x[n] = e^{-(n/40)}u[n] \qquad \text{and} \qquad x_m[n] = e^{-(n/40)} \sin\left(\frac{2\pi n}{8}\right) u[n] \qquad \textbf{(11.61)}$$

and draw pole–zero diagrams for $X(z)$ and $X_m(z)$.

■ **Solution**

Using

$$\alpha^n u[n] \xrightarrow{\;\;\mathcal{Z}\;\;} \frac{z}{z - \alpha} = \frac{1}{1 - \alpha z^{-1}} \qquad \textbf{(11.62)}$$

we get

$$e^{-(n/40)}u[n] \xrightarrow{\;\;\mathcal{Z}\;\;} \frac{z}{z - e^{-(1/40)}}. \qquad \textbf{(11.63)}$$

Therefore, $X(z) = \dfrac{z}{z - e^{-(1/40)}}$.

We can rewrite $x_m[n]$ as

$$x_m[n] = e^{-(n/40)} \frac{e^{j(2\pi n/8)} - e^{-j(2\pi n/8)}}{j2} u[n] \qquad \textbf{(11.64)}$$

or

$$x_m[n] = -\frac{j}{2} \left[ e^{-(n/40)} e^{j(2\pi n/8)} - e^{-(n/40)} e^{-j(2\pi n/8)} \right] u[n]. \qquad \textbf{(11.65)}$$

Then, starting with

$$e^{-(n/40)}u[n] \xrightarrow{\;\;\mathcal{Z}\;\;} \frac{z}{z - e^{-(1/40)}} \qquad \textbf{(11.66)}$$

and using the change-of-scale property

$$\alpha^n g[n] \xrightarrow{\;\;\mathcal{Z}\;\;} G\left(\frac{z}{\alpha}\right), \qquad \textbf{(11.67)}$$

we get

$$e^{j(2\pi n/8)} e^{-(n/40)}u[n] \xrightarrow{\;\;\mathcal{Z}\;\;} \frac{ze^{-j(2\pi/8)}}{ze^{-j(2\pi/8)} - e^{-(1/40)}} \qquad \textbf{(11.68)}$$

and

$$e^{-j(2\pi n/8)} e^{-(n/40)}u[n] \xrightarrow{\;\;\mathcal{Z}\;\;} \frac{ze^{j(2\pi/8)}}{ze^{j(2\pi/8)} - e^{-(1/40)}} \qquad \textbf{(11.69)}$$

and

$$-\frac{j}{2} \left[ e^{-(n/40)} e^{j(2\pi n/8)} - e^{-(n/40)} e^{-j(2\pi n/8)} \right] u[n]$$

$$\xrightarrow{\;\;\mathcal{Z}\;\;} -\frac{j}{2} \left[ \frac{ze^{-j(2\pi/8)}}{ze^{-j(2\pi/8)} - e^{-(1/40)}} - \frac{ze^{j(2\pi/8)}}{ze^{j(2\pi/8)} - e^{-(1/40)}} \right] \qquad \textbf{(11.70)}$$

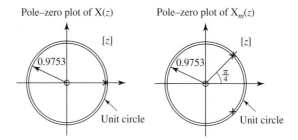

**Figure 11.12**
Pole–zero plots of X($z$) and X$_m$($z$).

or

$$X_m(z) = -\frac{j}{2}\left[\frac{ze^{-j(2\pi/8)}}{ze^{-j(2\pi/8)} - e^{-(1/40)}} - \frac{ze^{j(2\pi/8)}}{ze^{j(2\pi/8)} - e^{-(1/40)}}\right]$$

$$= \frac{ze^{-(1/40)}\sin(2\pi/8)}{z^2 - 2ze^{-(1/40)}\cos(2\pi/8) + e^{-(1/20)}} \tag{11.71}$$

or

$$X_m(z) = \frac{0.6896z}{z^2 - 1.3793z + 0.9512} = \frac{0.6896z}{(z - 0.6896 - j0.6896)(z - 0.6896 + j0.6896)} \tag{11.72}$$

(Figure 11.12). ∎

## INITIAL VALUE THEOREM

The initial value theorem is similar to its counterpart in the Laplace transform. If we take the limit as $z$ approaches infinity of the $z$ transform G($z$), of any function g[$n$], all the terms except the g[0]$z^0$ term approach zero leaving only the first term.

$$\lim_{z\to\infty} G(z) = \lim_{z\to\infty} \sum_{n=0}^{\infty} g[n]z^{-n} = \lim_{z\to\infty}\left[g[0] + \frac{g[1]}{z} + \frac{g[2]}{z^2} + \cdots\right] = g[0] \tag{11.73}$$

$$\boxed{g[0] = \lim_{z\to\infty} G(z)}. \tag{11.74}$$

## $z$-DOMAIN DIFFERENTIATION

Differentiation in the $z$ domain is related to a multiplication by $-n$ in the DT domain.

$$G(z) = \sum_{0}^{\infty} g[n]z^{-n} \tag{11.75}$$

$$\frac{d}{dz}G(z) = \frac{d}{dz}\sum_{n=0}^{\infty} g[n]z^{-n} = \sum_{n=0}^{\infty} g[n]\left(-nz^{-(n+1)}\right) = -z^{-1}\underbrace{\sum_{0}^{\infty} ng[n]z^{-n}}_{\mathcal{Z}(ng[n])}. \tag{11.76}$$

Therefore,

$$\boxed{-ng[n] \overset{\mathcal{Z}}{\longleftrightarrow} z\frac{d}{dz}G(z)}. \tag{11.77}$$

**EXAMPLE 11.6**

Using the $z$-domain differentiation property show that the $z$ transform of $n\mathrm{u}[n]$ is $\frac{z}{(z-1)^2}$.

■ **Solution**
Start with

$$\mathrm{u}[n] \overset{\mathcal{Z}}{\longleftrightarrow} \frac{z}{z-1}. \tag{11.78}$$

Then, using the $z$-domain differentiation property,

$$-n\mathrm{u}[n] \overset{\mathcal{Z}}{\longleftrightarrow} z\frac{d}{dz}\left(\frac{z}{z-1}\right) = -\frac{z}{(z-1)^2} \tag{11.79}$$

or

$$n\mathrm{u}[n] \overset{\mathcal{Z}}{\longleftrightarrow} \frac{z}{(z-1)^2}. \tag{11.80}$$

## CONVOLUTION IN DISCRETE TIME

We have seen in the Fourier and Laplace transforms that there is an important relationship between convolution in one domain and multiplication in the other domain. A similar relationship exists for the $z$ transform.

$$\mathrm{g}[n] * \mathrm{h}[n] = \sum_{m=-\infty}^{\infty} \mathrm{g}[m]\mathrm{h}[n-m]. \tag{11.81}$$

Taking the $z$ transform of both sides,

$$\mathcal{Z}(\mathrm{g}[n] * \mathrm{h}[n]) = \sum_{n=0}^{\infty} \sum_{m=-\infty}^{\infty} \mathrm{g}[m]\mathrm{h}[n-m]z^{-n}$$

$$\mathcal{Z}(\mathrm{g}[n] * \mathrm{h}[n]) = \sum_{m=-\infty}^{\infty} \mathrm{g}[m] \underbrace{\sum_{n=0}^{\infty} \mathrm{h}[n-m]z^{-n}}_{z^{-m}\mathrm{H}(z)} = \sum_{m=-\infty}^{\infty} \mathrm{g}[m]z^{-m}\mathrm{H}(z) \tag{11.82}$$

$$\mathcal{Z}(\mathrm{g}[n] * \mathrm{h}[n]) = \mathrm{H}(z)\sum_{m=0}^{\infty} \mathrm{g}[m]z^{-m} = \mathrm{H}(z)\mathrm{G}(z)$$

$$\boxed{\mathrm{g}[n] * \mathrm{h}[n] \overset{\mathcal{Z}}{\longleftrightarrow} \mathrm{H}(z)\mathrm{G}(z)}. \tag{11.83}$$

In words, convolution of two DT functions in the DT domain corresponds to multiplication of their $z$ transforms in the $z$ domain, exactly as was true for the Fourier and Laplace transforms. Consideration of the effect of multiplication of two DT-domain functions is beyond the scope of this text.

## DIFFERENCING

Differencing is the DT operation that is analogous to CT differentiation. The first backward difference of $\mathrm{g}[n]$ is

$$\Delta(\mathrm{g}[n-1]) = \mathrm{g}[n] - \mathrm{g}[n-1]. \tag{11.84}$$

Using the time-shifting property (for causal functions), the $z$ transform of the right side of (11.84) forms the pair,

$$g[n] - g[n-1] \xleftrightarrow{\ z\ } G(z) - z^{-1}G(z) = (1 - z^{-1})G(z). \qquad \textbf{(11.85)}$$

Therefore,

$$\boxed{g[n] - g[n-1] \xleftrightarrow{\ z\ } (1 - z^{-1})G(z)}. \qquad \textbf{(11.86)}$$

## ACCUMULATION

Accumulation is the DT operation which is analogous to CT integration, and the proof of the property can be done in an analogous manner. First realize that accumulation is equivalent to convolution with a unit sequence,

$$u[n] * g[n] = \sum_{m=-\infty}^{\infty} u[m]g[n-m] = \sum_{m=0}^{n} g[m]. \qquad \textbf{(11.87)}$$

The last summation in (11.87) has an upper limit of $n$ because $g[n]$ is assumed causal in (11.37).

$$\mathcal{Z}\left(\sum_{m=0}^{n} g[m]\right) = Z(u[n] * g[n]) = G(z)U(z) = \frac{z}{z-1}G(z) \qquad \textbf{(11.88)}$$

Therefore,

$$\boxed{\sum_{m=0}^{n} g[m] \xleftrightarrow{\ z\ } \frac{z}{z-1}G(z) = \frac{1}{1-z^{-1}}G(z)}. \qquad \textbf{(11.89)}$$

**EXAMPLE 11.7**

Using the accumulation property show that the $z$ transform of $nu[n]$ is $z/(z-1)^2$.

■ **Solution**
First express $nu[n]$ as an accumulation,

$$nu[n] = \sum_{m=0}^{n} u[m-1]. \qquad \textbf{(11.90)}$$

Then, using the time-shifting property, find the $z$ transform of $u[n-1]$,

$$u[n-1] \xleftrightarrow{\ z\ } z^{-1}\frac{z}{z-1} = \frac{1}{z-1}. \qquad \textbf{(11.91)}$$

Then applying the accumulation property,

$$nu[n] = \sum_{m=0}^{n} u[m-1] \xleftrightarrow{\ z\ } \left(\frac{z}{z-1}\right)\frac{1}{z-1} = \frac{z}{(z-1)^2}. \qquad \textbf{(11.92)}$$

## FINAL VALUE THEOREM

Begin the derivation of the final value theorem by considering the $z$ transform of the difference between a DT function and a shifted version of the same function (a first forward difference),

$$\mathcal{Z}(g[n+1] - g[n]) = \lim_{n\to\infty} \sum_{m=0}^{n} (g[m+1] - g[m])z^{-m}. \qquad \textbf{(11.93)}$$

Taking the $z$ transform and using the time-shifting property

$$z(G(z) - g[0]) - G(z) = \lim_{n\to\infty} \sum_{m=0}^{n} (g[m+1] - g[m])z^{-m}. \qquad \textbf{(11.94)}$$

Now take the limit as $z \to 1$ on both sides,

$$\lim_{z\to1}\{(z-1)G(z) - zg[0]\} = \lim_{z\to1}\left\{ \lim_{n\to\infty} \sum_{m=0}^{n} (g[m+1] - g[m])z^{-m} \right\}. \qquad \textbf{(11.95)}$$

Taking the $z$ limit first on the right side,

$$\lim_{z\to1}\{(z-1)G(z)\} - g[0] = \lim_{n\to\infty} \sum_{m=0}^{n} (g[m+1] - g[m])$$

$$= \lim_{n\to\infty} (g[1] - g[0] + g[2] - g[1] + \cdots + g[n+1] - g[n])$$

$$\textbf{(11.96)}$$

$$= \lim_{n\to\infty} (g[n+1] - g[0])$$

$$= \lim_{n\to\infty} g[n] - g[0]$$

*if* $\lim_{n\to\infty} g[n]$ exists. Rearranging and simplifying,

$$\boxed{\lim_{n\to\infty} g[n] = \lim_{z\to1}(z-1)G(z)} \qquad \textbf{(11.97)}$$

*only if* $\lim_{n\to\infty} g[n]$ exists. Corresponding to what we saw in the final value theorem of the Laplace transform, $\lim_{z\to1}(z-1)G(z)$ may exist even though $\lim_{n\to\infty} g[n]$ does not. For example, if

$$X(z) = \frac{z\sin(\Omega_0)}{z^2 - 2z\cos(\Omega_0) + 1}, \qquad \textbf{(11.98)}$$

then

$$\lim_{z\to1}(z-1)X(z) = \lim_{z\to1}(z-1)\frac{z\sin(\Omega_0)}{z^2 - 2z\cos(\Omega_0) + 1} = 0. \qquad \textbf{(11.99)}$$

But $x[n] = \sin(\Omega_0 n)$ and $\lim_{n\to\infty} x[n]$ does not exist. Therefore, the conclusion from (11.99) that the final value is zero, is wrong.

In a manner similar to the analogous proof for Laplace transforms, it can be shown that if there are any poles on or outside the unit circle, except for a single pole at $z = 1$, that the final value theorem does not apply.

## SUMMARY OF z-TRANSFORM PROPERTIES

Linearity $\qquad \alpha g[n] + \beta h[n] \overset{\mathcal{Z}}{\longleftrightarrow} \alpha G(z) + \beta H(z)$

Time shifting $\qquad g[n - n_0] \overset{\mathcal{Z}}{\longleftrightarrow} z^{-n_0} G(z) \qquad n_0 \geq 0$

$\qquad g[n + n_0] \overset{\mathcal{Z}}{\longleftrightarrow} z^{n_0} \left( G(z) - \sum_{m=0}^{n_0-1} g[m] z^{-m} \right) \qquad n_0 > 0$

Change of scale $\qquad \alpha^n g[n] \overset{\mathcal{Z}}{\longleftrightarrow} G\left(\dfrac{z}{\alpha}\right)$

Initial value theorem $\qquad g[0] = \lim_{z \to \infty} G(z)$

z-Domain differentiation $\qquad -n g[n] \overset{\mathcal{Z}}{\longleftrightarrow} z \dfrac{d}{dz} G(z)$

Convolution in discrete time $\qquad g[n] * h[n] \overset{\mathcal{Z}}{\longleftrightarrow} H(z) G(z)$

Differencing $\qquad g[n] - g[n - 1] \overset{\mathcal{Z}}{\longleftrightarrow} (1 - z^{-1}) G(z)$

Accumulation $\qquad \sum_{m=0}^{n} g[m] \overset{\mathcal{Z}}{\longleftrightarrow} \dfrac{z}{z - 1} G(z) = \dfrac{1}{1 - z^{-1}} G(z)$

Final value theorem $\qquad \lim_{n \to \infty} g[n] = \lim_{z \to 1} (z - 1) G(z)$

## 11.4 THE INVERSE z TRANSFORM

There is a direct formula for finding the inverse z transform. It is

$$ x[n] = \frac{1}{j2\pi} \oint_C X(z) z^{n-1} \, dz \qquad \textbf{(11.100)} $$

where $C$ is a closed circular contour traversed in the counterclockwise direction in the region of convergence. Since this text assumes that contour integration in the complex plane is beyond the reader's experience, we will not pursue this method of finding inverse z transforms.

There are two other methods of finding inverse z transforms that are the most common in practice, and each has its advantages and disadvantages. The first method is synthetic division of the z-domain expression. For example, let the z-domain expression be

$$ H(z) = \frac{z^2 \left(z - \frac{1}{2}\right)}{\left(z - \frac{2}{3}\right)\left(z - \frac{1}{3}\right)\left(z - \frac{1}{4}\right)} = \frac{z^3 - (z^2/2)}{z^3 - \frac{15}{12}z^2 + \frac{17}{36}z - \frac{1}{18}}. \qquad \textbf{(11.101)} $$

We can divide the denominator into the numerator yielding

$$\begin{array}{r} 1 + \dfrac{3}{4}z^{-1} + \dfrac{67}{144}z^{-2} + \cdots \\[2mm] z^3 - \dfrac{15}{12}z^2 + \dfrac{17}{36}z - \dfrac{1}{18} \overline{\smash{\big)}\, z^3 - \dfrac{z^2}{2} \quad\qquad} \end{array}$$

$$z^3 - \frac{15}{12}z^2 + \frac{17}{36}z - \frac{1}{18}$$

$$\frac{3}{4}z^2 - \frac{17}{36}z + \frac{1}{18}$$

$$\frac{3}{4}z^2 - \frac{45}{48}z + \frac{51}{144} - \frac{3}{72}z^{-1}$$

$$\frac{67}{144}z\cdots$$

$$\vdots$$

(11.102)

Comparing this to the definition of the $z$ transform

$$\mathrm{H}(z) = \sum_{n=0}^{\infty} \mathrm{h}[n]z^{-n} = \mathrm{h}[0] + \mathrm{h}[1]z^{-1} + \mathrm{h}[2]z^{-2} + \mathrm{h}[3]z^{-3} + \cdots, \qquad \textbf{(11.103)}$$

we see that

$$\mathrm{H}(z) = 1 + \frac{3}{4}z^{-1} + \frac{67}{144}z^{-2} + \cdots \qquad \textbf{(11.104)}$$

and, therefore,

$$\mathrm{h}[0] = 1, \quad \mathrm{h}[1] = \frac{3}{4}, \quad \mathrm{h}[2] = \frac{67}{144}, \ldots . \qquad \textbf{(11.105)}$$

So this technique yields the values of the DT function directly as a sequence. The advantage of this technique is that it is straightforward and will always yield the inverse $z$ transform of a ratio of polynomials in $z$. The disadvantage is that the inverse transform is not in closed form.

The second commonly used method of finding inverse $z$ transforms is very similar to the method for finding inverse Laplace transforms, by a partial-fraction expansion of the $z$-domain expression and identifying transform pairs using a table of transforms and the transform properties. Let the $z$-domain expression be the same as in the synthetic-division example,

$$\mathrm{H}(z) = \frac{z^2\left(z - \frac{1}{2}\right)}{\left(z - \frac{2}{3}\right)\left(z - \frac{1}{3}\right)\left(z - \frac{1}{4}\right)}. \qquad \textbf{(11.106)}$$

This is an improper fraction in $z$, and, therefore, we cannot directly expand it in partial fractions. However, if we view it as

$$\mathrm{H}(z) = z\mathrm{H}_1(z), \qquad \textbf{(11.107)}$$

we see that $\mathrm{H}_1(z)$ is a proper fraction in $z$ and we can expand it in partial fractions as

$$\mathrm{H}_1(z) = \frac{z\left(z - \frac{1}{2}\right)}{\left(z - \frac{2}{3}\right)\left(z - \frac{1}{3}\right)\left(z - \frac{1}{4}\right)} = \frac{\frac{4}{5}}{z - \frac{2}{3}} + \frac{2}{z - \frac{1}{3}} - \frac{\frac{9}{5}}{z - \frac{1}{4}}. \qquad \textbf{(11.108)}$$

Then, multiplying through by $z$,

$$H(z) = \frac{\frac{4}{5}z}{z - \frac{2}{3}} + \frac{2z}{z - \frac{1}{3}} - \frac{\frac{9}{5}z}{z - \frac{1}{4}}, \tag{11.109}$$

and, using

$$\alpha^n u[n] \overset{z}{\longleftrightarrow} \frac{z}{z - \alpha} = \frac{1}{1 - \alpha z^{-1}}, \tag{11.110}$$

we get

$$h[n] = \left[ \frac{4}{5}\left(\frac{2}{3}\right)^n + 2\left(\frac{1}{3}\right)^n - \frac{9}{5}\left(\frac{1}{4}\right)^n \right] u[n]. \tag{11.111}$$

Since this is the same function for which we found the inverse $z$ transform by the synthetic-division technique, this result must be equivalent to the previous one. Evaluating $h[n]$ for $n = 0, 1, 2, 3, \ldots$, we get

$$h[0] = 1, \quad h[1] = \frac{3}{4}, \quad h[2] = \frac{67}{144}, \cdots \tag{11.112}$$

which agrees with the previous result.

Complex pole-pairs and repeated poles are handled for the $z$ transform exactly as they are for the Laplace transform because the partial-fraction expansion methods are algebraically the same.

## EXAMPLE 11.8

Find the inverse $z$ transform of

$$X(z) = \frac{1}{(z^2 - 2z + 1)\left(z^2 - z + \frac{1}{2}\right)}. \tag{11.113}$$

■ **Solution**

The denominator can be factored, producing

$$X(z) = \frac{1}{(z - 1)^2\left(z - \frac{1}{2} - (j/2)\right)\left(z - \frac{1}{2} + (j/2)\right)}. \tag{11.114}$$

Since this fraction is proper in $z$, it can be expressed in partial fractions as

$$X(z) = \frac{2}{(z - 1)^2} - \frac{4}{z - 1} + \frac{2}{z - \frac{1}{2} - (j/2)} + \frac{2}{z - \frac{1}{2} + (j/2)} \tag{11.115}$$

or, to help in finding the inverse transforms using tables,

$$X(z) = z^{-1}\left( \frac{2z}{(z - 1)^2} - \frac{4z}{z - 1} + \frac{2z}{z - \frac{1}{2} - (j/2)} + \frac{2z}{z - \frac{1}{2} + (j/2)} \right). \tag{11.116}$$

We can now find the inverse $z$ transform directly in terms of complex exponentials or combine the last two terms into one term to yield a real function. Taking the first path to solution, the inverse $z$ transform is

$$x[n] = \left[ 2(n - 1) - 4 + 2\left(\frac{1}{2} + \frac{j}{2}\right)^{n-1} + 2\left(\frac{1}{2} - \frac{j}{2}\right)^{n-1} \right] u[n - 1] \tag{11.117}$$

or, combining complex exponentials,

$$x[n] = \left[ 2(n-1) - 4 + \frac{(1+j)^{n-1} + (1-j)^{n-1}}{2^{n-2}} \right] u[n-1]. \qquad \textbf{(11.118)}$$

Then using $1 \pm j = \sqrt{2} e^{\pm(j\pi/4)}$,

$$x[n] = 2\left[ n - 3 + \frac{2}{\left(\sqrt{2}\right)^{n-1}} \cos\left(\frac{\pi(n-1)}{4}\right) \right] u[n-1]. \qquad \textbf{(11.119)}$$

Taking the alternate route we can combine the two complex terms in (11.116),

$$X(z) = z^{-1} \left( \frac{2z}{(z-1)^2} - \frac{4z}{z-1} + \frac{2z(2z-1)}{z^2 - z + \frac{1}{2}} \right). \qquad \textbf{(11.120)}$$

We can rewrite the last term in a form that allows us to find the inverse transform directly from a table,

$$X(z) = z^{-1} \left( \frac{2z}{(z-1)^2} - \frac{4z}{z-1} + 4\frac{z^2 - \frac{1}{2}z}{z^2 - z + \frac{1}{2}} \right). \qquad \textbf{(11.121)}$$

Then the inverse $z$ transform is

$$x[n] = 2\left[ n - 3 + \frac{2}{\left(\sqrt{2}\right)^{n-1}} \cos\left(\frac{\pi(n-1)}{4}\right) \right] u[n-1] \qquad \textbf{(11.122)}$$

as before.  ∎

# 11.5 SOLUTION OF DIFFERENCE EQUATIONS WITH INITIAL CONDITIONS

The $z$ transform bears a relationship to *difference* equations analogous to the relationship of the Laplace transform to *differential* equations. A linear differential equation with initial conditions can be converted by the Laplace transform into an algebraic equation. The solution is then found in the Laplace domain and is inverse Laplace-transformed to find the CT-domain solution. A linear difference equation with initial conditions can be converted by the $z$ transform into an algebraic equation. Then it is solved, and the solution in the DT domain is found by an inverse $z$ transform.

**EXAMPLE 11.9**

Solve the difference equation

$$y[n+2] - \frac{3}{2}y[n+1] + \frac{1}{2}y[n] = \left(\frac{1}{4}\right)^n \qquad \text{for } n \geq 0 \qquad \textbf{(11.123)}$$

with initial conditions

$$y[0] = 10 \qquad \text{and} \qquad y[1] = 4. \qquad \textbf{(11.124)}$$

Initial conditions for a second-order differential equation usually consist of a specification of the initial value of the function and its first derivative. Initial conditions for a second-order difference equation usually consist of the specification of the initial two values of the function (in this case, y[0] and y[1]). To see the analogy, imagine that when the time between samples $T_s$ becomes arbitrarily small that the initial value of the derivative could be computed from the difference between the initial two values of the function (and the time between samples). In both cases the initial conditions account for everything that has happened up until time $t = 0$ or $n = 0$, the time at which the solution begins.

### ■ Solution

Taking the $z$ transform of both sides of the difference equation (using the time translation property of the $z$ transform),

$$z^2[Y(z) - y[0] - z^{-1}y[1]] - \frac{3}{2}z[Y(z) - y[0]] + \frac{1}{2}Y(z) = \frac{z}{z - \frac{1}{4}}. \qquad (11.125)$$

Solving for $Y(z)$,

$$Y(z) = \frac{z/\left(z - \frac{1}{4}\right) + z^2y[0] + zy[1] - \frac{3}{2}zy[0]}{z^2 - \frac{3}{2}z + \frac{1}{2}}$$

$$Y(z) = z\frac{z^2y[0] - z((7y[0]/4) - y[1]) - (y[1]/4) + (3y[0]/8) + 1}{\left(z - \frac{1}{4}\right)\left(z^2 - \frac{3}{2}z + \frac{1}{2}\right)} \qquad (11.126)$$

Substituting in the numerical values of the initial conditions,

$$Y(z) = z\frac{10z^2 - \frac{27}{2}z + \frac{15}{4}}{\left(z - \frac{1}{4}\right)\left(z - \frac{1}{2}\right)(z - 1)} \qquad (11.127)$$

The coefficient of the first $z$,

$$Y_1(z) = \frac{10z^2 - \frac{27}{2}z + \frac{15}{4}}{\left(z - \frac{1}{4}\right)\left(z - \frac{1}{2}\right)(z - 1)}, \qquad (11.128)$$

is a proper fraction in $z$ and can, therefore, be expanded in partial fractions,

$$Y(z) = zY_1(z) = z\left(\frac{\frac{16}{3}}{z - \frac{1}{4}} + \frac{4}{z - \frac{1}{2}} + \frac{\frac{2}{3}}{z - 1}\right). \qquad (11.129)$$

Then using

$$\alpha^n u[n] \overset{\mathcal{Z}}{\longleftrightarrow} \frac{z}{z - \alpha} \qquad (11.130)$$

and taking the inverse $z$ transform,

$$y[n] = \left[\frac{16}{3}\left(\frac{1}{4}\right)^n + 4\left(\frac{1}{2}\right)^n + \frac{2}{3}\right]u[n]. \qquad (11.131)$$

Evaluating this expression for $n = 0$ and 1 yields

$$y[0] = \frac{16}{3}\left(\frac{1}{4}\right)^0 + 4\left(\frac{1}{2}\right)^0 + \frac{2}{3} = 10 \qquad (11.132)$$

$$y[1] = \frac{16}{3}\left(\frac{1}{4}\right)^1 + 4\left(\frac{1}{2}\right)^1 + \frac{2}{3} = \frac{16}{12} + 2 + \frac{2}{3} = 4 \qquad (11.133)$$

which agree with the initial conditions. Substituting the solution into the difference equation,

$$\left\{\frac{16}{3}\left(\frac{1}{4}\right)^{n+2} + 4\left(\frac{1}{2}\right)^{n+2} + \frac{2}{3} - \frac{3}{2}\left[\frac{16}{3}\left(\frac{1}{4}\right)^{n+1} + 4\left(\frac{1}{2}\right)^{n+1} + \frac{2}{3}\right]\right.$$

$$\left. + \frac{1}{2}\left[\frac{16}{3}\left(\frac{1}{4}\right)^n + 4\left(\frac{1}{2}\right)^n + \frac{2}{3}\right]\right\} = \left(\frac{1}{4}\right)^n \qquad \text{for } n \geq 0 \qquad (11.134)$$

or

$$\frac{1}{3}\left(\frac{1}{4}\right)^n + \left(\frac{1}{2}\right)^n + \frac{2}{3} - 2\left(\frac{1}{4}\right)^n - 3\left(\frac{1}{2}\right)^n - 1 + \frac{8}{3}\left(\frac{1}{4}\right)^n$$

$$+ 2\left(\frac{1}{2}\right)^n + \frac{1}{3} = \left(\frac{1}{4}\right)^n \qquad \text{for } n \geq 0 \tag{11.135}$$

or

$$\left(\frac{1}{4}\right)^n = \left(\frac{1}{4}\right)^n \qquad \text{for } n \geq 0, \tag{11.136}$$

which proves that the solution does indeed solve the difference equation. ■

## 11.6 THE RELATIONSHIP BETWEEN THE $z$ AND LAPLACE TRANSFORMS

We explored in Chapters 5 and 7 important relationships between Fourier transform methods. In particular we showed that there is an information equivalence between a DT signal formed by sampling a CT signal,

$$x[n] = x(nT_s), \tag{11.137}$$

and a CT impulse signal formed by impulse sampling the same CT signal,

$$x_\delta(t) = x(t) f_s \, \text{comb}(f_s t), \tag{11.138}$$

where $f_s = 1/T_s$. We also derived the relationships between the DTFT of $x[n]$ and the CTFT of $x_\delta(t)$. Since the $z$ transform applies to a DT signal and is a generalization of the DTFT and a Laplace transform applies to a CT signal and is a generalization of the CTFT, we should expect a close relationship between them also.

Consider two systems, a DT system with impulse response $h[n]$ and a CT system with impulse response $h_\delta(t)$, and let them be related by

$$h_\delta(t) = \sum_{n=-\infty}^{\infty} h[n]\delta(t - nT_s). \tag{11.139}$$

This equivalence indicates that everything that happens to $x[n]$ in the DT system happens in a directly corresponding way to $x_\delta(t)$ in the CT system (Figure 11.13). Therefore,

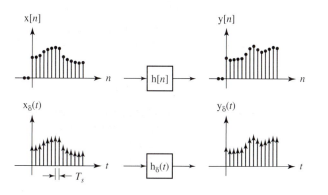

**Figure 11.13**
Equivalence of DT and CT systems.

it is possible to analyze DT systems using the Laplace transform with the strengths of CT impulses representing the values of the DT signals at equally spaced points in time. But it is notationally more convenient to use the *z* transform instead.

The transfer function of the DT system is

$$H(z) = \sum_{n=0}^{\infty} h[n]z^{-n},$$ 
(11.140)

and the transfer function of the CT system is

$$H_\delta(s) = \sum_{n=0}^{\infty} h[n]e^{-nT_s s}.$$ 
(11.141)

If the impulse responses are equivalent in the sense of (11.139), then the transfer functions must also be equivalent. The equivalence is seen in the relationship,

$$H_\delta(s) = H(z)|_{z \to e^{sT_s}}.$$ 
(11.142)

It is important at this point to consider some of the implications of the transformation $z \to e^{sT_s}$. One good way of seeing the relationship between the *s* and *z* complex planes is to *map* a contour or region in the *s* plane into a corresponding contour or region in the *z* plane. Consider first a very simple contour in the *s* plane, $s = j\omega = j2\pi f$, with $\omega$ and *f* representing real radian and cyclic frequency, respectively. This contour is the imaginary axis of the *s* plane. The corresponding contour in the *z* plane is $e^{j\omega T_s}$ or $e^{j2\pi f T_s}$, and, for any real value of $\omega$ and *f*, must lie on the unit circle. However, the mapping is not as simple as the last statement makes it sound. To illustrate the complication, map the segment of the imaginary axis in the *s* plane

$$-\frac{\pi}{T_s} < \omega < \frac{\pi}{T_s} \qquad \text{or} \qquad -\frac{1}{2T_s} < f < \frac{1}{2T_s}$$ 
(11.143)

into the corresponding contour in the *z* plane. As $\omega$ traverses the contour $-(\pi/T_s) \to \omega \to \pi/T_s$, *z* traverses the unit circle from $e^{-j\pi}$ to $e^{+j\pi}$ in the counterclockwise direction, making one complete traversal of the unit circle. Now if we let $\omega$ traverse the contour $\pi/T_s \to \omega \to 3\pi/T_s$, *z* traverses the unit circle from $e^{j\pi}$ to $e^{+j3\pi}$, which is exactly the same contour again because

$$e^{-j\pi} = e^{j\pi} = e^{j3\pi} = e^{j(2n+1)\pi} \qquad \text{where } n \text{ is any integer.}$$ 
(11.144)

Therefore, it is apparent that the transformation $z \to e^{sT_s}$ maps the imaginary axis of the *s* plane into the unit circle of the *z* plane *infinitely many times* (Figure 11.14).

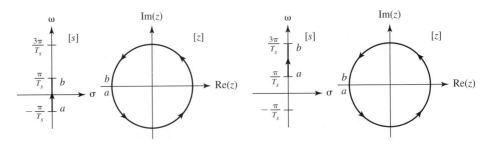

**Figure 11.14**
Mapping the $\omega$ axis of the *s* plane into the unit circle of the *z* plane.

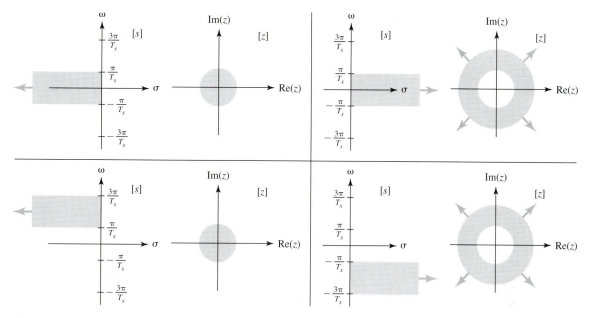

**Figure 11.15**
Mapping of the regions of the $s$ plane into regions in the $z$ plane.

This is another way of looking at the phenomenon of aliasing. All those segments of the imaginary axis of the $s$ plane of length $2\pi/T_s$ look exactly the same when translated into the $z$ plane because of the effects of sampling. So, for every point on the imaginary axis of the $s$ plane there is a corresponding unique point on the unit circle in the $z$ plane. But this unique correspondence does not work the other way. For every point on the unit circle in the $z$ plane there are infinitely many corresponding points on the imaginary axis of the $s$ plane.

Carrying the mapping idea one step farther, the left half-plane of the $s$ plane maps into the interior of the unit circle in the $z$ plane and the right half-plane of the $s$ plane maps into the exterior of the unit circle in the $z$ plane (infinitely many times in both cases). The corresponding ideas about stability and pole locations translate in the same way. A stable CT system has a transfer function with all its poles in the open left-half of the $s$ plane, and a stable DT system has a transfer function with all its poles in the open interior of the unit circle in the $z$ plane (Figure 11.15).

## 11.7 THE BILATERAL $z$ TRANSFORM

The unilateral $z$ transform applies to causal signals only. We began this chapter by defining the bilateral $z$ transform. We will now see how to find the responses of systems with noncausal excitations and/or noncausal impulse responses using the bilateral $z$ transform and a table of unilateral $z$ transforms.

The bilateral $z$ transform of $\mathrm{x}[n]$ is

$$\mathrm{X}(z) = \sum_{n=-\infty}^{\infty} \mathrm{x}[n]z^{-n} = \sum_{n=0}^{\infty} \mathrm{x}[n]z^{-n} + \sum_{n=-\infty}^{-1} \mathrm{x}[n]z^{-n}. \qquad \textbf{(11.145)}$$

By making the transformation $n \to -n$ in the second summation we get

$$X(z) = \sum_{n=0}^{\infty} x[n]z^{-n} + \sum_{n=1}^{\infty} x[-n]z^{n} \qquad \textbf{(11.146)}$$

or

$$X(z) = \sum_{n=0}^{\infty} x[n]z^{-n} - x[0] + \sum_{n=0}^{\infty} x[-n]z^{n}.$$

If we define

$$X_c(z) = \sum_{n=0}^{\infty} x[n]z^{-n} \qquad \text{and} \qquad X_{ac}(z) = \sum_{n=0}^{\infty} x[-n]z^{n}, \qquad \textbf{(11.147)}$$

then

$$X(z) = X_c(z) - x[0] + X_{ac}(z). \qquad \textbf{(11.148)}$$

If (11.147) defines $X_{ac}(z)$, then

$$X_{ac}\left(\frac{1}{z}\right) = \sum_{n=0}^{\infty} x[-n]z^{-n}, \qquad \textbf{(11.149)}$$

which is the unilateral $z$ transform of the discrete-time inverse of $x[n]$. Therefore, the bilateral $z$ transform of $x[n]$ is the sum of the $z$ transform of $x[n]u[n]$ and the $z$ transform of $x[-n]u[n]$, with $z$ replaced by $1/z$, minus $x[0]$.

The procedure for finding a bilateral $z$ transform using unilateral $z$ transforms is

1.  Find the unilateral $z$ transform $X_c(z)$ of the causal signal $x[n]u[n]$, along with its ROC, the open exterior of a circle whose radius is the distance of the farthest pole from the origin of the $z$ plane.

2.  Find the unilateral $z$ transform $X_{ac}(1/z)$ of the causal signal $x[-n]u[n]$, along with its ROC, the open exterior of a circle whose radius is the distance of the farthest pole from the origin of the $z$ plane.

3.  Make the change of variable $z \to 1/z$ in $X_{ac}(1/z)$ and in its ROC, yielding $X_{ac}(z)$, along with its ROC, the open interior of a circle whose radius is the distance of the closest pole to the origin of the $z$ plane.

4.  Add $X_c(z)$ to $X_{ac}(z)$ and subtract $x[0]$ to form $X(z)$. The ROC of $X(z)$ is the region of the $z$ plane which is common to the ROCs of $X_c(z)$ and $X_{ac}(z)$. If such a region does not exist, the bilateral $z$ transform of $x[n]$ does not exist either.

## PROPERTIES

As was true of the bilateral Laplace transform in comparison with the unilateral Laplace transform some of the properties of the bilateral $z$ transform are different than the corresponding properties of the unilateral $z$ transform.

| | |
|---|---|
| Linearity | $\alpha g[n] + \beta h[n] \xleftrightarrow{\mathcal{L}} \alpha G(z) + \beta H(z)$ |
| | $\text{ROC} = R_G \cap R_H$ |
| Time shifting | $g[n - n_0] \xleftrightarrow{\mathcal{L}} G(z)z^{-n_0}$ |
| | $\text{ROC} = R_G$ (except for possible pole–zero cancellation) |
| Change of scale | $\alpha^n g[n] \xleftrightarrow{\mathcal{Z}} G\left(\dfrac{z}{\alpha}\right) \qquad \text{ROC} = |a|R_G$ |
| Complex-frequency differentiation | $-n g[n] \xleftrightarrow{\mathcal{Z}} z\dfrac{d}{dz}G(z) \qquad \text{ROC} = R_G$ |

Convolution $\quad$ $g[n] * h[n] \xleftrightarrow{\mathcal{Z}} H(z)G(z)$

$\mathrm{ROC} = R_G \cap R_H$ at least

Differencing $\quad$ $g[n] - g[n-1] \xleftrightarrow{\mathcal{Z}} (1 - z^{-1})G(z)$

$\mathrm{ROC} = R_G \cap \{|z| > 0\}$ at least

Accumulation $\quad$ $\displaystyle\sum_{m=-\infty}^{n} g[m] \xleftrightarrow{\mathcal{Z}} \frac{z}{z-1}G(z) = \frac{1}{1 - z^{-1}}G(z)$

$\mathrm{ROC} = R_G \cap \{|z| > 1\}$ at least

**EXAMPLE 11.10**

Find the bilateral *z* transform of

$$x[n] = \alpha^{|n|} \tag{11.150}$$

(Figure 11.16).

**■ Solution**

Start with the unilateral *z* transform of $x[n]u[n]$,

$$X_c(z) = \mathcal{Z}(\alpha^{|n|}u[n]) = \mathcal{Z}(\alpha^n u[n]) = \frac{z}{z - \alpha} \qquad |z| > |\alpha|. \tag{11.151}$$

Then find the unilateral *z* transform of $x[-n]u[n]$,

$$X_{ac}\left(\frac{1}{z}\right) = \mathcal{Z}(\alpha^{|-n|}u[n]) = \mathcal{Z}(\alpha^n u[n]) = \frac{z}{z - \alpha} \qquad |z| > |\alpha|. \tag{11.152}$$

Convert (11.152) to the bilateral *z* transform

$$X_{ac}(z) = \frac{1/z}{(1/z) - \alpha} \qquad \left|\frac{1}{z}\right| > |\alpha| \tag{11.153}$$

or

$$X_{ac}(z) = \frac{1}{1 - \alpha z} \qquad |z| < \left|\frac{1}{\alpha}\right|. \tag{11.154}$$

Then combine the two *z* transforms into

$$X(z) = \frac{z}{z - \alpha} - \underbrace{x[0]}_{1} + \frac{1}{1 - \alpha z} \qquad |\alpha| < |z| < \left|\frac{1}{\alpha}\right| \tag{11.155}$$

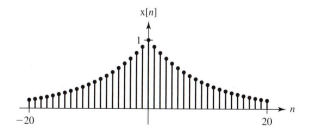

**Figure 11.16**
$x[n] = \alpha^{|n|}$ (with $\alpha = 0.9$).

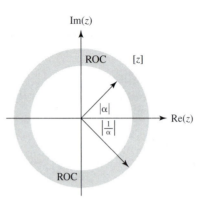

**Figure 11.17**
ROC of $X(z) = \frac{z}{z-\alpha} - \frac{z}{z-(1/\alpha)}$.

or

$$X(z) = \frac{z}{z - \alpha} - \frac{z}{z - (1/\alpha)} \qquad |\alpha| < |z| < \left|\frac{1}{\alpha}\right| \qquad \textbf{(11.156)}$$

(Figure 11.17). ∎

A bilateral $z$ transform which clearly illustrates why the ROC is so important in applying the bilateral $z$ transform is the bilateral $z$ transform of

$$x[n] = -\alpha^n u[-n - 1]. \qquad \textbf{(11.157)}$$

Since this signal is anticausal and $x[0] = 0$, we need only find the unilateral $z$ transform of

$$x[-n]u[n] = -\alpha^{-n}u[n - 1]u[n] = -\left(\frac{1}{\alpha}\right)^n u[n - 1], \qquad \textbf{(11.158)}$$

which is

$$X\left(\frac{1}{z}\right) = -\frac{1}{\alpha}z^{-1}\frac{z}{z - (1/\alpha)} = -\frac{\alpha^{-1}}{z - (1/\alpha)} \qquad |z| > \left|\frac{1}{\alpha}\right|. \qquad \textbf{(11.159)}$$

Then

$$X(z) = -\frac{\alpha^{-1}}{(1/z) - (1/\alpha)} = \frac{z}{z - \alpha} \qquad |z| < |\alpha|. \qquad \textbf{(11.160)}$$

The bilateral $z$ transform (and the unilateral $z$ transform) of the causal signal $\alpha^n u[n]$ is $z/(z - \alpha)$, $|z| > |\alpha|$. It is the same as the bilateral $z$ transform of the anticausal signal $-\alpha^n u[-n - 1]$, but the ROC is different. In fact, in this case, the two ROCs are mutually exclusive. That means that the bilateral $z$ transform of the sum of these two signals does not exist because no common ROC can be found for them. So in applying the bilateral $z$ transform, we must constantly keep the ROC in mind to reach a correct solution.

Although the method previously given for finding the bilateral $z$ transform in terms of unilateral transforms works, it is often easier, and less subject to error, to instead consult a table of bilateral $z$ transforms directly. This is especially true when trying to find *inverse* bilateral $z$ transforms. Appendix G is a table of $z$ transforms.

EXAMPLE 11.11

A DT system with an impulse response

$$h[n] = \left(\frac{3}{4}\right)^{|n|} \qquad (11.161)$$

is excited by

$$x[n] = \left(\frac{1}{2}\right)^{n} u[n] + \left(\frac{2}{3}\right)^{-n} u[-n-1] \qquad (11.162)$$

(Figure 11.18).

Find the system response.

■ **Solution**

From Example 11.10,

$$H(z) = \frac{z}{z - \frac{3}{4}} - \frac{z}{z - \frac{4}{3}} \qquad \frac{3}{4} < |z| < \frac{4}{3}. \qquad (11.163)$$

The bilateral $z$ transform of the excitation is

$$X(z) = \frac{z}{z - \frac{1}{2}} - \frac{z}{z - \frac{3}{2}} \qquad \frac{1}{2} < |z| < \frac{3}{2}. \qquad (11.164)$$

The $z$ transform of the system response is the product of the $z$ transforms of the impulse response and excitation,

$$Y(z) = \left(\frac{z}{z - \frac{3}{4}} - \frac{z}{z - \frac{4}{3}}\right)\left(\frac{z}{z - \frac{1}{2}} - \frac{z}{z - \frac{3}{2}}\right) \qquad \frac{3}{4} < |z| < \frac{4}{3}$$

$$\qquad (11.165)$$

$$Y(z) = \frac{z}{z - \frac{1}{2}}\frac{z}{z - \frac{3}{4}} - \frac{z}{z - \frac{3}{2}}\frac{z}{z - \frac{3}{4}} - \frac{z}{z - \frac{1}{2}}\frac{z}{z - \frac{4}{3}} + \frac{z}{z - \frac{3}{2}}\frac{z}{z - \frac{4}{3}} \qquad \frac{3}{4} < |z| < \frac{4}{3}.$$

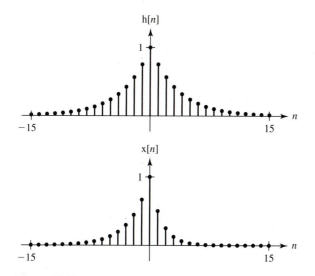

**Figure 11.18**
Impulse response and excitation.

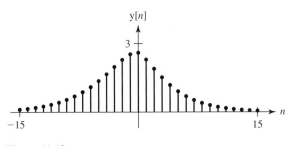

**Figure 11.19**
System response y[*n*].

Expanding in partial fractions and simplifying,

$$Y(z) = \frac{4z}{z - \frac{3}{4}} - \frac{7}{5}\frac{z}{z - \frac{1}{2}} - \frac{48}{5}\frac{z}{z - \frac{4}{3}} + \frac{7z}{z - \frac{3}{2}} \qquad \frac{3}{4} < |z| < \frac{4}{3}. \qquad \textbf{(11.166)}$$

Upon inverse *z* transformation, the terms with poles closer to the origin than the ROC produce causal signals, the terms with poles farther from the origin than the ROC produce anticausal signals, and the DT response is

$$y[n] = \left[4\left(\frac{3}{4}\right)^n - \frac{7}{5}\left(\frac{1}{2}\right)^n\right]u[n] + \left[\frac{48}{5}\left(\frac{4}{3}\right)^n - 7\left(\frac{3}{2}\right)^n\right]u[-n-1] \qquad \textbf{(11.167)}$$

(Figure 11.19). ∎

## 11.8  SUMMARY OF IMPORTANT POINTS

1.  Some signals which do not have a DTFT do have a *z* transform.
2.  Every *z* transform has an associated region of convergence in the *z* plane.
3.  An inverse *z* transform can be found by the direct inversion integral, iteration, or partial-fraction expansion. Use of the direct inversion integral is rare, and iteration does not provide a closed-form result. Therefore, partial-fraction expansion is usually preferred.
4.  The unilateral *z* transform can be used to solve difference equations with initial conditions.
5.  It is possible to perform an analysis of DT systems with the Laplace transform through the use of impulses to simulate discrete time. But the *z* transform is notationally more convenient.
6.  The bilateral *z* transform can be used to analyze noncausal signals and systems, and bilateral *z* transforms can be found using unilateral *z*-transform tables.

## EXERCISES WITH ANSWERS

1.   Using the definition of the *z* transform and/or the transform pairs,

$$\alpha^n u[n] \xleftrightarrow{\;\mathcal{Z}\;} \frac{z}{z - \alpha} = \frac{1}{1 - \alpha z^{-1}} \qquad |z| > |\alpha|$$

and

$$\sin(\Omega_0 n)\, u[n] \xrightarrow{\ z\ } \frac{z\sin(\Omega_0)}{z^2 - 2z\cos(\Omega_0) + 1} = \frac{\sin(\Omega_0)z^{-1}}{1 - 2\cos(\Omega_0)z^{-1} + z^{-2}}\quad |z| > 1,$$

find the $z$ transforms of these DT signals.

a.  $x[n] = u[n]$  b.  $x[n] = e^{-10n}u[n]$

c.  $x[n] = e^n \sin(n)\, u[n]$  d.  $x[n] = \delta[n]$

**Answers:**

1, all $z$;  $\dfrac{z}{z-1}$, $|z| > 1$;  $\dfrac{z}{z - e^{-10}}$, $|z| > e^{-10}$;

$\dfrac{ze\sin(1)}{z^2 - 2ez\cos(1) + e^2}$, $|z| > |0|$

2.  Sketch the region of convergence (if it exists) in the $z$ plane of the bilateral $z$ transform of these DT signals.

a.  $x[n] = u[n] + u[-n]$

b.  $x[n] = u[n] - u[n - 10]$

**Answers:**

Does not exist,  $|z| > 0$

3.  Using the time-shifting property, find the $z$ transforms of these signals.

a.  $x[n] = u[n - 5]$  b.  $x[n] = u[n + 2]$

c.  $x[n] = \left(\dfrac{2}{3}\right)^n u[n + 2]$

**Answers:**

$\dfrac{z^{-4}}{z-1}$,  $|z| > 1$;  $\dfrac{z}{z-1}$,  $|z| > 1$;  $\dfrac{z}{z - \frac{2}{3}}$,  $|z| > \dfrac{2}{3}$

4.  Draw system diagrams for these transfer functions.

a.  $H(z) = \dfrac{z^2}{z + \frac{1}{2}}$  b.  $H(z) = \dfrac{z}{z^2 + z + 1}$

**Answers:**

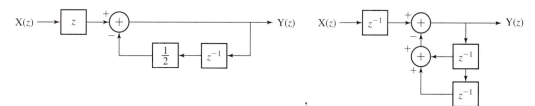

5.  Using the change-of-scale property, find the $z$ transform of

$$x[n] = \sin\left(\frac{2\pi n}{32}\right)\cos\left(\frac{2\pi n}{8}\right)u[n].$$

**Answer:**

$$z\frac{0.1379\,z^2 - 0.3827\,z + 0.1379}{z^4 - 2.7741\,z^3 + 3.8478\,z^2 - 2.7741\,z + 1}$$

6.  Using the z-domain differentiation property find the z transform of

$$x[n] = n\left(\frac{5}{8}\right)^n u[n].$$

**Answer:**

$$\frac{\frac{5}{8}z}{\left(z - \frac{5}{8}\right)^2}$$

7.  Using the convolution property, find the z transforms of these signals.
    a.  $x[n] = (0.9)^n u[n] * u[n]$
    b.  $x[n] = (0.9)^n u[n] * (0.6)^n u[n]$

**Answer:**

$$\frac{z^2}{z^2 - 1.9z + 0.9}, \quad \frac{z^2}{z^2 - 1.5z + 0.54}$$

8.  Using the differencing property and the z transform of the unit sequence, find the z transform of the DT unit impulse and verify your result by checking the z-transform table.

9.  Find the z transform of

$$x[n] = u[n] - u[n - 10]$$

and, using that result and the differencing property, find the z transform of

$$x[n] = \delta[n] - \delta[n - 10].$$

Compare this result with the z transform found directly by applying the time-shifting property to a DT impulse.

10.  Using the accumulation property, find the z transforms of these signals.
    a.  $x[n] = \text{ramp}[n]$
    b.  $x[n] = \sum_{m=0}^{n}(u[m] - u[m - 5])$

**Answers:**

$$\frac{z}{(z - 1)^2}, \quad \frac{z^2(1 - z^{-5})}{(z - 1)^2}$$

11.  Using the final-value theorem, find the final value of functions that are the inverse z transforms of these functions (if the theorem applies).

    a.  $X(z) = \dfrac{z}{z - 1}$

    b.  $X(z) = z\dfrac{2z - \frac{7}{4}}{z^2 - \frac{7}{4}z + \frac{3}{4}}$

**Answers:**

1, 1

**12.** Find the inverse $z$ transforms of these functions in series form by synthetic division.

$$\text{a.} \quad X(z) = \frac{z}{z - \frac{1}{2}} \qquad\qquad \text{b.} \quad X(z) = \frac{z - 1}{z^2 - 2z + 1}$$

**Answers:**

$$\frac{1}{z} + \frac{1}{z^2} + \frac{1}{z^3} + \cdots + \frac{1}{z^k} + \cdots = \frac{1}{z - 1},$$

$$1 + \frac{1}{2z} + \frac{1}{4z^2} + \cdots + \frac{1}{(2z)^k} + \cdots$$

**13.** Find the inverse $z$ transforms of these functions in closed form using partial-fraction expansions, a $z$-transform table, and the properties of the $z$ transform.

$$\text{a.} \quad X(z) = \frac{1}{z\left(z - \frac{1}{2}\right)}$$

$$\text{b.} \quad X(z) = \frac{z^2}{\left(z - \frac{1}{2}\right)\left(z - \frac{3}{4}\right)}$$

$$\text{c.} \quad X(z) = \frac{z^2}{z^2 + 1.8z + 0.82}$$

**Answers:**

$$\left(\frac{1}{2}\right)^{n-2} u[n - 2], \quad (0.9055)^n[\cos(3.031n) - 9.03\sin(3.031n)]u[n],$$

$$\left[3\left(\frac{3}{4}\right)^n - 2\left(\frac{1}{2}\right)^n\right]u[n]$$

**14.** Using the $z$ transform, find the total solutions to these difference equations with initial conditions, for discrete time $n \geq 0$.

$$\text{a.} \quad 2y[n + 1] - y[n] = \sin\left(\frac{2\pi n}{16}\right)u[n], \quad y[0] = 1$$

$$\text{b.} \quad 5y[n + 2] - 3y[n + 1] + y[n] = (0.8)^n u[n], \quad y[0] = -1, y[1] = 10$$

**Answers:**

$$0.2934\left(\frac{1}{2}\right)^{n-1}u[n - 1] + \left(\frac{1}{2}\right)^n u[n] - 0.2934\left[\cos\left(\frac{\pi}{8}(n - 1)\right)\right.$$

$$\left. - 2.812\sin\left(\frac{\pi}{8}(n - 1)\right)\right]u[n - 1],$$

$$y[n] = 0.4444(0.8)^n u[n] - \{\delta[n] - 9.5556(0.4472)^{n-1}[\cos(0.8355(n - 1))$$

$$+ 0.9325\sin(0.8355(n - 1))]u[n - 1]\}$$

**15.** From each block diagram, write the difference equation and find and sketch the response y[n] of the system for discrete time $n \geq 0$, assuming no initial energy storage in the system and impulse excitation $x[n] = \delta[n]$.

*a.*

*b.*

*c.*

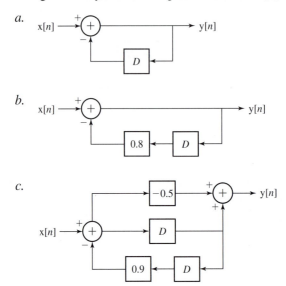

**Answers:**

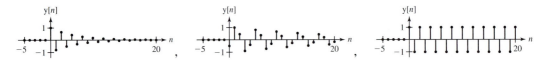

**16.** Sketch regions in the *z* plane corresponding to these regions in the *s* plane.

*a.* $0 < \sigma < \dfrac{1}{T_s}, 0 < \omega < \dfrac{\pi}{T_s}$

*b.* $-\dfrac{1}{T_s} < \sigma < 0, -\dfrac{\pi}{T_s} < \omega < 0$

*c.* $-\infty < \sigma < \infty, 0 < \omega < \dfrac{2\pi}{T_s}$

**Answers:**
The entire *z* plane,

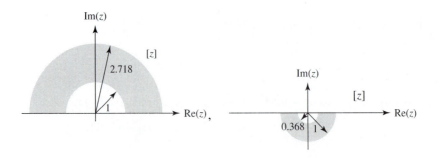

**17.** Find the bilateral $z$ transforms and ROCs of these signals.

    *a.*    $x[n] = u[-n]$              *b.*    $x[n] = \alpha^n u[-n]$

    *c.*    $x[n] = (0.5)^n u[-n] + (0.3)^n u[n]$

    *d.*    $x[n] = (-1.5)^n \cos\left(\dfrac{2\pi n}{8}\right) u[-n]$

**Answers:**

$$\dfrac{1}{1-z}, |z| < 1; \quad \dfrac{1}{1-\frac{z}{\alpha}}, |z| < \dfrac{1}{|\alpha|}; \quad -0.3143\,\dfrac{z - 1.414}{z^2 + 0.9427z + 0.4444}, |z| < \dfrac{2}{3};$$

$$\dfrac{2z^2 - 2z + 0.3}{2z^2 - 1.6z + 0.3}, 0.3 < |z| < \dfrac{1}{2}$$

# EXERCISES WITHOUT ANSWERS

**18.** Using the definition of the $z$ transform verify the $z$ transforms of the following functions:

    *a.*    $x[n] = u[n]$              *b.*    $x[n] = \dfrac{n^2}{2!} u[n]$

    *c.*    $x[n] = n\alpha^n u[n]$          *d.*    $x[n] = \alpha^n \sin(2\pi F_0 n)\, u[n]$

**19.** Sketch the region of convergence (if it exists) in the $z$ plane of the bilateral $z$ transform of these DT signals.

    *a.*    $x[n] = \left(\dfrac{1}{2}\right)^n u[n]$

    *b.*    $x[n] = \left(\dfrac{5}{4}\right)^n u[n] + \left(\dfrac{10}{7}\right)^n u[-n]$

**20.** Using the time-shifting property, find the $z$ transforms of these signals.

    *a.*    $x[n] = \left(\dfrac{2}{3}\right)^{n-1} u[n-1]$

    *b.*    $x[n] = \left(\dfrac{2}{3}\right)^n u[n-1]$

    *c.*    $x[n] = \sin\left(\dfrac{2\pi(n-1)}{4}\right) u[n-1]$

**21.** Draw system diagrams for these transfer functions.

    *a.*    $H(z) = \dfrac{z\left(z + \frac{2}{3}\right)}{z^2 + \frac{2}{3}z + \frac{3}{4}}$

    *b.*    $H(z) = \dfrac{z^2}{(z - 0.75)(z + 0.1)(z - 0.3)}$

**22.** If the $z$ transform of $x[n]$ is $X(z) = \dfrac{1}{z - \frac{3}{4}}$ and

$$Y(z) = j\left[X\left(e^{j(\pi/6)}z\right) - X\left(e^{-j(\pi/6)}z\right)\right],$$

what is $y[n]$?

23. Using the convolution property, find the *z* transforms of these signals.

    *a.*   $x[n] = \sin\left(\dfrac{2\pi n}{8}\right) u[n] * u[n]$

    *b.*   $x[n] = \sin\left(\dfrac{2\pi n}{8}\right) u[n] * (u[n] - u[n-8])$

24. Find the inverse *z* transforms of these functions in closed form using partial-fraction expansions, a *z*-transform table, and the properties of the *z* transform.

    *a.*   $X(z) = \dfrac{z-1}{z^2 + 1.8z + 0.82}$

    *b.*   $X(z) = \dfrac{z-1}{z(z^2 + 1.8z + 0.82)}$

    *c.*   $X(z) = \dfrac{z^2}{z^2 - z + \frac{1}{4}}$

25. Find the response of a system with impulse response

    $$h[n] = \text{rect}_{N_w}[n]$$

    to the excitation

    $$x[n] = \alpha^{|n|}\sin\left(\dfrac{2\pi n}{N_0}\right).$$

    For what range of values of $\alpha$ does the bilateral *z* transform of the excitation exist? What is the relationship between $N_w$ and $N_0$ which minimizes the signal energy of the response?

# *z*-Transform Analysis of Signals and Systems

## 12.1 INTRODUCTION AND GOALS

This chapter is an exploration of the application of *z*-transform methods to the analysis of DT signals and systems. Using the *z* transform we can find the total response of a discrete-time LTI system to any arbitrary excitation. Also, by inspection of the system transfer function, we can determine its stability and how it generally responds to different types of signals. Section 12.13 is the development of systematic state-space techniques for the analysis of multiple-input, multiple-output DT systems. These methods are analogous to state-space methods for CT systems.

### CHAPTER GOALS

1. To illustrate the application of the *z* transform to the design and analysis of DT systems through examples

2. To assess the stability of a DT system directly from its transfer function

3. To see how system responses to standard signals reveal system characteristics

4. To develop systematic methods of analysis for multiple-input, multiple-output DT systems

## 12.2 TRANSFER FUNCTIONS

The real power of the Laplace transform is in the analysis of the dynamic behavior of CT systems. In an analogous manner, the real power of the *z* transform is in the analysis of the dynamic behavior of DT systems. Most CT systems analyzed by engineers are described by differential equations, and most DT systems are described by difference equations. The general form of a difference equation describing a DT system with an excitation x[$n$] and a response y[$n$] is

$$\sum_{k=0}^{D} b_k y[n-k] = \sum_{k=0}^{N} a_k x[n-k]. \qquad \textbf{(12.1)}$$

If both x[$n$] and y[$n$] are causal, and we z-transform both sides, we get

$$\sum_{k=0}^{D} b_k z^{-k} Y(z) = \sum_{k=0}^{N} a_k z^{-k} X(z). \tag{12.2}$$

The transfer function H($z$) is the ratio of the response Y($z$) to the excitation X($z$),

$$H(z) = \frac{Y(z)}{X(z)} = \frac{\sum_{k=0}^{N} a_k z^{-k}}{\sum_{k=0}^{D} b_k z^{-k}} = \frac{a_0 + a_1 z^{-1} + \cdots + a_N z^{-N}}{b_0 + b_1 z^{-1} + \cdots + b_D z^{-D}}. \tag{12.3}$$

So the transfer function of a DT system described by a difference equation is a ratio of polynomials in $z$ just as the transfer function of a CT system described by a differential equation is a ratio of polynomials in $s$.

DT systems are conveniently described by block diagrams just as CT systems are, and transfer functions can be written directly from block diagrams of DT systems. Consider the DT system in Figure 12.1. The describing difference equation is

$$y[n] = 2x[n] - x[n-1] - \frac{1}{2}y[n-1]. \tag{12.4}$$

We can redraw the block diagram to make it a z-domain block diagram instead of a DT-domain block diagram (Figure 12.2). In the $z$ domain the describing equation is

$$Y(z) = 2\,X(z) - z^{-1}X(z) - \frac{1}{2}z^{-1}Y(z) \tag{12.5}$$

and the transfer function is

$$H(z) = \frac{Y(z)}{X(z)} = \frac{2 - z^{-1}}{1 + \frac{1}{2}z^{-1}} = \frac{2z - 1}{z + \frac{1}{2}}. \tag{12.6}$$

We can also find the overall transfer function for the system of Figure 12.2 by block-diagram reduction. The rules for block-diagram reduction of z-domain block diagrams are exactly the same as for s-domain block diagrams. The steps are illustrated in Figure 12.3.

The MATLAB control toolbox functions can, for the most part, also be used for DT systems. The overloaded operators * and + and the common commands tf, zpk, tfdata, zpkdata, step, impulse, feedback, and rlocus all work for DT systems. The command bode also works but with some small differences. DT systems are described in almost exactly the same way as CT systems are. The only real difference

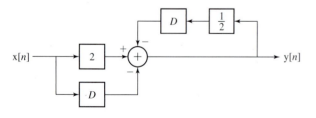

**Figure 12.1**
DT-domain block diagram of a DT system.

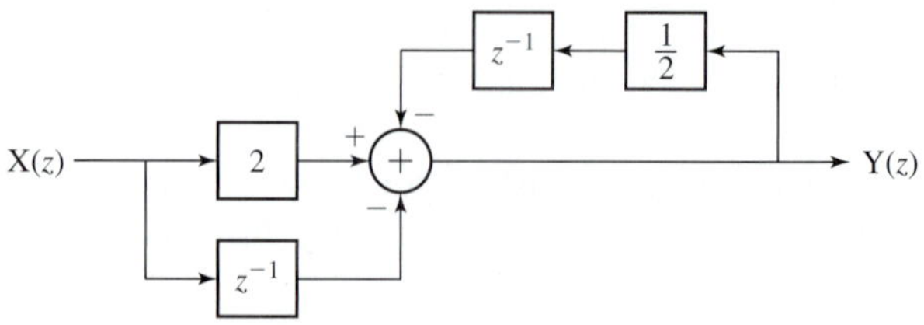

**Figure 12.2**
z-domain block diagram of a DT system.

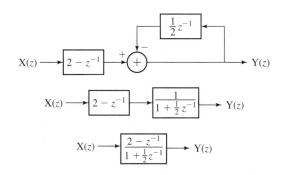

**Figure 12.3**
Block-diagram reduction.

is that in the system description, the time between samples must also be specified. For example, a system whose transfer function is

$$H(z) = \frac{z(z-1)}{z^2 + (z/2) + \frac{1}{2}}, \qquad (12.7)$$

with a sampling rate of 100 Hz, could be described by these MATLAB commands.

```
»fs = 100 ;
»Ts = 1/fs ;
»H = tf([1 -1 0],[1 1/2 1/2],Ts) ;
»H

Transfer function:
     z^2 - z
-----------------------
z^2 + 0.5 z + 0.5

Sampling time: 0.01
```

In fact, the way MATLAB knows that a system description is for a DT system is that $T_s$ is not zero. For CT systems the parameter $T_s$ is zero. For example,

```
»HCT = tf([1 -1 0],[1 1/2 1/2],0) ;
»HCT

Transfer function:
     s^2 - s
-----------------------
s^2 + 0.5 s + 0.5
```

MATLAB assumes with a sampling time of zero that the system is a CT system and returns a description as a function of $s$ instead of $z$.

## 12.3 SYSTEM STABILITY

A causal DT system is stable if its impulse response is absolutely summable, that is, if

$$\left| \sum_{n=0}^{\infty} h[n] \right| < \infty. \qquad (12.8)$$

For a system whose transfer function is a ratio of polynomials in $z$ of the form

$$H(z) = \frac{Y(z)}{X(z)} = \frac{\displaystyle\sum_{k=0}^{N} a_k z^{-k}}{\displaystyle\sum_{k=0}^{D} b_k z^{-k}} = \frac{a_0 + a_1 z^{-1} + \cdots + a_N z^{-N}}{b_0 + b_1 z^{-1} + \cdots + b_D z^{-D}}, \qquad (12.9)$$

with $N \leq D$ and all distinct poles, the transfer function can be written in the partial-fraction form

$$H(z) = \frac{K_1 z}{z - p_1} + \frac{K_2 z}{z - p_2} + \cdots + \frac{K_D z}{z - p_D} \qquad (12.10)$$

and the impulse response is then of the form

$$h[n] = K_1 p_1^n + K_2 p_2^n + \cdots + K_D p_D^n \qquad (12.11)$$

(some of the $p$'s may be complex). For stability, all the poles must satisfy the condition $|p_k| < 1$. The geometrical interpretation of this requirement is that in a DT system all the poles must lie in the open interior of the unit circle in the $z$ plane for system stability. This is directly analogous to the requirement in CT systems that all the poles lie in the open left half of the $s$ plane for system stability. This analysis was done for the most common case in which all the poles are distinct. If there are repeated poles, it can be shown that the requirement that all the poles lie in the open interior of the unit circle for system stability is unchanged.

## 12.4  PARALLEL, CASCADE, AND FEEDBACK CONNECTIONS

The transfer functions of components in the cascade, parallel, and feedback connections of DT systems combine in the same way they do in CT systems (Figures 12.4 through 12.6). We find the overall transfer function of a feedback system by the same technique used for CT systems, and the result is

$$H(z) = \frac{Y(z)}{X(z)} = \frac{H_1(z)}{1 + H_1(z)H_2(z)} = \frac{H_1(z)}{1 + T(z)}, \qquad (12.12)$$

where

$$T(z) = H_1(z)H_2(z) \qquad (12.13)$$

is the loop transfer function.

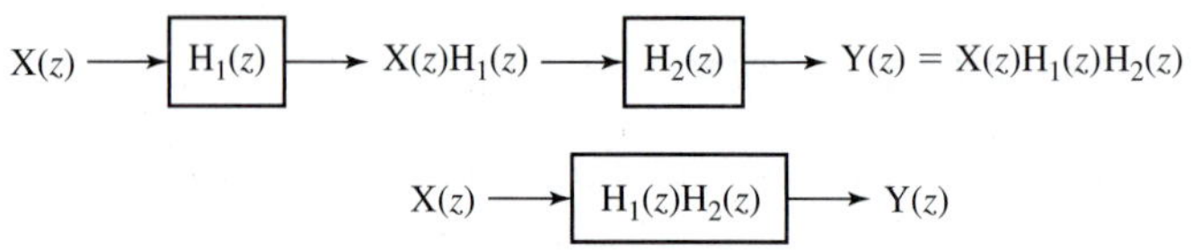

**Figure 12.4**
Cascade connection of systems.

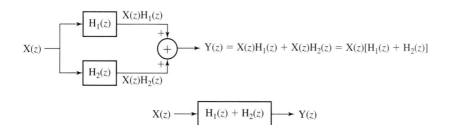

**Figure 12.5**
Parallel connection of systems.

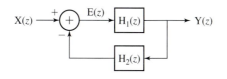

**Figure 12.6**
Feedback connection of systems.

## 12.5 SYSTEM RESPONSES TO STANDARD SIGNALS

The responses of DT systems to the unit sequence and to a suddenly applied DT sinusoid are good indicators of system dynamic performance.

### UNIT SEQUENCE RESPONSE

Let the transfer function of a DT system be

$$H(z) = \frac{N(z)}{D(z)}. \tag{12.14}$$

Then the unit sequence response of the system in the $z$ domain is

$$Y(z) = \frac{z}{z-1}\frac{N(z)}{D(z)}. \tag{12.15}$$

If the order of $N(z)$ is less than or equal to the order of $D(z)$, then the unit sequence response can be written in the partial-fraction form

$$Y(z) = z\left[\frac{N_1(z)}{D(z)} + \frac{H(1)}{z-1}\right] = z\frac{N_1(z)}{D(z)} + H(1)\frac{z}{z-1}. \tag{12.16}$$

If the system is stable and causal, the inverse $z$ transform of the term $z(N_1(z)/D(z))$ is a DT signal which decays with time (the transient response) and the inverse $z$ transform of the term $H(1)z/(z-1)$ is a unit sequence multiplied by the value of the transfer function at $z = 1$ (the steady-state response).

**EXAMPLE 12.1**

A DT system has a transfer function,

$$H(z) = \frac{100z}{z-\frac{1}{2}}. \tag{12.17}$$

Find and plot the unit sequence response.

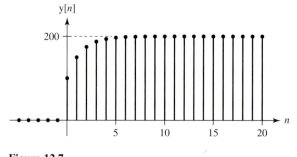

**Figure 12.7**
Unit sequence response.

■ **Solution**

In the $z$ domain the unit sequence response is

$$Y(z) = \frac{z}{z-1}\frac{100z}{z-\frac{1}{2}} = z\left[\frac{-100}{z-\frac{1}{2}} + \frac{200}{z-1}\right] = 100\left[\frac{2z}{z-1} - \frac{z}{z-\frac{1}{2}}\right]. \qquad \textbf{(12.18)}$$

The DT-domain, unit sequence response is the inverse $z$ transform which is

$$y[n] = 100\left[2 - \left(\frac{1}{2}\right)^n\right]u[n] \qquad \textbf{(12.19)}$$

(Figure 12.7). The final value which the unit sequence response approaches is 200, which is the same as $H(1)$. ■

In signal and system analysis, the two most commonly encountered systems are the one-pole and two-pole systems. The typical transfer function of a one-pole system is of the form

$$H(z) = \frac{Kz}{z-p}, \qquad \textbf{(12.20)}$$

where $p$ is the location of a real pole in the $z$ plane. Its $z$-domain response to a unit sequence excitation is

$$Y(z) = \frac{z}{z-1}\frac{Kz}{z-p} = \frac{K}{1-p}\left(\frac{z}{z-1} - \frac{pz}{z-p}\right), \qquad \textbf{(12.21)}$$

and its DT-domain response is

$$y[n] = \frac{K}{1-p}(1 - p^{n+1})u[n]. \qquad \textbf{(12.22)}$$

To simplify this expression and isolate effects, let the gain constant $K$ be $1 - p$. Then

$$y[n] = (1 - p^{n+1})u[n]. \qquad \textbf{(12.23)}$$

The steady-state response is $u[n]$ and the transient response is $-p^{n+1}u[n]$.

This is the DT counterpart of the classic unit step response of a one-pole CT system, and the speed of the response is determined by the pole location. For $0 < p < 1$, the system is stable and the closer $p$ is to 1, the slower the response is (Figure 12.8). For $p > 1$, the system is unstable.

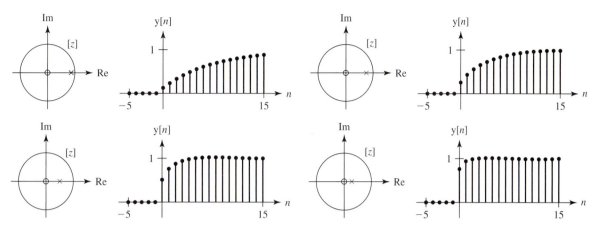

**Figure 12.8**
Response of a one-pole DT system to a unit sequence excitation as the pole location changes.

EXAMPLE 12.2

A DT system has a transfer function of the form

$$H(z) = K\frac{z^2}{z^2 - 2r_0\cos(\Omega_0)z + r_0^2}. \tag{12.24}$$

Find and plot the pole–zero diagrams and the unit sequence responses for

a.  $r_0 = \dfrac{1}{2}, \Omega_0 = \dfrac{\pi}{6}$

b.  $r_0 = \dfrac{1}{2}, \Omega_0 = \dfrac{\pi}{3}$

c.  $r_0 = \dfrac{3}{4}, \Omega_0 = \dfrac{\pi}{6}$

d.  $r_0 = \dfrac{3}{4}, \Omega_0 = \dfrac{\pi}{3}$

■ **Solution**

The poles of $H(z)$ lie at

$$p_{1,2} = r_0 e^{\pm j\Omega_0}. \tag{12.25}$$

If $r_0 < 1$, both poles lie inside the unit circle and the system is stable. The $z$ transform of the unit sequence response is

$$Y(z) = K\frac{z}{z - 1}\frac{z^2}{z^2 - 2r_0\cos(\Omega_0)z + r_0^2}. \tag{12.26}$$

For $\Omega_0 \neq \pm m\pi$, where $m$ is an integer, the partial fraction expansion of $Y(z)/Kz$ is

$$\frac{Y(z)}{Kz} = \frac{1}{1 - 2r_0\cos(\Omega_0) + r_0^2}\left[\frac{1}{z - 1} + \frac{(r_0^2 - 2r_0\cos(\Omega_0))z + r_0^2}{z^2 - 2r_0\cos(\Omega_0)z + r_0^2}\right] \tag{12.27}$$

Then

$$Y(z) = \frac{Kz}{1 - 2r_0\cos(\Omega_0) + r_0^2}\left[\frac{1}{z-1} + \frac{\left(r_0^2 - 2r_0\cos(\Omega_0)\right)z + r_0^2}{z^2 - 2r_0\cos(\Omega_0)z + r_0^2}\right] \qquad \textbf{(12.28)}$$

or

$$Y(z) = H(1)\left[\frac{z}{z-1} + z\frac{\left(r_0^2 - 2r_0\cos(\Omega_0)\right)z + r_0^2}{z^2 - 2r_0\cos(\Omega_0)z + r_0^2}\right] \qquad \textbf{(12.29)}$$

which can be written as

$$Y(z) = H(1)\left(\frac{z}{z-1} + r_0\left\{\begin{array}{l} [r_0 - 2\cos(\Omega_0)]\dfrac{z^2 - r_0\cos(\Omega_0)z}{z^2 - 2r_0\cos(\Omega_0)z + r_0^2} \\[4mm] + \dfrac{1 + [r_0 - 2\cos(\Omega_0)]\cos(\Omega_0)}{\sin(\Omega_0)}\dfrac{zr_0\sin(\Omega_0)}{z^2 - 2r_0\cos(\Omega_0)z + r_0^2}\end{array}\right\}\right) \qquad \textbf{(12.30)}$$

The inverse $z$ transform is

$$y[n] = H(1)\left(1 + r_0\left\{[r_0 - 2\cos(\Omega_0)]r_0^n\cos(n\Omega_0)\right.\right.$$
$$\left.\left. + \frac{1 + [r_0 - 2\cos(\Omega_0)]\cos(\Omega_0)}{\sin(\Omega_0)}r_0^n\sin(n\Omega_0)\right\}\right)u[n] \qquad \textbf{(12.31)}$$

This is the general solution for the unit-sequence response of a unity-gain, second-order DT system. If we let

$$K = 1 - 2r_0\cos(\Omega_0) + r_0^2 \qquad \textbf{(12.32)}$$

then the system has unity gain ($H(1) = 1$). In Figure 12.9 are the pole–zero diagrams and unit-sequence responses for the values of $r_0$ and $\Omega_0$ given above.

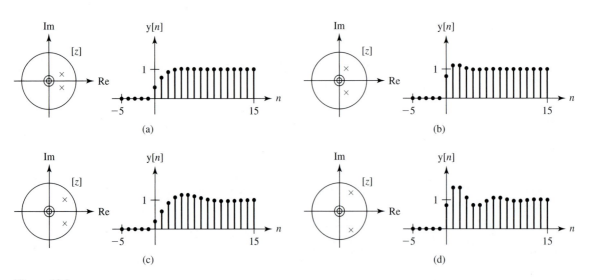

**Figure 12.9**
Pole–zero diagrams and unit sequence responses of a unity-gain, second-order DT system for four combinations of $r_0$ and $\Omega_0$.

As $r_0$ is increased, the response becomes more underdamped, ringing for a longer time. As $\Omega_0$ is increased, the speed of the ringing is increased. So we can generalize by saying that poles near the unit circle cause a more underdamped response than poles farther away from (and inside) the unit circle. We can also say that the rate of ringing of the response depends on the angle of the poles, being greater for a greater angle.

∎

## RESPONSE TO A SUDDENLY-APPLIED SINUSOID

The response of a DT system to a suddenly-applied unit amplitude DT cosine of DT radian frequency $\Omega_0$ is

$$Y(z) = \frac{N(z)}{D(z)} \frac{z[z - \cos(\Omega_0)]}{z^2 - 2z\cos(\Omega_0) + 1}. \tag{12.33}$$

The poles of this response are the poles of the transfer function plus the roots of

$$z^2 - 2z\cos(\Omega_0) + 1 = 0, \tag{12.34}$$

which are the complex-conjugate pair

$$p_1 = e^{j\Omega_0} \quad \text{and} \quad p_2 = e^{-j\Omega_0}. \tag{12.35}$$

Therefore,

$$p_1 = p_2^* \qquad p_1 + p_2 = 2\cos(\Omega_0) \qquad p_1 - p_2 = j2\sin(\Omega_0) \quad \text{and} \quad p_1 p_2 = 1. \tag{12.36}$$

Then if $\Omega_0 \neq m\pi$, where $m$ is an integer, and if there is no pole–zero cancellation, these poles are distinct and the response can be written in partial-fraction form as

$$Y(z) = z\left[\frac{N_1(z)}{D(z)} + \frac{1}{p_1 - p_2}\frac{H(p_1)(p_1 - \cos(\Omega_0))}{z - p_1}\right.$$

$$\left. + \frac{1}{p_2 - p_1}\frac{H(p_2)(p_2 - \cos(\Omega_0))}{z - p_2}\right] \tag{12.37}$$

or, after simplification,

$$Y(z) = z\left[\left\{\frac{N_1(z)}{D(z)} + \left[\frac{H_r(p_1)(z - p_{1r}) - H_i(p_1)p_{1i}}{z^2 - z(2p_{1r}) + 1}\right]\right\}\right], \tag{12.38}$$

where

$$p_1 = p_{1r} + jp_{1i} \quad \text{and} \quad H(p_1) = H_r(p_1) + jH_i(p_1). \tag{12.39}$$

This can be written in terms of the original parameters as

$$Y(z) = \left\{z\frac{N_1(z)}{D(z)} + \left[\text{Re}\left(H(\cos(\Omega_0) + j\sin(\Omega_0))\right)\frac{z^2 - z\cos(\Omega_0)}{z^2 - z(2\cos(\Omega_0)) + 1}\right.\right.$$

$$\left.\left. - \text{Im}(H(\cos(\Omega_0) + j\sin(\Omega_0)))\frac{z\sin(\Omega_0)}{z^2 - z(2\cos(\Omega_0)) + 1}\right]\right\}. \tag{12.40}$$

The inverse $z$ transform is

$$y[n] = \mathcal{Z}^{-1}\left(z\frac{N_1(z)}{D(z)}\right) + [\text{Re}\left(H(\cos(\Omega_0) + j\sin(\Omega_0))\right)\cos(\Omega_0 n)$$

$$- \text{Im}\left(H(\cos(\Omega_0) + j\sin(\Omega_0))\right)\sin(\Omega_0 n)]\,u[n] \tag{12.41}$$

or, using

$$\text{Re}(A)\cos(\Omega_0 n) - \text{Im}(A)\sin(\Omega_0 n) = |A|\cos(\Omega_0 n + \angle A), \qquad \textbf{(12.42)}$$

$$y[n] = \mathcal{Z}^{-1}\left(z\frac{N_1(z)}{D(z)}\right) + |H(\cos(\Omega_0) + j\sin(\Omega_0))|\cos(\Omega_0 n$$

$$+\angle H(\cos(\Omega_0) + j\sin(\Omega_0)))\,u[n] \qquad \textbf{(12.43)}$$

or

$$y[n] = \mathcal{Z}^{-1}\left(z\frac{N_1(z)}{D(z)}\right) + |H(p_1)|\cos(\Omega_0 n + \angle H(p_1))\,u[n]. \qquad \textbf{(12.44)}$$

If the system is stable, the term $\mathcal{Z}^{-1}(z(N_1(z)/D(z)))$ (the transient response) decays to zero with discrete time and the term $|H(p_1)|\cos(\Omega_0 n + \angle H(p_1))\,u[n]$ (the steady-state response) is equal to a sinusoid after discrete time $n = 0$ and persists forever.

---

## EXAMPLE 12.3

The DT system of Example 12.1 has a transfer function

$$H(z) = \frac{100z}{z - \frac{1}{2}}. \qquad \textbf{(12.45)}$$

Find and plot the response to the suddenly applied cosine

$$x[n] = \cos(\Omega_0 n)\,u[n] \qquad \Omega_0 = \frac{\pi}{4}. \qquad \textbf{(12.46)}$$

■ **Solution**

In the $z$ domain the response is of the form

$$Y(z) = \frac{Kz}{z - p}\frac{z\,[z - \cos(\Omega_0)]}{z^2 - 2z\cos(\Omega_0) + 1} = \frac{Kz}{z - p}\frac{z\,[z - \cos(\Omega_0)]}{(z - e^{j\Omega_0})(z - e^{-j\Omega_0})}, \qquad \textbf{(12.47)}$$

where $K = 100$, $p = \frac{1}{2}$, and $\Omega_0 = \pi/4$. This response can be written in the partial-fraction form

$$Y(z) = Kz\left[\frac{p\,[p - \cos(\Omega_0)]/(p - e^{j\Omega_0})(p - e^{-j\Omega_0})}{z - p} + \frac{Az + B}{z^2 - 2z\cos(\Omega_0) + 1}\right]. \qquad \textbf{(12.48)}$$

Solving for the constants,

$$A = \frac{1 - p\cos(\Omega_0)}{p^2 - 2p\cos(\Omega_0) + 1} \qquad \text{and} \qquad B = \frac{p - \cos(\Omega_0)}{p^2 - 2p\cos(\Omega_0) + 1} \qquad \textbf{(12.49)}$$

and

$$Y(z) = \frac{K}{p^2 - 2p\cos(\Omega_0) + 1}\left[p\,[p - \cos(\Omega_0)]\frac{z}{z - p}\right.$$

$$\left. + z\frac{[1 - p\cos(\Omega_0)]z + p - \cos(\Omega_0)}{z^2 - 2z\cos(\Omega_0) + 1}\right]. \qquad \textbf{(12.50)}$$

This can be written in the form

$$Y(z) = \frac{K}{p^2 - 2p\cos(\Omega_0) + 1}\left[p\left[p - \cos(\Omega_0)\right]\frac{z}{z - p}\right.$$

$$\left. + \left[1 - p\cos(\Omega_0)\right]\frac{z[z - \cos(\Omega_0)]}{z^2 - 2z\cos(\Omega_0) + 1} + p\sin(\Omega_0)\frac{z\sin(\Omega_0)}{z^2 - 2z\cos(\Omega_0) + 1}\right]. \quad \textbf{(12.51)}$$

The inverse $z$ transform is

$$y[n] = K\frac{p\left[p - \cos(\Omega_0)\right]p^n + \left[1 - p\cos(\Omega_0)\right]\cos(n\Omega_0) + p\sin(\Omega_0)\sin(n\Omega_0)}{p^2 - 2p\cos(\Omega_0) + 1}u[n]. \quad \textbf{(12.52)}$$

Substituting in numbers,

$$y[n] = 100\frac{-0.1035\left(\frac{1}{2}\right)^n + 0.6464\cos\left((\pi/4)n\right) + 0.3535\sin\left((\pi/4)n\right)}{0.5429}u[n] \quad \textbf{(12.53)}$$

or

$$y[n] = \left[-19.064\left(\frac{1}{2}\right)^n + 119.06\cos((\pi/4)n) + 65.113\sin((\pi/4)n)\right]u[n]. \quad \textbf{(12.54)}$$

Then, using

$$\text{Re}(A)\cos(\Omega_0 n) - \text{Im}(A)\sin(\Omega_0 n) = |A|\cos(\Omega_0 n + \angle A) \quad \textbf{(12.55)}$$

we can write

$$y[n] = \left[-19.064\left(\frac{1}{2}\right)^n + 135.7\cos\left(\frac{\pi}{4}n - 0.5\right)\right]u[n] \quad \textbf{(12.56)}$$

(Figure 12.10).

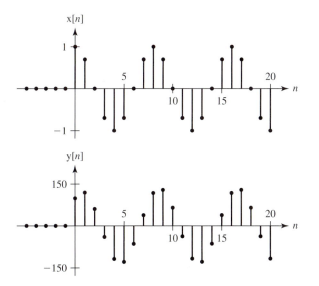

**Figure 12.10**
Suddenly applied DT cosine excitation and DT system response.

For comparison let's find the DT system response to a DT cosine (not suddenly applied) using the DTFT. The transfer function, expressed as a function of DT radian frequency $\Omega$ using the relationship $z = e^{j\Omega}$, is

$$H(j\Omega) = \frac{100e^{j\Omega}}{e^{j\Omega} - \frac{1}{2}}. \qquad (12.57)$$

The DTFT of the excitation $x[n]$ is

$$X(j\Omega) = \frac{1}{2}\left[ \text{comb}\left(\frac{\Omega - \Omega_0}{2\pi}\right) + \text{comb}\left(\frac{\Omega + \Omega_0}{2\pi}\right)\right]. \qquad (12.58)$$

Therefore, the response is

$$Y(j\Omega) = \frac{1}{2}\left[ \text{comb}\left(\frac{\Omega - \Omega_0}{2\pi}\right) + \text{comb}\left(\frac{\Omega + \Omega_0}{2\pi}\right)\right] \frac{100e^{j\Omega}}{e^{j\Omega} - \frac{1}{2}} \qquad (12.59)$$

or

$$Y(j\Omega) = 50\left[ \sum_{k=-\infty}^{\infty} \frac{e^{j\Omega}}{e^{j\Omega} - \frac{1}{2}}\delta\left(\frac{\Omega - \Omega_0}{2\pi} - k\right) + \sum_{k=-\infty}^{\infty} \frac{e^{j\Omega}}{e^{j\Omega} - \frac{1}{2}}\delta\left(\frac{\Omega + \Omega_0}{2\pi} - k\right)\right] \qquad (12.60)$$

$$Y(j\Omega) = 50 \sum_{k=-\infty}^{\infty}\left[ \frac{e^{j(\Omega_0+2\pi k)}}{e^{j(\Omega_0+2\pi k)} - \frac{1}{2}}\delta\left(\frac{\Omega - \Omega_0}{2\pi} - k\right) + \frac{e^{j(-\Omega_0+2\pi k)}}{e^{j(-\Omega_0+2\pi k)} - \frac{1}{2}}\delta\left(\frac{\Omega + \Omega_0}{2\pi} - k\right)\right].$$

Since $e^{j(\Omega_0+2\pi k)} = e^{j\Omega_0}$ and $e^{j(-\Omega_0+2\pi k)} = e^{-j\Omega_0}$ for integer values of $k$,

$$Y(j\Omega) = 50 \sum_{k=-\infty}^{\infty}\left[ \frac{e^{j\Omega_0}\delta((\Omega - \Omega_0)/2\pi - k)}{e^{j\Omega_0} - \frac{1}{2}} + \frac{e^{-j\Omega_0}\delta((\Omega + \Omega_0)/2\pi - k)}{e^{-j\Omega_0} - \frac{1}{2}}\right] \qquad (12.61)$$

or

$$Y(j\Omega) = 50\left[ \frac{e^{j\Omega_0}\text{comb}((\Omega - \Omega_0)/2\pi)}{e^{j\Omega_0} - \frac{1}{2}} + \frac{e^{-j\Omega_0}\text{comb}((\Omega + \Omega_0)/2\pi)}{e^{-j\Omega_0} - \frac{1}{2}}\right]. \qquad (12.62)$$

Finding a common denominator, applying Euler's identity, and simplifying,

$$Y(j\Omega) = \frac{50}{\frac{5}{4} - \cos(\Omega_0)}\left\{\left(1 - \frac{1}{2}\cos(\Omega_0)\right)\left[\text{comb}\left(\frac{\Omega - \Omega_0}{2\pi}\right) + \text{comb}\left(\frac{\Omega + \Omega_0}{2\pi}\right)\right]\right.$$
$$\left. + \frac{j}{2}\sin(\Omega_0)\left[\text{comb}\left(\frac{\Omega + \Omega_0}{2\pi}\right) - \text{comb}\left(\frac{\Omega - \Omega_0}{2\pi}\right)\right]\right\}. \qquad (12.63)$$

Finding the inverse DTFT,

$$y[n] = \frac{50}{\frac{5}{4} - \cos(\Omega_0)}\left\{\left[1 - \frac{1}{2}\cos(\Omega_0)\right]2\cos(\Omega_0 n) + \sin(\Omega_0)\sin(\Omega_0 n)\right\} \qquad (12.64)$$

or, since $\Omega_0 = \pi/4$,

$$y[n] = 119.06\cos\left(\frac{\pi}{4}n\right) + 65.113\sin\left(\frac{\pi}{4}n\right) = 135.7\cos\left(\frac{\pi}{4}n - 0.5\right). \qquad (12.65)$$

This is exactly the same as the steady-state part of the response to the suddenly applied cosine in (12.56).

## 12.6 POLE–ZERO DIAGRAMS AND THE GRAPHICAL CALCULATION OF FREQUENCY RESPONSE

The $z$ transform is a generalization of the DTFT, and the DTFT is a transform which describes DT signals in terms of a continuous function of DT frequency $\Omega$ or $F$. As we have just seen in Section 12.5, the response to a suddenly applied sinusoid is the sum of a transient response and a steady-state response. For stable systems, the transient response decays to zero as discrete time proceeds. The steady-state response persists after the transient response has decayed away. The steady-state response to a suddenly applied sinusoid, found using the $z$ transform, is the same as the response that would be found using the DTFT with a sinusoidal excitation of the same amplitude, phase, and frequency (Figure 12.11).

The generalization of the DTFT to the $z$ transform is based on the transformation $e^{j\Omega} \rightarrow z$. Therefore, to examine the frequency response of DT systems we can specialize the $z$ transform to the DTFT through the opposite transformation $z \rightarrow e^{j\Omega}$, with $\Omega$ being a real variable representing DT radian frequency. The fact that $\Omega$ is real means that in determining frequency response the only values of $z$ that we are now considering are those on the unit circle in the $z$ plane because $|e^{j\Omega}| = 1$ for any real $\Omega$. This is directly analogous to determining the frequency response of a CT system by examining the behavior of its $s$-domain transfer function as $s$ moves along the imaginary axis in the $s$ plane, and a similar graphical technique can be used.

Suppose the transfer function of a DT system is

$$H(z) = \frac{z}{z^2 - (z/2) + \frac{5}{16}} = \frac{z}{(z - p_1)(z - p_2)}, \tag{12.66}$$

where

$$p_1 = \frac{1 + j2}{4} \qquad \text{and} \qquad p_2 = \frac{1 - j2}{4}. \tag{12.67}$$

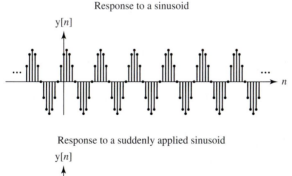

Response to a sinusoid

$y[n]$

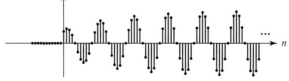

Response to a suddenly applied sinusoid

$y[n]$

**Figure 12.11**
Typical DT system responses to a true sinusoid and a suddenly applied sinusoid.

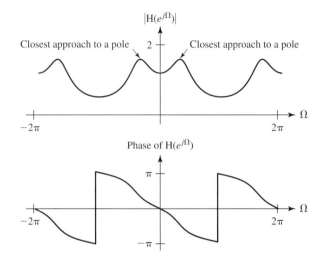

**Figure 12.13**
Magnitude and phase frequency response of the DT system whose transfer function is $H(z) = \frac{z}{z^2-(z/2)+\frac{5}{16}}$.

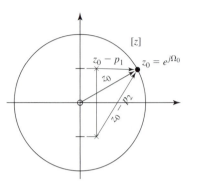

**Figure 12.12**
z-domain pole–zero diagram of a DT system transfer function.

The transfer function has a zero at zero and two complex-conjugate poles (Figure 12.12). The frequency response of the system at any particular DT radian frequency $\Omega_0$ is determined (to within a multiplicative constant) by the vectors from the poles and zeros of the transfer function to the z-plane point $z_0 = e^{j\Omega_0}$. The magnitude of the frequency response is the product of the magnitudes of the zero vectors divided by the product of the magnitudes of the pole vectors. In this case,

$$|H(e^{j\Omega})| = \frac{|e^{j\Omega}|}{|e^{j\Omega} - p_1|\,|e^{j\Omega} - p_2|}. \qquad (12.68)$$

It is apparent that as $e^{j\Omega}$ approaches a pole $p_1$, for example, the magnitude of the difference $e^{j\Omega} - p_1$ becomes small, making the magnitude of the denominator small and making the magnitude of the transfer function large. The opposite effect occurs when $e^{j\Omega}$ approaches a zero.

The phase of the frequency response is the sum of the angles of the zero vectors minus the sum of the angles of the pole vectors. In this case,

$$\angle H(e^{j\Omega}) = \angle e^{j\Omega} - \angle(e^{j\Omega} - p_1) - \angle(e^{j\Omega} - p_2) \qquad (12.69)$$

(Figure 12.13). The maximum magnitude frequency response occurs at approximately $z = e^{j\Omega_0} = e^{\pm j1.11}$, which are the points on the unit circle at the same angle as the poles of the transfer function and, therefore, the points on the unit circle at which the denominator factors $e^{j\Omega_0} - p_1$ and $e^{j\Omega_0} - p_2$ in (12.68) reach their minimum magnitudes.

**EXAMPLE 12.4**

Find and graph the pole–zero plot and frequency response for the system whose transfer function is

$$H(z) = \frac{z^2 - 0.96z + 0.9028}{z^2 - 1.56z + 0.8109}. \qquad (12.70)$$

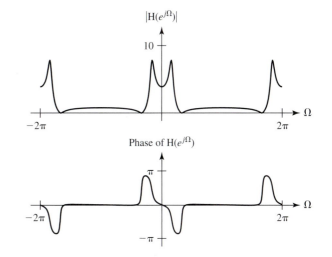

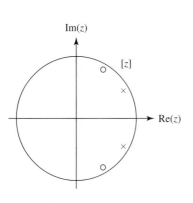

**Figure 12.14**
Pole–zero diagram of the transfer function $H(z) = \frac{z^2 - 0.96z + 0.9028}{z^2 - 1.56z + 0.8109}$.

**Figure 12.15**
Magnitude and phase frequency response of the system whose transfer function is $H(z) = \frac{z^2 - 0.96z + 0.9028}{z^2 - 1.56z + 0.8109}$.

■ **Solution**

The transfer function can be factored into

$$H(z) = \frac{(z - 0.48 + j0.82)(z - 0.48 + j0.82)}{(z - 0.78 + j0.45)(z - 0.78 - j0.45)}. \qquad \textbf{(12.71)}$$

The pole–zero diagram is shown in Figure 12.14. The magnitude and phase frequency responses of the system are illustrated in Figure 12.15. ■

**EXAMPLE 12.5**

Find and graph the pole–zero plot and frequency response for the system whose transfer function is

$$H(z) = \frac{0.0686}{(z^2 - 1.087z + 0.3132)(z^2 - 1.315z + 0.6182)}. \qquad \textbf{(12.72)}$$

■ **Solution**

This transfer function can be factored into

$H(z) =$

$$\frac{0.0686}{(z - 0.5435 + j0.1333)(z - 0.5435 - j0.1333)(z - 0.6575 + j0.4312)(z - 0.6575 - j0.4312)} \qquad \textbf{(12.73)}$$

The pole–zero diagram is illustrated in Figure 12.16. The magnitude and phase frequency responses of the system are illustrated Figure 12.17.

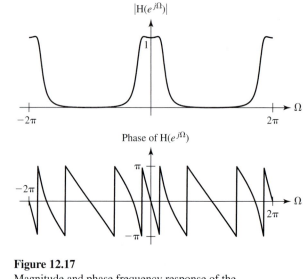

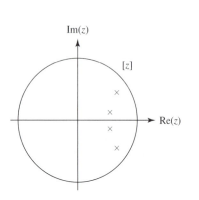

**Figure 12.16**
Pole–zero diagram for the transfer function
$$H(z) = \frac{0.0686}{(z^2-1.087z+0.3132)(z^2-1.315z+0.6182)}.$$

**Figure 12.17**
Magnitude and phase frequency response of the
system whose transfer function is
$$H(z) = \frac{0.0686}{(z^2-1.087z+0.3132)(z^2-1.315z+0.6182)}.$$

## 12.7 DISCRETE-TIME SYSTEMS WITH FEEDBACK

### THE JURY STABILITY TEST

The requirement for DT system stability is that all the poles of the system transfer function lie inside the unit circle of the $z$ plane. For the typical system whose transfer function is a ratio of polynomials in $z$, we can find the pole locations by factoring the denominator. But, as was true for CT systems, factoring the denominator is not necessary. For CT systems we had the Routh–Hurwitz test for system stability. It is possible to use the Routh–Hurwitz test for DT systems by making the change of variable

$$z = \frac{s+1}{s-1}, \tag{12.74}$$

which maps the interior of the unit circle in the $z$ plane into the left half of the $s$ plane. Once the change of variable has been made, the Routh–Hurwitz test proceeds normally. However, this modified form of the Routh–Hurwitz test is not as convenient for DT systems as it was for CT systems.

A more direct stability-checking method is the *Jury stability test*. The process is similar to the Routh–Hurwitz test. If the transfer function is of the form

$$H(z) = \frac{N(z)}{D(z)} \tag{12.75}$$

and the denominator is of the form

$$D(z) = a_D z^D + a_{D-1} z^{D-1} + \cdots + a_1 z + a_0 \qquad a_D \neq 0, \tag{12.76}$$

then a Jury array is formed from the coefficients in $D(z)$ (Figure 12.18). The first row is simply the coefficients of $z$ in the denominator, in ascending-subscript order. The

| 1 | $a_0$ | $a_1$ | $a_2$ | $\cdots$ | $a_{D-2}$ | $a_{D-1}$ | $a_D$ |
|---|---|---|---|---|---|---|---|
| 2 | $a_D$ | $a_{D-1}$ | $a_{D-2}$ | $\cdots$ | $a_2$ | $a_1$ | $a_0$ |
| 3 | $b_0$ | $b_1$ | $b_2$ | $\cdots$ | $b_{D-2}$ | $b_{D-1}$ | |
| 4 | $b_{D-1}$ | $b_{D-2}$ | $b_{D-3}$ | $\cdots$ | $b_1$ | $b_0$ | |
| 5 | $c_0$ | $c_1$ | $c_2$ | $\cdots$ | $c_{D-2}$ | | |
| 6 | $c_{D-2}$ | $c_{D-3}$ | $c_{D-4}$ | $\cdots$ | $c_0$ | | |
| $\vdots$ | $\vdots$ | $\vdots$ | $\vdots$ | $\cdots$ | | | |
| $2D-3$ | $s_0$ | $s_1$ | $s_2$ | | | | |

**Figure 12.18**
The Jury array.

second row is the same set of coefficients but in the reverse order. The third row is computed from the first two by

$$b_0 = \begin{vmatrix} a_0 & a_D \\ a_D & a_0 \end{vmatrix}, \quad b_1 = \begin{vmatrix} a_0 & a_{D-1} \\ a_D & a_1 \end{vmatrix}, \quad b_2 = \begin{vmatrix} a_0 & a_{D-2} \\ a_D & a_2 \end{vmatrix}, \quad \ldots, \quad b_{D-1} = \begin{vmatrix} a_0 & a_1 \\ a_D & a_{D-1} \end{vmatrix}.$$
(12.77)

The fourth row is the set of coefficients of the third row, but in reverse order. The $c$'s are computed from the $b$'s in exactly the same way the $b$'s are computed from the $a$'s. This process continues until only three entries appear. Then the system is stable if

$$D(1) > 0 \qquad \text{and} \qquad (-1)^D D(-1) > 0$$

and

$$a_D > |a_0|$$
$$|b_0| > |b_{D-1}|$$
$$|c_0| > |c_{D-2}|$$
$$\vdots$$
$$|s_0| > |s_2|.$$

**EXAMPLE 12.6**

Using the Jury stability test, find the value of $K$ in the feedback system of Figure 12.19 at which the system becomes unstable if

$$H_1(z) = \frac{z-2}{\left(z + \frac{1}{2}\right)\left(z + \frac{2}{3}\right)} \qquad \text{and} \qquad H_2(z) = \frac{z}{z - \frac{1}{4}}.$$
(12.78)

**■ Solution**

The overall transfer function for this system is

$$H(z) = \frac{K \dfrac{z-2}{\left(z + \frac{1}{2}\right)\left(z + \frac{2}{3}\right)}}{1 + K \dfrac{z-2}{\left(z + \frac{1}{2}\right)\left(z + \frac{2}{3}\right)} \dfrac{z}{z - \frac{1}{4}}} = K \frac{z^2 - \frac{9}{4}z + \frac{1}{2}}{z^3 + \left(K + \frac{11}{12}\right)z^2 + \left(\frac{1}{24} - 2K\right)z - \frac{1}{12}}.$$
(12.79)

and

$$D(z) = z^3 + \left(K + \frac{11}{12}\right)z^2 + \left(\frac{1}{24} - 2K\right)z - \frac{1}{12}.$$
(12.80)

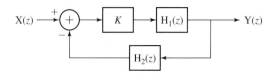

**Figure 12.19**
A DT feedback system.

The Jury array is

| 1 | $-\frac{1}{12}$ | $\frac{1}{24} - 2K$ | $K + \frac{11}{12}$ | 1 |
|---|---|---|---|---|
| 2 | 1 | $K + \frac{11}{12}$ | $\frac{1}{24} - 2K$ | $-\frac{1}{12}$ |
| 3 | $-\frac{143}{144}$ | $-\frac{265}{288} - \frac{5K}{6}$ | $\frac{23K}{12} - \frac{17}{144}$ | |

Therefore,

$$D(1) = 1 + K + \frac{11}{12} + \frac{1}{24} - 2K - \frac{1}{12} = -K + \frac{45}{24} = \frac{15}{8} - K$$

and, thus, $K$ must be less than $\frac{15}{8}$. Then

$$D(-1) = -1 + K + \frac{11}{12} - \frac{1}{24} + 2K - \frac{1}{12} = 3K - \frac{5}{24}.$$

Therefore, $(-1)^3 D(-1) = \frac{5}{24} - 3K$ and $K$ must be less than $\frac{5}{72}$. Checking the remaining criteria,

$$a_D > |a_0| \Rightarrow 1 > \frac{1}{12}$$

$$|s_0| > |s_2| \Rightarrow \frac{143}{144} > \frac{23K}{12} - \frac{17}{144} \Rightarrow K < 0.5797$$

Therefore, considering all criteria, $K$ must be less than $\frac{5}{72}$ for stability. ∎

## THE ROOT-LOCUS METHOD

Just as was true for CT feedback systems, a root locus can be drawn for a DT feedback system of the form illustrated in Figure 12.20. The procedure for drawing the root locus is exactly the same as for CT systems except that the loop transfer function

$$T(z) = K H_1(z) H_2(z) \tag{12.81}$$

is a function of $z$ instead of $s$. However, the interpretation of the root locus, after it is drawn, is a little different. For CT systems, the forward-path gain $K$ at which the root locus crosses into the right half-plane is the value at which the system becomes unstable. For DT systems, the statement is the same except that "right half-plane" is replaced with "exterior of the unit circle."

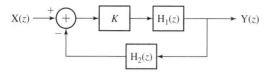

**Figure 12.20**
A typical DT feedback system.

EXAMPLE 12.7

Draw a root locus for the DT system whose forward-path transfer function is

$$H_1(z) = K\frac{z - 1}{z + \frac{1}{2}} \tag{12.82}$$

and whose feedback-path transfer function is

$$H_2(z) = \frac{z - \frac{2}{3}}{z + \frac{1}{3}}.$$

■ **Solution**

The loop transfer function is

$$T(z) = K\frac{z - 1}{z + \frac{1}{2}}\frac{z - \frac{2}{3}}{z + \frac{1}{3}}. \tag{12.83}$$

There are two zeros at $z = \frac{2}{3}$ and $z = 1$ and two poles at $z = -\frac{1}{2}$ and $z = -\frac{1}{3}$ (Figure 12.21). It is apparent from the root locus that this system is unconditionally stable for any finite positive $K$.

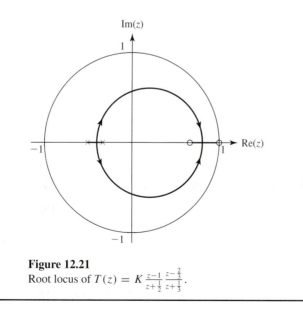

**Figure 12.21**
Root locus of $T(z) = K\frac{z-1}{z+\frac{1}{2}}\frac{z-\frac{2}{3}}{z+\frac{1}{3}}$.

**EXAMPLE 12.8**

Draw the root-locus plot for the system of Example 12.6 to confirm that at some value of $K$ the system does become unstable.

■ **Solution**

Since the root locus does go outside the unit circle, the system goes unstable if the gain constant $K$ is large enough (Figure 12.22).

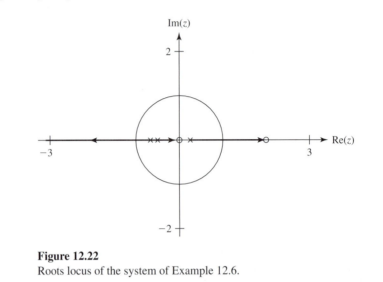

**Figure 12.22**
Roots locus of the system of Example 12.6.

## 12.8 SIMULATING CONTINUOUS-TIME SYSTEMS WITH DISCRETE-TIME SYSTEMS

In Chapter 7 we examined how CT signals are converted to DT signals by sampling. We found that, under certain conditions, the DT signal was a good representation of the CT signal in the sense that it preserved all or practically all of its information. A DT signal formed by properly sampling a CT signal in a sense simulates the CT signal. In Chapter 11 we examined the equivalence between a DT system with impulse response $h[n]$ and a CT system with impulse response

$$h_\delta(t) = \sum_{n=-\infty}^{\infty} h[n]\delta(t - nT_s). \tag{12.84}$$

The system whose impulse response is $h_\delta(t)$ is a very special type of CT system because its impulse response consists only of impulses. As a practical matter, this is impossible to achieve because the transfer function of such a system, being periodic, has a nonzero response at frequencies approaching infinity. No real CT system can have an impulse response that contains actual impulses, although in some cases that might be a good approximation for analysis purposes.

To simulate a CT system with a DT system we must first address the problem of forming a useful equivalence between a DT system, whose impulse response must be discrete, and a CT system, whose impulse response must be continuous. The most obvious and direct equivalence between a DT signal and a CT signal is to have the values

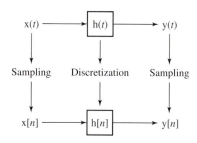

**Figure 12.23**
Signal sampling and system discretization.

of the CT signal at the sampling instants be the same as the values of the DT signal at the corresponding discrete times

$$x[n] = x(nT_s). \qquad \textbf{(12.85)}$$

So if the excitation of a DT system is a sampled version of an excitation of a CT system, we want the response of the DT system to be a sampled version of the response of the CT system (Figure 12.23).

The most natural choice for h[n] would be made in the same way as in (12.85),

$$h[n] = h(nT_s). \qquad \textbf{(12.86)}$$

This choice establishes an equivalence between the impulse responses of the two systems. With this choice of DT impulse response, if a unit CT impulse excites the CT system and a unit DT impulse of the same strength excites the DT system, the DT response y[n] is a sampled version of the CT response y($t$),

$$y[n] = y(nT_s). \qquad \textbf{(12.87)}$$

But it is important to realize that even though the two systems have equivalent impulse responses in the sense of (12.86) and (12.87), that does not mean that the system responses to other excitations will be equivalent in the same sense.

It is important to point out here that if we choose to make $h[n] = h(nT_s)$, and we excite both systems with unit impulses, that the responses are related by $y[n] = y(nT_s)$, but we cannot say that $x[n] = x(nT_s)$ as in Figure 12.23. This figure indicates that the DT excitation is formed by sampling the CT excitation. But for (12.86) and (12.87) to be valid, the CT excitation must be an impulse. Imagine trying to sample a CT impulse. We would have to say that $\delta[n] = \delta(nT_s)$, but this makes no sense because the amplitude of a CT impulse at its time of occurrence is not defined, so we cannot establish the corresponding strength of the DT impulse $\delta[n]$. A DT system design for which $h[n] = h(nT_s)$ is called an *impulse-invariant* design because of the equivalence of the system responses to unit impulses. This is only one type of equivalence between CT and DT systems. We will explore others in the sections to follow.

## 12.9 SAMPLED-DATA SYSTEMS

Because of the great increases in microprocessor speed and memory and large reductions in the cost of microprocessors, modern system design often uses DT subsystems to replace CT subsystems to save money, space, or power consumption and to increase the flexibility or reliability of the system. Aircraft autopilots, industrial chemical process

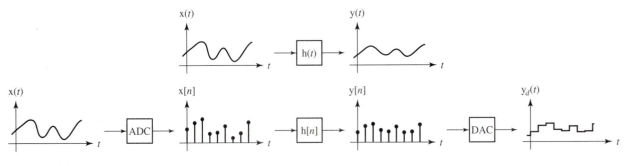

**Figure 12.24**
A common sampled-data simulation of a CT system.

control, manufacturing processes, automobile ignition, and fuel systems are examples. Systems which contain both DT and CT subsystems and mechanisms for converting between DT and CT signals are called *sampled-data* or, sometimes, *hybrid* systems.

The first type of sampled-data system used to replace a CT system, and still the most prevalent type, comes from a natural idea. We convert a CT signal to a DT signal with an analog-to-digital converter (ADC). We process the samples from the ADC in a DT system of some kind. Then we convert the DT response back to CT form using a digital-to-analog converter (DAC) (Figure 12.24). The desired design would have the response of the sampled-data system be very close to the response that would have come from the CT system. To do that we must choose h[$n$] properly and in order to do that, we must also understand the actions of the ADCs and DACs.

It is straightforward to model the action of the ADC. It simply acquires the value of its input signal at the sampling time and responds with a number proportional to that signal value. (It also quantizes the signal, but we will ignore that effect in this analysis.) The DT subsystem with impulse response h[$n$] is then designed to make the sampled-data system emulate the action of the CT system whose impulse response is h($t$). (We will attack the problem of how to determine h[$n$] soon.)

The action of the DAC is a little more complicated to model mathematically than that of the ADC. It is excited by a number from the DT subsystem, a DT impulse, and responds with a CT signal proportional to that number, which stays constant until the number changes to a new value. This can be modeled by thinking of the process as two steps. First let the DT impulse be converted to a CT impulse of the same strength. Then let the CT impulse excite a subsystem called a *zero-order hold* (first introduced in Chapter 9) with an impulse response which is rectangular with height, one, and width $T_s$ and which begins at time $t = 0$,

$$\mathrm{h_{zoh}}(t) = \begin{cases} 0 & t < 0 \\ 1 & 0 < t < T_s \\ 0 & t > T_s \end{cases} = \mathrm{rect}\left(\frac{t - (T_s/2)}{T_s}\right) \tag{12.88}$$

(Figure 12.25). The transfer function of the zero-order hold is the Laplace transform of its impulse response $\mathrm{h_{zoh}}(t)$,

$$\mathrm{H_{zoh}}(s) = \int_{0^-}^{\infty} \mathrm{h_{zoh}}(t) e^{-st}\, dt = \int_{0^-}^{T_s} e^{-st}\, dt = \left[\frac{e^{-st}}{-s}\right]_{0^-}^{T_s} = \frac{1 - e^{-sT_s}}{s}. \tag{12.89}$$

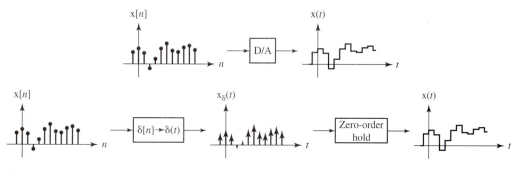

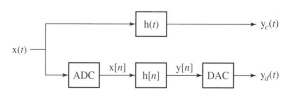

**Figure 12.25**
Equivalence of a DAC and a DT-to-CT impulse conversion followed by a zero-order hold.

**Figure 12.26**
Desired equivalence of CT and sampled-data systems.

The next design task is to make $h[n]$ emulate the action of $h(t)$ in the sense that the overall system responses will be as close as possible. If the CT system is excited by a signal $x(t)$ and produces a response $y_c(t)$, we would like to design the corresponding sampled-data system such that if we convert the CT excitation $x(t)$ to a DT excitation $x[n] = x(nT_s)$ with an ADC, process that with a DT system to produce the response $y[n]$, and then convert that to a CT response $y_d(t)$ with a DAC, that $y_d(t) = y_c(t)$ (Figure 12.26). Unfortunately this cannot be accomplished exactly (except in the theoretical limit in which the sampling rate approaches infinity). But we can establish conditions under which a good approximation can be made, one which gets better as the sampling rate is increased.

As a step toward determining the impulse response $h[n]$ of the DT subsystem, first consider the response of the CT system, not to the actual excitation $x(t)$ but rather to an impulse excitation $x_\delta(t)$, which is related to $x(t)$ by

$$x_\delta(t) = \sum_{n=-\infty}^{\infty} x(nT_s)\delta(t - nT_s) = x(t)f_s \, \text{comb}(f_s t). \qquad \textbf{(12.90)}$$

The response to $x_\delta(t)$ is

$$y(t) = h(t) * x_\delta(t) = h(t) * \sum_{m=-\infty}^{\infty} x[m]\delta(t - mT_s) = \sum_{m=-\infty}^{\infty} x[m]h(t - mT_s),$$

$$\textbf{(12.91)}$$

and the response at the $n$th multiple of $T_s$ is

$$y(nT_s) = \sum_{m=-\infty}^{\infty} x[m]h((n - m)T_s). \qquad \textbf{(12.92)}$$

Compare this to the response of a DT system with impulse response

$$h[n] = h(nT_s) \tag{12.93}$$

to the excitation

$$x[n] = x(nT_s), \tag{12.94}$$

which is

$$y[n] = x[n] * h[n] = \sum_{m=-\infty}^{\infty} x[m]h[n-m]. \tag{12.95}$$

By comparing (12.92) and (12.95) it is apparent that the response $y(t)$ of a CT system with impulse response $h(t)$ at the sampling instants $nT_s$ to a CT impulse excitation

$$x_\delta(t) = \sum_{n=-\infty}^{\infty} x(nT_s)\delta(t - nT_s) \tag{12.96}$$

can be found by finding the response of a DT system with impulse response $h[n] = h(nT_s)$ to the DT excitation $x[n] = x(nT_s)$ and making the equivalence, $y(nT_s) = y[n]$ (Figure 12.27).

Now, returning to our original CT and sampled-data systems, modify the CT system as illustrated in Figure 12.28. Using the equivalence in Figure 12.27, $y[n] = y(nT_s)$.

Now change both the CT system and DT subsystem impulse responses by multiplying them by the time between samples $T_s$ (Figure 12.29). In this modified system

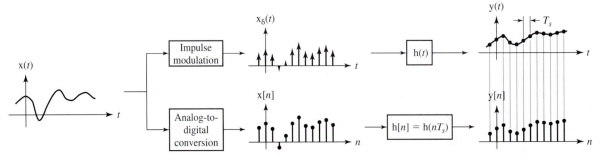

**Figure 12.27**
Equivalence, at continuous time $nT_s$ and corresponding discrete time $n$ of the responses of CT and DT systems excited by CT and DT signals derived from the same CT signal.

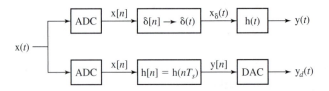

**Figure 12.28**
CT and sampled-data systems when the CT system is excited by $x_\delta(t)$ instead of $x(t)$.

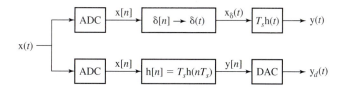

**Figure 12.29**
CT and sampled-data systems when their impulse responses are
multiplied by $T_s$.

we can still say that $y[n] = y(nT_s)$ where now

$$y(t) = x_\delta(t) * T_s h(t) = \left[ \sum_{n=-\infty}^{\infty} x(nT_s)\delta(t - nT_s) \right] * h(t)T_s = \sum_{n=-\infty}^{\infty} x(nT_s)h(t - nT_s)T_s,$$

(12.97)

$$y[n] = \sum_{m=-\infty}^{\infty} x[m]h[n - m] = \sum_{m=-\infty}^{\infty} x[m]T_s h((n - m)T_s),$$  (12.98)

$h[n] = T_s h(nT_s)$ is the new DT subsystem impulse response, and $h(t)$ still represents
the impulse response of the original CT system. Now in (12.97) let $T_s$ approach zero.
In that limit, the summation on the right-hand side becomes the convolution integral
first developed in the derivation of convolution in Chapter 3,

$$\lim_{T_s \to 0} y(t) = \lim_{T_s \to 0} \sum_{n=-\infty}^{\infty} x(nT_s)h(t - nT_s)T_s = \int_{-\infty}^{\infty} x(\tau)h(t - \tau)\, d\tau,$$  (12.99)

which is the CT signal $y_c(t)$, the response of the original CT system in Figure 12.26 to
the CT signal $x(t)$. Also, in that limit, $y[n] = y_c(nT_s)$. So, in the limit, the spacing between points $T_s$ approaches zero, the sampling instants $nT_s$ merge into a continuum $t$,
and there is a one-to-one correspondence between the DT signal values $y[n]$ and the
CT signal values $y_c(t)$. The response of the sampled-data system $y_d(t)$ will be indistinguishable from the response $y_c(t)$ of the original CT system to the CT signal $x(t)$.
Of course, in practice we can never sample at an infinite rate, so the correspondence
$y[n] = y_c(nT_s)$ can never be exact, but it does establish an approximate equivalence
between a CT and a sampled-data system.

There is another conceptual route to arriving at the same conclusion, $h[n] =
T_s h(nT_s)$. In the first development we formed a CT impulse signal

$$x_\delta(t) = \sum_{n=-\infty}^{\infty} x(nT_s)\delta(t - nT_s)$$  (12.100)

whose impulse strengths were equal to samples of the CT signal $x(t)$. Now, instead,
form a modified version of this CT impulse signal. Let the new correspondence between $x(t)$ and $x_\delta(t)$ be that the strength of an impulse at $nT_s$ is approximately the area
under $x(t)$ in the sampling interval $nT_s < t < (n + 1)T_s$, not the value at $nT_s$. That
is, the equivalence between $x(t)$ and $x_\delta(t)$ is based on (approximately) equal areas
(Figure 12.30).

The area under $x(t)$ is approximately $T_s x(nT_s)$ in each sampling interval. Therefore, the new CT impulse signal would be

$$x_\delta(t) = T_s \sum_{n=-\infty}^{\infty} x(nT_s)\delta(t - nT_s).$$  (12.101)

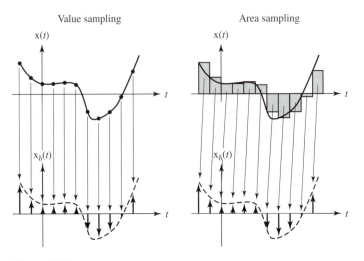

**Figure 12.30**
Comparison of value sampling and area sampling.

If we now apply this CT impulse signal to a system with impulse response h($t$), we get exactly the same response as in (12.97),

$$y(t) = \sum_{n=-\infty}^{\infty} x(nT_s)h(t - nT_s)T_s \qquad (12.102)$$

and, of course, the same result that y[$n$] = y$_c$($nT_s$) in the limit as the sampling rate approaches infinity. All we have done in this development is associate the factor $T_s$ with the excitation instead of with the impulse response. When the two are convolved, the result is the same. If we sampled signals setting impulse strengths equal to signal areas over a sampling interval, instead of setting them equal to signal values at sampling instants, then the correspondence h[$n$] = h($nT_s$) would be the correct design correspondence between a CT system and a sampled-data system which simulates it. (This is the way the impulse invariant design technique works.) But, since we don't sample that way (because most ADCs do not work that way), we instead associate the factor $T_s$ with the impulse response and form the correspondence h[$n$] = $T_s$h($nT_s$).

**EXAMPLE 12.9**

A CT system is characterized by a transfer function

$$H_s(s) = \frac{1}{s^2 + 40s + 300}. \qquad (12.103)$$

Design a sampled-data system of the form of Figure 12.24 to simulate this CT system. Do the design for two sampling rates $f_s$ = 10 and $f_s$ = 100 and compare step responses.

■ **Solution**
The impulse response of the CT system is

$$h(t) = \frac{1}{20}(e^{-10t} - e^{-30t})u(t). \qquad (12.104)$$

The DT-subsystem impulse response is then

$$
h[n] = \frac{T_s}{20}(e^{-10nT_s} - e^{-30nT_s})u[n],
\tag{12.105}
$$

and the corresponding $z$-domain transfer function is

$$
H_z(z) = \frac{T_s}{20}\left(\frac{z}{z - e^{-10T_s}} - \frac{z}{z - e^{-30T_s}}\right).
\tag{12.106}
$$

The step response of the CT system is

$$
y_c(t) = \frac{2 - 3e^{-10t} + e^{-30t}}{600}u(t).
\tag{12.107}
$$

The response of the DT subsystem to a unit sequence is

$$
y[n] = \frac{T_s}{20}\left[\frac{e^{-10T_s} - e^{-30T_s}}{(1 - e^{-10T_s})(1 - e^{-30T_s})} + \frac{e^{-10T_s}}{e^{-10T_s} - 1}e^{-10nT_s} - \frac{e^{-30T_s}}{e^{-30T_s} - 1}e^{-30nT_s}\right]u[n],
\tag{12.108}
$$

and the response of the DAC is

$$
y_d(t) = \sum_{n=0}^{\infty} y[n]\, \text{rect}\left(\frac{t - nT_s - T_s/2}{T_s}\right)
\tag{12.109}
$$

(Figure 12.31).

For the lower sampling rate the sampled-data system simulation is very poor. It approaches a steady-state response value which is about 78 percent of the steady-state response of the CT system. At the higher sampling rate the simulation is much better with a steady-state response approaching a value which is about 99 percent of the steady-state response of the CT system. Also, at the higher sampling rate, there is much less graininess in the simulation (the steps are smaller).

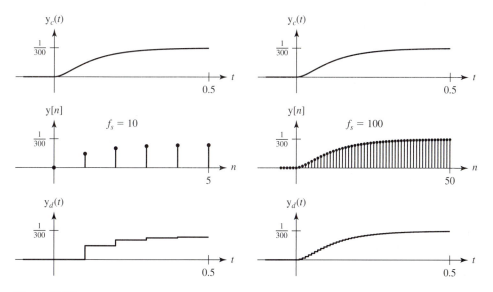

**Figure 12.31**
Comparison of the step responses of a CT system and two sampled-data systems which simulate it with different sampling rates.

We can see why the disparity between steady-state values exists by examining the expression

$$y[n] = \frac{T_s}{20} \left[ \frac{e^{-10T_s} - e^{-30T_s}}{(1 - e^{-10T_s})(1 - e^{-30T_s})} + \frac{e^{-10T_s}}{e^{-10T_s} - 1} e^{-10nT_s} - \frac{e^{-30T_s}}{e^{-30T_s} - 1} e^{-30nT_s} \right] u[n].$$

(12.110)

The steady-state response is

$$y_{ss} = \frac{T_s}{20} \frac{e^{-10T_s} - e^{-30T_s}}{(1 - e^{-10T_s})(1 - e^{-30T_s})}.$$

(12.111)

If we approximate the exponential functions by the first two terms in their series expansions, as $e^{-10T_s} \cong 1 - 10T_s$ and $e^{-30T_s} \cong 1 - 30T_s$, we get

$$y_{ss} = \frac{1}{300},$$

(12.112)

which is the correct steady-state response. However, if $T_s$ is not small enough, the approximation of the exponential function by the first two terms of its series expansion is not very good and actual and ideal steady-state values are significantly different. When $f_s = 10$, we get

$$e^{-10T_s} = 0.368 \qquad \text{and} \qquad 1 - 10T_s = 0$$

(12.113)

and

$$e^{-30T_s} = 0.0498 \qquad \text{and} \qquad 1 - 30T_s = -2,$$

(12.114)

which are terrible approximations. But when $f_s = 100$, we get

$$e^{-10T_s} = 0.905 \qquad \text{and} \qquad 1 - 10T_s = 0.9$$

(12.115)

and

$$e^{-30T_s} = 0.741 \qquad \text{and} \qquad 1 - 30T_s = 0.7,$$

(12.116)

which are much better approximations. ∎

## 12.10  DIGITAL FILTERS

### DIGITAL FILTER DESIGN METHODS

Probably the most important of the systems that electrical engineers design are filters. As we have already seen in earlier chapters, the analysis and design of CT filters is a large and important topic. An equally large and important topic (maybe even more important) is the design of *digital* filters (DT filters) which simulate some of the popular kinds of standard *analog* filters (CT filters). Nearly all DT systems are filters in a sense because they have frequency responses which are not constant with frequency.

There are many optimized standard filter design techniques for analog filters. One very popular way of designing digital filters is to simulate a proven analog filter design. All the commonly used standard analog filters have s-domain transfer functions which are ratios of polynomials in *s* and, therefore, have impulse responses that endure for an infinite time. This type of impulse response is called an *infinite-duration impulse response* (IIR). Many of the techniques which simulate the analog filter with a digital filter create a digital filter which also has an infinite-duration impulse response, and these types of digital filters are called IIR filters. Another popular design technique for digital filters produces filters with a finite-duration impulse response and these filters are called FIR filters.

Recall from Section 12.9 that if we excite a CT system whose impulse response is h($t$) with a CT impulse signal

$$x_\delta(t) = \sum_{n=-\infty}^{\infty} x[n]\delta(t - nT_s) = x(t)f_s \, \text{comb}(f_s t), \qquad \textbf{(12.117)}$$

we get a response y($t$). We showed there that we can find the response values y($nT_s$) at the sampling instants by exciting a DT system whose impulse response is h[$n$] = h($nT_s$) with a DT signal x[$n$] = x($nT_s$) and that the correspondence is y($nT_s$) = y[$n$].

One approach to the design of digital filters is the impulse-invariant method previously discussed. Another approach to establishing an equivalence between an analog and a digital filter is through a frequency-domain argument. Recall that the complex exponential is the eigenfunction of LTI systems, and any signal can be expressed as a linear combination of complex exponentials. If we apply the complex-exponential excitation $e^{st}$ to a continuous-time LTI system whose transfer function is H$_s(s)$, the response is H$_s(s)e^{st}$. Similarly, if we apply the complex-exponential excitation $z^n$ to a discrete-time LTI system whose transfer function is H$_z(z)$, the response is H$_z(z)z^n$. If we make the two complex-exponential excitations correspond through sampling at a rate $f_s$, then

$$z^n = e^{snT_s} \Rightarrow z = e^{sT_s} \qquad \textbf{(12.118)}$$

and the responses H$_s(s)e^{st}$ and H$_z(z)z^n$ are the same at the sampling instants if

$$H_s(s) = H_z(e^{sT_s}). \qquad \textbf{(12.119)}$$

The two responses are only guaranteed equal at the sampling instants, not between. Therefore, for equality at all times $t$, there must not be any "between." That is, the time between samples must approach zero, meaning the equivalence is only exact in the infinite-sampling-rate limit, as before.

Both of these methods of equating analog and digital filter responses depend, for perfect correspondence, on the sampling rate approaching infinity. This indicates that, generally speaking, as the sampling rate is increased, the performance of a digital filter more closely approaches that of the corresponding analog filter.

## IMPULSE- AND STEP-INVARIANT DESIGN

One approach to digital filter design is to try to make the digital filter response to a standard DT excitation a sampled version of the analog filter response to the corresponding standard CT excitation. This idea leads to the *impulse-invariant* and *step-invariant* design procedures. Impulse-invariant design makes the response of the DT system to a DT impulse a sampled version of the response of the CT system to a CT impulse. Step-invariant design makes the response of the DT system to a unit sequence a sampled version of the response of the CT system to a unit step. Each of these design processes produces an IIR filter (Figure 12.32).

Impulse-invariant design is probably the most straightforward technique for approximating an analog filter with a digital filter. If the transfer function of the analog filter in the Laplace domain is H$_s(s)$, then the impulse response is h($t$). If h($t$) is sampled, a discrete-time function h[$n$] is formed and the $z$ transform of h[$n$] can be taken to yield the desired $z$-domain transfer function H$_z(z)$.

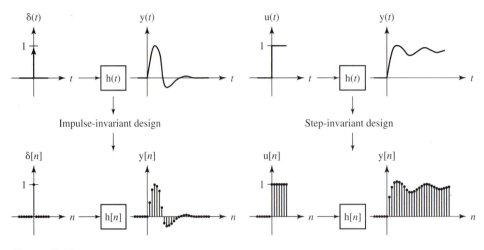

**Figure 12.32**
The impulse-invariant and step-invariant digital filter design techniques.

To illustrate the impulse-invariant design technique, let the CT filter transfer function be $H_s(s) = 1/(s + a)$. The inverse Laplace transform (CT impulse response) is $h(t) = e^{-at}u(t)$. That can be sampled with a time $T_s$ between samples forming the DT impulse response

$$h[n] = e^{-anT_s}u[n], \qquad (12.120)$$

and its $z$ transform is

$$H_z(z) = \frac{Y(z)}{X(z)} = \frac{z}{z - e^{-aT_s}} = \frac{1}{1 - e^{-aT_s}z^{-1}}. \qquad (12.121)$$

The actual design of the digital filter is the implementation of the recursion relation between the response y and the excitation x. In this case

$$Y(z) = X(z) + e^{-aT_s}z^{-1}Y(z) \qquad (12.122)$$

or, taking the inverse $z$ transform,

$$y[n] = x[n] + e^{-aT_s}y[n - 1]. \qquad (12.123)$$

The filter design is illustrated by the block diagram in Figure 12.33. The actual comparison between the two filters still depends, of course, on the choice of the sampling interval $T_s$. To make a concrete comparison, set $a$ to one and let $T_s = 0.1$. Then

$$y[n] = x[n] + 0.904y[n - 1]. \qquad (12.124)$$

The impulse response of the analog filter is

$$h(t) = e^{-t}u(t). \qquad (12.125)$$

The DT and CT impulse responses are illustrated in Figure 12.34.

It is instructive to check the step response of the filter in addition to the impulse response. Let the excitation be the unit sequence (Figure 12.35). Although the shape of the response looks about right, there is a scale factor of 10.4166 difference between the steady-state values of the responses. This points to what was mentioned earlier. Even though one aspect of the CT filter may be closely simulated, another may not. A simple

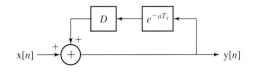

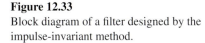

**Figure 12.33**
Block diagram of a filter designed by the impulse-invariant method.

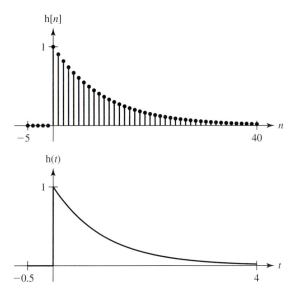

**Figure 12.34**
Impulse responses of an analog filter and a digital filter which simulates it through the impulse-invariant design technique.

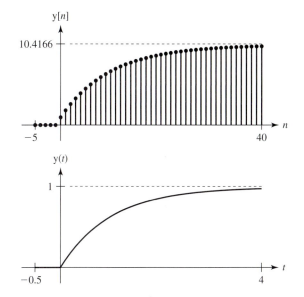

**Figure 12.35**
Comparison of the step responses of a CT filter and a DT filter which approximates it through the impulse-invariant design technique.

gain difference can easily be compensated for by a constant. A common way of doing that is to modify the $z$-domain transfer function so that its value at zero frequency is one. *Zero frequency* here means a value of $\Omega$ equal to zero. That translates through $e^{j\Omega} \to z$ into $z = 1$. At $z = 1$,

$$H_z(1) = \frac{1}{1 - e^{-aT_s}} = 10.4166 \tag{12.126}$$

and that accounts for the fact that the zero-frequency transfer function of this filter is 10.4166. One can always simply multiply $H_z(z)$ by $1 - e^{-aT_s}$. Then the zero-frequency transfer function will become one. That is, let

$$H_z(z) = \frac{Y(z)}{X(z)} = \frac{1 - e^{-aT_s}}{1 - e^{-aT_s} z^{-1}}. \tag{12.127}$$

(Of course the digital filter is no longer impulse invariant.) Now the frequency response of the filter will look about the same as that of the analog filter at low frequencies, but the approximation will become progressively worse as half the sampling rate

is approached (Figure 12.36). The left-hand graphs in the figure are the frequency response as a function of DT radian frequency $\Omega$, and the right-hand graphs are the same
frequency response except plotted versus CT cyclic frequency $f$ with the time between
samples being 0.1 s. As frequency progresses from zero upward, the initial magnitude
frequency response looks like a typical CT single-pole lowpass filter (Figure 12.37).
But the curve soon deviates significantly and at half the sampling rate turns around and
goes back up. This is inherent in digital filter design because of the aliasing caused by
sampling. All digital-filter frequency responses are periodic.

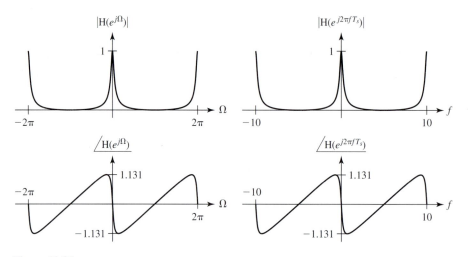

**Figure 12.36**
Frequency response of a DT filter, designed by the impulse-invariant technique and then normalized to a unity zero-frequency transfer function ($T_s = 0.1$).

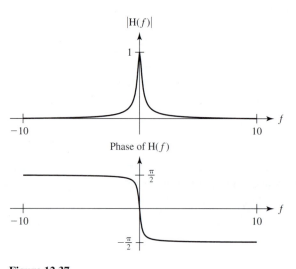

**Figure 12.37**
Frequency response of a CT filter whose transfer function
is $H_s(s) = \frac{1}{s+1}$.

The MATLAB signal toolbox has a command `impinvar` which does impulse-invariant digital design. The syntax is

$$[\text{bz,az}] = \text{impinvar(b,a,fs)}$$

where b = vector of coefficients of $s$ in numerator of $s$-domain transfer function
  a = vector of coefficients of $s$ in denominator of $s$-domain transfer function
  fs = sampling rate, Hz
  bz = vector of coefficients of $z$ in numerator of $z$-domain transfer function
  az = vector of coefficients of $z$ in denominator of $z$-domain transfer function

**EXAMPLE 12.10**

Using the impulse-invariant design method, design a digital filter to simulate the analog filter whose transfer function is

$$H_s(s) = \frac{s}{s^2 + 400s + 2 \times 10^5}. \tag{12.128}$$

Compare the frequency responses of the two filters.

■ **Solution**

To facilitate inverse transformation, this transfer function can be rearranged into

$$H_s(s) = \frac{s + 200}{(s + 200)^2 + 1.6 \times 10^5} - \frac{1}{2}\frac{400}{(s + 200)^2 + 1.6 \times 10^5} \tag{12.129}$$

and the inverse transform is

$$h(t) = e^{-200t}\left[\cos(400t) - \frac{1}{2}\sin(400t)\right]u(t). \tag{12.130}$$

This impulse response is an exponentially damped sinusoid with a time constant of 5 ms and a sinusoidal frequency of $400/2\pi \cong 63.7$ Hz. For a reasonably accurate simulation we should choose a sampling rate such that the sinusoid is oversampled and there are several samples of the exponential decay per time constant. Let the sampling rate $f_s$ be 1 kHz. Then the DT impulse response would be

$$h[n] = e^{-200nT_s}\left[\cos(400nT_s) - \frac{1}{2}\sin(400nT_s)\right]u(nT_s) \tag{12.131}$$

or

$$h[n] = (0.8187)^n\left[\cos(0.4n) - \frac{1}{2}\sin(0.4n)\right]u[n]. \tag{12.132}$$

The $z$ transform of this DT impulse response is the transfer function

$$H(z) = \frac{z(z - 0.8187\cos(0.4))}{z^2 - 2(0.8187)\cos(0.4)z + (0.819)^2} - \frac{1}{2}\frac{0.8187\sin(0.4)z}{z^2 - 2(0.8187)\cos(0.4)z + (0.8187)^2} \tag{12.133}$$

or

$$H(z) = \frac{Y(z)}{X(z)} = \frac{z(z - 0.9135)}{z^2 - 1.508z + 0.6703}. \tag{12.134}$$

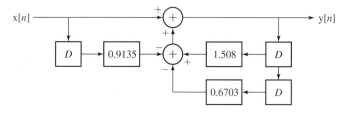

**Figure 12.38**
DT system with the transfer function $H(z) = \frac{z(z-0.9135)}{z^2-1.508\,z+0.6703}$.

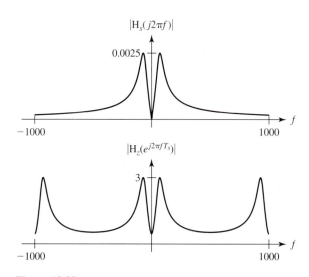

**Figure 12.39**
Magnitude frequency responses of the analog filter and its digital simulation by the impulse-invariant technique.

The corresponding recursion relation is

$$y[n] = x[n] - 0.9135\,x[n-1] + 1.508\,y[n-1] - 0.6703\,y[n-2] \qquad \textbf{(12.135)}$$

(Figure 12.38). The magnitude frequency responses of the analog and digital filters are compared in Figure 12.39.

Two things immediately stand out about this design. First the analog filter has a response of zero at $f = 0$ and the digital filter does not. Since this filter is probably intended as a bandpass filter, this is an undesirable design result. The gain of the digital filter is much greater than the gain of the analog filter. The gain could be made the same as the analog filter by a simple adjustment of the multiplication factor in the expression for $H_z(z)$. Also, although the frequency response does peak at the right frequency, the attenuation of the digital filter in the stopband is not very good. If we had used a higher sampling rate, this attenuation would have been better.    ■

A closely related design technique for digital filters is the step-invariant method. In this method the step response of the digital filter is designed to match the step response of the analog filter at the sampling instants. If an analog filter has a transfer function $H_s(s)$, the Laplace transform of its step response is $H_s(s)/s$. The step response itself is the inverse Laplace transform,

$$h_{-1}(t) = \mathcal{L}^{-1}\left(\frac{H_s(s)}{s}\right). \tag{12.136}$$

The equivalent DT unit sequence response is then

$$h_{-1}[n] = h_{-1}(nT_s). \tag{12.137}$$

Its $z$ transform is the product of the $z$-domain transfer function and the $z$ transform of a unit sequence,

$$\mathcal{Z}(h_{-1}[n]) = \frac{z}{z-1}H_z(z). \tag{12.138}$$

We can summarize by saying that, given an $s$-domain transfer function $H_s(s)$, we can find the corresponding $z$-domain transfer function $H_z(z)$ as

$$H_z(z) = \frac{z-1}{z}\mathcal{Z}\left(\mathcal{L}^{-1}\left(\frac{H_s(s)}{s}\right)_{(t)\to(nT_s)\to[n]}\right). \tag{12.139}$$

**EXAMPLE 12.11**

Using the step-invariant method, design a digital filter to approximate the analog filter whose transfer function is the same as in Example 12.10,

$$H_s(s) = \frac{s}{s^2 + 400s + 2\times10^5}, \tag{12.140}$$

with the same sampling rate $f_s = 1$ kHz.

■ **Solution**

The step response is

$$h_{-1}(t) = \mathcal{L}^{-1}\left(\frac{H_s(s)}{s}\right) = \mathcal{L}^{-1}\left(\frac{1}{s}\frac{s}{s^2 + 400s + 2\times10^5}\right) \tag{12.141}$$

or

$$h_{-1}(t) = \frac{1}{400}\mathcal{L}^{-1}\left(\frac{400}{(s+200)^2 + 1.6\times10^5}\right) = \frac{e^{-200t}\sin(400t)}{400}u(t). \tag{12.142}$$

The DT step response is

$$h_{-1}[n] = \frac{e^{-200nT_s}\sin(400nT_s)}{400}u[n]. \tag{12.143}$$

The $z$-domain transfer function is

$$H_z(z) = \frac{1}{400}\frac{z-1}{z}\frac{ze^{-200T_s}\sin(400T_s)}{z^2 - 2e^{-200T_s}z\cos(400T_s) + e^{-400T_s}} \tag{12.144}$$

or

$$H_z(z) = \frac{1}{400}\frac{e^{-0.2}\sin(0.4)(z-1)}{z^2 - 2e^{-0.2}z\cos(0.4) + e^{-0.4}} = 7.97\times10^{-4}\frac{(z-1)}{z^2 - 1.509z + 0.6708}. \tag{12.145}$$

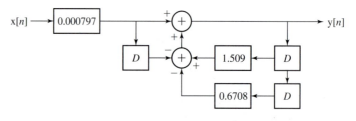

**Figure 12.40**
DT system with the transfer function $H_z(z) = 7.97 \times 10^{-4} \frac{(z-1)}{z^2 - 1.509\,z + 0.6708}$.

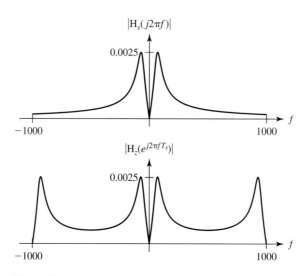

**Figure 12.41**
Magnitude frequency responses of the analog filter and its
digital simulation by the step-invariant technique.

The corresponding recursion relation is

$$y[n] = 7.97 \times 10^{-4}(x[n] - x[n-1]) + 1.509\,y[n-1] - 0.6708\,y[n-2] \qquad \textbf{(12.146)}$$

(Figure 12.40).

The magnitude frequency responses of the analog and digital filters are compared in Figure 12.41. In contrast with the impulse-invariant design, this digital filter has a response of zero at $f = 0$. Also, the filter gain is the same as the analog filter. However, its stopband attenuation is still not very good. As in the impulse-invariant design, that could be improved by using a higher sampling rate. ∎

## DIFFERENCE EQUATION APPROXIMATION TO DIFFERENTIAL EQUATIONS

A second technique commonly used in designing digital filters to simulate analog filters is to approximate the differential equation describing the linear system with a difference equation. The basic idea in this technique is to start with a desired transfer function in the $s$ domain $H_s(s)$ and find the integrodifferential equation corresponding to it in the time domain. Then derivatives and integrals in the CT domain are approximated by differences and accumulations in the DT domain and the resulting expression is a $z$-domain transfer function approximating the original $s$-domain transfer function. For example, suppose that

$$H_s(s) = \frac{1}{s+a}.$$    **(12.147)**

Since this is a transfer function, it is the ratio of the response $Y_s(s)$ to the excitation $X_s(s)$,

$$\frac{Y_s(s)}{X_s(s)} = \frac{1}{s+a}.$$    **(12.148)**

Then

$$Y_s(s)(s+a) = X_s(s).$$    **(12.149)**

Taking the inverse Laplace transform of both sides,

$$\frac{d}{dt}(y(t)) + a y(t) = x(t).$$    **(12.150)**

A derivative can be approximated by various finite-difference expressions, and each choice has a slightly different effect on the approximation of the digital filter to the analog filter. Let the derivative in this case be approximated by the *forward difference*,

$$\frac{d}{dt}(y(t)) \cong \frac{y[n+1] - y[n]}{T_s}.$$    **(12.151)**

Then the difference-equation approximation to the differential equation is

$$\frac{y[n+1] - y[n]}{T_s} + a y[n] = x[n]$$    **(12.152)**

and the corresponding recursion relation is

$$y[n+1] = x[n]T_s + (1 - aT_s)y[n].$$    **(12.153)**

The $z$-domain transfer function can be found by $z$-transforming the equation into

$$z(Y(z) - y[0]) = T_s X(z) + (1 - aT_s)Y(z).$$    **(12.154)**

Transfer functions are computed based on the assumption of no initial energy storage in a system. Therefore, $y[0] = 0$ and

$$H(z) = \frac{Y(z)}{X(z)} = \frac{T_s}{z - (1 - aT_s)}.$$    **(12.155)**

The DT filter could also have been based on a *backward-difference* approximation to the derivative,

$$\frac{d}{dt}(y(t)) \cong \frac{y[n] - y[n-1]}{T_s},$$    **(12.156)**

or a *central-difference* approximation to the derivative,

$$\frac{d}{dt}(y(t)) \cong \frac{y[n+1] - y[n-1]}{2T_s}.$$    **(12.157)**

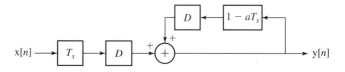

**Figure 12.42**
Block diagram of a digital filter designed by approximating a
differential equation with a difference equation using forward
differences.

The step, impulse, and frequency responses of the DT filters designed using different
approximations to a derivative are all slightly different. The block diagram of this filter is illustrated in Figure 12.42.

We can systematize this technique by realizing that every $s$ in an $s$-domain expression represents a corresponding differentiation in the time domain,

$$\frac{d}{dt}(\mathrm{x}(t)) \xleftarrow{\mathcal{L}} s\mathrm{X}(s). \tag{12.158}$$

We can approximate derivatives with forward, backward, or central differences,

$$\frac{d}{dt}(\mathrm{x}(t)) \cong \frac{\mathrm{x}(t + T_s) - \mathrm{x}(t)}{T_s}, \tag{12.159}$$

$$\frac{d}{dt}(\mathrm{x}(t)) \cong \frac{\mathrm{x}(t) - \mathrm{x}(t - T_s)}{T_s}, \tag{12.160}$$

or

$$\frac{d}{dt}(\mathrm{x}(t)) \cong \frac{\mathrm{x}(t + T_s) - \mathrm{x}(t - T_s)}{2T_s}. \tag{12.161}$$

The $z$ transforms of these differences are

$$\frac{\mathrm{x}(t + T_s) - \mathrm{x}(t)}{T_s} \xleftarrow{z} \frac{z - 1}{T_s}\mathrm{X}(z), \tag{12.162}$$

$$\frac{\mathrm{x}(t) - \mathrm{x}(t - T_s)}{T_s} \xleftarrow{z} \frac{1 - z^{-1}}{T_s}\mathrm{X}(z), \tag{12.163}$$

or

$$\frac{\mathrm{x}(t + T_s) - \mathrm{x}(t - T_s)}{2T_s} \xleftarrow{z} \frac{z - z^{-1}}{2T_s}\mathrm{X}(z). \tag{12.164}$$

Now we can replace every $s$ in an $s$-domain expression with the corresponding $z$-domain
expression. Then we can approximate the $s$-domain transfer function

$$\mathrm{H}_s(s) = \frac{1}{s + a} \tag{12.165}$$

with a forward-difference approximation to a derivative,

$$\mathrm{H}(z) = \left(\frac{1}{s + a}\right)_{s \to (z-1)/T_s} = \frac{1}{((z - 1)/T_s) + a} = \frac{T_s}{z - 1 + aT_s}, \tag{12.166}$$

which is exactly the same as (12.155). This avoids having to actually write the differential equation and substitute a finite difference for each derivative.

EXAMPLE 12.12

Using the difference-equation design method with a backward difference, design a digital filter to simulate the analog filter of Example 12.10 whose transfer function is

$$H_s(s) = \frac{s}{s^2 + 400s + 2 \times 10^5} \tag{12.167}$$

using the same sampling rate $f_s = 1$ kHz. Compare the frequency responses of the two filters.

■ **Solution**

The $z$-domain transfer function is

$$H(z) = \left( \frac{s}{s^2 + 400s + 2 \times 10^5} \right)_{s \to (1-z^{-1})/T_s} = \frac{\frac{1-z^{-1}}{T_s}}{((1 - z^{-1})/T_s)^2 + 400((1 - z^{-1})/T_s) + 2 \times 10^5}, \tag{12.168}$$

which can be simplified to

$$H(z) = \frac{zT_s(z - 1)}{\left(1 + 400T_s + (2 \times 10^5)T_s^2\right) z^2 - (2 + 400T_s)z + 1} \tag{12.169}$$

or

$$H(z) = T_s \frac{z(z - 1)}{\left(1 + 400T_s + (2 \times 10^5)T_s^2\right) z^2 - (2 + 400T_s)z + 1}. \tag{12.170}$$

Substituting in $T_s = 0.001$,

$$H(z) = 0.001 \frac{z(z - 1)}{1.6z^2 - 2.4z + 1} = 6.25 \times 10^{-4} \frac{z(z - 1)}{z^2 - 1.5z + 0.625} \tag{12.171}$$

and the recursion relation is

$$y[n] = (6.25 \times 10^{-4})(x[n] - x[n - 1]) + 1.5\, y[n - 1] - 0.625\, y[n - 2] \tag{12.172}$$

(Figure 12.43). The magnitude frequency responses of the analog and digital filters are compared in Figure 12.44. This result is very similar to the step-invariant result except that the gains are different.

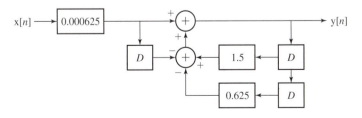

**Figure 12.43**
DT system with the transfer function $H(z) = 6.25 \times 10^{-4} \frac{z(z-1)}{z^2 - 1.5z + 0.625}$.

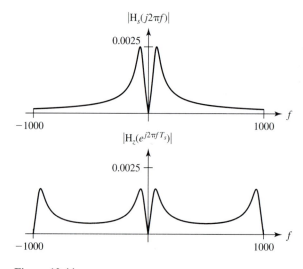

**Figure 12.44**
Magnitude frequency responses of the analog filter and its
digital simulation by the difference-equation technique.

## EXAMPLE 12.13

Using the difference-equation design method with a forward difference, design a digital filter to simulate the analog filter whose transfer function is

$$H_s(s) = \frac{1}{s^2 + 600s + 4 \times 10^5} \tag{12.173}$$

using a sampling rate $f_s = 500$ Hz.

**■ Solution**

The $z$-domain transfer function is

$$H(z) = \frac{1}{((z-1)/T_s)^2 + 600((z-1)/T_s) + 4 \times 10^5}$$

or

$$H(z) = \frac{T_s^2}{z^2 + (600T_s - 2)z + \left(1 - 600T_s + (4 \times 10^5)T_s^2\right)} \tag{12.174}$$

or

$$H(z) = \frac{4 \times 10^{-6}}{z^2 - 0.8z + 1.4}.$$

This result looks quite simple and straightforward, but there is a problem with this design. The poles of this $z$-domain transfer function are outside the unit circle, and the filter is, therefore, unstable, even though the $s$-domain transfer function is stable. This illustrates a disadvantage of the difference-equation method of digital filter design. Stable $s$-domain transfer functions can be converted into unstable $z$-domain transfer functions. Stability can be restored by increasing the sampling rate.

## DIRECT SUBSTITUTION AND THE MATCHED $z$ TRANSFORM

A third approach to the design of digital filters is to find a transformation from $s$ to $z$ which maps the $s$ plane into the $z$ plane, converts the poles and zeros of the $s$-domain transfer function into appropriate corresponding locations in the $z$ plane, and converts stable $s$-domain systems into stable $z$-domain systems. The most common techniques that use this idea are the matched $z$ transform, direct substitution, and the bilinear transformation. This type of design process produces an IIR filter (Figure 12.45).

The $z$ transform was derived by generalizing the DTFT by making the transformation $e^{j\Omega} \to z$. The relation between DT frequency and CT frequency (usually just called frequency) is

$$\Omega = \omega T_s, \tag{12.175}$$

where $T_s$ is the time between samples. Therefore, the transformation could be written as

$$e^{j\omega T_s} \to z. \tag{12.176}$$

This approach to designing a digital filter consists of two similar techniques, the so-called direct substitution and matched $z$-transform methods. These methods are based on the idea of simply mapping the poles and zeros of the $s$-domain transfer function into the $z$ domain through the relationship

$$z = e^{sT_s}, \tag{12.177}$$

where $s$ has taken the place of $j\omega$ in $e^{j\omega T_s} \to z$.

For example, to transform the CT filter frequency response

$$H_s(s) = \frac{1}{s + a}, \tag{12.178}$$

which has a pole at $s = -a$, we simply map the pole at $-a$ to the corresponding location in the $z$ plane. Then the $z$-domain pole location is given by $e^{-aT_s}$. The direct substitution method implements the transformation

$$s - a \to z - e^{aT_s}, \tag{12.179}$$

while the matched $z$-transform method implements the transformation

$$s - a \to 1 - e^{aT_s}z^{-1}. \tag{12.180}$$

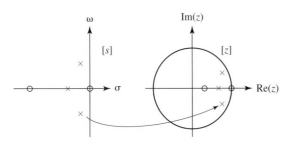

**Figure 12.45**
Mapping of poles and zeros from the $s$ plane to the $z$ plane.

The z-domain transfer functions which result (in this case) are as follows:
Direct substitution:

$$H_z(z) = \frac{1}{z - e^{-aT_s}} = \frac{z^{-1}}{1 - e^{-aT_s}z^{-1}} \qquad \text{with a pole at } z = e^{-aT_s} \text{ and no finite zeros}$$

(12.181)

Matched z transform:

$$H_z(z) = \frac{1}{1 - e^{-aT_s}z^{-1}} = \frac{z}{z - e^{-aT_s}} \qquad \text{with a pole at } z = e^{-aT_s} \text{ and a zero at } z = 0$$

(12.182)

Notice that the matched z-transform result is exactly the same as was obtained using the impulse-invariant method and the direct substitution result is the same except for a single sample delay due to the $z^{-1}$ factor. For more complicated s-domain transfer functions the results of these methods are not so similar. These transformation methods do not directly involve any time-domain analysis. The design is done entirely in the s and z domains. The transformations $s - a \rightarrow z - e^{aT}$ and $s - a \rightarrow 1 - e^{aT}z^{-1}$ both map a pole in the open left half of the s plane into a pole in the open interior of the unit circle in the z plane. Therefore, stable s-domain systems are transformed into stable z-domain systems.

## EXAMPLE 12.14

Using the matched z-transform design method, design a digital filter to simulate the analog filter of Example 12.10 whose transfer function is

$$H_s(s) = \frac{s}{s^2 + 400s + 2 \times 10^5}$$

(12.183)

using the same sampling rate $f_s = 1$ kHz. Compare the frequency responses of the two filters.

■ **Solution**
This transfer function has a zero at $s = 0$ and poles at $s = -200 \pm j400$. Using the mapping

$$s - a \rightarrow 1 - e^{aT}z^{-1},$$

(12.184)

we get a z-domain zero at $z = 1$, poles at

$$z = e^{(-200 \pm j400)T_s} = e^{-0.2 \pm j0.4} = 0.7541 \pm j0.3188,$$

(12.185)

and a z-domain transfer function,

$$H_z(z) = \frac{1 - z^{-1}}{(1 - (0.7541 + j0.3188)z^{-1})(1 - (0.7541 - j0.3188)z^{-1})}$$

(12.186)

or

$$H_z(z) = \frac{1 - z^{-1}}{1 - 1.509z^{-1} + 0.6708z^{-2}} = \frac{z(z - 1)}{z^2 - 1.509z + 0.6708}.$$

(12.187)

The recursion relation is

$$y[n] = x[n] - x[n-1] + 1.509\,y[n-1] - 0.6708\,y[n-2]$$

(12.188)

(Figure 12.46). The magnitude frequency responses of the analog and digital filters are compared Figure 12.47.

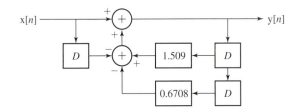

**Figure 12.46**
DT system with the transfer function
$H_z(z) = \frac{z(z-1)}{z^2 - 1.509\,z + 0.6708}$.

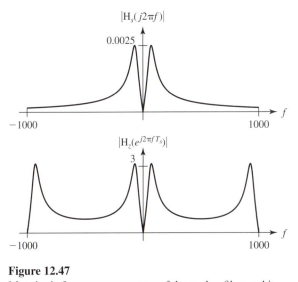

**Figure 12.47**
Magnitude frequency responses of the analog filter and its digital simulation by the matched $z$ transform technique.

## THE BILINEAR TRANSFORMATION

The impulse-invariant and step-invariant design techniques try to make the digital filter's DT-domain response match the corresponding analog filter's CT-domain response to a corresponding standard excitation. Another way to approach digital filter design is to try to make the frequency response of the digital filter match the frequency response of the analog filter. But, just as a DT-domain response can never exactly match a CT-domain response, the frequency response of a digital filter cannot exactly match the frequency response of an analog filter. One reason, mentioned earlier, for this is that the frequency response of a DT system is inherently periodic. As first mentioned in Chapter 2, when a sinusoidal CT signal is sampled to create a sinusoidal DT excitation, if the frequency of the CT signal is changed by an integer multiple of the sampling rate, the DT signal does not change at all. The DT system cannot tell the difference and responds the same way as it would respond to the original signal (Figure 12.48).

According to Shannon's sampling theorem, if a CT signal can be guaranteed never to have any frequency components outside the range $|f| < f_s/2$, then when it is sampled at the rate $f_s$, the DT signal so created contains all the information in the CT

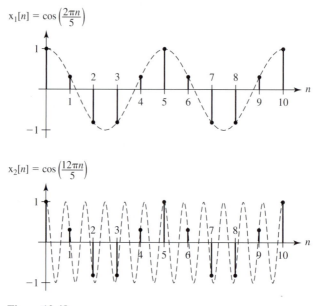

**Figure 12.48**
Two identical DT signals formed by sampling two different sinusoids.

signal. Then, when the DT signal excites a DT system, the response contains all the information in a corresponding CT signal. So the design process becomes a matter of making the digital filter frequency response match the analog filter frequency response only in the frequency range $|f| < f_s/2$, not outside it. In general this still cannot be done exactly, but it is often possible to make a good approximation. Of course, no signal is truly band-limited. Therefore, in practice we must arrange to have very little signal power beyond half the sampling rate instead of no signal power (Figure 12.49).

If a CT excitation does not have any frequency components outside the range $|f| < f_s/2$, any nonzero response of an analog filter outside that range would have no effect because it has nothing to filter. Therefore, in the design of a DT system to simulate a CT system, the sampling rate should be chosen such that the response of the analog filter at frequencies $|f| > f_s/2$ is approximately zero. Then all the filtering action will occur at frequencies in the range $|f| < f_s/2$. So the starting point for a frequency-domain design process is to specify the sampling rate such that

$$X(f) \cong 0 \quad \text{and} \quad H(f) \cong 0 \quad |f| > \frac{f_s}{2} \quad \text{(12.189)}$$

or

$$X(j\omega) \cong 0 \quad \text{and} \quad H(j\omega) \cong 0 \quad |\omega| > \pi f_s = \frac{\omega_s}{2}. \quad \text{(12.190)}$$

Now the problem is to find a DT transfer function which has approximately the same shape as the CT transfer function we are trying to simulate in the frequency range $|f| < f_s/2$. As discussed earlier, the straightforward method to accomplish this goal would be to use the transformation $e^{sT_s} \rightarrow z$ to convert a desired transfer function $H_s(s)$ into the corresponding $H_z(z)$,

$$H_z(z) = H_s(s)|_{s \rightarrow (1/T_s)\ln(z)}. \quad \text{(12.191)}$$

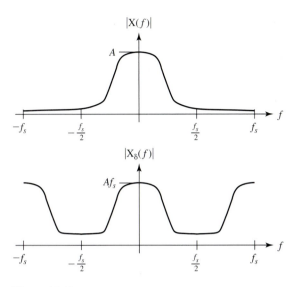

**Figure 12.49**
Magnitude spectrum of a CT signal and a DT signal
formed by impulse sampling the CT signal.

Although this development of the transformation technique is satisfying from a
theoretical point of view, it uses the functional transformation

$$s \rightarrow \frac{1}{T_s} \ln(z),$$ (12.192)

which transforms a CT system transfer function in the common form of the ratio of two
polynomials into a DT system transfer function which involves a ratio of polynomials,
not in $z$ but rather in $\ln(z)$, making the function transcendental with infinitely many
poles and zeros. So, although this process looks nice, it really does not lead to a very
practical DT-system transfer function design.

At this point it is common to make an approximation to simplify the form of the
DT-system transfer function. One such transformation arises from the series expression for the exponential function

$$e^x = 1 + x + \frac{x^2}{2!} + \frac{x^3}{3!} + \cdots = \sum_{k=0}^{\infty} \frac{x^k}{k!}.$$ (12.193)

We can apply that to the transformation

$$e^{sT_s} \rightarrow z,$$ (12.194)

yielding

$$1 + sT_s + \frac{(sT_s)^2}{2!} + \frac{(sT_s)^3}{3!} + \cdots \rightarrow z.$$ (12.195)

If we approximate this series by the first two terms, we get

$$1 + sT_s \rightarrow z$$ (12.196)

or

$$s \rightarrow \frac{z-1}{T_s}.$$ (12.197)

The equation $e^{sT_s} \cong 1 + sT_s$ is a good approximation if $T_s$ is small and gets better as $T_s$ gets smaller and, of course, as $f_s$ gets larger. That is, this approximation becomes very good at high sampling rates. (Notice that this is exactly the same as a forward-difference approximation to a derivative in the difference-equation method.)

Examine the transformation $s \rightarrow (z-1)/T_s$. A multiplication by $s$ in the $s$ domain corresponds to a differentiation with respect to $t$ of the corresponding function in the CT domain. A multiplication by $(z-1)/T_s$ in the $z$ domain corresponds to a forward difference divided by the sampling time $T_s$ of the corresponding function in the DT domain. This is a forward-difference approximation to a derivative. So, as mentioned in the difference-equation method, the two operations, multiplication by $s$ and by $(z-1)/T_s$, are analogous.

The transformation $s \rightarrow (z-1)/T_s$ is a mapping of the $s$ plane into the $z$ plane. For example, the contour $s = a + j\omega$, where $a$ is real, in the $s$ plane maps into the contour $z = (1 + aT_s) + j\omega T_s$, the line $\text{Re}(z) = 1 + aT_s$, in the $z$ plane (Figure 12.50).

If $a$ is negative, the contour $s = a + j\omega$ lies in the open left half-plane and the line $\text{Re}(z) = 1 + aT_s$ lies to the left of $\text{Re}(z) = 1$. But part of the contour $\text{Re}(z) = 1 + aT_s$ lies outside the unit circle. Therefore, a pole in the left half of the $s$ plane can map into a pole outside the unit circle in the $z$ plane. So this mapping is not generally satisfactory because it can transform a stable CT system into an unstable DT system. But if the sampling rate is high enough, the poles will be inside the unit circle and the digital filter will be stable.

A very clever modification of this transformation solves the problem of creating an unstable DT system from a stable CT system. We can write the transformation from the $s$ domain to the $z$ domain as

$$e^{sT_s} = \frac{e^{s(T_s/2)}}{e^{-s(T_s/2)}} \rightarrow z, \tag{12.198}$$

approximate *both* exponentials with an infinite series,

$$\frac{1 + \dfrac{sT_s}{2} + \dfrac{(sT_s/2)^2}{2!} + \dfrac{(sT_s/2)^3}{3!} + \cdots}{1 - \dfrac{sT_s}{2} + \dfrac{(sT_s/2)^2}{2!} - \dfrac{(sT_s/2)^3}{3!} + \cdots} \rightarrow z \tag{12.199}$$

and then truncate both series to two terms

$$\frac{1 + (sT_s/2)}{1 - (sT_s/2)} \rightarrow z \tag{12.200}$$

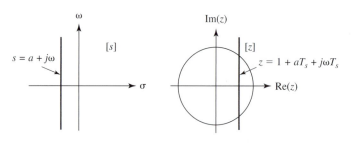

**Figure 12.50**
Corresponding contours in the $s$ and $z$ planes for the transformation $s \rightarrow \frac{z-1}{T_s}$.

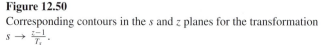

yielding

$$s \rightarrow \frac{2}{T_s} \frac{z-1}{z+1} \qquad \text{or} \qquad z \rightarrow \frac{2+sT_s}{2-sT_s}. \qquad \textbf{(12.201)}$$

This mapping from $s$ to $z$ is called the *bilinear z transform* because the numerator and denominator are both linear functions of $z$. (Don't get the terms *bilinear* and *bilateral z transform* confused.) The bilinear $z$ transform transforms any stable CT system into a stable DT system because it maps the entire open left half of the $s$ plane into the open interior of the unit circle in the $z$ plane. This was also true of the matched $z$ transform and direct substitution, but the correspondences are different. The mapping $z = e^{sT_s}$ maps any strip $\omega_0/T_s < \omega < (\omega_0 + 2\pi)/T_s$ of the $s$ plane into *the entire z plane*. The mapping from $s$ to $z$ is unique, but the mapping from $z$ to $s$ is not unique. The mapping $s \rightarrow (2/T_s)((z-1)/(z+1))$, maps each point in the $s$ plane into a unique point in the $z$ plane, and the inverse mapping $z \rightarrow (2+sT_s)/(2-sT_s)$ maps each point in the $z$ plane into a unique point in the $s$ plane. To see how the mapping works consider the contour $s = j\omega$ in the $s$ plane. Setting $z = (2+sT_s)/(2-sT_s)$ we get

$$z = \frac{2+j\omega T_s}{2-j\omega T_s} = 1\angle 2\tan^{-1}\left(\frac{\omega T_s}{2}\right) = e^{j2\tan^{-1}(\omega T_s/2)}. \qquad \textbf{(12.202)}$$

The corresponding contour (12.202) in the $z$ plane is the unit circle and, for $-\infty < \omega < \infty$, it is traversed exactly once. For the more general contour $s = \sigma_0 + j\omega$, where $\sigma_0$ is a constant, the corresponding contour is also a circle but with a different radius and centered on the $\text{Re}(z)$ axis such that as $\omega$ approaches $\pm\infty$, $z$ approaches $-1$ (Figure 12.51).

As the contours in the $s$ plane move to the left, the contours in the $z$ plane become smaller circles whose centers move closer to the $z = -1$ point. The mapping from $s$ to $z$ is a one-to-one mapping, but the distortion of regions becomes more and more severe as $s$ moves away from the origin. A higher sampling rate brings all poles and zeros in

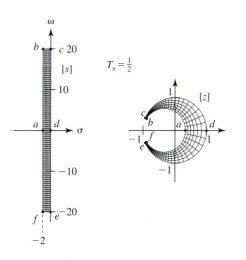

**Figure 12.51**
Mapping of an *s*-plane region into a corresponding *z*-plane region through the bilinear *z* transform.

the $s$ plane nearer to the $z = 1$ point in the $z$ plane where the distortion is minimal. That can be seen by taking the limit as $T_s$ approaches zero. In that limit, $z$ approaches $+1$.

The important difference between the bilinear $z$ transform method and the impulse-invariant or matched $z$-transform methods is that there is no aliasing using the bilinear $z$ transform because of the unique mapping between the $s$ and $z$ planes. However, there is a warping that occurs because of the way the $s = j\omega$ axis is mapped into the unit circle, $|z| = 1$, and vice versa. Letting $z = e^{j\Omega}$, where $\Omega$ is real, determines the unit circle in the $z$ plane. The corresponding contour in the $s$ plane is

$$s = \frac{2}{T_s}\frac{e^{j\Omega} - 1}{e^{j\Omega} + 1} = j\frac{2}{T_s}\tan\left(\frac{\Omega}{2}\right) \tag{12.203}$$

and, since $s = \sigma + j\omega$, $\sigma = 0$, and $\omega = (2/T_s)\tan(\Omega/2)$, or inverting the function, $\Omega = 2\tan^{-1}(\omega T_s/2)$ (Figure 12.52).

For small frequencies, the mapping is almost linear, but the distortion gets progressively worse as we increase frequency because we are forcing high frequencies $\omega$ in the $s$ domain to fit inside the range $-\pi < \Omega < \pi$ in the $z$ domain. This means that the asymptotic behavior of an analog filter as $f$ or $\omega$ approaches positive infinity occurs in the $z$ domain at $\Omega = \pi$, which, through $\Omega = \omega T_s = 2\pi f T_s$, is at $f = f_s/2$, half the sampling rate. Therefore, the warping forces the full infinite range of CT frequencies into the DT frequency range $-\pi < \Omega < \pi$ with a nonlinear invertible function, thereby avoiding aliasing.

The bilinear $z$ transform can also be derived by an entirely different route. Consider a first-order differential equation

$$y'(t) + ay(t) = bx(t). \tag{12.204}$$

Taking the Laplace transform of both sides (with the system initially at rest),

$$sY(s) + aY(s) = bX(s) \tag{12.205}$$

and the transfer function is

$$H_s(s) = \frac{Y(s)}{X(s)} = \frac{b}{s + a}. \tag{12.206}$$

The time function $y(t)$ is the integral of its derivative,

$$y(t) = \int_{t_0}^{t} y'(\lambda)\, d\lambda + y(t_0), \tag{12.207}$$

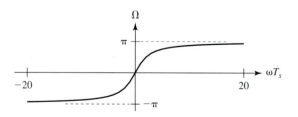

**Figure 12.52**
Frequency warping caused by the bilinear transformation.

and, in particular, over one sampling interval

$$y(nT_s) = \int_{(n-1)T_s}^{nT_s} y'(\lambda)\,d\lambda + y((n-1)T_s). \qquad \textbf{(12.208)}$$

We can approximate the derivative in this range of time by the average of its values at the two endpoints of the sampling interval,

$$y'(t) \cong \frac{y'(nT_s) + y'((n-1)T_s)}{2}. \qquad \textbf{(12.209)}$$

The approximation to an integral, (12.208), using (12.209) is called *trapezoidal rule* numerical integration. Then, performing the integral in (12.208) using this approximation,

$$y(nT_s) = \frac{y'(nT_s) + y'((n-1)T_s)}{2} \int_{(n-1)T_s}^{nT_s} d\lambda + y((n-1)T_s) \qquad \textbf{(12.210)}$$

or

$$y(nT_s) - y((n-1)T_s) = T_s \frac{y'(nT_s) + y'((n-1)T_s)}{2}. \qquad \textbf{(12.211)}$$

From the original differential equation (12.204),

$$y'(nT_s) = b\,x(nT_s) - a\,y(nT_s). \qquad \textbf{(12.212)}$$

Then in (12.211),

$$y(nT_s) - y((n-1)T_s) = T_s \frac{\overbrace{b\,x(nT_s) - a\,y(nT_s)}^{y'(nT_s)} + \overbrace{b\,x((n-1)T_s) - a\,y((n-1)T_s)}^{y'((n-1)T_s)}}{2}$$

$$\textbf{(12.213)}$$

and

$$y(nT_s) - y((n-1)T_s) = \frac{T_s}{2}\{b[x(nT_s) + x((n-1)T_s)] - a[y(nT_s) + y((n-1)T_s)]\}. \qquad \textbf{(12.214)}$$

Since (12.214) only depends on the values of the excitation and response at the sampling instants, we can write the difference equation in terms of DT signals as

$$y[n] - y[n-1] = \frac{T_s}{2}\{b(x[n] + x[n-1]) - a(y[n] + y[n-1])\} \qquad \textbf{(12.215)}$$

and then, $z$-transforming both sides,

$$Y(z) - z^{-1}Y(z) = \frac{T_s}{2}\{b[X(z) + z^{-1}X(z)] - a[Y(z) + z^{-1}Y(z)]\}. \qquad \textbf{(12.216)}$$

Solving for the ratio of $Y(z)$ to $X(z)$,

$$H_z(z) = \frac{Y(z)}{X(z)} = \frac{b(1 + z^{-1})(T_s/2)}{1 - z^{-1} + a(1 + z^{-1})(T_s/2)} \qquad \textbf{(12.217)}$$

or

$$H_z(z) = \frac{b}{(2/T_s)((z-1)/(z+1)) + a}.$$   **(12.218)**

Comparing this result with (12.206), it is apparent that

$$H_z(z) = H_s(s)|_{s \to (2/T_s)((z-1)/(z+1))},$$   **(12.219)**

and this is the bilinear $z$ transform in (12.201). So the bilinear transformation can be thought of as the $z$-transform equivalent of approximating integration by the trapezoidal rule.

The MATLAB signal toolbox has a command `bilinear` for designing a digital filter using the bilinear transformation. The syntax is

```
[zd,pd,kd] = bilinear(z,p,k,fs)
```

where  z = vector of $s$-domain zero locations
      p = vector of $s$-domain pole locations
      k = gain factor of $s$-domain filter
     fs = sampling rate, Hz
     zd = vector of $z$-domain zero locations
     pd = vector of $z$-domain pole locations
     kd = gain factor of $z$-domain filter

For example,

```
»z = [] ; p = -10 ; k = 1 ; fs = 4 ;
»[zd,pd,kd] = bilinear(z,p,k,fs) ;
»zd

zd =

    -1

»pd

pd =

    -0.1111

»kd

kd =

    0.0556
```

## EXAMPLE 12.15

Using the bilinear transformation, design a digital filter to approximate the analog filter whose transfer function is

$$H_s(s) = \frac{1}{s + 10}$$   **(12.220)**

and compare the frequency responses of the analog and digital filters for sampling rates of 4, 20, and 100 Hz.

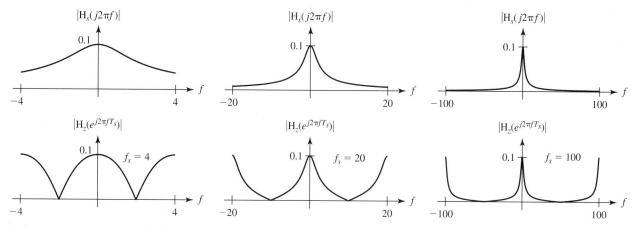

**Figure 12.53**
Magnitude frequency responses of the analog filter and three digital filters designed using the bilinear transform and three different sampling rates.

■ **Solution**

Using the transformation $s \rightarrow \dfrac{2}{T_s}\dfrac{(z-1)}{(z+1)}$,

$$H_z(z) = \frac{1}{(2/T_s)((z-1)/(z+1)) + 10} = \left(\frac{T_s}{2+10T_s}\right)\frac{z+1}{z - (2-10T_s)/(2+10T_s)}.$$
$$\tag{12.221}$$

For a 4-Hz sampling rate,

$$H_z(z) = \frac{1}{18}\frac{z+1}{z+\frac{1}{9}}.$$
$$\tag{12.222}$$

For a 20-Hz sampling rate,

$$H_z(z) = \frac{1}{50}\frac{z+1}{z-\frac{3}{5}}.$$
$$\tag{12.223}$$

For a 100-Hz sampling rate,

$$H_z(z) = \frac{1}{210}\frac{z+1}{z-\frac{19}{21}}.$$
$$\tag{12.224}$$

(Figure 12.53).                                                            ∎

## EXAMPLE 12.16

Using the bilinear z-transform design method, design a digital filter to simulate the analog filter of Example 12.10 whose transfer function is

$$H_s(s) = \frac{s}{s^2 + 400s + 2 \times 10^5}$$
$$\tag{12.225}$$

using the same sampling rate $f_s = 1$ kHz. Compare the frequency responses of the two filters.

■ **Solution**

Using the transformation $s \rightarrow (2/T_s)((z-1)/(z+1))$ and simplifying,

$$H(z) = \frac{z^2 - 1}{z^2 - 1.52z + 0.68}.$$
$$\tag{12.226}$$

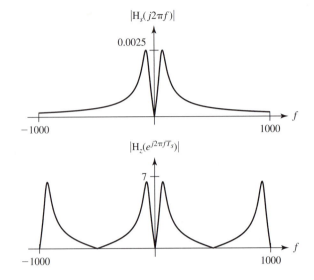

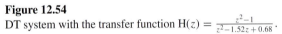

**Figure 12.54**
DT system with the transfer function $H(z) = \frac{z^2 - 1}{z^2 - 1.52z + 0.68}$.

**Figure 12.55**
Magnitude frequency responses of the analog filter and its digital simulation by the bilinear *z* transform technique.

The recursion relation is

$$y[n] = x[n] - x[n-2] + 1.52y[n-1] - 0.68y[n-2] \qquad \textbf{(12.227)}$$

(Figure 12.54). The magnitude frequency responses of the analog and digital filters are compared in Figure 12.55. ∎

## FIR FILTER DESIGN

Even though the commonly used analog filters have infinite-duration impulse responses, since they are stable systems their impulse responses approach zero as time *t* approaches positive infinity. Therefore, another way of simulating an analog filter is to sample the impulse response, as in the impulse-invariant design technique, but then to chop off the impulse response beginning at discrete time $n = N$, where it has fallen to some low level, creating a finite-duration impulse response (Figure 12.56). Digital filters which have finite-duration impulse responses are called FIR filters.

The technique of truncating an analog-filter impulse response can also be extended to approximating *noncausal* filters. The ideal lowpass, highpass, bandpass, and bandstop filters introduced in Chapter 6 all have impulse responses with nonzero values which extend over all time $-\infty < t < \infty$. They are, therefore, noncausal. But if the part of an ideal filter's impulse response which occurs before time $t = 0$ is insignificant in comparison with the part that occurs after time $t = 0$, then it can be truncated, forming an impulse response which is causal. It can also be truncated after some later time when the impulse response has fallen to a low value, as previously described (Figure 12.57). Of course, the truncation of an IIR to an FIR causes some difference between the impulse and frequency responses of the ideal analog and actual digital filters, but that is inherent in digital filter design. So the problem of digital filter design is still an approximation problem. The approximation is just done in a different way in this design method.

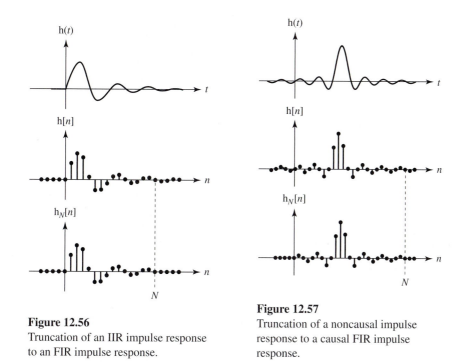

**Figure 12.56**
Truncation of an IIR impulse response to an FIR impulse response.

**Figure 12.57**
Truncation of a noncausal impulse response to a causal FIR impulse response.

Once the impulse response has been truncated and sampled, the design of a FIR filter is quite straightforward. The DT impulse response is in the form of a finite summation of DT impulses,

$$h_N[n] = \sum_{m=0}^{N-1} a_m \delta[n-m]. \tag{12.228}$$

This impulse response can be realized by a DT system of the form illustrated Figure 12.58. One essential difference between this type of filter design and all the others presented so far is that there is no feedback of the response to combine with the excitation to produce the next response. This type of filter has only feed-forward paths. Its transfer function is of the form

$$H_N(z) = \sum_{m=0}^{N-1} a_m z^{-m}. \tag{12.229}$$

This type of filter has a transfer function with $N-1$ poles, all located at $z = 0$, and, therefore, is guaranteed to be absolutely stable regardless of the choice of the coefficients $a$. (As it turns out, this type of filter is also less sensitive to quantization errors in real systems using ADCs to do the sampling and using finite-length, fixed-point arithmetic to do the computation.)

This type of digital filter is an approximation to an analog filter. It is obvious what the difference is between the two impulse responses; one is a truncated version of the other. But what effect does that have in the frequency domain? Truncation in the CT domain between time $t = 0$ and time $t = T$ can be described mathematically by

$$h_T(t) = \begin{cases} h(t) & 0 < t < T \\ 0 & \text{otherwise} \end{cases} = h(t)w(t), \tag{12.230}$$

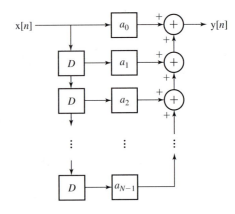

**Figure 12.58**
Prototypical FIR filter.

where $w(t)$ is a window function, in this case

$$w(t) = \text{rect}\left(\frac{t - (T/2)}{T}\right). \tag{12.231}$$

Multiplying by a window function is called *windowing*. Since multiplication in the time domain corresponds to convolution in the frequency domain,

$$H_T(f) = H(f) * W(f). \tag{12.232}$$

If the function $w(t)$ is the time-shifted rectangle in (12.234), its CTFT is a sinc function with a linear phase shift,

$$W(f) = T \, \text{sinc}(Tf)e^{-j\pi fT}. \tag{12.233}$$

Convolution with a sinc function causes a ripple effect on the magnitude frequency response. As the truncation time $T$ increases, the sinc function approaches an impulse. Therefore, the greater $T$ is, the smaller the ripple will be. To illustrate the effect let the ideal transfer function be

$$H(f) = \text{rect}\left(\frac{f}{2B}\right)e^{-j\pi fT}, \tag{12.234}$$

where $B$ is the bandwidth of an ideal lowpass filter. Then the impulse response would be

$$h(t) = 2B \, \text{sinc}\left(2B\left(t - \frac{T}{2}\right)\right). \tag{12.235}$$

Let the truncated impulse response be

$$h_T(t) = 2B \, \text{sinc}\left(2B\left(t - \frac{T}{2}\right)\right) \text{rect}\left(\frac{t - (T/2)}{T}\right). \tag{12.236}$$

Then the corresponding transfer function is

$$H_T(f) = \text{rect}\left(\frac{f}{2B}\right)e^{-j\pi fT} * T \, \text{sinc}(Tf)e^{-j\pi fT} \tag{12.237}$$

(Figure 12.59).

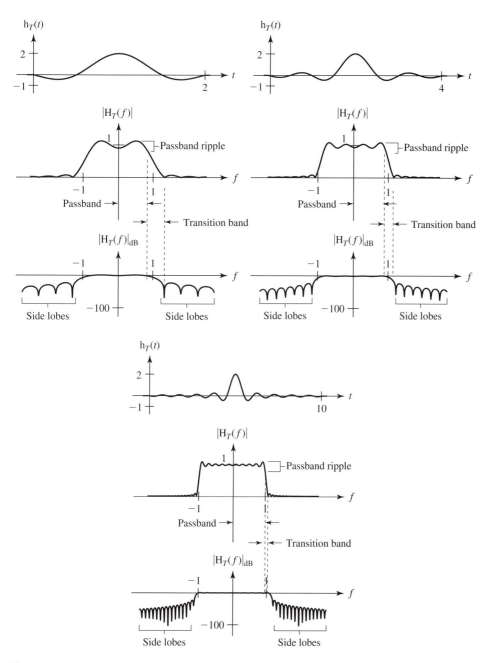

**Figure 12.59**
Three different truncated ideal-lowpass-filter CT impulse responses and their associated magnitude frequency responses.

As the nonzero length of the truncated impulse increases, the frequency response approaches the ideal. The similarity in appearance to the convergence of a CTFS is not accidental. A truncated CTFS exhibits the Gibbs phenomenon in the reconstructed signal. In this case, the truncation occurs in the CT domain and the ripple, which is the equivalent of the Gibbs phenomenon, occurs in the frequency domain. This phenomenon causes the effects marked as "passband ripple" and "side lobes" in Figure 12.59.

The amplitude of the passband ripple does not diminish as the truncation time increases, but it is more and more confined to the region near the cutoff frequency.

The effects of truncation in the DT domain are similar to the effects in the CT domain. The truncated impulse response is

$$
h_N[n] = \begin{cases} h[n] & 0 \le n < N \\ 0 & \text{otherwise} \end{cases} = h[n]w[n],
\tag{12.238}
$$

and the DTFT is

$$
H_N(j\Omega) = H(j\Omega) \circledast W(j\Omega)
\tag{12.239}
$$

(Figure 12.60).

We can also reduce the ripple effect in the frequency domain, without using a longer truncation time, by using a softer truncation in the time domain. That is, instead of windowing the original impulse response with a rectangular function, we could use a differently shaped window function which does not cause such large discontinuities in the truncated impulse response. There are many window shapes whose Fourier transforms have less ripple than a rectangular window's Fourier transform. Some of the most popular are

1.  von Hann or Hanning

$$
w[n] = \frac{1}{2}\left[1 - \cos\left(\frac{2\pi n}{N-1}\right)\right] \qquad 0 \le n < N
\tag{12.240}
$$

2.  Bartlett

$$
w[n] = \begin{cases} \dfrac{2n}{N-1} & 0 \le n \le \dfrac{N-1}{2} \\ 2 - \dfrac{2n}{N-1} & \dfrac{N-1}{2} \le n < N \end{cases}
\tag{12.241}
$$

3.  Hamming

$$
w[n] = 0.54 - 0.46\cos\left(\frac{2\pi n}{N-1}\right) \qquad 0 \le n < N
\tag{12.242}
$$

4.  Blackman

$$
w[n] = 0.42 - 0.5\cos\left(\frac{2\pi n}{N-1}\right) + 0.08\cos\left(\frac{4\pi n}{N-1}\right) \qquad 0 \le n < N
\tag{12.243}
$$

5.  Kaiser

$$
w[n] = \frac{I_0\left(\omega_a\sqrt{((N-1)/2)^2 - (n - (N-1)/2)^2}\right)}{I_0(\omega_a(N-1)/2)}
\tag{12.244}
$$

where $I_0$ is the modified zeroth-order Bessel function of the first kind and $\omega_a$ is a parameter which can be adjusted to trade off between transition-band width and side-lobe amplitude (Figure 12.61). The transforms of these window functions determine how the truncated impulse response will be affected. The magnitudes of the transforms of these common window functions are illustrated in Figure 12.62.

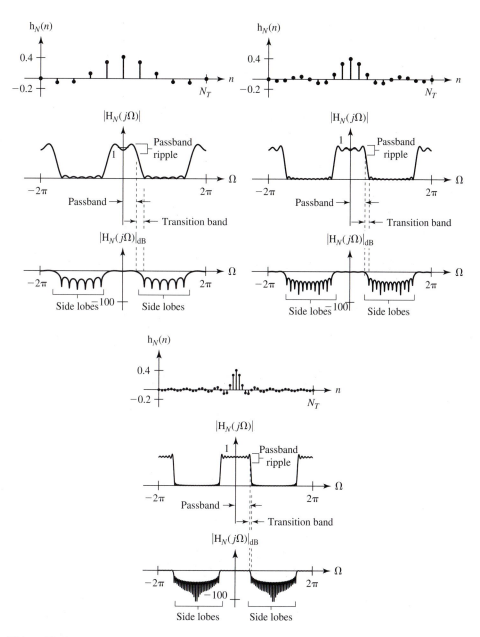

**Figure 12.60**
Three truncated ideal-lowpass-filter DT impulse responses and their associated magnitude
frequency responses.

Looking at the magnitudes of the transforms of the window functions it is apparent that, for a fixed $N$, two design goals are in conflict. When approximating ideal filters with FIR filters, we usually desire to have a very narrow transition band and very
high attenuation in the stopband. The transfer function of the FIR filter is the convolution of the ideal filter's transfer function with the transform of the window function. So
the ideal window function would have a transform that is an impulse, and the corresponding window function would be an infinite-width rectangle. That is impossible, so

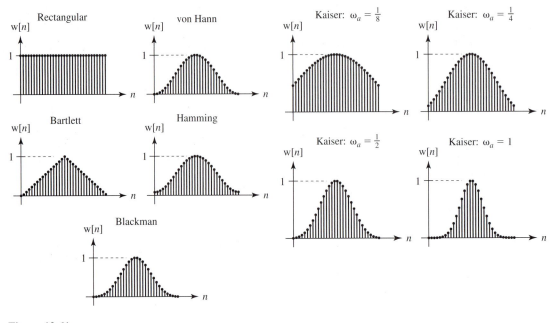

**Figure 12.61**
Window functions ($N = 32$).

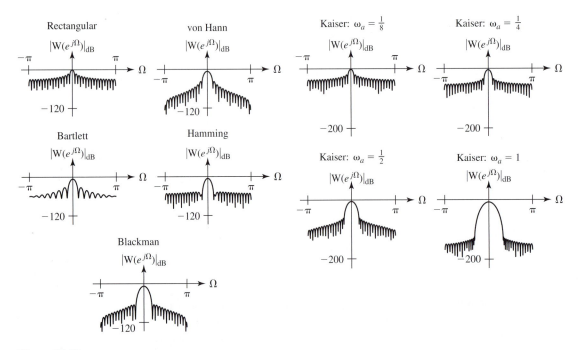

**Figure 12.62**
Magnitudes of $z$ transforms of window functions ($N = 32$).

we must compromise. If we use a finite-width rectangle, the transform is the sinc function and we get the transform illustrated in Figure 12.62 for a rectangle, which makes a relatively fast transition from the peak of its central lobe to its first null, but then the sinc function rises again to a peak which is only about 13 dB below the maximum. In other words, when we convolve it with an ideal lowpass filter frequency response, the transition band is narrow (compared to the other windows), but the stopband attenuation is not very good. Contrast that with the Blackman window. The central lobe width of its transform magnitude is more than twice that of the sinc function, so the transition band will not be as narrow. But once the magnitude goes down, it stays more than 60 dB down. So its stopband attenuation is much better.

One other feature of an FIR filter that makes it attractive is that it can be designed to have a linear phase response. The general form of the FIR impulse response is

$$\mathrm{h}[n] = \mathrm{h}[0]\delta[n] + \mathrm{h}[1]\delta[n-1] + \cdots + \mathrm{h}[N-1]\delta[n-(N-1)], \quad \textbf{(12.245)}$$

its $z$ transform is

$$\mathrm{H}(z) = \mathrm{h}[0] + \mathrm{h}[1]z^{-1} + \cdots + \mathrm{h}[N-1]z^{-(N-1)}, \quad \textbf{(12.246)}$$

and the corresponding frequency response is

$$\mathrm{H}(e^{j\Omega}) = \mathrm{h}[0] + \mathrm{h}[1]e^{-j\Omega} + \cdots + \mathrm{h}[N-1]e^{-j(N-1)\Omega}. \quad \textbf{(12.247)}$$

The length $N$ can be even or odd. First, let $N$ be even, and let the coefficients be chosen such that

$$\mathrm{h}[0] = \mathrm{h}[N-1], \quad \mathrm{h}[1] = \mathrm{h}[N-2], \quad \ldots, \quad \mathrm{h}\left[\frac{N}{2}-1\right] = \mathrm{h}\left[\frac{N}{2}\right] \quad \textbf{(12.248)}$$

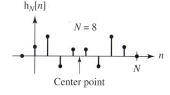

**Figure 12.63**
Example of a symmetric DT impulse response for $N = 8$.

(Figure 12.63). This type of impulse response is symmetric about its center point. Then we can write the frequency response as

$$\mathrm{H}(e^{j\Omega}) = \left\{ \mathrm{h}[0] + \mathrm{h}[0]e^{-j(N-1)\Omega} + \mathrm{h}[1]e^{-j\Omega} + \mathrm{h}[1]e^{-j(N-2)\Omega} + \cdots \right.$$

$$\left. + \mathrm{h}\left[\frac{N}{2}-1\right]e^{-j((N/2)-1)\Omega} + \mathrm{h}\left[\frac{N}{2}-1\right]e^{-j(N/2)\Omega} \right\} \quad \textbf{(12.249)}$$

or

$$\mathrm{H}(e^{j\Omega}) = e^{-j((N-1)/2)\Omega} \left\{ \mathrm{h}[0] \left( e^{j((N-1)/2)\Omega} + e^{-j((N-1)/2)\Omega} \right) \right.$$

$$\left. + \mathrm{h}[1] \left( e^{j((N-3)/2)\Omega} + e^{-j((N-3)/2)\Omega} \right) + \cdots + \mathrm{h}\left[\frac{N}{2}-1\right] \left( e^{-j\Omega} + e^{j\Omega} \right) \right\} \quad \textbf{(12.250)}$$

or

$$\mathrm{H}(e^{j\Omega}) = 2e^{-j((N-1)/2)\Omega} \left\{ \mathrm{h}[0] \cos\left( \left(\frac{N-1}{2}\right)\Omega \right) + \mathrm{h}[1] \cos\left( \left(\frac{N-3}{2}\right)\Omega \right) + \cdots \right.$$

$$\left. + \mathrm{h}\left[\frac{N}{2}-1\right]\cos(\Omega) \right\}. \quad \textbf{(12.251)}$$

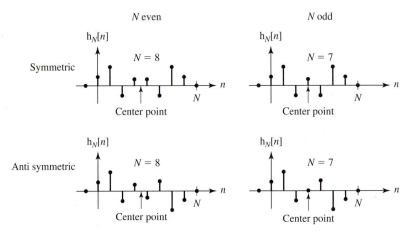

**Figure 12.64**
Examples of symmetric and antisymmetric DT impulse responses for *N* even and *N* odd.

This frequency response consists of the product of a factor $e^{-j((N-1)/2)\Omega}$, which has a linear phase shift with frequency and some other factors which have real values for all $\Omega$. Therefore, the overall frequency response is linear with frequency (except for jumps of $\pi$ rad at frequencies at which the sign of the real part changes). In a similar manner it can be shown that if the filter coefficients are antisymmetric, meaning

$$\text{h}[0] = -\text{h}[N-1], \quad \text{h}[1] = -\text{h}[N-2], \quad \ldots, \quad \text{h}\left[\frac{N}{2}-1\right] = -\text{h}\left[\frac{N}{2}\right], \tag{12.252}$$

then the phase shift is also linear with frequency. If *N* is odd, the results are similar. If the coefficients are symmetric,

$$\text{h}[0] = \text{h}[N-1], \quad \text{h}[1] = \text{h}[N-2], \quad \ldots, \quad \text{h}\left[\frac{N-3}{2}\right] = \text{h}\left[\frac{N+1}{2}\right], \tag{12.253}$$

or antisymmetric,

$$\text{h}[0] = -\text{h}[N-1], \quad \text{h}[1] = -\text{h}[N-2], \quad \ldots, \quad \text{h}\left[\frac{N-3}{2}\right] = -\text{h}\left[\frac{N+1}{2}\right],$$

$$\text{h}\left[\frac{N-1}{2}\right] = 0, \tag{12.254}$$

the phase frequency response is linear. Notice that in the case where *N* is odd there is a center point and, if the coefficients are antisymmetric, the center coefficient $\text{h}[(N-1)/2]$ must be zero (Figure 12.64).

## EXAMPLE 12.17

Using the FIR method, design a digital filter to approximate a single-pole lowpass analog filter whose transfer function is

$$\text{H}_s(s) = \frac{a}{s+a}. \tag{12.255}$$

Truncate the analog filter impulse response at three time constants and sample the truncated impulse response with a time between samples which is one-fourth of the time constant, forming a DT function. Then divide that DT function by $a$ to form the DT impulse response of the digital filter.

a.  Find and plot the magnitude frequency response of the digital filter versus DT radian frequency $\Omega$.

b.  Repeat part (a) for a truncation time of five time constants and a sampling rate of 10 samples per time constant.

■ **Solution**

a.  The impulse response is

$$\mathrm{h}(t) = a e^{-at} \mathrm{u}(t). \tag{12.256}$$

The time constant is $1/a$. Therefore, the truncation time is $3/a$, the time between samples is $1/4a$, and the samples are taken at discrete times $0 \le n < 12$. The FIR impulse response is then

$$\mathrm{h}[n] = a e^{-(n/4)}(\mathrm{u}[n] - \mathrm{u}[n - 12]) = a \sum_{m=0}^{11} e^{-(m/4)} \delta[n - m]. \tag{12.257}$$

The $z$-domain transfer function is

$$\mathrm{H}_z(z) = a \sum_{m=0}^{11} e^{-(m/4)} z^{-m} \tag{12.258}$$

or in terms of DT radian frequency

$$\mathrm{H}_z(e^{j\Omega}) = a \sum_{m=0}^{11} e^{-(m/4)}(e^{j\Omega})^{-m} = a \sum_{m=0}^{11} e^{-m((1/4)+j\Omega)}. \tag{12.259}$$

b.  The truncation time is $5/a$, the time between samples is $1/10a$, and the samples are taken at discrete times $0 \le n < 50$. The FIR impulse response is then

$$\mathrm{h}[n] = a e^{-(n/10)}(\mathrm{u}[n] - \mathrm{u}[n - 50]) = a \sum_{m=0}^{49} e^{-(m/4)} \delta[n - m]. \tag{12.260}$$

The $z$-domain transfer function is

$$\mathrm{H}_z(z) = a \sum_{m=0}^{49} e^{-(m/10)} z^{-m} \tag{12.261}$$

or in terms of DT radian frequency

$$\mathrm{H}_z(e^{j\Omega}) = a \sum_{m=0}^{49} e^{-(m/10)}(e^{j\Omega})^{-m} = a \sum_{m=0}^{49} e^{-m((1/10)+j\Omega)} \tag{12.262}$$

(Figure 12.65). The effects of truncation of the impulse response are visible as the ripple in the frequency response of the first FIR design with the lower sampling rate and shorter truncation time.

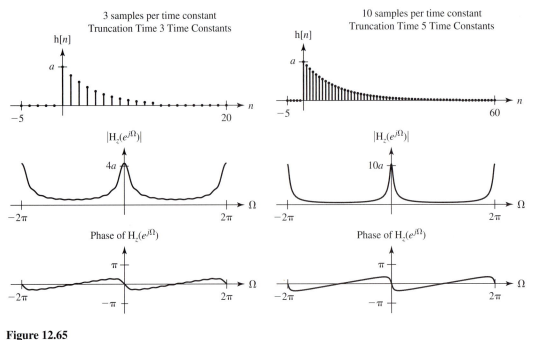

**Figure 12.65**
Impulse responses and frequency responses for the two FIR designs.

## EXAMPLE 12.18

A range of frequencies between 900 and 905 MHz is divided into 20 equal-width channels in which wireless signals may be transmitted. To transmit in any of the channels, a transmitter must send a signal whose amplitude spectrum fits within the constraints of Figure 12.66. The transmitter operates by modulating a sinusoidal carrier whose frequency is the center frequency of one of the channels, with the baseband signal. Before modulating the carrier, the baseband signal, which has an approximately flat spectrum, is prefiltered by a FIR filter which ensures that the transmitted signal meets the constraints of Figure 12.66. Assuming a sampling rate of 2 MHz, design the filter.

### ■ Solution

We know the shape of the ideal baseband CT lowpass filter's impulse response. It is of the form

$$\text{h}(t) = 2Af_m \, \text{sinc}(2f_m(t - t_0)). \tag{12.263}$$

The sampled impulse response is

$$\text{h}[n] = 2Af_m \, \text{sinc}(2f_m(nT_s - t_0)). \tag{12.264}$$

We can set the corner frequency of the ideal lowpass filter to about halfway between 100 and 125 kHz, say 115 kHz or 5.75 percent of the sampling rate. Let the gain constant $A$ be one. The time between samples is 0.5 $\mu s$. The filter will approach the ideal as its length approaches infinity. As a first try set the mean-squared difference between the filter's impulse response and the ideal filter's impulse response to be less than 1 percent and use a rectangular window. We can iteratively determine how long the filter must be by computing the mean-squared difference between the filter and a very long filter. Enforcing a mean-squared error of less than 1 percent sets a filter length of 108 or more. This design yields the frequency responses shown in Figure 12.67.

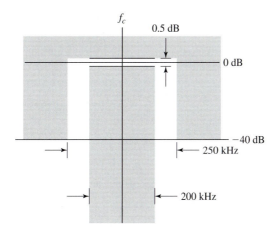

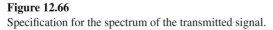

**Figure 12.66**
Specification for the spectrum of the transmitted signal.

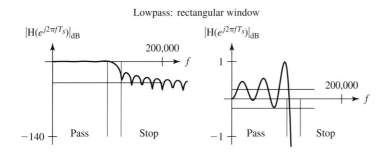

**Figure 12.67**
Frequency response of an FIR filter with a rectangular window and less
than 1 percent error in impulse response.

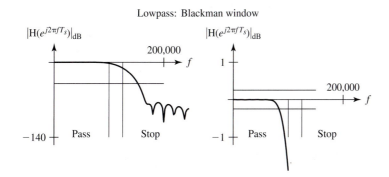

**Figure 12.68**
Frequency response of an FIR filter with a Blackman window and less than
1 percent error in impulse response.

This design is not good enough. The passband ripple is too large and the stopband attenuation is not great enough. We can reduce the ripple by using a window. Let's try a Blackman
window with every other parameter the same (Figure 12.68).

This design is also inadequate. We need to make the mean-squared error smaller. Making
the mean-squared error less than 0.25 percent sets a filter length of 210 and yields the magnitude

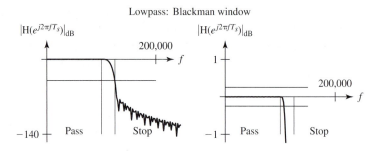

**Figure 12.69**
Frequency response of an FIR filter with a Blackman window and less than 0.25 percent error in impulse response.

frequency response in Figure 12.69. This filter meets specifications. The stopband attenuation just barely meets the specification, and the passband ripple easily meets specification. This design is by no means unique. Many other designs with slightly different corner frequencies, mean-squared errors, or windows could also meet the specification. ∎

## 12.11 DIGITAL FILTER DESIGN AND IMPLEMENTATION WITH MATLAB

In addition to the MATLAB features already mentioned in earlier chapters and in earlier sections of this chapter, there are many other commands and functions in MATLAB which can help in the design of digital filters. Probably the most generally useful function is `filter`. This is a function which actually digitally filters a vector of data representing a finite-time piece of a DT signal. The syntax is

```
y = filter(b,a,x),
```

where `x` is the vector of data to be filtered and `b` and `a` are vectors of coefficients in the recursion relation for the filter. The recursion relation is of the form

```
a(1)*y(n) = b(1)*x(n) + b(2)*x(n-1) + ... + b(nb+1)*x(n-nb)
                      - a(2)*y(n-1) - ... - a(na+1)*y(n-na).
```

[written in MATLAB syntax which uses (·) for arguments of all functions without making a distinction between CT and DT functions]. A related function is `filtfilt`. It operates exactly like `filter` except that it filters the data vector in the normal sense and then filters the resulting data vector backward. This makes the phase shift of the overall filtering operation identically zero at all frequencies and doubles the magnitude effect of the filtering operation.

There are four related functions which design a digital filter. The function `butter` designs an *N*th-order lowpass Butterworth digital filter through the syntax

```
[b,a] = butter[N,wn]
```

where `N` is the filter order and `wn` is the corner frequency expressed as a fraction of half the sampling rate. The function returns filter coefficients `b` and `a`, which can be used directly with `filter` or `filtfilt` to filter a vector of data. This function can also design a bandpass Butterworth filter simply by making `wn` a row vector of two corner frequencies of the form `[w1,w2]`. The passband of the filter is then `w1 < w < w2` in the same sense of being fractions of half the sampling rate. By adding a string `'high'` or `'stop'`, this function can also design highpass and bandstop digital filters.

Here are some examples.

| | |
|---|---|
| `[b,a] = butter[3,0.1]` | Lowpass third-order Butterworth filter, corner frequency of $0.05 f_s$ |
| `[b,a] = butter[4,[0.1 0.2]]` | Bandpass fourth-order Butterworth filter, corner frequencies of $0.05 f_s$ and $0.1 f_s$ |
| `[b,a] = butter[4,0.02,'high']` | Highpass fourth-order Butterworth filter, corner frequency of $0.01 f_s$ |
| `[b,a] = butter[2,[0.32 0.34],'stop']` | Bandstop second-order Butterworth filter, corner frequencies of $0.16 f_s$ and $0.17 f_s$ |

(There are also alternate syntaxes for `butter`. Type `help butter` for details. It can also be used to do analog filter design.)

The other three related digital filter design functions are `cheby1`, `cheby2`, and `ellip`. They design Chebyshev and elliptical filters. Chebyshev and elliptical filters have a narrower transition region for the same filter order than Butterworth filters but do so at the expense of passband and/or stopband ripple. Their syntax is similar except that maximum allowable ripple must also be specified in the passband and/or stopband.

Several standard window functions are available for use with FIR filters. They are `bartlett`, `blackman`, `boxcar` (rectangular), `chebwin` (Chebyshev), `hamming`, `hanning` (von Hann), `kaiser`, and `triang` (similar to, but not identical to, `bartlett`).

The function `freqz` finds the frequency response of a digital filter in a manner similar to the operation of the function `freqs` for analog filters. The syntax of `freqz` is

```
[H,w] = freqz(num,den,N) ;
```

where H = complex frequency response of filter

w = vector of DT frequencies, in rad (not rad/s because it is a DT frequency) at which H is computed

num, den = vectors of coefficients of numerator and denominator of digital filter transfer function

N = number of points

The function `upfirdn` changes the sampling rate of a signal by upsampling, FIR filtering, and downsampling. Its syntax is

```
y = upfirdn(x,h,p,q) ;
```

where y = signal resulting from change of sampling rate

x = signal whose sampling rate is to be changed

h = impulse response of FIR filter

p = factor by which signal is upsampled by zero insertion before filtering

q = factor by which signal is downsampled (decimated) after filtering

These are by no means all the digital signal processing capabilities of MATLAB. Type `help signal` for other functions.

---

### EXAMPLE 12.19

Digitally filter the DT signal

$$x[n] = u[n] - u[n - 10] \qquad (12.265)$$

with a third-order highpass digital Butterworth filter whose corner DT frequency is $\pi/6$ rad.

#### ■ Solution

```
%    Use 30 points to represent the excitation, x, and the response, y.

N = 30 ;

%    Generate the excitation signal.

n = 0:N-1 ; x = uDT(n) - uDT(n-10) ;

%    Design a third-order highpass digital Butterworth filter.

[num,den] = butter(3,1/6,'high') ;

%  Filter the signal.

y = filter(num,den,x) ;
```

The excitation and response are illustrated in Figure 12.70.

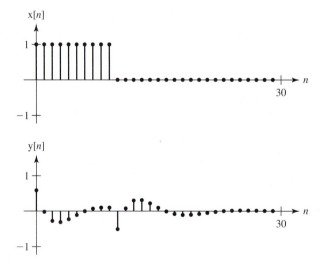

**Figure 12.70**
Excitation and response of a third-order highpass digital
Butterworth filter.

## 12.12 STANDARD REALIZATIONS OF SYSTEMS

The realization of DT systems very closely parallels the realization of CT systems. The same general techniques apply, and the same types of realizations result. The canonical realization is directly formed from the transfer function of the form

$$H(z) = \frac{Y(z)}{X(z)} = \frac{\sum_{k=0}^{N} b_k z^k}{\sum_{k=0}^{N} a_k z^k} = \frac{b_N z^N + b_{N-1} z^{N-1} + \cdots + b_1 z + b_0}{z^N + a_{N-1} z^{N-1} + \cdots + a_1 z + a_0} \qquad a_N = 1,$$

$$\text{(12.266)}$$

which can be separated into the cascade of two subsystem transfer functions,

$$H_1(z) = \frac{Y_1(z)}{X(z)} = \frac{1}{z^N + a_{N-1} z^{N-1} + \cdots + a_1 z + a_0} \qquad \text{(12.267)}$$

and

$$H_2(z) = \frac{Y(z)}{Y_1(z)} = b_N z^N + b_{N-1} z^{N-1} + \cdots + b_1 z + b_0. \qquad \text{(12.268)}$$

From (12.267),

$$z^N Y_1(z) = X(z) - [a_{N-1} z^{N-1} Y_1(z) + \cdots + a_1 z Y_1(z) + a_0 Y_1(z)] \qquad \text{(12.269)}$$

(Figure 12.71). Then, adding the second subsystem $H_2(z)$, we get the canonical realization of the overall system (Figure 12.72).

We can also realize the system in a cascade or parallel form from the factored form of the transfer function,

$$H(z) = A \frac{z - z_1}{z - p_1} \frac{z - z_2}{z - p_2} \cdots \frac{z - z_N}{z - p_N} \frac{1}{z - p_{N+1}} \frac{1}{z - p_{N+2}} \cdots \frac{1}{z - p_D} \qquad \text{(12.270)}$$

(Figure 12.73), or the partial-fraction form,

$$H(z) = \frac{K_1}{z - p_1} + \frac{K_2}{z - p_2} + \cdots + \frac{K_D}{z - p_D} \qquad \text{(12.271)}$$

(Figure 12.74).

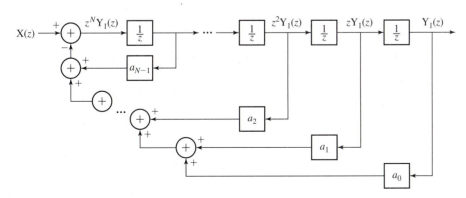

**Figure 12.71**
Realization of $H_1(z)$.

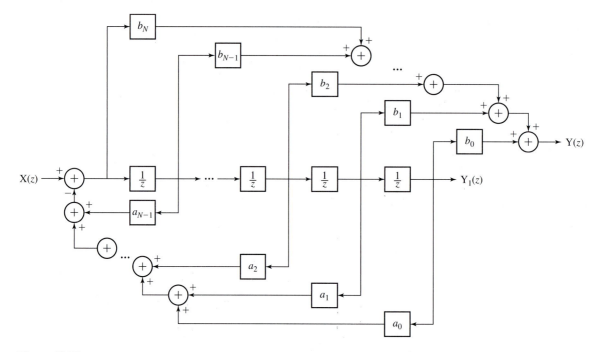

**Figure 12.72**
Overall canonical system realization.

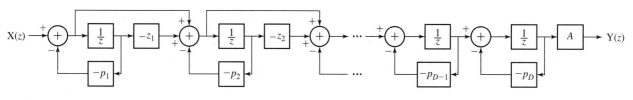

**Figure 12.73**
Overall cascade system realization.

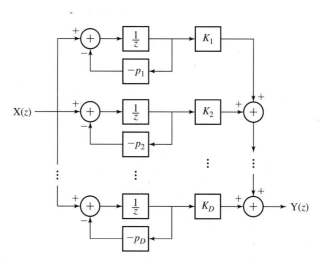

**Figure 12.74**
Overall parallel system realization.

DT systems are actually built using digital hardware (logic gates, flip-flops, counters, multiplexers, adders, etc.). In these systems the signals are all in the form of binary numbers with a finite number of bits. The operations are usually performed in *fixed-point* arithmetic which means that all the signals are quantized to a finite number of possible values and, therefore, are not exact representations of the ideal signals. This type of design usually leads to the fastest and most efficient system, but the round-off error between the ideal signals and the actual signals must be managed to avoid noisy, or in some cases even unstable, system operation. The analysis of such errors is beyond the scope of this text, but, generally speaking, the cascade and parallel realizations are more tolerant and forgiving of such errors than the canonical realization.

## 12.13 STATE-SPACE SIGNAL AND SYSTEM ANALYSIS

As is true for CT systems, the analysis of large DT systems is best done using a systematic technique like state-space analysis. This analysis directly parallels state-space analysis of CT systems. We still need to identify a number of state variables, which equals the order of the system. We begin with an example system (Figure 12.75).

In CT system state-space realization the derivatives of the state variables are set equal to a linear combination of the state variables and the excitations. In DT system state-space realization the *next* state-variable values are equated to a linear combination of the *present* state-variable values and the present excitations. The system and output equations are

$$
\begin{aligned}
\mathbf{q}[n + 1] &= \mathbf{A}\mathbf{q}[n] + \mathbf{B}\mathbf{x}[n] \\
\mathbf{y}[n] &= \mathbf{C}\mathbf{q}[n] + \mathbf{D}\mathbf{x}[n].
\end{aligned}
\tag{12.272}
$$

We choose the state variables in the simplest way, as the responses of the delay blocks. Then the state variables and matrices are

$$
\mathbf{q}[n] = \begin{bmatrix} q_1[n] \\ q_2[n] \end{bmatrix} \qquad
\mathbf{A} = \begin{bmatrix} \frac{1}{3} & \frac{1}{4} \\ \frac{1}{2} & 0 \end{bmatrix} \qquad
\mathbf{B} = \begin{bmatrix} 1 & 0 \\ 0 & 1 \end{bmatrix} \qquad \text{and} \qquad
\mathbf{x}[n] = \begin{bmatrix} x_1[n] \\ x_2[n] \end{bmatrix}
\tag{12.273}
$$

$$
\mathbf{y}[n] = [y[n]] \qquad \mathbf{C} = \begin{bmatrix} 2 & 3 \end{bmatrix} \qquad \mathbf{D} = \begin{bmatrix} 0 & 0 \end{bmatrix}.
\tag{12.274}
$$

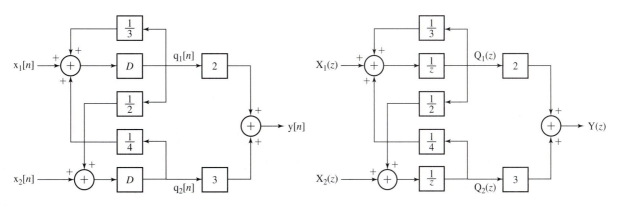

**Figure 12.75**
An example DT system.

**Table 12.1** States and response found by recursion

| $n$ | $q_1[n]$ | $q_2[n]$ | $y[n]$ |
|---|---|---|---|
| 0 | 0 | 0 | 0 |
| 1 | 1 | 1 | 5 |
| 2 | 1.5833 | 0.5 | 4.667 |
| 3 | 1.6528 | 0.7917 | 5.681 |
| $\vdots$ | $\vdots$ | $\vdots$ | $\vdots$ |

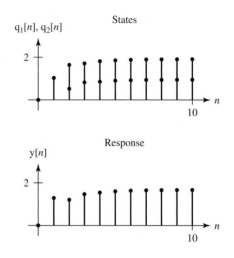

**Figure 12.76**
The states and response of the DT system.

A direct way of solving the state equations is by recursion. To illustrate the process let the excitation vector be

$$\mathbf{x}[n] = \begin{bmatrix} \mathrm{u}[n] \\ \delta[n] \end{bmatrix} \tag{12.275}$$

and let the system be initially at rest, $\mathbf{q}[0] = [0]$. Then doing the recursion directly from (12.272), we get the values in Table 12.1 and the plot in Figure 12.76.

We can generalize the recursion process. From (12.272),

$$\mathbf{q}[1] = \mathbf{A}\mathbf{q}[0] + \mathbf{B}\mathbf{x}[0]$$

$$\mathbf{q}[2] = \mathbf{A}\mathbf{q}[1] + \mathbf{B}\mathbf{x}[1] = \mathbf{A}^2\mathbf{q}[0] + \mathbf{A}\mathbf{B}\mathbf{x}[0] + \mathbf{B}\mathbf{x}[1]$$

$$\mathbf{q}[3] = \mathbf{A}\mathbf{q}[2] + \mathbf{B}\mathbf{x}[2] = \mathbf{A}^3\mathbf{q}[0] + \mathbf{A}^2\mathbf{B}\mathbf{x}[0] + \mathbf{A}\mathbf{B}\mathbf{x}[1] + \mathbf{B}\mathbf{x}[2] \tag{12.276}$$

$$\vdots$$

$$\mathbf{q}[n] = \mathbf{A}^n\mathbf{q}[0] + \mathbf{A}^{n-1}\mathbf{B}\mathbf{x}[0] + \mathbf{A}^{n-2}\mathbf{B}\mathbf{x}[1] + \cdots + \mathbf{A}^1\mathbf{B}\mathbf{x}[n-2] + \mathbf{A}^0\mathbf{B}\mathbf{x}[n-1]$$

and

$$\mathbf{y}[1] = \mathbf{C}\mathbf{q}[1] + \mathbf{D}\mathbf{x}[1] = \mathbf{C}\mathbf{A}\mathbf{q}[0] + \mathbf{C}\mathbf{B}\mathbf{x}[0] + \mathbf{D}\mathbf{x}[1]$$

$$\mathbf{y}[2] = \mathbf{C}\mathbf{q}[2] + \mathbf{D}\mathbf{x}[2] = \mathbf{C}\mathbf{A}^2\mathbf{q}[0] + \mathbf{C}\mathbf{A}\mathbf{B}\mathbf{x}[0] + \mathbf{C}\mathbf{B}\mathbf{x}[1] + \mathbf{D}\mathbf{x}[2]$$

$$\mathbf{y}[3] = \mathbf{C}\mathbf{q}[3] + \mathbf{D}\mathbf{x}[3] = \mathbf{C}\mathbf{A}^3\mathbf{q}[0] + \mathbf{C}\mathbf{A}^2\mathbf{B}\mathbf{x}[0] + \mathbf{C}\mathbf{A}\mathbf{B}\mathbf{x}[1] + \mathbf{C}\mathbf{B}\mathbf{x}[2] + \mathbf{D}\mathbf{x}[3]$$

$$\vdots$$

$$\mathbf{y}[n] = \mathbf{C}\mathbf{A}^n\mathbf{q}[0] + \mathbf{C}\mathbf{A}^{n-1}\mathbf{B}\mathbf{x}[0] + \mathbf{C}\mathbf{A}^{n-2}\mathbf{B}\mathbf{x}[1] + \cdots + \mathbf{C}\mathbf{A}^0\mathbf{B}\mathbf{x}[n-1] + \mathbf{D}\mathbf{x}[n].$$

$$\tag{12.277}$$

These can be written in the forms,

$$\mathbf{q}[n] = \mathbf{A}^n \mathbf{q}[0] + \sum_{m=0}^{n-1} \mathbf{A}^{n-m-1} \mathbf{B}\mathbf{x}[m] \tag{12.278}$$

and

$$\mathbf{y}[n] = \mathbf{C}\mathbf{A}^n \mathbf{q}[0] + \mathbf{C} \sum_{m=0}^{n-1} \mathbf{A}^{n-m-1} \mathbf{B}\mathbf{x}[m] + \mathbf{D}\mathbf{x}[n]. \tag{12.279}$$

In (12.278) the term $\mathbf{A}^n \mathbf{q}[0]$ is the the zero-input response caused by the initial state of the system $\mathbf{q}[0]$. The matrix $\mathbf{A}^n$ is called the *state transition* matrix and is often denoted by the symbol $\boldsymbol{\phi}[n]$. The name comes from the idea that the transition from one state to another is controlled by the dynamics of the system as characterized by the matrix $\mathbf{A}^n = \boldsymbol{\phi}[n]$. The second term $\sum_{m=0}^{n-1} \mathbf{A}^{n-m-1} \mathbf{B}\mathbf{x}[m]$ is the zero-state response of the system. This term is equivalent to the DT convolution sum $\mathbf{A}^{n-1}\mathrm{u}[n-1] * \mathbf{B}\mathbf{x}[n]\mathrm{u}[n]$ or, under the usual assumption in state-variable analysis that $\mathbf{x}[n]$ is zero for negative discrete time,

$$\sum_{m=0}^{n-1} \mathbf{A}^{n-m-1} \mathbf{B}\mathbf{x}[m] = \mathbf{A}^{n-1}\mathrm{u}[n-1] * \mathbf{B}\mathbf{x}[n]. \tag{12.280}$$

Then we can rewrite (12.278) as

$$\mathbf{q}[n] = \underbrace{\boldsymbol{\phi}[n]\mathbf{q}[0]}_{\substack{\text{zero-excitation} \\ \text{response}}} + \underbrace{\boldsymbol{\phi}[n-1]\mathrm{u}[n-1] * \mathbf{B}\mathbf{x}[n]}_{\substack{\text{zero-state} \\ \text{response}}} \tag{12.281}$$

By a similar process we can rewrite (12.279) as

$$\mathbf{y}[n] = \mathbf{C}\boldsymbol{\phi}[n]\mathbf{q}[0] + \mathbf{C}\boldsymbol{\phi}[n-1]\mathrm{u}[n-1] * \mathbf{B}\mathbf{x}[n] + \mathbf{D}\mathbf{x}[n]. \tag{12.282}$$

The last two results (12.281) and (12.282) are the DT-domain solutions for the states and the responses of the system.

We can also solve the state equations by using the $z$ transform. Transforming both sides of the system equation in (12.272)

$$z\mathbf{Q}(z) - z\mathbf{q}[0] = \mathbf{A}\mathbf{Q}(z) + \mathbf{B}\mathbf{X}(z). \tag{12.283}$$

We can solve for the state-variable vector as

$$\mathbf{Q}(z) = [z\mathbf{I} - \mathbf{A}]^{-1}[\mathbf{B}\mathbf{X}(z) + z\mathbf{q}[0]] = \underbrace{[z\mathbf{I} - \mathbf{A}]^{-1}\mathbf{B}\mathbf{X}(z)}_{\substack{\text{zero-state} \\ \text{response}}} + \underbrace{z[z\mathbf{I} - \mathbf{A}]^{-1}\mathbf{q}[0]}_{\substack{\text{zero-excitation} \\ \text{response}}}. \tag{12.284}$$

A comparison of (12.284) with (12.281) shows that

$$\boldsymbol{\phi}[n] \overset{\mathcal{Z}}{\longleftrightarrow} z[z\mathbf{I} - \mathbf{A}]^{-1}. \tag{12.285}$$

Therefore, it is consistent and logical to define the $z$ transform of the state transition matrix as

$$\Phi(z) = z[z\mathbf{I} - \mathbf{A}]^{-1}. \tag{12.286}$$

Notice the similarity to the corresponding result from CT state-space analysis, $\Phi(s) = [s\mathbf{I} - \mathbf{A}]^{-1}$.

To demonstrate a numerical solution, let the excitation vector again be

$$\mathbf{x}[n] = \begin{bmatrix} \mathrm{u}[n] \\ \delta[n] \end{bmatrix} \tag{12.287}$$

and let the system again be initially at rest $\mathbf{q}[0] = [0]$. Then

$$\mathbf{Q}(z) = \begin{bmatrix} z - \frac{1}{3} & -\frac{1}{4} \\ -\frac{1}{2} & z \end{bmatrix}^{-1} \begin{bmatrix} 1 & 0 \\ 0 & 1 \end{bmatrix} \begin{bmatrix} z/(z-1) \\ 1 \end{bmatrix} \tag{12.288}$$

or

$$\mathbf{Q}(z) = \begin{bmatrix} \dfrac{z^2 + (z/4) - \frac{1}{4}}{z^3 - \frac{4}{3}z^2 + \frac{5}{24}z + \frac{1}{8}} \\[3mm] \dfrac{z^2 - \frac{5}{6}z + \frac{1}{3}}{z^3 - \frac{4}{3}z^2 + \frac{5}{24}z + \frac{1}{8}} \end{bmatrix} = \begin{bmatrix} \dfrac{z^2 + (z/4) - \frac{1}{4}}{(z-1)(z-0.5575)(z+0.2242)} \\[3mm] \dfrac{z^2 - \frac{5}{6}z + \frac{1}{3}}{(z-1)(z-0.5575)(z+0.2242)} \end{bmatrix}. \tag{12.289}$$

Expanding in partial fractions,

$$\mathbf{Q}(z) = \begin{bmatrix} \dfrac{1.846}{z-1} - \dfrac{0.578}{z-0.5575} - \dfrac{0.268}{z+0.2242} \\[3mm] \dfrac{0.923}{z-1} - \dfrac{0.519}{z-0.5575} + \dfrac{0.596}{z+0.2242} \end{bmatrix} \tag{12.290}$$

Taking the inverse z transform

$$\mathbf{q}[n] = \begin{bmatrix} 1.846 - 0.578(0.5575)^{(n-1)} - 0.268(-0.2242)^{(n-1)} \\ 0.923 - 0.519(0.5575)^{(n-1)} + 0.596(-0.2242)^{(n-1)} \end{bmatrix} \mathrm{u}[n-1] \tag{12.291}$$

After finding the solution for the state-variable vector, finding the response vector is easy. It is

$$\mathbf{y}[n] = \left[ 6.461 - 2.713(0.5575)^{(n-1)} + 1.252(-0.2242)^{(n-1)} \right] \mathrm{u}[n-1] \tag{12.292}$$

Substituting values for $n$ into (12.291) and (12.292) we get Table 12.2 which agrees exactly with Table 12.1, verifying that the two solution methods, using recursion and the z transform, yield the same result.

From the state-space equations we can find the matrix transfer function relating all responses to all excitations. Starting with (12.283),

$$z\mathbf{Q}(z) - z\mathbf{q}[0] = \mathbf{AQ}(z) + \mathbf{BX}(z), \tag{12.293}$$

and setting the initial state to zero (which it must be for a transfer function to be defined), we can solve for $\mathbf{Q}(z)$ as

$$\mathbf{Q}(z) = [z\mathbf{I} - \mathbf{A}]^{-1}\mathbf{BX}(z) = z^{-1}\Phi(z)\mathbf{BX}(z), \tag{12.294}$$

The response $\mathbf{Y}(z)$ is

$$\mathbf{Y}(z) = \mathbf{CQ}(z) + \mathbf{DX}(z) = z^{-1}\mathbf{C}\Phi(z)\mathbf{BX}(z) + \mathbf{DX}(z), \tag{12.295}$$

**Table 12.2** States and response found from closed-form solutions

| $n$ | $q_1[n]$ | $q_2[n]$ | $y[n]$ |
|---|---|---|---|
| 0 | 0 | 0 | 0 |
| 1 | 1 | 1 | 5 |
| 2 | 1.5833 | 0.5 | 4.667 |
| 3 | 1.6528 | 0.7917 | 5.681 |
| $\vdots$ | $\vdots$ | $\vdots$ | $\vdots$ |

and the transfer function which is the ratio of the response to the excitation is

$$\mathbf{H}(z) = z^{-1}\mathbf{C}\Phi(z)\mathbf{B} + \mathbf{D} = \mathbf{C}[z\mathbf{I} - \mathbf{A}]^{-1}\mathbf{B} + \mathbf{D}. \qquad (12.296)$$

Everything derived in CT state-space analysis about transformation from one set of state variables to another set applies exactly to DT state-space analysis. If $\mathbf{q}_2[n] = \mathbf{T}\mathbf{q}_1[n]$ and $\mathbf{q}_1[n+1] = \mathbf{A}_1\mathbf{q}_1[n] + \mathbf{B}_1\mathbf{x}[n]$, then $\mathbf{q}_2[n+1] = \mathbf{A}_2\mathbf{q}_2[n] + \mathbf{B}_2\mathbf{x}[n]$, where $\mathbf{A}_2 = \mathbf{T}\mathbf{A}_1\mathbf{T}^{-1}$ and $\mathbf{B}_2 = \mathbf{T}\mathbf{B}_1$, and $\mathbf{y}[n] = \mathbf{C}_2\mathbf{q}_2[n] + \mathbf{D}_2\mathbf{x}[n]$, where $\mathbf{C}_2 = \mathbf{C}_1\mathbf{T}^{-1}$ and $\mathbf{D}_2 = \mathbf{D}_1$.

### EXAMPLE 12.20

Find the responses of the system in Figure 12.77 which is initially at rest to the excitations

$$x_1[n] = u[n] \qquad \text{and} \qquad x_2[n] = -u[n-2]. \qquad (12.297)$$

**■ Solution**

The system equations are

$$q_1[n+1] = q_2[n] - x_1[n]$$
$$q_2[n+1] = x_2[n] - \left[\frac{4}{5}(q_2[n] - x_1[n]) + \frac{2}{3}(q_1[n] + x_1[n])\right], \qquad (12.298)$$

and the output equations are

$$y_1[n] = \frac{4}{5}(q_2[n] - x_1[n]) + \frac{2}{3}(q_1[n] + x_1[n]) \qquad (12.299)$$
$$y_2[n] = q_1[n] + x_1[n].$$

In standard matrix form,

$$\mathbf{q}[n+1] = \mathbf{A}\mathbf{q}[n] + \mathbf{B}\mathbf{x}[n], \qquad (12.300)$$

where

$$\mathbf{A} = \begin{bmatrix} 0 & 1 \\ -\frac{2}{3} & -\frac{4}{5} \end{bmatrix} \qquad \text{and} \qquad \mathbf{B} = \begin{bmatrix} -1 & 0 \\ \frac{2}{15} & 1 \end{bmatrix}, \qquad (12.301)$$

and

$$\mathbf{y}[n] = \mathbf{C}\mathbf{q}[n] + \mathbf{D}\mathbf{x}[n], \qquad (12.302)$$

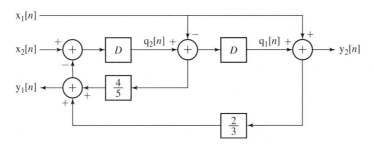

**Figure 12.77**
A DT system.

where

$$
\mathbf{C} = \begin{bmatrix} \frac{2}{3} & \frac{4}{5} \\ 1 & 0 \end{bmatrix} \quad \text{and} \quad \mathbf{D} = \begin{bmatrix} -\frac{2}{15} & 0 \\ 1 & 0 \end{bmatrix}. \tag{12.303}
$$

The system is initially at rest, so we can use the transfer function to find the responses. The matrix transfer function is

$$
\mathbf{H}(z) = \mathbf{C}[z\mathbf{I} - \mathbf{A}]^{-1}\mathbf{B} + \mathbf{D} = \begin{bmatrix} \frac{2}{3} & \frac{4}{5} \\ 1 & 0 \end{bmatrix} \begin{bmatrix} z & -1 \\ \frac{2}{3} & z + \frac{4}{5} \end{bmatrix}^{-1} \begin{bmatrix} -1 & 0 \\ \frac{2}{15} & 1 \end{bmatrix} + \begin{bmatrix} -\frac{2}{15} & 0 \\ 1 & 0 \end{bmatrix} \tag{12.304}
$$

or

$$
\mathbf{H}(z) = \frac{1}{z^2 + \frac{4}{5}z + \frac{2}{3}} \begin{bmatrix} \frac{2}{3} & \frac{4}{5} \\ 1 & 0 \end{bmatrix} \begin{bmatrix} z + \frac{4}{5} & 1 \\ -\frac{2}{3} & z \end{bmatrix} \begin{bmatrix} -1 & 0 \\ \frac{2}{15} & 1 \end{bmatrix} + \begin{bmatrix} -\frac{2}{15} & 0 \\ 1 & 0 \end{bmatrix} \tag{12.305}
$$

or

$$
\mathbf{H}(z) = \frac{1}{z^2 + \frac{4}{5}z + \frac{2}{3}} \begin{bmatrix} -\frac{42}{75}z + \frac{4}{45} & \frac{2}{3} + \frac{4}{5}z \\ -z - \frac{2}{3} & 1 \end{bmatrix} + \begin{bmatrix} -\frac{2}{15} & 0 \\ 1 & 0 \end{bmatrix}. \tag{12.306}
$$

The *z* transform of the excitation vector is

$$
\mathbf{X}(z) = \begin{bmatrix} \dfrac{z}{z - 1} \\ -\dfrac{z^{-1}}{z - 1} \end{bmatrix}. \tag{12.307}
$$

Then the *z*-domain response vector is

$$
\mathbf{Y}(z) = \mathbf{H}(z)\mathbf{X}(z) = \left( \frac{1}{z^2 + \frac{4}{5}z + \frac{2}{3}} \begin{bmatrix} -\frac{42}{75}z + \frac{4}{45} & \frac{2}{3} + \frac{4}{5}z \\ -z - \frac{2}{3} & 1 \end{bmatrix} + \begin{bmatrix} -\frac{2}{15} & 0 \\ 1 & 0 \end{bmatrix} \right) \begin{bmatrix} \frac{z}{z-1} \\ -\frac{z^{-1}}{z-1} \end{bmatrix} \tag{12.308}
$$

or

$$
\mathbf{Y}(z) = \frac{1}{z^2 + \frac{4}{5}z + \frac{2}{3}} \begin{bmatrix} \left(-\frac{42}{75}z + \frac{4}{45}\right)\frac{z}{z-1} - \left(\frac{2}{3} + \frac{4}{5}z\right)\frac{z^{-1}}{z-1} \\ \left(-z - \frac{2}{3}\right)\frac{z}{z-1} - \frac{z^{-1}}{z-1} \end{bmatrix} + \begin{bmatrix} -\frac{2}{15}\frac{z}{z-1} \\ \frac{z}{z-1} \end{bmatrix} \tag{12.309}
$$

or

$$
\mathbf{Y}(z) = \frac{z^{-1}}{z^2 + \frac{4}{5}z + \frac{2}{3}} \begin{bmatrix} -\frac{-\frac{42}{75}z^3 - \frac{4}{45}z^2 + \frac{4}{5}z + \frac{2}{3}}{z-1} \\ -\frac{z^3 + \frac{2}{3}z^2 + 1}{z-1} \end{bmatrix} + \begin{bmatrix} -\frac{2}{15}\frac{z}{z-1} \\ \frac{z}{z-1} \end{bmatrix}
$$

or

$$
\mathbf{Y}(z) = -z^{-1} \begin{bmatrix} \dfrac{\frac{42}{75}z^3 - \frac{4}{45}z^2 + \frac{4}{5}z + \frac{2}{3}}{(z-1)\left(z^2 + \frac{4}{5}z + \frac{2}{3}\right)} \\ \dfrac{z^3 + \frac{2}{3}z^2 + 1}{(z-1)\left(z^2 + \frac{4}{5}z + \frac{2}{3}\right)} \end{bmatrix} + \begin{bmatrix} -\frac{2}{15}\frac{z}{z-1} \\ \frac{z}{z-1} \end{bmatrix}.
$$

Expanding in partial fractions,

$$\mathbf{Y}(z) = -z^{-1} \begin{bmatrix} 0.56 + \frac{0.7856}{z-1} - \frac{0.7625\,z+0.5163}{z^2+\frac{4}{5}z+\frac{2}{3}} \\[2mm] 1 + \frac{1.081}{z-1} - \frac{0.2144\,z+0.9459}{z^2+\frac{4}{5}z+\frac{2}{3}} \end{bmatrix} + \begin{bmatrix} -\frac{2}{15}\frac{z}{z-1} \\[2mm] \frac{z}{z-1} \end{bmatrix} \qquad (12.310)$$

or

$$\mathbf{Y}(z) = -z^{-1} \begin{bmatrix} 0.56 + \frac{0.7856}{z-1} - \frac{0.7625}{0.7118}\frac{0.7118\,z}{z^2+\frac{4}{5}z+\frac{2}{3}} - z^{-1}\frac{0.5163}{0.7118}\frac{0.7118\,z}{z^2+\frac{4}{5}z+\frac{2}{3}} \\[2mm] 1 + \frac{1.081}{z-1} - \frac{0.2144}{0.7118}\frac{0.7118\,z}{z^2+\frac{4}{5}z+\frac{2}{3}} - z^{-1}\frac{0.9459}{0.7118}\frac{0.7118\,z}{z^2+\frac{4}{5}z+\frac{2}{3}} \end{bmatrix} + \begin{bmatrix} -\frac{2}{15}\frac{z}{z-1} \\[2mm] \frac{z}{z-1} \end{bmatrix}.$$

Taking the inverse $z$ transform,

$$\mathbf{y}[n] = - \begin{bmatrix} (0.56\delta[n-1] + 0.7856\,\mathrm{u}[n-2] \\ \quad -1.071(0.8165)^{n-1}\sin(2.083(n-1))\,\mathrm{u}[n-1] \\ \quad - 0.7253(0.8165)^{n-2}\sin(2.083(n-2))\,\mathrm{u}[n-2]) \\ (\delta[n-1] + 1.081\mathrm{u}[n-2] \\ \quad - 0.3012(0.8165)^{n-1}\sin(2.083(n-1))\,\mathrm{u}[n-1] \\ \quad - 1.329(0.8165)^{n-2}\sin(2.083(n-2))\,\mathrm{u}[n-2]) \end{bmatrix} + \begin{bmatrix} -\frac{2}{15}\mathrm{u}[n] \\[2mm] \mathrm{u}[n] \end{bmatrix}. \qquad (12.311)$$

The MATLAB system-object concept includes DT state-space models of systems just as it does for CT systems. The fundamental function is `ss`, and its syntax is

```
sys = ss(A,B,C,D,Ts) ;
```

where `A`, `B`, `C`, and `D` are the state-space-representation matrices of the same names and `Ts` is the time between samples. The function `ssdata` extracts state-space matrices from a system description in a manner analogous to `zpkdata` and `tfdata`. The function `ss2ss` transforms a state-space model to another state-space model. The syntax is

```
sys = ss2ss(sys,T) ;
```

where `T` is the transformation matrix.

## 12.14 SUMMARY OF IMPORTANT POINTS

1. Block-diagram reduction for DT systems follows exactly the same rules as block-diagram reduction for CT systems.
2. The poles of a stable DT system transfer function all lie inside the unit circle of the $z$ plane.
3. If a sinusoid is applied to a stable DT system at time $n = 0$, the response will eventually be the same as the response of that DT system to the same sinusoid applied for all time.
4. The dynamic behavior and frequency response of a DT system are completely determined by its pole–zero diagram except for a frequency-independent gain.
5. The Jury stability test for DT systems is analogous to the Routh–Hurwitz test for stability for CT systems.
6. A root locus can be drawn for DT systems just as it is drawn for CT systems, but the interpretation is that when the locus moves outside the unit circle the system becomes unstable.

7. CT systems can be approximated by sampled-data systems and, as the sampling rate is increased, the approximation becomes better.
8. Digital filters can be designed by multiple methods to approximate analog filters. Each method yields a different design, but all methods approach the same design as the sampling rate is increased.
9. Digital filters are generally classified as infinite-duration impulse response (IIR) or finite-duration impulse response (FIR).
10. An FIR digital filter is inherently stable and can be designed to achieve linear phase in its passband.
11. The standard DT system realizations are analogous to the corresponding standard CT system realizations.
12. State-space analysis methods are very useful in the analysis of multiple-excitation, multiple-response DT systems and have many similarities to state-space analysis of CT systems.

## EXERCISES WITH ANSWERS

1. Find the transfer functions for these systems by block-diagram reduction.

    *a.*

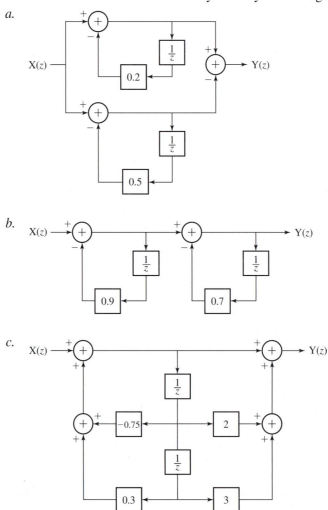

    *b.*

    *c.*

**Answers:**

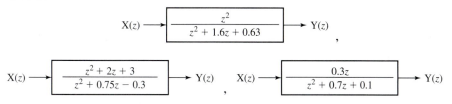

,

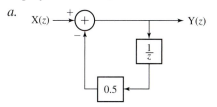          $X(z) \longrightarrow \boxed{\dfrac{0.3z}{z^2 + 0.7z + 0.1}} \longrightarrow Y(z)$

2. Evaluate the stability of the systems with each of these transfer functions.

   *a.* $H(z) = \dfrac{z}{z - 2}$

   *b.* $H(z) = \dfrac{z}{z^2 - \frac{7}{8}}$

   *c.* $H(z) = \dfrac{z}{z^2 - \frac{3}{2}z + \frac{9}{8}}$

   *d.* $H(z) = \dfrac{z^2 - 1}{z^3 - 2z^2 + 3.75z - 0.5625}$

**Answers:**
Three unstable and one stable

3. A feedback DT system has a transfer function

$$H(z) = \frac{K}{1 + K(z/(z - 0.9))}.$$

   For what range of $K$'s is this system stable?

**Answer:**
$K > -0.1$ or $K < -1.9$

4. Find the overall transfer functions of these systems in the form of a single ratio of polynomials in $z$.

   *a.*

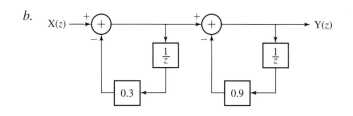

   *b.*

**Answers:**

$$\frac{z}{z + 0.3}, \quad \frac{z^2}{z^2 + 1.2z + 0.27}$$

5.   Find the DT-domain responses y[n] of the systems with these transfer functions
     to the unit sequence excitation x[n] = u[n].

   *a.*  $H(z) = \dfrac{z}{z-1}$                    *b.*  $H(z) = \dfrac{z-1}{z-\frac{1}{2}}$

**Answers:**

$\left(\dfrac{1}{2}\right)^n u[n]$,   ramp$[n+1]$

6.   Find the DT-domain responses y[n] of the systems with these transfer functions
     to the excitation $x[n] = \cos(2\pi n/8)\, u[n]$. Then show that the steady-state
     response is the same as would have been obtained by using DTFT analysis
     with an excitation $x[n] = \cos(2\pi n/8)$.

   *a.*  $H(z) = \dfrac{z}{z-0.9}$                    *b.*  $H(z) = \dfrac{z^2}{z^2 - 1.6z + 0.63}$

**Answers:**

$$y[n] = \left\{ 0.03482\,(0.7)^n + 1.454(0.9)^n + 1.9293 \cos\left(\dfrac{2\pi n}{8} - 1.3145\right) \right\} u[n],$$

$$0.3232(0.9)^n u[n] + 1.3644 \cos\left(\dfrac{\pi}{4}n - 1.0517\right) u[n]$$

7.   Sketch the magnitude frequency response of these systems from their pole–zero
     diagrams.

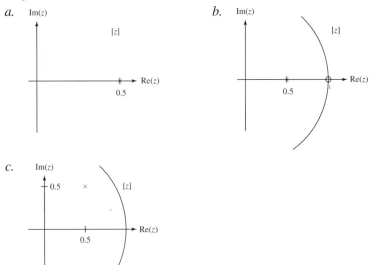

**Answers:**

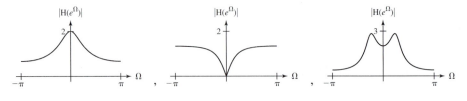

8. Use the Jury stability test to determine which of these transfer functions are for unstable systems.

   a. $H(z) = \dfrac{z^2 - z}{z^3 - 0.25z^2 - 0.6528z + 0.2083}$

   b. $H(z) = \dfrac{z - 1}{z^4 - 0.9z^3 - 0.65z^2 + 0.873z}$

   c. $H(z) = \dfrac{z}{z^4 - 1.5z^3 + 0.5z^2 + 0.25z - 0.25}$

**Answers:**

Marginally stable,   stable,   unstable

9. Draw a root locus for each system with the given forward- and feedback-path transfer functions.

   a. $H_1(z) = K\dfrac{z - 1}{z + \frac{1}{2}}, \quad H_2(z) = \dfrac{4z}{z - 0.8}$

   b. $H_1(z) = K\dfrac{z - 1}{z + \frac{1}{2}}, \quad H_2(z) = \dfrac{4}{z - 0.8}$

   c. $H_1(z) = K\dfrac{z}{z - \frac{1}{4}}, \quad H_2(z) = \dfrac{z + \frac{1}{5}}{z - \frac{3}{4}}$

   d. $H_1(z) = K\dfrac{z}{z - \frac{1}{4}}, \quad H_2(z) = \dfrac{z + 2}{z - \frac{3}{4}}$

   e. $H_1(z) = K\dfrac{1}{z^2 - \frac{1}{3}z - \frac{2}{9}}, \quad H_2(z) = 1$

**Answers:**

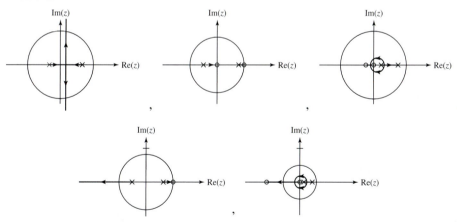

10. Using the impulse-invariant design method, design a DT system to approximate the CT systems with these transfer functions at the sampling rates specified. Compare the impulse and unit step (or sequence) responses of the CT and DT systems.

    a. $H(s) = \dfrac{6}{s + 6}, \quad f_s = 4$ Hz

    b. $H(s) = \dfrac{6}{s + 6}, \quad f_s = 20$ Hz

### Answers:

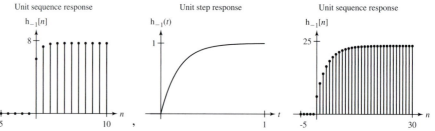

11. Using the impulse-invariant and step-invariant design methods, design digital filters to approximate analog filters with these transfer functions. In each case choose a sampling frequency which is 10 times the magnitude of the distance of the farthest pole or zero from the origin of the $s$ plane. Graphically compare the step responses of the digital and analog filters.

    *a.*   $H(s) = \dfrac{2}{s^2 + 3s + 2}$

    *b.*   $H(s) = \dfrac{6s}{s^2 + 13s + 40}$

    *c.*   $H(s) = \dfrac{250}{s^2 + 10s + 250}$

### Answers:

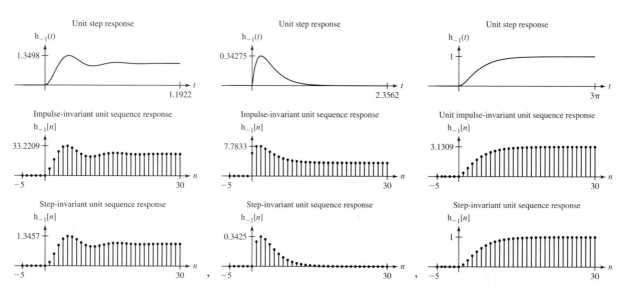

12. Using the difference-equation method and all backward differences, design digital filters to approximate analog filters with these transfer functions. If a sampling frequency is not specified, choose a sampling frequency which is 10 times the magnitude of the distance of the farthest pole or zero from the

origin of the $s$ plane. Graphically compare the step responses of the digital and analog filters.

a. $H(s) = s$, $\quad f_s = 1$ MHz

b. $H(s) = \dfrac{1}{s}$, $\quad f_s = 1$ kHz

c. $H(s) = \dfrac{2}{s^2 + 3s + 2}$

**Answers:**

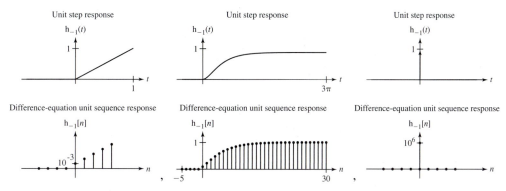

13. Using the matched $z$-transform method, design digital filters to approximate analog filters with these transfer functions. In each case choose a sampling frequency which is 10 times the magnitude of the distance of the farthest pole or zero from the origin of the $s$ plane (unless all poles or zeros are at the origin, in which case the sampling rate will not matter, in this method). Graphically compare the step responses of the digital and analog filters.

a. $H(s) = s$

b. $H(s) = \dfrac{1}{s}$

c. $H(s) = \dfrac{2s}{s^2 + 10s + 25}$

**Answers:**

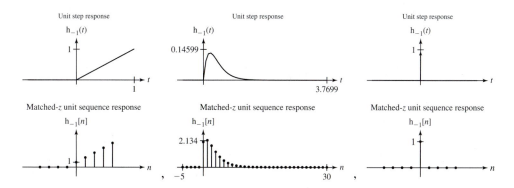

14. Using the bilinear $z$-transform method, design digital filters to approximate analog filters with these transfer functions. In each case choose a sampling frequency which is 10 times the magnitude of the distance of the farthest pole

or zero from the origin of the $s$ plane. Graphically compare the step responses of the digital and analog filters.

*a.*   $H(s) = \dfrac{s - 10}{s + 10}$          *b.*   $H(s) = \dfrac{10}{s^2 + 11s + 10}$

*c.*   $H(s) = \dfrac{3s}{s^2 + 11s + 10}$

### Answers:

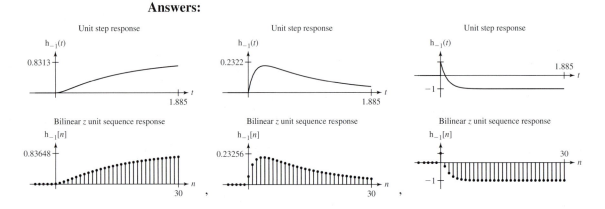

15.   Design a digital-filter approximation to each of these ideal analog filters by sampling a truncated version of the impulse response and using the specified window. In each case choose a sampling frequency which is 10 times the highest frequency passed by the analog filter. Choose the delays and truncation times such that no more than 1 percent of the signal energy of the impulse response is truncated. Graphically compare the magnitude frequency responses of the digital and ideal analog filters using a decibel magnitude scale versus linear frequency.

*a.*   Lowpass, $f_c = 1$ Hz, rectangular window
*b.*   Lowpass, $f_c = 1$ Hz, von Hann window

### Answers:

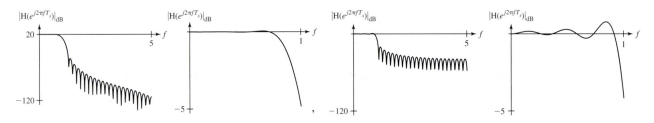

16.   Draw a canonical-form block diagram for each of these system transfer functions.

*a.*   $H(z) = \dfrac{z(z - 1)}{z^2 + 1.5z + 0.8}$

*b.*   $H(z) = \dfrac{z^2 - 2z + 4}{\left(z - \frac{1}{2}\right)(2z^2 + z + 1)}$

**Answers:**

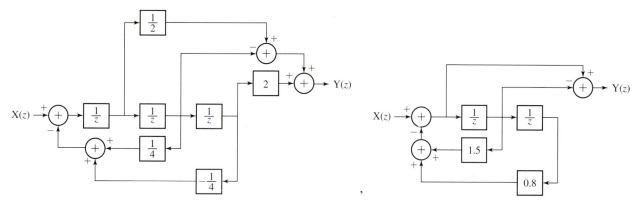

**17.** Draw a cascade-form block diagram for each of these system transfer functions.

$$a. \quad H(z) = \frac{z}{\left(z + \frac{1}{3}\right)\left(z - \frac{3}{4}\right)} \qquad b. \quad H(z) = \frac{z - 1}{4z^3 + 2z^2 + 2z + 3}$$

**Answers:**

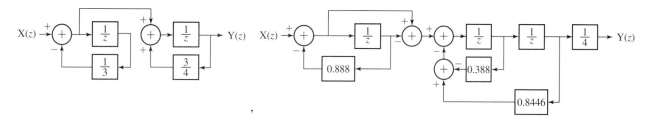

**18.** Draw a parallel-form block diagram for each of these system transfer functions.

$$a. \quad H(z) = \frac{z}{\left(z + \frac{1}{3}\right)\left(z - \frac{3}{4}\right)} \qquad b. \quad H(z) = \frac{8z^3 - 4z^2 + 5z + 9}{7z^3 + 4z^2 + z + 2}$$

**Answers:**

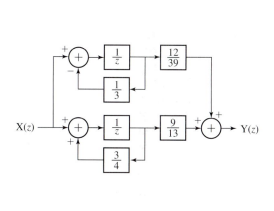

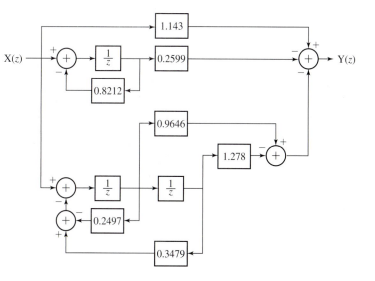

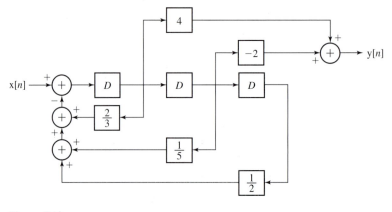

**Figure E19**
A DT system.

19.   For the system in Figure E19 write state equations and output equations.

**Answer:**

$$
\begin{bmatrix} q_1[n+1] \\ q_2[n+1] \\ q_3[n+1] \end{bmatrix} = \begin{bmatrix} 0 & 1 & 0 \\ 0 & 0 & 1 \\ -\frac{1}{2} & -\frac{1}{5} & -\frac{2}{3} \end{bmatrix} \begin{bmatrix} q_1[n] \\ q_2[n] \\ q_3[n] \end{bmatrix} + \begin{bmatrix} 0 \\ 0 \\ 1 \end{bmatrix} x[n],
$$

$$
y[n] = [\,0 \quad -2 \quad 4\,] \begin{bmatrix} q_1[n] \\ q_2[n] \\ q_3[n] \end{bmatrix}
$$

20.   Write a set of state equations and output equations corresponding to these transfer functions.

   *a.*   $H(z) = \dfrac{0.9z}{z^2 - 1.65z + 0.9}$

   *b.*   $H(z) = \dfrac{4(z-1)}{(z-0.9)(z-0.7)}$

**Answers:**

$$
\begin{bmatrix} q_1[n+1] \\ q_2[n+1] \end{bmatrix} = \begin{bmatrix} 0 & 1 \\ -0.63 & 1.6 \end{bmatrix} \begin{bmatrix} q_1[n] \\ q_2[n] \end{bmatrix} + \begin{bmatrix} 0 \\ 1 \end{bmatrix} x[n],
$$

$$
y[n] = [\,-4 \quad 4\,] \begin{bmatrix} q_1[n] \\ q_2[n] \end{bmatrix};
$$

$$
\begin{bmatrix} q_1[n+1] \\ q_2[n+1] \end{bmatrix} = \begin{bmatrix} 0 & 1 \\ -0.9 & 1.65 \end{bmatrix} \begin{bmatrix} q_1[n] \\ q_2[n] \end{bmatrix} + \begin{bmatrix} 0 \\ 0.9 \end{bmatrix} x[n],
$$

$$
y[n] = [\,0 \quad 1\,] \begin{bmatrix} q_1[n] \\ q_2[n] \end{bmatrix}
$$

21.   Convert the difference equation

$$
10y[n] + 4y[n-1] + y[n-2] + 2y[n-3] = \cos\left(\frac{2\pi n}{16}\right) u[n]
$$

   into a set of state equations and output equations.

**Answer:**

$$\begin{bmatrix} q_1[n+1] \\ q_2[n+1] \\ q_3[n+1] \end{bmatrix} = \begin{bmatrix} -0.4 & -0.1 & -0.2 \\ 1 & 0 & 0 \\ 0 & 1 & 0 \end{bmatrix} \begin{bmatrix} q_1[n] \\ q_2[n] \\ q_3[n] \end{bmatrix}$$
$$+ \begin{bmatrix} 1 & 0 & 0 \\ 0 & 0 & 0 \\ 0 & 0 & 0 \end{bmatrix} \begin{bmatrix} 0.1\cos\left(\frac{2\pi n}{16}\right) u[n+1] \\ 0 \\ 0 \end{bmatrix},$$

$$y[n] = \begin{bmatrix} 1 & 0 & 0 \end{bmatrix} \begin{bmatrix} q_1[n] \\ q_2[n] \\ q_3[n] \end{bmatrix}$$

22.  Convert the state equations and output equation

$$\begin{bmatrix} q_1[n+1] \\ q_2[n+1] \end{bmatrix} = \begin{bmatrix} -2 & -5 \\ 1 & 0 \end{bmatrix} \begin{bmatrix} q_1[n] \\ q_2[n] \end{bmatrix} + \begin{bmatrix} 1 & 0 \\ 0 & 0 \end{bmatrix} \begin{bmatrix} \left(\frac{1}{3}\right)^n u[n] \\ 0 \end{bmatrix}$$

$$y[n] = \begin{bmatrix} 1 & 0 \end{bmatrix} \begin{bmatrix} q_1[n] \\ q_2[n] \end{bmatrix}$$

into a single difference equation.

**Answer:**

$$y[n] + 2y[n-1] + 5y[n-2] = \left(\frac{1}{3}\right)^{n-1} u[n-1]$$

23.  Find the responses of the system described by this set of state equations and output equations. (Assume the system is initially at rest.)

$$\begin{bmatrix} q_1[n+1] \\ q_1[n+1] \end{bmatrix} = \begin{bmatrix} 3 & 1 \\ 0 & -2 \end{bmatrix} \begin{bmatrix} q_1[n] \\ q_1[n] \end{bmatrix} + \begin{bmatrix} 4 \\ 3 \end{bmatrix} u[n]$$

$$\begin{bmatrix} y_1[n] \\ y_1[n] \end{bmatrix} = \begin{bmatrix} 1 & -1 \\ 2 & 0 \end{bmatrix} \begin{bmatrix} q_1[n] \\ q_1[n] \end{bmatrix}$$

**Answer:**

$$y[n] = \begin{bmatrix} 2.3(3)^n + 1.2(-2)^n - 3.5 \\ 4.6(3)^n + 0.4(-2)^n - 5 \end{bmatrix} u[n]$$

# EXERCISES WITHOUT ANSWERS

24.  Find the overall transfer functions of these systems in the form of a single ratio of polynomials in $z$.

*a.*

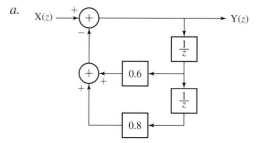

*b.*

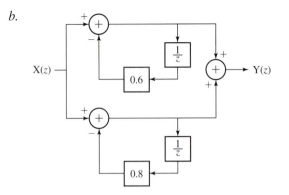

25. Find the DT-domain responses y[n] of the systems with these transfer functions to the unit sequence excitation x[n] = u[n].

    *a.*   $H(z) = \dfrac{z}{z^2 - 1.8z + 0.82}$     *b.*   $H(z) = \dfrac{z^2 - 1.932z + 1}{z(z - 0.95)}$

26. Sketch the magnitude frequency responses of these systems from their pole–zero diagrams.

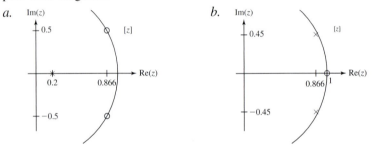

27. Using the impulse-invariant design method, design a DT system to approximate the CT systems with these transfer functions at the sampling rates specified. Compare the impulse and unit step (or sequence) responses of the CT and DT systems.

    *a.*   $H(s) = \dfrac{712s}{s^2 + 46s + 240}$,   $f_s = 20$ Hz

    *b.*   $H(s) = \dfrac{712s}{s^2 + 46s + 240}$,   $f_s = 200$ Hz

28. Using the impulse-invariant and step-invariant design methods, design digital filters to approximate analog filters with these transfer functions. In each case choose a sampling frequency which is 10 times the magnitude of the distance of the farthest pole or zero from the origin of the *s* plane. Graphically compare the step responses of the digital and analog filters.

    *a.*   $H(s) = \dfrac{16s}{s^2 + 10s + 250}$     *b.*   $H(s) = \dfrac{s + 4}{s^2 + 12s + 32}$

    *c.*   $H(s) = \dfrac{s^2 + 4}{s(s^2 + 12s + 32)}$

29. Using the difference-equation method and all backward differences, design digital filters to approximate analog filters with these transfer functions. In each case choose a sampling frequency which is 10 times the magnitude of the distance of the farthest pole or zero from the origin of the $s$ plane. Graphically compare the step responses of the digital and analog filters.

   a. $H(s) = \dfrac{s^2}{s^2 + 3s + 2}$

   b. $H(s) = \dfrac{s + 60}{s^2 + 120s + 2000}$

   c. $H(s) = \dfrac{16s}{s^2 + 10s + 250}$

30. Using the direct substitution method, design digital filters to approximate analog filters with these transfer functions. In each case choose a sampling frequency which is 10 times the magnitude of the distance of the farthest pole or zero from the origin of the $s$ plane (unless all poles or zeros are at the origin, in which case the sampling rate will not matter, in this method). Graphically compare the step responses of the digital and analog filters.

   a. $H(s) = \dfrac{s^2}{s^2 + 1100s + 10^5}$

   b. $H(s) = \dfrac{s^2 + 100s + 5000}{s^2 + 120s + 2000}$

   c. $H(s) = \dfrac{s^2 + 4}{s(s^2 + 12s + 32)}$

31. Using the bilinear $z$-transform method, design digital filters to approximate analog filters with these transfer functions. In each case choose a sampling frequency which is 10 times the magnitude of the distance of the farthest pole or zero from the origin of the $s$ plane. Graphically compare the step responses of the digital and analog filters.

   a. $H(s) = \dfrac{s^2}{s^2 + 100s + 250{,}000}$

   b. $H(s) = \dfrac{s^2 + 100s + 5000}{s^2 + 120s + 2000}$

   c. $H(s) = \dfrac{s^2 + 4}{s^2 + 12s + 32}$

32. Design a digital-filter approximation to each of these ideal analog filters by sampling a truncated version of the impulse response and using the specified window. In each case choose a sampling frequency which is 10 times the highest frequency passed by the analog filter. Choose the delays and truncation times such that no more than 1 percent of the signal energy of the impulse response is truncated. Graphically compare the magnitude frequency responses of the digital and ideal analog filters using a decibel magnitude scale versus linear frequency.

   a. Bandpass, $f_{low} = 10$ Hz,   $f_{high} = 20$ Hz, rectangular window

   b. Bandpass, $f_{low} = 10$ Hz,   $f_{high} = 20$ Hz, Blackman window

33. Draw a canonical-form block diagram for each of these system transfer functions.

    a.   $H(z) = \dfrac{z^2}{2z^4 + 1.2z^3 - 1.06z^2 + 0.08z - 0.02}$

    b.   $H(z) = \dfrac{z^2(z^2 + 0.8z + 0.2)}{(2z^2 + 2z + 1)(z^2 + 1.2z + 0.5)}$

34. Draw a cascade-form block diagram for each of these system transfer functions.

    a.   $H(z) = \dfrac{z^2}{z^2 - 0.1z - 0.12} + \dfrac{z}{z - 1}$

    b.   $H(z) = \dfrac{z/(z - 1)}{1 + (z/(z - 1))\left(z^2/\left(z^2 - \frac{1}{2}\right)\right)}$

35. Draw a parallel-form block diagram for each of these system transfer functions.

    a.   $H(z) = (1 + z^{-1})\dfrac{18}{(z - 0.1)(z + 0.7)}$

    b.   $H(z) = \dfrac{z/(z - 1)}{1 + (z/(z - 1))\left(z^2/\left(z^2 - \frac{1}{2}\right)\right)}$

36. Write a set of state equations and output equations corresponding to these transfer functions (which are for DT Butterworth filters).

    a.   $H(z) = \dfrac{0.06746z^2 + 0.1349z + 0.06746}{z^2 - 1.143z + 0.4128}$

    b.   $H(z) = \dfrac{0.0201z^4 - 0.0402z^2 + 0.0201}{z^4 - 2.5494z^3 + 3.2024z^2 - 2.0359z + 0.6414}$

37. For the system in Figure E37 write state equations and output equations.

38. Find the response of the system in Exercise E37 to the excitation $x[n] = u[n]$. (Assume that the system is initially at rest.)

39. A DT system is excited by a unit sequence, and the response is

$$y[n] = \left(8 + 2\left(\frac{1}{2}\right)^{n-1} - 9\left(\frac{3}{4}\right)^{n-1}\right)u[n - 1].$$

    Write state equations and output equations for this system.

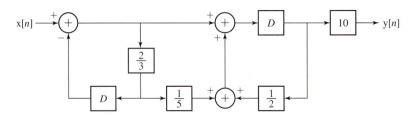

**Figure E37**
A DT system.

**40.** Define new states which transform this set of state equations and output equations into a set of diagonalized state equations and output equations, and write the new state equations and output equations.

$$
\begin{bmatrix} q_1[n+1] \\ q_2[n+1] \\ q_3[n+1] \end{bmatrix} = \begin{bmatrix} -0.4 & -0.1 & -0.2 \\ 0.3 & 0 & -0.2 \\ 1 & 0 & -1.3 \end{bmatrix} \begin{bmatrix} q_1[n] \\ q_2[n] \\ q_3[n] \end{bmatrix} + \begin{bmatrix} 2 & -0.5 \\ 1 & 0 \\ 0 & 3 \end{bmatrix}
$$

$$
\times \begin{bmatrix} 0.1\cos(2\pi n/16)u[n] \\ \left(\frac{3}{4}\right)^n u[n] \end{bmatrix}
$$

$$
\begin{bmatrix} y_1[n+1] \\ y_2[n+1] \end{bmatrix} = \begin{bmatrix} 1 & 0 & -1 \\ 0 & 0.3 & 0.7 \end{bmatrix} \begin{bmatrix} q_1[n] \\ q_2[n] \\ q_3[n] \end{bmatrix}
$$

**41.** Find the response of the system described by this set of state equations and output equations. (Assume the system is initially at rest.)

$$
\begin{bmatrix} q_1[n+1] \\ q_2[n+1] \end{bmatrix} = \begin{bmatrix} -\frac{1}{2} & -\frac{1}{5} \\ 0 & \frac{7}{10} \end{bmatrix} \begin{bmatrix} q_1[n] \\ q_2[n] \end{bmatrix} + \begin{bmatrix} 2 & -3 \\ 1 & 1 \end{bmatrix} \begin{bmatrix} u[n] \\ \left(\frac{3}{4}\right)^n u[n] \end{bmatrix}
$$

$$
y[n] = \begin{bmatrix} 4 & -1 \end{bmatrix} \begin{bmatrix} q_1[n] \\ q_2[n] \end{bmatrix} + \begin{bmatrix} 1 & 0 \end{bmatrix} \begin{bmatrix} u[n] \\ \left(\frac{3}{4}\right)^n u[n] \end{bmatrix}
$$

# A APPENDIX

# Useful Mathematical Relations

$$e^x = 1 + x + \frac{x^2}{2!} + \frac{x^3}{3!} + \frac{x^4}{4!} + \cdots$$

$$\sin(x) = x - \frac{x^3}{3!} + \frac{x^5}{5!} - \frac{x^7}{7!} + \cdots$$

$$\cos(x) = 1 - \frac{x^2}{2!} + \frac{x^4}{4!} - \frac{x^6}{6!} + \cdots$$

$$\cos(x) = \cos(-x) \quad \text{and} \quad \sin(x) = -\sin(-x)$$

$$e^{jx} = \cos(x) + j\sin(x)$$

$$\sin^2(x) + \cos^2(x) = 1$$

$$\cos(x)\cos(y) = \frac{1}{2}[\cos(x - y) + \cos(x + y)]$$

$$\sin(x)\sin(y) = \frac{1}{2}[\cos(x - y) - \cos(x + y)]$$

$$\sin(x)\cos(y) = \frac{1}{2}[\sin(x - y) + \sin(x + y)]$$

$$\cos(x + y) = \cos(x)\cos(y) - \sin(x)\sin(y)$$

$$\sin(x + y) = \sin(x)\cos(y) + \cos(x)\sin(y)$$

$$A\cos(x) + B\sin(x) = \sqrt{A^2 + B^2}\cos\left(x - \tan^{-1}\left(\frac{B}{A}\right)\right)$$

$$\frac{d}{dx}[\tan^{-1}(x)] = \frac{1}{1 + x^2}$$

$$\int u\, dv = uv - \int v\, du$$

$$\int x^n \sin(x)\, dx = -x^n \cos(x) + n \int x^{n-1} \cos(x)\, dx$$

$$\int x^n \cos(x)\, dx = x^n \sin(x) - n \int x^{n-1} \sin(x)\, dx$$

$$\int x^n e^{ax}\, dx = \frac{e^{ax}}{a^{n+1}}[(ax)^n - n(ax)^{n-1} + n(n-1)(ax)^{n-2} + \cdots$$

$$+ (-1)^{n-1} n!(ax) + (-1)^n n!] \qquad n \geq 0$$

$$\int e^{ax} \sin(bx)\, dx = \frac{e^{ax}}{a^2 + b^2}[a \sin(bx) - b \cos(bx)]$$

$$\int e^{ax} \cos(bx)\, dx = \frac{e^{ax}}{a^2 + b^2}[a \cos(bx) + b \sin(bx)]$$

$$\int \frac{dx}{a^2 + (bx)^2} = \frac{1}{ab} \tan^{-1}\left(\frac{bx}{a}\right)$$

$$\int \frac{dx}{(x^2 \pm a^2)^{1/2}} = \ln|x + (x^2 \pm a^2)^{1/2}|$$

$$\int_0^\infty \frac{\sin(mx)}{x}\, dx = \left\{ \begin{array}{ll} \frac{\pi}{2} & m > 0 \\ 0 & m = 0 \\ -\frac{\pi}{2} & m < 0 \end{array} \right\} = \frac{\pi}{2} \operatorname{sgn}(m)$$

$$|Z|^2 = Z Z^*$$

$$\sum_{n=0}^{N-1} r^n = \left\{ \begin{array}{ll} \dfrac{1 - r^N}{1 - r} & r \neq 1 \\ N & r = 1 \end{array} \right.$$

$$\sum_{n=0}^{\infty} r^n = \frac{1}{1 - r} \qquad |r| < 1$$

$$\sum_{n=k}^{\infty} r^n = \frac{r^k}{1 - r} \qquad |r| < 1$$

$$\sum_{n=0}^{\infty} n r^n = \frac{r}{(1 - r)^2} \qquad |r| < 1$$

If $T_0/2w$ is not an integer,

$$\operatorname{sinc}\left(\frac{t}{w}\right) * f_0 \operatorname{comb}(f_0 t) = w f_0(2M + 1) \operatorname{drcl}(f_0 t, 2M + 1)$$

(where $M$ is the greatest integer in $T_0/2w$ and $T_0 = 1/f_0$).

If $T_0/2w$ is an integer,

$$\text{sinc}\left(\frac{t}{w}\right) * f_0 \, \text{comb}(f_0 t) = w f_0 \left[\cos\left(\frac{\pi t}{w}\right) + \left(\frac{T_0}{w} - 1\right) \text{drcl}\left(f_0 t, \frac{T_0}{w} - 1\right)\right].$$

$$\frac{e^{j\pi n}}{e^{j\pi(n/N_0)}} \text{drcl}\left(\frac{n}{N_0}, N_0\right) = \text{comb}_{N_0}[n] \qquad \text{where } n \text{ and } N_0 \text{ are integers}$$

$$\text{drcl}\left(\frac{n}{2m + 1}, 2m + 1\right) = \text{comb}_{2m+1}[n] \qquad \text{where } n \text{ and } m \text{ are integers}$$

APPENDIX B

# Introduction to MATLAB

M ATLAB is a high-level computer-based mathematical tool for performing complex or repetitive calculations under program control. MATLAB can generate or read numerical data and store or display results of data analysis. This tutorial will introduce the features of MATLAB most often used for signal and system analysis. It is by no means an exhaustive exploration of all the capabilities of MATLAB.

The easiest way to become familiar with MATLAB is to sit at a console and experiment with various operators, functions, and commands until their properties are understood and then later to progress to writing MATLAB scripts and functions to execute a sequence of instructions to accomplish an analytical goal.

A logical progression in learning how to use MATLAB to solve signal and system problems is to understand

1. What kinds of numbers can be represented
2. How variables are named and how values are assigned to them
3. What mathematical operators are built into MATLAB and how they operate on numbers and variables
4. What mathematical functions are built into MATLAB and how to get descriptions of their syntax and use
5. How to write a script file which is a list of instructions which are executed sequentially or according to flow-control commands
6. How to write function files which contain user-defined functions which can be used just like built-in functions
7. How to display and format results of calculations in a graphical manner that aids quick comprehension of relationships among variables

## B.1 NUMBERS, VARIABLES, AND MATRICES

MATLAB is vector and matrix oriented. That is, everything in MATLAB is a matrix. A row vector with $m$ elements is a $1 \times m$ matrix, and a column vector with $m$ elements is an $m \times 1$ matrix. A scalar is a $1 \times 1$ matrix. MATLAB can handle real numbers or complex numbers. For example, the real number 7.89 is represented as simply `7.89` in MATLAB. The real number $15.8 \times 10^{-11}$ can be written in MATLAB as `15.8e-11`. The complex number whose real part is 8 and whose imaginary part is 3 can be represented either as `8+3*i` or as `8+j*3` because the two letters `i` and `j` are both preassigned by MATLAB to be equal to $\sqrt{-1}$. Other constants preassigned by MATLAB are `pi` for $\pi$, `inf` for infinity, and `NaN` for "not a number." Any of these predefined constants can be redefined by the user, but, unless the user has a very good reason to change them, they should be left as defined by MATLAB to avoid confusion.

## B.2 OPERATORS

The most commonly used built-in mathematical operators in MATLAB are

| | | | |
|---|---|---|---|
| = | assignment | – | subtraction and unary minus |
| + | addition | .* | array multiplication |
| * | matrix multiplication | .^ | array power |
| ^ | matrix power | ./ | array division |
| / | division | & | logical AND |
| <> | relational operators | ~ | logical NOT |
| \| | logical OR | ' | transpose |
| == | equality | .' | nonconjugated transpose |

In the examples to follow, which illustrate MATLAB features, the boldface text is what the user types in and the lightface text is the MATLAB response. The » character is the MATLAB prompt indicating MATLAB is waiting for user instructions.

Suppose we type in the assignment statement

```
»a = 2 ;
```

This assigns the value 2 to the scalar variable named `a`. The semicolon `;` terminates the instruction and suppresses display of the result of the operation. This statement could also be written as

```
»a=2 ;
```

The extra spaces in **a = 2** compared with **a=2** have no effect. They are ignored by MATLAB. The syntax for assigning values to a row vector is illustrated with the following two assignment statements which are typed on a single line and both terminated by a semicolon.

```
»b = [1 3 2] ; c = [2,5,1] ;
```

The elements of a row vector can be separated by either a space or a comma. Given the assignments already made, if we now simply type the name of a variable, MATLAB displays the value.

```
»a
a =
    2
»b
b =
    1    3    2
»c
c =
    2    5    1
```

Vectors and matrices can be multiplied by a scalar. The result is a vector or matrix whose elements have each been multiplied by that scalar.

```
»a*b
ans =
       2     6     4
```

`ans` is the name given to a result when the user does not assign a name.

```
»a*c
ans =
       4    10     2
```

A scalar can also be added to a vector or matrix.

```
»a + b
ans =
       3     5     4
```

Addition of a scalar to a vector or matrix simply adds the scalar to every element of the vector or matrix. Subtraction is similarly defined.

```
»a + c
ans =
       4     7     3
»a - b
ans =
       1    -1     0
»a - c
ans =
       0    -3     1
```

Vectors and matrices add (or subtract) in the way normally defined in mathematics. That is, the two vectors or matrices must have the same shape to be added (or subtracted). (Unless one of them is a $1 \times 1$ matrix, a scalar, as previously illustrated.)

```
»b + c
ans =
       3     8     3
»c - b
ans =
       1     2    -1
```

Vectors and matrices can be multiplied according to the usual rules of linear algebra.

```
»b*c
??? Error using ==> *
Inner matrix dimensions must agree.
```

This result illustrates a common error in MATLAB. Matrices must be *commensurate* to be multiplied using the * operator. Premultiplication of a $1 \times 3$ row vector like c by a $1 \times 3$ row vector like b is not defined. But if c were transposed to a $3 \times 1$ column vector, the multiplication would be defined. Transposition is done with the ' operator.

```
»c'
ans =
       2
       5
       1
```

```
»b*c'
ans =
     19
```

This is the product $bc^T$.

```
»b'*c
ans =
     2      5      1
     6     15      3
     4     10      2
```

This is the product $b^Tc$.

Often it is very useful to multiply two vectors or matrices of the same shape element by element instead of using the usual rules of matrix multiplication. That kind of multiplication is called *array multiplication* in MATLAB and is done using the . * operator.

```
»b.*c
ans =
     2     15      2
```

Now define a 3 by 3 matrix A.

```
»A = [3 5 1 ; 9 -1 2 ; -7 -4 3] ;
```

The ; operator puts the next entry in the next row of the matrix. So the ; operator has a dual use, to terminate instructions and to enter the next row when specifying a matrix.

```
»A
A =
      3      5      1
      9     -1      2
     -7     -4      3
```

Two-dimensional matrices are displayed in rows and columns. As previously indicated we can multiply a matrix by a scalar.

```
»a*A
ans =
      6     10      2
     18     -2      4
    -14     -8      6
```

We can also multiply a matrix and a vector, if they are commensurate.

```
»A*b
??? Error using ==> *
Inner matrix dimensions must agree.
»A*b'
ans =
     20
     10
    -13
```

Now define two more $3 \times 3$ matrices.

```
»B = [3 2 7 ; 4 1 2 ; -1 3 1] ;
»C = [4 5 5 ; -1 -3 2 ; 8 3 1] ;
»B
B =
        3       2       7
        4       1       2
       -1       3       1
»C
C =
        4       5       5
       -1      -3       2
        8       3       1
»B + C
ans =
        7       7      12
        3      -2       4
        7       6       2
»B*C
ans =
       66      30      26
       31      23      24
        1     -11       2
»B.*C
ans =
       12      10      35
       -4      -3       4
       -8       9       1
```

Another important operator in MATLAB is the $\wedge$ operator for raising a number or variable to a power. We can raise a scalar to a power.

```
»a^2
ans =
        4
```

We can raise a matrix to a power according to the rules of linear algebra.

```
»A^2
ans =
       47       6      16
        4      38      13
      -78     -43      -6
```

We can raise each element of a matrix to a power.

```
»A.^2
ans =
        9      25       1
       81       1       4
       49      16       9
```

Sometimes it is desired to know what variables are currently defined and have values stored by MATLAB. The who command accomplishes that.

```
»who
Your variables are:
A            C            ans            c
B            a            b
```

If you want to undefine a variable (take it out of the list MATLAB recognizes), the clear command accomplishes that.

```
»clear A
»A
??? Undefined function or variable. Symbol in question ==> A
```

We can redefine A by a matrix multiplication.

```
»A = B*C
A =
      66        30        26
      31        23        24
       1       -11         2
```

Another important operation is division. We can divide a scalar by a scalar.

```
»a/3
ans =
    0.6667
```

We can divide a vector by a scalar.

```
»b/a
ans =
    0.5000    1.5000    1.0000
```

We can divide a matrix by a scalar.

```
»A/10
ans =
    6.6000    3.0000    2.6000
    3.1000    2.3000    2.4000
    0.1000   -1.1000    0.2000
```

If we divide anything by a vector or matrix, we must choose which operator to use. The / operator can be used to divide one matrix by another in the sense that the MATLAB operation A/B is equivalent to the mathematical operation $\mathbf{AB}^{-1}$, where $\mathbf{B}^{-1}$ is the matrix inverse of $\mathbf{B}$. Of course, $\mathbf{A}$ and $\mathbf{B}^{-1}$ must be commensurate for A/B to work properly.

```
»A/B
ans =
   -2.6875   19.8125    5.1875
    0.2969    8.6719    4.5781
    1.3594   -1.7656   -3.9844
```

We can also do *array* division in which each element of the numerator is divided by the corresponding element of the denominator.

```
»A./B
ans =
      22.0000    15.0000     3.7143
       7.7500    23.0000    12.0000
      -1.0000    -3.6667     2.0000
```

The array division operator ./ can also be used to divide a scalar by a matrix in the sense that a new matrix is formed, each element of which is the scalar divided by the corresponding element of the dividing matrix.

```
»10./A
ans =
       0.1515     0.3333     0.3846
       0.3226     0.4348     0.4167
      10.0000    -0.9091     5.0000
```

There is an array of *relational* and *logical* operators to compare numbers and to make logical decisions. The == operator compares two numbers and returns a logical result which is 1 if the two numbers are the same and 0 if they are not. This operator can be used on scalars, vectors, or matrices; on scalars with vectors; or on scalars with matrices. If two vectors are compared or if two matrices are compared, they must have the same shape. The comparison is done on an element-by-element basis.

```
»1 == 2
ans =
       0

»5 == 5
ans =
       1

»a == b
ans =
       0     0     1

»a == B
ans =
       0     1     0
       0     0     1
       0     0     0

»B == C
ans =
       0     0     0
       0     0     1
       0     1     1
```

The ~= operator compares two numbers and returns a 1 if they are not equal and a 0 if they are equal.

The > operator compares two numbers and returns a 1 if the first number is greater than the second and a 0 if it is not.

The < operator compares two numbers and returns a 1 if the first number is less than the second and a 0 if it is not.

The >= operator compares two numbers and returns a 1 if the first number is greater than or equal to the second and a 0 if it is not.

The <= operator compares two numbers and returns a 1 if the first number is less than or equal to the second and a 0 if it is not.

All of these relational operators can be used with scalars, vectors, and matrices in the same way that == can.

```
»-3 ~= 7
ans =
     1

»B ~= C
ans =
     1     1     1
     1     1     0
     1     0     0

»c > 2
ans =
     0     1     0

»B < 2
ans =
     0     0     0
     0     1     0
     1     0     1

»B <= 2
ans =
     0     1     0
     0     1     1
     1     0     1

»B >= 2
ans =
     1     1     1
     1     0     1
     0     1     0

»B > 2
ans =
     1     0     1
     1     0     0
     0     1     0

»b < c
ans =
     1     1     0

»b < B
??? Error using ==> <
Matrix dimensions must agree.
```

MATLAB has three logical operators, & (logical AND), | (logical OR), and ~ (logical NOT). Logical operators operate on any numbers. The number 0 is treated as a logical 0 and any other number is treated as a logical 1 for purposes of the logical operation. The result of any logical operation is either the number 0 or the number 1.

The & operator returns a 1 if both operands are nonzero and a 0 otherwise.

The | operator returns a 0 if both operands are 0 and a 1 otherwise.

The ~ operator is a unary operator which returns a 1 if its operand is a 0, and a 0 if its operand is nonzero.

```
»0 & 1
ans =
     0

»1 & 1
ans =
     1

»1 & [1 0 -3 22]
ans =
     1    0    1    1

»0 | [3 ; 0 ; -18]
ans =
     1
     0
     1

»~[0 ; 1 ; 0.3]
ans =
     1
     0
     0

»(1 < 2 ) | (-5 > -1)
ans =
     1

»(1 < 2 ) & (-5 > -1)
ans =
     0
```

One of the most powerful operators in MATLAB is the : operator. This operator can be used to generate sequences of numbers, and it can also be used to select only certain rows and/or columns of a matrix. When used in the form a:b, where a and b are scalars, this operator generates a sequence of numbers from a to b separated by 1.

```
»3:8
ans =
     3    4    5    6    7    8
```

When used in the form a:b:c, where a, b, and c are scalars, it generates a sequence of numbers from a to c separated by b.

```
»-4:3:17
ans =
     -4    -1    2    5    8    11    14    17
»-4:3:16
ans =
     -4    -1    2    5    8    11    14
»(2:-3:-11)'
ans =
      2
     -1
     -4
     -7
    -10
```

If the increment b is positive, the sequence terminates at the last value that is less than or equal to the specified end value c. If the increment b is negative, the sequence terminates at the last value that is greater than or equal to the specified end value c.

Another use of the : operator is to form a column vector consisting of all the elements of a matrix. For example, the instruction A(:) forms a column vector of the elements in A.

```
»A
A =
     66     30    26
     31     23    24
      1    -11     2
»A(:)
ans =
     66
     31
      1
     30
     23
    -11
     26
     24
      2
```

The : operator can also be used to extract a submatrix from a matrix. For example, A(:,2) forms a matrix which is the second column of A.

```
»A(:,2)
ans =
     30
     23
    -11
```

The instruction A(3,:) forms a matrix which is the third row of A.

```
»A(3,:)
ans =
      1    -11     2
```

We can also extract partial rows and partial columns or combinations of those.

```
»A(1:2,3)
ans =
    26
    24

»A(1:2,:)
ans =
    66    30    26
    31    23    24

»A(1:2,2:3)
ans =
    30    26
    23    24
```

For a complete listing of MATLAB operators consult a MATLAB reference manual.

# B.3  SCRIPTS AND FUNCTIONS

The real power of any programming language lies in writing sequences of instructions to implement some algorithm. There are two types of programs in MATLAB, scripts and functions. Both types of programs are stored on disk as .m files, so called because the extension is .m. A script is what is normally called a *program*. It is a sequence of instructions that execute sequentially except when flow-control statements change the sequence. For example,

```
    .
    .
    .
f01 = 6 ; f02 = 9 ; f0 = gcd(f01,f02) ; T0 = 1/f0 ;
T01 = 1/f01 ; T02 = 1/f02 ; T0min = min(T01,T02) ; dt = T0min/24 ;
nPts = 2*T0/dt ; t = dt*[0:nPts]' ;
th1 = (rand(1,1)-0.5)*2*pi ; x1 = cos(2*pi*f01*t + th1) ;
th2 = (rand(1,1)-0.5)*2*pi ; x2 = cos(2*pi*f02*t + th2) ;
    .
    .
    .
```

Any text editor can be used to write a script file or a function file.

A *function* is a modular program which accepts arguments and returns results and is usually intended to be reusable in a variety of situations. The arguments passed to the function can be scalars, vectors, or matrices, and the results returned can also be scalars, vectors, or matrices.

```
%    Chi-squared probability density function
%
%    N = degrees of freedom
%    xsq = chi-squared
```

```
function y = pchisq(N,xsq)
     y = xsq.^(N/2-1).*exp(-xsq/2).*u(xsq)/(2^(N/2)*gamma(N/2)) ;


%      Rectangular to polar conversion
%
%      x and y are the rectilinear components of a vector.
%      r and theta are the magnitude and angle of the vector.
function [r,theta] = rect2polar(x,y)
     r = sqrt(x^2 + y^2) ; theta = atan2(y,x) ;
```

The first executable line in a function must begin with the keyword `function`. The variable representing the value returned by the function is next, in this case `y`. The name of the function is next, in this case `pchisq`. Then any parameters to be passed to the function are included in parentheses and separated by commas, in this case `N,xsq`.

Any line in a script or function which begins with the character `%` is a *comment* line and is ignored by MATLAB in executing the program. It is there for the benefit of anyone who reads the program to help in following the algorithm the program implements.

## B.4  MATLAB FUNCTIONS AND COMMANDS

MATLAB has a long list of built-in functions and commands to perform common mathematical tasks. In this tutorial the word *function* will be used for those MATLAB entities which are, or are similar to, mathematical functions. The word *command* will be used for other operations like plotting, input-output, formatting, etc. The word *instruction* will refer to a sequence of functions, commands, and/or operations that work as a unit and are terminated by a `;`. The word *operation* will refer to what is done by operators like $+$, $-$, $*$, $/$, and $\wedge$. The functions and commands are divided into groups. The exact groups and the number of groups varies slightly between platforms and between versions of MATLAB, but the following groups are common to all recent versions of MATLAB.

| | |
|---|---|
| general | General-purpose commands |
| ops | Operators and special characters |
| lang | Programming language constructs |
| elmat | Elementary matrices and matrix manipulation |
| elfun | Elementary math functions |
| specfun | Specialized math functions |
| matfun | Matrix functions and numerical linear algebra |
| datafun | Data analysis and Fourier transforms |
| polyfun | Interpolation and polynomials |
| funfun | Function functions and ordinary differential equation (ODE) solvers |
| sparfun | Sparse matrices |
| graph2d | Two-dimensional graphs |
| graph3d | Three-dimensional graphs |
| specgraph | Specialized graphs |

| graphics | Handle graphics |
|----------|-----------------|
| uitools | Graphical user-interface tools |
| strfun | Character strings |
| iofun | File input-output |
| timefun | Time and dates |
| datatypes | Data types and structures |

In addition to these groups there are *toolboxes* available from MATLAB which supply additional special-purpose functions. Three of the most common toolboxes are symbolic, signal, and control. The *symbolic* toolbox adds capabilities for MATLAB to do symbolic, as opposed to numerical, operations. Examples would be differentiation, integration, solving systems of equations, and integral transforms. The *signal* toolbox adds functions and commands which do common signal-processing operations like waveform generation, filter design and implementation, numerical transforms, statistical analysis, windowing, and parametric modeling. The control toolbox adds functions and commands which do common control-system operations like creation of LTI system models, state-space system description and analysis, responses to standard signals, pole–zero diagrams, frequency-response plots, and root locus. The "ops" group of operators has already been discussed. This tutorial will cover only those function and command groups and the functions and commands within those groups that are the most important in elementary signal and system analysis.

## General-Purpose Commands

The `help` command may be the most important in MATLAB. It is a convenient way of getting documentation on all MATLAB commands. If the user simply types `help`, she gets a list of the function and command groups available. For example,

```
»help

HELP topics:

Toolbox:symbolic   - Symbolic Math Toolbox.
Toolbox:signal     - Signal Processing Toolbox.
Toolbox:control    - Control System Toolbox.
matlab:general     - General purpose commands.
matlab:ops         - Operators and special characters.
matlab:lang        - Programming language constructs.
matlab:elmat       - Elementary matrices and matrix manipulation.
matlab:elfun       - Elementary math functions.
matlab:specfun     - Specialized math functions.
matlab:matfun      - Matrix functions - numerical linear algebra.
matlab:datafun     - Data analysis and Fourier transforms.
matlab:polyfun     - Interpolation and polynomials.
matlab:funfun      - Function functions and ODE solvers.
matlab:sparfun     - Sparse matrices.
matlab:graph2d     - Two dimensional graphs.
matlab:graph3d     - Three dimensional graphs.
matlab:specgraph   - Specialized graphs.
matlab:graphics    - Handle Graphics.
```

```
matlab:uitools       - Graphical user interface tools.
matlab:strfun        - Character strings.
matlab:iofun         - File input/output.
matlab:timefun       - Time and dates.
matlab:datatypes     - Data types and structures.
matlab:demos         - Examples and demonstrations.
Toolbox:matlab       - (No table of contents file)
Toolbox:local        - Preferences.
MATLAB 5:bin         - (No table of contents file)
MATLAB 5:extern      - (No table of contents file)
MATLAB 5:help        - (No table of contents file)

For more help on directory/topic, type "help topic".
```

The exact list may vary somewhat between platforms and MATLAB versions. Then if the user types `help` followed by the name of a group, he gets a list of functions and commands in that group. For example,

```
»help elfun

Elementary math functions.

Trigonometric.
    sin       - Sine.
    sinh      - Hyperbolic sine.
    asin      - Inverse sine.
    asinh     - Inverse hyperbolic sine.
    cos       - Cosine.
    cosh      - Hyperbolic cosine.
    acos      - Inverse cosine.
    acosh     - Inverse hyperbolic cosine.
    tan       - Tangent.
    tanh      - Hyperbolic tangent.
    atan      - Inverse tangent.
    atan2     - Four quadrant inverse tangent.
    atanh     - Inverse hyperbolic tangent.
    sec       - Secant.
    sech      - Hyperbolic secant.
    asec      - Inverse secant.
    asech     - Inverse hyperbolic secant.
    csc       - Cosecant.
    csch      - Hyperbolic cosecant.
    acsc      - Inverse cosecant.
    acsch     - Inverse hyperbolic cosecant.
    cot       - Cotangent.
    coth      - Hyperbolic cotangent.
    acot      - Inverse cotangent.
    acoth     - Inverse hyperbolic cotangent.

Exponential.
    exp       - Exponential.
    log       - Natural logarithm.
    log10     - Common (base 10) logarithm.
```

```
        log2        - Base 2 logarithm and dissect floating point number.
        pow2        - Base 2 power and scale floating point number.
        sqrt        - Square root.
        nextpow2    - Next higher power of 2.

    Complex.
        abs         - Absolute value.
        angle       - Phase angle.
        conj        - Complex conjugate.
        imag        - Complex imaginary part.
        real        - Complex real part.
        unwrap      - Unwrap phase angle.
        isreal      - True for real array.
        cplxpair    - Sort numbers into complex conjugate pairs.

    Rounding and remainder.
        fix         - Round towards zero.
        floor       - Round towards minus infinity.
        ceil        - Round towards plus infinity.
        round       - Round towards nearest integer.
        mod         - Modulus (signed remainder after division).
        rem         - Remainder after division.
        sign        - Signum.
```

Then if the user types `help` followed by a function or command name, she gets a description of its use. For example,

```
»help abs
```

```
 ABS  Absolute value.
    ABS(X) is the absolute value of the elements of X. When
    X is complex, ABS(X) is the complex modulus (magnitude) of
    the elements of X.

    See also SIGN, ANGLE, UNWRAP.

 Overloaded methods
    help sym/abs.m
```

The `who` and `clear` commands were covered earlier. The `who` command returns a list of the currently defined variables, and the `clear` command can clear any or all of the variables currently defined.

## Programming Language Flow Control

As in most programming languages, instructions in a MATLAB program are executed in sequence unless the flow of execution is modified by certain flow-control commands.

The `if...then...else` construct allows the programmer to make a logical decision based on variable values and branch to one of two choices based on that decision. A typical `if` decision structure might look like

```
if a > b then
    Handle the case of a > b
    .
    .
```

```
else
        Handle the case of a <= b
        .
        .
end
```

The `else` choice is optional. It could be omitted to form

```
if a > b then
        Handle the case of a > b
        .
        .
end
```

The decision criterion between `if` and `then` can be anything that evaluates to a logical value. A logical 1 sends the program flow to the first group of statements between `if` and `else` (or `if` and `end`, if `else` is omitted), and a logical 0 sends the program flow to the second group of statements between `else` and `end` (or past `end`, if `else` is omitted).

The `for` statement provides a way to specify that a group of instructions will be performed some number of times with a variable having a defined value each time.

```
for n = 1:15,
        .
        .
        .
end
```

This notation means set `n` to `1`, execute the instructions between the `for` and `end` statements, then set `n` to `2` and repeat the instructions, and keep doing that up through `n = 15` and then go to the next instruction after `end`. The more general syntax is

```
for n = N,
        .
        .
        .
end
```

where `N` is a vector of values of `n`. The vector `N` could be, for example,

`1:3:22,  15:-2:6`  or  `[0,-4,8,27,-11,19]`.

```
»N = 1:10 ; x = [] ;
for n = N,
        x = [x,n^2] ;
end
x

x =

      1    4    9   16   25   36   49   64   81   100
```

The variable `n` is simply cycled through the values in the vector, in sequence, and the instructions in the `for` loop are executed each time.

The `while` flow-control command also defines a loop between the `while` command and an `end` statement.

```
while x > 32,
      .
      .
      .
end
```

If $x$ is initially greater than 32, the instructions in this loop will execute sequentially once and then recheck the value of $x$. If it is still greater than 32, the loop will execute again. It will keep executing until $x$ is not greater than 32. Whenever the condition $x > 32$ is no longer true, flow proceeds to the next instruction after the `end` statement. So it is important that something inside the loop change the value of $x$. Otherwise the loop will never terminate execution! The condition $x > 32$ is just an example. More generally the condition can be anything that evaluates to a logical value. While the value is a logical 1, the loop executes. When the value changes to a logical 0, the loop passes execution to the next instruction after `end`.

The `switch` command is a kind of generalization of the `if...then...else` flow control. The syntax is

```
switch expression
    case value1,
       ...
    case value2,
       ...
       .
       .
       .
    case valueN
       ...
end
```

The `expression` is evaluated. If its value is `value1`, the instructions inside `case value1` (all instructions up to the `case value2` statement) are executed and control is passed to the instruction following the `end` statement. If its value is `value2`, the instructions inside `case value2` are executed and control is passed to the instruction following the `end` statement, etc. All `case`'s are checked in sequence.

The `input` statement is very useful. It allows the user to enter a variable value while the program is running. The value entered can be a number or a sequence of characters (a string). The general syntax for supplying a number is

```
n = input('message') ;
```

where `'message'` is any sequence of characters the programmer wants to appear to prompt the user of the program to enter a value. The general syntax for entering a string is

```
s = input('message','s') ;
```

where `'message'` is any sequence of characters the programmer wants to appear to prompt the user to enter the string and the `'s'` indicates to MATLAB to interpret whatever is entered by the user as a string instead of as a number.

The `pause` statement allows the programmer to cause the program to interrupt execution for a specified time and then resume execution. The syntax is

```
pause(n) ;
```

where $n$ is the number of seconds to pause or simply

```
pause ;
```

which causes the program to wait for the user to press a key. These commands are useful for giving the user time to look at a graphical (or numerical) result before proceeding to the next set of instructions.

## Elementary Matrices and Matrix Manipulation

There are several useful commands for generating matrices of various types. The most commonly used ones in signal and system analysis are `zeros`, `ones`, `eye`, `rand`, and `randn`. The commands `zeros` and `ones` are similar. Each one generates a matrix of a specified size filled with either 0s or 1s.

```
»zeros(1,3)
ans =
        0     0     0

»ones(5,2)
ans =
        1     1
        1     1
        1     1
        1     1
        1     1

»zeros(3)
ans =
        0     0     0
        0     0     0
        0     0     0
```

The command `eye` generates an identity matrix ("eye" for the "i" in identity).

```
»eye(4)
ans =
        1     0     0     0
        0     1     0     0
        0     0     1     0
        0     0     0     1
```

The commands `rand` and `randn` are used to generate a matrix of a specified size filled with random numbers taken from a distribution of random (actually pseudorandom) numbers. The command `rand` takes its numbers from a uniform distribution between 0 and 1. The command `randn` takes its numbers from a normal distribution (gaussian with zero mean and unit variance).

```
»rand(2,2)
ans =
    0.9501    0.6068
    0.2311    0.4860

»rand(2,2)
ans =
    0.8913    0.4565
    0.7621    0.0185
```

```
»randn(3,2)
ans =
    -0.4326     0.2877
    -1.6656    -1.1465
     0.1253     1.1909
»randn(3)
ans =
     1.1892     0.1746    -0.5883
    -0.0376    -0.1867     2.1832
     0.3273     0.7258    -0.1364
```

Two handy functions for determining the size of a matrix or the length of a vector are `size` and `length`. The function `size` returns a vector containing the number of rows, columns, etc., of a matrix.

```
»b
b =
     1     3     2
»size(b)
ans =
     1     3
»c
c =
     2     5     1
»size(c')
ans =
     3     1
»B
B =
     3     2     7
     4     1     2
    -1     3     1
»size(B)
ans =
     3     3
```

The function `length` returns the length of a vector.

```
»length(b)
ans =
     3
»length(c')
ans =
     3
»D
D =
     1     2
     3     6
    -2     9
```

```
»length(D)
ans =
    3
```

Notice that MATLAB returned a *length* for the 3 × 2 matrix D. In this case the matrix D is interpreted as a column vector of row vectors and the length is the number of elements in that column vector, the number of row vectors in D.

The keyword eps is reserved to always hold information about the precision of calculations in MATLAB. Its value is the distance from the number 1.0 to the next largest number representable by MATLAB. Therefore, it is a measure of the precision of number representation in MATLAB.

```
»eps
ans =
    2.2204e-16
```

This indicates that numbers near 1 are represented with a maximum error of about two parts in $10^{16}$ or about $2.22 \times 10^{-14}$ percent. In some numerical algorithms knowing the precision of representation can be important in deciding on when to terminate calculations.

## Elementary Math Functions

All the mathematical functions that are important in signal and system analysis are available in MATLAB. Some common ones are the trigonometric functions, sine, cosine, tangent. They all accept arguments in radians, and they will all accept vector or matrix arguments.

```
»sin(2)
ans =
    0.9093

»cos([2 4])
ans =
   -0.4161   -0.6536

»tan(B)
ans =
   -0.1425   -2.1850    0.8714
    1.1578    1.5574   -2.1850
   -1.5574   -0.1425    1.5574
```

Two other common functions are the exponential and logarithm functions and the square root function.

```
»exp([3 -1 j*pi/2])
ans =
   20.0855            0.3679            0 + 1.0000i

»log([3 ; 6])
ans =
    1.0986
    1.7918
```

```
»sqrt([3 ; j*2 ; -1])
ans =
    1.7321
    1.0000 + 1.0000i
         0 + 1.0000i
```

Five functions are very useful with complex numbers, absolute value, angle, conjugate, real part, and imaginary part.

```
»abs(1+j)
ans =
    1.4142
```

```
»angle([1+j ; -1+j])
ans =
    0.7854
    2.3562
```

```
»conj([1 j*3 -2-j*6])
ans =
    1.0000          0 - 3.0000i   -2.0000 + 6.0000i
```

```
»real([1+j ; exp(j*pi/2) ; (2-j)^2])
ans =
    1.0000
         0
    3.0000
```

```
»imag([1+j ; exp(j*pi/2) ; (2-j)^2])
ans =
    1.0000
    1.0000
   -4.0000
```

Some other useful functions are fix, floor, ceiling, round, modulo, remainder, and signum. The fix function rounds a number to the nearest integer toward zero. The floor function rounds a number to the nearest integer toward minus infinity. The ceil function rounds a number to the nearest integer toward plus infinity. The round function rounds a number to the nearest integer. The mod function implements the operation

```
mod(x,y) = x - y.*floor(x./y)
```

if y is not zero. The rem function implements the function

```
rem(x,y) = x - y.*fix(x./y)
```

if y is not zero. The sign function returns a 1 if its argument is positive, a 0 if its argument is zero, and a −1 if its argument is negative.

```
»fix([pi 2.6 -2.6 -2.4])
ans =
     3     2    -2    -2
```

```
»floor([pi 2.6 -2.6 -2.4])
ans =
     3     2    -3    -3
```

```
»ceil([pi 2.6 -2.6 -2.4])
ans =
      4      3     -2     -2
»round([pi 2.6 -2.6 -2.4])
ans =
      3      3     -3     -2
»mod(-5:5,3)
ans =
      1     2     0     1     2     0     1     2     0     1     2
»rem(-5:5,3)
ans =
     -2    -1     0    -2    -1     0     1     2     0     1     2
»sign([15 0 pi -2.9 -1e-18])
ans =
      1     0     1    -1    -1
»sign(C)
ans =
      1     1     1
     -1    -1     1
      1     1     1
```

## Specialized Math Functions

MATLAB provides many specialized and advanced functions. Four that are sometimes useful in signal and system analysis are the error function, the complementary
error function, the least-common-multiple function and the greatest-common-divisor
function. The error function $\text{erf}$ is defined mathematically by

$$\text{erf}(x) = \frac{2}{\sqrt{\pi}} \int_0^x e^{-u^2} \, du,$$

and the complementary error function $\text{erfc}$ is defined mathematically by

$$\text{erfc}(x) = \frac{2}{\sqrt{\pi}} \int_x^\infty e^{-u^2} \, du = 1 - \text{erf}(x).$$

These functions appear in the calculations of the probabilities of events which are
gaussian distributed. The least-common-multiple function $\text{lcm}$ accepts as arguments
two matrices of positive integers and returns a matrix of the smallest integer into which
corresponding elements in both arguments divide an integer number of times. The
greatest-common-divisor function $\text{gcd}$ accepts as arguments two matrices of nonnegative integers and returns a matrix of the greatest integer that will divide into the corresponding elements in both arguments an integer number of times. These functions
can be used to determine the period or frequency of a sum of periodic functions.

```
»erf([-1 ; 0.5])
ans =
    -0.8427
     0.5205
```

```
»erfc([-1 ; 0.5])
ans =
    1.8427
    0.4795

»lcm([6 6],[9 7])
ans =
    18    42

»gcd([19 8],[41 12])
ans =
    1     4
```

## Matrix Functions and Numerical Linear Algebra

Two specialized matrix operations are common in signal and system analysis, matrix inversion and finding eigenvalues. The function, `inv` inverts a matrix. The matrix must be square and its determinant must not be zero.

```
»inv(A)
ans =
    0.0315   -0.0351    0.0124
   -0.0039    0.0108   -0.0789
   -0.0369    0.0767    0.0597

»inv(A)*A
ans =
    1.0000    0.0000    0.0000
    0.0000    1.0000    0.0000
    0.0000    0.0000    1.0000
```

The command `eig` finds the eigenvalues of a square matrix.

```
»eig(B)
ans =
    6.5101
   -0.7550 + 3.0432i
   -0.7550 - 3.0432i
```

## Data Analysis and Fourier Transforms

There are several functions in this group that are often useful. The function `max` can be used with either one argument or two. If one vector is passed to it, it returns the value of the largest element in that vector. If one matrix is passed to it, it returns a row vector of the largest element values in each column. If two matrices of the same shape are passed to `max`, it returns a matrix of the greater of the two element values for each pair of corresponding elements in the two matrices. The function `min` operates exactly like `max` except that the minimum value or values are returned instead of the maximum.

```
»C
C =
    4     5     5
   -1    -3     2
    8     3     1
```

```
»max(C)
ans =
      8      5      5

»b
b =
      1      3      2

»max(b)
ans =
      3

»min(b)
ans =
      1

»c
c =
      2      5      1

»min(b,c)
ans =
      1      3      1
```

There are two statistical functions that are very useful, mean and standard deviation. The function `mean` returns the mean value of a vector, and for a matrix it returns a row vector containing the mean values of the columns of the vector. The function `std` returns the standard deviation of a vector, and for a matrix it returns a row vector containing the standard deviations of the columns of the vector.

```
»mean(A)
ans =
   32.6667     14.0000     17.3333

»std(B)
ans =
    2.6458      1.0000      3.2146

»mean(b)
ans =
      2
```

The function `sort` rearranges the elements of a vector into ascending order and rearranges the elements of each column of a matrix into ascending order.

```
»C
C =
      4      5      5
     -1     -3      2
      8      3      1

»sort(C)
ans =
     -1     -3      1
      4      3      2
      8      5      5
```

There are five functions, sum, product, difference, cumulative sum, and cumulative product that are useful, especially with discrete-time functions. The function, sum, returns the sum of the element values in a vector and returns a row vector of the sums of the element values in the columns of a matrix. The function prod returns the product of the element values in a vector and returns a row vector of the products of the element values in the columns of a matrix. For a vector the function diff returns a vector for which each element is the difference between the corresponding element in the input vector and the previous element in the input vector. The returned vector is always one element shorter than the input vector. For a matrix, diff returns a matrix of difference vectors of the columns of the input matrix calculated in the same way. For vectors, the function cumsum returns a vector for which each element is the sum of the previous element values in the input vector, and for matrices, it returns a matrix in which each column contains the cumulative sum of the corresponding columns of the input matrix. The function cumprod operates in the same way that cumsum does except that it returns products instead of sums.

```
»sum(b)
ans =
      6

»sum(B)
ans =
      6      6     10

»prod(b)
ans =
      6

»prod(B)
ans =
    -12      6     14

»diff(A)
ans =
    -35     -7     -2
    -30    -34    -22

»diff(c)
ans =
      3     -4

»cumsum(C)
ans =
      4      5      5
      3      2      7
     11      5      8

»cumprod(C)
ans =
      4      5      5
     -4    -15     10
    -32    -45     10
```

Four more useful functions and commands are histogram, covariance, convolution, and fast Fourier transform. The command `hist` graphs a histogram of a vector of data. It can also be used as a function to return data which can be used to graph a histogram. The function `cov` finds the variance of a vector or the covariance matrix of a matrix, considering each row to represent one observation and each column to contain data from one variable. The function `conv` convolves two vectors. The length of the returned vector is one less than the sum of the lengths of the vectors convolved. The function `fft` computes the discrete Fourier transform of a vector. The command

```
»hist(randn(1000,1))
```

produces the graph shown in Figure B.1.

```
»cov(randn(5,5))
ans =
      0.9370     -0.0487     -1.3801      0.3545      0.1551
     -0.0487      1.2900     -0.8799      0.5086     -0.1737
     -1.3801     -0.8799      4.8511     -0.0141     -0.5036
      0.3545      0.5086     -0.0141      0.8204     -0.3830
      0.1551     -0.1737     -0.5036     -0.3830      0.5912
```

```
»conv(ones(1,5),ones(1,5))
ans =
     1      2      3      4      5      4      3      2      1
```

```
»fft(cos(2*pi*(0:15)'/4))
ans =
   -0.0000
   -0.0000 -  0.0000i
    0.0000 -  0.0000i
   -0.0000 -  0.0000i
    8.0000 -  0.0000i
    0.0000 -  0.0000i
   -0.0000 -  0.0000i
    0.0000 -  0.0000i
    0.0000
    0.0000 +  0.0000i
   -0.0000 +  0.0000i
    0.0000 +  0.0000i
    8.0000 +  0.0000i
   -0.0000 +  0.0000i
    0.0000 +  0.0000i
   -0.0000 +  0.0000i
```

## Interpolation and Polynomials

In signal and system analysis it is often important to find the roots of equations. The function `roots` finds all the roots of an algebraic equation. The equation is assumed to be of the form,

$$a_N x^N + a_{N-1} x^{N-1} + \cdots + a_2 x^2 + a_1 x + a_0 = 0,$$

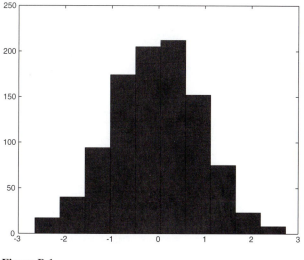

**Figure B.1**

and the vector a with the coefficients in descending order $a_N \cdots a_0$ is the argument sent to roots. The roots are returned in a vector of length $N - 1$.

```
»roots([3 9 8 1])
ans =
  -1.4257 + 0.4586i
  -1.4257 - 0.4586i
  -0.1486
```

Another function that is useful, especially in the study of random variables, is polyfit. The name is a contraction of *polynomial curve fitting*. This function accepts an X vector containing the values of an independent variable, a Y vector containing the values of a dependent variable, and a scalar N, which is the degree of the polynomial used to fit the two sets of data. The function returns the coefficients, in ascending order, of a polynomial of degree N, which has the minimum mean-squared error between its computed values of the dependent variable and the actual values of the dependent variable.

```
»X = 1:10 ;

»Y = 5 + 3*X + randn(1,10)*2 ;

»polyfit(X,Y,2)
ans =
    0.0170   2.4341   6.4467
```

## Two-Dimensional Graphs

One of the most important capabilities of MATLAB for signal and system analysis is its graphical data-plotting. Two-dimensional plotting is the plotting of one vector versus another in a two-dimensional coordinate system. The most commonly used plotting command is, as one might expect, plot. It accepts two vector arguments plus,

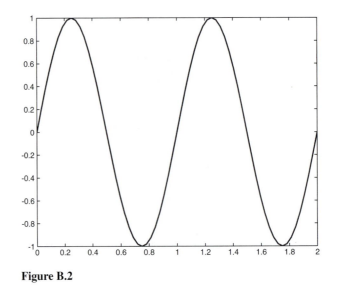

**Figure B.2**

optionally, some formatting commands and plots the second vector vertically versus the first horizontally. The instruction sequence

```
»t = 0:1/32:2 ; x = sin(2*pi*t) ;
»plot(t,x) ;
```

produces the plot shown in Figure B.2 (with a blue curve although it appears black in the figure).

If a third argument is added after the second vector, it controls the plotting color and style. It is a string which determines the color of the line, the style of the line, and the symbols (if any) used to mark points. Below is a listing of the characters and what they control.

| Color | | Marker | | Line style | |
|---|---|---|---|---|---|
| y | yellow | . | point | – | solid |
| m | magenta | o | circle | : | dotted |
| c | cyan | x | x-mark | – . | dash-dot |
| r | red | + | plus | – – | dashed |
| g | green | * | star | | |
| b | blue | s | square | | |
| w | white | d | diamond | | |
| k | black | v | triangle (down) | | |
| | | ^ | triangle (up) | | |
| | | < | triangle (left) | | |
| | | > | triangle (right) | | |
| | | p | pentagram | | |
| | | h | hexagram | | |

The instruction

```
»plot(t,x,'ro') ;
```

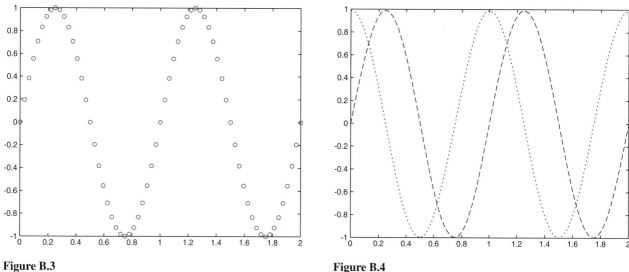

**Figure B.3**

**Figure B.4**

produces the plot shown in Figure B.3 (with red circles although they appear black in the figure), and the instructions

```
»y = cos(2*pi*t) ;
»plot(t,x,'k--',t,y,'r:') ;
```

produce the plot shown in Figure B.4.

The function `linspace` is a convenient tool for producing a vector of equally spaced independent variable values. It accepts either two or three scalar arguments. If two scalars d1 and d2 are provided, `linspace` returns a row vector of 100 equally spaced values between d1 and d2 (including both endpoints). If three scalars d1, d2, and N are provided, `linspace` returns a row vector of N equally spaced values between d1 and d2 (including both endpoints).

```
»linspace(0,5,11)
ans =
  Columns 1 through 7
         0    0.5000   1.0000   1.5000   2.0000   2.5000   3.0000
  Columns 8 through 11
    3.5000    4.0000   4.5000   5.0000
```

The function `logspace` is a convenient tool for producing a vector of equally logarithmically spaced independent variable values. It also accepts either two or three scalar arguments. If two scalars d1 and d2 are provided, `logspace` returns a row vector of 50 equally logarithmically spaced values between $10^{d1}$ and $10^{d2}$ (including both endpoints). If three scalars d1, d2, and N are provided, `logspace` returns a row vector of N equally logarithmically spaced values between $10^{d1}$ and $10^{d2}$ (including both endpoints).

```
»logspace(1,2,10)
ans =
  Columns 1 through 7
    10.0000   12.9155   16.6810   21.5443   27.8256   35.9381   46.4159
  Columns 8 through 10
    59.9484 77.4264 100.0000
```

The command `loglog` also plots one vector versus another, but with a logarithmic scale both vertically and horizontally. The syntax of formatting commands is exactly like `plot`. The instruction sequence

```
»f = logspace(1,4) ;
»H = 1./(1+j*f/100) ;
»loglog(f,abs(H)) ;
```

produces the plot shown in Figure B.5 (with a blue curve although it appears black in the figure).

The command `semilogx` produces a plot with a logarithmic horizontal scale and a linear vertical scale, and the command `semilogy` produces a plot with a logarithmic vertical scale and a linear horizontal scale. The syntax of formatting commands is exactly like `plot`. The instruction,

```
»semilogx(f,20*log10(abs(H))) ;
```

produces the plot shown in Figure B.6 (with a blue curve although it appears black in the figure), and the instruction

```
»semilogy(f,abs(H)) ;
```

produces the plot shown in Figure B.7 (with a blue curve although it appears black in the figure).

The command `stem` is used to plot discrete-time functions. Each data point is indicated by a small circle at the end of a line connecting it to the horizontal axis. The instruction sequence

```
»t = 0:1/32:2 ; x = sin(2*pi*t) ;
»stem(t,x) ;
```

produces the plot shown in Figure B.8 (with blue stems and circles although they appear black in the figure).

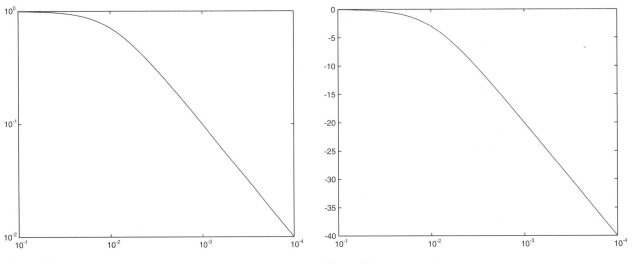

**Figure B.5**                                        **Figure B.6**

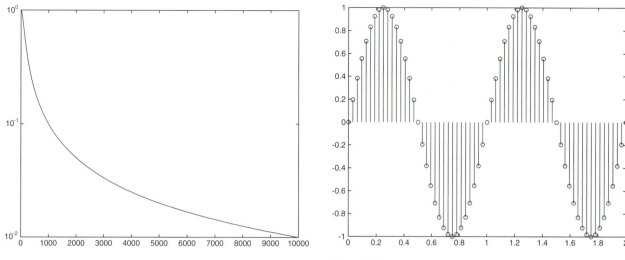

**Figure B.7**

**Figure B.8**

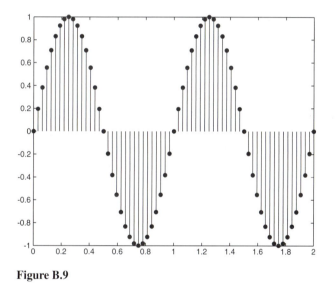

**Figure B.9**

This plot can be modified to look more like a conventional stem plot (see Figure B.9) by using two formatting commands.

```
»stem(t,x,'k','filled') ;
```

Often it is useful to display multiple plots simultaneously on the screen. The command subplot allows that. Plots are arranged in a matrix. Three arguments are passed to subplot. The first is the number of rows of plots, the second is the number of columns of plots, and the third is the number designating which of the plots is to be used by the next plot command. It is not necessary to actually use all the plots indicated by the number of rows and columns of any subplot command. Different subplot commands using different numbers of rows and columns can be used to

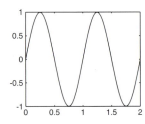

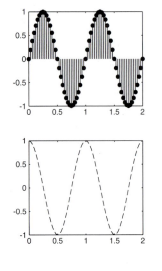

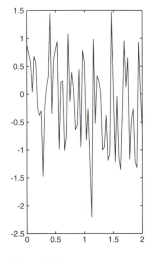

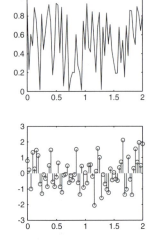

**Figure B.10**                                    **Figure B.11**

place plots on one screen display. The instruction sequence

```
»subplot(2,2,1) ; plot(t,x,'k') ;
»subplot(2,2,2) ; stem(t,x,'k','filled') ;
»x = cos(2*pi*t) ;
»subplot(2,2,4) ; plot(t,x,'k--') ;
```

produces the plot shown in Figure B.10, while the instruction sequence

```
»subplot(1,2,1) ; plot(t,randn(length(t),1)) ;
»subplot(2,2,2) ; plot(t,rand(length(t),1)) ;
»subplot(2,2,4) ; stem(t,randn(length(t),1)) ;
```

produces the plot shown in Figure B.11.

There are several useful commands which control the appearance of plots. The axis command allows the user to set the scale ranges arbitrarily instead of accepting the ranges assigned by MATLAB by default. The command axis takes an argument which is a four-element vector containing, in this sequence, the minimum x (horizontal) value, the maximum x value, the minimum y (vertical) value, and the maximum y value. The instruction

```
»plot(t,x) ; axis([-0.5,2.5,-1.5,1.5]) ;
```

produces the plot shown in Figure B.12.

The command zoom allows the user to interactively zoom in and out of the plot to examine fine detail. Zooming is accomplished by clicking the mouse at the center of the area to be magnified or by dragging a selection box around the area to be magnified. A double click returns the plot to its original scale, and a shift click zooms out instead of in. The plot shown in Figure B.13 was made by zooming in on the plot shown in Figure B.12. The command zoom enables zooming if it is currently disabled and disables it if it is currently enabled. The command zoom on enables zooming unconditionally, and the command zoom off disables zooming unconditionally.

The command grid puts a set of grid lines on the current plot. The command grid turns the grid on if it is currently off and off if it is currently on. The command

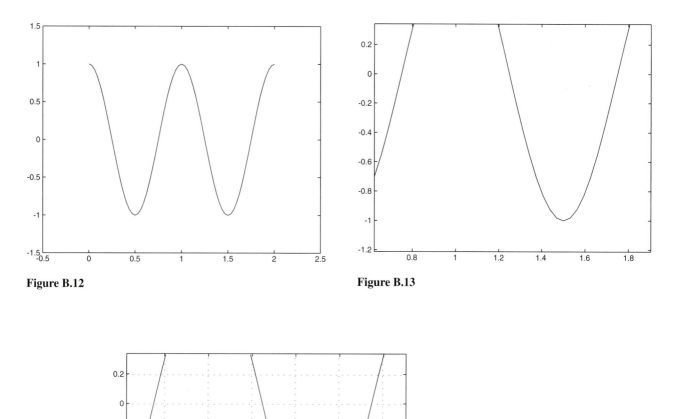

**Figure B.12**                    **Figure B.13**

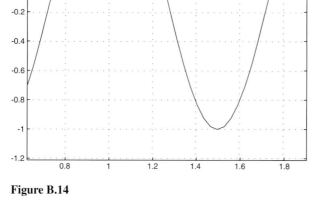

**Figure B.14**

`grid on` unconditionally turns the grid on, and the command `grid off` unconditionally turns the grid off. The instruction

`»grid on ;`

produces the plot shown in Figure B.14.

Sometimes it is desirable to plot one curve and later plot another on the same scale and the same set of axes. That can be done using the `hold` command. If a plot is made and then the command `hold on` is executed, the next time a `plot` command is executed the plot will be on the same set of axes. All subsequent plots will be on that same set of axes until a `hold off` command is executed. The instruction sequence

```
»plot(t,x) ;
»hold on ;
»plot(t/2,x/2) ;
»hold off ;
```

produces the plot shown in Figure B.15.

There are four commands for annotating plots, `title`, `xlabel`, `ylabel`, and `text`. As the names suggest, `title`, `xlabel`, and `ylabel` are used to place a title on the plot, label the *x* axis and label the *y* axis. Their syntax is the same in each case. Each requires a string argument to be displayed. The `text` command is used to place a text string at an arbitrary position on the plot. Its arguments are an *x* position, a *y* position, and a string. In addition, for each command, the font type, size, and style, as well as how the text is positioned and other formatting can be set with optional extra arguments. (Type `help title`, `xlabel`, `ylabel`, or `text` for more detail.) The instruction sequence

```
»plot(t,x)
»title('Plot demonstrating annotation commands') ;
»xlabel('x Label') ;
»ylabel('y Label') ;
»text(1,0,'Text label') ;
```

produces the plot shown in Figure B.16.

### Three-Dimensional Graphs

Sometimes it is useful to plot a curve in three-dimensional space. The command, `plot3` plots a curve in three-dimensional space determined by its three vector arguments. The

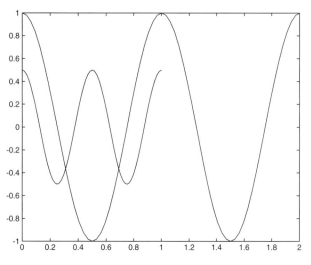

**Figure B.15**

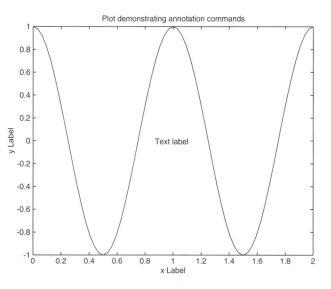

**Figure B.16**

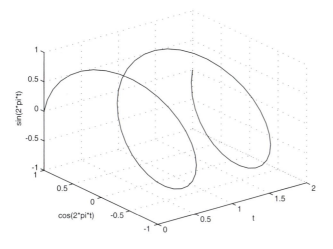

**Figure B.17**

instruction sequence

```
»plot3(t,cos(2*pi*t),sin(2*pi*t)) ;
»grid on ;
»xlabel('t') ; ylabel('cos(2*pi*t)'); zlabel('sin(2*pi*t)') ;
```

produces the plot shown in Figure B.17. Notice the use of `zlabel` which is directly analogous to `xlabel` and `ylabel`.

The other common type of three-dimensional plot is a function of two independent variables. The command `mesh` plots a view of a function of two independent variables which is geometrically a surface above a plane determined by two orthogonal axes representing those two independent variables. The arguments are a vector or matrix containing values of the x variable, a vector or matrix containing values of the y variable, and a matrix containing values of the function of x and y, z. If x is a vector of length $L_x$, and y is a vector of length $L_y$, then z must be an $L_y \times L_x$ matrix. The other choice is to make x, y, and z all matrices of the same shape. A related function which aids in setting up x and y for three-dimensional plotting is `meshgrid`. This command takes two vector arguments and returns two matrices. If x and y are the two vector arguments and X and Y are the two returned matrices, the rows of X are copies of the vector x and the columns of Y are copies of the vector y. The instruction sequence

```
»x = 0:0.5:2 ; y = -1:0.5:1 ;
»[X,Y] = meshgrid(x,y) ;
»X
X =
          0    0.5000    1.0000    1.5000    2.0000
          0    0.5000    1.0000    1.5000    2.0000
          0    0.5000    1.0000    1.5000    2.0000
          0    0.5000    1.0000    1.5000    2.0000
          0    0.5000    1.0000    1.5000    2.0000
```

```
»Y
Y =
    -1.0000   -1.0000   -1.0000   -1.0000   -1.0000
    -0.5000   -0.5000   -0.5000   -0.5000   -0.5000
          0         0         0         0         0
     0.5000    0.5000    0.5000    0.5000    0.5000
     1.0000    1.0000    1.0000    1.0000    1.0000
»x = 0:0.1:2 ; y = -1:0.1:1 ;
»[X,Y] = meshgrid(x,y) ;
»z = (X.^2).*(Y.^3) ;
»mesh(x,y,z) ;
»xlabel('x') ; ylabel('y') ; zlabel('x^2y^3') ;
```

produces the plot shown in Figure B.18.

## Specialized Graphs

The command `stairs` is a special plotting command. It accepts two arguments in the same way `plot` and `stem` do but graphs a stair-step plot. The instruction sequence

```
»t = 0:1/32:2 ; x = cos(2*pi*t) ;
»stairs(t,x) ;
```

produces the plot shown in Figure B.19.

The command `image` accepts a matrix and interprets its elements as specifying the intensity or color of an image. The argument can be an $m \times n$ matrix or an $m \times n \times 3$ matrix. If the argument is an $m \times n$ matrix, the values of the elements are interpreted as colors according to the current `colormap`. A `colormap` is an $m \times 3$ array containing specifications of the red, blue and green intensities of colors. The value of an element of the matrix argument is used as an index into the current `colormap` array to determine what color is displayed on the screen. For example, the `colormap`, `gray`, is $64 \times 3$. So an element value of one would select the first color in the `gray colormap`, and an element value of 64 would select the last color

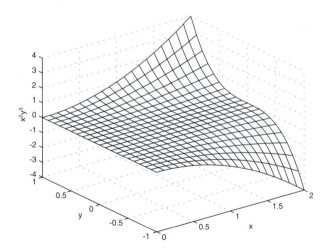

**Figure B.18**

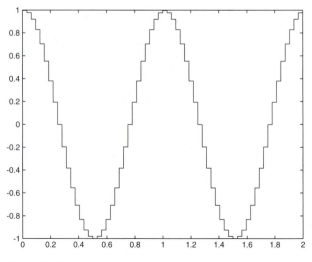

**Figure B.19**

in the `gray colormap`. If the argument passed to `image` is an $m \times n \times 3$ matrix, the three element values at each row and column position are interpreted as the specification of the intensity of red, green, and blue components of a color at that position. The instruction sequence

```
»x = 0:0.025:2 ; y = 0:0.025:2 ;
» [X,Y] = meshgrid(x,y) ;
»colormap(gray) ;
»z = sin(4*pi*X).*cos(4*pi*Y) ;
»minz = min(min(z)) ; maxz = max(max(z)) ;
»z = (z - minz)*64/(maxz-minz) ; z = uint8(z) ;
»image(x,y,z) ;
```

produces the plot shown in Figure B.20.

## Handle Graphics

In MATLAB every graph or plot is an object whose appearance can be modified by using the `set` command to change its properties. Each time a graph is created with the `plot`, `stem`, or `stairs` commands (or many other commands), a graphic object is created. That object can be modified by reference to its "handle". In the instruction,

```
»h = plot(t,sin(2*pi*t)) ;
```

h is a handle to the plot created by the `plot` command, creating the plot shown in Figure B.21. Then we can use a command like

```
»set(h,'LineWidth',2) ;
```

to change the plot to that shown in Figure B.22.

Graphic objects have many properties. Their properties and some possible values of the properties can be seen by using the `set` command.

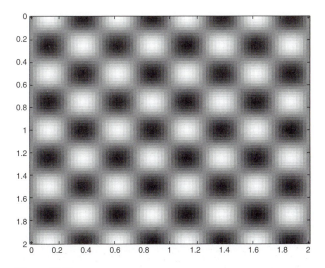

**Figure B.20**

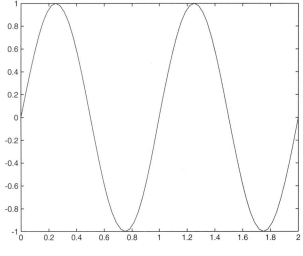

**Figure B.21**

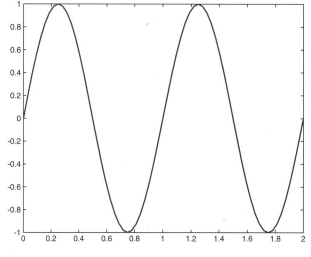

**Figure B.22**

```
»set(h)
    Color
    EraseMode: [ {normal} | background | xor | none ]
    LineStyle: [ {-} | -- | : | -. | none ]
    LineWidth
    Marker: [ + | o | * | . | x | square | diamond | v | ^ | > | < |
            pentagram | hexagram | {none} ]
    MarkerSize
    MarkerEdgeColor: [ none | {auto} ] -or- a ColorSpec.
    MarkerFaceColor: [ {none} | auto ] -or- a ColorSpec.
    XData
    YData
    ZData

    ButtonDownFcn
    Children
    Clipping: [ {on} | off ]
    CreateFcn
    DeleteFcn
    BusyAction: [ {queue} | cancel ]
    HandleVisibility: [ {on} | callback | off ]
    HitTest: [ {on} | off ]
    Interruptible: [ {on} | off ]
    Parent
    Selected: [ on | off ]
    SelectionHighlight: [ {on} | off ]
    Tag
    UIContextMenu
    UserData
    Visible: [ {on} | off ]
```

The command `get` lists the properties of an object and the current value of each property.

```
»get(h)
     Color = [0 0 1]
     EraseMode = normal
     LineStyle = -
     LineWidth = [2]
     Marker = none
     MarkerSize = [6]
     MarkerEdgeColor = auto
     MarkerFaceColor = none
     XData = [ (1 by 65) double array]
     YData = [ (1 by 65) double array]
     ZData = []

     ButtonDownFcn =
     Children = []
     Clipping = on
     CreateFcn =
     DeleteFcn =
     BusyAction = queue
     HandleVisibility = on
     HitTest = on
     Interruptible = on
     Parent = [3.00134]
     Selected = off
     SelectionHighlight = on
     Tag =
     Type = line
     UIContextMenu = []
     UserData = []
     Visible = on
```

The number of properties that can be modified is much too great to explore in this limited tutorial. (For more information consult a MATLAB manual.)

When a plotting command like `plot` or `stem` is used, a `figure` window with default properties is automatically created by MATLAB and the plot is put into that `figure`. The user can also create a figure window and specify its properties using the `figure` command. A figure window can be removed from the screen by closing it with the `close` command.

Not only is the graph an object, the axes on which the graph is drawn are themselves an object and the figure window is an object. The command `gcf` (get current figure) returns a handle to the current figure, and the command `gca` (get current axes) returns a handle to the current axes.

```
»set(gcf)
     BackingStore: [ {on} | off ]
     CloseRequestFcn
     Color
     Colormap
     CurrentAxes
```

```
CurrentObject
CurrentPoint
Dithermap
DithermapMode: [ auto | {manual} ]
IntegerHandle: [ {on} | off ]
InvertHardcopy: [ {on} | off ]
KeyPressFcn
MenuBar: [ none | {figure} ]
MinColormap
Name
NextPlot: [ {add} | replace | replacechildren ]
NumberTitle: [ {on} | off ]
PaperUnits: [ {inches} | centimeters | normalized | points ]
PaperOrientation: [ {portrait} | landscape ]
PaperPosition
PaperPositionMode: [ auto | {manual} ]
PaperType: [ {usletter} | uslegal | A0 | A1 | A2 | A3 | A4 | A5 | B0
        | B1 | B2 | B3 | B4 | B5 | arch-A | arch-B | arch-C | arch-D |
        arch-E | A | B | C | D | E | tabloid ]
Pointer: [ crosshair | fullcrosshair | {arrow} | ibeam | watch | topl
        | topr | botl | botr | left | top | right | bottom | circle |
        cross | fleur | custom ]
PointerShapeCData
PointerShapeHotSpot
Position
Renderer: [ {painters} | zbuffer | OpenGL ]
RendererMode: [ {auto} | manual ]
Resize: [ {on} | off ]
ResizeFcn
ShareColors: [ {on} | off ]
Units: [ inches | centimeters | normalized | points | {pixels} |
        characters ]
WindowButtonDownFcn
WindowButtonMotionFcn
WindowButtonUpFcn
WindowStyle: [ {normal} | modal ]

ButtonDownFcn
Children
Clipping: [ {on} | off ]
CreateFcn
DeleteFcn
BusyAction: [ {queue} | cancel ]
HandleVisibility: [ {on} | callback | off ]
HitTest: [ {on} | off ]
Interruptible: [ {on} | off ]
Parent
Selected: [ on | off ]
SelectionHighlight: [ {on} | off ]
Tag
UIContextMenu
```

```
        UserData
        Visible: [ {on} | off ]
»set(gca)
        AmbientLightColor
        Box: [ on | {off} ]
        CameraPosition
        CameraPositionMode: [ {auto} | manual ]
        CameraTarget
        CameraTargetMode: [ {auto} | manual ]
        CameraUpVector
        CameraUpVectorMode: [ {auto} | manual ]
        CameraViewAngle
        CameraViewAngleMode: [ {auto} | manual ]
        CLim
        CLimMode: [ {auto} | manual ]
        Color
        ColorOrder
        DataAspectRatio
        DataAspectRatioMode: [ {auto} | manual ]
        DrawMode: [ {normal} | fast ]
        FontAngle: [ {normal} | italic | oblique ]
        FontName
        FontSize
        FontUnits: [ inches | centimeters | normalized | {points} | pixels ]
        FontWeight: [ light | {normal} | demi | bold ]
        GridLineStyle: [ - | -- | {:} | -. | none ]
        Layer: [ top | {bottom} ]
        LineStyleOrder
        LineWidth
        NextPlot: [ add | {replace} | replacechildren ]
        PlotBoxAspectRatio
        PlotBoxAspectRatioMode: [ {auto} | manual ]
        Projection: [ {orthographic} | perspective ]
        Position
        TickLength
        TickDir: [ {in} | out ]
        TickDirMode: [ {auto} | manual ]
        Title
        Units: [ inches | centimeters | {normalized} | points | pixels |
              characters ]
        View
        XColor
        XDir: [ {normal} | reverse ]
        XGrid: [ on | {off} ]
        XLabel
        XAxisLocation: [ top | {bottom} ]
        XLim
        XLimMode: [ {auto} | manual ]
        XScale: [ {linear} | log ]
        XTick
```

```
XTickLabel
XTickLabelMode: [ {auto} | manual ]
XTickMode: [ {auto} | manual ]
YColor
YDir: [ {normal} | reverse ]
YGrid: [ on | {off} ]
YLabel
YAxisLocation: [ {left} | right ]
YLim
YLimMode: [ {auto} | manual ]
YScale: [ {linear} | log ]
YTick
YTickLabel
YTickLabelMode: [ {auto} | manual ]
YTickMode: [ {auto} | manual ]
ZColor
ZDir: [ {normal} | reverse ]
ZGrid: [ on | {off} ]
ZLabel
ZLim
ZLimMode: [ {auto} | manual ]
ZScale: [ {linear} | log ]
ZTick
ZTickLabel
ZTickLabelMode: [ {auto} | manual ]
ZTickMode: [ {auto} | manual ]

ButtonDownFcn
Children
Clipping: [ {on} | off ]
CreateFcn
DeleteFcn
BusyAction: [ {queue} | cancel ]
HandleVisibility: [ {on} | callback | off ]
HitTest: [ {on} | off ]
Interruptible: [ {on} | off ]
Parent
Selected: [ on | off ]
SelectionHighlight: [ {on} | off ]
Tag
UIContextMenu
UserData
Visible: [ {on} | off ]
```

There are two very useful primitive graphics commands `line` and `patch`. The command `line` draws a straight line between any two points according to the scale of the current axes. The command `patch` draws a filled polygon according to the scale of the current axes.

```
»subplot(2,1,1) ; line(randn(10,1),randn(10,1)) ;
»subplot(2,1,2) ; patch(rand(10,1),rand(10,1),'r') ;
```

(Figure B.23).

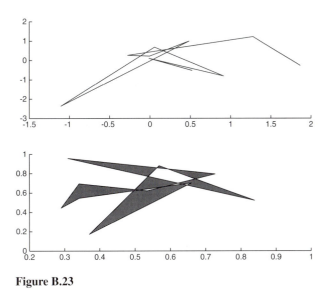

**Figure B.23**

## Graphical User Interface Tools

There is a set of tools in MATLAB which a programmer can use to develop a graphical user interface. They allow the programmer to create windows, buttons, menus, etc., and link program execution to them. A discussion of these features is outside the scope of this tutorial.

## Character Strings

String handling is often important and MATLAB has a complete set of string manipulation functions. The function `char` converts positive integers to the corresponding character according to the American Standard Code for Information Interchange (ASCII).

```
»char([40:52])
ans =
()*+,-./01234

»char([[65:90] ; [97:122]])
ans =
ABCDEFGHIJKLMNOPQRSTUVWXYZ
abcdefghijklmnopqrstuvwxyz
```

A *string* is a vector of characters. Several MATLAB functions operate on strings. The function `strcat` concatenates two strings.

```
»s1 = char([65:90]) ;
»s2 = char([97:122]) ;
»s3 = strcat(s1,s2)
s3 =
ABCDEFGHIJKLMNOPQRSTUVWXYZabcdefghijklmnopqrstuvwxyz
```

The function `strcmp` compares two strings. If they are the same, it returns a logical 1; otherwise it returns a logical 0.

```
»strcmp(s1,s2)
ans =
     0
```

```
»strcmp('ABCDEFGHIJKLMNOPQRSTUVWXYZ',s1)
ans =
     1
```

The function `findstr` finds one string within another. If the shorter of the two input strings is found within the longer, `findstr` returns the index of the first character in the longer string. If the shorter string is not found within the longer string, an empty matrix is returned.

```
»findstr(s1,'MNO')
ans =
     13
```

```
»findstr('tuv',s2)
ans =
     20
```

```
»findstr('tuv',s1)
ans =
     []
```

There are also functions to convert from strings to numbers, and vice versa. The function, `num2str` converts a number to a string.

```
»n = 25.3 ;
```

```
»n
n =
   25.3000
```

```
»length(n)
ans =
     1
```

```
»ns = num2str(n) ;
```

```
»ns
ns =
25.3
```

```
»length(ns)
ans =
     4
```

```
»ns(2)
ans =
5
```

The function `str2num` converts a string to a number.

```
»ns = '368.92' ;
```

```
»length(ns)
ans =
     6
```

```
»n = str2num(ns) ;

»n
n =
   368.9200

»length(n)
ans =
      1
```

## File Input-Output

Another important function of any programming language is to read data from a file and to store data in a file. In MATLAB a file is opened for reading or writing by the `fopen` command. The argument of the `fopen` command is a string specifying the file name and, optionally, the path to the file name, and it returns a positive integer file identifier which is used in future references to that file. If the file should already exist and cannot be found, `fopen` returns a $-1$ to indicate that the file was not found. If a file is opened for write access, it need not already exist; MATLAB will create it. When all interaction with the file is complete, it is closed by the `fclose` command. The argument of `fclose` is the file identifier number.

Files can contain data encoded in different ways. Two very common formats are binary and text. The two commands `fread` and `fwrite` read and write, respectively, binary data to and from a file. The arguments of these functions are the file identification number, the number of data to be read, and a string which is a precision indicator specifying how to read the data. The precision argument indicates how many bits represent a number. The available formats are

| MATALAB | C or Fortran | Description |
|---------|--------------|-------------|
| char    | char*1       | Character, 8 bits |
| uchar   | unsigned char | Unsigned character, 8 bits |
| schar   | signed char  | Signed character, 8 bits |
| int8    | integer*1    | Integer, 8 bits |
| int16   | integer*2    | Integer, 16 bits |
| int32   | integer*4    | Integer, 32 bits |
| int64   | integer*8    | Integer, 64 bits |
| uint8   | integer*1    | Unsigned integer, 8 bits |
| uint16  | integer*2    | Unsigned integer, 16 bits |
| uint32  | integer*4    | Unsigned integer, 32 bits |
| uint64  | integer*8    | Unsigned integer, 64 bits |
| float32 | real*4       | Floating point, 32 bits |
| float64 | real*8       | Floating point, 64 bits |

If the file is a text file, it can be read as text by the `fgetl` command, which reads in one line of text (up to the first line terminator) and returns that string without the terminator. The next invocation of `fgetl` reads in the next line of text in the same way.

The command `save` saves all the current variables to a disk file. If a file name is provided, the data are put into a file of that name with a `.mat` extension. If a file name is not provided, the data are put into a file named `matlab.mat`. The command `load` retrieves variables saved with the `save` command. There are also options with each command which allow the saving and retrieval of a selected set of variables instead of all of them.

## Time and Dates

MATLAB has a group of functions which return the current time or date or allow timing of the execution speed of a program. They are in the group `timefun`.

## Data Types and Structures

MATLAB allows the creation of structures and cell arrays to handle large groups of disparate data as a unit and to convert data from one precision to another. These features are described in the group `datatypes`.

# Method for Finding Least Common Multiples

Finding the least common multiple of any set of integers proceeds as follows:

1.  Find the prime factors of all the integers.
2.  For each prime factor that occurs find the number of occurrences in each of the integers.
3.  Form the product of the prime factors, with each prime factor occurring the maximum number of times that it occurred in the integers.

The answer obtained in step 3 is the least common multiple.

As an example, we will find the least common multiple of the integers 1 through 10.

1.  Find the prime factors:

$$1 = 1, 2 = 2, 3 = 3, 4 = 2 \times 2, 5 = 5, 6 = 2 \times 3,$$
$$7 = 7, 8 = 2 \times 2 \times 2, 9 = 3 \times 3, 10 = 2 \times 5.$$

2.  The maximum number of occurrences of each of the prime factors is summarized in the table.

| Prime factor | Number of occurrences |
|:---:|:---:|
| 1 | 1 |
| 2 | 3 |
| 3 | 2 |
| 5 | 1 |
| 7 | 1 |

3.  Therefore, the least common multiple is

$$2 \times 2 \times 2 \times 3 \times 3 \times 5 \times 7 = (2)^3(3)^2(5)(7) = 2520.$$

If the numbers are not integers, premultiply them by some common factor to make them all integers (if possible), perform each step, and then divide the result by that same factor. For example, we can find the least common multiple of $2/\pi$, $3/2\pi$, $5/7\pi$.

If we premultiply by $14\pi$, we get the integers 28, 21, and 10. The prime factors are $2 \times 2 \times 7$, $3 \times 7$, and $2 \times 5$. Therefore, the least common multiple of 28, 21, and 10 is $2 \times 2 \times 3 \times 5 \times 7 = 420$, and the least common multiple of $2/\pi, 3/2\pi, 5/7\pi$ is $420/14\pi = 30/\pi$. The fact that this is a common multiple can be confirmed by finding the ratios,

$$\frac{30/\pi}{2/\pi} = 15 \qquad \frac{30/\pi}{3/2\pi} = 20 \qquad \frac{30/\pi}{5/7\pi} = 42$$

which are all integers.

# Convolution Properties

## D.1 DT CONVOLUTION PROPERTIES

### Commutativity

The commutativity of DT convolution can be proven by starting with the definition of convolution

$$x[n] * h[n] = \sum_{k=-\infty}^{\infty} x[k]h[n-k] \tag{D.1}$$

and letting $q = n - k$. Then we have

$$x[n] * h[n] = \sum_{q=-\infty}^{\infty} x[n-q]h[q] = \sum_{q=-\infty}^{\infty} h[q]x[n-q] = h[n] * x[n]. \tag{D.2}$$

### Associativity

If we convolve $g[n] = x[n] * y[n]$ with $z[n]$, we get

$$g[n] * z[n] = (x[n] * y[n]) * z[n] = \underbrace{\left( \sum_{k=-\infty}^{\infty} x[k]y[n-k] \right)}_{g[n]} * z[n] \tag{D.3}$$

or

$$g[n] * z[n] = \sum_{q=-\infty}^{\infty} \underbrace{\left( \sum_{k=-\infty}^{\infty} x[k]y[q-k] \right)}_{g[q]} z[n-q]. \tag{D.4}$$

Exchanging the order of summation,

$$(x[n] * y[n]) * z[n] = \sum_{k=-\infty}^{\infty} x[k] \sum_{q=-\infty}^{\infty} y[q-k]z[n-q]. \tag{D.5}$$

Let $n - q = m$ and let $h[n] = y[n] * z[n]$. Then

$$(x[n] * y[n]) * z[n] = \sum_{k=-\infty}^{\infty} x[k] \underbrace{\sum_{m=-\infty}^{\infty} z[m]y[(n - k) - m]}_{z[n]*y[n-k]=y[n-k]*z[n]=h[n-k]} \tag{D.6}$$

or

$$(x[n] * y[n]) * z[n] = \underbrace{\sum_{k=-\infty}^{\infty} x[k]h[n - k]}_{x[n]*h[n]} = x[n] * \left( \underbrace{y[n] * z[n]}_{h[n]} \right). \tag{D.7}$$

### Distributivity

If we convolve $x[n]$ with the sum of $y[n]$ and $z[n]$, we get

$$x[n] * (y[n] + z[n]) = \sum_{k=-\infty}^{\infty} x[k] (y[n - k] + z[n - k]) \tag{D.8}$$

or

$$x[n] * (y[n] + z[n]) = \underbrace{\sum_{k=-\infty}^{\infty} x[k]y[n - k]}_{=x[n]*y[n]} + \underbrace{\sum_{k=-\infty}^{\infty} x[k]z[n - k]}_{=x[n]*z[n]}. \tag{D.9}$$

Therefore,

$$x[n] * (y[n] + z[n]) = x[n] * y[n] + x[n] * z[n]. \tag{D.10}$$

## D.2  CT CONVOLUTION PROPERTIES

### Commutativity

By making the change of variable $\lambda = t - \tau$ in one form of the definition of CT convolution

$$x(t) * h(t) = \int_{-\infty}^{\infty} x(\tau)h(t - \tau) \, d\tau, \tag{D.11}$$

it becomes

$$x(t) * h(t) = -\int_{\infty}^{-\infty} x(t - \lambda)h(\lambda) \, d\lambda = \int_{-\infty}^{\infty} h(\lambda)x(t - \lambda) \, d\lambda = h(t) * x(t), \tag{D.12}$$

proving that convolution is commutative.

### Associativity

Associativity can be proven by considering the two operations

$$[x(t) * y(t)] * z(t) \qquad \text{and} \qquad x(t) * [y(t) * z(t)]. \tag{D.13}$$

Using the definition of convolution

$$x(t) * h(t) = \int_{-\infty}^{\infty} x(\tau) h(t - \tau)\, d\tau \tag{D.14}$$

we get

$$[x(t) * y(t)] * z(t) = \left[ \int_{-\infty}^{\infty} x(\tau_{xy}) y(t - \tau_{xy})\, d\tau_{xy} \right] * z(t) \tag{D.15}$$

or

$$[x(t) * y(t)] * z(t) = \int_{-\infty}^{\infty} \left[ \int_{-\infty}^{\infty} x(\tau_{xy}) y(\tau_{yz} - \tau_{xy})\, d\tau_{xy} \right] z(t - \tau_{yz})\, d\tau_{yz} \tag{D.16}$$

and

$$x(t) * [y(t) * z(t)] = x(t) * \left[ \int_{-\infty}^{\infty} y(\tau_{yz}) z(t - \tau_{yz})\, d\tau_{yz} \right] \tag{D.17}$$

or

$$x(t) * [y(t) * z(t)] = \int_{-\infty}^{\infty} x(\tau_{xy}) \left[ \int_{-\infty}^{\infty} y(\tau_{yz}) z(t - \tau_{xy} - \tau_{yz})\, d\tau_{yz} \right] d\tau_{xy}.$$

Then the proof consists of showing that

$$\int_{-\infty}^{\infty} \int_{-\infty}^{\infty} x(\tau_{xy}) y(\tau_{yz} - \tau_{xy}) z(t - \tau_{yz})\, d\tau_{xy}\, d\tau_{yz}$$

$$= \int_{-\infty}^{\infty} \int_{-\infty}^{\infty} x(\tau_{xy}) y(\tau_{yz}) z(t - \tau_{xy} - \tau_{yz})\, d\tau_{xy}\, d\tau_{yz}. \tag{D.18}$$

In the right-hand $\tau_{xy}$ integration make the change of variable

$$\lambda = \tau_{xy} + \tau_{yz} \qquad \text{and} \qquad d\lambda = d\tau_{xy}. \tag{D.19}$$

Then

$$\int_{-\infty}^{\infty} \int_{-\infty}^{\infty} x(\tau_{xy}) y(\tau_{yz} - \tau_{xy}) z(t - \tau_{yz})\, d\tau_{xy}\, d\tau_{yz}$$

$$= \int_{-\infty}^{\infty} \int_{-\infty}^{\infty} x(\lambda - \tau_{yz}) y(\tau_{yz}) z(t - \lambda)\, d\lambda\, d\tau_{yz}. \tag{D.20}$$

Next, in the right-hand $\tau_{xy}$ integration make the change of variable

$$\eta = \lambda - \tau_{yz} \qquad \text{and} \qquad d\eta = -d\tau_{yz}. \tag{D.21}$$

Then

$$\int_{-\infty}^{\infty} \int_{-\infty}^{\infty} x(\tau_{xy}) y(\tau_{yz} - \tau_{xy}) z(t - \tau_{yz}) \, d\tau_{xy} \, d\tau_{yz} = -\int_{-\infty}^{\infty} \int_{\infty}^{-\infty} x(\eta) y(\lambda - \eta) z(t - \lambda) \, d\lambda \, d\eta$$

(D.22)

or

$$\int_{-\infty}^{\infty} \int_{-\infty}^{\infty} x(\tau_{xy}) y(\tau_{yz} - \tau_{xy}) z(t - \tau_{yz}) \, d\tau_{xy} \, d\tau_{yz} = \int_{-\infty}^{\infty} \int_{-\infty}^{\infty} x(\eta) y(\lambda - \eta) z(t - \lambda) \, d\lambda \, d\eta.$$

(D.23)

Except for the names of the variables of integration, the two integrals (D.22) and (D.23) are the same; therefore, the integrals are equal and the associativity of convolution is proven.

**Distributivity**

Convolution is also distributive,

$$x(t) * [h_1(t) + h_2(t)] = x(t) * h_1(t) + x(t) * h_2(t)$$

$$= \int_{-\infty}^{\infty} x(t)[h_1(t - \tau) + h_2(t - \tau)] \, d\tau$$

$$= \int_{-\infty}^{\infty} x(t)h_1(t - \tau) \, d\tau + \int_{-\infty}^{\infty} x(t)h_2(t - \tau) \, d\tau$$

$$= x(t) * h_1(t) + x(t) * h_2(t).$$

(D.24)

# APPENDIX E

## Table of Fourier Pairs

### E.1  FOURIER SERIES

**Continuous-Time Fourier Series (CTFS)**

The following Fourier pairs are for a periodic CT function represented over the period $T_F$.

$$x(t) = \sum_{k=-\infty}^{\infty} X[k]e^{j2\pi(kf_F)t} \xleftrightarrow{\mathcal{FS}} X[k] = \frac{1}{T_F} \int_{T_F} x(t)e^{-j2\pi(kf_F)t} \, dt$$

$$x(t) = \sum_{k=-\infty}^{\infty} X[k]e^{j(k\omega_F)t} \xleftrightarrow{\mathcal{FS}} X[k] = \frac{1}{T_F} \int_{T_F} x(t)e^{-j(k\omega_F)t} \, dt$$

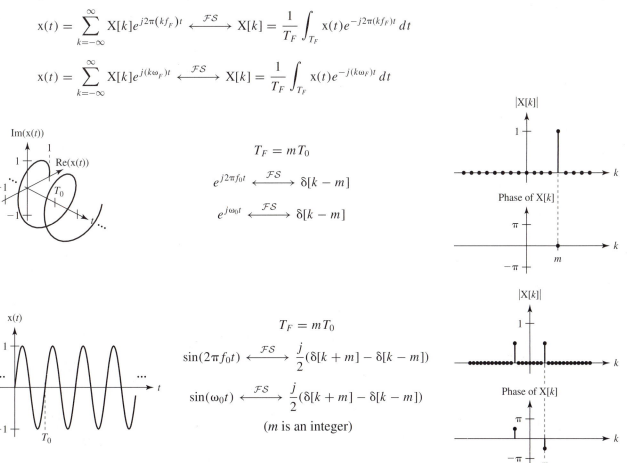

$$T_F = mT_0$$

$$e^{j2\pi f_0 t} \xleftrightarrow{\mathcal{FS}} \delta[k - m]$$

$$e^{j\omega_0 t} \xleftrightarrow{\mathcal{FS}} \delta[k - m]$$

$$T_F = mT_0$$

$$\sin(2\pi f_0 t) \xleftrightarrow{\mathcal{FS}} \frac{j}{2}(\delta[k + m] - \delta[k - m])$$

$$\sin(\omega_0 t) \xleftrightarrow{\mathcal{FS}} \frac{j}{2}(\delta[k + m] - \delta[k - m])$$

($m$ is an integer)

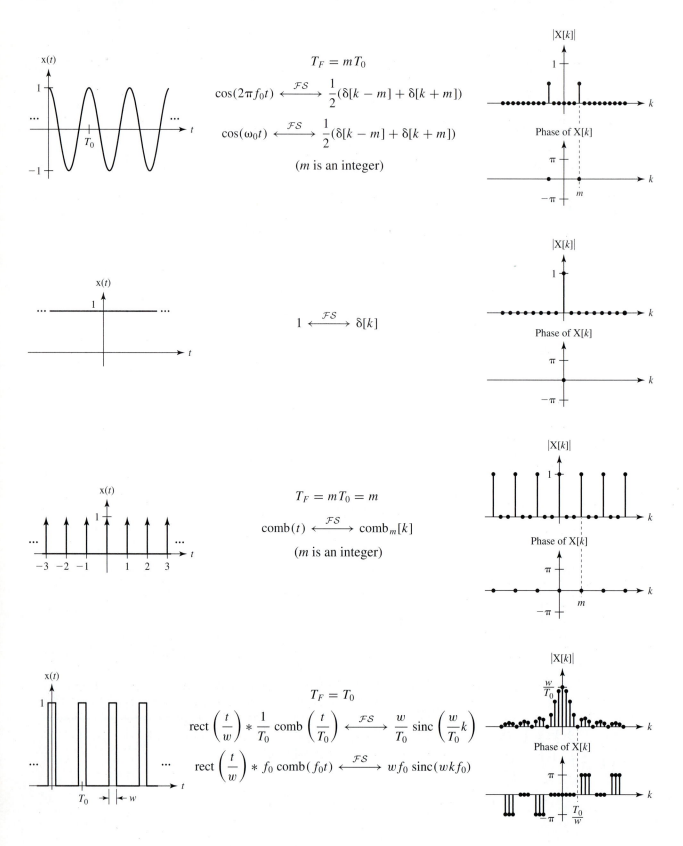

$$T_F = mT_0$$

$$\cos(2\pi f_0 t) \xleftrightarrow{\mathcal{FS}} \frac{1}{2}(\delta[k-m] + \delta[k+m])$$

$$\cos(\omega_0 t) \xleftrightarrow{\mathcal{FS}} \frac{1}{2}(\delta[k-m] + \delta[k+m])$$

*(m is an integer)*

$$1 \xleftrightarrow{\mathcal{FS}} \delta[k]$$

$$T_F = mT_0 = m$$

$$\text{comb}(t) \xleftrightarrow{\mathcal{FS}} \text{comb}_m[k]$$

*(m is an integer)*

$$T_F = T_0$$

$$\text{rect}\left(\frac{t}{w}\right) * \frac{1}{T_0}\text{comb}\left(\frac{t}{T_0}\right) \xleftrightarrow{\mathcal{FS}} \frac{w}{T_0}\text{sinc}\left(\frac{w}{T_0}k\right)$$

$$\text{rect}\left(\frac{t}{w}\right) * f_0\,\text{comb}(f_0 t) \xleftrightarrow{\mathcal{FS}} wf_0\,\text{sinc}(wkf_0)$$

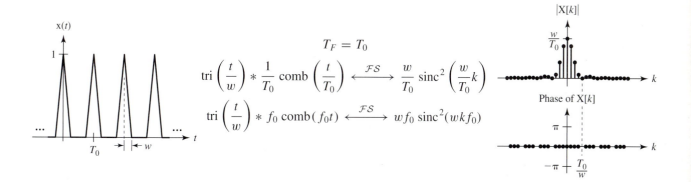

$$T_F = T_0$$

$$\text{tri}\left(\frac{t}{w}\right) * \frac{1}{T_0} \text{comb}\left(\frac{t}{T_0}\right) \xleftrightarrow{\mathcal{FS}} \frac{w}{T_0} \text{sinc}^2\left(\frac{w}{T_0}k\right)$$

$$\text{tri}\left(\frac{t}{w}\right) * f_0 \text{comb}(f_0 t) \xleftrightarrow{\mathcal{FS}} w f_0 \text{sinc}^2(w k f_0)$$

$$T_F = T_0$$

$$\text{sinc}\left(\frac{t}{w}\right) * \frac{1}{T_0} \text{comb}\left(\frac{t}{T_0}\right) \xleftrightarrow{\mathcal{FS}} \frac{w}{T_0} \text{rect}\left(\frac{w}{T_0}k\right)$$

$$\text{sinc}\left(\frac{t}{w}\right) * f_0 \text{comb}(f_0 t) \xleftrightarrow{\mathcal{FS}} w f_0 \text{rect}(w k f_0)$$

$$w f_0 (2M + 1) \text{drcl}(f_0 t, 2M + 1) \xleftrightarrow{\mathcal{FS}}$$
$$w f_0 \text{rect}(w k f_0)$$

$$\left(M \text{ is the greatest integer in } \frac{T_0}{2w}\right)$$

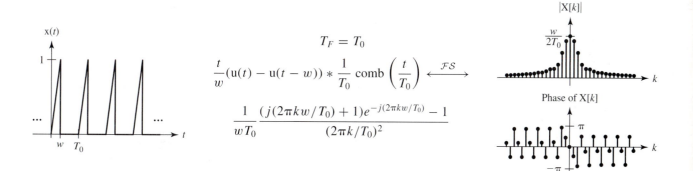

$$T_F = T_0$$

$$\frac{t}{w}(\text{u}(t) - \text{u}(t - w)) * \frac{1}{T_0} \text{comb}\left(\frac{t}{T_0}\right) \xleftrightarrow{\mathcal{FS}}$$

$$\frac{1}{wT_0} \frac{(j(2\pi k w / T_0) + 1)e^{-j(2\pi k w / T_0)} - 1}{(2\pi k / T_0)^2}$$

## Discrete-Time Fourier Series (DTFS)

The following Fourier pairs are for a periodic DT function represented over the period $N_F$.

$$\text{x}[n] = \sum_{k=\langle N_F \rangle} \text{X}[k] e^{j2\pi(kn/N_F)} \xleftrightarrow{\mathcal{FS}} \text{X}[k] = \frac{1}{N_F} \sum_{n=\langle N_F \rangle} \text{x}[n] e^{-j2\pi(kn/N_F)}$$

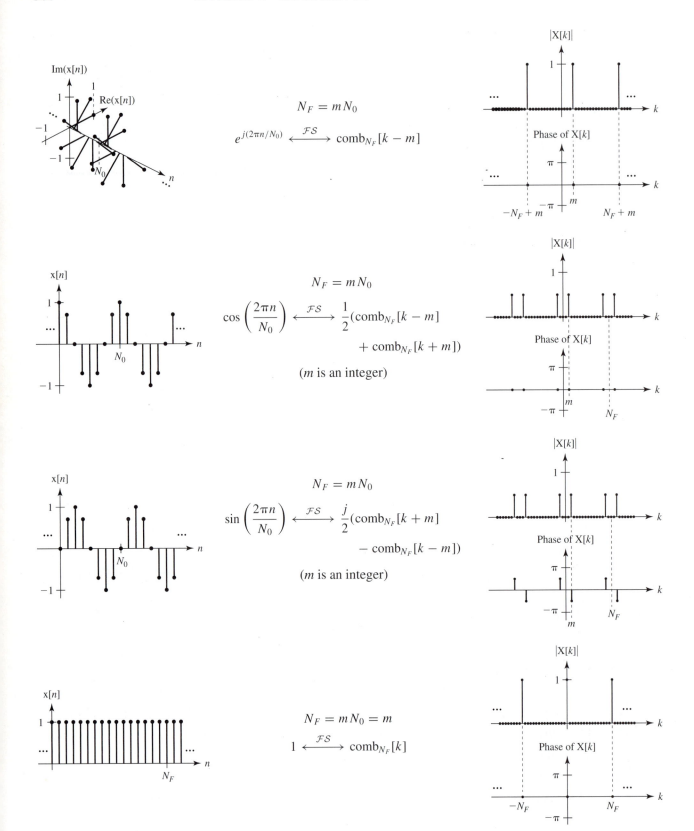

$$N_F = m N_0$$

$$e^{j(2\pi n/N_0)} \xleftarrow{\mathcal{FS}} \text{comb}_{N_F}[k - m]$$

$$N_F = m N_0$$

$$\cos\left(\frac{2\pi n}{N_0}\right) \xleftarrow{\mathcal{FS}} \frac{1}{2}(\text{comb}_{N_F}[k - m]$$
$$+ \text{comb}_{N_F}[k + m])$$

($m$ is an integer)

$$N_F = m N_0$$

$$\sin\left(\frac{2\pi n}{N_0}\right) \xleftarrow{\mathcal{FS}} \frac{j}{2}(\text{comb}_{N_F}[k + m]$$
$$- \text{comb}_{N_F}[k - m])$$

($m$ is an integer)

$$N_F = m N_0 = m$$

$$1 \xleftarrow{\mathcal{FS}} \text{comb}_{N_F}[k]$$

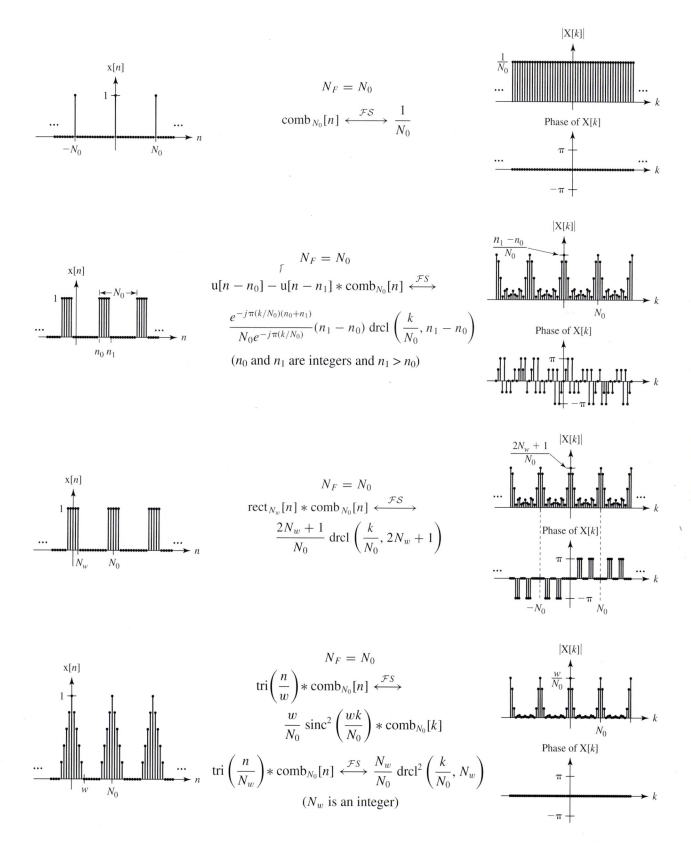

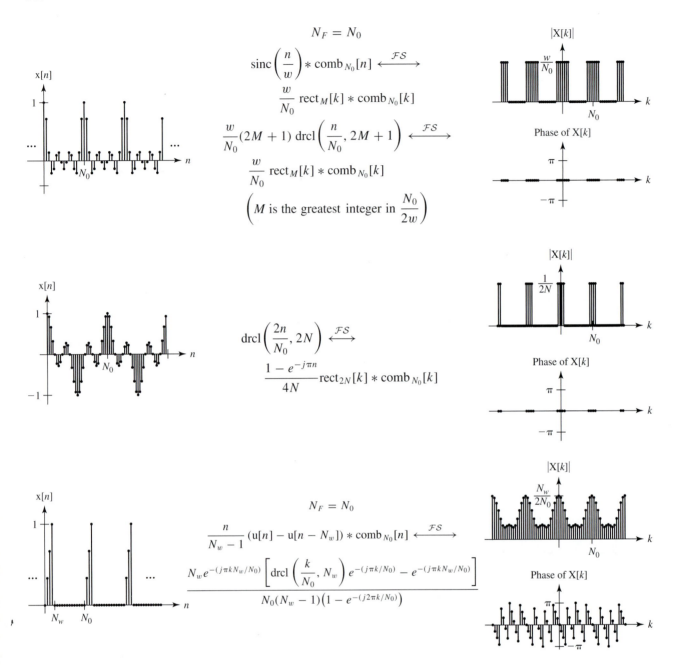

## E.2 FOURIER TRANSFORM

### Continuous-Time Fourier Transform (CTFT)

$$x(t) = \int_{-\infty}^{\infty} X(f)e^{+j2\pi ft}\, df \xleftrightarrow{\mathcal{F}} X(f) = \int_{-\infty}^{\infty} x(t)e^{-j2\pi ft}\, dt$$

$$x(t) = \frac{1}{2\pi}\int_{-\infty}^{\infty} X(j\omega)e^{+j\omega t}\, d\omega \xleftrightarrow{\mathcal{F}} X(j\omega) = \int_{-\infty}^{\infty} x(t)e^{-j\omega t}\, dt$$

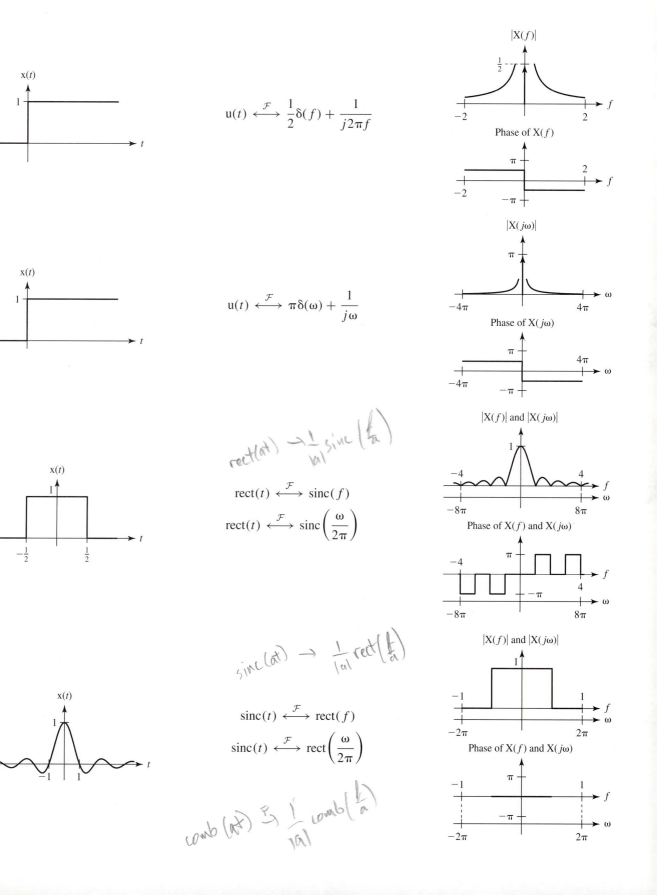

$$u(t) \overset{\mathcal{F}}{\longleftrightarrow} \frac{1}{2}\delta(f) + \frac{1}{j2\pi f}$$

$$u(t) \overset{\mathcal{F}}{\longleftrightarrow} \pi\delta(\omega) + \frac{1}{j\omega}$$

$$\mathrm{rect}(at) \to \frac{1}{|a|}\mathrm{sinc}\left(\frac{f}{a}\right)$$

$$\mathrm{rect}(t) \overset{\mathcal{F}}{\longleftrightarrow} \mathrm{sinc}(f)$$

$$\mathrm{rect}(t) \overset{\mathcal{F}}{\longleftrightarrow} \mathrm{sinc}\left(\frac{\omega}{2\pi}\right)$$

$$\mathrm{sinc}(at) \to \frac{1}{|a|}\mathrm{rect}\left(\frac{f}{a}\right)$$

$$\mathrm{sinc}(t) \overset{\mathcal{F}}{\longleftrightarrow} \mathrm{rect}(f)$$

$$\mathrm{sinc}(t) \overset{\mathcal{F}}{\longleftrightarrow} \mathrm{rect}\left(\frac{\omega}{2\pi}\right)$$

$$\mathrm{comb}(at) \overset{\mathcal{F}}{\to} \frac{1}{|a|}\mathrm{comb}\left(\frac{f}{a}\right)$$

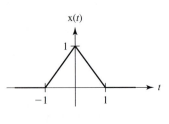

$$\text{tri}(t) \xleftrightarrow{\mathcal{F}} \text{sinc}^2(f)$$

$$\text{tri}(t) \xleftrightarrow{\mathcal{F}} \text{sinc}^2\left(\frac{\omega}{2\pi}\right)$$

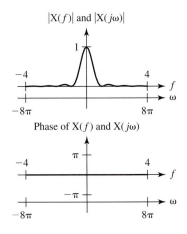

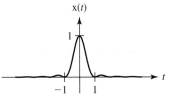

$$\text{sinc}^2(t) \xleftrightarrow{\mathcal{F}} \text{tri}(f)$$

$$\text{sinc}^2(t) \xleftrightarrow{\mathcal{F}} \text{tri}\left(\frac{\omega}{2\pi}\right)$$

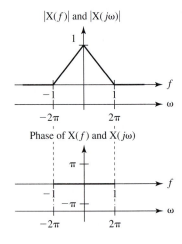

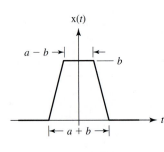

$$\frac{a+b}{2}\,\text{tri}\left(\frac{2t}{a+b}\right) - \frac{a-b}{2}\,\text{tri}\left(\frac{2t}{a-b}\right) \xleftrightarrow{\mathcal{F}}$$
$$|ab|\,\text{sinc}(af)\,\text{sinc}(bf)$$

$$\frac{a+b}{2}\,\text{tri}\left(\frac{2t}{a+b}\right) - \frac{a-b}{2}\,\text{tri}\left(\frac{2t}{a-b}\right) \xleftrightarrow{\mathcal{F}}$$
$$|ab|\,\text{sinc}\left(\frac{a\omega}{2\pi}\right)\text{sinc}\left(\frac{b\omega}{2\pi}\right)$$

$$(a > b > 0)$$

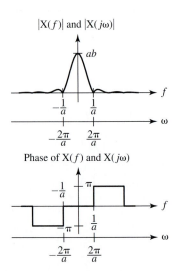

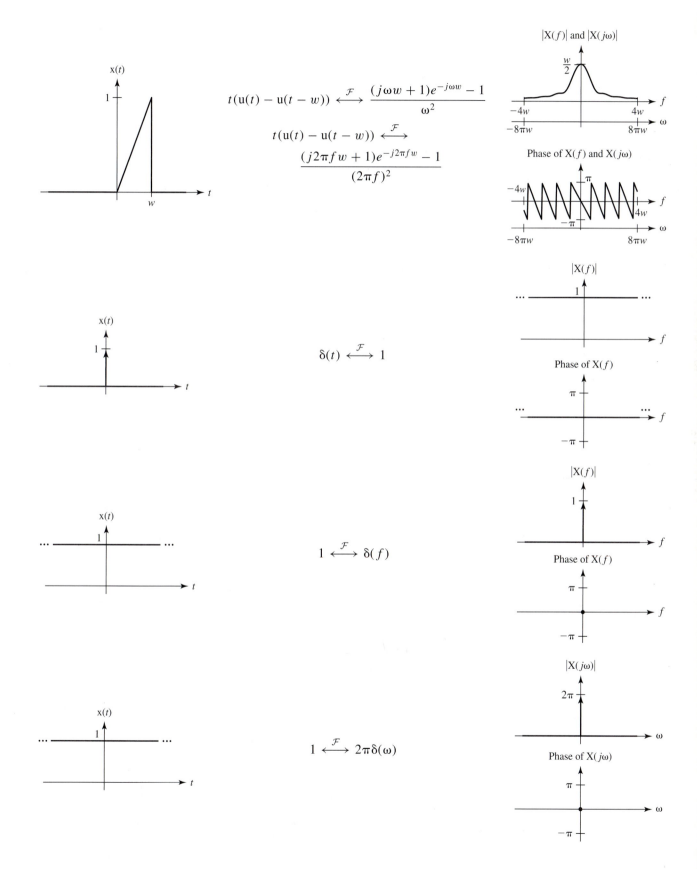

$$t(\mathrm{u}(t) - \mathrm{u}(t - w)) \stackrel{\mathcal{F}}{\longleftrightarrow} \frac{(j\omega w + 1)e^{-j\omega w} - 1}{\omega^2}$$

$$t(\mathrm{u}(t) - \mathrm{u}(t - w)) \stackrel{\mathcal{F}}{\longleftrightarrow} \frac{(j2\pi f w + 1)e^{-j2\pi f w} - 1}{(2\pi f)^2}$$

$$\delta(t) \stackrel{\mathcal{F}}{\longleftrightarrow} 1$$

$$1 \stackrel{\mathcal{F}}{\longleftrightarrow} \delta(f)$$

$$1 \stackrel{\mathcal{F}}{\longleftrightarrow} 2\pi\delta(\omega)$$

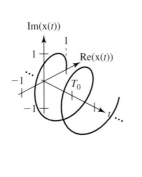

$$e^{j2\pi f_0 t} \overset{\mathcal{F}}{\longleftrightarrow} \delta(f - f_0)$$

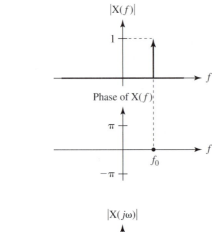

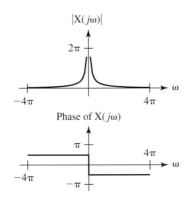

$$e^{j\omega_0 t} \overset{\mathcal{F}}{\longleftrightarrow} 2\pi\delta(\omega - \omega_0)$$

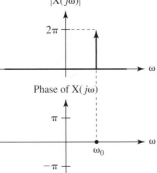

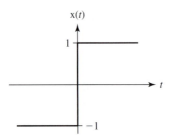

$$\mathrm{sgn}(t) \overset{\mathcal{F}}{\longleftrightarrow} \frac{1}{j\pi f}$$

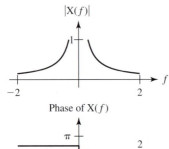

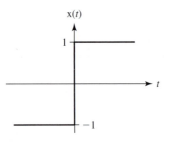

$$\mathrm{sgn}(t) \overset{\mathcal{F}}{\longleftrightarrow} \frac{2}{j\omega}$$

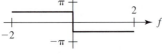

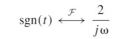

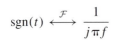

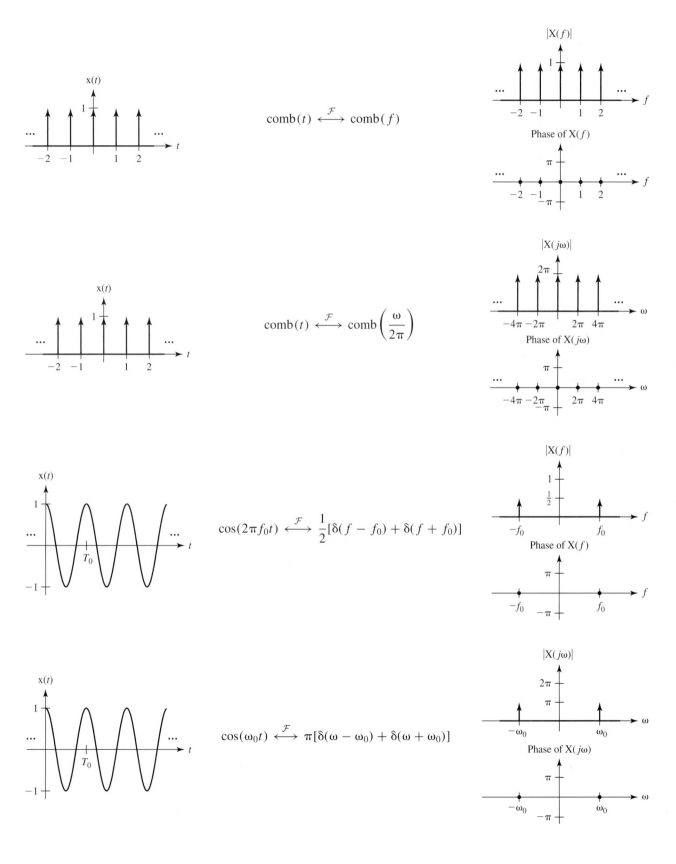

$$\text{comb}(t) \overset{\mathcal{F}}{\longleftrightarrow} \text{comb}(f)$$

$$\text{comb}(t) \overset{\mathcal{F}}{\longleftrightarrow} \text{comb}\left(\frac{\omega}{2\pi}\right)$$

$$\cos(2\pi f_0 t) \overset{\mathcal{F}}{\longleftrightarrow} \frac{1}{2}[\delta(f - f_0) + \delta(f + f_0)]$$

$$\cos(\omega_0 t) \overset{\mathcal{F}}{\longleftrightarrow} \pi[\delta(\omega - \omega_0) + \delta(\omega + \omega_0)]$$

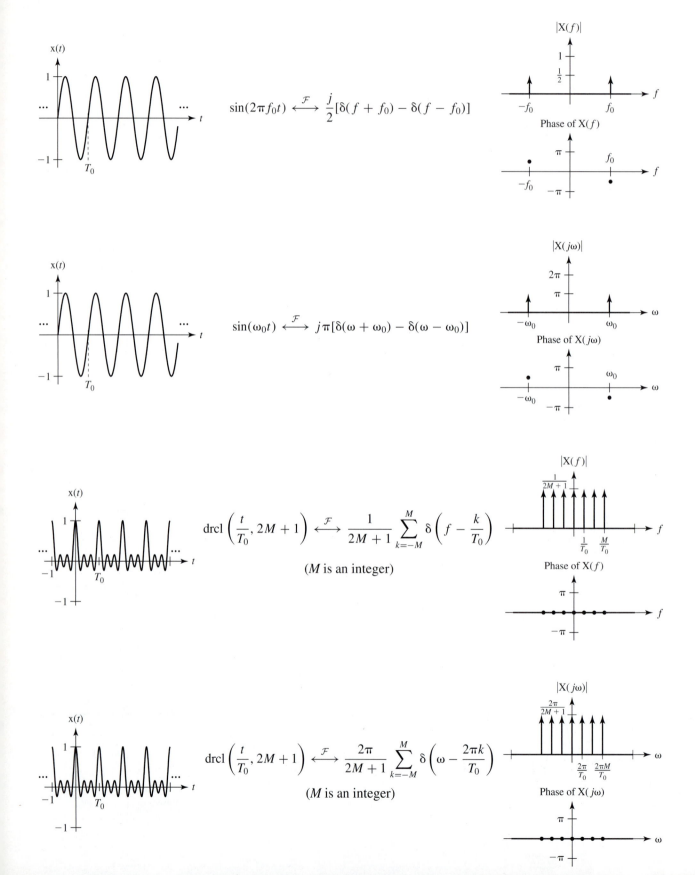

$$\sin(2\pi f_0 t) \xrightarrow{\mathcal{F}} \frac{j}{2}[\delta(f + f_0) - \delta(f - f_0)]$$

$$\sin(\omega_0 t) \xrightarrow{\mathcal{F}} j\pi[\delta(\omega + \omega_0) - \delta(\omega - \omega_0)]$$

$$\mathrm{drcl}\left(\frac{t}{T_0}, 2M + 1\right) \xrightarrow{\mathcal{F}} \frac{1}{2M + 1}\sum_{k=-M}^{M}\delta\left(f - \frac{k}{T_0}\right)$$

($M$ is an integer)

$$\mathrm{drcl}\left(\frac{t}{T_0}, 2M + 1\right) \xrightarrow{\mathcal{F}} \frac{2\pi}{2M + 1}\sum_{k=-M}^{M}\delta\left(\omega - \frac{2\pi k}{T_0}\right)$$

($M$ is an integer)

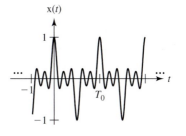

$$\text{drcl}\left(\frac{2t}{T_0}, 2M\right) \overset{\mathcal{F}}{\longleftrightarrow}$$

$$\frac{1}{4M} \sum_{k=-2M}^{2M} (1 - e^{-j\pi k})\, \delta\left(f - \frac{k}{T_0}\right)$$

(*M* is an integer)

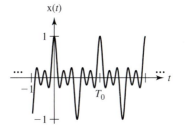

$$\text{drcl}\left(\frac{2t}{T_0}, 2M\right) \overset{\mathcal{F}}{\longleftrightarrow}$$

$$\frac{\pi}{2M} \sum_{k=-2M}^{2M} (1 - e^{-j\pi k})\, \delta\left(\omega - \frac{2\pi k}{T_0}\right)$$

(*M* is an integer)

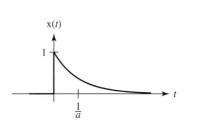

$$e^{-at}\mathrm{u}(t) \overset{\mathcal{F}}{\longleftrightarrow} \frac{1}{a + j\omega}$$

$$e^{-at}\mathrm{u}(t) \overset{\mathcal{F}}{\longleftrightarrow} \frac{1}{a + j2\pi f}$$

$$[\mathrm{Re}(a) > 0]$$

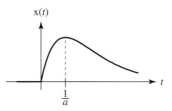

$$te^{-at}\mathrm{u}(t) \overset{\mathcal{F}}{\longleftrightarrow} \frac{1}{(a + j\omega)^2}$$

$$te^{-at}\mathrm{u}(t) \overset{\mathcal{F}}{\longleftrightarrow} \frac{1}{(a + j2\pi f)^2}$$

$$[\mathrm{Re}(a) > 0]$$

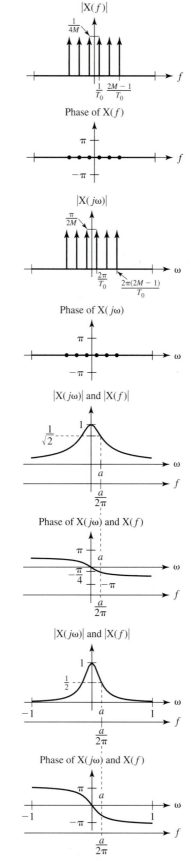

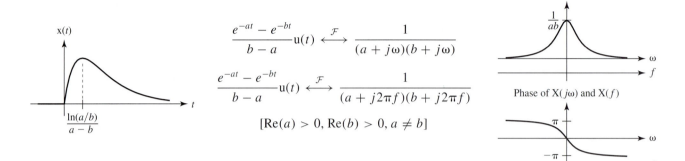

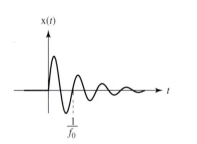

$$e^{-at}\sin(bt)\,u(t) \overset{\mathcal{F}}{\longleftrightarrow} \frac{b}{(j\omega + a)^2 + b^2}$$

$$e^{-\zeta\omega_0 t}\sin\left(\omega_0\sqrt{1 - \zeta^2}\,t\right)u(t) \overset{\mathcal{F}}{\longleftrightarrow}$$

$$\frac{\omega_0\sqrt{1 - \zeta^2}}{(j\omega)^2 + j\omega(2\zeta\omega_0) + \omega_0^2}$$

$$\left(b = \omega_0\sqrt{1 - \zeta^2},\ a = \zeta\omega_0\right)$$

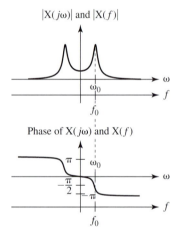

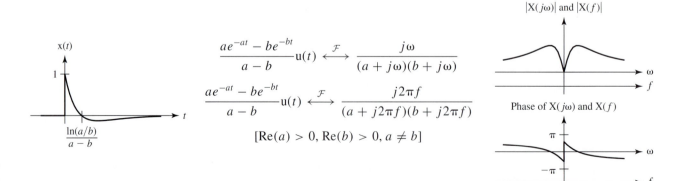

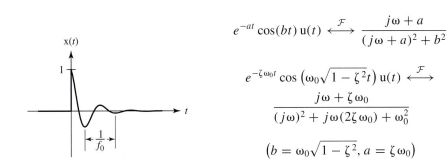

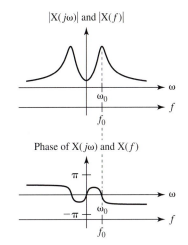

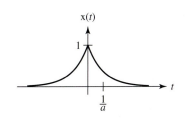

$$e^{-a|t|} \overset{\mathcal{F}}{\longleftrightarrow} \frac{2a}{a^2 + \omega^2}$$

$$e^{-a|t|} \overset{\mathcal{F}}{\longleftrightarrow} \frac{2a}{a^2 + (2\pi f)^2}$$

$$[\text{Re}(a) > 0]$$

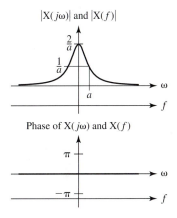

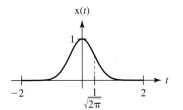

$$e^{-\pi t^2} \overset{\mathcal{F}}{\longleftrightarrow} e^{-\pi f^2}$$

$$e^{-\pi t^2} \overset{\mathcal{F}}{\longleftrightarrow} e^{-(\omega^2/4\pi)}$$

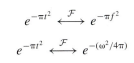

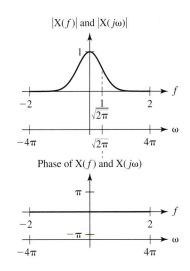

## Discrete Time Fourier Transform (DTFT)

$$x[n] = \int_1 X(F)e^{j2\pi Fn}\,dF \overset{\mathcal{F}}{\longleftrightarrow} X(F) = \sum_{n=-\infty}^{\infty} x[n]e^{-j2\pi Fn}$$

$$x[n] = \frac{1}{2\pi}\int_{2\pi} X(j\Omega)e^{j\Omega n}\,dF \overset{\mathcal{F}}{\longleftrightarrow} X(j\Omega) = \sum_{n=-\infty}^{\infty} x[n]e^{-j\Omega n}$$

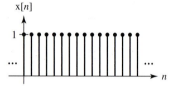

$$1 \overset{\mathcal{F}}{\longleftrightarrow} \text{comb}(F)$$

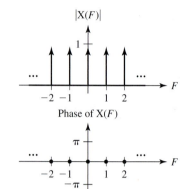

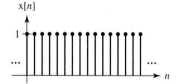

$$1 \overset{\mathcal{F}}{\longleftrightarrow} \text{comb}\left(\frac{\Omega}{2\pi}\right)$$

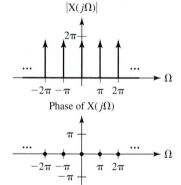

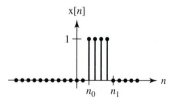

$$u[n-n_0] - u[n-n_1] \overset{\mathcal{F}}{\longleftrightarrow}$$
$$\frac{e^{-j\pi F(n_0+n_1)}}{e^{-j\pi F}}(n_1-n_0)\,\text{drcl}(F, n_1-n_0)$$

$$u[n-n_0] - u[n-n_1] \overset{\mathcal{F}}{\longleftrightarrow}$$
$$\frac{e^{-j(\Omega/2)(n_0+n_1)}}{e^{-j(\Omega/2)}}(n_1-n_0)\,\text{drcl}\left(\frac{\Omega}{2\pi}, n_1-n_0\right)$$

($n_0$ and $n_1$ are integers and $n_1 > n_0$)

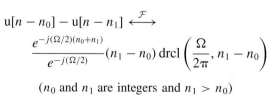

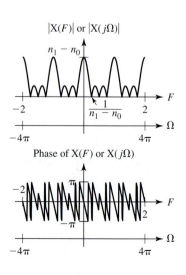

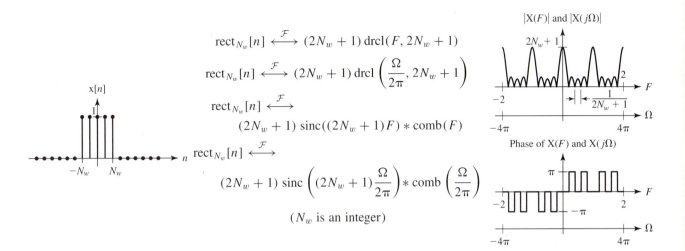

$$\text{rect}_{N_w}[n] \xleftrightarrow{\mathcal{F}} (2N_w + 1)\,\text{drcl}(F, 2N_w + 1)$$

$$\text{rect}_{N_w}[n] \xleftrightarrow{\mathcal{F}} (2N_w + 1)\,\text{drcl}\left(\frac{\Omega}{2\pi}, 2N_w + 1\right)$$

$$\text{rect}_{N_w}[n] \xleftrightarrow{\mathcal{F}}$$
$$(2N_w + 1)\,\text{sinc}((2N_w + 1)F) * \text{comb}(F)$$

$$\text{rect}_{N_w}[n] \xleftrightarrow{\mathcal{F}}$$
$$(2N_w + 1)\,\text{sinc}\left((2N_w + 1)\frac{\Omega}{2\pi}\right) * \text{comb}\left(\frac{\Omega}{2\pi}\right)$$

$$(N_w \text{ is an integer})$$

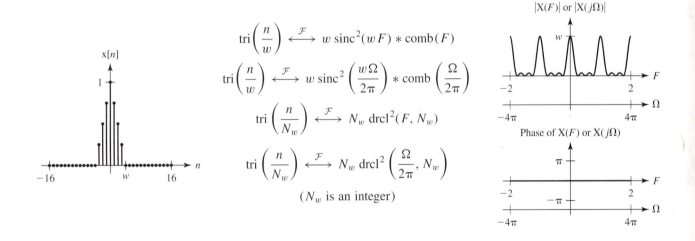

$$\text{tri}\left(\frac{n}{w}\right) \xleftrightarrow{\mathcal{F}} w\,\text{sinc}^2(wF) * \text{comb}(F)$$

$$\text{tri}\left(\frac{n}{w}\right) \xleftrightarrow{\mathcal{F}} w\,\text{sinc}^2\left(\frac{w\Omega}{2\pi}\right) * \text{comb}\left(\frac{\Omega}{2\pi}\right)$$

$$\text{tri}\left(\frac{n}{N_w}\right) \xleftrightarrow{\mathcal{F}} N_w\,\text{drcl}^2(F, N_w)$$

$$\text{tri}\left(\frac{n}{N_w}\right) \xleftrightarrow{\mathcal{F}} N_w\,\text{drcl}^2\left(\frac{\Omega}{2\pi}, N_w\right)$$

$$(N_w \text{ is an integer})$$

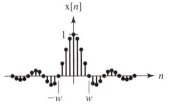

$$\text{sinc}\left(\frac{n}{w}\right) \xleftrightarrow{\mathcal{F}} w\,\text{rect}(wF) * \text{comb}(F)$$

$$\text{sinc}\left(\frac{n}{w}\right) \xleftrightarrow{\mathcal{F}} w\,\text{rect}\left(\frac{w\Omega}{2\pi}\right) * \text{comb}\left(\frac{\Omega}{2\pi}\right)$$

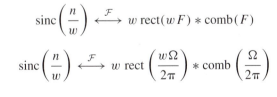

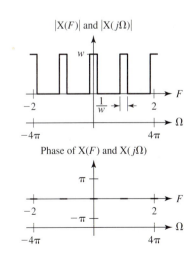

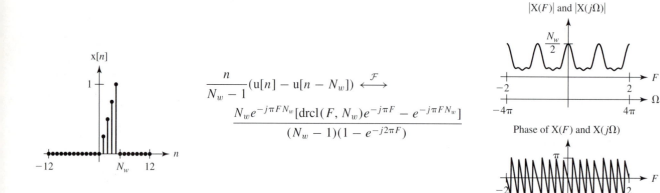

$$\frac{n}{N_w - 1}(\mathrm{u}[n] - \mathrm{u}[n - N_w]) \overset{\mathcal{F}}{\longleftrightarrow}$$

$$\frac{N_w e^{-j\pi F N_w}[\mathrm{drcl}(F, N_w)e^{-j\pi F} - e^{-j\pi F N_w}]}{(N_w - 1)(1 - e^{-j2\pi F})}$$

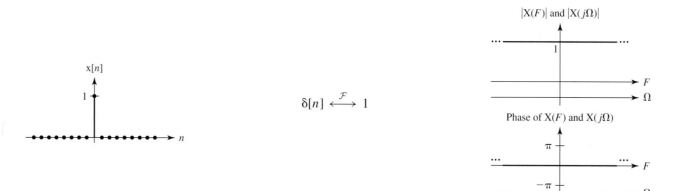

$$\delta[n] \overset{\mathcal{F}}{\longleftrightarrow} 1$$

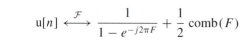

$$\mathrm{u}[n] \overset{\mathcal{F}}{\longleftrightarrow} \frac{1}{1 - e^{-j2\pi F}} + \frac{1}{2}\,\mathrm{comb}(F)$$

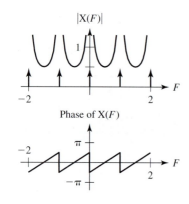

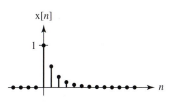

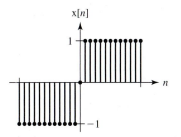

$$u[n] \overset{\mathcal{F}}{\longleftrightarrow} \frac{1}{1 - e^{-j\Omega}} + \frac{1}{2} \, \mathrm{comb}\left(\frac{\Omega}{2\pi}\right)$$

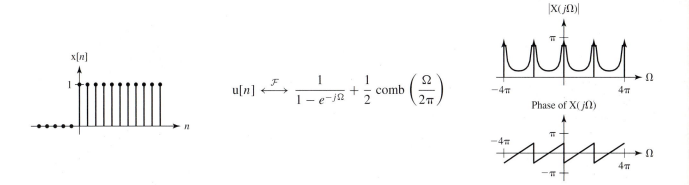

$$\alpha^n u[n] \overset{\mathcal{F}}{\longleftrightarrow} \frac{1}{1 - \alpha e^{-j\Omega}}$$

$$\alpha^n u[n] \overset{\mathcal{F}}{\longleftrightarrow} \frac{1}{1 - \alpha e^{-j2\pi F}}$$

$$(|\alpha| < 1)$$

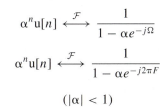

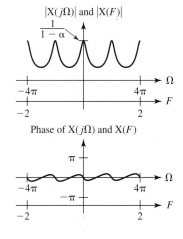

$$\mathrm{sgn}(n) \overset{\mathcal{F}}{\longleftrightarrow} j \frac{\sin(\Omega)}{\cos(\Omega) - 1}$$

$$\mathrm{sgn}(n) \overset{\mathcal{F}}{\longleftrightarrow} j \frac{\sin(2\pi F)}{\cos(2\pi F) - 1}$$

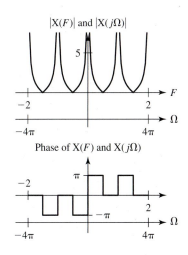

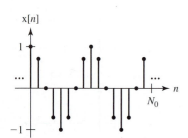

$$\text{comb}_{N_0}[n] \overset{\mathcal{F}}{\longleftrightarrow} \text{comb}(N_0 F)$$

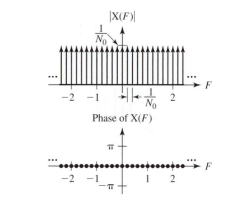

$$\text{comb}_{N_0}[n] \overset{\mathcal{F}}{\longleftrightarrow} \text{comb}\left(N_0 \frac{\Omega}{2\pi}\right)$$

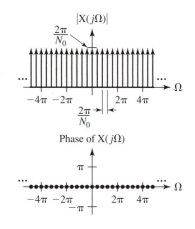

$$\cos(2\pi F_0 n) \overset{\mathcal{F}}{\longleftrightarrow} \frac{1}{2}[\text{comb}(F - F_0) + \text{comb}(F + F_0)]$$

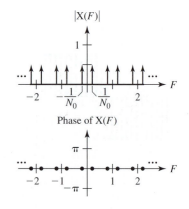

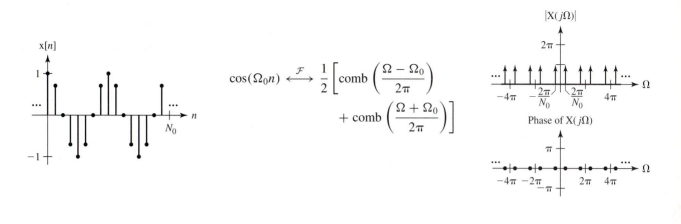

$$\cos(\Omega_0 n) \xleftrightarrow{\mathcal{F}} \frac{1}{2}\left[\text{comb}\left(\frac{\Omega - \Omega_0}{2\pi}\right) + \text{comb}\left(\frac{\Omega + \Omega_0}{2\pi}\right)\right]$$

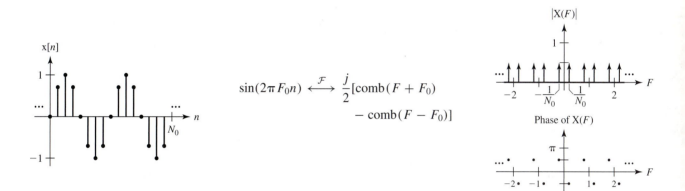

$$\sin(2\pi F_0 n) \xleftrightarrow{\mathcal{F}} \frac{j}{2}[\text{comb}(F + F_0) - \text{comb}(F - F_0)]$$

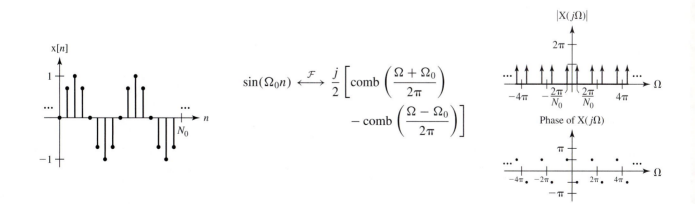

$$\sin(\Omega_0 n) \xleftrightarrow{\mathcal{F}} \frac{j}{2}\left[\text{comb}\left(\frac{\Omega + \Omega_0}{2\pi}\right) - \text{comb}\left(\frac{\Omega - \Omega_0}{2\pi}\right)\right]$$

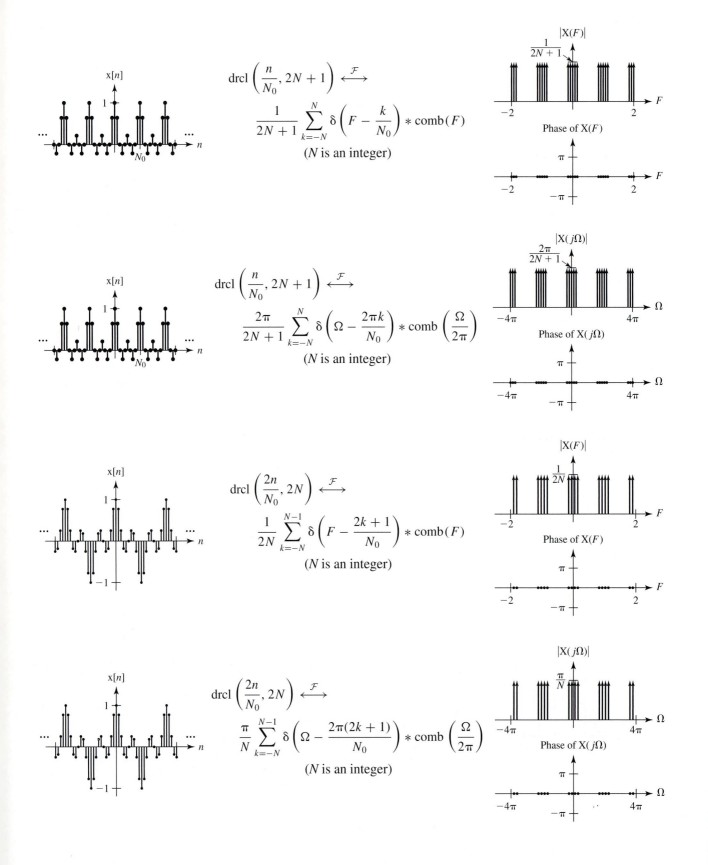

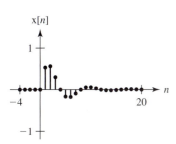

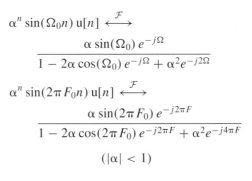

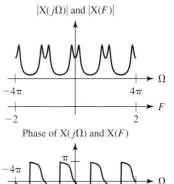

$$\alpha^n \sin(\Omega_0 n)\, u[n] \overset{\mathcal{F}}{\longleftrightarrow}$$

$$\frac{\alpha \sin(\Omega_0)\, e^{-j\Omega}}{1 - 2\alpha \cos(\Omega_0)\, e^{-j\Omega} + \alpha^2 e^{-j2\Omega}}$$

$$\alpha^n \sin(2\pi F_0 n)\, u[n] \overset{\mathcal{F}}{\longleftrightarrow}$$

$$\frac{\alpha \sin(2\pi F_0)\, e^{-j2\pi F}}{1 - 2\alpha \cos(2\pi F_0)\, e^{-j2\pi F} + \alpha^2 e^{-j4\pi F}}$$

$$(|\alpha| < 1)$$

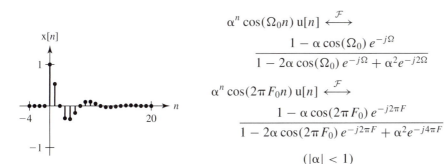

$$\alpha^n \cos(\Omega_0 n)\, u[n] \overset{\mathcal{F}}{\longleftrightarrow}$$

$$\frac{1 - \alpha \cos(\Omega_0)\, e^{-j\Omega}}{1 - 2\alpha \cos(\Omega_0)\, e^{-j\Omega} + \alpha^2 e^{-j2\Omega}}$$

$$\alpha^n \cos(2\pi F_0 n)\, u[n] \overset{\mathcal{F}}{\longleftrightarrow}$$

$$\frac{1 - \alpha \cos(2\pi F_0)\, e^{-j2\pi F}}{1 - 2\alpha \cos(2\pi F_0)\, e^{-j2\pi F} + \alpha^2 e^{-j4\pi F}}$$

$$(|\alpha| < 1)$$

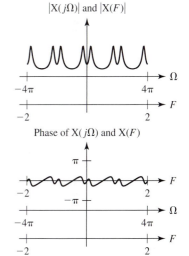

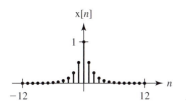

$$\alpha^{|n|} \overset{\mathcal{F}}{\longleftrightarrow} \frac{1 - \alpha^2}{1 - 2\alpha \cos(2\pi F) + \alpha^2}$$

$$\alpha^{|n|} \overset{\mathcal{F}}{\longleftrightarrow} \frac{1 - \alpha^2}{1 - 2\alpha \cos(\Omega) + \alpha^2}$$

$$(|\alpha| < 1)$$

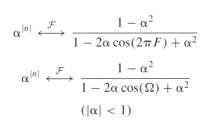

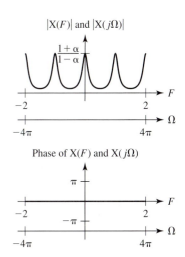

# F APPENDIX

## Table of Laplace Transform Pairs

### F.1  CAUSAL FUNCTIONS

$$\delta(t) \overset{\mathcal{L}}{\longleftrightarrow} 1 \qquad \text{all } s$$

$$u(t) \overset{\mathcal{L}}{\longleftrightarrow} \frac{1}{s} \qquad \text{Re}(s) > 0$$

$$u_{-n}(t) = \underbrace{u(t) * \cdots * u(t)}_{(n-1)\,\text{convolutions}} \overset{\mathcal{L}}{\longleftrightarrow} \frac{1}{s^n} \qquad \text{Re}(s) > 0$$

$$t\,u(t) \overset{\mathcal{L}}{\longleftrightarrow} \frac{1}{s^2} \qquad \text{Re}(s) > 0$$

$$e^{-\alpha t}\,u(t) \overset{\mathcal{L}}{\longleftrightarrow} \frac{1}{s + \alpha} \qquad \text{Re}(s) > -\alpha$$

$$t^n\,u(t) \overset{\mathcal{L}}{\longleftrightarrow} \frac{n!}{s^{n+1}} \qquad \text{Re}(s) > 0$$

$$t e^{-\alpha t}\,u(t) \overset{\mathcal{L}}{\longleftrightarrow} \frac{1}{(s + \alpha)^2} \qquad \text{Re}(s) > -\alpha$$

$$t^n e^{-\alpha t}\,u(t) \overset{\mathcal{L}}{\longleftrightarrow} \frac{n!}{(s + \alpha)^{n+1}} \qquad \text{Re}(s) > -\alpha$$

$$\sin(\beta t)\,u(t) \overset{\mathcal{L}}{\longleftrightarrow} \frac{\beta}{s^2 + \beta^2} \qquad \text{Re}(s) > 0$$

$$\cos(\beta t)\,u(t) \overset{\mathcal{L}}{\longleftrightarrow} \frac{s}{s^2 + \beta^2} \qquad \text{Re}(s) > 0$$

$$e^{-\alpha t}\sin(\beta t)\,u(t) \overset{\mathcal{L}}{\longleftrightarrow} \frac{\beta}{(s + \alpha)^2 + \beta^2} \qquad \text{Re}(s) > -\alpha$$

$$e^{-\alpha t}\cos(\beta t)\,u(t) \overset{\mathcal{L}}{\longleftrightarrow} \frac{s + \alpha}{(s + \alpha)^2 + \beta^2} \qquad \text{Re}(s) > -\alpha$$

$$e^{-\alpha t}\left[A\cos(\beta t) + \left(\frac{B - A\alpha}{\beta}\right)\sin(\beta t)\right]u(t) \overset{\mathcal{L}}{\longleftrightarrow} \frac{As + B}{(s + \alpha)^2 + \beta^2}$$

$$e^{-\alpha t}\left[\sqrt{A^2 + \left(\frac{B - A\alpha}{\beta}\right)^2}\cos\left(\beta t - \tan^{-1}\left(\frac{B - A\alpha}{A\beta}\right)\right)\right]u(t) \overset{\mathcal{L}}{\longleftrightarrow} \frac{As + B}{(s + \alpha)^2 + \beta^2}$$

$$e^{-(C/2)t}\left[A\cos\left(\sqrt{D - \left(\frac{C}{2}\right)^2}\,t\right) + \frac{2B - AC}{\sqrt{4D - C^2}}\sin\left(\sqrt{D - \left(\frac{C}{2}\right)^2}\,t\right)\right]u(t)$$

$$\overset{\mathcal{L}}{\longleftrightarrow} \frac{As + B}{s^2 + Cs + D}$$

$$e^{-(C/2)t}\left[\sqrt{A^2 + \left(\frac{2B - AC}{\sqrt{4D - C^2}}\right)^2}\cos\left(\sqrt{D - \left(\frac{C}{2}\right)^2}\,t - \tan^{-1}\left(\frac{2B - AC}{A\sqrt{4D - C^2}}\right)\right)\right]u(t)$$

$$\overset{\mathcal{L}}{\longleftrightarrow} \frac{As + B}{s^2 + Cs + D}$$

## F.2  ANTICAUSAL FUNCTIONS

$$-u(-t) \overset{\mathcal{L}}{\longleftrightarrow} \frac{1}{s} \qquad \mathrm{Re}(s) < 0$$

$$-e^{-\alpha t}u(-t) \overset{\mathcal{L}}{\longleftrightarrow} \frac{1}{s + \alpha} \qquad \mathrm{Re}(s) < -\alpha$$

$$-t^n u(-t) \overset{\mathcal{L}}{\longleftrightarrow} \frac{n!}{s^{n+1}} \qquad \mathrm{Re}(s) < 0$$

## F.3  NONCAUSAL FUNCTIONS

$$e^{-\alpha|t|} \overset{\mathcal{L}}{\longleftrightarrow} \frac{1}{s + \alpha} - \frac{1}{s - \alpha} \qquad -\alpha < \mathrm{Re}(s) < \alpha$$

$$\mathrm{rect}(t) \overset{\mathcal{L}}{\longleftrightarrow} \frac{e^{s/2} - e^{-(s/2)}}{s} \qquad \text{all } s$$

$$\mathrm{tri}(t) \overset{\mathcal{L}}{\longleftrightarrow} \left(\frac{e^{s/2} - e^{-(s/2)}}{s}\right)^2 \qquad \text{all } s$$

# G

**APPENDIX**

# Table of $z$ Transforms

## G.1 CAUSAL FUNCTIONS

$$\delta[n] \overset{z}{\longleftrightarrow} 1 \qquad \text{all } z$$

$$u[n] \overset{z}{\longleftrightarrow} \frac{z}{z-1} = \frac{1}{1-z^{-1}} \qquad |z| > 1$$

$$\alpha^n u[n] \overset{z}{\longleftrightarrow} \frac{z}{z-\alpha} = \frac{1}{1-\alpha z^{-1}} \qquad |z| > |\alpha|$$

$$n u[n] \overset{z}{\longleftrightarrow} \frac{z}{(z-1)^2} = \frac{z^{-1}}{(1-z^{-1})^2} \qquad |z| > 1$$

$$\frac{n^2}{2!} u[n] \overset{z}{\longleftrightarrow} \frac{z(z+1)}{2(z-1)^3} = \frac{1+z^{-1}}{2z(1-z^{-1})} \qquad |z| > 1$$

$$n\alpha^n u[n] \overset{z}{\longleftrightarrow} \frac{z\alpha}{(z-\alpha)^2} = \frac{\alpha z^{-1}}{(1-\alpha z^{-1})^2} \qquad |z| > |\alpha|$$

$$n^m \alpha^n u[n] \overset{z}{\longleftrightarrow} (-z)^m \frac{d^m}{dz^m}\left(\frac{z}{z-\alpha}\right) \qquad |z| > |\alpha|$$

$$\frac{n(n-1)(n-2)\cdots(n-m+1)}{m!}\alpha^{n-m} u[n] \overset{z}{\longleftrightarrow} \frac{z}{(z-\alpha)^{m+1}} \qquad |z| > |\alpha|$$

$$\sin(\Omega_0 n)\, u[n] \overset{z}{\longleftrightarrow} \frac{z\sin(\Omega_0)}{z^2 - 2z\cos(\Omega_0) + 1} = \frac{\sin(\Omega_0)\, z^{-1}}{1 - 2\cos(\Omega_0)\, z^{-1} + z^{-2}} \qquad |z| > 1$$

$$\cos(\Omega_0 n)\, u[n] \overset{z}{\longleftrightarrow} \frac{z[z-\cos(\Omega_0)]}{z^2 - 2z\cos(\Omega_0) + 1} = \frac{1 - \cos(\Omega_0)\, z^{-1}}{1 - 2\cos(\Omega_0)\, z^{-1} + z^{-2}} \qquad |z| > 1$$

$$\alpha^n \sin(\Omega_0 n)\, u[n] \overset{z}{\longleftrightarrow} \frac{z\alpha \sin(\Omega_0)}{z^2 - 2\alpha z\cos(\Omega_0) + \alpha^2}$$

$$= \frac{\alpha \sin(\Omega_0)\, z^{-1}}{1 - 2\alpha \cos(\Omega_0)\, z^{-1} + \alpha^2 z^{-2}} \qquad |z| > |\alpha|$$

$$\alpha^n \cos(\Omega_0 n)\, u[n] \overset{\mathcal{Z}}{\longleftrightarrow} \frac{z[z - \alpha \cos(\Omega_0)]}{z^2 - 2\alpha z \cos(\Omega_0) + \alpha^2}$$

$$= \frac{1 - \alpha \cos(\Omega_0)\, z^{-1}}{1 - 2\alpha \cos(\Omega_0)\, z^{-1} + \alpha^2 z^{-2}} \qquad |z| > |\alpha|$$

## G.2 ANTICAUSAL FUNCTIONS

$$-u[-n-1] \overset{\mathcal{Z}}{\longleftrightarrow} \frac{z}{z-1} \qquad |z| < 1$$

$$-\alpha^n u[-n-1] \overset{\mathcal{Z}}{\longleftrightarrow} \frac{z}{z-\alpha} \qquad |z| < |\alpha|$$

$$-n\alpha^n u[-n-1] \overset{\mathcal{Z}}{\longleftrightarrow} \frac{\alpha z}{(z-\alpha)^2} \qquad |z| < |\alpha|$$

## G.3 NONCAUSAL FUNCTIONS

$$\alpha^{|n|} \overset{\mathcal{Z}}{\longleftrightarrow} \frac{z}{z-\alpha} - \frac{z}{z-(1/\alpha)} \qquad |\alpha| < |z| < \left|\frac{1}{\alpha}\right|$$

# H APPENDIX

# Complex Numbers and Complex Functions

## H.1  BASIC PROPERTIES OF COMPLEX NUMBERS

In the history of mathematics there is a progressive broadening of the concept of numbers. The first numbers were the natural counting numbers, 1, 2, 3, . . . Next were zero and the negative numbers completing the set we now call *integers*. Fractions (ratios of integers) filled in some of the points between integers, and later irrational numbers filled in all the gaps between fractions to form what we now call the *real* numbers, an infinite continuum of one-dimensional numbers.

In trying to solve quadratic equations of the form $ax^2 + bx + c = 0$, real solutions can always be found if $b^2 - 4ac$ is greater than or equal to zero. But if $b^2 - 4ac$ is less than zero, no real solution can be found. The essence of the problem is in trying to solve the equation $x^2 = -1$, for $x$. None of the real numbers can be the solution of this equation. The proposal that an *imaginary* number could be the solution to this equation led to a whole new field of mathematics, complex variables. The idea of complex numbers seemed artificial and abstract at first, but as mathematical and physical theory has developed, the usefulness of complex numbers for solving real problems has been conclusively shown. The square root of $-1$ has been given the symbol $j$ and, therefore, $j^2 = -1$.

> Different authors use different symbols to indicate the square root of $-1$. A commonly used symbol is $i$. This is used in many mathematics and physics books. The symbol $j$ is preferred in most electrical engineering books to avoid confusion because the symbol $i$ is usually reserved for electric current.

A complex number $z$ can be expressed as the sum of a real number $x$ and an imaginary number $jy$, where $y$ is also a real number. In the complex number $z = x + jy$, $x$ is the real part and $y$ is the imaginary part. (Notice that, although it sounds strange, the imaginary part of a complex number is a real number.) Two complex numbers are equal if, and only if, their real and imaginary parts are equal separately. Let $z_1 = x_1 + jy_1$ and $z_2 = x_2 + jy_2$. Then, if $z_1 = z_2$, that implies that $x_1 = x_2$ and $y_1 = y_2$. In the following material the symbol $z$ will represent some arbitrary complex number and the symbols $x$ and $y$ will represent the real and imaginary parts of $z$, respectively.

The sum and product of two complex numbers are defined as

$$z_1 + z_2 = (x_1 + jy_1) + (x_2 + jy_2) = x_1 + x_2 + j(y_1 + y_2) \qquad \textbf{(H.1)}$$

and

$$z_1 z_2 = (x_1 + jy_1)(x_2 + jy_2) = x_1 x_2 - y_1 y_2 + j(x_1 y_2 + x_2 y_1). \qquad \textbf{(H.2)}$$

From (H.1),

$$z + 0 = z + (0 + j0) = (x + 0) + j(y + 0) = x + jy = z,$$

proving that the number, zero, is the additive identity for complex numbers just as it is for real numbers. From (H.2),

$$z(1) = z(1 + j0) = (x(1) - y(0)) + j(x(0) + y(1)) = x + jy = z, \qquad \textbf{(H.3)}$$

proving that the real number, one, is the multiplicative identity for complex numbers, just as it is for real numbers.

By a straightforward extension of the law of addition, subtraction is defined by

$$z_1 - z_2 = x_1 - x_2 + j(y_1 - y_2). \qquad \textbf{(H.4)}$$

Division can be derived from multiplication, and the result is

$$\frac{z_1}{z_2} = \frac{x_1 x_2 + y_1 y_2}{x_2^2 + y_2^2} + j\frac{x_2 y_1 - x_1 y_2}{x_2^2 + y_2^2} = x_3 + jy_3 = z_3. \qquad \textbf{(H.5)}$$

From (H.5) it follows that

$$\frac{z}{z} = 1, \qquad \frac{z_1}{z_2} = z_1\left(\frac{1}{z_2}\right) \qquad \text{and} \qquad \frac{1}{z_1 z_2} = \frac{1}{z_1}\frac{1}{z_2} \qquad (z_1 \neq 0,\ z_2 \neq 0).$$

<hr>

**EXAMPLE H.1**

MATLAB handles complex numbers just as easily as real numbers. The following four numerical calculations can be done by MATLAB directly at the computer console in the interactive mode.

$$(3 + j2) + (-1 + j6) = 2 + j8 \qquad (3 + j2) - (-1 + j6) = 4 - j4$$

$$(3 + j2)(-1 + j6) = -15 + j16 \qquad \frac{3 + j2}{-1 + j6} = \frac{9}{37} - j\frac{20}{37}$$

Below is a copy of a MATLAB session doing these calculations.

```
»A = 3 + j*2 ; B = -1+j*6 ;
»A + B
ans =
    2.0000 + 8.0000i
»A - B
ans =
    4.0000 - 4.0000i
»A*B
ans =
  -15.0000 +16.0000i
»A/B
```

```
ans =
    0.2432 - 0.5405i
```

The square root of $-1$ is predefined in MATLAB and is the default value of the variables $i$ and $j$.

The commutativity and associativity of complex numbers under addition and multiplication

$$z_1 + z_2 = z_2 + z_1 \qquad z_1 + (z_2 + z_3) = (z_1 + z_2) + z_3 \tag{H.6}$$

and

$$z_1 z_2 = z_2 z_1 \qquad z_1(z_2 z_3) = (z_1 z_2)z_3 \tag{H.7}$$

and the distributivity of complex numbers

$$z_1(z_2 + z_3) = z_1 z_2 + z_1 z_3 \tag{H.8}$$

can be proven from the definition of complex numbers and the commutativity, associativity, and distributivity of real numbers. These properties, (H.6), (H.7), and (H.8), lead to the results

$$\frac{z_1 + z_2}{z_3} = \frac{z_1}{z_3} + \frac{z_2}{z_3} \qquad \text{and} \qquad \frac{z_1 z_2}{z_3 z_4} = \frac{z_1}{z_3} \frac{z_2}{z_4} \qquad (z_3 \neq 0, z_4 \neq 0). \tag{H.9}$$

Just as any real number can be geometrically represented as a point in a one-dimensional space (the real line), any complex number can be represented geometrically by a point in a two-dimensional space, the complex plane (Figure H.1). The complex plane has two orthogonal axes, the real axis and the imaginary axis. Any particular complex number $z_0$ is defined by its real and imaginary parts $x_0$ and $y_0$.

A vector from the origin of the complex plane to a point can also be used to represent a complex number. The sum and difference of two complex numbers can be found by the usual rules of vector addition and subtraction. In Figure H.2 are two examples of the addition of two complex numbers.

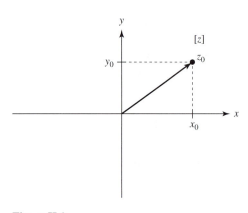

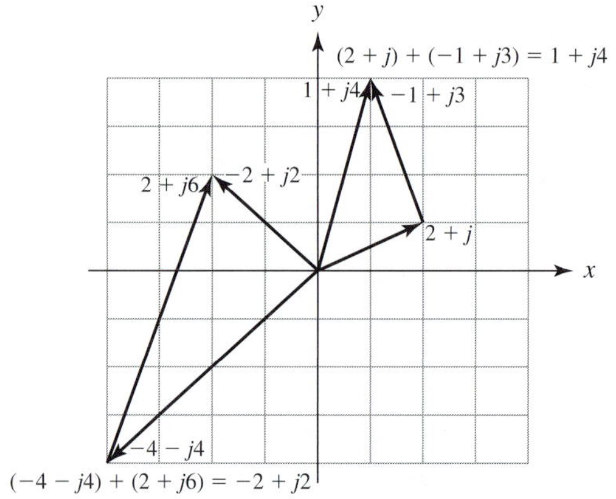

**Figure H.2**
Graphical addition of complex numbers by vector addition.

**Figure H.1**
The complex plane.

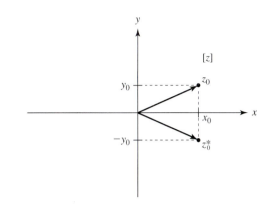

**Figure H.3**
Complex conjugates.

The complex conjugate of a complex number is found by negating its imaginary part. It is indicated by the addition of a * to the number. If $z_0 = x_0 + jy_0$, the complex conjugate of $z_0$ is $z_0^* = x_0 - jy_0$. The complex conjugate of a complex number is its reflection in the real axis of the complex plane (Figure H.3). Some properties of conjugates that can be derived from earlier properties of complex numbers are

$$(z_1 + z_2)^* = z_1^* + z_2^* \qquad (z_1 - z_2)^* = z_1^* - z_2^* \qquad (z_1 z_2)^* = z_1^* z_2^* \qquad \left(\frac{z_1}{z_2}\right)^* = \frac{z_1^*}{z_2^*}.$$

**(H.10)**

Also the sum of any complex number and its conjugate is real, and the difference between any complex number and its conjugate is imaginary.

The *absolute value* $|z|$ (or magnitude or modulus) of a complex number $z = x + jy$ is the length of the vector in the complex plane that represents $z$, which is (from the Pythagorean theorem) $|z| = \sqrt{x^2 + y^2}$. By extension, the distance between any two complex numbers $z_1$ and $z_2$ in the complex plane is

$$|z_1 - z_2| = \sqrt{(x_1 - x_2)^2 + (y_1 - y_2)^2}.$$

**(H.11)**

Pythagoras of Samos,
569–475 B.C.

The magnitude of a complex number is a real number,

$$|z| = \sqrt{x^2 + y^2} + j0.$$

**(H.12)**

A handy relation in the study of complex variables and functions of a complex variable is

$$|z|^2 = zz^* = x^2 + y^2.$$

**(H.13)**

Also,

$$|z_1 z_2| = |z_1||z_2| \qquad \left|\frac{z_1}{z_2}\right| = \frac{|z_1|}{|z_2|}.$$

**(H.14)**

## H.2 THE POLAR FORM

It is often convenient in analysis to represent a complex number in polar form. Instead of specifying its real and imaginary parts, we specify its magnitude $r$ and the angle $\theta$ that its vector representation in the complex plane makes with the positive real axis,

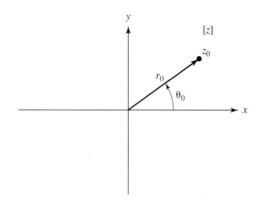

**Figure H.4**
The polar form of a complex number.

with the counterclockwise direction being positive. The relations are

$$x = r\cos(\theta) \qquad y = r\sin(\theta) \qquad z = r[\cos(\theta) + j\sin(\theta)]. \qquad \textbf{(H.15)}$$

The length of the vector $r$ is the magnitude of the complex number, $r = |z|$, and the angle or phase $\theta$ is related to $x$ and $y$ by $\tan(\theta) = y/x$ (Figure H.4). There is more than one value of $\theta$ that satisfies $\tan(\theta) = y/x$; therefore, the angle or phase of a complex number is *multiple valued*. If $\theta$ is a solution, so is $\theta + 2n\pi$, where $n$ is any integer. One special case is worthy of note: $x = y = 0$. In this case, the ratio $y/x$ is undefined. That means that $\tan(\theta) = y/x$ and, by implication, $\theta$ are also undefined. The phase of a complex number whose magnitude is zero is undefined. This should not be cause for alarm. If the magnitude is zero, the vector from the origin of the complex plane to the complex number is a zero-length vector, or a point, the origin. Geometrically the angle from the positive real axis to this vector has no meaning because the vector has no length and, therefore, no direction. Also, since the real and imaginary parts of the complex number are found from $x = r\cos(\theta)$ and $y = r\sin(\theta)$, if $r$ is zero, $x$ and $y$ are also zero regardless of the value of $\theta$. So the mathematics is telling us something logical (as it usually does). If the magnitude is zero, phase has no meaning!

The product of two complex numbers, written in polar form, is

$$z_1 z_2 = r_1[\cos(\theta_1) + j\sin(\theta_1)]r_2[\cos(\theta_2) + j\sin(\theta_2)]$$

$$z_1 z_2 = r_1 r_2\{\cos(\theta_1)\cos(\theta_2) - \sin(\theta_1)\sin(\theta_2) + j[\cos(\theta_1)\sin(\theta_2) + \sin(\theta_1)\cos(\theta_2)]\}$$

$$z_1 z_2 = \frac{r_1 r_2}{2}\{\cos(\theta_1 - \theta_2) + \cos(\theta_1 + \theta_2) - [\cos(\theta_1 - \theta_2) - \cos(\theta_1 + \theta_2)]$$
$$+ j[\sin(\theta_2 - \theta_1) + \sin(\theta_2 + \theta_1) + \sin(\theta_1 - \theta_2) + \sin(\theta_1 + \theta_2)]\}$$

$$z_1 z_2 = r_1 r_2[\cos(\theta_1 + \theta_2) + j\sin(\theta_2 + \theta_1)]. \qquad \textbf{(H.16)}$$

The *magnitude of the product* of two complex numbers is the product of their magnitudes, and the *angle of the product* of two complex numbers is the sum of their angles. Applying this idea to the product of multiple complex numbers leads to de Moivre's theorem,

$$z^n = r^n[\cos(n\theta) + j\sin(n\theta)]. \qquad \textbf{(H.17)}$$

It also follows that the *magnitude of the quotient* of two complex numbers is the quotient of their magnitudes and the *angle of the quotient* of two complex numbers is the

Abraham de Moivre,
5/26/1667–11/27/1754

difference of their angles.

$$\frac{z_1}{z_2} = \frac{r_1}{r_2}[\cos(\theta_1 - \theta_2) + j\sin(\theta_1 - \theta_2)] \qquad r_2 \neq 0. \tag{H.18}$$

The following are examples of complex numbers represented in various ways and some operations on complex numbers.

$$9 + j7 = 11.4[\cos(0.661) + j\sin(0.661)] = 11.4\angle 0.661 = 11.4\angle 37.87°$$

$$-4 + j8 = 8.94[\cos(2.034 + j\sin(2.034))] = 8.94\angle 2.034 = 8.94\angle 116.57°$$

$$(9 + j7)(-4 + j8) = 11.4[\cos(0.661) + j\sin(0.661)] \times 8.94[\cos(2.034) + j\sin(2.034)]$$

$$(9 + j7)(-4 + j8) = 11.4 \times 8.94[\cos(0.661 + 2.034) + j\sin(0.661 + 2.034)]$$

$$(9 + j7)(-4 + j8) = 101.98[\cos(2.695) + j\sin(2.695)] = -92 + j44$$

$$\frac{9 + j7}{-4 + j8} = \frac{11.4[\cos(0.661) + j\sin(0.661)]}{8.94[\cos(2.034) + j\sin(2.034)]}$$
$$= \frac{11.4}{8.94}[\cos(0.661 - 2.034) + j\sin(0.661 - 2.034)]$$
$$= 1.275[\cos(-1.373) + j\sin(-1.373)] = \frac{1}{4} - j\frac{5}{4}$$

In MATLAB,

```
»A = 9+j*7 ; B = -4+j*8 ;
»abs(A)
ans =
   11.4018
»angle(A)
ans =
   0.6610
»abs(B)
ans =
   8.9443
»angle(B)
ans =
   2.0344
»A*B
ans =
   -92.0000 +44.0000i
»abs(A*B)
ans =
  101.9804
»angle(A*B)
ans =
   2.6955
»A/B
ans =
   0.2500 - 1.2500i
```

```
»abs(A/B)
ans =
    1.2748
»angle(A/B)
ans =
    -1.3734
```

∎

The $n$th root $z_0$ of a complex number $z$ is the solution of the equation $z_0^n = z$, where $n$ is a positive integer. In polar form,

$$z = r[\cos(\theta) + j \sin(\theta)] \quad \text{and} \quad z_0 = r_0[\cos(\theta_0) + j \sin(\theta_0)]. \quad \textbf{(H.19)}$$

Then

$$r_0^n[\cos(n\theta_0) + j \sin(n\theta_0)] = r[\cos(\theta) + j \sin(\theta)] \quad \textbf{(H.20)}$$

and the solutions for $r_0$ and $\theta_0$ are

$$r_0 = \sqrt[n]{r} \qquad \theta_0 = \frac{\theta + 2k\pi}{n}, \quad \textbf{(H.21)}$$

where $k$ is an integer and there are exactly $n$ distinct values. The $n$ distinct $n$th roots of the real number $+1$ are

$$1^{1/n} = \cos\left(\frac{2k\pi}{n}\right) + j \sin\left(\frac{2k\pi}{n}\right) \qquad k = 0, 1, \ldots, n - 1. \quad \textbf{(H.22)}$$

Notice that each of the $n$th roots of any complex number lies on a circle in the complex plane. The circle is centered at the origin, the radius of the circle is the positive real $n$th root of the magnitude of the complex number, and the $n$ distinct $n$th roots are spaced at equal angular intervals of $2\pi/n$ rad (Figure H.5). Therefore, in any problem of finding roots of a complex number, if we can find one root, the others are easily found by putting them in a symmetrical array of complex numbers with the same magnitude and the proper angular spacing. Usually the easiest root to find first is the one whose magnitude is the positive real $n$th root of the magnitude of the complex number and whose angle is the angle of the complex number divided by $n$.

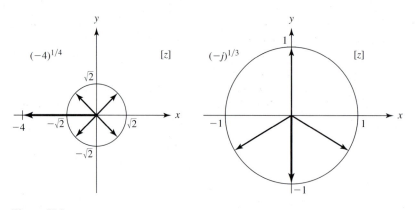

**Figure H.5**
Roots of complex numbers.

MATLAB has a function for finding the roots of an equation, `roots`. If the equation is of the form

$$a_n z^n + a_{n-1} z^{n-1} + \cdots + a_2 z^2 + a_1 z + a_0 = 0, \tag{H.23}$$

then the MATLAB command `roots([a_n a_{n-1} ... a_2 a_1 a_0])`, `roots([a_n, a_{n-1}, ... a_2, a_1, a_0])`, or `roots([a_n;a_{n-1}; ... a_2;a_1;a_0])` returns all the distinct roots of (H.23). For example,

```
»roots([3 2])

ans =

 -0.6667
```

or

```
»roots([9,1,-3,6])

ans =

-1.0432
  0.4661 + 0.6495i
  0.4661 - 0.6495i
```

or

```
»roots([2 ; j*3 ; 1 ; -9])

ans =

-0.6573 - 2.1860i

-0.7464 + 1.1162i
  1.4037 - 0.4301i
```

## H.3  FUNCTIONS OF A COMPLEX VARIABLE

In the study of signals and systems, probably the most important of the functions of a complex variable is the exponential function, defined as

$$\exp(z) = e^x[\cos(y) + j\sin(y)]. \tag{H.24}$$

Common nomenclature in signal and system analysis is that the exponential function of a complex variable is called a *complex exponential*. Note that if $y = 0$ in (H.24), $z$ becomes real and this definition collapses to the more familiar definition of the exponential function for real variables,

$$\exp(z) = \exp(x) = e^x. \tag{H.25}$$

If $z$ is purely imaginary, $x = 0$ and

$$\exp(jy) = \cos(y) + j\sin(y). \tag{H.26}$$

Equation (H.26) is known as Euler's (pronounced "oilers") identity after Leonhard Euler, one of the great early mathematicians.

The most common occurrences of the complex exponentials in signal and system theory are with either time $t$, cyclic frequency $f$, or radian frequency $\omega$ as the independent

Leonhard Euler,
4/15/1707–9/18/1783

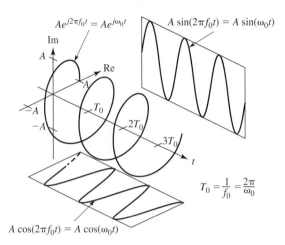

**Figure H.6**
Relation between a complex sinusoid and a real sine and
a real cosine.

variable; for example,

$$x(t) = 24e^{(\sigma_0 - j\omega_0)t} \qquad \text{or} \qquad X(f) = 5e^{-j2\pi f t_0} \qquad \text{or} \qquad X(j\omega) = -3e^{j4\omega},$$

$$\textbf{(H.27)}$$

where $\sigma_0$, $\omega_0$, and $t_0$ are real constants. When the argument of the exponential function is purely imaginary, the resulting complex exponential is called a *complex sinusoid* because it contains a cosine and a sine as its real and imaginary parts as illustrated in Figure H.6 for a complex sinusoid in time. The projection of the complex sinusoid onto a plane parallel to the plane containing the real and $t$ axes is the cosine function, and the projection onto a plane parallel to the plane containing the imaginary and $t$ axes is the sine function.

Other important properties of the exponential function are

$$\frac{\exp(z_1)}{\exp(z_2)} = \exp(z_1 - z_2) \qquad\qquad \textbf{(H.28)}$$

$$[\exp(z)]^n = \exp(nz) \qquad\qquad \textbf{(H.29)}$$

$$[\exp(z)]^{m/n} = \exp\left[\frac{m}{n}(z + j2k\pi)\right] \qquad k = 0, 1, \dots, n - 1. \qquad \textbf{(H.30)}$$

The exponential function is periodic with period $j2\pi$. That is, it is periodic in the imaginary dimension. This is shown from the definition (H.24) by substituting $z + j2n\pi$ for $z$,

$$\exp(z + j2n\pi) = e^x[\cos(y + 2n\pi) + j\sin(y + 2n\pi)]$$
$$= e^x[\cos(y) + j\sin(y)] = \exp(z), \qquad \textbf{(H.31)}$$

where $n$ is an integer. Also $\exp(z^*) = [\exp(z)]^*$. Lastly, if a particular complex number $z$ is represented by the polar form

$$z = r[\cos(\theta) + j\sin(\theta)], \qquad\qquad \textbf{(H.32)}$$

then from Euler's identity (H.26) one can write

$$z = r \exp(j\theta) = re^{j\theta}, \tag{H.33}$$

which is a convenient way of representing a complex number in many types of analysis. From Euler's identity (H.26), one can form

$$e^{-j\theta} = \cos(\theta) - j\sin(\theta). \tag{H.34}$$

Adding $e^{j\theta} = \cos(\theta) + j\sin(\theta)$ and $e^{-j\theta} = \cos(\theta) - j\sin(\theta)$,

$$e^{j\theta} + e^{-j\theta} = 2\cos(\theta) \Rightarrow \cos(\theta) = \frac{e^{j\theta} + e^{-j\theta}}{2}. \tag{H.35}$$

Similarly,

$$\sin(\theta) = \frac{e^{j\theta} - e^{-j\theta}}{j2}. \tag{H.36}$$

These two results are important because they show again the intimate relationship between sines, cosines, and complex exponentials. It is important to note that if $\theta$ is a real number, then the function $\cos(\theta)$ is real-valued. But the equality

$$\cos(\theta) = \frac{e^{j\theta} + e^{-j\theta}}{2} \tag{H.37}$$

expresses this real-valued function of a real variable in terms of a combination of complex-valued functions. This only works because $e^{j\theta}$ and $e^{-j\theta}$ are complex conjugates (for real $\theta$) and when a number is added to its complex conjugate the sum is real.

Other properties of trigonometric functions of a complex variable are summarized here.

$$\frac{d}{dz}(\sin(z)) = \cos(z) \text{ and } \frac{d}{dz}(\cos(z)) = -\sin(z) \tag{H.38}$$

$$\tan(z) = \frac{\sin(z)}{\cos(z)} \tag{H.39}$$

$$\cos(z) = \frac{e^y + e^{-y}}{2}\cos(x) - j\frac{e^y - e^{-y}}{2}\sin(x) \tag{H.40}$$

$$\sin^2(z) + \cos^2(z) = 1 \tag{H.41}$$

$$\sin(z_1 + z_2) = \sin(z_1)\cos(z_2) + \cos(z_1)\sin(z_2) \tag{H.42}$$

$$\cos(z_1 + z_2) = \cos(z_1)\cos(z_2) - \sin(z_1)\sin(z_2) \tag{H.43}$$

$$\sin(-z) = -\sin(z) \quad \text{and} \quad \cos(-z) = \cos(z) \tag{H.44}$$

$$\sin\left(\frac{\pi}{2} - z\right) = \cos(z) \tag{H.45}$$

$$\sin(2z) = 2\sin(z)\cos(z) \quad \text{and} \quad \cos(2z) = \cos^2(z) - \sin^2(z) \tag{H.46}$$

MATLAB implements all the exponential and trigonometric functions. The exponential function is `exp`, the sine function is `sin`, the cosine function is `cos`, the tangent function is `tan`, etc. In the trigonometric functions, the argument is always interpreted as an angle in radians. For example,

```
»exp(1)
ans =
    2.7183
```

```
»exp(-j*pi)
ans =
      -1
»cos(3*pi/4)
ans =
      -0.7071
»tan(-pi/4)
ans =
      -1.0000
```

## H.4 COMPLEX FUNCTIONS OF A REAL VARIABLE

In transform analysis there are many examples of complex functions of a real variable. Since the function value is complex, it cannot be as simply graphed as a single plot of a real function of a real variable. There are several methods of plotting functions like this, and each has its advantages and disadvantages. We can plot the real and imaginary parts separately as functions of the real independent variable, plot the magnitude and phase separately as functions of the real independent variable, or plot the real and imaginary parts versus the independent variable in one three-dimensional isometric plot. As an illustrative example, suppose we want to plot the function

$$x(t) = e^{j2\pi t}. \tag{H.47}$$

Figures H.7 through H.9 illustrate the three types of plots of this function.

Although plots of the real and imaginary parts are sometimes useful, for most analysis purposes, separate plots of the magnitude and phase of a complex function of a real variable are preferred. In the study of transform methods, the independent variable will often be frequency $f$ or $\omega$ instead of time $t$.

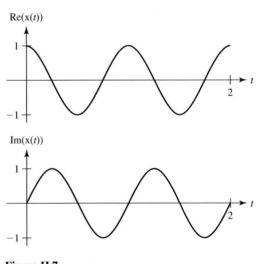

**Figure H.7**
Real and imaginary parts plotted separately versus the independent variable.

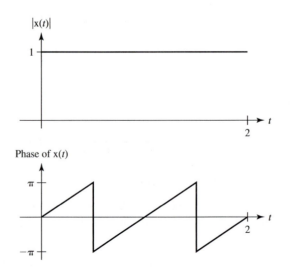

**Figure H.8**
Magnitude and phase plotted separately versus the independent variable.

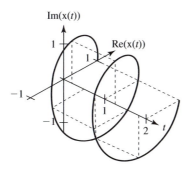

**Figure H.9**
Real and imaginary parts plotted
together versus the independent
variable in a three-dimensional
isometric plot.

Consider a complex function of radian frequency $\omega$,

$$H(j\omega) = \frac{1}{1 + j\omega}. \qquad \textbf{(H.48)}$$

How would we plot its magnitude and phase? The square of the magnitude of any complex number is the product of the number and its complex conjugate. Therefore, the magnitude of $H(j\omega)$ is

$$|H(j\omega)| = \sqrt{H(j\omega)H^*(j\omega)}. \qquad \textbf{(H.49)}$$

In this case

$$|H(j\omega)| = \sqrt{\frac{1}{1 + j\omega}\frac{1}{1 - j\omega}} = \sqrt{\frac{1}{1 + \omega^2}} = \frac{1}{\sqrt{1 + \omega^2}}. \qquad \textbf{(H.50)}$$

The phase of a complex number is the inverse tangent of the ratio of its imaginary part to its real part. The real and imaginary parts of $H(j\omega)$ are

$$\text{Re}(H(j\omega)) = \text{Re}\left(\frac{1}{1 + j\omega}\frac{1 - j\omega}{1 - j\omega}\right) = \text{Re}\left(\frac{1 - j\omega}{1 + \omega^2}\right) = \frac{1}{1 + \omega^2} \qquad \textbf{(H.51)}$$

and

$$\text{Im}(H(j\omega)) = \text{Im}\left(\frac{1}{1 + j\omega}\frac{1 - j\omega}{1 - j\omega}\right) = \text{Im}\left(\frac{1 - j\omega}{1 + \omega^2}\right) = -\frac{j\omega}{1 + \omega^2}. \qquad \textbf{(H.52)}$$

Therefore, the phase of $H(j\omega)$ is

$$\angle H(j\omega) = \tan^{-1}\left(\frac{\text{Im}(H(j\omega))}{\text{Re}(H(j\omega))}\right) = \tan^{-1}\left(\frac{-(j\omega/(1 + \omega^2))}{1/(1 + \omega^2)}\right) = \tan^{-1}(-j\omega). \qquad \textbf{(H.53)}$$

It should be noted here that the inverse tangent function is *multiple-valued*. This means that, strictly speaking, there is a countable infinity of correct values of $\angle H(j\omega)$ at any arbitrary value of $\omega$. (There is also an uncountable infinity of incorrect values!) If $\theta$ is any correct value of $\angle H(j\omega)$, then $\theta + 2n\pi$, where $n$ is any integer, is also a correct value of $\theta$ because the sines of $\theta$ and $\theta + 2n\pi$ are identical and the cosines of $\theta$ and $\theta + 2n\pi$ are identical. Therefore, the real part of $H(j\omega)$,

$$|H(j\omega)|\cos(\theta + 2n\pi),\qquad\qquad\text{(H.54)}$$

is the same for any integer value of $n$ and the imaginary part of $H(j\omega)$,

$$|H(j\omega)|\sin(\theta + 2n\pi),\qquad\qquad\text{(H.55)}$$

is also the same for any integer value of $n$. To avoid any needless confusion caused by the multiple-valued nature of the inverse tangent function it is conventional to restrict plots of phase to lie in some range of angles for which the inverse tangent function is single-valued, for example, $-\pi < \theta \le \pi$. This simply means that when we evaluate the inverse tangent function we choose a correct value that lies in that range. Since any correct phase is as good as any other, this causes no problems. Using this convention, the magnitude and phase of $H(j\omega)$ versus frequency are illustrated in Figure H.10. These plots were made using MATLAB so that they would be very accurate. But it is important to develop quick, approximate methods to visualize and sketch the magnitude and phase of complex functions of a real variable. This is a skill that helps an engineer in the design and analysis of systems. Look again at the function

$$H(j\omega) = \frac{1}{1 + j\omega}.\qquad\qquad\text{(H.56)}$$

We can get a very good quick indication of the general shape of the magnitude and phase by finding the magnitude and phase at some extreme points, $\omega$ approaching zero from above or below and $\omega$ approaching plus or minus infinity.

For $\omega$ equal to zero, the denominator $1 + j\omega$ of $H(j\omega)$ is simply one and $H(j\omega)$ obviously equals one, the real number one whose magnitude is one and whose phase is zero. For $\omega$ approaching zero from above (from positive values) the phase of $H(j\omega)$ is the phase of the numerator (which is zero) minus the phase of the denominator. The phase of the denominator for small positive $\omega$ is a small positive phase. Therefore, the phase of $H(j\omega)$ is zero minus a small positive phase, that is, a small negative phase.

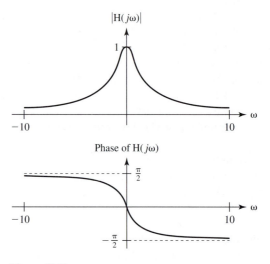

**Figure H.10**
Magnitude and phase of $H(j\omega) = \frac{1}{1+j\omega}$.

This shows that as ω approaches zero from above, the phase is negative and approaching zero. By similar reasoning as ω approaches zero from below (from negative values), the phase is positive and approaching zero. This analysis is confirmed by the phase plot in Figure H.10.

As ω approaches positive infinity, the denominator $1 + j\omega$ becomes infinite in magnitude and, since the numerator is finite, the magnitude of H($j\omega$) approaches zero. Also the one in the denominator $1 + j\omega$ of H($j\omega$) becomes negligible in comparison with $j\omega$, and the phase of H($j\omega$) approaches zero minus $\pi/2$, which is $-(\pi/2)$ rad. As ω approaches negative infinity, the phase of H($j\omega$) approaches zero minus $(-(\pi/2))$, which is $\pi/2$ rad. These limits are also confirmed by Figure H.10. For a function as simple as this example, we can sketch a fairly accurate magnitude and phase plot very quickly just using these simple principles.

Now let's try a somewhat more complicated example, a complex function of cyclic frequency, $f$,

$$\text{H}(f) = \frac{1 - f^2}{1 - f^2 + jf}. \qquad \textbf{(H.57)}$$

Using the quick-approximation ideas just presented, at $f = 0$, H($f$) is one. For $f$ approaching zero from above, the phase is the phase of $1 - f^2$, which for small $f$ is zero, minus the phase of $1 - f^2 + jf$, which for small positive $f$ is a small positive phase. Therefore, the phase for $f$ approaching zero from above is a small negative phase approaching zero. Similarly for $f$ approaching zero from below, the phase is a small positive phase approaching zero.

For $f$ approaching either positive or negative infinity, the $f^2$ terms in the numerator $1 - f^2$ and in the denominator $1 - f^2 + jf$ dominate and the ratio of numerator to denominator approaches $-f^2/-f^2$, which is one. So the magnitude approaches one and the phase approaches zero in that limit. So far we see that for very small or very large values of $f$ the magnitude approaches one and the phase approaches zero. We might be inclined to assume that the magnitude is one for all frequencies. But consider the case $f = \pm 1$. At those values of $f$, the magnitude of H($f$) is zero. Therefore the magnitude must begin at one for $f = 0$, go to zero at $f = \pm 1$, and approach one for $f$ approaching $\pm\infty$. Also, as $f$ approaches $+1$ from below, the numerator is a small positive real number with a phase of zero, and the denominator is a small positive number plus an imaginary number approaching $j$. The denominator phase is approaching $\pi/2$, so the phase of H($f$) is approaching $-(\pi/2)$. As $f$ approaches zero from above, the numerator is a small negative real number with a phase of $\pi$, the phase of the denominator approaches $\pi/2$, and the phase of H($f$) approaches $\pi/2$. So as $f$ moves from just below $+1$ to just above $+1$; the phase changes discontinuously from $-(\pi/2)$ to $\pi/2$. Notice that this discontinuity in the phase occurs where the magnitude is exactly zero. Figure H.11 is a plot generated by MATLAB of the magnitude and phase of H($f$), and it confirms all these observations about the magnitude and phase.

We have already explored the multiple-valued nature of the inverse tangent function. There is one more wrinkle in the computation of phase that is important. We will illustrate it by finding the phase of the complex number $z = -1 + j$. If we take a simple direct approach using a handheld calculator, we might calculate the phase as

$$\text{Phase of } z = \tan^{-1}\left(\frac{1}{-1}\right) = \tan^{-1}(-1) = -\frac{\pi}{4}. \qquad \textbf{(H.58)}$$

However, a plot of $z$ in the complex plane (Figure H.12) shows that this answer is wrong. The plot shows that $z$ lies in the second quadrant. The calculator result indicates

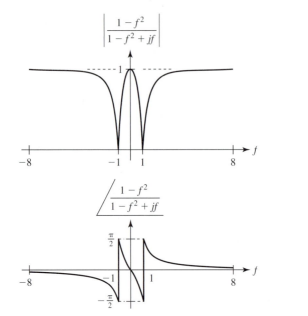

**Figure H.11**
Magnitude and phase of a complex function of a real frequency.

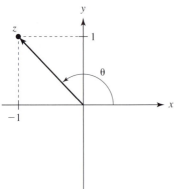

**Figure H.12**
Location of $-1 + j$ in the complex plane.

that $z$ lies in the fourth quadrant. Instead of simply evaluating the inverse tangent of a complex number or function, we should evaluate the *four-quadrant* inverse tangent using our knowledge of the real and imaginary parts separately instead of knowledge of their ratio alone. This enables us to locate the quadrant in which the number lies and eliminate a false answer which lies at $\pi$ rad from the correct answer in the diagonally opposite quadrant. The problem in using the simple inverse tangent function without thinking is that

$$\text{Phase of } (-1 + j) = \tan^{-1}\left(\frac{1}{-1}\right) = \tan^{-1}(-1) = -\frac{\pi}{4} \qquad \textbf{(H.59)}$$

and

$$\text{Phase of } (1 - j) = \tan^{-1}\left(\frac{-1}{1}\right) = \tan^{-1}(-1) = -\frac{\pi}{4}. \qquad \textbf{(H.60)}$$

A handheld calculator typically returns a value $\theta$, using the inverse tangent function, in the range $-(\pi/2) < \theta \leq \pi/2$. The exact location of the complex number in the complex plane is lost when the ratio of the imaginary to real part is taken. Therefore, any four-quadrant inverse tangent must take two arguments, the real and imaginary parts separately, rather than their ratio. (MATLAB has a function `angle` which finds the four-quadrant angle or phase of a complex number.) Using a four-quadrant inverse tangent, the correct answer in (H.59) would be

$$\text{Phase of } (-1 + j) = \tan^{-1}\left(\frac{1}{-1}\right) = \frac{3\pi}{4}. \qquad \textbf{(H.61)}$$

What is the phase of the real number $-1$? Using the four-quadrant tangent, the answer is $\pi$ (plus or minus any integer multiple of $2\pi$). Therefore, when plotting the magnitude and phase of a real-valued function, the magnitude is always nonnegative and the phase switches back and forth between zero and $\pi$ (or $-\pi$) as the function values go through zero. Of course, plotting the magnitude and phase of a real function is a little silly since it can be plotted with positive and negative values on a single plot. But, as previously illustrated, as soon as the function becomes complex, magnitude and phase plots are one of the best ways to graphically represent the function.

**EXAMPLE H.3**

The magnitude and phase plot of the function

$$X(f) = \frac{1 - f^2}{1 - f^2 + jf} \tag{H.62}$$

versus $f$, which appears in Figure H.11, can be easily plotted using the `fplot` command in MATLAB. The following sequence of MATLAB commands produces the plot of the magnitude and phase of this function in Figure H.13.

```
subplot(2,1,1) ; fplot('abs((1-f^2)/(1-f^2+j*f))',[-8,8],'k') ;
xlabel('Frequency, f (Hz)') ; ylabel('|X(f)|') ;
subplot(2,1,2) ; fplot('angle((1-f^2)/(1-f^2+j*f))',[-8,8],'k') ;
xlabel('Frequency, f (Hz)') ; ylabel('Phase of X(f)') ;
```

Although this plot is easy to generate, it does not allow as much user control over formatting and scaling as a somewhat more involved plotting technique. We are plotting a continuous function of the variable $f$. MATLAB does not know what a continuous function is. MATLAB can only draw straight lines. Therefore, we must formulate the problem so as to get a plot that looks like the continuous function using numerical calculations and plotting with straight lines (Figure H.14).

When we use `fplot`, MATLAB decides how to assign the values of the independent variable at which the function values will be calculated. Although the algorithm used by MATLAB

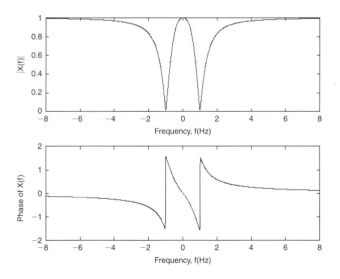

**Figure H.13**
Magnitude and phase of $X(f) = \frac{1-f^2}{1-f^2+jf}$.

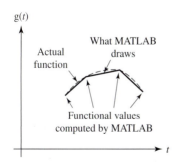

**Figure H.14**
Illustration of how MATLAB plots an approximation to a function.

is generally very good, we can have more control over the plotting of the function and formatting of the plot if we generate the independent-variable values ourselves and use the `plot` command instead. The `plot` command is more rudimentary than the `fplot` command, but it also allows the programmer more options in plotting.

In order to get a smooth-looking curve we must be sure that the points in $f$ are close enough together that when we draw straight lines between points the result looks like the actual curved contours of $X(f)$. The plot in Figure H.11 covers the range $-8 < f < 8$. How many points do we need to make the plot look smooth? The main requirement on how close the points should be is to resolve the region around $f = \pm 1$ where there are some sharp corners in the function's magnitude and phase. Let's try a spacing between points of $\frac{1}{10}$. The MATLAB program then might look like the following code.

```
%     Program to plot the function (1-f^2)/(1-f^2+j*f).
%-------------------------------------------------------------
%     This section actually calculates values of the function.
%-------------------------------------------------------------
df = 1/10 ;                      %  "df" - spacing between frequencies
fmin = -8 ; fmax = 8 ;           %  "fmin" & "fmax" - beginning and ending
                                 %     frequencies
f = fmin:df:fmax ;               %  "f" - vector of frequencies for
                                 %     plotting function with straight lines
                                 %     between points
X = (1-f.^2)./(1-f.^2+j*f)  ;    %  "X" - vector of function values

%-------------------------------------------------------------
%      This section displays the results and formats the plots.
%-------------------------------------------------------------

subplot(2,1,1) ;                 % Plot two plots, one on top and one on
                                 %    the bottom.
                                 % First draw the top plot.
p = plot(f,abs(X),'k') ;         % Plot |X(f)| with black lines between
                                 %    points.
set(p,'LineWidth',2) ;           % Make the plot line heavier.
xlabel('Frequency' f (Hz)') ;    % Label the "f" axis.
ylabel('|X(f)|') ;               % Label the "|X(f)|" axis.
title('Plot of (1-f.^2)./(1-f.^2+j*f)') ; %    Title the plots.
subplot(2,1,2) ;                 % Draw the second plot.
p = plot(f,angle(X),'k') ;       %  Plot the phase (angle) of X(f) with
                                 %    black lines between points.
set(p,'LineWidth',2) ;           % Make the plot line heavier.
xlabel('Frequency, f (Hz)')      % Label the "f" axis.
ylabel('Phase of X(f)') ;        %  Label the "Phase of X(f)" axis.
```

The actual MATLAB graph is displayed in Figure H.15. Although this plot looks very much like Figure H.11, it is not exactly the same. The jumps in the phase plot are not as nearly vertical as in Figure H.11. That is because the spacing between points is not quite small enough. Figure H.16 is the phase plot redone with the point locations indicated by small dots. Try a smaller spacing and see what your plot looks like.

Why is there a jump in the phase? Does the phase of $X(f)$ really change discontinuously at $f = \pm 1$? Notice that the size of the jump is exactly $\pi$ rad. One way to grasp what is really happening near $f = \pm 1$ is to graph the imaginary part of $X(f)$ versus the real part of $X(f)$ in the

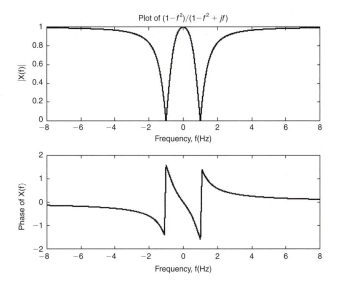

**Figure H.15**
MATLAB plots of the magnitude and phase of $X(f) = \frac{1-f^2}{1-f^2+jf}$.

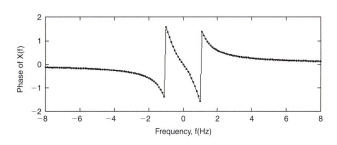

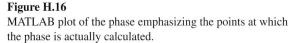

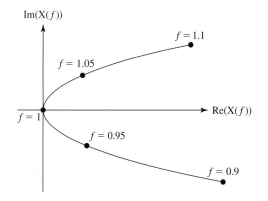

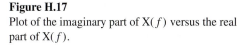

**Figure H.16**
MATLAB plot of the phase emphasizing the points at which the phase is actually calculated.

**Figure H.17**
Plot of the imaginary part of $X(f)$ versus the real part of $X(f)$.

complex plane, for a succession of $f$'s near $f = 1$ (Figure H.17). At $f = 1$, the plot goes through the origin of the complex plane, tangent to the imaginary axis. Therefore, the angle of a vector from the origin to the complex value of $X(f)$ approaches $-(\pi/2)$ just before reaching the origin, and as it passes through the origin the angle changes suddenly to $+(\pi/2)$, agreeing with the plot in Figure H.15. So the phase is discontinuous, even though the complex value of $X(f)$ is continuous! This can only happen where the complex value of $X(f)$ passes through zero. At any other point in the complex plane a phase discontinuity would cause a discontinuity in the complex value of $X(f)$, unless the size of the discontinuity of phase is exactly an integer multiple of $2\pi$ rad. (In this case the phase discontinuity is only apparent, not real, because we can always replace that phase with one which is continuous, making the phase plot again continuous.) ∎

# EXERCISES WITH ANSWERS

1. Find all the solutions of
   a. $z^2 + 8 = 2$
   b. $z^2 - 2z + 10 = 0$
   c. $7z^2 + 3z + 8 = 5$

**Answers:**
$-0.2143 \pm j0.6186, \quad \pm j\sqrt{6}, \quad 1 \pm j3$

2. If $z_1 = 3 - j6$, $z_2 = 2 + j8$, and $z = x + jy$, find $x$ and $y$ in each case.
   a. $z = z_1 + z_2$          b. $z = z_1 - z_2$
   c. $z = z_2 - z_1$          d. $z = z_1 z_2$
   e. $z = \dfrac{z_1}{z_2}$          f. $z = \dfrac{z_2}{z_1}$
   g. $z = \dfrac{1}{z_1}$          h. $z = \dfrac{1}{z_2}$

**Answers:**

$\dfrac{1}{34} - j\dfrac{2}{17}, \quad -1 + j14, \quad \dfrac{1}{15} + j\dfrac{2}{15}, \quad 5 + j2, \quad 54 + j12,$

$-\dfrac{14}{15} + j\dfrac{4}{5}, \quad 1 - j14, \quad -\dfrac{21}{34} - j\dfrac{9}{17}$

3. If $z_1 = (1 + j2)/5$, $z_2 = j(4 - j3)$, and $z = x + jy$, find $x$ and $y$ in each case.
   a. $z = z_1 + z_2$          b. $z = z_1^* + z_2$
   c. $z = z_2^*$          d. $z = z_2 + z_2^*$
   e. $z = z_2 - z_2^*$          f. $z = z_1 z_1^*$
   g. $z = \dfrac{z_1}{z_2^*}$

**Answers:**

$\dfrac{16}{5} + j\dfrac{18}{5}, \quad 3 - j4, \quad \dfrac{1}{5}, \quad 6, \quad \dfrac{16}{5} + j\dfrac{22}{5}, \quad j8, \quad -\dfrac{1}{25} + j\dfrac{2}{25}$

4. If $z_1 = (j - 3)^*$ and $z_2 = \dfrac{3 - j2}{-4 - j}$, find $|z|$ in each case.
   a. $z = z_1$          b. $z = z_2$
   c. $z = z_1 z_1^*$          d. $z = z_2 z_2^*$
   e. $z = z_1 + z_2^*$          f. $z = z_2 z_1^*$
   g. $z = \dfrac{z_1 + z_2}{z_1}$          h. $z = \dfrac{z_1}{z_1^*}$
   i. $z = \dfrac{z_2}{z_2^*}$

**Answers:**

$\dfrac{\sqrt{4505}}{17}, \quad 1, \quad \dfrac{13}{17}, \quad 10, \quad 1, \quad \sqrt{\dfrac{130}{17}}, \quad \sqrt{10}, \quad \dfrac{\sqrt{3757}}{17\sqrt{10}}, \quad \sqrt{\dfrac{13}{17}}$

**5.** Find the magnitude and angle of these complex numbers.

    *a.*   $z = 1 + j$                     *b.*   $z = 1 - j$

    *c.*   $z = 3 - j3$                   *d.*   $z = -4 + j3$

    *e.*   $z = (1 + j)(-1 - j)$           *f.*   $z = \dfrac{1}{1 + j}$

    *g.*   $z = \dfrac{2 - j}{1 + j3}$             *h.*   $z = \left(\dfrac{2 - j}{1 + j3}\right)^{*}$

**Answers:**

$$\sqrt{2}\angle\frac{\pi}{4} \pm 2n\pi, \quad \frac{1}{\sqrt{2}}\angle-\frac{\pi}{4}, \quad 5\angle2.498, \quad \frac{1}{\sqrt{2}}\angle1.713, \quad 2\angle-\frac{\pi}{2},$$

$$\frac{1}{\sqrt{2}}\angle-1.713, \quad \sqrt{2}\angle-\frac{\pi}{4} \pm 2n\pi, \quad 3\sqrt{2}\angle-\frac{\pi}{4} \pm 2n\pi,$$

**6.** Find all the distinct solutions to these equations.

    *a.*   $z^2 = j$                     *b.*   $z^3 = j$

    *c.*   $z^5 = -1$                   *d.*   $z^4 - 3 = j$

    *e.*   $z^3 - 8 = 0$

**Answers:**

$$1\angle\frac{\pi}{4}, 1\angle-\frac{3\pi}{4}; \quad 1\angle\frac{\pi}{6}, 1\angle\frac{5\pi}{6}, 1\angle-\frac{\pi}{2}; \quad 2\angle0, 2\angle\frac{2\pi}{3}, 2\angle-\frac{2\pi}{3};$$

$$1\angle\frac{\pi}{5}, 1\angle\frac{3\pi}{5}, -1, 1\angle-\frac{\pi}{5}, 1\angle-\frac{3\pi}{5};$$

$$1.333\angle0.0804, 1.333\angle1.651, 1.333\angle-3.061, 1.333\angle-1.490$$

**7.** Evaluate these exponential functions.

    *a.*   $e^{j\pi}$                      *b.*   $e^{j(\pi/2)}$

    *c.*   $e^{-j(\pi/2)}$                   *d.*   $e^{j(3\pi/2)}$

    *e.*   $e^{j\pi} + e^{-j\pi}$           *f.*   $e^{j(\pi/2)} + \left(e^{j(\pi/2)}\right)^{*}$

    *g.*   $e^{\pi} + e^{-\pi}$             *h.*   $e^{\pi/2} + (e^{\pi/2})^{*}$

**Answers:**

$9.621, \quad j, \quad -2, \quad -j, \quad 23.184, \quad 0, \quad -1, \quad -j$

**8.** If $z = x + jy = Ae^{j\theta}$, find $x$, $y$, $A$, and $\theta$.

    *a.*   $z = 4e^{j(\pi/2)}$

    *b.*   $z = 4e^{1-j(\pi/2)}$

    *c.*   $z = \left(4e^{1-j(\pi/2)}\right)\left(j2e^{-1-j(\pi/2)}\right)$

    *d.*   $z = \left(-10e^{j(3\pi/2)}\right)^{3}$

    *e.*   $z = \left(e^{j(3\pi/2)}\right)^{3/2}$

**Answers:**

$$x = 0, y = 4, A = 4, \theta = \frac{\pi}{2} + 2n\pi;$$

$$x = 0, y = -1000, A = 1000, \theta = -\frac{\pi}{2} + 2n\pi;$$

$$x = 0, \ y = -8, \ A = 8, \ \theta = -\frac{\pi}{2} + 2n\pi;$$

$$x = \frac{1}{\sqrt{2}}, \ y = \frac{1}{\sqrt{2}}, \ A = 1, \ \theta = \frac{\pi}{4} + 2n\pi \ \text{or}$$

$$x = -\frac{1}{\sqrt{2}}, \ y = -\frac{1}{\sqrt{2}}, \ A = 1, \ \theta = \frac{5\pi}{4} + 2n\pi;$$

$$x = 0, \ y = -10.87, \ A = 10.87, \ \theta = -\frac{\pi}{2} + 2n\pi$$

9.  Using MATLAB plot the magnitude and phase of the following complex functions of the real independent variable $t$, $f$, or $\omega$ over the range indicated.

   a.   $x(t) = 2e^{-j4\pi t}, \quad -1 < t < 1$

   b.   $x(t) = 2e^{(-1+j4\pi)t}, \quad -4 < t < 4$

   c.   $X(f) = \dfrac{1}{1 + j2\pi f}, \quad -2 < f < 2$

   d.   $X(j\omega) = \dfrac{j\omega}{1 + j\omega}, \quad -4\pi < \omega < 4\pi$

**Answers:**

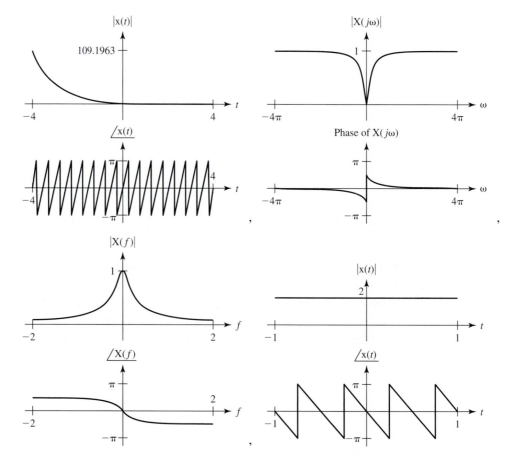

# EXERCISES WITHOUT ANSWERS

**10.** Convert these complex numbers to the polar form $Ae^{j\theta}$ with an angle in the range $-\pi < \theta \le \pi$.

    *a.*   $1 + j$                    *b.*   $3 - j2$

    *c.*   $-j$                      *d.*   $-j + 1$

**11.** Convert these complex numbers to the rectangular form $x + jy$.

    *a.*   $e^{j\pi}$                     *b.*   $4\angle 45°$

    *c.*   $3e^{2-j(\pi/4)}$          *d.*   $10e^{-j(11\pi/4)}$

**12.** Find the numerical value of $z$ in both the rectangular and polar forms.

    *a.*   $z = 2e^{-1+j(\pi/2)} + 4 - j2$       *b.*   $z = (1 - j)(4 + j5)^2$

    *c.*   $z = |(-j)^3|$                *d.*   $z = \dfrac{2e^{j\pi}}{(1 + j)^4}$

    *e.*   $z = \dfrac{2e^{-j\pi}}{(1 - j)^4}$         *f.*   $z = e^{1+j} - e^{1-j}$

**13.** Using MATLAB, plot graphs of the magnitude and phase of the following functions of the real variable $f$ over the range indicated.

    *a.*   $X(f) = \dfrac{10}{1 + j(f/100)}, \quad -400 < f < 400$

    *b.*   $X(f) = \dfrac{j10f}{1 + j(f/100)}, \quad -400 < f < 400$

    *c.*   $X(f) = e^{j\pi f} - e^{j2\pi f}, \quad -8 < f < 8$

    *d.*   $X(f) = \dfrac{5}{1 - f^2 + j(f/4)}, \quad -4 < f < 4$

    *e.*   $X(j\omega) = \dfrac{1}{\left(j\omega - e^{j(3\pi/4)}\right)\left(j\omega - e^{j(5\pi/4)}\right)\left(j\omega - e^{-j(3\pi/4)}\right)\left(j\omega - e^{-j(5\pi/4)}\right)},$

                                                      $-2\pi < \omega < 2\pi$

    *f.*   $X(j\omega) = \dfrac{(j\omega)^4}{\left(j\omega - e^{j(3\pi/4)}\right)\left(j\omega - e^{j(5\pi/4)}\right)\left(j\omega - e^{-j(3\pi/4)}\right)\left(j\omega - e^{-j(5\pi/4)}\right)},$

                                                       $-2\pi < \omega < 2\pi$

**14.**  *a.*   Show that the magnitude of the complex function $e^{jx}$, where $x$ is a real number, is one, regardless of the value of $x$.

    *b.*   Find the simplest expression you can for the phase of $e^{jx}$ as a function of $x$.

    *c.*   Graph the phase of $e^{jx}$ by hand and then write a MATLAB program to graph the same phase. The graphs should extend over values of $x$ in the range $-4\pi < x < 4\pi$. If the graphs are different, explain why.

# I
## A P P E N D I X

# Differential and Difference Equations

## I.1 INTRODUCTION

*Differential* equations are those in which an equality is expressed in terms of a function of one or more independent variables and derivatives of the function with respect to one or more of those independent variables. *Difference* equations are those in which an equality is expressed in terms of a function of one or more independent variables and finite differences of the function. Differential equations are important in signal and system analysis because they describe the dynamic behavior of continuous-time (CT) physical systems. Difference equations are important in signal and system analysis because they describe the dynamic behavior of discrete-time (DT) systems. *Discrete time* is simply equally spaced points in time, separated by some time difference $\Delta t$. In DT signals and systems the behavior of a signal and the action of a system are known only at discrete points in time and are not defined between those discrete points in time.

Differential equations have several properties by which they are classified: linear and nonlinear, ordinary and partial, homogeneous and inhomogeneous. They are also classified by their *order,* which is the highest order of a derivative in the equation after it is put into a standard form, and by the coefficients of the derivatives, which may either be constants or functions of the independent variable. Difference equations are classified in a similar manner in which the order of the difference equation is the highest-order difference after being put into standard form. Fortunately the great majority of systems are described (at least approximately) by the types of differential or difference equations that are easiest to solve: ordinary, linear differential, or difference equations with constant coefficients. This appendix covers only equations of this type.

## I.2 HOMOGENEOUS CONSTANT-COEFFICIENT LINEAR DIFFERENTIAL EQUATIONS

Let us begin with an example of the simplest differential equation, a homogeneous first-order linear ordinary differential equation,

$$2\,\frac{d\mathrm{y}\,(t)}{dt} + 7\mathrm{y}\,(t) = 0. \tag{I.1}$$

We can streamline the notation by indicating differentiation by

$$\frac{dy(t)}{dt} = y', \quad \frac{d^2y(t)}{dt^2} = y'', \quad \frac{d^3y(t)}{dt^3} = y''', \quad \dots \tag{I.2}$$

In (I.2) $t$ is the independent variable and y is the dependent variable, a function of $t$. Rewriting (I.1) in the streamlined notation,

$$2y' + 7y = 0. \tag{I.3}$$

A homogeneous linear ordinary differential equation is a linear combination of the dependent variable and its derivatives, set equal to zero. We can rearrange (I.3) into

$$y' = -\frac{7}{2}y. \tag{I.4}$$

This equation must be satisfied for any arbitrary value of the independent variable $t$. That means that y, which is a function of $t$, must have the same functional form as y'. The only function which has that property is the exponential function because

$$\frac{d}{dt}(e^{\lambda t}) = \lambda e^{\lambda t}. \tag{I.5}$$

Therefore, the functional form of the solution of (I.1) is $y(t) = e^{\lambda t}$, where $\lambda$ is a constant, as yet undetermined. The exponential function is unique to, or characteristic of, this type of differential equation because it is the only functional form that can solve it. The *characteristic function* of a differential equation is commonly referred to as the *eigenfunction,* after the German word *Eigenfunktion* meaning "characteristic function." To check the validity of this solution form, we put $y(t) = e^{\lambda t}$ into (I.1) and perform the indicated operations,

$$2y' + 7y = 2\lambda e^{\lambda t} + 7e^{\lambda t} = 0 \tag{I.6}$$

or

$$2\lambda + 7 = 0. \tag{I.7}$$

This equation is sometimes referred to as the *characteristic equation* associated with the differential equation. It is an algebraic equation and is satisfied if

$$\lambda = -\frac{7}{2}. \tag{I.8}$$

Then $y(t) = e^{\lambda t}$ is a solution of (I.1) if $\lambda = -\frac{7}{2}$. But it is not the most general solution. We can multiply by an arbitrary constant $K$ to get a solution form

$$y(t) = Ke^{\lambda t}. \tag{I.9}$$

When we put it into (I.3) we get

$$2y' + 7y = 2K\lambda e^{\lambda t} + 7Ke^{\lambda t} = 0 \tag{I.10}$$

or

$$2\lambda + 7 = 0 \tag{I.11}$$

as before. This is the most general form of solution. The particular value of $\lambda$ which solves the characteristic equation is called an *eigenvalue,* and the solution works for any arbitrary value of $K$.

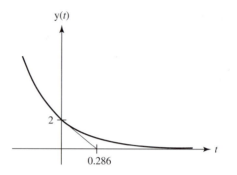

**Figure I.1**
The complete solution of the first-order, linear,
constant-coefficient, ordinary differential
equation with boundary conditions.

An exact value of $K$ can only be specified by using more information than is contained in the differential equation itself. It is found by applying *boundary conditions*. In order to specify $K$ one must know the value of y or its first derivative at some particular value of $t$. Suppose it is known that when $t$ is zero, y is two. Then

$$2 = Ke^{-(7/2)(0)} = K. \tag{I.12}$$

We see now that the arbitrary constant $K$ is needed to satisfy both the equation and the boundary condition. Then the full numerical solution of (I.3) with boundary conditions is

$$y(t) = 2e^{-(7/2)t} \tag{I.13}$$

(Figure I.1). In this case the boundary condition was given at $t = 0$ and, if $t$ represents time, this type of boundary condition is called an *initial* condition. By analogy to the procedures followed in this example, the solution of any equation of the form

$$y' = ay \tag{I.14}$$

can be found.

### Inhomogeneous Constant-Coefficient Linear Differential Equations

The next step up in equation complexity is the *inhomogeneous* first-order linear ordinary differential equation. An inhomogeneous linear ordinary differential equation is a linear combination of the dependent variable and its derivatives set equal to a function of the independent variable which is often called the *forcing function*. For example,

$$2y' + 7y = 4\cos(3t). \tag{I.15}$$

We can rewrite (I.15) as

$$y' = -\frac{7}{2}y + 2\cos(3t). \tag{I.16}$$

(This form, in which the first derivative of the function is on the left side of the equation and the right side has the function itself followed by the forcing function, is a standard way of writing differential equations which is used in systems of multiple differential equations.) We need to find some function y($t$) for which (I.16) is satisfied for any arbitrary $t$. If we choose y to be a cosine function of $t$, that yields a cosine on the right side, but the derivative will be a sine function of $t$ which will appear on the left side and that does not work. We could choose a sine function of $t$ and that would make the derivative have the right form, but the function itself would not. But if we choose a linear combination of a sine and a cosine, maybe we can arrange to have the two sine functions cancel somehow and leave just a cosine function. Let's try a solution of the form

$$y(t) = K_1 \sin(3t) + K_2 \cos(3t). \tag{I.17}$$

Substituting that into (I.15)

$$2[3K_1 \cos(3t) - 3K_2 \sin(3t)] + 7[K_1 \sin(3t) + K_2 \cos(3t)] = 4 \cos(3t). \tag{I.18}$$

For this to be a solution for any arbitrary $t$, since cosines and sines are different functions of $t$, the cosine parts on each side must be equal and the sine parts on each side must be equal, independently. That is,

$$6K_1 \cos(3t) + 7K_2 \cos(3t) = 4 \cos(3t) \tag{I.19}$$

$$-6K_2 \sin(3t) + 7K_1 \sin(3t) = 0 \tag{I.20}$$

or

$$6K_1 + 7K_2 = 4 \qquad \text{and} \qquad -6K_2 + 7K_1 = 0. \tag{I.21}$$

Solving,

$$K_1 = \frac{24}{85} \qquad \text{and} \qquad K_2 = \frac{28}{85}. \tag{I.22}$$

Therefore, $y(t) = \frac{24}{85} \sin(3t) + \frac{28}{85} \cos(3t)$ is one solution to (I.15).

---

The solution form

$$y(t) = K_1 \sin(3t) + K_2 \cos(3t)$$

could have been written in an equivalent form by using the trigonometric identity

$$\cos(x + y) = \cos(x) \cos(y) - \sin(x) \sin(y).$$

Then

$$y(t) = K \cos(3t + \theta),$$

where

$$K = \sqrt{K_1^2 + K_2^2} \qquad \text{and} \qquad \tan(\theta) = -\frac{K_1}{K_2}.$$

Sometimes this form is more convenient or appropriate.

---

The method used to find the coefficients $K_1$ and $K_2$ is called the *method of undetermined coefficients,* and it can be applied to find a solution to any linear ordinary inhomogeneous differential equation with constant coefficients. It is important to point out here that this is one solution to the equation but not the total solution. This solution is called the *particular* solution and will be denoted here by $y_p(t)$. The total solution is the sum of the particular solution and the solution of the homogeneous form of the equation $y_h(t)$, which was found in the previous example. The sum of the two is also a solution of

$$2y' + 7y = 4\cos(3t) \tag{I.23}$$

because $2y'_h + 7y_h = 0$. If the total solution is $y = y_h + y_p$ and that is substituted into (I.15), we get

$$2(y_h + y_p)' + 7(y_h + y_p) = 2(y'_h + y'_p) + 7(y_h + y_p) = 4\cos(3t) \tag{I.24}$$

or

$$\underbrace{2y'_h + 7y_h}_{=0} + 2y'_p + 7y_p = 4\cos(3t), \tag{I.25}$$

which is the same as the original inhomogeneous equation (I.15). Therefore, the total solution to (I.15) is

$$y(t) = Ke^{-(7/2)t} + \frac{24}{85}\sin(3t) + \frac{28}{85}\cos(3t) \tag{I.26}$$

where, again, $K$ must be found by matching boundary conditions, but this time with the total solution, not just the homogeneous solution. (In the previous example the homogeneous solution was the total solution.) Suppose y is zero when $t$ is $\pi/3$. Then

$$0 = Ke^{-(7/6)\pi} + \frac{24}{85}\sin(\pi) + \frac{28}{85}\cos(\pi) \Rightarrow K = -12.86. \tag{I.27}$$

We can always add the homogeneous solution to the particular solution to get a total solution because, when we substitute it into the differential equation and do the differentiations, it always adds to zero! The homogeneous solution is also called the *transient* solution because for stable physical systems described by this kind of equation, the homogeneous solution *decays* away with time or called the natural response because its form indicates the nature of the system described by the differential equation. The particular solution is also called the *steady-state* solution because it is the solution which persists after the transient solution has died away or called the *forced* response because it is the part of the solution that is forced to exist by the action of the forcing function.

We found the solution (I.26) by assuming a particular solution in a form consisting of a linear combination of sines and cosines,

$$y(t) = K_1\sin(3t) + K_2\cos(3t). \tag{I.28}$$

Since sines and cosines can be expressed in terms of complex exponentials, we could change this solution form to

$$y(t) = \frac{K_2 - jK_1}{2}e^{j3t} + \frac{K_2 + jK_1}{2}e^{-j3t} \tag{I.29}$$

or

$$y(t) = K'_1 e^{j3t} + K'_2 e^{-j3t}, \tag{I.30}$$

where

$$K'_1 = \frac{K_2 - jK_1}{2} \quad \text{and} \quad K'_2 = \frac{K_2 + jK_1}{2}. \tag{I.31}$$

Then, substituting this solution form into the original equation

$$2y' + 7y = 4\cos(3t) = 2(e^{j3t} + e^{-j3t}), \tag{I.32}$$

we get

$$j6K_1'e^{j3t} - j6K_2'e^{-j3t} + 7K_1'e^{j3t} + 7K_2'e^{-j3t} = 2(e^{j3t} + e^{-j3t}). \qquad \textbf{(I.33)}$$

Notice that when we equate like functional forms on both sides we get

$$j6K_1'e^{j3t} + 7K_1'e^{j3t} = 2e^{j3t} \qquad \text{and} \qquad -j6K_2'e^{-j3t} + 7K_2'e^{-j3t} = 2e^{-j3t} \qquad \textbf{(I.34)}$$

or

$$(7 + j6)K_1' = 2 \qquad \text{and} \qquad (7 - j6)K_2' = 2. \qquad \textbf{(I.35)}$$

For (I.35) to be satisfied $K_1' = (K_2')^*$, a requirement we have already seen in (I.31). Solving the left-hand equation in (I.35),

$$K_1' = \frac{14 - j12}{85}. \qquad \textbf{(I.36)}$$

Then

$$K_2' = \frac{14 + j12}{85} \qquad \textbf{(I.37)}$$

and

$$y_p(t) = \frac{14 - j12}{85}e^{j3t} + \frac{14 + j12}{85}e^{-j3t}. \qquad \textbf{(I.38)}$$

This solution form can be converted into

$$y_p(t) = \frac{28\cos(3t) + 24\sin(3t)}{85}, \qquad \textbf{(I.39)}$$

which is exactly the same as the previous solution form. Since, in this solution method the undetermined coefficients must always occur in complex-conjugate pairs, we can abbreviate the solution process by finding the solution to

$$2y' + 7y = 4e^{j3t}, \qquad \textbf{(I.40)}$$

which is

$$y_p(t) = \frac{4}{7 + j6}e^{j3t} = \frac{28 - j24}{85}(\cos(3t) + j\sin(3t)) \qquad \textbf{(I.41)}$$

or

$$y_p(t) = \frac{28\cos(3t) + 24\sin(3t) - j24\cos(3t) + j28\sin(3t)}{85}. \qquad \textbf{(I.42)}$$

Observe that the real part of this solution is the same as the solution of the original equation.

We found a particular solution for a particular forcing function $4\cos(3t)$. Of course, if the numerical coefficients changed, we could still find a solution by the same technique. But what happens if the functional form of the forcing function changes? Then the functional form of the solution must also change. There are many possible functional forms of forcing functions, but the most common ones that occur in engineering practice are ones which are at least piecewise continuous and differentiable with a finite number of unique functional forms of the derivatives. Commonly occurring functions which have these properties are

$$g(t) = A_0 + A_1 t + A_2 t^2 + \cdots + A_N t^N \tag{I.43}$$

$$g(t) = A e^{at} \tag{I.44}$$

$$g(t) = A \cos(at) \tag{I.45}$$

$$g(t) = A \sin(at) \tag{I.46}$$

and sums of products of these functions, for example,

$$g(t) = A e^{at} \cos(bt) + B e^{ct} \sin(dt). \tag{I.47}$$

As long as we restrict ourselves to forcing functions of these forms, we can always find a particular solution by proposing a solution form containing the forcing function form and all its unique derivatives. This may sound unnecessarily restrictive. After all, do all real systems have forcing functions of these few forms? No. But, as it turns out, all forcing functions with any engineering usefulness can be expressed as linear combinations of functions of these forms. In fact linear combinations of complex sinusoids are sufficient to describe any forcing function with engineering usefulness.

The next step up in complexity is to a higher-order differential equation. For example,

$$y'' + 5y' + 3y = 6t^2. \tag{I.48}$$

The total solution is the sum of the homogeneous and particular solutions. By reasoning similar to that for the first-order linear constant-coefficient differential equation, any solution of the homogeneous equation

$$y'' + 5y' + 3y = 0 \tag{I.49}$$

must be of the functional form $e^{\lambda t}$. Substituting into (I.49) and solving,

$$\lambda^2 e^{\lambda t} + 5\lambda e^{\lambda t} + 3 e^{\lambda t} = 0 \tag{I.50}$$

or

$$\lambda^2 + 5\lambda + 3 = 0 \tag{I.51}$$

The characteristic equation (I.51), is quadratic, and there are two solutions,

$$\lambda_1 = -0.6972 \qquad \text{and} \qquad \lambda_2 = -4.303. \tag{I.52}$$

So for this second-order differential equation there are two eigenvalues. Which one should we choose? Can we use both? We could propose the solution

$$y_{1h}(t) = K_{1h} e^{\lambda_1 t} \tag{I.53}$$

as we did in solving the first-order equation. This solution satisfies the equation but so does the solution

$$y_{2h}(t) = K_{2h} e^{\lambda_2 t}. \tag{I.54}$$

We would like to find the most general solution possible. If we put the sum of these two solutions into the homogeneous differential equation, we get

$$\lambda_1^2 K_{1h} e^{\lambda_1 t} + 5\lambda_1 K_{1h} e^{\lambda_1 t} + 3 K_{1h} e^{\lambda_1 t} + \lambda_2^2 K_{2h} e^{\lambda_2 t} + 5\lambda_2 K_{2h} e^{\lambda_2 t} + 3 K_{2h} e^{\lambda_2 t} = 0. \tag{I.55}$$

Then, substituting in the eigenvalues,

$$\underbrace{[(-0.6972)^2 + 5(-0.6972) + 3]}_{=0} K_{1h}e^{\lambda_1 t} + \underbrace{[(-4.303)^2 + 5(-4.303) + 3]}_{=0} K_{2h}e^{\lambda_2 t} = 0$$

(I.56)

and the homogeneous equation is also satisfied by the sum of these two solutions,

$$y_h(t) = K_1 e^{-0.6972t} + K_2 e^{-4.303t},$$

(I.57)

which is the most general possible solution of the homogeneous equation. This result can be generalized to a differential equation of any order. The solution of the homogeneous equation is a linear combination of the eigenfunctions, one for each unique eigenvalue.

> However, when any two eigenvalues are the same, the corresponding two eigenfunctions are not independent and can be combined into one eigenfunction. This happens only very rarely in practice, so it is not of great practical significance. In such a case the needed extra eigenfunction has the functional form $te^{\lambda t}$.

A particular solution of (I.48) can be found by proposing a solution which is a linear combination of the forcing function and all its unique derivatives of the form

$$y_p(t) = At^2 + Bt + C.$$

(I.58)

Substituting into (I.48)

$$2A + 5(2At + B) + 3(At^2 + Bt + C) = 6t^2$$

(I.59)

and solving,

$$A = 2 \qquad B = -\frac{20}{3} \qquad C = \frac{88}{9}.$$

(I.60)

Therefore, the total solution is

$$y(t) = y_h(t) + y_p(t) = K_1 e^{-0.6972t} + K_2 e^{-4.303t} + 2t^2 - \frac{20}{3}t + \frac{88}{9}$$

(I.61)

and the two remaining constants $K_1$ and $K_2$ must be found by applying two independent boundary conditions. (The number of boundary conditions needed is always equal to the order of the differential equation.)

## Systems of Linear Differential Equations

So far we have only considered the solution of a single differential equation for a single unknown function. A very common situation in signal and system analysis is the solution of systems of differential equations. Most systems of interest are described by more than one differential equation. An example that will introduce some solution methods is the relatively simple two-differential-equation system

$$y_1' + 5y_1 + 2y_2 = 10$$
$$y_2' + 3y_2 + y_1 = 0$$

(I.62)

with initial conditions $y_1(0) = 1$ and $y_2(0) = 0$.

We will solve this system of equations using two different methods. The first method will be an ad hoc method in which we combine the two first-order equations, each in two functions, into two second-order equations, each in only one function, and then solve these equations by the methods of the previous section. The second method is a more systematic technique of solving the two original first-order equations simultaneously. The second method is a natural stepping stone to techniques for solving systems of arbitrary numbers of first-order equations using matrix methods.

We can rearrange the second equation in (I.62) to form

$$y_1 = -y_2' - 3y_2. \tag{I.63}$$

Then, substituting (I.63) into the first equation in (I.62) we get

$$-y_2'' - 3y_2' + 5(-y_2' - 3y_2) + 2y_2 = 10 \tag{I.64}$$

or

$$y_2'' + 8y_2' + 13y_2 = -10. \tag{I.65}$$

We can solve this equation using the techniques of the previous section and, when we do, we get the characteristic equation

$$\lambda^2 + 8\lambda + 13 = 0, \tag{I.66}$$

the eigenvalues $\lambda_1 = -4 + \sqrt{3}$ and $\lambda_2 = -4 - \sqrt{3}$, the particular solution $y_p = -\frac{10}{13}$, and the total solution

$$y_2 = K_1 e^{-\lambda_1 t} + K_2 e^{-\lambda_2 t} - \frac{10}{13}. \tag{I.67}$$

We can apply the two initial conditions. The condition $y_2(0) = 0$ applied to (I.67) yields

$$K_1 + K_2 = \frac{10}{13}, \tag{I.68}$$

and we can use (I.63) with (I.67) and the initial condition $y_1(0) = 1$ to form

$$(\lambda_1 + 3)K_1 + (\lambda_2 + 3)K_2 = \frac{17}{13}. \tag{I.69}$$

Solving, we get

$$K_1 = 0.9841 \qquad \text{and} \qquad K_2 = -0.2151 \tag{I.70}$$

and a total numerical solution

$$y_2(t) = 0.9841 e^{-2.268t} - 0.2151 e^{-5.732t} - 0.769. \tag{I.71}$$

Now we can use (I.63) along with (I.71) to find $y_1(t)$,

$$y_1(t) = -0.720 e^{-2.268t} - 0.588 e^{-5.732t} + 2.308. \tag{I.72}$$

Notice that because of the relationship between the two solution functions $y_1(t)$ and $y_2(t)$ in (I.62), that the eigenvalues for the two functions are the same. Only the constants multiplying the eigenfunctions and the particular solutions are different. The fact that the eigenvalues are the same is a result of the fact that the two first-order differential equations are coupled and, therefore, not independent.

Now we will solve the system (I.62) using a different technique. Since each equation in (I.62) is first-order, we assume the solution forms

$$y_{1h}(t) = K_{1h} e^{\lambda t} \qquad \text{and} \qquad y_{2h}(t) = K_{2h} e^{\lambda t}. \tag{I.73}$$

Then, substituting these forms into (I.62) and simplifying we get two first-order coupled characteristic equations,

$$(\lambda + 5)K_{1h} + 2K_{2h} = 0$$
$$K_{1h} + (\lambda + 3)K_{2h} = 0. \tag{I.74}$$

This is a system of two equations in three unknowns, so we should not expect to be able to find a unique solution, but in this case, because of the form of the equations, we can solve for $\lambda$. Rearranging (I.74),

$$\frac{K_{1h}}{K_{2h}} = -\frac{2}{\lambda + 5}$$

$$\frac{K_{1h}}{K_{2h}} = -(\lambda + 3). \tag{I.75}$$

Then, equating the two equations in (I.75),

$$\frac{2}{\lambda + 5} = \lambda + 3 \tag{I.76}$$

or

$$\lambda^2 + 8\lambda + 13 = 0. \tag{I.77}$$

This is exactly the same characteristic equation, (I.66), we got in the previous solution and the eigenvalues are again

$$\lambda_1 = -4 + \sqrt{3} \qquad \text{and} \qquad \lambda_2 = -4 - \sqrt{3}. \tag{I.78}$$

So there are two values of $\lambda$ for which (I.74) can be satisfied [and they are the same two eigenvalues for which (I.66) has a solution]. This is a little different from the ordinary experience of solving two simultaneous equations. Even though we have three unknowns we are still able to solve for one of them. But, when we do, we get two possible values for that unknown instead of one. We have not found unique values, but we have narrowed the field of possible values.

Now let's do what one would ordinarily do in solving algebraic equations, choose one eigenvalue $\lambda_1$ and put it into the equations in (I.74) and try to find the arbitrary constants $K_{1h}$ and $K_{2h}$ [with numerical values substituted into (I.74)].

$$K_{1h} = -\frac{2}{\sqrt{3} + 1}K_{2h} \tag{I.79}$$

and

$$\left[ -\frac{2}{\sqrt{3} + 1} - 1 + \sqrt{3} \right]K_{2h} = 0. \tag{I.80}$$

Equation (I.80) can be satisfied if $K_{2h}$ is zero or the coefficient of $K_{2h}$ equals zero, or both. In other words, to find a nontrivial solution $K_{2h} \neq 0$, the coefficient $[-(2/(\sqrt{3} + 1)) - 1 + \sqrt{3}]$ must be zero. Simplifying (I.80) we get

$$(0)K_{2h} = 0. \tag{I.81}$$

Therefore, for this eigenvalue, we know it is possible for $K_{2h}$ to be nonzero, but we don't yet know what it is. This should have been expected because we are trying to solve a system of two equations in three unknowns and there is no unique solution. If we do the same thing with the other eigenvalue, we get the same result. But we can say one more thing that is useful about the arbitrary constants $K_{1h}$ and $K_{2h}$. From (I.75)

$$\frac{K_{1h}}{K_{2h}} = -(\lambda + 3).  \tag{I.82}$$

So, for the first eigenvalue,

$$\frac{K_{1h}}{K_{2h}} = 1 - \sqrt{3} = -0.732.  \tag{I.83}$$

Similarly, for the second eigenvalue,

$$\frac{K_{1h}}{K_{2h}} = 1 + \sqrt{3} = 2.732.  \tag{I.84}$$

We have not yet found unique solutions for all three unknowns, but we have established certain relationships among them. To accommodate these results for the two eigenvalues we need homogeneous solutions which are linear combinations of the two eigenfunctions corresponding to the two eigenvalues. So we assume solutions of the forms

$$y_{1h}(t) = K_{11h}e^{\lambda_1 t} + K_{12h}e^{\lambda_2 t}  \quad \text{and} \quad  y_{2h}(t) = K_{21h}e^{\lambda_1 t} + K_{22h}e^{\lambda_2 t},  \tag{I.85}$$

where

$$\frac{K_{11h}}{K_{21h}} = -0.732  \quad \text{and} \quad  \frac{K_{12h}}{K_{22h}} = 2.732.  \tag{I.86}$$

We need something to establish the exact values of the constants instead of just their ratios and that requires using the initial conditions. But before applying the initial conditions we need a particular solution to complete the total solution. Since the forcing function is a constant, we can assume particular solutions

$$y_{1p}(t) = K_{1p}  \quad \text{and} \quad  y_{2p}(t) = K_{2p}.  \tag{I.87}$$

Doing that and solving, we get

$$K_{1p} = \frac{30}{13}  \quad \text{and} \quad  K_{2p} = -\frac{10}{13}.  \tag{I.88}$$

So the total solutions are of the forms

$$y_1(t) = K_{11h}e^{\lambda_1 t} + K_{12h}e^{\lambda_2 t} + \frac{30}{13}  \quad \text{and} \quad  y_2(t) = K_{21h}e^{\lambda_1 t} + K_{22h}e^{\lambda_2 t} - \frac{10}{13}.  \tag{I.89}$$

Since we know two relations, (I.86), among the four arbitrary constants, there are actually only two unknowns, so we need only two initial conditions. We can now use $y_1(0) = 1$ and $y_2(0) = 0$. Applying the initial conditions we get

$$y_1(0) = K_{11h} + K_{12h} + \frac{30}{13} = 1  \quad \text{and} \quad  y_2(0) = K_{21h} + K_{22h} - \frac{10}{13} = 0.  \tag{I.90}$$

These two equations, together with (I.83) and (I.84), lead to the final numerical solution

$$y_1(t) = -0.720e^{-2.268t} - 0.588e^{-5.732t} + 2.308  \tag{I.91}$$

and

$$y_2(t) = 0.9841 e^{-2.268t} - 0.2151 e^{-5.732t} - 0.769. \qquad \textbf{(I.92)}$$

This was only the next step up in differential equation complexity, a two-differential-equation system, and a simple one at that! Try to imagine what it would be like to solve more complicated systems of differential equations. Fortunately, systematic techniques have been developed to solve systems of equations like this. Although these techniques don't actually reduce the total amount of calculation, they do arrange the computations in a way that makes them easy to program on a computer and, therefore, makes the solutions much easier in practice for humans to obtain.

MATLAB can solve systems of differential equations. Here is the MATLAB `help` message for the command DSOLVE.

```
DSOLVE Symbolic solution of ordinary differential equations.
    DSOLVE('eqn1','eqn2', ...) accepts symbolic equations representing
    ordinary differential equations and initial conditions. Several
    equations or initial conditions may be grouped together, separated
    by commas, in a single input argument.

    By default, the independent variable is 't'. The independent
    variable may be changed from 't' to some other symbolic variable by
    including that variable as the last input argument.

    The letter 'D' denotes differentiation with respect to the
    independent variable, i.e. usually d/dt. A "D" followed by a digit
    denotes repeated differentiation; e.g., D2 is d^2/dt^2. Any
    characters immediately following these differentiation operators are
    taken to be the dependent variables; e.g., D3y denotes the third
    derivative of y(t). Note that the names of symbolic variables should
    not contain the letter "D".

    Initial conditions are specified by equations like 'y(a) = b' or
    'Dy(a) = b' where y is one of the dependent variables and a and b
    are constants. If the number of initial conditions given is less
    than the number of dependent variables, the resulting solutions will
    obtain arbitrary constants, C1, C2, etc.

    Examples:
        dsolve('Dx = -a*x') returns

        ans = exp(-a*t)*C1

      x = dsolve('Dx = -a*x','x(0) = 1','s') returns

        x = exp(-a*s)

      y = dsolve('(Dy)^2 + y^2 = 1','y(0) = 0') returns

        y =
        [ sin(t)]
        [ -sin(t)]

      S = dsolve('Df = f + g','Dg = -f + g','f(0) = 1','g(0) = 2')
      returns a structure S with fields
```

```
    S.f = exp(t)*cos(t)+2*exp(t)*sin(t)
    S.g = -exp(t)*sin(t)+2*exp(t)*cos(t)
dsolve('Df = f + sin(t)', 'f(pi/2) = 0')
dsolve('D2y = -a^2*y', 'y(0) = 1, Dy(pi/a) = 0')
S = dsolve('Dx = y', 'Dy = -x', 'x(0)=0', 'y(0)=1')
S = dsolve('Du=v, Dv=w, Dw=-u','u(0)=0, v(0)=0, w(0)=1')
w = dsolve('D3w = -w','w(0)=1, Dw(0)=0, D2w(0)=0')
```

See also SOLVE, SUBS.

## I.3  LINEAR ORDINARY DIFFERENCE EQUATIONS

### Finite-Difference Approximations to a Derivative

To illustrate a connection between difference equations and differential equations, let us begin with (I.1), the homogeneous first-order constant-coefficient ordinary differential equation in Section I.2,

$$2\frac{dy(t)}{dx} + 7y(t) = 0 \tag{I.93}$$

and approximate it by a difference equation. We can do this by approximating derivatives by finite differences. Recall these definitions of a derivative,

$$\frac{dy(t)}{dt} = \lim_{\Delta t \to 0} \frac{y(t + \Delta t) - y(t)}{\Delta t}, \tag{I.94}$$

$$\frac{dy(t)}{dt} = \lim_{\Delta t \to 0} \frac{y(t) - y(t - \Delta t)}{\Delta t}, \tag{I.95}$$

and

$$\frac{dy(t)}{dt} = \lim_{\Delta t \to 0} \frac{y(t + \Delta t) - y(t - \Delta t)}{2\Delta t}. \tag{I.96}$$

At any point at which y(t) is differentiable, any of these definitions of a derivative yield exactly the same result when the limit is taken. A derivative in continuous time can be approximated by a finite difference in discrete time by

$$\frac{y((n+1)\,\Delta t) - y(n\,\Delta t)}{\Delta t}. \tag{I.97}$$

This is called a *forward difference* because it uses the present or current value of y, y(n Δt), and the next or future value of y, y((n + 1) Δt). Similarly

$$\frac{y(n\,\Delta t) - y((n-1)\,\Delta t)}{\Delta t} \tag{I.98}$$

is a *backward difference* and

$$\frac{y((n+1)\,\Delta t) - y((n-1)\,\Delta t)}{2\,\Delta t} \tag{I.99}$$

is a *central difference.* In the limit as Δt approaches zero these are all the same, but in discrete time, Δt is fixed and is not zero and these three approximations to a continuous-time derivative are, in general, different.

As an illustration we will convert the differential equation (I.93) to a difference equation by using a forward-difference approximation

$$2\frac{y((n+1)\,\Delta t) - y(n\,\Delta t)}{\Delta t} + 7y(n\,\Delta t) = 0. \qquad \textbf{(I.100)}$$

To simplify the notation let

$$y[n] = y(n\,\Delta t), \qquad \textbf{(I.101)}$$

where the square brackets [ ] distinguish a function of discrete time from a function of continuous time which is indicated by using parentheses ( ). In this notation, *time* is not explicitly indicated, but since the time between consecutive discrete-time values of the function y is always $\Delta t$, we do not need to explicitly indicate time. Using the simplified notation, (I.100) becomes

$$2\frac{y[n+1] - y[n]}{\Delta t} + 7y[n] = 0 \qquad \textbf{(I.102)}$$

or

$$2(y[n+1] - y[n]) + 7\,\Delta t\,y[n] = 0 \qquad \textbf{(I.103)}$$

which is a homogeneous difference equation.

Some authors use the notation $y_n = y(n\Delta t)$ for the $n$th value of y. This is an exact equivalent of $y[n]$.

First differences are analogous to first derivatives. In finite-difference mathematics we can denote a first difference of a discrete-time function $x[n]$ by the use of the operator $\Delta(\ )$ and define that operation by

$$\Delta(x[n]) = x[n+1] - x[n]. \qquad \textbf{(I.104)}$$

This is the first forward difference of $x[n]$. Then, consistent with that definition, a first backward difference of $x[n]$ would be the first forward difference of $x[n-1]$ or

$$\Delta(x[n-1]) = x[n] - x[n-1]. \qquad \textbf{(I.105)}$$

These operations are called *differencing* and are analogous to the operation of differentiation for continuous-time functions. Where it is convenient and unambiguous, we can use a shorthand notation for a difference just like the shorthand notation for a derivative of a continuous-time function

$$\Delta(x[n]) = x'[n]. \qquad \textbf{(I.106)}$$

There is a set of rules for differencing which exactly parallels the analogous rules for differentiation (Table I.1). In case you are asking why the exponential function was left out of the table, the answer is, it wasn't. It is hiding inside the power function. In the formula

$$\Delta(C^n) = C^n(C - 1), \qquad \textbf{(I.107)}$$

$C$ is a constant, possibly complex. Therefore, it could be represented by

$$C = e^\beta, \qquad \textbf{(I.108)}$$

**Table I.1** Rules for differences and derivative

| Function | Difference | Derivative |
|---|---|---|
| Constant $C$ | $\Delta(C) = 0$ | $\dfrac{dC}{dt} = 0$ |
| Constant times a function | $\Delta(Cx[n]) = Cx'[n]$ | $\dfrac{d}{dt}(Cx(t)) = Cx'(t)$ |
| Sum of functions | $\Delta(x[n] + y[n]) = x'[n] + y'[n]$ | $\dfrac{d}{dt}(x(t) + (t)) = x'(t) + y'(t)$ |
| Product of functions | $\Delta(x[n]y[n]) = x[n]y'[n] + y[n+1]x'[n]$ <br> (Notice the $[n+1]$.) | $\dfrac{d}{dt}(x(t)y(t)) = x(t)y'(t) + x'(t)y(t)$ |
| Quotient of functions | $\Delta\left(\dfrac{x[n]}{y[n]}\right) = \dfrac{y[n]x'[n] - x[n]y'[n]}{y[n+1]y[n]}$ <br> (Notice the $[n+1]$.) | $\dfrac{d}{dt}\left(\dfrac{x(t)}{y(t)}\right) = \dfrac{y(t)x'(t) - x(t)y'(t)}{y^2(t)}$ |
| Power function | $\Delta(C^n) = C^n(C - 1)$ | $\dfrac{d}{dt}(t^n) = nt^{n-1}$ |
| Cosine | $\Delta(\cos(n)) = -2\,\sin\left(\dfrac{1}{2}\right)\sin\left(n + \dfrac{1}{2}\right)$ | $\dfrac{d}{dt}(\cos(t)) = -\sin(t)$ |
| Sine | $\Delta(\sin(n)) = 2\,\sin\left(\dfrac{1}{2}\right)\cos\left(n + \dfrac{1}{2}\right)$ | $\dfrac{d}{dt}(\sin(t)) = \cos(t)$ |

where $\beta$ is an appropriately chosen constant, also possibly complex. Then the power function difference becomes

$$\Delta((e^\beta)^n) = \Delta(e^{\beta n}) = e^{\beta n}(e^\beta - 1). \tag{I.109}$$

Equation (I.103) is a finite-difference approximation to (I.93). Written with the new shorthand notation, (I.103) becomes

$$2y'[n] + 7\,\Delta t\,y[n] = 0. \tag{I.110}$$

Notice the similarity of (I.110) to the first-order differential equation it approximates,

$$2y'(t) + 7y(t) = 0. \tag{I.111}$$

To solve (I.110) we can rewrite it in recursion form,

$$y[n+1] = \frac{2 - 7\,\Delta t}{2}y[n]. \tag{I.112}$$

In recursion form, the difference equation expresses the next value of y, $y[n+1]$, in terms of the present value of y, $y[n]$. In words, the $[n+1]$th value of y is a multiple of the $n$th value of y, for any $n$. Equation (I.112) can be turned around to form

$$y[n] = \frac{2}{2 - 7\,\Delta t}y[n+1], \tag{I.113}$$

which expresses the present value of y in terms of the next value of y. Also, from (I.112) we can write

$$y[n] = \frac{2 - 7\,\Delta t}{2}y[n-1], \tag{I.114}$$

in which the present value is written in terms of the immediate past value. Therefore, by using (I.113) and (I.114) if we know any particular value in the sequence of y[$n$] values, we can find all the rest and we can express the entire sequence in terms of any single value in the sequence. For example, if we know y[0],

$$\vdots$$

$$y[-1] = \left(\frac{2 - 7\,\Delta t}{2}\right)^{-1} y[0]$$

$$y[0] = \left(\frac{2 - 7\,\Delta t}{2}\right)^{0} y[0]$$

$$y[1] = \left(\frac{2 - 7\,\Delta t}{2}\right)^{1} y[0] \tag{I.115}$$

$$y[2] = \left(\frac{2 - 7\,\Delta t}{2}\right)^{2} y[0]$$

or, more succinctly,

$$\vdots$$

$$y[n] = y[0] \left(\frac{2 - 7\,\Delta t}{2}\right)^{n} \tag{I.116}$$

implying that we need the value of y[0] to exactly determine the sequence values. Of course, any particular value of y is sufficient, so a more general form of the solution is

$$y[n] = K \left(\frac{2 - 7\Delta t}{2}\right)^{n}. \tag{I.117}$$

Suppose we know that y[0] $= 2$ as in the differential equation example. Then the exact solution is determined just as it was in the case of the solution of differential equations.

The nature of the solution of the difference equation and how closely it approximates the exact solution of the differential equation depends on the choice of $\Delta t$. For a good approximation, $\Delta t$ should be small compared to the time constant of the system which, in this case, is $\frac{2}{7}$ s. Figure I.2 illustrates the effect of different choices of $\Delta t$ by graphing the solution of the differential equation and the solution of the difference equation approximation to it for four different choices of $\Delta t$. Notice that when $\Delta t = 0.5$, y[$n$] is not only inaccurate, it is totally different in character from y($t$). It alternates sign with $n$ because the quantity $(2 - 7\,\Delta t)/2$ is negative. At a value of $\Delta t = \frac{2}{7}$, which is the time constant of the solution of the original differential equation, the quantity $(2 - 7\,\Delta t)/2$ is zero. For $\Delta t$ less than $\frac{2}{7}$ the sign of y[$n$] no longer alternates with $n$, but if the quantity $(2 - 7\,\Delta t)/2$ is near zero, the solution y[$n$] is still very inaccurate. As $\Delta t$ is made smaller, the solution y[$n$] approaches samples of the exact solution y($t$) as can be seen in the bottom two plots.

MATLAB has a function for finding differences, `diff`. It operates on a vector of length $N$ and returns a vector of forward differences of length $N - 1$. For example,

```
»diff([4 1 -9 3 -4 8])

ans =

 -3 -10 12 -7 12
```

The MATLAB code

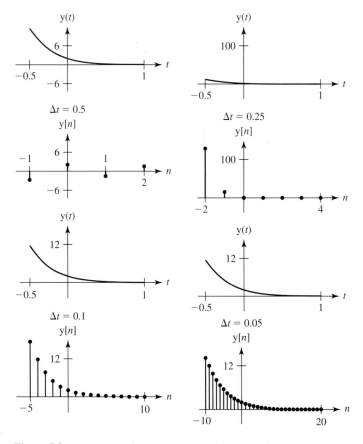

**Figure I.2**

Solution of the first-order, linear, constant-coefficient, ordinary difference equation approximating a differential equation for four different choices of $\Delta t$.

```
dt = 1/16 ; N = 16 ; n = 0:N ; x = sin(2*pi*n*dt) ;
subplot(2,1,1) ;
p = stem(n,x,'k','filled') ;
set(p,'LineWidth',2,'MarkerSize',4) ;
xlabel('n') ; ylabel('x[n]') ; axis([0 16,-1,1]) ;
subplot(2,1,2) ; nd = 0:N-1 ;
p = stem(nd,diff(x),'k','filled') ;
set(p,'LineWidth',2,'MarkerSize',4) ;
xlabel('n') ; ylabel('\Delta(x[n])') ; axis([0 16,-1,1]) ;
```

produces the graphs in Figure I.3.

## Homogeneous Linear Constant-Coefficient Difference Equations

The previous example was introduced as a way of approximating the solution of a differential equation. Methods like this are used in numerical analysis for exactly that purpose. However, it is important to point out that the solution of difference equations is more than just a way of approximating the solution of differential equations. There are systems which are inherently discrete-time and are not described by differential

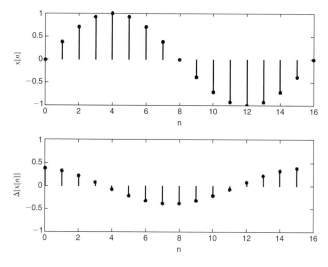

**Figure I.3**
A DT function and its first forward difference.

equations. In those situations the solution of the difference equation is the exact solution because the system is inherently discrete-time.

One classical example of an inherently discrete-time system is a financial system in which interest on an investment is accrued at discrete times. Suppose there is an initial investment of $P$ dollars (the principal) and it earns interest at an annual percentage interest rate $r$ compounded annually. Then $\Delta t$ is one year and $n$ is the number of the present year. Let the beginning of year zero be the time at which the money was invested, and let A$[n]$ be the amount of money in the account at the beginning of the $n$th year. Then the difference equation describing this DT system is

$$\mathrm{A}[n] = \left(1 + \frac{r}{100}\right)\mathrm{A}[n-1] \tag{I.118}$$

or

$$\mathrm{A}[n] - \left(1 + \frac{r}{100}\right)\mathrm{A}[n-1] = 0 \tag{I.119}$$

and the initial condition is A$[0] = P$. This can also be written in the form

$$\mathrm{A}'[n-1] - \frac{r}{100}\mathrm{A}[n-1] = 0, \tag{I.120}$$

which emphasizes its similarity to a first-order homogeneous differential equation. This is a linear constant-coefficient homogeneous difference equation. The solution of (I.119) is of the form

$$\mathrm{A}[n] = K\left(1 + \frac{r}{100}\right)^n, \tag{I.121}$$

and the constant $K$ is obviously $P$ in this case. Therefore, the exact solution of (I.119) is

$$\mathrm{A}[n] = P\left(1 + \frac{r}{100}\right)^n. \tag{I.122}$$

We have just solved a particular linear constant-coefficient homogeneous difference equation. We can generalize to any linear constant-coefficient homogeneous difference equation. Just as the exponential function of the form $Ae^{\lambda t}$, where $A$ and $\lambda$ are constants (possibly complex), is the eigenfunction for linear constant-coefficient homogeneous differential equations of the form

$$y^{(n)}(t) + a_{n-1}y^{(n-1)}(t) + \cdots + a_2y''(t) + a_1y'(t) + a_0y(t) = 0, \qquad \textbf{(I.123)}$$

the eigenfunction for linear constant-coefficient homogeneous difference equations of the form

$$y[n+k] + a_{n+k-1}y[n+k-1] + \cdots + a_2y[n+2] + a_1y[n+1] + a_0y[n] = 0$$

$$\textbf{(I.124)}$$

is a function of the form

$$y[n] = A\alpha^n, \qquad \textbf{(I.125)}$$

where $A$ and $\alpha$ are constants (possibly complex). To illustrate a similarity to the eigenfunctions of differential equations, we can express $\alpha$ as an exponential,

$$\alpha = e^{\beta}, \qquad \textbf{(I.126)}$$

where $\beta = \ln(\alpha)$. Then the form of the eigenfunction is

$$y[n] = A\alpha^n = A(e^{\beta})^n = Ae^{\beta n}, \qquad \textbf{(I.127)}$$

which is very similar to the form of the exponential eigenfunctions for differential equations. The difference is that $t$ can have any real value and $n$ can only have real integer values.

## Inhomogeneous Linear Constant-Coefficient Difference Equations

Most investment is not done as described in the previous example. It is much more common for an investor to begin an investment program with an initial investment and to add to the account with regular contributions. Let's now modify the previous example to include this effect. Let the yearly contribution, made at the end of each year, be $C$. The new difference equation is

$$A[n] = \left(1 + \frac{r}{100}\right)A[n-1] + C. \qquad \textbf{(I.128)}$$

This can be rewritten as

$$A[n] - \left(1 + \frac{r}{100}\right)A[n-1] = C \qquad \textbf{(I.129)}$$

or

$$A'[n-1] - \frac{r}{100}A[n-1] = C. \qquad \textbf{(I.130)}$$

Equation (I.129) is an inhomogeneous linear constant-coefficient difference equation. We already know the solution

$$A_h[n] = K_h\left(1 + \frac{r}{100}\right)^n \qquad \textbf{(I.131)}$$

of the corresponding *homogeneous* difference equation (I.119), which is denoted here with a subscript $h$ to distinguish it from the particular solution. Now we need the particular solution of the difference equation. Since the forcing function is a constant $C$, let's try to find a solution of that form,

$$A_p[n] = K_p. \tag{I.132}$$

Substituting into (I.129) we get

$$K_p - \left(1 + \frac{r}{100}\right)K_p = C. \tag{I.133}$$

Solving for $K_p$,

$$K_p = -100\frac{C}{r}. \tag{I.134}$$

Then the total solution of (I.128) is

$$A[n] = K_h\left(1 + \frac{r}{100}\right)^n - 100\frac{C}{r}. \tag{I.135}$$

Applying the initial condition,

$$A[0] = P = K_h - 100\frac{C}{r} \Rightarrow K_h = P + 100\frac{C}{r} \tag{I.136}$$

and

$$A[n] = \left(P + 100\frac{C}{r}\right)\left(1 + \frac{r}{100}\right)^n - 100\frac{C}{r}. \tag{I.137}$$

To make the example concrete let the parameter values be $P = \$10,000$, $C = \$1000$, and $r = 6$ percent. Figure I.4 shows the accumulation of account value over a 40-year

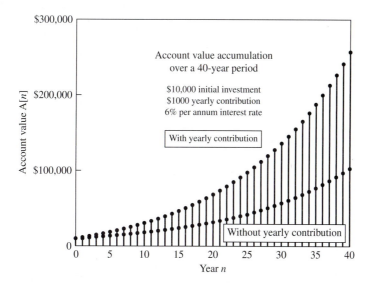

**Figure I.4**
Account value accumulation over a 40-year period with $10,000 initial investment at 6 percent interest compounded annually, with and without $1000 per year contribution.

period. Albert Einstein was once asked by a reporter what the most powerful force in the universe was. Looking at the graph of accumulation of wealth over time in Figure I.4 one can well understand why his answer was "compound interest."

As with differential equations, different forcing functions produce different particular solution forms. Some commonly occurring forcing functions and particular solution forms, where $A$, $K$, $a$, and $k$ are constants, are given in the following table.

| Forcing function | Particular solution form |
|---|---|
| Constant $A$ | Constant $K$ |
| $Aa^n$ | $Ka^n$ |
| $An^k$ | $K_0 + K_1 n + K_2 n^2 + \cdots + K_k n^k$ |
| $\sin(An)$    or    $\cos(An)$ | $K_1 \sin(An) + K_2 \cos(An)$ |
| $n^k a^n$ | $a^n(K_0 + K_1 n + K_2 n^2 + \cdots + K_k n^k)$ |
| $a^n \sin(An)$    or    $a^n \cos(An)$ | $a^n[K_1 \sin(An) + K_2 \cos(An)]$ |

## Systems of Linear Difference Equations

As was true of differential equations, a very common situation in signal and system analysis is the solution of systems of difference equations. Consider the two-difference-equation system

$$3y_1[n] + 2y_1[n-1] + y_2[n] = 0$$
$$4y_2[n] + 2y_2[n-1] + y_1[n] = 5 \tag{I.138}$$

with initial conditions $y_1[0] = 0$ and $y_2[0] = 2$. The functional form of the solutions is the eigenfunction $\alpha^n$. The homogeneous solutions are of the forms

$$y_1[n] = K_{h1}\alpha^n \qquad \text{and} \qquad y_2[n] = K_{h2}\alpha^n. \tag{I.139}$$

Substituting the eigenfunction form into (I.138) and simplifying we get the characteristic equations

$$(3\alpha + 2)K_{h1} + K_{h2}\alpha = 0$$
$$K_{h1}\alpha + (4\alpha + 2)K_{h2} = 0. \tag{I.140}$$

We can combine the two equations in (I.140) to form

$$\frac{\alpha}{3\alpha + 2} = -\frac{K_{h1}}{K_{h2}} = \frac{4\alpha + 2}{\alpha} \tag{I.141}$$

or

$$11\alpha^2 + 14\alpha + 4 = 0. \tag{I.142}$$

Therefore, the two eigenvalues are

$$\alpha_1 = -0.4331 \qquad \text{and} \qquad \alpha_2 = -0.8396. \tag{I.143}$$

The forms of the homogeneous solutions are

$$y_{1h}[n] = K_{h11}\alpha_1^n + K_{h12}\alpha_2^n \qquad \text{and} \qquad y_{2h}[n] = K_{h21}\alpha_1^n + K_{h22}\alpha_2^n \tag{I.144}$$

and, enforcing (I.141),

$$\frac{K_{h11}}{K_{h21}} = 0.618 \quad \text{and} \quad \frac{K_{h12}}{K_{h22}} = -1.618. \tag{I.145}$$

Since the forcing functions are constants, we can assume particular solutions,

$$y_{1p}[n] = K_{p1} \quad \text{and} \quad y_{2p}[n] = K_{p2}. \tag{I.146}$$

Doing that and solving, we get

$$K_{p1} = -0.1724 \quad \text{and} \quad K_{p2} = 0.8621. \tag{I.147}$$

So the total solution is of the form

$$y_1[n] = K_{h11}\alpha_1^n + K_{h12}\alpha_2^n - 0.1724 \quad \text{and} \quad y_2[n] = K_{h21}\alpha_1^n + K_{h22}\alpha_2^n + 0.8621. \tag{I.148}$$

We can now use $y_1[0] = 0$ and $y_2[0] = 2$. Applying the initial conditions we get

$$y_1[0] = K_{h11} + K_{h12} - 0.1724 = 0 \quad \text{and} \quad y_2[0] = K_{h21} + K_{h22} + 0.8621 = 2. \tag{I.149}$$

These two equations together with (I.145) lead to the final numerical solution

$$\begin{aligned} K_{h11} &= 0.557 & K_{h12} &= -0.3841 \\ K_{h21} &= 0.9005 & K_{h22} &= 0.2374, \end{aligned} \tag{I.150}$$

$$y_1[n] = 0.557(-0.4331)^n - 0.3841(-0.8396)^n - 0.1724, \tag{I.151}$$

and

$$y_2[n] = 0.9005(-0.4331)^n + 0.2374(-0.8396)^n + 0.8621. \tag{I.152}$$

Using vectors and matrices there is a more systematic way of solving systems of differential and difference equations (see Appendix J).

## EXERCISES WITH ANSWERS

1. Find the solutions of these differential equations with the boundary conditions indicated.

   *a* $\quad y' = -10y, \quad y(0) = 1$ $\qquad$ *b.* $\quad 3y' - 4y = 0, \quad y(2) = -1$

   *c.* $\quad \dfrac{y'}{2} + y = 0, \quad \dfrac{d}{dt}(y(t))\Big|_{t=0} = 4$

**Answers:**
$y(t) = -0.069e^{(4/3)t}, \quad y(t) = e^{-10t}, \quad y(t) = -2e^{-2t}$

2. Find the solutions of these differential equations with the boundary conditions indicated.

   *a.* $\quad y' + 10y = 5, \quad y(0) = 0$

   *b.* $\quad 3y' - 4y = 10\cos(20\pi t), \quad y(0) = 0$

   *c.* $\quad y'' + 10y' + 100y = e^{-5t}, \quad y(0), = 10, \quad \dfrac{d}{dt}(y(t))\Big|_{t=0} = -1$

    *d.*  $4y'' + 10y' + 8y = \sin(10\pi t)$,   $y(0), = 0$,   $\left.\dfrac{d}{dt}(y(t))\right|_{t=0} = 0$

    *e.*  $y''(t) + 5y'(t) + 10y(t) = 4$,   $y(0) = 1$,   $\left.\dfrac{d}{dt}(y(t))\right|_{t=0} = -3$

**Answers:**

$$y(t) = (10.05 \times 10^{-6} - j0.003013)e^{-\frac{5+j\sqrt{7}}{4}t} + (10.05 \times 10^{-6} + j0.003013)e^{-\frac{5-j\sqrt{7}}{4}t}$$

$$- 0.2522 \times 10^{-3}\sin(10\pi t) - 20.1 \times 10^{-6}\cos(10\pi t), \quad y(t) = \frac{1}{2}(1 - e^{-10t}),$$

$$y(t) = (4.993 - j2.83)e^{((-10+j\sqrt{300})/2)t} + (4.993 + j2.83)e^{((-10-j\sqrt{300})/2)t} + \frac{e^{-5t}}{75},$$

$$y(t) = 0.00113\,e^{(4/3)t} - 0.00113\cos(20\pi t) + 0.053\sin(20\pi t),$$

$$y(t) = e^{-(5/2)t}\left[0.6\cos\left(\frac{\sqrt{15}}{2}t\right) - \frac{3}{\sqrt{15}}\sin\left(\frac{\sqrt{15}}{2}t\right)\right] + 0.4$$

  **3.**  Solve the system of differential equations

$$y_1' + 2y_1 + 8y_2 = 0$$
$$y_2' + y_2 + 5y_1 = -4$$

      with initial conditions $y_1(0) = 3$ and $y_2(0) = -6$.

**Answer:**

$$y_1(t) = -1.844e^{-7.844t} + 5.686e^{4.844t} - 0.8421$$
$$y_2(t) = -1.347e^{-7.844t} - 4.864e^{4.844t} + 0.2105$$

  **4.**  Find the first derivative and then approximations to the first derivative, using the first forward, backward, and central differences of the function $x(t) = e^{-t}$, at time $t = 1$, using $\Delta t = 1, 0.1$ and $0.01$.

**Answers:**

|         | $-0.3697$, | $-0.3684$, | $-0.4323$, | $-0.36788$, | $-0.366$, |
|---------|-----------|-----------|-----------|------------|----------|
| $-0.35$, | $-0.387$, | $-0.36788$, | $-0.632$, | $-0.2325$  |          |

  **5.**  Convert the differential equation with an initial condition

$$4y'(t) + 8y(t) = 0 \qquad y(0) = -10$$

      to a difference equation using backward differences, with $\Delta t = 0.05$; solve the resulting difference equation; and plot a graph of the discrete-time solution $y[n]$ versus $n$ for $0 \le n < 40$. Compare the solution to samples of the solution of the differential equation

$$y(t) = -10e^{-2t}.$$

  **6.**  Repeat Exercise 5 using forward differences instead of backward differences and compare the solutions.

  **7.**  Find the total solution to the difference equation

$$4y[n + 1] + y[n] = 0$$

      with the initial condition $y[0] = -5$.

**Answer:**

$$y[n] = -5\left(-\frac{1}{4}\right)^n$$

8. Find the total solution to the difference equation

$$y[n] + y[n-1] + 2y[n-2] = 0$$

with the initial conditions $y[0] = 5$ and $y[1] = 3$.

**Answer:**

$$y[n] = 6.5(1.414)^n \cos(-0.694 + 1.932n)$$

9. Find the total solution of the difference equation

$$y[n] - 4y[n-1] = e^{-2n}$$

subject to the initial conditions $y[0] = 0$.

**Answer:**

$$y[n] = 0.035[(4)^n - e^{-2n}]$$

10. Find the total solution of the difference equation

$$y[n] + 10y[n-1] = 8$$

subject to the initial condition

$$y[0] = 2.$$

**Answer:**

$$y[n] = \frac{14(-10)^n + 8}{11}$$

11. Find the total solution of the difference equation

$$3y[n] + 8y[n-1] + 4y[n-2] = 2\cos\left(\frac{2\pi n}{7}\right)$$

subject to the initial conditions $y[0] = -1$ and $y[1] = 0$.

**Answer:**

$$y[n] = -2.237\left(-\frac{2}{3}\right)^n + 0.903(-2)^n + 0.1355\sin\left(\frac{2\pi n}{7}\right) + 0.334\cos\left(\frac{2\pi n}{7}\right)$$

12. Find the total solution of the difference equation

$$y[n] + 4y[n-1] + 2y[n-2] = \cos\left(\frac{n}{8}\right)$$

subject to the initial conditions $y[0] = 3$, $y[1] = 0$.

**Answer:**

$$y[n] = 3.385\left(-2 + \sqrt{2}\right)^n - 0.523\left(-2 - \sqrt{2}\right)^n + 0.1651\cos\left(\frac{n}{8}\right) + 0.1412\sin\left(\frac{n}{8}\right)$$

13. Solve the two-difference-equation system

$$5y_1[n] + 2y_1[n-1] - y_2[n-1] = 2$$

$$-6y_2[n] + 3y_2[n-1] + 3y_1[n-1] = 0$$

with initial conditions $y_1[0] = -4$ and $y_2[0] = 1$.

**Answer:**

$$y_1[n] = \frac{45}{54}\alpha_1^n - \frac{279}{54}\alpha_2^n + \frac{1}{3}, \quad y_2[n] = \frac{5}{54}\alpha_1^n + \frac{31}{54}\alpha_2^n + \frac{1}{3}$$

## EXERCISES WITHOUT ANSWERS

14.  After time $t = 0$, a circuit is described by the differential equation

$$Ri(t) + Li'(t) = A,$$

where $i(t)$ is the current through the series combination of a resistor $R$, an inductor $L$, and a voltage source and the voltage source is a constant voltage $A$. Assume that there is initially no stored energy in the circuit. Then the initial current must be zero because of the inductor. That is,

$$i(0^+) = 0.$$

If the resistance $R$ is 10 $\Omega$, the inductance $L$ is 2 H, and $A$ is 10 V, find the total numerical solution $i(t)$ for all time.

15.  A current source $5\sin(20{,}000\,\pi t)$ is suddenly applied at time $t = 0$ to the parallel combination of an inductance of 10 mH, a resistance of 20 $\Omega$, and a capacitance of 5 $\mu$F. There is no energy stored in the circuit before time $t = 0$. Find the total numerical solution for the voltage across the four parallel elements for all time.

16.  Write and solve a differential equation for the voltage $v_C(t)$ in the circuit given in Figure E16 for time $t > 0$, and then find an expression for the current $i(t)$ for time $t > 0$.

17.  Find the first derivative and then approximations to the first derivative, using the first forward, backward, and central differences, of the function $x(t) = \cos(8\pi t)$ at time, $t = \frac{1}{32}$, using $\Delta t = 0.25, 0.1$, and 0.01.

18.  Approximate the differential equation for the voltage $v_C(t)$ found in Exercise 9 with a difference equation, using a $\Delta t$ of one-tenth of the circuit's time constant, and numerically solve for the voltage $v_C(t)$, over a time span of five time constants beginning at time $t = 0$ by iteration using MATLAB. Graph the exact solution $v_C(t)$ and the numerical solution $v_C[n\Delta t]$ on the same time scale for comparison.

19.  A second derivative is a first derivative of a first derivative. Then a second difference should be a first difference of a first difference. Show that a second difference can be written as

$$\Delta^2 x[n] = \Delta(\Delta x[n]) = x[n+2] - 2x[n+1] + x[n]$$

and that a third difference can be written as

$$\Delta^3 x[n] = \Delta(\Delta(\Delta x[n])) = x[n+3] - 3x[n+2] + 3x[n+1] - x[n].$$

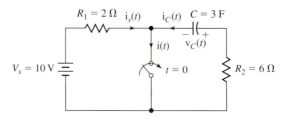

**Figure E16**

20. Mortgage loan payments are usually based on a system of payments in which the debt is retired by making equal monthly payments over the life of the loan. Every month interest accumulates on the unpaid balance. Some of each monthly payment pays interest costs and the rest is applied to the principal. Assign these variables:

$A$ = loan amount

$r$ = annual interest rate in percent

$M$ = total payment at the end of the $n$th month, a constant

$I[n]$ = interest payment at the end of the $n$th month

$P[n]$ = principal payment at the end of the $n$th month

$N$ = total number of months to pay the loan

Write difference equations expressing the relationships between these quantities. Let the time increment be one month. Find formulas for the monthly interest payment, the monthly principal payment, and the overall monthly payment. Then find the monthly payment for a $100,000 loan for 30 years at a 10 percent annual interest rate. (The formula for the summation of a finite geometric series

$$\sum_{n=0}^{N-1} \beta^n = \begin{cases} N & \beta = 1 \\ \dfrac{1 - \beta^N}{1 - \beta} & \beta \neq 1 \end{cases}$$

may prove useful.)

21. A water tank is filled by an inflow $x(t)$ and is emptied by an outflow $y(t)$ The outflow is controlled by a valve which offers resistance $R$ to the flow of water out of the tank. The water height in the tank is $h(t)$, and the surface area of the water is $A$ and is independent of height (cylindrical tank). The outflow is related to the water height (head) by

$$y(t) = \frac{h(t)}{R}.$$

a. Write the differential equation for the water height.

b. If the valve resistance is 10 s/m² and the inflow is 0.05 m³/s, at what water height will the inflow and outflow rates be equal, making the water height constant?

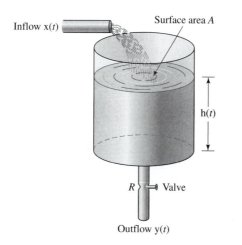

Inflow $x(t)$     Surface area $A$

$h(t)$

$R$ ⊳ ⊨ Valve

Outflow $y(t)$

   *c.* Let the tank's full height be 1.5 m, let the initial water height at time $t = 0$ be 0.5 m, and let the tank diameter be 1 m. If the inflow is a constant 0.2 m³/s, when will the tank start to overflow?

22. Pharmacokinetics is the study of how drugs are absorbed into, distributed through, metabolized by, and excreted from the human body. Some drug processes can be approximately modeled by a one-compartment model of the body in which $V$ is the volume of the compartment, $C_p$ is the drug concentration in that compartment, $k_{el}$ is a rate constant for excretion of the drug from the compartment, and $k_0$ is the infusion rate at which the drug enters the compartment. These can be combined into a single differential equation for the drug concentration as a function of time,

$$V \frac{d}{dt}(C_p(t)) = k_0 - V k_{el} C_p(t).$$

Let the parameter values be $k_{el} = 0.4\,\text{h}^{-1}$, $V = 20$ L, and $k_0 = 100$ mg/h. If the initial drug concentration is $C_p(0) = 10$ mg/L, plot the drug concentration as a function of time for 10 h.

23. At the beginning of the year 2000, the country Freedonia had a population $p$ of 100 million people. The birth rate is 4 percent per annum and the death rate is 2 percent per annum, compounded daily. That is, the births and deaths occur every day at a uniform fraction of the current population, and the next day the number of births and deaths changes because the population changed the previous day. For example, every day the number of people who die is the fraction 0.02/365 of the total population at the end of the previous day (neglect leap-year effects). Every year 100,000 immigrants enter Freedonia, the same number every day.

   *a.* Write a difference equation for the population at the beginning of the *n*th day after January 1, 2000.

   *b.* What will the population of Freedonia be at the beginning of the year 2050?

24. The decay of radioactive substances is governed by the principle that the rate of decay is proportional to the amount of the substance remaining. If the decay rate constant $k$ of radium is $-1.4 \times 10^{-11}\,\text{s}^{-1}$, what is its half-life in years? (Half-life is the time in which half of the original amount of a substance has decayed.)

25. An automobile ignition operates on the general principle of storing energy in an inductor (the spark coil) and then releasing the energy into the spark plug in such a way as to produce a high voltage across the spark gap, causing it to arc and thereby ignite the fuel–air mixture in the combustion chamber. The circuit including the spark coil (transformer) and condensor is first connected to a voltage source through a small resistance so that a current builds up in the primary winding of the coil. Then the circuit is disconnected from the voltage source. Up to the time the spark gap arcs the circuit can be analyzed by linear circuit analysis. Let the circuit be represented by the model given in Figure E25. Write and solve a system of two differential equations for the capacitor voltage and the transformer primary current. If the spark gap arcs at 50 kV, how long after the switching occurs does the arc occur? (Assume the transformer windings are completely coupled, $M = \sqrt{L_p L_s}$.)

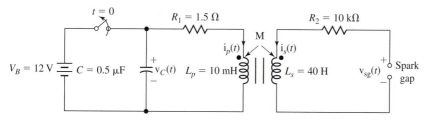

**Figure E25**

26. An aluminum block is heated to a temperature of 100°C. It is then dropped into a flowing stream of water which is held at a constant temperature of 10°C. After 1 min the temperature of the block is 60°C. (Aluminum is such a good heat conductor that its temperature is essentially uniform throughout its volume during the cooling process.) Assuming that the rate of cooling is proportional to the temperature difference between the block and the water, find the time required for the block to reach a temperature of 15°C.

27. A well-stirred vat has been fed for a long time by two streams of liquid, freshwater at 0.2 m³/s and concentrated blue dye at 0.1 m³/s. The vat contains 10 m³ of this mixture, and the mixture is being drawn from the vat at a rate of 0.3 m³/s to maintain the volume. The blue dye is suddenly changed to red dye at the same flow rate. At what time after the switch does the mixture drawn from the vat contain a ratio of red to blue dye of 99:1?

# J APPENDIX

# Vectors and Matrices

## J.1 DEFINITIONS AND OPERATIONS

A very common situation in linear system analysis is finding the solution to multiple simultaneous equations which describe complicated systems. Suppose we have $N$ linear equations in the $N$ unknowns $q_1, q_2, \ldots, q_N$ of the form,

$$
\begin{array}{ccccc}
a_{11}q_1 & a_{12}q_2 & \cdots & a_{1N}q_N & = x_1 \\
a_{21}q_1 & a_{21}q_2 & \cdots & a_{2N}q_N & = x_2 \\
\vdots & \vdots & \ddots & \vdots & \vdots \\
a_{N1}q_1 & a_{N2}q_2 & \cdots & a_{NN}q_N & = x_N
\end{array}
\tag{J.1}
$$

When $N$ becomes large, this formulation of the equations becomes clumsy and finding a solution becomes tedious. The use of vectors and matrices is a way of compacting the notation of systems of equations and leads to very good techniques of systematically solving them. Formulating systems of equations in matrix form allows us to appreciate the "forest," the overall system of equations, without being distracted by all the detailed relationships between the "trees," the individual equations.

A vector is a one-dimensional ordered array of numbers, variables, or functions, given a single name and manipulated as a group. For example, the variables $q_1, q_2, \ldots, q_N$ in (J.1) can be written as a single vector variable as

$$
\mathbf{q} = \begin{bmatrix} q_1 \\ q_2 \\ \vdots \\ q_N \end{bmatrix} \qquad \text{or} \qquad \mathbf{q} = [\, q_1 \quad q_2 \quad \cdots \quad q_N \,].
\tag{J.2}
$$

The first form is called a *column* vector, and the second form is called a *row* vector. The variables (or numbers or functions) $q_1, q_2, \ldots, q_N$ are called the *elements* of the vector. Boldface type distinguishes a vector (and later a matrix) from a scalar variable written in lightface. A vector can be conceived as a location in a space. The number of dimensions of the space is the same as the number of elements in the vector. For example, the vector

$$
\mathbf{q} = \begin{bmatrix} q_1 \\ q_2 \end{bmatrix} = \begin{bmatrix} -2 \\ 3 \end{bmatrix}
\tag{J.3}
$$

identifies a position in a two-dimensional space and can be illustrated graphically as in Figure J.1. A three-dimensional vector can also be represented graphically, but higher-dimensional vectors have no convenient graphical representation, even though they are just as real and useful as two- and three-dimensional vectors. Therefore, an $N$-dimensional vector is often spoken of as a location in an $N$ space (short for an $N$-dimensional space).

A matrix is a two-dimensional array of numbers, variables, or functions given a single name and manipulated as a group. For example, the array of coefficients in (J.1) forms the matrix

$$\mathbf{A} = \begin{bmatrix} a_{11} & a_{12} & \cdots & a_{1N} \\ a_{21} & a_{22} & \cdots & a_{2N} \\ \vdots & \vdots & \ddots & \vdots \\ a_{N1} & a_{N2} & \cdots & a_{NN} \end{bmatrix}. \tag{J.4}$$

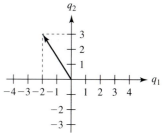

**Figure J.1**
A two-dimensional vector.

This is called an $N \times N$ matrix because it has $N$ rows and $N$ columns. By that same terminology, the vector

$$\mathbf{q} = \begin{bmatrix} q_1 \\ q_2 \\ \vdots \\ q_N \end{bmatrix} \tag{J.5}$$

is an $N \times 1$ matrix and the vector

$$\mathbf{q} = [\, q_1 \quad q_2 \quad \cdots \quad q_N \,] \tag{J.6}$$

is a $1 \times N$ matrix. In any specification of the size of a matrix, the notation $m \times n$ means a matrix with $m$ rows and $n$ columns. The notation $m \times n$ is conventionally spoken as "$m$ by $n$." Sometimes an alternate notation $(a_{rc})$ is used to indicate a matrix like $\mathbf{A}$. In any reference to a single element of a matrix in the form $a_{rc}$, the first subscript $r$ indicates the row and the second subscript $c$ indicates the column of that element.

If a matrix has the same number of rows and columns, it is called a *square matrix* and the number of rows (or columns) is called its *order*. A vector or matrix with only one element is a scalar. The diagonal of a square matrix $\mathbf{A}$ of order $N$ containing the elements $a_{11}, a_{22}, \ldots, a_{NN}$ is called the *principal diagonal*. If all the elements of the matrix which are not on the principal diagonal are zero, the matrix is called a *diagonal matrix,* for example,

$$\mathbf{A} = \begin{bmatrix} a_{11} & 0 & \cdots & 0 \\ 0 & a_{22} & \cdots & 0 \\ \vdots & \vdots & \ddots & \vdots \\ 0 & 0 & \cdots & a_{NN} \end{bmatrix}. \tag{J.7}$$

If, in addition, the elements on the principal diagonal are all one, the matrix is called a *unit matrix* or an *identity matrix* and is usually indicated by the symbol $\mathbf{I}$,

$$\mathbf{I} = \begin{bmatrix} 1 & 0 & \cdots & 0 \\ 0 & 1 & \cdots & 0 \\ \vdots & \vdots & \ddots & \vdots \\ 0 & 0 & \cdots & 1 \end{bmatrix}. \tag{J.8}$$

A matrix whose elements are all zero is called a *zero matrix* or a *null matrix*.

Two matrices are equal if, and only if, every element of one matrix is equal to the corresponding element in the same row and column in the other. Let two matrices **A** and **B** be defined by

$$
\mathbf{A} = \begin{bmatrix} a_{11} & a_{12} & \cdots & a_{1n} \\ a_{21} & a_{22} & \cdots & a_{2n} \\ \vdots & \vdots & \ddots & \vdots \\ a_{m1} & a_{m2} & \cdots & a_{mn} \end{bmatrix} \quad \text{and} \quad \mathbf{B} = \begin{bmatrix} b_{11} & b_{12} & \cdots & b_{1n} \\ b_{21} & b_{22} & \cdots & b_{2n} \\ \vdots & \vdots & \ddots & \vdots \\ b_{m1} & b_{m2} & \cdots & b_{mn} \end{bmatrix}. \quad \textbf{(J.9)}
$$

Then if $\mathbf{A} = \mathbf{B}$, that implies that $a_{rc} = b_{rc}$ for any $r$ and $c$. Obviously the two matrices must have the same number of rows and columns to be equal. The sum of these two matrices is defined by

$$
\mathbf{A} + \mathbf{B} = \begin{bmatrix} a_{11} + b_{11} & a_{12} + b_{12} & \cdots & a_{1n} + b_{1n} \\ a_{21} + b_{21} & a_{22} + b_{22} & \cdots & a_{2n} + b_{2n} \\ \vdots & \vdots & \ddots & \vdots \\ a_{m1} + b_{m1} & a_{m2} + b_{m2} & \cdots & a_{mn} + b_{mn} \end{bmatrix} = (a_{rc} + b_{rc}). \quad \textbf{(J.10)}
$$

In words, each element in the sum of two matrices is the sum of the corresponding elements in the two matrices being added. Subtraction follows directly from addition,

$$
\mathbf{A} - \mathbf{B} = \begin{bmatrix} a_{11} - b_{11} & a_{12} - b_{12} & \cdots & a_{1n} - b_{1n} \\ a_{21} - b_{21} & a_{22} - b_{22} & \cdots & a_{2n} - b_{2n} \\ \vdots & \vdots & \ddots & \vdots \\ a_{m1} - b_{m1} & a_{m2} - b_{m2} & \cdots & a_{mn} - b_{mn} \end{bmatrix} = (a_{rc} - b_{rc}). \quad \textbf{(J.11)}
$$

A matrix can be multiplied by a scalar. Let

$$
\mathbf{A} = \begin{bmatrix} a_{11} & a_{12} & \cdots & a_{1n} \\ a_{21} & a_{22} & \cdots & a_{2n} \\ \vdots & \vdots & \ddots & \vdots \\ a_{m1} & a_{m2} & \cdots & a_{mn} \end{bmatrix} \quad \textbf{(J.12)}
$$

represent any $m \times n$ matrix and let $c$ represent any scalar. Then

$$
c\mathbf{A} = \mathbf{A}c = \begin{bmatrix} ca_{11} & ca_{12} & \cdots & ca_{1n} \\ ca_{21} & ca_{22} & \cdots & ca_{2n} \\ \vdots & \vdots & \ddots & \vdots \\ ca_{m1} & ca_{m2} & \cdots & ca_{mn} \end{bmatrix}. \quad \textbf{(J.13)}
$$

In the special case of $c = -1$, $c\mathbf{A} = \mathbf{A}c = (-1)\mathbf{A} = -\mathbf{A}$, and $\mathbf{B} + (-\mathbf{A}) = \mathbf{B} - \mathbf{A}$. Using the laws of addition, subtraction, and multiplication by a scalar it is easy to show that

$$
\mathbf{A} + \mathbf{B} = \mathbf{B} + \mathbf{A} \quad \textbf{(J.14)}
$$

$$
\mathbf{A} + (\mathbf{B} + \mathbf{C}) = (\mathbf{A} + \mathbf{B}) + \mathbf{C} \quad \textbf{(J.15)}
$$

$$
c(\mathbf{A} + \mathbf{B}) = c\mathbf{A} + c\mathbf{B} \quad \textbf{(J.16)}
$$

$$
(a + b)\mathbf{A} = a\mathbf{A} + b\mathbf{A} \quad \textbf{(J.17)}
$$

and

$$
a(b\mathbf{A}) = (ab)\mathbf{A}. \quad \textbf{(J.18)}
$$

An operation that comes up often in matrix manipulation is the *transpose*. The transpose of a matrix $\mathbf{A}$ is indicated by the notation $\mathbf{A}^T$. If

$$\mathbf{A} = \begin{bmatrix} a_{11} & a_{12} & \cdots & a_{1n} \\ a_{21} & a_{22} & \cdots & a_{2n} \\ \vdots & \vdots & \ddots & \vdots \\ a_{m1} & a_{m2} & \cdots & a_{mn} \end{bmatrix}, \tag{J.19}$$

then

$$\mathbf{A}^T = \begin{bmatrix} a_{11} & a_{21} & \cdots & a_{m1} \\ a_{12} & a_{22} & \cdots & a_{m2} \\ \vdots & \vdots & \ddots & \vdots \\ a_{1n} & a_{2n} & \cdots & a_{mn} \end{bmatrix}. \tag{J.20}$$

In words, the transpose is the matrix formed by making rows into columns and columns into rows. If $\mathbf{A}$ is an $m \times n$ matrix, then $\mathbf{A}^T$ is an $n \times m$ matrix.

The real power of matrix methods in linear system analysis comes in the use of the product of two matrices. The product of two matrices is defined in a way that allows us to represent a system of multiple equations as one matrix equation. If a matrix $\mathbf{A}$ is $m \times n$ and another matrix $\mathbf{B}$ is $n \times p$, then the matrix product $\mathbf{C} = \mathbf{AB}$ is an $m \times p$ matrix whose elements are given by

$$c_{rc} = a_{r1}b_{1c} + a_{r2}b_{2c} + \cdots + a_{rn}b_{nc} = \sum_{k=1}^{n} a_{rk}b_{kc}. \tag{J.21}$$

The product $\mathbf{AB}$ can be described as $\mathbf{A}$ *postmultiplied* by $\mathbf{B}$ or as $\mathbf{B}$ *premultiplied* by $\mathbf{A}$. To be able to multiply two matrices the number of columns in the premultiplying matrix ($\mathbf{A}$ in $\mathbf{AB}$) must equal the number of rows in the postmultiplying matrix ($\mathbf{B}$ in $\mathbf{AB}$). The process of computing the $rc$ element of $\mathbf{C}$ can be conceived geometrically as the sum of the products of the elements in the $r$th row of $\mathbf{A}$ with the corresponding elements in the $c$th column of $\mathbf{B}$. Consider the simple case of the premultiplication of a column vector by a row vector. Let

$$\mathbf{A} = [\, a_1 \quad a_2 \quad \cdots \quad a_N \,] \qquad \text{and} \qquad \mathbf{B} = \begin{bmatrix} b_1 \\ b_2 \\ \vdots \\ b_N \end{bmatrix}. \tag{J.22}$$

Then

$$\mathbf{AB} = [\, a_1 \quad a_2 \quad \cdots \quad a_N \,] \begin{bmatrix} b_1 \\ b_2 \\ \vdots \\ b_N \end{bmatrix} = a_1 b_1 + a_2 b_2 + \cdots + a_N b_N. \tag{J.23}$$

$\mathbf{A}$ is $1 \times N$ and $\mathbf{B}$ is $N \times 1$, so the product is $1 \times 1$, a scalar. This special case in which a row vector premultiplies a column vector of the same length is called a *scalar product* because the product is a scalar. For contrast consider the product **BA**,

$$\mathbf{BA} = \begin{bmatrix} b_1 \\ b_2 \\ \vdots \\ b_N \end{bmatrix} [\, a_1 \quad a_2 \quad \cdots \quad a_N \,] = \begin{bmatrix} b_1 a_1 & b_1 a_2 & \cdots & b_1 a_N \\ b_2 a_1 & b_2 a_2 & \cdots & b_2 a_N \\ \vdots & \vdots & \ddots & \vdots \\ b_N a_1 & b_N a_2 & \cdots & b_N a_N \end{bmatrix}. \tag{J.24}$$

This result is very different, an $N \times N$ matrix. Obviously matrix multiplication is not (generally) commutative. If **A** and **B** are more general than simply vectors, the process of matrix multiplication can be broken down into multiple scalar products. In the matrix product $\mathbf{C} = \mathbf{AB}$, the $rc$ element of **C** is simply the scalar product of the $r$th row of **A** with the $c$th column of **B**. Even though matrix multiplication is not commutative, it is associative and distributive,

$$(\mathbf{AB})\mathbf{C} = \mathbf{A}(\mathbf{BC}), \tag{J.25}$$

$$\mathbf{A}(\mathbf{B} + \mathbf{C}) = \mathbf{AB} + \mathbf{AC}, \tag{J.26}$$

and

$$(\mathbf{A} + \mathbf{B})\mathbf{C} = \mathbf{AC} + \mathbf{BC}. \tag{J.27}$$

The product of any matrix **A** and the identity matrix **I** (in either order) is the matrix **A**.

$$\mathbf{AI} = \mathbf{IA} = \mathbf{A}. \tag{J.28}$$

One other useful multiplication rule is that

$$(\mathbf{AB})^T = \mathbf{B}^T \mathbf{A}^T. \tag{J.29}$$

In words, the transpose of a product of two matrices in a given order is the product of the transposes of those matrices in the reverse of that order.

Using the rules of matrix multiplication, a system of $N$ linear equations in $N$ unknowns can be written in matrix form as the single matrix equation

$$\mathbf{Aq} = \mathbf{x}, \tag{J.30}$$

where

$$\mathbf{A} = \begin{bmatrix} a_{11} & a_{12} & \cdots & a_{1N} \\ a_{21} & a_{22} & \cdots & a_{2N} \\ \vdots & \vdots & \ddots & \vdots \\ a_{N1} & a_{N2} & \cdots & a_{NN} \end{bmatrix} \quad \mathbf{q} = \begin{bmatrix} q_1 \\ q_2 \\ \vdots \\ q_N \end{bmatrix} \quad \mathbf{x} = \begin{bmatrix} x_1 \\ x_2 \\ \vdots \\ x_N \end{bmatrix}. \tag{J.31}$$

MATLAB has a rich complement of commands and functions to handle almost any vector or matrix manipulation. The standard arithmetic operators +, −, *, and ^ are all defined for both scalar and vector operands. That is, the multiplication operator * performs a true matrix multiplication and the power operator ^ performs a true matrix power operation. These can be modified to perform an array operation instead of a matrix operation by the addition of a period before the operator, as in the .* and .^ operators. In these forms the operators simply operate on corresponding pairs of elements in the two matrix operands. (See the MATLAB tutorial in Appendix B for more detail.) In addition, the ' operator transposes a matrix. (If the matrix elements are complex, this operator also *conjugates* the elements as it transposes the matrix.) The .' operator transposes without conjugating. Some other useful commands and functions are

| | |
|---|---|
| `zeros` | Zeros array |
| `ones` | Ones array |
| `rand` | Uniformly distributed random numbers |
| `randn` | Normally distributed random numbers |
| `linspace` | Linearly spaced vector |
| `logspace` | Logarithmically spaced vector |
| `size` | Size of matrix |
| `length` | Length of vector |
| `find` | Find indices of nonzero elements |

For example, let two matrices A and B be defined by

```
»A = round(4*randn(3,3))

A =

    -2     1     5
    -7    -5     0
     1     5     1

»B = round(4*randn(3,3))

B =

     1    -2     0
    -1     9     4
     3    -1     0
```

Then

```
»A+B

ans =

    -1    -1     5
    -8     4     4
     4     4     1

»A-B

ans =

    -3     3     5
    -6   -14    -4
    -2     6     1

»A*B

ans =

    12     8     4
    -2   -31   -20
    -1    42    20

»A^2
```

```
ans =

        2      18      -5
       49      18     -35
      -36     -19       6

»A.*B

ans =

       -2      -2       0
        7     -45       0
        3      -5       0

»A.^2

ans =

        4       1      25
       49      25       0
        1      25       1

»A'

ans =

       -2      -7       1
        1      -5       5
        5       0       1
```

## J.2  DETERMINANTS, CRAMER'S RULE, AND THE MATRIX INVERSE

Matrices and related concepts can be used to systematically solve systems of linear equations. Consider first a system of two equations and two unknowns,

$$a_{11}q_1 + a_{12}q_2 = x_1$$
$$a_{21}q_1 + a_{22}q_2 = x_2$$
or  $\mathbf{Aq} = \mathbf{x}.$   **(J.32)**

Using nonmatrix methods we can solve for the unknowns by any convenient method, and the answers are found to be

$$q_1 = \frac{a_{22}x_1 - a_{12}x_2}{a_{11}a_{22} - a_{21}a_{12}} \qquad \text{and} \qquad q_2 = \frac{a_{11}x_2 - a_{21}x_1}{a_{11}a_{22} - a_{21}a_{12}}. \qquad \textbf{(J.33)}$$

Notice that the two denominators in (J.33) are the same. This denominator which is common to both solutions is called the *determinant* of this system of equations and is conventionally indicated by the notation

$$\Delta_{\mathbf{A}} = \begin{vmatrix} a_{11} & a_{12} \\ a_{21} & a_{22} \end{vmatrix} = |\mathbf{A}| . \qquad \textbf{(J.34)}$$

Thus, the notations $\Delta_{\mathbf{A}}$ and $|\mathbf{A}|$ mean the same thing and can be used interchangeably. The determinant of a $2 \times 2$ system of equations written in the standard form of (J.34) is a scalar found by forming the product of the elements with subscripts 11 and 22 in the matrix of coefficients $\mathbf{A}$ and subtracting from that the product of the elements with subscripts 21 and 12. The numerators in (J.33) can be interpreted as determinants also,

$$\Delta_1 = x_1 a_{22} - x_2 a_{12} = \begin{vmatrix} x_1 & a_{12} \\ x_2 & a_{22} \end{vmatrix} \quad \text{and} \quad \Delta_2 = a_{11} x_2 - a_{21} x_1 = \begin{vmatrix} a_{11} & x_1 \\ a_{21} & x_2 \end{vmatrix}.$$

$$(\textbf{J.35})$$

Then, using this notation, the solutions of the two linear equations can be written in a very compact form,

$$q_1 = \frac{\Delta_1}{\Delta_A} \quad \text{and} \quad q_2 = \frac{\Delta_2}{\Delta_A} \qquad \Delta_A \neq 0. \qquad (\textbf{J.36})$$

As indicated in (J.36), these solutions only exist if the determinant is not zero. If the determinant is zero, that is an indication that the equations are not independent.

In preparation for extending this technique to larger systems of equations we will formalize and generalize the process of finding a determinant by defining the terms minor and cofactor. In any square matrix, the *minor* of any element of the matrix is defined as the determinant of the matrix found by eliminating all the elements in the same row and the same column as the element in question. For example, the minor of element $a_{11}$ in $\mathbf{A}$ is the determinant of the single-element matrix $a_{22}$, a scalar. The determinant of a scalar is just the scalar itself. The *cofactor* of any particular element $a_{rc}$ of a matrix is the product of the minor of that element and the factor $(-1)^{r+c}$. So the cofactor of the element $a_{12}$ is $-a_{21}$. The determinant $\Delta$ can be found by choosing any row or column of the matrix and, for each element in that row or column, forming the product of the element and the determinant of its cofactor and then adding all such products for that row or column. For example, expanding along the second column of $\mathbf{A}$ we would calculate the determinant to be

$$\Delta_A = a_{12}(-a_{21}) + a_{22}a_{11}, \qquad (\textbf{J.37})$$

which is the same determinant we got before. Expanding along the bottom row,

$$\Delta_A = a_{21}(-a_{12}) + a_{22}a_{11}. \qquad (\textbf{J.38})$$

This general procedure can be extended to larger systems of equations. Applying this technique to a $3 \times 3$ system,

$$\begin{aligned} a_{11}q_1 + a_{12}q_2 + a_{13}q_3 &= x_1 \\ a_{21}q_1 + a_{22}q_2 + a_{23}q_3 &= x_2 \qquad \text{or} \qquad \mathbf{Aq} = \mathbf{x} \\ a_{31}q_1 + a_{32}q_2 + a_{33}q_3 &= x_3 \end{aligned} \qquad (\textbf{J.39})$$

$$q_1 = \frac{\Delta_1}{\Delta_A} \qquad q_2 = \frac{\Delta_2}{\Delta_A} \qquad q_3 = \frac{\Delta_3}{\Delta_A}, \qquad (\textbf{J.40})$$

where one way (among many equivalent ways) of expressing the determinant is

$$\Delta_A = |\mathbf{A}| = \begin{vmatrix} a_{11} & a_{12} & a_{13} \\ a_{21} & a_{22} & a_{23} \\ a_{31} & a_{32} & a_{33} \end{vmatrix} = a_{11} \begin{vmatrix} a_{22} & a_{23} \\ a_{32} & a_{33} \end{vmatrix} - a_{12} \begin{vmatrix} a_{21} & a_{23} \\ a_{31} & a_{33} \end{vmatrix} + a_{13} \begin{vmatrix} a_{21} & a_{22} \\ a_{31} & a_{32} \end{vmatrix}$$

$$(\textbf{J.41})$$

and

$$\Delta_1 = \begin{vmatrix} x_1 & a_{12} & a_{13} \\ x_2 & a_{22} & a_{23} \\ x_3 & a_{32} & a_{33} \end{vmatrix} \qquad \Delta_2 = \begin{vmatrix} a_{11} & x_1 & a_{13} \\ a_{21} & x_2 & a_{23} \\ a_{31} & x_3 & a_{33} \end{vmatrix} \qquad \Delta_3 = \begin{vmatrix} a_{11} & a_{12} & x_1 \\ a_{21} & a_{22} & x_2 \\ a_{31} & a_{32} & x_3 \end{vmatrix}. \qquad (\textbf{J.42})$$

This method of finding solutions to systems of linear equations is called *Cramer's rule*. It is very handy, especially in symbolic solutions of systems of equations. In the actual numerical solution of systems of equations on a computer, Cramer's rule is considerably less efficient than other techniques, like gaussian elimination for example.

Here are some other properties of determinants that are sometimes useful in vector and matrix analysis of signals and systems:

1.  If any two rows or columns of a matrix are exchanged, the determinant changes sign (but not magnitude).

$$
\begin{vmatrix}
\vdots & \vdots & \vdots & \cdots & \vdots \\
a_{k1} & a_{k2} & a_{k3} & \cdots & a_{kn} \\
\vdots & \vdots & \vdots & \cdots & \vdots \\
a_{q1} & a_{q2} & a_{q3} & \cdots & a_{qn} \\
\vdots & \vdots & \vdots & \cdots & \vdots
\end{vmatrix}
= -
\begin{vmatrix}
\vdots & \vdots & \vdots & \cdots & \vdots \\
a_{q1} & a_{q2} & a_{q3} & \cdots & a_{qn} \\
\vdots & \vdots & \vdots & \cdots & \vdots \\
a_{k1} & a_{k2} & a_{k3} & \cdots & a_{kn} \\
\vdots & \vdots & \vdots & \cdots & \vdots
\end{vmatrix}
\tag{J.43}
$$

2.  The determinant of the identity matrix is one.

$$
\begin{vmatrix}
1 & 0 & \cdots & 0 \\
0 & 1 & \cdots & 0 \\
\vdots & \vdots & \ddots & \vdots \\
0 & 0 & \cdots & 1
\end{vmatrix}
= 1
\tag{J.44}
$$

3.  If any two rows or two columns of a matrix are equal, the determinant is zero.

$$
\begin{vmatrix}
\vdots & \vdots & \vdots & \cdots & \vdots \\
a_{k1} & a_{k2} & a_{k3} & \cdots & a_{kn} \\
\vdots & \vdots & \vdots & \cdots & \vdots \\
a_{k1} & a_{k2} & a_{k3} & \cdots & a_{kn} \\
\vdots & \vdots & \vdots & \cdots & \vdots
\end{vmatrix}
= 0
\tag{J.45}
$$

4.  A matrix with a row or column of zeros has a determinant of zero.

$$
\begin{vmatrix}
\vdots & 0 & \cdots & \vdots \\
\vdots & 0 & \cdots & \vdots \\
\vdots & \vdots & \vdots & \vdots \\
\vdots & 0 & \vdots & \vdots
\end{vmatrix}
= 0
\tag{J.46}
$$

5.  The determinant of the product of two matrices is the product of the determinants.

$$
|\mathbf{AB}| = |\mathbf{A}||\mathbf{B}|
\tag{J.47}
$$

6.  Transposing a matrix does not change its determinant.

$$
|\mathbf{A}| = |\mathbf{A}^T|
\tag{J.48}
$$

One other important operation is the inverse of a matrix. The inverse of a matrix $\mathbf{A}$ is defined as the matrix $\mathbf{A}^{-1}$, which when premultiplied or postmultiplied by $\mathbf{A}$ yields the identity matrix

$$\mathbf{A}\mathbf{A}^{-1} = \mathbf{A}^{-1}\mathbf{A} = \mathbf{I}. \tag{J.49}$$

The inverse of a matrix can be found by multiple methods. One formula for the inverse of a matrix is

$$\mathbf{A}^{-1} = \frac{\begin{bmatrix} A_{11} & A_{12} & \cdots & A_{1N} \\ A_{21} & A_{22} & \cdots & A_{2N} \\ \vdots & \vdots & \ddots & \vdots \\ A_{N1} & A_{N2} & \cdots & A_{NN} \end{bmatrix}^{T}}{|\mathbf{A}|}, \tag{J.50}$$

where $A_{rc}$ is the cofactor of the element $a_{rc}$ in the matrix $\mathbf{A}$. In words, the inverse of a matrix is the transpose of the matrix of cofactors of the elements of $\mathbf{A}$, divided by the determinant of $\mathbf{A}$. A term that is useful here is the *adjoint* of a matrix. The adjoint of $\mathbf{A}$ (adj $\mathbf{A}$) is the transpose of the matrix of cofactors. Therefore, the inverse of a matrix is the adjoint of the matrix, divided by the determinant of the matrix,

$$\mathbf{A}^{-1} = \frac{\text{adj } \mathbf{A}}{|\mathbf{A}|}. \tag{J.51}$$

Of course, if the determinant is zero, the inverse of $\mathbf{A}$ is undefined. In that case it does not have an inverse. One use of the inverse of a matrix can be seen in the solution of a system of equations written in matrix form as

$$\mathbf{A}\mathbf{q} = \mathbf{x}. \tag{J.52}$$

If we premultiply both sides of this matrix equation by $\mathbf{A}^{-1}$, we get

$$\mathbf{A}^{-1}(\mathbf{A}\mathbf{q}) = (\mathbf{A}^{-1}\mathbf{A})\mathbf{q} = \mathbf{I}\mathbf{q} = \mathbf{q} = \mathbf{A}^{-1}\mathbf{x}. \tag{J.53}$$

So a straightforward way of solving a matrix equation of the form $\mathbf{A}\mathbf{q} = \mathbf{x}$ is to premultiply both sides by the inverse of $\mathbf{A}$. This directly yields the solution $\mathbf{q} = \mathbf{A}^{-1}\mathbf{x}$, if $|\mathbf{A}| \neq 0$. The determinant of the inverse of a matrix is the reciprocal of the determinant of the matrix.

$$|\mathbf{A}^{-1}| = \frac{1}{|\mathbf{A}|}. \tag{J.54}$$

If we have a scalar equation

$$aq = 0, \tag{J.55}$$

we know that either $a$ or $q$ or both must be zero to satisfy the equation. If $a$ is a constant and $q$ is a variable and we want to find a nonzero value of $q$ that satisfies the equation, $a$ must be zero. A very common situation in matrix analysis of systems of differential or difference equations is a matrix equation of the form

$$\mathbf{A}\mathbf{q} = \mathbf{0}, \tag{J.56}$$

where $\mathbf{A}$ is the $N \times N$ matrix of coefficients of $N$ independent differential or difference equations and $\mathbf{q}$ is an $N \times 1$ vector of variables. This equation has nonzero solutions for **q** only if the determinant of *A* is zero. If **A** is the zero matrix, its determinant is zero and any vector **q** will satisfy the equation. But **A** need not be a zero matrix; it must only have a determinant of zero. If **A** is not the zero matrix and its determinant is zero, then there are only certain particular vectors **q** that can solve the equation. If **q** is a solution to **Aq** = **0**, then for any scalar *c*, *c***q** is also a solution. If **q**$_1$ and **q**$_2$ are both solutions to **Aq** = **0**, then any linear combination of **q**$_1$ and **q**$_2$ is also a solution. These properties will be significant when we come to eigenvalues and eigenvectors.

MATLAB can find the determinant of a matrix and the inverse of a matrix, using the `det` and `inv` functions. For example,

```
»A = round(4*randn(3,3))

A =

      0     -5     -3
     -3      3      3
      1      6      5

»det(A)

ans =

    -27

»inv(A)

ans =

    0.1111    -0.2593     0.2222
   -0.6667    -0.1111    -0.3333
    0.7778     0.1852     0.5556

»inv(A)*A

ans =

    1.0000    -0.0000    -0.0000
         0     1.0000          0
         0    -0.0000     1.0000
```

## J.3 DERIVATIVES AND DIFFERENCES

The derivative of a matrix is simply the matrix of the derivatives of the corresponding elements of the matrix. For example, if

$$\mathbf{A} = \begin{bmatrix} t^2 & yt \\ \sin(t) & e^{5t} \end{bmatrix}, \tag{J.57}$$

then

$$\frac{d}{dt}(\mathbf{A}) = \begin{bmatrix} 2t & y \\ \cos(t) & 5e^{5t} \end{bmatrix}. \tag{J.58}$$

Some common differentiation rules which follow from this definition are

$$\frac{d}{dt}(\mathbf{A} + \mathbf{B}) = \frac{d}{dt}(\mathbf{A}) + \frac{d}{dt}(\mathbf{B}) \tag{J.59}$$

$$\frac{d}{dt}(c\mathbf{A}) = c\frac{d}{dt}(\mathbf{A}) \tag{J.60}$$

$$\frac{d}{dt}(\mathbf{AB}) = \mathbf{A}\frac{d}{dt}(\mathbf{B}) + \mathbf{B}\frac{d}{dt}(\mathbf{A}), \tag{J.61}$$

which are formally just like their scalar counterparts.

In an analogous manner the first forward difference of a matrix is the matrix of the first forward differences of the corresponding elements. If

$$\mathbf{A} = \begin{bmatrix} \alpha^n & e^{-(n/2)} \\ n^2 + 3n & \cos(2\pi n/N_0) \end{bmatrix}, \tag{J.62}$$

then

$$\Delta(\mathbf{A}) = \begin{bmatrix} \alpha^{n+1} - \alpha^n & e^{-(n+1)/2} - e^{-(n/2)} \\ (n+1)^2 + 3(n+1) - (n^2 + 3n) & \cos\left(\frac{2\pi(n+1)}{N_0}\right) - \cos\left(\frac{2\pi n}{N_0}\right) \end{bmatrix}, \tag{J.63}$$

or

$$\Delta(\mathbf{A}) = \begin{bmatrix} \alpha^n(\alpha - 1) & e^{-(n/2)}\left(e^{-(1/2)} - 1\right) \\ 2n + 4 & \cos\left(\frac{2\pi(n+1)}{N_0}\right) - \cos\left(\frac{2\pi n}{N_0}\right) \end{bmatrix}.$$

> Don't confuse $\Delta_\mathbf{A}$, the determinant of the matrix $\mathbf{A}$, with $\Delta(\mathbf{A})$, the first forward difference of the matrix $\mathbf{A}$. The difference is usually clear in context.

## J.4 EIGENVALUES AND EIGENVECTORS

It is always possible to write a system of linear independent constant-coefficient ordinary differential equations as a single matrix equation of the form

$$\mathbf{q}' = \mathbf{Aq} + \mathbf{Bx}. \tag{J.64}$$

where $\mathbf{q}$ = vector of solution functions
$\mathbf{x}$ = vector of forcing functions
$\mathbf{A}, \mathbf{B}$ = coefficient matrices

(This is the form that is used in state-variable analysis.) Thus (J.64) is a linear constant-coefficient ordinary first-order matrix differential equation. Using an example from Appendix I, the system of differential equations

$$\begin{aligned} y_1' + 5y_1 + 2y_2 &= 10 \\ y_2' + 3y_2 + y_1 &= 0 \end{aligned} \tag{J.65}$$

with initial conditions $y_1(0) = 1$ and $y_2(0) = 0$ can be written as $\mathbf{q}' = \mathbf{A}\mathbf{q} + \mathbf{B}\mathbf{x}$, where

$$\mathbf{A} = \begin{bmatrix} -5 & -2 \\ -1 & -3 \end{bmatrix} \qquad \mathbf{q} = \begin{bmatrix} y_1(t) \\ y_2(t) \end{bmatrix} \qquad \mathbf{B} = \begin{bmatrix} 1 \\ 0 \end{bmatrix} \qquad \mathbf{x} = 10 \qquad \textbf{(J.66)}$$

and

$$\mathbf{q}' = \begin{bmatrix} y_1'(t) \\ y_2'(t) \end{bmatrix} = \frac{d}{dt}\begin{bmatrix} y_1(t) \\ y_2(t) \end{bmatrix} = \frac{d}{dt}(\mathbf{q}) \qquad \textbf{(J.67)}$$

with an initial-condition vector $\mathbf{q}_0 = \begin{bmatrix} y_1(0) \\ y_2(0) \end{bmatrix} = \begin{bmatrix} 1 \\ 0 \end{bmatrix}$. The corresponding homogeneous equation is

$$\mathbf{q}' = \mathbf{A}\mathbf{q}, \qquad \textbf{(J.68)}$$

and we know that the solution of the homogeneous equation is a linear combination of solutions of the form

$$\mathbf{q}_h = \mathbf{K}_h e^{\lambda t}, \qquad \textbf{(J.69)}$$

where $\mathbf{K}_h$ is a $2 \times 1$ vector of arbitrary constants instead of the single arbitrary constant we would have if we were solving a first-order scalar differential equation. Therefore, we know that

$$\mathbf{q}_h' = \mathbf{K}_h \lambda e^{\lambda t} = \lambda \mathbf{q} \qquad \textbf{(J.70)}$$

and, equating (J.70) and (J.68), that the solution of the homogeneous system of equations is the solution of the matrix equation

$$\mathbf{A}\mathbf{q} = \lambda \mathbf{q}. \qquad \textbf{(J.71)}$$

This can be rearranged into

$$\mathbf{A}\mathbf{q} - \lambda \mathbf{q} = (\mathbf{A} - \lambda \mathbf{I})\mathbf{q} = 0. \qquad \textbf{(J.72)}$$

For a nontrivial solution, $\mathbf{q} \neq \mathbf{0}$,

$$|\mathbf{A} - \lambda \mathbf{I}| = 0 \qquad \textbf{(J.73)}$$

or

$$\left|\begin{bmatrix} -5 & -2 \\ -1 & -3 \end{bmatrix} - \lambda \begin{bmatrix} 1 & 0 \\ 0 & 1 \end{bmatrix}\right| = \begin{vmatrix} -5 - \lambda & -2 \\ -1 & -3 - \lambda \end{vmatrix} = 0 \qquad \textbf{(J.74)}$$

or

$$(-5 - \lambda)(-3 - \lambda) - 2 = 0, \qquad \textbf{(J.75)}$$

which is equivalent to the system of characteristic equations

$$\begin{aligned} (\lambda + 5)K_{h1} + 2K_{h2} &= 0 \\ K_{h1} + (\lambda + 3)K_{h2} &= 0. \end{aligned} \qquad \textbf{(J.76)}$$

The formulation $|\mathbf{A} - \lambda \mathbf{I}| = 0$ is the *matrix characteristic equation* for a matrix differential or difference equation. From (J.76) we get

$$\lambda^2 + 8\lambda + 13 = 0. \qquad \textbf{(J.77)}$$

The eigenvalues are completely determined by the coefficient matrix $\mathbf{A}$. For each eigenvalue there is a corresponding eigenvector $\mathbf{q}$, which, together with the eigenvalue, solves (J.71). For $\lambda_1$ we have the equality

$$[\mathbf{A} - \lambda_1\mathbf{I}]\mathbf{q} = \mathbf{0}. \tag{J.78}$$

Any row of $\mathbf{A} - \lambda_1\mathbf{I}$ can be used to determine the direction of an eigenvector. For example, using the first row,

$$(-5 - \lambda_1)q_1 - 2q_2 = 0 \tag{J.79}$$

or

$$\frac{q_1}{q_2} = \frac{y_1(t)}{y_2(t)} = \frac{K_{h1}e^{\lambda_1 t}}{K_{h2}e^{\lambda_1 t}} = \frac{K_{h1}}{K_{h2}} = -\frac{2}{5 + \lambda_1}. \tag{J.80}$$

This sets the ratio of the components of the eigenvector and, therefore, its direction (but not any unique magnitude, yet) in the $N$ space, where, in this case, $N = 2$. Therefore, this eigenvector would be

$$\mathbf{q}_1 = K_1 \begin{bmatrix} -\dfrac{2}{5 + \lambda_1} \\ 1 \end{bmatrix} e^{\lambda_1 t} = K_1 \begin{bmatrix} -0.732 \\ 1 \end{bmatrix} e^{-2.268t}, \tag{J.81}$$

where $K_1$ is an arbitrary constant. If we had used the second row, we would have gotten exactly the same vector direction. Using $\lambda_2$ we would get

$$\mathbf{q}_2 = K_2 \begin{bmatrix} -\dfrac{2}{5 + \lambda_2} \\ 1 \end{bmatrix} e^{\lambda_2 t} = K_2 \begin{bmatrix} 2.732 \\ 1 \end{bmatrix} e^{-5.732t}. \tag{J.82}$$

Notice that we have arbitrarily set the $q_2$ component of $\mathbf{q}$ to one in both cases. This is just a convenience. We can do it because we only know the ratio of the two components of $\mathbf{q}$, not their exact values (yet). Since the exact values of the two vector components are not known, only their ratio, they are often chosen so as to make the length of $\mathbf{q}$ equal to one, making $\mathbf{q}$ a unit vector. Writing the vectors as unit vectors, we get

$$\mathbf{q}_1 = \begin{bmatrix} -0.5907 \\ 0.807 \end{bmatrix} e^{-2.268t} \quad \text{and} \quad \mathbf{q}_2 = \begin{bmatrix} 0.9391 \\ 0.3437 \end{bmatrix} e^{-5.732t}. \tag{J.83}$$

The most general homogeneous solution is a linear combination of the eigenvectors of the form

$$\mathbf{q} = K_{h1}\mathbf{q}_1 + K_{h2}\mathbf{q}_2 = [\,\mathbf{q}_1 \quad \mathbf{q}_2\,] \begin{bmatrix} K_{h1} \\ K_{h2} \end{bmatrix}, \tag{J.84}$$

where the two arbitrary constants $K_{h1}$ and $K_{h2}$ must be chosen to satisfy initial conditions.

The next solution step is to find the particular solution $\mathbf{q}_p = \begin{bmatrix} y_{1p}(t) \\ y_{2p}(t) \end{bmatrix}$. Since the forcing function is a constant $\mathbf{x} = 10$, the particular solution is a vector of constants $\mathbf{q}_p = \begin{bmatrix} K_{p1} \\ K_{p2} \end{bmatrix}$. Substituting into (J.64),

$$\mathbf{q}_p' = \mathbf{A}\,\mathbf{q}_p + \mathbf{B}\mathbf{x}. \tag{J.85}$$

Since $\mathbf{q}_p$ is a vector of constants, $\mathbf{q}_p' = 0$ and

$$\mathbf{A}\mathbf{q}_p = -\mathbf{B}\mathbf{x}. \tag{J.86}$$

Solving,

$$\mathbf{q}_p = -\mathbf{A}^{-1}\mathbf{B}\mathbf{x}. \tag{J.87}$$

The inverse of $\mathbf{A}$ is

$$\mathbf{A}^{-1} = \frac{1}{13}\begin{bmatrix} -3 & 2 \\ 1 & -5 \end{bmatrix}. \tag{J.88}$$

Therefore,

$$\mathbf{q}_p = -\frac{1}{13}\begin{bmatrix} -3 & 2 \\ 1 & -5 \end{bmatrix}\begin{bmatrix} 1 \\ 0 \end{bmatrix}10 = -\frac{1}{13}\begin{bmatrix} -3 \\ 1 \end{bmatrix}10 = \frac{1}{13}\begin{bmatrix} 30 \\ -10 \end{bmatrix}. \tag{J.89}$$

Now we know that the total solution is

$$\mathbf{q} = [\mathbf{q}_1 \quad \mathbf{q}_2]\begin{bmatrix} K_{h1} \\ K_{h2} \end{bmatrix} + \mathbf{q}_p \tag{J.90}$$

or

$$\mathbf{q} = \begin{bmatrix} -0.5907e^{-2.268t} & 0.9391e^{-5.732t} \\ 0.807e^{-2.268t} & 0.3437e^{-5.732t} \end{bmatrix}\begin{bmatrix} K_{h1} \\ K_{h2} \end{bmatrix} + \begin{bmatrix} 2.308 \\ -0.769 \end{bmatrix}. \tag{J.91}$$

The only task left is to solve for the arbitrary constants $K_{h1}$ and $K_{h2}$. The vector of initial conditions (at time $t = 0$) is

$$\mathbf{q}_0 = \begin{bmatrix} 1 \\ 0 \end{bmatrix}. \tag{J.92}$$

Then

$$\begin{bmatrix} -0.5907 & 0.9391 \\ 0.807 & 0.3437 \end{bmatrix}\begin{bmatrix} K_{h1} \\ K_{h2} \end{bmatrix} + \begin{bmatrix} 2.308 \\ -0.769 \end{bmatrix} = \begin{bmatrix} 1 \\ 0 \end{bmatrix}. \tag{J.93}$$

Solving,

$$\begin{bmatrix} K_{h1} \\ K_{h2} \end{bmatrix} = \begin{bmatrix} -0.5907 & 0.9391 \\ 0.807 & 0.3437 \end{bmatrix}^{-1}\begin{bmatrix} -1.308 \\ 0.769 \end{bmatrix} = \begin{bmatrix} 1.2194 \\ -0.6258 \end{bmatrix} \tag{J.94}$$

and, finally,

$$\mathbf{q} = \begin{bmatrix} y_1(t) \\ y_2(t) \end{bmatrix} = \begin{bmatrix} -0.5907e^{-2.268t} & 0.9391e^{-5.732t} \\ 0.807e^{-2.268t} & 0.3437e^{-5.732t} \end{bmatrix}\begin{bmatrix} 1.2194 \\ -0.6258 \end{bmatrix} + \begin{bmatrix} 2.308 \\ -0.769 \end{bmatrix} \tag{J.95}$$

or

$$\mathbf{q} = \begin{bmatrix} y_1(t) \\ y_2(t) \end{bmatrix} = \begin{bmatrix} -0.7203e^{-2.268t} - 0.5877e^{-5.732t} \\ 0.9841e^{-2.268t} - 0.2151e^{-5.732t} \end{bmatrix} + \begin{bmatrix} 2.308 \\ -0.769 \end{bmatrix}.$$

Just as was true with differential equations, it is always possible to write a system of linear independent constant-coefficient ordinary difference equations as a single matrix equation of the form

$$\mathbf{q}[n + 1] = \mathbf{A}\mathbf{q}[n] + \mathbf{B}\mathbf{x}[n], \tag{J.96}$$

where $\mathbf{q} =$ vector of solution functions

$\mathbf{x} =$ vector of forcing functions

$\mathbf{A}, \mathbf{B} =$ coefficient matrices

Thus (J.96) is a linear constant-coefficient ordinary first-order matrix difference equation. Consider the example from Appendix I, the system of difference equations

$$3y_1[n] + 2y_1[n-1] + y_2[n] = 0$$
$$4y_2[n] + 2y_2[n-1] + y_1[n] = 5 \tag{J.97}$$

with initial conditions $y_1[0] = 0$ and $y_2[0] = 2$. It can be rearranged into the form

$$3y_1[n] + y_2[n] = -2y_1[n-1]$$
$$4y_2[n] + y_1[n] = -2y_2[n-1] + 5 \tag{J.98}$$

or the equivalent form

$$3y_1[n+1] + y_2[n+1] = -2y_1[n]$$
$$y_1[n+1] + 4y_2[n+1] = -2y_2[n] + 5. \tag{J.99}$$

There seems to be a problem here. How do we arrange this equation into the form $\mathbf{q}[n+1] = \mathbf{A}\mathbf{q}[n] + \mathbf{B}\mathbf{x}[n]$? The answer lies in redefining the functions we are solving for. Let

$$q_1[n] = 3y_1[n] + y_2[n] \quad \text{and} \quad q_2[n] = y_1[n] + 4y_2[n] \tag{J.100}$$

implying that

$$y_1[n] = \frac{4q_1[n] - q_2[n]}{11} \quad \text{and} \quad y_2[n] = -\frac{q_1[n] - 3q_2[n]}{11}. \tag{J.101}$$

Then (J.99) can be written as

$$q_1[n+1] = \frac{-8q_1[n] + 2q_2[n]}{11}$$
$$q_2[n+1] = \frac{2q_1[n] - 6q_2[n]}{11} + 5. \tag{J.102}$$

We can express (J.102) in the standard matrix form

$$\mathbf{q}[n+1] = \mathbf{A}\mathbf{q}[n] + \mathbf{B}\mathbf{x}[n], \tag{J.103}$$

where

$$\mathbf{A} = \frac{1}{11}\begin{bmatrix} -8 & 2 \\ 2 & -6 \end{bmatrix} \quad \mathbf{q} = \begin{bmatrix} q_1[n] \\ q_2[n] \end{bmatrix} \quad \mathbf{B} = \begin{bmatrix} 0 \\ 1 \end{bmatrix} \quad \mathbf{x} = 5 \tag{J.104}$$

with an initial-condition vector

$$\mathbf{q}_0 = \begin{bmatrix} 3y_1[0] + y_2[0] \\ y_1[0] + 4y_2[0] \end{bmatrix} = \begin{bmatrix} 2 \\ 8 \end{bmatrix}. \tag{J.105}$$

The corresponding homogeneous equation is

$$\mathbf{q}[n+1] = \mathbf{A}\mathbf{q}[n], \tag{J.106}$$

and we know that the solution of the homogeneous equation is a linear combination of solutions of the form

$$\mathbf{q}_h = \mathbf{K}_h \alpha^n, \tag{J.107}$$

where $\mathbf{K}_h$ is a $2 \times 1$ vector of arbitrary constants. Therefore, we know that

$$\mathbf{q}_h[n+1] = \mathbf{K}_h \alpha^{n+1} = \alpha \mathbf{q}_h \tag{J.108}$$

and, substituting (J.108) into (J.106), that the solution of the homogeneous system of equations is the solution of the matrix equation

$$\mathbf{Aq} = \alpha\mathbf{q}, \tag{J.109}$$

which can be rearranged into

$$\mathbf{Aq} - \alpha\mathbf{q} = (\mathbf{A} - \alpha\mathbf{I})\,\mathbf{q} = \mathbf{0}. \tag{J.110}$$

For a nontrivial solution $\mathbf{q} \neq \mathbf{0}$,

$$|\mathbf{A} - \alpha\mathbf{I}| = 0 \tag{J.111}$$

or

$$\left| \frac{1}{11} \begin{bmatrix} -8 & 2 \\ 2 & -6 \end{bmatrix} - \alpha \begin{bmatrix} 1 & 0 \\ 0 & 1 \end{bmatrix} \right| = \frac{1}{11} \begin{vmatrix} -8 - 11\alpha & 2 \\ 2 & -6 - 11\alpha \end{vmatrix} = 0 \tag{J.112}$$

or

$$(8 + 11\alpha)(6 + 11\alpha) - 4 = 0. \tag{J.113}$$

The formulation $|\mathbf{A} - \alpha\mathbf{I}| = 0$ is the matrix characteristic equation for a matrix difference equation. From (J.113), we get

$$121\alpha^2 + 154\alpha + 44 = 0 \tag{J.114}$$

and the eigenvalues are $\alpha_1 = -0.4331$ and $\alpha_2 = -0.8396$ as we found in the previous solution of this system of difference equations. Notice that the redefinition of the functions we are solving for did not change the eigenvalues. The eigenvalues are completely determined by the coefficient matrix $\mathbf{A}$. For each eigenvalue there is a corresponding eigenvector $\mathbf{q}$ which, together with the eigenvalue, solves (J.109). For $\alpha_1$ we have the equality

$$[\mathbf{A} - \alpha_1\mathbf{I}]\mathbf{q} = \mathbf{0}. \tag{J.115}$$

Any row of $\mathbf{A} - \alpha_1\mathbf{I}$ can be used to determine the direction of an eigenvector. For example, using the first row,

$$(-8 - 11\alpha_1)q_1 + 2q_2 = 0 \tag{J.116}$$

or

$$\frac{q_1[n]}{q_2[n]} = \frac{K_{h1}\alpha_1^n}{K_{h2}\alpha_1^n} = \frac{K_{h1}}{K_{h2}} = \frac{2}{8 + 11\alpha_1}. \tag{J.117}$$

Therefore, a unit eigenvector would be

$$\mathbf{q}_1 = K_1 \begin{bmatrix} \dfrac{2}{8 + 11\alpha_1} \\ 1 \end{bmatrix} \alpha_1^n = \begin{bmatrix} 0.5257 \\ 0.8507 \end{bmatrix} (-0.4331)^n. \tag{J.118}$$

Using $\alpha_2$ we would get the other unit eigenvector,

$$\mathbf{q}_2 = \begin{bmatrix} \dfrac{2}{8 + 11\alpha_2} \\ 1 \end{bmatrix} \alpha_2^n = \begin{bmatrix} -0.8507 \\ 0.5257 \end{bmatrix} (-0.8396)^n. \tag{J.119}$$

Again the most general homogeneous solution is a linear combination of the eigenvectors of the form

$$\mathbf{q} = K_{h1}\mathbf{q}_1 + K_{h2}\mathbf{q}_2 = [\,\mathbf{q}_1 \quad \mathbf{q}_2\,]\begin{bmatrix} K_{h1} \\ K_{h2} \end{bmatrix},\tag{J.120}$$

where the two arbitrary constants $K_{h1}$ and $K_{h2}$ must be chosen to satisfy initial conditions.

The next solution step is to find the particular solution $\mathbf{q}_p = \begin{bmatrix} q_{1p}[n] \\ q_{2p}[n] \end{bmatrix}$. Since the forcing function is a constant $\mathbf{x} = 5$, the particular solution is a vector of constants $\mathbf{q}_p = \begin{bmatrix} K_{p1} \\ K_{p2} \end{bmatrix}$. Substituting into (J.102),

$$\mathbf{q}_p[n+1] = \mathbf{A}\mathbf{q}_p[n] + \mathbf{B}\mathbf{x}[n].\tag{J.121}$$

Since $\mathbf{q}_p$ is a vector of constants, $\mathbf{q}_p[n+1] = \mathbf{q}_p[n]$ and

$$(\mathbf{I} - \mathbf{A})\mathbf{q}_p[n] = \mathbf{B}\mathbf{x}[n].\tag{J.122}$$

Solving,

$$\mathbf{q}_p[n] = (\mathbf{I} - \mathbf{A})^{-1}\mathbf{B}\mathbf{x}[n].\tag{J.123}$$

The inverse of $(\mathbf{I} - \mathbf{A})$ is

$$(\mathbf{I} - \mathbf{A})^{-1} = \begin{bmatrix} 0.5862 & 0.0690 \\ 0.0690 & 0.6552 \end{bmatrix}.\tag{J.124}$$

Therefore,

$$\mathbf{q}_p = \begin{bmatrix} 0.5862 & 0.0690 \\ 0.0690 & 0.6552 \end{bmatrix}\begin{bmatrix} 0 \\ 1 \end{bmatrix}5 = \begin{bmatrix} 0.345 \\ 3.276 \end{bmatrix}.\tag{J.125}$$

Now we know that the total solution is

$$\mathbf{q} = [\,\mathbf{q}_1 \quad \mathbf{q}_2\,]\begin{bmatrix} K_{h1} \\ K_{h2} \end{bmatrix} + \mathbf{q}_p\tag{J.126}$$

or

$$\mathbf{q} = \begin{bmatrix} 0.5257\,(-0.4331)^n & -0.8507\,(-0.8396)^n \\ 0.8507\,(-0.4331)^n & 0.5257\,(-0.8396)^n \end{bmatrix}\begin{bmatrix} K_{h1} \\ K_{h2} \end{bmatrix} + \begin{bmatrix} 0.345 \\ 3.276 \end{bmatrix}.$$

The only task left is to solve for the arbitrary constants $K_{h1}$ and $K_{h2}$. The vector of initial conditions (at time $n = 0$) is

$$\mathbf{q}_0 = \begin{bmatrix} 2 \\ 8 \end{bmatrix}.\tag{J.127}$$

Then

$$\begin{bmatrix} 0.5257 & -0.8507 \\ 0.8507 & 0.5257 \end{bmatrix}\begin{bmatrix} K_{h1} \\ K_{h2} \end{bmatrix} + \begin{bmatrix} 0.345 \\ 3.276 \end{bmatrix} = \begin{bmatrix} 2 \\ 8 \end{bmatrix}.\tag{J.128}$$

Solving,

$$\begin{bmatrix} K_{h1} \\ K_{h2} \end{bmatrix} = \begin{bmatrix} 0.5257 & -0.8507 \\ 0.8507 & 0.5257 \end{bmatrix}^{-1}\begin{bmatrix} 1.655 \\ 4.724 \end{bmatrix} = \begin{bmatrix} 4.8885 \\ 1.0754 \end{bmatrix}\tag{J.129}$$

and, finally, combining (J.126) and (J.129),

$$\mathbf{q} = \begin{bmatrix} 2.57(-0.4331)^n - 0.915(-0.8396)^n \\ 4.1585(-0.4331)^n + 0.5655(-0.8396)^n \end{bmatrix} + \begin{bmatrix} 0.345 \\ 3.276 \end{bmatrix}. \qquad \textbf{(J.130)}$$

This solution should be equivalent to the previous solution of this system of difference equations in Appendix I,

$$y_1[n] = 0.557(-0.4331)^n - 0.3841(-0.8396)^n - 0.1724 \qquad \textbf{(J.131)}$$

and

$$y_2[n] = 0.9005(-0.4331)^n + 0.2374(-0.8396)^n + 0.8621. \qquad \textbf{(J.132)}$$

Using (J.101)

$$y_1[n] = \frac{4q_1[n] - q_2[n]}{11} \qquad \text{and} \qquad y_2[n] = -\frac{q_1[n] - 3q_2[n]}{11}, \qquad \textbf{(J.133)}$$

with (J.130) we get

$$y_1[n] = 0.557(-0.4331)^n - 0.3841(-0.8396)^n - 0.1724$$
$$y_2[n] = 0.9005(-0.4331)^n + 0.2374(-0.8396)^n + 0.8621 \qquad \textbf{(J.134)}$$

confirming that the two solution techniques agree.

## EXERCISES WITH ANSWERS

1. Graph these $\mathbf{q}$ vectors in the $q_1$-$q_2$ plane.

   a. $\mathbf{q}_a = \begin{bmatrix} 2 \\ -3 \end{bmatrix}$

   b. $\mathbf{q}_b = \begin{bmatrix} 3 \\ 2 \end{bmatrix}$

   c. $\mathbf{q}_c = \begin{bmatrix} 0 \\ 1 \end{bmatrix}$

   d. $\mathbf{q}_d = \mathbf{q}_a + \mathbf{q}_b$

   e. $\mathbf{q}_e = \mathbf{q}_a - \mathbf{q}_c$

   **Answers:**

   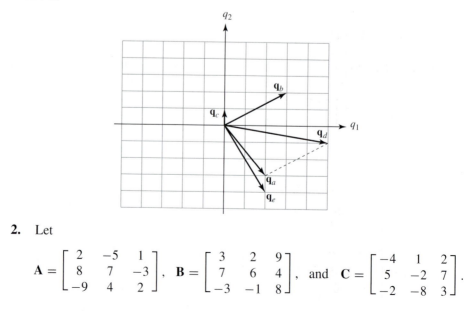

2. Let

$$\mathbf{A} = \begin{bmatrix} 2 & -5 & 1 \\ 8 & 7 & -3 \\ -9 & 4 & 2 \end{bmatrix}, \quad \mathbf{B} = \begin{bmatrix} 3 & 2 & 9 \\ 7 & 6 & 4 \\ -3 & -1 & 8 \end{bmatrix}, \quad \text{and} \quad \mathbf{C} = \begin{bmatrix} -4 & 1 & 2 \\ 5 & -2 & 7 \\ -2 & -8 & 3 \end{bmatrix}.$$

Find these matrices.

a.  **A + B**          b.  **A + C**          c.  **B − C**

**Answers:**

$$\begin{bmatrix} 7 & 1 & 7 \\ 2 & 8 & -3 \\ -1 & 7 & 5 \end{bmatrix}, \quad \begin{bmatrix} 5 & -3 & 10 \\ 15 & 13 & 1 \\ -12 & 3 & 10 \end{bmatrix}, \quad \begin{bmatrix} -2 & -4 & 3 \\ 13 & 5 & 4 \\ -11 & -4 & 5 \end{bmatrix}$$

**3.**  Using the **A, B,** and **C** matrices of Exercise 2, find these matrices.

a.  **−A**          b.  **3B + 2C**          c.  **C − 2B**

**Answers:**

$$\begin{bmatrix} 1 & 8 & 31 \\ 31 & 14 & 26 \\ -13 & -19 & 30 \end{bmatrix}, \quad \begin{bmatrix} -10 & -3 & -16 \\ -9 & -14 & -1 \\ 4 & -6 & -13 \end{bmatrix}, \quad \begin{bmatrix} -2 & 5 & -1 \\ -8 & -7 & 3 \\ 9 & -4 & -2 \end{bmatrix}$$

**4.**  Find the transposes of these matrices.

a.  $\mathbf{A} = \begin{bmatrix} 2 \\ -1 \\ 3 \\ 1 \end{bmatrix}$          b.  $\mathbf{B} = \begin{bmatrix} 2 & -1 \\ 1 & 3 \end{bmatrix}$

c.  $\mathbf{C} = [\,4 \quad 9 \quad -3 \quad -1\,]$          d.  $\mathbf{D} = \mathbf{A} + \mathbf{C}^T$

e.  $\mathbf{E} = \mathbf{B} + \mathbf{B}^T$

**Answers:**

$$\begin{bmatrix} 4 \\ 9 \\ -3 \\ -1 \end{bmatrix}, \quad [\,2 \quad -1 \quad 3 \quad 1\,], \quad \begin{bmatrix} 4 & 0 \\ 0 & 6 \end{bmatrix}, \quad \begin{bmatrix} 2 & 1 \\ -1 & 3 \end{bmatrix}, \quad \begin{bmatrix} 6 \\ 8 \\ 0 \\ 0 \end{bmatrix}$$

**5.**  Find these matrix products.

a.  $\mathbf{A} = [\,2 \quad 3\,] \begin{bmatrix} 3 \\ -2 \end{bmatrix}$          b.  $\mathbf{B} = \begin{bmatrix} 3 \\ -2 \end{bmatrix} [\,2 \quad 3\,]$

c.  $\mathbf{C} = \begin{bmatrix} 3 & -4 \\ 1 & 9 \end{bmatrix} \begin{bmatrix} 1 \\ 3 \end{bmatrix}$          d.  $\mathbf{D} = \mathbf{BB}^T$

e.  $\mathbf{E} = \mathbf{C}^T \mathbf{C}$          f.  $\mathbf{F} = \begin{bmatrix} 1 \\ 3 \end{bmatrix}^T \begin{bmatrix} 3 & -4 \\ 1 & 9 \end{bmatrix}^T$

g.  $\mathbf{G} = \begin{bmatrix} 2 & -5 & 1 \\ -2 & 20 & 10 \\ 7 & -1 & 1 \end{bmatrix} \begin{bmatrix} -2 & 1 & 5 \\ -7 & -5 & 0 \\ 1 & 5 & 1 \end{bmatrix}$

h.  $\mathbf{H} = \begin{bmatrix} -2 & -7 & 1 \\ 1 & -5 & 5 \\ 5 & 0 & 1 \end{bmatrix} \begin{bmatrix} 2 & -2 & 7 \\ -5 & 20 & -1 \\ 1 & 10 & 1 \end{bmatrix}$

**Answers:**

$$\begin{bmatrix} 32 & 32 & 11 \\ -126 & -52 & 0 \\ -6 & 17 & 36 \end{bmatrix}, \quad \begin{bmatrix} 32 & -126 & -6 \\ 32 & -52 & 17 \\ 11 & 0 & 36 \end{bmatrix}, \quad \begin{bmatrix} 81 & -252 \\ -252 & 784 \end{bmatrix}, \quad \mathbf{0}, \quad \begin{bmatrix} -9 \\ 28 \end{bmatrix},$$

$$\begin{bmatrix} -9 & 28 \end{bmatrix}, \quad \begin{bmatrix} 117 & -78 \\ -78 & 52 \end{bmatrix}, \quad \begin{bmatrix} 6 & 9 \\ -4 & -6 \end{bmatrix}$$

**6.** Solve these systems of equations using Cramer's rule, if it is possible. If it is not possible, state why.

    *a.*   $2q_1 + 7q_2 = -4$

           $-q_1 - 4q_2 = 9$

    *b.*  $\begin{bmatrix} -1 & -12 & 0 \\ -7 & 6 & 8 \\ 3 & 15 & 11 \end{bmatrix} \begin{bmatrix} q_1 \\ q_2 \\ q_3 \end{bmatrix} = \begin{bmatrix} -8 \\ -7 \\ 3 \end{bmatrix}$

    *c.*  $\begin{bmatrix} -4 & 12 & 0 & -10 \\ 6 & 6 & -1 & 13 \\ 7 & 11 & -14 & -7 \\ 6 & -11 & 2 & 5 \end{bmatrix} \begin{bmatrix} q_1 \\ q_2 \\ q_3 \\ q_4 \end{bmatrix} = \begin{bmatrix} 1 \\ -3 \\ -7 \\ 0 \end{bmatrix}$

**Answers:**

$$q_1 = -\frac{2544}{16,254}, \; q_2 = -\frac{861}{16,254}, \; q_3 = \frac{6999}{16,254}, \; q_4 = -\frac{1641}{16,254};$$

$$q_1 = 47, \; q_2 = -14;$$

$$q_1 = \frac{780}{1158}, \; q_2 = \frac{707}{1158}, \; q_3 = -\frac{861}{1158}$$

**7.** Invert these matrices, if it is possible. If it is not possible, state why.

    *a.*  $\mathbf{A} = \begin{bmatrix} 3 & 1 \\ -9 & -3 \end{bmatrix}$          *b.*  $\mathbf{B} = \begin{bmatrix} -1 & -6 \\ 0 & 22 \end{bmatrix}$

    *c.*  $\mathbf{C} = \begin{bmatrix} -9 & 15 & 3 \\ 6 & 5 & -9 \\ 5 & -6 & 0 \end{bmatrix}$      *d.*  $\mathbf{D} = \begin{bmatrix} -1 & 3 \\ 7 & 2 \\ -8 & -3 \end{bmatrix}$

    *e.*  $\mathbf{E} = \begin{bmatrix} 0 & 1 \\ 1 & 0 \end{bmatrix}$             *f.*  $\mathbf{F} = \mathbf{AB}$

    *g.*  $\mathbf{G} = \mathbf{D}^T \mathbf{D}$

**Answers:**

Not invertible, not invertible, $\begin{bmatrix} -1 & -\frac{3}{11} \\ 0 & \frac{1}{22} \end{bmatrix}$, $-\frac{1}{372} \begin{bmatrix} -54 & -18 & -150 \\ -45 & -15 & -63 \\ -61 & 21 & -135 \end{bmatrix}$,

not invertible, $\frac{1}{1283} \begin{bmatrix} 22 & -35 \\ -35 & 114 \end{bmatrix}$, $\begin{bmatrix} 0 & 1 \\ 1 & 0 \end{bmatrix}$

**8.** Where possible solve the equations of Exercise 6 using a matrix inverse.

**Answers:**

See Exercise 6.

**9.** Solve these systems of differential or difference equations.

a. $\begin{matrix} y_1'(t) = -3y_1(t) + 2y_2(t) + 5 \\ y_2'(t) = 4y_1(t) + y_2(t) - 2 \end{matrix}$, $y_1(0) = 3$, $y_2(0) = -1$

b. $\begin{matrix} y_1'(t) = 4y_1(t) + 10y_2'(t) + 5e^{-t} \\ y_2'(t) = -y_1(t) + 3y_2(t) - 7e^{-3t} \end{matrix}$, $y_1(0) = 0$, $y_2(0) = 9$

c. $\begin{matrix} y_1[n+1] = 4y_1[n] - 12y_2[n] + 3 \\ y_2[n+1] = -y_1[n] + 7y_2[n] - 1 \end{matrix}$, $y_1[0] = 1$, $y_2[0] = 0$

d. $y_1[n+1] = 6y_1[n] + 8y_2[n] - \left(\frac{1}{2}\right)^n$

$y_2[n+1] = 5y_1[n+1] - 3y_2[n] + \left(\frac{3}{4}\right)^n$, $y_1[0] = -6$, $y_2[0] = 1$

**Answers:**

$$\begin{bmatrix} y_1(t) \\ y_2(t) \end{bmatrix} = \begin{bmatrix} 1.642e^{-4.4641t} + 0.539e^{2.4641t} \\ -1.202e^{-4.4641t} + 1.475e^{2.4641t} \end{bmatrix} + \frac{1}{11}\begin{bmatrix} 9 \\ -14 \end{bmatrix},$$

$$\begin{bmatrix} y_1(t) \\ y_2(t) \end{bmatrix} = e^{-1.5t}\begin{bmatrix} -15.5\cos(3.1225t) + 74.37\sin(3.1225t) \\ 5.4166\cos(3.1225t) + 12.7688\sin(3.1225t) \end{bmatrix}$$
$$+ \begin{bmatrix} -2e^{-t} + 17.5e^{-3t} \\ -0.5e^{-t} + 4.0833e^{-3t} \end{bmatrix},$$

$$\begin{bmatrix} y_1[n] \\ y_2[n] \end{bmatrix} = \begin{bmatrix} -2.651(0.5422)^n - 0.9488(-0.0922)^n \\ -7.746(0.5422)^n + 3.246(-0.0922)^n \end{bmatrix} + \begin{bmatrix} 4.6 \\ 4.5 \end{bmatrix},$$

$$\begin{bmatrix} y_1[n] \\ y_2[n] \end{bmatrix} = \begin{bmatrix} -0.7147(43.4146)^n - 5.214(43.4146)^n \\ -3.3425(43.4146)^n + 4.171(-0.4146)^n \end{bmatrix}$$
$$+ \begin{bmatrix} 0.0892\left(\frac{1}{2}\right)^n - 0.1610\left(\frac{3}{4}\right)^n \\ 0.0638\left(\frac{1}{2}\right)^n + 0.1057\left(\frac{3}{4}\right)^n \end{bmatrix}$$

# BIBLIOGRAPHY

## Analog Filters

Huelsman, L., and Allen, P., *Introduction to the Theory and Design of Active Filters,* New York, McGraw-Hill, 1980.

Van Valkenburg, M., *Analog Filter Design,* New York, Holt-Rinehart-Winston, 1982.

## Basic Linear Signals and Systems

Brown, R., and Nilsson, J., *Introduction to Linear Systems Analysis,* New York, John Wiley & Sons, 1966.

Chen, C., *Linear System Theory and Design,* New York, Holt-Rinehart-Winston, 1984.

Cheng, D., *Analysis of Linear Systems,* Reading, MA, Addison-Wesley, 1961.

ElAli, T., and Karim, M., *Continuous Signals and Systems with MATLAB,* Boca Raton, FL, CRC Press, 2001.

Gardner, M., and Barnes, J., *Transients in Linear Systems,* New York, John Wiley & Sons, 1947.

Gaskill, J., *Linear Systems, Fourier Transforms and Optics,* New York, John Wiley & Sons, 1978.

Haykin, S., and VanVeen, B., *Signals and Systems,* New York, John Wiley & Sons, 1999.

Jackson, L., *Signals, Systems and Transforms,* Reading, MA, Addison-Wesley, 1991.

Kamen, E., and Heck, B., *Fundamentals of Signals and Systems,* Upper Saddle River, NJ, Prentice-Hall, 1997.

Lathi, B., *Signal Processing and Linear Systems,* Carmichael, CA, Berkeley-Cambridge, 1998.

Lindner, D., *Introduction to Signals and Systems,* New York, McGraw-Hill, 1999.

Neff, H., *Continuous and Discrete Linear Systems,* New York, Harper & Row, 1984.

Oppenheim, A., and Willsky, A., *Signals and Systems,* Upper Saddle River, NJ, Prentice-Hall, 1997.

Phillips, C., and Parr, J., *Signals, Systems, and Transforms,* Upper Saddle River, NJ, Prentice-Hall, 1999.

Schwartz, R., and Friedland, B., *Linear Systems,* New York, McGraw-Hill, 1965.

Sherrick, J., *Concepts in System and Signals,* Upper Saddle River, NJ, Prentice-Hall, 2001.

Soliman, S., and Srinath, M., *Continuous and Discrete Signals and Systems,* Englewood Cliffs, NJ, Prentice-Hall, 1990.

Ziemer, R., Tranter, W., and Fannin, D., *Signals and Systems Continuous and Discrete,* Upper Saddle River, NJ, Prentice-Hall, 1998.

## Circuit Analysis

Dorf, R., and Svoboda, J., *Introduction to Electric Circuits,* New York, John Wiley & Sons, 2001.

Hayt, W., Kemmerly, J., and Durbin, S., *Engineering Circuit Analysis,* New York, McGraw-Hill, 2002.

Irwin, D., *Basic Engineering Circuit Analysis,* New York, John Wiley & Sons, 2002.

Nilsson, J., and Riedel, S., *Electric Circuits,* Upper Saddle River, NJ, Prentice-Hall, 2000.

Paul, C., *Fundamentals of Electric Circuit Analysis,* New York, John Wiley & Sons, 2001.

Thomas, R., and Rosa, A., *The Analysis and Design of Linear Circuits,* New York, John Wiley & Sons, 2001.

## Communication Systems

Couch, L., *Digital and Analog Communication Systems,* Upper Saddle River, NJ, Prentice-Hall, 1997.

Lathi, B., *Modern Digital and Analog Communication Systems,* New York, Holt-Rinehart-Winston, 1983.

Roden, M., *Analog and Digital Communication Systems,* Upper Saddle River, NJ, Prentice-Hall, 1996.

Shenoi, K., *Digital Signal Processing in Telecommunications,* Upper Saddle River, NJ, Prentice-Hall, 1995.

Stremler, F., *Introduction to Communication Systems,* Reading, MA, Addison-Wesley, 1982.

Thomas, J., *Statistical Communication Theory,* New York, John Wiley & Sons, 1969.

Ziemer, R., and Tranter, W., *Principles of Communications,* New York, John Wiley & Sons, 1988.

## Discrete-Time Signals and Systems and Digital Filters

Bose, N., *Digital Filters: Theory and Applications,* New York, North-Holland, 1985.

Cadzow, J., *Discrete-Time Systems,* Englewood Cliffs, NJ, Prentice-Hall, 1973.

Childers, D., and Durling, A., *Digital Filtering and Signal Processing,* St. Paul, MN, West, 1975.

DeFatta, D., Lucas, J., and Hodgkiss, W., *Digital Signal Processing: A System Design Approach,* New York, John Wiley & Sons, 1988.

Gold, B., and Rader, C., *Digital Processing of Signals,* New York, McGraw-Hill, 1969.

Hamming, R., *Digital Filters,* Englewood Cliffs, NJ, Prentice-Hall, 1989.

Ifeachor, E., and Jervis, B., *Digital Signal Processing,* Harlow, England, Prentice-Hall, 2002.

Kuc, R., *Introduction to Digital Signal Processing,* New York, McGraw-Hill, 1988.

Kuo, B., *Analysis and Synthesis of Sampled-Data Control Systems,* Englewood Cliffs, NJ, Prentice-Hall, 1963.

Ludeman, L., *Fundamentals of Digital Signal Processing,* New York, John Wiley & Sons, 1987.

Oppenheim, A., *Applications of Digital Signal Processing,* Englewood Cliffs, NJ, Prentice-Hall, 1978.

Oppenheim, A., and Shafer, R., *Digital Signal Processing,* Englewood Cliffs, NJ, Prentice-Hall, 1975.

Peled, A., and Liu, B., *Digital Signal Processing: Theory Design and Implementation,* New York, John Wiley & Sons, 1976.

Rabiner, L., and Gold, B., *Theory and Application of Digital Signal Processing,* Englewood Cliffs, NJ, Prentice-Hall, 1975.

Roberts, R., and Mullis, C., *Digital Signal Processing,* Reading, MA, Addison-Wesley, 1987.

Shenoi, K., *Digital Signal Processing in Telecommunications,* Upper Saddle River, NJ, Prentice-Hall, 1995.

Stanley, W., *Digital Signal Processing,* Reston, VA, Reston Publishing Company, 1975.

Strum, R., and Kirk, D., *Discrete Systems and Digital Signal Processing,* Reading, MA, Addison-Wesley, 1988.

Young, T., *Linear Systems and Digital Signal Processing,* Englewood Cliffs, NJ, Prentice-Hall, 1985.

### The Fast Fourier Transform

Brigham, E., *The Fast Fourier Transform,* Englewood Cliffs, NJ, Prentice-Hall, 1974.

Cooley, J., and Tukey, J., "An Algorithm for the Machine Computation of the Complex Fourier Series," *Mathematics of Computation,* vol. 19, April 1965, pp. 297–301.

### Fourier Optics

Gaskill, J., *Linear Systems, Fourier Transforms and Optics,* New York, John Wiley & Sons, 1978.

Goodman, J., *Introduction to Fourier Optics,* New York, McGraw-Hill, 1968.

### Random Signals and Statistics

Bendat, J., and Piersol, A., *Random Data: Analysis and Measurement Procedures,* New York, John Wiley & Sons, 1986.

Cooper, G., and McGillem, C., *Probabilistic Methods of Signal and System Analysis,* New York, Oxford University Press, 1999.

Davenport, W., and Root, W., *Introduction to the Theory of Random Signals and Noise,* New York, John Wiley & Sons, 1987.

Fante, R., *Signal Analysis and Estimation,* New York, John Wiley & Sons, 1988.

Leon-Garcia, A., *Probability and Random Processes for Electrical Engineering,* Reading, MA, Addison-Wesley, 1994.

Mix, D., *Random Signal Processing,* Englewood Cliffs, NJ, Prentice-Hall, 1995.

Papoulis, A., *Probability, Random Variables and Stochastic Processes,* New York, McGraw-Hill, 1991.

Thomas, J., *Statistical Communication Theory,* New York, Wiley-IEEE Press, 1996.

### Related Mathematics

Abramowitz, M., and Stegun, I., *Handbook of Mathematical Functions,* New York, Dover, 1970.

Churchill, R., Brown, J., and Pearson, C., *Complex Variables and Applications,* New York, McGraw-Hill, 1990.

Churchill, R., *Operational Mathematics,* New York, McGraw-Hill, 1958.

Craig, E., *Laplace and Fourier Transforms for Electrical Engineers,* New York, Holt-Rinehart-Winston, 1964.

Goldman, S., *Laplace Transform Theory and Electrical Transients,* New York, Dover, 1966.

Jury, E., *Theory and Application of the z-Transform Method,* Malabar, FL, R. E. Krieger, 1982.

Kreyszig, E., *Advanced Engineering Mathematics,* New York, John Wiley & Sons, 1998.

Matthews, J., and Walker, R., *Mathematical Methods of Physics,* New York, W. A. Benjamin, Inc., 1970.

Noble, B., *Applied Linear Algebra,* Englewood Cliffs, NJ, Prentice-Hall, 1969.

Scheid, F., *Numerical Analysis,* New York, McGraw-Hill, 1968.

Sokolnikoff, I., and Redheffer, R., *Mathematics of Physics and Modern Engineering,* New York, McGraw-Hill, 1966.

Spiegel, M., *Complex Variables,* New York, McGraw-Hill, 1968.

Strang, G., *Introduction to Linear Algebra,* Wellesley, MA, Wellesley-Cambridge Press, 1993.

### Specialized Related Topics

DeRusso, P., Roy, R., and Close, C., *State Variables for Engineers,* New York, John Wiley & Sons, 1998.